OPERATIONS RESEARCH AND MANAGEMENT SCIENCE HANDBOOK

The Operations Research Series

Series Editor: A. Ravi Ravindran
Dept. of Industrial & Manufacturing Engineering
The Pennsylvania State University, USA

Integer Programming: Theory and Practice
John K. Karlof

Operations Research: A Practical Approach
Michael W. Carter and Camille C. Price

Operations Research and Management Science Handbook
A. Ravi Ravindran

Operations Research Calculations Handbook
Dennis Blumenfeld

Forthcoming Titles

Applied Nonlinear Optimization in Modeling Environments
Janos D. Pinter

Operations Research Calculations Handbook, Second Edition
Dennis Blumenfeld

Probability Models in Operations Research
Richard C. Cassady and Joel A. Nachlas

OPERATIONS RESEARCH AND MANAGEMENT SCIENCE HANDBOOK

EDITED BY
A. RAVI RAVINDRAN

CRC Press is an imprint of the
Taylor & Francis Group, an **informa** business

CRC Press
Taylor & Francis Group
6000 Broken Sound Parkway NW, Suite 300
Boca Raton, FL 33487-2742

© 2008 by Taylor & Francis Group, LLC
CRC Press is an imprint of Taylor & Francis Group, an Informa business

No claim to original U.S. Government works
Printed in the United States of America on acid-free paper
10 9 8 7 6 5 4 3 2 1

International Standard Book Number-13: 978-0-8493-9721-9 (Hardcover)

This book contains information obtained from authentic and highly regarded sources. Reprinted material is quoted with permission, and sources are indicated. A wide variety of references are listed. Reasonable efforts have been made to publish reliable data and information, but the author and the publisher cannot assume responsibility for the validity of all materials or for the consequences of their use.

Except as permitted under U.S. Copyright Law, no part of this book may be reprinted, reproduced, transmitted, or utilized in any form by any electronic, mechanical, or other means, now known or hereafter invented, including photocopying, microfilming, and recording, or in any information storage or retrieval system, without written permission from the publishers.

For permission to photocopy or use material electronically from this work, please access www.copyright.com (http://www.copyright.com/) or contact the Copyright Clearance Center, Inc. (CCC) 222 Rosewood Drive, Danvers, MA 01923, 978-750-8400. CCC is a not-for-profit organization that provides licenses and registration for a variety of users. For organizations that have been granted a photocopy license by the CCC, a separate system of payment has been arranged.

Trademark Notice: Product or corporate names may be trademarks or registered trademarks, and are used only for identification and explanation without intent to infringe.

Library of Congress Cataloging-in-Publication Data

Operations research and management science handbook / editor, A. Ravi Ravindran.
    p. cm. -- (Operations research series)
  Includes bibliographical references and index.
  ISBN 978-0-8493-9721-9 (alk. paper)
  1. Operations research--Handbooks, manuals, etc. I. Ravindran, Ravi. II. Title. III. Series.

T57.6.A32 1982
658.4'034--dc22
                                                    2007019976

Visit the Taylor & Francis Web site at
http://www.taylorandfrancis.com

and the CRC Press Web site at
http://www.crcpress.com

Dedication

This book is dedicated to the memory of Professor G. V. Loganathan, author of Chapter 24, who was killed in the Virginia Tech campus tragedy on April 16, 2007.

Contents

Preface xiii
Acknowledgments xv
Editor xvii
Contributors xix
History of Operations Research and Management Science xxi

I OR/MS Models and Methods

1 Linear Programming
Katta G. Murty **1-1**
1.1 Brief History of Algorithms for Solving Linear Equations, Linear Inequalities, and LPs 1-1
1.2 Applicability of the LP Model: Classical Examples of Direct Applications 1-4
1.3 LP Models Involving Transformations of Variables 1-12
1.4 Intelligent Modeling Essential to Get Good Results, an Example from Container Shipping 1-18
1.5 Planning Uses of LP Models 1-22
1.6 Brief Introduction to Algorithms for Solving LP Models ... 1-26
1.7 Software Systems Available for Solving LP Models 1-31
1.8 Multiobjective LP Models 1-31

2 Nonlinear Programming
Theodore B. Trafalis and Robin C. Gilbert **2-1**
2.1 Introduction 2-1
2.2 Unconstrained Optimization 2-3
2.3 Constrained Optimization 2-15
2.4 Conclusion 2-19

3 Integer Programming
Michael Weng **3-1**
3.1 Introduction 3-1
3.2 Formulation of IP Models 3-3
3.3 Branch and Bound Method 3-7
3.4 Cutting Plane Method 3-12
3.5 Other Solution Methods and Computer Solution 3-15

4 Network Optimization
Mehmet Bayram Yildirim **4-1**
4.1 Introduction 4-1
4.2 Notation 4-2

4.3	Minimum Cost Flow Problem	4-3
4.4	Shortest Path Problem	4-4
4.5	Maximum Flow Problem	4-8
4.6	Assignment Problem	4-13
4.7	Minimum Spanning Tree Problem	4-14
4.8	Minimum Cost Multicommodity Flow Problem	4-18
4.9	Conclusions	4-19

5 Multiple Criteria Decision Making
Abu S. M. Masud and A. Ravi Ravindran **5-1**

5.1	Some Definitions	5-3
5.2	The Concept of "Best Solution"	5-4
5.3	Criteria Normalization	5-5
5.4	Computing Criteria Weights	5-6
5.5	Multiple Criteria Methods for Finite Alternatives	5-8
5.6	Multiple Criteria Mathematical Programming Problems	5-15
5.7	Goal Programming	5-19
5.8	Method of Global Criterion and Compromise Programming	5-27
5.9	Interactive Methods	5-29
5.10	MCDM Applications	5-34
5.11	MCDM Software	5-35
5.12	Further Readings	5-35

6 Decision Analysis
Cerry M. Klein **6-1**

6.1	Introduction	6-1
6.2	Terminology for Decision Analysis	6-2
6.3	Decision Making under Risk	6-3
6.4	Decision Making under Uncertainty	6-17
6.5	Practical Decision Analysis	6-21
6.6	Conclusions	6-28
6.7	Resources	6-29

7 Dynamic Programming
José A. Ventura **7-1**

7.1	Introduction	7-1
7.2	Deterministic Dynamic Programming Models	7-3
7.3	Stochastic Dynamic Programming Models	7-19
7.4	Conclusions	7-24

8 Stochastic Processes
Susan H. Xu **8-1**

8.1	Introduction	8-1
8.2	Poisson Processes	8-7
8.3	Discrete-Time Markov Chains	8-14
8.4	Continuous-Time Markov Chains	8-27

8.5	Renewal Theory	**8-39**
8.6	Software Products Available for Solving Stochastic Models	**8-46**

9 Queueing Theory
Natarajan Gautam **9-1**

9.1	Introduction	**9-1**
9.2	Queueing Theory Basics	**9-2**
9.3	Single-Station and Single-Class Queues	**9-8**
9.4	Single-Station and Multiclass Queues	**9-21**
9.5	Multistation and Single-Class Queues	**9-28**
9.6	Multistation and Multiclass Queues	**9-34**
9.7	Concluding Remarks	**9-37**

10 Inventory Control
Farhad Azadivar and Atul Rangarajan **10-1**

10.1	Introduction	**10-1**
10.2	Design of Inventory Systems	**10-4**
10.3	Deterministic Inventory Systems	**10-9**
10.4	Stochastic Inventory Systems	**10-20**
10.5	Inventory Control at Multiple Locations	**10-27**
10.6	Inventory Management in Practice	**10-34**
10.7	Conclusions	**10-35**
10.8	Current and Future Research	**10-36**

11 Complexity and Large-Scale Networks
Hari P. Thadakamalla, Soundar R.T. Kumara, and Réka Albert **11-1**

11.1	Introduction	**11-1**
11.2	Statistical Properties of Complex Networks	**11-6**
11.3	Modeling of Complex Networks	**11-11**
11.4	Why "Complex" Networks	**11-16**
11.5	Optimization in Complex Networks	**11-18**
11.6	Conclusions	**11-26**

12 Simulation
Catherine M. Harmonosky **12-1**

12.1	Introduction	**12-1**
12.2	Basics of Simulation	**12-3**
12.3	Simulation Languages and Software	**12-15**
12.4	Simulation Projects—The Bigger Picture	**12-20**
12.5	Summary	**12-22**

13 Metaheuristics for Discrete Optimization Problems
Rex K. Kincaid **13-1**

13.1	Mathematical Framework for Single Solution Metaheuristics	**13-3**
13.2	Network Location Problems	**13-3**
13.3	Multistart Local Search	**13-5**
13.4	Simulated Annealing	**13-6**
13.5	Plain Vanilla Tabu Search	**13-8**
13.6	Active Structural Acoustic Control (ASAC)	**13-10**

	13.7	Nature Reserve Site Selection	**13-13**
	13.8	Damper Placement in Flexible Truss Structures	**13-21**
	13.9	Reactive Tabu Search	**13-29**
	13.10	Discussion	**13-35**

14 Robust Optimization
H. J. Greenberg and Tod Morrison **14-1**
14.1	Introduction	**14-1**
14.2	Classical Models	**14-2**
14.3	Robust Optimization Models	**14-10**
14.4	More Applications	**14-16**
14.5	Summary	**14-30**

II OR/MS Applications

15 Project Management
Adedeji B. Badiru **15-1**
15.1	Introduction	**15-1**
15.2	Critical Path Method	**15-3**
15.3	PERT Network Analysis	**15-18**
15.4	Statistical Analysis of Project Duration	**15-23**
15.5	Precedence Diagramming Method	**15-26**
15.6	Software Tools for Project Management	**15-34**
15.7	Conclusion	**15-37**

16 Quality Control
Qianmei Feng and Kailash C. Kapur **16-1**
16.1	Introduction	**16-1**
16.2	Quality Control and Product Life Cycle	**16-2**
16.3	New Trends and Relationship to Six Sigma	**16-5**
16.4	Statistical Process Control	**16-7**
16.5	Process Capability Studies	**16-16**
16.6	Advanced Control Charts	**16-18**
16.7	Limitations of Acceptance Sampling	**16-20**
16.8	Conclusions	**16-20**

17 Reliability
Lawrence M. Leemis **17-1**
17.1	Introduction	**17-1**
17.2	Reliability in System Design	**17-3**
17.3	Lifetime Distributions	**17-10**
17.4	Parametric Models	**17-17**
17.5	Parameter Estimation in Survival Analysis	**17-22**
17.6	Nonparametric Methods	**17-32**
17.7	Assessing Model Adequacy	**17-36**
17.8	Summary	**17-40**

18 Production Systems
Bobbie L. Foote and Katta G. Murty **18-1**
- 18.1 Production Planning Problem **18-1**
- 18.2 Demand Forecasting **18-2**
- 18.3 Models for Production Layout Design **18-12**
- 18.4 Scheduling of Production and Service Systems **18-20**

19 Energy Systems
C. Randy Hudson and Adedeji B. Badiru **19-1**
- 19.1 Introduction **19-1**
- 19.2 Definition of Energy **19-2**
- 19.3 Harnessing Natural Energy **19-3**
- 19.4 Mathematical Modeling of Energy Systems **19-3**
- 19.5 Linear Programming Model of Energy Resource Combination **19-4**
- 19.6 Integer Programming Model for Energy Investment Options **19-5**
- 19.7 Simulation and Optimization of Distributed Energy Systems **19-11**
- 19.8 Point-of-Use Energy Generation **19-11**
- 19.9 Modeling of CHP Systems **19-12**
- 19.10 Economic Optimization Methods **19-13**
- 19.11 Design of a Model for Optimization of CHP System Capacities **19-16**
- 19.12 Capacity Optimization **19-21**
- 19.13 Implementation of the Computer Model **19-24**
- 19.14 Other Scenarios **19-27**

20 Airline Optimization
Jane L. Snowdon and Giuseppe Paleologo **20-1**
- 20.1 Introduction **20-1**
- 20.2 Schedule Planning **20-5**
- 20.3 Revenue Management **20-14**
- 20.4 Aircraft Load Planning **20-21**
- 20.5 Future Research Directions and Conclusions **20-23**

21 Financial Engineering
Aliza R. Heching and Alan J. King **21-1**
- 21.1 Introduction **21-1**
- 21.2 Return **21-3**
- 21.3 Estimating an Asset's Mean and Variance **21-4**
- 21.4 Diversification **21-6**
- 21.5 Efficient Frontier **21-8**
- 21.6 Utility Analysis **21-10**
- 21.7 Black–Litterman Asset Allocation Model **21-13**
- 21.8 Risk Management **21-18**
- 21.9 Options **21-22**
- 21.10 Valuing Options **21-24**
- 21.11 Dynamic Programming **21-28**
- 21.12 Pricing American Options Using Dynamic Programming **21-29**
- 21.13 Comparison of Monte Carlo Simulation and Dynamic Programming **21-33**
- 21.14 Multi-Period Asset Liability Management **21-33**
- 21.15 Conclusions **21-36**

22 Supply Chain Management
Donald P. Warsing 22-1
- 22.1 Introduction 22-1
- 22.2 Managing Inventories in the Supply Chain 22-6
- 22.3 Managing Transportation in the Supply Chain 22-24
- 22.4 Managing Locations in the Supply Chain 22-38
- 22.5 Managing Dyads in the Supply Chain 22-48
- 22.6 Discussion and Conclusions 22-58

23 E-Commerce
Sowmyanarayanan Sadagopan 23-1
- 23.1 Introduction 23-1
- 23.2 Evolution of E-Commerce 23-3
- 23.3 OR/MS and E-Commerce 23-5
- 23.4 OR Applications in E-Commerce 23-7
- 23.5 Tools–Applications Matrix 23-20
- 23.6 Way Forward 23-21
- 23.7 Summary 23-21

24 Water Resources
G.V. Loganathan 24-1
- 24.1 Introduction 24-1
- 24.2 Optimal Operating Policy for Reservoir Systems 24-4
- 24.3 Water Distribution Systems Optimization 24-10
- 24.4 Preferences in Choosing Domestic Plumbing Materials 24-18
- 24.5 Stormwater Management 24-21
- 24.6 Groundwater Management 24-23
- 24.7 Summary 24-25

25 Military Applications
Jeffery D. Weir and Marlin U. Thomas 25-1
- 25.1 Introduction 25-1
- 25.2 Background on Military OR 25-2
- 25.3 Current Military Applications of OR 25-3
- 25.4 Concluding Remarks 25-10

26 Future of OR/MS Applications: A Practitioner's Perspective
P. Balasubramanian 26-1
- 26.1 Past as a Guide to the Future 26-2
- 26.2 Impact of the Internet 26-6
- 26.3 Emerging Opportunities 26-8

Index I-1

Preface

Operations Research (OR), which began as an interdisciplinary activity to solve complex problems in the military during World War II, has grown in the last 50 years to a full-fledged academic discipline. Now OR is viewed as a body of established mathematical models and methods to solve complex management problems. OR provides a quantitative analysis of the problem from which management can make an objective decision. OR has drawn upon skills from mathematics, engineering, business, computer science, economics, and statistics to contribute to a wide variety of applications in business, industry, government, and military. Operations Research and Management Science (OR/MS) methodologies and their applications continue to grow and flourish in a number of decision-making fields.

The objective of this handbook is to provide a comprehensive overview of OR/MS models, methods, and applications in a single volume. The handbook is not an OR textbook or a research monograph. The intent is that the handbook becomes the first resource a practitioner would reach for when faced with an OR/MS problem or question. The key features of the handbook are the following:

- Single source guide in OR/MS
- Comprehensive resource, but concise
- Covers all topics in OR/MS
- Quick reference guide to students, researchers, and practitioners
- Bridges theory and practice
- References to computer software availability
- Designed and edited with non-experts in mind
- Unified and up-to-date coverage ideal for ready reference

The handbook is divided into two parts. Part I contains 14 chapters that cover the fundamental OR/MS models and methods. Each chapter gives an overview of a particular OR/MS model and its solution methods, illustrates successful applications, and provides references to computer software availability. Each chapter in Part I is devoted to a topic listed below:

- Linear programming
- Nonlinear programming
- Integer programming
- Network optimization
- Multiple criteria decision making
- Decision analysis
- Dynamic programming
- Stochastic processes
- Queueing theory
- Inventory control
- Complexity and large-scale networks
- Simulation
- Metaheuristics
- Robust optimization

Part II of the handbook contains 11 chapters discussing the OR/MS applications in specific areas. They include:

- Airlines
- E-commerce
- Energy systems
- Finance
- Military
- Production systems
- Project management
- Quality control
- Reliability
- Supply chain management
- Water resources

Part II ends with a chapter on the future of OR/MS applications.

The handbook is an ideal reference book for OR/MS practitioners in business, industry, government, and academia. It can also serve as a supplemental text in undergraduate and graduate OR/MS courses at the university level.

A. Ravi Ravindran
University Park, Pennsylvania

Acknowledgments

First and foremost I would like to thank the authors, who have worked diligently in writing the various handbook chapters that are comprehensive, concise, and easy to read, bridging the gap between theory and practice. The development and evolution of this handbook have also benefited substantially from the advice and counsel of my colleagues and friends in academia and industry, who are too numerous to acknowledge individually. They helped me identify the key topics to be included in the handbook, suggested chapter authors, and served as reviewers of the manuscripts.

I express my sincere appreciation to Atul Rangarajan, an industrial engineering doctoral student at Penn State University, for serving as my editorial assistant and for his careful review of the page proofs returned by the authors. Several other graduate students also helped me with the handbook work, in particular, Ufuk Bilsel, Ajay Natarajan, Richard Titus, Vijay Wadhwa, and Tao Yang. Special thanks go to Professor Prabha Sharma at the Indian Institute of Technology, Kanpur, for her careful review of several chapter manuscripts. I also acknowledge the pleasant personality and excellent typing skills of Sharon Frazier during the entire book project.

I thank Cindy Carelli, Senior Acquisitions Editor, and Jessica Vakili, project coordinator at CRC Press, for their help from inception to publication of the handbook. Finally, I wish to thank my dear wife, Bhuvana, for her patience, understanding, and support when I was focused completely on the handbook work.

A. Ravi Ravindran

Editor

A. Ravi Ravindran, Ph.D., is a professor and the past department head of Industrial and Manufacturing Engineering at the Pennsylvania State University. Formerly, he was a faculty member at the School of Industrial Engineering at Purdue University for 13 years and at the University of Oklahoma for 15 years. At Oklahoma, he served as the director of the School of Industrial Engineering for 8 years and as the associate provost of the university for 7 years, with responsibility for budget, personnel, and space for the academic area. He holds a B.S. in electrical engineering with honors from the Birla Institute of Technology and Science, Pilani, India. His graduate degrees are from the University of California, Berkeley, where he received an M.S. and a Ph.D. in industrial engineering and operations research.

Dr. Ravindran's area of specialization is operations research with research interests in multiple criteria decision-making, financial engineering, health planning, and supply chain optimization. He has published two major textbooks (*Operations Research: Principles and Practice* and *Engineering Optimization: Methods and Applications*) and more than 100 journal articles on operations research. He is a fellow of the Institute of Industrial Engineers. In 2001, he was recognized by the Institute of Industrial Engineers with the Albert G. Holzman Distinguished Educator Award for significant contributions to the industrial engineering profession by an educator. He has won several Best Teacher awards from IE students. He has been a consultant to AT&T, General Motors, General Electric, IBM, Kimberly Clark, Cellular Telecommunication Industry Association, and the U.S. Air Force. He currently serves as the Operations Research Series editor for Taylor & Francis/CRC Press.

Contributors

Réka Albert
Pennsylvania State University
University Park, Pennsylvania

Farhad Azadivar
University of Massachusetts–Dartmouth
North Dartmouth, Massachusetts

Adedeji B. Badiru
Air Force Institute of Technology
Dayton, Ohio

P. Balasubramanian
Theme Work Analytics
Bangalore, India

Qianmei Feng
University of Houston
Houston, Texas

Bobbie L. Foote
U.S. Military Academy (Retd.)
West Point, New York

Natarajan Gautam
Texas A&M University
College Station, Texas

Robin C. Gilbert
University of Oklahoma
Norman, Oklahoma

H. J. Greenberg
University of Colorado at Denver
and
Health Sciences Center
Denver, Colorado

Catherine M. Harmonosky
Pennsylvania State University
University Park, Pennsylvania

Aliza R. Heching
IBM T. J. Watson Research Center
Yorktown Heights, New York

C. Randy Hudson
Oak Ridge National Laboratory
Oak Ridge, Tennessee

Kailash C. Kapur
University of Washington
Seattle, Washington

Rex K. Kincaid
College of William and Mary
Williamsburg, Virginia

Alan J. King
IBM T. J. Watson Research Center
Yorktown Heights, New York

Cerry M. Klein
University of Missouri–Columbia
Columbia, Missouri

Soundar R. T. Kumara
Pennsylvania State University
University Park, Pennsylvania

Lawrence M. Leemis
College of William and Mary
Williamsburg, Virginia

G. V. Loganathan
Virginia Tech
Blacksburg, Virginia

Abu S. M. Masud
Wichita State University
Wichita, Kansas

Tod Morrison
University of Colorado at Denver
and
Health Sciences Center
Denver, Colorado

Katta G. Murty
University of Michigan
Ann Arbor, Michigan

Giuseppe Paleologo
IBM T. J. Watson Research Center
Yorktown Heights, New York

Atul Rangarajan
Pennsylvania State University
University Park, Pennsylvania

A. Ravi Ravindran
Pennsylvania State University
University Park, Pennsylvania

Sowmyanarayanan Sadagopan
Indian Institute of Information Technology
Bangalore, India

Jane L. Snowdon
IBM T. J. Watson Research Center
Yorktown Heights, New York

Hari P. Thadakamalla
Pennsylvania State University
University Park, Pennsylvania

Marlin U. Thomas
Air Force Institute of Technology
Wright-Patterson AFB, Ohio

Theodore B. Trafalis
University of Oklahoma
Norman, Oklahoma

José A. Ventura
Pennsylvania State University
University Park, Pennsylvania

Donald P. Warsing
North Carolina State University
Raleigh, North Carolina

Jeffery D. Weir
Air Force Institute of Technology
Wright-Patterson AFB, Ohio

Michael Weng
University of South Florida
Tampa, Florida

Susan H. Xu
Pennsylvania State University
University Park, Pennsylvania

Mehmet Bayram Yildirim
Wichita State University
Wichita, Kansas

History of Operations Research and Management Science

A. Ravi Ravindran
Pennsylvania State University

Origin of Operations Research

To understand what operations research (OR) is today, one must know something of its history and evolution. Although particular models and techniques of OR can be traced back to much earlier origins, it is generally agreed that the discipline began during World War II. Many strategic and tactical problems associated with the Allied military effort were simply too complicated to expect adequate solutions from any one individual, or even a single discipline. In response to these complex problems, groups of scientists with diverse educational backgrounds were assembled as special units within the armed forces. These teams of scientists started working together, applying their interdisciplinary knowledge and training to solve such problems as deployment of radars, anti-aircraft fire control, deployment of ships to minimize losses from enemy submarines, and strategies for air defense. Each of the three wings of Britain's armed forces had such interdisciplinary research teams working on military management problems. As these teams were generally assigned to the commanders in charge of military operations, they were called *operational research* (OR) *teams*. The nature of their research came to be known as *operational research* or *operations research*.

The work of these OR teams was very successful and their solutions were effective in military management. This led to the use of such scientific teams in other Allied nations, in particular the United States, France, and Canada. At the end of the war, many of the scientists who worked in the military operational research units returned to civilian life in universities and industries. They started applying the OR methodology to solve complex management problems in industries. Petroleum companies were the first to make use of OR models for solving large-scale production and distribution problems. In the universities advancements in OR techniques were made that led to the further development and applications of OR. Much of the postwar development of OR took place in the United States.

An important factor in the rapid growth of operations research was the introduction of electronic computers in the early 1950s. The computer became an invaluable tool to the *operations researchers*, enabling them to solve large problems in the business world.

The Operations Research Society of America (ORSA) was formed in 1952 to serve the professional needs of these operations research scientists. Due to the application of OR in industries, a new term called management science (MS) came into being. In 1953, a national society called The Institute of Management Sciences (TIMS) was formed in the United States to promote scientific knowledge in the understanding and practice of management. The journals of these two societies, *Operations Research* and *Management Science*, as well as the joint conferences of their members, helped to draw together the many diverse results into some semblance of a coherent body of knowledge. In 1995, the two societies, ORSA and TIMS, merged to form the Institute of Operations Research and Management Sciences (INFORMS).

Another factor that accelerated the growth of operations research was the introduction of OR/MS courses in the curricula of many universities and colleges in the United States. Graduate programs leading to advanced degrees at the master's and doctorate levels were introduced in major American universities. By the mid-1960s many theoretical advances in OR techniques had been made, which included linear programming, network analysis, integer programming, nonlinear programming, dynamic programming, inventory theory, queueing theory, and simulation. Simultaneously, new applications of OR emerged in service organizations such as banks, health care, communications, libraries, and transportation. In addition, OR came to be used in local, state, and federal governments in their planning and policy-making activities.

It is interesting to note that the modern perception of OR as a body of established models and techniques—that is, a discipline in itself—is quite different from the original concept of OR as an *activity,* which was preformed by interdisciplinary teams. An evolution of this kind is to be expected in any emerging field of scientific inquiry. In the initial formative years, there are no experts, no traditions, no literature. As problems are successfully solved, the body of specific knowledge grows to a point where it begins to require specialization even to know what has been previously accomplished. The pioneering efforts of one generation become the standard practice of the next. Still, it ought to be remembered that at least a portion of the record of success of OR can be attributed to its ecumenical nature.

Meaning of Operations Research

From the historical and philosophical summary just presented, it should be apparent that the term "operations research" has a number of quite distinct variations of meaning. To some, OR is that certain body of problems, techniques, and solutions that has been accumulated under the name of OR over the past 50 years and we apply OR when we recognize a problem of that certain genre. To others, it is an activity or process, which by its very nature is applied. It would also be counterproductive to attempt to make distinctions between "operations research" and the "systems approach." For all practical purposes, they are the same.

How then can we define operations research? The Operational Research Society of Great Britain has adopted the following definition:

> Operational research is the application of the methods of science to complex problems arising in the direction and management of large systems of men, machines, materials and money in industry, business, government, and defense. The distinctive approach is to develop a scientific model of the system, incorporating measurement of factors such as chance and risk, with which to predict and compare the outcomes of alternative decisions, strategies or controls. The purpose is to help management determine its policy and actions scientifically.

The Operations Research Society of America has offered a shorter, but similar, description:

> Operations research is concerned with scientifically deciding how to best design and operate man–machine systems, usually under conditions requiring the allocation of scarce resources.

In general, most of the definitions of OR emphasize its methodology, namely its unique approach to problem solving, which may be due to the use of interdisciplinary teams or due to the application of scientific and mathematical models. In other words, each problem may be analyzed differently, though the same basic approach of operations research is employed. As more research went into the development of OR, the researchers were able to classify to some extent many of the important management problems that arise in practice. Examples of such problems are those relating to allocation, inventory, network, queuing, replacement, scheduling, and so on. The theoretical research in OR concentrated on developing appropriate mathematical models and techniques for analyzing these problems under different conditions. Thus, whenever a management problem is identified as belonging to a particular class, all the models and techniques available for that class can be used to study that problem. In this context, one could view OR as a collection of mathematical models and techniques to solve complex management problems. Hence, it is very common to find OR courses in universities emphasizing different mathematical techniques of operations research such as mathematical programming, queueing theory, network analysis, dynamic programming, inventory models, simulation, and so on.

For more on the early activities in operations research, see Refs. 1–5. Readers interested in the timeline of major contributions in the history of OR/MS are referred to the excellent review article by Gass [6].

References

1. Haley, K.B., War and peace: the first 25 years of OR in Great Britain, *Operations Research*, 50, Jan.–Feb. 2002.
2. Miser, H.J., The easy chair: what OR/MS workers should know about the early formative years of their profession, *Interfaces*, 30, March–April 2000.
3. Trefethen, F.N., A history of operations research, in *Operations Research for Management*, J.F. McCloskey and F.N. Trefethen, Eds., Johns Hopkins Press, Baltimore, MD, 1954.
4. Horner, P., History in the making, *ORMS Today*, 29, 30–39, 2002.
5. Ravindran, A., Phillips, D.T., and Solberg, J.J., *Operations Research: Principles and Practice*, Second Edition, John Wiley & Sons, New York, 1987 (Chapter 1).
6. Gass, S.I., Great moments in histORy, *ORMS Today*, 29, 31–37, 2002.

I

OR/MS Models and Methods

1. **Linear Programming** *Katta G. Murty* 1-1
   Brief History of Algorithms for Solving Linear Equations, Linear Inequalities, and LPs • Applicability of the LP Model: Classical Examples of Direct Applications • LP Models Involving Transformations of Variables • Intelligent Modeling Essential to Get Good Results, an Example from Container Shipping • Planning Uses of LP Models • Brief Introduction to Algorithms for Solving LP Models • Software Systems Available for Solving LP Models • Multiobjective LP Models

2. **Nonlinear Programming** *Theodore B. Trafalis and Robin C. Gilbert*... 2-1
   Introduction • Unconstrained Optimization • Constrained Optimization • Conclusion

3. **Integer Programming** *Michael Weng* 3-1
   Introduction • Formulation of IP Models • Branch and Bound Method • Cutting Plane Method • Other Solution Methods and Computer Solution

4. **Network Optimization** *Mehmet Bayram Yildirim* 4-1
   Introduction • Notation • Minimum Cost Flow Problem • Shortest Path Problem • Maximum Flow Problem • Assignment Problem • Minimum Spanning Tree Problem • Minimum Cost Multicommodity Flow Problem • Conclusions

5. **Multiple Criteria Decision Making** *Abu S. M. Masud and A. Ravi Ravindran* 5-1
   Some Definitions • The Concept of "Best Solution" • Criteria Normalization • Computing Criteria Weights • Multiple Criteria Methods for Finite Alternatives • Multiple Criteria Mathematical Programming Problems • Goal Programming • Method of Global Criterion and Compromise Programming • Interactive Methods • MCDM Applications • MCDM Software • Further Readings

6. **Decision Analysis** *Cerry M. Klein* 6-1
   Introduction • Terminology for Decision Analysis • Decision Making under Risk • Decision Making under Uncertainty • Practical Decision Analysis • Conclusions • Resources

7. **Dynamic Programming** *José A. Ventura* 7-1
   Introduction • Deterministic Dynamic Programming Models • Stochastic Dynamic Programming Models • Conclusions

8. **Stochastic Processes** *Susan H. Xu* 8-1
   Introduction • Poisson Processes • Discrete-Time Markov Chains • Continuous-Time Markov Chains • Renewal Theory • Software Products Available for Solving Stochastic Models

9 **Queueing Theory** *Natarajan Gautam* **9**-1
   Introduction • Queueing Theory Basics • Single-Station and Single-Class Queues
   • Single-Station and Multiclass Queues • Multistation and Single-Class Queues •
   Multistation and Multiclass Queues • Concluding Remarks

10 **Inventory Control** *Farhad Azadivar and Atul Rangarajan* **10**-1
   Introduction • Design of Inventory Systems • Deterministic Inventory Systems •
   Stochastic Inventory Systems • Inventory Control at Multiple Locations • Inventory
   Management in Practice • Conclusions • Current and Future Research

11 **Complexity and Large-Scale Networks** *Hari P. Thadakamalla,*
   Soundar R. T. Kumara and Réka Albert **11**-1
   Introduction • Statistical Properties of Complex Networks • Modeling of Complex
   Networks • Why "Complex" Networks • Optimization in Complex Networks •
   Conclusions

12 **Simulation** *Catherine M. Harmonosky* **12**-1
   Introduction • Basics of Simulation • Simulation Languages and Software •
   Simulation Projects—The Bigger Picture • Summary

13 **Metaheuristics for Discrete Optimization Problems**
   Rex K. Kincaid **13**-1
   Mathematical Framework for Single Solution Metaheuristics • Network Location
   Problems • Multistart Local Search • Simulated Annealing • Plain Vanilla Tabu
   Search • Active Structural Acoustic Control (ASAC) • Nature Reserve Site Selection
   • Damper Placement in Flexible Truss Structures • Reactive Tabu Search •
   Discussion

14 **Robust Optimization** *H. J. Greenberg and Tod Morrison* **14**-1
   Introduction • Classical Models • Robust Optimization Models • More Applications
   • Summary

1
Linear Programming

Katta G. Murty
University of Michigan

1.1	Brief History of Algorithms for Solving Linear Equations, Linear Inequalities, and LPs	**1-1**
1.2	Applicability of the LP Model: Classical Examples of Direct Applications	**1-4**
	Product Mix Problems • Blending Problems • The Diet Problem • The Transportation Model • Multiperiod Production Planning, Storage, Distribution Problems	
1.3	LP Models Involving Transformations of Variables	**1-12**
	Min–Max, Max–Min Problems • Minimizing Positive Linear Combinations of Absolute Values of Affine Functions	
1.4	Intelligent Modeling Essential to Get Good Results, an Example from Container Shipping	**1-18**
1.5	Planning Uses of LP Models	**1-22**
	Finding the Optimum Solutions • Infeasibility Analysis • Values of Slack Variables at an Optimum Solution • Marginal Values, Dual Variables, and the Dual Problem, and Their Planning Uses • Evaluating the Profitability of New Products	
1.6	Brief Introduction to Algorithms for Solving LP Models	**1-26**
	The Simplex Method • Interior Point Methods for LP	
1.7	Software Systems Available for Solving LP Models	**1-31**
1.8	Multiobjective LP Models	**1-31**
	References	**1-34**

1.1 Brief History of Algorithms for Solving Linear Equations, Linear Inequalities, and LPs

The study of mathematics originated with the construction of linear equation models for real world problems several thousand years ago. As an example we discuss an application that leads to a model involving a system of simultaneous linear equations from Murty (2004).

Example 1.1: Scrap Metal Blending Problem

A steel company has four different types of scrap metal (SM-1 to SM-4) with the following compositions (Table 1.1).

The company needs to blend these four scrap metals into a mixture for which the composition by weight is: Al—4.43%, Si—3.22%, C—3.89%, and Fe—88.46%. How should they prepare this mixture? To answer this question, we need to determine the proportions of the four scrap metals SM-1 to SM-4 in the blend to be prepared. ∎

The most fundamental idea in mathematics that was discovered more than 5000 years ago by the Chinese, Indians, Iranians, Babylonians, and Greeks is to represent the quantities that we wish to determine by symbols, usually letters of the alphabet like x, y, z, and then express the relationships between the quantities represented by these symbols in the form of equations, and finally use these equations as tools to find out the true values represented by the symbols. The symbols representing the unknown quantities to be determined are nowadays called *unknowns* or *variables* or *decision variables*. The process of representing the relationships between the variables through equations or other functional relationships is called *modeling* or *mathematical modeling*.

This process gradually evolved into *algebra*, one of the chief branches of mathematics. Even though the subject originated more than 5000 years ago, the name algebra itself came much later; it is derived from the title of an Arabic book *Al-Maqala fi Hisab al-jabr w'almuqabalah* written by Al-Khawarizmi around 825 AD. The term "al-jabr" in Arabic means "restoring" in the sense of solving an equation. In Latin translation the title of this book became *Ludus Algebrae*, the second word in this title surviving as the modern word "algebra" for the subject, and Al-Khawarizmi is regarded as the father of algebra. The earliest algebraic systems constructed are systems of linear equations.

In the scrap metal blending problem, the decision variables are: $x_j =$ proportion of SM-j by weight in the mixture, for $j = 1$–4. Then the percentage by weight of the element Al in the mixture will be $5x_1 + 7x_2 + 2x_3 + x_4$, which is required to be 4.43. Arguing the same way for the elements Si, C, and Fe, we find that the decision variables x_1 to x_4 must satisfy each equation in the following *system of linear equations* to lead to the desired mixture:

$$5x_1 + 7x_2 + 2x_3 + x_4 = 4.43$$
$$3x_1 + 6x_2 + x_3 + 2x_4 = 3.22$$
$$4x_1 + 5x_2 + 3x_3 + x_4 = 3.89$$
$$88x_1 + 82x_2 + 94x_3 + 96x_4 = 88.46$$
$$x_1 + x_2 + x_3 + x_4 = 1$$

The last equation in the system stems from the fact that the sum of the proportions of various ingredients in a blend must always be equal to 1. This system of equations is the mathematical model for our scrap metal blending problem; it consists of five equations

TABLE 1.1 Scrap Metal Composition Data

	% of element by weight, in type			
Type	Al	Si	C	Fe
SM-1	5	3	4	88
SM-2	7	6	5	82
SM-3	2	1	3	94
SM-4	1	2	1	96

in four variables. It is clear that a solution to this system of equations makes sense for the blending application only if all the variables in the system have nonnegative values in it. The nonnegativity restrictions on the variables are *linear inequality constraints*. They cannot be expressed in the form of linear equations, and as nobody knew how to handle linear inequalities at that time, they ignored them.

Linear algebra dealing with methods for solving systems of linear equations is the classical subject that initiated the study of mathematics a long time ago. The most effective method for solving systems of linear equations, called the *elimination method*, was discovered by the Chinese and the Indians over 2500 years ago and this method is still the leading method in use today. This elimination method was unknown in Europe until the nineteenth century when the German mathematician Gauss rediscovered it while calculating the orbit of the asteroid Ceres based on recorded observations in tracking it. The asteroid was lost from view when the Sicilian astronomer Piazzi tracking it fell ill. Gauss used the method of least squares to estimate the values of the parameters in the formula for the orbit. It led to a system of 17 linear equations in 17 unknowns that he had to solve, which is quite a large system for mental computation. Gauss's accurate computations helped in relocating the asteroid in the skies in a few months' time, and his reputation as a mathematician soared. Another German, Wilhelm Jordan, popularized the algorithm in a late nineteenth-century book that he wrote. From that time, the method has been popularly known as the Gauss–Jordan elimination method. Another version of this method, called the Gaussian elimination method, is the most popular method for solving systems of linear equations today.

Even though linear equations were resolved thousands of years ago, systems of linear inequalities remained unsolved until the middle of the twentieth century. The following theorem (Murty, 2006) relates systems of linear inequalities to systems of linear equations.

THEOREM 1.1 *Consider the system of linear inequalities in variables x*

$$A_{i.}x \geq b_i, \quad i = 1, \ldots, m \tag{1.1}$$

where $A_{i.}$ is the coefficient vector for the i-th constraint. If this system has a feasible solution, then there exists a subset $\boldsymbol{P} = \{p_1, \ldots, p_s\} \subset \{1, \ldots, m\}$ such that every solution of the system of equations: $A_{i.}x = b_i$, $i \in \boldsymbol{P}$ is also a feasible solution of the original system of linear inequalities (Equation 1.1).

This theorem can be used to generate a finite enumerative algorithm to find a feasible solution to a system of linear constraints containing inequalities, based on solving subsystems in each of which a subset of the inequalities are converted into equations and the other inequality constraints are eliminated. However, if the original system has m inequality constraints, in the worst case this enumerative algorithm may have to solve 2^m systems of linear equations before it either finds a feasible solution of the original system, or concludes that it is infeasible. The effort required grows exponentially with the number of inequalities in the system in the worst case.

In the nineteenth century, Fourier generalized the classical elimination method for solving linear equations into an elimination method for solving systems of linear inequalities. The method called *Fourier elimination*, or the *Fourier–Motzkin elimination method*, is very elegant theoretically. However, the elimination of each variable adds new inequalities to the remaining system, and the number of these new inequalities grows exponentially as more and more variables are eliminated. So this method is also not practically viable for large problems.

The simplex method for linear programming developed by Dantzig (1914–2005) in the mid-twentieth century (Dantzig, 1963) is the first practically and computationally viable method for solving systems of linear inequalities. This has led to the development of *linear programming* (LP), a branch of mathematics developed in the twentieth century as an extension of linear algebra to solve systems of *linear inequalities*. The development of LP is a landmark event in the history of mathematics and its applications that brought our ability to solve general systems of linear constraints (including linear equations, inequalities) to a state of completion.

A general system of linear constraints in decision variables $x = (x_1, \ldots, x_n)^T$ is of the form: $Ax \geq b$, $Dx = d$, where the coefficient matrices A, D are given matrices of orders $m \times n$, $p \times n$, respectively. The inequality constraints in this system may include sign restrictions or bounds on individual variables.

A general LP is the problem of finding an optimum solution for the problem of minimizing (or maximizing) a given linear objective function $z = cx$ say, subject to a system of linear constraints.

Suppose there is no objective function to optimize, and only a feasible solution of a system of linear constraints is to be found. When there are inequality constraints in the system, the only practical method to even finding a feasible solution is to solve a linear programming formulation of it as a Phase I linear programming problem. Dantzig developed this Phase I formulation as part of the simplex method for LPs that he developed in the mid-twentieth century.

1.2 Applicability of the LP Model: Classical Examples of Direct Applications

LP has now become a dominant subject in the development of efficient computational algorithms, the study of convex polyhedra, and in algorithms for decision making. But for a short time in the beginning, its potential was not well recognized. Dantzig tells the story of how when he gave his first talk on LP and his simplex method for solving it at a professional conference, Hotelling (a burly person who liked to swim in the sea; the popular story about him was that when he does, the level of the ocean rises perceptibly) dismissed it as unimportant since everything in the world is nonlinear. But Von Neumann came to the defense of Dantzig saying that the subject will become very important (Dantzig and Thapa, 1997, vol. 1, p. xxvii). The preface in this book contains an excellent account of the early history of LP from the inventor of the most successful method in OR and in the mathematical theory of polyhedra.

Von Neumann's early assessment of the importance of LP turned out to be astonishingly correct. Today, the applications of LP in almost all areas of science are numerous. The LP model is suitable for modeling a real world decision-making problem if

- All the decision variables are continuous variables
- There is a single objective function that is required to be optimized
- The objective function and all the constraint functions defining the constraints in the problem are linear functions of the decision variables (i.e., they satisfy the usual *proportionality* and *additivity assumptions*)

There are many applications in which the reasonableness of the linearity assumptions can be verified and an LP model for the problem constructed by direct arguments. We present some classical applications like this in this section; this material is from Murty (1995, 2005b).

Linear Programming

In all these applications you can judge intuitively that the assumptions needed to handle them using an LP model are satisfied to a reasonable degree of approximation.

Of course LP can be applied to a much larger class of problems. Many important applications involve optimization models in which a nonlinear objective function that is piecewise linear and convex is to be minimized subject to linear constraints. These problems can be transformed into LPs by introducing additional variables. These transformations are discussed in the next section.

1.2.1 Product Mix Problems

This is an extremely important class of problems that manufacturing companies face. Normally the company can make a variety of products using the raw materials, machinery, labor force, and other resources available to them. The problem is to decide how much of each product to manufacture in a period, to maximize the total profit subject to the availability of needed resources.

To model this, we need data on the units of each resource necessary to manufacture one unit of each product, any bounds (lower, upper, or both) on the amount of each product manufactured per period, any bounds on the amount of each resource available per period, the expected demand for each product, and the cost or net profit per unit of each product manufactured.

Assembling this type of reliable data is one of the most difficult jobs in constructing a product mix model for a company, but it is very worthwhile. The process of assembling all the needed data is sometimes called the *input–output analysis* of the company. The coefficients, which are the resources necessary to make a unit of each product, are called *input–output (I/O) coefficients*, or *technology coefficients*.

Example 1.2: The Fertilizer Product Mix Problem

As an example, consider a fertilizer company that makes two kinds of fertilizers called Hi-phosphate (Hi-ph) and Lo-phosphate (Lo-ph). The manufacture of these fertilizers requires three raw materials called RM 1, RM 2, and RM 3. At present their supply of these raw materials comes from the company's own quarry that is only able to supply maximum amounts of 1500, 1200, 500 tons/day, respectively, of RM 1, RM 2, and RM 3. Even though there are other vendors who can supply these raw materials if necessary, at the moment they are not using these outside suppliers.

They sell their output of Hi-ph and Lo-ph fertilizers to a wholesaler who is willing to buy any amount that they can produce, so there are no upper bounds on the amounts of Hi-ph and Lo-ph manufactured daily.

At the present rates of operation their Cost Accounting Department estimates that it is costing the quarry $50, $40, and $60/ton respectively to produce and deliver RM 1, RM 2, and RM 3 at the fertilizer plant. Also, at the present rates of operation, all other production costs (for labor, power, water, maintenance, depreciation of plant and equipment, floor space, insurance, shipping to the wholesaler, etc.) come to $7/ton to manufacture Hi-ph or Lo-ph and deliver to the wholesaler.

The sale price of the manufactured fertilizers to the wholesaler fluctuates daily, but their averages over the last one month have been $222 and $107 per ton, respectively, for Hi-Ph and Lo-ph fertilizers. We will use these prices to construct the mathematical model. ∎

The Hi-ph manufacturing process needs as inputs 2 tons RM 1 and 1 ton each of RM 2 and RM 3 for each ton of Hi-ph manufactured. Similarly, the Lo-ph manufacturing process needs as inputs 1 ton RM 1 and 1 ton of RM 2 for each ton of Lo-ph manufactured.

So, the net profit/ton of fertilizer manufactured is $(222 - 2 \times 50 - 1 \times 40 - 1 \times 60 - 7) = 15$ and $(107 - 1 \times 50 - 1 \times 40 - 7) = 10$, respectively, for Hi-ph and Lo-ph.

There are clearly two decision variables in this problem; these are: $x_1 =$ the tons of Hi-ph produced per day, x_2 the tons of Lo-ph produced per day. Associated with each variable in the problem is an *activity* that the decision maker can perform. The activities in this example are: Activity 1: to make 1 ton of Hi-ph, Activity 2: to make 1 ton of Lo-ph. The variables in the problem just define the *levels* at which these activities are carried out.

As all the data are given on a per ton basis, they provide an indication that the linearity assumptions are quite reasonable in this problem. Also, the amount of each fertilizer manufactured can vary continuously within its present range. So, LP is an appropriate model for this problem.

Each raw material leads to a constraint in the model. The amount of RM 1 used is $2x_1 + x_2$ tons, and it cannot exceed 1500, leading to the constraint $2x_1 + x_2 \leq 1500$. As this inequality compares the amount of RM 1 used to the amount available, it is called a *material balance inequality*. All goods that lead to constraints in the model for the problem are called *items*. The material balance equations or inequalities corresponding to the various items are the constraints in the problem. When the objective function and all the constraints are obtained, the formulation of the problem as an LP is complete. The LP formulation of the fertilizer product mix problem is given below.

$$\begin{array}{lrll}
\text{Maximize } p(x) = 15x_1 + 10x_2 & & & \text{Item} \\
\text{subject to} & 2x_1 + x_2 & \leq 1500 & \text{RM 1} \\
& x_1 + x_2 & \leq 1200 & \text{RM 2} \\
& x_1 & \leq 500 & \text{RM 3} \\
& x_1 \geq 0, x_2 \geq & 0 &
\end{array} \quad (1.2)$$

Real world product mix models typically involve large numbers of variables and constraints, but their structure is similar to that in this small example.

1.2.2 Blending Problems

This is another large class of problems in which LP is applied heavily. Blending is concerned with mixing different materials called the *constituents* of the mixture (these may be chemicals, gasolines, fuels, solids, colors, foods, etc.) so that the mixture conforms to specifications on several properties or characteristics.

To model a blending problem as an LP, the *linear blending assumption* must hold for each property or characteristic. This implies that the value for a characteristic of a mixture is the weighted average of the values of that characteristic for the constituents in the mixture, the weights being the proportions of the constituents. As an example, consider a mixture consisting of four barrels of fuel 1 and six barrels of fuel 2, and suppose the characteristic of interest is the octane rating (Oc.R). If linear blending assumption holds, the Oc.R of the mixture will be equal to (4 times the Oc.R of fuel 1 + 6 times the Oc.R of fuel 2)/(4 + 6).

The linear blending assumption holds to a reasonable degree of precision for many important characteristics of blends of gasolines, crude oils, paints, foods, and so on. This makes it possible for LP to be used extensively in optimizing gasoline blending, in the manufacture of paints, cattle feed, beverages, and so on.

The decision variables in a blending problem are usually either the quantities or the proportions of the constituents in the blend. If a specified quantity of the blend needs to be made, then it is convenient to take the decision variables to be the quantities of the

Linear Programming

various constituents blended; in this case one must include the constraint that the sum of the quantities of the constituents is equal to the quantity of the blend desired.

If there is no restriction on the amount of blend made, but the aim is to find an optimum composition for the mixture, it is convenient to take the decision variables to be the proportions of the various constituents in the blend; in this case one must include the constraint that the sum of all these proportions is 1.

We provide a gasoline blending example. There are more than 300 refineries in the United States processing a total of more than 20 million barrels of crude oil daily. Crude oil is a complex mixture of chemical components. The refining process separates crude oil into its components that are blended into gasoline, fuel oil, asphalt, jet fuel, lubricating oil, and many other petroleum products. Refineries and blenders strive to operate at peak economic efficiencies, taking into account the demand for various products. To keep the example simple, we consider only one characteristic of the mixture, the Oc.R of the blended fuels in this example. In actual application there are many other characteristics to be considered also.

A refinery takes four raw gasolines and blends them to produce three types of fuel. The company sells raw gasoline not used in making fuels at \$38.95/barrel if its Oc.R is >90, and at \$36.85/barrel if its Oc.R is ≤90. The cost of handling raw gasolines purchased and blending them into fuels or selling them as is is estimated to be \$2 per barrel by the Cost Accounting Department. Other data are given in Table 1.2.

The problem is to determine how much raw gasoline of each type to purchase, the blend to use for the three fuels, and the quantities of these fuels to make to maximize total daily net profit.

We will use the quantities of the various raw gasolines in the blend for each fuel as the decision variables, and we assume that the linear blending assumption holds for the Oc.R. Let

RG_i = raw gasoline type i to purchase/day, $i = 1$–4

$x_{ij} = \begin{cases} \text{barrels of raw gasoline type } i \text{ used in making fuel} \\ \text{type } j \text{ per day, } i = 1 \text{ to } 4, \quad j = 1, 2, 3 \end{cases}$

y_i = barrels of raw gasoline type i sold as is/day

F_j = barrels of fuel type j made/day, $j = 1, 2, 3$

So, the total amount of fuel type 1 made daily is $F_1 = x_{11} + x_{21} + x_{31} + x_{41}$. If this is >0, by the linear blending assumption its Oc.R will be $(68x_{11} + 86x_{21} + 91x_{31} + 99x_{41})/F_1$. This is required to be ≥ 95. So, the Oc.R. constraint on fuel type 1 can be represented by the linear constraint: $68x_{11} + 86x_{21} + 91x_{31} + 99x_{41} - 95F_1 \geq 0$. Proceeding in a similar manner, we obtain the following LP formulation for this problem.

TABLE 1.2 Data for the Fuel Blending Problem

Raw gas type	Octane rating (Oc.R)	Available daily (barrels)	Price per barrel
1	68	4000	\$31.02
2	86	5050	33.15
3	91	7100	36.35
4	99	4300	38.75

Fuel type	Minimum Oc.R	Selling price (\$) (barrel)	Demand
1	95	47.15	At most 10,000 barrels/day
2	90	44.95	No limit
3	85	42.99	At least 15,000 barrels/day

$$\text{Maximize} \quad 47.15F_1 + 44.95F_2 + 42.99F_3 + y_1(36.85 - 31.02)$$
$$+ y_2(36.85 - 33.15) + y_3(38.95 - 36.35) + y_4(38.95$$
$$- 38.75) - (31.02 + 2)RG_1 - (33.15 + 2)RG_2$$
$$- (36.35 + 2)RG_3 - (38.75 + 2)RG_4$$

subject to
$$RG_i = x_{i1} + x_{i2} + x_{i3} + y_i, \quad i = 1,\ldots,4$$
$$0 \leq (RG_1, RG_2, RG_3, RG_4) \leq (4000, 5050, 7100, 4300)$$
$$Fj = x_{1j} + x_{2j} + x_{3j} + x_{4j}, \quad j = 1, 2, 3$$
$$0 \leq F_1 \leq 10,000$$
$$F_3 \leq 15,000$$
$$68x_{11} + 86x_{21} + 91x_{31} + 99x_{41} - 95F_1 \geq 0$$
$$68x_{12} + 86x_{22} + 91x_{32} + 99x_{42} - 90F_2 \geq 0$$
$$68x_{13} + 86x_{23} + 91x_{33} + 99x_{43} - 85F_3 \geq 0$$
$$F_2 \geq 0, \quad x_{ij}, y_i \geq 0, \quad \text{for all } i, j$$

Blending models are economically significant in the petroleum industry. The blending of gasoline is a very popular application. A single grade of gasoline is normally blended from about 3 to 10 individual components, none of which meets the quality specifications by itself. A typical refinery might have 20 different components to be blended into four or more grades of gasoline and other petroleum products such as aviation gasoline, jet fuel, and middle distillates, differing in Oc.R and properties such as pour point, freezing point, cloud point, viscosity, boiling characteristics, vapor pressure, and so on, by marketing region.

1.2.3 The Diet Problem

A *diet* has to satisfy many constraints; the most important is that it should be palatable (i.e., be tasty) to the one eating it. This is a very difficult constraint to model mathematically, particularly if the diet is for a human individual. So, early publications on the diet problem have ignored this constraint and concentrated on meeting the minimum daily requirement (MDR) of each nutrient identified as being important for the individual's well-being. Also, these days most of the applications of the diet problem are in the farming sector, and farm animals and birds are usually not very fussy about what they eat.

The diet problem is one among the earliest problems formulated as an LP. The first paper on it was by Stigler (1945). Those were the war years, food was expensive, and the problem of finding a minimum cost diet was of more than academic interest. Nutrition science was in its infancy in those days, and after extensive discussions with nutrition scientists, Stigler identified nine essential nutrient groups for his model. His search of the grocery shelves yielded a list of 77 different available foods. With these, he formulated a diet problem that was an LP involving 77 nonnegative decision variables subject to 9 inequality constraints.

Stigler did not know of any method for solving his LP model at that time, but he obtained an approximate solution using a trial and error search procedure that led to a diet meeting the MDR of the nine nutrients considered in the model at an annual cost of $39.93 at 1939 prices! After Dantzig developed the simplex method for solving LPs in 1947, Stigler's diet problem was one of the first nontrivial LPs to be solved by the simplex method on a computer, and it gave the true optimum diet with an annual cost of $39.67 at 1939 prices. So, the trial and error solution of Stigler was very close to the optimum.

The Nobel prize committee awarded the 1982 Nobel prize in economics to Stigler for his work on the diet problem and later work on the functioning of markets and the causes and effects of public regulation.

Linear Programming

TABLE 1.3 Data on the Nutrient Content of Grains

Nutrient	Nutrient units/kg of grain type		MDR of nutrient in units
	1	2	
Starch	5	7	8
Protein	4	2	15
Vitamins	2	1	3
Cost ($/kg.) of food	0.60	0.35	

The data in the diet problem consist of a list of nutrients with the MDR for each; a list of available foods with the price and composition (i.e., information on the number of units of each nutrient in each unit of food) of every one of them; and the data defining any other constraints the user wants to place on the diet. As an example we consider a very simple diet problem in which the nutrients are starch, protein, and vitamins as a group; the foods are two types of grains with data given in Table 1.3.

The activities and their levels in this model are: activity j: to include 1 kg of grain type j in the diet, associated level $= x_j$, for $j = 1, 2$. The items in this model are the various nutrients, each of which leads to a constraint. For example, the amount of starch contained in the diet x is $5x_1 + 7x_2$, which must be ≥ 8 for feasibility. This leads to the formulation given below.

$$\begin{array}{ll}
\text{Minimize } z(x) = 0.60x_1 + 0.35x_2 & \underline{\text{Item}} \\
\text{subject to} \quad 5x_1 + 7x_2 \geq 8 & \text{Starch} \\
\quad 4x_1 + 2x_2 \geq 15 & \text{Protein} \\
\quad 2x_1 + x_2 \geq 3 & \text{Vitamins} \\
\quad x_1 \geq 0, \quad x_2 \geq 0 &
\end{array}$$

Nowadays almost all the companies in the business of making feed for cattle, other farm animals, birds, and the like use LP extensively to minimize their production costs. The prices and supplies of various grains, hay, and so on are constantly changing, and feed makers solve the diet model frequently with new data values, to make their buy-decisions and to formulate the optimum mix for manufacturing the feed. ■

Once I met a farmer at a conference discussing commercial LP software systems. He operates reasonable size cattle and chicken farms. He was carrying his laptop with him. He told me that in the fall harvest season, he travels through agricultural areas extensively. He always has his laptop with LP-based diet models for the various cattle and chicken feed formulations inside it. He told me that before accepting an offer from a farm on raw materials for the feed, he always uses his computer to check whether accepting this offer would reduce his overall feed costs or not, using a sensitivity analysis feature in the LP software in his computer. He told me that this procedure has helped him save his costs substantially.

1.2.4 The Transportation Model

An essential component of our modern life is the shipping of goods from where they are produced to markets worldwide. Nationally, within the United States alone transportation of goods is estimated to cost over 1 trillion/year. The aim of this problem is to find a way of carrying out this transfer of goods at minimum cost. Historically, it was among the first LPs to be modeled and studied. The Russian economist L. V. Kantorovitch studied this problem in the 1930s and developed the dual simplex method for solving it, and published a book on it, *Mathematical Methods in the Organization and Planning of Production*, in

TABLE 1.4 Data for the Transportation Problem

	c_{ij} (cents/ton)			Availability at
	$j=1$	2	3	mine (tons) daily
Mine $i=1$	11	8	2	**800**
2	7	5	4	**300**
Requirement at plant (tons) daily	**400**	**500**	**200**	

Russian in 1939. In the United States, (Hitchcock, 1941) developed an algorithm similar to the primal simplex algorithm for finding an optimum solution to the transportation problem. And (Koopmans, 1949) developed an optimality criterion for a basic solution to the transportation problem in terms of the dual basic solution (discussed later on). The early work of L. V. Kantorovitch and T. C. Koopmans in these publications was part of their effort for which they received the 1975 Nobel prize for economics.

The classical single commodity transportation problem is concerned with a set of nodes or places called *sources* that have a commodity available for shipment, and another set of places called *sinks* or *demand centers* or *markets* that require this commodity. The data consists of the *availability* at each source (the amount available there to be shipped out), the *requirement* at each market, and the cost of transporting the commodity per unit from each source to each market. The problem is to determine the quantity to be transported from each source to each market so as to meet the requirements at minimum total shipping cost.

As an example, we consider a small problem where the commodity is iron ore, the sources are mines 1 and 2 that produce the ore, and the markets are three steel plants that require the ore. Let c_{ij} = cost (cents per ton) to ship ore from mine i to steel plant j, $i = 1, 2, j = 1, 2, 3$. The data are given in Table 1.4. To distinguish between different data elements, we show the cost data in normal size letters, and the supply and requirement data in bold face letters.

The decision variables in this model are: x_{ij} = ore (in tons) shipped from mine i to plant j. The items in this model are the ore at various locations. We have the following LP formulation for this problem.

$$\text{Minimize } z(x) = 11x_{11} + 8x_{12} + 2x_{13} + 7x_{21} + 5x_{22} + 4x_{23} \quad \text{Item}$$

$$\text{Ore at}$$

$$\text{subject to} \quad x_{11} + x_{12} + x_{13} = 800 \quad \text{mine 1}$$
$$x_{21} + x_{22} + x_{23} = 300 \quad \text{mine 2}$$
$$x_{11} + x_{21} = 400 \quad \text{plant 1}$$
$$x_{12} + x_{22} = 500 \quad \text{plant 2}$$
$$x_{13} + x_{23} = 200 \quad \text{plant 3}$$

$$x_{ij} \geq 0 \quad \text{for all } i = 1, 2, \ j = 1, 2, 3$$

Let G denote the directed network with the sources and sinks as nodes, and the various routes from each source to each sink as the arcs. Then this problem is a single commodity minimum cost flow problem in G. So, the transportation problem is a special case of single commodity minimum cost flow problems in directed networks. Multicommodity flow problems are generalizations of these problems involving two or more commodities.

The model that we presented for the transportation context is of course too simple. Real world transportation problems have numerous complicating factors, both in the constraints to be satisfied and the objective functions to optimize, that need to be addressed. Starting

Linear Programming

1.2.5 Multiperiod Production Planning, Storage, Distribution Problems

The LP model finds many applications for making production allocation, planning, storage, and distribution decisions in companies. Companies usually like to plan ahead; when they are planning for one period, they usually like to consider also a few periods into the future. This leads to multiperiod planning problems.

To construct a mathematical model for a multiperiod horizon, we need reliable data on the expected production costs, input material availabilities, production capacities, demand for the output, selling prices, and the like in each period. With economic conditions changing rapidly and unexpectedly these days, it is very difficult to assemble reliable data on such quantities beyond a few periods from the present. That is why multiperiod models used in practice usually cover the current period and a few periods following it, for which data can be estimated with reasonable precision.

For example, consider the problem of planning the production, storage, and marketing of a product whose demand and selling price vary seasonally. An important feature in this situation is the profit that can be realized by manufacturing the product in seasons during which the production costs are low, storing it, and putting it in the market when the selling price is high. Many products exhibit such seasonal behavior, and companies and businesses take advantage of this feature to augment their profits. A linear programming formulation of this problem has the aim of finding the best production-storage-marketing plan over the planning horizon, to maximize the overall profit. For constructing a model for this problem we need reasonably good estimates of the demand and the expected selling price of the product in each period of the planning horizon; availability and cost of raw materials, labor, machine times, etc. necessary to manufacture the product in each period; and the availability, and cost of storage space.

As an example, we consider the simple problem of a company making a product subject to such seasonal behavior. The company needs to make a production plan for the coming year, divided into six periods of 2 months each, to maximize net profit (= sales revenue − production and storage costs). Relevant data are in Table 1.5. The production cost there includes the cost of raw material, labor, machine time, and the like, all of which fluctuate from period to period. And the production capacity arises due to limits on the availability of raw material and hourly labor.

Product manufactured during a period can be sold in the same period, or stored and sold later on. Storage costs are $2/ton of product from one period to the next. Operations begin in period 1 with an initial stock of 500 tons of the product in storage, and the company would like to end up with the same amount of the product in storage at the end of period 6.

TABLE 1.5 Data for the 6-Period Production Planning Problem

Period	Total Production cost ($/ton)	Production capacity (tons)	Demand (tons)	Selling price ($/ton)
1	20	1500	1100	180
2	25	2000	1500	180
3	30	2200	1800	250
4	40	3000	1600	270
5	50	2700	2300	300
6	60	2500	2500	320

The decision variables in this problem are, for period $j = 1$–6

x_j = product made (tons) during period j
y_j = product left in storage (tons) at the end of period j
z_j = product sold (tons) during period j

In modeling this problem the important thing to remember is that inventory equations (or material balance equations) must hold for the product for each period. For period j this equation expresses the following fact.

$$\left. \begin{array}{l} \text{Amount of product in storage} \\ \text{at the beginning of period } j + \\ \text{the amount manufactured} \\ \text{during period } j \end{array} \right\} = \left\{ \begin{array}{l} \text{Amount of product sold during} \\ \text{period } j + \text{the amount left in} \\ \text{storage at the end of period } j \end{array} \right.$$

The LP model for this problem is given below:

$$\begin{aligned}
\text{Maximize } & 180(z_1 + z_2) + 250z_3 + 270z_4 + 300z_5 + 320z_6 \\
& - 20x_1 - 25x_2 - 30x_3 - 40x_4 - 50x_5 - 60x_6 \\
& - 2(y_1 + y_2 + y_3 + y_4 + y_5 + y_6)
\end{aligned}$$

subject to $x_j, y_j, z_j \geq 0$ for all $j = 1$ to 6
$x_1 \leq 1500, x_2 \leq 2000, x_3 \leq 2200, x_4 \leq 3000, x_5 \leq 2700, x_6 < 2500$
$z_1 \leq 1100, z_2 \leq 1500, z_3 \leq 1800, z_4 \leq 1600, z_5 \leq 2300, z_6 \leq 2500$

$$500 + x_1 - (y_1 + z_1) = 0$$
$$y_1 + x_2 - (y_2 + z_2) = 0$$
$$y_2 + x_3 - (y_3 + z_3) = 0$$
$$y_3 + x_4 - (y_4 + z_4) = 0$$
$$y_4 + x_5 - (y_5 + z_5) = 0$$
$$y_5 + x_6 - (y_6 + z_6) = 0$$
$$y_6 = 500$$

Many companies have manufacturing facilities at several locations, and usually make and sell several different products. Production planning at such companies involves production allocation decisions (what products will each facility manufacture) and transportation-distribution decisions (plan to ship the output of each facility to each market) in addition to the material balance constraints of the type discussed in the example above for each product and facility.

1.3 LP Models Involving Transformations of Variables

In this section, we will extend the range of application of LP to include problems that can be modeled as those of optimizing a convex piecewise linear objective function subject to linear constraints. These problems can be transformed easily into LPs in terms of additional variables. This material is from Murty (under preparation).

Let $\theta(\lambda)$ be a real valued function of a single variable $\lambda \in R^1$. $\theta(\lambda)$ is said to be a *piecewise linear* (PL) *function* if it is continuous and if there exists a partition of R^1 into intervals

Linear Programming

TABLE 1.6 The PL Function $\theta(\lambda)$

Interval for λ	Slope	Value of $\theta(\lambda)$
$\lambda \leq \lambda_1$	c_1	$c_1 \lambda$
$\lambda_1 \leq \lambda \leq \lambda_2$	c_2	$\theta(\lambda_1) + c_2(\lambda - \lambda_1)$
$\lambda_2 \leq \lambda \leq \lambda_3$	c_3	$\theta(\lambda_2) + c_3(\lambda - \lambda_2)$
$\vdots$	$\vdots$	$\vdots$
$\lambda \geq \lambda_r$	c_{r+1}	$\theta(\lambda_r) + c_{r+1}(\lambda - \lambda_r)$

of the form $[-\infty, \lambda_1] = \{\lambda \leq \lambda_1\}$, $[\lambda_1, \lambda_2], \ldots, [\lambda_{r-1}, \lambda_r]$, $[\lambda_r, \infty]$ (where $\lambda_1 < \lambda_2 < \cdots < \lambda_r$ are the breakpoints in this partition) such that inside each interval the slope of $\theta(\lambda)$ is a constant. If these slopes in the various intervals are $c_1, c_2, \ldots, c_{r+1}$, the values of this function at various values of λ are tabulated in Table 1.6.

This PL function is said to be *convex* if its slope is monotonic increasing with λ, that is, if $c_1 < c_2 \cdots < c_{r+1}$. If this condition is not satisfied it is nonconvex. Here are some numerical examples of PL functions of the single variable λ (Tables 1.7 and 1.8).

Example 1.3: PL Function $\theta(\lambda)$

TABLE 1.7 PL Convex Function $\theta(\lambda)$

Interval for λ	Slope	Value of $\theta(\lambda)$
$\lambda \leq 10$	3	3λ
$10 \leq \lambda \leq 25$	5	$30 + 5(\lambda - 10)$
$\lambda \geq 25$	9	$105 + 9(\lambda - 25)$

Example 1.4: PL Function $g(\lambda)$

TABLE 1.8 Nonconvex PL Function $\theta(\lambda)$

Interval for λ	Slope	Value of $\theta(\lambda)$
$\lambda \leq 100$	10	10λ
$100 \leq \lambda \leq 300$	5	$1000 + 5(\lambda - 100)$
$300 \leq \lambda \leq 1000$	11	$2000 + 11(\lambda - 300)$
$\lambda \geq 1000$	20	$9700 + 20(\lambda - 1000)$

Both functions $\theta(\lambda)$, $g(\lambda)$ are continuous functions and PL functions. $\theta(\lambda)$ is convex because its slope is monotonic increasing, but $g(\lambda)$ is not convex as its slope is not monotonic increasing with λ.

A PL function $h(\lambda)$ of the single variable $\lambda \in R^1$ is said to be a *PL concave function* iff $-h(\lambda)$ is a PL convex function; that is, iff the slope of $h(\lambda)$ is monotonic decreasing as λ increases.

PL Functions of Many Variables

Let $f(x)$ be a real valued function of $x = (x_1, \ldots, x_n)^T$. $f(x)$ is said to be a PL (piecewise linear) function of x if there exists a partition of R^n into convex polyhedral regions $K_1, \ldots, K_r$ such that $f(x)$ is linear within each K_t, for $t = 1$ to r; and a *PL convex function* if it is also convex. It can be proved mathematically that $f(x)$ is a PL convex function iff there exists a finite number, r say, of linear (more precisely affine) functions $c_0^t + c^t x$, (where c_0^t, and $c^t \in R^n$ are the given coefficient vectors for the t-th linear function) $t = 1$ to r such that for each $x \in R^n$

$$f(x) = \text{Maximum}\{c_0^t + c^t x : t = 1, \ldots, r\} \tag{1.3}$$

A function $f(x)$ defined by Equation 1.3 is called the *pointwise maximum (or supremum) function* of the linear functions $c_0^t + c^t x$, $t = 1$ to r. PL convex functions of many variables that do not satisfy the additivity hypothesis always appear in this form (Equation 1.3) in real world applications.

Similarly, a PL function $h(x)$ of $x = (x_1, \ldots, x_n)^T$ is said to be a *PL concave function* if there exist a finite number s of affine functions $d_0^t + d^t x$, $t = 1$ to s, such that $h(x)$ is their *pointwise infimum*, that is, for each $x \in R^n$

$$h(x) = \text{Minimum}\{d_0^t + d^t x : t = 1, \ldots, s\}$$

Now we show how to transform various types of problems of minimizing a PL convex function subject to linear constraints into LPs, and applications of these transformations.

Minimizing a Separable PL Convex Function Subject to Linear Constraints

A real valued function $z(x)$ of variables $x = (x_1, \ldots, x_n)^T$ is said to be *separable* if it satisfies the additivity hypothesis, that is, if it can be written as the sum of n functions, each one involving only one variable as in: $z(x) = z_1(x_1) + z_2(x_2) + \cdots + z_n(x_n)$. Consider the following general problem of this type:

$$\begin{aligned}
\text{Minimize} \quad & z(x) = z_1(x_1) + \cdots + z_n(x_n) \\
\text{subject to} \quad & Ax = b \\
& x \geq 0
\end{aligned} \tag{1.4}$$

where each $z_j(x_j)$ is a PL convex function with slopes in intervals as in Table 1.9, $r_j + 1$ is the number of different slopes $c_{j1} < c_{j2} < \cdots < c_{j,r_j+1}$ of $z_j(x_j)$, and $\ell_{j1}, \ell_{j2}, \ldots, \ell_{j,r_j+1}$ are the lengths of the various intervals in which these slopes apply.

As the objective function to be minimized does not satisfy the proportionality assumption, this is not an LP. However, the convexity property can be used to transform this into an LP by introducing additional variables. This transformation expresses each variable x_j as a sum of $r_j + 1$ variables, one associated with each interval in which its slope is constant. Denoting these variables by $x_{j1}, x_{j2}, \ldots, x_{j,r_j+1}$, the variable x_j becomes $= x_{j1} + \cdots + x_{j,r_j+1}$ and $z_j(x_j)$ becomes the linear function $c_{j1} x_{j1} + \cdots + c_{j,r_j+1} x_{j,r_j+1}$ in terms of the new variables. The reason for this is that as the slopes are monotonic (i.e., $c_{j1} < c_{j2} < \cdots < c_{j,r_j+1}$), for any value of $\bar{x}_j \geq 0$, if $(\bar{x}_{j1}, \bar{x}_{j2}, \ldots, \bar{x}_{j,r_j+1})$ is an optimum

$$\begin{aligned}
\text{Minimize} \quad & c_{j1} x_{j1} + \cdots + c_{j,r_j+1} x_{j,r_j+1} \\
\text{Subject to} \quad & x_{j1} + \cdots + x_{j,r_j+1} = \bar{x}_j \\
& 0 \leq x_{jt} < \ell_{jt}, \quad t = 1, \ldots, r_j + 1
\end{aligned}$$

TABLE 1.9 The PL Function $Z_j(x_j)$

Interval	Slope in interval	Value of $z_j(x_j)$	Length of interval
$0 \leq x_j \leq k_{j1}$	c_{j1}	$c_{j1} x_j$	$\ell_{j1} = k_{j1}$
$k_{j1} \leq x_j \leq k_{j2}$	c_{j2}	$z_j(k_{j1}) + c_{j2}(x_j - k_{j1})$	$\ell_{j2} = k_{j2} - k_{j1}$
$\vdots$	$\vdots$	$\vdots$	$\vdots$
$k_{j,r_j} \leq x_j$	c_{j,r_j+1}	$z_j(k_{j,r_j}) + c_{j,r_j+1}(x_j - k_{j,r_j})$	$\ell_{j,r_j+1} = \infty$

Linear Programming

$\bar{x}_{j,t+1}$ will not be positive unless $\bar{x}_{jk} = \ell_{jk}$ for $k=1$ to t, for each $t=1$ to r_j. Hence the optimum objective value in this problem will be equal to $z_j(x_j)$. This shows that our original problem (Equation 1.4) is equivalent to the following transformed problem which is an LP.

$$\text{Minimize} \quad \sum_{j=1}^{j=n} \sum_{t=1}^{t=r_j+1} c_{jt} x_{jt}$$

$$\text{subject to} \quad x_j = \sum_{t=1}^{t=r_j+1} x_{jt}, \quad j = 1, \ldots, n$$

$$Ax = b$$

$$0 \leq x_{jt} \leq \ell_{jt}, \quad t = 1, \ldots, r_j + 1; \quad j = 1, \ldots, n$$

$$x \geq 0.$$

The same type of transformation can be used to transform a problem involving the maximization of a separable PL concave function subject to linear constraints into an LP.

Example 1.5

A company makes products P_1, P_2, P_3 using limestone (LI), electricity (EP), water (W), fuel (F), and labor (L) as inputs. Labor is measured in man hours, and other inputs in suitable units. ∎

Each input is available from one or more sources. The company has its own quarry for LI, which can supply up to 250 units/day at a cost of \$20/unit. Beyond that, LI can be purchased in any amounts from an outside supplier at \$50/unit. EP is only available from the local utility. Their charges for EP are: \$30/unit for the first 1000 units/day, \$45/unit for up to an additional 500 units/day beyond the initial 1000 units/day, \$75/unit for amounts beyond 1500 units/day. Up to 800 units/day of water is available from the local utility at \$6/unit; beyond that they charge \$7/unit of water/day. There is a single supplier for F who can supply at most 3000 units/day at \$40/unit; beyond that there is currently no supplier for F. From their regular workforce they have up to 640 man hours of labor/day at \$10/man hour; beyond that they can get up to 160 man hours/day at \$17/man hour from a pool of workers.

They can sell up to 50 units of P_1 at \$3000/unit/day in an upscale market; beyond that they can sell up to 50 more units/day of P_1 to a wholesaler at \$250/unit. They can sell up to 100 units/day of P_2 at \$3500/unit. They can sell any quantity of P_3 produced at a constant rate of \$4500/unit.

Data on the inputs needed to make the various products are given in Table 1.10. Formulate the product mix problem to maximize the net profit/day at this company.

Maximizing the net profit is the same as minimizing its negative, which is = (the costs of all the inputs used/day) − (sales revenue/day). We verify that each term in this sum

TABLE 1.10 I/O Data

	Input units/unit made				
Product	LI	EP	W	F	L
P_1	$\frac{1}{2}$	3	1	1	2
P_2	1	2	$\frac{1}{4}$	1	1
P_3	$\frac{3}{2}$	5	2	3	1

is a PL convex function. So, we can model this problem as an LP in terms of variables corresponding to each interval of constant slope of each of the input and output quantities.

Let LI, EP, W, F, L denote the quantities of the respective inputs used/day; and P_1, P_2, P_3 denote the quantities of the respective products made and sold/day. Let LI_1 and LI_2 denote the units of limestone used daily from own quarry and outside supplier. Let EP_1, EP_2, and EP_3 denote the units of electricity used/day at \$30, 45, 75/unit, respectively. Let W_1 and W_2 denote the units of water used/day at rates of \$6, 7/unit, respectively. Let L_1 and L_2 denote the man hours of labor used/day from regular workforce, pool, respectively. Let P_{11} and P_{12} denote the units of P_1 sold at the upscale market and to the wholesaler, respectively.

Then the LP model for the problem is:

Minimize $\quad z = 20LI_1 + 50LI_2 + 30EP_1 + 45EP_2 + 75EP_3 + 6W_1 + 7W_2 + 40F + 10L_1 + 17L_2 - 3000P_{11} - 250P_{12} - 3500P_2 - 4500P_3$

subject to

$$(1/2)P_1 + P_2 + (3/2)P_3 = LI$$
$$3P_1 + 2P_2 + 5P_3 = EP$$
$$P_1 + (1/4)P_2 + 2P_3 = W$$
$$P_1 + P_2 + 3P_3 = F$$
$$2P_1 + P_2 + P_3 = L$$
$$LI_1 + LI_2 = LI, \quad W_1 + W_2 = W$$
$$EP_1 + EP_2 + EP_3 = EP$$
$$L_1 + L_2 = L, \quad P_{11} + P_{12} = P_1, \quad \text{all variables} \geq 0$$

$$(LI_1, EP_1, EP_2, W_1) \leq (250, 1000, 500, 800)$$
$$(F, L_1, L_2) \leq (3000, 640, 160)$$
$$(P_{11}, P_{12}, P_2) \leq (50, 50, 100).$$

1.3.1 Min–Max, Max–Min Problems

As discussed above, a PL convex function in variables $x = (x_1, \ldots, x_n)^T$ can be expressed as the pointwise maximum of a finite set of linear functions. Minimizing a function like that is appropriately known as a min–max problem. Similarly, a PL concave function in x can be expressed as the pointwise minimum of a finite set of linear functions. Maximizing a function like that is appropriately known as a max–min problem. Both min–max and max–min problems can be expressed as LPs in terms of just one additional variable.

If the PL convex function $f(x) = \min\{c_0^t + c^t x : t = 1, \ldots, r\}$, then $-f(x) = \max\{-c_0^t - c^t x : t = 1, \ldots, r\}$ is PL concave and conversely. Using this, any min–max problem can be posed as a max–min problem and vice versa. So, it is sufficient to discuss max–min problems. Consider the max–min problem

$$\max z(x) = \min\{c_0^1 + c^1 x, \ldots, c_0^r + c^r x\}$$
$$\text{subject to} \quad Ax = b$$
$$x \geq 0$$

To transform this problem into an LP, introduce the new variable x_{n+1} to denote the value of the objective function $z(x)$ to be maximized. Then the equivalent LP with additional

Linear Programming

linear constraints is:

$$\begin{array}{ll} \max & x_{n+1} \\ \text{subject to} & x_{n+1} \leq c_0^1 + c^1 x \\ & x_{n+1} \leq c_0^2 + c^2 x \\ & \quad \vdots \\ & x_{n+1} \leq c_0^r + c^r x \\ & Ax = b \\ & x \geq 0 \end{array}$$

The fact that x_{n+1} is being maximized and the additional constraints together imply that if $(\bar{x}, \bar{x}_{n+1})$ is an optimum solution of this LP model, then $\bar{x}_{n+1} = \min\{c_0^1 + c^1 \bar{x}, \ldots, c_0^r + c^r \bar{x}\} = z(\bar{x})$, and that $\bar{x}_{n+1}$ is the maximum value of $z(x)$ in the original max–min problem.

Example 1.6: Application in Worst Case Analysis

Consider the fertilizer maker's product mix problem with decision variables x_1 and x_2 (Hi-ph, Lo-ph fertilizers to be made daily in the next period) discussed in Example 1.2, Section 1.2. There we discussed the case where the net profit coefficients c_1 and c_2 of these variables are estimated to be $15 and $10, respectively. In reality, the prices of fertilizers are random variables that fluctuate daily. Because of unstable conditions, and new agricultural research announcements, suppose market analysts have only been able to estimate that the expected net profit coefficient vector (c_1, c_2) is likely to be one of $\{(15, 10), (10, 15), (12, 12)\}$ without giving a single point estimate. So, here we have three possible scenarios. In scenario 1, $(c_1, c_2) = (15, 10)$, expected net profit $= 15x_1 + 10x_2$; in scenario 2, $(c_1, c_2) = (10, 15)$, expected net profit $= 10x_1 + 15x_2$; in scenario 3 $(c_1, c_2) = (12, 12)$, expected net profit $= 12x_1 + 12x_2$. Suppose the raw material availability data in the problem is expected to remain unchanged. The important question is: which objective function to optimize for determining the production plan for the next period. ∎

Irrespective of which of the three possible scenarios materializes, at the worst the minimum expected net profit of the company will be $p(x) = \min\{15x_1 + 10x_2, 10x_1 + 15x_2, 12x_1 + 12x_2\}$ under the production plan $x = (x_1, x_2)^T$. *Worst case analysis* is an approach that advocates determining the production plan to optimize this worst case net profit $p(x)$ in this situation. This leads to the max–min model: maximize $p(x) = \min\{15x_1 + 10x_2, 10x_1 + 15x_2, 12x_1 + 12x_2\}$ subject to the constraints in Equation 1.2. The equivalent LP model corresponding to this is:

$$\begin{array}{ll} \max & p \\ \text{subject to} & p \leq 15x_1 + 10x_2 \\ & p \leq 10x_1 + 15x_2 \\ & p \leq 12x_1 + 12x_2 \\ & 2x_1 + x_2 \leq 1500 \\ & x_1 + x_2 \leq 1200 \\ & x_1 \leq 500, \quad x_1, x_2 \geq 0 \end{array}$$

1.3.2 Minimizing Positive Linear Combinations of Absolute Values of Affine Functions

Consider the problem

$$\begin{array}{ll} \min & z(x) = w_1 |c_0^1 + c^1 x| + \cdots + w_r |c_0^r + c^r x| \\ \text{subject to} & Ax \geq b \end{array}$$

where the weights $w_1, \ldots, w_r$ are all strictly positive. In this problem the objective function to be minimized, $z(x)$, is a PL convex function; hence this problem can be transformed into an LP. To transform, define for each $t = 1$ to r two new nonnegative variables $u_t^+ = \max\{0, c_0^t + c^t x\}$, $u_t^- = -\min\{0, c_0^t + c^t x\}$. u_t^+ is called the *positive part* of $c_0^t + c^t x$, and u_t^- its *negative part*. It can be verified that $(u_t^+)(u_t^-)$ is zero by definition, and because of this we have: $c_0^t + c^t x = (u_t^+) - (u_t^-)$ and $|c_0^t + c^t x| = (u_t^+) + (u_t^-)$. Using this, we can transform the above problem into the following LP:

$$\begin{aligned}
\min \quad & w_1[(u_1^+) + (u_1^-)] + \cdots + w_r[(u_r^+) + (u_r^-)] \\
\text{subject to} \quad & c_0^1 + c^1 x = (u_t^+) - (u_t^-) \\
& \quad \vdots \qquad \vdots \\
& c_0^r + c^r x = (u_r^+) - (u_r^-) \\
& Ax \geq b \\
& (u_t^+), (u_t^-) \geq 0, \quad t = 1, \ldots, r
\end{aligned}$$

Using the special structure of this problem it can be shown that the condition $(u_t^+)(u_t^-) = 0$ for all $t = 1$ to r will hold at all optimum solutions of this LP. This shows that this transformation is correct.

An application of this transformation is discussed in the next section. This is an important model that finds many applications.

1.4 Intelligent Modeling Essential to Get Good Results, an Example from Container Shipping

To get good results from a linear programming application, it is very important to develop a good model for the problem being solved. There may be several ways of modeling the problem, and it is very important to select the one most appropriate to model it intelligently to get good results. Skill in modeling comes from experience; unfortunately there is no theory to teach how to model intelligently. We will now discuss a case study of an application carried out for routing trucks inside a container terminal to minimize congestion. Three different ways of modeling the problem have been tried. The first two approaches lead to (1) an integer programming model and (2) a large-scale multicommodity flow LP model, respectively. Both these models gave very poor results. The third and final model developed uses a substitute objective function technique; that is, it optimizes another simpler objective function that is highly correlated to the original, because that other objective function is much easier to control. This approach led to a small LP model, and gives good results.

Today most of the nonbulk cargo is packed into steel boxes called containers (typically of size $40 \times 8 \times 9$ in feet) and transported in oceangoing vessels. A container terminal in a port is the place where these vessels dock at berths for unloading of inbound containers and loading of outbound containers. The terminals have storage yards for the temporary storage of these containers. The terminal's internal trucks (TIT) transport containers between the berth and the storage yard (SY). The SY is divided into rectangular areas called blocks, each served by one or more cranes (rubber tired gantry cranes, or RTGC) to unload/load containers from/to trucks. Customers bring outbound containers into the terminal in their own trucks (called external trucks, or XT), and pick up from the SY and take away their inbound containers on these XT. Each truck (TIT or XT) can carry only one container at a time.

The example (from Murty et al., 2005a,b) deals with the mathematical modeling of the problem of routing the trucks inside the terminal to minimize congestion. We represent the terminal road system by a directed network $G = (\mathcal{N}, \mathcal{A})$ where $\mathcal{N}$ is the set of nodes (each block, berth unloading/loading position, road intersection, terminal gate is a node), $\mathcal{A}$ is the set of arcs (each lane of a road segment joining a pair of nodes is an arc). Each (berth unloading/loading position, block), (block, berth unloading/loading position), (gate, block), (block, gate) is an origin–destination pair for trucks that have to go from the origin to the destination; they constitute a separate commodity that flows in G. Let T denote the number of these commodities. Many terminals use a 4-hour planning period for their truck routing decisions.

Let $f = (f_{ij}^r)$ denote the flow vector of various commodities on G in the planning period, where $f_{ij}^r =$ expected number of trucks of commodity r passing through arc (i, j) in the planning period for $r = 1$ to T, and $(i, j) \in \mathcal{A}$. Let $\theta = \max \left\{ \sum_{r=1}^{T} f_{ij}^r : (i, j) \in \mathcal{A} \right\}$, $\mu = \min \left\{ \sum_{r=1}^{T} f_{ij}^r : (i, j) \in \mathcal{A} \right\}$. Then either θ or $\theta - \mu$ can be used as measures of congestion on G during the planning period, to optimize.

As storage space allocation to arriving containers directly determines how many trucks travel between each origin–destination pair, the strategy used for this allocation plays a critical role in controlling congestion. This example deals with mathematical modeling of the problem of storage space allocation to arriving containers to minimize congestion.

Typically, a block has space for storing 600 containers, and a terminal may have 100 (some even more) blocks. At the beginning of the planning period, some spaces in the SY would be occupied by containers already in storage, and the set of occupied storage positions changes every minute; it is very difficult to control this change. Allocating a specific open storage position to each container expected to arrive in the planning period has been modeled as a huge integer program, which takes a long time to solve. In fact, even before this integer programming model is entered into the computer, the data change. So these traditional integer programming models are not only impractical but also inappropriate for the problem.

So a more practical way is to break up the storage space allocation decision into two stages: Stage 1 determines only the *container quota* x_i, for each block i, which is the number of newly arriving containers that will be dispatched to block i for storage during the planning period. Stage 1 will not determine which of the specific arriving containers will be stored in any block; that decision is left to Stage 2, which is a dispatching policy that allocates each arriving container to a specific block for storage at the time of its arrival, based on conditions prevailing at that time. So, Stage 2 makes sure that by the end of the planning period the number of new containers sent for storage to each block is its quota number determined in Stage 1, while minimizing congestion at the blocks and on the roads.

Our example deals with the Stage 1 problem. The commonly used approach is based on a batch-processing strategy. Each batch consists of all the containers expected to arrive/leave at each node during the planning period. At the gate, this is the number of outbound containers expected to arrive for storage. At a block it is the number of stored containers expected to be retrieved and sent to each berth or the gate. At each berth it is the number of inbound containers expected to be unloaded to be sent for storage to SY. With this data, the problem can be modeled as a multicommodity network flow problem. It is a large-scale LP with many variables and thousands of constraints. However, currently available LP software systems are fast; this model can be solved using them in a few minutes of computer time.

But the output from this model turned out to be poor, as the model is based solely on the total estimated workload during the planning period. Such a model gives good results for the real problem only if the workload in the terminal (measured in number of containers handled/unit time) is distributed more or less uniformly over time during the planning period. In reality the workload at terminals varies a lot over time. At the terminal where we did this work, the number of containers handled per hour varied from 50 to 400 in a 4-hour planning period.

Let $f_i(t)$ denote the fill ratio in block i at time point t, which is equal to (number of containers in storage in block i at time point t)/(number of storage spaces in block i). We observed that the fill ratio in a block is highly positively correlated with the number of containers being moved in and out of the block/minute. So, maintaining fill ratios in all the blocks nearly equal, along with a good dispatching policy, will ensure that the volumes of traffic in the neighborhoods of all the blocks are nearly equal, thus ensuring equal distribution of traffic on all the terminal roads and hence minimizing congestion. This leads to a substitute-objective-function technique for controlling congestion indirectly. For the planning period, we define the following:

$x_i =$ The container quota for block $i =$ number of containers arriving in this period to be dispatched for storage to block i, a decision variable.

$a_i =$ The number of stored containers that will remain in block i at the end of this period if no additional containers are sent there for storage during this period, a data element.

$N =$ The number of new containers expected to arrive at the terminal in this period for storage, a data element.

$B, A =$ The total number of blocks in the storage yard, the number of storage positions in each block, data elements.

The fill-ratio equalization policy determines the decision variables x_i to make sure that the fill ratios in all the blocks are as nearly equal as possible at one time during the period, namely the end of the period. The fill ratio in the whole yard at the end of this period will be $F = (N + \sum_i a_i)/(A \times B)$. If the fill ratios in all the blocks at the end of this period are all equal, they will all be equal to F. Thus, this policy determines x_i to guarantee that the fill ratio in each block will be as close to F as possible by the end of this period. Using the least sum of absolute deviations measure, this leads to the following model to determine x_i.

$$\text{Minimize} \quad \sum_{i=1}^{B} |a_i + x_i - AF|$$
$$\text{subject to} \quad \sum_{i=1}^{B} x_i = N$$
$$x_i \geq 0 \quad \text{for all } i$$

Transforming this we get the following LP model to determine x_i

$$\text{Minimize} \quad \sum_{i=1}^{B} (u_i^+ + u_i^-)$$
$$\text{subject to} \quad \sum_{i=1}^{B} x_i = N$$
$$a_i + x_i - AF = u_i^+ - u_i^- \quad \text{for all } i$$
$$x_i, u_i^+, u_i^- \geq 0 \quad \text{for all } i$$

This is a much simpler and smaller LP model with only $B+1$ constraints. Using its special structure, it can be verified that its optimum solution can be obtained by the following

Linear Programming

combinatorial scheme: Rearrange the blocks in increasing order of a_i from top to bottom. Then begin at the top and determine x_i one after the other to bring $a_i + x_i$ to the level AF or as close to it as possible until all the N new containers expected to arrive are allocated.

We will illustrate with a small numerical example of an SY with $B = 9$ blocks, $A = 600$ spaces in each block, with $N = 1040$ new containers expected to arrive during the planning period. Data on a_i already arranged in increasing order are given in the following table. So, the fill ratio in the whole yard at the end of the planning period is expected to be $F = (N + \sum_i a_i)/(AB) = 3547/5400 \approx 0.67$, and so the average number of containers in storage/block will be $AF \approx 400$. So, the LP model for determining x_i for this planning period to equalize fill ratios is

$$\text{Minimize} \quad \sum_{i=1}^{9} (u_i^+ + u_i^-)$$

$$\text{subject to} \quad \sum_{i=1}^{9} x_i = 1040$$

$$a_i + x_i - u_i^+ + u_i^- = 400 \quad \text{for all } i$$

$$x_i, u_i^+, u_i^- \geq 0 \quad \text{for all } i$$

The optimum solution (x_i) of this model obtained by the above combinatorial scheme is given in Table 1.11. $a_i + x_i$ is the expected number of containers in storage in block i at the end of this planning period; it can be verified that its values in the various blocks are nearly equal.

Stage 1 determines only the container quota numbers for the blocks, not the identities of containers that will be stored in each block. The storage block to which each arriving container will be sent for storage is determined by the dispatching policy discussed in Stage 2. Now we describe Stage 2 briefly.

Regardless of how we determine the container quota numbers x_i, if we send a consecutive sequence of arriving container trucks to the same block in a short time interval, we will create congestion at that block. To avoid this possibility, the dispatching policy developed in Stage 2 ensures that the yard crane in that block has enough time to unload a truck we send there before we send another. For this we had to develop a system to monitor continuously over time: $w_i(t) =$ the number of trucks waiting in block i to be served by the yard cranes there at time point t. As part of our work on this project, the terminal where we did this work developed systems to monitor $w_i(t)$ continuously over time for each block i.

TABLE 1.11 Optimum Solution x_i

Block i	a_i	x_i	$a_i + x_i$
1	100	300	400
2	120	280	400
3	150	250	400
4	300	100	400
5	325	75	400
6	350	35	385
7	375	0	375
8	400	0	400
9	450	0	450
Total	2570	1040	

They also developed a dispatching cell that has the responsibility of dispatching each truck in the arriving stream to a block; this cell gets this $w_i(t)$ information continuously over time.

As time passes during the planning period, the dispatching cell also keeps track of how many containers in the arriving stream have already been sent to each block for storage. When this number becomes equal to the container quota number for that block, they will not send any more containers for storage to that block during the planning period. For each block i, let $x_i^R(t) = x_i -$ (number of new containers sent to block i for storage up to time t in the planning period) = remaining container quota number for block i at time t in the planning period.

This policy dispatches each truck arriving (at the terminal gate, and at each berth) at time point t in the period to a block i satisfying: $w_i(t) = \text{Min}\{w_j(t) : j \text{ satisfying } x_i^R(t) > 0\}$, that is, a block with a remaining positive quota that has the smallest number of trucks waiting in it.

This strategy for determining the quota numbers for blocks, x_i described above, along with the dynamic dispatching policy to dispatch arriving container trucks using real time information on how many trucks are waiting in each block, turned out to be highly effective. It reduced congestion and helped reduce truck turnaround time by over 20%.

In this work, our first two mathematical models for the problem turned out to be ineffective; the third one was not only the simplest but a highly effective one. This example shows that to get good results in real world applications, it is necessary to model the problems intelligently. Intelligent modeling + information technology + optimization techniques is a powerful combination for solving practical problems.

1.5 Planning Uses of LP Models

When LP is the appropriate technique to model a real world problem, and the LP model is constructed, we discuss here what useful information can be derived using the model (from [Murty, 1995, 2005b]).

1.5.1 Finding the Optimum Solutions

Solving the model gives an optimum solution, if one exists. The algorithms can actually identify the set of all the optimum solutions if there are alternate optimum solutions. This may be helpful in selecting a suitable optimum solution to implement (one that satisfies some conditions that may not have been included in the model, but which may be important).

Solving the fertilizer product mix problem, we find that the unique optimum solution for it is to manufacture 300 tons Hi-ph and 900 tons Lo-ph, leading to a maximum daily profit of $13,500.

1.5.2 Infeasibility Analysis

We may discover that the model is *infeasible* (i.e., it has no feasible solution). If this happens, there must be a subset of constraints that are mutually contradictory in the model (maybe we promised to deliver goods without realizing that our resources are inadequate to manufacture them on time). In this case the algorithms can indicate how to modify the constraints to make the model feasible. For example, suppose the system of constraints in the original LP is: $Ax = b$, $x \geq 0$, where A is an $m \times n$ matrix and $b = (b_i) \in R^m$, and the equality constraints are recorded so that $b \geq 0$. The Phase I problem for finding an initial feasible solution for this problem is

$$\text{Minimize} \quad w(t) = \sum_{i=1}^{m} t_i$$
$$\text{subject to} \quad Ax + It = b$$
$$x, t \geq 0$$

where I is the unit matrix of order m, and $t = (t_1, \ldots, t_m)^T$ is the vector of artificial variables. Suppose the optimum solution of this Phase I problem is $(\bar{x}, \bar{t})$. If $\bar{t} = 0$, $\bar{x}$ is a feasible solution of the original LP that can now be solved using it as the initial feasible solution. If $\bar{t} \neq 0$, the original LP is infeasible. Mathematically there is nothing more that can be done on the original model. But the real world problem for which this model is constructed does not go away; it has to be tackled somehow. So, we have to investigate what practically feasible changes can be carried out on the model to modify it into a feasible system. Infeasibility analysis is the study of such changes (see Murty, 1995; Murty et al., 2000; Brown and Graves, 1977; Chinnek and Dravineks, 1991).

Sometimes it may be possible to eliminate some constraints to make the model feasible. But the most commonly used technique to make the model feasible is to modify some data elements in the model. Making changes in the technology coefficient matrix A involves changing the technological processes used in the system; hence these changes are only considered rarely in practice. Data elements in the RHS constants vector b represent things like resource quantities made available, delivery commitments made, and so on; these can be modified relatively easily. That's why most often it is the RHS constants in the model that are changed to make the model feasible.

One simple modification that will make the model feasible is to change the RHS constants vector b into $\bar{b} = b - \bar{t}$. Then the constraints in the modified model are $Ax = \bar{b}$, $x \geq 0$; $\bar{x}$ is a feasible solution for it. Starting with $\bar{x}$, the modified model can be solved. As an example, consider the system

$$\begin{aligned} 2x_1 + 3x_2 + x_3 - x_4 &= 10 \\ x_1 + 2x_2 - x_3 + x_5 &= 5 \\ x_1 + x_2 + 2x_3 &= 4 \\ x_j \geq 0 \quad \text{for all } j \end{aligned}$$

The Phase I problem to find a feasible solution of this system is

$$\begin{aligned} \text{Minimize} \quad & t_1 + t_2 + t_3 \\ \text{subject to} \quad & 2x_1 + 3x_2 + x_3 - x_4 + t_1 = 10 \\ & x_1 + 2x_2 - x_3 + x_5 + t_2 = 5 \\ & x_1 + x_2 + 2x_3 + t_3 = 4 \\ & x_j, t_i \geq 0 \quad \text{for all } j, i \end{aligned}$$

An optimum solution of this Phase I problem is $(\bar{x}, \bar{t})$, where $\bar{x} = (3, 1, 0, 0, 0)^T$, $\bar{t} = (1, 0, 0)^T$, so the original system is infeasible. To make the original system feasible we can modify the RHS constants vector in the original model $b = (10, 5, 4)^T$ to $b - \bar{t} = (9, 5, 4)^T$. For the modified system, $\bar{x}$ is a feasible solution.

This modification only considers reducing the entries in the RHS constants vector in the original model; also it gives the decision maker no control on which RHS constants b_i are changed to make the system feasible. Normally there are costs associated with changing the value of b_i, and these may be different for different i. To find a least costly modification of the b-vector to make the system feasible, let

c_i^+, c_i^- = cost per unit increase, decrease respectively in the value of b_i

p_i, q_i = maximum possible increase, decrease allowed in the value of b_i

Then the model to minimize the total cost of all the changes to make the model feasible is the LP

$$\text{Minimize} \quad \sum_{i=1}^{m}(c_i^+ u_i^+ + c_i^- u_i^-)$$
$$\text{subject to} \quad Ax + Iu^+ - Iu^- = b$$
$$u^+ \leq q, u^- \leq p$$
$$x, u^+, u^- \geq 0$$

where $u^+ = (u_1^+, \ldots, u_m^+)^T$, $u^- = (u_1^-, \ldots, u_m^-)^T$, $p = (p_i)$, $q = (q_i)$, and I is the unit matrix of order m. If $(\bar{x}, \bar{u}^+, \bar{u}^-)$ is an optimum solution of this LP, then $b' = b - \bar{u}^+ + \bar{u}^-$ is the optimum modification of the original RHS vector b under this model; and $\bar{x}$ is a feasible solution of the modified model (Brown and Graves, 1977; Chinnek and Dravineks, 1991).

1.5.3 Values of Slack Variables at an Optimum Solution

The values of the slack variables corresponding to inequality constraints in the model provide useful information on which supplies and resources will be left unused and in what quantities, if that solution is implemented.

For example, in the fertilizer product mix problem, the optimum solution is $\hat{x} = (300, 900)$. At this solution, RM 1 slack is $\hat{x}_3 = 1500 - 2\hat{x}_1 - \hat{x}_2 = 0$, RM 2 slack is $\hat{x}_4 = 1200 - \hat{x}_1 - \hat{x}_2 = 0$, and RM 3 slack $\hat{x}_5 = 500 - \hat{x}_1 = 200$ tons.

Thus, if this optimum solution is implemented, the daily supply of RM 1 and RM 2 will be completely used up, but 200 tons of RM 3 will be left unused. This shows that the supplies of RM 1 and RM 2 are very critical to the company, and that there is currently an oversupply of 200 tons of RM 3 daily that cannot be used in the optimum operation of the Hi-ph and Lo-ph fertilizer processes.

This also suggests that it may be worthwhile to investigate if the maximum daily profit can be increased by lining up additional supplies of RM 1 and RM 2 from outside vendors or if additional capacity exists in the Hi-ph, Lo-ph manufacturing processes. A useful planning tool for this investigation is discussed next.

1.5.4 Marginal Values, Dual Variables, and the Dual Problem, and Their Planning Uses

Each constraint in an LP model is the material balance constraint of some item, the RHS constant in that constraint being the availability or the requirement of that item. The *marginal value* of that item (also called the marginal value corresponding to that constraint) is defined to be the rate of change in the optimum objective value of the LP, per unit change in the RHS constant in the constraint. This marginal value associated with a constraint is also called the *dual variable* corresponding to that constraint.

Associated with every LP there is another LP called its *dual problem*; both share the same data. In this context, the original problem is called the *primal problem*. The variables in the dual problem are the marginal values or dual variables defined above; each of these variables is associated with a constraint in the primal. Given the primal LP, the derivation of its dual problem through marginal economic arguments is discussed in many LP books; for example, see Dantzig (1963), Gale (1960), and Murty (1995, 2005b). For an illustration, let the primal be

$$\begin{aligned} \text{Minimize} \quad & z = cx \\ \text{subject to} \quad & Ax = b \\ & x \geq 0 \end{aligned}$$

where $x = (x_1, \ldots, x_n)^T$ is the vector of primal variables, and A is of order $m \times n$. Denoting the dual variable associated with the i-th constraint in $Ax = b$ by π_i, the vector of dual variables associated with these constraints is the row vector $\pi = (\pi_1, \ldots, \pi_m)$. Let the dual variable associated with the nonnegativity restriction $x_j \geq 0$ be denoted by $\bar{c}_j$ for $j = 1$ to n, and let $\bar{c}$ = the row vector $(\bar{c}_1, \ldots, \bar{c}_n)$. Then the dual problem is

$$\begin{aligned} \text{Maximize} \quad & v = \pi b \\ \text{subject to} \quad & \pi A + \bar{c} = c \\ & \bar{c} \geq 0 \end{aligned}$$

The dual variables $\bar{c}$ associated with the nonnegativity constraints $x \geq 0$ in the primal are called *relative* or reduced *cost coefficients* of the primal variables. Given π we can get $\bar{c}$ from $\bar{c} = c - \pi A$. Hence, commonly people omit $\bar{c}$, and refer to π itself as the dual solution.

When the optimum solution of the dual problem is unique, it is the vector of marginal values for the primal problem. All algorithms for linear programming have the property that when they obtain an optimum solution of the primal, they also produce an optimum solution of the dual; this is explained in Section 1.6. Also, most software packages for LP provide both the primal and dual optimum solutions when they solve an LP model.

For the fertilizer product mix problem (Equation 1.2) discussed in Example 1.2 (Section 1.2), the dual optimum solution $\pi = (\pi_1, \pi_2, \pi_3) = (5, 5, 0)$ is unique; hence it is the vector of marginal values for the problem. As the objective function in the problem is in units of net profit dollars, this indicates that the marginal values of raw materials RM 1 and RM 2 are both \$5/ton in net profit dollars. As the current price of RM 1 delivered to the company is \$50/ton, this indicates that as long as the price charged by an outside vendor per ton of RM 1 delivered is $\leq \$50 + 5 = 55$/ton, it is worth getting additional supplies of RM 1 from that vendor. \$55/ton delivered is the breakeven price for acquiring additional supplies of RM 1 for profitability.

In the same way, as the current price of RM 2 delivered to the company is \$40/ton, we know that the breakeven price for acquiring additional supplies of RM 2 for profitability is \$40 + 5 = \$45/ton delivered.

Also, since the marginal value of RM 3 is zero, there is no reason to get additional supplies of RM 3, as no benefit will accrue from it.

This type of analysis is called *marginal analysis*. It helps companies to determine what their most critical resources are and how the requirements or resource availabilities can be modified to arrive at much better objective values than those possible under the existing requirements and resource availabilities.

1.5.5 Evaluating the Profitability of New Products

Another major use of marginal values is in evaluating the profitability of new products. It helps to determine whether they are worth manufacturing, and if so at what level they should be priced so that they are profitable in comparison with existing product lines.

We will illustrate this again using the fertilizer product mix problem. Suppose the company's research chemist has come up with a new fertilizer that he calls *lushlawn*. Its manufacture requires per ton, as inputs, 3 tons of RM 1, 2 tons of RM 2, and 2 tons of RM 3.

At what rate/ton should lushlawn be priced in the market, so that it is competitive in profitability with the existing Hi-ph and Lo-ph that the company currently makes?

To answer this question, we computed the marginal values RM 1, RM 2, RM 3 to be $\pi_1 = 5$, $\pi_2 = 5$, $\pi_3 = 0$.

So, the input packet of $(3, 2, 2)^T$ tons of (RM 1, RM 2, RM 3)^T needed to manufacture one ton of lushlawn has value to the company of $3\pi_1 + 2\pi_2 + 2\pi_3 = 3 \times 5 + 2 \times 5 + 2 \times 0 = \$25$ in terms of net profit dollars. On the supply side, the delivery cost of this packet of raw materials is $3 \times 50 + 2 \times 40 + 2 \times 60 = \$350$.

So, clearly, for lushlawn to be competitive with Hi-ph and Lo-ph, its selling price in the market/ton should be $\geq \$25 + 350 +$ (its production cost/ton). The company can conduct a market survey and determine whether the market will accept lushlawn at a price $\geq$ this breakeven level. Once this is known, the decision whether to produce lushlawn would be obvious.

By providing this kind of valuable planning information, the LP model becomes a highly useful decision making tool.

1.6 Brief Introduction to Algorithms for Solving LP Models

1.6.1 The Simplex Method

The celebrated simplex method developed by George B. Dantzig in 1947 is the first computationally viable method for solving LPs and systems of linear inequalities (Dantzig, 1963). Over the years the technology for implementing the simplex method has gone through many refinements, with the result that even now it is the workhorse behind many of the successful commercial LP software systems.

Before applying the simplex method, the LP is transformed into a standard form through simple transformations (like introducing slack variables corresponding to inequality constraints, etc.). The general step in the simplex method is called a *pivot step*. We present the details of it for the LP in the most commonly used standard form, which in matrix notation is:

$$\begin{aligned} \text{Minimize} \quad & z \\ \text{subject to} \quad & Ax = b \\ & cx - z = 0 \\ & x \geq 0 \end{aligned} \quad (1.5)$$

where A is a matrix of order $m \times n$ of full rank m. A basic vector for this problem is a vector of m variables among the x_j, and then $-z$, of the form $(x_B, -z)$ where $x_B = (x_{j1}, \ldots, x_{jm})$, satisfying the property that the submatrix B consisting of the column vectors of these basic variables is a basis, that is, a nonsingular square submatrix of order $m + 1$. Let x_D denote the remaining vector of nonbasic variables. The primal basic solution corresponding to this basic vector is given by

$$x_D = 0, \begin{pmatrix} x_B \\ -z \end{pmatrix} = B^{-1} \begin{pmatrix} b \\ 0 \end{pmatrix} = \begin{pmatrix} \bar{b} \\ \vdots \\ \bar{b}_m \\ -\bar{z} \end{pmatrix} \quad (1.6)$$

The basic vector $(x_B, -z)$ is said to be a primal feasible basic vector if the values of the basic variables in x_B in the basic solution in Equation 1.6 satisfy the nonnegativity restrictions on these variables, primal infeasible basic vector otherwise. The basic solution in Equation 1.6 is called a basic feasible solution (BFS) for Equation 1.5 if $(x_B, -z)$ is a primal feasible basic vector.

There is also a dual basic solution corresponding to the basic vector $(x_B, -z)$. If that dual basic solution is π, then the last row of B^{-1} will be $(-\pi, 1)$, so the dual basic solution corresponding to this basic vector can be obtained from the last row of B^{-1}.

A pivot step in the simplex method begins with a feasible basic vector $(x_B, -z)$, say associated with the basis B, dual basic solution π, and primal BFS given in Equation 1.6. $\bar{c} = c - \pi A$ is called the vector of relative cost coefficients of (x_j) corresponding to this basic vector. The optimality criterion is: $\bar{c} \geq 0$; if it is satisfied the BFS in Equation 1.6 and π are optimum solutions of the primal LP and its dual; and the method terminates.

If $\bar{c} \not\geq 0$, any variable x_j associated with a $\bar{c}_j < 0$ will be a nonbasic variable which is called a variable eligible to enter the present basic vector to obtain a better solution than the present BFS. One such eligible variable, x_s say, is selected as the entering variable. Its updates column: $(\bar{a}_{1s}, \ldots, \bar{a}_{ms}, \bar{c}_s)^T = B^{-1}$ (column vector of x_s in Equation 1.5) is called the pivot column for this pivot step. The minimum ratio in this pivot step is defined to be

$$\theta = \min\{\bar{b}_i/\bar{a}_{is} : 1 \leq i \leq m \text{ such that } \bar{a}_{is} > 0\}$$

where $(\bar{b}_i)$ are the values of the basic variables in the present BFS. If the minimum ratio is attained by $i = r$, then the r-th basic variable in $(x_B, -z)$ will be the dropping variable to be replaced by x_s to yield the next basic vector. The basis inverse corresponding to the new basic vector is obtained by performing a Gauss–Jordan pivot step on the columns of the present B^{-1} with the pivot column and row r as the pivot row. The method goes to the next step with the new basic vector, and is continued the same way until termination occurs.

We will illustrate with the fertilizer problem (Equation 1.2) formulated in Example 1.2. Introducing slack variables x_3, x_4, x_5 corresponding to the three inequalities, and putting the objective function in minimization form, the standard form for the problem in a detached coefficient tableau is as shown in Table 1.12.

It can be verified that $(x_1, x_3, x_4, -z)$ is a feasible basic vector for the problem. The corresponding basis B and its inverse B^{-1} are

$$B = \begin{pmatrix} 2 & 1 & 0 & 0 \\ 1 & 0 & 1 & 0 \\ 1 & 0 & 0 & 0 \\ -15 & 0 & 0 & 1 \end{pmatrix}, \quad B^{-1} = \begin{pmatrix} 0 & 0 & 1 & 0 \\ 1 & 0 & -2 & 0 \\ 0 & 1 & -1 & 0 \\ 0 & 0 & 15 & 1 \end{pmatrix}$$

So the corresponding primal BFS is given by: $x_2 = x_5 = 0$, $(x_1, x_3, x_4, -z)^T = B^{-1}(1500, 1200, 500, 0)^T = (500, 700, 500, 7500)^T$.

From the last row of B^{-1} we see that the corresponding dual basic solution is $\pi = (0, 0, -15)$.

TABLE 1.12 Original Tableau for Fertilizer Problem

x_1	x_2	x_3	x_4	x_5	$-z$	RHS
2	1	1	0	0	0	1500
1	1	0	1	0	0	1200
1	0	0	0	1	0	500
−15	−10	0	0	0	1	0

$x_j \geq 0$ for all j, min z

TABLE 1.13 Pivot Step to Update B^{-1}

B^{-1}				PC
0	0	1	0	0
1	0	−2	0	1
0	1	−1	0	$\boxed{1}$
0	0	15	1	−10
0	0	1	0	0
1	−1	−1	0	0
0	1	−1	0	1
0	10	5	1	0

So, the relative cost vector is $\bar{c} = (0, 0, 15, 1)$ (the matrix consisting of columns of x_1 to x_5 in Table 1.12) $= (0, -10, 0, 0, 0) \not\geq 0$. So, x_2 is the only eligible variable to enter; we select it as the entering variable. The pivot column = updated column of $x_2 = B^{-1}$(orginal column of x_2) $= (0, 1, 1, -10)^T$.

The minimum ratio $\theta = \min \{700/1, 500/1\} = 500$, attained for $r = 3$. So the third basic variable in the present basic vector, x_4, is the dropping variable to be replaced by x_2. So, the next basic vector will be $(x_1, x_3, x_2, -z)$. To update the basis inverse we perform the pivot step with the pivot element enclosed in a box. PC = pivot column (Table 1.13).

So the bottom matrix on the left is the basis inverse associated with the new basic vector $(x_1, x_3, x_2, -z)$. It can be verified that the BFS associated with this basic vector is given by: $x_4 = x_5 = 0$, $(x_1, x_3, x_2, -z) = (500, 200, 500, 12,500)$.

Hence, in this pivot step the objective function to be minimized in this problem, z, has decreased from −7500 to −12,500. The method now goes to another step with the new basic vector, and repeats until it terminates.

The method is initiated with a known feasible basic vector. If no feasible basic vector is available, a Phase I problem with a known feasible basic vector is set up using artificial variables; solving the Phase I problem by the same method either gives a feasible basic vector for the original problem, or concludes that it is infeasible.

1.6.2 Interior Point Methods for LP

Even though a few isolated papers have discussed some interior point approaches for LP as early as the 1960s, the most explosive development of these methods was triggered by the pioneering paper by Karmarkar (1984). In it he developed a new interior point method for LP, proved that it is a polynomial time method, and outlined compelling reasons why it has the potential to be also practically efficient and likely to beat the simplex method for large-scale problems. In the tidal wave that followed, many different interior point methods were developed. Computational experience has confirmed that some of them do offer an advantage over the simplex method for solving large-scale sparse problems.

We will briefly describe a popular method known as the *primal-dual path following interior point method* from Wright (1996). It considers the primal LP: minimize $c^T x$, subject to $Ax = b$, $x \geq 0$; and its dual in which the constraints are: $A^T y + s = c$, $s \geq 0$, where A is a matrix of order $m \times n$ and rank m (in Section 1.5, we used the symbol $\bar{c}$ to denote s). The system of primal and dual constraints put together is:

$$\begin{aligned} Ax &= b \\ A^T y + s &= c \\ (x, s) &\geq 0 \end{aligned} \quad (1.7)$$

In LP literature, a feasible solution (x, y, s) to Equation 1.7 is called an *interior feasible solution* if $(x, s) > 0$. Let $\mathcal{F}$ denote the set of all feasible solutions of Equation 1.7, and $\mathcal{F}^0$ the

Linear Programming

set of all interior feasible solutions. For any $(x, y, s) \in \mathcal{F}^0$ define $X = \text{diag}(x_1, \ldots, x_n)$, the square diagonal matrix of order n with diagonal entries $x_1, \ldots, x_n$; and $S = \text{diag}(s_1, \ldots, s_n)$.

The Central Path

This path, $\mathcal{C}$, is a nonlinear curve in $\mathcal{F}^0$ parametrized by a positive parameter $\tau > 0$. For each $\tau > 0$, the point $(x^\tau, y^\tau, s^\tau) \in \mathcal{C}$ satisfies: $(x^\tau, s^\tau) > 0$ and

$$A^T y^\tau + s^\tau = c^T$$
$$A x^\tau = b$$
$$x_j^\tau s_j^\tau = \tau, \quad j = 1, \ldots, n$$

If $\tau = 0$, the above equations define the optimality conditions for the LP. For each $\tau > 0$, the solution (x^τ, y^τ, s^τ) is unique, and as τ decreases to 0 the central path converges to the center of the optimum face of the primal, dual pair of LPs.

Optimality Conditions

For the primal, dual pair of LPs under discussion, an $(x^r = (x_j^r); y^r = (y_i^r), s^r = (s_j^r))$ of primal and dual feasible solutions is an optimum solution pair for the two problems iff $x_j^r s_j^r = 0$ for all j. These conditions are called complementary slackness optimality conditions.

We will use the symbol e to denote the column vector consisting of all 1s in R^n. From optimality conditions, solving the LP is equivalent to finding a solution (x, y, s) satisfying $(x, s) \geq 0$, to the following system of $2n + m$ equations in $2n + m$ unknowns.

$$F(x, y, s) = \begin{bmatrix} A^T y + s - c \\ Ax - b \\ XSe \end{bmatrix} = 0 \tag{1.8}$$

This is a nonlinear system of equations because of the last equation.

The General Step in the Method

The method begins with an interior feasible solution to the problem. If no interior feasible solution is available to initiate the method, it could be modified into an equivalent problem with an initial interior feasible solution by introducing artificial variables.

Starting with an interior feasible solution, in each step the method computes a direction to move at that point, and moves in that direction to the next interior feasible solution, and continues from there the same way.

Consider the step in which the current interior feasible solution at the beginning is $(\bar{x}, \bar{y}, \bar{s})$. So, $(\bar{x}, \bar{s}) > 0$. Also, the variables in y are unrestricted in sign in the problem.

Once the direction to move from the current point $(\bar{x}, \bar{y}, \bar{s})$ is computed, we may move from it only a small step length in that direction, and since $(\bar{x}, \bar{s}) > 0$, such a move in any direction will take us to a point that will continue satisfying $(x, s) > 0$. So, in computing the direction to move at the current point, the nonnegativity constraints $(x, s) \geq 0$ can be ignored. The only remaining conditions to be satisfied for attaining optimality are the equality conditions (Equation 1.8). So the direction finding routine concentrates only on trying to satisfy Equation 1.8 more closely.

Equation 1.8 is a square system of nonlinear equations; $(2n + m)$ equations in $(2n + m)$ unknowns. It is nonlinear because the third condition in Equation 1.8 is nonlinear. Experience in nonlinear programming indicates that the best directions to move in algorithms for solving nonlinear equations are either the Newton direction or some modified Newton direction. So, this method uses a modified Newton direction to move. To define that, two parameters are used: μ (an average complementary slackness property violation measure) = $\bar{x}^T \bar{s}/n$, and $\sigma \in [0,1]$ (a centering parameter). Then the direction for the move denoted by $(\Delta x, \Delta y, \Delta s)$ is the solution to the following system of linear equations

$$\begin{pmatrix} 0 & A^T & I \\ A & 0 & 0 \\ S & 0 & X \end{pmatrix} \begin{pmatrix} \Delta x \\ \Delta y \\ \Delta s \end{pmatrix} = \begin{pmatrix} 0 \\ 0 \\ -XSe + \sigma\mu e \end{pmatrix}$$

where 0 in each place indicates the appropriate matrix or vector of zeros, I the unit matrix of order n, and e indicates the column vector of order n consisting of all 1s. If $\sigma = 1$, the direction obtained will be a centering direction, which is a Newton direction toward the point (x^μ, y^μ, s^μ) on $\mathcal{C}$ at which the products $x_j s_j$ of all complementary pairs in this primal, dual pair of problems are $=\mu$. Many algorithms choose σ from the open interval (0,1) to trade off between twin goals of reducing μ and improving centrality.

Then take the next point to be $(\hat{x}, \hat{y}, \hat{s}) = (\bar{x}, \bar{y}, \bar{s}) + \alpha(\Delta x, \Delta y, \Delta s)$, where α is a positive step length selected so that $(\hat{x}, \hat{s})$ remains > 0.

With $(\hat{x}, \hat{y}, \hat{s})$ as the new current interior feasible solution, the method now goes to the next step.

It has been shown that the sequence of interior feasible solutions obtained in this method converges to a point in the optimum face.

A Gravitational Interior Point Method for LP

This is a new type of interior point method discussed recently in Murty (2006). We will describe the main ideas in this method briefly. It considers LP in the form: minimize $z(x) = cx$ subject to $Ax \geq b$, where A is a matrix of order $m \times n$. Let K denote the set of feasible solutions, and K^0 its interior $= \{x : Ax > b\}$. Let $A_{i.}$ denote the i-th row vector of A; assume $||c|| = ||A_{i.}|| = 1$ for all i.

This method also needs an interior feasible solution for initiating the method. Each iteration in the method consists of two steps. We will discuss each of these steps.

Step 1: *Centering step*: Let x^0 be the current interior feasible solution. The orthogonal distance of the point x^0 to the boundary hyperplane defined by the equation $A_{i.}x = b_i$ is $A_{i.}x^0 - b_i$ for $i = 1$ to n. The minimum value of these orthogonal distances is the radius of the largest sphere that can be constructed within K with x^0 as center; hence its radius is $\delta^0 = \min\{A_{i.}x^0 - b_i : i = 1, \ldots, m\}$.

This step tries to move from x^0 to another interior feasible solution on the objective plane $H^0 = \{x : cx = cx^0\}$ through x^0, to maximize the radius of the sphere that can be constructed within K with the center in $H^0 \cap K^0$. That leads to another LP: max δ subject to $\delta \leq A_{i.}x - b_i$, $i = 1, \ldots, m$, and $cx = cx^0$.

In the method, this new LP is solved approximately using a series of line searches on $H^0 \cap K^0$ beginning with x^0. The directions considered for the search are orthogonal projections of normal directions to the facetal hyperplanes of K on the hyperplane $\{x : cx = 0\}$, which form the set: $P = \{P_{.i} = (I - c^T c)A_i^T : i = 1, \ldots, m\}$. There are other line search directions that can be included in this search, but this set of directions has given excellent results in the limited computational testing done so far.

Let x^r be the current center. At x^r, $P_{.i} \in P$ is a profitable direction to move (i.e., this move leads to a better center) only if all the dot products $A_{t.}P_{.i}$ for $t \in T$ have the same sign, where $T = \{t : t \text{ ties for the minimum in } \{A_{q.}x^r - b_q : q = 1, \ldots, m\}\}$. If $P_{.i}$ is a profitable direction to move at x^r, the optimum step length is the optimum α in the following LP in two variables θ, α: Max θ subject to $\theta - \alpha A_{t.}P_{.i} \leq A_{t.}x^r - b_t$, $t = 1, \ldots, m$, which can be solved very efficiently by a special algorithm.

The line searches are continued either until a point is reached where none of the directions in P are profitable, or the improvement in the radius of the sphere in each line search becomes small.

Step 2: *Descent Step*: Let $\bar{x}$ be the final center in $H^0 \cap K^0$ selected in the centering step. At $\bar{x}$ two descent directions are available, $-c^T$ and $\bar{x} - \tilde{x}$, where $\tilde{x}$ is the center selected in the previous iteration. Move from $\tilde{x}$ in the direction among these two that gives the greatest decrease in the objective value, to within $\in$ of the boundary, where $\in$ is a tolerance for interiorness.

If x^* is the point obtained after the move, go to the next iteration with x^* as the new current interior feasible solution.

It has been proved that this method takes no more than $O(m)$ iterations if the centering step is carried out exactly in each iteration. The method with the approximate centering strategy is currently undergoing computational tests. The biggest advantage of this method over the others is that it needs no matrix inversions, and hence offers a fast descent method to solve LPs whether they are sparse or not. Also, the method is not affected by any redundant constraints in the model.

1.7 Software Systems Available for Solving LP Models

There are two types of software. *Solver software* takes an instance of an LP model as input and applies one or more solution methods and outputs the results. *Modeling software* does not incorporate solution methods; it offers a computer modeling language for expressing LP models, features for reporting, model management, and application development, in addition to a translator for the language. Most modeling systems offer a variety of bundled solvers.

The most commonly talked about commercial solvers for LP are CPLEX, OSL, MATLAB, LINDO, LOQO, EXCEL, and several others. The most common modeling software systems are AMPL, GAMS, MPL, and several others.

Detailed information about these systems and their capabilities and limitations can be obtained from the paper by Fourer (2005), and the Web sites for the various software systems and their vendors are also given there.

Also, users can submit jobs to the NEOS server maintained by Argonne National Laboratory and retrieve job results. See the Web site http://www-neos.mcs.anl.gov/for details. You can see a complete list of currently available solvers and detailed information on each of them at the Optimization Software Guide Web site (http://www-fp.mcs.anl.gov/otc/Guide/SoftwareGuide/Categories/linearprog.html).

1.8 Multiobjective LP Models

So far we have discussed LP models for problems in which there is a well-defined single objective function to optimize. But many real world applications involve several objective functions simultaneously. For example, most manufacturing companies are intensely interested in attaining large values for several things, such as the company's net profit, its market share, and the public's recognition of the company as a progressive organization.

In all such applications, it is extremely rare to have one feasible solution that simultaneously optimizes all of the objective functions. Typically, optimizing one of the objective functions has the effect of moving another objective function away from its most desirable value. These are the usual conflicts among the objective functions in multiobjective models. Under such *conflicts*, a multiobjective problem is not really a mathematically well-posed problem unless information on how much value of one objective function can be sacrificed for unit gain in the value of another is given. Such *tradeoff information* is usually not available, but when it is available, it makes the problem easier to analyze.

Because of the conflicts among the various objective functions, there is no well-accepted concept of *optimality* in multiobjective problems. Concepts like Pareto optima, or nondominated solutions, or efficient points, or equilibrium solutions have been defined, and mathematical algorithms to enumerate all such solutions have been developed. Usually there are many such solutions, and there are no well-accepted criteria to select one of them for implementation, with the result that all this methodology remains unused in practice.

The most practical approach, and one that is actually used by practitioners is the *goal programming approach*, originally proposed by Charnes and Cooper (1961, 1977), which we will now discuss. This material is based on the section on goal programming in Murty (2005b and manuscript under preparation).

Let $c^1 x, \ldots, c^k x$ be the various objective functions to be optimized over the set of feasible solutions of $Ax = b$, $x \geq 0$. The goal programming approach has the added conveniences that different objective functions can be measured in different units, and that it is not necessary to have all the objective functions in the same (either maximization or minimization) form. So, some of the objective functions among $c^1 x, \ldots, c^k x$ may have to be minimized, others maximized.

In this approach, instead of trying to optimize each objective function, the decision maker is asked to specify a *goal* or *target value* that realistically is the most desirable value for that function. For $r = 1$ to k, let g_r be the specified goal for $c^r x$.

At any feasible solution x, for $r = 1$ to k, we express the deviation in the r-th objective function from its goal, $c^r x - g_r$, as a difference of two nonnegative variables

$$c^r x - g_r = u_r^+ - u_r^-$$

where u_r^+, u_r^- are the *positive* and *negative parts of the deviation* $c^r x - g_r$, that is,

$$u_r^+ = \begin{cases} 0 & \text{if } c^r x - g_r \leq 0 \\ c^r x - g_r & \text{if } c^r x - g_r > 0 \end{cases}$$
$$u_r^- = \begin{cases} 0 & \text{if } c^r x - g_r \geq 0 \\ -(c^r x - g_r) & \text{if } c^r x - g_r < 0 \end{cases}$$

The goal programming model for the original multiobjective problem will be a single objective problem in which we try to minimize a *linear penalty function* of these deviation variables of the form $\sum_1^k (\alpha_r u_r^+ + \beta_r u_r^-)$, where $\alpha_r, \beta_r \geq 0$ for all r. We now explain how the coefficients α_r, β_r are to be selected.

If the objective function $c^r x$ is one which is desired to be maximized, then feasible solutions x which make $u_r^- = 0$ and $u_r^+ \geq 0$ are desirable, while those which make $u_r^+ = 0$ and $u_r^- > 0$ become more and more undesirable as the value of u_r^- increases. In this case u_r^+ measures the (desirable) excess in this objective value over its specified target, and u_r^- measures the (undesirable) shortfall in its value from its target. To guarantee that the algorithm seeks solutions in which u_r^- is as small as possible, we associate a positive penalty coefficient β_r with u_r^-, and include a term of the form $\alpha_r u_r^+ + \beta_r u_r^-$ (where $\alpha_r = 0, \beta_r > 0$) in the penalty function that the goal programming model tries to minimize. $\beta_r > 0$ measures the loss or penalty per unit shortfall in the value of $c^r x$ from its specified goal of g_r. The value of β_r should reflect the importance attached by the decision maker for attaining the specified goal on this objective function (higher values of β_r represent greater importance). The coefficient α_r of u_r^+ is chosen to be 0 in this case because our desire is to see u_r^+ become positive as far as possible.

If the objective function $c^r x$ is one which is desired to be minimized, then positive values for u_r^- are highly desirable, whereas positive values for u_r^+ are undesirable. So, for these r we include a term of the form $\alpha_r u_r^+ + \beta_r u_r^-$, where $\alpha_r > 0$ and $\beta_r = 0$ in the penalty function that goal programming model minimizes. Higher values of α_r represent greater importance attached by the decision maker to the objective functions in this class.

There may be some objective functions $c_r(x)$ in the original multiobjective problem for which both positive and negative deviations are considered undesirable. For objective functions in this class we desire values that are as close to the specified targets as possible. For each such objective function we include a term $\alpha_r u_r^+ + \beta_r u_r^-$, with both α_r and $\beta_r > 0$, in the penalty function that the goal programming model minimizes.

So, the goal programming model is the following single objective problem.

$$\text{Minimize} \quad \sum_{1}^{k} (\alpha_r u_r^+ + \beta_r u_r^-)$$
$$\text{subject to} \quad c^r x - g_r = u_r^+ - u_r^-, \quad r = 1 \text{ to } k$$
$$Ax = b$$
$$x, u_r^+, u_r^- \geq 0, \quad r = 1 \text{ to } k$$

As this model is a single objective function linear program it can be solved by the algorithms discussed earlier. Also, as all α_r and $\beta_r \geq 0$, and from the manner in which the values for α_r, β_r are selected, $(u_r^+)(u_r^-)$ will be zero for all r in an optimum solution of this model; that is, $u_r^+ = \text{maximum}\{c^r x - g_r, 0\}$, $u_r^- = \text{minimum}\{0, -(c^r x - g_r)\}$ will hold for all r. Hence solving this model will try to meet the targets set for each objective function, or deviate from them in the desired direction as far as possible.

The optimum solution of this goal programming model depends critically on the goals selected and on the choice of the penalty coefficients $\alpha = (\alpha_1, \ldots, \alpha_k)$, $\beta = (\beta_1, \ldots, \beta_k)$. Without any loss of generality we can assume that the vectors α, β are scaled so that $\sum_{r=1}^{k} (\alpha_r + \beta_r) = 1$. Then the larger an α_r or β_r, the more the importance the decision maker places on attaining the goal set for $c_r(x)$. Again, there may not be universal agreement among all the decision makers involved on the penalty coefficient vectors α, β to be used; it has to be determined by negotiations among them. Once α and β are determined, an optimum solution of the goal programming model is the solution to implement. Solving it with a variety of penalty vectors α and β and reviewing the various optimum solutions obtained may make the choice in selecting one of them for implementation easier. One can also solve the goal programming model with different sets of goal vectors for the various objective functions. This process can be repeated until at some stage, an optimum solution obtained for it seems to be a reasonable one for the original multiobjective problem. Exploring with the optimum solutions of this model for different goal and penalty coefficient vectors in this manner, one can expect to get a practically satisfactory solution for the multiobjective problem.

Example 1.7

As an example, consider the problem of the fertilizer manufacturer to determine the best values for x_1, x_2, the tons of Hi-ph and Lo-ph fertilizer to manufacture daily, discussed in

Example 1.2 (Section 1.2). There we considered only the objective of maximizing the net daily profit, $c^1 x = \$(15x_1 + 10x_2)$.

Suppose the fertilizer manufacturer is also interested in maximizing the market share of the company, which is usually measured by the total daily fertilizer sales of the company, $c^2 x = (x_1 + x_2)$ tons.

In addition, suppose the fertilizer manufacturer is also interested in maximizing the public's perception of the company as being one on the forefront of technology, measured by the Hi-ph tonnage sold by the company daily, $c^3 x = x_1$ tons. So, we now have three objective functions $c^1 x$, $c^2 x$, $c^3 x$, all to be maximized simultaneously, subject to the constraints in Equation 1.2.

Suppose the company has decided to set a goal of $g_1 = \$13,000$ for daily net profit; $g_2 = 1150$ tons for total tonnage of fertilizer sold daily; and $g_3 = 400$ tons for Hi-ph tonnage sold daily. Also, suppose the penalty coefficients associated with shortfalls in these goals are required to be 0.5, 0.3, 0.2, respectively. With this data, the goal programming formulation of this problem, after transferring the deviation variables to the left-hand side, is

$$
\begin{aligned}
\text{Minimize} \quad & 0.5 u_1^- + 0.3 u_2^- + 0.2 u_3^- \\
\text{subject to} \quad & 15 x_1 + 10 x_2 + u_1^- - u_1^+ = 13{,}000 \\
& x_1 + x_2 + u_2^- - u_2^+ = 1150 \\
& x_1 + u_3^- - u_3^+ = 400 \\
& 2 x_1 + x_2 \le 1500 \\
& x_1 + x_2 \le 1200 \\
& x_1 \le 500 \\
& x_1, x_2, u_1^-, u_1^+, u_2^-, u_2^+, u_3^-, u_3^+ \ge 0
\end{aligned}
$$

An optimum solution of this problem is $\hat{x} = (\hat{x}_1, \hat{x}_2)^T = (350, 800)^T$. $\hat{x}$ attains the goals set for net daily profit and total fertilizer tonnage sold daily, but falls short of the goal on the Hi-ph tonnage sold daily by 50 tons. $\hat{x}$ is the solution for this multiobjective problem obtained by goal programming, with the goals and penalty coefficients given above.

References

1. Brown, G. and Graves, G. 1977, "Elastic Programming," Talk given at an ORSA-TIMS Conference.
2. Charnes, A. and Cooper, W. W. 1961, *Management Models and Industrial Applications of LP*, Wiley, New York.
3. Charnes, A. and Cooper, W. W. 1977, "Goal Programming and Multiple Objective Optimizations, Part I," *European Journal of Operations Research*, 1, 39–54.
4. Chinnek, J. W. and Dravineks, E. W. 1991, "Locating Minimal Infeasible Sets in Linear Programming," *ORSA Journal on Computing*, 3, 157–168.
5. Dantzig, G. B. 1963, *Linear Programming and Extensions*, Princeton University Press, Princeton, NJ.
6. Dantzig, G. B. and Thapa, M. N. 1997, *Linear Programming, 1. Introduction, 2. Theory and Extensions*, Springer-Verlag, New York.
7. Fourer, R. 2005, "Linear Programming Software Survey," *ORMS Today*, 32(3), 46–55.
8. Gale, D. 1960, *The Theory of Linear Economic Models*, McGraw-Hill, New York.
9. Hitchcock, F. L. 1941, "The Distribution of a Product from Several Sources to Numerous Localities," *Journal of Mathematics and Physics*, 20, 224–230.

10. Karmarkar, N. K. 1984, "A New Polynomial-Time Algorithm for Linear Programming," *Combinatorica*, 4, 373–395.
11. Koopmans, T. C. 1949, "Optimum Utilization of the Transportation System," *Econometrica*, 17, Supplement.
12. Murty, K. G. 1995, *Operations Research—Deterministic Optimization Models*, Prentice Hall, Englewood Cliffs, NJ.
13. Murty, K. G., Kabadi, S. N., and Chandrasekaran, R. 2000, "Infeasibility Analysis for Linear Systems, A Survey," *The Arabian Journal for Science and Engineering*, 25(1C), 3–18.
14. Murty, K. G. 2004, *Computational and Algorithmic Linear Algebra and n-Dimensional Geometry*, Freshman-Sophomore level linear algebra book available as download at: http://ioe.engin.umich.edu/people/fac/books/murty/algorithmic_linear_algebra/.
15. Murty, K. G. 2005a, "A Gravitational Interior Point Method for LP," *Opsearch*, 42(1), 28–36.
16. Murty, K. G. 2005b, *Optimization Models for Decision Making: Volume 1 (Junior Level)*, at http://ioe.engin.umich.edu/people/fac/books/murty/opti_model/.
17. Murty, K. G. *Optimization Models for Decision Making: Volume 2 (Masters Level)*, under preparation.
18. Murty, K. G., Wan, Y.-W., Liu, J., Tseng, M., Leung, E., Kam, K., Chiu, H., 2005a, "Hongkong International Terminals Gains Elastic Capacity Using a Data-Intensive Decision-Support System," *Interfaces*, 35(1), 61–75.
19. Murty, K. G., Liu, J., Wan, Y.-W., and Linn, R. 2005b, "A DSS (Decision Support System) for Operations in a Container Terminal," *Decision Support Systems*, 39(3), 309–332.
20. Murty, K. G. 2006, "A New Practically Efficient Interior Point Method for LP," *Algorithmic Operations Research, Dantzig Memorial Issue*, 1, 1 (January 2006)3–19; paper can be seen at the website: http://journals.hil.unb.ca/index.php/AOR/index.
21. Stigler, G. J. 1945, "The Cost of Subsistence," *Journal of Farm Economics*, 27.
22. Wright, S. J. 1996, *Primal-Dual Interior-Point Methods*, SIAM, Philadelphia.

2
Nonlinear Programming

Theodore B. Trafalis and
Robin C. Gilbert
University of Oklahoma

2.1 Introduction **2-1**
   Problem Statement • Optimization Techniques
2.2 Unconstrained Optimization **2-3**
   Line Searches • Multidimensional Optimization
2.3 Constrained Optimization **2-15**
   Direction Finding Methods • Transformation Methods
2.4 Conclusion **2-19**
References **2-20**

2.1 Introduction

Nonlinear programming is the study of problems where a nonlinear function has to be minimized or maximized over a set of values in $\mathbb{R}^n$ delimited by several nonlinear equalities and inequalities. Such problems are extremely frequent in engineering, science, and economics. Since World War II, mathematicians have been engaged to solve problems of resource allocation, optimal design, and industrial planning involving nonlinear objective functions as well as nonlinear constraints. These problems required the development of new algorithms that have been benefited with the invention of digital computers. This chapter is concentrated on presenting the fundamentals of deterministic algorithms for nonlinear programming. Our purpose is to present a summary of the basic algorithms of nonlinear programming. A pseudocode for each algorithm is also presented. We believe that our presentation in terms of a fast cookbook for nonlinear programming algorithms can benefit the practitioners of operations research and management science.

The chapter is organized as follows. In Section 2.1.1, the definition of the general nonlinear programming problem is discussed. In Section 2.1.2, optimization techniques are discussed. In Section 2.2, nonlinear deterministic techniques for unconstrained optimization are discussed. Constrained optimization techniques are discussed in Section 2.3. Finally, Section 2.4 concludes the chapter.

2.1.1 Problem Statement

Let $X \subseteq \mathbb{R}^n$ with $n \in \mathbb{N}^*$. Consider the following *optimization problem* with equality and inequality constraints:

$$\begin{aligned}
\underset{\mathbf{x} \in X}{\text{minimize}} \quad & \mathbf{f}(\mathbf{x}) \\
\text{subject to} \quad & \mathbf{g}(\mathbf{x}) \leq \mathbf{0} \\
& \mathbf{h}(\mathbf{x}) = \mathbf{0}
\end{aligned} \qquad (2.1)$$

where $\mathbf{f}\colon \mathrm{X}\to\mathbb{R}^m$, $\mathbf{g}\colon \mathrm{X}\to\mathbb{R}^p$, and $\mathbf{h}\colon \mathrm{X}\to\mathbb{R}^q$ with $(m,p,q)\in\mathbb{N}^3$. The vector function $\mathbf{f}$ is called the *objective function*, the vector function $\mathbf{g}$ is called the *inequality constraint function*, and the vector function $\mathbf{h}$ is called the *equality constraint function*. The vector $\mathbf{x}\in \mathrm{X}$ is called the *optimization variable*. If $\mathbf{x}$ is such that the constraints of problem (2.1) hold, then $\mathbf{x}$ is called a *feasible solution*. Some problems generalize the above formulation and consider $\mathbf{f}$, $\mathbf{g}$, and $\mathbf{h}$ to be random vectors. In this case, we have a *stochastic optimization problem*. By opposition we may say that problem (2.1) is a *deterministic optimization problem*.

The categories of problems vary with the choices of X, m, p, and q. If $p=q=0$ then we say that we have an *unconstrained optimization problem*. Conversely, if one of q or p is nonzero then we say that we have a *constrained optimization problem*. The categories of problems when m varies are the following:

- If $m=0$ we have a *feasibility problem*.
- If $m=1$ we have a classical optimization problem. The space of feasible solutions is called the *feasible set*. If a vector $\bar{\mathbf{x}}$ is a feasible solution such that $f(\bar{\mathbf{x}})\le f(\mathbf{x})$ for every $\mathbf{x}\in \mathrm{X}$, then $\bar{\mathbf{x}}$ is called *optimal solution* and $f(\bar{\mathbf{x}})$ is called *optimal value*. If X is continuous and if the objective function is convex then we have a *convex optimization problem* that belongs to a very important category of optimization problems.
- If $m\ge 2$ then the above problem is a *multiobjective optimization problem* where the space of feasible solutions is called *decision space*. The optimality notion in the case $m=1$ is replaced by the *Pareto optimality* and "optimal" solutions are said to be *Pareto optimal*. A feasible solution $\bar{\mathbf{x}}$ is Pareto optimal if for every $\mathbf{x}$ in the decision space we have $f_i(\bar{\mathbf{x}})\le f_i(\mathbf{x})$ for every $i\in [\![1,m]\!]$.

If $\mathrm{X}\subseteq \mathbb{Z}^n$ then we have an *integer optimization problem*. If only some of the components of the optimization variables are integers then we have a *mixed-integer optimization problem*. In the other cases we have a *continuous optimization problem*.

Most of the optimization problems found in the literature deal with the case $m=1$ for deterministic optimization problems. From now on, we will always consider this case in all of the following discussions. In this class of optimization problems we have many subcategories that depend on the choice of f and the choice of the constraint functions. The most common cases are:

- The objective function f and the constraint functions $\mathbf{g}$ and $\mathbf{h}$ are linear. In this case we have a *linear optimization problem*.
- The objective function f is quadratic and the constraint functions $\mathbf{g}$ and $\mathbf{h}$ are linear. In this case we have a *quadratic optimization problem*.
- The objective function f is quadratic and the constraint functions $\mathbf{g}$ and $\mathbf{h}$ are quadratic. In this case we have a *quadratically constrained quadratic optimization problem*.
- The objective function f is linear and subject to second-order cone constraints (of the type $\|\mathbf{A}\mathbf{x}+\mathbf{b}\|_2 \le \langle \mathbf{c}\cdot\mathbf{x}\rangle + d$). In this case we have a *second-order cone optimization problem*.
- The objective function f is linear and subject to a matrix inequality. In this case we have a *semidefinite optimization problem*.

- The objective function f and the constraint functions **g** and **h** are nonlinear. In this case we have a *nonlinear optimization problem*.

We will be dealing with the latter case in this chapter.

2.1.2 Optimization Techniques

There are three major groups of optimization techniques in nonlinear optimization. These techniques are summarized in the following list:

Deterministic techniques: These methods are commonly used in convex optimization. Convex nonlinear optimization problems have been extensively studied in the last 60 years and the literature is now thriving with convergence theorems and optimality conditions for them. The backbone of most of these methods is the so-called descent algorithm. The steepest descent method, the Newton method, the penalty and barrier methods, and the feasible directions methods fall in this group.

Stochastic techniques: These methods are based on probabilistic meta-algorithms. They seek to explore the feasibility region by moving from feasible solutions to feasible solutions in directions that minimize the objective value. They have been proved to converge toward local minima by satisfying certain conditions. Simulated annealing is an example of a stochastic technique (see Refs. [30,36,47]).

Heuristic strategies: These methods are based on heuristics for finding good feasible solutions to very complicated optimization problems. They may not work for some problems and their behavior is not yet very well understood. They are used whenever the first two groups of techniques fail to find solutions or if these techniques cannot be reasonably used to solve given problems. Local searches and swarm intelligence are some of the numerous heuristic techniques introduced in the past few years (see Refs. [9,12,20,21,47]).

2.2 Unconstrained Optimization

In this section, several common algorithms for continuous nonlinear convex optimization are reviewed for the cases where no constraints are present. Section 2.2.1 reviews the most usual line searches and Section 2.2.2 presents methods for optimizing convex functions on $\mathbb{R}^n$.

2.2.1 Line Searches

The purpose of line searches is to locate a minimum of real-valued functions over an interval of $\mathbb{R}$. They are extremely fundamental procedures that are frequently used as subroutines in other optimization algorithms based on descent directions. The objective function f is required to be strictly unimodal over a specified interval $[a, b]$ of $\mathbb{R}$; that is, there exists a $\bar{x}$ that minimizes f over $[a, b]$ and for every $(x_1, x_2) \in [a, b]^2$ such that $x_1 < x_2$ we have that $\bar{x} \leq x_1 \Rightarrow f(x_1) < f(x_2)$ and $x_2 \leq \bar{x} \Rightarrow f(x_2) < f(x_1)$. Depending on the line search method, f may need to be differentiable.

There are at least three types of line searches:

- Line searches based on region elimination. These methods exploit the assumption of strict unimodality to sequentially shrink the initial interval until it reaches a negligible length. The Fibonacci search and the bisection search are very successful line search methods (see the sections "Fibonacci Search" and "Bisection Search").
- Lines searches based on curve fitting. These methods try to exploit the shape of the objective function to accelerate the bracketing of a minimum. The quadratic fit line search and the cubic fit line search are very popular curve fitting line searches (see the section "Quadratic Fit Search").
- Approximated line searches. Contrary to the above line searches, these methods try to find acceptable "minimizers" with the minimum function evaluations possible. This is often necessary when it is computationally difficult to evaluate the objective function. The backtracking search is an approximated line search that is adapted with a different set of stopping rules (notably the popular Armijo's rule). See Algorithm 2.5.

Bracketing of the Minimum

Line searches often require an initial interval containing a minimum to be provided before starting the search. The following technique provides a way to bracket the minimum of a strictly unimodal function defined over an open interval of $\mathbb{R}$. It starts from an initial point belonging to the domain of definition of the objective function and exploits the strict unimodality assumption to gradually expand the length of an inspection interval by following the decreasing slope of the function. It stops if the initial point does not belong to the domain of definition or whenever the slope of the function increases.

Algorithm 2.1 is a modified version of a method credited to Swann [52]. This variant can handle strictly unimodal functions that are defined on an open interval of $\mathbb{R}$ with the supplementary assumption that their values are ∞ outside this interval. Such modification is quite useful when this bracketing technique is used with barrier functions (see the section "Sequential Unconstrained Minimization Techniques"). This method requires only functional evaluations and a positive step parameter δ. It starts from an initial point x, determines the decreasing slope of the function (lines 1–11), and then detects a change in the slope by jumping successively on the domain of definition by a step length of δ times a power of 2 (lines 12–18). If the current test point is feasible, then the step length at the next iteration will be doubled. Otherwise the step length will be halved.

Fibonacci Search

The Fibonacci line search is a derivative-free line search based on the region elimination scheme that finds the minimizer of a strictly unimodal function on a closed bounded interval of $\mathbb{R}$. This method is credited to Kiefer [29]. As the objective function is strictly unimodal, it is possible to eliminate successively parts of the initial interval by knowing the objective function value at four different points of the interval. The ways to manage the locations of the test points vary with the region-elimination-based line searches but the Fibonacci search is specially designed to reduce the initial interval to a given length with the minimum

Algorithm 2.1: Modified Swann's bracketing method.

Input: function f, starting point x, step parameter $\delta > 0$.
Output: interval bracketing the minimum.
1 $k \leftarrow 1$, fail $\leftarrow k$, $\delta_a \leftarrow \delta$, $\delta_b \leftarrow \delta$, **if** $f(x) = \infty$ **then return** $\emptyset$;
2 **repeat**
3    $a_k \leftarrow x - \delta_a$, **if** $f(a_k) = \infty$ **then** $\delta_a \leftarrow \delta_a/2$, fail $\leftarrow k+1$;
4 **until** $f(a_k) < \infty$;
5 **repeat**
6    $b_k \leftarrow x + \delta_b$, **if** $f(b_k) = \infty$ **then** $\delta_b \leftarrow \delta_b/2$, fail $\leftarrow k+1$;
7 **until** $f(b_k) < \infty$;
8 **if** $f(x) \leq f(a_k)$ and $f(x) \leq f(b_k)$ **then return** $[a_k, b_k]$;
9 $a_{k+1} \leftarrow x$;
10 **if** $f(x) \leq f(b_k)$ **then** $b_{k+1} \leftarrow a_k$, $\delta \leftarrow \delta_a$, $\sigma \leftarrow -1$;
11 **else** $b_{k+1} \leftarrow b_k$, $\delta \leftarrow \delta_b$, $\sigma \leftarrow 1$;
12 **while** $f(b_{k+1}) < f(a_{k+1})$ **do**
13    $k \leftarrow k+1$, $a_{k+1} \leftarrow b_k$, **if** fail $= k$ **then** $\delta \leftarrow \delta/2$ **else** $\delta \leftarrow 2\delta$;
14   **repeat**
15      $b_{k+1} \leftarrow b_k + \sigma\delta$, **if** $f(b_{k+1}) = \infty$ **then** $\delta \leftarrow \delta/2$, fail $\leftarrow k+1$;
16   **until** $f(b_{k+1}) < \infty$;
17 **end**
18 **if** $\sigma > 0$ **then return** $[a_k, b_{k+1}]$ **else return** $[b_{k+1}, a_k]$;

number of functional evaluations. The Fibonacci search is notably slightly better than the golden section search for reducing the search interval.

As its name suggests, the method is based on the famous Fibonacci sequence $(F_n)_{n \in \mathbb{N}}$ defined by $F_0 = F_1 = 1$ and $F_{n+2} = F_{n+1} + F_n$ for $n \in \mathbb{N}$. The test points are chosen such that the information about the function obtained at the previous iteration is used in the next. In this way the search requires only a single functional evaluation per iteration in its main loop. Contrary to the golden section search with which the Fibonacci search is almost identical, the reduction interval ratio at each iteration varies.

Algorithm 2.2 implements the Fibonacci search. The method requires only functional evaluations and an initial interval $[a, b]$. It is assumed that a termination scalar $\varepsilon > 0$ that represents the maximum length of the final bracketing interval is given. It is suggested that ε should be at least equal to the square root of the spacing of floating point numbers [46]. The algorithm determines first the needed number of iterations (line 1) of its main loop (lines 3–10). Then it returns the mid-point of the final bracketing interval as an estimation of the minimizer of the objective function (line 11). The region-elimination scheme is loosely the following: if the functional values are greater on the left of the current interval (line 5), then the leftmost part is deleted and a new test point is created on the right part (line 6). Conversely, if the functional values are greater on the right (line 7) then the rightmost part is deleted and a new test point on the left is created (line 8). Naturally this scheme works only for strictly unimodal functions.

Bisection Search

Like the Fibonacci search, the bisection search is based on the region elimination scheme. Unlike the Fibonacci search, it requires the objective function to be differentiable and performs evaluations of the derivative of f. The bisection search, which is sometimes referred

Algorithm 2.2: Fibonacci search.

    Input: function f, initial interval $[a,b]$.
    Output: approximated minimum.
    Data: termination parameter ε.
1   $k \leftarrow 0$, determine $n > 2$ such that $F_n > (b-a)/\varepsilon$;
2   $u \leftarrow a + (F_{n-2}/F_n)(b-a)$, $v \leftarrow a + (F_{n-1}/F_n)(b-a)$;
3   **while** $k < n - 2$ **do**
4       $k \leftarrow k+1$;
5   **if** $f(u) > f(v)$ **then**
6           $a \leftarrow u$, $u \leftarrow v$, $v \leftarrow a + (F_{n-k-1}/F_{n-k})(b-a)$;
7   **else**
8           $b \leftarrow v$, $v \leftarrow u$, $u \leftarrow a + (F_{n-k-2}/F_{n-k})(b-a)$;
9   **end**
10 **end**
11 **if** $f(u) > f(u + \varepsilon/2)$ **then return** $(u+b)/2$ **else return** $(a+u)/2$;

---

to as the Bolzano search, is one of the oldest line searches. It finds the minimizer of a pseudoconvex function on a closed bounded interval $[a,b]$ of $\mathbb{R}$. A differentiable function is said to be pseudoconvex on an open set $\mathbf{X}$ of $\mathbb{R}^n$ if for every $(\mathbf{x}_1, \mathbf{x}_2) \in \mathbf{X}^2$ such that $0 \leq \nabla f(\mathbf{x}_1)^t (\mathbf{x}_2 - \mathbf{x}_1)$, we have $f(\mathbf{x}_1) \leq f(\mathbf{x}_2)$. A pseudoconvex function is strictly unimodal. In our case $n=1$, therefore, a real-valued function f is pseudoconvex if for every $(x_1, x_2) \in (a,b)^2$ such that $0 \leq df(x_1)(x_2 - x_1)$ we have $f(x_1) \leq f(x_2)$.

Algorithm 2.3 implements the bisection search. This search uses the information conveyed by the derivative function to sequentially eliminate portions of the initial bracketing interval $[a,b]$. The algorithm determines first the needed number of iterations (line 1) of its main loop (lines 2–8). Then it returns the mid-point of the final bracketing interval as an estimation of the minimizer of the objective function (line 9). The main loop works like the main loop of the Fibonacci search: if the derivative value at the mid-point is strictly positive then the rightmost part of the current interval is deleted; otherwise if the derivative value is strictly negative then the leftmost part is deleted (line 6). In the case where the derivative value at the mid-point is null then the algorithm stops as it has reached optimality (line 4).

Algorithm 2.3: Bisection search.

    Input: derivative function df, initial interval $[a,b]$.
    Output: approximated minimum.
    Data: termination parameter ε.
1   $k \leftarrow 0$, determine $n \geq 0$ such that $2^n \geq (b-a)/\varepsilon$;
2   **while** $k < n$ **do**
3       $k \leftarrow k+1$, $u \leftarrow (a+b)/2$;
4   **if** $df(u) = 0$ **then return** u;
5   **else**
6   **if** $df(u) > 0$ **then** $b \leftarrow u$ **else** $a \leftarrow u$;
7   **end**
8   **end**
9   **return** $(a+b)/2$;

Quadratic Fit Search

The quadratic fit search tries to interpolate the location of the minimum of a strictly unimodal function at each iteration and refines in this way the location of the true minimizer. In a way, this method tries to guess the shape of the function to fasten the search for the minimizer. This method requires only functional evaluations. Nevertheless, it is sometimes used in conjunction with the bisection search when the derivative of the objective function is available. Another curve fitting line search based on the cubic interpolation of f is also available. However, this method requires f to be differentiable and may sometimes face severe ill-conditioning effects.

The implementation of the quadratic fit search is shown in Algorithm 2.4. Using three starting points $x_1 < x_2 < x_3$ with $f(x_2) \leq f(x_1)$ and $f(x_2) \leq f(x_3)$, the method finds a minimizer of the quadratic interpolation function (lines 2–4), then corrects its position (lines 5–8) and updates the values of x_1, x_2, and x_3 with some sort of region-elimination scheme (lines 9–13). Once the main loop is completed (lines 1–14), the minimizer is estimated by x_2 (line 15).

Algorithm 2.4: Quadratic fit search.

Input: function f, starting points $x_1 < x_2 < x_3$ with $f(x_2) \leq f(x_1)$ and $f(x_2) \leq f(x_3)$.
Output: approximated minimum.
Data: termination parameter ε.

1 **while** $x_3 - x_1 > \varepsilon$ **do**
2      $a_1 \leftarrow x_2^2 - x_3^2$, $a_2 \leftarrow x_3^2 - x_1^2$, $a_3 \leftarrow x_1^2 - x_2^2$;
3      $b_1 \leftarrow x_2 - x_3$, $b_2 \leftarrow x_3 - x_1$, $b_3 \leftarrow x_1 - x_2$;
4      $\overline{x} \leftarrow (a_1 f(x1) + a_2 f(x_2) + a_3 f(x_3))/(2(b_1 f(x_1) + b_2 f(x_2) + b_3 f(x_3)))$;
5   **if** $\overline{x} = x_2$ **then**
6   **if** $x_3 - x_2 \leq x_2 - x_1$ **then** $\overline{x} \leftarrow x_2 - \varepsilon/2$ **else** $\overline{x} \leftarrow x_2 + \varepsilon/2$;
7   **end**
8   **if** $\overline{x} > x_2$ **then**
9   **if** $f(\overline{x}) \geq f(x_2)$ **then** $x_3 \leftarrow \overline{x}$ **else** $x_1 \leftarrow x_2$, $x_2 \leftarrow \overline{x}$;
10   **else**
11   **if** $f(\overline{x}) \geq f(x_2)$ **then** $x_1 \leftarrow \overline{x}$ **else** $x_3 \leftarrow x_2$, $x_2 \leftarrow \overline{x}$;
12   **end**
13 **end**
14 **return** x_2;

Backtracking Search

The backtracking search is an approximated line search that aims to find acceptable (and *positive*) scalars that will give low functional values while trying to perform the least number of functional evaluations possible. It is often used when a single functional evaluation requires non-negligible computational resources. This line search does not need to be started with an initial interval bracketing the minimum. Instead it just needs a (positive) guess value x_1 that belongs to the domain of definition of the objective function. It is important to notice that this approach will look for acceptable scalars that have the same sign of the initial guess.

The backtracking line search has many different stopping criteria. The most famous one is Armijo's condition [1] that uses the derivative of f to check that the current test scalar

x is giving "enough" decrease in f. Armijo's condition is sometimes coupled with a supplementary curvature condition on the slope of f at x. These two conditions are better known as the *Wolfe conditions* (see Ref. [39]). There exist many refinements for the backtracking search, notably safeguard steps that ensure that the objective function is well defined at the current test scalar. These safeguard steps are useful whenever barrier functions are used (see the section "Sequential Unconstrained Minimization Techniques").

Algorithm 2.5 is a straightforward implementation of the backtracking search with Armijo's condition. Depending on whether Armijo's condition is satisfied or not (line 2), the initial guess is sequentially doubled until Armijo's condition does not hold any longer (line 3). This loop actually tries to avoid the solution to be too close to zero. Once the loop is over, the previous solution is returned (line 4). Conversely, if Armijo's condition does not hold for the initial guess, the current scalar is sequentially halved until Armijo's condition is restored (line 6). It is important to note that the sequence generated by this algorithm consists only of positive scalars.

Algorithm 2.5: Backtracking search with Armijo's condition.

Input: function f, derivative df, initial guess $x_1 > 0$.
Output: acceptable step length.
Data: parameters $0 < c < 1$ (usually $c = 0.2$).
1 $k \leftarrow 0$;
2 **if** $f(x_{k+1}) \leq f(0) + cx_{k+1}\, df(0)$ **then**
3   **repeat** $k \leftarrow k+1$, $x_{k+1} \leftarrow 2x_k$ **until** $f(x_{k+1}) > f(0) + cx_{k+1}\, df(0)$;
4   **return** x_k;
5 **else**
6   **repeat** $k \leftarrow k+1$, $x_{k+1} \leftarrow x_k/2$ **until** $f(x_{k+1}) \leq f(0) + cx_{k+1}\, df(0)$;
7   **return** x_{k+1};
8 **end**

2.2.2 Multidimensional Optimization

There exist several strategies for optimizing a convex function f over $\mathbb{R}^n$. Some of them require derivatives, while others do not. However, almost all of them require a line search inside their main loop. They can be categorized by the following:

- Methods without derivatives:
  - Simplex Search method or S^2 method: see the section "Simplex Search Method."
  - Pattern search methods: see the section "The Method of Hooke and Jeeves."
  - Methods with adaptive search directions: see the section "The Method of Rosenbrock" and "The Method of Zangwill."
- Methods with derivatives:
  - Steepest descent method: see the section "Steepest Descent Method."
  - Conjugate directions methods: see the section "Method of Fletcher and Reeves" and "Quasi-Newton Methods."
  - Newton-based methods: see the section "Levenberg–Marquardt Method."

Concerning recent works on methods without derivatives, the reader might be interested in the paper of Lewis et al. [33] and the article of Kolda et al. [31].

Simplex Search Method or S² Method

The original simplex search method (which is different from the simplex method used in linear programming) is credited to Spendley et al. [51]. They suggested sequentially transforming a nondegenerate simplex in $\mathbb{R}^n$ by reflecting one of its vertices through the centroid at each iteration and by re-initializing this simplex if it remains unchanged for a certain number of iterations. In this way it takes at most $n+1$ iterations to find a downhill direction. Nelder and Mead [37,38] proposed a variant where operations like expansion and contraction are allowed.

Algorithm 2.6 implements the simplex search of Nelder and Mead. First, with the help of a scalar $c > 0$, a simplex S is created from a base point $\mathbf{x}_1 \in \mathbb{R}^n$ (lines 1–3). Then while the size of the simplex S is greater than some critical value (line 4), the extremal functional values at the vertex of S are taken (line 5). The vertex with the maximum functional value is identified as $\mathbf{x}_M$ and the vertex with the lowest functional value is identified as $\mathbf{x}_m$. Then the centroid $\overline{\mathbf{x}}$ of the polyhedron $S \setminus \{\mathbf{x}_M\}$ is computed and the reflexion of $\mathbf{x}_m$ through $\overline{\mathbf{x}}$, denoted by $\mathbf{x}_r$, is computed (line 6). If the functional value at the reflexion point is smaller than the smallest functional value previously computed (line 7), then the expansion of the centroid $\overline{\mathbf{x}}$ through the reflexion point is computed and stored in $\mathbf{x}_e$. Then the vertex $\mathbf{x}_M$ is replaced by the expansion point if the functional value at the expansion point is less than the functional value at the reflexion point. Otherwise the vertex $\mathbf{x}_M$ is replaced by the reflexion point (line 8). If line 7 is incorrect then the vertex $\mathbf{x}_M$ is replaced by the reflexion point if the functional value at that point is smaller than the greatest functional value at $S \setminus \{\mathbf{x}_M\}$ (line 10). Otherwise a contraction point $\mathbf{x}_c$ is computed (line 12) and the simplex S

Algorithm 2.6: Nelder and Mead simplex search method.

Input: function f, starting point $\mathbf{x}_1 \in \mathbb{R}^n$, simplex size $c > 0$.
Output: approximated minimum.
Data: termination parameter ε, coefficients $\alpha > 0$ (reflexion), $0 < \beta < 1$ (contraction), $\gamma > 1$ (expansion) ($\alpha = 1$, $\beta = 0.5$, $\gamma = 2$).

1 $a \leftarrow c(\sqrt{n+1} + n - 1)/(n\sqrt{2})$, $b \leftarrow c(\sqrt{n+1} - 1)/(n\sqrt{2})$;
2 **for** $i \in [\![1, n]\!]$ **do** $\mathbf{d}_i \leftarrow b\mathbf{1}_n$, $(\mathbf{d}_i)_i \leftarrow a$, $\mathbf{x}_{i+1} \leftarrow \mathbf{x}_1 + \mathbf{d}_i$;
3 $S = \{\mathbf{x}_1, \ldots, \mathbf{x}_{n+1}\}$;
4 **while** size(S) $\geq$ critical_value(ε) **do**
5     $\mathbf{x}_m = \mathrm{argmin}_{\mathbf{x} \in S} f(\mathbf{x})$, $\mathbf{x}_M = \mathrm{argmax}_{\mathbf{x} \in S} f(\mathbf{x})$;
6     $\overline{\mathbf{x}} \leftarrow \frac{1}{n}\sum_{i=1, i \neq M}^{n+1} \mathbf{x}_i$, $\mathbf{x}_r \leftarrow \overline{\mathbf{x}} + \alpha(\overline{\mathbf{x}} - \mathbf{x}_M)$;
7   **if** $f(\mathbf{x}_m) > f(\mathbf{x}_r)$ **then**
8       $\mathbf{x}_e \leftarrow \overline{\mathbf{x}} + \gamma(\mathbf{x}_r - \overline{\mathbf{x}})$, **if** $f(\mathbf{x}_r) > f(\mathbf{x}_e)$ **then** $\mathbf{x}_M \leftarrow \mathbf{x}_e$ **else** $\mathbf{x}_M \leftarrow \mathbf{x}_r$;
9   **else**
10   **if** $\max_{i \in [\![1, n+1]\!], i \neq M} f(\mathbf{x}_i) \geq f(\mathbf{x}_r)$ **then** $\mathbf{x}_M \leftarrow \mathbf{x}_r$;
11   **else**
12        $\hat{\mathbf{x}} = \mathrm{argmin}_{\mathbf{x} \in \{\mathbf{x}_r, \mathbf{x}_M\}} f(\mathbf{x})$, $\mathbf{x}_c \leftarrow \overline{\mathbf{x}} + \beta(\hat{\mathbf{x}} - \overline{\mathbf{x}})$;
13   **if** $f(\mathbf{x}_c) > f(\hat{\mathbf{x}})$ **then for** $i \in [\![1, n+1]\!]$ **do** $\mathbf{x}_i \leftarrow \mathbf{x}_i + (\mathbf{x}_m - \mathbf{x}_i)/2$;
14   **else** $\mathbf{x}_M \leftarrow \mathbf{x}_c$;
15   **end**
16   **end**
17 **end**
18 **return** $\frac{1}{n+1}\sum_{i=1}^{n+1} \mathbf{x}_i$;

is either contracted (line 13) or the vertex $\mathbf{x}_M$ is replaced by the contraction point (line 14). Once the main loop is completed, the centroid of the simplex is returned as an estimation of the minimum of f (line 18).

The Method of Hooke and Jeeves

The method of Hooke and Jeeves [27] is a globally convergent method that embeds a pattern search that tries to guess the shape of the function f to locate an efficient downhill direction.

Algorithm 2.7 implements a variant of the method of Hooke and Jeeves, found in Ref. [2], which uses line searches that are used at different steps. Line 3 implements the pattern search: after calculating a downhill direction $\mathbf{d}$, a line search minimizes the functional value from the current test point $\mathbf{x}_{k+1}$ along the direction $\mathbf{d}$. The result is stored at the point $\mathbf{z}_1$ (line 4). Then, starting from $\mathbf{z}_1$, a cyclic search along each coordinate direction iteratively locates a minimizer of f and updates the elements of the sequence $(\mathbf{z}_i)_{i \in [\![1, n+1]\!]}$ (line 5). This step is an iteration of the cyclic coordinate method (see Ref. [2]). The Hooke and Jeeves method is a combination of a pattern search (line 3) with the cyclic coordinate method (lines 4–6).

Algorithm 2.7: Method of Hooke and Jeeves.

Input: function f, starting point $\mathbf{x}_1 \in \mathbb{R}^n$.
Output: approximated minimum.
Data: termination parameter ε, coordinate directions $(\mathbf{e}_1, \ldots, \mathbf{e}_n)$.

1 $k \leftarrow 0$, $\mathbf{x}_k \leftarrow \mathbf{0}$;
2 **repeat**
3    $\mathbf{d} \leftarrow \mathbf{x}_{k+1} - \mathbf{x}_k$, **if** $k \neq 0$ **then** $\overline{\lambda} \leftarrow \mathrm{argmin}_{\lambda \in \mathbb{R}}\, f(\mathbf{x}_{k+1} + \lambda \mathbf{d})$ **else** $\overline{\lambda} \leftarrow 0$;
4    $k \leftarrow k+1$, $\mathbf{z}_1 \leftarrow \mathbf{x}_k + \overline{\lambda}\mathbf{d}$;
5   **for** $i \in [\![1,n]\!]$ **do** $\overline{\lambda} \leftarrow \mathrm{argmin}_{\lambda \in \mathbb{R}} f(\mathbf{z}_i + \lambda \mathbf{e}_i)$, $\mathbf{z}_{i+1} \leftarrow \mathbf{z}_i + \overline{\lambda}\mathbf{e}_i$;
6    $\mathbf{x}_{k+1} \leftarrow \mathbf{z}_{n+1}$;
7 **until** $\|\mathbf{x}_{k+1} - \mathbf{x}_k\|_\infty < \varepsilon$;
8 **return** $\mathbf{x}_{k+1}$;

The Method of Rosenbrock

The method of Rosenbrock [48] bears a similarity to the method of Hooke and Jeeves: it uses the scheme of the cyclic coordinate method in conjunction with a step that optimizes the search directions. At each step the set of search directions is rotated to best fit the shape of function f. Like the method of Hooke and Jeeves, the method of Rosenbrock is globally convergent in convex cases. This method is also considered to be a good minimization technique in many cases.

Algorithm 2.8 implements a variant of the method of Rosenbrock, found in Ref. [2], which uses line searches. Lines 11–13 are almost identical to lines 4–6 of the method of Hooke and Jeeves: this is the cyclic coordinate method. Lines 3–10 implement the Gram–Schmidt orthogonalization procedure to find a new set of linearly independent and orthogonal directions adapted to the shape of the function at the current point.

The Method of Zangwill

The method of Zangwill [53] is a modification of the method of Powell [42], which originally suffered from generating dependent search directions in some cases. The method of Zangwill

---
Algorithm 2.8: Method of Rosenbrock.

Input: function f, starting point $\mathbf{x}_1 \in \mathbb{R}^n$.
Output: approximated minimum.
Data: termination parameter ε, coordinate directions $(\mathbf{e}_1, \ldots, \mathbf{e}_n)$.
1 $k \leftarrow 0$;
2 **repeat**
3   **if** $k \neq 0$ **then**
4   **for** $i \in [\![1, n]\!]$ **do**
5   **if** $\lambda_i = 0$ **then** $\mathbf{a}_i \leftarrow \mathbf{e}_i$ **else** $\mathbf{a}_i \leftarrow \sum_{\ell=i}^{n} \lambda_\ell \mathbf{e}_\ell$;
6   **if** $i = 1$ **then** $\mathbf{b}_i \leftarrow \mathbf{a}_i$ **else** $\mathbf{b}_i \leftarrow \mathbf{a}_i - \sum_{\ell=1}^{i-1} (\mathbf{a}_i^t \mathbf{d}_\ell) \mathbf{d}_\ell$;
7        $\mathbf{d}_i \leftarrow \mathbf{b}_i / \|\mathbf{b}_i\|$;
8   **end**
9   **for** $i \in [\![1, n]\!]$ **do** $\mathbf{e}_i \leftarrow \mathbf{d}_i$;
10   **end**
11    $k \leftarrow k+1$, $\mathbf{z}_1 \leftarrow \mathbf{x}_k$;
12   **for** $i \in [\![1, n]\!]$ **do** $\lambda_i \leftarrow \operatorname{argmin}_{\lambda \in \mathbb{R}} f(\mathbf{z}_i + \lambda \mathbf{e}_i)$, $\mathbf{z}_{i+1} \leftarrow \mathbf{z}_i + \lambda_i \mathbf{e}_i$;
13    $\mathbf{x}_{k+1} \leftarrow \mathbf{z}_{n+1}$;
14 **until** $\|\mathbf{x}_{k+1} - \mathbf{x}_k\|_\infty < \varepsilon$;
15 **return** $\mathbf{x}_{k+1}$;
---

uses steps of the method of Hooke and Jeeves in its inner loop and, like the method of Rosenbrock, it is sequentially modifying the set of search directions in order to best fit the shape of the function f at the current point. This method is globally convergent in the convex cases and is considered to be a quite good minimization method.

Algorithm 2.9 is implementing the method of Zangwill. In the main loop (lines 2–15), a set of n linearly independent search directions are generated at each iteration (lines 4–13). To generate a new search direction, a pattern search similar to the one of the method of Hooke and Jeeves is completed (lines 5–6). The first search direction is discarded and the pattern direction is inserted in the set of search directions (lines 8–9). Then to avoid the generation of dependent search directions, the starting point of the next pattern search is moved away by one iteration of the cyclic coordinate method (lines 9–11). After n iterations of the inner loop (lines 4–13), a set of n independent search directions adapted to the shape of f is generated and the last point generated by the last pattern search is the starting point of a new iteration of the main loop (line 14).

Steepest Descent Method

The steepest descent method is a globally convergent method and an elementary minimization technique that uses the information carried by the derivative of f to locate the steepest downhill direction. Algorithm 2.10 implements the steepest descent method. Note that the line search requires the step length to be positive. In this case approximate line searches with Wolfe's conditions can be used.

The steepest descent method is prone to severe ill-conditioning effects. Its convergence rate can be extremely slow in some badly conditioned cases. It is known for producing zigzagging trajectories that are the source of its bad convergence rate. To get rid of the drawback of the steepest descent method, the other methods based on the derivative of f correct the values of the gradient of f at the current point of the iteration by a linear operation (also called *deflection* of the gradient).

Algorithm 2.9: Method of Zangwill.

Input: function f, starting point $\mathbf{x}_1 \in \mathbb{R}^n$.
Output: approximated minimum.
Data: termination parameter ε, coordinate directions $(e_1, \ldots, e_n)$.
1 $k \leftarrow 0$, **for** $i \in [\![1, n]\!]$ **do** $\mathbf{d}_i \leftarrow e_i$;
2 **repeat**
3     $k \leftarrow k+1$, $\mathbf{y}_1 \leftarrow \mathbf{x}_k$, $\mathbf{z}_1 \leftarrow \mathbf{x}_k$;
4   **for** $i \in [\![1, n]\!]$ **do**
5   **for** $j \in [\![1, n]\!]$ **do** $\overline{\lambda} \leftarrow \mathrm{argmin}_{\lambda \in \mathbb{R}} f(\mathbf{z}_j + \lambda \mathbf{d}_j)$, $\mathbf{z}_{j+1} \leftarrow \mathbf{z}_j + \overline{\lambda}\mathbf{d}_j$;
6         $\mathbf{d} \leftarrow \mathbf{z}_{n+1} - \mathbf{z}_1$, $\overline{\lambda} \leftarrow \mathrm{argmin}_{\lambda \in \mathbb{R}} f(\mathbf{z}_{n+1} + \lambda \mathbf{d})$, $\mathbf{y}_{i+1} \leftarrow \mathbf{z}_{n+1} + \overline{\lambda}\mathbf{d}$;
7   **if** $i < n$ **then**
8   **for** $j \in [\![1, n-1]\!]$ **do** $\mathbf{d}_j \leftarrow \mathbf{d}_{j+1}$;
9             $\mathbf{d}_n \leftarrow \mathbf{d}$, $\mathbf{z}_1 \leftarrow \mathbf{y}_{i+1}$;
10   **for** $j \in [\![1, n]\!]$ **do** $\overline{\lambda} \leftarrow \mathrm{argmin}_{\lambda \in \mathbb{R}} f(\mathbf{z}_j + \lambda e_j)$, $\mathbf{z}_{j+1} \leftarrow \mathbf{z}_j + \overline{\lambda} e_j$;
11            $\mathbf{z}_1 \leftarrow \mathbf{z}_{n+1}$;
12   **end**
13 **end**
14 $\mathbf{x}_{k+1} \leftarrow \mathbf{y}_{n+1}$;
15 **until** $\|\mathbf{x}_{k+1} - \mathbf{x}_k\|_\infty < \varepsilon$;
16 **return** $\mathbf{x}_{k+1}$;

Algorithm 2.10: Steepest descent method.

Input: function f, gradient ∇f, starting point $\mathbf{x}$.
Output: approximated minimum.
Data: termination parameter ε.
1 **while** $\|\nabla f(\mathbf{x})\| \geq \varepsilon$ **do**
2     $\mathbf{d} \leftarrow -\nabla f(\mathbf{x})$, $\overline{\lambda} \leftarrow \mathrm{argmin}_{\lambda \in \mathbb{R}_+} f(\mathbf{x} + \lambda \mathbf{d})$, $\mathbf{x} \leftarrow \mathbf{x} + \overline{\lambda}\mathbf{d}$;
3 **end**
4 **return** $\mathbf{x}$;

The Method of Fletcher and Reeves

The method of the conjugate gradient of Fletcher and Reeves [19] is based on the mechanisms of the steepest descent method and implements a technique of gradient deflection. It was derived from a method proposed by Hestenes and Stiefel [26].

This method uses the notion of conjugate directions. In other words, given an $n \times n$ symmetric positive definite matrix $\mathbf{H}$, the n linearly independent directions $\mathbf{d}_1, \ldots, \mathbf{d}_n$ of $\mathbb{R}^n$ are $\mathbf{H}$-conjugate if $\mathbf{d}_i^t \mathbf{H} \mathbf{d}_j = 0$ for every $i \neq j$. In the quadratic case, if $\mathbf{d}_1, \ldots, \mathbf{d}_n$ are conjugate directions with the Hessian matrix $\mathbf{H}$, the function f at any point of $\mathbb{R}^n$ can be decomposed into the summation of n real-valued functions on $\mathbb{R}$. Then n line searches along the conjugate directions are sufficient to obtain a minimizer of f from any point of $\mathbb{R}^n$. The method of Fletcher and Reeves iteratively constructs conjugate directions while at the same time it deflects the gradient. In the quadratic case, the method should stop in at most n iterations. Nevertheless, while this technique may not be as efficient as others, it requires modest memory resources and therefore it is suitable for large nonlinear problems.

Nonlinear Programming

The method of Fletcher and Reeves is implemented in Algorithm 2.11. While the increase of f at the current point is still non-negligible (line 2), a procedure that has some similarities with the cyclic coordinate method is repeated (lines 3–8). The method goes on by producing a sequence of points that minimize f along each conjugate direction (line 3) and the next conjugate direction that deflects the gradient is computed at line 5. As in the steepest descent method, the line search at line 3 requires the step length to be positive. In this case approximate line searches with Wolfe's conditions can be used. Then whenever n consecutive conjugate directions are generated, the method is restarted (line 7).

Algorithm 2.11: Method of the conjugate gradient of Fletcher and Reeves.

Input: function f, gradient ∇f, starting point $\mathbf{x}$.
Output: approximated minimum.
Data: termination parameter ε.

1 $i \leftarrow 1$, $\mathbf{z}_i \leftarrow \mathbf{x}$, $\mathbf{d}_i \leftarrow -\nabla f(\mathbf{z}_i)$;
2 **while** $\|\nabla f(\mathbf{z}_i)\| \geq \varepsilon$ **do**
3 $\quad \bar{\lambda} \leftarrow \operatorname{argmin}_{\lambda \in \mathbb{R}_+} f(\mathbf{z}_i + \lambda \mathbf{d}_i)$, $\mathbf{z}_{i+1} \leftarrow \mathbf{z}_i + \bar{\lambda} \mathbf{d}_i$;
4 $\quad$ **if** $i < n$ **then**
5 $\quad\quad \alpha \leftarrow \|\nabla f(\mathbf{z}_{i+1})\|^2 / \|\nabla f(\mathbf{z}_i)\|^2$, $\mathbf{d}_{i+1} \leftarrow -\nabla f(\mathbf{z}_{i+1}) + \alpha \mathbf{d}_i$, $i \leftarrow i+1$;
6 $\quad$ **else**
7 $\quad\quad \mathbf{x} \leftarrow \mathbf{z}_{n+1}$, $i \leftarrow 1$, $\mathbf{z}_i \leftarrow \mathbf{x}$, $\mathbf{d}_i \leftarrow -\nabla f(\mathbf{z}_i)$;
8 $\quad$ **end**
9 **end**
10 **return x**;

Other conjugate gradient methods have been derived from Hestenes and Stiefel. Variants of the method of Fletcher and Reeves have different restart procedures (and, consequently, a slightly different gradient deflection). The reader should refer to Beale [3] and Powell [44] for further details on restart procedures. There exist different gradient updates, notably in the method of Polak and Ribiere [40], in the method of Polyak [41], and in the method of Sorenson [50]. The Polak and Ribiere update is usually deemed to be more efficient than the Fletcher and Reeves update. Using the above notations, the update is:

$$\alpha \leftarrow \frac{(\nabla f(\mathbf{z}_{i+1}) - \nabla f(\mathbf{z}_i))^{\mathrm{t}} \nabla f(\mathbf{z}_{i+1})}{\nabla f(\mathbf{z}_i)^{\mathrm{t}} \nabla f(\mathbf{z}_i)}$$

Quasi-Newton Methods

Quasi-Newton methods, like the method of Fletcher and Reeves, are based on the notion of conjugate directions. However, the deflection of the gradient is made differently. The original method is credited to Davidson [11]. It has been improved by Fletcher and Powell [18], and simultaneously refined by Broyden [6,7], Fletcher [15], Goldfarb [22], and Shanno [49]. In these methods, the negative of the gradient at $\mathbf{x}$, $-\nabla f(\mathbf{x})$, is deflected by multiplying it by a positive definite matrix $\mathbf{A}$ that is updated at each iteration (this matrix evolves toward the inverse of the Hessian matrix in the quadratic case). These methods are like the steepest descent method with an affine scaling. Those are referred as quasi-Newton methods and, like the nonglobally convergent Newton method, perform an affine scaling to deflect the gradient.

Note that Newton's method uses the inverse of the Hessian at the current iteration point (assuming that the Hessian exists and is positive definite).

Algorithm 2.12 implements a quasi-Newton method with both Davidson-Fletcher-Powell (DFP) updates and Broyden-Fletcher-Goldfarb-Shanno (BFGS) updates. The user may switch between these two updates by choosing the value of a parameter ϕ. If $\phi = 0$ the method is using DFP updates that produce sometimes numerical problems. If $\phi = 1$ the method is using BFGS updates that are deemed to exhibit a superior behavior. Other values for ϕ are fine too; however, some negative values may lead to degenerate cases. As in the steepest descent method, the line search at line 3 requires the step length to be positive. In this case approximate line searches with Wolfe's conditions can be used. However, inexact line searches may corrupt the positive definiteness of the matrices $\mathbf{A}_i$ and some quasi-Newton methods are proposing a varying ϕ that counters this phenomenon. Sometimes the matrices $\mathbf{A}_i$ are multiplied by a strictly positive scalar updated at each iteration to avoid some ill-conditioning effects when minimizing nonquadratic functions. For further information, the reader should refer to Fletcher [17].

Algorithm 2.12: A Quasi-Newton method.

Input: function f, gradient ∇f, starting point $\mathbf{x}$.
Output: approximated minimum.
Data: termination parameter ε, parameter ϕ, positive definite matrix $\mathbf{A}_1$ (or $\mathbf{A}_1 = \mathbf{I}$).
1 $i \leftarrow 1$, $\mathbf{z}_i \leftarrow \mathbf{x}$;
2 **while** $\|\nabla f(\mathbf{z}_i)\| \geq \varepsilon$ **do**
3      $\mathbf{d} \leftarrow -\mathbf{A}_i \nabla f(\mathbf{z}_i)$, $\bar{\lambda} \leftarrow \mathrm{argmin}_{\lambda \in \mathbb{R}_+} f(\mathbf{z}_i + \lambda \mathbf{d})$, $\mathbf{z}_{i+1} \leftarrow \mathbf{z}_i + \bar{\lambda}\mathbf{d}$;
4   **if** $i < n$ **then**
5          $\mathbf{u} \leftarrow \bar{\lambda}\mathbf{d}$, $\mathbf{v} \leftarrow \nabla f(\mathbf{z}_{i+1}) - \nabla f(\mathbf{z}_i)$;
6          $\mathbf{B}_1 \leftarrow \mathbf{u}\mathbf{u}^t/\mathbf{u}^t\mathbf{v} - \mathbf{A}_i \mathbf{v}\mathbf{v}^t \mathbf{A}_i / \mathbf{v}^t \mathbf{A}_i \mathbf{v}$;
7          $\mathbf{B}_2 \leftarrow ((1 + \mathbf{v}^t \mathbf{A}_i \mathbf{v}/\mathbf{u}^t\mathbf{v})\mathbf{u}\mathbf{u}^t - \mathbf{A}_i \mathbf{v}\mathbf{u}^t - \mathbf{u}\mathbf{v}^t\mathbf{A}_i)/\mathbf{u}^t\mathbf{v}$;
8          $\mathbf{A}_{i+1} \leftarrow \mathbf{A}_i + (1-\phi)\mathbf{B}_1 + \phi \mathbf{B}_2$, $i \leftarrow i+1$;
9   **else** $\mathbf{x} \leftarrow \mathbf{z}_{n+1}$, $i \leftarrow 1$, $\mathbf{z}_i \leftarrow \mathbf{x}$;
10 **end**
11 **return** $\mathbf{x}$;

Levenberg–Marquardt Method

The Levenberg–Marquardt method [32,35] is a globally convergent modification of Newton's method. Like the quasi-Newton methods, the Levenberg–Marquardt method is an adaptation of the steepest descent method with an affine scaling of the gradient as a deflection technique. The positive definite matrix that multiplies the gradient is the summation of the Hessian of f at the current point (this is the approach of the Newton method) and a scaled identity matrix $c\mathbf{I}$ (almost like in the steepest descent method) that enforces the positive definiteness.

Algorithm 2.13 implements the Levenberg–Marquardt method. While the increase of f at the current point is still non-negligible (line 2), a Choleski decomposition of $c_k \mathbf{I} + \mathbf{H}(\mathbf{x}_k)$ is repeated until success (lines 3–11). The decomposition gives $c_k \mathbf{I} + \mathbf{H}(\mathbf{x}_k) = \mathbf{L}\mathbf{L}^t$. Lines 12–13 solve the linear system $\mathbf{L}\mathbf{L}^t\mathbf{d} = -\nabla f(\mathbf{x}_k)$ and the descent direction $\mathbf{d}$ is obtained. Line 14 performs a line search and the next point $\mathbf{x}_{k+1}$ is computed. As in the steepest descent method, the line search requires the step length to be positive. In this case approximate line searches with Wolfe's conditions can be used. Line 15 computes the ratio ρ of the actual decrease over the predicted decrease. If the actual decrease is not enough, then the scale

Algorithm 2.13: Levenberg–Marquardt method.

Input: function f, gradient ∇f, Hessian $\mathbf{H}$, starting point $\mathbf{x}_1 \in \mathbb{R}^n$.
Output: approximated minimum.
Data: termination parameter ε, parameter c_1 (usually $c_1 \in \{0.25, 0.75, 2, 4\}$),
parameters $0 < \rho_1 < \rho_2 < 1$ (usually $\rho_1 = 0.25$ and $\rho_2 = 0.75$).

1 $k \leftarrow 1$;
2 **while** $\|\nabla f(\mathbf{x}_k)\| \geq \varepsilon$ **do**
3   **repeat**
4   fail $\leftarrow 0$, $A \leftarrow c_k \mathbf{I} + \mathbf{H}(\mathbf{x}_k)$;
5   **for** $i \in [\![1, n]\!]$, $j \in [\![i, n]\!]$ **do**
6        $S \leftarrow a_{ji} - \sum_{\ell=1}^{i-1} a_{j\ell} a_{i\ell}$;
7   **if** $i = j$ **then**
8   **if** $S \leq 0$ **then** $c_k \leftarrow 4 c_k$, fail $\leftarrow 1$, **break else** $l_{ii} \leftarrow \sqrt{S}$;
9   **else** $l_{ji} \leftarrow S/l_{ii}$;
10   **end**
11   **until** fail $= 0$;
12   **for** $i \in [\![1, n]\!]$ **do** $S \leftarrow -(\nabla f(\mathbf{x}_k))_i - \sum_{\ell=1}^{i-1} l_{i\ell} y_\ell$, $y_i \leftarrow S/l_{ii}$;
13   **for** $i \in [\![0, n-1]\!]$ **do** $S \leftarrow y_{n-i} - \sum_{\ell=n-i+1}^{n} l_{\ell, n-i} d_\ell$, $d_{n-i} \leftarrow S/l_{ii}$;
14    $\overline{\lambda} \leftarrow \mathrm{argmin}_{\lambda \in \mathbb{R}_+} f(\mathbf{x}_k + \lambda \mathbf{d})$, $\mathbf{x}_{k+1} \leftarrow \mathbf{x}_k + \overline{\lambda} \mathbf{d}$;
15    $\rho \leftarrow (f(\mathbf{x}_{k+1}) - f(\mathbf{x}_k))/(\overline{\lambda} \nabla f(\mathbf{x}_k)^t \mathbf{d} + \overline{\lambda}^2 \mathbf{d}^t \mathbf{H}(\mathbf{x}_k) \mathbf{d}/2)$;
16   **if** $\rho < \rho_1$ **then** $c_{k+1} \leftarrow 4 c_k$;
17   **else**
18   **if** $\rho_2 < \rho$ **then** $c_{k+1} \leftarrow c_k/2$ **else** $c_{k+1} \leftarrow c_k$;
19   **end**
20   **if** $\rho \leq 0$ **then** $\mathbf{x}_{k+1} \leftarrow \mathbf{x}_k$;
21    $k \leftarrow k + 1$;
22 **end**
23 **return** $\mathbf{x}_k$;

parameter is quadrupled (line 16). Otherwise the scale parameter is halved if the actual decrease is too big (line 18). If the ratio ρ is negative, then the method is re-initialized (20). The operations at lines 15–20 are similar to the operations of the trust region methods.

2.3 Constrained Optimization

In this section, we review some techniques of minimization in the case where equality and inequality constraints are present. A lot of algorithms are specialized in cases where only one kind of constraint is present or in cases where the constraints are linear. We will review globally convergent algorithms for the general convex case. There exist at least two types of approaches that solve this kind of problem:

- The transformation methods. These techniques transform the objective function to incorporate the constraints and therefore transform a constrained problem into a sequence of unconstrained problems. Thus any unconstrained minimization algorithm in Section 2.2 can be used to solve the constrained problem.
- The direction finding methods. These techniques converge toward a minimizer by moving from feasible points to feasible points and determining feasible directions

and feasible step lengths at each iteration. The methods described in this section are globally convergent and use first or second order approximations of f, $\mathbf{g}$, and $\mathbf{h}$.

2.3.1 Direction Finding Methods

Penalty Successive Linear Programming Method

The penalty successive linear programming (PSLP) method is credited to Zhang et al. [54]. It is the first globally convergent form of the successive linear programming (SLP) approach introduced by Griffith and Stewart [23]. A variant of the SLP approach that includes quadratic approximations has been proposed by Fletcher [16]. At each iteration, the method finds feasible directions by solving a linear programming problem based on the first-order approximations of the objective function and constraint functions, and on some constraints on the direction components. In this method, the linear equality and inequality constraints $\mathbf{Ax} \leq \mathbf{b}$ are separated and treated differently than the purely nonlinear constraints $\mathbf{g}(\mathbf{x}) \leq \mathbf{0}$ and $\mathbf{h}(\mathbf{x}) = \mathbf{0}$. This method will converge quadratically if the optimal solution is close to a vertex of the feasible region. Otherwise the convergence can be very slow.

The PSLP method is implemented in Algorithm 2.14. It needs a trust region vector $\mathbf{d}$ greater than a vector $\lambda > \mathbf{0}$ that gives the maximum admissible values for the components of the search directions. Line 12 uses a linear programming solver to obtain an optimal solution

---

Algorithm 2.14: PSLP method.

Input: function f, nonlinear constraint functions $\mathbf{g}$ and $\mathbf{h}$, gradients ∇f, ∇g_i for $i \in [\![1, p]\!]$ and ∇h_i for $i \in [\![1, q]\!]$, feasible starting point $\mathbf{x}_1$ for the linear constraints $\mathbf{Ax} \leq \mathbf{b}$.
Output: approximated minimum.
Data: lower bound vector $\mathbf{0} < \lambda \in \mathbb{R}^n$, trust region vector $\mathbf{d} \geq \lambda$, scalars $0 < \rho_0 < \rho_1 < \rho_2 < 1$ (usually $\rho_0 = 10^{-6}$, $\rho_1 = 0.25$, $\rho_2 = 0.75$), multiplier α (usually $\alpha = 2$), penalty parameters $\mathbf{0} < \mu \in \mathbb{R}^p$, $\mathbf{0} < \eta \in \mathbb{R}^q$; function $D_{\mathbf{x}_k} f : \mathbf{x} \mapsto f(\mathbf{x}_k) + \nabla f(\mathbf{x}_k)^{\mathrm{t}}(\mathbf{x} - \mathbf{x}_k)$; function π: $(\mathbf{x}, \mu, \eta) \mapsto \sum_{i=1}^{p} \mu_i$ max $(g_i(\mathbf{x}), 0) + \sum_{i=1}^{q} \eta_i |h_i(\mathbf{x})|$; function $\delta_{\mathbf{x}_k} \pi$: $(\mathbf{x}, \mu, \eta) \mapsto \sum_{i=1}^{p} \mu_i$ max $(D_{\mathbf{x}_k} g_i(\mathbf{x}), 0) + \sum_{i=1}^{q} \eta_i |D_{\mathbf{x}_k} h_i(\mathbf{x})|$ function θ:$(\mathbf{x}, \mu, \eta) \mapsto f(\mathbf{x}) + \pi\,(\mathbf{x}, \mu, \eta)$; function $\delta_{\mathbf{x}_k} \theta$: $(\mathbf{x}, \mu, \eta) \mapsto D_{\mathbf{x}_k} f(\mathbf{x}) + \delta_{\mathbf{x}_k} \pi(\mathbf{x}, \mu, \eta)$.

1   $k \leftarrow 1$;
2   **repeat**
3   **if** $k \neq 1$ **then**
4         $\rho \leftarrow (\theta(\mathbf{x}_k, \mu, \eta) - \theta\,(\overline{\mathbf{x}}, \mu, \eta))/(\theta(\mathbf{x}_k, \mu, \eta) - \delta_{\mathbf{x}_k} \theta(\mathbf{x}_{k+1}, \mu, \eta))$;
5   **if** $\rho < \rho_0$ **then** $\mathbf{d} \leftarrow \max(\mathbf{d}/\alpha, \lambda)$;
6   **else**
7   **if** $0 \leq \rho < \rho_1$ **then** $\mathbf{d} \leftarrow \mathbf{d}/\alpha$;
8   **if** $\rho_2 < \rho$ **then** $\mathbf{d} \leftarrow \alpha \mathbf{d}$;
9            $\mathbf{d} \leftarrow \max(\mathbf{d}, \lambda)$, $\mathbf{x}_{k+1} \leftarrow \overline{\mathbf{x}}$, $k \leftarrow k + 1$;
10   **end**
11   **end**
12      $\overline{\mathbf{x}} \leftarrow \mathrm{argmin}_{\mathbf{x} \in \mathbb{R}^n} \{\delta_{\mathbf{x}_k} \theta(\mathbf{x}, \mu, \eta) : \mathbf{Ax} \leq \mathbf{b}, -\mathbf{d} \leq \mathbf{x} - \mathbf{x}_k \leq \mathbf{d}\}$;
13 **until** $\overline{\mathbf{x}} = \mathbf{x}_k$;
14 **return** $\overline{\mathbf{x}}$;

Nonlinear Programming

of the first order approximation problem. It can be initialized by $\mathbf{x}_k$ if $\mathbf{x}_k$ is feasible for the linear constraints; otherwise a feasibility step must be introduced. Line 4 calculates the ratio of the decrease of the penalty function over the predicted decrease of the penalty function. If the ratio ρ is negligible or negative, then the trust region is contracted and the method is re-initialized (line 5). Otherwise the region is contracted if ρ is too small, and expanded if ρ is large (lines 7–8). This part of the algorithm is similar to the trust region methods.

Merit Successive Quadratic Programming Method

The following method was presented in Ref. [2] and is based on the same ideas as the PSLP method. Bazaraa et al. give credits to Han [24] and Powell [45] for this method. At each iteration, the method finds feasible directions by solving a quadratic programming problem based on the second-order approximations of the objective function and the first-order approximations of the constraint functions. This method is quite demanding in terms of computational power; however, it does not have the drawbacks of the PSLP method.

The merit successive quadratic programming method is implemented in Algorithm 2.15. Since the function θ at line 4 is not differentiable, only an exact line search can be used to find an optimal $\overline{\lambda}$. Line 7 uses a quadratic programming solver to determine the next descent direction. It can be initialized by $\mathbf{x}_k$ if $\mathbf{x}_k$ is feasible; otherwise a feasibility step must be introduced. The matrix $\mathbf{B}$ can be updated at line 5, provided that it always remains positive definite. As suggested by Bazaraa et al. [2], an update of $\mathbf{B}$ based on a quasi-Newton scheme can be used but it is not necessary.

Algorithm 2.15: Merit successive quadratic programming method.

Input: function f, nonlinear constraint functions $\mathbf{g}$ and $\mathbf{h}$, gradients ∇f, ∇g_i for $i \in [\![1, p]\!]$ and ∇h_i for $i \in [\![1, q]\!]$, feasible starting point $\mathbf{x}_1$, positive definite matrix $\mathbf{B}$.

Output: approximated minimum.

Data: penalty parameters $\mathbf{0} < \mu \in \mathbb{R}^p_+, \mathbf{0} < \eta \in \mathbb{R}^q$; function
$\pi : (\mathbf{x}, \mu, \eta) \mapsto \sum_{i=1}^p \mu_i \max(g_i(\mathbf{x}), 0) + \sum_{i=1}^q \eta_i |h_i(\mathbf{x})|$; function
$\theta : (\mathbf{x}, \mu, \eta) \mapsto f(\mathbf{x}) + \pi(\mathbf{x}, \mu, \eta)$; function $D_{\mathbf{x}_k} f : \mathbf{x} \mapsto f(\mathbf{x}_k) + \nabla f(\mathbf{x}_k)^t (\mathbf{x} - \mathbf{x}_k)$; function $Q_{\mathbf{x}_k} : \mathbf{x} \mapsto D_{\mathbf{x}_k} f(\mathbf{x}) + (\mathbf{x} - \mathbf{x}_k)^t \mathbf{B} (\mathbf{x} - \mathbf{x}_k)/2$.

1 $k \leftarrow 1$;
2 **repeat**
3   **if** $k \neq 1$ **then**
4         $\overline{\lambda} \leftarrow \operatorname{argmin}_{\lambda \in \mathbb{R}_+} \theta(\mathbf{x}_k + \lambda(\overline{\mathbf{x}} - \mathbf{x}_k), \mu, \eta)$;
5         $\mathbf{x}_{k+1} \leftarrow \mathbf{x}_k + \overline{\lambda}(\overline{\mathbf{x}} - \mathbf{x}_k), k \leftarrow k+1$;
6   **end**
7     $\overline{\mathbf{x}} \leftarrow \operatorname{argmin}_{\mathbf{x} \in \mathbb{R}^n} \{ Q_{\mathbf{x}_k}(\mathbf{x}) : D_{\mathbf{x}_k} g_i(\mathbf{x}) \leq 0 |_{i=1}^p, D_{\mathbf{x}_k} h_i(\mathbf{x}) = 0 |_{i=1}^q \}$;
8 **until** $\overline{\mathbf{x}} = \mathbf{x}_k$;
9 **return** $\overline{\mathbf{x}}$;

2.3.2 Transformation Methods

Sequential Unconstrained Minimization Techniques

The sequential unconstrained minimization techniques (SUMT) were developed by Fiacco and McCormick [13,14] to solve constrained nonlinear convex optimization problems by transforming them into a sequence of unconstrained problems using penalty and barrier functions. The introduction of penalty functions for solving constrained problems is credited to Courant [10], while the use of barrier functions is credited to Caroll [8].

Algorithms 2.16 and 2.17 implement two different SUMT. Algorithm 2.16 is a parametrized SUMT in which parameter $\mu > 0$ is progressively increased at each iteration to give more weight to the penalty part of the unconstrained objective function. As μ increases, the solution $\bar{\mathbf{x}}$ moves toward a feasible point. For μ sufficiently large, $\bar{\mathbf{x}}$ becomes close enough to an optimal solution of the constrained problem. However, it is not recommended to start this method with a large μ as it might slow down the convergence and trigger ill-conditioning effects. Algorithm 2.17 tries to overcome these difficulties by removing the parameter μ. However, a scalar $a < \inf\{f(\mathbf{x}) : \mathbf{g}(\mathbf{x}) < \mathbf{0}, \mathbf{h}(\mathbf{x}) = \mathbf{0}\}$ must be determined before the computation starts.

Algorithm 2.16: SUMT.

Input: see unconstrained optimization method (starting point $\bar{\mathbf{x}}$).
Output: approximated minimum.
Data: termination parameter ε, penalty parameter $\mu > 0$, scalar $\alpha > 1$, penalty function $\pi : \mathbf{x} \mapsto \sum_{i=1}^{p} \pi_g(g_i(\mathbf{x})) + \sum_{i=1}^{q} \pi_h(h_i(\mathbf{x}))$ with π_g and π_h continuous, $\pi_g(z) = 0$ if $z \leq 0$, $\pi_g(z) > 0$ otherwise, and $\pi_h(z) = 0$ if $z = 0$, $\pi_h(z) > 0$ otherwise.

1 while $\mu\pi(\bar{\mathbf{x}}) \geq \varepsilon$ do
2    $\bar{\mathbf{x}} \leftarrow \operatorname{argmin}_{\mathbf{x} \in \mathbb{R}^n} \{f(\mathbf{x}) + \mu\pi(\mathbf{x}) : \bar{\mathbf{x}} \text{ starting point}\}, \mu \leftarrow \alpha\mu$;
3 end
4 return $\bar{\mathbf{x}}$;

Algorithm 2.17: Parameter-free SUMT.

Input: see unconstrained optimization method (starting point $\bar{\mathbf{x}}$).
Output: approximated minimum.
Data: termination parameter ε, scalar $a < \inf\{f(\mathbf{x}) : \mathbf{g}(\mathbf{x}) < \mathbf{0}, \mathbf{h}(\mathbf{x}) = \mathbf{0}\}$, $\pi : (\mathbf{x}, a) \mapsto \max^2(f(\mathbf{x}) - a, 0) + \sum_{i=1}^{p} \max^2(g_i(\mathbf{x}), 0) + \|\mathbf{h}(\mathbf{x})\|_2^2$.

1 while $\pi(\bar{\mathbf{x}}, a) \geq \varepsilon$ do
2    $\bar{\mathbf{x}} \leftarrow \operatorname{argmin}_{\mathbf{x} \in \mathbb{R}^n} \{\pi(\mathbf{x}, a) : \bar{\mathbf{x}} \text{ starting point}\}, a \leftarrow f(\bar{\mathbf{x}})$;
3 end
4 return $\bar{\mathbf{x}}$;

Algorithm 2.18 implements a method proposed by Fiacco and McCormick in 1968 [14] that mixes a penalty function with a barrier function. The penalty function π is aimed for the equality constraints and the barrier function β is aimed for the inequality constraints. Note that the line searches that are used by the unconstrained optimization subroutines must provide a safeguard technique to avoid the computational difficulties induced by the domain of definition of barrier functions. Lines 1–8 compute a feasible point for the inequality constraints. If such a point cannot be found, the algorithm is stopped and no solution is returned (line 7). Using the feasible starting point $\bar{\mathbf{x}}$, the mixed penalty-barrier solves the constrained problem (lines 10–13).

Nonlinear Programming

Algorithm 2.18: Mixed penalty-barrier method.

Input: see unconstrained optimization method (starting point $\bar{\mathbf{x}}$).
Output: approximated minimum.
Data: termination parameter ε, barrier parameter $\mu_0 > 0$, penalty parameter $\eta > 0$, scalars $0 < c_1 < 1$ and $c_2 > 1$; penalty function $\pi : \mathbf{x} \mapsto \sum_{i=1}^{q} \pi_h(h_i(\mathbf{x}))$ with π_h continuous, $\pi_h(z) = 0$ if $z = 0$, $\pi_h(z) > 0$ otherwise; barrier function $\beta_I : \mathbf{x} \mapsto \sum_{i \in I} \beta_g(g_i(\mathbf{x}))$ with $I \subseteq [\![1, p]\!]$, β_g continuous over $\mathbb{R}_-^*$, $\beta_g(z) \geq 0$ if $z < 0$ and $\lim_{z \to 0^+} \beta_g(z) = \infty$. Note: $\beta = \beta_{[\![1,p]\!]}$.

1 $I \leftarrow \{i \in [\![1, p]\!] : g_i(\bar{\mathbf{x}}) < 0\}$;
2 **while** $I \neq [\![1, p]\!]$ **do**
3   Select $j \in [\![1, p]\!] \setminus I$, $\mu \leftarrow \mu_0$;
4   **while** $\mu \beta_I / (\bar{\mathbf{x}}) \geq \varepsilon$ **do**
5        $\bar{\mathbf{x}} \leftarrow \operatorname{argmin}_{\mathbf{x} \in \mathbb{R}^n} \{g_j(\mathbf{x}) + \mu \beta_I(\mathbf{x}) : \bar{\mathbf{x}} \text{ starting point}\}$, $\mu \leftarrow c_1 \mu$;
6   **end**
7   **if** $g_j(\bar{\mathbf{x}}) \geq 0$ **then return** $\emptyset$ **else** $I \leftarrow \{i \in [\![1, p]\!] : g_i(\bar{\mathbf{x}}) < 0\}$;
8 **end**
9 $\mu \leftarrow \mu_0$;
10 **while** $\mu \beta(\bar{\mathbf{x}}) + \eta \pi(\bar{\mathbf{x}}) \geq \varepsilon$ **do**
11     $\bar{\mathbf{x}} \leftarrow \operatorname{argmin}_{\mathbf{x} \in \mathbb{R}^n} \{f(\mathbf{x}) + \mu \beta(\mathbf{x}) + \eta \pi(\mathbf{x}) : \bar{\mathbf{x}} \text{ starting point}\}$;
12     $\mu \leftarrow c_1 \mu$, $\eta \leftarrow c_2 \eta$;
13 **end**
14 **return** $\bar{\mathbf{x}}$;

Method of Multipliers

The method of multipliers (MOM) was introduced independently by Hestenes [25] and Powell [43] in 1969. It uses Lagrange multipliers with a penalty function in a SUMT scheme.

Algorithm 2.19 implements the MOM. The algorithm iterates until the violation of the constraints is small enough (line 2). In line 3, the penalty function is minimized with an unconstrained optimization method and the result is stored at point $\mathbf{x}_{k+1}$. Then the Lagrange multipliers are updated if the violation function shows some improvements at the new iterate $\mathbf{x}_{k+1}$; k is incremented and a new iteration begins (lines 4–12). Otherwise the penalty parameters are increased and the method is restarted (lines 13–18).

2.4 Conclusion

In this chapter, we have discussed several common algorithms on deterministic optimization for convex nonlinear problems. However, formal explanations on the construction and the convergence of the mentioned methods have not been discussed. Useful references were given for each algorithm, and the most important features were briefly reviewed. For further details on nonlinear optimization, the reader might consult Bazaraa et al. [2], Bertsekas [4], Boyd and Vandenberghe [5], Corne et al. [9], Fletcher [17], Horst et al. [28], Luenberger [34], and Reeves [47]. Owing to limitations of space we did not discuss heuristic and evolutionary optimization techniques such as tabu search [21]. Nonlinear programming techniques are used in several areas of engineering and are powerful tools in the arsenal of operations research.

Please note that the reader of this chapter should be warned about the different ways of implementing each algorithm. It is important to note that within the scope of this quick introduction we did not assess problems related to numerical instabilities that can arise

Algorithm 2.19: MOM.

Input: see unconstrained optimization method (starting point $\mathbf{x}_1$).
Output: approximated minimum.
Data: termination parameter ε, penalty parameters $\mathbf{0} < \mu \in \mathbb{R}^p$, $\mathbf{0} < \eta \in \mathbb{R}^q$, initial multipliers $\mathbf{0} \leq \mathbf{u} \in \mathbb{R}^p$ and $\mathbf{v} \in \mathbb{R}^q$, scalars $0 < c < 1$, α_μ, $\alpha_\eta > 1$ (usually $c = 0.25$, $\alpha_\mu = \alpha_\eta = 10$), violation function $\varphi : \mathbf{x} \mapsto \max(\|\mathbf{h}(\mathbf{x})\|_\infty, \max_{i \in [\![1,p]\!]}(g_i(\mathbf{x}), 0))$, penalty functions π_g and π_h such that $\pi_g(\mathbf{x}, \mathbf{u}, \mu) = \sum_{i=1}^{p} \left(\mu_i \max^2 \left(g_i(\mathbf{x}) + \frac{u_i}{2\mu_i}, 0 \right) - \frac{u_i^2}{4\mu_i} \right)$ and $\pi_h(\mathbf{x}, \mathbf{v}, \eta) = \sum_{i=1}^{q} (v_i h_i(\mathbf{x}) + \eta_i h_i^2(\mathbf{x}))$.

1 $k \leftarrow 1$;
2 **while** $\varphi(\mathbf{x}_k) \geq \varepsilon$ **do**
3    $\mathbf{x}_{k+1} \leftarrow \text{argmin}_{\mathbf{x} \in \mathbb{R}^n} \{f(\mathbf{x}) + \pi_g(\mathbf{x}, \mathbf{u}, \mu) + \pi_h(\mathbf{x}, \mathbf{v}, \eta) : \mathbf{x}_k \text{ starting point}\}$;
4   **if** $k = 1$ **then**
5   **for** $i \in [\![1, p]\!]$ **do** $u_i \leftarrow u_i + \max(2\mu_i g_i(\mathbf{x}_{k+1}), -u_i)$;
6   **for** $i \in [\![1, q]\!]$ **do** $v_i \leftarrow v_i + 2\eta_i h_i(\mathbf{x}_{k+1})$;
7       $k \leftarrow k + 1$;
8   **else**
9   **if** $\varphi(\mathbf{x}_{k+1}) \leq c\varphi(\mathbf{x}_k)$ **then**
10   **for** $i \in [\![1, p]\!]$ **do** $u_i \leftarrow u_i + \max(2\mu_i g_i(\mathbf{x}_{k+1}), -u_i)$;
11   **for** $i \in [\![1, q]\!]$ **do** $v_i \leftarrow v_i + 2\eta_i h_i(\mathbf{x}_{k+1})$;
12          $k \leftarrow k + 1$
13   **else**
14   **for** $i \in [\![1, p]\!]$ **do**
15   **if** $\max(g_i(\mathbf{x}_{k+1}), 0) > c\varphi(\mathbf{x}_k)$ **then** $\mu_i \leftarrow \alpha_\mu \mu_i$;
16   **end**
17   **for** $i \in [\![1, q]\!]$ **do if** $|h_i(\mathbf{x}_{k+1})| > c\varphi(\mathbf{x}_k)$ **then** $\eta_i \leftarrow \alpha_\eta \eta_i$;
18   **end**
19   **end**
20 **end**
21 **return** $\mathbf{x}_k$;

from round-off errors and truncation errors in every implementation. For further discussion about the algorithmic properties of these techniques, the reader may refer to the NEOS Guide which is an online portal to the optimization community that provides links to stable solvers for different platforms. The NEOS Guide can be found at the following address: http://www-fp.mcs.anl.gov/otc/Guide/index.html.

References

1. L. Armijo. Minimization of functions having Lipschitz continuous first-partial derivatives. *Pacific Journal of Mathematics*, 16(1):1–3, 1966.
2. M. Bazaraa, H. Sherali, and C. Shetty. *Nonlinear Programming: Theory and Algorithms*. Third edition, Wiley, New York, 2006.
3. E. Beale. A derivation of conjugate gradients. In F. Lootsma, editor, *Numerical Methods for Nonlinear Optimization*, pages 39–43, Academic Press, London, 1972.

4. D. Bertsekas. *Nonlinear Programming.* Second edition, Athena Scientific, Nashua, NH, 1999.
5. S. Boyd and L. Vandenberghe. *Convex Optimization.* Cambridge University Press, Cambridge, UK, 2004.
6. C. Broyden. Quasi-Newton methods and their application to function minimization. *Mathematics of Computation*, 21(99):368–381, 1967.
7. C. Broyden. The convergence of a class of double rank minimization algorithms 2. The new algorithm. *Journal of the Institute of Mathematics and Its Applications*, 6:222–231, 1970.
8. C. Carroll. The created response surface technique for optimizing nonlinear restrained systems. *Operations Research*, 9:169–184, 1961.
9. D. Corne, M. Dorigo, and F. Glover. *New Ideas in Optimization.* McGraw-Hill, New York, 1999.
10. R. Courant. Variational methods for the solution of problems of equilibrium and vibrations. *Bulletin of the American Mathematical Society*, 49:1–23, 1943.
11. W. Davidson. Variable metric method for minimization. Technical Report ANL-5990, AEC Research and Development, 1959.
12. M. Dorigo and T. Stiitzle. *Ant Colony Optimization.* MIT Press, Cambridge, MA, 2004.
13. A. Fiacco and G. McCormick. The sequential unconstrained minimization technique for nonlinear programming, a primal-dual method. *Management Science*, 10:360–366, 1964.
14. A. Fiacco and G. McCormick. *Nonlinear Programming: Sequential Unconstrained Minimization Techniques.* Wiley, New York, 1968.
15. R. Fletcher. A new approach to variable metric algorithms. *Computer Journal*, 13:317–322, 1970.
16. R. Fletcher. Numerical experiments with an Li exact penalty function method. In O. Mangasarian, R. Meyer, and S. Robinson, editors, *Nonlinear Programming, 4*, Academic Press, New York, 1981.
17. R. Fletcher. *Practical Methods of Optimization*, Second edition, Wiley, New York, 1987.
18. R. Fletcher and M. Powell. A rapidly convergent descent method for minimization. *Computer Journal*, 6(2):163–168, 1963.
19. R. Fletcher and C. Reeves. Function minimization by conjugate gradients. *Computer Journal*, 7:149–154, 1964.
20. F. Glover and G. Kochenberger. *Handbook of Metaheuristics.* Kluwer Academic Publishers, New York, 2002.
21. F. Glover and M. Laguna. *Tabu Search.* Kluwer Academic Publishers, New York, 1997.
22. D. Goldfarb. A family of variable metric methods derived by variational means. *Mathematics of Computation*, 24:23–26, 1970.
23. R. Griffith and R. Stewart. A nonlinear programming technique for the optimization of continuous process systems. *Management Science*, 7:379–392, 1961.
24. S. Han. A globally convergent method for nonlinear programming. Technical Report TR 75–257, Computer Science, Cornell University, Ithaca, NY, 1975.
25. M. Hestenes. Multiplier and gradient methods. *Journal of Optimization Theory and Applications*, 4:303–320, 1969.
26. M. Hestenes and E. Stiefel. Methods of conjugate gradients for solving linear systems. *Journal of Research of the National Bureau of Standards*, 49(6): 409–436, 1952.
27. R. Hooke and T. Jeeves. Direct search solution of numerical and statistical problems. *Journal of the Association of Computer Machinery*, 8(2):212–229, 1961.
28. R. Horst, P. Pardalos, and N. Thoai. *Introduction to Global Optimization*, Second edition, Springer, New York, 2000.
29. J. Kiefer. Sequential minimax search for a maximum. *Proceedings of the American Mathematical Society*, 4:502–506, 1953.

30. S. Kirkpatrick, C. Gelatt, and M. Vecchi. Optimization by simulated annealing. *Science*, 220(4598):671–680, 1983.
31. T. Kolda, R. Lewis, and V. Torczon. Optimization by direct search: new perspectives on some classical and modern methods. *SIAM Review*, 45(3):385–482, 2003.
32. K. Levenberg. A method for the solution of certain problems in least squares. *Quarterly Journal of Applied Mathematics*, 2:164–168, 1944.
33. R. Lewis, V. Torczon, and M. Trosset. Direct search methods: then and now. In *Numerical Analysis 2000*, pages 191–207, Elsevier, New York, 2001.
34. D. Luenberger. *Linear and Nonlinear Programming*, Second edition, Springer, New York, 2003.
35. D. Marquardt. An algorithm for least-squares estimation of nonlinear parameters. *Journal of the Society for Industrial and Applied Mathematics*, 11(2):431–441, 1963.
36. Z. Michalewicz. *Genetic Algorithms + Data Structures = Evolution Programs*. Third edition, Springer, New York, 1996.
37. J. Nelder and R. Mead. A simplex method for function minimization. *Computer Journal*, 7(4):308–313, 1964.
38. J. Nelder and R. Mead. A simplex method for function minimization—errata. *Computer Journal*, 8:27, 1965.
39. J. Nocedal and S. Wright. *Numerical Optimization*. Springer, New York, 1999.
40. E. Polak and G. Ribiere. Note sur la convergence de méthodes de directions conjugées. *Revue Française d'Informatique et de Recherche Opérationelle*, 16:35–43, 1969.
41. B. Polyak. The conjugate gradient method in extremal problems. *U.S.S.R. Computational Mathematics and Mathematical Physics*, 9(4):94–112, 1969.
42. M. Powell. An efficient method for finding the minimum of a function of several variables without calculating derivatives. *Computer Journal*, 7:155–162, 1964.
43. M. Powell. A method for nonlinear constraints in minimization problems. In R. Fletcher, editor, *Optimization*, pages 283–298, Academic Press, London and New York, 1969.
44. M. Powell. Restart procedures for the conjugate gradient method. *Mathematical Programming*, 12:241–254, 1977.
45. M. Powell. A fast algorithm for nonlinearly constrained optimization calculations. In G. Watson, editor, *Numerical Analysis Proceedings, Biennial Conference, Dundee, Lecture Notes in Mathematics (630)*, pages 144–157, Springer, Berlin, 1978.
46. W. Press, B. Flannery, S. Teukolsky, and W. Vetterling. *Numerical Recipes in C*. Second edition, Cambridge University Press, Cambridge, UK, 1992.
47. C. Reeves. *Modern Heuristic Techniques for Combinatorial Problems*. McGraw-Hill, New York, 1995.
48. H. Rosenbrock. An automatic method for finding the greatest or least value of a function. *Computer Journal*, 3:175–184, 1960.
49. D. Shanno. Conditioning of quasi-Newton methods for function minimizations. *Mathematics of Computation*, 24:641–656, 1970.
50. H. Sorenson. Comparison of some conjugate direction procedures for function minimization. *Journal of the Franklin Institute*, 288:421–441, 1969.
51. W. Spendley, G. Hext, and F. Himsworth. Sequential application of simplex designs in optimization and evolutionary operation. *Technometrics*, 4(4):441–461, 1962.
52. W. Swann. Report on the development of a new direct search method of optimization. Technical Report 64/3, Imperial Chemical Industries Ltd. Central Instrument Research Laboratory, London, 1964.
53. W. Zangwill. Minimizing a function without calculating derivatives. *Computer Journal*, 10(3):293–296, 1967.
54. J. Zhang, N. Kim, and L. Lasdon. An improved successive linear programming algorithm. *Management Science*, 31(10):1312–1331, 1985.

3
Integer Programming

3.1	Introduction..	**3**-1
3.2	Formulation of IP Models............................	**3**-3
	Capital Budgeting Problem • The Fixed-Charge Problem • Either-Or Constraints • If-Then Constraints • Functions with N Possible Values	
3.3	Branch and Bound Method..........................	**3**-7
	Branching • Computing Lower and Upper Bounds • Fathoming • Search Strategies • A Bound-First B&B Algorithm for Minimization (Maximization) • An Example	
3.4	Cutting Plane Method.................................	**3**-12
	Dual Fractional Cut (The Gomory Cuts) for Pure IPs • Dual Fractional Cutting-Plane Algorithm for ILP • An Example • Dual Fractional Cut (The Gomory Cuts) for MILPs • Dual Fractional Cutting-Plane Algorithm for MILP	
3.5	Other Solution Methods and Computer Solution	**3**-15
References...		**3**-16

Michael Weng
University of South Florida

3.1 Introduction

An integer programming (IP) problem is a mathematical (linear or nonlinear) programming problem in which some or all of the variables are restricted to assume only integer or discrete values. If all variables take integer values, then the problem is called a pure IP. On the other hand, if both integer and continuous variables coexist, the problem is called a mixed integer program (MIP). In addition, if the objective function and all constraints are linear functions of all variables, such a problem is referred to as a pure integer linear programming (ILP) or mixed integer linear programming (MILP) model. The term linear may be omitted unless it is necessary to contrast these models with nonlinear programming problems.

With the use of the matrix notation, an ILP problem can be written as follows:

$$\text{(ILP)} \quad \text{minimize} \quad \mathbf{dy}$$
$$\text{subject to} \quad \mathbf{By} \geq \mathbf{b}$$
$$\mathbf{y} \geq \mathbf{0} \quad \text{and integer}$$

where $\mathbf{y}$ represents the vector of integer variables, $\mathbf{d}$ is the coefficient vector of the objective function, and $\mathbf{B}$ and $\mathbf{b}$ are the coefficient matrix and right-hand-side vector of the constraints, respectively. All elements of $\mathbf{d}$, $\mathbf{B}$, and $\mathbf{b}$ are known constants. Some constraints may be equations and of the type "$\leq$." For an ILP with k variables and m constraints, $\mathbf{y} = k \times 1$ vector, $\mathbf{d} = 1 \times k$ vector, $\mathbf{B} = m \times k$ matrix, and $\mathbf{b} = m \times 1$ vector.

Similarly, an MILP is generally written as follows

$$\text{(MILP)} \quad \text{minimize} \quad \mathbf{cx} + \mathbf{dy}$$
$$\text{subject to} \quad \mathbf{Ax} + \mathbf{By} \geq \mathbf{b}$$
$$\mathbf{x} \geq \mathbf{0}$$
$$\mathbf{y} \geq \mathbf{0}, \quad \text{and integer}$$

where the additional notation $\mathbf{x}$ represents the vector of continuous variables, $\mathbf{c}$ is the coefficient vector for $\mathbf{x}$ in the objective function, and $\mathbf{A}$ the coefficient matrix for $\mathbf{x}$ in the constraints.

Note that any variable in $\mathbf{y}$ above can assume any nonnegative integer, as long as all constraints are satisfied. One special case is that all integer variables are restricted to be binary (i.e., either 0 or 1). Such an IP problem is referred to as a binary ILP (BILP) or binary MILP (BMILP). In this case, the constraint set ($\mathbf{y} \geq \mathbf{0}$, and integer) is replaced by $\mathbf{y} \in \{0,1\}$.

Although several solution methods have been developed for ILP or MILP, none of these methods is totally reliable in view of computational efficiency, particularly as the number of integer variables increases. In fact, most methods can be classed as either enumeration techniques, cutting-plane techniques, or a combination of these. Unlike LP, where problems with millions of variables and thousands of constraints can be solved in a reasonable time, computational experience with ILP or MILP remains elusive.

If all the integer restrictions on all the variables are omitted (i.e., allow an integer variable to assume continuous values), an ILP becomes an LP, called its LP relaxation. It is obvious that the feasible region of an ILP is a subset of the feasible region of its LP relaxation, and therefore, the optimal objective value of an ILP is always no better than that of its LP relaxation. In general, the optimal solution to the LP relaxation will not satisfy all the integer restrictions. There is one exception: For an ILP, if matrix B is totally unimodular, then the optimal solution to its LP relaxation is guaranteed to be integer-valued and therefore, is optimal to the ILP. Unfortunately, this is generally not the case for most ILPs. For a minimization ILP, the optimal objective value of its LP relaxation is a lower bound for the ILP optimal objective. Due to the computational difficulty associated with solving an ILP or MILP, LP relaxation is often used in developing solution techniques for solving them.

Once the optimal solution to the LP relaxation is obtained, one may round the noninteger values of integer variables to the closest integers. However, there is no guarantee in such rounding that the resulting rounded solution would be feasible to the underlying ILP. In fact, it is generally true that if the original ILP has one or more equality constraints, the rounded solution will not satisfy all the constraints and would be infeasible to the ILP. The infeasibility created by rounding may be tolerated in two aspects. One aspect is that, in general, the (estimated) parameters of the problems may not be exact. But there are typical equality constraints in integer problems where the parameters are exact. For example, the parameters in the multiple-choice constraint $y_1 + y_2 + \cdots + y_n = p$ (where p is the integer, and $y_j = 0$ or 1, for all j) are exact. The other aspect is that an integer variable can assume large values and rounding up or rounding down will not lead to a significant economic or social impact. It is most likely unacceptable, however, to use rounding as an approximation if an integer variable represents, for example, a one-time purchase of major objects such as large ships and jumbo jets, or a decision to finance or not to finance a major project.

Before discussing solution techniques for solving an ILP or MILP in detail, we first describe the formulation of several typical IP problems.

3.2 Formulation of IP Models

Similar to LP, there are three basic steps in formulating an IP model: (1) identifying and defining all integer and continuous decision variables; (2) identifying all restrictions and formulating all corresponding constraints in terms of linear equations or inequalities; and (3) identifying and formulating the objective as a linear function of the decision variables to be optimized (either minimized or maximized). The remainder of this section illustrates a variety of modeling techniques for constructing ILPs or MILPs via some sample problems.

3.2.1 Capital Budgeting Problem

Six projects are being considered for execution over the next 3 years. The expected returns in net present value and yearly expenditures of each project as well as funds available per year are tabulated below (all units are million dollars):

	Annual Expenditures			
Project	Year 1	Year 2	Year 3	Returns
1	5	2	8	20
2	9	7	9	40
3	4	6	3	15
4	7	4	4	21
5	3	5	2	12
6	8	4	9	28
Available Funds	26	22	25	

The problem seeks to determine which projects should be executed over the next 3 years such that at the end of the 3-year period, the total returns of the executed projects are at the maximum, subject to the availability of funds each year.

1. Identify the decision variables. Each project is either executed or rejected. Therefore, the problem reduces to a "yes-no" decision for each project. Such a decision can be represented as a binary variable, where the value 1 means "yes" and 0 means "no." That is,

    $y_j = 1$, if project j is selected to execute; and 0, otherwise, for $j = 1, 2, \ldots, 6$.

2. Formulate the constraints. In this problem, the constraints are that total annual expenditures of the selected projects do not exceed funds available for each year. These constraints can be mathematically expressed as follows:

    $$5y_1 + 9y_2 + 4y_3 + 7y_4 + 3y_5 + 8y_6 \leq 26, \text{ for year 1}$$
    $$2y_1 + 7y_2 + 6y_3 + 4y_4 + 5y_5 + 4y_6 \leq 22, \text{ for year 2}$$
    $$8y_1 + 9y_2 + 3y_3 + 4y_4 + 2y_5 + 9y_6 \leq 25, \text{ for year 3}$$

3. Formulate the objective function. The objective of this problem is to maximize the total returns of the selected projects at the end of the 3-year planning period. The total returns are given mathematically as

    $$Z = 20y_1 + 40y_2 + 15y_3 + 21y_4 + 12y_5 + 28y_6$$

The ILP model for this capital budgeting problem is

$$\text{maximize} \quad Z = 20y_1 + 40y_2 + 15y_3 + 21y_4 + 12y_5 + 28y_6$$
$$\text{subject to} \quad 5y_1 + 9y_2 + 4y_3 + 7y_4 + 3y_5 + 8y_6 \leq 26$$
$$2y_1 + 7y_2 + 6y_3 + 4y_4 + 5y_5 + 4y_6 \leq 22$$
$$8y_1 + 9y_2 + 3y_3 + 4y_4 + 2y_5 + 9y_6 \leq 25$$
$$y_1, \quad y_2, \quad y_3, \quad y_4, \quad y_5, \quad y_6 = 0 \text{ or } 1$$

Using matrix notation, this ILP can be written as

$$(\text{ILP}) \quad \text{minimize} \quad \mathbf{dy}$$
$$\text{subject to} \quad \mathbf{By} \geq \mathbf{b}$$
$$\mathbf{y} \geq \mathbf{0}, \text{ and binary}$$

where $\mathbf{d} = (20\ 40\ 15\ 21\ 12\ 28)$, $\mathbf{y} = (y_1\ y_2\ y_3\ y_4\ y_5\ y_6)^T$,

$$\mathbf{B} = \begin{bmatrix} 5 & 9 & 4 & 7 & 3 & 8 \\ 2 & 7 & 6 & 4 & 5 & 4 \\ 8 & 9 & 3 & 4 & 2 & 9 \end{bmatrix}, \text{ and } \mathbf{b} = \begin{pmatrix} 26 \\ 22 \\ 25 \end{pmatrix}$$

The optimal solution to the LP relaxation, obtained by imposing the upper bounds $x_j \leq 1$, for all j, is $y_1 = y_2 = y_5 = 1$, $y_3 = 0.7796$, $y_4 = 0.7627$, and $y_6 = 0.0678$ with an objective value of \$101.61 million. This solution has no meaning to the ILP, and rounding to the closest integer values leads to an infeasible solution. The optimal integer solution is $y_1 = y_2 = y_3 = y_4 = 1$ and $y_5 = y_6 = 0$ with $Z = \$96$ million.

The capital budgeting problem has a special case: the set-covering problem. There are k potential sites for new facilities and the cost associated with selecting a facility at site j is d_j. Each facility can service (or cover) a subset of m areas. For example, a facility may represent a fire station and the areas represent all sections of a city. The objective is to select the least-cost subset of all the potential sites such that each and every area is covered by at least one selected facility. Then the corresponding ILP can be written as above, except that all elements in matrix $\mathbf{B}$ are either 0 or 1.

If the planning horizon is reduced to a one-year period, then there is a single constraint, and the one-constraint capital budgeting problem is called the 0–1 knapsack problem. In addition, if each integer variable can assume any nonnegative discrete values, we get the so-called general knapsack problem.

3.2.2 The Fixed-Charge Problem

In a typical production planning problem involving n products, the production cost for product j may consist of a variable per-unit cost c_j, and a fixed cost (charge) $K_j(>0)$, which occurs only if product j is produced. Thus, if x_j = the production level of product j, then its production cost $C_j(x_j)$ is

$$C_j(x_j) = \begin{cases} K_j + c_j x_j, & x_j > 0 \\ 0, & x_j = 0 \end{cases}$$

This cost function is depicted in Figure 3.1. The objective would be to minimize $Z = \sum_j C_j(x_j)$. This objective function is nonlinear in variables x_j due to the discontinuity at the origin, and can be converted into a linear function by introducing additional

Integer Programming

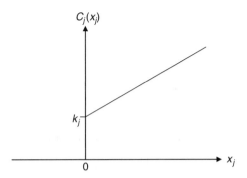

FIGURE 3.1 Fixed-charge cost function.

binary variables as follows. Define

$$y_j = \begin{cases} 1, & x_j > 0 \\ 0, & x_j = 0 \end{cases}$$

This condition can then be expressed as a single linear constraint as

$$x_j \leq M y_j$$

where $M(>0)$ is sufficiently large to guarantee that constraint $x_j \leq M$ is redundant whenever $y_j = 1$. On the other hand, the minimization of the linear objective function $Z = \sum_j (c_j x_j + K_j y_j)$ assures that $y_j = 0$, whenever $x_j = 0$.

3.2.3 Either-Or Constraints

There are practical instances when at least one of two constraints must be satisfied, and it may be impossible to satisfy both constraints simultaneously. For example, consider the following restriction

$$|x_j - 50| \geq 5$$

This restriction means that x_j has to deviate from 50 by at least 5. This constraint is nonlinear and can be replaced by the following two conflicting linear constraints: $x_j - 50 \geq 5$ (i.e., $55 - x_j \leq 0$), and $50 - x_j \geq 5$ (i.e., $x_j - 45 \leq 0$). In this case, it is obvious that both cannot be satisfied simultaneously. This conflict can be resolved by defining a binary variable as follows:

$$y_j = \begin{cases} 1, & x_j \geq 55 \\ 0, & x_j \leq 45 \end{cases}$$

Then the restriction $|x_j - 50| \geq 5$ can be replaced by

$$55 - x_j \leq M(1 - y)$$
$$x_j - 45 \leq My$$

where $M(>0)$ is a sufficiently large constant. These two constraints guarantee that either $x_j \geq 55$ or $x_j \leq 45$ (but not both) will hold.

Another typical either-or application concerns job sequencing on a processor: either job A proceeds job B, or job B proceeds job A, but not both.

In general, the either-or constraints can be described as follows: at least one of the two constraints $g(\mathbf{x}) \leq b$ and $h(\mathbf{x}) \leq e$ must be satisfied and the other constraint may or may not be satisfied. This can be modeled by using a binary variable y as

$$g(\mathbf{x}) \leq b + M(1-y)$$
$$h(\mathbf{x}) \leq e + My$$

The either-or constraints can be extended to the case of *satisfying at least p out of $m(>p)$* constraints. Suppose that p of the following m constraints must be satisfied.

$$g_1(\mathbf{x}) \leq b_1$$
$$g_2(\mathbf{x}) \leq b_2$$
$$\vdots$$
$$g_m(\mathbf{x}) \leq b_m$$

Define binary variable y_i for constraint $g_i(\mathbf{x}) \leq b_i$ as follows.

$$y_j = \begin{cases} 1, & \text{constraint i } is \text{ } satisfied \\ 0, & otherwise \end{cases}$$

Then the following guarantees that *at least p constraints* will be satisfied.

$$g_1(\mathbf{x}) \leq b_1 + M(1-y_1)$$
$$g_2(\mathbf{x}) \leq b_2 + M(1-y_2)$$
$$\vdots$$
$$g_m(\mathbf{x}) \leq b_m + M(1-y_m)$$
$$\sum_i y_i \geq p$$

Replacing the last constraint $\sum_i y_i \geq p$ by $\sum_i y_i = p$ and $\sum_i y_i \leq p$ models, respectively, exactly and at most p out of m constraints must be satisfied.

In many investment, production, or distribution problems, there might be minimum purchase or production requirements that must be met. For example, an investment opportunity might require a minimum investment of $100,000, or the introduction of a new product might be required to produce a minimum of 2000 units. Let x represent the continuous decision variable and C the minimum requirement. Then either $x = 0$ or $x \geq C$, and both cannot be met simultaneously. Define a binary variable y, and this restriction can be modeled simply by the following two constraints.

$$x \leq My$$
$$x \geq Cy$$

3.2.4 If-Then Constraints

There are situations where one constraint must be met if another constraint is to be satisfied. For example, two projects, 1 and 2, are considered. If project 1 is selected, then project 2 must also be selected. This case can easily be handled as follows. Let $y_j = 1$ if project j is selected, and 0 if not, $j = 1$ and 2. Then constraint $y_2 \geq y_1$ does the trick. Note that, if project 1 is not selected, there is no restriction on the selection of project 2 (i.e., project 2 may or may not be selected).

Integer Programming

In general, the if-then constraints can be described as follows. If constraint $g_1(\mathbf{x}) \leq b$ is met, then constraint $g_2(\mathbf{x}) \leq e$ must also be satisfied. But if constraint $g_1(\mathbf{x}) \leq b$ is not met, there is no restriction on the satisfaction of constraint $g_2(\mathbf{x}) \leq e$. This can be modeled by using two binary variables y_1 and y_2 as follows:

$$g_1(\mathbf{x}) \leq b + M(1 - y_1)$$
$$g_2(\mathbf{x}) \leq e + M(1 - y_2)$$
$$y_2 \geq y_1$$

If $y_1 = 1$, y_2 must also be 1, and the first two constraints above reduce to their original forms, which means that the if-then requirement is satisfied.

On the other hand, if $y_1 = 0$, $g_1(\mathbf{x}) \leq b + M$ will hold for any $\mathbf{x}$, since M is so large. This effectively eliminates the original constraint $g_1(\mathbf{x}) \leq b$. In addition, $y_2 \geq y_1$ reduces to $y_2 \geq 0$, and, therefore, $g_2(\mathbf{x}) \leq e$ may or may not be satisfied.

3.2.5 Functions with N Possible Values

Consider the situation that a function $f(\mathbf{x})$ is required to take on exactly one of N given values. That is, $f(\mathbf{x}) = b_1, b_2, \ldots, b_{N-1}$, or b_N. This requirement can be modeled by

$$f(\mathbf{x}) = \sum b_j y_j$$
$$\sum y_j = 1$$
$$y_j \text{ binary}, \quad \text{for } j = 1, 2, \ldots, N$$

3.3 Branch and Bound Method

In this section, we discuss in detail the branch and bound (B&B) method. Without loss of generality, it is assumed that the objective function is to be minimized. That is, the ILP is a minimization problem.

As mentioned earlier, total enumeration is not practical to solve ILPs with a large number of variables. However, the B&B method, which is an implicit enumeration approach, is the most effective and widely used technique for solving large ILPs. The B&B method theoretically allows one to solve any ILP by solving a series of LP relaxation problems (called subproblems).

The B&B method starts with solving the LP relaxation (problem). If the optimal solution to the relaxed LP is integer-valued, the optimal solution to the LP relaxation is also optimal to the ILP, and we are done. However, it is most likely that the optimal solution to the LP relaxation does not satisfy all the integrality restrictions. See, for instance, the optimal solution to the relaxed LP of the capital budgeting problem in Section 3.2.1. In this case, we partition the ILP into a number of subproblems that are generally smaller in size or easier to solve than the original problem. This process of partitioning any given problem into two or more smaller or easier subproblems is commonly called branching, and each subproblem is called a branch. The B&B method is then repeated for each subproblem.

As mentioned earlier, the optimal solution to the LP relaxation is a lower bound on the optimal solution to the corresponding ILP. Let Z be the best known objective function value for the original ILP. For any given subproblem, let Z_{relax} be the optimal objective value to its LP relaxation. This means that the optimal integer solution to this subproblem is no better than Z_{relax}. If $Z_{\text{relax}} \geq Z$, then this subproblem (branch) cannot yield a feasible integer solution better than Z and can be eliminated from further consideration. This process

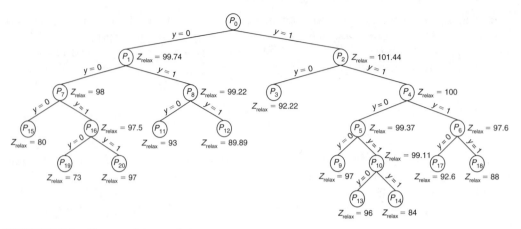

FIGURE 3.2 The complete search tree.

of eliminating a branch from further consideration is usually called fathoming. However, if $Z_{\text{relax}} < Z$, then a conclusion cannot be reached and further branching from this subproblem is needed. This process continues until all branches are fathomed, and the best integer solution is optimal to the original ILP.

The B&B method can be represented by a search tree that has many levels (Figure 3.2). The original ILP is the root, the first level consists of all the subproblems branched from the original ILP, the second level consists of all the subproblems branched from all the first level subproblems that have not been fathomed, and so on. We now discuss in more detail each of the three key components of the B&B method: branching, computing bounds, and fathoming.

3.3.1 Branching

For convenience, let P_0 denote a given ILP or MILP. The set of all subproblems branched from P_0 must represent all of P_0 to find an optimal solution to P_0. In addition, it is an approximation that any two subproblems are mutually exclusive. For example, let $q_1, q_2, \ldots, q_r$ be all the possible values that integer variable y_j can assume. Let P_i represent the subproblem obtained by fixing y_j at q_i, for $i = 1, 2, \ldots, r$. If P_0 is partitioned into $P_1, P_2, \ldots, P_r$, then

$$\{P_0\} = \{P_1\} \cup \{P_2\} \cup \ldots \cup \{P_r\}, \text{ and}$$
$$\{P_i\} \cap \{P_s\} = \emptyset, \quad \text{for all } i \neq s$$

Note that each subproblem P_i has one fewer integer variable since y_j now is a fixed constant.

If y_j is a binary variable, there are only two obvious branches with $y_j = 0$ and $y_j = 1$. Among all binary variables, y_j should be selected as one that does not equal to 0 or 1 in the optimal solution to the LP relaxation. In the capital budgeting problem given in Section 3.2.1, any of y_3, y_4, and y_5 can be chosen as y_j.

If y_j is a general integer variable, it may require a lot of effort to find all its possible feasible integer values. A more effective way is to derive two branches: $y_j \leq t$ and $y_j \geq t+1$, where t is a nonnegative integer. Again, one of the variables whose values are not an integer in the optimal solution to the LP relaxation should be selected and t is the largest integer smaller than the corresponding variable value.

As one can see, the branching process is essentially to add additional restrictions to form subproblems. Therefore, the optimal solution to any subproblem is no better than the

Integer Programming 3-9

branching subproblem. In particular, for minimization, the optimal solution to any subproblem is always greater than or equal to the optimal solution to the branching subproblem and thus to the original ILP. As we move further from the root in the search tree, the optimal objective values of subproblems increase or remain the same for minimization ILPs.

3.3.2 Computing Lower and Upper Bounds

Let Z_{best} be the objective function value of a known integer solution. Then the optimal objective value Z^* is at least as good as Z_{best}. That is, Z_{best} is an upper bound on Z^* for a minimization ILP. If a feasible integer solution with an objective value strictly better than Z_{best} is found, then replace Z_{best} by the new solution. In doing so, the upper bound Z_{best} remains the smallest upper bound and the corresponding feasible integer solution is called the incumbent solution (i.e., the best known integer solution). If no integer solutions have been identified yet, Z_{best} is set to be ∞.

For any subproblem P_j in the B&B search, we attempt to find a lower bound Z_L on the optimal objective value of P_j. That is, the optimal solution of P_j cannot be better than Z_L. If $Z_L > Z_{\text{best}}$ (the objective value of the incumbent integer solution), then the subproblem P_j can be discarded from further consideration, and the corresponding branch is then fathomed. However, if $Z_L \leq Z_{\text{best}}$, then a conclusion cannot be reached and the subproblem P_j needs to be branched further.

In general, it is no easy task to find a lower bound for an optimization problem. Recall, however, that for any ILP, the optimal objective value associated with its LP relaxation provides a lower bound on the optimal objective value of the ILP.

3.3.3 Fathoming

In the B&B search, the ILP (the root) is branched into two or more level-1 subproblems; each level-1 subproblem is either fathomed or branched into two or more level-2 subproblems; and each level-2 subproblem is either fathomed or further branched into two or more level-3 subproblems, and so on. If all subproblems (branches) are fathomed, the search stops and the incumbent integer solution is optimal to the ILP.

A subproblem may be fathomed in one of the following three ways:

1. An optimal integer solution is found. In this case, further branching from this subproblem is unnecessary and the incumbent solution may be updated.
2. The subproblem is found to be infeasible.
3. The optimal objective value Z_L to its LP relaxation is strictly greater than Z_{best}.

3.3.4 Search Strategies

During the B&B searching process, there are generally many subproblems (at different levels) that remain to be further branched. The question is which remaining subproblem should be selected to branch next? That is, what search strategy should be used?

Width-first and depth-first searches are commonly used search strategies. In width-first search, all subproblems at a level are examined before any subproblem at the next level will be considered, and the search always moves forward from one level to the next. In depth-first search, an arbitrary subproblem at the newest level is examined first, and the search may move backwards. Since depth-first search can yield a feasible solution faster, which can be used to fathom, it is computationally more efficient than width-first search. However, width-first strategy generally requires less computer memory. Another commonly used strategy is best-bound first, where the sub-problem with the best bound is branched first. For minimization, the subproblem with the smallest lower bound is branched. In doing so, it is likely to generate a good integer solution early in the searching process and therefore speed up the search process.

In the next section, we present a B&B algorithm using the bound-first search strategy.

3.3.5 A Bound-First B&B Algorithm for Minimization (Maximization)

Step 1: Solve the LP relaxation. If the optimal solution happens to be integer-valued, then stop; and this is the optimal solution to the ILP. Otherwise, set $Z_{\text{best}} = \infty (= -\infty)$, define the ILP as the candidate problem P, and go to Step 2.

Step 2: Let y_j be a variable whose value in the optimal solution to the LP relaxation of P is not integer, and let t be the noninteger value of y_j in the solution. Branch P into two subproblems by adding two constraints $y_j \leq \text{int}(t)$ and $y_j \geq \text{int}(t) + 1$, where $\text{int}(t)$ is the integer part of t. If y_j is binary, let $y_j = 0$ and $y_j = 1$. Solve the LP relaxation problems. If a subproblem is infeasible, fathom it. Otherwise, proceed to Step 3.

Step 3: If an integer solution is found, fathom the corresponding subproblem, and replace Z_{best} by the new feasible objective value if it is strictly smaller (greater) than Z_{best}.

Step 4: If a noninteger solution is found, fathom it if the optimal objective value Z_{relax} is strictly greater (smaller) than the current solution, Z_{best}.

Step 5: If all subproblems are fathomed, then stop; an optimal integer solution has been found with the optimal objective value Z_{best}.

Step 6: Replace the candidate subproblem P by the remaining subproblem with the smallest (largest) bound Z_{relax}, and go to Step 2.

3.3.6 An Example

This section presents an example to illustrate the use of the B&B method. For simplicity, a BILP problem is considered. In particular, we use the capital budgeting problem presented in Section 3.2.1. We denote this ILP P_0.

Since this is a maximization ILP, set $Z_{\text{best}} = -\infty$. The optimal solution to the LP relaxation of P_0 is $y_1 = y_2 = y_5 = 1$, $y_3 = 0.7796$, $y_4 = 0.7627$, and $y_6 = 0.0678$ with an objective value $Z_{\text{relax}} = 101.61$. This is a non-integer solution. Choose, for example, y_3 as y_j. Fixing y_3 at 0 and 1 respectively yields the following two sub-problems that have only five variables.

P_1: maximize $\quad Z = 20y_1 + 40y_2 + 21y_4 + 12y_5 + 28y_6$
subject to
$$5y_1 + 9y_2 + 7y_4 + 3y_5 + 8y_6 \leq 26$$
$$2y_1 + 7y_2 + 4y_4 + 5y_5 + 4y_6 \leq 22$$
$$8y_1 + 9y_2 + 4y_4 + 2y_5 + 9y_6 \leq 25$$
$$y_1, \quad y_2, \quad y_4, \quad y_5, \quad y_6 = 0 \text{ or } 1$$

P_2: maximize $\quad Z = 20y_1 + 40y_2 + 21y_4 + 12y_5 + 28y_6 + 15$
subject to
$$5y_1 + 9y_2 + 7y_4 + 3y_5 + 8y_6 \leq 22$$
$$2y_1 + 7y_2 + 4y_4 + 5y_5 + 4y_6 \leq 16$$
$$8y_1 + 9y_2 + 4y_4 + 2y_5 + 9y_6 \leq 22$$
$$y_1, \quad y_2, \quad y_4, \quad y_5, \quad y_6 = 0 \text{ or } 1$$

Integer Programming

The optimal relaxed LP solutions are

P_1: $y_2 = y_4 = y_5 = 1, y_1 = 0.8947,$ and $y_6 = 0.3158$ with an objective value $Z_{\text{relax}} = 99.74$.

P_2: $y_1 = y_2 = 1, y_4 = 0.767, y_5 = 0.7476,$ and $y_6 = 0.0485$ with an objective value $Z_{\text{relax}} = 101.44$.

Since P_2 has the largest lower bound $Z_{\text{relax}} = 101.44$, it is selected to branch from the next. Let us choose to set $y_4 = 0$ and $y_4 = 1$. This will lead to two subproblems with only four decision variables $y_1, y_2, y_5,$ and y_6: $P_3(y_4 = 0)$ and $P_4(y_4 = 1)$. The two relaxed LP solutions are

P_3: $y_2 = y_6 = 1, y_1 = 0.2778,$ and $y_5 = 0.8889$ with $Z_{\text{relax}} = 92.22$.

P_4: $y_1 = y_2 = 1, y_6 = 0,$ and $y_5 = 0.3333$ with $Z_{\text{relax}} = 100$.

Of the three remaining subproblems P_1, P_3, and P_4, P_4 is selected to branch next by setting $y_5 = 0(P_5)$ and $y_5 = 1(P_6)$.

P_5: $y_2 = 1, y_1 = 0.9474,$ and $y_6 = 0.1579$ with $Z_{\text{relax}} = 99.37$.

P_6: $y_1 = 1, y_2 = 0.6,$ and $y_6 = 0.2$ with $Z_{\text{relax}} = 97.6$.

Among the four remaining subproblems P_1, P_3, P_5, and P_6, P_1 is branched next by fixing $y_1 = 0(P_7)$ and $y_1 = 1(P_8)$.

P_7: $y_2 = y_5 = y_6 = 1,$ and $y_4 = 0.8571$ with $Z_{\text{relax}} = 98$.

P_8: $y_2 = y_4 = y_5 = 1,$ and $y_6 = 0.2222$ with $Z_{\text{relax}} = 99.22$.

Of P_3, P_5, P_6, P_7, and P_8, the subproblem P_5 is the next one to branch from. Choose y_1 to be fixed at 0 and 1.

P_9: $y_2 = 1, y_6 = 0.75$ with $Z_{\text{relax}} = 97$.

P_{10}: $y_2 = 1, y_6 = 0.1111$ with $Z_{\text{relax}} = 99.11$.

We next branch from subproblem P_8 and set y_6 equal to 0 and 1.

P_{11}: $y_2 = y_4 = y_5 = 1$ with $Z_{\text{relax}} = 93$. This is an integer solution. Since $Z_{\text{relax}} > Z_{\text{best}} = -\infty$, reset $Z_{\text{best}} = 93$. P_{11} becomes the first incumbent solution. Fathom P_{11}.

P_{12}: $y_4 = y_5 = 1,$ and $y_2 = 0.2222$ with $Z_{\text{relax}} = 89.89$.

Since $Z_{\text{relax}} < Z_{\text{best}}$ for subproblems P_3 and P_{12}, fathom both P_3 and P_{12}. Of the four remaining subproblems P_6, P_7, P_9, and P_{10}, we branch from P_{10} next and fix y_6.

P_{13}: $y_2 = 1$ with $Z_{\text{relax}} = 96$. Again, this is an integer solution. Since $Z_{\text{relax}} > Z_{\text{best}} = 93$, reset $Z_{\text{best}} = 96$, replace the current incumbent solution P_{11} by P_{13}, and fathom P_{13}.

P_{14}: $y_2 = 0$ with $Z_{\text{relax}} = 84$. This solution is all integer-values, and simply fathom P_{14} since $Z_{\text{relax}} < Z_{\text{best}} = 96$.

There are three subproblems P_6, P_7, and P_9 that remain to be examined. We now branch from P_7 and fix y_4 at 0 and 1.

P_{15}: $y_2 = y_5 = y_6 = 1$ with $Z_{\text{relax}} = 80$. Fathom P_{15}.

P_{16}: $y_2 = y_5 = 1$ and $y_6 = 0.875$ with $Z_{\text{relax}} = 97.5$. Fathom P_{15}.

There still are three subproblems P_6, P_9, and P_{16} left. Branch from P_6 by fixing y_2 at 0 and 1.

P_{17}: $y_1 = 1$ and $y_6 = 0.875$ with $Z_{\text{relax}} = 92.6$. P_{17} is fathomed since $Z_{\text{relax}} < Z_{\text{best}} = 96$.

P_{18}: $y_1 = 0$ and $y_6 = 0$ with $Z_{\text{relax}} = 88$. P_{18} is fathomed since the solution is all integer-valued with $Z_{\text{relax}} < Z_{\text{best}} = 96$.

There are only two subproblems remaining. Branch from P_{16} leads to the following:

P_{19}: $y_2 = y_5 = 1$ with $Z_{\text{relax}} = 73$. Fathom P_{19}.

P_{20}: $y_2 = 1$ and $y_5 = 2/3$ with $Z_{\text{relax}} = 97$.

Both the remaining subproblems P_9 and P_{20} have the same $Z_{\text{relax}} = 97$. Branching from both leads to four subproblems that will be fathomed since their relaxed LP solutions are all strictly worse than Z_{best}. Now all subproblems have been fathomed, and the B&B search stops with the current incumbent solution $P_{13}(y_1 = y_2 = y_3 = y_4 = 1$ and $y_5 = y_6 = 0$ with $Z_{\text{best}} = 96$) being the optimal solution to the ILP.

This ILP has six binary variables. Since each variable can assume either 0 or 1, the total enumeration will have to evaluate a total of $2^6 = 64$ solutions (many of them may be infeasible). In the above B&B search, a total of only 25 subproblems are evaluated—over 60% reduction of computational effort. As the number of variables increases, computational effort reduction can become more significant. This is especially true when the number of variables that equal 1 in an optimal integer solution is a small portion of all the binary integer variables.

In the B&B search, the number of possible subproblems at least doubles as the level increases by 1. At the final level, for an ILP with k integer variables, there are at least 2^k possible sub-problems. For a small $k = 20$, $2^k > 1$ million, and $k = 30$, $2^k > 1$ billion. Too many subproblems! One way to reduce the search effort is to limit the number of subproblems to branch further to some constant (e.g., 50). Such a search method is called a beam search, and the constant is the beam size. Beam search may quickly find an integer solution that may be good, but its optimality is not guaranteed.

3.4 Cutting Plane Method

Recall that the feasible region of any LP is a convex set, and each solution found by the simplex method corresponds to an extreme point of the feasible region. In other words, the simplex method only searches the extreme points of the feasible region in identifying an optimal solution. In general, the optimal extreme point associated with the optimal solution may not be all integer-valued. If the coefficient matrix **B** in an ILP is totally unimodular, then each extreme point of the feasible region is guaranteed to have all integer variable values, and therefore, the relaxed LP solution is guaranteed to be all integer-valued. But this is generally not true for most ILPs. However, if only the optimal extreme point of the relaxed LP feasible region is all integer-valued, then the LP relaxation solution is also optimal to the ILP. This is the underlying motivation for a cutting-plane method.

The basic idea is to change the boundaries of the convex set of the relaxed LP feasible region by adding "cuts" (additional linear constraints) so that the optimal extreme point becomes all-integer when all such cuts are added. The added cuts will slice off some off the feasible region of the LP relaxation (around the current noninteger optimal extreme point), but will not cut off any feasible integer solutions. That is, the areas sliced off from the feasible region of the LP relaxation do not include any of the integer solutions feasible

Integer Programming

to the original ILP. When enough such cuts are added, the new optimal extreme point of the sliced feasible region becomes all-integer, and thus is optimal to the ILP.

We first describe in detail the dual fractional cutting-plane method for pure ILPs and then extend it to MILPs.

3.4.1 Dual Fractional Cut (The Gomory Cuts) for Pure IPs

Consider a maximization ILP, where all coefficients are integer and all constraints are equalities. Its LP relaxation is

$$\text{(LP Relaxation)} \quad \text{maximize} \quad \mathbf{dy}$$
$$\text{subject to} \quad \mathbf{By} = \mathbf{b}$$
$$\mathbf{y} \geq \mathbf{0}$$

The optimal simplex solution to the LP relaxation can be expressed as follows:

$$\mathbf{y}_B + \mathbf{S}^{-1}\mathbf{N}\mathbf{y}_N = \mathbf{S}^{-1}\mathbf{b}$$

where $\mathbf{y}_N = \mathbf{0}$, and $\mathbf{y}_B = \mathbf{S}^{-1}\mathbf{b}(\geq \mathbf{0})$ are respectively the vectors of non-basic and basic variables, and $\mathbf{S}$ and $\mathbf{N}$ consist of columns of coefficients in the constraints corresponding to basic and non-basic variables, respectively. The square matrix $\mathbf{S}$ is commonly called a basis.

Let $\mathbf{a}_j$ denote the column of constraint coefficients of variable y_j. Then the above equation can be rewritten as follows:

$$\mathbf{y}_B + \sum_{\text{non-basic } j} \mathbf{u}_j y_j = \mathbf{v}$$

where $\mathbf{u} = \mathbf{S}^{-1}\mathbf{a}_j$ and $\mathbf{v} = \mathbf{S}^{-1}\mathbf{b}$. If $\mathbf{v}$ is all-integer, then this solution is also optimal to the ILP. Otherwise, $\mathbf{v}$ has at least one noninteger element. Let v_r be a noninteger element in $\mathbf{v}$. A cut will be constructed based on the rth equation in the above vector equation, which can be explicitly written as follows:

$$y_{B,r} + \sum_{\text{non-basic } j} u_{r,j} y_j = v_r$$

Since v_r is noninteger and positive, it can be expressed as $v_r = \text{int}(v_r) + f_r$, where $\text{int}(v_r)$ is the integer rounded down from v_r, and f_r is the non-integer part of v_r. Clearly, $0 < f_r < 1$. Similarly, we can write $u_{r,j} = \text{int}(v_{r,j}) + f_{r,j}$, where $0 \leq f_{r,j} < 1$, for all non-basic variable y_j. Substituting v_r and $u_{r,j}$ into the above equation yields

$$y_{B,r} + \sum_{\text{non-basic } j} \Big[\text{int}(u_{r,j}) + f_{r,j}\Big] y_j = \text{int}(v_r) + f_r$$

The above equation can be rewritten as

$$\sum_{\text{non-basic } j} f_{r,j} y_j - f_r = \text{int}(v_r) - \Bigg[y_{B,r} + \sum_{\text{non-basic } j} \text{int}(u_{r,j}) y_j \Bigg]$$

Since the right-hand side of this equation is always integer for any feasible integer solution, the left-hand side must also be integer-valued for all feasible integer solutions. In addition, the first term $\sum_{\text{non-basic } j} f_{u,j} y_j$ is nonnegative for any nonnegative solutions. Therefore,

$$\sum_{\text{non-basic } j} f_{u,j} y_j - f_r = y$$

where y is some nonnegative integer. Adding this as an additional constraint to the ILP will not reduce any feasible solution. This cut is referred to as a Gomory cut (Gomory, 1960). In addition, the ILP remains a pure IP after adding this cut.

Resolve the LP relaxation of the ILP with the added cuts. If an all-integer solution is obtained, it is optimal to the original ILP and therefore stop. Otherwise, repeat.

The following algorithm outlines the basic steps of the dual fractional cutting-plane approach.

3.4.2 Dual Fractional Cutting-Plane Algorithm for ILP

Step 1: Start with an all-integer simplex tableau and solve it as an LP (i.e., the LP relaxation). If it is infeasible, so is the ILP and stop. If the optimal solution is all-integer, it is also optimal to the ILP, and stop. Otherwise, proceed to Step 2.

Step 2: In the optimal simplex tableau, identify a row (say, row r) associated with a non-integer basic variable. Use this row to construct a cut as follows.

$$\sum_{\text{non-basic } j} f_{u,j} y_j - f_r = y$$

where y is an additional nonnegative integer variable. Add this new constraint to the bottom of the current optimal simplex tableau, and go to Step 3.

Step 3: Reoptimize the new LP relaxation, using the dual simplex method. If the new LP relaxation is infeasible, so is the original ILP, and stop. If the optimal solution to the new LP relaxation is all-integer, it is also optimal to the original ILP, and stop. Otherwise, go to Step 2.

3.4.3 An Example

The following example is considered to illustrate the use of Gomory cuts.

$$\begin{aligned}
\text{maximize} \quad & Z = 7y_1 + 9y_2 \\
\text{subject to} \quad & -y_1 + 3y_2 + y_3 = 6 \\
& 7y_1 + y_2 + y_4 = 35 \\
& y_1, y_2, y_3, y_4 \geq 0, \quad \text{and integer}
\end{aligned}$$

The optimal solution to the LP relaxation is $(y_1, y_2, y_3, y_4) = (9/2, 7/2, 0, 0)$ with $Z = 63$. The equation associated with noninteger basic variable y_2 in the final optimal simplex tableau is $y_2 + (7/22)y_3 + (1/22)y_4 = 7/2$ which can be rewritten as

$$y_2 + \left(0 + \frac{7}{22}\right) y_3 + \left(0 + \frac{1}{22}\right) y_4 = 3 + \frac{1}{2}$$

Thus, the resulting Gomory cut is

$$\frac{7}{22} y_3 + \frac{1}{22} y_4 - \frac{1}{2} = y_5$$

Solve the LP relaxation and we get the optimal solution $\mathbf{y} = (32/7, 3, 11/7, 0, 0)$ with $Z = 59$. The equation associated with non-basic variable y_1 is $y_1 + (1/7)y_4 + (-1/7)y_5 = 32/7$; this yields the following Gomory cut:

$$\frac{1}{7} y_4 + \frac{6}{7} y_5 - \frac{4}{7} = y_6$$

Again, solving the new LP relaxation yields the optimal solution $\mathbf{y} = (4, 3, 1, 4, 0, 0)$ with $Z = 55$. This solution is all integer-valued. Therefore, it is also optimal to the original ILP, and stop.

3.4.4 Dual Fractional Cut (The Gomory Cuts) for MILPs

The cutting-plane algorithm for ILPs presented in Section 3.4.1 can be extended to deal with MILPs. Again, the intent is to whittle the feasible region down to an optimal extreme point that has integer values for all of the integer variables by adding cuts. Let row r correspond to an integer variable whose value is noninteger in the optimal solution to the LP relaxation. Then the cut takes the following form:

$$\sum_{\text{non-basic } j} g_{u,j} y_j - f_r = y$$

where y is a nonnegative integer variable, and $g_{u,j} =$

1. $u_{r,j}$, if $u_{r,j} > 0$ and y_j is a continuous variable,
2. $[f_r/(f_r - 1)]u_{v,j}$, if $u_{v,j} < 0$ and y_j is a continuous variable,
3. $f_{r,j}$, if $f_{r,j} \leq f_r$ and y_j is an integer variable,
4. $[f_r/(1 - f_r)](1 - f_{r,j})$, if $f_{r,j} \leq f_r$ and y_j is an integer variable.

3.4.5 Dual Fractional Cutting-Plane Algorithm for MILP

Step 1: Solve the LP relaxation. If it is infeasible, so is the MILP and stop. If the optimal solution satisfies all integer requirements, it is also optimal to the MILP, and stop. Otherwise, proceed to Step 2.

Step 2: In the optimal simplex tableau, identify a row (say, row r) which contains an integer basic variable whose value is not integer. Use this row to construct a cut as follows:

$$\sum_{\text{non-basic } j} g_{u,j} y_j - f_r = y$$

where y is an additional nonnegative integer variable. Add this new constraint to the bottom of the current optimal simplex tableau, and go to Step 3.

Step 3: Reoptimize the new LP relaxation, using the dual simplex method. If the new LP relaxation is infeasible, so is the original MILP, and stop. If the optimal solution to the new LP relaxation satisfies all integer restrictions, it is also optimal to the original MILP, and stop. Otherwise, go to Step 2.

3.5 Other Solution Methods and Computer Solution

In LP, the simplex method is based on recognizing that the optimum occurs at an extreme point of the convex feasible region defined by the linear constraints. This powerful result reduces the search for the optimum from an infinite number of many solutions to a finite number of extreme point solutions. On the other hand, pure ILPs with a bounded feasible region are guaranteed to have just a finite number of feasible solutions. It may seem that ILPs should be relatively easy to solve. After all, LPs can be solved extremely efficiently and the only difference is that ILPs have far fewer solutions to be considered.

Unfortunately, this is not the case. While LPs are polynomially solvable via the interior point methods (Bazarra et al., 1990), ILPs are NP-hard. The NP-hardness of ILPs can be established since every instance of the generalized assignment problem which is NP-hard is an ILP (for more discussion on computational complexity, refer to Garey and Johnson, 1979). Therefore, ILPs are, in general, much more difficult to solve than LPs. The integer nature of the variables makes it difficult to devise an efficient algorithm that searches directly among the integer points of the feasible region. In view of this difficulty, researchers have developed solution procedures (e.g., B&B search and cutting-plane methods) that are based on exploiting the tremendous success in solving LPs. A more detailed discussion of B&B search and cutting-plane methods can be found in, for example, Salkin and Mathur (1989) and Schrijver (1986). Other classical methods include, for example, partitioning algorithms (Benders, 1962) and group theoretic algorithms (Gomory, 1967).

More recently, local search methods have been developed to find heuristic solutions to ILPs. Glover and Laguna (1997) give excellent discussion of tabu search in integer programming. Heuristic search algorithms based on simulated annealing and genetic algorithms can be found in Aarts and Korst (1990) and Rayward et al. (1996).

There are many sophisticated software packages for ILPs or MILPs that build on recent improvements in integer programming. For example, the developers of the powerful mathematical programming package CPLEX have an ongoing project to further develop a fully state-of-the-art IP module. The inclusion of the Solver in Microsoft EXCEL has certainly revolutionized the practical use of computer software in solving ILPs. Refer to Fourer (2005) for an excellent survey of commercial software packages.

References

1. Aarts, E. and Korst, J., *Simulated Annealing and Boltzmann Machines*, John Wiley & Sons, New York, 1990.
2. Bazarra, M. S., Jarvis, J. J., and Sherali, H. D., *Linear Programming and Network Flows*, second edition, John Wiley & Sons, New York, 1990.
3. Benders, J., "Partitioning Procedures for Solving Mixed-Variables Programming Problems," *Numerische Mathematik*, **4**, pp. 238–252, 1962.
4. Fourer, R., "Linear Programming Software Survey," *ORMS Today*, June 2005.
5. Garey, M. R. and Johnson, D. S., *Computers and Intractability: A Guide to the Theory of NP-Completeness*, W. H. Freeman, San Francisco, 1979.
6. Glover, F. and Laguna, M., *Tabu Search*, Kluwer Academic Publishers, Boston, 1997.
7. Gomory, R. E., "An Algorithm for the Mixed Integer Problem," Research Memorandum RM-2597, The RAND Corporation, Santa Monica, CA, 1960.
8. Gomory, R. E., "Faces of an Integer Polyhedron," *Proceedings of the National Academy of Science*, **57** (1), 16–18, 1967.
9. Rayward-Smith, V. J., Osman, I. H., Reeves, C. R., and Smith, G. D. (Eds.), *Modern Heuristic Search Methods*, John Wiley & Sons, New York, 1996.
10. Salkin, H. M. and Mathur, K., *Foundations of Integer Programming*, North Holland, New York, 1989.
11. Schrijver, A., *Theory of Linear and Integer Programming*, John Wiley & Sons, New York, 1986.

4
Network Optimization

4.1	Introduction	4-1
4.2	Notation	4-2
4.3	Minimum Cost Flow Problem	4-3
	Linear Programming Formulation	
4.4	Shortest Path Problem	4-4
	Linear Programming Formulation • Dijkstra's Algorithm	
4.5	Maximum Flow Problem	4-8
	Linear Programming Formulation • Augmenting Path Algorithm	
4.6	Assignment Problem	4-13
	Linear Programming Formulation • The Hungarian Algorithm • Application of the Algorithm Through an Example	
4.7	Minimum Spanning Tree Problem	4-14
	Linear Programming Formulation • Kruskal's Algorithm	
4.8	Minimum Cost Multicommodity Flow Problem	4-18
	Linear Programming Formulation	
4.9	Conclusions	4-19
	References	4-19

Mehmet Bayram Yildirim
Wichita State University

4.1 Introduction

Communication on the Internet or via phones, satellites, or landlines; distribution of goods over a supply chain; transmission of blood through veins; movement of traffic on roads, connection of towns by roads, connection of yards by railroads, flight of planes between airports, and the like, have something in common: all can be modeled as networks. Taha (2002) reported that as much as 70% of the real-world mathematical programming problems can be represented by network-related models. These models are widely utilized in real life for many reasons. First, networks are excellent visualization tools and easy to understand as problems can be presented pictorially over networks. Second, network models have several computationally efficient and easy-to-implement algorithms that take advantage of the special structure of networks, thus providing a key advantage for researchers to handle large-size real-life combinatorial problems. Furthermore, network flow problems appear as subproblems when mathematical programming techniques such as decomposition, column generation, or Lagrangian relaxation are utilized to solve large-scale complex mathematical models. This is a key advantage because using efficient network algorithms results in faster convergence when the above-mentioned techniques are utilized. For example, shortest path problems, one of the simplest network flow models that can be solved quite efficiently, appear in many networks: transportation, communication, logistics, supply chain management, Internet routing, molecular biology, physics, sociology, and so on.

In this chapter, our goal is not to focus on the details of network flow theory, but instead to describe the general problem and provide a mathematical formulation for each problem. For most of the fundamental network problems presented in this chapter, we will describe an algorithm and illustrate how it can be applied to a sample network flow problem. We will also provide a variety of problems for some of the network models presented.

The organization of this chapter is as follows: First, the notation utilized in this chapter is presented. Next, the minimum cost flow problem is introduced. The special cases of this problem, the shortest path problem, maximum flow problem, and the assignment problem are then presented. The multicommodity flow problem and the minimum spanning tree problem are also studied in this chapter.

4.2 Notation

A *directed network* (or a *directed graph*) is defined as $G = (N, A)$, where N is the *node set* (i.e., $N = \{1, 2, 3, 4, 5, 6, 7\}$), and A is the *arc set* whose elements are ordered pairs of distinct nodes (i.e., $A = \{(1,2), (1,3), (2,3), (2,4), (3,6), (4,5), (4,7), (5,2), (5,3), (5,7), (6,7)\}$). An *arc* (i, j) starts with a *tail* node i and ends with a *head* node j, where nodes i and j are the *endpoints* of arc (i, j). In Figure 4.1, arcs (1,2) and (1,4) are directed arcs. An *arc adjacency list* $A(i)$ of a node i is the set of arcs emanating from node i, that is, $A(i) = \{(i, j) : (i, j) \in A\}$. For example, in the figure below, $A(4) = \{(4, 5), (4,7)\}$. The *indegree* of a node is the number of incoming arcs of that node, and the *outdegree* of a node is the number of outgoing arcs. The *degree* of a node is the sum of its *indegree* and *outdegree*. For example, node 5 has an indegree of 1 and outdegree of 3. As a result, the degree of node 5 is 4. Consequently, $\sum_{i \in N} |A(i)| = |N|$.

A *walk* in a directed graph G is a subgraph of G consisting of a sequence of nodes and arcs. This also implies that the subgraph is connected. A *directed walk* is an "oriented" walk such that for any two consecutive nodes i_k and i_{k+1}, (i_k, i_{k+1}) is an arc in A. For example, 1,2,5,7,4,2 is a walk but not a directed walk. 1,2,4,7 is a directed walk. A *path* is a walk without any repetition of nodes (e.g., 1,2,5,7), and a *directed path* is a directed walk without any repetition of nodes. For example, in the directed graph below, 1,2,4,5 is a directed path, but 1,2,4,5,2 is not. A cycle is a closed path that begins and ends at the same node (e.g., 4,5,7,4). A *directed cycle* is a directed closed path that begins and ends at the same node (e.g., 2,4,5,2). A network is called *acyclic* if it does not contain any directed cycle. A tree is a connected acyclic graph. A *spanning tree* on a given undirected graph is a subgraph of G that is a tree and spans (touches) all nodes. A graph $G = (N, A)$ is a *bipartite graph* if we

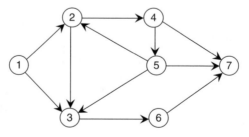

FIGURE 4.1 A directed network.

can partition the nodes in this graph into two disjoint subsets N_1 and N_2 so that for every arc $(i,j) \in A, i \in N_1$ and $j \in N_2$, or vice versa.

An *undirected network* is defined as $G = (N, A)$, where N is the node set and A is the arc set whose elements are unordered pairs of distinct nodes. An arc in an undirected network is similar to a two-way street (i.e., flow is allowed in both directions), while an arc in a directed network allows flow in one direction only. An undirected network can be transformed into a directed network by replacing an undirected arc (i,j) with two directed arcs (i,j) and (j,i). Consequently, node i is adjacent to node j, and node j is adjacent to node i. Thus, $\sum_{i \in N} |A(i)| = 2|N|$. The definitions above hold for undirected networks, except that there is no distinction between a cycle and a directed cycle, a path and a directed path, and a walk and a directed walk.

4.3 Minimum Cost Flow Problem

The minimum cost flow problem is one of the most fundamental network flow problems, where the goal is to send flow from supply nodes to demand nodes using arcs with capacities and involve the minimum total cost of transportation given availability of supply and demand in a directed network (if the network is undirected, an undirected arc between nodes i and j is replaced with two directed arcs, (i,j) and (j,i), with the same cost and capacity as the undirected arc to obtain a directed network).

A minimum cost flow problem has several applications: distribution of a product from manufacturing plants to warehouses, or from warehouses to retailers; flow of raw materials and intermediate goods through stations in a production line; routing of automobiles through a street network; and routing of calls through a telephone system.

4.3.1 Linear Programming Formulation

Let x_{ij} denote the amount of flow on arc (i,j). The minimum cost flow problem can be formulated as follows:

$$\text{Minimize:} \quad \sum_{(i,j) \in A} c_{ij} x_{ij} \quad (4.1)$$

$$\text{Subject to:} \quad \sum_{\{j:(i,j) \in A\}} x_{ij} - \sum_{\{j:(j,i,) \in A\}} x_{ji} = b_i \quad i \in N \quad (4.2)$$

$$l_{ij} \leq x_{ij} \leq u_{ij} \quad (i,j) \in A \quad (4.3)$$

$$x_{ij} \geq 0 \quad (i,j) \in A \quad (4.4)$$

where c_{ij} is the cost of arc $(i,j) \in A$ and b_i is the supply/demand at node i. If node i is a supply node, then $b_i > 0$, whereas $b_i < 0$ for a demand node, and $b_i = 0$ for a transshipment node. To have a feasible solution, $\sum_{i \in N} b_i = 0$. The objective here is to minimize the cost of transporting the commodity from supply nodes to demand nodes. Equation 4.2 is the mass balance (flow balance or conservation of flow) constraints: the difference between the total flow emanating from node i (*outflow*, the first term in Equation 4.2) and entering node i (*inflow*, the second term in Equation 4.2) is equal to the demand/supply at that node. Equation 4.3 states that flow on any arc (i,j) should be between the allowable range, that is, between the lower (l_{ij}) and upper bounds (u_{ij}) of flow on arc (i,j), where $l_{ij} \leq 0$ or $l_{ij} > 0$, and $u_{ij} < \infty$. When $u_{ij} = \infty$ for all arcs (i.e., there is no upper bound on the arc capacity), the problem becomes an *uncapacitated network flow problem*. Note that the above formulation has $|A|$ nonnegative variables, lower and upper bound constraints, and $|N|$ mass balance constraints.

The basic feasible solution of the equation system defined by Equations 4.2 to 4.3 are integer-valued if all b_i's are integer-valued. In other words, when the right-hand side for all constraints (i.e., the supply and demand) is an integer, the network flow models provide integer solutions.

The above problem can be solved using linear programming techniques such as the simplex algorithm. For network problems, calculations of the simplex tableau values become easier. The specialized simplex algorithm to solve network problems is defined as the *network simplex method*. In the network simplex algorithm, a feasible spanning tree structure is successively transformed into an improved spanning tree structure until optimality is achieved.

Minimum cost flow problems have several special cases. For example, in *transportation problems*, the network is bipartite. Each node is either a supply or a demand node. Supply nodes are connected to demand nodes via arcs without capacities. The goal is to minimize the overall transportation cost. In a transportation problem, when the supply and demand at each node is equal to one, and the number of supply nodes is equal to the number of demand nodes, an *assignment problem* is obtained. Other special cases of minimum cost flow problems include shortest path problems and maximum flow problems.

4.4 Shortest Path Problem

Shortest path problems involve a general network structure in which the only relevant parameter is cost. The goal is to find the shortest path (the path with minimum total distance) from the origin to the destination. Computationally, finding the shortest path from an origin to all other nodes (often also known as a shortest path tree problem) is not any more difficult than determining the shortest path from an origin to a single destination. Note that a shortest path problem is a specialized minimum cost flow problem, where the origin ships *one* unit of flow to every other node on the network (thus, a total of $m-1$ units of flow).

Shortest path problems arise in a variety of practical settings, both as stand-alone problems or as subproblems of more complex settings. Applications include finding a path of minimum time, minimum length, minimum cost, or maximum reliability. The shortest path problem arises in a transportation network where the goal is to travel the shortest distance between two locations, or in a telecommunication problem where a message must be sent between two nodes in the quickest way possible. Other applications include equipment replacement, project scheduling, project management, cash flow management, workforce planning, inventory planning, production planning, DNA sequencing, solving certain types of differential equations, and approximating functions. Shortest path problems appear as subproblems in traffic assignment problems, in multicommodity flow problems, and network design problems. More detailed descriptions of some of the problems that can be modeled as shortest path problem are provided below:

The *maximum reliability path problem* determines a directed path of maximum reliability from the source node to every other node in the network where each arc is associated with a reliability measure or probability of that arc being operational. The reliability of a path can be calculated by the product of reliabilities of each arc on that path. To solve this problem using a shortest path algorithm, one can take the cost of each arc as the logarithm of its reliability and convert the maximization problem into a minimization problem by multiplying the objective function by -1.

In an *equipment replacement problem*, the input is the total cash inflow/outflow for purchasing the equipment using the equipment for a certain period of time, and finally selling the equipment. This data can be transformed into a network structure by assuming that the nodes represent the timeline and that the arcs represent an equipment replacement decision. The cost of an arc between two nodes separated by k *years* is the total cost of buying, keeping, and then selling the equipment for k *years*. The goal in an equipment replacement problem is to determine the best equipment replacement policy over a T-*year* planning horizon, which is equivalent to solving a shortest path problem over the network described above.

4.4.1 Linear Programming Formulation

The objective in a shortest path problem is to minimize the total cost of travel from an origin node s (i.e., a source node) to all other nodes in a directed network where the cost of traversing arc (i, j) is given by c_{ij}. The shortest path problem can be considered a minimum cost flow problem with the goal of sending one unit of flow from the source node s to every other node in the network. Let x_{ij} be the amount of flow on arc (i, j). The shortest path problem can be formulated as follows:

$$\text{Minimize:} \quad \sum_{(i,j) \in A} c_{ij} x_{ij} \tag{4.5}$$

$$\text{Subject to:} \quad \sum_{\{j:(i,j) \in A\}} x_{ij} - \sum_{\{j:(j,i,) \in A\}} x_{ji} = (n-1), \quad i = s \tag{4.6}$$

$$\sum_{\{j:(i,j) \in A\}} x_{ij} - \sum_{\{j:(j,i,) \in A\}} x_{ji} = -1 \quad i = N - \{s\} \tag{4.7}$$

$$x_{ij} \geq 0 \quad (i,j) \in A \tag{4.8}$$

In the above formulation, the objective function, Equation 4.5, minimizes the total cost of sending $(n-1)$ units of flow node s to all other nodes on the network. Equation 4.6 is the mass balance constraint (node balance, conservation of flow constraint) for the origin. Equation 4.7 is the mass balance constraint for all other nodes. The above formulation has $|A|$ variables and $|N|$ constraints. The shortest path problem can be solved utilizing several specialized, very efficient algorithms. We will describe one of the most famous network optimization algorithms, Dijkstra's algorithm, to solve a shortest path problem in the following. In the presence of negative cycles, the optimal solution of the shortest path problem is unbounded.

4.4.2 Dijkstra's Algorithm

Dijkstra's algorithm is a widely used, simple-to-implement algorithm to solve shortest path problems. Dijkstra's algorithm is a label-setting algorithm: Initially, all nodes are assigned tentative distance labels (temporary shortest path distances), and then iteratively, the shortest path distance to a node or set of nodes at each step is determined. In the Dijkstra's implementation described below, $d(j)$ is the (temporary) distance label of node j (shortest path distance, or minimum cost directed path from the source node s to node j). Distance labels are upper bounds on shortest path distances. $Pred(j)$ is the immediate predecessor of node j on the shortest path tree. LIST is a data structure in which candidate nodes that have been assigned temporary shortest distances are stored. $A(j)$ is the arc adjacency list. The efficiency of the algorithm depends on the structure of the network (e.g., acyclic networks, arcs with nonnegative lengths, integer-valued arc costs, etc.) and how the LIST data structure is implemented (e.g., Dial's implementation, and heap implementations such as Fibonacci heap, radix heap, d-heap, etc.). Below is a description of the Dijkstra's algorithm:

Step 1: For the source node, let $d(s)=0$ and $pred(s)=0$. For all other nodes (i.e., $j \in N - \{s\}$), $d(j) = \infty$, $pred(j) = |N| + 1$. Set LIST $:= \{s\}$;

Step 2: Permanently label the node i with $d(i) = \min\{d(j) : j \in LIST\}$. Remove node i from the LIST;

Step 3: For each $(i, j) \in A(i)$, do a distance update: if $d(j) > d(i) + c_{ij}$, then $d(j) := d(i) + c_{ij}$, $pred(j) := i$, and if $j \notin LIST$, then add j to LIST;

Step 4: If LIST $\neq \emptyset$, then go to Step 2.

In the above implementation of Dijkstra's algorithm, we assumed that the network is directed. In an undirected network, to satisfy this assumption, each undirected arc (i, j) with cost c_{ij} can be replaced by two directed arcs, arc (i, j) and arc (j, i) with costs c_{ij}. It is also assumed that the network contains a directed path from the origin (the source node) to every other node in the network, and the cost c_{ij} for arc $(i, j) \in A$ is nonnegative.

We apply the Dijkstra's algorithm to the network depicted in Figure 4.2. The iterations of the algorithm are presented in Table 4.1. In the initialization stage, all distance labels other than the origin $(d(s) = 0)$ are set to infinity. At each iteration, a node is permanently labeled (i.e., the distance from the origin, node s, to that node is determined) and distance labels of all nodes that are accessible from the permanently labeled node are updated. In Table 4.1, shaded columns indicate nodes that have been permanently labeled. Below, we will describe a couple of iterations of the Dijkstra's algorithm.

At the first iteration, initially, the LIST contains only node 1 (the origin). Thus, node 1, the node with the shortest distance from the origin, is permanently labeled and removed from the LIST. Nodes 2 and 3 are added to the list, as the distance for those nodes through node 1 is less than their current temporary distance labels of infinity. In the second iteration, node 2 is the node with the shortest distance from the origin. Thus, node 2 is permanently labeled and removed from the LIST. Distance labels of nodes 4, 5, and 6 are updated and added to the LIST. The predecessor of these nodes is node 2. Dijkstra's algorithm takes six iterations to permanently label all nodes on the network. The shortest path from node 1 to node 6 is 1-2-4-5-6, with a total distance of 60. In Table 4.1, the distance labels and predecessors of each node are presented. The final shortest path tree is presented in Figure 4.3.

For more generalized networks (including networks with negative arc lengths), shortest path problems can be solved using label-correcting algorithms. Label-correcting algorithms determine the shortest path distance at the time the algorithm terminates (i.e., all of the

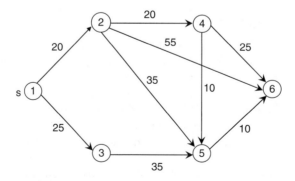

FIGURE 4.2 Shortest path example: the goal is to determine the shortest path from node 1 to all other nodes.

TABLE 4.1 Dijkstra's Algorithm Implementation

Initialization LIST = {1}

	\multicolumn{6}{c}{Node}					
	1	2	3	4	5	6
$d(j)$	0	∞	∞	∞	∞	∞
$Pred(j)$	0	7	7	7	7	7

Iteration 1 Permanently labeled node = 1 LIST = { }

	Node					
	1	2	3	4	5	6
$d(j)$	0	20	25	∞	∞	∞
$Pred(j)$	0	1	1	7	7	7

LIST = {2,3}

Iteration 2 Permanently labeled node = 2 LIST = {3}

	Node					
	1	2	3	4	5	6
$d(j)$	0	20	25	40	55	75
$Pred(j)$	0	1	1	2	2	2

LIST = {3,4,5,6}

Iteration 3 Permanently labeled node = 3 LIST = {4,5,6}

	Node					
	1	2	3	4	5	6
$d(j)$	0	20	25	40	55	75
$Pred(j)$	0	1	1	2	2	2

LIST = {4,5,6}

Iteration 4 Permanently labeled node = 4 LIST = {5,6}

	Node					
	1	2	3	4	5	6
$d(j)$	0	20	25	40	50	65
$Pred(j)$	0	1	1	2	4	4

LIST = {5,6}

Iteration 5 Permanently labeled node = 5 LIST = {6}

	Node					
	1	2	3	4	5	6
$d(j)$	0	20	25	40	50	60
$Pred(j)$	0	1	1	2	4	5

LIST = {6}

Iteration 6 Permanently labeled node = 6 LIST = { } **TERMINATE**

	Node					
	1	2	3	4	5	6
$d(j)$	0	20	25	40	50	60
$Pred(j)$	0	1	1	2	4	5

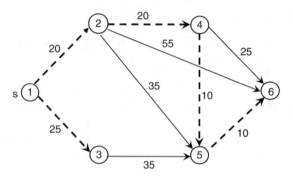

FIGURE 4.3 Shortest path tree for the network in Figure 4.2.

distance labels are temporary until the termination). Label-setting algorithms can also be viewed as a special case of label-correcting algorithms.

Some of the generalizations of the shortest path problems can be listed as follows: The *constrained shortest path problem* finds the shortest distance from an origin to all other nodes while keeping the traversal time of that path below a predetermined amount. The traversal time between nodes can be different from the distance between the given nodes. In the *k shortest path problem*, one determines not only the shortest path but also the 2nd, 3rd,..., $(k-1)$st, and kth shortest paths from an origin to every other node. The *multi-criteria shortest path problem* intends to determine a path that minimizes simultaneously all the criteria (i.e., distance, time, cost, reliability) under consideration. The multicriteria shortest path problem finds *nondominated paths*, that is, paths for which there is no other path with better values for all criteria. In the *all pair shortest path problem*, the shortest path between all pairs of nodes is found.

4.5 Maximum Flow Problem

In a capacitated network, often, sending a maximum amount of flow from an origin (source) node to a destination (sink) node might be advantageous. In many situations, links on the network can be considered as having capacity that limits the quantity of a product that may be shipped through the arc. Often in these situations, knowledge of the overall capacity of the network might provide an advantage over competitors. This can be achieved by finding the maximum flow that can be shipped on the network. The objective in a maximum flow problem is to determine a feasible pattern of flow through the network that maximizes the total flow from the supply node (source, origin) to the destination node (sink). The maximum flow problem is a special case of a minimum cost flow problem. Any maximum flow problem can be transformed into a minimum cost flow problem using the following transformation: add an arc with -1 cost and ∞ capacity between the sink node and source node (i.e., add arc (d, s)). All other arcs on the network have a cost of zero. As the objective in a minimum cost flow problem is to minimize total cost, the maximum possible flow is delivered to the sink node when the minimum cost flow problem is solved for the new network.

Maximum flow problems have several applications. For example, consider a pipeline that transports water from lakes (sources) to a residential area (sinks) via a pipeline (arcs on the network). Water has to be transported via pipes with different capacities. Intermediate pumping stations (transshipment nodes) are installed at appropriate distances. Before being available for consumption, the water has to be treated at treatment plants. The intermediate pumping stations and treatment plants are transshipment nodes. The goal is to determine the maximum capacity of this pipeline (i.e., the maximum amount that can be pumped from sources to sinks). Similar problems exist in crude oil and natural gas pipelines. Other examples of maximum flow problems are staff scheduling, airline scheduling, tanker scheduling, and so on. In summary, maximal flow problems play a vital role in the design and operation of water, gas, petroleum, telecommunication, information, electricity, and computer networks like the Internet and company intranets.

4.5.1 Linear Programming Formulation

Let f represent the amount of flow in the network from source node s to sink node d (this is equivalent to the flow on arc (d, s)). Then the maximal flow problem can be stated as follows:

$$\text{Maximize:} \quad f \tag{4.9}$$

$$\text{Subject to:} \quad \sum_{\{j:(i,j)\in A\}} x_{ij} = f \quad i = s \tag{4.10}$$

$$\sum_{\{j:(i,j)\in A\}} x_{ij} - \sum_{\{j:(j,i)\in A\}} x_{ji} = 0 \quad i = N\backslash\{s,d\} \tag{4.11}$$

$$-\sum_{\{j:(j,i)\in A\}} x_{ji} = -f \quad i = d \tag{4.12}$$

$$x_{ij} \leq u_{ij} \quad (i,j) \in A \tag{4.13}$$

$$x_{ij} \geq 0 \quad (i,j) \in A \tag{4.14}$$

The objective is to maximize the total flow sent from origin (node s) to destination node d. Equations 4.10 through 4.12 are conservation of flow (mass balance) constraints. At the origin and destination, the net inflow and outflow are f and $-f$, respectively. The mathematical program has $|A|+1$ variables, $|A|$ capacity constraints, and $|N|$ node balance constraints. Maximum flow problems can be solved using the network simplex method. However, algorithms such as the augmented path algorithm, the preflow-push algorithm, and excess scaling algorithms can be utilized to solve large-scale maximum flow problems efficiently. Below we present an augmenting path algorithm.

A special case of the maximum flow problem (thus, the minimum cost flow problem) is the *circulation problem*, in which the objective is to determine if the network flow problem is feasible, that is, if there is a solution in which the flow on each arc is between the lower and upper bounds of that arc. The circulation problem does not have any supply or demand nodes (i.e., all nodes are transshipment nodes, thus, $b_i = 0$ for all $i \in N$). A circulation problem can be converted to a minimum cost flow problem by adding an arc with -1 cost and ∞ capacity between the sink node and source node (i.e., add arc (d, s)). All other arcs on the network have a zero cost.

4.5.2 Augmenting Path Algorithm

In an augmenting path algorithm, the flow along the paths from source node to destination node is incrementally augmented. This algorithm maintains mass balance constraints at every node of the network other than the origin and destination. Furthermore, instead of the original network, a residual network is utilized to determine the paths from origin (supply node) to destination (demand node) where there can be a positive flow. The residual network is defined as follows: For each arc (i, j), an additional arc (j, i) is defined. When there is a flow of e_{ij} on arc (i, j), the remaining capacity of arc (i, j) is $c_{ij} - e_{ij}$, whereas the remaining capacity of arc (j,i) is increased from 0 to e_{ij}. Whenever some amount of flow is added to an arc, that amount is subtracted from the residual capacity in the same direction and added to the capacity in the opposite direction. An augmented path between an origin–destination pair is a directed path for which every arc on the path has strictly positive residual capacity. The minimum of these residual capacities of the augmented path is the residual capacity (c^*) of the residual path. Then the remaining capacities on arcs $((i,j), (j,i))$ are modified along the path to either $(c_{ij} - e_{ij} - c^*, e_{ij} + c^*)$ or $(c_{ij} - e_{ij} + c^*, e_{ij} - c^*)$, depending on whether the flow is on arc (i,j) or (j,i). The augmenting path algorithm can be stated as follows:

Step 1: Modify the original network to obtain the residual network.

Step 2: Identify an augmenting path (i.e., a directed path from source node to sink node) with positive residual capacity.

Step 3: Determine the maximum amount of flow that can be sent over the augmenting path (suppose this is equal to c^*). Increase the flow on this path by c^*, i.e., total flow = total flow + c^*.

Step 4: Decrease the *residual capacity* on each arc of the augmenting path by c^*, and increase *residual capacity* on each arc in the opposite direction on the augmenting path by c^*.

Step 5: If there is an augmented path from source to sink with positive residual capacity, then go to step 3.

We apply the augmenting path algorithm to the network shown in Figure 4.4a. The residual network is obtained after initialization. In Figure 4.4b, the residual capacity of each arc is listed above/below the header (arrow/orientation) of each arc. For example, initially the remaining capacity on arc (2,4) is 20 (that is the capacity of this arc) and (4,2) is 0. The augmented paths and updates in flows are given in Table 4.2. Initially, all of the newly added reverse arcs carry zero flow. At iteration 1, 10 units of flow are sent on path 1-2-5-6 (the minimum residual arc capacity on this path). The shaded cells are the arcs on this path for which there is a change in the residual arc capacity. For example, arc (5,6) can no longer carry any flow, while its reverse arc, (6,5), has 10 units of residual capacity. In iteration 2, the bottleneck (the arc with minimum capacity) occurs at arc (5,2). As a result, 10 units of flow are transferred on path 1-3-5-2-4-6. Finally, after 10 units of flow are sent on path 1-2-4-6, there is no more augmenting path with strictly positive residual capacity from node 1 to node 6 (see Figure 4.4c for the flows on paths and the residual arc capacities). As a result, the maximum flow on this network is 30 units.

The maximum flow problem can also be solved using the *minimum cut algorithm*. A *cut* in a connected network defines the set of directed arcs, which, if removed from the network, would make it impossible to travel from the source/origin to the sink/destination. In other words, a cut should include at least one arc from every directed path from the source node to the sink node to prevent flow going from origin to destination. The *capacity of a cut* is the sum of the capacities of the arcs in the cut. The goal in the minimum cut problem is to determine the cut with the lowest capacity. To determine a cut, we first partition all nodes into two subsets where origin and destination cannot be in the same subset, that is, $N = N_1 \cup N_2$, where origin $\in N_1$ and destination $\in N_2$. The cut is a set of all forward arcs that connects the subset that contains the origin and the other subset with destination. For example, suppose $N_1 = \{1, 2\}$ and $N_2 = \{3, 4, 5, 6\}$. Then the cut includes arcs (1,3), (2,4), (2,5) and (2,6) and the cut capacity is $u_{13} + u_{24} + u_{25} + u_{26} = 135$. The minimum cut problem is the dual of the maximum flow problem. Thus, any feasible solution to the minimum cut problem provides an upper bound to the maximum flow problem and

Network Optimization

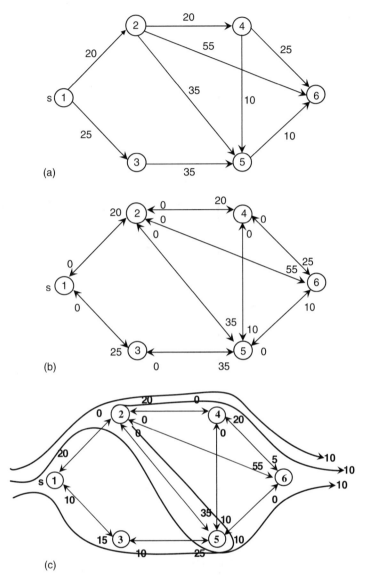

FIGURE 4.4 Application of the augmenting path algorithm on a network. (a) Original network. (b) Transformed network with residual arcs. (c) Paths and flows on the network after the augmenting flow algorithm is run to determine the maximum flows.

at optimality the minimum cut capacity is equal to the maximum flow. As a result, the maximum flow problem can be solved using the following algorithm:

Step 1: Identify all the cuts of the network.
Step 2: Determine the capacities of the cuts.
Step 3: Select the cut with the minimum capacity.

Note that if all possible cuts are not identified, then this algorithm may not identify the maximum flow. For the network in Figure 4.4a, if the nodes are partitioned as $N_1 = \{1, 3, 5\}$

TABLE 4.2 Application of Augmenting Path Algorithm to the Network in Figure 4.4a

			Residual Arc Capacity																	
Iteration	Path	Flow	(1,2)	(2,1)	(1,3)	(3,1)	(2,4)	(4,2)	(2,5)	(5,2)	(2,6)	(6,2)	(3,5)	(5,3)	(4,5)	(5,4)	(4,6)	(6,4)	(5,6)	(6,5)
0			20	0	25	0	20	0	35	0	55	0	35	0	10	0	25	0	10	0
1	1-2-5-6	10	10	10	25	0	20	0	25	10	55	0	35	0	10	0	25	0	0	10
2	1-3-5-2-4-6	10	10	10	15	10	10	10	35	0	55	0	25	10	10	0	15	10	0	10
3	1-2-4-6	10	0	20	15	10	0	20	35	0	55	0	25	10	10	0	5	20	0	10

and $N_2=\{2,4,6\}$, then the cut capacity is $u_{12}+u_{56}=20+10=30$, which is equal to the maximum flow in the network.

4.6 Assignment Problem

In the assignment problem, on a weighted bipartite graph $G=(N,A)$, where for every $(i,j)\in A$, $i\in N_1$ and $j\in N_2$, or vice versa, one seeks to pair nodes in N_1 and N_2 in such a way that the total cost of this pairing is minimized. In other words, given two sets of objects, the best, most efficient, least-cost pairing is sought.

There are several applications of assignment problems: For example, assigning professors to classes; matching people to jobs, rooms, events, machines, projects, or to each other; assigning crews to flights, jobs to machines, and so on. Each assignment has a value, and we want to make the assignments to minimize the sum of these values.

4.6.1 Linear Programming Formulation

Given a weighted bipartite network $G=(N_1\cup N_2, A)$ with $|N_1|=|N_2|$ and arc weights of c_{ij}, the assignment problem can be formulated as follows:

$$\text{Minimize:} \quad \sum_{(i,j)\in A} c_{ij} x_{ij} \tag{4.15}$$

$$\text{Subject to:} \quad \sum_{\{j:(i,j)\in A\}} x_{ij} = 1 \quad i \in N_1 \tag{4.16}$$

$$\sum_{\{i:(i,j)\in A\}} x_{ij} = 1 \quad j \in N_2 \tag{4.17}$$

$$x_{ij} \geq 0 \quad (i,j) \in A \tag{4.18}$$

where x_{ij} takes a value of 1 if node i and node j are paired. The objective is to achieve the minimum cost pairing. Equation 4.16 guarantees that each node in N_1 is assigned to exactly one node in N_2. Similarly, Equation 4.17 assures that any node in N_2 is assigned exactly to one node in N_1. The mathematical program has $|N_1|^2$ variables and $2|N_1|$ constraints.

Note that the assignment problem is a special case of a minimum cost flow problem on a bipartite graph, with N_1 and N_2 as disjoint node sets. Supply at each node in N_1 is one unit, and demand at each node in N_2 is one unit. The cost on each arc is c_{ij}.

4.6.2 The Hungarian Algorithm

The assignment problem can be solved using the Hungarian algorithm. The input for the Hungarian algorithm is an $|N_1|\times|N_1|$ cost matrix. The output is the optimal pairing of each element. The Hungarian algorithm can be described as follows:

Step 1: Find the minimum element in each row, and subtract the minimums from the cells of each row to obtain a new matrix. For the new matrix, find the minimums in each column. Construct a new matrix (called the reduced cost matrix) by subtracting the minimums from the cells of each column.

TABLE 4.3 Setup Time in Hours

	Job 1	Job 2	Job 3
Machine 1	5	1	1
Machine 2	1	3	7
Machine 3	1	5	3

Step 2: Draw the minimum number of lines (horizontal or vertical) that are needed to cover all zeroes in the reduced cost matrix. If m lines are required, an optimal solution is available among the covered zeroes in the matrix. Go to step 4. If fewer than m lines are needed, then the current solution is not optimal, and proceed to step 3.

Step 3: Find the smallest uncovered element (say, k) in the reduced cost matrix. Subtract k from each uncovered element in the reduced cost matrix, and add k to each element that is covered by two lines. Return to step 2.

Step 4: When the optimal solution is found in step 2, the next step is determining the optimal assignment. While making the assignments, start with the rows or columns that contain a single zero-valued cell. Once a cell with 0 value is chosen, eliminate the row and the column that the cell belongs to from further considerations. If no row or column is found with only one cell with 0 value, then check if there are rows/columns with a successively higher number of cells with 0 values, and choose arbitrarily from the available options. Note that, if there are several zeros at multiple locations, alternative optimal solutions might exist.

4.6.3 Application of the Algorithm Through an Example

EFGH Manufacturing has three machines and three jobs to be completed. Each machine must complete one job. The time required to set up each machine for completing each job is shown in Table 4.3. EFGH Manufacturing wants to minimize the total setup time required to complete all three jobs by assigning each job to a machine.

The Hungarian algorithm is applied to the EFGH example as follows: After the row and column minimums are subtracted from each cell in the cost matrix, two lines are required to cover all zeros in the modified matrix in Table 4.4.c, which implies the nonoptimality of the current solution. The minimum value of all uncovered cells is 2. As a result, subtract 2 from each uncovered cell, and add 2 to the cell that is covered by two lines (i.e., first row, first column) to obtain the matrix in Figure 4.5d. For this modified matrix, the minimum number of lines required to cover is three, and thus, the optimal solution can be obtained. The optimal assignment is job 1 to machine 3, job 2 to machine 2, and job 3 to machine 1, with a total setup time (cost) of 5.

4.7 Minimum Spanning Tree Problem

In a connected, undirected graph, a spanning tree, T, is a subgraph that connects all nodes on the network. Given that each arc has a weight c_{ij}, the length (or cost) of the spanning tree is equal to the sum of the weights of all arcs on that tree, that is, $\sum_{(i,j) \in T} c_{ij}$. The minimum spanning tree problem identifies the spanning tree with the minimum length. In other words, a minimum spanning tree is a spanning tree with a length less than or equal to the length of every other spanning tree. The minimum spanning tree problem is similar to the shortest path problem discussed in Section 4.4, with the following differences: In shortest

TABLE 4.4 Solving an Assignment Problem Using the Hungarian Method

Row Minimum

5	1	1	1
1	3	7	1
1	5	3	1

4	0	0
0	2	6
0	4	2

Column Minimum 0 0 0

(a) Subtract the row minimum
(b) Subtract the column minimum

4	0	0
0	2	6
0	4	2

Uncovered minimum = 2

6	0	0
2	0	4
0	2	0

(c) Minimum number of lines to cover zeros is two, thus not optimal
(d) Minimum number of lines to cover zeros is three, thus optimal

6	0	**0**
2	**0**	4
0	2	0

5	1	**1**
1	**3**	7
1	5	3

(e) Assignment of jobs to machines on the modified cost matrix
(f) Assignment of jobs to machines on the orignal cost matrix, total cost = 5

path problems, the objective is to determine the minimum total cost of a set of links (or arc) that connect source to sink. In minimum spanning tree problems, the objective is to select a minimum total cost set of links (or arcs) such that all the nodes are connected and there is no source or sink node. The optimal shortest path tree from an origin to all other nodes is a spanning tree, but not necessarily a minimum spanning tree.

Minimum spanning tree problems have several applications, specifically in network design. For example, the following problems can be optimized by minimum spanning tree algorithms: connecting different buildings on a university campus with phone lines or high-speed Internet lines while minimizing total installation costs; connecting different components on a printed circuit board to minimize the length of wires, capacitance, and delay line effects; and constructing a pipeline network connecting a number of towns to minimize the total length of pipeline.

4.7.1 Linear Programming Formulation

Let $A(S) \subseteq A$ denote the set of arcs induced by the node set S, that is, if $(i,j) \in A(S)$ then $i \in S$ and $j \in S$. The linear programming formulation can be stated as follows:

$$\text{Minimize:} \quad \sum_{(i,j) \in A} c_{ij} x_{ij} \quad (4.19)$$

$$\text{Subject to:} \quad \sum_{(i,j) \in A} x_{ij} = |N| - 1 \quad (4.20)$$

$$\sum_{(i,j) \in A(S)} x_{ij} = |S| - 1 \quad S \subseteq N \quad (4.21)$$

$$x_{ij} \in \{0,1\} \quad (i,j) \in A \quad (4.22)$$

In this formulation, x_{ij} is the decision variable that identifies if arc (i,j) should be included in the spanning tree. The objective is to minimize the total cost of arcs included in the spanning tree. In a network with $|N|$ nodes, $|N|-1$ arcs should be included in the tree (Equation 4.20). Using Equation 4.21, we ensure that there are no cycles on the minimum spanning tree subgraph. Note that when $A(S) = A$, Equations 4.20 and 4.21 are equivalent.

4.7.2 Kruskal's Algorithm

Kruskal's algorithm builds an optimal spanning tree by adding one arc at a time. By defining LIST as the set of arcs that is chosen as part of a minimum spanning tree, Kruskal's algorithm can be stated as follows:

Step 1: Sort all arcs in non-decreasing order of their costs to obtain the set A'.
Step 2: Set LIST $= \emptyset$.
Step 3: Select the arc with the minimum cost, and remove this arc from A'.
Step 4: Add this arc to the LIST if its addition does not create a cycle.
Step 5: If the number of arcs in the list is $|\text{LIST}| = |N| - 1$, then stop; arcs in the LIST form a minimum spanning tree. Otherwise, go to step 3.

In the example below, a network design problem is considered. In this problem, Wichita State University has six buildings that are planned to be connected via a gigabit network. The cost of connecting the buildings (in thousands) for which there can be a direct connection is given in Figure 4.5a. Using Kruskal's algorithm, the minimum cost design of the gigabit network for Wichita State University is determined. First, all arcs are sorted in nondecreasing order of their costs: (5,6), (4,5), (1,2), (2,4), (4,6), (1,3), (2,5), (3,5), and (2,6). In the first four iterations, arcs (5,6), (4,5), (1,2), and (2,4) are added, that is, the LIST = {(5,6), (4,5), (1,2), (2,4)} (see Figure 4.5b–e). In the fifth iteration, arc (4,6) cannot be added as it creates the cycle 4-5-6 (see Figure 4.5f). Finally, when arc (1,3) is added to the LIST, all nodes on the network are connected, that is, the minimum spanning tree is obtained. As shown in Figure 4.5g, the minimum spanning tree is LIST = {(5,6), (4,5), (1,2), (2,4),(1,3)}, and the total cost of the minimum cost tree is $10 + 10 + 20 + 20 + 25 = 85$.

Other types of minimum spanning tree problems (Garey and Johnson 1979) are as follows: A *degree-constrained spanning tree* is a spanning tree where the maximum vertex degree is limited to a certain constant k. *The degree-constrained spanning tree problem* is used to determine if a particular network has such a spanning tree for a particular k. In the *maximum leaf spanning tree problem*, the goal is to determine if the network has at least

Network Optimization

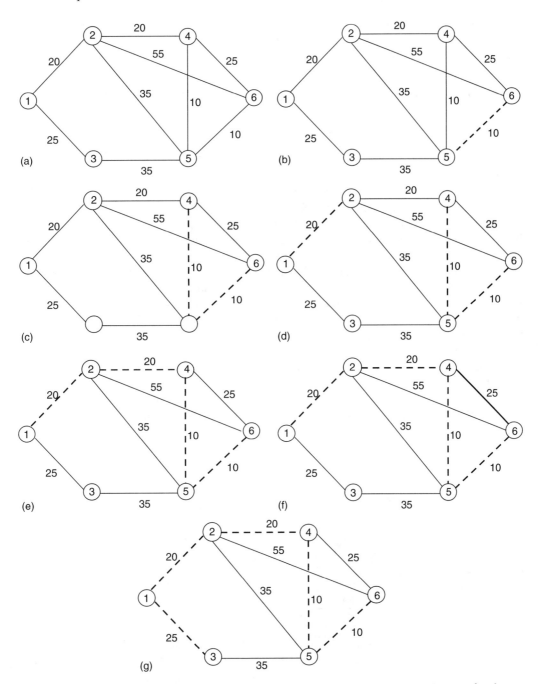

FIGURE 4.5 Application of Kruskal's algorithm to determine the minimum spanning tree for the network shown in Figure 4.5a. (a) Original network. (b) Arc (5,6) is added. (c) Arc (4,5) is added. (d) Arc (1,2) is added. (e) Arc (2,4) is added. (f) Arc (4,6) cannot be added as it creates a cycle. (g) Arc (1,3) is added. Algorithm terminates.

k nodes having a degree 1, where $k < |N|$. In the shortest total path length spanning tree problem, the goal is to ensure that the distance between any two nodes on the minimum spanning tree is less than a constant B.

4.8 Minimum Cost Multicommodity Flow Problem

The minimum cost multicommodity network flow problem (MCMNF) is a generalization of a minimum cost flow problem that considers a single commodity. In MCMNF, the goal is to minimize the shipping cost of several commodities from supply nodes to demand nodes while sharing a fixed arc capacity on the network. The commodities in a multicommodity flow problem can be differentiated by physical characteristics or by origin/destination pairs. The decision of assigning a percentage of each arc's capacity to different commodities to minimize the overall cost increases the complexity of this problem significantly when compared with other network flow problems such as the shortest path and maximum flow problems.

Multicommodity network flow problems have several applications: in communication networks, packets from different origin destination pairs share the same capacity. Furthermore, the same network might transfer several types of data, such as cable TV, phone, and video conferencing, which might have different service requirements simultaneously. A similar observation can be made for transportation networks if commodities are defined with respect to origin–destination pairs or different types of transportation such as cars, busses, trucks, and the like. In distribution networks, commodities can be defined as different types of products that must be transferred from plants to warehouses and then from warehouses to customers.

In aviation networks, passengers from different origins are transferred through hubs to travel to their destinations. Passengers must travel on planes with different types of capacity. Furthermore, airports have limited capacity, which restricts the number of planes that can park at the airport at any time. The goal is to minimize the total cost of transportation on this network.

4.8.1 Linear Programming Formulation

In minimum cost multicommodity flow problems, the objective is to minimize the overall cost of transportation of $|K|$ commodities from origins to destinations. Let c_{ij}^k be the cost of transporting a unit of commodity k on arc (i,j). The decision variable x_{ij}^k is the amount of commodity k transported on arc (i,j). The mathematical programming formulation for a minimum cost multicommodity flow problem can be stated as follows:

$$\text{Minimize:} \quad \sum_{k \in K} \sum_{(i,j) \in A} c_{ij}^k x_{ij}^k \qquad (4.23)$$

$$\text{Subject to:} \quad \sum_{\{j:(i,j) \in A\}} x_{ij}^k - \sum_{\{j:(j,i,) \in A\}} x_{ji}^k = b_i^k \quad i \in N, k \in K \qquad (4.24)$$

$$\sum_{k \in K} x_{ij}^k \leq u_{ij} \qquad (i,j) \in A \qquad (4.25)$$

$$x_{ij}^k \geq 0 \qquad (i,j) \in A \qquad (4.26)$$

In the above formulation, the mass balance constraint for each commodity (Equation 4.24) ensures the conservation of flow at each node for that commodity. Equation 4.25 is the capacity constraint on each arc. This mathematical model has $|K||A|$ variables, $|K||N|$

node balance constraints, and $|A|$ capacity constraints. Note that all of the network models described in this chapter, except the minimum spanning tree problem, are a special case of the minimum cost multicommodity flow problem. In the literature, this problem has been solved using methods like Lagrangian relaxation, column generation approaches, and Dantzig–Wolfe decomposition.

4.9 Conclusions

Network optimization is a very popular tool to solve large-scale real-life problems. Because of the availability of very fast network optimization methods, researchers can tackle very complex real-life problems efficiently. All of the network flow models presented in this chapter, except the minimum spanning tree problem, assume that some sort of flow must be sent through a network; thus, one has to decide how this can be achieved using the algorithms presented in this chapter to optimize the operations in a network. On the other hand, the minimum spanning tree problem can be used to optimize the design of a network.

This chapter presents some very basic models and solution procedures. More detailed analysis of these network models can be found in the network flow literature. The bibliography below provides a selected list of books in operations research and network optimization. The introduction to operations research texts by Hillier and Lieberman (2005), Jensen and Bard (2003), and Winston (2005) contain overviews of network flow models similar to those presented here at the elementary level. The two best comprehensive references on network models are by Ahuja, Magnanti, and Orlin (1993) and Bertsekas (1991). The notation used in this chapter is similar to the former.

Below is a short list of websites that contain either freely available software or datasets related to network optimization:

- NEOS Guide, a repository about optimization
  http://www-fp.mcs.anl.gov/otc/GUIDE/
- Andrew Goldberg's codes for shortest path and minimum cost flow algorithms
  http://www.avglab.com/andrew/#library
- Dmitri Bertsekas Network Optimization Codes
  http://web.mit.edu/dimitrib/www/noc.htm
- GOBLIN-C++ Object for network optimization
  http://www.math.uni-augsburg.de/opt/goblin.html
- J E Beasley OR library: data sets for OR problems
  http://people.brunel.ac.uk/~mastjjb/jeb/info.html
- C++ code for multicommodity flow problems
  http://www.di.unipi.it/di/groups/optimize/

Network problems can also be solved using professional LP/MIP solvers. For example, ILOG/CPLEX, SUNSET SOFTWARE/XA, and DASH/XPRESS are powerful solvers that take into account the network structure of large-scale problems to determine a solution in a reasonable amount of time.

References

1. Ahuja, R. K., Magnanti, T. L., and Orlin, J. B., *Network Flows: Theory, Algorithms, and Applications*, Prentice-Hall, Englewood Cliffs, NJ, 1993.

2. Balakrishnan, V., *Network Optimization*, First Edition, Chapman & Hall/CRC, Boca Raton, FL, 1995.
3. Bazaraa, M. S., Jarvis, J. I., and Sherali, H. D., *Linear Programming and Network Flows*, Third Edition, Wiley, New York, 2005.
4. Bertsekas, D. P., *Linear Network Optimization*, MIT Press, Cambridge, MA, 1991.
5. Evans, J. R. and Minieka, E., *Optimization Algorithms for Networks and Graphs*, Second Edition, Marcel Dekker, New York, 1992.
6. Garey, M. R. and Johnson, D. S., *Computers and Intractability: A Guide to the Theory of NP-Completeness*, W. H. Freeman, New York, 1979.
7. Glover, F., Klingman, D., and Phillips, N. V., *Network Models in Optimization and Their Applications in Practice*, Wiley, New York, 1992.
8. Hillier, F. S. and Lieberman, G. J., *Introduction to Operations Research*, Eighth Edition, McGraw Hill, New York, 2005.
9. Jensen, P. A. and Bard, J. F., *Operation Research Models and Methods*, Wiley, New York, 2003.
10. Jensen, P. A. and Barnes, J. W., *Network Flow Programming*, Wiley, New York, 1980.
11. Lawler, E. L., *Combinatorial Optimization: Networks and Matroids*, Holt, Rinehart and Winston, New York, 1976.
12. Murty, K. G., *Network Programming*, Prentice-Hall, Englewood Cliffs, NJ, 1992.
13. Papadimitriou, C. H. and Steiglitz, K., *Combinatorial Optimization: Algorithms and Complexity*, Dover, Mineola, NY, 1998.
14. Philips D. T. and Garcia-Diaz, A., *Fundamentals of Network Analysis*, Prentice-Hall, Englewood Cliffs, NJ, 1981.
15. Sheffi, Y., *Urban Transportation Networks: Equilibrium Analysis with Mathematical Programming Methods*, Prentice-Hall, Englewood Cliffs, NJ, 1985.
16. Taha, H. A., *Operations Research: An Introduction*, Seventh Edition, Prentice Hall, Upper Saddle River, NJ, 2002.
17. Winston, W. L., *Operations Research, Applications and Algorithms*, Fourth Edition, Thompson, Belmont, CA, 2005.

5
Multiple Criteria Decision Making

5.1	Some Definitions	5-3
5.2	The Concept of "Best Solution"	5-4
5.3	Criteria Normalization.............................	5-5
	Linear Normalization • Vector Normalization • Use of 10 Raised to Appropriate Power • Use of Range Equalization Factor	
5.4	Computing Criteria Weights........................	5-6
	Weights from Ranks • Rating Method • Ratio Weighing Method	
5.5	Multiple Criteria Methods for Finite Alternatives	5-8
	Max–Min Method • Min–Max (Regret) Method • Compromise Programming • TOPSIS Method • ELECTRE Method • Analytic Hierarchy Process • PROMETHEE Method	
5.6	Multiple Criteria Mathematical Programming Problems	5-15
	Definitions • Determining an Efficient Solution [12] • Test for Efficiency • Classification of MCMP Methods	
5.7	Goal Programming...................................	5-19
	Goal Programming Formulation • Partitioning Algorithm for Preemptive Goal Programs • Other Goal Programming Models	
5.8	Method of Global Criterion and Compromise Programming	5-27
	Method of Global Criterion • Compromise Programming	
5.9	Interactive Methods..................................	5-29
	Classification of Interactive Methods • Inconsistency of the DM • Computational Studies	
5.10	MCDM Applications.................................	5-34
5.11	MCDM Software	5-35
5.12	Further Readings....................................	5-35
	References	5-36

Abu S. M. Masud
Wichita State University

A. Ravi Ravindran
Pennsylvania State University

Decision problems often exhibit these characteristics: the presence of multiple, conflicting criteria for judging the alternatives and the need for making compromises or trade-offs regarding the outcomes of alternate courses of action. This chapter covers some of the practical methods available for helping make better decisions for these types of problems.

A multiple criteria decision making (MCDM) problem can be represented by the following generalized model:

$$\text{Maximize } [C_1(\mathbf{x}), C_2(\mathbf{x}), \ldots, C_k(\mathbf{x})] \tag{5.1}$$
$$\mathbf{x} \in \mathbf{X}$$

where
$\mathbf{x}$ is any specific alternative,
$\mathbf{X}$ is a set representing the feasible region or available alternatives, and
C_l is the lth evaluation criterion.

MCDM problems can be broadly classified as "selection problems" or "mathematical programming problems." The focus of multiple criteria selection problems (MCSP) is on selecting the best or preferred alternative(s) from a finite set of alternatives, and the alternatives are usually known *a priori*. The MCDM methods that help in identifying the "best" alternative for such problems will be referred to as the multiple criteria methods for finite alternatives (MCMFA). The MCSP is also referred to in the literature as multiple attribute decision making (MADM) problem [1]. MCSP are often represented in terms of a pay-off table. Table 5.1 shows a general format of a pay-off table, where θ_{ij} is the outcome of alternative i with respect to evaluation criterion j.

The focus of multiple criteria mathematical programming (MCMP) problems is to fashion or create an alternative when the possible number of alternatives is high (or infinite) and all alternatives are not known *a priori*. MCMP problems are usually modeled using explicit mathematical relationships, involving decision variables incorporated within constraints and objectives. The MCDM methods for identifying the "best" alternative in a MCMP will be referred to in this book as the multiple criteria mathematical programming methods (MCMPM). The MCMP problem is also known in the literature as a multiple objective decision making (MODM) problem or a vector optimization problem. MCMP problems are usually formulated as

$$\text{Max}[f_1(\mathbf{x}), f_2(\mathbf{x}), \ldots, f_k(\mathbf{x})] \tag{5.2}$$
$$\text{Subject to: } g_j(\mathbf{x}) \leq 0, j = 1, 2, \ldots, m \text{ and } \mathbf{x} = \{x_i|\ i = 1, 2, \ldots, n\}$$

where
$f_l(\mathbf{x}) = l$th objective function, $\quad l = 1, 2, \ldots, k$
$g_j(\mathbf{x}) = j$th constraint function, $\quad j = 1, 2, \ldots, m$

TABLE 5.1 Pay-Off Table

Alternatives	Criteria				
	C_1	C_2	$\ldots$	C_j	$\ldots$ C_k
$A_1 = \mathbf{x}^1$	θ_{11}	θ_{12}		θ_{1j}	θ_{1k}
$A_2 = \mathbf{x}^2$	θ_{21}	θ_{22}		θ_{2j}	θ_{2k}
$\ldots$					
$A_i = \mathbf{x}^i$	θ_{i1}	θ_{i2}		θ_{ij}	θ_{ik}
$\ldots$					
$A_m = \mathbf{x}^m$	θ_{m1}	θ_{m2}		θ_{mj}	θ_{mk}

5.1 Some Definitions

To provide a common understanding of the MCDM problem and its solution methods, we provide definitions of critical terms and concepts used in this chapter. In the literature of MCDM, these terms have special meaning and some are used interchangeably. Most of the definitions provided here are based on those given in Refs. [1,2].

Alternatives: Alternatives are the possible courses of action in a decision problem.

Alternatives are at the heart of decision making. In many decision situations, particularly for MCSP, alternatives can be prespecified. In such cases, it is important that every attempt is made for the development of all alternatives. Failure to do so may result in selecting an alternative for implementation that is inferior to other unexplored ones. In other situations, usually in MCMP, prespecification is not possible. In problems where prespecification of alternatives is not possible, alternatives are defined implicitly through mathematical relationships between decision variables. The challenge here is to define appropriate decision variables and to develop the mathematical relations involving these variables.

Attributes: These are the traits, characteristics, qualities, or performance parameters of the alternatives.

For example, if the decision situation is one of choosing the "best" car to purchase, then the attributes could be color, gas mileage, attractiveness, size, and the like. Attributes, from the decision making point of view, are the descriptors of the alternatives. For MCSP, attributes usually form the evaluation criteria.

Objectives: These are the directions of improvement or to do better, as perceived by the decision maker (DM).

For example, considering the same example of choosing the "best" car, an objective may be to "maximize gas mileage." This objective indicates that the DM prefers higher gas mileage, the higher the better. For MCMP, objectives form the evaluation criteria.

Goals: These are specific (or desired) status of attributes or objectives. Goals are targets or thresholds of objective or attribute values that are expected to be attained by the "best" alternative.

For example, in choosing the "best" car, a goal may be to buy a car that achieves an average gas mileage of 20 mpg or more; another example, buy a "4-door car."

(Evaluation) Criteria: These are the rules of acceptability or standards of judgment for the alternatives. Therefore, criteria encompass attributes, goals, and objectives.

Criteria may be *true* or *surrogate*. When a criterion is directly measurable, it is called a *true* criterion. An example of a true criterion is "the cost of the car," which is directly measured by its dollar price. When a criterion is not directly measurable, it may be substituted by one or more surrogate criteria. A *surrogate* criterion is used in place of one or more others that are more expressive of the DM's underlying values but are more difficult to measure directly. An example may be using "headroom for back seat passengers" as a surrogate for "passenger comfort." "Passenger comfort" is more expressive as a criterion, but is very difficult to measure. "Headroom for back seat passengers" is, however, easier to measure and can be used as one of the surrogate criteria for representing "passenger comfort."

5.2 The Concept of "Best Solution"

In single criterion decision problems, the "best" solution is defined in terms of an "optimum solution" for which the criterion value is maximized (or minimized) when compared to any other alternative in the set of all feasible alternatives. In MCDM problems, however, as the optimum of each criterion do not usually point to the same alternative, a conflict exists. The notion of an "optimum solution" does not usually exist in the context of conflicting, multiple criteria. Decision making in a MCDM problem is usually equivalent to choosing the best compromise solution. The "best solution" of an MCDM problem may be the "preferred (or best compromise) solution" or a "satisficing solution." In the absence of an optimal solution, the concepts of dominated and nondominated solutions become relevant. In the MCDM literature, the terms "nondominated solution," "Pareto optimal solution," and "efficient solution" are used interchangeably. In addition, concepts of "ideal solution" and "anti-ideal solution" are relevant in many MCDM methods.

Satisficing Solution: It is a feasible solution that meets or exceeds the DM's minimum expected level of achievement (or outcomes) of criteria values.

Nondominated Solution: A feasible solution (alternative) $\mathbf{x}^1$ dominates another feasible solution (alternative) $\mathbf{x}^2$ if $\mathbf{x}^1$ is at least as good as (i.e., as preferred as) $\mathbf{x}^2$ with respect to all criteria and is better than (i.e., preferred to) $\mathbf{x}^2$ with respect to at least one criterion. A nondominated solution is a feasible solution that is not dominated by any other feasible solution. That is, for a nondominated solution an increase in the value of any one criterion is not possible without some decrease in the value of at least one other criterion. Mathematically, a solution $\mathbf{x}^1 \in \mathbf{X}$ is nondominated if there is no other $\mathbf{x} \in \mathbf{X}$ such that $C_i(\mathbf{x}) \geq C_i(\mathbf{x}^1)$, $i = 1, 2, \ldots, k$, and $C_i(\mathbf{x}) \neq C_i(\mathbf{x}^1)$.

Ideal Solution: An ideal solution $\mathbf{H}^*$ is an artificial solution, defined in criterion space, each of whose elements is the maximum (or optimum) of a criterion's value for all $\mathbf{x} \in \mathbf{X}$. That is, the ideal solution consists of the upper bound of the criteria set.

The ideal solution is also known as positive-ideal solution or the utopia solution. Mathematically, the ideal solution for Equation 5.1 is obtained by

$$\mathbf{H}^* = \{H_i^* = \text{Max } C_i(\mathbf{x}) | \mathbf{x} \in \mathbf{X}, \quad i = 1, 2, \ldots, k\} \tag{5.3}$$

In all nontrivial MCDM problems, the ideal solution is an infeasible solution.

Anti-Ideal Solution: The anti-ideal solution, $\mathbf{L}_*$ consists of the lower bound of the criteria set.

This solution is also known as the negative-ideal solution or the nadir solution. One mathematical definition of the anti-ideal solution for Equation 5.1 is

$$\mathbf{L}_* = \{L_{i*} = \text{Min } C_i(\mathbf{x}_j^*) | \ j = 1, 2, \ldots, m; \quad i = 1, 2, \ldots, k\} \tag{5.4}$$

That is, the minimum values in each column of the pay-off table constitute the anti-ideal solution. This definition works well for problems where all the alternatives are prespecified. In problem (5.2) where all alternatives are never identified, this can cause a problem. This situation can be avoided by solving the following $i = 1, 2, \ldots, k$ single criterion minimization problems:

$$\text{Min} \quad f_i(\mathbf{x}) \tag{5.5}$$
$$\text{Subject to} \quad \mathbf{x} \in \mathbf{X}$$

Multiple Criteria Decision Making

For simplicity, the pay-off table is commonly used to identify the anti-ideal solution. However, with this simplification, we run the risk of being far off from the real anti-ideal.

5.3 Criteria Normalization

A common problem in multiple criteria decision making with the use of differing units of evaluation measures is that relative rating of alternatives may change merely because the units of measurement have changed. This issue can be addressed by normalization. Normalization allows intercriterion comparison. In the following discussion of normalization, assume that a benefit criterion is one in which the DM prefers more of it (i.e., more is better) and a cost criterion is one in which the DM prefers less of it (i.e., less is better). In general, a cost criterion can be transformed mathematically to an equivalent benefit criterion by multiplying by -1 or by taking the inverse of it.

5.3.1 Linear Normalization

Linear normalization converts a measure to a proportion of the way along the allowed range, where the allowed range is transformed to that between 0 and 1. A measure $C_j(\mathbf{x}^i)$, outcome in criterion j for alternative $\mathbf{x}^i$, is normalized to r_{ij} as follows:

$$r_{ij} = \frac{C_j(\mathbf{x}^i) - L_{j*}}{H_j^* - L_{j*}} \text{ (for benefit criterion)}$$
$$r_{ij} = \frac{L_{j*} - C_j(\mathbf{x}^i)}{L_{j*} - H_j^*} \text{ (for cost criterion)} \qquad (5.6)$$

where

$L_{j*} = \text{Max } \{C_j(\mathbf{x}^i), \quad i = 1, 2, \ldots, m\}$ for cost criterion j and Min $\{C_j(\mathbf{x}^i),$
$\quad i = 1, 2, \ldots, m\}$ for benefit criterion j

$H_j^* = \text{Min } \{C_j(\mathbf{x}^i), \quad i = 1, 2, \ldots, m\}$ for cost criterion j and Max $\{C_j(\mathbf{x}^i),$
$\quad i = 1, 2, \ldots, m\}$ for benefit criterion j

Note that, after normalization, all criteria are transformed to benefit criteria (i.e., to be maximized).

A less common form of linear normalization works as follows:

$$r_{ij} = \frac{H_j^* - C_j(\mathbf{x}^i)}{H_j^*} \qquad (5.7)$$

This form of normalization considers the distance of the ideal value of a criterion from the origin as a unit distance. Note that, after such normalization, all the criteria are transformed to cost criteria (i.e., they are to be minimized).

5.3.2 Vector Normalization

In vector normalization, each criterion outcome $C_j(\mathbf{x}^i)$ is divided by a norm L_{pj} as defined below:

$$r_{ij} = \frac{C_j(\mathbf{x}^i)}{L_{pj}} \qquad (5.8)$$

$$L_{pj} = \left(\sum_{i=1}^{i=n} |C_j(\mathbf{x}^i)|^p \right)^{1/p}, \quad 1 \leq p \leq \infty \qquad (5.9)$$

The usual values of p are 1, 2, or ∞ and the corresponding L_p norms are:

$$L_{1j} = \sum_{i=1}^{i=n} |C_j(\mathbf{x}^i)| \tag{5.10}$$

$$L_{2j} = \left(\sum_{i=1}^{i=n} |C_j(\mathbf{x}^i)|^2\right)^{1/2} \tag{5.11}$$

$$L_{\infty j} = \text{Max}\{|C_j(\mathbf{x}^i)|, \quad i = 1, 2, \ldots, n\} \tag{5.12}$$

5.3.3 Use of 10 Raised to Appropriate Power

This method, suggested by Steuer [3], rescales outcomes across all criteria to make them comparable. This is achieved by multiplying each $C_j(\mathbf{x}^i)$, $i = 1, 2, \ldots, n$, by 10 raised to an appropriate power that makes all outcomes of the same order of magnitude:

$$r_{ij} = C_j(\mathbf{x}^i) \times 10^{q_j} \tag{5.13}$$

where q_j is an appropriate number that will make criterion j outcomes similar in magnitude to the other criteria outcomes.

5.3.4 Use of Range Equalization Factor

This approach, also suggested by Steuer [3], tries to make the range of variation of the achievement values for all criteria comparable. This is done by first computing a range equalization factor, π_j, for criterion j and then multiplying each jth criterion outcome by this factor:

$$\pi_j = \frac{1}{\Delta_j} \left(\sum_{l=1}^{l=k} \frac{1}{\Delta_l}\right)^{-1} \tag{5.14}$$

$$r_{ij} = C_j(x^i) \times \pi_j \tag{5.15}$$

where $\Delta_j = |H_j^* - L_{j*}|$.

5.4 Computing Criteria Weights

Many MCDM methods require the use of relative importance weights of criteria. Many of these methods require ratio-scaled weights proportional to the relative value of unit changes in criteria value functions.

5.4.1 Weights from Ranks

This is a simple and commonly used method in which only the rank order of the criteria is used for developing the weights. First, the DM ranks the criteria in order of increasing relative importance; the highest ranked criterion gets a rank of 1. Let r_i represent the rank of the ith criterion. Next, determine criterion weight, λ_i, as follows:

$$\lambda_i = \frac{k - r_i + 1}{\sum_{j=1}^{j=k}(k - r_j + 1)} \tag{5.16}$$

where k is the number of criteria. This method produces an ordinal scale but does not guarantee the correct type of criterion importance because ranking does not capture the strength of preference information.

When a large number of criteria are considered, it may be easier for the DM to provide pairwise ranking instead of complete ranking. Assuming consistency in pairwise ranking information, $k(k-1)/2$ such comparisons can be used to derive the complete rank order. The number of times criterion i is ranked higher than all other criteria (in pairwise comparisons) is used to generate the rank order. If c_{ij} indicates the comparison of criterion C_i with criterion C_j, then $c_{ij} = 1$ if C_i is preferred over C_j and $c_{ij} = 0$ if C_i is not preferred over C_j. Note: $c_{ii} = 1$. Next, find criterion totals, $t_i = \sum_{j=1}^{j=k} c_{ij}$ and criteria weights $\lambda_i = \frac{t_i}{\sum_{i=1}^{i=k} t_i}$.

5.4.2 Rating Method

The method works as follows: first, an appropriate rating scale is agreed to (e.g., from 0 to 10). The scale should be clearly understood to be used properly. Next, using the selected scale, the DM provides rating for each criterion, r_i, judgmentally. Finally, normalize the ratings to determine weights:

$$\lambda_i = \frac{r_i}{\sum_{j=1}^{j=k} r_j} \qquad (5.17)$$

This method fails to assure a ratio scale and may not even provide the appropriate importance.

5.4.3 Ratio Weighing Method

In this method, originally proposed by Saaty [4], the DM compares two criteria at a time and indicates a_{ij}, which is the "number of times criterion C_i, is more important than criterion C_j." At least $(k-1)$ pairwise evaluations are needed. As inconsistency is expected in a large number of pairwise comparisons, many such methods require more than $(k-1)$ comparisons. Saaty has proposed a method for determining criteria weights based on the principal eigenvector of the pairwise comparison matrix. Let matrix A represent the pairwise comparisons (note that $a_{ii} = 1$).

$$A = \begin{bmatrix} 1 & a_{12} & \cdots & a_{1k} \\ a_{21} & 1 & \cdots & a_{2k} \\ & & & \\ a_{k1} & a_{k2} & \cdots & 1 \end{bmatrix}$$

If $a_{ij} = 1/a_{ji}$, $a_{ij} = a_{il} \times a_{lj}$, and $a_{ij} > 0$, then A is called a positive reciprocal matrix. With $(k-1)$ comparisons, the rest of the matrix can be filled by using the above relations. The principal eigenvector of A, π, can be found by finding the largest eigenvalue, $\alpha_{\max}$, for the following set of equations:

$$A\pi = \alpha_{\max} \pi \qquad (5.18)$$

Weights are simply the normalized principal eigenvector values:

$$\lambda_i = \frac{\pi_i}{\sum_{j=1}^{j=k} \pi_j} \qquad (5.19)$$

An easy way of computing the principal eigenvector and the weights is given by Harker [5]. He recursively computes the ith estimate π^i as follows:

$$\pi^i = \frac{A^i e}{e^T A^i e} \qquad (5.20)$$

Note: $A^1 = A$, $A^i = (A^{i-1} A)$ and $e^T = (1, 1, \ldots, 1)$. Saaty also provides a measure for checking the consistency of A; see the section on AHP for details. This method assures a ratio scale.

5.5 Multiple Criteria Methods for Finite Alternatives

All methods discussed in this section are appropriate for the following general MCDM problem:

$$\text{Max}[C_1(\mathbf{x}), C_2(\mathbf{x}), \ldots, C_k(\mathbf{x})] \tag{5.21}$$
$$\text{Subject to} \quad \mathbf{x} \in \mathbf{X}$$

In the context of an MCSP,

$C_l(\mathbf{x})$ = the lth attribute for alternative $\mathbf{x}$, $\theta_l(\mathbf{x})$
$\mathbf{x} \in \mathbf{X}$ = the set of available alternatives
$\quad \mathbf{x}$ = any alternative
$\theta_{jl} = \theta_l(\mathbf{x}^j)$ = lth attribute value for jth alternative

5.5.1 Max–Min Method

This method is based on the assumption that the DM is very pessimistic in his/her outlook and wants to maximize, over the decision alternatives, the achievement in the weakest criterion. Alternatives are characterized by the minimum achievement among all of its criterion values. It, thus, uses only a portion of the available information by ignoring all other criterion values. To make intercriterion comparison possible, all criterion values are normalized. Geometrically, this method maximizes the minimum normalized distance from the anti-ideal solution along each criterion for all available alternatives.

Mathematically, this method works as follows (assuming linear normalization):

$$\text{Max}\left[\text{Min}\left\{\frac{C_l(\mathbf{x}) - L_{l*}}{H_l^* - L_{l*}}\right\}, l = 1, 2, \ldots, k\right] \tag{5.22}$$
$$\text{Subject to} \quad \mathbf{x} \in \mathbf{X}$$

where
H_l^* = ideal solution value for the lth criterion
L_{l*} = anti-ideal solution value for the lth criterion

5.5.2 Min–Max (Regret) Method

In this method, it is assumed that the DM wants to minimize the maximum opportunity loss. Opportunity loss is defined as the difference between the ideal (solution) value of a criterion and the achieved value of that criterion in an alternative. Thus, the Min–Max method tries to identify a solution that is close to the ideal solution. Geometrically, this method finds a solution that minimizes the maximum normalized distance from the ideal solution along each criterion for all available alternatives.

Mathematically, this method is represented by the following problem,

$$\text{Min}\left[\text{Max}\left\{\frac{H_l^* - C_l(\mathbf{x})}{H_l^* - L_{l*}}\right\}, l = 1, 2, \ldots, k\right] \tag{5.23}$$
$$\text{Subject to} \quad \mathbf{x} \in \mathbf{X}$$

where
H_l^* = ideal solution value for the lth criterion
L_{l*} = anti-ideal solution value for the lth criterion

5.5.3 Compromise Programming

Compromise programming (CP) identifies the preferred solution (alternative) that is as close to the ideal solution as possible. That is, it identifies the solution whose distance from the ideal solution is minimum. Distance is measured with one of the metrics, M_p, defined below. Distance is usually normalized to make it comparable across criteria units. Note that CP is also known as the global criterion method. Mathematically, compromise programming involves solving the following problem:

$$\text{Min } M_p(\mathbf{x}) \tag{5.24}$$
$$\text{Subject to } \mathbf{x} \in \mathbf{X}$$

where the metric M_p is defined (using linear normalization) as follows:

$$M_p(\mathbf{x}) = \left(\sum_{i=1}^{i=k} \left| \frac{H_i^* - C_i(\mathbf{x})}{H_i^* - L_{i*}} \right|^p \right)^{1/p}, \quad 1 \leq p \leq \infty \tag{5.25}$$

As $H_i^* \geq C_i(\mathbf{x})$ is always true for Equation 5.25, the CP problem formulation can be restated as follows:

$$\text{Min } M_p(\mathbf{x}) = \left(\sum_{i=1}^{i=k} \left(\frac{H_i^* - C_i(\mathbf{x})}{H_i^* - L_{i*}} \right)^p \right)^{1/p}, \quad 1 \leq p \leq \infty \tag{5.26}$$
$$\text{Subject to } \mathbf{x} \in \mathbf{X}$$

Using Equation 5.26 is simpler and it is the form commonly used in compromise programming.

Note that geometrically, the distance measures in a CP problem have different meanings depending on the value of p chosen. For $p = 1$, $M_1(\mathbf{x})$ measures the "city-block" or "Manhattan block" distance (sum of distances along all axes) from H^*; for $p = 2$, $M_2(\mathbf{x})$ measures the straight-line distance from H^*; for $p = \infty$, $M_\infty(\mathbf{x})$ measures the maximum of the axial distances from H^*.

5.5.4 TOPSIS Method

TOPSIS (technique for order preference by similarity to ideal solution) was originally proposed by Hwang and Yoon [2] for the MCSP. TOPSIS operates on the principle that the preferred solution (alternative) should simultaneously be closest to the ideal solution, H^*, and farthest from the negative-ideal solution, L_*. TOPSIS does not require the specification of a value (utility) function but it assumes the existence of monotonically increasing value (utility) function for each (benefit) criterion. The method uses an index that combines the closeness of an alternative to the positive-ideal solution with its remoteness from the negative-ideal solution. The alternative that maximizes this index value is the preferred alternative. In TOPSIS, the pay-off matrix is first normalized as follows:

$$r_{ij} = \frac{\theta_{ij}}{\left[\left(\sum_i \theta_{ij}^2 \right) \right]^{1/2}} \quad i = 1, \ldots, m; \; j = 1, \ldots, k \tag{5.27}$$

Next, the weighted pay-off matrix, Q, is computed:

$$q_{ij} = \lambda_j r_{ij} \quad i = 1, 2, \ldots, m; \; j = 1, 2, \ldots, k \tag{5.28}$$

where λ_j is the relative importance weight of the jth attribute; $\lambda_j \geq 0$ and $\sum \lambda_j = 1$.

Using the weighted pay-off matrix, ideal and negative-ideal solutions (H^* and L_*) are identified as follows:

$$H^* = \{q_j^*, j = 1, 2, \ldots, k\} = \{\text{Max } q_{ij}, \text{for all } i; j = 1, 2, \ldots, k\}$$
$$L_* = \{q_{*j}, j = 1, 2, \ldots, k\} = \{\text{Min } q_{ij}, \text{for all } i; j = 1, 2, \ldots, k\} \quad (5.29)$$

Based on these solutions, separation measures for each solution (alternative) are calculated:

$$P_i^* = \left[\sum_j (q_{ij} - q_j^*)^2\right]^{1/2}, \quad i = 1, 2, \ldots, m$$
$$P_{*i} = \left[\sum_j (q_{ij} - q_{*j})^2\right]^{1/2}, \quad i = 1, 2, \ldots, m \quad (5.30)$$

where P_i^* is the distance of the ith solution (alternative) from the ideal solution and P_{*i} is the distance of the same solution from the negative-ideal solution. TOPSIS identifies the preferred solution by minimizing the similarity index, D, defined below. Note that all the solutions can be ranked by their index values; a solution with a higher index value is preferred over that with index values smaller than its value.

$$D_i = P_{*i}/(P_i^* + P_{*i}), \quad i = 1, 2, \ldots, m \quad (5.31)$$

Note that $0 \leq D_i \leq 1$; $D_i = 0$ when the ith alternative is the negative-ideal solution and $D_i = 1$ when the ith alternative is the ideal solution.

5.5.5 ELECTRE Method

ELECTRE method, developed by Roy [6], falls under the category called outranking methods. It compares two alternatives at a time (i.e., uses pairwise comparison) and attempts to build an outranking relationship to eliminate alternatives that are dominated using the outranking relationship. Six successive models of this method have been developed over time. They are: ELECTRE I, II, III, IV, Tri, and IS. Excellent overviews of the history and foundations of ELECTRE methods are given by Roy [6,7] and Rogers et al. [8]. We will explain only ELECTRE I in this section. The outcome of ELECTRE I is a (smaller than original) set of alternatives (called the kernel) that can be presented to the DM for the selection of "best solution." Complete rank ordering of the original set of alternatives is possible with ELECTRE II.

An alternative A_i outranks another alternative A_j (i.e., $A_i \to A_j$) when it is realistic to accept the risk of regarding A_i as at least as good as (or not worse than) A_j, even when A_i does not dominate A_j mathematically. This outranking relationship is not transitive. That is, it is possible to have $A_p \to A_q$, $A_q \to A_r$ but $A_r \to A_p$. Each pair of alternatives (A_i, A_j) is compared with respect to two indices: a concordance index, $c(i,j)$, and a discordance index, $d(i,j)$. The concordance index $c(i,j)$ is a weighted sum of the number of criteria in which A_i is better than A_j. The discordance index $d(i,j)$ is the maximum weighted difference in criterion levels among criteria for which A_i is worse than A_j.

Let θ_{ip} = pth criterion achievement level for alternative A_i
 λ_p = relative importance weight of criterion p; $\sum_{p=1}^{k} \lambda_p = 1$ and $\lambda_p > 0$
 r_{ip} = pth criterion achievement level (normalized) for A_i; use any appropriate normalization scheme to derive r_{ip} from θ_{ip} (see previous section)
 $q_{ip} = \lambda_p \, r_{ip}$

Multiple Criteria Decision Making

g is the criterion index for which $q_{ip} > q_{jp}$
l is the criterion index for which $q_{ip} < q_{jp}$
e is the criterion index for which $q_{ip} = q_{jp}$
s is the index of all criteria ($s = g + l + e$)

Then,

$$c(i,j) = \sum_{p \in g} \lambda_p + \sum_{p \in e} \varphi_p \lambda_p \tag{5.32}$$

$$d(i,j) = \frac{\text{Max}|q_{ip} - q_{jp}|, \forall l}{\text{Max}|q_{ip} - q_{jp}|, \forall s}$$

where φ_p is usually set equal to 0.5.

ELECTRE assumes that criterion levels are measurable on an interval scale for the discordance index. Ordinal measures are acceptable for the concordance index. Weights should be ratio scaled and represent relative importance to unit changes in criterion values. Two threshold values, α and β, are used and these are set by the DM. Sensitivity analysis with respect to α and β is needed to test the stability of the outranking relationship.

Alternative A_i outranks alternative A_j iff $c(i,j) \geq \alpha$ and $d(i,j) \leq \beta$. Based on the outranking relation developed, the preferred set of alternatives, that is, a kernel (K), is defined by the following conditions:

1. Each alternative in K is not outranked by any other alternative in K.
2. Every alternative not in K is outranked by at least one alternative in K.

5.5.6 Analytic Hierarchy Process

The analytic hierarchy process (AHP) method was first proposed by Saaty [4,9]. AHP is applicable only for MCSP. With AHP, value (utility) function does not need to be evaluated, nor does it depend on the existence of such a function. To use this method, the decision problem is first structured in levels of a hierarchy. At the top level is the goal or overall purpose of the problem. The subsequent levels represent criteria, subcriteria, and so on. The last level represents the decision alternatives.

After the problem has been structured in the form of a hierarchy, the next step is to seek value judgments concerning the alternatives with respect to the next higher level subcriteria. These value judgments may be obtained from available measurements or, if measurements are not available, from pairwise comparison or preference judgments. The pairwise comparison or preference judgments can be provided using any appropriate ratio scale. Saaty has proposed the following scale for providing preference judgment.

Scale value	Explanation
1	Equally preferred (or important)
3	Slightly more preferred (or important)
5	Strongly more preferred (or important)
7	Very strongly more preferred (or important)
9	Extremely more preferred (or important)
2, 4, 6, 8	Used to reflect compromise between scale values

After the value judgments of alternatives with respect to subcriteria and relative importances (or priorities) of the sub-criteria and criteria have been received (or computed), composite values indicating overall relative priorities of the alternative are then determined by finding weighted average values across all levels of the hierarchy.

AHP is based on the following set of four axioms. The description of the axioms is based on Harker [5].

Axiom 1: Given two alternatives (or subcriteria) A_i and A_j, the DM can state θ_{ij}, the pairwise comparison (or preference judgment), with respect to a given criterion from a set of criteria such that $\theta_{ji} = 1/\theta_{ij}$ for all i and j. Note that θ_{ij} indicates how strongly alternative A_i is preferred to (or better than) alternative A_j.

Axiom 2: When judging alternatives A_i and A_j, the DM never judges one alternative to be infinitely better than another, that is, $\theta_{ij} \neq \infty$ with respect to any criterion.

Axiom 3: One can formulate the decision problem as a hierarchy.

Axiom 4: All criteria and alternatives that impact the decision problem are represented in the hierarchy (i.e., it is complete).

When relative evaluations of subcriteria or alternatives are obtained through pairwise comparison, Saaty [9] has proposed a methodology (the eigenvector method) for computing the relative values of alternatives (and relative weights of subcriteria). With this method, the principal eigenvector is computed as follows:

$$\boldsymbol{\theta}\mathbf{v} = \lambda_{\max}\mathbf{v} \tag{5.33}$$

where $\mathbf{v}$ = vector of relative values (weights) and $\lambda_{\max}$ = maximum eigenvalue.

According to Harker [5], the principal eigenvector can be determined by raising the matrix θ to increasing powers k and then normalizing the resulting system:

$$\mathbf{v} = \lim_{k \to \infty} \frac{(\theta^k e)}{(e^T \theta^k e)} \tag{5.34}$$

where $e^T = (1, 1, \ldots, 1, 1)$. The $\mathbf{v}$ vector is then normalized to the $\mathbf{w}$ vector, such that $\sum_{i=1}^{i=n} w_i = 1$. See Equation 5.20 for an easy heuristic proposed by Harker [5] for estimating $\mathbf{v}$. Once the $\mathbf{w}$ vector has been determined, $\lambda_{\max}$ can be determined as follows:

$$\lambda_{\max} = \frac{\left(\sum_{j=1}^{j=n} \theta_{1j} w_j\right)}{w_1} \tag{5.35}$$

As there is scope for inconsistency in judgments, the AHP method provides for a measure of such inconsistency. If all the judgments are perfectly consistent, then $\lambda_{\max} = n$ (where n is the number of subcriteria or alternatives under consideration in the current computations); otherwise, $\lambda_{\max} > n$. Saaty defines consistence index (CI) as follows:

$$\text{CI} = \frac{(\lambda_{\max} - n)}{(n - 1)} \tag{5.36}$$

For different sizes of comparison matrix, Saaty conducted experiments with randomly generated judgment values (using the 1–9 ratio scale discussed before). Against the means of

Multiple Criteria Decision Making

the CI values of these random experiments, called random index (RI), the computed CI values are compared, by means of the consistency ratio (CR):

$$\text{CR} = \frac{\text{CI}}{\text{RI}} \tag{5.37}$$

As a rule of thumb, CR ≤ 0.10 indicates acceptable level of inconsistency.

The experimentally derived RI values are:

n	1	2	3	4	5	6	7	8	9	10	11	12	13	14	15
RI	0.00	0.00	0.58	0.90	1.12	1.24	1.32	1.41	1.45	1.49	1.51	1.48	1.56	1.57	1.59

After the relative value vectors, **w**, for different sub-elements of the hierarchy have been computed, the next step is to compute the overall (or composite) relative values of the alternatives. A linear additive function is used to represent the composite relative evaluation of an alternative. The procedure for determining the composite evaluation of alternatives is based on maximizing the "overall goal" at the top of the hierarchy. When multiple DMs are involved, one may take geometric mean of the individual evaluations at each level. For more on AHP, see Saaty [9]. An excellent tutorial on AHP is available in Forman and Gass [10].

5.5.7 PROMETHEE Method

The preference ranking organization method of enrichment evaluations (PROMETHEE) methods have been developed by Brans and Mareschal [11] for solving MCSP. PROMETHEE I generates a partial ordering on the set of possible alternatives, while PROMETHEE II generates a complete ordering of the alternatives. The PROMETHEE methods seek to enrich the usual dominance relation to generate better solutions for the general selection type problem. Only PROMETHEE I will be summarized in this section.

We assume that there exists a set of n possible alternatives, $\mathbf{A} = [A_1, A_2, \ldots, A_n]$, and k criteria, $\mathbf{C} = [C_1, C_2, \ldots, C_k]$, each of which is to be maximized. In addition, we assume that the relative importance weights, $\boldsymbol{\lambda} = [\lambda_1, \lambda_2, \ldots, \lambda_k]$, associated with the k criteria, are known in advance. We further assume that the criteria achievement matrix θ is normalized, using any appropriate method, to $\mathbf{R}$ so as to eliminate all scaling effects.

Traditionally, alternative A_1 is said to dominate alternative A_2 iff $\theta_{1j} \geq \theta_{2j}$, $\forall j$ and $\theta_{1j} > \theta_{2j}$, for at least one j. However, this definition does not work very well in situations where A_1 is better than A_2 with respect to the first criteria by a very wide margin while A_2 is better than A_1 with respect to the second criteria by a very narrow margin or A_1 is better than A_2 with respect to criterion 1 by a very narrow margin and A_2 is better than A_1 with respect to criterion 2, again by a very narrow margin, or A_1 is marginally better than A_2 with respect to both criteria.

To overcome such difficulties associated with the traditional definition of dominance, the PROMETHEE methods take into consideration the amplitudes of the deviations between the criteria. For each of the k criteria, consider all pairwise comparisons between alternatives. Let us define the amplitude of deviation, d_i, between alternative a and alternative b with respect to criterion i as

$$d_i = C_i(a) - C_i(b) = \theta_{ai} - \theta_{bi}$$

The following preference structure summarizes the traditional approach:

If $d_i > 0$ then a is preferred to b and we write aPb.
If $d_i = 0$ then a is indifferent to b and we write aIb.
If $d_i < 0$ then b is preferred to a and we write bPa.

In PROMETHEE a preference function for criteria i, $P_i(a, b)$, is introduced to indicate the intensity of preference of alternative a over alternative b with respect to criterion i. (Note: $P_i(b, a)$ gives the intensity of preference of alternative b over alternative a.) $P_i(a, b)$ is defined such that $0 \leq P_i(a, b) \leq 1$ and

$P_i(a, b) = 0$ if $d_i \leq 0$ (equal to or less than 0), indicating "no preference" between a and b
$P_i(a, b) \approx 0$ if $d_i > 0$ (slightly greater than 0), indicating "weak preference" of a over b
$P_i(a, b) \approx 1$ if $d_i \gg 0$ (much greater than 0), indicating "strong preference" of a over b
$P_i(a, b) = 1$ if $d_i \ggg 0$ (extremely greater than 0), indicating "strict preference" of a over b

Next function $E_i(a,b)$ is defined, which can be of six forms. The most commonly used form is the Gaussian (or normal distribution); we will use this in this section. For each criterion, the decision maker and the analyst must cooperate to determine the parameter s where $0 < s < 1$. This parameter is a threshold delineating the weak preference area from the strong preference area. Once this parameter is established the values are calculated as follows:

$$E_i(a, b) = \begin{cases} 1 - e^{-\frac{(d_i)^2}{2s^2}}, & d_i \geq 0 \\ 0, & d_i < 0 \end{cases} \tag{5.38}$$

If $E_i(a, b) > 0$, then a is preferred to b with respect to criterion i. $\pi(a, b)$, preference index function, is defined next as follows:

$$\pi(a, b) = \sum_{j=1}^{j=k} \lambda_j E_j(a, b) \tag{5.39}$$

where λ_j is the weight of criterion j. The preference index function expresses the global preference of alternative a over alternative b. Note:

$\pi(a, a) = 0$
$0 \leq \pi(a, b) \leq 1$
$\pi(a, b) \approx 0$ implies a weak global preference of a over b
$\pi(a, b) \approx 1$ implies a strong global preference of a over b
$\pi(a, b)$ expresses intensity of dominance of a over b
$\pi(b, a)$ expresses intensity of dominance of b over a

Using $\pi(a, b)$, positive-outranking flow, $\varphi^+(a)$, and negative-outranking flow, $\varphi^-(a)$, are calculated as follows. Positive-outranking expresses how alternative "a" outranks all other alternatives and, therefore, the higher the value the better the alternative is. The negative-outranking expresses how "a" is outranked by all other alternatives and, therefore, the lower the value the better the alternative is.

$$\varphi^+(a) = \sum_{i=1}^{i=n} \pi(a, A_i) \tag{5.40}$$

$$\varphi^-(a) = \sum_{i=1}^{i=n} \pi(A_i, a) \tag{5.41}$$

Multiple Criteria Decision Making

The following function is defined to assist in ordering the preference of the alternatives:

$$\begin{cases} a\ S^+ b, & \varphi^+(a) > \varphi^+(b) \\ a\ I^+ b, & \varphi^+(a) = \varphi^+(b) \end{cases} \quad (5.42)$$

$$\begin{cases} a\ S^- b, & \varphi^-(a) < \varphi^-(b) \\ a\ I^- b, & \varphi^-(a) = \varphi^-(b) \end{cases} \quad (5.43)$$

The PROMETHEE I partial relation is the intersection of these two pre-orders. There are three possible conclusions when making pairwise comparisons between alternatives.

Conclusion I: a outranks b

$$a\ P^I\ b \text{ if } \begin{cases} a\ S^+\ b \text{ and } a\ S^-\ b \\ a\ S^+\ b \text{ and } a\ I^-\ b \\ a\ I^+\ b \text{ and } a\ S^-\ b \end{cases} \quad (5.44)$$

In this case the positive flow exceeds or is equal to the positive flow of b and the negative flow of b exceeds or is equal to the negative flow of a. The flows agree and the information is sure.

Conclusion II: a indifferent to b

$$a\ I^I\ b \quad \text{if } a\ I^+\ b \text{ and } a\ I^-\ b \quad (5.45)$$

In this case the positive and negative flows of the two alternatives are equal. The alternatives are concluded to be roughly equivalent.

Conclusion III: a incomparable to b
This will generally occur if alternative a performs well with respect to a subset of the criteria for which b is weak while b performs well on the criteria for which a is weak.

5.6 Multiple Criteria Mathematical Programming Problems

In the previous sections, our focus was on solving MCDM problems with a *finite* number of alternatives, where each alternative is measured by several conflicting criteria. These MCDM problems were called multiple criteria selection problems (MCSP). The methods we discussed earlier helped in identifying the best alternative or rank order all the alternatives from the best to the worst.

In this and the subsequent sections, we will focus on MCDM problems with an *infinite number of alternatives*. In other words, the feasible alternatives are not known *a priori* but are represented by a set of mathematical (linear/nonlinear) constraints. These MCDM problems are called *multicriteria mathematical programming* (MCMP) problems.

MCMP Problem

$$\begin{aligned} \text{Max} \quad & \mathbf{F}(\mathbf{x}) = \{f_1(\mathbf{x}), f_2(\mathbf{x}), \ldots, f_k(\mathbf{x})\} \\ \text{Subject to} \quad & g_j(\mathbf{x}) \leq 0 \quad \text{for } j = 1, \ldots, m \end{aligned} \quad (5.46)$$

where $\mathbf{x}$ is an n-vector of *decision variables* and $f_i(\mathbf{x})$, $i = 1, \ldots, k$ are the k criteria/objective functions.

Let $S = \{\mathbf{x}/g_j(\mathbf{x}) \leq 0, \text{ for all "}j\text{"}\}$
$Y = \{\mathbf{y}/\mathbf{F}(\mathbf{x}) = \mathbf{y} \text{ for some } \mathbf{x} \in S\}$
S is called the *decision space* and Y is called the *criteria or objective space* in MCMP.

A solution to MCMP is called a *superior solution* if it is feasible and maximizes all the objectives simultaneously. In most MCMP problems, superior solutions do not exist as the objectives conflict with one another.

5.6.1 Definitions

Efficient, Non-Dominated, or Pareto Optimal Solution: A solution $\mathbf{x}^\circ \in S$ to MCMP is said to be *efficient* if $f_k(\mathbf{x}) > f_k(\mathbf{x}^\circ)$ for some $\mathbf{x} \in S$ implies that $f_j(\mathbf{x}) < f_j(\mathbf{x}^\circ)$ for at least one other index j. More simply stated, an efficient solution has the property that an improvement in any one objective is possible only at the expense of at least one other objective.

A Dominated Solution is a feasible solution that is not efficient.

Efficient Set: The set of all efficient solutions is called the *efficient set* or *efficient frontier*.

Note: Even though the solution of MCMP reduces to finding the efficient set, it is not practical because there could be an infinite number of efficient solutions.

Example 5.1

Consider the following bi-criteria linear program:

$$\text{Max } Z_1 = 5x_1 + x_2$$

$$\text{Max } Z_2 = x_1 + 4x_2$$

$$\text{Subject to: } x_1 \leq 5$$

$$x_2 \leq 3$$

$$x_1 + x_2 \leq 6$$

$$x_1, x_2 \geq 0$$

The decision space and the objective space are given in Figures 5.1 and 5.2, respectively. Corner Points C and D are efficient solutions whereas corner points A, B, and E are dominated. The set of all efficient solutions is given by the line segment CD in both figures.

An ideal solution is the vector of individual optima obtained by optimizing each objective function separately ignoring all other objectives.

In Example 5.1, the maximum value of Z_1, ignoring Z_2, is 26 and occurs at point D. Similarly, maximum Z_2 of 15 is obtained at point C. Thus the ideal solution is (26,15) but is *not* feasible or achievable.

Note: One of the popular approaches to solving MCMP problems is to find an efficient solution that comes "as close as possible" to the ideal solution. We will discuss these approaches later. ∎

Multiple Criteria Decision Making

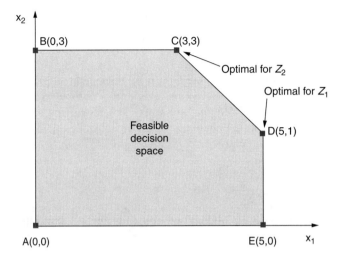

FIGURE 5.1 Decision space (Example 5.1).

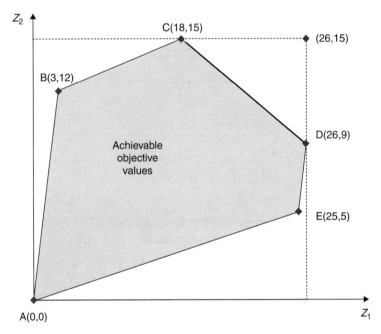

FIGURE 5.2 Objective space (Example 5.1).

5.6.2 Determining an Efficient Solution [12]

For the MCMP problem given in Equation 5.46, consider the following single objective optimization problem, called the P_λ problem.

$$\text{Max } Z = \sum_{i=1}^{k} \lambda_i f_i(\mathbf{x})$$

Subject to: $\mathbf{x} \in S$

$$\sum_{i=1}^{k} \lambda_i = 1 \qquad (5.47)$$

$$\lambda_i \geq 0$$

THEOREM 5.1 (Sufficiency) Let $\lambda_i > 0$ for all i be specified. If $\mathbf{x}^\circ$ is an optimal solution for the P_λ problem (Equation 5.47), then $\mathbf{x}^\circ$ is an efficient solution to the MCMP problem.

In Example 5.1, if we set $\lambda_1 = \lambda_2 = 0.5$ and solve the P_λ problem, the optimal solution will be at D, which is an efficient solution.

The P_λ problem (Equation 5.47) is also known as the weighted objective problem.
Warning: Theorem 5.1 is only a sufficient condition and is not necessary. For example, there could be efficient solutions to MCMP that could not be obtained as optimal solutions to the P_λ problem. Such situations occur when the objective space is not a convex set. However, for MCMP problems, where the objective functions, and constraints are linear, Theorem 5.1 is both necessary and sufficient.

5.6.3 Test for Efficiency

Given a feasible solution $\overline{\mathbf{x}} \in S$ for MCMP, we can test whether it is efficient by solving the following single objective problem.

$$\text{Max} \quad W = \sum_{i=1}^{k} d_i$$

$$\text{Subject to:} \quad f_i(\mathbf{x}) \geq f_i(\overline{\mathbf{x}}) + d_i \quad \text{for } i = 1, 2, \ldots, k$$

$$\mathbf{x} \in S$$

$$d_i \geq 0$$

THEOREM 5.2

1. If Max $W > 0$, then $\overline{x}$ is a dominated solution.
2. If Max $W = 0$, then $\overline{x}$ is an efficient solution.

Note: If Max $W > 0$, then at least one of the d_i's is positive. This implies that at least one objective can be improved without sacrificing on the other objectives.

5.6.4 Classification of MCMP Methods

In MCMP problems, often there are an infinite number of efficient solutions and they are not comparable without the input from the DM. Hence, it is generally assumed that the DM has a real-valued *preference function* defined on the values of the objectives, but it is not known explicitly. With this assumption, the primary objective of the MCMP solution methods is to find the *best compromise solution*, which is an efficient solution that maximizes the DM's preference function.

In the last two decades, most MCDM research have been concerned with developing solution methods based on different assumptions and approaches to measure or derive the DM's preference function. Thus, the MCMP methods can be categorized by the basic assumptions made with respect to the DM's preference function as follows:

1. When *complete* information about the preference function is available from the DM.
2. When *no* information is available.
3. Where *partial* information is obtainable progressively from the DM.

Multiple Criteria Decision Making

In the following sections we will discuss MCMP methods such as goal programming, compromise programming and interactive methods as examples of categories 1, 2, and 3 type approaches.

5.7 Goal Programming

One way to treat multiple criteria is to select one criterion as primary and the other criteria as secondary. The primary criterion is then used as the optimization objective function, while the secondary criteria are assigned acceptable minimum or maximum values depending on whether the criterion is maximum or minimum and are treated as problem constraints. However, if careful consideration is not given while selecting the acceptable levels, a feasible design that satisfies all the constraints may not exist. This problem is overcome by goal programming, which has become a practical method for handling multiple criteria. Goal programming [13] falls under the class of methods that use completely prespecified preferences of the decision maker in solving the MCMP problem.

In goal programming, all the objectives are assigned target levels for achievement and relative priority on achieving these levels. Goal programming treats these targets as goals to aspire for and not as absolute constraints. It then attempts to find an optimal solution that comes as "close as possible" to the targets in the order of specified priorities. In this section, we shall discuss how to formulate goal programming models and their solution methods.

Before we discuss the formulation of goal programming problems, we should discuss the difference between the terms *real constraints and goal constraints* (or simply goals) as used in goal programming models. The real constraints are absolute restrictions on the decision variables, whereas the goals are conditions one would like to achieve but are not mandatory. For instance, a real constraint given by

$$x_1 + x_2 = 3$$

requires all possible values of $x_1 + x_2$ to always equal 3. As opposed to this, a goal requiring $x_1 + x_2 = 3$ is not mandatory, and we can choose values of $x_1 + x_2 \geq 3$ as well as $x_1 + x_2 \leq 3$. In a goal constraint, positive and negative deviational variables are introduced as follows:

$$x_1 + x_2 + d_1^- - d_1^+ = 3 \quad d_1^+, d_1^- \geq 0$$

Note that if $d_1^- > 0$, then $x_1 + x_2 < 3$, and if $d_1^+ > 0$, then $x_1 + x_2 > 3$.

By assigning suitable weights w_1^- and w_1^+ on d_1^- and d_1^+ in the objective function, the model will try to achieve the sum $x_1 + x_2$ as close as possible to 3. If the goal were to satisfy $x_1 + x_2 \geq 3$, then only d_1^- is assigned a positive weight in the objective, while the weight on d_1^+ is set to zero.

5.7.1 Goal Programming Formulation

Consider the general MCMP problem given in Section 5.6 (Equation 5.46). The assumption that there exists an optimal solution to the MCMP problem involving multiple criteria implies the existence of some preference ordering of the criteria by the DM. The goal programming (GP) formulation of the MCMP problem requires the DM to specify an acceptable level of achievement (b_i) for each criterion f_i and specify a weight w_i (ordinal or cardinal) to be associated with the deviation between f_i and b_i. Thus, the GP model of an MCMP problem becomes:

$$\text{Minimize } Z = \sum_{i=1}^{k}(w_i^+ d_i^+ + w_i^- d_i^-) \tag{5.48}$$

$$\text{Subject to: } f_i(\mathbf{x}) + d_i^- - d_i^+ = b_i \quad \text{for } i = 1, \ldots, k \tag{5.49}$$

$$g_j(\mathbf{x}) \leq 0 \quad \text{for } j = 1, \ldots, m \tag{5.50}$$

$$x_j, d_i^-, d_i^+ \geq 0 \quad \text{for all } i \text{ and } j \tag{5.51}$$

Equation 5.48 represents the objective function of the GP model, which minimizes the weighted sum of the deviational variables. The system of equations (Equation 5.49) represents the goal constraints relating the multiple criteria to the goals/targets for those criteria. The variables, d_i^- and d_i^+, in Equation 5.49 are called deviational variables, representing the under achievement and over achievement of the ith goal. The set of weights (w_i^+ and w_i^-) may take two forms:

1. Prespecified weights (cardinal)
2. Preemptive priorities (ordinal).

Under prespecified (cardinal) weights, specific values in a relative scale are assigned to w_i^+ and w_i^- representing the DM's "trade-off" among the goals. Once w_i^+ and w_i^- are specified, the goal program represented by Equations. 5.48 to 5.51 reduces to a single objective optimization problem. The cardinal weights could be obtained from the DM using any of the methods discussed in Sections 5.1 to 5.5 including the AHP method. However, for this method to work, the criteria values have to be scaled or normalized using the methods given in Section 5.3.

In reality, goals are usually incompatible (i.e., incommensurable) and some goals can be achieved only at the expense of some other goals. Hence, preemptive goal programming, which is more common in practice, uses ordinal ranking or preemptive priorities to the goals by assigning incommensurable goals to different priority levels and weights to goals at the same priority level. In this case, the objective function of the GP model (Equation 5.48) takes the form

$$\text{Minimize } Z = \sum_{p} P_p \sum_{i}(w_{ip}^+ d_i^+ + w_{ip}^- d_i^-) \tag{5.52}$$

where P_p represents priority p with the assumption that P_p is much larger then P_{p+1} and w_{ip}^+ and w_{ip}^- are the weights assigned to the ith deviational variables at priority p. In this manner, lower priority goals are considered only after attaining the higher priority goals. Thus, preemptive goal programming is essentially a sequential single objective optimization process, in which successive optimizations are carried out on the alternate optimal solutions of the previously optimized goals at higher priority.

The following example illustrates the formulation of a preemptive GP problem.

Example 5.2

Suppose a company has two machines for manufacturing a product. Machine 1 makes 2 units per hour, while machine 2 makes 3 units per hour. The company has an order for 80 units. Energy restrictions dictate that only one machine can operate at one time. The company has 40 hours of regular machining time, but overtime is available. It costs $4.00 to run

Multiple Criteria Decision Making

machine 1 for 1 hour, while machine 2 costs $5.00/hour. The company's goals, in order of importance, are as follows:

1. Meet the demand of 80 units exactly.
2. Limit machine overtime to 10 hours.
3. Use the 40 hours of normal machining time.
4. Minimize costs. ∎

Formulation. Letting x_j represent the number of hours machine j is operating, the goal programming model is

$$\text{Minimize } Z = P_1(d_1^- + d_1^+) + P_2 d_3^+ + P_3(d_2^- + d_2^+) + P_4 d_4^+$$

$$\text{Subject to: } 2x_1 + 3x_2 + d_1^- - d_1^+ = 80 \tag{5.53}$$

$$x_1 + x_2 + d_2^- - d_2^+ = 40 \tag{5.54}$$

$$d_2^+ + d_3^- - d_3^+ = 10 \tag{5.55}$$

$$4x_1 + 5x_2 + d_4^- - d_4^+ = 0 \tag{5.56}$$

$$x_i, d_i^-, d_i^+ \geqq 0 \quad \text{for all } i$$

where P_1, P_2, P_3, and P_4 represent the preemptive priority factors such that $P_1 \gg P_2 \gg P_3 \gg P_4$. Note that the target for cost is set at an unrealistic level of zero. As the goal is to minimize d_4^+, this is equivalent to minimizing the total cost of production.

In this formulation, Equation 5.55 does not conform to the general model given by Equations 5.49 to 5.51, where no goal constraint involves a deviational variable defined earlier. However, if $d_2^+ > 0$, then $d_2^- = 0$, and from Equation 5.54 we get

$$d_2^+ = x_1 + x_2 - 40$$

which, when substituted into Equation 5.55, yields

$$x_1 + x_2 + d_3^- - d_3^+ = 50 \tag{5.57}$$

Thus Equation 5.57 can replace Equation 5.55 and the problem fits the general model, where each deviational variable appears in only one goal constraint and has at most one positive weight in the objective function.

5.7.2 Partitioning Algorithm for Preemptive Goal Programs

Linear Goal Programs

Linear goal programming problems can be solved efficiently by the partitioning algorithm developed by Arthur and Ravindran [14,15]. It is based on the fact that the definition of preemptive priorities implies that higher order goals must be optimized before lower order goals are even considered. Their procedure consists of solving a series of linear programming subproblems by using the solution of the higher priority problem as the starting solution for the lower priority problem.

The partitioning algorithm begins by solving the smallest subproblem S_1, which is composed of those goal constraints assigned to the highest priority P_1 and the corresponding

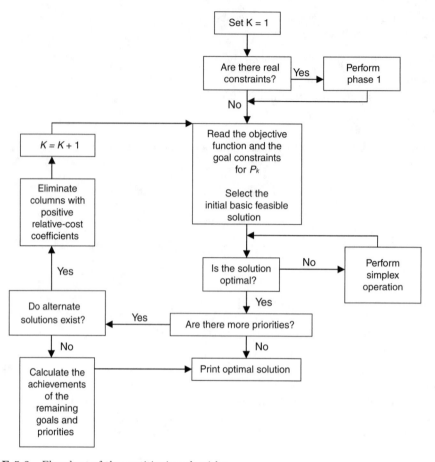

FIGURE 5.3 Flowchart of the partitioning algorithm.

terms in the objective function. The optimal tableau for this subproblem is then examined for alternate optimal solutions. If none exist, then the present solution is optimal for the original problem with respect to all the priorities. The algorithm then substitutes the values of the decision variables into the goal constraints of the lower priorities to calculate their attainment levels, and the problem is solved. However, if alternate optimal solutions do exist, the next set of goal constraints (those assigned to the second highest priority) and their objective function terms are added to the problem. This brings the algorithm to the next largest subproblem in the series, and the optimization resumes. The algorithm continues in this manner until no alternate optimum exists for one of the subproblems or until all priorities have been included in the optimization. The linear dependence between each pair of deviational variables simplifies the operation of adding the new goal constraints to the optimal tableau of the previous subproblem without the need for a dual-simplex iteration.

When the optimal solution to the subproblem S_{k-1} is obtained, a variable elimination step is preformed prior to the addition of goal constraints of priority k. The elimination step involves deleting all nonbasic columns that have a positive relative cost ($\overline{C_j} > 0$) in the optimal tableau of S_{k-1} from further consideration. This is based on the well-known linear programming (LP) result that a nonbasic variable with a positive relative cost in an optimal tableau cannot enter the basis to form an alternate optimal solution. Figure 5.3 gives a flowchart of the partitioning algorithm.

TABLE 5.2 Solution to Subproblem S_1

$P_1 : c_j$		0	0	1	1	
c_B	Basis	x_1	x_2	d_1^-	d_1^+	b
1	d_1^+	2	3	1	-1	80
$P_1 : \bar{c}$	Row	-2	-3	0	2	$Z_1 = 80$
0	x_2	$\frac{2}{3}$	1	$\frac{1}{3}$	$-\frac{1}{3}$	$\frac{80}{3}$
$P_1 : \bar{c}$	Row	0	0	1	1	$Z_1 = 0$

TABLE 5.3 Solution to Subproblem S_2

$P_2 : c_j$		0	0	0	1	
c_B	Basis	x_1	x_2	d_3^-	d_3^+	b
0	x_2	$\frac{2}{3}$	1	0	0	$\frac{80}{3}$
0	d_3^-	$\frac{1}{3}$	0	1	-1	$\frac{70}{3}$
$P_2 : \bar{c}$	Row	0	0	0	1	$Z_2 = 0$

We now illustrate the partitioning algorithm using Example 5.2. The subproblem S_1 for priority P_1 to be solved initially is given below:

$$S_1: \quad \text{Minimize} \quad Z_1 = d_1^- + d_1^+$$

$$\text{Subject to:} \quad 2x_1 + 3x_2 + d_1^- - d_1^+ = 80$$

$$x_1, x_2, d_1^-, d_1^+ \geq 0$$

The solution to subproblem S_1 by the simplex method is given in Table 5.2. However, alternate optima exist to subproblem S_1 (the nonbasic variable x_1 has a relative cost of zero). As the relative costs for d_1^+ and d_1^- are positive, they cannot enter the basis later; else they destroy the optimality achieved for priority 1. Hence, they are eliminated from the tableau from further consideration.

We now add the goal constraint assigned to the second priority (Equation 5.57):

$$x_1 + x_2 + d_3^- - d_3^+ = 50$$

Since x_2 is a basic variable in the present optimal tableau (Table 5.2), we perform a row operation on the above equation to eliminate x_2, and we get

$$\frac{1}{3}x_1 + d_3^- - d_3^+ = \frac{70}{3} \tag{5.58}$$

Equation 5.58 is now added to the present optimal tableau after deleting the columns corresponding to d_1^- and d_1^+. This is shown in Table 5.3. The objective function of subproblem S_2 is given by

$$\text{Minimize } Z_2 = d_3^+$$

As the right-hand side of the new goal constraint (Equation 5.57) remained nonnegative after the row reduction (Equation 5.58), d_3^- was entered as the basic variable in the new tableau (Table 5.3). If, on the other hand, the right-hand side had become negative, the row would be multiplied by -1 and d_3^+ would become the new basic variable. Table 5.3 indicates

TABLE 5.4 Solution to Subproblem S_3

c_B	$P_3 : c_j$ Basis	0 x_1	0 x_2	0 d_3^-	1 d_2^-	1 d_2^+	b
0	x_2	$\frac{2}{3}$	1	0	0	0	$\frac{80}{3}$
0	d_3^-	$\frac{1}{3}$	0	1	0	0	$\frac{70}{3}$
1	d_2^-	$\frac{1}{3}$	0	0	1	-1	$\frac{40}{3}$
$P_3 : \bar{c}$	Row	$-\frac{1}{3}$	0	0	0	2	
0	x_2	0	1	0	-2	2	0
0	d_3^-	0	0	1	-1	1	10
0	x_1	1	0	0	3	3	40
$P_3 : \bar{c}$	Row	0	0	0	1	1	$Z_3 = 0$

that we have found an optimal solution to S_2. As alternate optimal solutions exist, we add the goal constraint and objective corresponding to priority 3. We also eliminate the column corresponding to d_3^+. The goal constraint assigned to P_3, given by

$$x_1 + x_2 + d_2^- - d_2^+ = 40$$

is added after elimination of x_2. This is shown in Table 5.4. Now x_1 can enter the basis to improve the priority 3 goal, while maintaining the levels achieved for priorities 1 and 2. Then d_2^- is replaced by x_1 and the next solution becomes optimal for subproblem S_3 (see Table 5.4). Moreover, the solution obtained is unique. Hence, it is not possible to improve the goal corresponding to priority 4, and we terminate the partitioning algorithm. It is only necessary to substitute the values of the decision variables ($x_1 = 40$ and $x_2 = 0$) into the goal constraint for P_4 (Equation 5.56) to get $d_4^+ = 160$. Thus, the cost goal is not achieved and the minimum cost of production is $160.

Integer Goal Programs

Arthur and Ravindran [16] show how the partitioning algorithm for linear GP problems can be extended with a modified branch and bound strategy to solve both pure and mixed integer GP problems. The variable elimination scheme used in the PAGP algorithm is not applicable for integer goal programs. They demonstrate the applicability of the branch and bound algorithm with constraint partitioning for integer goal programs with a multiple objective nurse scheduling problem [17].

Nonlinear Goal Programs

Saber and Ravindran [18] present an efficient and reliable method called the partitioning gradient based (PGB) algorithm for solving nonlinear GP problems. The PGB algorithm uses the partitioning technique developed for linear GP problems and the generalized reduced gradient (GRG) method to solve single objective nonlinear programming problems. The authors also present numerical results by comparing the PGB algorithm against a modified pattern search method for solving several nonlinear GP problems. The PGB algorithm found the optimal solution for all test problems proving its robustness and reliability, whereas the pattern search method failed in more than half the test problems by converging to a nonoptimal point.

Multiple Criteria Decision Making

Kuriger and Ravindran [19] have developed three intelligent search methods to solve nonlinear GP problems by adapting and extending the simplex search, complex search, and pattern search methods to account for multiple criteria. These modifications were largely accomplished by using partitioning concepts of goal programming. The paper also includes computational results with several test problems.

5.7.3 Other Goal Programming Models

In addition to the preemptive and non-preemptive goal programming models, other approaches to solving MCMP problems using goal programming have been proposed. In both preemptive and non-preemptive GP models, the DM has to specify the targets or goals for each objective. In addition, in the preemptive GP models, the DM specifies a preemptive priority ranking on the goal achievements. In the non-preemptive case, the DM has to specify relative weights for goal achievements.

To illustrate, consider the following bi-criteria linear program (BCLP):

Example 5.3: BCLP

$$\text{Max } f_1 = x_1 + x_2$$
$$\text{Max } f_2 = x_1$$
$$\text{Subject to: } 4x_1 + 3x_2 \leq 12$$
$$x_1, x_2 \geq 0$$

Maximum f_1 occurs at $\mathbf{x} = (0, 4)$ with $(f_1, f_2) = (4, 0)$. Maximum f_2 occurs at $\mathbf{x} = (3, 0)$ with $(f_1, f_2) = (3, 3)$. Thus the ideal values of f_1 and f_2 are 4 and 3, respectively, and the bounds on (f_1, f_2) on the efficient set will be:

$$3 \leq f_1 \leq 4$$
$$0 \leq f_2 \leq 3$$

Let the DM set the goals for f_1 and f_2 as 3.5 and 2, respectively. Then the GP model becomes:

$$x_1 + x_2 + d_1^- - d_1^+ = 3.5 \tag{5.59}$$

$$x_1 + d_2^- - d_2^+ = 2 \tag{5.60}$$

$$4x_1 + 3x_2 \leq 12 \tag{5.61}$$

$$x_1, x_2, d_1^-, d_1^+, d_2^-, d_2^+ \geq 0 \tag{5.62}$$

Under the preemptive GP model, if the DM indicates that f_1 is much more important than f_2, then the objective function will be

$$\text{Min } Z = P_1 d_1^- + P_2 d_2^-$$

subject to the constraints (Equations 5.59 to 5.62), where P_1 is assumed to be much larger than P_2.

Under the non-preemptive GP model, the DM specifies relative weights on the goal achievements, say w_1 and w_2. Then the objective function becomes

$$\text{Min } Z = w_1 d_1^- + w_2 d_2^-$$

subject to the same constraints (Equations 5.59 to 5.62).

Tchebycheff (Min–Max) Goal Programming

In this GP model, the DM only specifies the goals/targets for each objective. The model minimizes the maximum deviation from the stated goal. To illustrate, the objective function for the Tchebycheff goal program becomes:

$$\text{Min Max } (d_1^-, d_2^-) \quad (5.63)$$

subject to the same constraints (Equations 5.59 to 5.62).

Equation 5.63 can be reformulated as a linear objective by setting

$$\text{Max } (d_1^-, d_2^-) = M \geq 0$$

Then Equation 5.63 is equivalent to

$$\text{Min} \quad Z = M$$
$$\text{Subject to: } M \geq d_1^-$$
$$M \geq d_2^-$$

and the constraints given by Equations 5.59 to 5.62.

The advantage of the Tchebycheff goal program is that there is no need to get preference information (priorities or weights) about goal achievements from the DM. Moreover, the problem reduces to a single objective optimization problem. The disadvantages are (1) the scaling of goals is necessary (as required in nonpreemptive GP) and (2) outliers are given more importance and could produce poor solutions.

Fuzzy Goal Programming

Fuzzy goal programming uses the ideal values as targets and minimizes the maximum normalized distance from the ideal solution for each objective. The lower and upper bounds on the objectives are used for scaling the objectives. This is similar to the Min–Max (Regret) method discussed in Section 5.5.2.

To illustrate, consider Example 5.3 again. Using the ideal values of $(f_1, f_2) = (4, 3)$ and the bounds

$$3 \leq f_1 \leq 4$$
$$0 \leq f_2 \leq 3$$

the fuzzy goal programming formulation becomes:

$$\text{Min} \quad Z = M$$
$$\text{Subject to: } M \geq \frac{4 - (x_1 + x_2)}{4 - 3}$$
$$M \geq \frac{3 - x_1}{3 - 0}$$
$$4x_1 + 3x_2 \leq 12$$
$$x_1, x_2 \geq 0$$

For additional readings on the variants of fuzzy GP models, the reader is referred to Ignizio and Cavalier [20], Tiwari et al. [21,22], Mohammed [23], and Hu et al. [24].

An excellent source of reference for goal programming methods and applications is the textbook by Schniederjans [25].

5.8 Method of Global Criterion and Compromise Programming

5.8.1 Method of Global Criterion

The method of global criterion and compromise programming [1,26,27] fall under the class of MCMP methods that do not require any preference information from the DM.

Consider the MCMP problem given by Equation 5.46. Let

$$S = \{\mathbf{x}/g_j(\mathbf{x}) \leq 0, \quad \text{for all } j\}$$

Let the ideal values of the objectives $f_1, f_2, \ldots, f_k$ be $f_1^*, f_2^*, \ldots, f_k^*$. The method of global criterion finds an efficient solution that is "closest" to the ideal solution in terms of the L_p distant metric. It also uses the ideal values to normalize the objective functions. Thus the MCMP reduces to:

$$\text{Minimize} \quad Z = \sum_{i=1}^{k} \left(\frac{f_i^* - f_i}{f_i^*} \right)^p$$

Subject to: $\mathbf{x} \in S$

The values of f_i^* are obtained by maximizing each objective f_i subject to the constraints $\mathbf{x} \in S$, but ignoring the other objectives. The value of p can be $1, 2, 3, \ldots$, etc. Note that $p = 1$ implies equal importance to all deviations from the ideal. As p increases larger deviations have more weight.

Example 5.4: [1]

$$\text{Max } f_1 = 0.4x_1 + 0.3x_2$$

$$\text{Max } f_2 = x_1$$

Subject to: $x_1 + x_2 \leq 400$ (5.64)

$2x_1 + x_2 \leq 500$ (5.65)

$x_1, x_2 \geq 0$ (5.66)

Let S represent the feasible region constrained by Equations 5.64 to 5.66. Figure 5.4 illustrates the feasible region. The method of global criterion has two steps:

Step 1: Obtain the ideal point.
Step 2: Obtain the preferred solutions by varying the value of $p = 1, 2, \ldots$.

Step 1: *Ideal Point*: Maximizing $f_1(\mathbf{x}) = 0.4x_1 + 0.3x_2$ subject to $\mathbf{x} \in S$ gives the point B as the optimal solution, with $\mathbf{x}^*$ (100, 300) and $f_1^* = 130$ (see Figure 5.4).

Minimizing $f_2(\mathbf{x}) = x_1$ subject to $\mathbf{x} \in S$ gives the point C as the optimal solution, with x^* (250, 0) and $f_2^* = 250$ (see Figure 5.4). Thus, the ideal point $(f_1^*, f_2^*) = (130, 250)$.

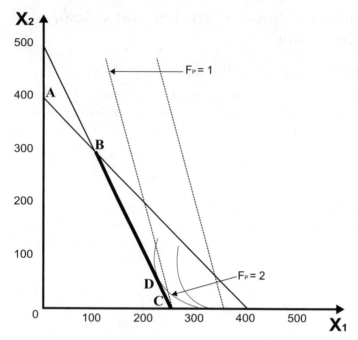

FIGURE 5.4 Illustration of the method of global criterion (Example 5.4).

Step 2: *Obtain Preferred Solutions*: A preferred solution is a nondominated or efficient solution, which is a point on the line segment BC. In Figure 5.4 all points on the line segment BC are efficient solutions.

Case 1: $p=1$

$$\text{Min } Z \left(\frac{130 - (.4x_1 + .3x_2)}{130}\right) + \left(\frac{250 - x_1}{250}\right)$$

$$= 2 - .00708x_1 - .00231x_2$$

subject to: $\mathbf{x} \in S$

The optimal solution to the LP problem when $p=1$ is given by

$$x_1 = 250, x_2 = 0, f_1 = 100, f_2 = 250$$

This is point C in Figure 5.4.

Case 2: $p=2$

$$\text{Min } Z \left(\frac{130 - (.4x_1 + .3x_2)}{130}\right)^2 + \left(\frac{250 - x_1}{250}\right)^2$$

subject to: $\mathbf{x} \in S$

This is a quadratic programming problem and the optimal solution is given by

$$x_1 = 230.7, x_2 = 38.6, f_1 = 103.9, f_2 = 230.7$$

This is point D in Figure 5.4.

The DM's preferred solution should be one of the efficient points on BC (Figure 5.4). It is possible that the solutions obtained for $p=1$ and $p=2$ may not satisfy the DM at all!

5.8.2 Compromise Programming

Compromise programming [26,28] is similar in concept to the one discussed in Section 5.5.3 for MCSP and the method of global criterion. It finds an efficient solution by minimizing the L_p distance metric from the ideal point as given below.

$$\text{Min } L_p = \left[\sum_{i=1}^{k} \lambda_i^p (f_i^* - f_i)^p \right]^{1/p} \qquad (5.67)$$

Subject to $\mathbf{x} \in S$ and $p = 1, 2, \ldots, \infty$

where λ_i's have to be specified or assessed subjectively. Note that λ_i could be set to $1/f_i^*$.

THEOREM 5.3 *Any point x^* that minimizes L_p (Equation 5.67) for $\lambda_i > 0$ for all i, $\sum \lambda_i = 1$ and $1 \leq p < \infty$ is called a compromise solution. Zeleny [26] has proved that these compromise solutions are non-dominated. As $p \to \infty$, Equation 5.67 becomes*

$$\text{Min } L_\infty = \text{Min } \text{Max}_i [\lambda_i (f_i^* - f_i)]$$

and is known as the Tchebycheff Metric.

5.9 Interactive Methods

Interactive methods for MCMP problems rely on the progressive articulation of preferences by the DM. These approaches can be characterized by the following procedure.

Step 1: Find a solution, preferably feasible and efficient.
Step 2: Interact with the DM to obtain his/her reaction or response to the obtained solution.
Step 3: Repeat Steps 1 and 2 until satisfaction is achieved or until some other termination criterion is met.

When interactive algorithms are applied to real-world problems, the most critical factor is the functional restrictions placed on the objective functions, constraints, and the unknown preference function. Another important factor is preference assessment styles (hereafter called interaction styles). Typical interaction styles are:

a. *Binary pairwise comparison*—the DM must compare a pair of two dimensional vectors at each interaction.
b. *Pairwise comparison*—the DM must compare a pair of p-dimensional vectors and specify a preference.
c. *Vector comparison*—the DM must compare a set of p-dimensional vectors and specify the best, the worst, or the order of preference (note that this can be done by a series of pairwise comparisons).
d. *Precise local trade-off ratio*—the DM must specify precise values of local trade-off ratios at a given point. It is the marginal rate of substitution between objectives f_i and f_j: in other words, trade-off ratio is how much the DM is willing to give up in objective j for a unit increase in objective i at a given efficient solution.

e. *Interval trade-off ratio*—the DM must specify an interval for each local trade-off ratio.

f. *Comparative trade-off ratio*—the DM must specify his preference for a given trade-off ratio.

g. *Index specification and value trade-off*—the DM must list the indices of objectives to be improved or sacrificed, and specify the amount.

h. *Aspiration levels* (or reference point)—the DM must specify or adjust the values of the objectives that indicate his/her optimistic wish concerning the outcomes of the objectives.

Shin and Ravindran [29] provide a detailed survey of MCMP interactive methods. The survey includes

- A classification scheme for all interactive methods.
- A review of methods in each category based on functional assumptions, interaction style, progression of research papers from the first publication to all its extensions, solution approach, and published applications.
- A rating of each category of methods in terms of the DM's cognitive burden, ease of use, effectiveness, and handling inconsistency.

5.9.1 Classification of Interactive Methods

Shin and Ravindran [29] classify the interactive methods as follows:

1. Feasible region reduction methods
2. Feasible direction methods
3. Criterion weight space methods
4. Trade-off cutting plane methods
5. Lagrange multiplier methods
6. Visual interactive methods using aspiration levels
7. Branch-and-bound methods

We will describe each of the above approaches briefly here.

Feasible Region Reduction Methods

Each iteration of this approach generally consists of three phases: a calculation phase; a decision phase; and a feasible region reduction phase. In the calculation phase, an efficient solution that is nearest to the ideal solution, in the minimax sense for given weights, is obtained. In the decision phase, the DM interacts with the method and his/her responses are used to construct additional constraints in the feasible region reduction phase. The method continues to perform the three phase iterations until the DM considers the current solution to be the best compromise solution. Advantages of this method are that it can terminate at a nonextreme point and extensions to integer and nonlinear cases are only dependent on the single objective optimization method. However, the method may present a dominated solution to the DM and it is an *ad hoc* procedure, as no preference function concept is utilized. In addition, most methods in this category require a series of index specifications and value trade-offs from the DM.

The first interactive approach in this category is the STEP method (STEM). It was originally described as the progress orientation procedure in Benayoun et al. [30] and later elaborated by Benayoun et al. [31]. STEM guarantees convergence in no more than p (number of objectives) iterations. Fichefet [32] combined the basic features of goal programming to form the goal programming STEM method. GPSTEM guides the DM to reach a compromise solution in fewer iterations than STEM. It also considers bimatrix games to allow multiple DMs in the solution process.

Feasible Direction Methods

This approach is a direct extension of the feasible direction methods developed for solving single objective nonlinear programming problems. It starts with a feasible solution and iteratively performs two major steps: (1) to find a "usable direction" (along which the preference of the DM appears to increase), called the direction finding step; and (2) to determine the step-size from the current solution along the usable direction, called the line search. In the direction finding step, the DM provides information about his preferences in the neighborhood of the current solution by specifying values of local trade-offs among criteria. This, when translated into an approximation of the gradient of the preference function at that point, guides the selection of a new solution with higher preference.

The pioneering method in this category is the GDF procedure by Geoffrion et al. [33]. They modified the Frank–Wolfe method for solving single objective convex programs and applied the GDF procedure to an academic planning problem. The major drawback of the GDF procedure is the difficulty in providing precise local trade-off ratios and the necessity of the line search.

Sadagopan and Ravindran [34] have extended the GDF procedure to nonlinear constraints using the generalized reduced gradient (GRG) method for solving single objective problems. They use "reduced gradients" to overcome the nonlinearity of the constraints. They also employ interval trade-off estimates rather then precise local trade-offs and eliminate the line search.

Criterion Weight Space Methods

When the decision space is a compact convex set and the objective functions to be maximized are concave, an efficient solution can be obtained by solving a single objective problem in which the objectives are combined using weights. (Recall the P_λ problem discussed in Section 5.6.2.) The domain of the weights is defined as the criterion weight space, and this approach reaches the best compromise solution by either searching for the optimal weight space or successively reducing the space. This approach has been popular in practice, but most methods are applicable to only multiple objective linear programming (MOLP) problems.

The Zionts–Wallenius (ZW) method [35] is a typical criterion weight space method. The method optimizes an LP problem for a given arbitrary set of weights on the objectives. Then a set of trade-offs (reduced costs of the objectives), which are associated with the optimal solution of the current LP problem, is presented to the DM. The DM's responses include a preference/nonpreference for a trade-off or indifference. The DM's responses are then used in reducing the criterion weight space by additional constraints on weights and generating a new efficient solution. If the DM is satisfied with the new solution (i.e., he does not want any trade-off), the procedure terminates. Malakooti and Ravindran [36] improve the ZW method by employing the paired comparisons of alternatives and strength of preference. They claim some advantages over the ZW method through a computational study. They also propose an interesting approach to handle the problem of the DM's inconsistent responses.

Steuer [37] proposes a procedure with interval criterion weights. Rather than obtain a single efficient extreme point with fixed weights, a cluster of efficient extreme points is generated. By widening or narrowing the intervals corresponding to each objective, different subsets of efficient extreme points can be generated. At each iteration, after interacting with the DM, the intervals are successively reduced until the best compromise solution is reached.

Trade-Off Cutting Plane Methods

This approach can be viewed as another variation of the feasible direction method. The methods of this class are unique in the way they isolate the best compromise solution. They iteratively reduce the objective space (equivalent to the reduction of the feasible space) by cutting planes and the line search is eliminated. The methods are applicable to general nonlinear problems, but they require precise local trade-off ratios. This approach transforms the MCMP problem through one-to-one functional mapping from decision space to objective space.

Musselman and Talavage [38] use local trade-offs to develop a cutting plane that progressively eliminates a portion of the objective space. They use the "method of centers" to locate the new solution point near the center of the remaining objective space, thus eliminating the line search requirement. The method was applied to storm drainage problem. Shin and Ravindran [39] combined the direction finding step of Sadagopan and Ravindran [34] and the cutting plane concept. The line search is optional in the method and a computational study using GRG2 was performed. Sadagopan and Ravindran [40] present a paired comparison method (PCM) and a comparative trade-off method (CTM) for solving the bicriterion problems. These methods also eliminate a certain portion of the objective space at each iteration via interactions with the DM. The PCM was applied successfully to a cardiovascular disease control problem in the U.S. Air Force [41].

Lagrange Multiplier Methods

The interactive methods that use Lagrange multipliers belong in this class. The interactive surrogate worth trade-off method (ISWTM) of Chankong and Haims [42] is a Lagrange multiplier method. It maximizes one objective subject to a varying set of bounded objectives and uses the resulting Lagrange multipliers of the constraints for interacting with the DM. The method first generates efficient solutions that form the trade-off functions in the objective surface derived from the generalized Lagrangian problem, and then searches for a preferred solution by interacting with the DM to generate shadow prices of all bounded objectives. In the generation of a shadow price, the DM is asked interactively to assess the indifference bounds to define a surrogate worth function. It must be pointed out that the amount of work to determine a shadow price (surrogate worth function) might be cumbersome, considering that it needs to be repeated at each step and for each objective.

Visual Interactive Methods

Most of the aforementioned approaches assume that the unknown DM's preference function remains unchanged during the interactive process. In an effort to relax the assumptions concerning the DM's behavior, Korhonen and Laakso [43] presented a graphic-aided interactive approach. Theoretically, this method can be seen as an extension of the GDF procedure in that the line search used is analogous to that on the GDF procedure. However, it determines new search directions using reference directions suggested by Wierzbicki [44], which reflect the DM's preference. At each iteration, the method generates a picture representing a subset of the efficient frontier for interacting with the DM. By modifying the method, "Pareto race" was developed by Korhonen and Wallenius [45] and it has been applied in pricing alcoholic beverages [46]. Pareto Race enables the DM to control the entire efficient frontier through the interactive process. With the rapid advances in personal computer technology, this type of approach is becoming popular.

Branch-and-Bound Method

The branch-and-bound method developed by Marcotte and Soland [47] divides the objective space into subsets. Each subset is a branch from the original objective space. They further branch into even smaller subsets if a branch is promising. At each subset they determine an ideal solution and use this ideal solution to form an upper bound for each subset. Note that the ideal solution of a subset dominates all the efficient points of that subset. Each subset forms a node and at each node solutions are compared against the incumbent solution by interacting with the DM. A node is fathomed if the ideal solution pertaining to that subset is not preferred to the incumbent solution by the DM. This method is also applicable to the discrete case and has a number of good properties, such as vector comparisons, nondependence on the preference function, and termination at an efficient solution.

Raman [48] developed a branch-and-bound interactive method for solving bicriteria linear integer programs. The author used the Tchebycheff's norm for generating efficient solutions. Eswaran et al. [49] implemented a weighted Tchebycheff's norm to develop an interactive method for solving bicriteria nonlinear integer programs with applications to quality control problems in acceptance sampling.

5.9.2 Inconsistency of the DM

Interactive methods require the DM to respond to a series of questions at each iteration. Therefore the consistency of the DM is one of the most important factors in the success of the methods. Because DMs are very subjective, different starting solutions may lead to different best compromise solutions. Moreover, different methods use different types of questions or interaction styles and could guide the DM to different solutions. Nearly all interactive methods require consistent responses from the DM to be successful. Thus, the assumption of the DM's consistency usually draws severe criticism and some researchers underrate the abilities of the interactive methods due to this drawback.

There are generally two ways to reduce the DM's inconsistency: (1) testing consistency during the procedure; and (2) minimizing the DM's cognitive burden. Testing whether the DM's responses are consistent with those to the previous questions has been used in trade-off assessment schemes. Also, tests for recognizing the DM's inconsistency have been developed by Malakooti and Ravindran [36]. The DM's inconsistency can be reduced by presenting easy questions to the DM. In this way the DM has little confusion with questions and less chance for making mistakes. Reducing the DM's cognitive burden is one of the motivations for new algorithmic developments in this area.

5.9.3 Computational Studies

There are only a few computational studies on interactive methods. Wallenius [50] describes an evaluation of several MCMP methods, such as STEM and the GDF procedure, and reports that none of the methods has been highly successful in practice. Klein et al. [51] claim that interactive procedures are easier to use and achieve more satisfactory solutions when compared to utility measurement methods. Their conclusions are based on simulated study of a quality control problem with several students serving as DMs.

Most methods are generally presented without computational studies, probably due to the difficulty of interacting with real DMs. Even if simulated preference functions are used, more computational study is needed to assess the usefulness of these methods. Also, the preferences of practitioners with regard to interaction styles and methodological approaches must be disclosed by comparative studies. This will help managers apply the interactive methods to real problems.

5.10 MCDM Applications

One of the most successful applications of multi-criteria decision making has been in the area of portfolio selection, an important problem faced by individual investors and financial analysts in investment companies. A portfolio specifies the amount invested in different securities that may include bonds, common stocks, mutual funds, bank CDs, Treasury notes, and others. Much of the earlier investment decisions were made by seat-of-the-pants approaches. Markowitz [52] in the 1950s pioneered the development of the modern portfolio theory, which uses bi-criteria mathematical programming models to analyze the portfolio selection problem. By quantifying the trade-offs between risks and returns, he showed how an investor can diversity portfolios such that the portfolio risk can be reduced without sacrificing returns. Based on Markowitz's work, Sharpe [53] introduced the concept of the market risk, and developed a bi-criteria linear programming model for portfolio analysis. For their pioneering work in modern portfolio theory, both Markowitz and Sharpe shared the 1990 Nobel Prize in Economics. The Nobel award was the catalyst for the rapid use of the modern portfolio theory by Wall Street firms in the 1990s. A detailed discussion of the mean-variance model of Markowitz is available in Part II of this handbook (Chapter 21).

Among the MCDM models and methods, the goal programming models have seen the most applications in industry and government. Chapter 4 of the textbook by Schniederjan [25] contains an extensive bibliography (666 citations) on goal programming applications categorized by areas—accounting, agriculture, economics, engineering, finance, government, international, management, and marketing. Zanakis and Gupta [54] also have a categorized bibliographic survey of goal programming applications.

Given below is a partial list of MCDM applications in practice:

- Academic planning [30,55–57]
- Accounting [58]
- Environment [59–61]
- Forest management [62–64]
- Health planning [17,40,41,65–67]
- Investment planning [52,53,68–70]
- Manpower planning [71–73]
- Metal cutting [74–76]
- Production planning and scheduling [77–82]
- Quality control [49,51,83–87]
- Reliability [88]
- Supply chain management [89–97]
- Transportation [98–101]
- Waste disposal [102]
- Water resources [38,103–107]

5.11 MCDM Software

One of the problems in applying MCDM methods in practice is the lack of commercially available software implementing these methods. There is some research software available. Two good resources for these are:

http://www.terry.uga.edu/mcdm/
http://www.sal.hut.fi/

The first is the Web page of the International Society on multiple criteria decision making. It has links to MCDM software and bibliography. A number of this software is available free for research and teaching use. The second link is to the research group at Helsinki University of Technology. It has links to some free downloads software, again for research and instructional use.

Following are some links to commercially available MCDM software:

Method(s)	Software
Utility/ValueTheory	LogicalDecisions http://www.logicaldecisions.com/
AHP	ExpertChoice http://www.expertchoice.com/software/
AHP/SMART	CriteriumDecisionPlus http://www.hearne.co.uk/products/decisionplus/
GoalProgramming	LindoSystems http://www.lindo.com
PROMETHEE	Decision Lab http://www.visualdecision.com/dlab.htm
ELECTRE	LAMSADE Web page http://l1.lamsade.dauphine.fr/english/software.html/

5.12 Further Readings

While a need for decision making in the context of multiple conflicting criteria has existed for a very long time, the roots of the discipline of multiple criteria decision making (MCDM) go back about half a century only. Development of the discipline has taken place along two distinct tracks. One deals in problems with a relatively small number of alternatives, often in an environment of uncertainty. A groundbreaking book that has influenced research and application in this area is by Keeney and Raiffa [108]. It is called *Decision Analysis* or *Multi Attribute Decision Making*. Chapter 6 of this handbook discusses decision analysis in detail. The other track deals with problems where the intent is to determine the "best" alternative utilizing mathematical relationships among various decision variables, called multiple criteria mathematical programming (MCMP) or commonly MCDM problems.

MCDM Text Books
Two of the earliest books describing methods for MCDM problems are Charnes and Cooper [109] and Ijiri [58]. Other books of significance dealing with the exposition of theories, methods, and applications in MCDM include Lee [110], Cochran and Zeleny [111], Hwang and Masud [1], Hwang and Yoon [2], Zeleny [25], Hwang and Lin [112], Steuer [3], Saaty [8],

Kirkwood [113], Ignizio [114], Gal et al. [115], Miettinen [116], Olson [117], Vincke [118], Figueira et al. [119], Ehrgott and Gandibleux [120], and Ehrgott [121].

MCDM Journal Articles
Recent MCDM survey articles of interest include Geldermann and Zhang [122], Kaliszewski [123], Leskinen et al. [124], Osman et al. [125], Tamiz et al., [126], Vaidya and Kumar [127], Alves and Climaco [128], Steuer and Na [129], and Zapounidis and Doumpos [130].

MCDM Internet Links
- Homepage of the International Society on Multiple Criteria Decision Making
  http://www.terry.uga.edu/mcdm
- EURO Working Group on Multi-criteria Decision Aids
  http://www.inescc.pt/~ewgmcda/index.html
- EMOO web page by Carlos Coello
  http://www.lania.mx/~ccoello/EMOO
- Decision Lab 2000
  http://www.visualdecision.com
- Kaisa Miettinen's website
  http://www.mit.jyu.fi/miettine/lista.html
- Decisionarium
  http://www.decisionarium.hut.fi
- Vincent Mousseau's MCDA database
  http://www.lamsade.dauphine.fr/mcda/biblio/

References

1. Hwang, C.L. and Masud, A., *Multiple Objective Decision Making—Methods and Applications*, Springer-Verlag, New York, 1979.
2. Hwang, C.L. and Yoon, K.S., *Multiple Attribute Decision Making—Methods and Applications*, Springer-Verlag, New York, 1981.
3. Steuer, R.E., *Multiple Criteria Optimization: Theory, Computation and Application*, John Wiley, New York, 1986.
4. Saaty, T.L., A scaling method for priorities in hierarchical structures, *J Math Psychol*, 15, 234–281, 1977.
5. Harker, P.T., The art and science of decision making: The analytic hierarchy process, in *The Analytic Hierarchy Process: Applications and Studies*, B.L. Golden, E.W. Wasil and P.T. Harker (Eds.), Springer-Verlag, New York, 1987.
6. Roy, B., The outranking approach and the foundations of Electre methods, in *Readings in Multiple Criteria Decision Aid*, C.A. Bana e Costa (Ed.), Springer-Verlag, New York, 1990.
7. Roy, B., *Multicriteria Methodology for Decision Making*, Kluwer Academic Publishers, Norwell, MA, 1996.
8. Rogers, M., Bruen, M., and Maystre, L.Y., *Electre and Decision Support*, Kluwer Publishers, Norwell, MA, 2000.
9. Saaty, T.L., *The Analytic Hierarchy Process*, 2nd ed., RWS Publications, Pittsburgh, PA, 1990.
10. Forman, E.H. and Gass, S., AHP—An Exposition, *Oper Res*, 49, 469–486, 2001.
11. Brans, J.P. and Mareschal, B., The PROMTHEE methods for MCDM, in *Readings in Multiple Criteria Decision Aid*, C.A. Bana e Costa (Ed.), Springer-Verlag, New York, 1990.

12. Geoffrion, A., Proper efficiency and theory of vector maximum, *J Math Anal Appl*, 22, 618–630, 1968.
13. Ravindran, A., Ragsdell, K.M., and Reklaitis, G.V., *Engineering Optimization: Methods and Applications*, 2nd ed., John Wiley, New York, 2006 (Chapter 11).
14. Arthur, J.L. and Ravindran, A., An efficient goal programming algorithm using constraint partitioning and variable elimination, *Manage Sci*, 24(8), 867–868, 1978.
15. Arthur, J.L. and Ravindran, A., PAGP—Partitioning algorithm for (linear) goal programming problems, *ACM T Math Software*, 6, 378–386, 1980.
16. Arthur, J.L. and Ravindran, A., A branch and bound algorithm with constraint partitioning for integer goal programs, *Eur J Oper Res*, 4, 421–425, 1980.
17. Arthur, J.L. and Ravindran, A., A multiple objective nurse scheduling model, *IIE Trans*, 13, 55–60, 1981.
18. Saber, H.M. and Ravindran, A., A partitioning gradient based (PGB) algorithm for solving nonlinear goal programming problem, *Comput Oper Res*, 23, 141–152, 1996.
19. Kuriger, G. and Ravindran, A., Intelligent search methods for nonlinear goal programs, *Inform Syst OR*, 43, 79–92, 2005.
20. Ignizio, J.M. and Cavalier, T.M., *Linear Programming*, Prentice Hall, Englewood Cliffs, NJ, 1994 (Chapter 13).
21. Tiwari, R.N., Dharmar S., and Rao J.R., Priority structure in fuzzy goal programming, *Fuzzy Set Syst*, 19, 251–259, 1986.
22. Tiwari, R.N., Dharmar, S., and Rao, J.R., Fuzzy goal programming—An additive model, *Fuzzy Set Syst*, 24, 27–34, 1987.
23. Mohammed, R.H., The relationship between goal programming and fuzzy programming, *Fuzzy Set Syst*, 89, 215–222, 1997.
24. Hu, C.F., Teng, C.J., and Li, S.Y., A fuzzy goal programming approach to multi objective optimization problem with priorities, *Euro J Oper Res*, 171, 1–15, 2006.
25. Schniederjans, M., *Goal Programming: Methodology and Applications*, Kluwer Academic, Dordrecht, 1995.
26. Zeleny, M., *Multiple Criteria Decision Making*, McGraw Hill, New York, 1982.
27. Tabucannon, M., *Multiple Criteria Decision Making in Industry*, Elsevier, Amsterdam, 1988.
28. Yu, P.-L., *Multiple Criteria Decision Making*, Plenum Press, New York, 1985.
29. Shin, W.S. and Ravindran, A., Interactive multi objective optimization: survey I—continuous case, *Comput Oper Res*, 18, 97–114, 1991.
30. Benayoun, R., Tergny, J., and Keuneman, D., Mathematical programming with multi-objective functions: a solution by P.O.P. (progressive orientation procedure), *Metric IX*, 279–299, 1965.
31. Benayoun, R., de Montgolfier, J., Tergny, J., and Larichev, O., Linear programming with multiple objective functions: step method (STEM), *Math Program*, 1, 366–375, 1971.
32. Fichefet, J., GPSTEM: an interactive multiobjective optimization method, *Prog Oper Res*, 1, 317–332, 1980.
33. Geoffrion, A.M., Dyer, J.S., and Fienberg, A., An interactive approach for multi-criterion optimization with an application to the operation of an academic department, *Manage Sci*, 19, 357–368, 1972.
34. Sadagopan, S. and Ravindran, A., Interactive algorithms for multiple criteria nonlinear programming problems, *Euro J Oper Res*, 25, 247–257, 1986.
35. Zionts, S. and Wallenius, J., An interactive programming method for solving the multiple criteria problem. *Manage Sci*, 22, 652–663, 1976.
36. Malakooti, B. and Ravindran, A., An interactive paired comparison simplex method for MOLP problems, *Ann Oper Res*, 5, 575–597, 1985/86.

37. Steuer, R.E., Multiple objective linear programming with interval criterion weights, *Manage Sci*, 23, 305–316, 1976.
38. Musselman, K. and Talavage, J., A tradeoff cut approach to multiple objective optimizations, *Oper Res*, 28, 1424–1435, 1980.
39. Shin, W.S. and Ravindran, A., An interactive method for multiple objective mathematical programming (MOMP) problems, *J Optimiz Theory App*, 68, 539–569, 1991.
40. Sadagopan, S. and Ravindran, A., Interactive solution of bicriteria mathematical programming, *Nav Res Log Qtly*, 29, 443–459, 1982.
41. Ravindran, A. and Sadagopan, S., Decision making under multiple criteria—A case study, *IEEE T Eng Manage*, EM-34, 127–177, 1987.
42. Chankong, V. and Haimes, Y.Y., An interactive surrogate worth tradeoff (ISWT) method for multiple objective decision making, in *Multiple Criteria Problem Solving*, S. Zionts (Ed.), Springer, New York, 1978.
43. Korhonen, P. and Laakso, L., A visual interactive method for solving the multiple criteria problem, *Euro J Oper Res*, 24, 277–278, 1986.
44. Wierzbicki, A.P., The use of reference objectives in multiobjective optimization, in *Multiple Criteria Decision Making*, G. Fandel and T. Gal (Eds.), Springer, New York, 1980.
45. Korhonen, P. and Wallenius, J., A Pareto race, *Nav Res Log Qtly*, 35, 615–623, 1988.
46. Korhonen, P. and Soismaa, M., A multiple criteria model for pricing alcoholic beverages, *Euro J Oper Res*, 37, 165–175, 1988.
47. Marcotte, O. and Soland, R.M., An interactive branch-and-bound algorithm for multiple criteria optimization, *Manage Sci*, 32, 61–75, 1985.
48. Raman, T., A branch and bound method using Tchebycheff Norm for solving bi-criteria integer problems, M.S. Thesis, University of Oklahoma, Norman, 1997.
49. Eswaran, P.K., Ravindran, A., and Moskowitz, H., Interactive decision making involving nonlinear integer bicriterion problems, *J Optimiz Theory Appl*, 63, 261–279, 1989.
50. Wallenius, J., Comparative evaluation of some interactive approaches to multicriterion optimization. *Manage Sci*, 21, 1387–1396, 1976.
51. Klein, G., Moskowitz, H., and Ravindran, A., Comparative evaluation of prior versus progressive articulation of preference in bicriterion optimization, *Nav Res Log Qtly*, 33, 309–323, 1986.
52. Markowitz, H., *Portfolio Selection: Efficient Diversification of Investments*, Wiley, New York, 1959.
53. Sharpe, W.F., A simplified model for portfolio analysis, *Manage Sci*, 9, 277–293, 1963.
54. Zanakis, S.H. and Gupta, S.K., A categorized bibliographic survey of goal programming, *Omega: Int J Manage Sci*, 13, 211–222, 1995.
55. Beilby, M.H. and T.H. Mott, Jr., Academic library acquisitions allocation based on multiple collection development goals, *Comput Oper Res*, 10, 335–344, 1983.
56. Lee, S.M. and Clayton, E.R., A goal programming model for academic resource allocation, *Manage Sci*, 17, 395–408, 1972.
57. Dinkelbach, W. and Isermann, H., Resource allocation of an academic department in the presence of multiple criteria-some experience with a modified STEM method, *Comput Oper Res*, 7, 99–106, 1980.
58. Ijiri, Y., *Management Goals and Accounting for Control*, Rand-McNally, Chicago, 1965.
59. Charnes, A., Cooper, W.W., Harrald, J., Karwan, K.R., and Wallace, W.A., A goal interval programming model for resource allocation on a marine environmental protection program, *J Environ Econ Manage*, 3, 347–362, 1976.
60. Sakawa, M., Interactive multiobjective decision making for large-scale systems and its application to environmental systems, *IEEE T Syst Man Cyb*, 10, 796–806, 1980.

61. Stewart, T.J., Experience with prototype multicriteria decision support systems for pelagic fish quota determination, *Nav Res Log Qtly*, 35, 719–731, 1988.
62. Schuler, A.T., Webster, H.H., and Meadows, J.C., Goal programming in forest management, *J Forestry*, 75, 320–324, 1977.
63. Steuer, R.E. and Schuler, A.T., An interactive multiple objective linear programming approach to a problem in forest management, *Oper Res*, 25, 254–269, 1978.
64. Harrison, T.P. and Rosenthal, R.E., An implicit/explicit approach to multiobjective optimization with an application to forest management planning, *Decision Sci*, 19, 190–210, 1988.
65. Musa, A.A. and Saxena, U., Scheduling nurses using goal programming techniques, *IIE Transactions*, 16, 216–221, 1984.
66. Trivedi, V.M., A mixed-integer goal programming model for nursing service budgeting, *Oper Res*, 29, 1019–1034, 1981.
67. Beal, K., Multicriteria optimization applied to radiation therapy planning, M.S. Thesis, Pennsylvania State University, University Park, 2003.
68. Lee, S.M. and Lerro, A.J., Optimizing the portfolio selection for mutual funds, *J Finance*, 28, 1087–1101, 1973.
69. Nichols, T. and Ravindran, A., Asset allocation using goal programming, *Proceedings of the 34th International Conference on Computers and IE*, San Francisco, November 2004.
70. Erlandson, Sarah, Asset allocation models for portfolio optimization, M.S. Thesis, Pennsylvania State University, University Park, 2002.
71. Charnes, A., Cooper, W.W., and Niehaus, R.J., Dynamic multi-attribute models for mixed manpower systems, *Nav Res Log Qtly*, 22, 205–220, 1975.
72. Zanakis, S.H. and Maret, M.W., A Markovian goal programming approach to aggregate manpower planning, *J Oper Res Soc*, 32, 55–63, 1981.
73. Henry, T.M. and Ravindran, A., A goal programming application for army officer accession planning, *Inform Syst OR J*, 43, 111–120, 2005.
74. Philipson, R.H. and Ravindran, A., Applications of mathematical programming to metal cutting, *Math Prog Stud*, 19, 116–134, 1979.
75. Philipson, R.H. and Ravindran, A., Application of goals programming to machinability data optimization, *J Mech Design: Trans ASME*, 100, 286–291, 1978.
76. Malakooti, B. and Deviprasad, J., An interactive multiple criteria approach for parameter selection in metal cutting, *Oper Res*, 37, 805–818, 1989.
77. Arthur, J.L. and Lawrence, K.D., Multiple goal production and logistics planning in a chemical and pharmaceutical company, *Comput Oper Res*, 9, 127–237, 1982.
78. Deckro, R.J., Herbert, J.E., and Winkofsky, E.P., Multiple criteria job-shop scheduling, *Comput Oper Res*, 9, 279–285, 1984.
79. Lawrence, K.D. and Burbridge, J.J., A multiple goal linear programming model for coordinated and logistic planning, *Int J Prod Res*, 14, 215–222, 1976.
80. Lee, S.M. and Moore, J.L., A practical approach to production scheduling, *Prod Inventory Manage*, 15, 79–92, 1974.
81. Lee, S.M., Clayton, E.R., and Taylor, B.W., A goal programming approach to multi-period production line scheduling, *Comput Oper Res*, 5, 205–211, 1978.
82. Lashine, S.H., Foote, B.L., and Ravindran, A., A nonlinear mixed integer goal programming model for the two-machine closed flow shop, *Euro J Oper Res*, 55, 57–70, 1991.
83. Moskowitz, H., Ravindran, A., Klein, G., and Eswaran, P.K., A bicriteria model for acceptance sampling in quality control, *TIMS Special Issue on Optim Stat*, 19, 305–322, 1982.
84. Moskowitz, H., Plante, R., Tang, K., and Ravindran, A., Multiattribute Bayesian acceptance sampling plans for screening and scrapping rejected lots, *IIE Transactions*, 16, 185–192, 1984.

85. Sengupta, S., A goal programming approach to a type of quality control problem, *J Oper Res Soc*, 32, 207–212, 1981.
86. Ravindran, A., Shin, W., Arthur, J., and Moskowitz, H., Nonlinear integer goal programming models for acceptance sampling, *Comput Oper Res*, 13, 611–622, 1986.
87. Klein, G., Moskowitz, H., and Ravindran, A., Interactive multiobjective optimization under uncertainty, *Manage Sci*, 36, 58–75, 1990.
88. Mohamed, A., Ravindran, A., and Leemis, L., An interactive availability allocation algorithm, *Int J Oper Quant Manage*, 8, 1–19, 2002.
89. Thirumalai, R., Multi criteria multi decision maker inventory models for serial supply chains, PhD Thesis, Pennsylvania State University, University Park, 2001.
90. Gaur, S. and Ravindran, A., A bi-criteria model for inventory aggregation under risk pooling, *Comput Ind Eng*, 51, 482–501, 2006.
91. DiFilippo, A.M., Multi criteria supply chain inventory models with transportation costs, M.S. Thesis, Pennsylvania State University, University Park, 2003.
92. Attai, T.D., A multi objective approach to global supply chain design, M.S. Thesis, Pennsylvania State University, University Park, 2003.
93. Wadhwa, V. and Ravindran, A., A multi objective model for supplier selection, *Comput Oper Res*, 34, 3725–3737, 2007.
94. Mendoza, A., Santiago, E., and Ravindran, A., A three phase multicriteria method to the supplier selection problem, Working Paper, Industrial Engineering, Pennsylvania State University, University Park, August 2005.
95. Lo-Ping, C., Fuzzy goal programming approach to supplier selection, M.S. Thesis, Pennsylvania State University, University Park, 2006.
96. Natarajan, A., Multi criteria supply chain inventory models with transportation costs, Ph.D. Thesis, Pennsylvania State University, University Park, 2006.
97. Yang, T., Multi objective optimization models for management supply risks in supply chains, Ph.D. Thesis, Pennsylvania State University, University Park, 2006.
98. Cook, W.D., Goal programming and financial planning models for highway rehabilitation, *J Oper Res Soc*, 35, 217–224, 1984.
99. Lee, S.M. and Moore, L.J., Multi-criteria school busing models, *Manage Sci*, 23, 703–715, 1977.
100. Moore, L.J, Taylor, B.W., and Lee, S.M., Analysis of a transshipment problem with multiple conflicting objectives, *Comput Oper Res*, 5, 39–46, 1978.
101. Sinha, K.C., Muthusubramanyam, M., and Ravindran, A., Optimization approach for allocation of funds for maintenance and preservation of the existing highway system, *Transport Res Rec*, 826, 5–8, 1981.
102. Mogharabi, S. and Ravindran, A., Liquid waste injection planning via goal programming, *Comput Ind Eng*, 22, 423–433, 1992.
103. Haimes, Y.Y., Hall, W.A., and Freedman, H.T., *Multiobjective Optimization in Water Resources Systems*, Elsevier, Amsterdam, 1975.
104. Johnson, L.E. and Loucks, D.P., Interactive multiobjective planning using computer graphics, *Comput Oper Res*, 7, 89–97, 1980.
105. Haimes, Y.Y., Loparo, K.A., Olenik, S.C., and Nanda, S.K., Multiobjective statistical method for interior drainage systems, *Water Resour Res* 16, 465–475, 1980.
106. Loganathan, G.V. and Sherali, H.D., A convergent interactive cutting-plane algorithm for multiobjective optimization, *Oper Res*, 35, 365–377, 1987.
107. Monarchi, D.E., Kisiel, C.C., and Duckstein, L., Interactive multiobjective programming in water resources: A case study, *Water Resour Res*, 9, 837–850, 1973.
108. Keeny, R.L. and Raiffa, H., *Decision with Multiple Objectives: Preferences and Value Trade Offs*, Wiley, New York, 1976.

109. Charnes, A. and Cooper, W.W., *Management Models and Industrial Applications of Linear Programming*, Wiley, New York, 1961.
110. Lee, S.M., *Goal Programming for Decision Analysis*, Auerbach Publishers, Philadelphia, PA, 1972.
111. Cochran, J.L. and Zeleny, M. (Eds.), *Multiple Criteria Decision Making*, University of South Carolina Press, Columbia, SC, 1973.
112. Hwang, C.L. and Lin, M.J., *Group Decision Making Under Multiple Criteria*, Springer-Verlag, New York, 1987.
113. Kirkwood, C.W., *Strategic Decision Making*, Wadsworth Publishing Co., Belmont, CA, 1997.
114. Ignizio, J.P., *Goal Programming and Its Extensions*, Heath, Lexington, MA, 1976.
115. Gal, T., Stewart T.J., and Hanne, T. (Eds.), *Multiple Criteria Decision Making: Advances in MCDM Models*, Kluwer Academic Publishers, Norwell, MA, 1999.
116. Miettinen, K.M., *Nonlinear Multiobjective Optimization*, Kluwer Academic Publishers, Norwell, MA, 1999.
117. Olson, D.L., *Decision Aids for Selection Problems*, Springer-Verlag, New York, 1996.
118. Vincke, P., *Multicriteria Decision Aid*, Wiley, Chichester, England, 1992.
119. Figueira, J.S., Greco, S., and Ehrgott, M. (Eds.), *Multiple Criteria Decision Analysis: State of the Art Surveys*, Springer, New York, 2005.
120. Ehrgott, M. and Gandibleux, X., *Multiple Criteria Optimization: State of the Art Annotated Bibliographic Surveys*, Kluwer Academic, Dordrecht, 2002.
121. Ehrgott, M., *Multicriteria Optimization*, Springer, New York, 2005.
122. Geldermann, J. and Zhang, K., Software review: Decision lab 2000, *J Multi-Criteria Decision Anal*, 10, 317–323, 2001.
123. Kaliszewski, I., Out of the mist towards decision-maker-friendly multiple criteria decision making, *Euro J Oper Res*, 158, 293–307, 2004.
124. Leskinen, P., Kangas, A.S., and Kangas, J., Rank-based modeling of preferences in multi-criteria decision making, *Euro J Oper Res*, 158, 721–733, 2004.
125. Osman, M., Kansan, S., Abou-El-Enien, T., and Mohamed, S., Multiple criteria decision making theory applications and software: a literature review, *Adv Model Anal B*, 48, 1–35, 2005.
126. Tamiz, M., Jones, D., and Romero, C., Goal programming for decision making: An overview of the current state-of-the-art, *Euro J Oper Res*, 111, 569–581, 1998.
127. Vaidya, O.S. and Kumar, S., Analytic hierarchy process: an overview of applications, *Euro J Oper Res*, 169, 1–29, 2006.
128. Alves, M.J. and Climaco, J., A review of interactive methods for multiobjective integer and mixed-integer programming, *Euro J Oper Res*, 180, 99–115, 2007.
129. Steuer, R.E. and Na, P., Multiple criteria decision making combined with finance: A categorized bibliographic study, *Euro J Oper Res*, 150, 496–515, 2003.
130. Zopounidis, C. and Doumpos, M., Multi-criteria decision aid in financial decision making: methodologies and literature review, *J Multi-Criteria Decision Anal*, 11, 167–186, 2002.

6
Decision Analysis

6.1	Introduction..	**6**-1
6.2	Terminology for Decision Analysis...................	**6**-2
	Terminology • An Investment Example	
6.3	Decision Making under Risk..........................	**6**-3
	Maximum Likelihood • Expected Value under Uncertainty • Expected Opportunity Loss or Expected Regret • Expected Value of Perfect Information • Decision Trees • Posterior Probabilities • Utility Functions	
6.4	Decision Making under Uncertainty................	**6**-17
	Snack Production Example • Maximin (Minimax) Criterion • Maximax (Minimin) Criterion • Hurwicz Criterion • Laplace Criterion or Expected Value • Minimax Regret (Savage Regret)	
6.5	Practical Decision Analysis	**6**-21
	Problem Definition • Objectives • Alternatives • Consequences • Tradeoffs • Uncertainty • Risk Tolerance • Linked Decisions	
6.6	Conclusions	**6**-28
6.7	Resources..	**6**-29
	References	**6**-30

Cerry M. Klein
University of Missouri–Columbia

6.1 Introduction

Everyone engages in the process of making decisions on a daily basis. Some of these decisions are quite easy to make and almost automatic. Other decisions can be very difficult to make and almost debilitating. Likewise, the information needed to make a good decision varies greatly. Some decisions require a great deal of information whereas others much less. Sometimes there is not much if any information available and hence the decision becomes intuitive, if not just a guess. Many, if not most, people make decisions without ever truly analyzing the situation and the alternatives that exist. There is a subjective and intrinsic aspect to all decision making, but there are also systematic ways to think about problems to help make decisions easier. The purpose of decision analysis is to develop techniques to aid the process of decision making, not replace the decision maker.

Decision analysis can thus be defined as the process and methodology of identifying, modeling, assessing, and determining an appropriate course of action for a given decision problem. This process often involves a wide array of tools and the basic approach is generally to break the problem down into manageable and understandable parts that the decision maker can comprehend and handle. It is then necessary to take these smaller elements and reconstitute them into a proper solution for the larger original problem. Through this process, the decision maker should also gain insight into the problem and the implementation of the solution. This should enable the decision maker to analyze complex situations and choose a course of action consistent with their basic values and knowledge [1].

When analyzing the decision making process, the context or environment of the decision to be made allows for a categorization of the decisions based on the nature of the problem or the nature of the data or both. There are two broad categories of decision problems: decision making under certainty and decision making under uncertainty. Some break decision making under uncertainty down further in terms of whether the problem can be modeled by probability distributions (risk) or not (uncertainty). There is also a break down of decision type based on whether one is choosing an alternative from a list of alternatives (attribute based) or allocating resources (objective based), or negotiating an agreement [2]. Decision analysis is almost always in the context of choosing one alternative from a set of alternatives and is the approach taken here.

Decision making under certainty means that the data are known deterministically or at least at an estimated level the decision maker is comfortable with in terms of variation. Likewise, the decision alternatives can be well defined and modeled. The techniques used for these problem types are many and include much of what is included in this handbook such as linear programming, nonlinear programming, integer programming, multiobjective optimization, goal programming, analytic hierarchy process, and others.

Decision making under risk means that there is uncertainty in the data, but this uncertainty can be modeled probabilistically. Note that there are some that do not use this designation as they believe all probability is subjective; hence all decisions not known with certainty are uncertain. However, we will use the common convention of referring to probabilistic models as decision making under risk.

Decision making under uncertainty means the probability model for the data is unknown or cannot be modeled probabilistically and hence the data are imprecise or vague.

Decisions made under risk and uncertainty are the focus of this chapter.

6.2 Terminology for Decision Analysis

There are many excellent works on decision analysis. This is a well-studied area with contributions in a wide variety of fields including operations research, operations management, psychology, public policy, leadership, and so on. With the wide variety of backgrounds and works, it is necessary to have a common language. To that end and to help structure the discussion the following terminology, adapted from Ravindran, Phillips, and Solberg [1, Chapter 5], is used.

6.2.1 Terminology

Decision Maker The entity responsible for making the decision. This may be a single person, a committee, company, and the like. It is viewed here as a single entity, not a group.

Alternatives A finite number of possible decision alternatives or courses of action available to the decision maker. The decision maker generally has control over the specification and description of the alternatives. Note that one alternative to always include is the action of doing nothing, which is maintaining the status quo.

States of Nature The scenarios or states of the environment that may occur but are not under control of the decision maker. These are the circumstances under which a decision is made. The states of nature are mutually exclusive events and

Decision Analysis

exhaustive. This means that one and only one state of nature is assumed to occur and that all possible states are considered.

Outcome Outcomes are the measures of net benefit, or payoff, received by the decision maker. This payoff is the result of the decision and the state of nature. Hence, there is a payoff for each alternative and outcome pair. The measures of payoff should be indicative of the decision maker's values or preferences. The payoffs are generally given in a payoff matrix in which a positive value represents net revenue, income, or profit and a negative value represents net loss, expenses, or costs. This matrix yields all alternative and outcome combinations and their respective payoff and is used to represent the decision problem.

6.2.2 An Investment Example

As an example, consider the common decision of determining where to invest money [3,4]. The example will be limited to choosing one form of investment, but similar approaches exist for considering a portfolio. Assume you have $10,000 to invest and are trying to decide between a speculative stock (SS, one that has high risk, but can generate substantial returns), a conservative stock (CS, one that will perform well in any environment, but doesn't have the potential for large returns), bonds (B), and certificates of deposit (CD). Data from the last several years have been collected and analyzed and estimated rates of return for each investment have been determined. These rates, however, are dependent on the state of the economy. From the analysis it has also been determined that there are three basic states of the economy to consider; *Strong*—the economy is growing at a rate greater than 5%, *Stable*—the economy is growing at a rate of 3%–4%, and *Weak*—the economy is growing at a rate of 0%–2%.

The information available indicates that SS has an estimated rate of return of 18% if the economy is strong, a rate of return of 10% if the economy is stable, and a rate of return of −5% if the economy is weak. The CS has an estimated rate of return of 13% if the economy is strong, a rate of return of 8% if the economy is stable, and a rate of return of 1% if the economy is weak. Bonds have an estimated return of 4% if the economy is strong, a rate of 5% if the economy is stable, and a rate of return of 6% if the economy is weak. CD have an estimated rate of return of 7% if the economy is strong, a rate of return of 3% if the economy is stable, and a rate of return of 2% if the economy is weak. Lastly, there is also the alternative of DN, that is of not investing the money, and that would yield 0% return regardless of the state of the economy.

Note that for each combination of decision alternative and state of nature there is a corresponding payoff. The payoff in this example is the expected rate of return. In general, the payoff is some quantitative value used to measure a possible outcome. This measure is generally given in terms of a monetary value, but any measure can be used. Likewise, the value used is often a statistical estimate of the possible payoff. As the payoffs result from a combination of alternatives and states of nature, they are easily represented by a *payoff matrix* (Table 6.1). The payoff matrix for this example is given in Table 6.1.

6.3 Decision Making under Risk

Every business situation, as well as most life situations, involves a level of uncertainty. The modeling of the uncertainty yields different approaches to the decision problem. One way to deal with the uncertainty is to make the uncertain more certain. This can be done by using probability to represent the uncertainty. That is, uncontrollable factors are modeled

TABLE 6.1 Payoff Matrix for Investment Problem

Alternative	State of Nature		
	Strong (%)	Stable (%)	Weak (%)
Speculative Stock (SS)	18	10	−5
Conservative Stock (CS)	13	8	1
Bonds (B)	4	5	6
Certificates of Deposit (CD)	7	3	2
Do Nothing (DN)	0	0	0

and estimated probabilistically. When this is possible, the uncertainty is characterized by a probability distribution.

There is a wide spectrum of uncertainty to consider when analyzing a problem. At one end of the spectrum is complete certainty of the data. At the other end is complete uncertainty. What lies between is varying degrees of uncertainty in which the payoff for each alternative and state of nature can be described by some probability distribution. This is what is termed as decision making under risk in this chapter.

It should be noted that there are inherent difficulties with any approach related to uncertainty. Uncertainty results from a lack of perfect information and a decision maker will often need to determine whether more information is needed before a decision can be made. This occurs at a cost, and may not yield a better decision. Additionally, the probability models themselves may not truly reflect the situation or may be difficult to obtain. Therefore, the decision maker must always keep in mind that the use of the probability models is to help the decision maker avoid adverse decisions and to help better understand the risk involved in any decision made. The probability models are decision support tools, not exact methods for giving a solution. Human input is always needed.

One way to estimate probabilities is to use prior probabilities for given events. The prior probabilities come from existing information about the possible states of nature that can be translated into a probability distribution if the states of nature are assumed to be random. These prior probabilities are often subjective and dependent on an individual's experience and need to be carefully determined. For more specific information on determining prior probabilities see Hillier and Lieberman [5, chapter 15] and Merkhofer [6].

In the given investment example, assume that a probability distribution for the possible states of nature can be determined based on the past data related to economic growth. These prior probabilities are given in Table 6.2.

The advantage of prior probabilities is that they can be used to determine expected values for different criteria. Expected values often give an acceptable estimate as to what is most likely to happen and therefore give a good basis on which to help make a decision. Several approaches based on expected value have been developed and are outlined below. Note, however, that expected value is not always the best indicator and the decision maker's preferences and knowledge always need to be included in the process.

TABLE 6.2 Investment Problem with Prior Probabilities

Alternative	State of Nature		
	Strong	Stable	Weak
SS	18%	10%	−5%
CS	13%	8%	1%
B	4%	5%	6%
CD	7%	3%	2%
DN	0%	0%	0%
Prior Probability of State of Nature	0.1	0.6	0.3

Decision Analysis

TABLE 6.3 Expected Values of Each Alternative

Alternative	State of Nature			
	Strong	Stable	Weak	Expected Value
SS	18%	10%	−5%	6.3%
CS	13%	8%	1%	6.4%
B	4%	5%	6%	5.2%
CD	7%	3%	2%	3.1%
DN	0%	0%	0%	0%
Prior Probability	0.1	0.6	0.3	

Another concept that can be included here is the concept of dominance. Dominance implies that an alternative will never be chosen because there is another alternative that is always better regardless of the state of nature. Hence, there is no reason to ever consider the alternative that is dominated; it cannot give a reasonable course of action. In the investment example the do nothing (DN) alternative is a dominated alternative. That is, there is at least one other alternative that would always be chosen regardless of the state of nature over do nothing. From Table 6.1 it can be seen that the alternatives, CS, B, and CD each dominate DN. It is then reasonable to exclude the alternative DN from further consideration. However, to illustrate this concept and to show that this alternative is never chosen, it is left in the problem for now.

6.3.1 Maximum Likelihood

The idea behind maximum likelihood is that good things always happen. If the decision maker is very optimistic about the future then why not choose the best possible outcome assuming the best possible state of nature will occur.

To find the best choice by maximum likelihood, first find the state of nature with the largest probability of occurring and then choose the alternative for that state of nature with the maximum payoff.

In the investment example given by Table 6.2, the state of nature with the largest probability is stable growth. For that state of nature, the SS has the largest rate of return. Therefore, based on this criterion the decision would be to invest in SS.

6.3.2 Expected Value under Uncertainty

For a more balanced approach, a decision maker can assume that the prior probabilities give an accurate representation of the chance of occurrence. Therefore, instead of being overly optimistic the decision maker can compute the expected value for each alternative over the states of nature and then choose based on those expected values.

To implement this approach, for each alternative determine its expected value based on the probability of the state of nature and the payoff for that alternative and state of nature. That is, for each row of the payoff matrix take the sum of each payoff times the corresponding state of nature probability.

From Table 6.2, for the alternative SS, the expected value would be $(0.1 \times 18\%) + (0.6 \times 10\%) + (0.3 \times (-5\%)) = 6.3\%$. The expected value for each alternative of the investment example is given in Table 6.3.

Based on expected value the best decision would be to invest in CS.

Note that this type of analysis can easily be done in a spreadsheet, which affords the opportunity to do a what-if analysis. As the prior probabilities are the most questionable of the data, in a spreadsheet, it would be possible to adjust these values to see the impact of different values for the prior probabilities. Likewise, it is also possible to do a sensitivity analysis of the prior probabilities and determine ranges for the prior probabilities for which each alternative would be chosen. There exists software designed to help with this type of analysis. One such package is called *SensIt* and is described in Ref. [5]. The Web address for this software is given later.

6.3.3 Expected Opportunity Loss or Expected Regret

There are times when the actual payoffs and their expected values are not sufficient for a decision maker. Many times when dealing with uncertainty, decision makers feel better about a decision if they know they have not made an error that has cost them a great deal in terms of opportunity that has been lost. This lost opportunity is generally called regret and can be used to determine an appropriate decision.

After a decision is made and the actual state of nature becomes known, the opportunity missed by the decision made can be determined. That is, for the investment example, if the actual state of nature that occurred was stable growth, but it was decided that B were the best investment, then the opportunity of going with SS was missed. Therefore, the missed opportunity of having a 10% rate of return instead of the 5% chosen, results in a missed opportunity (regret) of $10 - 5 = 5\%$.

To avoid the possibility of having missed a large opportunity, expected opportunity loss (EOL) looks to minimize the opportunity loss or regret. To do this, the possible opportunity loss for each state of nature is determined for each alternative. This is done by taking the largest payoff value in each column (state of nature) and then for each alternative subtracting the payoff for that alternative from the largest payoff in the column. For the investment example, the opportunity loss is computed for each alternative and state of nature in Table 6.4. Once the opportunity losses are known, the EOL is determined for each alternative (row) by using the prior probabilities. For example, in Table 6.4 the EOL for alternative CS is given by $(0.1 \times 5\%) + (0.6 \times 2\%) + (0.3 \times 5\%) = 3.2\%$.

The criterion then is to minimize the EOL. In this example, the best decision based on this criterion would be to select CS.

6.3.4 Expected Value of Perfect Information

All the methods presented thus far were dependent on the given prior information. The question then arises as to whether it would be advantageous to acquire additional information to help in the decision making process. As information is never completely reliable, it is not known if the additional information would be beneficial. Therefore, the question becomes, what is the value of additional information? The expected value of perfect information (EVPI) criterion gives a way to answer this question by measuring the improvement of a decision based on new information.

The idea behind EVPI is that if the state of nature that will occur is known with certainty, then the best alternative can be determined with certainty as well. This would give the best value for the decision, which can then be compared to the value expected under current

TABLE 6.4 Expected Opportunity Loss (EOL) of Each Alternative

Alternative	State of Nature			EOL
	Strong	Stable	Weak	
SS	$18 - 18 = 0\%$	$10 - 10 = 0\%$	$6 - (-5) = 11\%$	3.3%
CS	$18 - 13 = 5\%$	$10 - 8 = 2\%$	$6 - 1 = 5\%$	3.2%
B	$18 - 4 = 14\%$	$10 - 5 = 5\%$	$6 - 6 = 0\%$	4.4%
CD	$18 - 7 = 11\%$	$10 - 3 = 7\%$	$6 - 2 = 4\%$	6.5%
DN	$18 - 0 = 18\%$	$10 - 0 = 10\%$	$6 - 0 = 6\%$	9.6%
Prior Probability	0.1	0.6	0.3	

EVPI and Expected Value under Uncertainty

This approach gives a straightforward approach for computing EVPI. The value of EPVI is just simply the expected value under certainty minus the expected value under uncertainty. To compute the expected value under certainty simply take the best payoff under each state of nature and multiply it by its prior probability and sum these.

For the investment example the expected value under certainty would be $(0.1 \times 18\%) + (0.6 \times 10\%) + (0.3 \times 6\%) = 9.6\%$. The expected value under uncertainty is the best alternative value, which in this example comes from alternative CS that has an expected value of 6.4% from Table 6.3. With this information EVPI is computed as $\text{EVPI} = 9.6\% - 6.4\% = 3.2\%$.

Therefore, if an investment of $10,000 is being planned, the most additional information would be worth is $\$10,000 \times 3.2\% = \$320$.

EVPI and EOL

If the state of nature is known with certainty then there would be no opportunity loss. That is, under certainty the best alternative would always be chosen so there would be no regret or opportunity loss, it would always be zero. Hence, under uncertainty EOL gives the cost of uncertainty that could be eliminated with perfect information and therefore, $\text{EPVI} = \text{EOL}$. From Table 6.4 of EOL values, the best alternative is the one with the smallest EOL that is given by the alternative CS with an EOL of 3.2%. This is also the value for EPVI found above. Note that for each alternative its expected value plus its EOL is 9.6%, the expected value under certainty found above.

6.3.5 Decision Trees

To help better understand the decision process it can be represented graphically by a combination of lines and nodes called a decision tree. The purpose of the tree is to pictorially depict the sequence of possible actions and outcomes [1,3–5,7]. There are two types of nodes used in a decision tree. A square represents a decision point or fork, which is the action (alternative) taken by the decision maker and a circle represents an event or chance fork, which is the state of nature. The branches (lines) in the tree represent the decision path related to alternatives and states of nature.

Decision trees are generally most helpful when a sequence of decisions must be made, but they can also be used to illustrate a single decision. The decision tree in Figure 6.1 is for the investment example. Note that each square node denotes the alternative chosen and each circular node represents the state of nature and the numbers at the end of each decision path (lines) are the payoffs for that course of action.

Figure 6.1 gives a pictorial view of the information contained in Table 6.2, the payoff matrix. Notice that the same computations for expected value can be performed within the tree. Starting at the right hand side of the tree, for each circle (level) the expected value of that circle is computed by taking the probability on each branch times the payoff associated with each branch. For example, the top circle would have an expected value of $(0.1 \times 18\%) + (0.6 \times 10\%) + (0.3 \times (-5\%)) = 6.3\%$. The expected value can then be added to the tree as a value at each circle. Once the values are known at each circle, the value for the square is given by taking the maximum value of the circles. The decision tree with

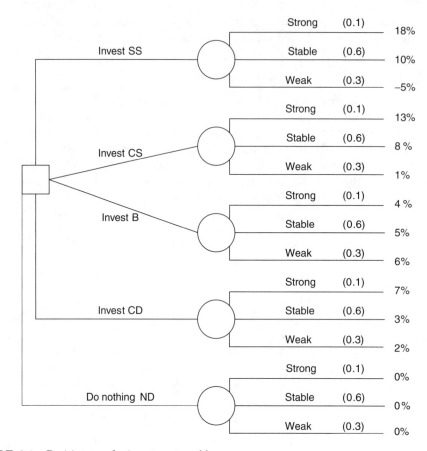

FIGURE 6.1 Decision tree for investment problem.

this additional expected value information is given in Figure 6.2 and is the same as the information given in Table 6.3.

This, however, represents a single decision. Decision trees are much more useful when dealing with a sequence of alternatives and states of nature in which a series of decisions must be made. These types of problems cannot be represented easily in matrix form and thus decision trees become a powerful tool. Decision trees are useful for a wide variety of scenarios and the interested reader is referred to Refs. [3–5,7] for additional examples.

6.3.6 Posterior Probabilities

As mentioned before, there is the possibility of gaining new knowledge to help in better defining the probability of an occurrence of a state of nature. This can be done through additional experimentation or sampling. The improved estimates of the probabilities are called posterior probabilities or Bayes' probabilities due to the use of Bayes' theorem in their computation.

To find posterior probabilities, additional information about the states of nature must be acquired. This can be done by experimentation, which in this sense can refer to the use of experts, analysts, or consultants as well as actual additional experimentation. Therefore, for any of these forms, the additional information is viewed as being obtained by an experiment.

The experimentation will have a possible set of outcomes that will help determine which alternative to select. Based on the outcomes of the experimentation and their probability

Decision Analysis

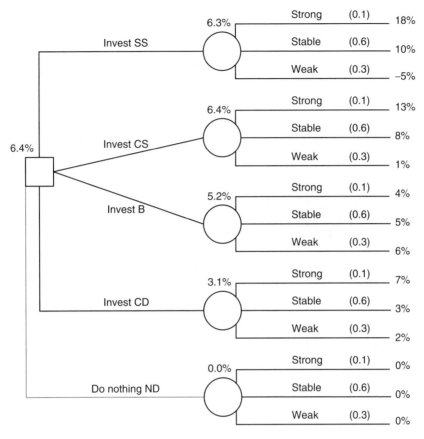

FIGURE 6.2 Decision tree for investment problem with expected values.

of occurrence, Bayes' theorem yields the posterior probability of a state of nature. The posterior probability of the state of nature given the test outcome equals the prior probability of the state of nature times the conditional probability of the outcome given the state of nature divided by the sum of the conditional probabilities for all the states of nature and outcomes. To state this mathematically, let S_i be the state of nature i and O_j be outcome j of the additional experimentation. Also let $P(S_i)$ be the prior probability of the state of nature i. Then the posterior probability of the state of nature i given test outcome j is $P(S_i|O_j)$. This is defined as:

$$P(S_i|O_j) = \frac{P(S_i)P(O_j|S_i)}{\sum_k P(S_k)P(O_j|S_k)}$$

To help illustrate how to compute these values, assume for our investment problem that financial analysts who spent 20 years working on Wall Street have been hired to give their opinion on what investment to make. They state their opinion in terms of the possible state of nature as a probability and their willingness to invest. This opinion is viewed as a test outcome. That is, they might state that based on the past 20 years of investing they have done, when the economic growth is strong (greater than 5%) they have invested 70% of the time, 10% of the time they have been neutral on investing, and 20% of the time they have not invested. This could be written as $P(Invest \mid Strong) = 0.7$, $P(Neutral \mid Strong) = 0.1$ and $P(Do\ Not\ Invest \mid Strong) = 0.2$. The financial analysts would also give

TABLE 6.5 Conditional Probabilities for Financial Analysts

Investment Opinion	Given State of Nature		
	Strong	Stable	Weak
Invest	0.7	0.6	0.3
Neutral	0.1	0.3	0.3
Don't Invest	0.2	0.1	0.4

similar probabilities for each possible state of nature in terms of investing. Table 6.5 gives the conditional probabilities given by the analysts for the investment problem.

With these conditional probabilities and with the prior probabilities from Table 6.2, the posterior probabilities are determined as follows:

$$P(Strong\,|\,Invest) = (0.1 \times 0.7)/((0.1 \times 0.7) + (0.6 \times 0.6) + (0.3 \times 0.3)) = 0.135$$
$$P(Strong\,|\,Neutral) = (0.1 \times 0.1)/((0.1 \times 0.1) + (0.6 \times 0.3) + (0.3 \times 0.3)) = 0.036$$
$$P(Strong\,|\,Don't\,Invest) = (0.1 \times 0.2)/((0.1 \times 0.2) + (0.6 \times 0.1) + (0.3 \times 0.4)) = 0.1$$
$$P(Stable\,|\,Invest) = (0.6 \times 0.6)/((0.1 \times 0.7) + (0.6 \times 0.6) + (0.3 \times 0.3)) = 0.692$$
$$P(Stable\,|\,Neutral) = (0.6 \times 0.3)/((0.1 \times 0.1) + (0.6 \times 0.3) + (0.3 \times 0.3)) = 0.643$$
$$P(Stable\,|\,Don't\,Invest) = (0.6 \times 0.1)/((0.1 \times 0.2) + (0.6 \times 0.1) + (0.3 \times 0.4)) = 0.3$$
$$P(Weak\,|\,Invest) = (0.3 \times 0.3)/((0.1 \times 0.7) + (0.6 \times 0.6) + (0.3 \times 0.3)) = 0.173$$
$$P(Weak\,|\,Neutral) = (0.3 \times 0.3)/((0.1 \times 0.1) + (0.6 \times 0.3) + (0.3 \times 0.3)) = 0.321$$
$$P(Weak\,|\,Don't\,Invest) = (0.3 \times 0.4)/((0.1 \times 0.2) + (0.6 \times 0.1) + (0.3 \times 0.4)) = 0.6$$

Additionally, the probability of a given opinion from the analyst is given by the denominator of the posterior probability related to that opinion. Hence, $P(Invest) = (0.1 \times 0.7) + (0.6 \times 0.6) + (0.3 \times 0.3) = 0.52$, $P(Neutral) = (0.1 \times 0.1) + (0.6 \times 0.3) + (0.3 \times 0.3) = 0.28$, $P(Don't\,Invest) = (0.1 \times 0.2) + (0.6 \times 0.1) + (0.3 \times 0.4) = 0.2$.

Note that once the prior and conditional probabilities are known, the process of finding the posterior probabilities can easily be done within a spreadsheet [5].

With the posterior probabilities, it is now possible to analyze the decision in terms of this new information. This is done through the use of the decision tree by adding a chance node at the beginning of the tree (left side) representing the opinion of the analyst and then replacing the prior probabilities on each branch with the corresponding posterior probabilities. Likewise, for each opinion the probability of that opinion can also be added to the tree for those branches. The tree for the investment problem with posterior probabilities is given in Figure 6.3. Note, however, that to simplify the tree the option of DN is not included as it is a dominated alternative and would never be chosen.

For this tree, the analysis is done as before to determine expected values except now the posterior probabilities are used and at each decision node the decision is based on the best expected value for that node.

To do the analysis start at the right side of the tree and for each circle determine its expected value. For example, for the neutral branch the expected value for the circle corresponding to invest in SS is $(0.036 \times 18\%) + (0.643 \times 10\%) + (0.321 \times (-5\%)) = 5.47\%$. The expected value for the circle corresponding to invest in CS is $(0.036 \times 13\%) + (0.643 \times 8\%) + (0.321 \times 1\%) = 5.93\%$. The expected value for the circle corresponding to invest in bonds is $(0.036 \times 4\%) + (0.643 \times 5\%) + (0.321 \times 6\%) = 5.28\%$. Lastly, the expected value for the circle corresponding to invest in CD is $(0.036 \times 7\%) + (0.643 \times 3\%) + (0.321 \times 2\%) = 2.82\%$. Therefore, the decision at the square on the neutral branch is given by the maximum

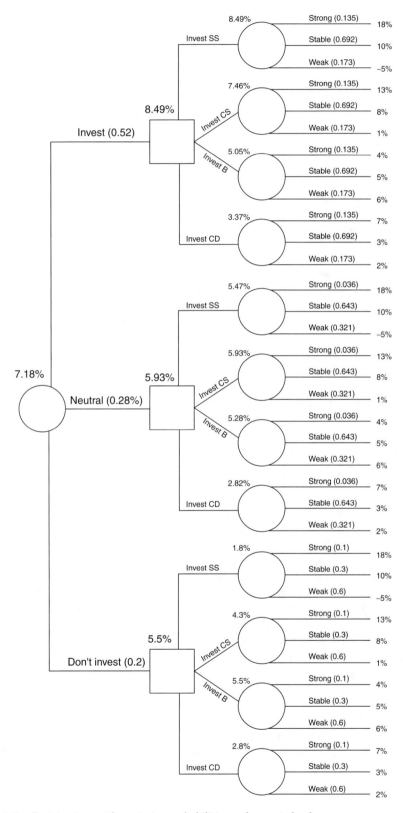

FIGURE 6.3 Decision tree with posterior probabilities and expected values.

expected value of the four circles associated with that decision square. The maximum value is 5.93%, which corresponds to invest in CS.

From the decision tree the expected payoff at each decision (square) is given which yields the decision of which stock to invest in. That is, if the analyst says to invest, then the best investment choice would be in SS. If the analyst is neutral then the best investment would be in CS and if the analyst says not to invest then the best investment would be in bonds.

Lastly, from the determined probabilities of each opinion the expected payoff for the entire tree can be computed as $(0.52 \times 8.49\%) + (0.28 \times 5.93\%) + (0.2 \times 5.5\%) = 7.18\%$. This value yields what one would expect the return on the investment to be if they employed the services of the analyst. Note that this is higher than the expected return of 6.4% from Figure 6.2, when an analyst is not used.

6.3.7 Utility Functions

In the preceding discussion and example, it was assumed that the payoff was monetary even though rates of return were used. For the investment example the return on investment could easily be converted to monetary terms. The investment SS under strong (greater than 5%) growth had a return of 18%. As $10,000 was being invested, the value $10,000 $\times$ 18% = $1800 could have been used instead of the rate of return. Monetary values could have been used in place of each rate of return and then the expected payoffs would have been in dollars as well. Either approach is correct and the choice is dependent on the preference of the decision maker.

Monetary value, however, is not always the best way to model a decision as the value of money to a decision maker can fluctuate depending on the circumstances. For instance, let us say you were going to play the lottery. The big payoff this week is $42 million. It costs you just $1 to purchase a ticket. If you have $100 in your wallet you might think spending $1 for a chance at $42 million is a good investment. Now, if you only have $5 in your wallet and are hungry you might think playing the lottery is not a good idea as $5 will by you a meal but $4 will not. That $1 for the ticket is viewed differently now than it was when you had $100. This varying value of money is called the *utility* of the money. Utility is given in terms of a *utility function* as the utility changes based on circumstances and time.

Another way of looking at this concept is to consider the following choice or lottery [5]: (1) Flip a coin and if it comes up heads you win $10,000 and if it comes up tails you win nothing or (2) accept $3,000. Which would you choose? Many people would choose the $3000. This concept is used in many popular game shows. The question though is why would one choose the $3000 when the expected payoff for the coin flip, $(0.5 \times \$10,000) + (0.5 \times \$0) = \$5000$, is more? The answer is that choices are not always made solely on expected monetary gain, but also on the potential for loss and other circumstances. An individual's view of money may also change over time and with the different situations they face. That is the utility of money.

When discussing utility, decision makers are usually classified into one of three classes: risk-averse, risk-neutral, and risk-seeking. The risk-averse person has a decreasing marginal utility for money. This means that the less money they have the more it is worth to them (the more utility it has). That is, assume for $1 its utility is 10 and for $2 its utility is 15. The unit increase in the amount of money did not have an equivalent increase in the utility. The utility of the additional dollar was only 5 instead of 10. As the number of dollars increases, each additional dollar will have less utility for the decision maker. This can be easily seen by graphing the utility of money versus the amount of money. This graph is called the utility function. Let $u(M)$ be the function representing the utility of M. For a

Decision Analysis

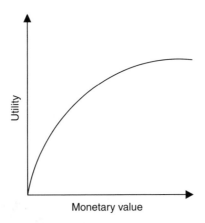

FIGURE 6.4 Risk-averse utility function.

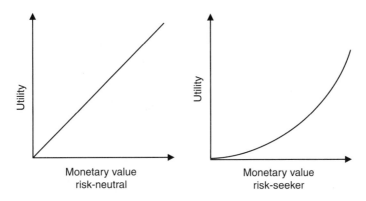

FIGURE 6.5 Risk-neutral and risk-seeker utility functions.

risk-averse person this function will always have a concave shape, representing decreasing marginal utility, as shown in Figure 6.4.

In contrast, the risk-neutral person views money the same regardless of the amount or circumstances. That is, if $1 has a utility of 10 then $2 will have a utility of 20; each dollar added will have the same utility. This would result in a linear function for utility. When the utility of money is not considered in a problem it is the same as if the decision maker is assumed to be risk neutral. The risk-seeking person will have an increasing marginal utility for money. That is, the more they have the more it is worth. The utility of each additional dollar increases. If $1 has a utility of 10 then $2 could have a utility of 25. The additional $1 had a higher utility (15) than the previous $1 (10). These decision makers are willing to make risky decisions for the opportunity of more money. A risk-seeker would gladly give up the $3000 for a 50–50 chance at $10,000. The utility function for a risk-seeker would be a convex function, representing increasing marginal utility. Figure 6.5 gives utility functions for both the risk-neutral and risk-seeking individual.

Most people do not always fit into one of the above three classifications. They tend to be a mixture of the three depending on the circumstances and will generally change over time. However, for a given point in time, it is assumed that a utility function for the decision maker can be determined. Note that utility functions are unique to an individual decision maker. Therefore, one might have two different decision makers looking at the same problem and each will make a different decision based on their utility.

Once a utility function for a decision maker has been determined, then the analysis of the problem proceeds as it does when using monetary value. The only difference is that the utility function value of the payoff amount is used in place of the monetary value and the expected utility is found instead of the expected value. Decisions are then based on the expected utility values.

How to determine or construct the utility function for a decision maker is then the fundamental aspect of applying utility functions. The keys to estimating a utility function for an individual are the following two fundamental properties [5,7].

PROPERTY 6.1 Under the assumptions of utility theory, the decision maker's utility function for money has the property that the decision maker is indifferent between two alternative courses of action if the two alternatives have the same expected utility.

PROPERTY 6.2 The utility of the worst outcome is always zero and the utility of the best outcome is always one.

From these two properties it can be seen that the determination of an individual's utility function is based on the comparison of equivalent lotteries.

A lottery is just simply a choice between outcomes with a given probability. Lotteries are generally denoted by trees where the branches of the tree represent outcomes. For example, a lottery in which there is a 50–50 chance of receiving $5000 and losing $1000 would be given by:

$$\begin{array}{c} \$5000\ (0.5) \\ -\$1000\ (0.5) \end{array}$$

A lottery that is certain has a probability of 1 and is represented by a single line.

$$\$3000\ (1)$$

To illustrate how to find a utility function, consider the following adaptation of the investment problem. You are to determine how to invest $5000. There are two stocks from which to choose. Stock HR is a high risk stock and will yield a return of $1000 if there is strong economic growth. Stock LR is a low risk stock and will yield a return of $300 if there is strong economic growth. However, if the economic growth is weak then the HR stock will lose $500 and the LR stock will only yield $100. The payoff matrix for the two stocks HR and LR and the states of nature, strong growth and weak growth, are given in Table 6.6. The prior probabilities are a forecast of the possible state of the economy based on past economic data.

TABLE 6.6 Payoff Matrix for HR and LR Stocks

	States of Nature	
Alternative	Strong	Weak
HR	1,000	−500
LR	300	100
Prior Probability	0.4	0.6

Decision Analysis

For this decision problem the best possible outcome is to make $1000 and the worst possible outcome is to lose $500. Let $u(x)$ be the utility function. Then from properties 1 and 2, the utility values are $u(1000) = 1$ and $u(-500) = 0$.

To construct the utility function, additional values of utility must be determined. This is done by determining a lottery with a known utility value and then finding an equivalent lottery. The equivalent lottery then has the same utility value.

In this example, $1000 and −$500 have known utility values. Therefore, the first step in constructing the rest of the utility function is to find the value x for which the decision maker is indifferent between the two lotteries below.

$$\underline{x\ (1.0)} \quad \text{and} \quad \begin{array}{|c|} \hline \$1000\ (0.5) \\ \hline -\$500\ (0.5) \\ \hline \end{array}$$

Note that the lottery on the right has an expected utility of $1 \times (0.5) + 0 \times (0.5) = 0.5$. As a result, the two lotteries are indifferent (equivalent) and the utility of x must be $u(x) = 0.5$.

Assume the decision maker states that the value x that makes these lotteries indifferent is $x = -\$200$. Hence, $u(-200) = 0.5$. With this value we can now compute the expected utility of a 50–50 lottery of $u(-200)$ and $u(-500)$ which is $0.5 \times (0.5) + 0 \times (0.5) = 0.25$. Therefore, any lottery indifferent to this 50–50 lottery will have a utility of 0.25. That is, the value x that makes the following lotteries indifferent has utility 0.25.

$$\underline{x\ (1.0)} \quad \text{and} \quad \begin{array}{|c|} \hline -\$200\ (.05) \\ \hline -\$500\ (0.5) \\ \hline \end{array}$$

Let $x = -400$ be the value that makes the lotteries indifferent. Then $u(-400) = 0.25$. This gives four points of the utility function. To approximate the utility function, several more points can be determined by repeating this process. This is given below.

Find x such that the following are indifferent.

The lottery on the right has expected utility $1.0 \times (0.5) + 0.5 \times (0.5) = 0.75$. Let $x = 300$ be the value that makes the lotteries indifferent. Then $u(300) = 0.75$.

Next, find x such that the following are indifferent.

The lottery on the right has expected utility $1.0 \times (0.5) + 0.75 \times (0.5) = 0.875$. Let $x = 600$ be the value that makes the lotteries indifferent. Then $u(600) = 0.875$.

Lastly, find x such that the following are indifferent.

The lottery on the right has expected utility $0.75 \times (0.5) + 0.5 \times (0.5) = 0.625$. Let $x = 100$ be the value that makes the lotteries indifferent. Then $u(100) = 0.625$.

From these values an approximate utility function can be plotted based on the points $(x, u(x))$. For the example, the points are $(-500, 0)$, $(-400, 0.25)$, $(-200, 0.5)$, $(100, 0.625)$, $(300, 0.75)$, $(600, 0.875)$, $(1000, 1)$. The utility function's approximate graph is given in Figure 6.6.

Figure 6.7 gives the decision tree for this problem based on the information in Table 6.6 and the utility values determined above. The values at each circle are the expected utilities and the value at the decision square is the expected utility of the tree, which indicates the best choice would be to invest in LR stock.

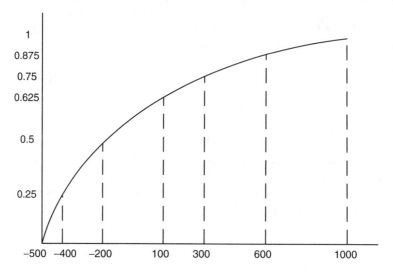

FIGURE 6.6 Approximate utility function.

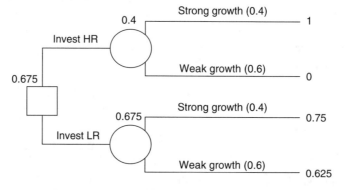

FIGURE 6.7 Decision tree for HR and LR stocks using expected utility.

TABLE 6.7 Expected Payoff and Utility for HR and LR Stock

	\$	u(x)	\$	u(x)	Monetary	Utility
	States of Nature					
	Strong		Weak		Expected Value	
Alternative	\$	u(x)	\$	u(x)	Monetary	Utility
HR	1000	1	−500	0	100	0.4
LR	300	0.75	100	0.625	180	0.675
Prior Probability		0.4		0.6		

Table 6.7 has the computations for both the expected utility and the expected monetary payoff for comparison. Note that for this problem they give the same decision; however, this is not always the case. It is quite possible to get different results when using utility values as opposed to monetary values. For example, if in this investment problem the utility values remained the same but the HR stock only lost \$300 instead of \$500 when the economy was weak then HR's expected monetary value would have been $(0.4 \times 1000) + (0.6 \times (-300)) = 220$, which is greater than the expected monetary value of 180 for the LR stock. For this scenario, stock HR would then be the decision made based on expected monetary value which would be different than the decision of stock LR based on expected utility values.

Utility theory gives a way to more realistically model a problem based on the decision maker's preferences and attitudes toward risk. The determination of utility functions may be difficult, but there are existing utility functions that one can potentially use to help in this process [5,7,8].

There can be drawbacks, however, to using utility functions. One is determining the utility function and making sure the decision maker's preferences satisfy the Von Neumann–Morgenstern axioms of utility theory [7,9]. There is also the possibility of prospecting and framing [7,10]. In both these cases, decision makers will make a decision that goes against what the expected utility would suggest. Prospect theory says that a decision maker will not treat the probabilities given for a problem as valid. They will "distort" the probabilities. This generally occurs at the boundaries as most people are more sensitive to changes in probabilities close to 0 or close to 1. Framing refers to how choices are presented to a decision maker. The framing of the question and its context can result in different outcomes. For a more detailed discussion of the utility theory see Refs. [7–13].

In this section, the determination of the "best" decision is dependent on expected values, which implies that a probability distribution can be determined for the states of nature. If this is possible, then decision trees can also be employed to help illustrate the decision process and to organize the data. There has been a variety of software developed to help in this process. Some of the software is very sophisticated and requires a great deal of background knowledge to use properly, and some of the software is very simple to use. For a thorough analysis of existing software see Maxwell [14,15].

6.4 Decision Making under Uncertainty

When it is not possible to find a probability distribution for a decision problem, or none exists, that problem is said to be uncertain. In this situation, it is not possible to determine expected values and one must find other factors on which to make a decision. The payoff matrix is still employed for problems under uncertainty, but without any prior probabilities.

There are several methods to deal with this kind of uncertainty. These methods differ mainly in how the decision maker views risk and the state of nature. The different methods could be viewed as being similar to aspects of utility theory in that they try to take into

TABLE 6.8 Payoff Matrix for Snack Production

	State of Nature			
Alternative	Decrease Slightly	Steady	Marginal Increase	Significant Increase
A1	−1500	−400	1100	2150
A2	−450	200	500	500
A3	−850	−75	450	1100
A4	−200	300	300	300
A5	−150	−250	−450	−850

account the decision maker's preferences and attitudes. These methods are discussed in the context of the following problem.

6.4.1 Snack Production Example

A local food processing plant has developed a low calorie snack that has taken the market by storm. After only being on the market for 3 months sales have outpaced forecasts by more than 60% and the plant cannot meet demand. It is running 24/7 and is lagging behind demand, at the moment, by 25%. To help alleviate this problem, management must decide how to increase production. Marketing was asked to develop new demand forecasts but has not been able to determine whether or not sales will continue to increase, hold steady, or decrease. After several different surveys it was found that sales could hold steady, increase marginally (somewhere between 10% and 30%), increase significantly (somewhere between 30% and 70%) or decrease slightly (somewhere between 1% and 10%), but marketing cannot give a distribution for the possible outcomes. After researching possible solutions the following alternatives have been put forth by the engineering team: (A1) Build an additional plant that would be able to increase the total production of the product by 120%; (A2) Add an identical additional line to the current plant that would increase production by 30%; (A3) Expand the current plant and replace current line with new technologies that would allow total production to increase by 60%; (A4) Hire a full time operations analyst to increase efficiency with an estimated increase in production of 15%; and (A5) Do Nothing and maintain the status quo. An economic analysis of each alternative has been carried out and the net profit gained from possible increased sales minus implementation costs is used as the payoff for the alternative. For the DN alternative this would simply be lost sales. Table 6.8 gives the payoff matrix, and the values are in $1000.

6.4.2 Maximin (Minimax) Criterion

This criterion is a conservative approach to managing the unknown risk. The intent is to determine the worst that can happen with each alternative and then pick the alternative that gives the best worst result. When payoffs are profit then the maximin (maximize the minimum value possible) criterion is used and when the payoff is cost then the minimax (minimize the maximum value possible) criterion is used.

Implementation of this criterion is quite simple. For maximin, identify the minimum in each row and then select the alternative with the largest row minimum. This is given in Table 6.9. The best choice under maximin would be alternative A4.

Under this approach, the decision maker is guaranteed they will never lose more than $200,000 but at the same time they will never make more than $300,000, whereas a different alternative such as A2 would guarantee they would never lose more than $450,000 but could make $500,000 but never any more. By choosing A4, they relinquish the opportunity to make more money if the state of nature is anything other than a decrease in demand.

Decision Analysis

TABLE 6.9 Maximin for Snack Production

	State of Nature				
Alternative	Decrease Slightly	Steady	Marginal Increase	Significant Increase	Row Minimum
A1	−1500	−400	1100	2150	−1500
A2	−450	200	500	500	−450
A3	−850	−75	450	1,100	−850
A4	−200	300	300	300	−200*
A5	−150	−250	−450	−850	−850

*represents best alternative

TABLE 6.10 Maximax for Snack Production

	State of Nature				
Alternative	Decrease Slightly	Steady	Marginal Increase	Significant Increase	Row Maximum
A1	−1500	−400	1100	2150	2150*
A2	−450	200	500	500	500
A3	−850	−75	450	1100	1100
A4	−200	300	300	300	300
A5	−150	−250	−450	−850	−150

*represents best alternative

6.4.3 Maximax (Minimin) Criterion

This criterion is the opposite of maximin in that it is very optimistic and risk-seeking. For this approach it is assumed that the best scenario possible will happen. Therefore, for each row the maximum is chosen and then the alternative with the maximum row maximum is selected. Table 6.10 illustrates this.

The alternative selected by this approach is A1, which yields the largest possible pay-off. However, just as with the maximin approach this leaves the decision maker open to significant losses. For A1 two states of nature yield a loss, one of which is quite significant.

6.4.4 Hurwicz Criterion

The two previous criteria were at the extremes, one very pessimistic and the other very optimistic. This criterion is designed to mitigate the extremes and to allow for a range of attitudes of the decision maker. The basis of this approach is an index of optimism given by α, such that $0 \leq \alpha \leq 1$. The more certain a decision maker is that the better states of nature will occur, the larger the value of α, the less certain the smaller the value of α. For a given value of α, the criterion is implemented by taking for each row (alternative), $((\alpha \times \text{row max}) - (1 - \alpha) \times |\text{row min}|)$, and then selecting the alternative with the largest value. For the example, let $\alpha = 0.48$, which indicates not a strong regard for optimism or pessimism, but with a hint of pessimism. For the alternatives, the computations are

$$\text{For A1} \quad 0.48 \times (2150) - 0.52 \times (1500) = 44$$
$$\text{For A2} \quad 0.48 \times (500) - 0.52 \times (450) = 6$$
$$\text{For A3} \quad 0.48 \times (1100) - 0.52 \times (850) = 85$$
$$\text{For A4} \quad 0.48 \times (300) - 0.52 \times (200) = 40$$
$$\text{For A5} \quad 0.48 \times (-150) - 0.52 \times (850) = -514$$

The best alternative under this criterion is A3.

TABLE 6.11 Laplace for Snack Production

Alternative	State of Nature				Expected Value
	Decrease Slightly	Steady	Marginal Increase	Significant Increase	
A1	−1500	−400	1100	2150	338*
A2	−450	200	500	500	188
A3	−850	−75	450	1100	156
A4	−200	300	300	300	175
A5	−150	−250	−450	−850	−425

* represents best alternative

TABLE 6.12 Minimax Regret for Snack Production

Alternative	State of Nature				Row Maximum
	Decrease Slightly	Steady	Marginal Increase	Significant Increase	
A1	1350	700	0	0	1350
A2	300	100	600	1650	1650
A3	700	375	650	1050	1050*
A4	50	0	800	1850	1850
A5	0	550	1550	3000	3000

* represents best alternative

6.4.5 Laplace Criterion or Expected Value

The Laplace criterion is another more optimistic approach to the problem. The basis of this approach is that since the probabilities are not known for the states of nature and there is no reason to think otherwise, each state of nature should be viewed as equally likely. This gives a probability distribution for the states of nature, which then allows the expected value to be determined. The alternative with the best expected value is selected.

For the snack production example, as there are four states of nature, the probability for each state would be 0.25. The expected value for each alternative is determined by taking the sum of the payoff for each state of nature times 0.25. Table 6.11 gives the Laplace (expectation) values and A1 is the best choice under this criterion.

6.4.6 Minimax Regret (Savage Regret)

The last approach is based on the concept of regret discussed previously. The idea is to not look at payoffs, but instead at lost opportunities (regret). The regret is based on each state of nature and is determined by looking at the best outcome of that state against the other possible outcomes. Therefore, regret is computed for each column of the payoff matrix by taking the maximum value in that column and replacing each payoff value in the column with the maximum value minus the payoff value. The regret matrix for the snack production problem is given in Table 6.12. Once the regret is known, the minimax criterion is applied that says to take the maximum value of each row and then choose the minimum of the row maximums. For this criterion the best alternative is A3.

Making decisions under uncertainty is a difficult task. The decision maker's attitude toward risk will affect the approach taken and the possibilities for large losses or gains as well as regrets. Generally, when the probabilities of the states of nature are truly unknown the decision maker will make more conservative decisions.

6.5 Practical Decision Analysis

In this chapter, a number of approaches for different types of decision making problems have been covered. The methods and procedures presented should be viewed as decision support tools and not as the final solutions to the problems. To help make better decisions in practice there are structured approaches to decision making that can be employed that use the discussed techniques as a part of the process. The intent of this section is to provide a practical, well thought-out approach to solving a problem in which the techniques of this chapter are but a part. The structure being proposed and the majority of the material in this section are based on the work of Hammond et al. [16], which should be consulted for a more complete discussion.

There are many approaches to solving problems, but most use a basic framework that involves defining the problem and alternatives. However, oftentimes the greatest value added to the decision process from a structured approach is the enhanced understanding and insight gained by the decision maker from going through the process. This in itself will lead to better decisions as most decisions made in practice are based on intuition, experience, and gut reactions without much analysis. This is not to say that these cannot lead to good outcomes, but this approach in the long run will produce more poor decisions than good ones.

As an example, consider the following Hospital Problem that you have been asked to solve. As the director of a private hospital you have been tasked by the governing board to increase profits through additional services. The hospital is located in a prosperous city that is home to a major university, several other health care facilities, good infrastructure, and many local and nearby attractions. The hospital is well run and well respected and especially noted for its geriatric care unit, its burn unit, its cardiac unit, its pharmacy, and its pediatrics unit. The major problems within the facility are related to staffing issues, but the board desires to increase profits without any drastic cost savings accomplished through staff attrition or layoffs. Some of the board feel that the hospital is already understaffed.

Any procedure that can be used to help structure the decision making process for this problem or any problem is very helpful in practice. The practical framework given here was developed by Hammond et al. [16], based on years of experience and research in the decision analysis field. They present a framework containing eight essential aspects of decision making that pertain to all decisions. These eight factors constitute the body of the process of making effective decisions. These eight factors are presented below.

6.5.1 Problem Definition

The first and key aspect of decision making is to identify the real problem. That is, to make sure you are solving the correct problem. "A good solution to a well-posed decision problem is always better than an excellent solution to a poorly posed one" [16].

To define the decision problem and pose it well requires practice and effort. To begin, one should understand what caused the need for the decision in the first place. For example, you might be asked to determine the best conveyor system to replace your current one, but the actual problem may not be that the current conveyor system needs replacing. The actual problem might be a cycle time issue or it might be that you need to determine what the best material handling system for your overall production facility is without limiting yourself to only conveyors. Time and thought need to go into why the apparent problem has arisen and into what the actual components are that are driving the problem. Too many mistakes are made by solving the wrong problem.

As the problem is defined, there are usually constraints that are identified that help narrow the alternatives. These constraints may be self-imposed or imposed from outside.

For example, a company might desire to maintain a zero defect quality control policy (self-imposed) and also must meet federal regulations on hazardous waste disposal. All constraints should be questioned to make sure they are legitimate, do not hinder you from seeing the best decision, or that they are not just a symptom of defining the wrong problem.

Next, to make sure "you are focused on the right goal" [16], it is essential that the correct elements of the problem are identified as well as what other decisions will be affected by the decision for this problem. How this plays into the decision process should also be determined. Use this to help refine the scope of the problem definition. That is, all aspects of the problem should be considered, but you do not want to make the problem so large or complicated that a decision cannot be made.

Lastly, re-examine the problem definition as the process goes along. Time and information may change your perspective or give new insights into the problem that will allow for the problem to be restated and better defined. Always look for opportunities to question the problem statement and to redefine it. This redefining of the problem statement helps focus the problem better, helps determine information needed, and tends to lead to better decisions.

Consider the hospital problem introduced at the beginning of this section. The problem seems at first to be to increase profit so as to please shareholders. This could be done through cost-cutting measures and staffing issues but that does not address the real issue and is against the board's wishes. The profit margins for the hospital have been increasing slightly each year for the last 5 years and it has been determined that this facility is one of the most profitable in the state. As you ask questions of the board, you find out that their desire to raise profits is related to wanting more positive exposure for the hospital at the corporate level and to wanting to attract better doctors, as well as improving the shareholders' positions. After much discussion and research you settle on the problem being that you need to increase profits without cutting cost while enhancing the reputation of the hospital. This will impact several areas of concern, such as staffing issues and attracting quality physicians. Therefore, you develop the following problem statement: Increase the profit margin of the hospital through the expansion of services that meet the needs of the community.

6.5.2 Objectives

With the problem well defined, you now need some way to determine the best decision. This is done by identifying what it is you really want to achieve. That is, how will you know if your decision is a good one? How will you measure the success of the decision? These criteria are called objectives. Hence, objectives tell you what you want to achieve and give a way of assessing the decision you have made. One of the keys in this process is to make sure all the objectives for the problem are identified. This helps one think through the problem and in this process of analysis and objective identification it might be possible to even identify possible solutions. This also helps one develop and identify alternatives for the problem.

Objectives should help guide the entire process of decision making. They tell you what information is needed, what decision should be made, why a decision is a good one, and the importance of the decision in terms of time and effort. Hammond et al. [16] give five steps to help identify objectives.

1. Write down all concerns you hope to address through your decision.
2. Convert your concerns into succinct objectives.
3. Separate ends from means to establish your fundamental objectives. Means objectives are milestones on the way to the end. The end or fundamental objective is what is needed for its own sake. It is the reason for the decision. Means objectives

Decision Analysis 6-23

help generate alternatives and give insight to the problem, but fundamental objectives are what are used to evaluate and compare alternatives.

4. Clarify what you mean by each objective. For example, what does improve quality really mean?
5. Test your objectives to see if they capture your interests. That is, based on the derived objectives test several alternatives to see what decision would be made and whether or not that decision is one you could live with.

When identifying your fundamental objectives keep in mind that objectives are personal, different people can have different objectives for the same problem, objectives should not be limited by what is easy or hard, fundamental objectives should remain relatively stable over time, and if a potential solution makes you uncomfortable, you may have missed an objective [16]. The process of identifying the objectives is very important to the overall decision process. It is imperative to take the time to get them right and to specify them fully and completely.

Note that in many problems there are multiple objectives to be considered. In the previous sections of this chapter, the methods presented were for a single objective. Often, when there appear to be multiple objectives, there is still only one fundamental or end objective and the methods developed here would be applicable to that objective. If there truly are multiple fundamental objectives, then one would need to apply different techniques such as multiple objective decision making (see Chapter 5).

For the hospital problem there is one overriding fundamental objective and that is to increase profits. This is easily measured. Likewise, as part of meeting this objective it is possible to measure its impact on the other stated objectives of enhancing reputation and attracting high quality physicians. The enhancement of reputation is more difficult to measure, but can be determined indirectly from survey results, newspaper articles, and increased patient demand. The attraction of high quality physicians is easily measured.

6.5.3 Alternatives

Alternatives are the potential choices one can make to solve the given problem. The dilemma, though, is that your solution will only be as good as your set of alternatives. Therefore, it is crucial to have a good set of alternatives that are not limited in scope. Do not think that just because alternatives to a similar problem are known, those alternatives will suffice for the new problem or that the apparent alternatives are all that is needed. Oftentimes, bad decisions are made because the only alternatives considered are the apparent ones or ones that are only incrementally better than the current solution. Realize that regardless of how well you generate alternatives, there still may be a better one out there. However, do not fall into the trap of trying to find the "best" alternative. This may lead to spending so much time looking for the alternative that no decision is made or that a decision is forced on you by the delay. Find a set of alternatives with which you are satisfied, and stop there, realizing you can adapt and consider new alternatives as you go through the process.

Generating a good set of alternatives takes time and effort. To help in the process consider the following techniques by Hammond et al. [16] to generate good alternatives.

1. Based on your objectives, ask how they can be achieved. Do this for means as well as fundamental objectives.
2. Challenge the constraints. Make sure a constraint is real and is needed. Assume there are no constraints and develop alternatives and then see how they fit the constraints or can be adapted to the constraints. Ask why the constraint is there

and what can be done to meet it or overcome it. Are there ways to nullify the constraint? Think creatively.

3. Set high aspirations. Force yourself to meet this high expectation by thinking more out of the box.
4. Do your own thinking first. Develop alternatives before you ask others for input. This helps keep you from having only alternatives that might have been biased from input from others.
5. Learn from experience. Have you made similar decisions? What was right or wrong in that decision? Use past experience to help generate alternatives, but do not fall into the trap of just repeating past mistakes or not considering other alternatives since you did not before.
6. Ask others for suggestions and input. This is done only after you have carefully thought about the problem and developed your own alternatives. This can be viewed as a sort of a brainstorming session. Keep an open mind and consider the suggestions and see what other alternatives they might trigger for you.
7. Never stop looking for alternatives. As you go through the decision process you might see other alternatives or get other ideas. Never think it is too late to consider different alternatives.
8. Know when to quit looking for alternatives. Realize that the perfect solution usually does not exist so spending a great deal of time looking for it is not usually productive. If you have thought hard about your alternatives and have a set of alternatives that give you a diversity of options and at least one you would be satisfied with, then you could stop. However, remember to always consider additional alternatives as they present themselves through the decision process.
9. Don't forget to always include the do nothing or status quo alternative. It is possible that the current state is the best alternative available.

For the hospital problem you have spent much time researching what other hospitals are doing, analyzing your current strengths, market trends, and the population demographics of your area. You have developed several alternatives based on this research. One is the development of an outpatient clinic focused on orthopedics. You believe there is great potential here based on the aging population, your already strong geriatrics unit, and the increasing number of sports-related injuries occurring at the local high schools and university. Another is the development of a complementary pediatrics unit that is designed similar to the obstetrics unit that allows for family visitors and birthing rooms. This unit will allow a family member to stay in the room and the rooms will be furnished more like a bedroom than a hospital room. These will cost more but will allow for more family input and attention. It is believed as well that through this process you can also address some of the staffing issues by having family members in the room to help monitor progress and other basic services. Another alternative would be to upgrade the cardiac unit to allow for more state-of-the-art surgeries and techniques. This would help attract physicians and open up the hospital to more specialized and profitable surgeries. Lastly, you could maintain the *status quo* and see if profits increase on their own.

After the determination of these alternatives you hold two brainstorming sessions, one with the staff and one with the physicians. In the session with the staff someone mentions the growing number of elder care facilities in the area and the lack of trained staff to check pharmaceuticals and prescriptions. This leads to the alternative of using your pharmacy and pharmacists as a pharmaceutical distribution center. You will meet the needs of the

Decision Analysis

surrounding elder care facilities by providing pharmaceutical services, quality and safety checks on prescriptions, door-to-door delivery, and expedited services. Another idea that came up was the development of a research unit dedicated to state-of-the-art genetic testing and automated surgeries on the assumption this would increase reputation as well as draw highly qualified physicians, which would result in increased profits. Due to the large capital outlay to equip such a laboratory, it was eliminated from consideration at this time. At this point no new alternatives or alternatives that meet the constraints are presented, so you move forward with the four alternatives: outpatient orthopedics, family pediatric unit, upgrade cardiac unit, and the pharmaceutical distribution network.

6.5.4 Consequences

Now that the problem has been defined, objectives determined, and alternatives developed you need to be able to determine which alternative is the "best." This is done by looking at the consequences of the alternatives related to the objectives and determining what the payoff for each alternative will be. In the case of a single objective, the techniques described in this chapter can be applied to find the best alternative. The key is to make sure that as you look at each alternative the measure of the consequences is consistent and is a measure you are willing to use. The key to making sure you have the right consequences for the alternatives is to build a consequence table. This is very similar to the payoff matrix, and generally the payoff matrix is part of the consequences table. The difference is that the consequences table includes more measures than just the single payoff measure and is used to help identify the main measure that is consistent across the alternatives.

The alternatives of the hospital problem are analyzed in terms of the demographics, potential profits, costs of implementation, staffing issues, and other pertinent factors. Several possible states of nature are also formulated related to the economy, demographics, health care costs, supply and demand of nurses, and household income. Based on this information a consequence table is developed where the main measure is potential profit increases. Additionally, within the consequence table are factors related to reputation and physician applications and recruitment. From this consequence table it is determined that the best alternative is to develop a pharmaceutical distribution network for local elder care facilities.

6.5.5 Tradeoffs

The concept of tradeoffs is most commonly applied to multiobjective problems, but the basic idea also has applicability to single objective problems.

Even with applicable techniques and appropriate consequences it may be that there is more than one possible alternative to consider. Recall from the earlier discussion of the different techniques applied to the investment and snack problem that each technique gave a different alternative. The question then becomes: which alternative to choose.

To help in this process one can look at possible tradeoffs within the alternatives. First, one should eliminate any alternative that is dominated by another alternative. As discussed earlier, that is when there is an alternative that is better than another alternative no matter the state of nature or the objective. The dominated alternative is eliminated from further consideration. For the other alternatives, consider what the tradeoffs are of choosing one over another. For example, in the snack production problem alternative A2 was never chosen by any of the methods. However, if the decision maker is conservative in nature they may be willing to choose A2 over A4 by viewing the tradeoff of losses in the first two states of nature compared to the gains in the other two states of nature as acceptable for this other conservative solution.

In the case of a single objective, when two alternatives are "equal" then other considerations will need to come into play. This might include subjective considerations or additional objectives not considered initially.

As an example, in the hospital problem there was actually one alternative that gave a higher expected profit than the distribution alternative. That alternative was the orthopedics unit. However, in analyzing the two alternatives it was seen that the difference in profit increases was not substantial, but the difference in reputation enhancement was quite large in favor of the distribution network. Likewise, the distribution network integrated well with the geriatrics unit and could possibly increase demand for this service substantially in the future. Hence, the distribution network was chosen.

6.5.6 Uncertainty

Virtually all problems deal with uncertainty in some form or another. The methods presented in this chapter are designed for dealing with uncertainty. The key is to realize uncertainty exists and that it will need to be dealt with appropriately. This involves determining the possible states of nature and their probability of occurrence, if possible, and appropriate courses of action based on those states of nature. The decision trees presented earlier are an excellent tool to help in the understanding of uncertainty and risk and its impact on the decision problem.

It is also important to understand that when uncertainty exists it is possible to make a good decision and still have the consequences of that decision turn out poorly. Likewise, it is possible to make a bad decision and still have the consequences turn out good. That is the nature of risk and uncertainty. Therefore, under uncertainty and risk, the consequences of the decision might not be the best way to assess the decision made or the process used to arrive at that decision. What truly needs to be assessed is the decision process itself. A poor process may occasionally yield good consequences, but over the long run it will be detrimental. Likewise, a good process may occasionally yield poor consequences, but over the long run this process will be beneficial.

When developing the problem statement uncertainty must be included. This is done through the use of a risk profile. To determine the risk profile, Hammond et al. [16] suggest answering the following four questions:

1. What are the key uncertainties?
2. What are the possible outcomes of these uncertainties?
3. What are the chances of occurrence of each possible outcome?
4. What are the consequences of each outcome?

Answering these four questions should lead to a payoff matrix with prior probabilities. The payoff matrix can then be used to generate solutions and to develop a corresponding decision tree.

As part of the consequence table for the hospital problem uncertainties were taken into account and probabilities determined. The key uncertainties were related to health care costs, demographics, economic conditions, and potential competition from other area health care providers. For example, it was determined that if the demographics indicated a downturn in the percentage of the population over 60 and the economy was weak then

Decision Analysis

elder care facilities would lose business. The consequence of this for the distribution network alternative would be an expected 2% decline in profits, for the orthopedic unit it would be an expected 3% decline in profits, for the family pediatric unit it would result in no change in profit, and for the cardiac unit it would represent an expected 1% decline in profits.

6.5.7 Risk Tolerance

Once the risk profile has been determined it is necessary to understand how you as the decision maker view risk. This is the process of determining your utility function. Are you risk-averse, risk-neutral, or risk-seeking? Taking this into account can help you make a better decision for yourself. Here it is important to remember that the measure of success is the decision process and not the consequences of the decision. A risk-seeker and someone who is risk-averse will come to two completely different decisions, but if the process is sound then those decisions are good decisions for them.

As the decision maker for the hospital problem you have determined that you are risk-averse, preferring more conservative approaches that give a better chance of smaller profits than riskier high-return alternatives. This is part of the reason why the family pediatric unit and the cardiac unit did not look promising to you.

6.5.8 Linked Decisions

Very few decisions are truly made independently. Current decisions have an impact on decisions that will need to be made later. These are called linked decisions. To make a good decision, at this point in time, requires you to think about the impact of the current decision on later decisions. A poor decision now may lead to limited alternatives for later decisions. "The essence of making smart linked decisions is planning ahead" [16].

Linked decisions tend to always contain certain components. These components are: (1) a basic decision must be made now; (2) the alternatives to choose from are affected by uncertainty; (3) desirability of an alternative is also affected by the possible future decisions that will need to be made based on the current decision; (4) the typical decision making pattern is to decide, then learn, then decide, then learn, then decide, and so on.

Considering and making linked decisions is really just another decision problem. The process outlined in this section can be used to help make the linked decisions. There are six basic steps in dealing with linked decisions [16]:

1. Understand the basic decision problem. This includes the steps outlined here of defining the problem, specifying objectives, generating alternatives, determining consequences, and identifying uncertainties. The uncertainties are what make linked decisions difficult, so it is important to spend time correctly identifying and acknowledging them.
2. Identify ways to reduce critical uncertainties. This generally involves obtaining additional information. Once obtained, then it might be possible to determine posterior probabilities.
3. Identify future decisions linked to the basic decision. In doing this, it is important to find an appropriate time horizon to consider. If the horizon is too long, you might be considering future possibilities that will not actually come into play. A rule of thumb is often the current decision and two future linked decisions.

4. Understand relationships in the linked decisions. Often a decision tree can be used here to help illustrate relationships and to discover unseen relationships.
5. Decide what to do in the basic decision. Again, this is where a decision tree can be most helpful in keeping the linked decisions before you. Look at what will be happening and then work backward from there determining the consequences of each choice. In the decision tree this is equivalent to starting at the right hand side and working back through the branches.
6. Treat later decisions as new decision problems.

Recognition of linked decisions is the first critical aspect. It is important to always view decisions in terms of short-range as well as long-range impacts. A more holistic view allows the decision maker to better gauge the impact of decisions on current objectives as well as possible unintended consequences and later effects. The consideration of the linking of a decision to future considerations leads to better decisions. A truly good decision will be viewed as such not only now but also in the future when its effects are experienced.

For the hospital problem, part of the reason the distribution network was chosen was its linkage with future decisions. The growing demographics of the region indicated that elder care would become a central component of the health care system for that region. Focusing on the distribution network would serve as a marketing device for the hospital in this demographic as well as a way to identify potential patients with concerns other than pharmaceuticals. It might then be possible to build the elder care reputation of the hospital to the point where it is the first choice among the elderly, which would then allow the hospital to go forward with its other alternative of orthopedics targeted primarily at the elderly. This in turn could open up other alternatives related to elder care not yet considered and make the hospital the dominant health care delivery entity in the region. This would then attract high quality physicians and enhance the reputation even further.

6.6 Conclusions

In this chapter, a number of approaches to different types of decision problems under risk and uncertainty have been presented. The intent of theses methodologies is to give tools to the decision maker to assist in the decision making process. They should not be viewed as a shortcut to good decisions or as a panacea for making decisions. To be a consistently good decision maker requires a systematic and thorough approach to decision making. The goal is to have a sound decision making process and framework that allows a decision maker to approach any problem with confidence. The worst scenario for a decision maker is to have decisions made for them due to their slowness in responding to a situation or their inability to decide. A decision maker's intent should be to make the best decision possible with the information given at the current point in time. If the process is well thought out and sound then the decisions made should generally be good.

The purpose of this chapter has been to help decision makers develop their decision making process and to give tools to help in that process. The following basic systems approach to decision making summarizes the approach taken in this chapter [16]:

1. Address the right decision problem.
2. Clarify your real objectives and recognize means and ends.
3. Develop a range of creative alternatives.
4. Understand the consequences of the decisions (short and long term).

Decision Analysis 6-29

5. Make appropriate tradeoffs between conflicting objectives.
6. Deal appropriately with uncertainties.
7. If there are multiple objectives, make appropriate tradeoffs.
8. Take account of you risk attitude.
9. Plan ahead for decisions linked over time.
10. Make your decision, analyze possible other solutions through the use of different methodologies, and perform sensitivity analysis.

There is no guarantee that a decision will always be good when uncertainties are present, but the chances of a decision being good increase significantly when the decision process is good. Making good decisions takes time and effort but the rewards are worth the investment. This is true for decisions in everyday life as well as those one wrestles with in their work. To help one make good decisions consistently, a decision maker needs to develop a good process, apply the process to all decisions, be flexible, adjust decisions as time and information become available, and enjoy what they are doing; then good decisions will occur.

6.7 Resources

Decision analysis is a well-studied field with many resources available. In addition to the information given here and the stated references thus far, there are many other works to consider. Hillier and Lieberman [5, p. 717] give a good listing of actual applications of decision analysis as presented in the practitioner-oriented journal *Interfaces*. Many other articles and books exist, which cannot all be detailed here. To help one begin the process of a deeper study of decision analysis, Refs. [17–32] are recommended as initial sources.

A quick search of the Web is also suggested in that there are many sites devoted to decision analysis. These include academic sites as well as professional and consulting sites. Some good places to start are Decision Analysis Society of INFORMS: http://faculty.fuqua.duke.edu/daweb/ and International Society on Multiple Criteria Decision Making: http://www.terry.uga.edu/mcdm/.

There is also a large selection of software available at many different levels to help in decision analysis. Maxwell [14,15] has done excellent surveys of decision analysis software. The latest survey is available on the Web at http://www.lionhrtpub.com/orms/orms-10-04/frsurvey.html.

Additionally, there is also a decision analysis software survey from *OR/MS Today* available at http://www.lionhrtpub.com/orms/surveys/das/das.html.

Along with these surveys, there are a number of sites dedicated to software. Some of this software is discussed in Ref. [5], which also gives some examples and brief tutorials. The software mentioned in this chapter related to decision trees can be found at http://www.treeplan.com/ and http://www.usfca.edu/~middleton/.

Some additional software sites include, but are not limited to, http://www.palisade.com/precisiontree/, http://www.lumina.com/, and http://www.vanguardsw.com/decisionpro/jgeneral.htm.

Remember, though, that the software is still just a decision support tool, not a decision making tool. It will give you the ability to analyze different scenarios and to do sensitivity analysis, but the decision must still be made by the decision maker.

References

1. Ravindran, A., Phillips, D.T., and Solberg, J.J., *Operations Research: Principles and Practice*, 2nd ed., John Wiley & Sons, 1987, chap. 5.
2. Watson, S.R. and Buede, D.M., *Decision Synthesis: The Principles and Practice of Decision Analysis*, Cambridge University Press, 1987.
3. Taha, H.A., *Operations Research: An Introduction*, 8th ed., Prentice Hall, 2007, chap. 13.
4. Arsham, H., Tools for Decision Analysis: Analysis of Risky Decisions, http://home.ubalt.edu/ntsbarsh/opre640a/partIX.htm.
5. Hillier, F.S. and Lieberman, G.J., *Introduction to Operations Research*, 8th ed., McGraw Hill, 2005, chap. 15.
6. Merkhofer, M.W., Quantifying Judgmental Uncertainty: Methodology, Experiences and Insights, *IEEE Transactions on Systems, Man, and Cybernetics*, 17:5, 741–752, 1987.
7. Winston, W.L., *Operations Research: Applications and Algorithms*, 4th ed., Brooks/Cole, 2004, chap. 13.
8. Pennings, J.M.E. and Smidts, A., The Shape of Utility Functions and Organizational Behavior, *Management Science*, 49:9, 1251–1263, 2003.
9. Keeney, R.L. and Raiffa, H., *Decisions with Multiple Objectives*, Wiley, 1976.
10. Tversky, A. and Kahneman, D., The Framing of Decisions and the Psychology of Choice, *Science*, 211:4481, 453–458, 1981.
11. Golub, A.L., *Decision Analysis: An Integrated Approach*, Wiley, 1997.
12. Biswas, T., *Decision Making under Uncertainty*, St. Martin's Press, 1997.
13. Gass, S.I. and Harris, C.M., Eds., *Encyclopedia of Operations Research and Management Science*, Kluwer Academic Publishers, 1996.
14. Maxwell, D.T., Software Survey: Decision Analysis, *OR/MS Today*, 29:3, 44–51, June 2002.
15. Maxwell, D.T., Decision Analysis: Aiding Insight VII, *OR/MS Today*, October 2004, http://www.lionhrtpub.com/orms/orms-10-04/frsurvey.html.
16. Hammond, J.S., Keeney, R.L., and Raiffa, H., *Smart Choices: A Practical Guide to Making Better Decisions*, Harvard Business School Press, 1999.
17. Ben-Haim, Y., *Information-Gap Decision Theory: Decisions under Severe Uncertainty*, Academic Press, 2001.
18. Clemen, R.T. and Reilly, T., *Making Hard Decisions with Decision Tools*, Duxbury Press, 2001.
19. Connolly, T., Arkes, H.R., and Hammond, K.R., Eds., *Judgment and Decision Making: An Interdisciplinary Reader*, Cambridge University Press, 2000.
20. Daellenbach, H.G., *Systems and Decision Making: A Management Science Approach*, Wiley, 1994.
21. Eiser, J., *Attitudes and Decisions*, Routledge, 1988.
22. Flin, R., et al., (Ed.), *Decision Making under Stress: Emerging Themes and Applications*, Ashgate Pub., 1997.
23. George, C., *Decision Making under Uncertainty: An Applied Statistics Approach*, Praeger Pub., 1991.
24. Goodwin, P. and Wright, G., *Decision Analysis for Management Judgment*, 2nd ed., Wiley, 1998.
25. Grünig, R., Kühn, R., and Matt, M., Eds., *Successful Decision-Making: A Systematic Approach to Complex Problems*, Springer, 2005.
26. Kirkwood, C.W., *Strategic Decision Making: Multiobjective Decision Analysis with Spreadsheets*, Duxbury Press, 1997.
27. Keeney, R.L., Foundations for Making Smart Decisions, *IIE Solutions*, 31:5, 24–30, May 1999.

28. Pratt, J.W., Raiffa, H., and Schlaifer, R., *Introduction to Statistical Decision Theory*, The MIT Press, 1995.
29. Van Asselt, M., *Perspectives on Uncertainty and Risk: The Prima Approach to Decision Support*, Kluwer Academic Publishers, 2000.
30. Van Gigch, J.P., *Metadecisions: Rehabilitating Epistemology*, Kluwer Academic Publishers, 2002.
31. Vose, D., *Risk Analysis: A Quantitative Guide*, 2nd ed., John Wiley & Sons, 2000.
32. Wickham, P., *Strategic Entrepreneurship: A Decision-making Approach to New Venture Creation and Management*, Pitman, 1998.

7
Dynamic Programming

7.1	Introduction...	**7-1**
7.2	Deterministic Dynamic Programming Models	**7-3**
	Capacity Expansion Problem [11,13] • Capacity Expansion Problem with Discounting [5,11,13] • Equipment Replacement Problem [3,8,10] • Simple Production Problem • Dynamic Inventory Planning Problem [11,15] • Reliability System Design [11]	
7.3	Stochastic Dynamic Programming Models	**7-19**
	Stochastic Equipment Replacement Problem [3,8,10] • Investment Planning [11]	
7.4	Conclusions	**7-24**
	References	**7-24**

José A. Ventura
Pennsylvania State University

7.1 Introduction

Dynamic programming (DP) is an optimization procedure that was developed by Richard Bellman in 1952 [1,2]. DP converts a problem with multiple decisions and limited resources into a sequence of interrelated subproblems arranged in stages, so that each subproblem is more tractable than the original problem. A key aspect of this procedure is that the decision in one stage cannot be made in isolation due to the resource constraints. The best decision must optimize the objective function with respect to the current and prior stages, or the current and future stages.

A wide variety of deterministic and stochastic optimization problems can be solved by DP. Some of them are multi-period planning problems, such as production planning, equipment replacement, and capital investment problems, in which the stages of the DP model correspond to the various planning periods. Other problems involve the allocation of limited resources to various activities or jobs. The latter case includes all variations of the knapsack and cargo loading problems.

The difficulty of DP is in the development of the proper model to represent a particular situation. Like an artist experience in model development is essential to be able to manage complex problems and establish recurrence relations between interrelated subproblems in consecutive stages. For this reason, this chapter introduces DP by analyzing different applications, where decision variables may be integer or continuous, objective functions and constraints may be either linear or nonlinear, and the data may be deterministic or random variables with known probability distribution functions. Below, the basic components of a

DP model are introduced and applied to the following nonlinear integer program with a single constraint:

$$\text{maximize } z = \sum_{i=1}^{n} c_i(x_i)$$

$$\text{subject to } \sum_{i=1}^{n} a_i x_i \leq b$$

$$x_i \geq 0, \text{ integer}, \quad i = 1, \ldots, n$$

where b and a_i, $i = 1, \ldots, n$, are positive real values. This optimization model represents the allocation of b units of a resource to n different activities. The objective is to maximize the total profit, given that function $f_i(x_i)$, $i = 1, \ldots, n$, defined for $x_i \in \{0, 1, 2, \ldots, \lfloor b/a_i \rfloor\}$, shows the profit for activity i given that x_i units of the activity are employed.

The *terminology* used to define a DP model includes the following main elements [2,3,4,6,7,10,11,14,16]:

Stage (i): the original problem is divided into n stages. There is an initial stage (stage n) and a terminating stage (stage 1). Index i represents a given stage, $i = 1, 2, \ldots, n$.

State (s_i): each stage has a number of states associated with it. The states are the various possible conditions in which the system might be at each particular stage of the problem.

Decision variable (x_i): there is one decision variable or a subset of decision variables for each stage of the problem.

Contribution function ($c_i(x_i)$): this function provides the value at stage i given that the decision is x_i.

Optimal value function ($f_i(s_i)$): best total function value from stage i to stage n, given that the state at stage i is s_i.

Optimal policy ($p_i(s_i) = x_i^*$): optimal decision at a particular stage depends on the state. The DP procedure is designed to find an optimal decision at each stage for all possible states.

Transformation function ($t_i(s_i, x_i)$): this function shows how the state for the next stage changes based on the current state, stage, and decision.

Example $\quad s_{i+1} = t_i(s_i, x_i) = s_i - a_i x_i$

Recurrence relation: this is an equation that identifies the optimal policy (decision) at stage i, given that the optimal policy at stage $i+1$ is available.

Example $\quad f_i(s_i) = \max_{x_i = 0, 1, \ldots, \lfloor s_i/a_i \rfloor} \{c_i(x_i) + f_{i+1}(s_i - a_i x_i)\}, \quad s_i = 0, \ldots, b;$

$$i = 1, \ldots, n-1$$

Boundary conditions: these are the initial conditions at stage n and obvious values of the optimal value function.

Example $\quad f_n(s_n) = \max_{x_n = 0, 1, \ldots, \lfloor s_n/a_n \rfloor} \{c_n(x_n)\}, \quad s_n = 0, \ldots, b$

Answer: the global optimal solution of the problem is determined in the terminating stage (stage 1).

Example $\quad f_1(b)$

Dynamic Programming

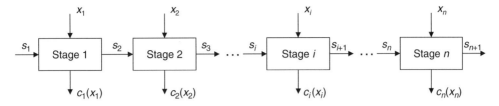

FIGURE 7.1 Graphical illustration of the stages of a DP formulation.

In this DP model, the recurrent process begins in stage n, moves backwards, and terminates in stage 1 (backward formulation). A similar formulation could be developed in reverse order from stage 1 to stage n (forward formulation). In many applications, both the two formulations are equivalent and require the same computational effort to solve a particular instance of the problem. In other applications, depending on the initial conditions and the size of the state space, one of the formulations may be more efficient than the other.

Figure 7.1 provides a graphical illustration equivalent to the above DP formulation. Each box corresponds to a stage. The state at a particular stage is predetermined by the prior decision. Based on the state and the decision made at this stage, an outcome represented by the contribution function and the value of the state at the following stage is obtained.

Two significant advantages of DP can be observed in the above formulation. One is that it transforms a problem with n decision variables into n single-variable subproblems. The second advantage over virtually all other optimization methods is that it finds the global maxima or minima rather than just local optima. The key limitation of DP is the dimensionality of the state space. In simple terms, if the model includes several state variables, then difficulties concerning the storage of information and time required to perform the computation may appear.

The justification of the DP procedure relies on the *Principle of Optimality*. Roughly, the principle states that, if $(x_1^*, x_2^*, \ldots, x_n^*)$ is an optimal policy to a given problem and s_i^* is the optimal state in stage i, then $(x_i^*, x_{i+1}^*, \ldots, x_n^*)$ is an optimal policy for the subproblem defined between stages i and n with s_i^* as the initial state. The correctness of a DP model can be proved by showing that all possible states are considered and the recurrence relation satisfies the principle of optimality.

The selection of the state variables for the DP model is critical in the sense that the states in the state space must satisfy the *Markov Property*. That is, the optimal policy from any stage i to stage n depends only on the entering state (s_i) and not in any other way on previous decisions.

DP models can be classified as deterministic and stochastic, depending on the type of data that are available to solve a problem. Obviously, if the data are known for a particular situation, a deterministic DP model will be used to find the best solution of the problem. If some of the data are probabilistic, then a stochastic DP model will be developed to optimize an expected value. Below, five applications of deterministic DP and two applications of stochastic DP are presented in Sections 7.2 and 7.3, respectively.

7.2 Deterministic Dynamic Programming Models

7.2.1 Capacity Expansion Problem [11,13]

A supplier of electronic components for the automotive industry has decided to expand its capacity by building nine new manufacturing facilities closer to customer locations in the next 4 years. A builder has made an interesting offer to build the facilities. The annual

TABLE 7.1 Annual Facility Requirements and Costs (in Millions of Dollars)

Year (i)	Required Facilities (r_i)	Fixed Cost (a_i)	Variable Cost (b_i)
1	2	1.2	4.3
2	4	1	4.7
3	7	1.4	4.4
4	9	1.2	5.1

construction cost will consist of a fixed cost plus a variable cost per facility to be built. The fixed cost does not have to be paid if no facilities are built in a given year. The maximum number of facilities that can be built in a single year is four. Table 7.1 shows the desired number of facilities that need to be finished by the end of each year along with the annual fixed and variable costs.

DP Formulation and Solution

This example can be formulated as a four-stage DP model where the stages correspond to the four planning years. The construction cost in year i, defined as $c_i(x_i)$, is a function of the number of facilities x_i to be built during the year:

$$c_i(x_i) = \begin{cases} 0, & \text{if } x_i = 0, \\ a_i + b_i x_i, & \text{if } x_i = 1, \ldots, 4 \end{cases}$$

At the beginning of a year (stage), the state of the system can be specified by the total number of facilities built in the prior years or the number of remaining facilities to be built. In the proposed model, the state of the system, s_i, is characterized by the number of facilities already built. At each stage i, it is possible to determine the range of feasible states. For example, for $i = 3$, $s_3 \in \{4, \ldots, 8\}$, because 4 is the minimum requirement of completed facilities by the end of year 2 and 8 is the most that can be built in 2 years. For $i = 4$, $s_4 \in \{7, 8, 9\}$. The lower bound 7 is the requirement for year 3 and 9 is the overall requirement for the entire planning horizon (4 years). The transformation function computes the number of facilities by the beginning of year $i + 1$ as a function of the number of facilities at the beginning of year i and the facilities built during the year:

$$s_{i+1} = s_i + x_i$$

The optimal value function, $f_i(s_i)$, is defined as the total minimum cost for years i to 4, given that s_i facilities have already been built by the beginning of year i. Then, the recurrence relation is as follows:

$$f_i(s_i) = \min_{x_i = \lambda_i, \ldots, u_i} \{c_i(x_i) + f_{i+1}(s_i + x_i)\}, \quad s_i = r_{i-1}, \ldots, \min\{4(i-1), 9\}; \quad i = 1, \ldots, 3$$

where $\lambda_i = \max\{0, r_i - s_i\}$ and $u_i = \max\{4, 9 - s_i\}$ defines the range of the decision variable x_i, based on the minimum number of required facilities by the end of year i, and the annual construction capacity or the facility requirements at the end of the 4-year planning horizon. Note that the lower bound on s_i is the minimum number of required facilities by the end of year $i - 1$, and the upper bound is the minimum of the maximum production capacity in the first $i - 1$ years, $4(i - 1)$, and the total number of facilities needed by the end of year 4.

The numerical solution for this problem is provided by the tables below. Each table summarizes the solutions of the various subproblems solved for all feasible values of the state

Dynamic Programming

variable in one particular stage. The second column of each table shows the computations of the values of the optimal value function for each possible decision. For example, when $s_3 = 6$ in stage 3, five values of the decision variable are considered. The decision $x_3 = 0$ is infeasible because at least seven facilities must be completed by the end of the year ($s_3 + x_3 = 6 + 0 < 7$). The case $x_3 = 4$ is non-optimal because only nine facilities are required by the end of year 4 ($s_3 + x_3 = 6 + 4 > 9$). The other three decisions are feasible and the corresponding function values need to be determined to select the optimal decision that provides minimum construction cost for years 3 and 4. In this subproblem, the optimal decision is $x_3^* = 3$ and the corresponding optimal value function is

$$f_3(6) = c_3(3) + f_4(6+3) = (1.4 + 3 \times 4.4) + 0 = 14.6.$$

The last three columns of the table in stage i summarize the optimal solutions for the subproblems in that stage, including the optimal value function, $f_i^*(s_i)$, the optimal decision, x_i^*, and the initial state at the next stage, s_{i+1}. The optimal value function is obtained by selecting the minimum of the function values calculated in the second column. The minimum value corresponds to a specific decision that is optimal for the subproblem. The last column provides the next state that is determined by using the transformation function. It is important to point out that the actual calculations in the tables can be performed very efficiently. The function value of any cell in the second column of Tables 7.1 to 7.3 is the sum of two values, $c_i(x_i) + f_{i+1}(s_i - x_i)$. The first value is constant for all the cells in the same subcolumn corresponding to decision x_i. The second value comes from column $f_{i+1}(x_{i+1})$ in stage $i+1$.

Stage 4: (Initialization)

	$c(x_4)$					
s_4	$x_4 = 0$	$x_4 = 1$	$x_4 = 2$	$f_4^*(s_4)$	x_4^*	s_5
7	Infeasible	Infeasible	$1.2 + 2 \times 5.1 = 11.4$	11.4	2	9
8	Infeasible	$1.2 + 5.1 = 6.3$	Non-optimal	6.3	1	9
9	0	Non-optimal	Non-optimal	0	0	9

Stage 3:

	$c_3(x_3) + f_4(s_3 + x_3)$							
s_3	$x_3 = 0$	$x_3 = 1$	$x_3 = 2$	$x_3 = 3$	$x_3 = 4$	$f_3(s_3)$	x_3^*	s_4
4	Infeasible	Infeasible	Infeasible	$14.6 + 11.4 = 26.0$	$19.0 + 6.3 = 25.3$	25.3	4	8
5	Infeasible	Infeasible	$10.2 + 11.4 = 21.6$	$14.6 + 6.3 = 20.9$	$19.0 + 0 = 19.0$	19.0	4	9
6	Infeasible	$5.8 + 11.4 = 17.2$	$10.2 + 6.3 = 16.5$	$14.6 + 0 = 14.6$	Non-optimal	14.6	3	9
7	$0 + 11.4 = 11.4$	$5.8 + 6.3 = 12.1$	$10.2 + 0 = 10.2$	Non-optimal	Non-optimal	10.2	2	9
8	$0 + 6.3 = 6.3$	$5.8 + 0 = 5.8$	Non-optimal	Non-optimal	Non-optimal	5.8	1	9

Stage 2:

	$c_2(x_2) + f_3(s_2 + x_2)$							
s_2	$x_2 = 0$	$x_2 = 1$	$x_2 = 2$	$x_2 = 3$	$x_2 = 4$	$f_2(s_2)$	x_2^*	s_3
2	Infeasible	Infeasible	$10.4 + 25.3 = 35.7$	$15.1 + 19.0 = 34.1$	$19.8 + 14.6 = 34.4$	34.1	3	5
3	Infeasible	$5.7 + 25.3 = 31.0$	$10.4 + 19.0 = 29.4$	$15.1 + 14.6 = 29.7$	$19.8 + 10.2 = 30.0$	29.4	2	5
4	$0 + 25.3 = 25.3$	$5.7 + 19.0 = 24.7$	$10.4 + 14.6 = 25.0$	$15.1 + 10.2 = 25.3$	$19.8 + 5.8 = 25.6$	24.7	1	5

Stage 1:

			$c_1(x_1) + f_2(s_1+x_1)$					
s_1	$x_1 = 0$	$x_1 = 1$	$x_1 = 2$	$x_1 = 3$	$x_2 = 4$	$f_1(s_1)$	x_1^*	s_2
0	Infeasible	Infeasible	$9.8 + 34.1 = 43.9$	$14.1 + 29.4 = 43.5$	$18.4 + 24.7 = 43.1$	43.1	4	4

Optimal Solution

The optimal solution, in this case unique, can be found by backtracking, starting at the last solved stage (stage 1) and terminating at the stage solved initially (stage 4). Using the transformation function, it is possible to determine the state at stage $i+1$, s_{i+1}, from the state in stage i, s_i, and the corresponding optimal decision, x_i^*. Details of this process are provided below.

$$s_1 = 0, \qquad x_1^*(0) = 4 \text{ plants to be built in year 1,}$$
$$s_2 = t_1(0,2) = 0 + 4 = 4, \qquad x_2^*(4) = 1 \text{ plant to be built in year 2,}$$
$$s_3 = t_2(3,3) = 4 + 1 = 5, \qquad x_3^*(5) = 4 \text{ plants to be built in year 3,}$$
$$s_4 = t_3(6,0) = 5 + 4 = 9, \qquad x_4^*(9) = 0 \text{ plants to be built in year 4.}$$

$f_1(0) = \$43.1$ millions (minimum construction cost for all 4 years).

7.2.2 Capacity Expansion Problem with Discounting [5,11,13]

The solution of the capacity expansion problem discussed in the prior section gives equal weight to the cost of a facility paid in year 1 and in year 4. However, in reality, the money spent in year 4 can be invested in year 1 to earn interest for the following 3 years. This situation can be resolved by either using the present values of all cost data within the original DP model or extending the DP model to take into account the time value of money.

If money can be safely invested at an annual interest rate r, then future expenditures can be invested at the same rate. The effect of that is that capital grows by a factor of $(1+r)$ each year. Similarly, assuming that all annual costs are always effective at the beginning of the corresponding year, a dollar spent next year has the same value as β dollars today, where $\beta = 1/(1+r)$ is called the discount rate.

The capacity expansion problem with a discount factor $\beta = 0.9$ can be solved using the original formulation with the discounted costs shown in Table 7.2. Note that the costs in year i have been multiplied by β^{i-1} to determine their present value at the beginning of stage 1. For example, the discounted fixed cost in year 3 becomes $1.4 \cdot 0.9^{3-1} = 1.26$ and the corresponding variable cost is $4.4 \cdot 0.9^{3-1} = 3.56$.

Alternatively, the original DP formulation can be extended by redefining the optimal value function and including β in the recurrence relation.

TABLE 7.2 Annual Facility Requirements and Discounted Costs (in Millions of Dollars)

Year (i)	Required Facilities (r_i)	Fixed Cost (a_i)	Variable Cost (b_i)
1	2	1.20	4.30
2	4	0.90	4.23
3	7	1.26	3.56
4	9	0.97	3.72

Dynamic Programming   7-7

Optimal value function $(f_i(s_i))$: present value of the total minimum cost at the beginning of year i for years i to 4 given that s_i facilities have already been built by the beginning of year i.

Recurrence relation:

$$f_i(s_i) = \min_{x_i = \lambda_i, \ldots, u_i} \{c_i(x_i) + \beta \cdot f_{i+1}(s_i + x_i)\}, \quad s_i = r_{i-1}, \ldots, \min\{4(i-1), 9\}; \quad i = 1, \ldots, 3$$

The optimal solution of the example in Section 7.2.1 with discounting is the following:

$$x_1^*(0) = 4 \text{ plants to be built in year 1},$$
$$x_2^*(4) = 0 \text{ plants to be built in year 2},$$
$$x_3^*(4) = 4 \text{ plants to be built in year 3},$$
$$x_4^*(8) = 1 \text{ plant to be built in year 4}.$$

The present value of the optimal construction cost can be determined using the revised recurrence relation that takes into account the time value of money:

$$f_4(8) = c_4(1) = 1.2 + 5.1 \cdot 1 = 6.3,$$
$$f_3(4) = c_3(4) + \beta \cdot f_4(4+4) = (1.4 + 4.4 \cdot 4) + 0.9 \cdot 6.3 = 24.67,$$
$$f_2(4) = c_2(0) + \beta \cdot f_3(4+0) = 0 + 0.9 \cdot 24.67 = 22.20,$$
$$f_1(0) = c_3(4) + \beta \cdot f_2(0+4) = (1.2 + 4.3 \cdot 4) + 0.9 \cdot 22.20 = 38.38.$$

Note that the optimal policy for the discounted model differs from that in the original model. While both policies require four facilities to be built in year 1 and four additional facilities in year 3, the ninth facility is built in year 2 in the original policy and in year 4 in the discounted policy. The present value of the total construction cost at the beginning of year 1 for the optimal policy is $38.38 million.

7.2.3 Equipment Replacement Problem [3,8,10]

A company needs to own a certain type of machine for the next n years. As the machine becomes older, annual operating costs increase so much that the machine must be replaced by a new one. The price of a new machine in year i is $a(i)$. The annual expenses for operating a machine of age j is $r(j)$. Whenever the machine is replaced, the company receives some compensation for the old one, as trade-in value. Let $t(j)$ be the trade-in value for a machine that has age j. At the end of year n, the machine can be salvaged for $v(j)$ dollars, where j is the age of the machine. Given that at the beginning of the first year the company owns a machine of age y, we need to determine the replacement policy for the machine that minimizes the total cost for the next n years, given that replacement decisions can only be made at the beginning of each year. It is assumed that annual operating costs and trade-in and salvage values are stationary and only depend on the age of the machine.

DP Formulation and Solution

The DP formulation for this problem involves n stages corresponding to the n planning periods (years). At the beginning of stage i, the state of the system can be completely defined by the age of the machine that has been used in the prior year. At the beginning of a stage, two possible decisions can be made: keep the current machine or replace the machine. If we decide to keep the machine, then the only cost in the current stage will be the cost of operating the machine. If the decision is to replace the machine, the annual cost will include the price of a new machine, minus the trade-in value received for the current machine, plus the operating cost of a new machine. The cost in stage i, $c_i(j)$, depends on the age of the current machine and the decision:

$$c_i(j) = \begin{cases} a(i) - t(j) + r(0), & \text{if we buy a new machine} \\ r(j), & \text{if we keep the current machine} \end{cases}$$

The remaining DP formulation is provided below.

Optimal value function ($f_i(s_i)$): minimum cost of owning a machine from the beginning of year i to the end of year n (or beginning of year $n+1$), starting year i with a machine that just turned age s_i.

Optimal policy ($p_i(s_i)$): "buy" or "keep."

Transformation function ($t_i(s_i, p_i)$): this function shows how the state for the next stage changes based on the current state, stage, and decision.

$$s_{i+1} = t_i(s_i, p_i) = \begin{cases} 1, & \text{if } p_i = \text{"buy"} \\ s_i + 1, & \text{if } p_i = \text{"keep"} \end{cases}$$

Recurrence relation:

$$f_i(s_i) = \min \begin{cases} a(i) - t(s_i) - r(0) + f_{i+1}(1), & \text{if } p_i = \text{"buy"} \\ r(s_i) + f_{i+1}(s_i + 1), & \text{if } p_i = \text{"keep"} \end{cases} \quad s_i = 1, 2, \ldots, i-1, y+i-1$$

Boundary conditions: $f_{n+1}(s_i) = -v(s_i)$, $\quad s_n = 1, 2, \ldots, n-1, y+n-1$

Answer: $f_1(y)$.

Example 7.1

This DP formulation is illustrated by solving a numerical problem for a 4-year planning horizon. The age of the machine at the beginning of the first year is 2. The cost of a new machine in year 1 is \$58,000 and increases by \$2,000 every year. Annual operating costs along with trade-in and salvage values are time independent and provided in Table 7.3.

TABLE 7.3 Cost Data (in Thousands) for the Equipment Replacement Example

Machine Age (s_i) in Years	Annual Operating Cost ($r_i(s_i)$)	Trade-in Value ($t_i(s_i)$)	Salvage Value ($v_i(s_i)$)
0	12	—	—
1	15	35	30
2	25	25	20
3	35	15	10
4	60	10	5
5	80	5	0
6	100	0	0

Solution 7.1

$$f_5(1) = -30, \quad f_5(2) = -20, \quad f_5(3) = -10, \quad f_5(4) = -5, \quad f_5(6) = 0$$

$$f_4(1) = \min \begin{Bmatrix} (58+6) - 35 + 12 + (-30) \\ 15 + (-20) \end{Bmatrix} = -5, \quad p_4(1) = \text{keep}$$

$$f_4(2) = \min \begin{Bmatrix} (58+6) - 25 + 12 + (-30) \\ 25 + (-10) \end{Bmatrix} = 15, \quad p_4(2) = \text{keep}$$

$$f_4(3) = \min \begin{Bmatrix} (58+6) - 15 + 12 + (-30) \\ 35 + (-5) \end{Bmatrix} = 30, \quad p_4(3) = \text{buy}$$

$$f_4(5) = \min \begin{Bmatrix} (58+6) - 5 + 12 + (-30) \\ 80 + 0 \end{Bmatrix} = 41, \quad p_4(5) = \text{buy}$$

$$f_3(1) = \min \begin{Bmatrix} (58+4) - 35 + 12 + (-5) \\ 15 + 15 \end{Bmatrix} = 30, \quad p_3(1) = \text{keep}$$

$$f_3(2) = \min \begin{Bmatrix} (58+4) - 25 + 12 + (-5) \\ 25 + 30 \end{Bmatrix} = 44, \quad p_3(2) = \text{buy}$$

$$f_3(4) = \min \begin{Bmatrix} (58+4) - 15 + 12 + (-5) \\ 60 + 30 \end{Bmatrix} = 54, \quad p_3(4) = \text{buy}$$

$$f_2(1) = \min \begin{Bmatrix} (58+2) - 35 + 12 + 30 \\ 15 + 44 \end{Bmatrix} = 59, \quad p_2(1) = \text{keep}$$

$$f_2(3) = \min \begin{Bmatrix} (58+2) - 15 + 12 + 54 \\ 35 + 54 \end{Bmatrix} = 99, \quad p_2(3) = \text{keep}$$

$$f_1(2) = \min \begin{Bmatrix} 58 - 25 + 12 + 59 \\ 25 + 99 \end{Bmatrix} = 104, \quad p_1(2) = \text{buy}$$

The feasible states for stage 5 are 1 to 4, and 6. Note that, if the original machine is not replaced at all, it will have age 6 by the beginning of year 5, and if the machine is replaced at least one time, it will be at most 4 years old by the beginning of year 5. Thus, state 5 is infeasible.

The total minimum cost for the 4-year planning horizon is \$104,000 based on an optimal replacement policy that includes buying a new machine at the beginning of years 1, 3, and 4.

7.2.4 Simple Production Problem

A company needs to produce at least d units of a product during the next n periods. The production cost in period i is a quadratic function of the quantity produced. If x_i units are produced in period i, the production cost is $c_i(x_i) = w_i x_i^2$, where w_i is a known positive coefficient, $i = 1, \ldots, n$. The objective of this problem is to determine the optimal production quantities that minimize the total cost for the n periods. Note that the production quantities are not restricted to be integer. This problem can be formulated as a nonlinear program with a quadratic objective function and a linear constraint:

$$\text{Minimize } z = \sum_{i=1}^{n} w_i x_i^2$$

$$\text{subject to} \quad \sum_{i=1}^{n} x_i \geq d$$

$$x_i \geq 0, \quad i = 1, \ldots, n$$

This model illustrates a situation in which decision variables are continuous. Thus, subproblems cannot be solved by checking all feasible solutions as the feasible region is not finite. In general, when decision variables are continuous, an optimization procedure must be used to find an optimal solution. In this particular problem, the single-variable subproblems can be solved by setting the first derivative of the optimal value function to zero and solving the resulting linear equation.

DP Formulation and Solution

The proposed DP model is based on a forward formulation in the sense that the boundary conditions are given for stage 1 and the recurrent process moves from any stage i to stage $i+1$, $i=1,\ldots,n-1$, where stages match the production periods. The state of the system at a given stage i, s_i, is defined by the number of units produced in the first i periods. The DP formulation is presented below.

Optimal value function ($f_i(s_i)$): minimum production cost for the first i periods given that s_i units are produced in these periods.

Optimal policy ($p_i(s_i) = x_i^*$): units produced in period i.

Transformation function ($t_i(s_i, x_i)$): it finds the number of units produced by the end of period $i-1$ as a function of the number of units produced in the first i periods and the number of units produced in period i.

$$s_{i-1} = t_i(s_i, x_i) = s_i - x_i$$

Recurrence relation:

$$f_i(s_i) = \min_{0 \le x_i \le s_i} \{w_i x_i^2 + f_{i-1}(s_i - x_i)\}, \quad 0 \le s_i \le d; \quad i = 1, \ldots, n-1$$

Boundary conditions:

$$f_1(s_1) = w_1 x_1^2, \quad \text{where } x_1 = s_1, \ 0 \le s_i \le d$$

Answer: $f_n(d)$.

The optimal value function in stage 1 is given in closed-form; that is, $f_1(s_1) = w_1 s_1^2$. This function can be used as an input to stage 2 to determine the corresponding optimal value function in closed-form:

$$f_2(s_2) = \min_{0 \le x_2 \le s_2} \{w_2 x_2^2 + f_1(s_2 - x_2)\} = \min_{0 \le x_2 \le s_2} \{w_2 x_2^2 + w_1(s_2 - x_2)^2\}$$

Figure 7.2 shows the single-variable quadratic function $w_2 x_2^2 + w_1(s_2 - x_2)^2$, which is convex. The convexity of the function is proved below by showing that the second derivative is positive.

$$\frac{\partial \{w_2 x_2^2 + w_1(s_2 - x_2)^2\}}{\partial x_2} = 2x_2(w_1 + w_2) - 2s_2 w_1$$

$$\frac{\partial^2 \{w_2 x_2^2 + w_1(s_2 - x_2)^2\}}{\partial x_2^2} = 2(w_1 + w_2) > 0, \quad \text{since } w_1, w_2 > 0$$

The unbounded minimum is determined by setting the first derivative to zero. If this minimum falls between bounds, $0 \le x_2 \le s_2$, it is feasible and solves the subproblem; otherwise, the minimum in the feasible range would be one of the two boundary points.

$$2x_2(w_1 + w_2) - 2s_2 w_1 = 0 \Rightarrow x_2 = \frac{w_1}{w_1 + w_2} s_2 = \frac{\frac{1}{w_2}}{\frac{1}{w_1} + \frac{1}{w_2}} s_2$$

Dynamic Programming

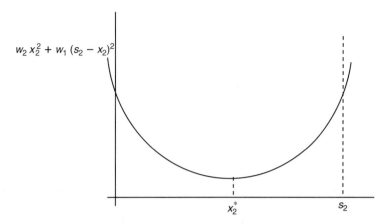

FIGURE 7.2 Function to be minimized in stage 2.

As the above solution is feasible (between bounds), it solves the subproblem in stage 2. This solution can be incorporated into the recurrence relation for stage 2 to compute the following optimal value function in closed-form:

$$f_2(s_2) = \frac{s_2^2}{\frac{1}{w_1} + \frac{1}{w_2}}$$

THEOREM 7.1 *In general, the optimal policy and value function for the simple production problem is given by the following equations:*

$$x_i = \frac{\frac{1}{w_i}}{\sum_{j=1}^{i} \frac{1}{w_j}} s_i \quad \text{and} \quad f_i(s_i) = \frac{s_i^2}{\sum_{j=1}^{i} \frac{1}{w_j}}, \quad i = 1, \ldots, n-1$$

PROOF 7.1 (By induction) Since the result has already been proved for stages 1 and 2, now we only need to show that, if it is true for stage i, it must be true for stage $i+1$. Note that in stage $i+1$,

$$f_{i+1}(s_{i+1}) = \min_{0 \le x_{i+1} \le s_{i+1}} \{w_{i+1} x_{i+1}^2 + f_i(s_{i+1} - x_{i+1})\}$$

Now, the term $f_i(s_{i+1} - x_{i+1})$ can be replaced by its closed-form expression, which is assumed to be correct in stage i.

$$f_{i+1}(s_{i+1}) = \min_{0 \le x_{i+1} \le s_{i+1}} \left\{ w_{i+1} x_{i+1}^2 + \frac{(s_{i+1} - x_{i+1})^2}{\sum_{j=1}^{i} \frac{1}{w_j}} \right\}$$

The minimum is obtained by setting the first derivative to zero.

$$\frac{\partial}{\partial x_{i+1}}\left\{w_{i+1}x_{i+1}^2 + \frac{(s_{i+1}-x_{i+1})^2}{\sum_{j=1}^{i}\frac{1}{w_j}}\right\} = \frac{2}{\sum_{j=1}^{i}\frac{1}{w_j}}\left[\left(w_{i+1}\sum_{j=1}^{i}\frac{1}{w_j}+1\right)x_{i+1}-s_{i+1}\right] = 0$$

$$\Rightarrow x_{i+1} = \frac{\frac{1}{w_{i+1}}}{\sum_{j=1}^{i+1}\frac{1}{w_j}}s_{i+1}$$

which proves the closed-form solution for the optimal policy. This value is the unique minimum because the second derivative can be shown to be positive. Now, by replacing this expression for x_{i+1} into $f_{i+1}(s_{i+1})$, the closed-form solution for the optimal value function can be derived:

$$f_{i+1}(s_{i+1}) = w_{i+1}\left(\frac{\frac{1}{w_{i+1}}}{\sum_{j=1}^{i+1}\frac{1}{w_j}}s_{i+1}\right)^2 + \frac{\left(s_{i+1}-\frac{\frac{1}{w_{i+1}}}{\sum_{j=1}^{i+1}\frac{1}{w_j}}s_{i+1}\right)^2}{\sum_{j=1}^{i}\frac{1}{w_j}}$$

which can be simplified to

$$f_{i+1}(s_{i+1}) = \frac{s_{i+1}^2}{\sum_{j=1}^{i+1}\frac{1}{w_j}} \quad \blacksquare$$

In this application, DP has been used to derive a simple closed-form solution for the problem. Below, the closed-form solution is applied to solve a numerical example.

Example 7.2

In this example, we solve a three-period production planning problem for a total demand of 9 units. The coefficients of the objective function are: $w_1 = 2$, $w_2 = 3$, and $w_3 = 6$.

Solution 7.2

The closed-form solution can be applied starting at stage 3 and going backwards to stages 2 and 1.

Stage $i=3$:

$$s_3 = d = 9 \Rightarrow x_3^* = \frac{\frac{1}{6} \cdot 9}{\frac{1}{2}+\frac{1}{3}+\frac{1}{6}} = 1.5$$

Stage $i=2$:

$$s_2 = 9-1.5 = 7.5 \Rightarrow x_2^* = \frac{\frac{1}{3} \cdot 7.5}{\frac{1}{2}+\frac{1}{3}} = 3$$

Stage $i=1$:

$$s_1 = 7.5 - 3 = 4.5 \Rightarrow x_1^* = 4.5$$

$$f_3(9) = \frac{9^2}{\frac{1}{2}+\frac{1}{3}+\frac{1}{6}} = 81$$

7.2.5 Dynamic Inventory Planning Problem [11,15]

One of the important applications of DP for many decades has been in the area of inventory planning and control over a finite number of time periods, where demand is known but may change from period to period. Let r_i be the demand in period i, $i = 1, \ldots, n$. At the beginning of each period the company needs to decide the number of units to be ordered to an external supplier or produced in the manufacturing floor in that period. If an order in period i is submitted, the ordering cost consist of a fixed setup cost, K_i, and a unit variable cost, c_i, which must be multiplied by the order quantity. Inventory shortages are not allowed. If inventory is carried from period i to period $i+1$, a holding cost is incurred in that period. The holding cost in period i is determined by multiplying the unit holding cost, h_i, by the number of units of inventory by the end of period i, which are carried from period i to period $i+1$. In the following DP model the initial inventory at the beginning of period 1 and the final inventory at the end of period n are assumed to be zero. If this is not the case, the original problem can be slightly modified by subtracting the initial inventory in period 1 from the demand in that period, and adding the final required inventory in period n to the demand of that period.

The proposed model belongs to the class of *periodic review models*, because the company reviews the inventory level at the beginning of each period, which can be a week or a month, and then the ordering decision is made. This model is an alternative to the continuous review model in which the company keeps track of the inventory level at all times and an order can be submitted at any time.

DP Formulation and Solution

Similar to the model in Section 7.2.3, the stages of this DP formulation correspond to the n production periods. The cost at any stage i, $i = 1, 2, \ldots, n$, is a function of the inventory on hand at the beginning of the stage before ordering, x_i, and the quantity ordered in that stage, z_i:

$$c_i(x_i, z_i) = \begin{cases} h_i(x_i - r_i), & \text{if } z_i = 0 \\ K_i + c_i z_i + h_i(x_i + z_i - r_i), & \text{if } z_i > 0 \end{cases}$$

Note that to avoid shortages, $z_i \geq \max\{0, r_i - x_i\}$, $i = 1, 2, \ldots, n$, which means that the initial inventory plus the order quantity must be greater than or equal to the demand in period i.

The DP approach presented here takes advantage of the following optimality condition, which is also known as the *zero inventory ordering property* (Wagner and Whitin, 1957): "For an arbitrary demand requirement and concave costs (e.g., a fixed setup cost, and linear production and holding costs), there exists an optimal policy that orders (or produces) only when the inventory level is zero." This condition implies that $x_i^* \cdot z_i^* = 0$, $i, = 1, 2, \ldots, n$. Based on this property, the only order quantities that need to be considered in each stage (period) i are either 0, r_i, $r_i + r_{i+1}, \ldots, r_i + r_{i+1} + \cdots + r_n$. By taking advantage of this property, a simple DP model can be developed with only one state in each stage, assuming that the initial inventory on hand, before ordering, is zero. Thus, given that an order must be submitted in stage i, the order quantity must correspond to the sum of demands of a certain number of periods ahead, $r_i + r_{i+1} + \cdots + r_j$, where j is the last period whose demand will be served by this order. Therefore, $j \in \{i, i+1, \ldots, n\}$ is the decision variable. An efficient DP formulation is provided below.

Optimal value function (f_i): the total cost of the best policy from the beginning of period i to the end of period n, given that the inventory on hand, before ordering, in period i is zero.

Optimal policy ($p_i = j^*$): given that the inventory on hand is zero, the optimal policy indicates the last period to be served from the order issued in stage i.

Transformation function ($t_i(j) = j + 1$): it shows the next stage $j+1$ in which an order will be issued, given that a production order is submitted in stage i.

Recurrence relation: The general recurrence relation is

$$f_i = \min_{j=i,i+1,\ldots,n} \{K_i + c_i(r_i + r_{i+1} + \cdots + r_j) + h_i(r_{i+1} + \cdots + r_j) \\ + h_{i+1}(r_{i+2} + \cdots + r_j) + \cdots + h_{j-1}r_j + f_{j+1}\}, \quad i = 1, \ldots, n$$

If the cost parameters are stationary, that is, $K_i = K$, $c_i = c$, $h_i = h$, $i = 1, \ldots, n$, the recurrence relation can be slightly simplified as follows:

$$f_i = \min_{j=i,i+1,\ldots,n} \{K + c(r_i + r_{i+1} + \cdots + r_j) + h[r_{i+1} + 2r_{i+2} + 3r_{i+3} + \cdots \\ + (j-i)r_j] + f_{j+1}\}, \quad i = 1, \ldots, n$$

Boundary conditions: $f_{n+1} = 0$.

Answer: f_1.

Example 7.3

In this example, we consider a five-period dynamic inventory problem with stationary cost data: $K = \$40$, $c = \$10/\text{unit}$, and $h = \$3/\text{unit}/\text{period}$. The known demand for each period is the following: $r_1 = 2$ units, $r_2 = 4$ units, $r_3 = 2$ units, $r_4 = 2$ units, and $r_5 = 3$ units.

Solution 7.3

Stage 6: (Initialization)

$$f_6 = 0$$

Stage 5: ($r_5 = 3$)

j	$K + cr_5 + f_6$	f_5	j^*
5	$40 + 10 \times 3 + 0 = 70$	70	5

Stage 4: ($r_4 = 2$)

j	$K + c(r_4 + \cdots + r_j) + h[r_5 + \cdots + (j-4)r_j] + f_{j+1}$	f_4	j^*
4	$40 + 10 \times 2 + 0 + 70 = 130$		
5	$40 + 10 \times (2+3) + 3 \times 3 + 0 = 99$	99	5

Dynamic Programming

Stage 3: ($r_3 = 2$)

j	$K + c(r_3 + r_4 + \cdots + r_j) + h[r_4 + 2r_5 + \cdots + (j-3)r_j] + f_{j+1}$	f_3	j^*
3	$40 + 10 \times 2 + 0 + 99 = 159$		
4	$40 + 10 \times (2+2) + 3 \times 2 + 70 = 156$	134	5
5	$40 + 10 \times (2+2+3) + 3 \times (2 + 2 \times 3) + 0 = 134$		

Stage 2: ($r_2 = 4$)

j	$K + c(r_2 + r_3 + \cdots + r_j) + h[r_3 + 2r_4 + \cdots + (j-2)r_j] + f_{j+1}$	f_2	j^*
2	$40 + 10 \times 4 + 0 + 134 = 214$		
3	$40 + 10 \times (4+2) + 3 \times 2 + 99 = 205$	195	5
4	$40 + 10 \times (4+2+2) + 3 \times (2 + 2 \times 2) + 70 = 208$		
5	$40 + 10 \times (4+2+2+3) + 3 \times (2 + 2 \times 2 + 3 \times 3) + 0 = 195$		

Stage 1: ($r_1 = 2$)

j	$K + c(r_1 + r_2 + \cdots + r_j) + h[r_2 + 2r_3 + \cdots + (j-1)r_j] + f_{j+1}$	f_1	j^*
1	$40 + 10 \times 2 + 0 + 195 = 255$		
2	$40 + 10 \times (2+4) + 3 \times 4 + 134 = 246$		
3	$40 + 10 \times (2+4+2) + 3 \times (4 + 2 \times 2) + 99 = 243$	243	3
4	$40 + 10 \times (2+4+2+2) + 3 \times (4 + 2 \times 2 + 3 \times 2) + 70 = 252$		
5	$40 + 10 \times (2+4+2+2+3) + 3 \times (4 + 2 \times 2 + 3 \times 2 + 4 \times 3) + 0 = 248$		

Solution 7.4

Minimum Total Cost: $f_1 = \$243$.

Stage 1:

$$j^* = 3 \quad \Rightarrow \quad \text{Order } r_1 + r_2 + r_3 = 2 + 4 + 2 = 8 \text{ units in period 1.}$$

Stage 4:

$$j^* = 5 \quad \Rightarrow \quad \text{Order } r_4 + r_5 = 2 + 3 = 5 \text{ units in period 4.}$$

Graphical Solution

A directed graph is constructed with $n+1$ nodes. Each node i, $i = 1, 2, \ldots, n+1$, represents the beginning of a production period with zero units of inventory on hand. For each node i, an arc (i, j) is added to the graph, for $j = i+1, \ldots, n+1$. Arc $(i, j+1)$ represents the case where the inventory on hand at the beginning of period i is zero and an order of $r_i + \cdots + r_j$ units is submitted in that period, so that at the beginning of period $j+1$ the inventory level will become zero again. The cost associated with arc $(i, j+1)$, denoted as $c_{i,j+1}$, represents the total ordering and holding cost for periods i to j for this case:

$$c_{i,j+1} = K_i + c_i(r_i + r_{i+1} + \cdots + r_j) + h_i(r_{i+1} + \cdots + r_j) \\ + h_{i+1}(r_{i+2} + \cdots + r_j) + \cdots + h_{j-1}r_j$$

If the cost parameters are stationary, that is, $K_i = K$, $c_i = c$, $h_i = h$, $i = 1, \ldots, n$, the arc cost $c_{i,j+1}$ becomes

$$c_{i,j+1} = K + c(r_i + r_{i+1} + \cdots + r_j) + h[r_{i+1} + 2r_{i+2} + 3r_{i+3} + \cdots + (j-i)r_j]$$

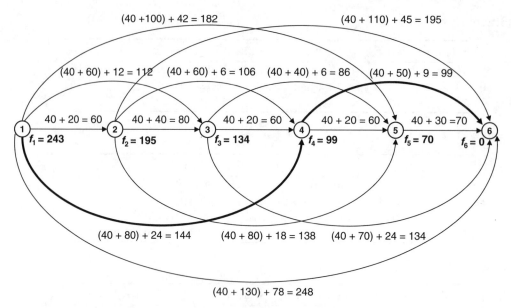

FIGURE 7.3 Graphical solution for the dynamic inventory example.

The optimal inventory policy is determined by finding the shortest path from node 1 to node $n+1$.

Figure 7.3 shows the 6-node graph for the example problem. The arc costs have been computed using the formula for the stationary cost parameters. Detailed calculations of two arc costs are shown below.

$$c_{1,6} = K + c(r_1 + r_2 + \cdots + r_5) + h[r_2 + 2r_3 + \cdots + 4r_5]$$
$$= 40 + 10 \times (2 + 4 + 2 + 2 + 3) + 3 \times (4 + 2 \times 2 + 3 + 2 + 4 \times 3) + 0 = 248$$
$$c_{2,4} = K + c(r_2 + r_3) + hr_3 = 40 + 10 \times (4 + 2 + 2) + 3 \times (2 + 2 \times 2) + 7 = 208$$

Once all arc costs are calculated, the shortest path from node 1 to node 6 can be found in a straightforward manner as the graph is acyclic. The computational effort required to find the shortest path is exactly the same than the effort required to solve all DP stages. Initially, the optimal value function at stage 6 is set to zero, $f_6 = 0$. At stage 5, the optimal value function (length of the shortest path from node (stage) 5 to node 6) is $f_5 = c_{5,6} + f_6 = 70 + 0 = 70$. The remaining optimal value functions can be computed recursively with the following simplified recurrence relation:

$$f_i = \min_{j=i,i+1,\ldots,n} \{c_{i,j} + f_{j+1}\}$$

7.2.6 Reliability System Design [11]

Figure 7.4 shows an illustration of an electromechanical device that contains three components in serial arrangement so that each component must work for the system to function. The reliability of the system can be improved by installing several parallel units in one or more of the components. Table 7.4 shows the reliability of each component as a function of the number of parallel units.

The reliability of the device is the product of the reliabilities of the three components. The cost of installing one, two, or three parallel units in each component (in thousands of dollars) is given in Table 7.5.

Dynamic Programming

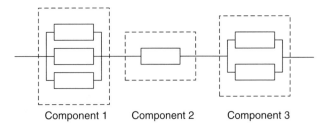

Component 1 Component 2 Component 3

FIGURE 7.4 Illustration of the electromechanical system.

TABLE 7.4 Reliability for Each Component

Parallel Units (x_i)	Probability of Functioning $p_i(x_i)$		
	Component $i=1$	Component $i=2$	Component $i=3$
1	0.6	0.7	0.8
2	0.7	0.8	0.9
3	0.9	0.9	0.95

TABLE 7.5 Cost of Each Component (in Thousands of Dollars)

Parallel Units (x_i)	Cost of the Component $c_i(x_i)$		
	Component $i=1$	Component $i=2$	Component $i=3$
1	3	2	1
2	5	4	4
3	6	5	5

Because of budget limitations, a maximum of $10,000 can be expended. We need to determine the number of units for each component so that the reliability of the system is maximized.

DP Formulation and Solution

This problem can be formulated as a three-stage DP model. At each stage i, a decision has to be made concerning the number of parallel units, x_i, to be assigned to the ith component. This decision depends on the remaining budget available for stages i to 3. Thus, the state of the system, s_i, is the remaining budget available at the beginning of that stage (in thousands of dollars). At each stage i, it is necessary to determine a range of feasible states. For example, for $i=3$, $s_3 \in \{1,\ldots,5\}$, because 1 is the cost of one unit of component 3 and 5 is the maximum available budget, given that at least one unit of component 1 and one unit of component 2 have to be bought, i.e., $10-(3+2)=5$. Similar arguments can be provided to establish the range of s_2: $s_2 \in \{3,\ldots,7\}$. The transformation function computes the budget available by the beginning of state $i+1$ as a function of the budget available at the beginning of stage i and the number of units assigned to component i:

$$s_{i+1} = t_i(s_i, x_i) = s_i - c_i(x_i)$$

where $c_i(x_i)$ is the cost of x_i units of component i (see Table 7.5).

The optimal value function, $f_i(s_i)$, gives the maximum probability of functioning of a subsystem defined by components $i=1,2,3$ in serial arrangement given that the budget available for these components is s_i. Based on this definition, the recurrent relation can be established as follows:

$$f_i(s_i) = \max_{x_i=1,2,3}\{p_i(x_i) \cdot f_{i+1}(s_i - c_i(x_i))\}, \quad s_i = l_i,\ldots,u_i; \quad i=1,2$$

where l_i and u_i are the lower and upper bounds on s_i, and $p_i(x_i)$ is the reliability of component i with x_i units in parallel. The boundary conditions in stage 3 are defined as follows:

$$f_3(s_3) = \max_{x_i=1,2,3} \{p_3(x_3) : c_3(x_3) \leq s_3\}$$

In this problem, $f_1(10)$ gives the maximum system reliability that can be obtained with a budget of \$10,000. The numerical solution for this problem is provided in the following tables.

Stage 3:

	$p_3(x_3)$					
s_3	$x_3=1$	$x_3=2$	$x_3=3$	$f_3^*(s_3)$	x_3^*	s_4
1	0.8	Infeasible	Infeasible	0.8	1	0
2	0.8	Infeasible	Infeasible	0.8	1	1
3	0.8	Infeasible	Infeasible	0.8	1	2
4	Non-optimal	0.9	Infeasible	0.9	2	0
5	Non-optimal	Non-optimal	0.95	0.95	3	0

Stage 2:

	$p_2(x_2) \cdot f_3(s_2 - c_2(x_2))$					
s_2	$x_2=1$	$x_2=2$	$x_2=3$	$f_2^*(s_2)$	x_2^*	s_3
3	$0.7 \times 0.8 = 0.56$	Infeasible	Infeasible	0.56	1	1
4	$0.7 \times 0.8 = 0.56$	Infeasible	Infeasible	0.56	1	2
5	$0.7 \times 0.8 = 0.56$	$0.8 \times 0.8 = 0.64$	Infeasible	0.64	2	1
6	$0.7 \times 0.9 = 0.63$	$0.8 \times 0.8 = 0.64$	Infeasible	0.64	2	2
7	$0.7 \times 0.95 = 0.665$	$0.8 \times 0.8 = 0.64$	$0.9 \times 0.8 = 0.72$	0.72	3	2

Stage 1:

	$p_1(x_1) \cdot f_2(s_1 - c_1(x_1))$					
s_1	$x_1=1$	$x_1=2$	$x_1=3$	$f_1^*(s_1)$	x_1^*	s_2
10	$0.6 \times 0.72 = 0.432$	$0.7 \times 0.64 = 0.448$	$0.9 \times 0.56 = 0.504$	0.504	3	4

Optimal Solution

The optimal solution is unique and can be obtained by backtracking, starting at stage 1 and moving to stages 2 and 3. Using the transformation function, it is possible to determine the state at stage $i+1$, s_{i+1}, from the state in stage i, s_i, and the corresponding optimal decision, x_i^*. Details of this process are provided below.

$s_1 = 10,$ $\qquad x_1^*(10) = 3$ units of component 1 are used

$s_2 = t_1(s_1, x_1) = 10 - 6 = 4,$ $\qquad x_2^*(4) = 1$ unit of component 2 is used

$s_3 = t_2(s_2, x_2) = 4 - 2 = 2,$ $\qquad x_3^*(1) = 1$ unit of component 3 is used

$f_1^*(10) = 0.504$ (system reliability)

7.3 Stochastic Dynamic Programming Models

Decision makers generally face situations in which they need to make decisions with many uncertainties concerning the consequences of these decisions. A manufacturer may produce items for inventory without knowing how many units customers will demand for each type of item. An investor may decide to buy a certain number of shares of a particular stock without knowing if the stock will go up or down in the future. Although the future may be uncertain, in general, it is still possible to estimate probability distributions for all possible outcomes associated with a decision [4,7,9,12].

All the DP applications discussed in the previous section have considered situations in which the consequences of all possible decisions were completely predetermined, including the system reliability problem in Section 7.2.6. Although the objective of that problem was to maximize the expected system reliability, it can be categorized as deterministic in the sense that, when a decision is made at any stage, the state resulting from that decision is precisely determined and known. The essential difference between the stochastic models to be discussed in this section and the prior deterministic models is that the state resulting from a decision is not predetermined, but can be described by a known probability distribution function that depends on the initial state and the decision taken.

In stochastic DP, let the pair (i,j) denote state j of stage i and $D(i,j)$ denote the set of all possible decisions associated with (i,j). If the system is observed to be in (i,j), some decision x in $D(i,j)$ must be selected. Assuming that stage i is not the last stage n, for example, $i < n$, decision x will cause a transition to some state k in stage $i+1$; the particular state k can be characterized by the following transition probability:

$P_{jk}^i(x) =$ probability that the state observed in stage $i+1$ will be k, given that the current state in stage i is j and decision x is made.

Note that, for a particular pair (i,j) and a specific decision $x \in D(i,j)$, $\sum_k P_{jk}^i(x) = 1$ and $P_{jk}^i(x) \geq 0$, for any state k in the state space. Two applications of stochastic DP using this type of probabilities are presented in this section.

Note that, in stochastic DP, a different set of states may occur for different replications of the same problem even though the same policy is applied. The simple explanation is that, for a given stage, state, and decision, the resulting state at the next stage is only known by a probability distribution. Therefore, in stochastic DP, an optimal policy is characterized by the best decision for each possible state at any stage.

7.3.1 Stochastic Equipment Replacement Problem [3,8,10]

First, we present an extension of the equipment replacement problem discussed in Section 7.2.3. Now, we assume that the annual operating cost of a machine of age j is a random variable with known probability distribution. In addition, it is assumed that the machine may suffer a severe failure at the end of any year and must be replaced. The probability of a severe failure only depends on the age of the machine at the beginning of the year. Finally, at the beginning of a year, we allow the option of salvaging the currently owned machine and leasing a new machine for a year. If the company has already leased a machine during the prior year, it can lease another machine or buy a new one at the beginning of the year. Assuming that the company owns a machine of age y by the beginning of the first year, and replacement or leasing decisions can only be made at the beginning of each year, we need to find the optimal replacement/leasing policy. The input data defining this problem are:

n = number of years to be considered

y = age of the machine at the beginning of year 1

$a(i)$ = purchasing price of a new machine at the beginning of year i

$\bar{r}(j)$ = expected annual operating cost for a machine of age j at the beginning of the year

$t_w(j)$ = trade-in value for a machine of age j that is in working condition

$v_w(j)$ = salvage value for a machine of age j that is in working condition

$t_b(j)$ = trade-in value for a machine of age j that is broken down

$v_b(j)$ = salvage value for a machine of age j that is broken down

$q(j)$ = probability that a machine of age j in working condition at the beginning of a year breaks down by the end of the year

λ = annual cost for leasing a machine

DP Formulation and Solution

In this problem, it is assumed that the company does not have to pay anything if a leased machine breaks down by the end of a year as the leasing cost includes insurance for the machine.

The following DP formulation uses two optimal value functions. One function provides the value of the best policy for years i to n, given that the company owns a machine at the beginning of year i. The second function gives the value of the best policy if a machine was leased in the prior year. In the case of the first function, a number of states need to be considered depending on the age of the owned machine at the beginning of year i. If the machine was leased, only one state exists because the company does not currently own any machine. In the two recurrence relations stated below, some states require two decisions. The first decision is applied at the beginning of year i and the second decision is taken only if the company decides to own a machine during the current year and the machine incurs a severe failure by the end of the year. If the decision taken at the beginning of the year is to lease a machine, then the second decision is unnecessary. The following notation is used to represent the decision making process:

B: "buy a new machine"

K: " keep the current machine"

and L: "lease a new machine for one year"

Stage (i): year i, $i = 1, \ldots, n$.

State: if the company owned a machine in year $i-1$, then the state of the system at the beginning of year i, denoted as s_i, is the age of the machine; if a machine was leased in year $i-1$, the only state is that the company does not own any machine at the beginning of year i.

Decision variable ($x_i = (x_{i1}, x_{i2})$): two decisions may have to be taken in each state. The first decision at the beginning of year i is $x_{i1} \in \{$"buy," "keep," "lease"$\}$,

Dynamic Programming

obviously the "keep" option is only available if the company owns a machine. The second decision may be necessary at the end of stage i, if an owned machine breaks down at the end of the year, $x_{i2} \in \{\text{"buy," "lease"}\}$.

Optimal value functions: the following two functions need to be evaluated in each stage:

$f_i(s_i) =$ minimum expected cost from the beginning of year i to the beginning of year $n+1$, given that the company owns a working machine of age s_i at the beginning of year i.

$g_i =$ minimum expected cost from the beginning of year i to the beginning of year $n+1$, given that the company does not own any machine at the beginning of year i as a machine was leased in the preceding year.

Optimal policy ($p_f(i,s_i)$ or $p_g(i) = x_i^*$): optimal replacement/leasing plan for year i.

Recurrence relation: For $i = n-1, \ldots, 2, 1$; $s_i = 1, 2, \ldots, i, y + (i-1)$

$$f_i(s_i) = \min \begin{cases} \text{B:} & p(i) - t_w(s_i) + \bar{r}(0) + (1 - q(0)) \cdot f_{i+1}(1) + q(0) \\ & \cdot \min \begin{cases} \text{B:} & p(i+1) - t_b(1) + f_{i+1}(0) \\ \text{L:} & \lambda - v_b(1) + \bar{r}(0) + g_{i+2} \end{cases} \\ \text{K:} & \bar{r}(s_i) + (1 - q(s_i)) \cdot f_{i+1}(s_i + 1) + q(s_i) \\ & \cdot \min \begin{cases} \text{B:} & p(i+1) - t_b(s_i + 1) + f_{i+1}(0) \\ \text{L:} & \lambda - v_b(s_i + 1) + \bar{r}(0) + g_{i+2} \end{cases} \\ \text{L:} & \lambda - v_w(s_i) + \bar{r}(0) + g_{i+1} \end{cases}$$

For $i = n-1, \ldots, 2, 1$; $s_i = 0$,

$$f_i(0) = \text{K:} \quad \bar{r}(0) + (1 - q(0)) \cdot f_{i+1}(1) + q(0) \cdot \min \begin{cases} \text{B:} & p(i+1) - t_b(1) + f_{i+1}(0) \\ \text{L:} & \lambda - v_b(1) + \bar{r}(0) + g_{i+2} \end{cases}$$

For $i = n-1, \ldots, 2, 1$,

$$g_i = \min \begin{cases} \text{B:} & p(i) + \bar{r}(0) + (1 - q(0)) \cdot f_{i+1}(1) + q(0) \\ & \cdot \min \begin{cases} \text{B:} & p(i+1) - t_b(1) + f_{i+1}(0) \\ \text{L:} & \lambda - v_b(1) + \bar{r}(0) + g_{i+2} \end{cases} \\ \text{L:} & \lambda + \bar{r}(0) + g_{i+1} \end{cases}$$

Boundary conditions: For $s_i = 1, 2, \ldots, n, y + (n-1)$,

$$f_n(s_i) = \min \begin{cases} \text{B:} & p(n) - t_w(s_i) + \bar{r}(0) + (1 - q(0)) \cdot v_w(1) + q(0) \cdot v_b(1) \\ \text{K:} & \bar{r}(s_i) + (1 - q(s_i)) \cdot v_w(s_i + 1) + q(s_i) \cdot v_b(s_i + 1) \\ \text{L:} & \lambda - v_w(s_i) + \bar{r}(0) \end{cases}$$

$$g_n = \min \begin{cases} \text{B:} & p(n) + \bar{r}(0) + (1 - q(0)) \cdot v_w(1) + q(0) \cdot v_b(1) \\ \text{L:} & \lambda + \bar{r}(0) \end{cases}$$

$g_{n+1} = 0$ (This case may be used in the computation of $f_{n-1}(s_i)$ and g_{n-1}.)

Answer: $f_1(y)$.

7.3.2 Investment Planning [11]

We have $10,000 to invest in the following 3 years. Investments can only be made for 1 year at the beginning of any year in multiples of $10,000. Two types of investment opportunities are available: A and B. Investment B is more conservative than investment A. If we invest in A in a given year, at the end of the year the money will be lost with probability 0.3 or will be doubled with probability 0.7. If the investment is made in B, at the end of the year the money will be returned with probability 0.9 or will be doubled with probability 0.1. The objective of this problem is to come up with an investment policy that maximizes the total expected return by the end of the third year. In order to simplify the solution procedure, it is assumed that only one type of investment can be selected in a given year.

DP Formulation and Solution

In the following formulation, the set of possible decisions depends on the amount of money available at the beginning of a year. If we have less than $10,000, the only possibility is to keep the money, but if the money available for investment is $10,000 or more, multiples of $10,000 can be invested in A or B.

Stage (i): year i, $i = 1, 2, 3$.
State (s_i): amount of money on hand at the beginning of year i. In year 1, $s_1 = \$10,000$.
Decision variable ($x_i = (x_{i1}, x_{i2})$): the decision in stage i consists of the amount of money to invest, x_{i1}, in multiples of $10,000, and the type of investment to be made, $x_{i2} \in \{0, A, B\}$, where 0 means that no investment is made.
Optimal value function ($f_i(s_i)$): maximum expected amount of money from the beginning of year i to the end of year 3 given that the money on hand is s_i at the beginning of year i.
Optimal policy ($p_i(s_i) = x_i^*$): optimal investments plan for year i, given that the money on hand is s_i.
Recurrent relation: For $0 \leq s_i < 10$, $x_i = (x_{i1}, x_{i2}) = (0, 0)$, $f_i(s_i) = s_i$.
For $s_i \geq 10$,

$$f_i(s_i) = \max \begin{cases} f_{i+1}(s_i) & \text{for } (x_{i1}, x_{i2}) = (0, 0) \\ \max_{x_{i1} \in A} \{0.3 \times f_{i+1}(s_i - x_{i1}) \\ \quad + 0.7 \times f_{i+1}(s_i + x_{i1})\} & \text{for } x_{i2} = A \\ \max_{x_{i1} \in A} \{0.9 \times f_{i+1}(s_i) + 0.1 \times f_{i+1}(s_i + x_{i1})\}, & \text{for } x_{i2} = B \end{cases}$$

where $A = \{a : a = 10, 20, \ldots; a \leq s_i\}$ is the set of possible investment quantities (in thousands of dollars).

Dynamic Programming

Boundary conditions: For $0 \leq s_3 < 10$, $x_3 = (x_{31}, x_{32}) = (0,0)$, $f_3(s_3) = s_3$.
For $s_3 \geq 10$,

$$f_3(s_3) = \max \begin{cases} s_3 & \text{for } (x_{31}, x_{32}) = (0,0) \\ \max_{x_{31} \in A} \{0.3 \times (s_3 - x_{31}) + 0.7 \times (s_3 + x_{31})\} & \text{for } x_{32} = A \\ \max_{x_{31} \in A} \{0.9 \times (s_3) + 0.1 \times (s_3 + x_{31})\} & \text{for } x_{32} = B \end{cases}$$

where $A = \{a : a = 10, 20, \ldots; a \leq s_3\}$.

Answer: $f_1(10)$.

Numerical solution

The tables below analyze all possible investment opportunities for each year (stage) depending on the money on hand at the beginning of the year (state). The decision that gives the maximum expected return by the end of the third year is selected.

Stage 3:

		s_3	$0.3 \times (s_3 - x_{31}) + 0.7 \times (s_3 + x_{31})$	$0.9 \times s_3 + 0.1 \times (s_3 + x_{31})$			
s_3	x_{31}	$x_{32} = 0$	$x_{32} = A$	$x_{32} = B$	x_{31}^*	x_{32}^*	$f_3(s_3)$
0	0	0	–	–	0	0	0
10	0	10	–	–			
	10	–	$0.3 \times (10-10) + 0.7 \times (10+10) = 14$	$0.9 \times 10 + 0.1 \times (10+10) = 11$	10	A	14
20	0	20	–	–			
	10	–	$0.3 \times (20-10) + 0.7 \times (20+10) = 24$	$0.9 \times 20 + 0.1 \times (20+10) = 21$			
	20	–	$0.3 \times (20-20) + 0.7 \times (20+20) = 28$	$0.9 \times 20 + 0.1 \times (20+20) = 22$	20	A	28
30	0	30	–	–			
	10	–	$0.3 \times (30-10) + 0.7 \times (30+10) = 34$	$0.9 \times 30 + 0.1 \times (30+10) = 31$			
	20	–	$0.3 \times (30-20) + 0.7 \times (30+20) = 38$	$0.9 \times 30 + 0.1 \times (30+20) = 32$			
	30	–	$0.3 \times (30-30) + 0.7 \times (30+30) = 42$	$0.9 \times 30 + 0.1 \times (30+30) = 33$	30	A	42
40	0	40	–	–			
	10	–	$0.3 \times (40-10) + 0.7 \times (40+10) = 44$	$0.9 \times 40 + 0.1 \times (40+10) = 41$			
	20	–	$0.3 \times (40-20) + 0.7 \times (40+20) = 48$	$0.9 \times 40 + 0.1 \times (40+20) = 42$			
	30	–	$0.3 \times (40-30) + 0.7 \times (40+30) = 52$	$0.9 \times 40 + 0.1 \times (40+30) = 43$			
	40	–	$0.3 \times (40-40) + 0.7 \times (40+40) = 56$	$0.9 \times 40 + 0.1 \times (40+40) = 44$	40	A	56

Stage 2:

		$f_3(s_2)$	$0.3 \times f_3(s_2 - x_{21}) + 0.7 \times f_3(s_2 + x_{21})$	$0.9 \times f_3(s_2) + 0.1 \times f_3(s_2 + x_{21})$			
s_2	x_{21}	$x_{22} = 0$	$x_{22} = A$	$x_{22} = B$	x_{21}^*	x_{22}^*	$f_2(s_2)$
0	0	$f_3(0) = 0$	–	–	0	0	0
10	0	$f_3(10) = 14$	–	–			
	10	–	$0.3 \times f_3(10-10) + 0.7 \times f_3(10+10)$ $= 0.3 \times 0 + 0.7 \times 28 = 19.6$	$0.9 \times f_3(10) + 0.1 \times f_3(10+10)$ $= 0.9 \times 14 + 0.1 \times 28 = 15.4$	10	A	19.6
20	0	$f_3(20) = 28$	–	–			
	10	–	$0.3 \times f_3(20-10) + 0.7 \times f_3(20+10)$ $= 0.3 \times 14 + 0.7 \times 42 = 33.6$	$0.9 \times f_3(20) + 0.1 \times f_3(20+10)$ $= 0.9 \times 28 + 0.1 \times 42 = 29.4$			
	20	–	$0.3 \times f_3(20-20) + 0.7 \times f_3(20+20)$ $= 0.3 \times 0 + 0.7 \times 56 = 39.2$	$0.9 \times f_3(20) + 0.1 \times f_3(20+20)$ $= 0.9 \times 28 + 0.1 \times 56 = 30.8$	20	A	39.2

Stage 1:

s_1	x_{11}	$f_2(s_1)$ $x_{12}=0$	$0.3 \times f_2(s_1 - x_{11}) + 0.7 \times f_2(s_1 + x_{11})$ $x_{12}=A$	$0.9 \times f_2(s_1) + 0.1 \times f_2(s_1 + x_{11})$ $x_{12}=B$	x_{11}^*	x_{12}^*	$f_1(s_1)$
10	0	$f_2(10) = 19.6$	—	—			
	10	—	$0.3 \times f_2(10-10) + 0.7 \times f_2(10+10)$ $= 0.3 \times 0 + 0.7 \times 39 = 27.3$	$0.9 \times f_2(10) + 0.1 \times f_2(10+10)$ $= 0.9 \times 19.6 + 0.1 \times 39.2 = 21.56$	10	A	27.3

Optimal Solution

$x_1^* = (x_{11}^*, x_{12}^*) = (10, A) \equiv$ Invest \$10,000 in A.

$f_1(10) = 27.3$ $\equiv$ The expected amount of money by the end of the third year will be \$27,300.

7.4 Conclusions

This chapter has introduced the reader to DP, which is a particular approach to solve optimization problems. In an optimization problem, we try to find the best solution from a set of alternatives. One of the main difficulties of DP is in the development of the mathematical formulation for a particular problem and the establishment of the recurrence relation that allows us to solve instances of the problem in stages in an efficient manner. For this reason, we have taken the approach of introducing DP by discussing the formulation of some of the important applications in the areas of industrial engineering and management science. These applications have been categorized in deterministic models, where data is known with certainty, and stochastic models, in which some of the information is uncertain and require the use of probability distribution functions.

In many real-world applications, the difficulty of DP is the large number of states that need to be considered to solve a problem. This difficulty was called the *curse of dimensionality* by Bellman (1952). The number of states can be reduced by making some additional assumptions to the problem, which may result in a model that does not capture the real-world setting. A DP model must be simple enough so that its behavior can be understood, but at the same time its robustness must be verifiable.

References

1. Bellman, R.E. (1952), "On the Theory of Dynamic Programming," *Proceedings of the National Academy of Sciences*, 38, pp. 716–719.
2. Bellman, R.E. (1957), *Dynamic Programming*, Princeton University Press, Princeton, NJ.
3. Bellman, R.E. and S.E. Dreyfus (1962), *Applied Dynamic Programming*, Princeton University Press, Princeton, NJ.
4. Bertsikas, D. (1976), *Dynamic Programming and Stochastic Control*, Academic Press, New York.
5. Blackwell, D. (1965), "Discounted Dynamic Programming," *Annals of Mathematical Statistics*, 36, pp. 226–235.
6. Cooper, L. and M.W. Cooper (1981), *Introduction to Dynamic Programming*, Pergamon Press, Elmsford, NY.
7. Denardo, E.V. (1982), *Dynamic Programming: Models and Applications*, Prentice Hall, Englewood Cliffs, NJ.

8. Derman, C. (1963), "On Optimal Replacement Rules when Changes of State are Markovian," in *Mathematical Optimization Techniques* (R.E. Bellman, Ed.), University of California Press, Berkeley, CA.
9. Derman, C. (1970), *Finite State Markovian Decision Processes*, Academic Press, New York.
10. Dreyfus, S. and A. Law (1976), *The Art of Dynamic Programming*, Academic Press, New York.
11. Hillier, F.S. and G.J. Lieberman (2005), *Introduction to Operations Research*, 8th Ed., McGraw Hill, Boston, MA.
12. Ross, S.M. (1983), *Introduction to Stochastic Dynamic Programming*, Academic Press, New York.
13. Manne, A.S. and A.F. Veinott, Jr. (1967), Chapter 11, in *Investments for Capacity Expansion: Size, Location, and Time-Phasing* (A.S. Manne, Ed.), MIT Press, Cambridge, MA.
14. Taha, H.A. (2003), *Operations Research: An Introduction*, 7th Ed., Prentice Hall, Upper Saddle River, NJ.
15. Wagner, H.M. and T. Whitin (1957), "Dynamic Problems in the Theory of the Firm," in *Theory of Inventory Management*, 2nd Ed. (T. Whitin, Ed.), Princeton University Press, Princeton, NJ.
16. Winston, W.L. (2004), *Operations Research: Applications and Algorithms*, 4th Ed., Brooks/Cole-Thomson, Belmont, CA.

8
Stochastic Processes

8.1	Introduction...	8-1
8.2	Poisson Processes	8-7
	The Exponential Distribution and Properties • Definition of Poisson Processes • Properties of Poisson Processes • Nonhomogeneous Poisson Processes and Compound Poisson Processes	
8.3	Discrete-Time Markov Chains......................	8-14
	Definition of Discrete-Time Markov Chains • Transient Analysis • Classification of States • Limiting Probabilities • Statistical Inference of Discrete-Time Markov Chains	
8.4	Continuous-Time Markov Chains..................	8-27
	Definition of Continuous-Time Markov Chains • Birth and Death Processes and Applications • Transient Analysis • Limiting Distribution • Statistical Inference of Continuous-Time Markov Chains	
8.5	Renewal Theory	8-39
	Renewal Processes • Renewal Reward Processes • Regenerative Processes • Semi-Markov Processes • Statistical Inference of Renewal Processes	
8.6	Software Products Available for Solving Stochastic Models	8-46
	References	8-47

Susan H. Xu
The Pennsylvania State University

8.1 Introduction

Deterministic models and stochastic models are two broad categories of mathematical models that aim at providing quantitative characterizations of a real system or a natural phenomenon under study. The salient difference between the two types of models is that, given a set of assumptions for each model, a deterministic model predicts a single outcome, whereas a stochastic model predicts a set of possible outcomes along with the likelihood or probability of each outcome. When a stochastic model is a more suitable choice for the purpose of investigation, it is often the case that the underlying system can be better represented by a collection or a family of random variables, indexed by a parameter such as time or space. Such a family of random variables is called a *stochastic process*. The field of stochastic processes represents a collection of models and methods used to depict the dynamic relationship of a family of random variables evolving in time or space.

The study of stochastic processes was started at the beginning of the twentieth century, and it has been an actively researched area ever since, doubtlessly because of its deep connections with practical problems. Today, stochastic processes are widely applied in different disciplines such as engineering, business, physics, biology, health care, and the military, to name a few. Stochastic processes can be used to understand the variability inherent in the underlying process, to make predictions about the system behavior, to gain insight on effective design and control of the system, and to aid in managerial decision making. The following examples illustrate the applications of stochastic processes in different fields.

Example 8.1: A brand switching model for consumer behavior

Suppose there are several brands of a product competing in a market. For example, those brands might be competing brands of soft drinks. Let us assume that every week a consumer buys one of the three brands, labeled as 1, 2, and 3. In each week, a consumer may either buy the same brand he bought the previous week or switch to a different brand. A consumer's preference can be influenced by many factors, such as brand loyalty and brand pressure (i.e., a consumer is persuaded to purchase the same brand; see Whitaker (1978)). To gauge consumer behavior, sample surveys are frequently conducted. Suppose that one of such surveys identifies the following consumer behavior:

	Following week		
Current week	Brand 1	Brand 2	Brand 3
Brand 1	0.51	0.35	0.14
Brand 2	0.12	0.80	0.08
Brand 3	0.03	0.05	0.92

For example, of those who currently bought brand 1, 51% buy the same brand, 35% switch to brand 2 and 14% to brand 3, in the next week. The brand choices of a consumer over different weeks can be represented by a stochastic process that can enter three different states, namely, 1, 2, and 3. The market share of a brand during a period is defined as the average proportion of people who buy the brand during the period. The questions of interest might be: What is the market share of a specific brand in a short run (say in 3 months) or in a long run (i.e., the average market share of the brand when the number of weeks observed is sufficiently large)? How does repeat business, due to brand loyalty and brand pressure, affect a company's market share and profitability? What is the expected number of weeks that a consumer stays with a particular brand? ∎

Example 8.2: Automobile insurance

Most insurers around the world use the Bonus Malus (Latin for good-bad) system in automobile liability insurance. Such a system gives a merit rating, represented by a positive integer-valued state, to each policyholder and determines the annual premium accordingly. A policyholder's state changes from year to year in response to the number of at-fault accidents made by the policyholder. The system penalizes at-fault accidents of a policyholder by increasing his state value (resulting in an annual premium surcharge) and rewards claim-free years by decreasing his state value (resulting in a premium discount). The following table describes a hypothetical Bonus Malus system having four states (a real Bonus Malus system usually has many more states):

		Next state if			
Current state	Annual premium	0 claims	1 claim	2 claims	≥ 3 claims
1	$500	1	2	3	4
2	$600	1	3	4	4
3	$800	2	4	4	4
4	$1000	3	4	4	4

For instance, the table indicates that if a policyholder in state 2 makes no claims this year, then the person's rating would change to state 1 the next year. Empirical data can be collected and analyzed so that a theoretical probability distribution on the number of yearly claims from a policyholder can be obtained. The collection of states visited by a policyholder, indexed by year, is a stochastic process. Based on the above table and the probability distribution of annual claims, an insurer can compute the probability that a policyholder changes from one state to another in successive years. From a model like this, an insurer can compute various performance measures, such as the long-run average premium received from a policyholder and its own insurance risk. ∎

Example 8.3: Reliability

The reliability of a system, possibly consisting of several parts, is defined as the probability that the system will function during its assigned mission time. The measure is determined mainly by the lifetime distributions of the constituting components and the structure function of the system. For example, during the mission time, a k-out-of-n system will function if and only if at least k components out of n components will function. Special cases are the series system, which is an n-out-of-n system, and the parallel system, which is a 1-out-of-n system. If we associate with each time t a binary random variable that equals 1 if the system functions at time t and 0 otherwise, then the collection of the binary random variables for different t is a stochastic process, representing the availability of the system over time. The reliability of the system can be determined by the properties of the lifetime distributions of the components and system structure. ∎

Example 8.4: ALOHA protocols

ALOHA was a multiaccess communication protocol first deployed at the University of Hawaii in 1970. While the original version of the protocol is no longer used, its core design concept has become the basis for the almost universal Ethernet.

It was quickly noticed that the first version of the ALOHA protocol was not stable. Its throughput was low and the number of backlogged packets was high, while a large portion of available bandwidths was being wasted. Since then, several versions of the protocol have been proposed. The following *slotted* and *unslotted* ALOHA models are based on Kulkarni (1995) (also see Gautam, 2003).

In the slotted version, there are N users transmitting messages (in the form of packets) via satellites. At time slots $n = 1, 2, \ldots$, each user independently transmits a packet with probability p. If only one user transmits a packet then the transmission is successful and the packet departs the system. However, if two or more users simultaneously transmit, then a collision occurs and their packets are garbled. Such packets are backlogged in the system and have to be re-sent later. A backlogged packet will, independent of all else, retransmit with probability r in a time slot. A user with a backlogged packet will not transmit a new packet until his backlogged packet is successfully re-sent.

In the unslotted version (again see Kulkarni, 1995, and Gautam, 2003), it is assumed that each of the N users, when not backlogged, transmits a packet after an exponential amount of time with rate λ. Each packet requires an exponential amount of time with rate μ to transmit. A collision occurs when a user attempts to transmit while another user is transmitting, which causes all transmissions to terminate instantaneously and the collided packets to be backlogged. It is assumed that a backlogged packet retransmits after an exponential amount of time with rate γ.

In either the slotted or unslotted version of the protocol, the number of backlogged packets over time slots $n = 1, 2, \ldots$ or over continuous-time $t \geq 0$ form a stochastic process. The typical performance measures arising in the efficient satellite communication of ALOHA include the system throughput, the bandwidth utilization, the long-run behavior of the number of backlogged packets, and the time needed to successfully transmit a packet. ∎

Example 8.5: A model of social mobility

A problem of interest in the study of social structure is about the transitions between the social status of successive generations in a family. Sociologists often assume that the social class of a son depends only on his parents' social class, but not on his grandparents'. A famous U.K. study of social mobility was conducted after World War II (Glass, 1954), which identified three social classes: *upper class* (executive, managerial, high administrative, professional), *middle class* (high grade supervisor, non-manual, skilled manual), and *lower class* (semi-skilled or unskilled). Each family in the society occupies one of the three social classes, and its occupation evolves across different generations. Glass (1954) analyzed a random sample of 3500 male residents in England and Wales in 1949 and estimated that the transitions between the social classes of successive generations in a family were as the following:

	Following generation		
Current generation	Upper class	Middle class	Lower class
Upper class	0.45	0.48	0.07
Middle class	0.05	0.70	0.25
Lower class	0.01	0.50	0.49

A dataset like this has enabled sociologists to answer questions such as: How many generations are necessary for a lower class family to become a higher class family? What is the distribution of a family's occupation in the long run? ∎

With the help of the above examples, we now define a stochastic process.

DEFINITION 8.1 *The collection of random variables $X = \{X(t),\ t \in T\}$ is called a stochastic process, where T is called the index set.*

The values assumed by process X are called the *states*, and the set of all possible values is called the *state space* and denoted by S. Stochastic processes are further classified into four broad classes by the nature of the index set T and state space S, where each of them can be either discrete or continuous. The index $t \in T$ often corresponds to discrete units of time, and the index set is $T = \{0, 1, 2, \ldots\}$. In this case, X is called a *discrete-time* stochastic process, and it is customary to represent X by $\{X_n, n \geq 1\}$. Take the brand switching problem as an example: X_n might represent the brand preference of a consumer in week n, which can take values 1, 2, or 3, with a discrete state space $S = \{1, 2, 3\}$. As another example, suppose that X_n represents the total amount of claims (for convenience sake we assume it can take a nonnegative, continuous value, although in reality it is countable) made by a policyholder in year n, then $\{X_n, n \geq 0\}$ is a discrete-time stochastic process with a continuous state space. When the index set T is an interval, X is called a *continuous-time* stochastic process. In most common physical systems time is a natural index parameter, so $T = (0, \infty)$. In this case, we follow the convention to write X as $\{X(t), t \geq 0\}$. In the unslotted ALOHA example, if $X(t)$ represents the number of backlogged packets at time t, $t \geq 0$, then $\{X(t), t \geq 0\}$ is a continuous-time stochastic process with a discrete state space $S = \{0, 1, 2, \ldots\}$. Finally, an example of the continuous-time stochastic process with a continuous state space might be the cumulated amount of rainfall in an area continuously monitored during a season, or the market price of a stock continuously observed during a trading session.

The random variables in a stochastic process often exhibit some sort of interdependence. For instance, the current rating of a policyholder may depend on his claim history, and a son's occupation can be affected by his ancestors' occupations. Without knowing the dependence structure of a random sequence, little can be said or done about a stochastic process. As such, the study of stochastic processes is mainly centered around the characterizations and solution methods of some prototypical processes that have certain types of dependence structure. In this chapter, we shall discuss several fundamental processes that are most frequently used in modeling real-world problems.

1. *Poisson Processes*: A stochastic process $\{N(t), t \geq 0\}$ is called a *counting process* if $N(t)$ represents the number of "events" that have occurred during the interval $[0, t]$. A counting process is called a *Poisson process* if the interarrival times of successive events are independently and identically distributed (iid) random variables that follow a common exponential distribution. The Poisson process inherits the *memoryless property* of the exponential distribution, which translates into the following *stationary* and *independent increments* property: at any time epoch t, the process from time t onward is independent of its history and has the same distribution as the original process. In essence, this property reduces the analysis of the Poisson process into that of a sequence of independent random variables, where those random variables represent the numbers of events occurring in non-overlapping intervals. The Poisson process is a key building block in stochastic modeling. For example, it has been used extensively to model the arrivals to a service system, the traffic flow on a highway, the number of defective items in a manufacturing process, and the number of replacements of a component. The subjects related to the Poisson process are covered in Section 8.2.

2. *Discrete-Time Markov Chains*: A *discrete-time Markov chain* (DTMC) is a discrete-time stochastic process defined on $S = \{0, 1, 2, \ldots\}$ that has the simplest type of dependence structure, known as the *Markov property*: given the present state of the process, the future evolution of the process is independent of its history. The dependence structures in the brand switching, automobile insurance, and social mobility examples are all of this type. As it turns out, a DTMC can be completely specified by the *transition probability matrix* (e.g., the probability tables given in the aforementioned examples) and the distribution of the initial state (e.g., the distribution of the merit rating of a policyholder at time 0), which greatly simplifies the analysis of a DTMC. Although simple, the DTMC

proves to be the most useful modeling tool in analyzing practical problems. We shall treat the topics relevant to DTMCs in Section 8.3.

3. *Continuous-Time Markov Chains*: In a DTMC, the process stays in a state for a unit of time and then possibly makes a transition. If we relax this assumption and allow the sojourn time of the process in each state to be independent and follow a state-dependent exponential distribution, the resultant process is called a *continuous-time Markov chain* (CTMC). For example, the unslotted ALOHA protocol given in Example 8.4 can be modeled as a CTMC. As in the case of a DTMC, a CTMC is a simple yet powerful modeling tool to treat real-world stochastic systems. The topics related to CTMCs will be examined in Section 8.4.

4. *Renewal Theory*: In a Poisson process, the times between successive events are iid *exponential* random variables. As an extension, a *renewal process* is a counting process whose interarrival times are iid random variables following a general distribution. In renewal theory, an event is called a *renewal*, the time instance when a renewal takes place is called a *renewal epoch*, and the time interval between two consecutive renewals is called a *renewal cycle*. A renewal process has a stronger dependence structure than a Poisson process: while the latter can probabilistically restart itself at any time, the former can only do so at a renewal epoch. Nevertheless, renewal epochs facilitate the partition of the interval $(0, \infty)$ into disjoint, independent renewal cycles, which, in turn, reduce the long-run analysis of a renewal process to that of a typical renewal cycle. When there is a reward associated with each renewal, the resultant process is called a *renewal reward process*. A large number of practical problems can be formulated as renewal reward processes. Renewal theory also forms a cornerstone for the development of other stochastic processes with more complex dependence structures. For example, the *semi-Markov process* can be roughly understood as a combination of a Markov chain and a renewal process: it assumes that the process changes states as a DTMC, but the sojourn time in each state can follow a state-dependent, but otherwise general, distribution. We shall deal with the renewal process and its variants in Section 8.5.

For each of the aforementioned processes, our discussion shall be focused on the following aspects of the process:

- *Process characterization*: First, we shall give the formal definition of the underlying process, identify its basic structure, and characterize the important properties.
- *Transient analysis:* Second, we shall consider how the underlying process behaves in the transient state, i.e., to derive the distribution of $X(t)$ or X_n for a finite t or n. Transient analysis helps to answer questions such as "what is the distribution of the number of backlogged packets in ALOHA at time $t = 10$" or "what is the probability that a 1-out-of-n system will function in the next 24 hours?"
- *Long-run analysis*: Third, we shall study the long-run behavior of the process, i.e, derive the limiting distribution of $X(t)$ or X_n, as t or n tends to infinity. The long-run analysis seeks to answer questions such as "what is the long-run market share of a brand" or "what proportion of families will be in the middle class in steady-state?"
- *Statistical inference*: To apply results of stochastic processes to a real-life situation, data from the actual process have to be observed and analyzed to fit the characteristics of a prototypical process. This brings us to the topic of statistical

inference of stochastic processes. We shall briefly introduce parameter estimation and hypothesis testing methods used in stochastic processes.

For each process, we shall illustrate the basic concepts and methodologies using several practical problems extracted from different application areas. Because of the page limitation, the results will be stated without proof. There are numerous stochastic processes textbooks where the reader can consult the proofs of our stated results. The References section of this chapter lists a sample of the texts at different levels. At the introductory level, the reader may consult Ross (2003), Kulkarni (1999), Kao (1997), and Taylor and Karlin (1994). The books for a more advanced level include Cinlar (1975), Ross (1996), Wolff (1989), Bhat & Miller (2002), Tijms (1994), and Kulkarni (1995). There are also many books dealing with stochastic processes in specialized fields and a few of them are listed here: Aven & Jensen (1999) and Barlow & Proschan (1981) for reliability, Buzacott & Shanthikumar (1993) for manufacturing, Bolch et al. (2006) for computer networks, Kleinrock (1976) for queueing systems, Rolski et al. (1999) and Kijima (2003) for finance and insurance, Zipkin (2000) and Porteus (2002) for inventory systems, Helbing & Calek (1995) and Bartholomew (1982) for social sciences, and Goel & Richter-Dyn (2004) for biology.

8.2 Poisson Processes

It appears that the Poisson process was first investigated in detail by physicians A. Einstein and M. Smolukhovsky in the context of Brownian processes. The first application of the Poisson process, published in 1910, was to describe radioactive decay occurring randomly. The Poisson process, however, was named after the French mathematician Simeon-Denis Poisson (1781–1840), who was credited for the introduction of the Poisson distribution, but not of the Poisson process. Nevertheless, he certainly earned the right to be named after this famous process connected with his distribution.

Both the Poisson process to be dealt with here and the continuous-time Markov chain to be considered in Section 8.4 are intimately connected with the *exponential* distribution. Exponential distribution plays a special role in those processes because it is mathematically amenable and often a good approximation to the actual distribution. In the next section, we review several salient properties of the exponential distribution.

8.2.1 The Exponential Distribution and Properties

A continuous random variable T is said to follow an exponential distribution with parameter (or rate) $\lambda > 0$, where $\lambda = 1/E[T]$, if it has the probability density function (PDF)

$$f(t) = \lambda e^{-\lambda t}, \quad t \geq 0 \tag{8.1}$$

The property that makes the distribution attractive mathematically is that it has no memory, or is *memoryless*. That is, the exponential distribution satisfies the condition

$$P(T > t + s | X > s) = P(T > t) \quad \text{for any } t, s > 0$$

The memoryless property bears the following interpretation: if T is the lifetime of a part, then the above condition states that the probability that an s-year-old part lasts another t years is the same as the probability that a new part lasts t years. To see that the property is also a realistic assumption for an actual distribution, imagine that T is the time interval between two successive accidents occurring on a highway, with a mean of 3 days per accident. Suppose that there were no accidents in the past 2 days. Then, because of the total randomness of accidents, the remaining time until the next accident to occur from this point on should be no different, probabilistically, from the original waiting time between two successive accidents. Thus, beyond the 2 accident-free days, we should expect to wait another 3 days for the next accident to occur.

Besides the memoryless property, the exponential distribution also has several other nice properties useful in stochastic modeling, as stated below.

Further Properties of the Exponential Distribution

1. The sum of a fixed number of iid exponential random variables follows a *Gamma* (or *Erlang*) *distribution*. Suppose that $S_n = T_1 + T_2 + \cdots + T_n$, where T_i are iid exponential random variables with rate λ. Then random variable S_n has the probability density function

$$f_{S_n}(t) = \lambda e^{-\lambda t} \frac{(\lambda t)^{n-1}}{(n-1)!}, \quad \text{for } t \geq 0$$

   For example, suppose that the occurrence times between successive accidents on a highway are iid exponential random variables with a mean of 3 days per accident. Then the waiting time for the 5th accident to occur is Erlang with mean $n\lambda = 5(3) = 15$ days.

2. The minimum of independent exponential random variables is still an exponential random variable. Let T_i be the time when the ith event occurs, where T_i's are independent exponential variables with respective rates λ_i, $i = 1, 2, \ldots, n$. Then the time when the first of the n events occurs, $T = \min(T_1, T_2, \ldots, T_n)$, is an exponential random variable with rate $\sum_{i=1}^{n} \lambda_i$. For example, suppose that there are two clerks who can serve customers at the exponential rates 3 and 5, respectively. Given both clerks are currently busy, the time until one of the clerks finishes service follows an exponential distribution with rate $3 + 5 = 8$.

3. The probability that the ith event is the first to occur among the n events is proportional to λ_i. Let T_i, $i = 1, 2, \ldots, n$, be independent exponential variables with respective rates λ_i, and $T = \min(T_1, T_2, \ldots, T_n)$. Then

$$P(T_i = T) = \frac{\lambda_i}{\sum_{i=1}^{n} \lambda_i}, \quad i = 1, 2, \ldots, n \quad (8.2)$$

Applying the above result to our two clerks example, the probability that the clerk with the exponential rate 3 is the first one to complete service is $\left(\frac{3}{(3+5)} = \frac{3}{8}\right)$.

8.2.2 Definition of Poisson Processes

A stochastic process $\{N(t), t \geq 0\}$ is called a *counting process* if $N(t)$ represents the number of "events" that have occurred during the interval $[0, t)$. Here, the "events" can be inbound phone calls to a call center, machine breakdowns in a production system, or customer orders for a product. Let us first examine two properties that are desirable for a counting process. A counting process is said to have *independent increments* if the numbers of events that occur in nonoverlapping intervals are independent. This means, for example, the defective items produced by a machine between 8:00 and 10:00 A.M. is independent of that produced between 12:00 and 3:00 P.M. The counting process is said to have *stationary increments* if the number of events that occur in an interval depends only on how long the interval is.

In other words, $N(s+t) - N(s)$ and $N(t)$ are governed by the same distribution for any $s \geq 0$ and $t > 0$. For example, if the aforementioned defective counting process has stationary increments, then the number of defective items produced during 8:00–10:00 A.M. will have the same distribution as that produced during 1:00–3:00 P.M., since both time periods are 2 h. Note that the stationary and independent increment properties together mean that if we partition the time interval $(0, \infty)$ into the subintervals of an equal length, then the number of events that occurred in those subintervals are iid random variables.

Now we are ready to formally define a Poisson process.

DEFINITION 8.2 *A Poisson process with rate (or intensity) λ is a counting process $\{N(t), t \geq 0\}$ for which*

a. $N(t) \geq 0$;
b. *the process has independent increments;*
c. *the number of events in any interval of length t follows the Poisson distribution with rate λt:*

$$P(N(s+t) - N(s) = n) = \frac{(\lambda t)^n e^{-\lambda t}}{n!}, \quad for\ any\ s \geq 0 \quad and \quad t > 0 \quad (8.3)$$

(Property (c), in fact, implies that the Poisson process has stationary increments.)

Conditions (a)–(c) imply that the numbers of events that occur in nonoverlapping intervals are independent Poisson random variables, and those Poisson events occur at a constant rate λ. The Poisson process plays a key role in stochastic modeling largely because of its mathematical tractability and practical realism. Theoretically, the law of common events asserts that the Poisson distribution is an approximation of the binomial distribution if the number of events is large and the probability of actual occurrence of each event is small (Ross, 2003; Taylor & Karlin, 1994). Empirically, it has been found that counting processes arising in numerous applications indeed exhibit such characteristics.

From conditions (a)–(c), we can identify additional characterizations of a Poisson process:

d. *the probability that there is exactly one event occurring in a very small interval $[s, s+h]$ is approximately λh;*
e. *the probability that there are at least two events occurring in a very small interval $[s, s+h]$ is negligible; and*
f. *the interarrival times between successive events are a sequence of iid exponential random variables with rate λ.*

Conditions (d) and (e) postulate that the Poisson process is a process of "rare" events, that is, events can only occur one at a time. In fact, conditions (d) and (e), together with stationary and independent increments, serve as an alternative definition of a Poisson process. Now, condition (f) means that the Poisson process is memoryless, that is, at any time epoch, the remaining time until the next event to occur follows the same exponential distribution, regardless of the time elapsed since the last event. Condition (f) serves as the third alternative definition of a Poisson process. Those definitions are equivalent in the sense that from the set of conditions for one definition one can derive the set of conditions for another definition.

Practitioners can use any of the three definitions as a yardstick to justify whether a Poisson process is an adequate representation of the actual arrival process. For example, if the actual arrival process shows a pattern of batch arrivals or the arrival rates at different time instances are different (e.g., the arrival rate in a rush-hour is larger than that in normal hours), then the Poisson process can be excluded outright as a candidate for modeling the actual arrival process. Nevertheless, the law of rare events suggests that the Poisson process is often a good approximation of the actual arrival process.

8.2.3 Properties of Poisson Processes

We often need to split a Poisson process into several subprocesses. For example, it might be beneficial to classify the customers arriving to a service system as the priority customers and non-priority customers and route them to different agents. Suppose that we associate with each arrival in the Poisson process a distribution $\{p_i : \sum_{i=1}^{n} p_i = 1\}$, and let the arrival be of type i with probability p_i, $i = 1, \ldots, n$, independent of all else. This mechanism decomposes the original Poisson process into n subprocesses. In some applications we need to merge several independent Poisson streams into a single arrival stream. The following theorem states that both the decomposed processes and the superposed process are still Poisson processes.

THEOREM 8.1

a. Let $\{N(t), t \geq 0\}$ be a Poisson process with rate λ. Each time an event occurs, independent of all else, it is classified as a type i event with probability p_i, $\sum_{i=1}^{n} p_i = 1$. Let $\{N_i(t), t \geq 0\}$ be the arrival process of type i. Then $\{N_i(t), t \geq 0\}$, $i = 1, \ldots, n$, are n independent Poisson processes with respective rates λp_i, $i = 1, \ldots, n$.

b. Let $\{N_i(t), t \geq 0\}$, $i = 1, \ldots, n$, be independent Poisson processes with respective rates λ_i, $i = 1, \ldots, n$. Then the composite process $\{\sum_{i=1}^{n} N_i(t), t \geq 0\}$ is a Poisson process with rate $\sum_{i=1}^{n} \lambda_i$.

Example 8.6: The case of meandering messages

This case is adapted from Nelson (1995) and is a simplified version of a real computer system. While the numbers used here are fictitious, they are consistent with actual data. A computer at the Ohio State University (called osu.edu) receives e-mail from the outside world, and distributes the mail to other computers on campus, including the central computer in the College of Engineering (called eng.ohio-state.edu). Besides the mail from osu.edu, eng.ohio-state.edu also receives e-mail directly without passing through osu.edu. Records maintained by ohio.edu show that, over the two randomly selected days, it received 88,322 messages on one day and 84,478 messages on the other, during the normal business hours 7:30 A.M.– 7:30 P.M. Historically, 20% of the mail goes to the College of Engineering, with the average message size about 12K bytes. The College of Engineering does not keep detailed records, but estimates that the direct messages to eng.ohio-state.edu is about two-and-a-half times the traffic it receives from ohio.edu. The College of Engineering plans to replace eng.ohio-state.edu with a newer computer and wants the new computer to have enough capacity to handle even extreme bursts of traffic.

Provided that the rate of arrivals is reasonably steady throughout the business day, a Poisson process is a plausible model for mailing directly to ohio.edu and directly to eng.ohio-state.edu, because it is the result of a large number of senders acting independently. However, the planner should be aware that the Poisson process model is not perfect, as it does not represent "bulk mail" that occurs when a single message is simultaneously sent to multiple users on a list. Also, because e-mail traffic is significantly lighter at night, our conclusions have to be restricted to business hours, to amend to the stationary increments requirement of the Poisson process.

Two business days, with 12 h per day, amounts to 86,400 s. So the estimated arrival rate is $\hat{\lambda}_{osu} = ((88{,}322 + 84{,}478)/86{,}400) \approx 2$ arrivals/s (see Section 8.2 for parameter estimation in a Poisson process). The standard error is only $\hat{se} = \sqrt{\hat{\lambda}_{osu}/86{,}400} \approx 0.005$ arrivals/s, indicating a quite precise estimation of $\hat{\lambda}_{osu}$. The overall arrival process to eng.ohio-state.edu is composed of the direct arrivals and those distributed by osu.edu. If we say that each arrival to osu.edu has probability $p = 0.2$ of being routed to eng.ohio-state.edu, then the arrivals to the machine through ohio.edu form a Poisson process with rate $\hat{\lambda}_{routed} = p\hat{\lambda}_{osu} = 0.4$ arrivals/s. Based on speculation, the direct arrival rate to eng.ohio-state.edu is $\hat{\lambda}_{direct} = (2.5)\hat{\lambda}_{routed} = 1$ arrival/s. We can assign no standard error to this estimate as it is not based on data. Thus, we may want to do a sensitivity analysis over a range of values for $\hat{\lambda}_{direct}$. The overall arrival process to eng.ohio-state.edu, a superposition of two independent Poisson processes, is a Poisson process with rate $\hat{\lambda}_{eng} = \hat{\lambda}_{routed} + \hat{\lambda}_{direct} = 1.4$ arrivals/s. Based on the model, we can do some rough capacity planning by looking at the probability of extreme bursts (a better model requires to model the system as a queueing process). For example, if the new machine is capable of processing 3 messages/s, then

$$P(\text{more than three arrivals/s}) = 1 - \sum_{i=0}^{3} \frac{e^{-\hat{\lambda}_{eng}} \left(\hat{\lambda}_{eng}\right)^i}{i!} \approx 0.05$$

If the processing time of a message also depends on its size, then the average number of bytes received by eng.ohio-state.edu is $(12K \text{bytes}) \hat{\lambda}_{eng} = 16.8K$ bytes/s, assuming the message size is independent of the arrival process. Unfortunately, we cannot make a probability statement about the number of bytes received, as we have no knowledge of the distribution of the message size except for its mean. ∎

The next result relates a Poisson process to a *uniform distribution*, which provides tool for computing the cost model of a Poisson process. Let S_n be the time of occurrence of the nth event in a Poisson process, $n = 1, 2, \ldots$. From property (1) of the exponential distribution given in Section 8.2, we know that S_n has an Erlang distribution with parameters (n, λ), $n \geq 1$. If we are told that exactly n events have occurred during interval $[0, t]$, that is, $N(t) = n$, how does this information alter the joint distribution of $S_1, S_2, \ldots, S_n$? This can be answered intuitively as follows: as Poisson events occur completely randomly over time, it postulates that any small interval of a fixed length is equally likely to contain a Poisson event. In other words, given $N(t) = n$, the occurrence times $S_1, \ldots, S_n$, considered as *unordered* random variables, behave like a random sample from a *uniform* distribution between $[0, t]$. This intuition can be formalized by the following theorem.

THEOREM 8.2 *Given $N(t) = n$, the occurrence times $S_1, \ldots, S_n$ have the same distribution as the order statistics of n independent uniform random variables in the interval $[0, t]$.*

Example 8.7: Sum quota sampling

We wish to estimate the expected interarrival time of a Poisson process. To do so, we have observed the process for a pre-assigned quota, t, and collected a sample of the interarrival times, $X_1, X_2, \ldots, X_{N(t)}$, during interval $[0, t]$. Note that the sample size is a random variable. In *sum quota sampling*, we use the sample mean

$$\bar{X}_{N(t)} = \frac{S_{N(t)}}{N(t)} = \frac{X_1 + X_2 + \cdots + X_{N(t)}}{N(t)}, \quad N(t) > 0$$

to estimate the expected waiting time, provided that $N(t) > 0$. An important statistical concern is whether $E[\bar{X}_{N(t)}|N(t)>0]$ is an unbiased estimator of the expected interarrival time, say $E(X_1)$. The key is to evaluate the conditional expectation $E[S_{N(t)}|N(t)=n]$. Let $(U_1, \ldots, U_n)$ be a sample from the uniform distribution in the interval $[0, t]$. Then by Theorem 8.2,

$$E(S_{N(t)} \mid N(t) = n) = E[\max(U_1, \ldots, U_n)] = \int_0^t \left[1 - \left(\frac{x}{t}\right)^n\right] dx = \frac{nt}{n+1}$$

Then we get

$$E[\bar{X}_{N(t)} \mid N(t) > 0] = \sum_{n=1}^{\infty} E\left[\frac{S_n}{n} \mid N(t) = n\right] P(N(t) = n \mid N(t) > 0)$$

$$= \sum_{n=1}^{\infty} \frac{nt}{(n+1)n} \left[\frac{(\lambda t)^n e^{-\lambda t}}{n!(1 - e^{-\lambda t})}\right] = \frac{1}{\lambda}\left(1 - \frac{\lambda t}{e^{-\lambda t} - 1}\right)$$

The fraction of the bias to the true mean, $E[X_1] = 1/\lambda$, is

$$\frac{E[\bar{X}_{N(t)} \mid N(t) > 0] - E[X_1]}{E[X_1]} = -\frac{\lambda t}{e^{\lambda t} - 1} = -\frac{E[N(t)]}{e^{E[N(t)]} - 1}$$

The following table computes numerically the bias due to the sum quota sampling:

$E[N(t)]$	Fraction bias	$E[N(t)]$	Fraction bias
5	−0.0339	20	−4.12E-8
10	−4.54E-04	25	−3.47E-10
15	−4.59E-06	30	−2.81E-12

This suggests that, although the sample mean generated by sample quota sampling is downward biased, the bias tends to zero rapidly. Even for a moderate average sample size of 10, the fraction of the bias is merely about 0.5%. ∎

8.2.4 Nonhomogeneous Poisson Processes and Compound Poisson Processes

In many applications, it is pertinent to allow the arrival rate of the Poisson process to be dependent on time t, that is, $\lambda = \lambda(t)$. The resultant process is called the *nonhomogeneous Poisson process* with rate function $\lambda(t)$. For such a process, the number of events in an interval $[s, s+t]$ follows a Poisson distribution with rate $\int_s^{s+t} \lambda(u)du$, and the number of events that occur on non-overlapping intervals is independent.

Stochastic Processes 8-13

Example 8.8: Emergency 911 calls

In a study of police patrol operations in the New York City Policy Department (Green and Kolesar, 1989), the data of 911 calls were analyzed. The study tabulated the total number of calls to Precinct 77 in fifteen-minute intervals during June–July, 1980.

The data exhibit a strong nonhomogeneous arrival pattern of the calls: the average number of calls decreases gradually from about 37 calls at 12:00 midnight to 9 calls at 6:00 A.M., and then steadily increases, though with certain periods held stable, to about 45 calls when it approaches 12:00 midnight. Based on the arrival counts, an estimated rate function $\lambda(t)$ was obtained. The hypothesis tests based on the counts and interarrival times showed that the nonhomogeneous Poisson process was an adequate description of the arrival process of the 911 calls. ∎

Another generalization of the Poisson process is to associate each event with a random variable (called *mark*): let $\{Y_n, n \geq 1\}$ be a sequence of iid random variables that is also independent of the Poisson process $\{N(t), t \geq 0\}$. Suppose that for the nth event we associate with it a random variable $Y_n, n \geq 1$, and define $X(t) = \sum_{n=1}^{N(t)} Y_n$, $t \geq 0$. Then the process $\{X(t), t \geq 0\}$ is call a *compound Poisson process*.

Example 8.9: Examples of compound Poisson processes

1. *The Poisson process*: This is a special case of the compound Poisson process since if we let $Y_n \equiv 1$, $n \geq 1$, then $X(t) = N(t)$.
2. *Insurance risk*: Suppose that insurance claims occur in accordance with a Poisson process, and suppose that the magnitude of the nth claim is Y_n. Then the cumulative amount of claims up to time t can be represented by $X(t)$.
3. *Stock prices*: Suppose that we observe the market price variations of a stock according to a Poisson process. Denote Y_n, $n \geq 1$, as the change in the stock price between the $(n-1)$st and nth observations. Based on the random work hypothesis of stock prices, which is a postulation developed in finance, Y_n, $n \geq 1$, can be modeled as iid random variables. Then $X(t)$ represents the total price change up to time t. ∎

Statistical Inference of Poisson Processes

To apply mathematical results in a real life process, empirical data have to be collected, analyzed, and fitted to a theoretical process of choice. Parameter estimation and hypothesis testing are two basic statistical inference tools to extract useful information from the dataset. This section briefly discusses statistical inference concerning the Poisson process.

Suppose that, for a fixed time interval $[0, t]$, we observed n Poisson events at times $0 = s_0 < s_1 < s_2 < \cdots < s_n < t$. Since the interarrival times $s_i - s_{i-1}$, $i = 1, \ldots, n$, form a random sample from an exponential distribution, the maximum likelihood function of the sample is given by

$$f(\lambda) = \prod_{i=1}^{n} (\lambda e^{-\lambda(s_i - s_{i-1})}) e^{-\lambda(t - s_n)} = \lambda^n e^{-\lambda t}$$

The log maximum likelihood function $L(\lambda) = \ln f(\lambda)$ is

$$L(\lambda) = \ln f(\lambda) = n \ln \lambda - \lambda t$$

Solving $dL(\lambda)/d\lambda = 0$ yields the maximum likelihood estimate of λ as

$$\hat{\lambda} = \frac{n}{t} \quad (8.4)$$

As the number of events n in $[0, t]$ follows a Poisson distribution with the mean and variance both equal to λt, $\hat{\lambda}$ is an unbiased estimator of λ, $E(\hat{\lambda}) = \lambda$, with $V(\hat{\lambda}) = \frac{\lambda}{t}$. Clearly, the longer the observation interval $[0, t]$, the more precise the estimator $\hat{\lambda}$, as $s\hat{e} = \sqrt{\hat{\lambda}/t}$ decreases in t.

To test whether a process is Poisson for the data collected over a fixed interval, we can use several well-developed goodness-of-fit tests based on the exponential distribution of the interarrival times (see Gnedenko et al., 1969). We discuss here another simple test (Bhat & Miller, 2002), using the relationship between the uniform distribution and the Poisson process (see Theorem 8.2). Again assume that we have observed n Poisson events at times $0 = s_0 < s_1 < s_2 < \cdots < s_n < t$. Recall that the s_i, considered as unordered random variables, are iid uniform random variables between $[0, t]$. Hence, $Y_n = \sum_{i=1}^{n} s_i$ is the sum of n independent uniform random variables between $[0, t]$, with $E(Y_n) = nt/2$ and $V(Y_n) = nt^2/12$. When n is sufficiently large,

$$Z = \frac{Y_n - \frac{nt}{2}}{\sqrt{\frac{nt^2}{12}}} \quad (8.5)$$

is a standard normal random variable, by the central limit theorem.

Example 8.10: Testing for a Poisson process

This example is based on Lewis (1986), and the dataset was reprinted in Hand (1994). The data were the number of times that 41 successive vehicles driving northbound on Route M1 in England passed a fixed point near Junction 13. The 40 time points (in seconds) are given below:

12	14	20	22	41	46	80	84	85	89
97	104	105	126	132	143	151	179	185	189
194	195	213	222	227	228	249	250	251	256
259	273	278	281	285	290	291	294	310	312

Denote s_i as the ith observation in the above list, where $t = s_{40} = 312$ corresponds to the passing time of the last vehicle. We wish to test the null hypothesis that this process is Poisson. We calculate $Y_{40} = \sum_{i=1}^{40} s_i = 7062$ and $Z = \frac{Y_n - nt/2}{\sqrt{nt^2/12}} = \frac{7062 - 6240}{\sqrt{324,480}} = 1.443$. The p-value for the test is $2P(Z \geq 1.443) = 0.149$. Therefore, the dataset did not constitute sufficient evidence to reject the null hypothesis that the underlying process is Poisson. ∎

8.3 Discrete-Time Markov Chains

Most real-world systems that evolve dynamically in time exhibit some sort of temporal dependence, that is, the outcomes of successive trials in the process are not independent random variables. Dealing with temporal dependence in such a process is a formidable task. As the simplest generalization of the probability model of successive trials with independent outcomes, the *discrete-time Markov chain* (DTMC) is a random sequence in which the outcome of each trial depends only on that of its immediate predecessor (known as the first order dependence).

Stochastic Processes

The notion of Markov chains originated from the Russian mathematician A.A. Markov (1856–1922), who studied sequences of random variables with certain dependence structures in his attempt to extend the weak law of large numbers and the central limit theorem. For illustrative purposes, Markov in his 1906 manuscript applied his chain to the distribution of vowels and consonants in Pushkin's poem "Eugeny Onegin." In the model, he assumed that the outcome of each trial depends only on its immediate predecessor. The model turned out to be a very accurate estimation of the frequency at which consonants occur in Pushkin's poem. Today, the Markov chain finds applications in diverse fields such as physics, biology, sociology, meteorology, reliability, and many others, and proves to be the most useful tool for analyzing practical problems.

8.3.1 Definition of Discrete-Time Markov Chains

Consider the discrete-time stochastic process $\{X_n, n=0,1,\ldots\}$ that assumes values in the discrete state space $S = \{0, 1, 2, \ldots\}$. We say the process is in state i at time n if $X_n = i$. The process is called a discrete-time Markov chain (DTMC) if for all $i_0, i_1, \ldots, i, j$, and n,

$$P(X_{n+1} = j | X_n = i, X_{n-1} = i_{n-1}, \ldots, X_1 = i_1, X_0 = i_0) = P(X_{n+1} = j | X_n = i) \quad (8.6)$$

The above expression describes the *Markov property*, which states that to predict the future of the process it is sufficient to know the most recently observed outcome of the process. The right side of Equation 8.6 is called the *one-step transition probability*, which represents the conditional probability that the chain undergoes a state transition from i to j in period n. If for all i, j, and n,

$$P(X_{n+1} = j | X_n = i) = p_{ij}$$

then the DTMC is said to be *stationary* or *time-homogeneous*. We shall limit our discussion to the stationary case. Following the convention, we arrange probabilities p_{ij} in a matrix form, $\mathbf{P} = \{p_{ij}\}$, and call $\mathbf{P}$ the *one-step transition matrix*. Matrix $\mathbf{P}$ is a *stochastic matrix* in the sense that

$$p_{ij} \geq 0 \text{ for all } i,j, \quad \text{and} \quad \sum_j p_{ij} = 1 \text{ for } i = 1, 2, \ldots \quad (8.7)$$

Note that a DTMC is completely specified by the transition matrix $\mathbf{P}$ and the *initial distribution* $a = \{a_i\}$, where $a_i = P(X_0 = i)$ is the probability that the chain starts in state i, $i = 0, 1, \ldots$.

Several motivating examples given in Section 8.1, for example, the brand switching, social mobility, automobile insurance, and slotted Aloha examples, are all DTMC models. Next, we revisit some of those examples and also introduce other well-known DTMC models.

Example 8.1: A brand switching model (continued)

Let us continue our discussion on customers' brand switching behavior, based on a model extracted from Whitaker (1978). Whitaker defines *brand loyalty* as the proportion of consumers who repurchase a brand on the next occasion without persuasion, and *purchasing pressure* as the proportion of consumers who are persuaded to purchase a brand on the next occasion. Denote w_i and d_i, respectively, as the values of brand loyalty and brand pressure for brand i, where both d_i and w_i are between 0 and 1 and $\sum_i d_i = 1$. To illustrate, consider the following three-brand case:

	Brand 1	Brand 2	Brand 3
Brand loyalty w_i	0.30	0.60	0.90
Purchasing pressure d_i	0.30	0.50	0.20

In Whitaker's Markov brand switching model, brand loyalty and brand pressure are combined to give brand switching probabilities as follows:

$$p_{ij} = \begin{cases} d_i + (1-d_i)w_j & i = j \\ (1-d_i)w_j & i \neq j \end{cases} \quad (8.8)$$

Here, p_{ii}, the proportion of consumers who repurchase brand i on two consecutive occasions, includes the proportion of loyal consumers d_i who stay with brand i without being influenced by purchasing pressure, and the proportion of disloyal consumers $(1-d_i)$ who remain with brand i due to purchasing pressure w_i. The proportion of switching customers, p_{ij}, $j \neq i$, consists of the proportion of disloyal consumers $(1-d_i)$ who are subjected to purchasing pressure w_j from brand j. Applying Equation 8.8 to our dataset yields

$$\mathbf{P} = \begin{pmatrix} 0.51 & 0.35 & 0.14 \\ 0.12 & 0.80 & 0.08 \\ 0.03 & 0.05 & 0.92 \end{pmatrix}$$

This explains how we arrive at the probability table given in Example 8.1. ∎

Example 8.11: A random walk model

Consider a DTMC defined on the state space $S = \{0, 1, 2, \ldots, N\}$, where N can be infinity, with a trigonal transition matrix:

$$\mathbf{P} = \begin{pmatrix} r_0 & p_0 & & & \\ q_1 & r_1 & p_1 & & \\ & q_2 & r_2 & p_1 & \\ & & \ddots & \ddots & \ddots \\ & & & r_N & p_N \end{pmatrix}$$

where $q_i, r_i, p_i \geq 0$ and $q_i + r_i + p_i = 1$, for $i = 1, \ldots, N-1$, and $r_0 + p_0 = 1$ and $r_N + _N = 1$. The key feature of this DTMC is that it can only move at most one position at each step. It has the designated name *random walk*, as in the original article of Karl Pearson of 1905, it was used to describe a path of a drunk who either moves one step forward, one step backward, or stays in the same location (a more general model allows the drunk to move to the negative integer points, or to move on the integer points of a two-dimensional space). Since then, the random work model has been used in physics to describe the trajectories of particle movements, in finance to depict stock price changes, in biology to study how epidemics spread, and in operations research to model the number of customers in discrete queues.

When $r_0 = 0$ (hence $p_0 = 1$), the random walk is said to have a reflecting barrier at 0, as whenever the process hits 0, it bounces back to state 1. When $r_0 = 1$ (hence $p_1 = 0$), the random walk is said to have an absorbing barrier at 0, as whenever the chain hits 0, it stays

Stochastic Processes

there forever. When $N < \infty$, $p_0 = r_N = 0$ (hence $r_0 = p_N = 1$), $p_i = p$, and $q_i = 1-p$ (hence $r_i = 0$) for $i = 1, 2, \ldots, N-1$, it becomes the well-known *gambler's ruin* model. The model assumes that a gambler at each play of the game either wins \$1 with probability p or loses \$1 with probability $q = 1 - p$. The gambler will quit playing either when he goes broke (i.e., hits 0) or attains a fortune of N, whichever occurs first. Then $\{X_n, n \geq 0\}$ represents the gambler's fortune over time. ∎

Example 8.12: Success runs

Consider a DTMC on a state space $\{0, 1, \ldots, N\}$, where N can be infinity, with the transition matrix of the form:

$$P = \begin{pmatrix} p_0 & q_0 & 0 & 0 & \cdots & 0 & 0 \\ p_1 & r_1 & q_1 & 0 & \cdots & 0 & 0 \\ p_2 & 0 & r_2 & q_2 & \cdots & 0 & 0 \\ \vdots & \vdots & \vdots & \vdots & & \vdots & \vdots \\ p_{N-1} & 0 & 0 & 0 & \cdots & r_{N-1} & q_{N-1} \\ p_N & 0 & 0 & 0 & \cdots & 0 & q_N \end{pmatrix}$$

where $p_i, r_i, q_i \geq 0$, $p_i + r_i + q_i = 1$ for $i = 1, \ldots, N-1$, and $p_0 + q_0 = p_N + q_N = 1$. To see why the name *success runs* is appropriate, consider a sequence of trials, each results in a success, a failure, or a draw. If there have been i consecutive successes, then, the probabilities of a failure, a draw, and a success in the next trial are p_i, r_i, and q_i, respectively. Whenever there is a failure, the process starts over with a new sequence of trials. Then $\{X_n, n = 1, 2, \ldots\}$ forms a Markov chain, where X_n denotes the length of the run of successes at trial n.

The success-run chain is a source of rich examples and we consider two here. Suppose that customer accounts receivable in a firm are classified each month into four categories: current (state 1), 30 to 60 days past due (state 2), 60 to 90 days past due (state 3), and over 90 days past due (state 4). The company estimates the month-to-month transition matrix of customer accounts to be

$$\mathbf{P} = \begin{pmatrix} 0.9 & 0.1 & 0 & 0 \\ 0.5 & 0 & 0.5 & 0 \\ 0.3 & 0 & 0 & 0.7 \\ 0.2 & 0 & 0 & 0.8 \end{pmatrix}$$

Then $\{X_n, n \geq 0\}$ is a success-run chain, where X_n is the status of an account in month n.

Another application of the success-run chain is in reliability. Let us assume that a component has a discrete lifetime Y with the distribution $P(Y = i) = a_i$, $i = 1, 2, \ldots$. Starting with a new component, suppose that the component in service will be replaced by a new component either when it fails or when its age surpasses a predetermined value T, whichever occurs first. These types of policies are called *age replacement policies*. The motivation of instituting age replacement policies is to reduce unplanned system failures, as unplanned replacements disrupt normal operations of the system and are more costly then planned replacements. Let X_n be the age of the component in service at time n, with state space $S = \{0, 1, \ldots, T\}$. We have

$$P(X_{n+1} = 0 \mid X_n = i) = P(Y = i+1 \mid Y \geq i+1) = \frac{a_{i+1}}{\sum_{k=i+1}^{\infty} a_k} \equiv p_i, \quad i = 0, 1, \ldots, T$$

where p_i represents that an i-period old component fails by the end of the period. Therefore,

$$P = \begin{pmatrix} p_0 & 1-p_0 & 0 & 0 & \cdots & 0 \\ p_1 & 0 & 1-p_1 & 0 & \cdots & 0 \\ p_2 & 0 & 0 & 1-p_2 & \cdots & 0 \\ \vdots & \vdots & \vdots & \vdots & & \vdots \\ 1 & 0 & 0 & 0 & \cdots & 0 \end{pmatrix}$$

It can be easily seen that this is a success-run chain. ∎

Example 8.4: A slotted ALOHA model (continued)

The slotted ALOHA described in Example 8.4 can be modeled as a DTMC $\{X_n, n \geq 0\}$ on $S = \{0, 1, \ldots, N\}$, where N is the total number of users and X_n is the number of backlogged users at the beginning of the nth slot. The transition probabilities are given by (see Kulkarni, 1995):

$$p_{i,i-1} = P(\text{None of } N - X_n \text{ users transmit, exactly one of } X_n \text{ users retransmits})$$
$$= (1-p)^{N-i} ir(1-r)^{i-1},$$

$$p_{i,i+1} = P(\text{Exactly one of } N - X_n \text{ users transmits, at least one of } X_n \text{ users retransmits})$$
$$= (N-i)p(1-p)^{N-i-1}(1-(1-r)^i),$$

$$p_{i,i+j} = P(\text{Exactly } j \text{ of } N - X_n \text{ users transmit})$$
$$= \frac{(N-i)!}{j!(N-i-j)!} p^j (1-p)^{N-i-j}, \quad 2 \leq j \leq N-i,$$

$$p_{i,i} = P\begin{pmatrix}\text{Exactly one of } N - X_n \text{ users transmits, none of } X_n \text{ users retransmit; or} \\ \text{none of } N - X_n \text{ users transmit, 0 or more than one of } X_n \text{ users retransmit}\end{pmatrix}$$
$$= (N-i)p(1-p)^{N-i-1}(1-r)^i + (1-p)^{N-i}(1-ir(1-r)^{i-1}). \quad\blacksquare$$

8.3.2 Transient Analysis

Transient analysis of a DTMC is concerned with the performance measures that are functions of the time index n. These performance measures are probability statements about the possible realization of a DTMC at time n. We present two such performance measures. The first one is the conditional probability that the chain goes from i to j in n transitions:

$$p_{ij}^{(n)} \equiv P(X_n = j \,|\, X_0 = i), \quad n = 1, 2, \ldots \tag{8.9}$$

They are called the *n-step transition probabilities*. In the matrix form, those probabilities are denoted by $\mathbf{P}^{(n)} = \{p_{ij}^{(n)}\}$, $n = 1, 2, \ldots$. For example, in the brand switching model of Example 8.1, $p_{13}^{(5)}$ is the conditional probability that a consumer will buy brand 3 in week 5, given he bought brand 1 in week 0. The second performance measure of interest is

$$a_j^{(n)} \equiv P(X_n = j), n = 1, 2, \ldots \tag{8.10}$$

which is the unconditional probability that the DTMC is in state j after n steps, without the knowledge of the initial state. For instance, in the brand switching model, $a_3^{(5)}$ is the unconditional probability that a customer will purchase brand 3 in week 5, regardless of his choice in week 0.

Stochastic Processes

Next, we present the formulas to calculate the performance measures given in Equations 8.9 and 8.10. The basic equations to evaluate Equation 8.9 are the Chapman-Kolmogorov (CK) equations:

$$p_{ij}^{(m+n)} = \sum_{k=0}^{\infty} p_{ik}^{(m)} p_{kj}^{(n)} \tag{8.11}$$

The equations state that from state i, for the chain to be in state j after $m+n$ steps, it must be in some intermediate state k after m steps and then move from k onto j during the remaining n steps. The CK equations allow us to construct the convenient relationship for the transition probabilities between any two periods in which the Markov property holds. From the theory of matrices we recognize that Equation 8.11 can be expressed as $\mathbf{P}^{(m+n)} = \mathbf{P}^{(m)} \cdot \mathbf{P}^{(n)}$. By iterating this formula, we obtain

$$\mathbf{P}^{(n)} = \mathbf{P}^{(n-1)} \cdot \mathbf{P} = \mathbf{P}^{(n-2)} \cdot \mathbf{P} \cdot \mathbf{P} = \cdots = \mathbf{P} \cdot \mathbf{P} \cdots \mathbf{P} = \mathbf{P}^n \tag{8.12}$$

In words, the n-step transition matrix $\mathbf{P}^{(n)}$ can be computed by multiplying the one-step transition matrix $\mathbf{P}$ by itself n times.

To obtain $a_j^{(n)}$ defined in Equation 8.10, we condition on the state of the process in period $n-1$:

$$a_j^{(n)} = P(X_n = j) = \sum_i P(X_n = j \mid X_{n-1} = i) P(X_{n-1} = i) = \sum_i p_{ij} a_i^{(n-1)} \tag{8.13}$$

or, equivalently, $\mathbf{a}^{(n)} = \mathbf{a}^{(n-1)} \mathbf{P}$. Iteratively using this relationship, we get

$$\mathbf{a}^{(n)} = \mathbf{a}^{(n-1)} \mathbf{P} = \mathbf{a}^{(n-2)} \mathbf{P}^2 = \cdots = \mathbf{a}^{(0)} \mathbf{P}^n \tag{8.14}$$

Example 8.1: A brand switching model (continued)

Let $\mathbf{a}^{(0)} = (0.30, 0.30, 0.40)$ be the brand shares in week 0. Then the brand shares in week 2, computed by Equation 8.14, are

$$\mathbf{a}^{(2)} = \mathbf{a}^{(0)} \mathbf{P}^2 = (0.30, 0.30, 0.40) \begin{pmatrix} 0.51 & 0.35 & 0.14 \\ 0.12 & 0.80 & 0.08 \\ 0.03 & 0.05 & 0.92 \end{pmatrix}^2 = (0.159, 0.384, 0.457)$$

For example, the share of brand 3 in week 2 is $a_3^{(2)} = 0.457$. To obtain $p_{23}^{(2)}$, the proportion of consumers who bought brand 2 in week 0 and buy brand 3 in week 2, we use Equation 8.12 and obtain:

$$\mathbf{P}^{(2)} = \mathbf{P}^2 = \begin{pmatrix} 0.51 & 0.35 & 0.14 \\ 0.12 & 0.80 & 0.08 \\ 0.03 & 0.05 & 0.92 \end{pmatrix}^2 = \begin{pmatrix} 0.306 & 0.466 & 0.228 \\ 0.160 & 0.686 & 0.154 \\ 0.049 & 0.096 & 0.855 \end{pmatrix} \tag{8.15}$$

Then $p_{23}^{(2)} = 0.154$. ■

Example 8.12: Success runs (continued)

Consider a special case of the success-run model where $N = \infty$ and

$$p_{i0} = p_i = p, \; p_{i,i+1} = q_i = q = 1 - p, \; i = 0, 1, \ldots, N, \; \text{ and } \; r_i = 0, \; i = 1, \ldots, N$$

Let us compute $p_{0j}^{(n)}$. Although doable, it would be cumbersome to first compute $\mathbf{P}^n$ and then identify the appropriate terms $p_{0j}^{(n)}$ from $\mathbf{P}^n$. A better way is to explore the special structure of the transition matrix. Toward this end, we directly use the CK equations 8.11 and get

$$p_{00}^{(n)} = \sum_{k=0}^{\infty} p_{0k}^{(n-1)} p_{k0} = p \sum_{k=0}^{\infty} p_{0k}^{(n-1)} = p$$

Thus $p_{00}^{(n)} = p$ for all $n \geq 1$! Using the similar idea,

$$p_{01}^{(n)} = \sum_{k=0}^{\infty} p_{0k}^{(n-1)} p_{k1} = p_{00}^{(n-1)} \cdot p_{01} = pq = \begin{cases} q & \text{if } n = 1 \\ pq & \text{if } n > 1 \end{cases}$$

Let us now compute $p_{0j}^{(n)}$, for $j > 1$. If $j > n$, then $p_{0j}^{(n)} = 0$, as it takes a minimum of j steps to move from 0 to j. For $j \leq n$, we can recursively compute

$$p_{0j}^{(n)} = \sum_{k=0}^{\infty} p_{0k}^{(n-1)} p_{kj} = q p_{0,j-1}^{(n-1)} = q^2 p_{0,j-2}^{(n-2)} = \cdots = q^{j-1} p_{01}^{(n-j+1)} = \begin{cases} q^j & \text{if } n = j \\ pq^j & \text{if } n > j \end{cases}$$

where in the last equality, we used the expression for $p_{01}^{(n)}$. To verify the correctness of the above expressions, we find that

$$\sum_{j=0}^{\infty} p_{0j}^{(n)} = \sum_{j=0}^{n-1} pq^j + q^n = \frac{p(1-q^n)}{1-q} + q^n = 1$$

8.3.3 Classification of States

It is often necessary to obtain the limiting distribution of a DTMC as $n \to \infty$, which tells us how the DTMC behaves in a long-run. Specifically, we consider the limiting distributions of $\mathbf{P}^{(n)}$, as $n \to \infty$. It turns out that the existence and uniqueness of such a limiting distribution depend on the types of DTMCs. To understand possible complications that may arise in the limiting behavior of a DTMC, we consider, in turn, the following two-state DTMCs on the state space $S = \{1, 2\}$:

$$\mathbf{P}_1 = \begin{pmatrix} 0.7 & 0.3 \\ 0.4 & 0.6 \end{pmatrix}, \quad \mathbf{P}_2 = \begin{pmatrix} 1 & 0 \\ 0 & 1 \end{pmatrix}, \quad \mathbf{P}_3 = \begin{pmatrix} 0 & 1 \\ 1 & 0 \end{pmatrix}, \quad \mathbf{P}_4 = \begin{pmatrix} 0.5 & 0.5 \\ 0 & 1 \end{pmatrix} \quad (8.16)$$

For the DTMC governed by matrix $\mathbf{P}_1$, simple matrix multiplications show that

$$\mathbf{P}_1^{(2)} = \mathbf{P}_1 \cdot \mathbf{P}_1 = \begin{pmatrix} 0.7 & 0.3 \\ 0.4 & 0.6 \end{pmatrix} \begin{pmatrix} 0.7 & 0.3 \\ 0.4 & 0.6 \end{pmatrix} = \begin{pmatrix} 0.61 & 0.39 \\ 0.52 & 0.48 \end{pmatrix}$$

$$\mathbf{P}_1^{(4)} = \mathbf{P}_1^{(2)} \cdot \mathbf{P}_1^{(2)} = \begin{pmatrix} 0.61 & 0.39 \\ 0.52 & 0.48 \end{pmatrix} \begin{pmatrix} 0.61 & 0.39 \\ 0.52 & 0.48 \end{pmatrix} = \begin{pmatrix} 0.575 & 0.425 \\ 0.567 & 0.433 \end{pmatrix}$$

$$\mathbf{P}_1^{(8)} = \mathbf{P}_1^{(4)} \cdot \mathbf{P}_1^{(4)} = \begin{pmatrix} 0.575 & 0.425 \\ 0.567 & 0.433 \end{pmatrix} \begin{pmatrix} 0.575 & 0.425 \\ 0.567 & 0.433 \end{pmatrix} = \begin{pmatrix} 0.572 & 0.428 \\ 0.570 & 0.430 \end{pmatrix}$$

Observe that $\mathbf{P}_1^{(4)}$ is almost identical to $\mathbf{P}_1^{(8)}$, and they have almost identical row values. This suggests that there exist values π_j, $j = 1, 2$, such that

$$\lim_{n \to \infty} p_{ij}^{(n)} = \pi_j, \quad j = 1, 2$$

In this case, the limiting probability $p_{ij}^{(n)}$ is independent of i. This DTMC is an example of the *regular* Markov chain, where a Markov chain is said to be *regular* if for some $n \geq 1$, $\mathbf{P}^{(n)}$ has all *positive* entries. It turns out that the limiting distribution exists for any regular DTMC.

Now, the DTMC associated with $\mathbf{P}_2$ always returns to the same initial state. Indeed, both states are *absorbing states* as once entered they are never left. Since $\mathbf{P}_2^{(n)} = \mathbf{P}_2$ for all n, the limit of $p_{ij}^{(n)}$, as n tends infinity, exists, but it depends on the initial state.

Next, observe that the DTMC governed by $\mathbf{P}_3$ alternates between states 1 and 2, for example, $p_{11}^{(n)} = 0$ when n is odd, and $p_{11}^{(n)} = 1$ when n is even. As both states in the DTMC are *periodic*, $\mathbf{P}_3^{(n)}$ does not converge to a limit as $n \to \infty$.

Finally, for the DTMC governed by transition matrix $\mathbf{P}_4$, we get

$$\lim_{n \to \infty} \mathbf{P}_4^{(n)} = \lim_{n \to \infty} \begin{pmatrix} 0.5^n & 1 - 0.5^n \\ 0 & 1 \end{pmatrix} = \begin{pmatrix} 0 & 1 \\ 0 & 1 \end{pmatrix}$$

Here, state 1 is *transient* and the chain is eventually absorbed by state 2. Intuitively, a state is transient if, starting from the state, the process will eventually leave the state and never return.

Those four transition matrices illustrate four distinct types of convergence behaviors of DTMCs. To sort out various cases, we must first be able to classify the states of a DTMC. It is worth noting that the calculations of the transient performance measures hold to all DTMCs, regardless of the classification of states.

Let us now classify states in a DTMC, using the transition matrices $\mathbf{P}_1$ to $\mathbf{P}_4$ given in Equation 8.16 to illustrate the concepts. State j is said to be *accessible* from state i if, starting from state i, there is a positive probability that j can be reached from i in a finite number of transitions, that is, $p_{ij}^{(n)} > 0$ for some $n \geq 0$. States i and j are said to *communicate* if they are accessible to each other. For instance, the states in $\mathbf{P}_1$ or $\mathbf{P}_3$ communicate since each state can be accessed by another in one transition: $p_{12} > 0$ and $p_{21} > 0$. However, in $\mathbf{P}_2$ or $\mathbf{P}_4$, the two states do not communicate.

We can partition the states of a Markov chain into *equivalent classes*, where each equivalent class contains those states that communicate with each other. For example, there is a single class in the DTMC governed by $\mathbf{P}_1$ or $\mathbf{P}_3$, whereas there are two classes, each with a single state, in the DTMC governed by $\mathbf{P}_2$ or $\mathbf{P}_4$. It is possible, as illustrated by $\mathbf{P}_4$, that the process starts in one class and enters another class. However, once the process leaves a class, it cannot return to that class, or else the two classes should be combined to form a single class. The Markov chain is said to be *irreducible* if all the states belong to a single class, as in the case of $\mathbf{P}_1$ or $\mathbf{P}_3$.

State i is said to be *transient* if $\lim_{n \to \infty} p_{ii}^{(n)} = 0$ and *recurrent* if $\lim_{n \to \infty} p_{ii}^{(n)} > 0$. Intuitively, a state is transient if, starting from the state, the process will eventually leave the state and never return. In other words, the process will only visit the state a finite number of times. A state is recurrent if, starting from the state, the process is guaranteed to return to the state again and again, in fact, infinitely many times. In the example of $\mathbf{P}_4$, state 1 is transient and state 2 is recurrent. It turns out that for a finite-state DTMC, at least one state must be recurrent, as seen from $\mathbf{P}_1$ to $\mathbf{P}_4$.

If a state is recurrent, then it is said to be *positive recurrent* if, starting from the state, the expected number of transitions until the chain return to the state is finite. It can be shown that in a finite-state DTMC all the recurrent states are also positive recurrent. In an infinite-state DTMC, however, it is possible that a recurrent state is not positive recurrent. Such a recurrent state is called *null recurrent*.

State i is said be *periodic* if $p_{ii}^{(n)} > 0$ only when $n = d, 2d, 3d, \ldots$, and $p_{ii}^{(n)} = 0$ otherwise. A state that is not periodic is called *aperiodic*. In $\mathbf{P}_3$, both states have period $d = 2$, since the chain can reenter each state only at steps 2, 4, 6, and so on. The states in the other three matrices are all aperiodic.

It can be shown that recurrence, transientness, and periodicity are all *class properties*; that is, if state i is recurrent (positive recurrent, null recurrent, transient, periodic), then all other states in the same class of state i inherit the same property. The claim is exemplified by matrices $\mathbf{P}_1$ to $\mathbf{P}_4$.

8.3.4 Limiting Probabilities

The basic limiting theorem of a DTMC can be stated as follows.

THEOREM 8.3 *For a DTMC that is irreducible and ergodic (i.e., positive recurrent and aperiodic), $\lim_{n \to \infty} p_{ij}^{(n)}$, $j \geq 0$, exists and is independent of initial state i. Let $\lim_{t \to \infty} p_{ij}^{(n)} = \pi_j$, $j \geq 0$. Then, $\pi = (\pi_0, \pi_1, \ldots)$ is the unique solution of*

$$\pi_j = \sum_{i=0}^{\infty} \pi_i p_{ij}, \quad j = 0, 1, \ldots \tag{8.17}$$

$$\sum_{j=0}^{\infty} \pi_j = 1 \tag{8.18}$$

Equation 8.17 can be intuitively understood by the CK equations: recall that

$$p_{ij}^{(n)} = \sum_k p_{ik}^{(n-1)} p_{kj}, \quad \text{for all } i, j, n$$

If $p_{ij}^{(n)}$ indeed converges to a value that is independent of i, then by letting n approach to infinity on both sides of the above equation, and assuming that the limit and summation operations are interchangeable, we arrive at Equation 8.17.

Of the four matrices $\mathbf{P}_1$–$\mathbf{P}_4$, only $\mathbf{P}_1$ represents an irreducible ergodic DTMC. Matrix $\mathbf{P}_2$ or $\mathbf{P}_4$ has two classes and hence is not irreducible, and matrix $\mathbf{P}_3$ is *periodic* and hence is not ergodic.

Denote μ_{jj} as the mean recurrent time of state j, which is defined as the expected number of transitions until the process revisits state j. For example, in the brand switching example, μ_{jj} is the expected number of weeks between successive purchases of brand j, $j = 1, 2, 3$. It turns out that μ_{jj} is intimately related to π_j via the equation

$$\mu_{jj} = \frac{1}{\pi_j}, \quad j \geq 0 \tag{8.19}$$

The relationship makes an intuitive sense: since, on average, the chain will spend one unit of time in state j on every μ_{jj} units of time, the proportion of time it stays in state j, π_j, must be $1/\mu_{jj}$.

Note that in a finite-state DTMC, one of the equations in 8.17 is redundant, as we only need N equations to obtain N unknowns. This is illustrated by the following two examples.

Stochastic Processes

Example 8.5: A model of social mobility (continued)

The DTMC is obviously irreducible and ergodic. Based on the estimated transition probabilities shown in the table of Example 8.5, the limiting probabilities (π_1, π_2, π_3) can be obtained by solving

$$\pi_1 = 0.45\pi_1 + 0.05\pi_2 + 0.01\pi_3$$
$$\pi_2 = 0.48\pi_1 + 0.70\pi_2 + 0.50\pi_3$$
$$\pi_1 + \pi_2 + \pi_3 = 1$$

The solution is $\pi_1 = 0.07, \pi_2 = 0.62, \pi_3 = 0.31$. This means, regardless of the current occupation of a family, in a long-run, 7% of its descendants holds upper-class jobs, 62% middle-class jobs, and 31% lower-class jobs. Now, given a family currently holds an upper-class job, it takes in average $\mu_{11} = 1/\pi_1 = 14.29$ generations for the family to hold an upper-class job again! ∎

Example 8.2: Automobile insurance (continued)

In the Bonus Malus system discussed in Example 8.2, suppose that the number of yearly claims by a policyholder follows Poisson with rate 0.5, that is, $P(Y=n) = \frac{e^{-0.5} n^{0.5}}{n!} \equiv \theta_n$. Then $\theta_0 = .6065, \theta_1 = .3033, \theta_2 = .0758$, and $\sum_{n=3}^{\infty} \theta_n = 0.0144$. Based on the Bonus Males table given in Example 8.2, we obtain the transition matrix

$$\mathbf{P} = \begin{pmatrix} .6065 & .3033 & .0758 & .0144 \\ .6065 & .0000 & .3033 & .0902 \\ .0000 & .6065 & .0000 & .3935 \\ .0000 & .0000 & .6065 & .3935 \end{pmatrix}$$

The chain is irreducible and ergodic. The limiting distribution satisfies the system of equations

$$\pi_1 = .6065\pi_1 + .6065\pi_2$$
$$\pi_2 = .3033\pi_1 + .6065\pi_3$$
$$\pi_3 = .0758\pi_1 + .3033\pi_2 + .6065\pi_4$$
$$\sum_{i=1}^{4} \pi_i = 1$$

The solution is $\pi_1 = 0.3692, \pi_2 = 0.2395, \pi_3 = 0.2103$, and $\pi_4 = 0.1809$. Based on the limiting distribution, we can compute the average annual premium paid by a policyholder as

$$500\pi_1 + 600\pi_2 + 800\pi_3 + 1000\pi_4 = \$677.44 \qquad \blacksquare$$

Unfortunately, not all Markov chains can satisfy the conditions given in Theorem 8.3. What can be said about the limiting behavior of such Markov chains? It turns out that it depends on different cases, as explained by the following remark.

Remark 8.1

1. *The DTMC with irreducible, positive recurrent, but periodic states*: In this case, π is still a unique nonnegative solution of Equations 8.17 and 8.18. But now π_j must be understood as the long-run proportion of time that the process is in state j. An example is the DTMC associated with $\mathbf{P}_3$; it has the unique solution $\pi = (1/2, 1/2)$.

The long-run proportions are also known as the *stationary distribution* of the DTMC. The name comes from the fact that if the initial distribution of the chain is chosen to be π, then the process is *stationary* in the sense that the distribution of X_n remains the same for all n. It can be shown that when a limiting distribution exists, it is also a stationary distribution.

2. *The DTMC with several closed, positive recurrent classes*: In this case, the transition matrix of the DTMC takes the form

$$\mathbf{P} = \begin{pmatrix} \mathbf{P}_A & 0 \\ 0 & \mathbf{P}_B \end{pmatrix}$$

where the subprocess associated with either $\mathbf{P}_A$ or $\mathbf{P}_B$ itself is a Markov chain. Matrix $\mathbf{P}_2$ is an example of such a case. A slightly more general example is

$$\mathbf{P} = \begin{pmatrix} 0.5 & 0.5 & 0 \\ 0.75 & 0.25 & 0 \\ 0 & 0 & 1 \end{pmatrix} \tag{8.20}$$

with $\mathbf{P}_A = \begin{pmatrix} 0.5 & 0.5 \\ 0.75 & 0.25 \end{pmatrix}$ and $\mathbf{P}_B = (1)$. It can be easily obtained that

$$\mathbf{P}^{(n)} = \begin{pmatrix} \mathbf{P}_A^{(n)} & 0 \\ 0 & \mathbf{P}_B^{(n)} \end{pmatrix} \tag{8.21}$$

which means that each class is *closed*, that is, once the process is in a class, it remains there thereafter. In effect, the transition matrix is reduced to two irreducible matrices $\mathbf{P}_A$ and $\mathbf{P}_B$. From Equation 8.21, it follows that, for the matrix given in Equation 8.20,

$$\lim_{n \to \infty} \mathbf{P}^{(n)} = \begin{pmatrix} \pi_1^A & \pi_2^A & 0 \\ \pi_1^A & \pi_2^A & 0 \\ 0 & 0 & \pi_3^B \end{pmatrix}$$

where we can obtain $(\pi_1^A, \pi_2^B) = (1/3, 2/3)$ in the usual way using Equations 8.17 and 8.18. We of course have $\pi_3^B = 1$. In contrast to the irreducible ergodic DTMC, where the limiting distribution is independent of the initial state, the DTMC with several closed, positive recurrent classes has the limiting distribution that is dependent on the initial state.

3. *The DTMC with both recurrent and transient classes*: The DTMC associated with $\mathbf{P}_4$ is an example of such a case. In general, there may be several transient and several recurrent classes. In this situation, we often seek the probabilities that the chain is eventually absorbed by different recurrent classes. We illustrate the method using the well-known *gambler's ruin problem* described in Example 8.11.

Example 8.13: The gambler's ruin

The DTMC associated with the gambler's ruin problem has the transition matrix:

$$\mathbf{P} = \begin{pmatrix} 1 & 0 & 0 & \cdot & \cdot & \cdot & 0 \\ q & 0 & p & 0 & \cdot & \cdot & 0 \\ 0 & q & 0 & p & 0 & \cdot & 0 \\ \vdots & \vdots & \vdots & \vdots & \vdots & \vdots & \vdots \\ 0 & 0 & \cdot & \cdot & q & 0 & p \\ 0 & 0 & \cdot & \cdot & \cdot & 0 & 1 \end{pmatrix} \tag{8.22}$$

Stochastic Processes

The DTMC has three classes, $C_1 = \{0\}$, $C_2 = \{N\}$, and $C_3 = \{1, 2, \ldots, N-1\}$, where C_1 and C_2 are recurrent classes and C_3 is a transient class. Starting from any state i, we want to compute f_i, the probability that the chain is absorbed by C_2 (i.e., the gambler attains a fortune of N before being ruined), for $i = 0, 1, \ldots, N$. Conditioning on the outcome of the first play, we obtain a system of linear equations for the f_i:

$$f_0 = 0$$
$$f_i = pf_{i+1} + qf_{i-1}, \quad i = 1, 2, \ldots, N-1$$
$$f_N = 1$$

The solution of the above system of linear equations is given by:

$$f_i = \begin{cases} \dfrac{1 - (q/p)^i}{1 - (q/p)^N} & p \neq 1/2 \\ \dfrac{i}{N} & p = 1/2 \end{cases} \quad i = 0, 1, \ldots, N$$

An application of the gambler's ruin model in drug testing is given by Ross (2003). ∎

4. The *irreducible DTMC with null recurrent or transient states*: This case is only possible when the state space is infinite, since any finite-state, irreducible DTMC must be positive recurrent. In this case, neither the limiting distribution nor the stationary distribution exists. A well-known example of this case is the random walk model discussed in Example 8.13. It can be shown (see, for example, Ross, 2003) that when $N = \infty$, $p_0 = 1$, and $p_i = q_i = 1/2$ for $i \geq 1$, the Markov chain is null recurrent. If $N = \infty$, $p_0 = 1$, and $p_i = p, > 1 - p = q = q_i$ for $i \geq 1$, then the Markov chain is transient.

8.3.5 Statistical Inference of Discrete-Time Markov Chains

In this section, we briefly discuss parameter estimation and hypothesis testing issues for DTMCs. Further references on the subject can be found in Bhat & Miller (2002), Basawa & Prakasa Rao (1980), and references therein.

We first consider how to estimate the entries in transition matrix $\mathbf{P} = \{p_{ij}\}$, based on data collected. Suppose we have observed a DTMC on a finite state space $S = \{0, 1, \ldots, N\}$ for n transitions. Let n_i be the number of periods that the process is in state i, and n_{ij} the number of transitions from i to j, with $n_i = \sum_{j=0}^{N} n_{ij}$, $i = 1, 2, \ldots, N$, and $\sum_{i=0}^{N} n_i = n$. For each state i, the transition counts from this state to other states, $(n_{i0}, n_{i1}, \ldots, n_{iN})$, can be regarded as a sample of size n_i from a *multinomial* distribution with probabilities $(p_{i0}, p_{i1}, \ldots, p_{iN})$, $i = 0, 1, \ldots, N$. To obtain the maximum likelihood function of $\mathbf{P}$, we also have to take into account that the values $(n_0, n_1, \ldots, n_N)$ are random variables. Whittle (1955) shows that the maximum likelihood function has to include a correction factor A, and takes the form

$$f(\mathbf{P}) = A \prod_{i=0}^{N} \frac{n_i!}{n_{i0}! n_{i1}! \cdots n_{iN}!} p_{i0}^{n_{i0}} p_{i1}^{n_{i1}} \cdots P_{iN}^{n_{iN}}$$

where A turns out to be independent of $\mathbf{P}$. Therefore, the log likelihood function $L(\mathbf{P})$ is given by

$$L(\mathbf{P}) = \ln f(\mathbf{P}) = \ln B + \sum_{i=0}^{N}\sum_{j=0}^{N} n_{ij} \ln p_{ij}$$

$$= \ln B + \sum_{i=0}^{N}\left[\sum_{j=0}^{N-1} n_{ij} \ln p_{ij} + n_{iN} \ln\left(1 - \sum_{j=0}^{N-1} p_{ij}\right)\right]$$

where B contains all the terms independent of $\mathbf{P}$. For each i, we take the derivatives of $L(\mathbf{P})$ with respect to p_{ij} for $j = 0, 1, \ldots, N-1$ and set the resultant equations to zero:

$$\frac{\partial L(P)}{\partial p_{ij}} = \frac{n_{ij}}{p_{ij}} - \frac{n_{iN}}{1 - \sum_{j=0}^{N-1} p_{ij}} = 0, \quad j = 0, 1, \ldots, N-1$$

Solving the above equations yields the maximum likelihood estimator of p_{ij} as

$$\hat{p}_{ij} = \frac{n_{ij}}{n_i}, \quad i, j = 0, 1, \ldots, N \tag{8.23}$$

Example 8.14: Voters' attitude changes

Anderson (1954) used the data collected by the Bureau of Applied Social Research in Erie County, Ohio, in 1940 (Lazarsfeld et al., 1948), in a study of voters' attitude changes. The Bureau interviewed 600 voters from May to August on their voting preferences, D (Democrat), R (republican), and DK (Do not know or other candidates). Among the 600 people interviewed, 445 people responded to all six interviews. In that year, the Republic Convention was held between the June and July interviews, and the Democratic Convention was held between the July and August interviews. The transition counts for pairs of three successive interviews are given below.

	May–June					June–July					July–August			
	R	D	DK	Total		R	D	DK	Total		R	D	DK	Total
R	125	5	16	146	R	124	3	16	143	R	146	2	4	153
D	7	106	15	128	D	6	109	14	127	D	6	111	4	121
DK	11	18	142	171	DK	22	9	142	173	DK	40	36	96	172
				445					445					445

Using Equation 8.23, we obtain the estimates of the three sets of transition probabilities as:

	May–June				June–July				July–August		
	R	D	DK		R	D	DK		R	D	DK
R	0.856	0.034	0.110	R	0.867	0.021	0.112	R	0.961	0.013	0.026
D	0.055	0.828	0.117	D	0.047	0.845	0.108	D	0.050	0.917	0.033
DK	0.064	0.105	0.831	DK	0.127	0.052	0.821	DK	0.233	0.209	0.558

One question pertinent in the study was whether the voters' attitudes had changed due to the political events during the period under consideration. If no attitude changes were

Stochastic Processes

detected, the Markov chain is stationary, and hence the three sets of transition counts should be pooled to give a single estimate of the transition matrix. We shall revisit this issue later. ∎

We now discuss the *likelihood ratio test* for the stationarity of transition matrices. Let $p_{ij}^t = P(X_{t+1} = j | X_t = i)$ be the one-step transition probability from state i to state j at time t. Our null hypothesis, H_0, is $p_{ij}^t = p_{ij}$, for $t = 1, \ldots, T$. To test H_0, denote n_{ij}^t as the transition count from i to j in period t. For example, in the voters' attitude dataset, let $t = 1$, 2, 3 represent respectively May, June, and July, and states 1, 2, 3 correspond to D, R, and DK. Then $n_{12}^1 = 5$ means that 5 people changed their voting preferences from a Republican candidate to a Democratic candidate from May to June. Under H_0, the likelihood ratio test statistic follows a χ^2 distribution with $(T-1)N(N-1)$ degrees of freedom (see Bhat & Miller, 2002, pp. 134–135):

$$\chi^2_{(T-1)N(N-1)} = 2 \sum_{t=1}^{T} \sum_{i=0}^{N} \sum_{j=0}^{N} n_{ij}^t \ln \frac{p_{ij}^t}{p_{ij}} \qquad (8.24)$$

As discussed, the maximum likelihood estimates of p_{ij}^t and p_{ij} are given by

$$\hat{p}_{ij}^t = \frac{n_{ij}^t}{\sum_{j=0}^{N} n_{ij}^t}, \quad t = 1, 2, \ldots, T, \quad \hat{p}_{ij} = \frac{\sum_{t=1}^{T} n_{ij}^t}{\sum_{t=1}^{T} \sum_{j=0}^{N} n_{ij}^t} = \frac{n_{ij}}{n_i}$$

Example 8.14: Voters' attitude changes (continued)

We wish to test the stationarity of the three transition matrices, which tells us whether the voters exhibited the same behavior across May to August. If indeed the process were stationary, the three sets of transition counts can be pooled to form one set of transition counts. The pooled transition counts and the estimate of the transition matrix for the pooled data are given in the following table:

n_{ij}	R	D	DK	Total	$\hat{p}_{ij}$	R	D	DK
R	395	10	36	441	R	0.896	0.023	0.081
D	19	326	33	378	D	0.050	0.862	0.088
DK	73	63	380	516	DK	0.141	0.122	0.737
				1335				

Under H_0, the χ^2 statistic given in Equation 8.24 has $(3-1) \cdot 3 \cdot (3-1) = 12$ degrees of freedom. Using the three sets of transition probabilities computed in Example 8.14 along with the above estimated transition probabilities for the pooled data, we obtain $\chi^2_{12} = 97.644$. As $P(\chi^2_{12} > 97.644) < 0.001$, we conclude that there was sufficient evidence to support the claim that the voters' attitudes had been affected by political events. ∎

8.4 Continuous-Time Markov Chains

In a DTMC, the process stays in a state for a unit of time and then possibly makes a transition. However, in many practical situations the process may change state at any point of time. One type of model that is powerful to analyze a continuous-time stochastic system is the *continuous-time Markov chain* (CTMC), which is similar to the DTMC but assumes that the sojourn time of the process in each state is a state-dependent *exponential* random variable, independent of all else.

8.4.1 Definition of Continuous-Time Markov Chains

We call a continuous-time stochastic process $\{X(t), t \geq 0\}$ with state space $S = \{0, 1, \ldots\}$ a continuous-time Markov chain if for all i, j, t, and s,

$$P(X(t+s) = j | X(s) = i, X(u) = x(u), 0 \leq u < s) = P(X(t+s) = j | X(s) = i) \quad (8.25)$$

As in the definition of the DTMC, Equation 8.25 expresses the Markov property, which states, to predict the future state of the process, it is sufficient to know the most recently observed state. Here, we only consider the stationary, or time-homogeneous, CTMC, that is, the CTMC with its transition probability $P(X(t+s) = j | X(s) = i)$ depending on t but not on s, for any i, j, s, and t. We write

$$p_{ij}(t) \equiv P(X(t+s) = j | X(s) = i)$$

We take the convention to arrange the probabilities $p_{ij}(t)$ in the form of a matrix, $\mathbf{P}(t) = \{p_{ij}(t)\}$, which shall be called the transition matrix. Note that $\mathbf{P}(t)$ depends on t, that is, a different t specifies a different transition matrix. For each t, $\mathbf{P}(t)$ is a stochastic matrix in the sense defined in Equation 8.7. As in the DTMC case, the transition matrix of the CTMC satisfies the Chapman–Kolmogorov (CK) equations:

$$\mathbf{P}(s+t) = \mathbf{P}(s) \cdot \mathbf{P}(t) \quad (8.26)$$

How do we check whether a stochastic process is a CTMC? Directly examining the Markov property Equation 8.25 of the system is not practical and yields little insight. A more direct construction of a CTMC is via a *jump* process with exponentially distributed sojourn times between successive jumps. For this, consider a stochastic process evolving in the state space $S = \{0, 1, \ldots\}$ as follows: (1) if the system enters state i, it stays there for an exponentially distributed time with rate ν_i, independent of all else; (2) when the system leaves state i, it makes a transition to state $j \neq i$ with probability p_{ij}, independent of how long the system has stayed in i. By convention, the transition from a state to itself is not allowed. Define $X(t)$ as the state of the system at time t in the jump process described above. Then it can be shown that $\{X(t), t \geq 0\}$ is a CTMC. As such, a CTMC can be described by two sets of parameters: exponential sojourn time rates $\{\nu_i\}$ and transition probabilities $\{p_{ij}, j \neq i\}$.

Another view of a CTMC as a jump process is as follows: when the process is in state i, it attempts to make a jump to state j after a sojourn time T_{ij}, $j \geq 0$, where T_{ij} follows an exponential distribution with rate q_{ij}, independent of all else. The process moves from i to j if T_{ij} happens to be the smallest among all the sojourn times T_{ik}, $k \neq i$, that is, $T_{ij} = T_i = \min_{k \neq i}\{T_{ik}\}$. Then, by properties (2) and (3) of the exponential distribution (see Section 8.2), T_i follows the exponential distribution with rate $\sum_{k \neq i} q_{ik}$, and $P(T_{ij} = T_i) = q_{ij} / \sum_{k \neq i} q_{ik}$.

Let us define

$$\nu_i = \sum_{k \neq i} q_{ik}, \quad p_{ij} = \frac{q_{ij}}{\sum_{k \neq i} q_{ik}}, \quad i = 0, 1, \ldots, j \neq i \quad (8.27)$$

The preceding expressions mean that a CTMC can also be represented by a set of exponential rates q_{ij}, $j \neq i$, which shall be referred to as the transition rates hereafter. For convenience, we define

$$q_{ii} = -\nu_i = -\sum_{k \neq i} q_{ik}, \quad i = 0, 1, \ldots$$

Stochastic Processes

Then we can arrange all the rates by a matrix, $\mathbf{Q} = \{q_{ij}\}$. Matrix $\mathbf{Q}$ is known as the *infinitesimal generator* (or simply, the generator) of a CTMC. Supplemented by the initial distribution of the process, namely, $\mathbf{a} = \{a_j\}$, where $a_j = P(X(0) = j)$, $j \geq 0$, $\mathbf{Q}$ completely specifies the CTMC.

Example 8.15: A two-component cold standby system

A system has two components, only one of which is used at any given time. The standby component will be put in use when the online component fails. The standby component cannot fail (thus the name *cold standby*). The uptime of each component, when in use, is an exponential random variable with rate λ. There is a single repairman who repairs a breakdown component at an exponential rate μ.

Let $X(t)$ denote the number of failed components at time t; then $X(t)$ can take values 0, 1, or 2. It is not difficult to see that $\{X(t), t \geq 0\}$ is a DTMC with the parameters

$$\nu_0 = \lambda, \quad \nu_1 = \lambda + \mu, \quad \nu_2 = \mu,$$

$$p_{01} = p_{21} = 1, \quad p_{10} = \frac{\mu}{\lambda + \mu}, \quad p_{12} = \frac{\lambda}{\lambda + \mu}$$

From Equation 8.27, we obtain the infinitesimal generator $\mathbf{Q}$ of the CTMC as

$$\mathbf{Q} = \begin{pmatrix} -\lambda & \lambda & 0 \\ \mu & -(\lambda + \mu) & \lambda \\ 0 & -\mu & \mu \end{pmatrix}$$

∎

8.4.2 Birth and Death Processes and Applications

The birth and death process is a CTMC with state space $S = \{0, 1, 2, \ldots\}$, where the state represents the current number of people in the system, and the state changes when either a birth or a death occurs. More specifically, when the population size is i, a new arrival (birth) enters the system at an exponential rate λ_i, $i = 0, 1, \ldots$, and a new departure (death) leaves the system at an exponential rate μ_i, $i = 1, 2, \ldots$. The infinitesimal generator of the birth and death process is

$$\mathbf{Q} = \begin{pmatrix} -\lambda_0 & \lambda_0 & 0 & 0 & 0 & \cdots \\ \mu_1 & -(\lambda_1 + \mu_1) & \lambda_1 & 0 & 0 & \cdots \\ 0 & \mu_2 & -(\lambda_1 + \mu_1) & \lambda_2 & 0 & \cdots \\ \vdots & \vdots & \vdots & \vdots & \vdots & \vdots \end{pmatrix} \quad (8.28)$$

As seen, the key feature of the birth and death process is that, at any state, the process can only jump to an adjacent state of that state. The Poisson process is an example of the birth and death process, in which births occur at a constant rate λ and death rates are all 0. The cold standby reliability system is another example. The birth and death process may also be viewed as the continuous counterpart of the random walk model discussed in Example 8.11.

The study of the *birth and death process* was started in the second decade of the twentieth century, with its roots in biology. The investigations by the Dutch scientist A.K. Erlang on traffic of telephone networks also contributed to the development of the theory of the birth and death process and its application in queueing theory. We shall illustrate the use of a fundamental result credited to Erlang, the *Erlang C formula*, in the following example.

Example 8.16: A model of tele-queue: the Erlang C formula

The fundamental challenge for efficient inbound call center operations is to strike a correct balance between supply (the numbers of agents and trunks) and demand (the random and time varying incoming calls). The call center needs to maximize the utilization of its agents, while keeping a caller waiting time to an acceptable minimum. Gans et al. (2003) give a comprehensive review of queueing models used in call center operations. One of the simplest yet widely used model to evaluate a call center's performance is the Erlang C model that originated from A.K. Erlang. When Erlang worked for the Copenhagen Phone Company, he used the model to calculate the fraction of callers in a small village who had to wait because all the lines were in use.

By the queueing notation, Erlang C is an $M/M/s$ queue, which is a birth and death process. The model assumes that the arrival process is Poisson with rate λ, service times are iid exponential with rate μ, and there are s agents to answer calls. Erlang C has its parameters specified as

$$\lambda_i = \lambda, \quad i \geq 0, \quad \mu_i = \begin{cases} i\mu & \text{if } 0 < i \leq s \\ s\mu & \text{if } i > s \end{cases} \tag{8.29}$$

The death rate μ_i is understood as: when $0 < i \leq s$, $i(\leq s)$ agents are busy, and a caller departs the center when one of the i agents finishes service. Hence the service rate is $i s$. When $i \geq s$, all s agents are busy and they serve at an aggregated rate $s\mu$. We will revisit the model and discuss its solution methods in the subsequent sections. ∎

Erlang C has two drawbacks. First, it assumes that the system has an infinite number of trunks so a caller who cannot be answered immediately would not be blocked. In reality, a call center has a finite number of lines, and a caller will be blocked (i.e., get busy signals) when all the lines are occupied. Second, it does not take customer abandonments into account. Empirical data show that a significant proportion of callers abandon their lines before being served (Gans et al., 2003). The next model, Erlang A, incorporates both busy signals and abandonment in model construction.

Example 8.17: The Erlang A model

In this model, the arrival process is assumed to be Poisson and service times are iid exponential. There are s agents and k trunks, $s \leq k$. The maximum number of calls the system can accommodate is k and an incoming call will get busy signals if all k lines are occupied. Patience is defined as the maximal amount of time that a caller is willing to wait for service; if not served within this time, the caller deserts the queue. The queueing model associated with Erlang A is the $M/M/s/k+M$ queue, where the $+M$ notation means that patience is iid exponentially distributed, say with rate θ. The $M/M/s/k+M$ model is again a birth and death process with the transition rates

$$\begin{aligned} \lambda_i &= \lambda & i = 0, \ldots, k-1 \\ \mu_i &= \min\{i, s\}\mu + \max\{i-s, 0\}\theta & i = 1, \ldots, k \end{aligned} \tag{8.30}$$

To understand the departure rate μ_i, note that the first term $\min\{i, s\}\mu$ is the service completion rate, bearing a similar explanation as in the Erlang C model. The second term $\max\{i-s, 0\}\theta$ is the abandonment rate when there are i callers in the system, as the number of callers waiting in queue is $\max\{i-s, 0\}$, and each caller in queue abandons the line at an exponential rate θ. ∎

Stochastic Processes

8.4.3 Transient Analysis

Transient analysis of a CTMC is concerned with the distribution of $X(t)$ for a finite t. Since if we know $\mathbf{P}(t) = \{p_{ij}(t)\}$, we can compute the distribution of $X(t)$ by

$$P(X(t) = j) = \sum_{i=0}^{\infty} p_{ij}(t) a_i$$

where $\{a_i\}$ is the distribution of X_0, it suffices to know $\{p_{ij}(t)\}$. There are several ways to compute $\{p_{ij}(t)\}$, and we present two of them, the *uniformization* and *Kolmogorov differential equation* methods. While the first one is straightforward for numerical computation, the second one is mathematically more structured. Let us consider them in turn. The uniformization method needs the condition that the ν_i in the CTMC is bounded. For example, the condition holds for a finite-state CTMC. Let ν be any finite number such that $\max_i \{\nu_i\} \leq \nu < \infty$. Intuitively, what this technique does is to insert fictitious transitions from a state to itself so that, with both the real and fictitious transitions, the CTMC changes states at a constant rate ν, regardless of the state it is in, where ν is a uniform upper bound on ν_i for any i. Let matrix $\hat{\mathbf{P}} = \{\hat{p}_{ij}\}$ be defined as

$$\hat{p}_{ij} = \begin{cases} 1 - \dfrac{\nu_i}{\nu} & \text{if } i = j \\ \dfrac{q_{ij}}{\nu} & \text{if } i \neq j \end{cases} \quad (8.31)$$

Clearly, $\hat{\mathbf{P}}$ is a stochastic matrix. The following is the key result, which gives the distribution $\mathbf{P}(t)$:

$$\mathbf{P}(t) = \sum_{k=0}^{\infty} e^{-\nu t} \frac{(\nu t)^k}{k!} \hat{\mathbf{P}}^k, \quad t \geq 0 \quad (8.32)$$

Example 8.18:

Suppose that the infinitesimal generator of a CTMC is

$$\mathbf{Q} = \begin{pmatrix} -5 & 2 & 3 & 0 \\ 4 & -6 & 2 & 0 \\ 0 & 2 & -4 & 2 \\ 1 & 0 & 3 & -4 \end{pmatrix}$$

Since $q_{ii} = -\nu_i$, we have $\nu_0 = 5$, $\nu_1 = 6$, $\nu_3 = 4$, and $\nu_2 = 4$. Hence, we can choose $\nu = \max\{5, 6, 4, 4\} = 6$. The uniformized transition matrix $\hat{\mathbf{P}}$ becomes

$$\hat{\mathbf{P}} = \begin{pmatrix} 1/6 & 1/3 & 1/2 & 0 \\ 2/3 & 0 & 1/3 & 0 \\ 0 & 1/3 & 1/3 & 1/3 \\ 1/6 & 0 & 1/2 & 1/3 \end{pmatrix}$$

Let us compute $\mathbf{P}(0.5)$. To numerically evaluate $\mathbf{P}(t)$ given in Equation 8.32, we must truncate the summation of infinite terms to that of the first K terms. We shall use $K = 13$ (algorithms are available to determine K that guarantees the error to be controlled within a desired level. See Nelson (1995) and Kulkarni (1995). From Equation 8.32, we get

$$P(0.5) = \sum_{k=0}^{13} e^{-6 \times 0.5} \frac{(6 \times 0.5)^k}{k!} \begin{pmatrix} 1/6 & 1/3 & 1/2 & 0 \\ 2/3 & 0 & 1/3 & 0 \\ 0 & 1/3 & 1/3 & 1/3 \\ 1/6 & 0 & 1/2 & 1/3 \end{pmatrix}^k$$

$$= \begin{pmatrix} 0.2506 & 0.2170 & 0.3867 & 0.1458 \\ 0.2531 & 0.2381 & 0.3744 & 0.1341 \\ 0.1691 & 0.1936 & 0.4203 & 0.2170 \\ 0.1580 & 0.1574 & 0.3983 & 0.2862 \end{pmatrix}$$

Similarly, we can compute $\mathbf{P}(1)$, using $K = 19$. We get

$$P(1) = \begin{pmatrix} 0.2083 & 0.2053 & 0.3987 & 0.1912 \\ 0.2083 & 0.2053 & 0.3979 & 0.1885 \\ 0.1968 & 0.1984 & 0.4010 & 0.2039 \\ 0.1920 & 0.1940 & 0.4015 & 0.2125 \end{pmatrix}$$

We can verify that $\mathbf{P}(1) = \mathbf{P}(0.5) \cdot \mathbf{P}(0.5)$, as we would expect from the CK equations. ∎

Example 8.19: A repairable reliability system

Consider a machine that alternates between two conditions: state 0 means the machine is in working condition, whereas state 1 means it is under repair. Suppose the working time and repair time are both exponentially distributed with respective rates λ and μ. We are interested in the distribution of the state of the machine at a fixed time t. It is easily seen that the process has the generator given by

$$Q = \begin{pmatrix} -\lambda & \lambda \\ \mu & -\mu \end{pmatrix}$$

It is convenient to let $\nu = \lambda + \mu$. From Equation 8.31 we obtain

$$\hat{P} = \begin{pmatrix} \frac{\mu}{\lambda+\mu} & \frac{\lambda}{\lambda+\mu} \\ \frac{\mu}{\lambda+\mu} & \frac{\lambda}{\lambda+\mu} \end{pmatrix}$$

It can be easily verified that $\hat{\mathbf{P}}^k = \hat{\mathbf{P}}$, for any $k \geq 1$. This property allows us to obtain Equation 8.32 in the closed form (for detailed derivation, see, for example, Kulkarni, 1995):

$$P(t) = \begin{pmatrix} \frac{\mu}{\lambda+\mu} & \frac{\lambda}{\lambda+\mu} \\ \frac{\mu}{\lambda+\mu} & \frac{\lambda}{\lambda+\mu} \end{pmatrix} + \begin{pmatrix} \frac{\lambda}{\lambda+\mu} & \frac{-\lambda}{\lambda+\mu} \\ \frac{-\mu}{\lambda+\mu} & \frac{\mu}{\lambda+\mu} \end{pmatrix} \cdot e^{-(\lambda+\mu)t} \quad (8.33)$$

From Equation 8.33, we can read that the probability that the machine, now up, will be up at time t is

$$p_{00}(t) = \frac{\mu}{\lambda+\mu} + \frac{\lambda}{\lambda+\mu} e^{-(\lambda+\mu)t} \quad \blacksquare$$

The second approach to evaluate $\mathbf{P}(t)$ is based on the systems of *Kolmogorov differential equations*, which assert that $\mathbf{P}(t)$ must satisfy

$$\frac{d\mathbf{P}(t)}{dt} = \mathbf{P}(t) \cdot \mathbf{Q} \quad \text{and} \quad \frac{d\mathbf{P}(t)}{dt} = \mathbf{Q} \cdot \mathbf{P}(t) \quad (8.34)$$

Stochastic Processes 8-33

where $d\mathbf{P}(t)/dt = \{dp_{ij}(t)/dt\}$. The first set of equations is called the Kolmogorov *forward* equations and the second set the Kolmogorov *backward* equations. Under the mild conditions, which are satisfied by most practical problems, the solutions of the forward and backward equations are identical. We illustrate the method using the previous example.

Example 8.19: A repairable reliability system (revisited)

We shall use the forward equations to obtain $\mathbf{P}(t)$. Using the generator $\mathbf{Q}$ given in Example 8.19, we can write Equation 8.34 as

$$\begin{pmatrix} \dfrac{dp_{00}(t)}{dt} & \dfrac{dp_{01}(t)}{dt} \\ \dfrac{dp_{10}(t)}{dt} & \dfrac{dp_{11}(t)}{dt} \end{pmatrix} = \begin{pmatrix} p_{00}(t) & p_{01}(t) \\ p_{10}(t) & p_{11}(t) \end{pmatrix} \cdot \begin{pmatrix} -\lambda & \lambda \\ \mu & -\mu \end{pmatrix} \quad (8.35)$$

Note that there are in total four differential equations in Equation 8.35. Suppose we are only interested in $p_{00}(t)$. Do we have to solve all four differential equations to get $p_{00}(t)$? The answer is fortunately no. In general, if we are interested in $p_{ij}(t)$, then we only need to solve the set of differential equations involved in the ith row of $\{dp_{ij}(t)/dt\}$. In our case, we end up with two differential equations:

$$\frac{dp_{00}(t)}{dt} = -\lambda p_{00}(t) + \mu p_{01}(t)$$
$$\frac{dp_{01}(t)}{dt} = \lambda p_{00}(t) - \mu p_{01}(t)$$

Recognizing $p_{00}(t) = 1 - p_{01}(t)$, we further reduce the two equations to one:

$$\frac{dp_{00}(t)}{dt} = \mu - (\lambda + \mu)p_{00}(t)$$

Using the standard technique to solve this linear differential equation, we obtain

$$p_{00}(t) = \frac{\mu}{\lambda + \mu} + \frac{\lambda}{\lambda + \mu}e^{-(\lambda+\mu)t}$$

As expected, this result is the same as the result obtained by the uniformization method. ∎

8.4.4 Limiting Distribution

We now turn our attention to the limiting behavior of the CTMC. As in the DTMC case, the existence of the limiting distribution of $\mathbf{P}(t)$, when $t \to \infty$, requires the CTMC under study to satisfy certain conditions. The sufficient conditions under which the limiting distribution exists and is independent of the initial state are:

1. the CTMC is irreducible, that is, all states communicate; and
2. the CTMC is positive recurrent, that is, starting from any state, the mean time to return to that state is finite.

Note that any finite-state, irreducible CTMC is positive recurrent, and hence, exists the limiting distribution that is independent of the initial state.

When the above conditions are satisfied for the CTMC, denote $\pi = \{\pi_0, \pi_1, \ldots\}$ as the limiting distribution of the process, where $\pi_j = \lim_{t \to \infty} p_{ij}(t)$, $j \geq 0$. To derive the set of equations that $\pi = \{\pi_0, \pi_1, \ldots\}$ satisfy, consider the forward equations given in Equation 8.34.

If for any given j, probability $p_{ij}(t)$ indeed converges to a constant π_j, then we must have $\lim_{t\to\infty} \dfrac{\mathrm{d}p_{ij}(t)}{\mathrm{d}t} = 0$. Hence, the forward equations, in the limit, tend to

$$0 = \pi \mathbf{Q} = (\pi_0, \pi_1, \pi_2, \ldots) \begin{pmatrix} -v_0 & q_{01} & q_{02} & q_{03} & \cdots \\ q_{10} & -v_1 & q_{12} & q_{13} & \cdots \\ q_{20} & q_{21} & -v_2 & q_{23} & \cdots \\ \vdots & \vdots & \vdots & \vdots & \vdots \end{pmatrix} \tag{8.36}$$

where $v_i = \Sigma_j q_{ij}$. The above equations are equivalent to

$$\pi_j v_j = \sum_{i \neq j} \pi_i q_{ij}, \quad \text{for all } j \tag{8.37}$$

Evidently, we also need the normalizing equation:

$$\sum_j \pi_j = 1 \tag{8.38}$$

Equations 8.37 are called the *balance equations* and have a nice interpretation: the left side term, $\pi_j \nu_j$, represents the long-run rate at which the process leaves state j, since, when in j, the process leaves at rate ν_j, and the long-run proportion of time it is in j is π_j. The right side term, $\Sigma_{i \neq j} \pi_i p_{ij}$, is the long-run rate at which the process enters j, since, when in state i, the process enters j at rate q_{ij}, and the proportion of time the process is in i is π_i, $i \neq j$. Hence, Equation 8.37 asserts, the long-run rate into any state must be balanced with the long-run rate out of that state.

Equations 8.37 and 8.38 together can be used to compute the limiting distribution π of the CTMC when it exists. It turns out that the solution is also unique. The limiting distribution π sometimes is called the *stationary distribution*. The name comes from the fact that if the distribution of the initial state X_0 is chosen to be π, then $X(t)$ will have the same distribution π for any given t. In other words, the process is stationary starting from time 0.

Example 8.20: The limiting distribution of the birth and death process

The infinitesimal generator $\mathbf{Q}$ of the birth and death process is given in Equation 8.28. Using Equations 8.37 and 8.38, we get

$$\lambda_0 \pi_0 = \mu_1 \pi_1$$
$$(\lambda_i + \mu_i)\pi_i = \lambda_{i-1}\pi_{i-1} + \mu_{i+1}\pi_{i+1}, \quad i > 0$$
$$\sum_j \pi_j = 1$$

The solution of the preceding equations is (see, for example, Ross, 2003)

$$\pi_0 = \left[1 + \sum_{i=1}^{\infty} \frac{\lambda_0 \lambda_1, \ldots, \lambda_{i-1}}{\mu_1 \mu_1, \ldots, \mu_i}\right]^{-1} \tag{8.39}$$

$$\pi_i = \frac{\lambda_0 \lambda_1, \ldots, \lambda_{i-1}}{\mu_1 \mu_1, \ldots, \mu_i} \pi_0, \quad i > 0 \tag{8.40}$$

Stochastic Processes

The necessary and sufficient condition for the limiting distribution to exist is

$$\sum_{i=1}^{\infty} \frac{\lambda_0 \lambda_1, \ldots, \lambda_{i-1}}{\mu_1 \mu_1, \ldots, \mu_i} < \infty$$

The above formulas will be used to derive the limiting distributions of Erlang models, which are special cases of the birth and death process. ∎

Example 8.16: The Erlang C formula (continued)

We start with computing the limiting distribution for Erlang C (i.e., the $M/M/s$ queue). The transition rates of Erlang C are given in Equation 8.29. Define $\rho = \lambda/s\mu$ as the traffic intensity of the system. We assume $\rho < 1$, which is the necessary and sufficient condition for the existence of the limiting distribution. Substituting the parameters into Equations 8.39 and 8.40, we get

$$\pi_0 = \left[\sum_{i=0}^{s} \frac{(s\rho)^i}{i!} + \frac{\rho^{s+1} s^s}{s!}(1-\rho)^{-1} \right]^{-1}$$

$$\pi_i = \begin{cases} \dfrac{(s\rho)^i}{i!} \pi_0 & \text{if } i \leq s \\ \dfrac{s^s \rho^i}{s!} \pi_0 & \text{if } i \geq s \end{cases} \quad i \geq 1$$

We now use π to compute key performance measures in a call center operation. One way to use the Erlang C result is to estimate the service level (SL), which is defined as the probability that a caller's waiting time, W_q, is no more than a predetermined number, say D:

$$SL = P(W_q \leq D) \tag{8.41}$$

Conditioning on whether a caller has to wait, we can write the expression of SL as

$$SL = P(W_q \leq D) = 1 - P(W_q > D | W_q > 0) P(W_q > 0) \tag{8.42}$$

To compute SL, we need the expressions of $P(W_q > 0)$ and $P(W_q > D | W_q > 0)$; each is of interest in its own right, as the former is the proportion of callers who must wait, and the latter is the proportion of the callers in queue who have to wait more than D units of time. As a caller has to wait in queue if and only if there are at least s callers present when he arrives,

$$P(Wq > 0) = 1 - \sum_{i=0}^{s-1} \pi_i = 1 - \sum_{i=0}^{s-1} \frac{(s\rho)^i}{i!} \pi_0 = \frac{(s\rho)^s \pi_0}{s!(1-\rho)} \tag{8.43}$$

To compute $P(W_q > D | W_q > 0)$, we reason as follows: since π_i has a geometric tail, that is, $\pi_{i+1} = \rho \pi_i$, for $i \geq s$, the random variable $W_q | W_q > 0$ has an exponential distribution with rate $(s\mu - \lambda)$. Thus,

$$P(W_q > D | W_q > 0) = e^{-(s\mu - \lambda)D} \tag{8.44}$$

Substituting Equations 8.43 and 8.44 into Equation 8.42, we obtain

$$SL = 1 - e^{-(s\mu - \lambda)D} \frac{(s\rho)^s \pi_0}{s!(1-\rho)} \tag{8.45}$$

The average waiting time of a caller can be computed by the formula

$$E(W_q) = E(W_q | W_q > 0) P(W_q > 0) = \frac{(s\rho)^s \pi_0}{(s\mu - \lambda)(1-\rho)s!} \tag{8.46}$$

Erlang C can also be used to estimate the number of agents needed to satisfy a desirable SL. Here, the objective is to find the minimum number of agents needed to achieve a given SL. For example, suppose that the arrival rate is $\lambda = 10$ calls/min, the average service time is 3 min/call ($\mu = 1/3$), the acceptable waiting time is $D = 5$ min, and the desirable service rate is 90%. If we denote $W_q(\lambda, \mu, s)$ as the waiting time of a caller with system parameters λ, μ, and s, then we need to find the minimum number of agents s^* such that

$$SL = P(W_q(10, 1/3, s^*) \leq 5) \geq 90\%$$

To ease computation, various telecommunication service providers offer free, online "calculators" to evaluate performance measures such as SL given in Equation 8.45. For example, performance analysis calculators for Erlang models are publicly available at www.4callcenters.com and www.math.vu.nl/obp/callcenters. The results presented in Table 8.1 use the calculator available at the second listed URL, which is an Excel add-in package called "CallCenterDSS.xla," offered by Professor G. Koole at Vrije University, Amsterdam. In our computation, we first compute SL for an acceptable waiting time D, for the fixed arrival rate λ, service rate μ, and number of agents s; we then compute the number of agents s necessary to achieve an acceptable SL, for the fixed arrival rate λ, service rate μ, and acceptable waiting time D.

Example 8.17: The Erlang A model (continued)

Recall that Erlang A is an $M/M/s/k + M$ queue, where s is the number of agents, k is the number of trunks, $s \leq k$, and $+M$ means that each caller has an independent patience time that is exponentially distributed with rate, say θ. In principle, since Erlang A is a birth and death process, its limiting distribution can be computed by Equations 8.37 and 8.38, using the transition rates given in Equation 8.30. The expression of the limiting probabilities, however, is complex, and shall not be given here. As mentioned, several Erlang calculators are publicly available, and some of them are able to treat Erlang A. The results presented in Table 8.2 use "CallCenterDSS.xla" from www.math.vu.nl/obp/callcenters. The table shows that when callers become less patient, that is, when θ increases, the SL increases (good), the percentage of abandonment increases (bad), the percentage of total calls lost (i.e., the sum of the percentages of abandonment and calls blocked) increases (bad), and the percentage of calls blocked decreases (good). When the number of trunks k increases,

TABLE 8.1 Calculating Performance Measures in Erlang C

Service Level		Number of Agents Needed	
System Parameters: $\lambda = 10$, $\mu = 1/3$, $s = 35$		System Parameters: $\lambda = 10$, $\mu = 1/3$, $D = 5$	
D : Acceptable Waiting Time (minute)	Service Level %	Required SL %	# of Agents Needed
2	73.08	75	34
4	74.53	80	35
6	75.91	85	36
8	77.21	90	38
10	78.44	95	40

TABLE 8.2 Calculating Performance Measures in Erlang A

System Parameters: $\lambda = 10$, $\mu = 1/3$, $s = 30$, $D = 5$					
θ: Patience Rate	k: # of Trunks	SL %	Abandonment %	Blocking %	Total calls lost %
3	32	87.79	5.90	5.46	11.36
6	32	89.18	8.46	3.36	11.82
3	35	80.24	10.35	0.50	10.85
6	35	85.38	11.54	0.09	11.63

Stochastic Processes 8-37

the SL decreases (bad), the percentage of calls blocked decreases (good), the percentage of total calls lost decreases (good), and the percentage of abandonment increases (bad). Those comparative statistics suggest that management must balance several competing criteria to achieve efficient call center operation. ∎

8.4.5 Statistical Inference of Continuous-Time Markov Chains

The maximum likelihood procedure has been extensively used in the CTMC for parameter estimation. We illustrate the idea by a special case of the birth and death process with $\lambda_0 = \lambda_I$, $\lambda_i = \lambda_B$, for $i \geq 1$ (i.e., the arrival rate when the system is idle is different from the arrival rate when the system is busy), and $\mu_i = \mu$, for $i \geq 1$ (i.e., the service rate is a constant regardless of the number of customers in the system). We want to estimate parameters λ_I, λ_B, and μ.

Suppose that we have observed this special birth and death process for a total time t. During the period the system was idle for a total time t_I and busy for a total time t_B, such that $t_I + t_B = t$. We also keep track of the following counts during the observation period $[0, t]$:

1. n_I: the number of arrivals who observe an empty system (i.e., the number of type $0 \to 1$ transitions). Note that n_I can be considered as a random sample of the number of Poisson events with rate λ_I during the observation time $[0, t_I]$;

2. n_B: the number of arrivals who observe a busy system (i.e., the number of type $n \to n+1$ transitions, $n \geq 1$). Note that n_B can be considered as a random sample of the number of Poisson events with rate λ_B during the observation time $[0, t_B]$;

3. n_d: The number of departures (i.e., the number of type $n \to n-1$ transitions). Note that n_d can be considered as a random sample of the number of Poisson events with rate μ during the observation time $[0, t_B]$.

Recall that we have shown in Section 8.2 that the maximum likelihood estimator of the arrival rate in a Poisson process is $\hat{\lambda} = n/t$, where t is the observation period and n is the number of events occurred during the period. Here, we again encounter the parameter estimation issue of a Poisson process. However, our problem is more complicated since t_I and t_B are random variables. Nevertheless, it can be shown that the maximum likelihood estimators of λ_I, λ_B, and μ can be obtained as if t_I and t_B were constants (Bhat & Miller, 2002):

$$\hat{\lambda}_I = \frac{n_I}{t_I}, \quad \hat{\lambda}_B = \frac{n_B}{t_B}, \quad \hat{\mu} = \frac{n_d}{t_B} \tag{8.47}$$

These estimators are indeed intuitively plausible. Note that the idea is to decompose the process into three Poisson processes: the birth process when the system is empty, the birth process when the system is busy, and the death process when the system is busy. Equation 8.47 then gives the estimate of the arrival rate in the underlying Poisson process. This idea can be used to estimate the transition rates in the more general birth and death process.

When $\lambda_I = \lambda_B = \lambda$, the above simple birth and death process becomes an $M/M/1$ queue, and the counts n_I and n_B can be pooled to arrive at a single estimate of λ:

$$\hat{\lambda} = \frac{n_I + n_B}{t} \tag{8.48}$$

Example 8.21

The manager of a supermarket needs to decide the staffing level of his store so that the average waiting time of his customers is no more than 1 min. He feels that it is reasonable to assume that customers arriving at his store follow a Poisson process, and each customer needs an exponential amount of time to be served at the checkout counter. To estimate the traffic intensity of his store, he took observations for a 2-h period, during which he found 63 customers arrived at the counter and 44 customers left the counter. He also observed that the checkout counter was completely empty for 9 min during the 2-h observation period.

The problem described above is an $M/M/1$ queue with arrival rate λ and service rate μ. Using the notation given earlier, we have

$$t = 120 \text{ min}, \quad t_B = 111 \text{ min}$$
$$n_I + n_B = 63, \quad n_d = 44$$

From Equations 8.47 and 8.48, it can be estimated that

$$\hat{\lambda} = \frac{n_I + n_B}{t} = \frac{63}{120} = 0.53, \quad \hat{\mu} = \frac{n_d}{t_B} = \frac{44}{111} = 0.40$$

The manager needs to determine the number of checkers to hire so that the average waiting time of his customer is 1 min. For this, he needs to find s so that $E[W_q] = 1$ min. From Equation 8.46, s should be the smallest integer such that

$$E[W_q] = \frac{1}{s\mu - \lambda} P(W_q > 0) \leq 1 \text{ min}$$

Substituting the estimated parameters $\hat{\lambda}$ and $\hat{\mu}$ into the above expression, it is found that the smallest integer to ensure the above condition is $s = 3$. ∎

Next, we consider the hypothesis testing method for the DTMC in the context of the simple birth and death process discussed earlier, based on the study of Billingsley (1961). Suppose that we have observed the process for a duration t and collected statistics for t_I, t_B, n_I, n_B, and n_d. We are reasonably sure that the process is a birth and death process and we wish to know whether there are convincing evidences to support the hypothesis $H_0: \lambda_I = \lambda_I^0$, $\lambda_B = \lambda 0_B$, $\mu = \mu^0$. The log maximum likelihood function under the null hypothesis is

$$L(\lambda_I^0, \lambda_B^0, \mu^0) = n_I \ln \lambda_I^0 + n_B \ln \lambda_B^0 + n_d \ln \mu^0 - \lambda_I^0 t_I - \lambda_B^0 t_B - \mu^0 t_b + C$$

where C is a term involving n_I, n_B, and n_d and independent of λ_I^0, λ_B^0, and μ^0. The test statistic for H_0 follows χ^2 distribution with 2 degrees of freedom:

$$2[\max L(\lambda_I, \lambda_B, \mu) - L(\lambda_I^0, \lambda_B^0, \mu^0)] - 2 \begin{bmatrix} n_I \ln \frac{n_I}{\lambda_I^0 t_I} + n_B \ln \frac{n_B}{\lambda_B^0 t_B} + n_d \ln \frac{n_d}{\mu^0 t_B} \\ -(n_I + n_B + n_d) + \lambda_I^0 t_I + \lambda_B^0 t_B + \mu^0 t_B \end{bmatrix} \quad (8.49)$$

Example 8.21 (continued)

Suppose the manager of the supermarket anticipates the average service time of his checkout counter is 2 min. Further, he has assumed that the arrival rate is 0.5. To test the hypothesis $H_0: \lambda = 0.5$ and $\mu = 0.5$, we use Equation 8.49 and find that the χ^2 statistic is 2.71, with 2 degrees of freedom. Since $P(\chi^2 \geq 2.71) = 0.26$ is large enough, it indicates that the manager's estimates of system parameters, $\lambda = 0.5$ and $\mu = 0.5$, cannot be rejected.

Stochastic Processes

8.5 Renewal Theory

The stochastic process we have discussed so far, including the Poisson process, the DTMC and the CTMC, all have the Markov property, that is, the future of the process from time n (for the discrete-time case) or t (for the continuous-time case) onward depends only on the state of the system at time n or t and is independent of the process's history prior to time n or t. The Markov property proves to be the key enabler to make the analysis of such a system possible.

In this section, we discuss the stochastic processes that do not have the Markov property at all observation points. In particular, we consider the stochastic processes that have the Markov property only at some, possibly random, time epochs $0 = T_0 \leq T_1 \leq T_2 \leq \ldots$. In other words, only at times T_n, $n = 0, 1, \ldots$, the future of the process can be predicted based on the state of the system at time T_n. It turns out that the transient analysis of such a process is much harder than the process with the Markov property. Therefore, in this section we focus on the long-run behavior of the process. The processes we shall cover in this section include *renewal* processes, *renewal reward* processes, *regenerative* processes, and *semi-Markov* processes.

8.5.1 Renewal Processes

Recall that a Poisson process is a counting process in which the interarrival times between successive events are iid random variables with an *exponential* distribution. As a generalization of the Poisson process, we define:

DEFINITION 8.3 *A counting process $\{N(t), t \geq 0\}$ is said to be a renewal process if the interarrival times between successive events, $\{X_1, X_2, \ldots\}$, are a sequence of iid random variables.*

As defined, X_1 is the arrival epoch of the first event, which follows some distribution F, X_2 is the time between the first and second events, which follows the same distribution F, and so on. Therefore, a renewal process depicts a sequence of events occurring randomly over time, and $N(t)$ represents the number of such events that occurred by time t. Let $S_0 = 0$ and

$$S_n = \sum_{j=1}^{n} X_j, \quad n = 1, 2, \ldots \tag{8.50}$$

be the waiting time until the occurrence of the nth event. We shall call S_n the *renewal epoch* of the nth event, $n \geq 1$, and the event that occurred at a renewal epoch a *renewal*. Clearly, the renewal process starts afresh, probabilistically, at renewal epochs S_n, $n \geq 0$. In practice, $\{N(t), t \geq 0\}$ and $\{S_n, n \geq 1\}$ are interchangeably called the renewal process.

It is often of interest to compute the expected number of renewals occuring by time t. The function $m(t) = E[N(t)]$ is known as the *renewal function*. For example, the renewal function of a Poisson process with rate λ is $m(t) = \lambda t$.

Example 8.22

The prototypical example of a renewal process is the successive replacements of light bulbs, with the times required for replacements ignored. Suppose a new light bulb, with lifetime X_1, is put in use at time 0. At time $S_1 = X_1$, it is replaced by a new light bulb with lifetime X_2. The second light bulb fails at time $S_2 = X_1 + X_2$ and is immediately replaced by the third one, and the process continues as such. It is natural to assume that the lifetimes of light bulbs follow the same distribution and are independent of each other. Now, $N(t)$ is the number of light bulbs replaced up to time t, with mean $m(t)$, and S_n is the time epoch for the nth replacement. ∎

Although in theory we can derive certain properties associated with $N(t)$, S_n, and $m(t)$ (Wolff, 1989), they are generally hard to evaluate for finite t or n unless the process has the Markov property. For example, it is difficult to know exactly the average number of light bulbs replaced during a 1-year period except when the a light bulb has an exponential lifetime. As a consequence, renewal theory is primarily concerned with the limiting behavior of the process as $t \to \infty$. We first investigate the limiting behavior of $N(t)/t$ as $t \to \infty$. Let μ be the reciprocal of the mean interarrival time:

$$E(X_j) = \frac{1}{\mu} \tag{8.51}$$

To avoid trivialities, we assume that $0 < E(X_j) < \infty$. This assumption can be used to establish the fact that $N(t)$ is finite for finite t and goes to infinity as t goes to infinity.

THEOREM 8.4: *The Elementary Renewal Theorem*

1. $\lim_{t \to \infty} \frac{N(t)}{t} = \mu$ *with probability* 1.
2. $\lim_{t \to \infty} \frac{m(t)}{t} = \mu$.

The elementary renewal theorem is intuitively plausible. Take the first expression as an example: the left side term, $\lim_{t \to \infty} N(t)/t$, gives the long-run number of renewals per unit time, and the right side term, μ, is the reciprocal of the expected time between renewals. It becomes obvious that if each renewal occurs in average $E(X_j)$ units of time, then the number of renewals per unit time, in a long run, must approach to $1/E[X_j] = \mu$. Because of the theorem, μ is called the *renewal rate*, since it can be thought of as the number of renewals occurred per unit time in a long run.

The elementary renewal theorem, though simple and intuitive, proves to be extremely powerful to analyze the long run behavior of non-Markov systems. We illustrate its applications with several examples.

Example 8.23: Age replacement policies

Consider an age replacement policy that calls for replacing an item when it fails or when its age reaches T, whichever occurs first. We shall call the policy *Policy T*. Suppose that the lifetimes of successive items are iid random variables $Y_1, Y_2, \ldots$, with distribution G. Under *policy T*, the time the ith item spent in service is $X_i = \min\{Y_i, T\}$, $i = 1, 2, \ldots$, with the expectation

$$E[X_i] = \int_0^\infty P(\min\{Y_i, T\} > t)\mathrm{d}t = \int_0^T P(Y_i > t)\mathrm{d}t$$

Theorem 8.4 then tells us that, in a long run, the replacements occur at the rate

$$\mu = \frac{1}{E[X_i]} = \frac{1}{\int_0^T P(Y_i > t)\mathrm{d}t}$$

Stochastic Processes

For example, suppose the lifetime is uniformly distributed between $[0, 1]$ and $T = 1/2$. Then

$$\mu = \frac{1}{E[X_i]} = \frac{1}{\int_0^{1/2} (1-t)dt} = \frac{3}{8}$$

■

Example 8.24: The $M/G/1/1$ queue

Suppose that patients arrive at a single-doctor emergency room (ER) in accordance with a Poisson process with rate λ. An arriving patient will enter the room if the doctor is available; otherwise the patient leaves the ER and seeks service elsewhere. Let the amount of time the doctor treats a patient be a random variable having distribution G. We are interested in the rate at which patients enter the ER and the service level of the ER, defined as the proportion of the arriving patients who are actually being treated.

To see how the system can be formulated as a renewal process, let us suppose that at time 0 a patient has just entered the ER. We say that a renewal occurs whenever a patient enters the ER. Let $1/\mu_G$ be the mean time needed by the doctor to treat a patient. Then, by the memoryless property of the exponential interarrival times, the mean time between successive entering patients is

$$\frac{1}{\mu} = \frac{1}{\mu_G} + \frac{1}{\lambda}$$

and the rate at which patients enter the ER is

$$\mu = \frac{\lambda \mu_G}{\lambda + \mu_G}$$

Since the patient arrival process is Poisson with rate λ, the service level of the ER is given by

$$\frac{\mu}{\lambda} = \frac{\mu_G}{\lambda + \mu_G}$$

■

8.5.2 Renewal Reward Processes

Consider a renewal process $\{N(t), t \geq 0\}$ with interarrival times $\{X_n, n \geq 1\}$. Suppose that associated with each interarrival time X_n is a reward R_n (or a profit, i.e., reward-cost), $n \geq 1$. We allow X_n and R_n to be dependent (they usually are, as a reward may depend on the length of the interval), but assume that the pairs (X_1, R_1), (X_2, R_2), ... are iid bivariate random variables. Let $R(t)$ be the total reward received by the system over the interval $[0, t]$:

$$R(t) = \sum_{n=1}^{N(t)} R_n \tag{8.52}$$

The reward process $\{R(t), t \geq 0\}$ is called the *renewal reward process* or *cumulative process*. The expected cumulative reward up to time t is denoted by $E[R(t)]$, $t \geq 0$. In general, it is very difficult to derive the distribution of $R(t)$ and to compute the expectation $E[R(t)]$. Fortunately, it is quite easy to compute the long-run reward rate, that is, the limit of $R(t)/t$ as $t \to \infty$.

THEOREM 8.5: The Renewal Reward Theorem.
Let $E[X_n] = E[X]$ and $E[R_n] = E[R]$. If $E[X] < \infty$ and $E[R] < \infty$, then

1. $\lim_{t \to \infty} \dfrac{R(t)}{t} = \dfrac{E[R]}{E[X]}$, with probability 1.
2. $\lim_{t \to \infty} \dfrac{E[R(t)]}{t} = \dfrac{E[R]}{E[X]}$.

We shall call renewal intervals *renewal cycles*. With this terminology, the renewal reward theorem states that the long-run average reward per unit time (or the expected reward per unit time) is the expected reward per cycle divided by the expected cycle length. This simple and intuitive result is very powerful to compute the long-run reward rate, or the long-run proportion of time spent in different states, as illustrated by the following examples.

Example 8.23: Age replacement policies (continued)

Let us assume that, under policy T, the cost for a planned replacement is c_r, and that for an unplanned replacement is c_f, with $c_f > c_r$. The cost of the ith replacement, R_i, depends on lifetime Y_i as follows:

$$R_i = \begin{cases} c_r & \text{if } Y_i < T \\ c_f & \text{if } Y_i \geq T \end{cases}$$

Clearly, R_i depends on the in-service time of the ith item. The expected cost per replacement is

$$E(R) = c_f P(Y_i \leq T) + c_r P(Y_i > T)$$

The length of a cycle, as computed in Example 8.23, is $E(X) = \int_0^T P(Y_i > t)dt$. We have the long-run cost per unit time under policy T as

$$\dfrac{E(R)}{E(X)} = \dfrac{c_f P(Y_i \leq T) + c_r P(Y_i > T)}{\int_0^T P(Y_i > t)dt}$$

Given the distribution of Y_i and cost parameters c_r and c_f, the optimal age replacement policy would be the value T that minimizes the preceding equation. For a numerical example, let Y_i be a uniform random variable over (0, 1) years, $c_r = 50$, and $c_f = 100$. Then

$$\dfrac{E(R)}{E(X)} = \dfrac{100T + 50(1-T)}{T - \frac{T^2}{2}}$$

Taking the derivative of the above equation and setting it to zero, we get $T^* = \sqrt{3} - 1 = 0.732$ years. The annual cost under the optimal policy T^* is $186.6/year.

How is policy T^* compared with the "replace upon failure" policy? In this case the cost rate is $\dfrac{E(R)}{E(Y)} = \dfrac{100}{0.5} = \$200$/year. Thus, even though the age replacement policy replaces the item more often than the "replace upon failure" policy, it is actually more economical than the latter. ∎

Example 8.25: A simple continuous-review inventory model

In a continuous-review inventory system, the inventory level is continuously monitored and information is updated each time a transaction takes place. In this setting, let us consider a simple inventory system where customers arrive according to a renewal process with a mean interarrival time $1/\lambda$, and each customer requests a single item. Suppose that the

Stochastic Processes 8-43

system orders a batch of Q items each time the inventory level drops to zero. For simplicity, the replenishment lead time is assumed to be zero. The operating costs incurred include a fixed set-up cost K each time an order is placed and a unit holding cost h for each unsold item. The manager of the system needs to determine the optimal order quantity Q^* that minimizes the long-run operating cost of the system.

To see how the problem can be formulated as a renewal reward process, define a cycle as the time interval between two consecutive orders. The expected length of a cycle is the expected time required to receive Q orders. Since the mean interarrival time of demand is $1/\lambda$,

$$E[\text{length of a cycle}] = \frac{Q}{\lambda}$$

Let Y_n be the time between the $(n-1)$st and nth demands in a cycle, $n = 1, \ldots, Q$. Since, during the period Y_n, the system holds $(Q - n + 1)$ units of the item, the cost per cycle is

$$E[\text{cost of a cycle}] = K + E\left[\sum_{n=1}^{Q} h(Q - n + 1)Y_n\right] = K + \frac{hQ(Q+1)}{2\lambda}$$

Hence, the long-run average cost of the system is

$$\frac{E[\text{cost of a cycle}]}{E[\text{length of a cycle}]} = \frac{\lambda K}{Q} + \frac{h(Q+1)}{2}$$

Treating Q as a continuous variable and using a calculus, then the above expression is minimized at

$$\hat{Q} = \sqrt{\frac{2\lambda K}{h}}$$

If $\hat{Q}$ is an integer, then it is the optimal batch size; otherwise, the optimal batch size is either the largest integer smaller than $\hat{Q}$ or the smallest integer larger than $\hat{Q}$, whichever yields a smaller value in the cost function. ∎

Example 8.26: Prorated warranty

This example is based on Example 7.11 in Kulkarni (1999). A tire company issues a 50,000 mile prorated warranty on its tires as follows: if the tire fails after its mileage life, denoted by L, exceeds 50,000 miles, then the customer has to pay $95 for a new tire. If the tire fails before it reaches 50,000 miles, that is, $L \leq 50,000$, the customer pays the price of a new tire according to the prorated formula $\$95 \times (L/50,000)$. Suppose that the customer continues to buy the same brand of tire after each failure. Let L follow the distribution

$$f(x) = 2 \times 10^{-10}x, \quad \text{for } 0 \leq x \leq 100,000 \text{ miles}$$

The customer has the option to buy the tire without warranty for $90. Should the consumer purchase the warranty (we assume that the customer either always gets the warranty or never gets the warranty)? First suppose that the customer never purchases the warranty. The cycle is defined as each purchase of a new tire, with the mean

$$E(L) = \int_0^{100,000} 2 \times 10^{-10} x^2 \, dx = 2 \times 10^{-10} \left.\frac{x^3}{3}\right|_0^{100,000} = \frac{2}{3} \times 10^5$$

Since each new tire costs the customer $90, the average cost rate per mile is

$$\frac{E(R)}{E(L)} = \frac{\$90}{\frac{2}{3} \times 10^5} = \$135 \times 10^{-5} \text{ per mile}$$

or \$1.35 per 1000 miles. Now consider the option of always purchasing the tires under the warranty. The cycle is the same as before. However, the prorated warranty implies $R = \$95 \times \frac{\min\{50{,}000, L\}}{50{,}000}$. Hence

$$E(R) = \int_0^{50{,}000} 2 \times 10^{-10} \times \frac{95x^2}{50{,}000} dx + \int_{50{,}000}^{100{,}000} 2 \times 10^{-10} \times 95x \, dx = \$87.08$$

The long-run cost under the warranty is

$$\frac{97.08}{\frac{2}{3} \times 10^5} = \$130.63 \times 10^{-5} \text{ per mile}$$

or \$1.31 per 1000 miles. This implies that the customer should buy the warranty. ∎

8.5.3 Regenerative Processes

Consider a stochastic process $\{Z(t), t \geq 0\}$ defined on $S = \{0, 1, \ldots\}$ having the property that the process starts afresh, probabilistically, at (possibly random) time epochs $T_n, n \geq 1$. This means, there exist time epochs $T_n, n \geq 1$, such that the evolution of the process from T_n onward follows the same probability law as the process that starts at time T_{n-1}. We call such a process a *regenerative process*, the time epochs $\{T_n, n \geq 1\}$ *regenerative epochs*, and $Y_n = T_n - T_{n-1}, n \geq 1$, *regenerative cycles*. One may envision that $\{T_n, n \geq 1\}$ constitute the arrival times of a renewal process, and $\{Y_n, n \geq 1\}$ the interarrival times. Indeed, a renewal process is an example of a regenerative process. Another example is a recurrent Markov chain, where T_1 represents the time of the first transition into the initial state.

A very useful result about the long-run behavior of a regenerative process states that

$$\lim_{t \to \infty} P(Z(t) = i) = \frac{E[\text{amount of time in state } i \text{ in a cycle}]}{E[\text{length of a cycle}]} \tag{8.53}$$

Example 8.27:

Suppose that an irreducible, positive recurrent CTMC $\{X(t), t \geq 0\}$ starts in state i. By the Markov property, the process starts over again each time it re-enters state i. Thus T_n, $n \geq 1$, where T_n denotes the nth time the CTMC returns to state i, constitute regenerative epochs. Let μ_{ii} be the mean recurrent time of state i. From Equation 8.53,

$$\lim_{t \to \infty} P(X(t) = i) = \frac{E[\text{amount of time in state } i \text{ during a recurrent time of state } i]}{\mu_{ii}}$$

As each time the CTMC visits state i, it stays there for an exponential amount of time with rate ν_i, we have

$$\lim_{t \to \infty} P(X(t) = i) = \frac{1/\nu_i}{\mu_{ii}}$$ ∎

Example 8.25: A simple continuous-review inventory model (continued)

Suppose that in the continuous-review inventory system discussed in Example 8.25, we are interested in the limiting distribution of $I(t)$, where $I(t)$ is the number of units on hand at time t. Suppose that at time 0 there are Q units on hand, and each time inventory on hand drops to 0 we order a new batch of size Q. Then the inventory process $\{I(t), t \geq 0\}$ is a regenerative process, which regenerates itself each time a new order is placed. Then, for $i = 1, 2, \ldots, Q$,

$$\lim_{t \to \infty} P(I(t) = i) = \frac{E[\text{amount of time } i \text{ units on hand during a reorder cycle}]}{E[\text{length of a reordering cycle}]}$$

$$= \frac{1/\lambda}{(1/\lambda)Q} = \frac{1}{Q}$$

That is, inventory on hand is a discrete uniform random variable between 1 and Q. ∎

8.5.4 Semi-Markov Processes

Consider a process $\{X(t), t \geq 0\}$ that can be in any of the finite states $\{0, 1, 2, \ldots, N\}$. Suppose that each time the process enters state i, it remains there for a random amount of time with rate μ_i and then makes a transition to state j with probability p_{ij}. The sojourn time in a state and the next state reached do not need to be independent. Such a process is called the *semi-Markov process* (SMP). Let $\{T_n, n \geq 1\}$, where $T_0 = 0$, be the sequence of epochs at which the process makes transitions (returning to the same state is allowed). An important property of a semi-Markov process is that the process at each of the transition epochs $T_n, n \geq 1$, has the Markov property. That is, for each T_n, the evolution of the process from T_n onward depends only on $X(T_n)$, the state of the process observed at transition time T_n. Clearly, if the sojourn time in each state is identically 1, then the SMP is just a DTMC, and if the sojourn time in each state is exponentially distributed, then the SMP becomes a CTMC.

Let $M_j(t)$ be the total time that the SMP spends in state j up to time t. We define

$$p_j = \lim_{t \to \infty} \frac{M_j(t)}{t}$$

as the long-run proportion of time that the SMP is in state j, $j = 0, 1, \ldots, N$. Let us find those proportions. To do this, we denote $X_n = X(T_n)$ as the nth state visited by the SMP, $n \geq 1$. Note that $\{X_n, n \geq 1\}$ is a DTMC governed by the transition matrix $\mathbf{P} = \{p_{ij}\}$. This DTMC shall be called the *embedded DTMC* of the SMP. For simplicity, let us assume that the embedded DTMC is irreducible so that the limiting (or stationary) distribution $\pi = \{\pi_0, \pi_1, \ldots, \pi_n\}$ of the embedded DTMC exists. From the result of Section 8.3, π will be the unique non-negative solution of

$$\pi_j = \sum_{i=0}^{N} \pi_i p_{ij} \tag{8.54}$$

$$\sum_{j=0}^{N} \pi_j = 1 \tag{8.55}$$

Now, since the proportion of transitions that the SMP enters state j is π_j, and it remains in state j for an average $1/\mu_j$ units of time in each of such transitions, it is intuitively plausible, and indeed can be shown formally, that p_j is given by

$$p_j = \frac{\pi_j/\mu_j}{\sum_j \pi_j/\mu_j}, \quad j = 0, 1, \ldots, N \tag{8.56}$$

Example 8.26: Machine maintenance

We now take the repair times of the reliability system under policy T into account. Suppose that under policy T, an item in service is repaired upon failure (emergency renewal) or at age T (preventive renewal). Suppose that the lifetime of the item is L, and emergency and preventive renewals take random times Z_e and Z_p, respectively.

The aim is to determine the long-run availability of the system. The system can be in one of the three states, 0 (up), 1 (emergency renewal), and 2 (preventive renewal). Then the system can be modeled as a SMP with its embedded DTMC having the transition probability matrix

$$P = \begin{pmatrix} 0 & P(L<T) & P(L \geq T) \\ 1 & 0 & 0 \\ 1 & 0 & 0 \end{pmatrix}$$

The limiting distribution of the embedded CTMC can be obtained as $\pi_0 = 1/2$, $\pi_1 = P(L<T)/2$, and $\pi_2 = P(L \geq T)/2$. The mean sojourn time of the system in each state is

$$\frac{1}{\mu_0} = E[\min(L,T)] = \int_0^T P(L>l)\mathrm{d}l, \frac{1}{\mu_1} = E[Z_e], \frac{1}{\mu_2} = E[Z_p]$$

According to Equation 8.56, the stationary availability of the system is

$$A(T) = \frac{\int_0^T P(L>l)\mathrm{d}l}{\int_0^T P(L>l)\mathrm{d}l + E[Z_\mathrm{e}]P(L<T) + E[Z_\mathrm{p}]P(L \geq T)}$$

It is practically important that the system availability does not depend on the distributions of repair times Z_e and Z_p, but only on their expected values. ∎

8.5.5 Statistical Inference of Renewal Processes

Let $\{N(t), t \geq 0\}$ be a counting process with the times between successive events denoted by $\{X_n, n \geq 1\}$. If $\{N(t), t \geq 0\}$ is a renewal process, we require that $\{X_n, n \geq 1\}$ is a sequence of iid random variables from some distribution F. Given that we have observed a sample of interarrival event times, $(x_1, x_2, \ldots, x_n)$, statistical inference problems become choosing an appropriate distribution function F and estimating its parameters, which can be done by using standard statistical theory. We will not elaborate on the details of those methods here. We would like, however, to provide the reader with several useful references for further reading. Parametric inferences were studied by Barlow & Proschan (1981), and nonparametric inferences were considered by Karr (1991). Miller & Bhat (1997) presented several sampling methods to estimate the parameters of F when the inter-event times are not directly observable. Basawa & Prakasa Rao (1980) and Bhat & Miller (2002) also contain useful materials on this subject.

8.6 Software Products Available for Solving Stochastic Models

Both special-purpose and general-purpose software products are available to solve stochastic models, either numerically or symbolically. The special-purpose stochastic modeling software products, listed below, focus on Markov process analyses and their applications in different areas such as queueing, reliability, and telecommunications. Most of them have simulation capabilities.

1. **Probabilistic Symbolic Model Checker (PRISM)**: PRISM is a free and open source software developed at the University of Birmingham. PRISM is a tool for modeling and analyzing probabilistic systems including DTMCs, CTMCs,

and Markov decision processes (MDPs). PRISM uses a module-based system description language for model inputs. It then translates the system descriptions into an appropriate model and provides performance statistics. Hinton et al. (2006) provide an overview of the main features of PRISM. The Web site of PRISM is www.cs.bham.ac.uk/dxp/prism/index.php.

2. **MCQueue**: This is a free educational software for Markov Chains and queues (no commercial use). This software package contains two modules. The first module is for the analysis of DTMCs and CTMCs up to 100 states. The other module calculates performance measures for basic queueing models (e.g., $M/G/1$, $M/M/s$, $M/D/s$, $G/M/s$, and $M/M/s/s+M$ queues). The algorithms in this package are based on the methods discussed in Tijms (2003). The package can be downloaded from http://staff.feweb.vu.nl/tijms/.

3. **MARkov Chain Analyzer (MARCA)**: MARCA is developed at North Carolina State University. It facilitates the generation of large Markov chain models. It also has a set of Fortran subroutines to support model construction. Ordering information is available at www.csc.ncsu.edu/faculty/stewart/MARCA/marca.html.

4. Queueing theory is an important application area of stochastic processes. Not surprisingly, quite a few software products are available for queueing analysis and applications in telecommunications. A list of queueing software packages has been compiled by Dr. M. Hlynka at the University of Windsor and is available at www2.uwindsor.ca/hlynka/qsoft.html.

5. Several software packages are available for reliability applications. For example, **Relex Markov** by Relex Software Corporation (www.relexsoftware.co.uk/products/markov.ph), **Markov Analysis** by ITEM Software (www.itemuk.com/markov.html), and **SMART** by University of California at Riverside (www.cs.ucr.edu/ ciardo/SMART/).

6. **@Risk** and **Crystal Balls** are **Excel** add-ins that analyze stochastic models by Monte Carlo simulation. **Excel**, with the aid of **VBA**, can also be used in Monte Carlo simulation. **Arena, ProModel**, and **Extend** are popular software packages designed for discrete-event simulations.

Several general-purpose, programming language based software products, although not specifically developed for the stochastic modeling purpose, are popular in solving such problems. These include the proprietary computational packages such as **MATLAB, Maple,** and **Mathematica** for general mathematical computation and open source software packages such as **R** (sometimes described as **GNU S**) for statistical computing and graphics. Some textbooks provide computer program codes (e.g., Kao, 1997, who uses MATLAB). Statistical packages such as **SAS** and **Minitab** are also powerful in analyzing stochastic models.

References

1. Anderson, T. W. (1954), "Probability models for analyzing time changes in attitudes." In P. E. Lazarsfeld (Ed.), *Mathematical Thinking in the Social Sciences*, Free Press, Glencoe, IL.
2. Aven, T. and Jensen, U. (1999), *Stochastic Models in Reliability*, Springer, New York.
3. Barlow, R. E. and Proschan, F. (1981), *Statistical Theory of Reliability and Life Testing*, To Begin With, Silver Spring, MD.

4. Bartholomew, D. J. (1982), *Stochastic Models for Social Processes*, Wiley Series in Probability and Statistics, John Wiley & Sons, New York.
5. Basawa, I. V. and Prakasa Rao, B. L. S. (1980), *Statistical Inference for Stochastic Processes*, Academic Press, New York.
6. Bhat, U. N. and Miller, G. K. (2002), *Elements of Applied Stochastic Processes*, Wiley Series in Probability and Statistics, 3rd ed., John Wiley & Sons, New York.
7. Billingsley, P. (1961), *Statistical Inference for Markov Processes*, University of Chicago Press, Chicago.
8. Bolch, G., Greiner, S., and de Meer, H. (2006), *Queueing Networks and Markov Chains: Modeling and Performance Evaluation with Computer Science Applications*, 2nd ed., Wiley-Interscience, New York.
9. Buzacott, J. A. and Shanthikumar, J. G. (1993), *Stochastic Models of Manufacturing Systems*, Prentice Hall, Englewood Cliffs, NJ.
10. Cinlar, E. (1975), *Introduction to Stochastic Processes*, Prentice Hall, Englewood Cliffs, NJ.
11. Gans, N., Koole, G., and Mandelbaum, A. (2003), "Telephone call centers: Tutorial, review, and research prospects," *Manufacturing & Service Operations Management* **5**, 79–141.
12. Gautam, N. (2003), *Stochastic Models in Telecommunications*, Handbook of Statistics 21—Stochastic Processes: Modeling and Simulation (D. N. Shanbhag and C. R. Rao (eds.)), North-Holland, Amsterdam.
13. Glass, D. F. (Ed.) (1954), *Social Mobility in Britain*, London: Routledge & Kegan Paul.
14. Gnedenko, B. V., Belyayev, Y. K., and Solovyev, A. D. (1969), *Mathematical Models in Reliability*, Academic Press, New York.
15. Goel, N. S. and Richter-Dyn, N. (2004), *Stochastic Models in Biology*, The Blackburn Press, Caldwell, NJ.
16. Green, L. V. and Kolesar, P. J. (1989), "Testing the validity of a queueing model of police control," *Management Science* **35**, 127–148.
17. Hand, D. J. (1994), *A Handbook of Small Data Sets*, D. J. Hand, Fergus Daly, D. Lunn, K. McConway and E. Ostrowski (Eds.), Chapman & Hall, London.
18. Helbing, D. and Calek, R. (1995), *Quantitative Sociodynamics: Stochastic Methods and Models of Social Interaction Processes (Theory and Decision Library B)*, Springer, New York.
19. Hinton, A., Kwiatkowska, M., Norman, G., and Parker, D. (2006), "PRISM: A tool for automatic verification of probabilistic systems." In Hermanns and Palsberg (ed.), Proceedings of the 12th International Conference on Tools and Algorithms for the Construction and Analysis of Systems (TACAS'06), Vol. 3920 of *LNCS*, Springer, New York, pp. 441–444.
20. Kao, E. P. C. (1997), *An Introduction to Stochastic Processes*, Duxbury Press, New York.
21. Karr, A. F. (1991), *Point Processes and Their Statistical Inference*, 2nd ed., Marcel Dekker, New York.
22. Kijima, M. (2003), *Stochastic Processes with Applications to Finance*, Chapman & Hall/CRC, Boca Raton, FL.
23. Kleinrock, L. (1976), *Queueing Systems*, John Wiley & Sons, New York.
24. Kulkarni, V. G. (1995), *Modeling and Analysis of Stochastic Systems*. Texts in Statistical Science Series, Chapman & Hall, London.
25. Kulkarni, V. G. (1999), *Modeling, Analysis, Design, and Control of Stochastic Systems*. Springer Texts in Statistics, Springer, New York.
26. Lazarsfeld, P. E., Berelson, B., and Gaudet, H. (1948), *The People's Choice*, Columbia University Press, New York.
27. Lewis, T. (1986), "M345 statistical methods, unit 2: Basic methods: Testing and estimation," Milton Keynes, Buckinghamshire, England: Open University Vol. 16.

28. Miller, G. K. and Bhat, U. N. (1997), "Estimation for renewal processes with unobservable gamma or erlang interarrival times," *Journal of Statistical Planning and Inference* **61**, 355–372.
29. Nelson, B. L. (1995), *Stochastic Modeling: Analysis and Simulation*, McGraw-Hill Series in Industrial Engineering and Management Science, New York.
30. Porteus, E. L. (2002), *Foundations of Stochastic Inventory Systems*, Stanford University Press, Stanford, CA.
31. Rolski, T., Schmidli, H., Schmidt, V., and Teugels, J. (1999), *Stochastic Processes for Insurance and Finance*, Wiley Series in Probability and Statistics, John Wiley & Sons.
32. Ross, S. M. (1996), *Stochastic Processes*, 2nd ed., John Wiley & Sons, New York.
33. Ross, S. M. (2003), *Probability Models*, 8th ed., Academic Press, London.
34. Taylor, H. M. and Karlin, S. (1994), *An Introduction to Stochastic Modeling*, 2nd ed., Academic Press, London.
35. Tijms, H. C. (1994), *Stochastic Models: An Algorithmic Approach*, John Wiley & Sons, New York.
36. Tijms, H. C. (2003), *The First Course in Stochastic Models*, John Wiley & Sons.
37. Whitaker, D. (1978), "The derivation of a measure of brand loyalty using a Markov brand switching model," *Journal of the Operational Research Society* **29**, 959–970.
38. Whittle, P. (1955), "Some distribution and moment formulae for the Markov chain," *Journal of the Royal Statistical Society* **B17**, 235–242.
39. Wolff, R. W. (1989), *Stochastic Modeling and the Theory of Queues*, Prentice-Hall International Series in Industrial and Systems Engineering, Prentice Hall, Englewood Cliffs, NJ.
40. Zipkin, P. (2000), *Foundations of Inventory Management*, McGraw-Hill/Irwin, New York.

9
Queueing Theory

9.1	Introduction..	9-1
9.2	Queueing Theory Basics............................	9-2
	Fundamental Queueing Relations • Preliminary Results for the $GI/G/s$ Queue	
9.3	Single-Station and Single-Class Queues............	9-8
	The Classic $M/M/1$ Queue: Main Results • The Multiserver System: $M/M/s$ • Finite Capacity $M/M/s/K$ System • The $M/M/1/K$ Queue • The $M/M/s/s$ Queue • The Infinite Server $M/M/\infty$ Queue • Finite Population Queues • Bulk Arrivals and Service • The $M/G/1$ Queue • The $G/M/1$ Queue • The $G/G/1$ Queue • The $G/G/m$ Queue	
9.4	Single-Station and Multiclass Queues..............	9-21
	Multiclass $G/G/1$ Queue: General Results • $M/G/1$ Queue with Multiple Classes	
9.5	Multistation and Single-Class Queues..............	9-28
	Open Queueing Networks: Jackson Network • Closed Queueing Networks (Exponential Service Times) • Algorithms for Nonproduct-Form Networks	
9.6	Multistation and Multiclass Queues	9-34
	Scenario • Notation • Algorithm	
9.7	Concluding Remarks.................................	9-37
	Acknowledgments	9-38
	References	9-38

Natarajan Gautam
Texas A&M University

9.1 Introduction

What do a fast food restaurant, an amusement park, a bank, an airport security check point, and a post office all have in common? Answer: you are certainly bound to wait in a line before getting served at all these places. Such types of queues or waiting lines are found everywhere: computer-communication networks, production systems, transportation services, and so on. To efficiently utilize manufacturing and service enterprises, it is critical to effectively manage queues. To do that, in this chapter, we present a set of analytical techniques collectively called queueing theory. The main objective of queueing theory is to develop formulae, expressions, or algorithms for performance metrics, such as the average number of entities in a queue, mean time spent in the system, resource availability, probability of rejection, and the like. The results from queueing theory can directly be used to solve design and capacity planning problems, such as determining the number of servers, an optimum queueing discipline, schedule for service, number of queues, system architecture, and

TABLE 9.1 Examples to Illustrate Various Types of Queueing Systems

	Single-class	Multiclass
Single-station	Post office	Multi-lingual call center
Multistation	Theme park	Multi-ward hospital

the like. Besides making such strategic design decisions, queueing theory can also be used for tactical as well as operational decisions and controls.

The objective of this chapter is to introduce fundamental concepts in queues, clarify assumptions used to derive results, motivate models using examples, and point to software available for analysis. The presentation in this chapter is classified into four categories depending on the types of customers (one or many) and number of stations (one or many). Examples of the four types are summarized in Table 9.1.

The results presented in this chapter are a compilation of several excellent books and papers on various aspects of queueing theory. In particular, the bulk of the single-station and single-class analysis (which forms over half the chapter) is from Gross and Harris [1], which arguably is one of the most popular texts in queueing theory. The book by Bolch et al. [2] does a fantastic job presenting algorithms, approximations, and bounds, especially for multistage queues (i.e., queueing networks). For multiclass queues, the foundations are borrowed from the well-articulated chapters of Wolff [3] as well as Buzacott and Shanthikumar [4]. The set of papers by Whitt [5] explaining the queueing network analyzer is used for the multistage and multiclass queues. Most of the notation used in this chapter and the fundamental results are from Kulkarni [6]. If one is interested in a single site with information about various aspects of queues (including humor!), the place to visit is the page maintained by Hlynka [7]. In fact, the page among other things illustrates various books on queueing, course notes, and a list of software. A few software tools would be pointed out in this chapter, but it would be an excellent idea to visit Hlynka's site [8] for an up-to-date list of queueing software. In there, software that run on various other applications (such as MATLAB, Mathematica, Excel, etc.) are explained and the most suitable one for the reader can be adopted.

This chapter is organized as follows. First, some basic results that are used throughout the chapter are explained in Section 9.2. The bulk of this chapter is Section 9.3, which lays the foundation for the other types of systems by initially considering the single-station and single-class queue. Then in Section 9.4, the results for single-station and multiple classes are presented. Following that, the chapter moves from analyzing a single station to a network of queues in Section 9.5 where only one class is considered. This is extended to the most general form (of which all the previous models are special cases) of multistation and multiclass queue in Section 9.6. Finally, some concluding remarks are made in Section 9.7.

9.2 Queueing Theory Basics

Consider a single-station queueing system as shown in Figure 9.1. This is also called a single-stage queue. There is a single waiting line and one or more servers. A typical example can be found at a bank or post office. Arriving customers enter the queueing system and wait in the waiting area if a server is not free (otherwise they go straight to a server). When a server becomes free, one customer is selected and service begins. Upon service completion, the customer departs the system. A few key assumptions are needed to analyze the basic queueing system.

Queueing Theory

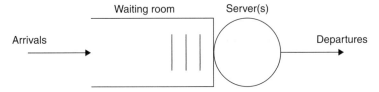

FIGURE 9.1 A single-station queueing system.

TABLE 9.2	Fields in the Kendall Notation
AP	M, G, E_k, H, PH, D, GI, etc.
ST	M, G, E_k, H, PH, D, GI, etc.
NS	denoted by s, typically $1, 2, \ldots, \infty$
Cap	denoted by k, typically $1, 2, \ldots, \infty$
	default: ∞
SD	FCFS, LCFS, ROS, SPTF, etc.
	default: FCFS

ASSUMPTION 9.1 *The customer interarrival times, that is, the time between arrivals, are independent and identically distributed (usually written as "iid"). Therefore, the arrival process is what is called a renewal process. All arriving customers enter the system if there is room to wait. Also, all customers wait till their service is completed in order to depart.*

ASSUMPTION 9.2 *The service times are independent and identically distributed random variables. Also, the servers are stochastically identical; that is, the service times are sampled from a single distribution. In addition, the servers adopt a work-conservation policy; that is, the server is never idle when there are customers in the system.*

The above assumptions can certainly be relaxed. There are a few models that do not require the above assumptions. However, for the rest of this chapter, unless explicitly stated otherwise, we will assume that Assumptions 9.1 and 9.2 hold.

To standardize description for queues, Kendall developed a notation with five fields: $AP/ST/NS/Cap/SD$. In the Kendall notation, AP denotes arrival process characterized by the interarrival distribution, ST denotes the service time distribution, NS is the number of servers in the system, Cap is the maximum number of customers in the whole system (with a default value of infinite), and SD denotes service discipline which describes the service order such as first come first served (FCFS), which is the default, last come first served (LCFS), random order of service (ROS), shortest processing time first (SPTF), and so on. The fields AP and ST can be specific distributions such as exponential (denoted by M which stands for memoryless or Markovian), Erlang (denoted by E_k), phase-type (PH), hyperexponential (H), deterministic (D), and so on. Sometimes, instead of a specific distribution, AP and ST fields could be G or GI, which denote general distribution (although GI explicitly says "general independent," G also assumes independence). Table 9.2 depicts values that can be found in the five fields of Kendall notation.

For example, $GI/H/4/6/LCFS$ implies that the arrivals are according to a renewal process with general distribution, service times are according to a hyperexponential distribution, there are four servers, a maximum of six customers are permitted in the system at a time (including four at the server), and the service discipline is LCFS. Also, $M/G/4/9$ implies that the interarrival times are exponential (whereby the arrivals are according to a Poisson process), service times are according to some general distribution, there are four servers, the system capacity is nine customers in total, and the customers are served according to FCFS. Finally, in an $M/M/1$ queue, the arrivals are according to a Poisson process, service times exponentially distributed, there is one server, the waiting space is infinite and the customers are served according to FCFS.

9.2.1 Fundamental Queueing Relations

Consider a single-station queueing system such as the one shown in Figure 9.1. Assume that this system can be described using Kendall notation. That means the interarrival time distribution, service time distribution, number of servers, system capacity, and service discipline are given. For such a system we now describe some parameters and measures of performance. Assume that customers (or entities) that enter the queueing system are assigned numbers with the nth arriving customer called customer-n. Most of the results presented in this section are available in Kulkarni [6] with possibly different notation.

In that light, let A_n denote the time when the nth customer arrives, and thereby $A_n - A_{n-1}$, an interarrival time. Let S_n be the service time for the nth customer. Let D_n be the time when the nth customer departs. We denote $X(t)$ as the number of customers in the system at time t, X_n as the number of customers in the system just after the nth customer departs, and X_n^* as the number of customers in the system just before the nth customer arrives. Although in this chapter we will not go into details, it is worthwhile mentioning that $X(t)$, X_n, and X_n^* are usually modeled as stochastic processes. We also define two other variables, which are usually not explicitly modeled but can be characterized in steady state. These are W_n, the waiting time of the nth customer and $W(t)$, the total remaining workload at time t (this is the total time it would take to serve all the customers in the system at time t). The above variables are described in Table 9.3 for easy reference, where customer n denotes the nth arriving customer.

It is usually very difficult to obtain distributions of the random variables $X(t)$, X_n, X_n^*, $W(t)$, and W_n. However, the corresponding steady-state values can be obtained, that is, the limiting distributions as n and t go to infinite. In that light, let p_j be the probability that there are j customers in the system in steady state, and let π_j and π_j^* be the respective probabilities that in steady state a departing and an arriving customer would see j other customers in the system. In addition, let $G(x)$ and $F(x)$ be the cumulative distribution functions of the workload and waiting time respectively in steady state. Finally, define L as the time-averaged number of customers in the system, and define W as the average waiting time (averaged across all customers). One of the primary objectives of queueing models is to obtain closed-form expressions for the performance metrics p_j, π_j, π_j^*, $G(x)$, $F(x)$, L,

TABLE 9.3 Variables and Their Mathematical and English Meanings

Variable	Mathematical Expression	Meaning
A_n		Arrival time of customer n
S_n		Service time of customer n
D_n		Departure time of customer n
$X(t)$		Number of customers in the system at time t
X_n	$X(D_n+)$	No. in system just after customer n's departure
X_n^*	$X(A_n-)$	No. in system just before customer n's arrival
W_n	$D_n - A_n$	Waiting time of customer n
$W(t)$		Total remaining workload at time t

and W. These performance metrics can be mathematically represented as follows:

$$p_j = \lim_{t \to \infty} P\{X(t) = j\}$$

$$\pi_j = \lim_{n \to \infty} P\{X_n = j\}$$

$$\pi_j^* = \lim_{n \to \infty} P\{X_n^* = j\}$$

$$G(x) = \lim_{t \to \infty} P\{W(t) \leq x\}$$

$$F(x) = \lim_{n \to \infty} P\{W_n \leq x\}$$

$$L = \lim_{t \to \infty} E[X(t)]$$

$$W = \lim_{n \to \infty} E[W_n]$$

Let $\overline{\lambda}$ be the average number of customers that enter the queueing system per unit time, referred to as the mean entering rate. Note that if the system capacity is finite, all arriving customers do not enter and therefore $\overline{\lambda}$ is specifically referred to as average rate of "entering" and not "arrival." The relation between L and W is given by Little's law (a result by Prof. John D.C. Little of MIT):

$$L = \overline{\lambda} W \tag{9.1}$$

It is important to note two things. First, for the finite capacity case W must be interpreted as the mean time in the system for customers that actually "enter" the system (and does not include customers that were turned away). Second, note that the average rate of departure from the system (if the system is stable) is also $\overline{\lambda}$. This is called conservation of customers, whereby customers are neither created nor destroyed; therefore the average customer entering rate equals average customer departure rate (if the system is stable).

We now focus our attention on infinite capacity queues in the next section. However, while we present results, if applicable, finite capacity queues' extensions will be explained. In addition, in future sections too we will mainly concentrate on infinite capacity queues (with some exceptions) due to issues of practicality and ease of analysis. From a practical standpoint, if a queue actually has finite capacity but the capacity is seldom reached, approximating the queue as an infinite capacity queue is reasonable.

9.2.2 Preliminary Results for the $GI/G/s$ Queue

Define the following for a single-stage $GI/G/s$ queue (interarrival times independent and identically distributed, service time any general distribution, s servers, infinite waiting room, FCFS service discipline):

- λ: Average arrival rate into the system (inverse of the average time between two arrivals); notice that as the capacity is finite, all customers that arrive, also *enter* the system.
- μ: Average service rate of a server (inverse of the average time to serve a customer); it is important that the units for λ and μ be the same; that is, both should be per second or both should be per minute, and so on.
- L_q: Average number of customers in the queue, not including ones in service (L defined earlier, includes the customers at the servers).

- W_q: Average time spent in the queue, not including in service (W defined earlier, includes customer service times). Note that, in units of $1/\mu$,

$$W = W_q + \frac{1}{\mu} \tag{9.2}$$

- $\rho = \frac{\lambda}{s\mu}$: the traffic intensity, which is a dimensionless quantity.

It is important to note that while extending the results to finite capacity queues, all the above definitions pertain only to customers that enter the system (and do not include those that were turned away when the system was full). However, the first result below is applicable only for infinite capacity queues as finite capacity queues are always stable.

RESULT 9.1 *A necessary condition for stability of a queueing system is*

$$\rho \leq 1$$

For most cases, the above condition is also sufficient (the sufficient condition actually is $\rho < 1$). However, in the case of queues with multiclass traffic that traverses through multi-station queues, this condition may not be sufficient.

Little's Law and Other Results Using Little's Law

As described in the previous section, we once again present Little's law, here $\overline{\lambda} = \lambda$.

RESULT 9.2 *For a GI/G/s queue,*

$$L = \lambda W \tag{9.3}$$

and

$$L_q = \lambda W_q \tag{9.4}$$

Notice that if we can compute one of L, L_q, W, or W_q, the other three can be obtained using the above relations. Little's law holds under very general conditions. In fact, even the service discipline does not have to be FCFS and the servers do not need to be work conserving. The result holds for any system with inputs and outputs. As an example, Equation 9.4 is nothing but using Little's law for the waiting space and not including the server. Note that if the system is stable, the output rate on an average is also λ. Using Little's law, some more interesting results can be obtained.

RESULT 9.3 *The probability that a particular server is busy p_b is given by*

$$p_b = \rho$$

which can be derived from $W = W_q + 1/\mu$ and Little's law via the relation $L = L_q + \lambda/\mu$. Also, for the special single server case of $s = 1$, that is, GI/G/1 queues, the probability that the system is empty, p_0 is

$$p_0 = 1 - \rho$$

Based on the definition of L, it can be written as

$$L = \sum_{j=0}^{\infty} j p_j$$

Queueing Theory

In a similar manner, let $L_{(k)}$ be the kth factorial moment of the number of customers in the system in steady state, that is,

$$L_{(k)} = \sum_{j=k}^{\infty} k! \binom{j}{k} p_j$$

Also, let $W^{(k)}$ be the kth moment of the waiting time in steady state, that is,

$$W^{(k)} = \lim_{n \to \infty} E\left[\{W_n\}^k\right]$$

Little's law can be extended for the $M/G/s$ queue in the following manner.

RESULT 9.4 *For an $M/G/s$ queue,*

$$L_{(k)} = \lambda^k W^{(k)} \tag{9.5}$$

Of course, the special case of $k=1$ is Little's law itself. However, the interesting result is that all moments of the queue lengths are related to corresponding moments of waiting times. Notice that from factorial moments it is easy to obtain actual moments.

Limiting Distributions of $X(t)$, X_n, and X_n^*

In some situations it may not be possible to obtain p_j easily. But it may be possible to get π_j or π_j^*. In that light, two results based on the limiting distributions (π_j, π_j^*, and p_j) will be presented. The first result relates the X_n and X_n^* processes in the limit (i.e., the relation between π_j and π_j^*). The second result illustrates the relation between the limiting distributions of $X(t)$ and X_n^* (i.e., p_j and π_j^*). They hold under very general cases beyond the cases presented here. However, one must be very careful while using the results in the more general situations.

RESULT 9.5 *Let π_j and π_j^* be as defined earlier as the limiting distributions of X_n and X_n^*, respectively. When either one of those limits exists, so does the other and*

$$\pi_j = \pi_j^* \quad \text{for all } j \geq 0$$

It can easily be shown that the limits described in the above result exists for queue length processes of $M/M/1$, $M/M/s$, $M/G/1$, and $G/M/s$ queueing systems. However, the limit for the more general $G/G/s$ case is harder to show but the result does hold. The result also holds for the $G/G/s/k$ case, whether we look at "entering" or "arriving" customers (in the "arriving" case, departing customers denote both the ones rejected as well as the ones that leave after service).

RESULT 9.6 *If the arrival process is Poisson (i.e., an $M/G/s$ queue), then the probability that an arriving customer in steady state will see the queueing system in state j is the probability that the system is in state j in the long run, that is,*

$$p_j = \lim_{t \to \infty} P\{X(t) = j\} = \pi_j^*$$

The above result is called PASTA (Poisson Arrivals See Time Averages). PASTA is a powerful result that can be used in situations beyond queueing. For example, if one is interested in computing an average over time, instead of observing the system continuously, it can be observed from time to time such that the interarrival times are exponentially distributed. In fact, one common mistake made by a lot of people is to compute averages by sampling at equally spaced intervals. In fact, sampling must be done in such a manner that the time between samples is exponentially distributed. Only then the averages obtained across such a sample will be equal to the average across time.

Therefore, when one out of L, W, L_q, or W_q is known, the other three can be computed. Also under certain conditions, when one out of p_j, π_j, or π_j^* is known, the others could be computed. In the next few sections, we will see how to compute "one" of those terms.

9.3 Single-Station and Single-Class Queues

In this section, we consider a single queue at a single station handling a single class of customers. We start with the simplest case of an $M/M/1$ queue and work our way through more complex cases. Note that all the results can be found in standard texts such as Gross and Harris [1] especially until Section 9.3.11.

9.3.1 The Classic $M/M/1$ Queue: Main Results

Consider a single-stage queueing system where the arrivals are according to a Poisson process with average arrival rate λ per unit time (which is written as $PP(\lambda)$), that is, the time between arrivals is according to an exponential distribution with mean $1/\lambda$. For this system the service times are exponentially distributed with mean $1/\mu$ and there is a single server.

The number in the system at time t in the $M/M/1$ queue, that is, $X(t)$, can be modeled as a continuous time Markov chain (CTMC), specifically a birth and death process. The condition for stability for the CTMC and subsequently the $M/M/1$ queue is that the traffic intensity ρ should be less than 1, that is, $\rho = \lambda/\mu < 1$. This means that the average arrival rate should be smaller than the average service rate. This is intuitive because the server would be able to handle all the arrivals only if the arrival rate is slower than the rate at which the server can process on an average.

The long-run probability that the number of customers in the system is j (when $\rho < 1$) is given by

$$p_j = \lim_{t \to \infty} P\{X(t) = j\} = (1-\rho)\rho^j \quad \text{for all } j \geq 0$$

Therefore, the long-run probability that there are more than n customers in the system is ρ^n. In addition, the key performance measures can also be obtained. Using p_j we have

$$L = \sum_{j=0}^{\infty} j p_j = \frac{\lambda}{\mu - \lambda}$$

and

$$L_q = 0 p_0 + \sum_{j=0}^{\infty} j p_{j+1} = \frac{\lambda^2}{\mu(\mu - \lambda)}$$

Recall that W_n is the waiting time of the nth arriving customer and $F(x) = \lim_{n \to \infty} P\{W_n \leq x\}$. Then

$$F(x) = 1 - e^{-(\mu - \lambda)x} \quad \text{for } x \geq 0$$

and therefore the waiting time for a customer arriving in steady-state is exponentially distributed with mean $1/(\mu - \lambda)$. Therefore

$$W = \frac{1}{\mu - \lambda}$$

which can also be derived using Little's law and the expression for L. In addition, using Little's law for L_q,

$$W_q = \frac{\lambda}{\mu(\mu - \lambda)}$$

$M/M/1$ Type Queues with Balking

When an arriving customer sees j other customers in the system, this customer joins the queue with probability α_j. In other words, this customer balks from the queueing system with probability $(1 - \alpha_j)$. This can be modeled as a CTMC, which is a birth and death process with birth parameters (i.e., rate for going from state n to $n+1$) $\lambda_{n+1} = \alpha_n \lambda$ for $n \geq 0$ and death parameters (i.e., rate for going from state n to $n-1$) $\mu_n = \mu$ for $n \geq 1$. It is not possible to obtain p_j in closed-form (and thereby L) except for some special cases. In general,

$$p_j = \frac{\prod_{k=0}^{j}(\lambda_k/\mu_k)}{1 + \sum_{n=1}^{\infty} \prod_{i=1}^{n} \frac{\lambda_i}{\mu_i}}$$

with $\lambda_0 = \mu_0 = 1$ and when the denominator exists. Also,

$$L = \sum_{j=0}^{\infty} j p_j$$

and using $\overline{\lambda} = \sum_{n=0}^{\infty} \lambda_{n+1} p_n$, W can be obtained as $L/\overline{\lambda}$.

$M/M/1$ Type Queues with Reneging

Every customer that joins a queue waits for an $\exp(\theta)$ amount of time before which if the service does not begin, the customer leaves the queueing system (which is called reneging from the queueing system). This can be modeled as a birth and death CTMC with birth parameters (see above for $M/M/1$ with balking for definition) $\lambda_{n+1} = \lambda$ for $n \geq 0$ and death parameters $\mu_n = \mu + (n-1)\theta$ for $n \geq 1$. It is not possible to obtain p_j in closed-form (and thereby L) except for some special cases. In general,

$$p_j = \frac{\prod_{k=0}^{j}(\lambda_k/\mu_k)}{1 + \sum_{n=1}^{\infty} \prod_{i=1}^{n} \frac{\lambda_i}{\mu_i}}$$

with $\lambda_0 = \mu_0 = 1$ and when the denominator exists. Also,

$$L = \sum_{j=0}^{\infty} j p_j$$

and using $\overline{\lambda} = \lambda$, it is possible to obtain W as L/λ. It is crucial to note that W is the time in the system for all customers, so it includes those customers that reneged as well as those that were served. A separate analysis must be performed to obtain the departure rate of customers after service. Using this departure rate as $\overline{\lambda}$, if W is obtained then it would be the average waiting time for customers that were served.

In case there is a queueing system with balking and reneging, then the analysis can be combined. However, it must be noted that if the reneging times are not exponential, then the analysis is a lot harder.

$M/M/1$ Queue with State-Dependent Service

Consider an $M/M/1$ type queue where the mean service rate depends on the state of the system. Many times when the number of customers waiting increases, the server starts working faster. This is typical when the servers are humans. Therefore, if there are n customers in the system, the mean service rate is μ_n. Note that in the middle of service if the number in service increases to $n+1$, the mean service rate also changes to μ_{n+1} (further, if it increases to $n+2$ then service rate becomes μ_{n+2}, and so on). This can also be modeled as a birth and death CTMC with birth parameters (defined in $M/M/1$ with balking) $\lambda_{n+1} = \lambda$ for $n \geq 0$ and death parameters μ_n for $n \geq 1$. It is not possible to obtain p_j in closed-form (and thereby L) except for some special cases. In general,

$$p_j = \frac{\prod_{k=0}^{j}(\lambda_k/\mu_k)}{1 + \sum_{n=1}^{\infty} \prod_{i=1}^{n} \frac{\lambda_i}{\mu_i}}$$

with $\lambda_0 = \mu_0 = 1$ and when the denominator exists. Also,

$$L = \sum_{j=0}^{\infty} j p_j$$

and using $\overline{\lambda} = \lambda$, W can be obtained as L/λ.

Note that if the mean service rate has to be picked and retained throughout the service of a customer, that system cannot be modeled as a birth and death process.

$M/M/1$ Queue with Processor Sharing

For p_j, L, and W it does not matter what the service discipline is (FCFS, LCFS, ROS, etc.) The results are the same as long as the customers are served one at a time. Now what if the customers are served using a processor sharing discipline? Customers arrive according to a Poisson process with mean arrival rate λ customers per unit time. The amount of work each customer brings is according to $\exp(\mu)$; that is, if each customer were served individually it would take $\exp(\mu)$ time for service. However, the processor is shared among all customers. So, if the system has i customers, each customer gets only an ith of the processing power. Therefore each of the i customers get a service rate of μ/i. However, the time for the first of the i to complete service is according to $\exp(i \times \mu/i)$. Therefore, the CTMC for the number of customers in the system is identical to that of an FCFS $M/M/1$ queue. And so, even the processor sharing discipline will have identical p_j, L, and W as that of the FCFS $M/M/1$ queue.

9.3.2 The Multiserver System: $M/M/s$

The description of an $M/M/s$ queue is similar to that of the classic $M/M/1$ queue with the exception that there are s servers. Note that by letting $s = 1$, all the results for the $M/M/1$ queue can be obtained. The number in the system at time t, $X(t)$, in the $M/M/s$ queue can be modeled as a CTMC, which again is a birth and death process. The condition for stability is $\rho = \lambda/(s\mu) < 1$ where ρ is called the traffic intensity. The long run probability that the number of customers in the system is j (when $\rho < 1$) is given by

$$p_j = \begin{cases} \dfrac{1}{j!}\left(\dfrac{\lambda}{\mu}\right)^j p_0 & \text{if } 0 \leq j \leq s-1 \\ \dfrac{1}{s! s^{j-s}}\left(\dfrac{\lambda}{\mu}\right)^j p_0 & \text{if } j \geq s \end{cases}$$

where $p_0 = \left[\sum_{n=0}^{s-1}\left\{\frac{1}{n!}(\lambda/\mu)^n\right\} + \frac{(\lambda/\mu)^s}{s!}\frac{1}{1-\lambda/(s\mu)}\right]^{-1}$. Thereby, using p_j, we can derive

$$L_q = \frac{p_0(\lambda/\mu)^s \lambda}{s!s\mu[1-\lambda/(s\mu)]^2}$$

Also, $W_q = L_q/\lambda$, $W = W_q + 1/\mu$, and $L = L_q + \lambda/\mu$. The steady-state waiting time for a customer has a cumulative distribution function (CDF) given by

$$F(x) = \frac{s(1-\rho)-w_0}{s(1-\rho)-1}(1-e^{-\mu x}) - \frac{1-w_0}{s(1-\rho)-1}(1-e^{-(s\mu-\lambda)x})$$

where $w_0 = 1 - \frac{\lambda^s p_0}{s!\mu^s(1-\rho)}$.

9.3.3 Finite Capacity $M/M/s/K$ System

In fact, this is one of the more general forms of the Poisson arrivals (with mean rate λ per unit time) and exponential service time (with mean $1/\mu$) queue. Using the results presented here, results for all the $M/M/\cdot/\cdot$ type queues can be obtained. For example, letting $s=1$ and $K=\infty$, the $M/M/1$ results can be obtained; $K=\infty$ would yield the $M/M/s$ results, $K=s$ would yield the $M/M/s/s$ results, $K=s=\infty$ would yield the $M/M/\infty$ results, and so on. The special cases are popular because (a) the results are available in closed-form and (b) insights can be obtained, especially while extending to the more general cases.

The number in the system at time t, $X(t)$, in the $M/M/s/K$ queue can be modeled as specifically a birth and death chain CTMC. Using the p_j values, one can derive

$$L_q = \frac{p_0(\lambda/\mu)^s \rho}{s!(1-\rho)^2}[1 - \rho^{K-s} - (K-s)\rho^{K-s}(1-\rho)]$$

where $\rho = \lambda/(s\mu)$ and $p_0 = \left[\sum_{n=0}^{s}\left\{\frac{1}{n!}(\lambda/\mu)^n\right\} + \frac{(\lambda/\mu)^s}{s!}\sum_{n=s+1}^{K}\rho^{n-s}\right]^{-1}$.

Caution: Since this is a finite capacity queue, ρ can be greater than 1. The probability that an arriving customer is rejected is p_K as is given by $p_K = \frac{(\lambda/\mu)^K}{s!s^{K-s}}p_0$. Therefore, the average entering rate $\overline{\lambda}$ is given by $\overline{\lambda} = (1-p_K)\lambda$. Hence W_q can be derived as $L_q/\overline{\lambda}$. Also, W and L can be obtained using $W = W_q + 1/\mu$ and $L = L_q + \overline{\lambda}/\mu$.

9.3.4 The $M/M/1/K$ Queue

The $M/M/1/K$ is another special case of the $M/M/s/K$ system with $s=1$. However, it is the most fundamental finite capacity queue example with Poisson arrivals (with mean rate λ) and exponential service times (with mean $1/\mu$). No more than K customers can be in the system at any time. The traffic intensity ρ (where $\rho = \lambda/\mu$) does not have to be less than one. Since the capacity is finite, the system cannot be unstable. However, we assume for now that $\rho \neq 1$. For the case $\rho = 1$, limits results from calculus can be used (such as L'Hospital's rule) to obtain the corresponding value. The number of customers in the system, not including any at the server, is

$$L_q = \frac{\rho}{1-\rho} - \frac{\rho(K\rho^K + 1)}{1-\rho^{K+1}}$$

To obtain W_q, we use
$$W_q = L/\overline{\lambda}$$
where $\overline{\lambda}$ is the average entering rate into the system and can be expressed as $\lambda(1-p_K)$ where
$$p_K = \frac{(\lambda/\mu)^K[1-\lambda/\mu]}{1-(\lambda/\mu)^{K+1}}$$

Also, W and L can be obtained using $W = W_q + 1/\mu$ and $L = L_q + \overline{\lambda}/\mu$.

9.3.5 The $M/M/s/s$ Queue

Although the $M/M/s/s$ queue is a special case of the $M/M/s/K$ system with $K=s$, there are several interesting aspects and unique applications for it. Customers arrive according to a Poisson process with mean rate λ per unit time. Essentially there are no queues. But there are s servers that can be thought of as s resources that customers hold on to for an exponential amount of time (with mean $1/\mu$). In fact, queueing theory started with such a system by a Danish Mathematician A.K. Erlang, who studied telephone switches with s lines. There is no waiting and if all s lines are being used, the customer gets a "busy" tone on their telephone. This system is also known as the Erlang loss system. However, there are many other applications for the $M/M/s/s$ queue such as: a rental agency with s items, gas stations where customers do not wait if a spot is not available among s possible spots, self-service area with maximum capacity s, and so on. In many of these systems there are no explicit servers.

The probability that there are j (for $j = 0, \ldots, s$) customers in the system in the long run is
$$p_j = \frac{\dfrac{(\lambda/\mu)^j}{j!}}{\sum_{i=0}^{s}\dfrac{(\lambda/\mu)^i}{i!}}$$

Therefore, the "famous" Erlang loss formula is the probability that an arriving customer is rejected (loss probability) and is given by
$$p_s = \frac{\dfrac{(\lambda/\mu)^s}{s!}}{\sum_{i=0}^{s}\dfrac{(\lambda/\mu)^i}{i!}}$$

A remarkable fact is that the above formula holds good even for the $M/G/s/s$ system with mean service time $1/\mu$. We will see that in the $M/G/s/s$ system explanation. We can derive
$$L = \frac{\lambda}{\mu}(1-p_s)$$

As the effective arrival rate is $\lambda(1-p_s)$, $W = 1/\mu$, which is obvious since there is no waiting, the average time in the system W is indeed the average service time. Clearly $L_q = 0$ and $W_q = 0$ for that same reason.

9.3.6 The Infinite Server $M/M/\infty$ Queue

This is identical to the $M/M/s/s$ system with $s = \infty$. In reality there are never infinite resources or servers. But when s is very large and a negligible number of customers are

Queueing Theory

rejected, the system can be assumed as $s = \infty$ as the results are expressed in closed-form. Systems such as the beach, grocery store (not counting the check out line), car rentals, and the like can be modeled as $M/M/\infty$ queues.

The probability that there are j customers in the system in the long run is $p_j = (\lambda/\mu)^j \frac{1}{j!} e^{-\lambda/\mu}$. Also, $L = \lambda/\mu$ and $W = 1/\mu$. Of course, $L_q = 0$ and $W_q = 0$.

9.3.7 Finite Population Queues

Until this point we assumed that there are an infinite number of potential customers and the arrival rates did not depend on the number of customers in the system. Now we look at the case where the arrivals are state-dependent. Consider a finite population of N customers. Each customer after completion of service returns to the queue after spending $\exp(\lambda)$ time outside the queueing system. There is a single server which serves customers in $\exp(\mu)$ amount of time. Clearly the arrival rate would depend on the number of customers in the system. If $X(t)$ denotes the number of customers in the system, then its limiting distribution is

$$p_j = \lim_{t \to \infty} P\{X(t) = j\} = \frac{\binom{N}{j} j!(\lambda/\mu)^j}{\sum_{i=0}^{N} \binom{N}{i} i!(\lambda/\mu)^i}$$

Clearly $L = \sum_{j=0}^{\infty} j p_j$; however, L_q, W, and W_q are tricky and need the effective arrival rate $\bar{\lambda}$. Using the fact that $\bar{\lambda} = \lambda(N - L)$ we can get

$$L_q = L - \lambda(N - L)/\mu, \quad W = \frac{L}{\lambda(N - L)}, \quad \text{and} \quad W_q = \frac{L_q}{\lambda(N - L)}$$

Notice that the arrivals are not according to a Poisson process (which requires that interarrival times be independent and identically distributed exponential random variables). Therefore, PASTA cannot be applied. However, it is possible to show that the probability that an arriving customer in steady state will see j in the system is

$$\pi_j^* = \frac{(N - j)p_j}{N - L}$$

9.3.8 Bulk Arrivals and Service

So far we have only considered the case of single arrivals and single service. In fact, in practice, it is not uncommon to see bulk arrivals and bulk service. For example, arrivals into theme parks are usually in groups, arrivals and service in restaurants are in groups, shuttle busses perform service in batches, and so on. We only present the cases where the interarrival times and service times are both exponentially distributed. However, unlike the cases seen thus far, here the CTMC models are not birth and death processes for the number in the system.

Bulk Arrivals Case: $M^{[X]}/M/1$ Queue

Arrivals occur according to $PP(\lambda)$ and each arrival brings a random number X customers into the system. A single server processes the customers one by one, spending $\exp(\mu)$ time with each customer. There is infinite waiting room and customers are processed according to FCFS. Such a system is denoted an $M^{[X]}/M/1$ queue. Let a_i be the probability that an arrival batch size is i, that is, $P\{X=i\}$ for $i>0$ (we do not allow batch size of zero). Let $E[X]$ and $E[X^2]$ be the first and second moments of X (where $E[X^2] = Var[X] + \{E[X]\}^2$). Define $\rho = \lambda E[X]/\mu$. The condition for stability is $\rho < 1$. We can derive the following results:

$$p_0 = 1 - \rho$$

$$L = \frac{\lambda\{E[X] + E[X^2]\}}{2\mu(1-\rho)}$$

Other p_j values can be computed in terms of λ, μ, and a_i. Note that the average entering rate $\bar{\lambda} = \lambda E[X]$. Therefore, using Little's law, $W = L/(\lambda E[X])$. Also, $W_q = W - 1/\mu$ and thereby $L_q = \lambda E[X] W_q$.

Bulk Service Case: $M/M^{[Y]}/1$ Queue

Single arrivals occur according to $PP(\lambda)$. The server processes a maximum of K customers at a time and any arrivals that take place during a service can join service (provided the number is less than K). There is a single server, infinite waiting room, and FCFS discipline. The service time for the entire batch is $\exp(\mu)$ whether the batch is of size K or not. In fact, this is also sometimes known as the $M/M^{[K]}/1$ queue (besides the $M/M^{[Y]}/1$ queue). Notice that this system is identical to a shuttle bus type system where customers arrive according to $PP(\lambda)$ and busses arrive with $\exp(\mu)$ as the inter-bus-arrival distribution. As soon as a bus arrives, the first K customers (if there are less than K, then all customers) instantaneously enter the bus and the bus leaves. Then the queue denotes the number of customers waiting for a shuttle bus.

To obtain the distribution of the number of customers waiting, let $(r_1, \ldots, r_{k+1})$ be the $K+1$ roots of the characteristic equation (with D as the variable)

$$\mu D^{K+1} - (\lambda + \mu)D + \lambda = 0$$

Let r_0 be the only root among the $K+1$ to be within 0 and 1. We can derive the following results:

$$p_n = (1 - r_0)r_0^n \quad \text{for } n \geq 0$$

$$L = \frac{r_0}{1 - r_0}, \quad L_q = L - \lambda/\mu$$

$$W = \frac{r_0}{\lambda(1 - r_0)}, \quad W_q = W - 1/\mu$$

9.3.9 The $M/G/1$ Queue

Consider a queueing system with $PP(\lambda)$ arrivals and general service times. The service times are iid with CDF $G(\cdot)$, mean $1/\mu$, and variance σ^2. Notice that in terms of S_n, the service time for any arbitrary customer n,

$$G(t) = P\{S_n \leq t\}$$

$$1/\mu = E[S_n]$$

$$\sigma^2 = Var[S_n]$$

Queueing Theory

There is a single server, infinite waiting room, and customers are served according to FCFS service discipline. It is important to note for some of the results, such as L and W, that the CDF $G(\cdot)$ is not required. However, for p_j and the waiting time distribution, the Laplace Steiltjes Transform (LST) of the CDF denoted by $\tilde{G}(s)$ and defined as

$$\tilde{G}(s) = E[e^{sS_n}] = \int_{t=0}^{\infty} e^{-st} dG(t)$$

is required.

Note that since the service time is not necessarily exponential, the number in the system for the $M/G/1$ queue cannot be modeled as a CTMC. However, notice that if the system was observed at the time of departure, a Markovian structure is obtained. Let X_n be the number of customers in the system immediately after the nth departure. Then it is possible to show that $\{X_n, n \geq 0\}$ is a discrete time Markov chain (DTMC) whose transition probability matrix can be obtained. The stability condition is $\rho < 1$ where $\rho = \lambda/\mu$. Let π_j be the limiting probability (under stability) that in the long run a departing customer sees j customers in the system, that is,

$$\pi_j = \lim_{n \to \infty} P\{X_n = j\}$$

Then, using PASTA, we have $p_j = \pi_j$ for all j. To obtain the π_j, consider the generating function $\phi(z)$ such that

$$\phi(z) = \sum_{j=0}^{\infty} \pi_j z^j$$

If the system is stable,

$$\pi_0 = 1 - \rho$$

$$\phi(z) = \frac{(1-\rho)(1-z)\tilde{G}(\lambda - \lambda z)}{\tilde{G}(\lambda - \lambda z) - z}$$

Although π_j values cannot be obtained in closed-form for the general case, they can be derived from $\phi(z)$ by repeatedly taking derivatives with respect to z and letting z go to zero.

However, L and W can be obtained in closed-form. The average number of customers in the system is

$$L = \rho + \frac{\lambda^2}{2} \frac{(\sigma^2 + 1/\mu^2)}{1 - \rho}$$

The average waiting time in the system is

$$W = 1/\mu + \frac{\lambda}{2} \frac{(\sigma^2 + 1/\mu^2)}{1 - \rho}$$

Also, $L_q = L - \rho$ and $W_q = W - 1/\mu$.

Recall that W_n is the waiting time of the nth arriving customer and $F(x) = \lim_{n \to \infty} P\{W_n \leq x\}$. Although it is not easy to obtain $F(x)$ in closed-form except for some special cases, it is possible to write it in terms of the LST, $\tilde{F}(s)$ defined as

$$\tilde{F}(s) = E[e^{sW_n}] = \int_{x=0}^{\infty} e^{-sx} dF(x)$$

in the following manner:

$$\tilde{F}(s) = \frac{(1-\rho)s\tilde{G}(s)}{s - \lambda(1 - \tilde{G}(s))}$$

It is important to realize that although inverting the LST in closed-form may not be easy, there are several software packages that can be used to invert it numerically. In addition, it is worthwhile to check that all the above results are true by trying exponential service times, as after all $M/M/1$ is a special case of the $M/G/1$ queue.

The $M/G/1$ Queue with Processor Sharing

Similar to the $M/M/1$ system, for the $M/G/1$ system as well, the expressions for the number in system, L, and waiting time, W, it does not matter what the service discipline is (FCFS, LCFS, ROS, etc). The results would be the same as long as the customers were served one at a time. Now what if the customers are served using a processor sharing discipline? Customers arrive according to a Poisson process with mean arrival rate λ customers per unit time. The amount of work each customer brings is according to some general distribution with CDF $G(\cdot)$ as described in the $M/G/1$ setting earlier. Also, if each customer were served individually it would take $1/\mu$ time for service on an average (and a variance of σ^2 for service time). However, for this case, the processor is shared among all customers. So if the system has i customers, each customer gets only an ith of the processing power. Therefore, each of the i customers get a service rate of $1/i$ of the server speed. For this system, it can be shown that

$$W = \frac{1}{\mu - \lambda}$$

The result indicates that the waiting time does not depend on the distribution of the service time but on the mean alone. Also, $L = \lambda W$, $W_q = 0$, and $L_q = 0$.

The $M/G/\infty$ Queue

Although this is an extension to the $M/M/\infty$ for the general service time case, the results are identical, indicating they are independent of the distribution of service time. The probability that there are j customers in the system in the long run is

$$p_j = e^{-\lambda/\mu} \frac{(\lambda/\mu)^j}{j!} \quad \text{for } j \geq 0$$

The departure process from the queue is $PP(\lambda)$. Also, $L = \lambda/\mu$ and $W = 1/\mu$. Of course $L_q = 0$ and $W_q = 0$.

The $M/G/s/s$ Queue

This is a queueing system where the arrivals are according to a Poisson process with mean arrival rate λ. The service times (also called holding times) are generally distributed with mean $1/\mu$. There are s servers but no waiting space. The results are identical to those of the $M/M/s/s$ queue. In fact, the Erlang loss formula was derived for this general case initially. For $0 \leq j \leq s$, the steady-state probability that there are j customers in the system is

$$p_j = \frac{\frac{(\lambda/\mu)^j}{j!}}{\sum_{k=0}^{s} \frac{(\lambda/\mu)^k}{k!}}$$

Queueing Theory

The departure process from the queue is $PP((1-p_s)\lambda)$. In addition, $L = \frac{\lambda}{\mu}(1-p_s)$, $W = 1/\mu$, $L_q = 0$, and $W_q = 0$.

9.3.10 The $G/M/1$ Queue

Consider a queueing system where the interarrival times are according to some given general distribution and service times are according to an exponential distribution with mean $1/\mu$. The interarrival times are iid with CDF $G(\cdot)$ and mean $1/\lambda$. This means that

$$G(t) = P\{A_{n+1} - A_n \le t\}$$

$$1/\lambda = E[A_{n+1} - A_n] = \int_0^\infty t\,dG(t)$$

Assume that $G(0) = 0$. Also, there is a single server, infinite waiting room, and customers are served according to FCFS service discipline. It is important to note for most of the results the LST of the interarrival time CDF denoted by $\tilde{G}(s)$ and defined as

$$\tilde{G}(s) = E[e^{s(A_{n+1} - A_n)}] = \int_{t=0}^\infty e^{-st}\,dG(t)$$

is required.

Similar to the $M/G/1$ queue, as all the random events are not necessarily exponentially distributed, the number in the system for the $G/M/1$ queue cannot be modeled as a CTMC. However, notice that if the system was observed at the time of arrivals, a Markovian structure is obtained. Let X_n^* be the number of customers in the system just before the nth arrival. Then it is possible to model the stochastic process $\{X_n^*, n \ge 0\}$ as a DTMC. The DTMC is ergodic if

$$\rho = \frac{\lambda}{\mu} < 1$$

which is the stability condition. Let π_j^* be the limiting probability that in the long run an arriving customer sees j other customers in the system, that is,

$$\pi_j^* = \lim_{x \to \infty} P\{X_n^* = j\}$$

If $\rho < 1$, we can show that

$$\pi_j^* = (1-\alpha)\alpha^j$$

where α is a unique solution in $(0,1)$ to

$$\alpha = \tilde{G}(\mu - \mu\alpha)$$

Using the notation W_n as the waiting time of the nth arriving customer under FCFS and $F(x) = \lim_{x \to \infty} P\{W_n \le x\}$, we have

$$F(x) = 1 - e^{-\mu(1-\alpha)x}$$

Therefore, under FCFS, the waiting time in the system in the long run is exponentially distributed with parameter $\mu(1-\alpha)$. Using that result, the average waiting time in the system is

$$W = \frac{1}{\mu(1-\alpha)}$$

Using Little's law, the average number of customers in the system is

$$L = \frac{\lambda}{\mu(1-\alpha)}$$

Note that we cannot use PASTA (as the arrivals are not Poisson, unlike the $M/G/1$ case). However, it is possible to obtain p_j using the following relation:

$$p_0 = 1 - \rho$$

$$p_j = \rho \pi^*_{j-1} \quad \text{when } j > 0$$

9.3.11 The $G/G/1$ Queue

Consider a single server queue with infinite waiting room where the interarrival times and service times are according to general distributions. The service discipline is FCFS. As the model is so general without a Markovian structure, it is difficult to model the number in the system as an analyzable stochastic process. Therefore, it is not possible to get exact expressions for the various performance measures. However, bounds and approximations can be derived. There are several of them and none are considered absolutely better than others. In this subsection, almost all the results are from Bolch et al. [2] unless otherwise noted.

Recall that W_n is the time in the system for the nth customer, S_n is the service time for the nth customer and A_n is the time of the nth arrival. To derive bounds and approximations for the $G/G/1$ queue, a few variables need to be defined. Define $I_{n+1} = \max(A_{n+1} - A_n - W_n, 0)$ and $T_{n+1} = A_{n+1} - A_n$. All the bounds and approximations are in terms of four parameters:

$$1/\lambda = E[T_n] \text{ average interarrival time}$$

$$C_a^2 = Var[T_n]/\{E[T_n]\}^2 \text{ SCOV of interarrival times}$$

$$1/\mu = E[S_n] \text{ average service time}$$

$$C_s^2 = Var[S_n]/\{E[S_n]\}^2 \text{ SCOV of service times}$$

where SCOV is the "squared coefficient of variation," that is, the ratio of the variance to the square of the mean (only for positive-valued random variables). Another parameter that is often used is $\rho = \lambda/\mu$, which is the traffic intensity.

Let random variables T, S, and I be the limiting values as $n \to \infty$ of T_n, S_n, and I_n, respectively. Although the mean and variance of T and S are known, $E[I]$ can be computed as $E(I) = E(T) - E(S)$, which requires $E(T) > E(S)$, that is, $\rho < 1$. It is possible to show that

$$W = \frac{E(S^2) - 2\{E(S)\}^2 - E(I^2) + E[T^2]}{2\{E(T) - E(S)\}}$$

Notice that the only unknown quantity above is $E(I^2)$. Therefore, approximations and bounds for W can be obtained through those of $E[I^2]$. As $L = \lambda W$ (using Little's law), bounds and approximations for L can also be obtained.

Since on many occasions, the departure process from a queue is the arrival process to another queue in a queueing network setting, it is important to study the mean and SCOV of the departure process of a $G/G/1$ queue. Let D_n be the time of departure of the nth customer. Define $\Delta_{n+1} = D_{n+1} - D_n$ as the interdeparture time with Δ being the interdeparture time in steady state. Then it is possible to show that under stability,

Queueing Theory

$E(\Delta) = E(I) + E(S) = E(T) = 1/\lambda$. Therefore, the departure rate equals arrival rate (conservation of flow of customers). In addition, let $C_d^2 = \text{Var}(\Delta)/\{E[\Delta]\}^2$; then

$$C_d^2 = C_a^2 + 2\rho^2 C_s^2 + 2\rho(1-\rho) - 2\lambda W(1-\rho)$$

Note that C_d^2 is in terms of W and hence approximations and bounds for W would yield the same for C_d^2.

Bounds for L, W, and C_d^2 for $G/G/1$ Queue

First some bounds on W:

$$W \leq \frac{C_a^2 + \rho^2 C_s^2}{2\lambda(1-\rho)} + E(S)$$

$$W \leq \frac{\rho(2-\rho)C_a^2 + \rho^2 C_s^2}{2\lambda(1-\rho)} + E(S)$$

$$W \geq \frac{\rho(C_a^2 - 1 + \rho) + \rho^2 C_s^2}{2\lambda(1-\rho)} + E(S) \quad \text{if } T \text{ is DFR}$$

$$W \leq \frac{\rho(C_a^2 - 1 + \rho) + \rho^2 C_s^2}{2\lambda(1-\rho)} + E(S) \quad \text{if } T \text{ is IFR}$$

where IFR and DFR are described subsequently. Note that $\rho(2-\rho) < 1$; therefore, the first bound is always inferior to the second. Also, IFR and DFR respectively denote increasing failure rate and decreasing failure rate random variables. Mathematically, the failure rate of a positive-valued random variable X is defined as $h(x) = f_X(x)/[1 - F_X(x)]$, where $f_X(x)$ and $F_X(x)$ are the probability density function and CDF of the random variable X. The reason they are called failure rate is because if X denotes the lifetime of a particular component, then $h(x)$ is the rate at which that component fails when it is x time units old. IFR and DFR imply that $h(x)$ is respectively the increasing and decreasing functions of x. Note that all random variables need not be IFR or DFR; they could be neither. Also, the exponential random variable has a constant failure rate.

We can also obtain bounds via the $M/G/1$ (disregarding C_a^2 and using Poisson arrival process with mean rate λ arrival process) and $G/M/1$ (disregarding C_s^2 and using exponentially distributed service times with mean $1/\mu$) results. It is important to note that to use the $G/M/1$ results, the distribution of the interarrival times is needed, not just the mean and SCOV. See the table below with LB and UB referring to lower and upper bounds:

C_a^2	C_s^2	$M/G/1$	$G/M/1$
>1	>1	LB	LB
>1	<1	LB	UB
<1	>1	UB	LB
<1	<1	UB	UB

That means (see second result above) if $C_a^2 > 1$ and $C_s^2 < 1$ for the actual $G/G/1$ system, then W using $M/G/1$ analysis would be a lower bound and correspondingly $G/M/1$ would yield an upper bound.

Next let us see what we have for L when T is DFR (although all bounds above for W can be used by multiplying by λ)

$$\frac{\rho(C_a^2 - 1 + \rho) + \rho^2 C_s^2}{2(1-\rho)} + \rho \leq L \leq \frac{\rho(2-\rho)C_a^2 + \rho^2 C_s^2}{2(1-\rho)} + \rho$$

Finally some bounds on C_d^2:

$$C_d^2 \geq (1-\rho)^2 C_a^2 + \rho^2 C_s^2$$

$$C_d^2 \leq (1-\rho)C_a^2 + \rho^2 C_s^2 + \rho(1-\rho) \text{ if } T \text{ is DFR}$$

$$C_d^2 \geq (1-\rho)C_a^2 + \rho^2 C_s^2 + \rho(1-\rho) \text{ if } T \text{ is IFR}$$

Approximations for L, W, and C_d^2 for $G/G/1$ Queue

The following are some approximations for L and C_d^2 taken from Buzacott and Shanthikumar [4]. Approximations for W can be obtained by dividing L by λ. There are several other approximations available in the literature, many of which are empirical. Only a few are presented here as follows:

Approx.	L	C_d^2
1	$\left(\dfrac{\rho^2(1+C_s^2)}{1+\rho^2 C_s^2}\right)\left(\dfrac{C_a^2 + \rho^2 C_s^2}{2(1-\rho)}\right) + \rho$	$(1-\rho^2)\left(\dfrac{C_a^2 + \rho^2 C_s^2}{1+\rho^2 C_s^2}\right) + \rho^2 C_s^2$
2	$\left(\dfrac{\rho^2(1+C_s^2)}{2-\rho+\rho C_s^2}\right)\left(\dfrac{\rho(2-\rho)C_a^2 + \rho^2 C_s^2}{2(1-\rho)}\right) + \rho$	$1 - \rho^2 + \rho^2 C_s^2 + (C_a^2 - 1)$
3	$\dfrac{\rho^2(C_a^2 + C_s^2)}{2(1-\rho)} + \dfrac{(1-C_a^2)C_a^2 \rho}{2} + \rho$	$\left(\dfrac{(1-\rho^2)(2-\rho) + \rho C_s^2(1-\rho)^2}{2-\rho+\rho C_s^2}\right)$ $(1-\rho)(1+\rho C_a^2)C_a^2 + \rho^2 C_s^2$

9.3.12 The $G/G/m$ Queue

Everything is similar to the $G/G/1$ queue explained before except that the number of servers is m. Getting closed-form expressions was impossible for $G/G/1$, so naturally for $G/G/m$ there is no question. However, several researchers have obtained bounds and approximations for the $G/G/m$ queue. In fact, letting $m = 1$ for the $G/G/m$ results would produce great results for $G/G/1$. Notice that the traffic intensity $\rho = \lambda/(m\mu)$. The random variables S and T, as well as parameters C_a^2 and C_s^2 used in the following bounds and approximations, have been defined in the $G/G/1$ system above.

- The Kingman upper bound:

$$W_q \leq \frac{Var(T) + Var(S)/m + (m-1)/(m^2\mu^2)}{2(1-\rho)}$$

- The Brumelle and Marchal lower bound:

$$W_q \geq \frac{\rho^2 C_s^2 - \rho(2-\rho)}{2\lambda(1-\rho)} - \frac{m-1}{m}\frac{(C_s^2+1)}{2\mu}$$

- Under heavy traffic conditions, for the $G/M/m$ systems,

$$W_q \approx \frac{Var(T) + Var(S)/m^2}{2(1-\rho)}\lambda$$

Queueing Theory

and waiting time in the queue is distributed approximately according to an exponential distribution with mean $1/W_q$. Note that "heavy traffic" implies that ρ is close to 1.

- And finally,

$$W_{G/G/m} \approx \frac{W_{M/M/m}}{W_{M/M/1}} W_{G/G/1} + E[S]$$

In the above approximation, the subscript for W denotes the type of queue. For example, $W_{M/M/m}$ implies the mean waiting time for the $M/M/m$ queue using the same λ and μ as the $G/G/m$ case.

- There are several approximations available in the literature, many of which are empirical. The most popular one is the following. Choose α_m such that

$$\alpha_m = \begin{cases} \frac{\rho^m + \rho}{2} & \text{if } \rho > 0.7 \\ \rho^{\frac{\rho+1}{2}} & \text{if } \rho < 0.7 \end{cases}$$

The waiting time in the queue is given by the approximation

$$W_q \approx \frac{\alpha_m}{\mu} \left(\frac{1}{1-\rho}\right) \left(\frac{C_a^2 + C_s^2}{2m}\right)$$

9.4 Single-Station and Multiclass Queues

In the models considered so far there was only a single class of customers in the system. However, there are several applications where customers can be differentiated into classes and each class has its own characteristics. For example, consider a hospital emergency room. The patients can be classified into emergency, urgent, and normal cases with varying arrival rates and service time requirements. Another example is a toll booth where the vehicles can be classified based on type (cars, buses, trucks, etc.) and each type has its own arrival rate and service time characteristics. There are several examples in production systems (routine maintenance versus breakdowns in repair shops) and communication systems (voice calls versus dial-up connection for Internet at a telephone switch) where entities must be classified due to the wide variability of arrival rates and service times.

Having made a case for splitting traffic in queues into multiple classes, it is also important to warn that unless absolutely necessary, due to the difficulty in analyzing such systems, one should not classify. There are two situations where it does make sense to classify. First, when the system has a natural classification where the various classes require their own performance measures (e.g., in a flexible manufacturing system, if a machine produces three types of parts and it is important to measure the in-process inventory of each of them individually, then it makes sense to model them as three classes). Second, when the service times are significantly different for the various classes that the distribution models would fit better, then it makes sense (e.g., if the service times have a bimodal distribution, then classifying into two classes with unimodal distribution for each class would possibly be better).

The next question to ask is how are the different classes of customers organized at the single station? There are two waiting line structures:

a. All classes of customers wait in the same waiting room. Examples: buffer in flexible manufacturing system, packets on a router interface, vehicles at a 1-lane road traffic light, and so on. Service scheduling policies are: FCFS, priority, ROS, LCFS, and so on.

b. Each class has a waiting room of its own and all classes of customers of a particular class wait in the same waiting room. Examples: robots handling several machine buffers, class-based queueing in routers, vehicles at a toll plaza (electronic payments, exact change, and full service), and the like. Service scheduling policies (especially when there is only a single server) across the classes typically are: priority, polling, weighted round-robin, and so on.

Within a class, the two waiting line structures use FCFS usually (LCFS and others are also possible). If the waiting room is of infinite capacity and there is no switch-over time from one queue to another, both (a) and (b) can be treated identically. However, in the finite waiting room case, they are different and in fact one of the design decisions is to figure out the buffer sizes, admission/rejection rules, and the like.

For the multiclass queues, several design decisions need to be made. These include:

- *Assigning classes*: how should the customers be classified? As alluded to before, it is critical, especially when there is no clear-cut classification, how customers should be classified and how many categories to consider.
- *Buffer sizing*: what should the size of the buffers be or how should a big buffer be partitioned for the various classes? These decisions can be made either one time (static) or changed as the system evolves (dynamic).
- *Scheduling rule*: how should the customers or entities be scheduled on the servers? For example, FCFS, shortest expected processing time first (FCFS within a class), round-robin across the K classes, priority-based scheduling, and so on. Sometimes these are "given" for the system and cannot be changed; other times these could be decisions that can be made.
- *Priority allocation*: if priority-based scheduling rule is used, then how should priorities be assigned? In systems like the hospital emergency room, the priorities are clear. However, in many instances one has to trade off cost and resources to determine priorities.
- *Service capacity*: how to partition resources such as servers (wholly or partially) among classes? For example, in a call-center handling customers that speak different languages and some servers being multilingual, it is important to allocate servers to appropriate queues. Sometimes these capacity allocations are made in a static manner and other times dynamically based on system state.

There are several articles in the literature that discuss various versions of the above design problems. In this chapter, we assume that the following are known or given: there are R classes already determined, infinite buffer size, scheduling rule already determined, and a single server that serves customers one at a time. For such a system, we first describe some general results for the $G/G/1$ case next and describe specific results for the $M/G/1$ case subsequently. Most of the results are adapted from Wolff [3] with possibly different notation.

9.4.1 Multiclass $G/G/1$ Queue: General Results

Consider a single-station queue with a single server that caters to R classes of customers. Customers belonging to class i ($i \in \{1, 2, \ldots, R\}$) arrive into the system at a mean rate λ_i and the arrival process is independent of other classes, but is also independent and identically distributed within a class. Customers belonging to class i ($i \in \{1, 2, \ldots, R\}$) require an average service time of $1/\mu_i$. Upon completion of service, the customers depart the system. We assume that the distribution of interarrival times and service times are known for each class. However, notice that the scheduling policy (i.e., service discipline) has not yet been specified. We describe some results that are invariant across scheduling policies (or at least a subset of policies).

For these systems, except for some special cases, it is difficult to obtain performance measures such as "distribution" of waiting time and queue length like in the single class cases. Therefore, we concentrate on obtaining average waiting time and queue length. Let L_i and W_i be the mean queue length and mean waiting time for class i customers. Irrespective of the scheduling policy, Little's law holds for each class, so for all $i \in [1, R]$,

$$L_i = \lambda_i W_i$$

That means that one can think of each class as a mini system in itself. Also, similar results can be derived for L_{iq} and W_{iq} which respectively denote the average number waiting in queue (not including customers at servers) and average time spent waiting before service. In particular, for all $i \in [1, R]$,

$$L_{iq} = \lambda_i W_{iq}$$

$$W_i = W_{iq} + \frac{1}{\mu_i}$$

$$L_i = L_{iq} + \rho_i$$

where

$$\rho_i = \frac{\lambda_i}{\mu_i}$$

In addition, L and W are the overall mean number of customers and mean waiting time averaged over all classes. Note that $L = L_1 + L_2 + \cdots + L_R$ and if $\lambda = \lambda_1 + \lambda_2 + \cdots + \lambda_R$, the net arrival rate, then $W = L/\lambda$. For the $G/G/1$ case with multiple classes, more results can be derived for a special class of scheduling policies called work-conserving disciplines that we describe next.

Work-Conserving Disciplines under $G/G/1$

We now concentrate on a subset of service-scheduling policies (i.e., service disciplines) called work-conserving disciplines where more results for the $G/G/1$ queue can be obtained. In fact, many of these results have not been explained in the single class in the previous sections, but by letting $R = 1$, they can easily be accomplished.

The essence of work-conserving disciplines is that the system workload at every instant of time remains unchanged over all work-conserving service scheduling disciplines. Intuitively this means that the server never idles and does not do any wasteful work. The server continuously serves customers if there are any in the system. For example, FCFS, LCFS, and ROS are work conserving. Certain priority policies that we will see later, such as non-preemptive and preemptive resume policies, are also work conserving. There are policies that are non-work-conserving, such as the preemptive repeat (unless the service times are exponential). Usually, when the server takes a vacation from service or if there is a switch-over time (or set-up time) during moving from classes, unless those can be explicitly accounted for in the service times, are non-work conserving.

To describe the results for the work-conserving disciplines, consider the notation used in Section 9.4.1. Define ρ, the overall traffic intensity, as

$$\rho = \sum_{i=1}^{R} \rho_i$$

An R-class $G/G/1$ queue with a work-conserving scheduling discipline is stable if

$$\rho < 1$$

In addition, when a $G/G/1$ system is work conserving, the probability that the system is empty is $1 - \rho$.

Let S_i be the random variable denoting the service time of a class i customer. Then we have the second moment of the overall service time as

$$E[S^2] = \frac{1}{\lambda} \sum_{i=1}^{R} \lambda_i E[S_i^2]$$

We now present two results that are central to work-conserving disciplines. These results were not presented for the single-class case (easily doable by letting $R = 1$).

RESULT 9.7 *If the $G/G/1$ queue is stable, then when the system is in steady state, the expected remaining service time at an arbitrary time in steady state is $\lambda E[S^2]/2$.*

As the total amount of work remains a constant across all work-conserving disciplines, and the above result represents the average work remaining for the customer at the server, the average work remaining due to all the customers waiting would also remain a constant across work-conserving discipline. That result is described below.

RESULT 9.8 *Let W_{iq} be the average waiting time in the queue (not including service) for a class i customer, then the expression*

$$\sum_{i=1}^{R} \rho_i W_{iq}$$

is a constant *over all work conserving disciplines.*

However, quantities such as L, W, L_i, and W_i (and the respective quantities with the q subscript) will depend on the service-scheduling policies. It is possible to derive these expressions in closed-form for $M/G/1$ queues that we describe next. The HOM software [9] can be used for numerical analysis of various scheduling policies for relatively general multiclass traffic.

9.4.2 $M/G/1$ Queue with Multiple Classes

Consider a special case of the $G/G/1$ queue with R classes where the arrival process is $PP(\lambda_i)$ for class i ($i = 1, 2, \ldots, R$). The service times are iid with mean $E[S_i] = 1/\mu_i$, second moment $E[S_i^2]$, and CDF $G_i(\cdot)$ for class i ($i = 1, 2, \ldots, R$) and $\rho_i = \lambda_i/\mu_i$. We present results for three work-conserving disciplines: FCFS, non-preemptive priority, and preemptive resume priority.

Multiclass $M/G/1$ with FCFS

In this service-scheduling scheme, the customers are served according to FCFS. None of the classes receive any preferential treatment. The analysis assumes that all the R classes

Queueing Theory

in some sense can be aggregated into one class as there is no differentiation. Hence, the net arrival process is $PP(\lambda)$ with $\lambda = \lambda_1 + \lambda_2 + \cdots + \lambda_R$. Let S be a random variable denoting the "effective" service time for an arbitrary customer. Then

$$G(t) = P(S \le t) = \frac{1}{\lambda} \sum_{i=1}^{R} \lambda_i G_i(t)$$

$$E[S] = \frac{1}{\mu} = \frac{1}{\lambda} \sum_{i=1}^{R} \lambda_i E[S_i]$$

$$E[S^2] = \sigma^2 + \frac{1}{\mu^2} = \frac{1}{\lambda} \sum_{i=1}^{R} \lambda_i E[S_i^2]$$

$$\rho = \lambda E[S]$$

Assume that the system is stable. Then, using standard $M/G/1$ results with $X(t)$ being the total number of customers in the system at time t, we get when $\rho < 1$,

$$L = \rho + \frac{1}{2} \frac{\lambda^2 E[S^2]}{1-\rho}$$

$$W = \frac{L}{\lambda}$$

$$W_q = W - \frac{1}{\mu}$$

$$L_q = \frac{1}{2} \frac{\lambda^2 E[S^2]}{1-\rho}$$

The expected number of class i customers in the system (L_i) as well as in the queue (L_{iq}) and the expected waiting time in the system for class i (W_i) as well as in the queue (L_{iq}) are given by:

$$W_{iq} = W_q = \frac{1}{2} \frac{\lambda E[S^2]}{1-\rho}$$

$$L_{iq} = \lambda_i W_{iq}$$

$$L_i = \rho_i + L_{iq}$$

$$W_i = W_{iq} + \frac{1}{\mu_i}$$

$M/G/1$ with Non-Preemptive Priority

Here we consider priorities among the various classes. For the following analysis assume that class 1 has highest priority and class R has the lowest. Service discipline within a class is FCFS. The server always starts serving a customer of the highest class among those waiting for service, and the first customer that arrived within that class. However, the server completes serving a customer before considering who to serve next. The meaning of non-preemptive priority is that a customer in service does not get preempted while in service by another customer of high priority (however preemption does occur while waiting). Assume that the system is stable.

Let $\alpha_i = \rho_1 + \rho_2 + \cdots + \rho_i$ with $\alpha_0 = 0$. Then we get the following results:

$$E[W_i^q] = \frac{\frac{1}{2}\sum_{j=1}^{R} \lambda_j E[S_j^2]}{(1-\alpha_i)(1-\alpha_{i-1})} \quad \text{for } 1 \leq i \leq R$$

$$E[L_i^q] = \lambda_i E[W_i^q]$$

$$W_i = E[W_i^q] + E[S_i]$$

$$L_i = E[L_i^q] + \rho_i$$

Sometimes performance measures for individual classes are required and other times aggregate performance measures across all classes. The results for the individual classes can also be used to obtain the overall or aggregate performance measures as follows:

$$L = L_1 + L_2 + \cdots + L_R$$

$$W = \frac{L}{\lambda}$$

$$W_q = W - \frac{1}{\mu}$$

$$L_q = \lambda W_q$$

Note: In the above analysis we assume we are given which class should get the highest priority, second highest, and so on. However, if we need to determine an optimal way of assigning priorities, one method is now provided. If you have R classes of customers and it costs the server C_j per unit time a customer of class j spends in the system (holding cost for class j customer), then to minimize the total expected cost per unit time in the long run, the optimal priority assignment is to give class i higher priority than class j if $C_i\mu_i > C_j\mu_j$. In other words, sort the classes in the decreasing order of the product $C_i\mu_i$ and assign first priority to the largest $C_i\mu_i$ and the last priority to the smallest $C_i\mu_i$ over all i. This is known as the $C\mu$ rule. Also note that if all the C_i values were equal, then this policy reduces to "serve the customer with the smallest expected processing time first."

$M/G/1$ with Preemptive Resume Priority

A slight modification to the $M/G/1$ non-preemptive priority considered above is to allow preemption during service. During the service of a customer, if another customer of higher priority arrives, then the customer in service is preempted and service begins for this new high priority customer. When the preempted customer returns to service, service resumes from where it was preempted. This is a work-conserving discipline (however, if the service has to start from the beginning which is called preemptive repeat, then it is not work conserving because the server wasted some time serving). Here, we consider the case where upon arrival, a customer of class i can preempt a customer of class j in service if $j > i$. Also, the total service time is unaffected by the interruptions, if any. Assume that the system is stable.

The waiting time of customers of class i is unaffected by customers of class j if $j > i$. Thus, class 1 customers face a standard single-class $M/G/1$ system with arrival rate λ_1 and service time distribution $G_1(\cdot)$. In addition, if only the first i classes of customers are considered, then the processing of these customers as a group is unaffected by the lower priority customers. The crux of the analysis is in realizing that the work content of this system (with only the top i classes) at all times is the same as an $M/G/1$ queue with FCFS and top i classes due to the work-conserving nature. Therefore, using the results for work-conserving systems, the performance analysis of this system is done.

Now consider an $M/G/1$ queue with only the first i classes and FCFS service. The net arrival rate is

$$\lambda(i) = \lambda_1 + \lambda_2 + \cdots + \lambda_i$$

the average service times is

$$\frac{1}{\mu(i)} = \sum_{j=1}^{i} \frac{\lambda_j E[S_j]}{\lambda(i)}$$

and the second moment of service times is

$$S^2(i) = \sum_{j=1}^{i} \frac{\lambda_j E[S_j^2]}{\lambda(i)}$$

Also let $\rho(i) = \lambda(i)/\mu(i)$. Let W_{jq}^{prp} be the waiting time in the queue for class j customers under preemptive resume policy. Using the principle of work conservation (see Result 9.8),

$$\sum_{j=1}^{i} \rho_j W_{jq}^{prp} = \rho(i) \frac{\lambda(i) S^2(i)}{2(1 - \lambda(i)/\mu(i))}$$

Notice that the left-hand side of the above expression is the first i classes under preemptive resume and the right-hand side being FCFS with only the first i classes of customers. Now, we can recursively compute W_{1q}, then W_{2q}, and so on till W_{Rq} via the above equations for $i = 1, 2, \ldots, R$.

Other average measures for the preemptive resume policy can be obtained as follows:

$$W_i^{prp} = W_{iq}^{prp} + E[S_i]$$

$$L_i^{prp} = \lambda_i W_i^{prp}$$

$$L_{iq}^{prp} = L_i^{prp} - \rho_i$$

Sometimes performance measures for individual classes are required and other times aggregate performance measures across all classes. The results for the individual classes can also be used to obtain the overall performance measures as follows:

$$L^{prp} = L_1^{prp} + L_2^{prp} + \cdots + L_R^{prp}$$

$$W^{prp} = \frac{L^{prp}}{\lambda}$$

$$W_q^{prp} = W^{prp} - \frac{1}{\mu}$$

$$L_q^{prp} = \lambda W_q^{prp}$$

9.5 Multistation and Single-Class Queues

So far we have only considered single-stage queues. However, in practice, there are several systems where customers go from one station (or stage) to other stations. For example, in a theme park the various rides are the different stations and customers wait in lines at each station and randomly move to other stations. Several engineering systems such as production, computer-communication, and transportation systems can also be modeled as queueing networks.

In this section, we only consider single-class queueing networks. The network is analyzed by considering each of the individual stations one by one. Therefore, the main technique would be to decompose the queueing network into individual queues or stations and develop characteristics of arrival processes for each individual station. Similar to the single-station case, here too we start with networks with Poisson arrivals and exponential service times, and then eventually move to more general cases. There are two types of networks: open queueing networks (customers enter and leave the networks) and closed queueing networks (the number of customers in the networks stays a constant).

9.5.1 Open Queueing Networks: Jackson Network

A Jackson network is a special type of open queueing network where arrivals are Poisson and service times are exponential. In addition, a queueing network is called a Jackson network if it satisfies the following assumptions:

1. It consists of N service stations (nodes).
2. There are s_i servers at node i ($1 \leq s_i \leq \infty$), $1 \leq i \leq N$.
3. Service times of customers at node i are iid $\exp(\mu_i)$ random variables. They are independent of service times at other nodes.
4. There is infinite waiting room at each node.
5. Externally, customers arrive at node i in a Poisson fashion with rate λ_i. All arrival processes are independent of each other and the service times. At least one λ_i must be nonzero.
6. When a customer completes service at node i, he or she or it departs the system with probability r_i or joins the queue at node j with probability p_{ij}. Here $p_{ii} > 0$ is allowed. It is required that $r_i + \sum_{j=1}^{N} p_{ij} = 1$ as all customers after completing service at node i either depart the system or join another node. The routing of a customer does not depend on the state of the network.
7. Let $P = [P_{ij}]$ be the routing matrix. Assume that $I - P$ is invertible, where I is an $N \times N$ identity matrix. The $I - P$ matrix is invertible if there is at least one node from where customers can leave the system.

To analyze the Jackson network, as mentioned earlier, we decompose the queueing network into the N individual nodes (or stations). The results are adapted from Kulkarni [6]. In steady state, the total arrival rate into node j (external and internal) is denoted by a_j and is given by

$$a_j = \lambda_j + \sum_{i=1}^{N} a_i p_{ij} \quad j = 1, 2, \ldots, N$$

Let $a = (a_1, a_2, \ldots, a_N)$. Then a can be solved as

$$a = \lambda(I - P)^{-1}$$

Queueing Theory

The following results are used to decompose the system: (a) the departure process from an $M/M/s$ queue is a Poisson process; (b) the superposition of Poisson process forms a Poisson process; and (c) Bernoulli (i.e., probabilistic) splitting of Poisson processes forms Poisson processes. Therefore, the resultant arrival into any node or station is Poisson. Then we can model node j as an $M/M/s_j$ queue with $PP(a_j)$ arrivals, $\exp(\mu_j)$ service, and s_j servers (if the stability condition at each node j is satisfied, i.e., $a_j < s_j\mu_j$). Hence, it is possible to obtain the steady-state probability of having n customers in node j as

$$\phi_j(n) = \begin{cases} \dfrac{1}{n!}\left(\dfrac{a_j}{\mu_j}\right)^n \phi_j(0) & \text{if } 0 \le n \le s_j - 1 \\ \dfrac{1}{s_j! s_j^{n-s_j}}\left(\dfrac{a_j}{\mu_j}\right)^n \phi_j(0) & \text{if } n \ge s_j \end{cases}$$

where $\phi_j(0) = \left[\sum_{n=0}^{s_j-1}\left\{\dfrac{1}{n!}(a_j/\mu_j)^n\right\} + \dfrac{(a_j/\mu_j)^s_j}{s_j!}\dfrac{1}{1 - a_j/(s_j\mu_j)}\right]^{-1}$

Now looking back into the network as a whole, let X_i be the steady-state number of customers in node i. Then it is possible to show that

$$P\{X_1 = x_1, X_2 = x_2, \ldots, X_N = x_N\} = \phi_1(x_1)\phi_2(x_2)\ldots\phi_N(x_N)$$

The above form of the joint distribution is known as product form. In steady state, the queue lengths at various nodes are independent random variables. Therefore, what this implies is that each node (or station) in the network behaves as if it is an independent $M/M/s$ queue. Hence, each node j can be analyzed as an independent system and performance measures can be obtained.

Specifically, it is possible to obtain performance measures at station j, such as the average number of customers (L_j), average waiting time (W_j), time in queue not including service (W_{jq}), number in queue not including service (L_{jq}), distribution of waiting time ($F_j(x)$), and all other measures using the single-station $M/M/s$ queue analysis in Section 9.3.2.

Besides the Jackson network, there are other product-form open queueing networks. The state-dependent service rate and the state-dependent arrival rate problems are two cases when product-form solution exists.

State-Dependent Service

Assume that the service rate at node i when there are n customers at that node is given by $\mu_i(n)$ with $\mu_i(0) = 0$. Also assume that the service rate does not depend on the states of the remaining nodes. Then define the following: $\phi_i(0) = 1$ and

$$\phi_i(n) = \prod_{j=1}^{n}\left(\dfrac{a_i}{\mu_i(j)}\right) \quad n \ge 1$$

where a_j is as before, the effective arrival rate into node j.

The steady-state probabilities are given by

$$P\{X_1 = x_1, X_2 = x_2, \ldots, X_N = x_N\} = c\prod_{i=1}^{N}\phi_i(x_i)$$

where the normalizing constant c is

$$c = \left\{\prod_{i=1}^{N}\left\{\sum_{n=0}^{\infty}\phi_i(n)\right\}\right\}^{-1}$$

Using the above joint distribution, it is possible to obtain certain performance measures. However, one of the difficulties is to obtain the normalizing constant. Once that is done, the marginal distribution at each node (or station) can be obtained. That can be used to get the distribution of the number of customers in the system as well as the mean (and even higher moments). Then, using Little's law, the mean waiting time can also be obtained.

State-Dependent Arrivals and Service

In a manner similar to the case of state-dependent service, the analysis of state-dependent arrivals and service can be extended. Let $\lambda(n)$ be the total arrival rate to the network as a whole when there are n customers in the entire network. Assume that u_i is the probability that an incoming customer joins node i, independently of other customers. Therefore, external arrivals to node i are at rate $u_i \lambda(n)$. The service rate at node i when there are n_i customers at that node is given by $\mu_i(n_i)$ with $\mu_i(0) = 0$.

Let b_i be the unique solution to

$$b_j = u_j + \sum_{i=1}^{N} b_i p_{ij}$$

Define the following: $\phi_i(0) = 1$ and

$$\phi_i(n) = \prod_{j=1}^{n} \left(\frac{b_i}{\mu_i(j)} \right) \quad \text{for } n \geq 1$$

Define $\hat{x} = \sum_{i=1}^{N} x_i$. The steady-state probabilities are given by

$$P\{X_1 = x_1, X_2 = x_2, \ldots, X_N = x_N\} = c \prod_{i=1}^{N} \phi_i(x_i) \prod_{j=1}^{\hat{x}} \lambda(j)$$

where the normalizing constant c is

$$c = \left\{ \sum_{x} \prod_{i=1}^{N} \phi_i(x_i) \prod_{j=1}^{\hat{x}} \lambda(j) \right\}^{-1}$$

9.5.2 Closed Queueing Networks (Exponential Service Times)

Closed queueing networks are networks where there are no external arrivals to the system and no departures from the system. They are popular in population studies, multiprogrammed computer systems, window flow control, Kanban, and so on. It is important to note that the number of customers being a constant is essentially what is required. This can happen if a new customer enters the network as soon as an existing customer leaves (a popular scheme in just-in-time manufacturing). Most of the results are adapted from Kulkarni [6]. We need a few assumptions to analyze these networks:

1. The network has N service stations and a total of C customers.
2. The service rate at node i, when there are n customers in that node, is $\mu_i(n)$ with $\mu_i(0) = 0$ and $\mu_i(n) > 0$ for $1 \leq n \leq C$.
3. When a customer completes service at node i, the customer joins node j with probability p_{ij}.

Queueing Theory

Let the routing matrix $P = [p_{ij}]$ be such that it is irreducible. That means it is possible to reach every node from every other node in one or more steps or hops. Define $\pi = (\pi_1, \pi_2, \ldots, \pi_N)$ such that

$$\pi = \pi P \quad \text{and} \quad \sum_{i=1}^{N} \pi_i = 1$$

Indeed, the P matrix is a stochastic matrix, which is a lot similar to the transition probability matrix of DTMCs. However, it is important to note that nothing is modeled as a DTMC.

Define the following: $\phi_i(0) = 1$ and

$$\phi_i(n) = \prod_{n=1}^{n} \left(\frac{\pi_i}{\mu_i(j)} \right) \quad n \geq 1$$

The steady-state probabilities are given by

$$P\{X_1 = x_1, X_2 = x_2, \ldots, X_N = x_N\} = G(C) \prod_{i=1}^{N} \phi_i(x_i)$$

where the normalizing constant $G(C)$ is chosen such that

$$\sum_{x_1, x_2, \ldots, x_N} P(X_1 = x_1, X_2 = x_2, \ldots, X_N = x_N) = 1$$

Note that for this problem, similar to the two previous product-form cases, the difficulty arises in computing the normalizing constant. In general it is not computationally trivial.

Some additional results can be obtained such as the *Arrival Theorem* explained below.

RESULT 9.9 *In a closed product-form queueing network, for any x, the probability that x jobs are seen at the time of arrival to node i when there are C jobs in the network is equal to the probability that there are x jobs at this node with one less job in the network (i.e., $C - 1$).*

This gives us the relationship between the arrival time probabilities and steady-state probabilities. Let $\pi_{ij}(C)$ denote the probability that in a closed-queueing network of C customers, an arriving customer into node i sees j customers ahead of him/her/it in steady state. Also, let $p_{ij}(C - 1)$ denote the probability that in a "hypothetical" closed-queueing network of $C - 1$ customers, there are j customers in node i in steady state. Result 9.9 states that

$$\pi_{ij}(C) = p_{ij}(C - 1)$$

Single-Server Closed Queueing Networks

Assume that for all i, there is a single server at node i with service rate μ_i. Then the mean performance measures can be computed without going through the computation of the normalizing constant. Define the following:

- $W_i(k)$: Average waiting time in node i when there are k customers in the network;
- $L_i(k)$: Average number in node i when there are k customers in the network;
- $\lambda(k)$: Overall throughput of the network when there are k customers in the network.

Initialize $L_i(0) = 0$ for $1 \leq i \leq N$. Then for $k = 1$ to C, iteratively compute for each i:

$$W_i(k) = \frac{1}{\mu_i}[1 + L_i(k-1)]$$

$$\lambda(k) = \frac{k}{\sum_{i=1}^{N} a_i W_i(k)}$$

$$L_i(k) = \lambda(k) W_i(k) a_i$$

The first of the above three equations comes from the Arrival Theorem (Result 9.9). The second and third come from Little's law applied to the network and a single node, respectively.

9.5.3 Algorithms for Nonproduct-Form Networks

The product form for the joint distribution enables one to analyze each node independently. When the interarrival times or service times are not exponential, then a nonproduct form emerges. We will now see how to develop approximations for these nonproduct-form networks. We present only one algorithm here, namely the diffusion approximation. However, the literature is rich with several others such as the maximum entropy method, QNA for single class, and so on. We now illustrate the diffusion approximation algorithm as described in Bolch et al. [2].

Diffusion Approximation: Open Queueing Networks

1. *Key Idea*: Substitute the discrete process $\{X_i(t), t \geq 0\}$ that counts the number in the node i, by a continuous diffusion process. Thus a product-form approximation can be obtained that works well under heavy traffic (i.e., traffic intensity in each node is above 0.95 at least).
2. *Assumptions*: Single server at each node. Service time at server i has mean $1/\mu_i$ and SCOV $C_{S_i}^2$. There is a single stream of arrivals into the network with interarrival times having a mean of $1/\lambda_i$ and SCOV C_A^2. There is a slight change of notations for the routing probabilities (consider the outside world as node 0):
   a. If $i > 0$ then p_{ij} is the probability of going from node i to node j upon service completion in node i;
   b. If $i = 0$ then p_{0j} is the probability that an external arrival joins node j;
   c. If $j = 0$ then p_{i0} is the probability of exiting the queueing network upon service completion in node i.
3. *The Algorithm*:
   a. Obtain visit ratios a_j for all $1 \leq j \leq N$ by solving

   $$a_j = \sum_{i=1}^{N} p_{ij} a_i + p_{0j}$$

   with $a_0 = 1$. Then for $1 \leq i, j \leq N$, if $P = [p_{ij}]$ an $N \times N$ matrix, then $a = [a_1, a_2, \ldots, a_N] = [p_{01}, p_{02}, \ldots, p_{0N}][I - P]^{-1}$.

b. For all $1 \leq i \leq N$, compute the following (assume $C_{S_0}^2 = C_A^2$):

$$C_{A_i}^2 = 1 + \sum_{j=0}^{N}(C_{S_j}^2 - 1)p_{ji}^2 a_j/a_i$$

$$\rho_i = \frac{\lambda a_i}{\mu_i}$$

$$\theta_i = \exp\left[\frac{-2(1-\rho_i)}{C_{A_i}^2 \rho_i + C_{S_i}^2}\right]$$

$$\phi_i(x_i) = \begin{cases} 1 - \rho_i & \text{if } x_i = 0 \\ \rho_i(1-\theta_i)\theta_i^{x_i-1} & \text{if } x_i > 0 \end{cases}$$

c. The steady-state joint probability is

$$p(x) = \prod_{i=1}^{N} \phi_i(x_i)$$

d. The mean number of customers in node i is

$$L_i = \frac{\rho_i}{1-\theta_i}$$

Diffusion Approximation: Closed Queueing Networks

All the parameters are identical to the open queueing network case. There are C customers in the closed queueing network. There are two algorithms, one for large C and the other for small C.

1. *Algorithm Bottleneck (for large C)*
   a. Obtain visit ratios a_j for all $1 \leq j \leq N$ by solving

   $$a_j = \sum_{i=1}^{N} p_{ij} a_i$$

   As there will not be a unique solution, one can normalize by $a_1 + a_2 + \cdots + a_N = 1$.
   b. Identify the bottleneck node b as the node with the largest a_i/μ_i value among all $i \in [1, N]$.
   c. Set $\rho_b = 1$. Using the relation $\rho_b = \lambda a_b/\mu_b$, obtain $\lambda = \mu_b/a_b$. Then for all $i \neq b$, obtain $\rho_i = \lambda a_i/\mu_i$.
   d. Follow the open queueing network algorithm now to obtain for all $i \neq b$, $C_{A_i}^2$, θ_i, and $\phi_i(x_i)$.

e. Then the average number of customers in node i ($i \neq b$) is

$$L_i = \frac{\rho_i}{1 - \theta_i}$$

and $L_b = C - \sum_{i \neq b} L_i$.

2. *Algorithm MVA (for small C):*

Consider MVA for product-form closed queueing networks (see Section 9.5.2). Use that analysis and iteratively compute for all $1 \leq k \leq C$, the quantities $W_i(k)$, $\lambda_i(k)$, and $L_i(k)$. Assume overall throughput $\lambda = \lambda(C)$. Then follow the open queueing network algorithm (in Section 9.5.3).

9.6 Multistation and Multiclass Queues

Consider an open queueing network with multiple classes where the customers are served according to FCFS. To obtain performance measures we use a decomposition technique. For that, we first describe the problem setting, develop some notation, and illustrate an algorithm.

9.6.1 Scenario

We first describe the setting, some of which involves underlying assumptions needed to carry out the analysis.

1. There are N service stations (nodes) in the open queueing network. The outside world is denoted by node 0 and the others $1, 2, \ldots, N$.
2. There are m_i servers at node i ($1 \leq m_i \leq \infty$), for $1 \leq i \leq N$.
3. The network has R classes of traffic and class switching is not allowed.
4. Service times of class r customers at node i are iid with mean $1/\mu_{i,r}$ and SCOV $C^2_{S_{i,r}}$.
5. The service discipline is FCFS.
6. There is infinite waiting room at each node.
7. Externally, customers of class r arrive at node i according to a general interarrival time with mean $1/\lambda_{0i,r}$ and SCOV $C^2_{A_{i,r}}$.
8. When a customer of class r completes service at node i, the customer joins the queue at node j ($j \in [0, N]$) with probability $p_{ij,r}$.

After verifying that the above scenario (and assumptions) is applicable, the next task is to obtain all the input parameters for the model described above, that is, for each $i \in [1, N]$ and $r \in [1, R]$, m_i, $1/\mu_{i,r}$, $C^2_{S_{i,r}}$, $1/\lambda_{0i,r}$, $C^2_{A_{i,r}}$, $p_{ij,r}$ (for $j \in [0, N]$).

9.6.2 Notation

Before describing the algorithm, some of the notations are in Table 9.4 for easy reference. A few of the notations are inputs to the algorithm (as described above) and others are derived in the algorithm. The algorithm is adapted from Bolch et al. [2], albeit with a different set of notations. The reader is also encouraged to refer to Bolch et al. [2] for further insights into the algorithm.

TABLE 9.4 Notation Used in Algorithm

N	Total number of nodes
Node 0	Outside world
R	Total number of classes
$\lambda_{ij,r}$	Mean arrival rate from node i to node j of class r
$\lambda_{i,r}$	Mean arrival rate to node i of class r (or mean departure rate from node i of class r)
$p_{ij,r}$	Fraction of traffic of class r that exit node i and join node j
λ_i	Mean arrival rate to node i
$\rho_{i,r}$	Utilization of node i due to customers of class r
ρ_i	Utilization of node i
μ_i	Mean service rate of node i
$C_{S_i}^2$	SCOV of service time of node i
$C_{ij,r}^2$	SCOV of time between two customers going from node i to node j
$C_{A_{i,r}}^2$	SCOV of class r interarrival times into node i
$C_{A_i}^2$	SCOV of interarrival times into node i
$C_{D_i}^2$	SCOV of inter-departure times from node i

9.6.3 Algorithm

The decomposition algorithm essentially breaks down the network into individual nodes and analyzes each node as an independent $GI/G/s$ queue with multiple classes (note that this is only FCFS and hence handling multiple classes is straightforward). For the $GI/G/s$ analysis, we require for each node and each class the mean arrival and service rates as well as the SCOV of the interarrival times and service times. The bulk of the algorithm in fact is to obtain them. There are three situations where this becomes hard: when multiple streams are merged (superposition), when traffic flows through a node (flow), and when a single stream is forked into multiple streams (splitting). For convenience, we assume that just before entering a queue, the superposition takes place, which results in one stream. Likewise, we assume that upon service completion, there is only one stream that gets split into multiple streams. There are three basic steps in the algorithm (a software developed by Kamath [10] uses the algorithm and refinements; it can be downloaded for free and used for analysis).

Step 1: Calculate the mean arrival rates, utilizations, and aggregate service rate parameters using the following:

$$\lambda_{ij,r} = \lambda_{i,r} p_{ij,r}$$

$$\lambda_{i,r} = \lambda_{0i,r} + \sum_{j=1}^{N} \lambda_{j,r} p_{ji,r}$$

$$\lambda_i = \sum_{r=1}^{R} \lambda_{i,r}$$

$$\rho_{i,r} = \frac{\lambda_{i,r}}{m_i \mu_{i,r}}$$

$$\rho_i = \sum_{r=1}^{R} \rho_{i,r} \quad \text{(condition for stability } \rho_1 < 1 \ \forall i)$$

$$\mu_i = \frac{1}{\sum_{r=1}^{R} \frac{\lambda_{i,r}}{\lambda_i} \frac{1}{m_i \mu_{i,r}}} = \frac{\lambda_i}{\rho_i}$$

$$C_{S_i}^2 = -1 + \sum_{r=1}^{R} \frac{\lambda_{i,r}}{\lambda_i} \left(\frac{\mu_i}{m_i \mu_{i,r}} \right)^2 (C_{S_{i,r}}^2 + 1)$$

Step 2: Iteratively calculate the coefficient of variation of interarrival times at each node. Initialize all $C_{ij,r} = 1$ for the iteration. Then until convergence perform i., ii., and iii. cyclically.

i. Superposition (aggregating customers from all nodes j and all classes r the SCOV of interarrival time into node i):

$$C_{A_{i,r}}^2 = \frac{1}{\lambda_{i,r}} \sum_{j=0}^{N} C_{ji,r}^2 \lambda_{j,r} p_{ji,r}$$

$$C_{A_i}^2 = \frac{1}{\lambda_i} \sum_{r=1}^{R} C_{A_{i,r}}^2 \lambda_{i,r}$$

ii. Flow (departing customers from node i have interdeparture time SCOV as a function of the arrival times, service times, and traffic intensity into node i):

$$C_{D_i}^2 = 1 + \frac{\rho_i^2 (C_{S_i}^2 - 1)}{\sqrt{m_i}} + (1 - \rho_i^2)(C_{A_i}^2 - 1)$$

iii. Splitting (computing the class-based SCOV for class r customers departing from node i and arriving at node j):

$$C_{ij,r}^2 = 1 + p_{ij,r}(C_{D_i}^2 - 1)$$

Note that the splitting formula is exact if the departure process is a renewal process. However, the superposition and flow formulae are approximations. Several researchers have provided expressions for the flow and superposition. The above is from Ward Whitt's QNA [5].

Step 3: Obtain performance measures such as mean queue length and mean waiting times using standard $GI/G/m$ queues. Treat each queue as independent. Choose α_{m_i} such that

$$\alpha_{m_i} = \begin{cases} \frac{\rho_i^{m_i} + \rho_i}{2} & \text{if } \rho_i > 0.7 \\ \rho_i^{\frac{m_i+1}{2}} & \text{if } \rho_i < 0.7 \end{cases}$$

Then the mean waiting time for class r customers in the queue (not including service) of node i is approximately

$$W_{iq} \approx \frac{\alpha_{m_i}}{\mu_i} \left(\frac{1}{1 - \rho_i} \right) \left(\frac{C_{A_i}^2 + C_{S_i}^2}{2 m_i} \right)$$

Notice that for all classes r at node i, W_{iq} is the waiting time in the queue. Other performance measures at node i and across the network can be obtained using standard relationships.

9.7 Concluding Remarks

In this chapter, we presented some of the fundamental scenarios and results for single as well as multiclass queueing systems and networks. However, this by no means does justice to the vast amount of literature available in the field, as the chapter has barely scratched the surface of queueing theory. But with this background it should be possible to read through relevant articles and books that model several other queueing systems. In a nutshell, queueing theory can be described as an analytical approach for system performance analysis. There are other approaches for system performance analysis such as simulations. It is critical to understand and appreciate situations when it is more appropriate to use queueing theory as well as situations where one is better off using simulations.

Queueing theory is more appropriate when: (a) several what-if situations need to be analyzed expeditiously, namely, what happens if the arrival rate doubles, triples, and so on; (b) insights into relationship between variables are required, namely, how is the service time related to waiting time; (c) to determine the best course of action for any set of parameters, namely, is it always better to have one queue with multiple servers than one queue for each server; (d) formulae are needed to plug into optimization routines, namely, to insert into a nonlinear program, the queue length must be written as a function to optimize service speed. Simulations, on the other hand, are more appropriate when: (a) system performance measures are required for a single set of numerical values; (b) performance of a set of given policies needs to be evaluated numerically; (c) assumptions needed for queueing models are unrealistic (which is the most popular reason for using simulations). Having said that, in practice it is not uncommon to use a simulation model to verify analytical results from queueing models or to use analytical models for special cases to verify simulations.

Another important aspect, especially for practitioners, is the tradeoff between using physics versus psychology. Queueing theory in general and this chapter in particular deals with the physics of waiting lines or queues. One should realize that the best solution is not necessarily one that uses physics of queues but maybe some psychological considerations. A classic example is a consultant who was approached by a hotel where customers were complaining about how long they waited to get to their rooms using the elevators. Instead of designing a new system with more elevators (and a huge cost thereby), the consultant simply advised placing mirrors near the elevator and inside the elevator. By doing that, although the actual time in the system does not improve, the perceived time surely does as the customers sometimes do not realize they are waiting while they busily staring at the mirrors!

From a research standpoint, there are several unsolved problems today, and a few of them are described below. For the single-class and single-station systems, issues such as long-range dependent arrivals and service, time-dependent arrival and service rates, nonidentical servers, and time-varying capacity and number of servers have received limited attention. For the multiclass and single-station queues, policies for scheduling customers, especially when some classes have heavy-tailed service times (and there are more than one servers), are being actively pursued from a research standpoint. For the single-class and multistation queues, the situation where arrival and service rates at a node depend on the states of some of the other nodes has not been explored. For the multiclass and multistation case, especially with re-entrant lines, performance analysis is being pursued for policies other than FCFS (such as preemptive and non-preemptive priority).

Acknowledgments

The author would like to thank the anonymous reviewers for their comments and suggestions that significantly improved the content and presentation of this book chapter. The author is also grateful to Prof. Ravindran for asking him to write this chapter.

References

1. D. Gross and C.M. Harris. *Fundamentals of Queueing Theory*, 3rd Ed. John Wiley & Sons, New York, 1998.
2. G. Bolch, S. Greiner, H. de Meer, and K.S. Trivedi. *Queueing Networks and Markov Chains*, 1st Ed. John Wiley & Sons, New York, 1998.
3. R.W. Wolff. *Stochastic Modeling and the Theory of Queues*. Prentice Hall, Englewood Cliffs, NJ, 1989.
4. J.A. Buzacott and J.G. Shanthikumar. *Stochastic Models of Manufacturing Systems*. Prentice-Hall, New York, 1992.
5. W. Whitt. *The Queueing Network Analyzer*. The Bell System Technical Journal, 62(9), 2779–2815, 1983.
6. V.G. Kulkarni. *Modeling and Analysis of Stochastic Systems*. Texts in Statistical Science Series. Chapman & Hall, London, 1995.
7. M. Hlynka. *Queueing Theory Page*. http://www2.uwindsor.ca/~hlynka/queue.html.
8. M. Hlynka. *List of Queueing Theory Software*. http://www2.uwindsor.ca/~hlynka/qsoft.html.
9. M. Moses, S. Seshadri, and M. Yakirevich. *HOM Software*. http://www.stern.nyu.edu/HOM.
10. M. Kamath. *Rapid Analysis of Queueing Systems Software*. http://www.okstate.edu/cocim/raqs/.

10
Inventory Control

10.1 Introduction... **10-1**
 On the Logic of Inventories • Types of Inventory • ABC Classification of Inventories
10.2 Design of Inventory Systems......................... **10-4**
 Factors in Favor of Higher Inventory Levels • Factors in Favor of Lower Inventory Levels • A Simple Classification of Inventory Management Models
10.3 Deterministic Inventory Systems.................... **10-9**
 The Economic Order Quantity Model • Uniform Demand with Replenishment Taking Place over a Period of Time • EOQ Model with Back Ordering • Time-Varying Demand: The Wagner–Whitin Model
10.4 Stochastic Inventory Systems....................... **10-20**
 Safety Stock • A Maximum–Minimum System with Safety Stock • A Simplified Solution Method for Stochastic Systems • Stochastic Inventory Model for One-Period Systems • Stochastic, Multi-Period Inventory Models: The (s, S) Policy
10.5 Inventory Control at Multiple Locations........... **10-27**
 Multi-Echelon Inventory Models: Preliminaries • Serial Supply Chain Inventory Models
10.6 Inventory Management in Practice **10-34**
 Inventory Management and Optimization Tools
10.7 Conclusions **10-35**
10.8 Current and Future Research....................... **10-36**
 Final Remarks
References **10-37**

Farhad Azadivar
University of Massachusetts Dartmouth

Atul Rangarajan
Pennsylvania State University

10.1 Introduction

Each product travels a path from where it consists of one or several raw materials and ends up in the hands of the consumer as a finished good. Along the way it goes through several manufacturing or other transformation processes. For example, consider a petroleum-based product, say gasoline, which starts as crude oil and is transported to a refinery where it is processed into its final form. Assume that the crude oil is transported to the refinery in tankers that arrive at certain intervals. Obviously, the refinery cannot work intermittently; it requires a continuous input of crude oil, forcing it to build holding spaces to store crude oil from tankers and then to use it at a continuous rate. Also, after producing gasoline the refinery cannot sell it at the exact rate at which it is produced. The refinery will need storage tanks to store its gasoline production and supply it to its users, that is, gas stations, according to demand. A little further down the path gas stations themselves cannot ask people needing 10 gallons of gasoline to wait until they can get this exact amount from the refinery. Thus, each gas station builds a storage tank that is filled by the supplies from the refinery and provides small amounts of gas to each car that comes to the station. The amount of raw material, semi-finished or finished product held at each node in the *supply chain* (SC) is referred to as the *inventory level* of that particular material at that node. Inventories are a part of our daily life; careful observation will show that we store (or hold inventories of) many items. One needs to look no further than the kitchen at home to see inventories of vegetables, milk, and the like. Running to the grocery store every time we want a glass of milk would be inefficient (and costly)!

10.1.1 On the Logic of Inventories

From the above petroleum example we see that one of the main functions of maintaining inventories is to provide for a smooth flow of product from its original raw form to its final form as a finished product in the hands of its ultimate users. However, even if all these processes could be arranged such that the flow could be kept moving smoothly without inventories, the variability involved with some of the processes would still create problems that holding inventories could resolve. For example, suppose that the crude oil could be pumped out at a rate exactly equal to the input rate to the refinery. Also assume that gasoline could be sent to gas stations as it is produced through a pipeline, and further, gas stations can supply cars at a continuous rate exactly equal to the output of the refinery. With all these unrealistic assumptions, the need for inventory still cannot be eliminated. For instance, if the pipeline carrying the crude oil is damaged as a result of an accident, customers of the disrupted gas stations will be affected almost immediately. Or, if there is a strike in the refinery, again the effect would be felt immediately. However, if inventories of crude oil and gasoline were held at the refinery and the corresponding gas stations, the effects may not be felt immediately or ever, as there would be enough supply to the end users while the pipelines are being fixed and the production gap could be filled by overtime production. In general, there are three main reasons to hold inventory:

- *Economies of scale*—Usually, placing an order has a certain cost that is independent of the quantities ordered. Thus, more frequent orders incur higher costs of ordering paper work. This may even cause higher transportation costs because the cost of transportation per unit is often smaller for larger orders. Economies of scale also play an important role in those cases where buying in large quantities results in a reduction of the unit price.
- *Uncertainties*—As mentioned above, as a product is converted from raw material to final product, variabilities in these transformational processes lead to losses. Inventories help mitigate the negative impact of uncertainties.
- *Customer service levels*—While variabilities in demand and supply are inevitable, inventories help buffer against these variations and ensure product availability. Consequently, delays in satisfying customer demand are reduced (when out of stock) and unnecessary loss of revenue is avoided while improving customer satisfaction.

There are a few reasons why *minimal* inventories should be held. First, inventories represent significant investment in capital and resources that may be employed to better purposes. Second, many products have limited life cycles after which they become obsolete. When the product is perishable (e.g., fresh food items), excess inventories may have to be thrown out after expiry. Even if the product is not perishable, markdowns may be required to dispose of excess inventory, leading to lower revenues or losses. Finally, inventories may hide inefficiencies in the system. Reduction of inventories coupled with better production processes form the basis of just-in-time systems (see Silver et al. [1, Chapter 16] for details on such systems).

Virtually all businesses hold inventory and inventory management remains one of the biggest business challenges. Many consider it to be a necessary evil; it helps ensure product availability for customers but consumes valuable resources to procure and maintain. The magnitude of inventory-related operations is staggering: in 1989, the total value of inventories held in the United States alone was close to $1 trillion [2] and had increased to about $1.1 trillion in total inventories held by private industry alone in the United States in 1999 [3]. Furthermore, new business models have emerged with inventory management as the strategic differentiator. Companies like Dell and Amazon.com have used supply chain management and inventory management to achieve great success. Between 1995 and 2005, Toyota carried 30 fewer days of inventory than General Motors, giving it immense advantage [4]. It is evident that these problems merit close attention. Inventory management problems are often extremely complex and mathematical models capture only a limited part of this complexity. Mathematical models from the fields of management science and operations research have emerged as strong contenders to help gain insight and provide scientific methods to address inventory management problems. Recent advances in computing and information technologies have enabled the development, solution, and implementation of many advanced models that are used in industry today.

Fundamentally, these inventory management models address two critical issues: *how often* should inventory be replenished and by *how much* each time? Models gain complexity as additional factors are incorporated, more constraints are added to the problem, and assumptions are made less rigid and more realistic. Research into mathematical inventory models may be traced back to more than nine decades but began to receive sustained interest from the 1950s onwards. While it would be impossible to comprehensively cover this vast body of literature, this chapter introduces the reader to the basic issues and models in inventory management. Consequently, the expository style used here is to bring out the main intuition and results and not toward theoretical rigor. The interested reader is referred to Axsäter [5] and Silver et al. [1] for excellent coverage of inventory management; see also Bramel and Simchi-Levi [6], Graves et al. [2], Hadley and Whitin [7], Porteus [8], and Zipkin [3] for more (mathematically) advanced treatments.

10.1.2 Types of Inventory

For production systems there is a need for basically three types of inventories. These are inventories of raw material, finished products, and work-in-process (wip). Any of these inventory systems could exist at any stage of the path of a product from its origin until it reaches the hands of the final users. Note that if various stages of production are performed at different locations, or sometimes even by different companies, work finished in one location could be considered the final product for that location and raw material for the next location downstream.

Inventory of raw materials: The need for keeping an inventory of raw materials arises because of several factors. As mentioned in the above example, it is often not possible or economical to procure raw materials exactly when the need for them arises. The cost factors involved include the cost of paperwork, transportation costs, economies of scale, and unforeseeable events. Unpredictable events such as international shortage of certain goods, a strike at the supplying plant, a dry season, and the like, may interrupt the production process if a sufficient inventory of raw materials is not on hand.

Inventory of work-in-process: This category of materials refers to semi-finished goods. These are the outputs of one stage of production that act as the input to the following stage. The uncertainties involved in this case are less than the raw material category but never nonexistent. The uncertainties are less because the capacities of each stage are known and controllable. Thus, the designer can match the capacities. However, uncertainties are still there because of problems in scheduling, unforeseeable breakdown of machinery, and so on.

Inventory of the finished product: The inventory of the final product involves even more variability than is experienced with raw material. The demand for the final product is often uncertain, and the manufacturer has little control over it. Another factor, which is important when considering final product inventories, is the cost involved in producing various sized lots. This sometimes favors production of larger than demanded lot sizes because of potential savings in the unit costs. All of these factors will be discussed in exploring inventory models.

Another important classification is to divide inventories into *cycle stocks, pipeline inventories*, and *safety stocks*. The amount of inventory on hand when ordering in batches is referred to as cycle stock. It is used to meet normal demand patterns. On the other hand, safety stocks refer to the average inventory on hand to buffer against uncertainties in supply and demand. Inventories (whether raw materials, wip, or finished products) spend quite some time in transit between various stages of the SC (say from factory to warehouse, warehouse to retailer, etc.) and are referred to as pipeline inventories. Such inventories are not physically accounted for though they have been ordered and shipped. Depending on the situation, either the buyer or the supplier must pay for these pipeline inventories and therefore they are an important type of inventory. Pipeline inventory costs are usually an integral part of inventory models that account for transportation costs.

10.1.3 ABC Classification of Inventories

Most firms manage inventories of more than one product (commonly known as stock keeping unit or SKU). Furthermore, these items may differ in their costs, demands, criticality, and so on. Management would like to monitor its most critical and high value items closely and expend lesser effort on other items. The ABC classification is a simple methodology to aid in this process. It has been observed in several large, multiple-item inventory systems, that a small percentage of items represent a significant portion of inventory investment. ABC analysis involves computing the annual value of items to the firm as follows: the product of the price and annual demand of each item is computed for all items. These are then ranked in descending order and the cumulative percentages of annual dollar values and number of SKUs are computed and plotted. An example of this is given in Figure 10.1a. The salient characteristics of each class are given in Figure 10.1b. While the procedure involves annual dollar value, we note that critical parts may be moved to class A or B irrespective of monetary value to avoid disruptions. As the adage goes, "for want of a horseshoe, the battle was lost." See Silver et al. [1, Chapter 3] for a detailed discussion on frameworks for inventory management.

10.2 Design of Inventory Systems

The discussion so far has indicated a need for holding inventories of raw material, wip, and final product to take advantage of economies of scale and to counter the effects of unexpected events. The design of inventory systems refers to the process of determining the amount to

Inventory Control

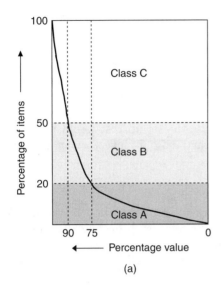

1. **Class A items:**
   - Few items ($\approx$ 20%) deserving close attention
   - Represent large investments ($\approx$ 80%)
   - May include critical components
   - Significant management by exception
2. **Class B items:**
   - Large number of items ($\approx$ 30%) deserving moderate attention (less critical items)
   - Represent moderate investments ($\approx$ 10%–20%)
   - Automated inventory management; occasional management by exception
3. **Class C items:**
   - Large number of items ($\approx$ 50%) deserving minimal attention
   - Represent moderate investments ($\approx$ 5%–10%)
   - Simple inventory management system

FIGURE 10.1 ABC analysis.

keep in inventory and the procedures and frequencies of replenishing consumed quantities, that is, answering the two fundamental questions of how often and how much to order.

A major criterion for deciding the size of an inventory is the cost of holding the inventory. One cannot increase the size of an inventory indefinitely to eliminate all disruptions of the production process, because the actual cost of keeping the goods in the inventory might exceed the potential loss due to the shortage of goods and materials. On the other hand, if lower inventory levels are maintained to reduce inventory holding costs, costs will go up due to more frequent ordering as well as losses that may result because of potential disruption of the production process. Thus, designing an inventory system requires finding an arrangement that is a compromise between the costs favoring a high level of inventory and those in favor of low levels. This, in fact, requires a system that minimizes the total costs. Some of the relevant costs are discussed below.

10.2.1 Factors in Favor of Higher Inventory Levels

Ordering Costs and Set-up Costs

When materials are purchased, they are ordered from a supplier. Ordering requires a series of actions with associated costs such as market research, bidding, and the like. Usually these costs are independent of the amount ordered. Sometimes even the transportation cost is independent of the size of the order. For example, if a company sends its own truck to pick up the purchased materials the truck may have to make the trip whether the quantity ordered is 1 ton or 10 tons. The ordering cost in this case would be the cost of making one trip, independent of the amount ordered. In the analysis provided here, any cost associated with ordering that is independent of the volume of the order is referred to as an ordering cost.

A fixed ordering cost implies that larger orders require fewer orders to be placed during a given period, thus incurring a lower total ordering cost for that period. For example, assume that a company requires 12,000 electric motors each year and there is a fixed ordering cost of \$60/order. If the company orders all the units in a single lot, it will order only once a year at an ordering cost of \$60/year. If the company orders 3000 units at a time, then it has to order $12{,}000/3000 = 4$ times a year at an annual ordering cost of $\$60 \times 4 = \$240$. Thus, fixed ordering cost favors higher quantities per order with a lower frequency of ordering.

For wip parts and final products, where the units are produced rather than purchased, the ordering cost is replaced by set-up costs. Set-up costs refer to the costs associated with changing the existing setup of the machinery and production capacity to the setup required for the next process. This involves stopping the work for a while, spending time and money to change the arrangement of machinery and schedule, and a temporary loss of efficiency due to the transition time required for the operators to get used to the new arrangements. All these costs can be evaluated and combined into a fixed set-up cost. Set-up costs are also independent of the number of units produced because the set-up cost would be the same for the production of either one unit or several thousand units. In inventory analysis set-up costs are dealt with in exactly the same manner as ordering costs.

Quantity Discounts

Sometimes ordering or producing in large quantities yields a reduction in the unit cost of purchased or produced items. Most suppliers will consider providing discounts for a customer who buys in large quantities. Besides that, sometimes one can save by ordering large shipments from a major supplier who normally does not sell in small quantities. The same is true in the case of producing a product. Producing larger quantities may make possible the use of more efficient equipment, thus reducing the cost per unit. Because of this, potential quantity discounts favor buying in larger volumes that in turn result in higher levels of inventory. Silver et al. [1, Chapter 5] cover basic quantity discount models; see also Munson and Rosenblatt [9].

Unreliable Sources of Supply

When supply sources are unreliable, management may decide to stock up its inventories to avoid the losses that could result from being out of stock. This also provides motivation to increase inventory levels.

10.2.2 Factors in Favor of Lower Inventory Levels

Inventory Holding Cost

Keeping the physical items in the inventory has certain costs. Some of these costs are

- The interest on the capital invested in the units retained.
- The cost of operating the physical warehousing facility where the items are held. These are costs such as depreciation of the building, insurance, utilities, record-keeping, and the like.
- The cost of obsolescence and spoilage. The item that becomes obsolete as a result of newer products in the market will lose some or all of its value. This loss in value is considered as a cost; so are the losses in values of perishable items such as foods or some chemicals and published items such as newspapers and magazines.
- Taxes that are based on the inventories on hand.

These costs, unlike ordering and set-up costs, are dependent upon the size of the inventory levels. The interest cost on the investment for each item can be easily found. Inventory holding cost per item is not as simple to estimate. This cost is usually estimated based on the average size of the storage place, which is usually proportional to the average size of the inventory being held. An average per unit cost of obsolescence and spoilage is even harder to evaluate, but it can be evaluated roughly as the cost of the average number of units lost divided by the average number of units in the inventory. The average figure is used here because the number of items that could be lost cannot be predicted with certainty.

10.2.3 A Simple Classification of Inventory Management Models

Consider the operation of a simplified inventory system as shown in Figure 10.2. An order of size Q units, called the *order quantity*, is placed with the supplier. The supplier may be either an upstream production stage or an external stocking/production location who ships this quantity (if available). This order is received after some (shipping) time delay. The time between when an order is placed and when it is received is known as the *lead time* and we denote it by L. The resulting inventory is used to satisfy customer demand (denoted by D).

Several variations in the supply and demand processes as well as the nature of the inventory we are trying to manage result in several possible models as seen in the literature on inventory management. Inventory models may be classified based on the following criteria (Table 10.1).

1. *Stocking location*: Often suppliers and customer locations are not explicitly considered; instead we look to control inventory at a single stocking location (*single location models*). When more than one stocking location is considered, such models are referred to as *multi-echelon inventory models* or SC inventory models in the literature.

2. *Control*: In multiechelon models, the multiple locations may be under the control of a single entity (allowing coordinated decisions) and are referred to as *centralized* SC inventory models [10]. Conversely, if multiple entities make inventory decisions independently, we refer to these as *decentralized* SC inventory models (see Refs. [11,14]). Centralized control models are usually harder to optimize while decentralized models pose challenges in coordination of the SC locations.

FIGURE 10.2 Basic inventory system flows.

TABLE 10.1 A Classification of Inventory Models

Factors			Single Location	Multi-Echelon
Control	Centralized			
	Decentralized			
Supply process	Deterministic	Zero/Fixed LT		
		OQ Based LT		
	Stochastic			
Demand process	Deterministic			
	Stochastic	Stationary		
		Non-Stationary		
		Unknown		
Capacity	Unlimited			
	Flexible			
	Fixed			
No. of items	Single Product			
	Multiple Products			
Sourcing options	Single Source			
	Multiple			
	Lateral Shipments			

3. *Supply process lead times*: Lead times are variable in reality. However, a common simplification is to assume that lead times are known and constant, that is, *deterministic*. In fact, lead times are often assumed to be zero; when $L > 0$ and deterministic, the model is usually analogous to the former case. Alternatively, the lead time is assumed to follow a known distribution (*stochastic lead time*).

4. *Demand process*: Similar to supply processes, customer demand (simply demand hereafter) may be *deterministic* and known or *stochastic* with known probability distribution. The demand pattern (i.e., the underlying stochastic process) may be time invariant or *stationary* with all demand parameters being constant over time. When the parameters of the process (say the mean, standard deviation, etc., of the distribution) change over time, the demand process is said to be *nonstationary*. Nonstationary demands represent a challenging class of problems; see Swaminathan and Tayur [15] for a review. Finally, there are many instances when complete details about the demand distribution are unknown (e.g., new product introductions, promotion periods, etc.). At best the estimates of the mean and variance of demand may be known; see Gallego et al. [16], Godfrey and Powell [17], and Moon and Gallego [18] for details.

5. *Capacities*:* Inventory models often assume that capacity restrictions may be ignored at every stocking location. This assumption may be relaxed to reflect capacity limits. Sometimes, capacity may be moderately flexible (e.g., outsourcing to subcontractors, using flexible manufacturing systems, etc.).

6. *Number of items*: Inventory models may look to manage the inventories of a *single product* or *multiple products* simultaneously. If demand for these multiple items is independent of the others and capacity restrictions do not exist, then the problem decomposes to several single-product problems. When capacity constraints or demand correlations exist, the problem becomes more complex. Bramel and Simchi-Levi [6] and Swaminathan and Tayur [15] provide references for the multiproduct case.

7. *Sourcing options*: Figure 10.2 shows a single upstream supplier. *Single sourcing* involves the use of one supplier for each component. When more than one supplier is used, the strategy is referred to as *multiple sourcing*. *Lateral shipments* refer to scenarios where some locations are allowed to become suppliers (temporarily) to other locations at the same stage in the SC. For example, a retailer may send an emergency shipment to another retail location to meet a surge in demand. We consider only single sourcing in this chapter; Minner [20] presents a recent review of multiple sourcing models.

Inventory Control Models

The objective of analyzing inventory systems is to design a system that is a compromise between factors favoring high levels of inventory and those in favor of low levels. This compromise is intended to result in the minimum cost of maintaining an inventory system. The cost of maintaining an inventory system is usually evaluated in terms of cost per unit time where the unit time can be any length of time, usually a year.

---

*The interplay between capacities, lead times, and inventory decisions is little researched but important. Karmarkar [19] remains an important reference on these issues.

Inventory Control

Inventory models are mathematical models of real systems. To evaluate inventory systems mathematically, these models make several assumptions that are sometimes restrictive. Generally, as the assumptions are relaxed, the mathematical analyses become more difficult. In this chapter some of the existing inventory models are discussed. First, it is assumed that no uncertainty exists and deterministic models are presented. Then some limited stochastic models will be presented.

10.3 Deterministic Inventory Systems

In deterministic inventory models it is assumed that the demand is fixed and known. Also it is assumed that when a lot is ordered, it arrives exactly at the time it is expected. Further, when the demand pattern remains constant over time, the inventory pattern is known exactly. Consequently, parameters like the maximum and minimum inventory levels are known exactly. Thus, these models may also be referred to as *maximum–minimum* inventory models.

10.3.1 The Economic Order Quantity Model

The economic order quantity model represents the firm inventory control model due to Harris [21]. This model can be applied to both raw material and finished goods. In the case of raw material inventory, the demand is represented by the amount used by the production process over a given period. The ordered quantities in this case are the amount ordered from the supplier. In the case of finished goods, the demand is the amount shipped to the retailer or consumer over a given period of time, and the orders are those sent to suppliers requesting new lots to be supplied. In this model the whole inventory process can be shown graphically. In a graph the amount of goods or products in the inventory is plotted against time. Let us explain this situation through an example.

Example 10.1

Consolidated Appliance Company maintains an inventory of electric motors that are used in the production of refrigerators. Figure 10.3a shows the graphical representation of the level of inventory as a function of time (with the y-axis being in 100 units scale). At the start there are 3000 motors on hand. The company uses 1000 motors a day; thus at the end of day 1, there will remain only 2000 units in the inventory. This situation is shown by point 1 with coordinates of 1 and 2000, which corresponds to units in the inventory at the end of day 1. At the end of day 2 there are 1000 units left, which is shown by point 2 with coordinates of 2 and 1000. At the end of day 3, the number of motors on hand should be zero, but the arrival of a new shipment of motors, which contains 7000 units, increases the inventory level to 7000 (point 3 in Figure 10.3a). In this way, the inventory level at the end of each day can be represented by a point. ∎

The points in Figure 10.3a show the number of units at the end of each period. However, if one wants to demonstrate a real variation of the level of inventory by time, one can observe the level at infinitesimal time differentials. This results in a curve rather than a series of points. As nothing is said about the pattern by which 1000 units are drawn from the inventory each day, the curve between two consecutive ends of the day points can have any shape. In the present model, however, it is assumed that the shape of the curve between the points is linear. Thus the demonstration in Figure 10.3a can be changed to the one shown in Figure 10.3b. From here on this linear graphical representation will be adopted for all

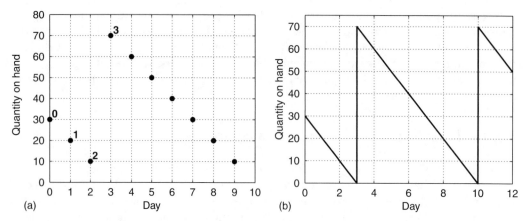

FIGURE 10.3 Example 1: (a) Inventory graph. (b) Linear representation of inventory system.

the deterministic cases. There are several concepts in the design of inventory systems that now can be explained using Figure 10.3b.

1. The system may be called a *maximum–minimum system* because the number of units in the system varies between a fixed maximum and a fixed minimum.
2. The minimum in this model corresponds to zero. As the decision maker is sure that what has been ordered will arrive exactly at the time expected, he will have nothing to worry about if he is out of stock for just a moment. However, as we will see later, this is not the case where uncertainties are involved in the time of arrival of the new shipment or the quantities of the demand.
3. As is seen in Figure 10.3b, it is assumed that everything ordered arrives at one point in time, thus instantly raising the level of inventory at days 3 and 10. Furthermore, it is clear that the maximum inventory in this case is equal to the amount ordered.

Formulating the Model

To design such an inventory system it is necessary to find several quantitative parameters with which the system can be uniquely defined. The values of these parameters should be such that they minimize the total cost of operating the inventory system. From Figure 10.3b it is clear that this system can be defined by three parameters: demand, the maximum level of inventory, and the minimum level of inventory. It is assumed that the demand is fixed and the designer has no control on this parameter. Also, as mentioned above, for this model the minimum inventory level is always zero. Thus, the only parameter to be determined in this case is the maximum inventory level. On the other hand, it was also mentioned that the maximum inventory level for this model is the quantity ordered. Then the design of the system can be defined as finding the ordering quantity that minimizes the total cost of maintaining the system. This best (optimal) order quantity is called the *economic order quantity* (EOQ).

To find the optimum values of parameters of an inventory system the system is formulated as an optimization problem according to the following procedure. First, a suitable period of analysis is selected. This period is usually 1 year. Then the total cost of maintaining the inventory over this period is evaluated as a function of the controllable parameters of the system. This function is then minimized by differentiation or any other applicable mathematical technique to obtain the optimum values for the controllable variables of the system. To formulate the EOQ model the following notation is utilized.

Inventory Control

Q = Order quantity
Q^* = Economic order quantity
D = Demand per time-period
B = Ordering cost, \$/order
E = Inventory holding costs, \$/unit/time-period
U = Unit cost, \$/unit
TC = Total cost per time-period

As was mentioned, the only controllable variable in this system is the order quantity. Thus the total cost must be stated in terms of this variable. The total cost for this system consists of two distinct types of expenses. These are inventory holding costs and ordering or set-up costs. When the unit price of items is affected by the quantity ordered (as in the case where quantity discounts are involved), the cost of units is also added to the total cost. In cases where the unit price is assumed constant, inclusion of this cost has no effect on the controllable variables of the system. In these situations, like the present model, this cost is not considered.

Inventory holding costs are always present and depend on the number of units in stock and the length of the time they are kept. In the electric motor example, suppose that keeping one motor in inventory for 1 day costs one cent. According to our notation $E = \$0.01/\text{unit}/\text{day}$. This means that if 1000 units are kept in the inventory, the company bears a cost of $1000 \times 0.01 = \$10/\text{day}$. If the amount of material kept in the inventory is constant over time, say 1000 units, one can say that the contribution of the holding cost to the total cost is \$10/day, but as the level of inventory varies from one day to another, so does the inventory holding cost. From Figure 10.3b, one can see that the inventory levels vary from 0 to 7000 units with the resulting holding cost varying from 0 to $7000 \times 0.01 = \$70/\text{day}$.

To find the average daily inventory cost, one may consider finding the inventory holding cost for each day and calculating their average. Another method to estimate the same value is to find the average number of units kept in the inventory and multiply that by the holding cost. As most of the time costs are calculated on an annual basis, it would be quite difficult to evaluate the holding cost for each of the 365 days in a year. As we will see in the following paragraph, evaluation of the average inventory and multiplying it by the annual inventory holding cost/unit is much easier. As we assume variations in inventory levels are linear, it is easy to see that the average inventory is equal to the average of the maximum and the minimum of these levels. Thus,

$$\text{Average Inventory} = \frac{\text{Maximum Inventory} + \text{Minimum Inventory}}{2} \quad (10.1)$$

However, we know that the minimum for our system is zero, and the maximum is just the order quantity. Then

$$\text{Average Inventory} = \frac{Q+0}{2} = \frac{Q}{2} \quad (10.2)$$

The average holding cost per unit time would be

$$\text{Average holding cost per unit time} = E\left(\frac{Q}{2}\right) = \frac{EQ}{2} \quad (10.3)$$

Now refer to the second part of inventory costs, which is ordering or set-up cost. This cost occurs whenever an order is placed, but this cannot be considered as the cost per day or per month. The best way to evaluate the contribution of this cost to the inventory cost for the evaluation period is to find how many times the order is placed in one evaluation period. For instance, if the costs are evaluated per year and four times a year an order is placed at a cost of $200, then the ordering cost per year would be $4 \times \$200 = \$800$. For our electric motor problem, if we assume a cost of $125/order, the annual ordering cost can be calculated as

$$\text{No. of orders per year} = \frac{\text{Annual Demand}}{\text{Quantity ordered each time}} \tag{10.4}$$

$$= \frac{365 \times 1000}{7000} = 52.14$$

$$\Rightarrow \text{Annual ordering costs} = \$125 \times 52.14 = \$6518$$

Similarly, if the analysis period is 1 day, then the daily cost of ordering would be based on ordering once a week. Thus,

$$\text{No. of orders per day} = \frac{1}{7}$$

$$\text{Daily ordering costs} = \$125 \times \frac{1}{7} = \$17.86$$

Now let us determine the frequency and the cost of ordering per period for a general model in terms of our only controllable variable involved, the order quantity. As the items are drawn from the inventory at a rate of D units per period, each order quantity lasts for Q/D units of time. The number of orders placed during one period is

$$\text{No. of orders per period} = \frac{1}{Q/D} = \frac{D}{Q} \tag{10.5}$$

Thus, paying $\$B$ per order, we have

$$\text{Ordering costs per period} = \frac{BD}{Q} \tag{10.6}$$

Now, returning to the total cost per period,

Total cost per period, TC = Inventory holding cost per period + Ordering cost per period

$$= \frac{EQ}{2} + \frac{BD}{Q} \tag{10.7}$$

To show that the addition of the acquisition cost of the units involved has no effect in optimizing the system, we add the cost of units to this total cost. Then,

TC = Inventory holding cost/period + Ordering cost/period + Cost of units/period

$$= \frac{EQ}{2} + \frac{BD}{Q} + UD \tag{10.8}$$

After stating the total cost as a function of the ordering quantity, we proceed to minimize the total cost. In this case, since there is only one variable involved, simple differentiation yields the result.

$$\frac{d(TC)}{dQ} = \frac{E}{2} - \frac{BD}{Q^2} = 0$$

$$\Rightarrow \text{Economic Order Quantity}, Q^* = \sqrt{\frac{2BD}{E}} \tag{10.9}$$

Inventory Control

As the above relationship shows, the cost of units involved does not play any role in this optimum value. Thus, for this example, the economic order quantity can be calculated as

$$Q^* = \sqrt{\frac{2 \times 125 \times 1000}{0.01}} = 5000 \text{ units}$$

Lead Time and Reorder Point

There is another parameter that is important in designing inventory systems. Usually orders are not received at the time they are placed. In inventory systems terminology, the length of time between when orders are placed and the time when they are received is called the *lead time* (denoted by L). In deterministic inventory models, it is assumed that the lead time is constant and known. In stochastic inventory systems, the lead time could be a random variable.

Going back to the design stage, we note that orders are not received immediately after they are placed. If one waits until the inventory level reaches zero and then orders, the inventory will be out of stock during the lead time. Thus orders have to be placed before the inventory level has reached zero. Two methods can be used to determine when an order should be placed. In the first method the time at which the inventory will reach zero is estimated and the order is placed a number of periods equal to the lead time earlier than that estimated time. For example, if the lead time is 4 days, one could estimate the time at which the inventory will reach the zero level and order 4 days before that time.

The second approach, which is used more often, is based on the level of inventory. In this approach, the order is placed whenever the inventory level reaches a level called the *reorder point* (ROP). This means that if the order is placed when the amount left in the inventory is equal to the reorder point, the inventory on hand will last until the new order arrives. Thus the reorder point is that quantity sufficient to supply the demand during the lead time. Now, if we assume that both the lead time and the demand are constant, the demand during the lead time is constant too. As the reorder point is equal to this demand, it can be calculated from the following relation:

$$ROP = L \times D \tag{10.10}$$

It is very important to keep in mind that the reorder point is not a point in time, but a quantity equal to the level of the inventory at the time the order is to be placed. Figure 10.4

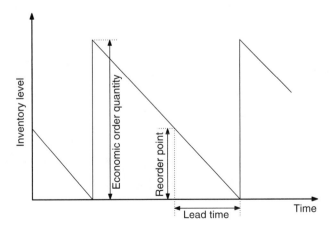

FIGURE 10.4 Graphical representation of economic order quantity, lead time, and reorder point.

shows the values of economic order quantity, lead time, and reorder point graphically for a simple maximum–minimum system.

10.3.2 Uniform Demand with Replenishment Taking Place over a Period of Time

In this model, the assumption that all of the ordered units arrive at the same time is relaxed, and is replaced by the assumption that units arrive at a uniform rate of P over a period of time. This model is especially useful for situations where the units are produced inside the plant and are stored to supply a uniform demand. Consequently, the model is often referred to as the *economic production quantity* (EPQ) model.

The first requirement to build up a stock is that the supply rate P be higher than the demand D during the production or supply period. During the supply period the inventory is replenished at a rate of P units per period and consumed at a rate of D units per period. Thus, in this period the inventory builds up at a rate of $(P - D)$ units per period. The graphical representation of this system is shown in Figure 10.5. The design parameter for this system is again the order quantity, but here the maximum inventory level and the order quantity are not the same. The reason is that it takes some time to receive all units of ordered quantity and during this time some of the units are consumed. Another way of defining this system is to specify the supply period instead of the order quantity. Knowing the supply rate and the supply period, the order quantity could be determined. In this chapter, the order quantity is specified as a design parameter.

Formulating the Model

This problem can be formulated very similarly to the basic model. The only difference here is that the average inventory level is not half of the order quantity. As the system is still linear and minimum inventory is zero, the average inventory is:

$$\text{Average Inventory} = \frac{\text{Maximum Inventory} + \text{Minimum Inventory}}{2} = \frac{\text{Maximum Inventory}}{2} \quad (10.11)$$

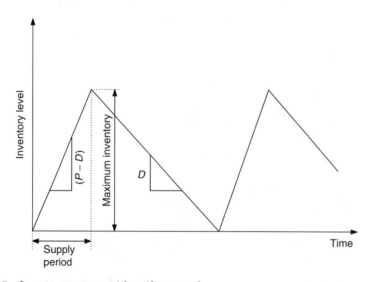

FIGURE 10.5 Inventory systems with uniform supply.

Inventory Control

Then if we can state the maximum inventory in terms of order quantity, the average cost of the inventory system can be found as a function of order quantity. The calculation for maximum inventory is done as follows. Let ST represent the length of the supply period. As the order quantity is Q and supply rate is P, then

$$ST = \frac{Q}{P}$$

During this supply period the inventory level is increased by $(P-D)$ units per unit time. As it is assumed that the supply starts arriving when the inventory level is zero, then the maximum inventory level is equal to the buildup during this period. Then,

$$\text{Maximum inventory, } I_{\max} = (P-D)ST = (P-D)\frac{Q}{P}$$

$$= Q\left(1 - \frac{D}{P}\right) \tag{10.12}$$

The average inventory (I_{avg}) is then given from Equations 10.11 and 10.12 as

$$I_{avg} = \frac{I_{\max}}{2} = \frac{Q}{2}\left(1 - \frac{D}{P}\right)$$

As a result, the inventory holding cost is $\dfrac{EQ}{2}\left(1 - \dfrac{D}{P}\right)$. The ordering cost is the same as before. Thus,

$$TC = \frac{EQ}{2}\left(1 - \frac{D}{P}\right) + \frac{BD}{Q} \tag{10.13}$$

As before, differentiating Equation 10.13 and equating to zero, we solve for the optimal Q, which is given as

$$Q^* = \sqrt{\frac{2BD}{E\left(1 - \frac{D}{P}\right)}} \tag{10.14}$$

Notes: In using Equation 10.14 it is important to note the following:

- If P is less than D, the term $(1 - D/P)$ becomes negative. This would result in a Q^* which is the square root of a negative value. This is due to insufficiency of the supply. As mentioned earlier, to build up an inventory the production rate must be higher than the consumption rate.
- The number of units produced and the number of units consumed over a long period should not be confused with the production rate and consumption rate. The production rate is higher, but production does not take place every day. To avoid problems of this nature, it is suggested that the consumption rate always be defined in terms of the period over which the production rate is defined. For example, if the production rate is given in terms of units per day and consumption in terms of units per month, the consumption rate should be converted into units per day.

10.3.3 EOQ Model with Back Ordering

In this model running out of stock is not forbidden, but a cost associated with back ordering of each unit is taken into account. The characteristics of this system can be explained using the graphical representation of this model as given in Figure 10.6. When the inventory level reaches zero, at point O, demand for units continues. When the demands cannot be met, they are recorded as back orders, meaning that the customer has to wait until the new shipment arrives. Back ordered units could be considered as negative inventory. When the new shipment arrives, the back orders are satisfied first. After supplying all back ordered units (which is assumed to take place instantly), the remainder of the units in the shipment raises the inventory level to the maximum of $I_{\max}$ represented by point M in Figure 10.6. There are two important points to be noted here. First, it is assumed that the new shipment is ordered such that when it arrives the back orders have reached their maximum of S. This changes the relation for the reorder point to

$$ROP = DL - S \tag{10.15}$$

For models where back orders are not allowed (as in the previous two models), $S=0$ and $ROP = DL$. The second point is that if the customer has to wait for his order, he may have to be compensated by some means, such as a discount. These costs, which can be considered as the penalty for late delivery, are represented by F and are stated as *back ordering cost* per unit per unit time.

Formulating the Model

To design this system one controllable variable is no longer sufficient to define it. Three variables can be defined for the system. These are the maximum inventory ($I_{\max}$), the

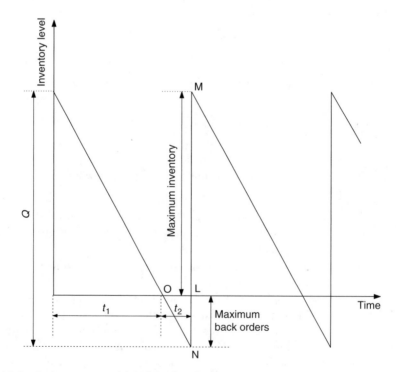

FIGURE 10.6 An inventory model with back ordering.

Inventory Control

maximum back orders (S), and the order quantity (Q). From Figure 10.6, see that

$$Q = LM + LN = I_{\max} + S \tag{10.16}$$

This indicates that any two variables out of the above three would be sufficient to define the system. The third variable can be obtained from the other two. Here, two variables, Q and S, are selected as the design variables of the system. Then, the maximum inventory can be obtained from Equation 10.16 as $I_{\max} = (Q - S)$.

As was shown earlier with a zero minimum inventory, the average inventory can be considered to be one-half of the maximum inventory. However, in this case it is important to note that at some period of time during each inventory cycle the level of inventory is zero. This period in Figure 10.6 is shown by t_2 as the time segment between points O and L. So,

Holding cost per unit time during interval $t_2 = 0$

Holding cost per unit time during interval $t_1 = E\dfrac{(Q-S)}{2}$

$$\text{Average holding cost/unit time} = \dfrac{\dfrac{E(Q-S)}{2}(t_1) + (0 \times t_2)}{(t_1 + t_2)}$$

$$= \dfrac{t_1}{t}\left[\dfrac{E(Q-S)}{2}\right] \tag{10.17}$$

where $t = t_1 + t_2$. The values of t_1 and t_2 can be found as follows.

t_2 is the time during which the back orders are built up to S at a rate of D units per unit time. Thus, $t_2 = S/D$. Further, $t = t_1 + t_2$ is the period during which Q units are spent at a rate of D units per unit time. Then $t = Q/D$ and $t_1 = t - t_2 = (Q-S)/D$. Then,

$$\text{Average holding cost/unit time} = \dfrac{(Q-S)/D}{Q/D}\left[\dfrac{E(Q-S)}{2}\right]$$

$$= \dfrac{E(Q-S)^2}{2Q}$$

The ordering cost does not change from its previous value, as the same ordering pattern continues and ordering cost/unit time $= BD/Q$.

For this model we have to add an additional cost that is called the *back ordering cost* and is represented by F, the cost per unit back ordered for a unit time. The number of units back ordered varies from 0 to S with an average of $S/2$, but again this occurs only during the time interval t_2. Then

Back ordering cost per unit time during $t_1 = 0$

Back ordering cost per unit time during $t_2 = F\dfrac{S}{2}$

Therefore,

$$\text{Average back ordering cost/unit time} = \dfrac{\left[t_1 \times 0 + t_2 \dfrac{FS}{2}\right]}{t}$$

$$= \dfrac{FS^2}{2Q}$$

Thus, the total inventory cost per unit time is

$$TC = \text{Holding cost/unit time} + \text{Ordering cost/unit time} + \text{Back ordering cost/unit time}$$
$$= \frac{E(Q-S)^2}{2Q} + \frac{BD}{Q} + \frac{FS^2}{2Q} \qquad (10.18)$$

There are two variables of Q and S to be determined. By partial differentiation with respect to Q and S and equating the results to zero, the values of Q and S can be found as follows.

$$\frac{\partial(TC)}{\partial Q} = \frac{E}{2}\left[\frac{2Q(Q-S) - (Q-S)^2}{Q^2}\right] - \frac{BD}{Q^2} - \frac{FS^2}{2Q^2} = 0$$

$$\frac{\partial(TC)}{\partial S} = \frac{E}{2}\left[\frac{-2(Q-S)}{Q}\right] + \frac{FS}{Q} = 0$$

Solving these two equations simultaneously yields the optimum values of Q and S as

$$Q^* = \sqrt{\frac{2BD}{E}} \times \sqrt{\frac{E+F}{F}} \qquad (10.19)$$

$$S^* = \sqrt{\frac{2BD}{E}} \times \sqrt{\frac{E^2}{F(E+F)}} \qquad (10.20)$$

Now that we have two parameters and we know that $I_{\max} = Q - S$, the optimal value for $I_{\max}$ may be shown to be

$$I_{\max}^* = \sqrt{\frac{2BD}{E}} \times \sqrt{\frac{F}{E+F}} \qquad (10.21)$$

10.3.4 Time-Varying Demand: The Wagner–Whitin Model

Thus far, we have assumed that demand is deterministic and constant over time (i.e., time-invariant or stationary). In this section, we relax this assumption of stationarity; the model is due to Wagner and Whitin [22]. The following assumptions are made.

1. Demand is deterministically known for the following N periods and given by the vector $D = \{D_1, D_2, \ldots, D_N\}$. Note that time-varying demand implies $D_i \neq D_j$ for at least one pair of i, j.
2. Let the inventory at the end of period j be given by I_j and let the starting inventory be $I_0 = 0$.
3. Supply is assumed to be instantaneous.
4. Shortages and back orders are not permitted.
5. Holding costs are incurred on the ending inventory in every period j, with h_j being the unit holding cost for period j. Then denote $H_j(x) = xh_j$. Similarly, order costs for period j are given by $C_j(x) = a_j\delta(x) + u_j x$ where a_j is the set-up costs in period j and u_j is the cost/unit in period j. Further, let $\delta(x) = 1$ if $x > 0$; 0 otherwise. We assume that H_j and C_j are concave functions of their respective arguments.

Our objective is to find the order quantities in each period that minimized total ordering and holding costs over the N periods. In other words, we want to find $Q = \{Q_1, Q_2, \ldots, Q_N\}$

Inventory Control

where all order quantities are non-negative. The model may be formulated as given below:

$$\min Z = \sum_{i=1}^{N} [C_i(Q_i) + H_i(I_i)] = \sum_{i=1}^{N} [a_i\delta_i(Q_i) + u_iQ_i + h_iI_i] \quad (10.22)$$

$$\text{subject to} \quad I_{i-1} + Q_i = D_i + I_i \quad \text{for all } i$$

$$Q_i, I_i \geq 0 \quad \text{for all } i$$

$$\delta_i(Q_i) \quad \text{binary for all } i \quad (10.23)$$

The above problem is a nonlinear, integer programming problem involving the *minimization* of a *concave* function (since H_j and C_j are concave functions). To solve the problem, we note the following [5,22]:

1. Inventory is held over a period if and only if the corresponding inventory costs are lower than the ordering costs. Suppose k units are left at the end of period $j-1$; then it would be optimal to hold these $\Leftrightarrow H_{j-1}(k) < C_j(k)$. If not, the previous replenishment should be reduced by these k units to reduce costs.
2. Inventory is replenished only when the inventory level is zero (*zero inventory ordering property*). Consequently, if an order is placed, it must cover demand for an integer number of future periods.

Using the above observations, Wagner and Whitin [22] show that the following must be true:

$$Q_{i-1}I_i = 0 \quad \text{for all } i \quad (10.24)$$

From the above properties and Equation 10.24, note that one of the following must be true as a consequence:

1. $I_{i-1} > 0 \Rightarrow Q_i = 0$; and $I_{i-1} > 0 \Rightarrow I_{i-1} > D_i$; $I_{i-1} \in \{D_i, D_i + D_{i+1}, \ldots, \sum_{k=i}^{N} D_k\}$. Essentially, this means that if the last period's ending inventory is positive, then this inventory level is at least the current period demand (D_i) and may be as much as all remaining periods' demand ($\sum_{k=i}^{N} D_k$).
2. $I_{i-1} = 0 \Rightarrow Q_i > 0$. In fact, $Q_i \in \{D_i, D_i + D_{i+1}, \ldots, \sum_{k=i}^{N} D_k\}$. In this case, if the ending inventory in period $i-1$ is zero, then an order must be placed in the current period and this order quantity (Q_i) must be, at a minimum, equal to the current period's demand with the maximum possible order quantity being equal to the total demand of all remaining periods.

Further, note that both I_{i-1} and Q_i cannot equal zero (unless $i = N + 1$, i.e., the last period demand has been met).

A simple method to "see" and solve the Wagner–Whitin model would be to represent it as an acyclic graph (Figure 10.7). To do so, let each node represent the beginning of a period $k = 1, 2, \ldots, N \Rightarrow$ node $k + 1$ represents the end of period k. The graph then has a total of

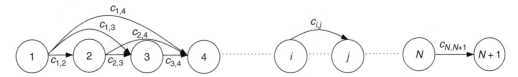

FIGURE 10.7 Graph representation of the Wagner–Whitin model.

$N+1$ nodes. The arc (i,j) represents ordering in period i with the next replenishment being in period j. Consequently, all demands in periods $i, i+1, \ldots, j-1$ are to be met from inventory. Further, let arc (i,j) exist only if $i < j$ $(i = 1, \ldots, N;\ j = 2, \ldots, N+1)$. The total number of arcs in the network is given by $\frac{N(N+1)}{2}$. Since the arc lengths (denoted by $c_{i,j}$) represent ordering and holding costs, we have

$$c_{i,j} = C_i \left(\sum_{k=i}^{j-1} D_k \right) + \sum_{k=i}^{j-2} \left[H_k \left(\sum_{r=k+1}^{j-1} D_r \right) \right] \tag{10.25}$$

The optimal solution to the Wagner–Whitin model is given by the *shortest path* from node 1 to node N and may be solved efficiently using Dijkstra [23] and Wagelmans et al. [24]. See Lee and Nahmias [25] and Silver et al. [1] for examples and references.

Comments: Zangwill [26] solved the case where back orders are allowed. In addition to these exact formulations, several heuristics like the Silver–Meal method, the Part–Period Balancing method, and the like have been developed to get near-optimal solutions efficiently [1,27]. Finally, a requirement of the Wagner–Whitin model is that there be a finite horizon (with known ending inventory level). Rangarajan and Ravindran [28] exploit this requirement to solve a SC inventory model efficiency.

10.4 Stochastic Inventory Systems

So far, in all the models studied, it was assumed that the demand is constant and known. Also, it was assumed that the units ordered will arrive exactly when they are expected. These assumptions eliminated uncertainties and allowed somewhat simple solutions for designing inventory systems. However, in the real world uncertainties almost always exist. Very few manufacturers can claim that they know exactly what their demand will be. Also, very seldom can they be certain that the supplier will deliver the ordered quantities exactly on time. Thus, applying deterministic models may result in some undesirable consequences that are discussed below.

Realistically, new orders do not necessarily arrive just when the last item in the inventory is used up. Let us consider an example. Suppose an automobile service center stocks car batteries. On an average, 10 cars that need new batteries are brought in every day. Let us assume that the service shop has used a deterministic model of inventory control and orders 100 batteries every 10 days. Also, let us assume that the lead time is 3 days, which results in a reorder point of 30 batteries. Suppose one day the shop operator notices that there are only 30 batteries left and orders a new shipment. If the demand is deterministic and is exactly 10 batteries a day, he would not have any problem while waiting 3 days for the order to arrive. His 30 batteries would be sufficient for these 3 days. However, the demand is usually not deterministic. During these 3 days, the total number of batteries demanded could be 40, 30, 20, or any other number. If the demand for these 3 days happens to be 20, there is no problem, but if it happens to be 40 or anything more than 30, the service shop will not have sufficient batteries to supply the demand. In this case, the service center is going to be out of stock for some period of time. In other words, an inventory system is out of stock when there is no unit available to meet all or part of the demand.

As another example, consider the electric motor case. The lead time in this situation was 3 days, and the reorder point was calculated to be 3000 motors. Now, if an order is placed at a point where there are 3000 units in stock and if a fixed consumption rate of 1000 units a day is assumed, the motors will be sufficient for the coming 3 days. However, if for some reason the supplier delivers the order on the fourth day instead of the third, the inventory

Inventory Control

will be zero for 1 day, and the demand for 1000 motors will be left unsatisfied. Again this inventory system will be out of stock but in this case for a different reason.

The undesirable consequences of being out of stock vary, depending on the nature of the operation. For example, in the case of the auto service center the customers may be willing to wait for one more day to get replacement batteries. In this case, being out of stock does not result in any problem. On the other hand, they may not be willing to wait and force the service center to do something about the situation, like buying the needed batteries from a neighboring supplier at a higher cost. It is also possible that the customers may take their business elsewhere. This would result in a loss of profit. Furthermore, those rejected customers may never come back to this shop for future repairs, which again causes the loss of goodwill and future revenue. In the case of the electric motors being used in manufacturing home appliances, being out of stock for 1 day may halt the whole manufacturing process, thus resulting in an appreciable loss.

The cost of being out of stock is called the shortage cost. The magnitude of this cost depends on the nature of the operation. Shortage costs can somewhat be avoided by maintaining some reserve stock called safety stock.

10.4.1 Safety Stock

In the auto service shop example, suppose the shop had ordered a new shipment at the time when there were 40 units in stock instead of 30. It would then have been prepared to meet a demand of up to 40 units during the lead time. On the other hand, if the demand had followed the usual pattern of 10 a day, at the end of the lead time when the new shipment was to arrive, 10 batteries would have still been left in stock. There also exists the possibility of the demand being less than the expected demand. This may result in say 20 units left in the inventory when the new shipment arrives.

To handle the uncertainties we will assume that the demand is a random variable with an average of 30 units for the lead time period. If we decide on the reorder point of 40 units, the units left in the inventory when the new shipment arrives would also be a random variable with an average of 10. This quantity of 10 units is called the safety stock for this inventory situation. The significance of safety stock is that it shows how many more units we should keep in addition to the expected demand to cope with uncertainties. The safety stock of 10 units for the above example may not solve all of the problems the auto service shop faces. For example, what happens if the demand during the lead time becomes 50? Again the shop will be 10 units short. One may suggest that the safety stock of 20 batteries might be better as it could respond to even greater uncertainties. In the extreme, to eliminate any chance of being out of stock, it may seem that a large safety stock would be necessary.

Although a large safety stock may eliminate the risk of being out of stock, it raises another problem. A large safety stock adds to the inventory holding costs. The capital spent on the items in inventory is tied up and the insurance and tax costs on holdings increase. This suggests that one should not increase the size of safety stock indefinitely to eliminate the risk of being out of stock. Rather, one has to establish a balance between the losses due to being out of stock and the cost of holding the additional units in stock.

The optimum value of the safety stock is the value for which the sum of the cost of being out of stock and the cost of holding the safety stock is a minimum. In stochastic inventory systems the demand or lead time or both are random. Thus, the shortage would be a random variable too. Then to evaluate shortage costs, some information about the probability distribution of the demand, especially demand during the lead time, must be given. In the derivations that follow these probability distributions will be assumed as given.

Before starting to study the stochastic inventory models, two points regarding safety stock must be made. First, when safety stock is considered, the procedure for calculating the reorder point changes. Without safety stock the reorder point was simply the product of lead time and the demand. With safety stock present, its value must be added to the previous reorder point to satisfy the variations in the demand. Thus,

$$ROP = LD + SS \tag{10.26}$$

where ROP, L, and D are as previously defined and SS is the safety stock.

It is very important at this point to examine this relationship carefully. As the factor directly affecting our decision is the reorder point rather than the safety stock, we usually determine the best reorder point before finding the best safety stock. Safety stock can be found from the reorder point by

$$SS = ROP - LD \tag{10.27}$$

Also, as safety stock is used when dealing with stochastic systems, the above relationship could be better defined by $SS = ROP -$ Expected value of the demand during the lead time.

The second point to keep in mind is that since the basis for ordering is the reorder point, the variation of the demand before reaching the reorder point does not create any problem. For example, in the auto service example suppose that a new order of 100 units arrives today. For the first 2 days assume that the demand has been 10 batteries a day. On the third day, when there are 80 batteries left, a demand of 20 batteries arrives and the stock is reduced to 60. As the reorder point of 30 has not yet been reached, the shop does not have to do anything. Suppose the next day the demand is 30 batteries, which will deplete the inventory to 30 units. Again, the shop has nothing to worry about because no customer has been turned away; the new reorder point has been reached, the new order is placed, and everything is in order. Now, if during the next 3 days the demands are 15, 15, and 10, respectively, the shop will be out of stock after 2 days, and even if it orders immediately, the order will not arrive on time. As a result, the shop will be 10 units short.

The above example shows that having two drastic demands of 20 and 30 units before reaching the reorder point did not create any problem, while two demands of 15 units each after the reorder point caused shortages. Thus we could conclude that the only time we have to worry about being out of stock is during the lead time, and the period of study for shortages should be restricted to this period.

To deal with uncertainties, like any other stochastic system, we would need some knowledge about the stochastic behavior of the system. In this case, the stochastic characteristics of interest would be the probability distribution of the demand during the lead time. The problem arises from the fact that given the probability distribution function of the demand per day or any other period, it is not always easy to determine the probability distribution of demand during the lead time, which consists of several periods. In other words, if the probability density function of demand per day is denoted by $f(x)$, the density function for demand during the lead time of n days is not always a simple function of $f(x)$. As a result of the above discussion, the first attempt in evaluating the safety stock must be to determine the distribution of the demand during the lead time. In the discussion and examples of this chapter only those cases will be presented for which this function can be easily evaluated.

10.4.2 A Maximum–Minimum System with Safety Stock

Here, a safety stock is added to the simple maximum–minimum inventory system. The general form of this model is shown in Figure 10.8. In this figure, when the ordered units

Inventory Control

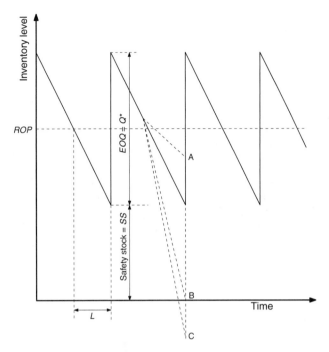

FIGURE 10.8 Inventory system with safety stock.

arrive, there are still a number of units equal to the safety stock SS left in the inventory. Of course, as was pointed out earlier, because of the random demand the number of units left when the new order arrives is a random variable. Three of the possible amounts left in the inventory when the new order arrives are shown in Figure 10.8 by points A, B, and C. SS is the expected value of these and all other possible values of this quantity. When new units are received, they are added to this safety stock. As a result the inventory level is built up to the maximum of

$$I_{\max} = SS + Q$$

For simplicity, it is assumed that the decline in the inventory level follows a linear path until it reaches its minimum level of $I_{\min} = SS$ where another order is received. Although it may seem unreasonable to assume a linear decline pattern when the demand is random, in the long run this assumption does not greatly affect the accuracy of the holding cost estimates. The major effect of the random demand on the cost occurs during the lead time where the possibility of shortage exists.

Now, having identified the maximum and the minimum levels of the inventory we can determine the total cost function as

Total Inventory Cost/Period = Inventory holding cost/period + Ordering cost/period
+ Shortage cost/period

From previous discussions we know that

Inventory holding cost/period = E (average inventory)
$$= E\left(\frac{Q + SS + SS}{2}\right) = E\left(\frac{Q}{2} + SS\right)$$

$$\text{Ordering cost/period} = \frac{BD}{Q}$$

$$\Rightarrow TC = E\left(\frac{Q}{2} + SS\right) + \frac{BD}{Q} + SC \tag{10.28}$$

where SC is the shortage cost per period. Equation 10.28 could be used to minimize the cost by finding the optimum values for the decision variables, Q and SS, provided that we could state the shortage cost as a function of these two variables. This is possible if we know the probability distribution of the random demand during the lead time. The demand during the lead time is considered because that is the only time during which the possibility of shortage exists. The calculation of shortage cost using that distribution is as follows.

Let the demand during the lead time be given by the random variable X with the density function of $f(x)$. Also let D and D_L be the expected values of the demand per period and the demand during the lead time, respectively. The shortages occur if the demand during the lead time is more than the reorder point (denoted ROP). The reorder point in this case is:

$$ROP = \text{Expected value of the demand during the lead time} + \text{Safety stock}$$

Expected value of the demand during lead time can be found as

$$D_L = \int_{-\infty}^{\infty} x f(x) \, dx$$

$$\Rightarrow ROP = D_L + SS \tag{10.29}$$

Now the only time we may have shortage is when the demand during lead time exceeds the reorder point. That is,

$$\text{Shortage cost/Lead time} = \begin{cases} 0 & \text{if } X \leq ROP \\ S \times \mathbb{E}[\text{No. of units short}] & \text{if } X > ROP \end{cases} \tag{10.30}$$

where S is the shortage cost per unit and $\mathbb{E}[\cdot]$ denotes the expected value. Notice that this assumption implies that the shortage cost per item is independent of the duration of shortage. In other words, if an item is short during the lead time, the cost of its shortage will be S whether this shortage happened 1 day before the new shipment arrived or 5 days.

However, the demand in excess of reorder point includes all values of demand greater than ROP. Now if the demand is x and reorder point R, the shortage would be a random variable that could be shown by $(x - R)$. The expected value of this random variable is

$$\mathbb{E}[\text{No. of units short}] = \int_R^{\infty} (x - R) f(x) \, dx$$

$$\text{and, Shortage cost/Lead time} = S \int_R^{\infty} (x - R) f(x) \, dx$$

As we have seen there are D/Q lead times per period

$$\text{Shortage cost/Period} = SC = \frac{DS}{Q} \int_R^{\infty} (x - R) f(x) \, dx \tag{10.31}$$

Although it is not explicitly stated, this cost is a function of Q and SS only. S is known and the term under the integral after integration will be stated in terms of the limits of

Inventory Control

the integral that are between ∞ and $R = D_L + SS$. Then the overall inventory cost (from Equations 10.28 and 10.31) will be

$$TC = E\left(\frac{Q}{2} + SS\right) + \frac{BD}{Q} + \frac{DS}{Q}\int_{D_L+SS}^{\infty}(x - D_L - SS)f(x)\mathrm{d}x \tag{10.32}$$

Now, if the density function can be integrated, the total cost will be in terms of Q and SS only. When the demand is discrete, the total cost function changes to

$$TC = E\left(\frac{Q}{2} + SS\right) + \frac{BD}{Q} + \frac{DS}{Q}\sum_{x_i > R}(x_i - R)p(x_i) \tag{10.33}$$

where the summation is for all demands that are greater than ROP and $p(x_i)$ denotes the probability mass function of the (discrete) demand distribution during the lead time. Interestingly, Lau et al. [29] suggest that under certain conditions, order-quantity, reorder-point models may yield impractical solutions or may be unsolvable. They suggest that cost parameters and service-level targets be chosen carefully to yield good solutions.

10.4.3 A Simplified Solution Method for Stochastic Systems

When the demand is discrete, Equation 10.33 does not produce a continuous function for differentiation. Also, sometimes even with continuous functions the resulting total cost function is complicated and cannot be optimized easily. In such situations, the following simplified method may result in a somewhat less accurate result, but the computations will be easier. Consider Equation 10.32 or 10.33. The terms of these relations can be divided into two categories. First, there are the terms associated with ordering and inventory holding costs. These terms, which are similar to those in the deterministic model, are

$$TC_1 = \frac{BD}{Q} + \frac{EQ}{2} \tag{10.34}$$

Second, there are the terms associated with holding the safety stock and shortage costs. These terms are

$$TC_2 = E \times SS + \frac{DS}{Q}\int_{D_L+SS}^{\infty}(x - D_L - SS)f(x)\mathrm{d}x$$
$$\text{or } TC_2 = E \times SS + \frac{DS}{Q}\sum_{x_i > R}(x_i - R)P(x_i) \tag{10.35}$$

In the simplified model it is assumed that these two cost figures can be minimized separately and independently from each other. This assumption is, of course, not correct because optimizing a function term by term does not necessarily optimize the entire function. However, the ease of evaluation obtained from this assumption sometimes compensates for the slight loss of accuracy of the results.

Thus, the order quantity is found based on the first two terms. Then this order quantity is used in the second group of the costs to determine the optimum value of the safety stock. The optimum order quantity from Equation 10.34 can be obtained as follows:

$$\frac{\mathrm{d}TC_1}{\mathrm{d}Q} = \frac{E}{2} - \frac{BD}{Q^3} = 0$$
$$\Rightarrow Q^* = \sqrt{\frac{2BD}{E}} = \sqrt{\frac{2B\,\mathbb{E}[D]}{E}} \tag{10.36}$$

Note that this is the relationship for the economic order quantity for the simple maximum–minimum inventory model (i.e., the EOQ model). The only difference here is that demand is represented by the expected value of the demand (as denoted explicitly in Equation 10.36). When demand is a continuous random variable, we need to evaluate the integral in Equation 10.35. Once this is done, the resulting function becomes a single variable function in terms of SS. Again, one has to be careful in minimizing such a function to make sure SS falls within the acceptable range. For discrete demands the cost could be evaluated using $ROP = DL - S$ for the range of acceptable values of SS, and the optimum value could be estimated as the one corresponding to the minimum cost.

10.4.4 Stochastic Inventory Model for One-Period Systems

This model applies to systems that operate for only one period and where items cannot be carried from one period to the next. Examples of this type are perishable products, fashion goods, or products that become obsolete or outdated after one period or selling season. Consider keeping an inventory of fruits and vegetables to supply a stochastic demand for a certain period. Obviously, at the end of the period, items left over are considered expired and must be thrown away or sold at a considerable discount. Another example is the procurement of newspapers and magazines. After a day or a week they become worthless. This model is commonly referred to as the *newsvendor model*. To design an inventory system for these situations one has to find the amount to procure such that the expected value of the profit is maximized as given below.

The newsvendor sells a single product over a single selling season with one procurement opportunity before the start of the season. Demand is stochastic with known distribution function $F(\cdot)$ and probability density function $f(\cdot)$. The product is procured at a cost of $\$c$/unit and sold at a price of $\$p$/unit. Any unsold inventory at the end of the selling period is salvaged (i.e., the firm receives this amount per unsold unit) at a value of $\$s$/unit ($p > c > s$). Back orders are not permitted and supply is assumed to be instantaneous with no set-up costs. The newsvendor is faced with the decision of how much to order at the start of the period so as to maximize *expected profit* (denoted by Π). As before, the order quantity is given by Q and demand by D. To develop the profit maximization model, note that

$$\mathbb{E}[\Pi] = \mathbb{E}[\text{profit when } D \in [0, Q]] + \mathbb{E}[\text{profit when } D \in [Q, \infty)] \quad (10.37)$$

When demand is less than or equal to Q, then all demand is met and a loss ($=(c-s)$/unit) is incurred for any excess inventory (first term on the right). When demand exceeds the available inventory, then all Q units are sold (indicated by the second term on the right in Equation 10.37). Let the random demand be given by X with known distribution. Then, we may write Equation 10.37 as

$$\mathbb{E}[\Pi] = \int_0^Q [(p-c)x - (c-s)(Q-x)] \mathrm{d}F(x) + \int_Q^\infty [(p-c)Q \, \mathrm{d}F(x)] \quad (10.38)$$

Using Leibniz's rule to differentiate Equation 10.38 and setting the derivative to zero, we can show that the optimal order quantity, Q^*, is given by

$$Q^* = F^{-1}\left(\frac{p-c}{p-s}\right) = F^{-1}(c_f) \quad (10.39)$$

where $c_f = (p-c)/(p-s)$ is known as the *critical fractile*. Now, note the following.

$$c_f = \frac{p-c}{p-s} = \frac{p-c}{(p-c)+(c-s)} = \frac{c_u}{c_u + c_o} \quad (10.40)$$

Inventory Control

where c_u, the *underage cost*, represents the profit/additional unit that is lost by being short of demand and c_o, the *overage cost*, represents the loss incurred/additional unit that is in excess of demand. See also Chapter 22 of this book for additional discussion and derivation of the newsvendor model result using marginal analysis.

Example 10.2

Let demand be random and uniformly distributed between 100 and 400 over the selling season. Further, let $p = 12\$/$unit, $c = 9\$/$unit, and $s = 0\$/$unit. From Equation 10.40 we have $c_f = (12-9)/(12-0) = 0.25$. The optimal order quantity, Q^*, is given by Equation 10.39. For a uniform random variable, $F(x) = \frac{x-a}{b-a}$; $x \in [a, b]$. Therefore, $Q^* = c_f(b-a) + a = 175$ units. ∎

Comments: (i) The case when demand follows a discrete distribution is analogous to the one above. (ii) Q^* is often called the *base stock level*. If the starting inventory is less than Q^*, then the optimal action is to bring the inventory level up to Q^* (i.e., order $Q = (Q^* - I_0)$ units, where I_0 is the inventory at the start before ordering). Such a policy is called a *base stock policy* in the literature and has wide applicability. Evidently, the maximum inventory level will equal the base stock level in this model. (iii) Several modifications of this model are possible [see 1,15,30]. The newsvendor model also forms the basis of many models addressing decisions in pricing [31–34], supply chain inventory models [35,36], production capacity investments [37,38], and so on.

10.4.5 Stochastic, Multi-Period Inventory Models: The (s, S) Policy

Given the simple solution to the single period problem above, it is obvious to ask if the analysis can be extended to a multi-period setting. Multiperiod inventory models are usually addressed using techniques from dynamic programming; see Chapter 7 of this book for an overview. Essentially, we are interested in finding the minimum expected cost over N periods. As such, two important cases are possible: the *finite horizon* case, where decisions are to be made over a finite number of periods ($N < \infty$) or the *infinite horizon* case ($N = \infty$). While the derivation of such models is beyond the scope of this chapter, it has been shown that in both cases [39,40] the optimal inventory control policy is of the form of (s, S) with $s < S$. The policy essentially implies the following ordering rule: *if inventory falls below the reorder level ($= s$), then place an order to bring the inventory level up to S*. To state this mathematically, let inventory at the start of a period be I and the decision variable be Q, the order quantity for that period. Then, (s, S) implies the following optimal ordering rule

$$Q = \begin{cases} (S-I) & \text{if } I < s \text{ [i.e., raise inventory to } S] \\ 0 & \text{if } I \geq s \text{ [i.e., do nothing]} \end{cases} \quad (10.41)$$

Evidently, the (s, S) policy is fairly simple in form and easy to implement. However, it must be noted that computation of the two parameters (s and S) is not always easy though several efficient heuristics are available. The (s, S) policy has also been shown to be optimal under fairly generic assumptions. For detailed development of the model, assumptions, extensions, and solution procedures, see Lee and Nahmias [25], Porteus [8], and Zipkin [3].

10.5 Inventory Control at Multiple Locations

In the models discussed thus far, we have looked exclusively at managing inventories at a single facility. In doing so, we have assumed that orders placed with an "external supplier" are assured to arrive after a suitable lead time (when $L > 0$). The astute reader may have

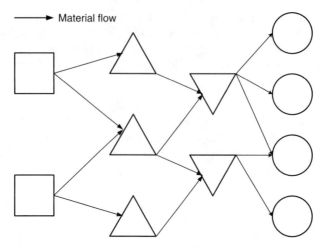

FIGURE 10.9 A general multi-echelon supply chain network.

reason to pause and ask: "What if the supplier is out of stock?" Furthermore, virtually all facilities holding inventories are part of a SC in reality. One therefore begins to see the need for looking at inventory management at multiple locations. Such models are referred to as SC inventory management models or as *multi-echelon inventory theory* in the research literature.*

As mentioned previously, inventory represents considerable investments and is considered one of the drivers of SC performance; see Chopra and Meindl [43] and Chapter 22 of this book for detailed discussions of the role of inventories in the SC. The core challenge of managing inventories at multiple facilities is the issue of *dependencies*. Consider a retailer who is supplied by a warehouse. In the models of Sections 10.3 and 10.4, we either assume that this supplier has infinite capacities (or inventories) or that such inventories arrive after a lead time that is either fixed or described by a probability distribution. However, this ignores the explicit consideration of the supplier inventory, the dependency of the retailer on the supplier's inventory and the need to perhaps control supplier inventories as well. To complicate matters further, inventories are usually held at multiple locations in the SC and the SC may be viewed as a network (Figure 10.9). In this section, we briefly overview the basics of multi-echelon inventory models and refer the reader to more detailed texts on this topic [2,5,6,44] for further details and references.

10.5.1 Multi-Echelon Inventory Models: Preliminaries

Quantitative models for complex SC structures as in Figure 10.9 are difficult to work with. Based on common network topologies observed in practice and to make models more tractable, most multi-echelon inventory models assume one of three common network structures. Before delving into this, we introduce some relevant terminology. In our

---

*Multi-echelon inventory theory emerged from the need to control inventories at multiple production stages in a facility and early literature reflects this "production" heritage. Clark [41] and Scarf [42], pioneers in the field, give some perspective on the early developments and their work. In this chapter, we focus more on SC inventory management that considers these locations to be distinct, reflecting contemporary practice.

Inventory Control

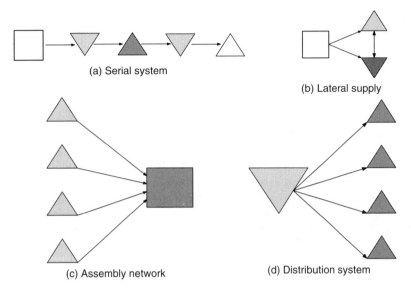

FIGURE 10.10 Common supply chain structures.

retailer–warehouse example, the retailer represents one *stage* or *echelon* of the SC and the warehouse is another stage of the SC. In general, an SC may have multiple stages and multiple facilities at each stage (as shown in Figure 10.9). In all cases, clear relationships exist between these stages; namely, material "flows" (or moves) from one stage to the next. As we move from the raw material suppliers toward the final stages that see end customer demand, we are said to be moving *downstream* in the SC and *upstream* when in the reverse direction. So, material flows from upstream suppliers to downstream retailers.

We can now look at the common SC structures used in many multi-echelon inventory models. These are represented graphically in Figure 10.10a–d.

1. *Serial system*, shown in Figure 10.10a, is an SC wherein every facility has at most one upstream supplier and one downstream customer and only one facility exists at each stage. This often represents the simplest structure.
2. *Assembly network* is one in which there is one downstream facility that is supplied by multiple upstream suppliers, who may in turn be supplied by multiple suppliers themselves. It is commonly assumed that each upstream stage has only one customer at the next downstream stage. Figure 10.10c gives an example of such a system.
3. *Distribution systems* are networks where upstream facilities supply multiple downstream stages. A special case of particular importance is when each facility has at most one supplier but may itself supply multiple downstream customers. Such a network is called an arborescent distribution network (Figure 10.10d).
4. *Lateral supply*—As mentioned in Section 10.2.3, in some cases, a facility may receive supplies from another facility at the same stage. For example, consider two Wal-Mart stores in a city. If one ships inventory to the other to make up for shortages at the latter facility, this is known as a lateral shipment. This is usually done as an emergency measure, though. Figure 10.10b represents this graphically.

Note that the serial system is a special case of both the assembly and distribution network types. In our discussions here, we will look at serial systems. See Graves et al. [2], Song and Yao [45], and Song and Zipkin [46] for details on SC structure and assembly systems.

Inventory Coordination Policies

The effect of dependencies between inventories in the SC has the potential to complicate operations. In the single warehouse, single retailer (serial) SC example, assume that the retailer orders every week and the warehouse every 5 days. Ideally, we would like these two reorder intervals to be such that the overall SC inventory costs are minimized; in other words, we would like the SC members to be *coordinated*. Inventory coordination schemes help smoothen operations. Coordination takes on special significance in a decentralized control setting where members in the SC operate independently, often leading to inefficiencies. Coordination is usually achieved in two ways: *quantity coordination* or *time coordination*. In quantity coordination, SC members change their order quantities so as to reduce system-wide costs. This is done using SC *contracts* that help induce this change and redistributes some of the savings between the SC members. In time coordination schemes, the reorder intervals of the members are changed to bring some regularity to operations (e.g., ordering weekly). Evidently, under deterministic settings, the two are equivalent in terms of the changes in the order quantities and reorder intervals (though contracts also look at the issue of sharing of cost savings). We discuss time coordination policies here for two reasons: (a) they are operational in nature, fairly simple, and do not involve issues like transfer pricing and compensation issues; and (b) optimal or near optimal policies are available even without the use of more complex contracts and this suffices for the discussion at hand. Cachon [47] and Tsay et al. [48] review the contracting literature.

The following notation is used in the discussions below. The term *facility* is used to denote an inventory stocking location. In general, we denote the first upstream facility as node 0 and the last down stream facility as node N in the serial system. The distribution system in Figure 10.10d is commonly referred to as the *single warehouse, multiple retailer* system, and we denote the warehouse as facility 0 and the retailers as facilities 1 through N. Let the review period length at each location be T_i, $i = 0, 1, \ldots, N$. The following are some common coordination policies [12,49].

- *Stationary Policy*: If each facility orders at equally spaced points in time and in equal quantities each time, then the policy is called a stationary policy.
- *Nested Policy*: If each facility orders every time any of its immediate suppliers does (and perhaps at other times as well), then it is known as a nested policy.
- *Stationary Nested*: A policy is stationary and nested if $T_i \leq T_0$ for all $i = 1, 2, \ldots, N$. Such a policy ensures that the order quantities at upstream stages are stationary.

Zero Inventory Ordering Property

Schwarz [50] considered a single warehouse, multi-retailer system and showed that an optimal policy may be fairly complex. He then defined a *basic policy* that has the following properties: (a) deliveries are made to the warehouse only when the warehouse and at least one retailer have zero inventory; (b) deliveries are made to a retailer only when the retailer has zero inventory; and (c) all deliveries made to any given retailer between successive deliveries to the warehouse are of equal size. This result is known as the *zero-inventory*

Inventory Control

ordering property. In fact, the result holds for many other systems also when demand is deterministic.

With this background, we now proceed to consider some basic multi-echelon inventory models facing deterministic end demand.

10.5.2 Serial Supply Chain Inventory Models

Consider a serial, two-stage SC consisting of a warehouse supplying a downstream retailer who faces demand of D units/time period. The warehouse and retailer have order costs of a_0, a_1, inventory holding costs of h_0, h_1, and the inventory at each facility is denoted by I_0, I_1 (in appropriate units), respectively. Further, it is assumed that $h_0 < h_1$ as unit costs and holding costs usually increase as we move downstream in the SC. We also assume that the SC follows a nested coordination policy wherein the warehouse reorder interval (T_0) is a positive integer multiple (m) of the retailer's reorder interval (T_1); that is,

$$T_0 = mT_1; \quad m = 1, 2, \ldots \tag{10.42}$$

Assumptions of the EOQ model are assumed to hold unless otherwise stated. Two control policies are possible (see Section 10.2.3), namely centralized and decentralized SC control and both are discussed below.

Decentralized Control of the Serial System

In this case, the warehouse and retailer make inventory decisions independent of the other subject to the nested policy constraint. So the retailer solves his/her inventory problem (using the EOQ model; Section 10.3.1) first and communicates the optimal reorder interval (T_1^*) to the warehouse. Thus, from Equation 10.9, we have

$$T_1^* = \sqrt{\frac{2a_1}{Dh_1}}; \quad Q_1^* = DT_1^* \tag{10.43}$$

The corresponding optimal cost is given by $Z_1^* = \sqrt{2a_1 D h_1}$. The inventory plot for the retailer is shown in the top panel of Figure 10.11. This is known as the retailer's *installation stock* and it follows the familiar "saw-tooth" pattern of the EOQ model.

Now, the warehouse sees a bulk demand of Q_1^* units every T_1^* time units. Thus, the warehouse inventory pattern is not a saw-tooth pattern and is represented by the step function in the middle panel of Figure 10.11. It is fairly easy to show that in this case, the average inventory at the warehouse (denoted $\bar{I}_0$) is given by

$$\bar{I}_0 = \frac{Q_1(m-1)}{2} \tag{10.44}$$

Then, the total variable costs at the warehouse is given by

$$Z_0(m) = \frac{a_0}{T_0} + \frac{h_0 Q_1^*(m-1)}{2} = \frac{a_0}{mT_1^*} + \frac{h_0 DT_1^*(m-1)}{2} \tag{10.45}$$

It is easy to show that Equation 10.45 is pointwise convex in the integer variable m and so the minimum value is attained when the first difference (Equation 10.46) equals zero or changes from negative to positive.

$$\Delta Z_0(m) = Z_0(m+1) - Z_0(m) \tag{10.46}$$

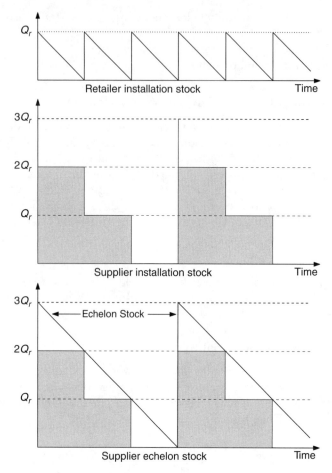

FIGURE 10.11 The concepts of installation and echelon inventories.

Substituting Equation 10.45 in 10.46 and simplifying, we can show that the optimal solution is given by identifying the smallest positive integer, m^*, such that the following is true.

$$\frac{2a_0}{h_0 D(T_1^*)^2} \leq m^*(m^*+1) \tag{10.47}$$

Note that the left hand side of Equation 10.47 is a constant. The problem is a single variable problem and efficient optimization techniques may be employed to find the optimal m^* to get Q_0^* and T_0^*. This completes the analysis of the decentralized SC.

Centralized Control of the Serial System

In this setting, we are looking to optimize the reorder intervals at both facilities simultaneously. It is common practice to assume that all information like costs, inventory levels, and the like at all facilities in the centralized network are known globally. We should note that this places greater control in the hands of the manager(s), though obtaining such detailed information in a timely manner may not be easy in practice. The optimization problem to minimize the system-wide total variable costs (i.e., the sum of the ordering and holding

Inventory Control

costs at both facilities) is given by

$$\text{Problem C: min } Z_c = \left(\frac{a_1}{T_1} + \frac{a_0}{T_0}\right) + \left[\frac{h_1 Q_1}{2} + \frac{h_0 Q_1 (m-1)}{2}\right]$$

$$= \frac{1}{T_1}\left(a_1 + \frac{a_0}{m}\right) + T_1\left[\frac{h_1 D}{2} + \frac{h_0 D(m-1)}{2}\right] \quad (10.48)$$

subject to

$$T_0 = mT_1; \quad m = 1, 2, \ldots$$
$$T_1 \geq 0$$

Problem C is a nonlinear, integer programming problem in two variables, m and T_1. Before proceeding, though, let us pause to consider the average inventory expression at the warehouse. Under more general SC structures, we may not get a nice expression for $\bar{I}_0$. Clark and Scarf [51] introduced the concept of echelon inventories, which is defined as the total inventory of a product at any particular stage in the SC and all subsequent downstream stages linked to it. From this definition, the echelon inventory at the retailer (stage N) is identical to its installation inventory. The echelon inventory helps us simplify the computation of $\bar{I}_0$ in many systems.

Consider Figure 10.11, where we assume that $T_0 = 3T_1 \Rightarrow Q_0 = 3Q_1$. Now, every T_1 time units, Q_1 is released from warehouse inventory to the retailer resulting in the step function shape of warehouse inventory level over time. However, note that this inventory merely moves down the SC to the retailer. In a centralized SC, information on this inventory is assumed to be known. So, by superimposing the two stages' inventory curves, we get the warehouse's echelon levels (bottom panel of Figure 10.11) which has the familiar saw-tooth pattern. Thus, the mean echelon inventory at the warehouse may be approximated in a fashion similar to that of the retailer.

To complete the analysis, we are given $h_0 < h_1$. Define $h'_0 = h_0$ and $h'_1 = h_1 - h_0$ as the echelon inventory holding rates at the warehouse and retailer, respectively. Further, let $g_i = \frac{h'_i D}{2}$; $i = 0, 1$. Substituting these in Equation 10.48 and simplifying gives us

$$Z_c = \frac{1}{T_1}\left(a_1 + \frac{a_0}{m}\right) + T_1(g_1 + mg_0) \quad (10.49)$$

Now, note that the form of Equation 10.49 is similar to the EOQ model objective function, but with two variables. To solve this for optimal values, the following method may be used: fix m to some value and compute the corresponding best value for the reorder interval given by

$$T'(m) = \sqrt{\frac{(a_1 + \frac{a_0}{m})}{(g_1 + mg_0)}} \quad (10.50)$$

Again using the first difference argument, we can show that the optimal integer multiple, m^*, is given by the smallest integer satisfying the following relation:

$$m^*(m^* + 1) \geq \frac{a_0/g_0}{a_1/g_1} \quad (10.51)$$

Equations 10.50 and 10.51 may be used to get the optimal reorder interval for the retailer (T_1^*) and this gives us the corresponding warehouse reorder interval ($T_0^* = m^* T_1^*$). The optimal variable costs for the serial SC is given below, completing the analysis.

$$Z_c^* = \sqrt{\left(a_1 + \frac{a_0}{m}\right)(g_1 + mg_0)}$$

Comments: (i) The above procedure may be generalized to a serial SC with N stages. (ii) Schwarz [50] proved that the nested policy is indeed optimal for a serial SC. See Muckstadt and Roundy [52] for further details.

The models above represent the simplest multi-echelon inventory models. Several extensions are possible and the reader is referred to [5,6,12,14,49,51,52] for more extensive discussions on multi-echelon inventory theory.

10.6 Inventory Management in Practice

One may wonder, given the many simplifications made in developing inventory management models, if the models are of value in practice. The short answer is a resounding "Yes!" We readily acknowledge that all the models are not applicable in all situations; real-world applications often use heuristics, simulations, and the like to manage inventories (in single and multi-echelon settings). However, most of these are based on theoretical models like the ones discussed in this chapter. Also, the focus in practice is often on getting good solutions rapidly (and often optimal solutions are not achievable due to complex constraints, etc.). In many situations, the models discussed previously can and have been applied in practice. A sample of applications are listed in Table 10.2. All these cases are taken from the *Interfaces* journal.*

TABLE 10.2 Examples of Inventory Management Applications in Practice

Reference	Industry	Company	Comments
Lee and Billington [53]	Computers	Hewlett-Packard (HP)	• Goal: Inventory management in decentralized SC for HP Printers • Inventory reduction of 10%–30%
Lin et al. [54]*	Computers	IBM	• Goal: Decision support system (DSS) for global SC (inventory) management • Approx. $750 million in inventory and markdown reductions
Koschat et al. [55]	Print media	Time Warner	• Goal: Optimize printing orders and distribution of magazines in three stage SC • Solutions based on the newsvendor model • $3.5 million increase in annual profits
Kapuscinski et al. [56]	Computers	Dell Inc.	• Goal: Identify inventory drivers in SC for better inventory management at Dell DCs • Expected savings of about $43 million; 67% increase in inventory turns; improved customer service
Bangash et al. [57]	Electronics	Lucent Technologies	• Goal: DSS tool for inventory management of multiple products • Solution based on (s, S) policies • $55 million in inventory reductions; fill rates increased by 30%
Bixby et al. [58]	Food	Swift & Co.	• Goal: Production management at beef products facilities; DSS tool for sales • Solution adapts production plans based on inventories and customer orders dynamically • $12.74 million in annual savings; better sales force utilization

*Franz Edelman Award winner.

*"*Interfaces*, a bimonthly journal of INFORMS, is dedicated to improving the practical application of OR/MS to decisions and policies in today's organizations and industries. Each article provides details of the completed application, along with the results and impact on the organization" (from the journal Web site; http://interfaces.pubs.informs.org). INFORMS also awards the *Franz Edelman Award* annually, from applications reported in *Interfaces*, the purpose of which is to "...call out, recognize and reward outstanding examples of management science and operations research practice in the world."

Inventory Control

TABLE 10.3 Inventory Management Software Vendors (sample)

Vendor	Web site
i2 Technologies	http://www.i2.com
Logility	http://www.logility.com
Manhattan Associates	http://www. manh.com
Oracle Corporation	http://www.oracle.com
SAP	http://www.sap.com

10.6.1 Inventory Management and Optimization Tools

At some point, most readers will want to implement some of the models discussed here and elsewhere. Several tools exist for managing inventories, including analytical and optimization applications. It is worth pointing out that several of the models described in this chapter (e.g., EOQ, Newsvendor, etc.) may be easily implemented using a spreadsheet program. This makes it easy to test out the models and perform what-if analysis. For large-scale implementations, specialized software tools exist.

Most models require good data inputs and the recent trend in industry is to use enterprise resource planning (ERP) systems to manage large amounts of data. Often, inventory management and SC management tools work in concert with ERP systems. Specialized inventory management and optimization applications also exist independent of such large (and expensive) ERP systems. In either case, many of these tools are capable of handling large problems of varying degree of complexity. Table 10.3 lists a few inventory and SC management software vendors that are publicly traded companies (in alphabetical order). For a more extensive list, see Foster [59] and McCrea [60].

In concluding this section, we would like to reiterate that sophisticated tools are not substitutes for sound analysis and good judgment. An understanding of the models being used, their strengths, limitations, and applicability is paramount to successfully implement a scientific inventory management program.

10.7 Conclusions

Although keeping a certain level of inventory might be necessary in many situations, one should always consider making changes in the production or business systems so that the optimum inventory level is held. Holding large amounts of inventory not only raises inventory costs, it may also hide the inefficiencies of the system's components. For instance, if an unreliable supplier supplies a company with raw material, the company's large inventory level may hide the fact that the supplier is unreliable and may prevent managers from looking for an alternative supplier. Conversely, if the inventory levels are too low, then the system may become vulnerable to disruptions, leading to increased losses of money and customer goodwill.

In this chapter, the objectives of maintaining inventories for raw material, work-in-process, and finished products were explained. It was indicated that the purpose of studying inventory systems is to design an optimum inventory policy for each particular case such that the overall cost of maintaining it is minimized. The concept of ABC analysis was presented as a simple method to classify inventories so as to ensure proper attention and resources are given to critical and high value inventories. We discussed the various costs relevant to holding inventories (inventory holding costs, ordering or set-up costs, and shortage costs) and the influence they have on inventories.

As mentioned at the outset, inventory management looks to answer the fundamental questions of how often and how much to order. Several models of inventory analysis were introduced to answer these questions under various settings. In each case, the relevant variables were specified and the method to compute the optimal values was given. Models for both deterministic and stochastic demand were presented. In reality, inventories are held across the SC and inventories at one location usually depend on inventories at other locations in the SC. We introduced multi-echelon inventory models that explicitly consider multiple stocking locations. It must be pointed out that the models described in this chapter are just a few representative cases of the existing models. Interested readers are referred to references given in the bibliography for more extensive coverage of inventory control models.

10.8 Current and Future Research

The development and use of operations research techniques and mathematical models to understand and control inventories continues. A very brief set of topics of current and future research efforts is given below with some references to relevant research literature.*

- *Inventory Models* continue to receive significant attention from researchers as they are of practical value and represent significant theoretical challenges. As noted previously, optimal policies for multi-echelon inventory models, even in the deterministic demand case are not known for all SC structures and identifying such policies would be of significant interest. In the case of stochastic multi-echelon inventory management, there are even fewer generic results and the search for better solutions continues.
- *Interface Between Inventory Management Models and Other Business Functions*: In some sense, the models covered in this chapter (with the exception of the newsvendor model, perhaps) represent "pure" inventory management models. However, inventory management is not done in a vacuum and is significantly influenced by other business functions like marketing, pricing, competition, and the like, and research in areas at the interface of inventory management and other business operations has been very active in recent times. For example, we have assumed that demand is not influenced by any action we take; in reality, promotions and markdowns have significant effect on the demand patterns. Elmaghraby and Keskinocak [61], Monahan et al. [32], Netessine [62], and Petruzzi and Dada [33] cover *pricing* issues in some depth. Another emerging area of interest is to actively incorporate the effect of competitors in the analysis. Models in this category usually use game theoretic methods in their analysis; see Cachon [63] and Cachon and Netessine [64] for reviews. Cachon and Netessine [65] provide an introduction to the methodologies of competitive analysis.
- *Supply Chain Management and Inventories*: Just as demand is influenced by pricing, inventory decisions are often made in concert with pricing offered by suppliers. Furthermore, in an attempt to coordinate SC members, *contracts* are often used, especially in the decentralized SC case. Such contracts often involve the use of (transfer) pricing to help compensate members in an equitable fashion and align SC members. For reviews, see Cachon [47] and Tsay et al. [48].

---

*Recent issues of many of the journals mentioned in the bibliography are excellent sources to identify current advances in inventory management.

We also have assumed that the inventories are placed at existing facilities whose locations are given. Facility locations, transportation modes, and the like can have a significant impact on inventory levels and ideally should be jointly optimized. Sarmiento and Nagi [66] give a review of *integrated SC models*. For an overview of the issues in transportation and inventory management, see Natarajan [67] and Chapter 22 of this book. The placement of safety stocks in the SC is also of strategic importance. However, the problem has been shown to be $\mathcal{NP}$-hard [68] and represents significant challenges; see Graves and Willems [69]. Billington et al. [70] present an application of some of this research to the HP SC. Finally, the advent of global SCs has seen inventories and facilities being spread worldwide. This has increased the vulnerability of the SC to disruptions and has begun to receive considerable interest, though only a limited number of models exist currently. Rangarajan and Guide [71] discuss the unique challenges presented by disruptions and review the relevant literature.

- *Reverse Logistics and Closed-Loop SCs*: We consider material flow to be from upstream stages to downstream stages in the SC. However, material may also flow in the reverse direction in the form of returns, defective parts, and the like. This "reverse flow" must be managed, sometimes due to laws to this effect. They can also represent a valuable income stream for a firm. The processes involved in harvesting value from these products can be very different from the manufacturing processes involved in forward flow of material. Research into managing inventories in the closed-loop SC is still in its infancy. The reader is referred to Fleischmann et al. [72], Guide and Van Wassenhove [73], and Guide et al. [74].

10.8.1 Final Remarks

In conclusion, mathematical models for inventory management are abstractions of real systems. Nevertheless, they are often sufficient to optimally control an inventory system or, at the very least, provide valuable insight into the behavior of the system and salient factors influencing it. As demonstrated by the examples presented earlier, these models, when suitably applied, can lead to significant improvements and increased profitability. Despite the long history and the many successes of inventory management methods, significant challenges (of both practical and theoretical interest) remain.

References

1. E.A. Silver, D.F. Pyke, and R. Peterson. *Inventory Management and Production Planning and Scheduling*. John Wiley & Sons, New York, 3rd edition, 1998.
2. S.C. Graves, A.H.G. Rinnooy Kan, and P.H. Zipkin, editors. *Logistics of Production and Inventory*, volume 4 of *Handbooks in Operations Research and Management Science*. North-Holland, Amsterdam, 1993.
3. P.H. Zipkin. *Foundations of Inventory Management*. McGraw-Hill, New York, 2000.
4. G. Cachon and M. Olivares. Drivers of finished goods inventory performance in the U.S. automobile industry. Working paper, The Wharton School, University of Pennsylvania, October 2006.
5. S. Axsäter. *Inventory Control*. Springer, 2nd edition, 2006.
6. J. Bramel and D. Simchi-Levi. *The Logic of Logistics: Theory, Algorithms and Applications for Logistics Management*. Springer-Verlag, New York, 1997.

7. G. Hadley and T. Whitin. *Analysis of Inventory Systems.* Prentice-Hall, Englewood Cliffs, NJ, 1963.
8. E.L. Porteus. *Foundations of Stochastic Inventory Theory.* Stanford University Press, Palo Alto, CA, 2002.
9. C. Munson and M. Rosenblatt. Theories and realities of quantity discounts: An exploratory study. *Production and Operations Management*, 7:352–359, 1998.
10. A. Federgruen. Centralized planning models for multi-echelon inventory systems under uncertainty. In volume 4 of *Handbooks in Operations Research and Management Science*, chapter 3. North-Holland, Amsterdam, 1993.
11. H.L. Lee and C. Billington. Material management in decentralized supply chains. *Operations Research*, 41(5):835–847, 1993.
12. A. Rangarajan. Inventory Models for Decentralized Supply Chains. Master's thesis, The Pennsylvania State University, University Park, December 2004.
13. C.C. Sherbrooke. METRIC: A multi-echelon technique for recoverable item control. *Operations Research*, 16:122–141, 1968.
14. C.C. Sherbrooke. *Optimal Inventory Modeling of Systems: Multi-Echelon Techniques.* Wiley, New York, 1992.
15. J.M. Swaminathan and S.R. Tayur. Tactical planning models for supply chain management. In A.G. de Kok and S.C. Graves, editors, *Supply Chain Management: Design, Coordination and Operation*, volume 11 of *Handbooks in Operations Research and Management Science*, chapter 8, pages 423–456. Elsevier, 2003.
16. G. Gallego, J.K. Ryan, and D. Simchi-Levi. Minimax analysis for finite-horizon inventory models. *IIE Transactions*, 33(10):861–874, 2001.
17. G.A. Godfrey and W.B. Powell. An adaptive, distribution-free algorithm for the newsvendor problem with censored demands, with applications to inventory and distribution. *Management Science*, 47(8):1101–1112, 2001.
18. I. Moon and G. Gallego. Distribution free procedures for some inventory models. *The Journal of the Operational Research Society*, 45(6):651–658, 1994.
19. U.S. Karmarkar. Manufacturing lead times, order release and capacity loading. In S.C. Graves, A.H.G. Rinnooy Kan, and P.H. Zipkin, editors, *Logistics of Production and Inventory*, volume 4 of *Handbooks in Operations Research and Management Science*, chapter 6. North-Holland, Amsterdam, 1993.
20. S. Minner. Multiple-supplier inventory models in supply chain management: A review. *International Journal of Production Economics*, 81–82:265–279, 2003.
21. F.W. Harris. How many parts to make at once. *Factory, The Magazine of Management*, 10(2):135–136, 1913.
22. H.M. Wagner and T.M. Whitin. Dynamic version of the economic lot size model. *Management Science*, 5:89–96, 1958.
23. E.W. Dijkstra. A note on two problems in connection with graphs. *Numerische Mathematik*, 1:269–271, 1959.
24. A. Wagelmans, S. Van Hoesel, and A. Kolen. Economic lot sizing: An O(nlogn) algorithm that runs in linear time in the Wagner-Whitin case. *Operations Research*, 40(S1):145–156, 1992.
25. H.L. Lee and S. Nahmias. Single-product, single-location models. In S.C. Graves, A.H.G. Rinnooy Kan, and P.H. Zipkin, editors, *Logistics of Production and Inventory*, volume 4 of *Handbooks in Operations Research and Management Science*, chapter 1, pages 3–55. North-Holland, Amsterdam, 1993.
26. W.I. Zangwill. A deterministic multi-period production scheduling model with backlogging. *Management Science*, 13:105–119, 1966.

27. S. Nahmias. *Production and Operations Analysis*. McGraw-Hill, Boston, MA, 4th edition, 2001.
28. A. Rangarajan and A.R. Ravindran. A base period inventory policy for decentralized supply chains. Working paper No. 05–04, Center for Supply Chain Research, Pennsylvania State University, 2005.
29. A. H-L. Lau, H-S. Lau, and D.F. Pyke. Degeneracy in inventory models. *Naval Research Logistics*, 49(7):686–705, 2002.
30. M. Khouja. The single-period (news-vendor) problem: Literature review and suggestions for future research. *Omega*, 27:537–553, 1999.
31. M.A. Lariviere and E.L. Porteus. Selling to the newsvendor: An analysis of price-only contracts. *Manufacturing & Services Operations Management*, 3(4):293–305, 2001.
32. G.E. Monahan, N.C. Petruzzi, and W. Zhao. The dynamic pricing problem from a newsvendor's perspective. *Manufacturing & Services Operations Management*, 6(1): 73–91, 2004.
33. N.C. Petruzzi and M. Dada. Pricing and the news vendor problem: A review with extensions. *Operations Research*, 47(2):183–194, 1999.
34. G. Raz and E.L. Porteus. A fractiles perspective to the joint price/quantity newsvendor model. *Management Science*, 52(11):1764–1777, 2006.
35. K.H. Shang and J-S. Song. Newsvendor bounds and heuristic for optimal policies in serial supply chains. *Management Science*, 49(5):618–638, 2003.
36. J.A. Van Van Mieghem and N. Rudi. Newsvendor networks: Inventory management and capacity investment with discretionary activities. *Manufacturing & Services Operations Management*, 4(4):313–335, 2002.
37. A. Burnetas and S. Gilbert. Future capacity procurements under unknown demand and increasing costs. *Management Science*, 47(7):979–992, 2001.
38. J.A. Van Mieghem. Capacity management, investment, and hedging: Review and recent developments. *Manufacturing & Service Operations Management*, 5(4):269–302, 2003.
39. D.L. Iglehart. Optimality of (s, S) policies in the infinite horizon dynamic inventory problem. *Management Science*, 9:259–267, 1963.
40. H.E. Scarf. The optimality of (s, S) policies in the dynamic inventory problem. In J.A. Kenneth, K. Samuel, and S. Patrick, editors. *Mathematical Methods in the Social Sciences*. Stanford University Press, Stanford, CA, 1960.
41. A.J. Clark. Multi-echelon inventory theory—a retrospective. *International Journal of Production Economics*, 35:271–275, 1994.
42. H.E. Scarf. Inventory theory. *Operations Research*, 50(1):186–191, 2002.
43. S. Chopra and P. Meindl. *Supply Chain Management: Strategy, Planning and Operation*. Prentice-Hall, Upper Saddle River, NJ, 2001.
44. S.R. Tayur, R. Ganeshan, and M.J. Magazine, editors. *Quantitative Models for Supply Chain Management*. Kluwer Academic Publishers, Norwell, MA, 1999.
45. J-S. Song and D.D. Yao, editors. *Supply Chain Structures: Coordination, Information and Optimization*. Kluwer Academic Publishers, Boston, MA, 2002.
46. J-S. Song and P. Zipkin. Supply chain operations: Assemble-to-order systems. In A.G. de Kok and S.C. Graves, editors, *Supply Chain Management: Design, Coordination and Operation*, volume 11 of *Handbooks in Operations Research and Management Science*, chapter 6, pages 561–596. Elsevier, 2003.
47. G.P. Cachon. Supply chain coordination with contracts. In A. G. de Kok and S. C. Graves, editors, *Supply Chain Management: Design, Coordination and Operation*, volume 11 of *Handbooks in Operations Research and Management Science*, chapter 6, pages 229–339. Elsevier, Boston, 2003.

48. A. Tsay, S. Nahmias, and N. Agarwal. Modeling supply chain contracts: A review, chapter 10. In *Quantitative Models for Supply Chain Management*, pages 299–336. Kluwer Academic, Dordrecht, 1999.
49. R. Roundy. 98%-effective integer-ratio lot-sizing for one-warehouse multi-retailer systems. *Management Science*, 31(11):1416–1430, 1985.
50. L.B. Schwarz. A simple continuous review deterministic one-warehouse n-retailer inventory problem. *Management Science*, 19:555–566, 1973.
51. A. Clark and H. Scarf. Optimal policies for a multi-echelon inventory problem. *Management Science*, 6:475–490, 1960.
52. J.A. Muckstadt and R.O. Roundy. *Analysis of Multistage Production Systems*, In S.C. Graves, A.H.G. Rinnooy Kan, and P.H. Zipkin, editors. *Logistics of Production and Inventory*, volume 4 of *Handbooks in Operations Research and Management Science*, chapter 2, pages 59–131. North-Holland, Amsterdam, 1993.
53. H.L. Lee and C. Billington. The evolution of supply-chain-management models and practice at Hewlett-Packard. *Interfaces*, 25(5):42–63, 1995.
54. G. Lin, M. Ettl, S. Buckley, S. Bagchi, D.D. Yao, B.L. Naccarato, R. Allan, K. Kim, and L. Koenig. Extended-enterprise supply-chain management at IBM Personal Systems Group and other divisions. *Interfaces*, 30(1):7–25, 2000.
55. M.A. Koschat, G.L. Berk, J.A. Blatt, N.M. Kunz, M.H. LePore, and S. Blyakher. Newsvendors tackle the newsvendor problem. *Interfaces*, 33(3):72–84, 2003.
56. R. Kapuscinski, R.Q. Zhang, P. Carbonneau, R. Moore, and B. Reeves. Inventory decisions in Dell's supply chain. *Interfaces*, 34(3):191–205, 2004.
57. A. Bangash, R. Bollapragada, R. Klein, N. Raman, H.B. Shulman, and D.R. Smith. Inventory requirements planning at Lucent Technologies. *Interfaces*, 34(5):342–352, 2004.
58. A. Bixby, B. Downs, and M. Self. A scheduling and capable-to-promise application for Swift & Company. *Interfaces*, 36(1):69–86, 2006.
59. T.A. Foster. The supply chain top 100 software vendors. *Supply Chain Management Review*, pages S2–S12, September/October 2006. Available at: http://www.scmr.com/info/59562.html.
60. B. McCrea. The state of the supply chain software industry. *Supply Chain Management Review*, pages S1–S4, October 2006. Available at: http://www.scmr.com/info/59562.html.
61. W. Elmaghraby and P. Keskinocak. Dynamic pricing in the presence of inventory considerations: Research overview, current practices, and future directions. *Management Science*, 49(10):1287–1309, 2003.
62. S. Netessine. Dynamic pricing of inventory/capacity with infrequent price changes. *European Journal of Operational Research*, 174(1):553–580, 2006.
63. G. Cachon. Competitive supply chain inventory management. In S. Tayur, R. Ganeshan, and M. Magazine, editors, *Quantitative Models for Supply Chain Management*. Kluwer, Boston, 1998.
64. G. Cachon and S. Netessine. Game theory in supply chain analysis. In D. Simchi-Levi, S.D. Wu, and Z-J. Shen, editors, *Handbook of Quantitative Supply Chain Analysis: Modeling in the eBusiness Era*, chapter 2. Kluwer, Boston, 2004.
65. G.P. Cachon and S. Netessine. Game theory in supply chain analysis. In Paul Gray, editor. *INFORMS TutORials in Operations Research*, pages 200–233. INFORMS, 2006.
66. A.M. Sarmiento and R. Nagi. A review of integrated analysis of production–distribution systems. *IIE Transactions*, 31(11):1061–1074, 1999.
67. A. Natarajan. Multi-Criteria Supply Chain Inventory Models with Transportation Costs. PhD thesis, The Pennsylvania State University, University Park, PA, May 2007.
68. E. Lesnaia. Optimizing Safety Stock Placement in General Network Supply Chains. Ph.D. thesis, Sloan School of Management, Massachusetts Institute of Technology, Cambridge, MA, September 2004. URL http://web.mit.edu/sgraves/www/LesnaiaThesis.pdf.

69. S.C. Graves and S.P. Willems. Supply chain design: Safety stock placement and supply chain configuration. In A.G. de Kok and S.C. Graves, editors, *Supply Chain Management: Design, Coordination and Operation*, volume 11 of *Handbooks in Operations Research and Management Science*, chapter 6, pages 95–132. Elsevier, Boston, 2003.
70. C. Billington, G. Callioni, B. Crane, J.D. Ruark, J.U. Rapp, T. White, and S.P. Willems. Accelerating the profitability of Hewlett-Packard's supply chains. *Interfaces*, 34(1):59–72, 2004.
71. A. Rangarajan and V.D.R. Guide, Jr. Supply chain risk management: An overview. Working paper, January 2006.
72. M. Fleischmann, J.M. Bloemhof-Ruwaard, R. Dekker, E. van der Laan, J.A.E.E. van Nunen, and L.N. Van Wassenhove. Quantitative models for reverse logistics: A review. *European Journal of Operational Research*, 103:1–17, 1997.
73. V.D.R. Guide, Jr. and L.N. Van Wassenhove. Closed-loop supply chains: An introduction to the feature issue. *Production and Operations Management*, 15(3 & 4), 2006.
74. V.D.R. Guide, Jr., T. Harrison, and L.N. Van Wassenhove. The challenge of closed-loop supply chains. *Interfaces*, 33(6):3–6, 2003.

11
Complexity and Large-Scale Networks

Hari P. Thadakamalla,
Soundar R. T. Kumara,
and Réka Albert
Pennsylvania State University

11.1 Introduction.. 11-1
11.2 Statistical Properties of Complex Networks....... 11-6
  Average Path Length and the Small-World Effect •
  Clustering Coefficient • Degree Distribution •
  Betweenness Centrality • Modularity and Community
  Structures • Network Resilience
11.3 Modeling of Complex Networks...................... 11-11
  Random Graphs • Small-World Networks • Scale-Free
  Networks
11.4 Why "Complex" Networks.......................... 11-16
11.5 Optimization in Complex Networks 11-18
  Network Resilience to Node Failures • Local Search •
  Other Topics
11.6 Conclusions 11-26
Acknowledgments 11-27
References 11-27

11.1 Introduction

In the past few decades, graph theory has been a powerful analytical tool for understanding and solving various problems in operations research (OR). The study of graphs (or networks) traces back to the solution of the Königsberg bridge problem by Euler in 1735. In Königsberg, the river Preger flows through the town dividing it into four land areas as shown in Figure 11.1a. These land areas are connected by seven different bridges. The Königsberg bridge problem is to find whether it is possible to traverse through the city on a route that crosses each bridge exactly once, and return to the starting point. Euler formulated the problem using a graph theoretical representation and proved that the traversal is not possible. He represented each land area as a vertex (or node) and each bridge as an edge between two nodes (land areas), as shown in Figure 11.1b. Then, he posed the question as whether there exists a path such that it passes every edge exactly once and ends at the start node. This path was later termed an Eulerian circuit. Euler proved that for a graph to have an Eulerian circuit, all the nodes in the graph need to have an even degree.

Euler's great insight lay in representing the Königsberg bridge problem as a graph problem with a set of vertices and edges. Later, in the twentieth century, graph theory developed into a substantial area of study that is applied to solve various problems in engineering and several other disciplines [1]. For example, consider the problem of finding the shortest route

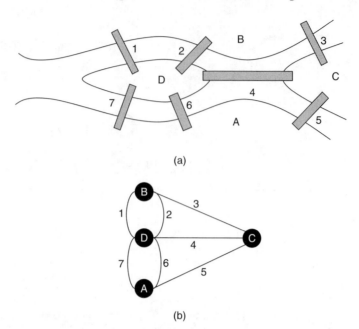

FIGURE 11.1 Königsberg bridge problem: (a) shows the river flowing through the town dividing it into four land areas A, B, C, and D. The land areas are connected by seven bridges numbered from 1 to 7. (b) Graph theoretical representation of the Königsberg bridge problem. Each node represents a land area and the edge between them represents the bridges connecting the land areas.

between two geographical points. The problem can be modeled as a shortest path problem on a network, where different geographical points are represented as nodes and they are connected by an edge if there exists a direct path between the two nodes. The weights on the edges represent the distance between the two nodes (see Figure 11.2). Let the network be $G(V, E)$ where V is the set of all nodes, E is the set of edges (i, j) connecting the nodes, and w is a function such that w_{ij} is the weight of the edge (i, j). The shortest path problem from node s to node t can be formulated as follows:

$$\text{minimize} \sum_{(i,j) \in \xi} w_{ij} x_{ij}$$

$$\text{subject to} \sum_{\{j|(i,j) \in \xi\}} x_{ij} - \sum_{\{j|(j,i) \in \xi\}} x_{ij} = \begin{cases} 1 & \text{if } i = s \\ -1 & \text{if } i = t \\ 0 & \text{otherwise} \end{cases}$$

$$x_{ij} \geq 0, \quad \forall (i,j) \in \xi$$

where $x_{ij} = 1$ or 0 depending on whether the edge from node i to node j belongs to the optimal path or not, respectively. Many algorithms have been proposed to solve the shortest path problem [1]. Using one such popular algorithm (Dijkstra's algorithm [1]), we find the shortest path from node 10 to node 30 as (10 - 1 - 3 - 12 - 30) (see Figure 11.2). Note that this problem and similarly other problems considered in traditional graph theory require finding the exact optimal path.

In the last few years there has been an intense amount of activity in understanding and characterizing large-scale networks, which led to the development of a new branch of science

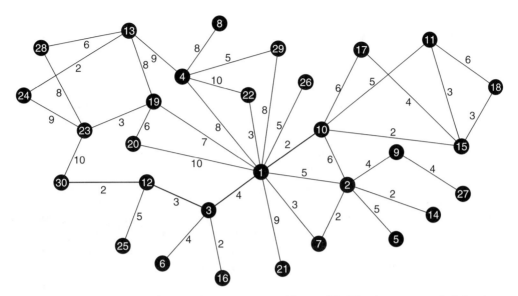

FIGURE 11.2 Illustration of a typical optimization problem in OR. The objective is to find the shortest path from node 10 to node 30. The values on the edges represent the distance between two nodes. Here, we use the exact distances between different nodes to calculate the shortest path 10 - 1 - 3 - 12 - 30.

called "*Network Science*" [2]. The scale of the size of these networks is substantially different from the networks considered in traditional graph theory. Further, these networks do not have any pre-specified structure/order or any design principles. Also, the problems posed in such networks are very different from traditional graph theory. These large-scale networks are referred to as *complex networks*, and we will discuss the reasons why they are termed "complex" networks later in Section 11.4. The following are examples of complex networks:

- *World Wide Web*: It can be viewed as a network where Web pages are the nodes and hyperlinks connecting one webpage to another are the directed edges. The World Wide Web is currently the largest network for which topological information is available. It had approximately *one billion* nodes at the end of 1999 [3] and is continuously growing at an exponential rate. A recent study [4] estimated the size to be 11.5 billion nodes as of January 2005.
- *Internet*: The Internet is a network of computers and telecommunication devices connected by wired or wireless links. The topology of the Internet is studied at two different levels [5]. At the router level, each router is represented as a node and physical connections between them as edges. At the domain level, each domain (autonomous system, Internet provider system) is represented as a node and inter-domain connections by edges. The number of nodes, approximately, at the router level was 150,000 in 2000 [6] and at the domain level was 4000 in 1999 [5].
- *Phone call network*: The phone numbers are the nodes and every completed phone call is an edge directed from the receiver to the caller. Abello et al. [7] constructed a phone call network from the long distance telephone calls made during a single day which had 53,767,087 nodes and over 170 million edges.
- *Power grid network*: Generators, transformers, and substations are the nodes and high-voltage transmission lines are the edges. The power grid network of the western United States had 4941 nodes in 1998 [8]. The North American power grid consisted of 14,099 nodes and 19,657 edges [9] in 2005.

- *Airline network*: Nodes are the airports and an edge between two airports represent the presence of a direct flight connection [10,11]. Barthelemy et al. [10] have analyzed the International Air Transportation Association database to form the world-wide airport network. The resulting network consisted of 3880 nodes and 18,810 edges in 2002.
- *Market graph*: Recently, Boginski et al. [12,13] represented the stock market data as a network where the stocks are nodes and two nodes are connected by an edge if their correlation coefficient calculated over a period of time exceeds a certain threshold value. The network had 6556 nodes and 27,885 edges for the U.S. stock data during the period 2000–2002 [13].
- *Scientific collaboration networks*: Scientists are represented as nodes and two nodes are connected if the two scientists have written an article together. Newman [14,15] studied networks constructed from four different databases spanning biomedical research, high-energy physics, computer science, and physics. One of these networks formed from the Medline database for the period from 1961 to 2001 had 1,520,251 nodes and 2,163,923 edges.
- *Movie actor collaboration network*: Another well-studied network is the movie actor collaboration network, formed from the Internet Movie Database [16], which contains all the movies and their casts since the 1890s. Here again, the actors are represented as nodes and two nodes are connected by an edge if two actors have performed together in a movie. This is a continuously growing network with 225,226 nodes and 13,738,786 edges in 1998 [8].

The above are only a few examples of complex networks pervasive in the real world [17–20]. Tools and techniques developed in the field of traditional graph theory involved studies that looked at networks of tens or hundreds or in extreme cases thousands of nodes. The substantial growth in the size of many such networks (see Figure 11.3) necessitates a different approach for analysis and design. The new methodology applied for analyzing complex networks is similar to the *statistical physics* approach to complex phenomena.

The study of large-scale complex systems has always been an active research area in various branches of science, especially in the physical sciences. Some examples are ferromagnetic properties of materials, statistical description of gases, diffusion, formation of crystals, and so on. For instance, let us consider a box containing one mole (6.022×10^{23}) of gas atoms as our system of analysis (see Figure 11.4a). If we represent the system with the microscopic properties of the individual particles such as their position and velocity, then it would be

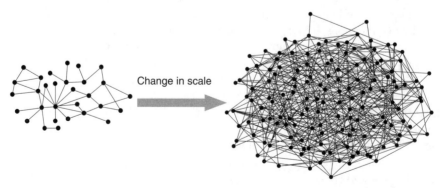

FIGURE 11.3 Pictorial description of the change in scale in the size of the networks found in many engineering systems. This change in size and lack of any order necessitates a change in the analytical approach.

Complexity and Large-Scale Networks

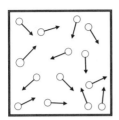

Individual properties: *Position and velocity*

Statistical properties: *Pressure, temperature*

$PV = nRT$

(a)

Individual properties: *Degree, neighbors, edge weights*

Statistical properties: *Degree, distribution, small-worldness*

(b)

FIGURE 11.4 Illustration of the analogy between a box of gas atoms and complex networks. (a) A mole of gas atoms (6.022×10^{23} atoms) in a box. (b) An example of a large-scale network. For analysis, we need to represent both the systems using statistical properties.

next to impossible to analyze the system. Rather, physicists use statistical mechanics to represent the system and calculate macroscopic properties such as temperature, pressure, and the like. Similarly, in networks such as the Internet and WWW, where the number of nodes is extremely large, we have to represent the network using macroscopic properties (such as degree distribution, edge-weight distribution, etc.), rather than the properties of individual entities in the network (such as the neighbors of a given node, the weights on the edges connecting this node to its neighbors, etc.) (see Figure 11.4b). Now let us consider the shortest path problem in such networks (for instance, WWW). We rarely require specific shortest path solutions such as from node A to node B (from Web page A to Web page B). Rather it is useful if we know the average distance (number of hops) taken from any node to any other node (any Web page to any other Web page) to understand dynamical processes (such as search in WWW). This new approach for understanding networked systems provides new techniques as well as challenges for solving conceptual and practical problems in this field. Furthermore, this approach has become feasible and has received a considerable boost by the availability of computers and communication networks that have made the gathering and analysis of large-scale datasets possible.

The objective of this chapter is to introduce this new direction of interdisciplinary research (Network Science) and discuss the new challenges for the OR community. During the last few years, there has been a tremendous amount of research activity dedicated to the study of these large-scale networks. This activity was mainly triggered by significant findings in real-world networks that we will elaborate later in the chapter. There was a revival of network modeling that gave rise to many path breaking results [17–20] and provoked vivid interest across different disciplines of the scientific community. Until now, a major part of this research was contributed by physicists, mathematicians, sociologists, and biologists. However, the ultimate goal of modeling these networks is to understand and optimize the dynamical processes taking place in the network. In this chapter, we address the urgent need and opportunity for the OR community to contribute to the fast-growing interdisciplinary research on Network Science. The methodologies and techniques developed until now will definitely aid the OR community in furthering this research.

The following is the outline of the chapter. In Section 11.2, we introduce different statistical properties that are prominently used for characterizing complex networks. We also present the empirical results obtained for many real complex networks that initiated a revival of network modeling. In Section 11.3, we summarize different evolutionary models proposed to explain the properties of real networks. In particular, we discuss Erdős–Rényi random graphs, small-world networks, and scale-free networks. In Section 11.4, we discuss briefly why these networks are called "complex" networks, rather than large-scale networks. We summarize typical behaviors of complex systems and demonstrate how the real networks have these behaviors. In Section 11.5, we discuss the optimization in complex networks by concentrating on two specific processes, robustness and local search, which are most relevant to engineering networks. We discuss the effects of statistical properties on these processes and demonstrate how they can be optimized. Further, we briefly summarize a few more important topics and give references for further reading. Finally, in Section 11.6, we conclude and discuss future research directions.

11.2 Statistical Properties of Complex Networks

In this section, we explain some of the statistical properties that are prominently used in the literature. These statistical properties help in classifying different kinds of complex networks. We discuss the definitions and present the empirical findings for many real networks.

11.2.1 Average Path Length and the Small-World Effect

Let $G(V, E)$ be a network where V is the collection of *entities* (or nodes) and E is the set of *arcs* (or edges) connecting them. A path between two nodes u and v in the network G is a sequence $[u = u_1, u_2, \ldots, u_n = v]$, where $u'_i s$ are the nodes in G and there exists an edge from u_{i-1} to u_i in G for all i. The path length is defined as the sum of the weights on the edges along the path. If all the edges are equivalent in the network, then the path length is equal to the number of edges (or hops) along the path. The average path length (l) of a connected network is the average of the shortest paths from each node to every other node in a network. It is given by

$$l \equiv \langle d(u,w) \rangle = \frac{1}{N(N-1)} \sum_{u \in V} \sum_{u \neq w \in V} d(u,w)$$

where N is the number of nodes in the network and $d(u, w)$ is the shortest path between u and w. Table 11.1 shows the values of l for many different networks. We observe that despite the large size of the network (w.r.t. the number of nodes), the average path length is small. This implies that any node can reach any other node in the network in a relatively small number of steps. This characteristic phenomenon, that most pairs of nodes are connected by a short path through the network, is called the *small-world effect*.

The existence of the small-world effect was first demonstrated by the famous experiment conducted by Stanley Milgram in the 1960s [24], which led to the popular concept of

TABLE 11.1 Average Path Length of Many Real Networks

Network	Size (number of nodes)	Average path length
WWW [21]	2×10^8	16
Internet, router level [6]	150,000	11
Internet, domain level [5]	4000	4
Movie actors [8]	212,250	4.54
Electronic circuits [22]	24,097	11.05
Peer-to-peer network [23]	880	4.28

Note that despite the large size of the network (w.r.t. the number of nodes), the average path length is very small.

six degrees of separation. In this experiment, Milgram randomly selected individuals from Wichita, Kansas, and Omaha, Nebraska, to pass on a letter to one of their acquaintances by mail. These letters had to finally reach a specific person in Boston, Massachusetts; the name and profession of the target was given to the participants. The participants were asked to send the letter to one of their acquaintances whom they judged to be closer (than themselves) to the target. Anyone who received the letter subsequently would be given the same information and asked to do the same until it reached the target person. Over many trials, the average length of these acquaintance chains for the letters that reached the targeted node was found to be approximately 6. That is, there is an acquaintance path of an average length 6 in the social network of people in the United States. Another interesting and even more surprising observation from this experiment is discussed in Section 11.5.2. Currently, Watts et al. [25] are involved in an Internet-based study to verify this phenomenon.

Mathematically, a network is considered to be small-world if the average path length scales logarithmically or slower with the number of nodes N ($\sim \log N$). For example, say the number of nodes in the network, N, increases from 10^3 to 10^6; then the average path length will increase approximately from 3 to 6. This phenomenon has critical implications on the dynamic processes taking place in the network. For example, if we consider the spread of information, computer viruses, or contagious diseases across a network, the small-world phenomenon implies that within a few steps it could spread to a large fraction of most of the real networks.

11.2.2 Clustering Coefficient

The clustering coefficient characterizes the local transitivity and order in the neighborhood of a node. It is measured in terms of the number of triangles (3-cliques) present in the network. Consider a node i that is connected to k_i other nodes. The number of possible edges between these k_i neighbors that form a triangle is $k_i(k_i - 1)/2$. The clustering coefficient of a node i is the ratio of the number of edges E_i that actually exist between these k_i nodes and the total number $k_i(k_i - 1)/2$ possible, that is,

$$C_i = \frac{2E_i}{k_i(k_i - 1)}$$

The clustering coefficient of the whole network (C) is then the average of $C_i's$ over all the nodes in the network, that is, $C = \frac{1}{n}\sum_i C_i$ (see Figure 11.5). The clustering coefficient is high for many real networks [17,20]. In other words, in many networks if node A is connected to node B and node C, then there is a high probability that node B and node C are also

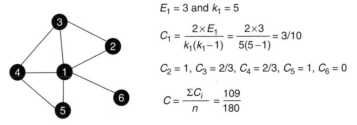

FIGURE 11.5 Calculating the clustering coefficient of a node and the network. For example, node 1 has degree 5 and the number of edges between the neighbors is 3. Hence, the clustering coefficient for node 1 is 3/10. The clustering coefficient of the entire network is the average of the clustering coefficients at each individual node (109/180).

connected. With respect to social networks, it means that it is highly likely that two friends of a person are also friends, a feature analyzed in detail in the so-called theory of balance [26].

11.2.3 Degree Distribution

The degree of a node is the number of edges incident on it. In a directed network, a node has both an in-degree (number of incoming edges) and an out-degree (number of outgoing edges). The degree distribution of the network is the function p_k, where p_k is the probability that a randomly selected node has degree k. Here again, a directed graph has both in-degree and out-degree distributions. It was found that most of the real networks including the World Wide Web [27–29], the Internet [5], metabolic networks [30], phone call networks [7,31], scientific collaboration networks [14,148], and movie actor collaboration networks [33–35] follow a power-law degree distribution ($p(k) \sim k^{-\gamma}$), indicating that the topology of the network is very heterogeneous, with a high fraction of small-degree nodes and a few large-degree nodes. These networks having power-law degree distributions are popularly known as *scale-free networks*. These networks were called scale-free networks because of the lack of a characteristic degree and the broad tail of the degree distribution. Figure 11.6 shows the empirical results for the Internet at the router level and co-authorship network of high-energy physicists. The following are the expected values and variances of the node degree in scale-free networks,

$$E[k] = \begin{cases} finite & \text{if } \gamma > 2 \\ \infty & \text{otherwise} \end{cases} \qquad V[k] = \begin{cases} finite & \text{if } \gamma > 3 \\ \infty & \text{otherwise} \end{cases}$$

where γ is the power-law exponent. Note that the variance of the node degree is infinite when $\gamma < 3$ and the mean is infinite when $\gamma < 2$. The power-law exponent (γ) of most of the networks lies between 2.1 and 3.0, which implies that there is high heterogeneity with respect to node degree. This phenomenon in real networks is critical because it was shown that the heterogeneity has a huge impact on the network properties and processes such as

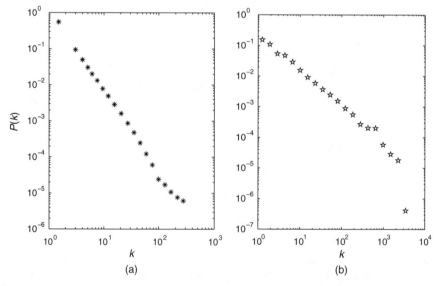

FIGURE 11.6 The degree distribution of real networks. (a) Internet at the router level. Data courtesy of Ramesh Govindan [6]. (b) Co-authorship network of high-energy physicists, after Newman [14].

network resilience [9,36], network navigation, local search [32], and epidemiological processes [37–41]. Later in this chapter, we will discuss the impact of this heterogeneity in detail.

11.2.4 Betweenness Centrality

The betweenness centrality (BC) of a node counts the fraction of shortest paths going through the node. The BC of a node i is given by

$$BC(i) = \sum_{s \neq i \neq t} \frac{\sigma_{st}(i)}{\sigma_{st}}$$

where σ_{st} is the total number of shortest paths from node s to t and $\sigma_{st}(i)$ is the number of these shortest paths passing through node i. If the BC of a node is high, it implies that this node is central and many shortest paths pass through this node. BC was first introduced in the context of social networks [42], and has been recently adopted by Goh et al. [43] as a proxy for the load (l_i) at a node i with respect to transport dynamics in a network. For example, consider the transportation of data packets in the Internet along the shortest paths. If many shortest paths pass through a node then the load on that node would be high. Goh et al. have shown numerically that the load (or BC) distribution follows a power-law, $P_L(l) \sim l^{-\delta}$ with exponent $\delta \approx 2.2$, and is insensitive to the detail of the scale-free network as long as the degree exponent (γ) lies between 2.1 and 3.0. They further showed that there exists a scaling relation $l \sim k^{(\gamma-1)/(\delta-1)}$ between the load and the degree of a node when $2 < \gamma \leq 3$. Later in this chapter, we discuss how this property can be utilized for local search in complex networks. Many other centrality measures exist in literature and a detailed review of these measures can be found in [44].

11.2.5 Modularity and Community Structures

Many real networks are found to exhibit a community structure (also called modular structure). That is, groups of nodes in the network have high density of edges within the group and lower density between the groups (see Figure 11.7). This property was first proposed in social networks [40], where people may divide into groups based on interests, age, profession, and the like. Similar community structures are observed in many networks which reflect the division of nodes into groups based on the node properties [20]. For example, in the WWW it reflects the subject matter or themes of the pages, in citation networks

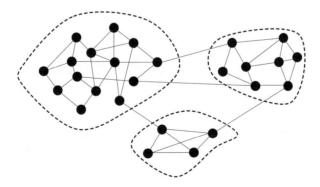

FIGURE 11.7 Illustration of a network with community structure. Communities are defined as a group of nodes in the network that have a higher density of edges within the group than between groups. In the above network, the group of nodes enclosed within a dotted loop is a community.

it reflects the area of research, in cellular and metabolic networks it may reflect functional groups [45,46].

In many ways, community detection is similar to a traditional graph partitioning problem (GPP). In GPP the objective is to divide the nodes of the network into k disjoint sets of specified sizes such that the number of edges between these sets is the minimum. This problem is NP-complete [47] and several heuristic methods [48–50] have been proposed to decrease the computation time. GPP arises in many important engineering problems that include mapping of parallel computations, laying out of circuits (VLSI design), and the ordering of sparse matrix computations [48]. Here, the number of partitions to be made is specified and the size of each partition is restricted. For example, in the mapping of parallel computations, the tasks have to be divided between a specified number of processors such that the communication between the processors is minimized and the loads on the processors are balanced. However, in real networks, we do not have any a priori knowledge about the number of communities into which we should divide or about the size of the communities. The goal is to find the naturally existing communities in the real networks rather than dividing the network into a pre-specified number of groups. As we do not know the exact partitions of network, it is difficult to evaluate the goodness of a given partition. Moreover, there is no unique definition of a community due to the ambiguity of how dense a group should be to form a community. Many possible definitions exist in the literature [42,51–54]. A simple definition given in [51,54] considers a subgraph as a community if each node in the subgraph has more connections within the community than with the rest of the graph. Newman and Girvan [52] have proposed another measure that calculates the fraction of links within the community minus the expected value of the same quantity in a randomized counterpart of the network. The higher this difference, the stronger is the community structure. It is important to note that in spite of this ambiguity, the presence of community structures is a common phenomenon across many real networks. Algorithms for detecting these communities are briefly discussed in Section 11.5.3.

11.2.6 Network Resilience

The ability of a network to withstand removal of nodes/edges is called network resilience or robustness. In general, the removal of nodes and edges disrupts the paths between nodes and can increase the distances thus making the communication between nodes harder. In more severe cases, an initially connected network can break down into isolated components that cannot communicate anymore. Figure 11.8 shows the effect of the removal of nodes/edges on a network. Observe that as we remove more nodes and edges, the network disintegrates into many components. There are different ways of removing nodes and edges to test the robustness of a network. For example, one can remove nodes at random with uniform probability or by selectively targeting certain classes of nodes, such as nodes with high degree. Usually, the removal of nodes at random is termed as random failures and the removal of nodes with higher degree is termed as targeted attacks; other removal strategies are discussed in detail in Ref. [55]. Similarly, there are several ways of measuring the degradation of the network performance after the removal. One simple way to measure it is to calculate the decrease in size of the largest connected component in the network. A connected component is a part of the network in which a path exists between any two nodes in that component and the largest connected component is the largest among the connected components. The less the decrease in the size of the largest connected component, the better the robustness of the network. In Figure 11.8, the size of the largest connected component decreases from 13 to 9 and then to 5. Another way to measure robustness is to calculate the increase of the average path length in the largest connected component. Malfunctioning of nodes/edges eliminates some

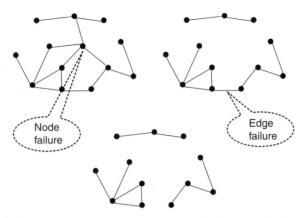

FIGURE 11.8 The effects of removing a node or an edge in the network. Observe that as we remove more nodes and edges the network disintegrates into small components/clusters.

existing paths and generally increases the distance between the remaining nodes. Again, the less the increase, the better the robustness of the network. We discuss more about network resilience and robustness with respect to optimization in Section 11.5.1.

11.3 Modeling of Complex Networks

In this section, we give a brief summary of different models for complex networks. Most of the modeling efforts focused on understanding the underlying process involved during the network evolution and capture the above-mentioned properties of real networks. In specific, we concentrate on three prominent models, namely, the Erdős–Rényi random graph model, the Watts–Strogatz small-world network model, and the Barabási–Albert scale-free network model.

11.3.1 Random Graphs

One of the earliest theoretical models for complex networks was given by Erdős and Rényi [56–58] in the 1950s and 1960s. They proposed uniform random graphs for modeling complex networks with no obvious pattern or structure. The following is the evolutionary model given by Erdős and Rényi:

- Start with a set of N isolated nodes.
- Connect each pair of nodes with a connection probability p.

Figure 11.9 illustrates two realizations for the Erdős–Rényi random graph model (ER random graphs) for different connection probabilities. Erdős and Rényi have shown that at $p_c \simeq 1/N$, the ER random graph abruptly changes its topology from a loose collection of small clusters to one which has a giant connected component. Figure 11.10 shows the change in size of the largest connected component in the network as the value of p increases, for $N = 1000$. We observe that there exists a threshold $p_c = 0.001$ such that when $p < p_c$, the network is composed of small isolated clusters and when $p > p_c$ a giant component suddenly appears. This phenomenon is similar to the *percolation transition*, a topic well studied in both mathematics and statistical mechanics [17].

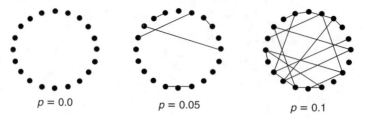

FIGURE 11.9 An Erdős–Rényi random graph that starts with $N=20$ isolated nodes and connects any two nodes with a probability p. As the value of p increases the the number of edges in the network increase.

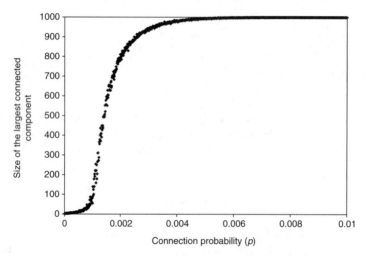

FIGURE 11.10 Illustration of percolation transition for the size of the largest connected component in the Erdős–Rényi random graph model. Note that there exists $p_c = 0.001$ such that when $p < p_c$, the network is composed of small isolated clusters and when $p > p_c$ a giant component suddenly appears.

In an ER random graph, the mean number of neighbors at a distance (number of hops) d from a node is approximately $<k>^d$, where $<k>$ is the average degree of the network. To cover all the nodes in the network, the distance (l) should be such that $<k>^l \sim N$. Thus, the average path length is given by $l = \frac{\log N}{\log <k>}$, which scales logarithmically with the number of nodes N. This is only an approximate argument for illustration; a rigorous proof can be found in Ref. [59]. Hence, ER random graphs are small-world. The clustering coefficient of the ER random graphs is found to be low. If we consider a node and its neighbors in an ER random graph, then the probability that two of these neighbors are connected is equal to p (probability that two randomly chosen neighbors are connected). Hence, the clustering coefficient of an ER random graph is $p = \frac{<k>}{N}$, which is small for large sparse networks. Now, let us calculate the degree distribution of the ER random graphs. The total number of edges in the network is a random variable with an expected value of $pN(N-1)/2$, and the number of edges incident on a node (the node degree) follows a binomial distribution with parameters $N-1$ and p,

$$p(k_i = k) = C_{N-1}^k p^k (1-p)^{N-1-k}$$

This implies that in the limit of large N, the probability that a given node has degree k approaches a Poisson distribution, $p(k) = \frac{<k>^k e^{-<k>}}{k!}$. Hence, ER random graphs are

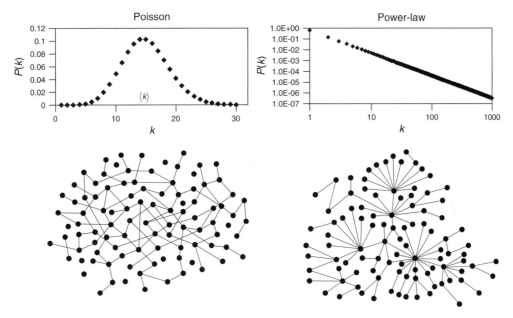

FIGURE 11.11 Comparison of networks with Poisson and power-law degree distribution of the same size. Note that the network with Poisson distribution is homogenous in node degree. Most of the nodes in the network have the same degree, which is close to the average degree of the network. However, the network with power-law degree distribution is highly heterogenous in node degree. There are a few nodes with large degree and many nodes with small degree.

statistically homogenous in node degree as the majority of the nodes have a degree close to the average, and significantly small and large node degrees are exponentially rare.

ER random graphs were used to model complex networks for a long time [59]. The model was intuitive and analytically tractable; moreover, the average path length of real networks is close to the average path length of an ER random graph of the same size [17]. However, recent studies on the topologies of diverse large-scale networks found in nature indicated that they have significantly different properties from ER random graphs [17–20]. It has been found [8] that the average clustering coefficient of real networks is significantly larger than the average clustering coefficient of ER random graphs with the same number of nodes and edges, indicating a far more ordered structure in real networks. Moreover, the degree distribution of many large-scale networks is found to follow a power-law $p(k) \sim k^{-\gamma}$. Figure 11.11 compares two networks with Poisson and power-law degree distributions. We observe that there is a remarkable difference between these networks. The network with Poisson degree distribution is more homogenous in node degree, whereas the network with power-law distribution is highly heterogenous. These discoveries along with others related to the mixing patterns of complex networks [17–20] initiated a revival of network modeling in the past few years.

Nonuniform random graphs are also studied [31,60–64] to mimic the properties of real-world networks, in specific, power-law degree distribution. Typically, these models specify either a degree sequence, which is a set of N values of the degrees k_i of nodes $i = 1, 2, \ldots, N$ or a degree distribution $p(k)$. If a degree distribution is specified then the sequence is formed by generating N random values from this distribution. This can be thought of as giving each node i in the network k_i "stubs" sticking out of it and then pairs of these stubs are connected randomly to form complete edges [64]. Molloy and Reed [62] have proved that for a random graph with a degree distribution $p(k)$, a giant connected component emerges almost surely when $\sum_{k\geq 1} k(k-2)p(k) > 0$, provided that the maximum degree is less than $N^{1/4}$. Later, Aiello et al. [31,60] introduced a two-parameter random graph model $P(\alpha, \gamma)$ for power-law graphs with exponent γ described as follows: Let n_k be the number of nodes with degree k, such that n_k and k satisfy $\log n_k = \alpha - \gamma \log k$. The total number of nodes in the network can be computed, noting that the maximum degree of a node in the network is $e^{\alpha/\gamma}$. Using the results from Molloy and Reed [62], they showed that there is almost surely a unique giant connected component if $\gamma < \gamma_0 = 3.47875\ldots$, whereas, there is no giant connected component almost surely when $\gamma > \gamma_0$.

Newman et al. [64] have developed a general approach to random graphs by using a generating function formalism [65]. The generating function for the degree distribution p_k is given by $G_0(x) = \sum_{k=0}^{\infty} p_k x^k$. This function captures all the information present in the original distribution since $p_k = \frac{1}{k!} \frac{d^k G_0}{dx^k}|_{x=0}$. The average degree of a randomly chosen node would be $<k> = \sum_k k p(k) = G_0'(1)$. Further, this formulation helps in calculating other properties of the network [64]. For instance, we can approximately calculate the relation for the average path length of the network. Let us consider the degree of the node reached by following a randomly chosen edge. If the degree of this node is k, then we are k times more likely to reach this node than a node of degree 1. Thus, the degree distribution of the node arrived by a randomly chosen edge is given by kp_k and not p_k. In addition, the distribution of the number of edges from this node (one less than the degree) q_k, is $\frac{(k+1)p_{k+1}}{\sum_k k p_k} = \frac{(k+1)p_{k+1}}{<k>}$. Thus, the generating function for q_k is given by $G_1(x) = \frac{\sum_{k=0}^{\infty}(k+1)p_{k+1}x^k}{k} = \frac{G_0'(x)}{G_0'(1)}$. Note that the distribution of the number of first neighbors of a randomly chosen node (degree of a node) is $G_0(x)$. Hence, the distribution of the number of second neighbors from the same randomly chosen node would be $G_0(G_1(x)) = \sum_k p_k [G_1(x)]^k$. Here, the probability that any of the second neighbors is connected to the first neighbors or to one another scales as N^{-1} and can be neglected in the limit of a large N. This implies that the average number of second neighbors is given by $\left[\frac{\partial}{\partial x} G_0(G_1(x))\right]_{x=1} = G_0'(1) G_1'(1)$. Extending this method of calculating the average number of nearest neighbors, we find that the average number of mth neighbors, z_m, is $[G_1'(1)]^{m-1} G_0'(1) = \left[\frac{z_2}{z_1}\right]^{m-1} z_1$. Now, let us start from a node and find the number of first neighbors, second, third $\ldots$ mth neighbors. Assuming that all the nodes in the network can be reached within l steps, we have $1 + \sum_{m=1}^{l} z_m = N$. As for most graphs $N \gg z_1$ and $z_2 \gg z_1$, we obtain the average path length of the network $l = \frac{N/z_1}{z_2/z_1} + 1$. The generating function formalism can further be extended to include other features such as directed graphs, bipartite graphs, and degree correlations [20].

Another class of random graphs that are especially popular in modeling social networks is the exponential random graphs models (ERGMs) or p* models [66–70]. The ERGM consists of a family of possible networks of N nodes in which each network G appears with probability $P(G) = \frac{1}{Z} \exp(-\sum_i \theta_i \epsilon_i)$, where the function Z is $\sum_G \exp(-\sum_i \theta_i \epsilon_i)$. This is similar to the Boltzmann ensemble of statistical mechanics with Z as the partition function [20]. Here, $\{\epsilon_i\}$ is the set of observables or measurable properties of the network such as number of nodes with a certain degree, number of triangles, and so on. $\{\theta_i\}$ are an adjustable set of parameters for the model. The ensemble average of a property ϵ_i is given as $\langle \epsilon_i \rangle = \sum_G \epsilon_i(G) P(G) = \frac{1}{Z} \epsilon_i \exp(-\sum_i \theta_i \epsilon_i) = \frac{\partial f}{\partial \theta_i}$. The major advantage of these models is that they can represent any kind of structural tendencies such as dyad and triangle formations. A detailed review of the parameter estimation techniques can be found in Refs. [66,71]. Once the parameters $\{\theta_i\}$ are specified, the networks can be generated by using the Gibbs or Metropolis–Hastings sampling methods [71].

11.3.2 Small-World Networks

Watts and Strogatz [8] presented a small-world network model to explain the existence of high clustering and small average path length simultaneously in many real networks, especially social networks. They argued that most of the real networks are neither completely regular nor completely random, but lie somewhere between these two extremes. The Watts–Strogatz model starts with a regular lattice on N nodes and each edge is rewired with a certain probability p. The following is the algorithm for the model:

- Start with a regular ring lattice on N nodes where each node is connected to its first k neighbors.
- Randomly rewire each edge with a probability p such that one end remains the same and the other end is chosen uniformly at random. The other end is chosen without allowing multiple edges (more than one edge joining a pair of nodes) and loops (edges joining a node to itself).

The resulting network is a regular network when $p=0$ and a random graph when $p=1$, since all the edges are rewired (see Figure 11.12). The above model is inspired from social networks where people are friends with their immediate neighbors such as neighbors on the street, colleagues at work, and so on (the connections in the regular lattice). Also, each person has a few friends who are a long way away (long-range connections attained by random rewiring). Later, Newman [72] proposed a similar model where instead of edge rewiring, new edges are introduced with probability p. The clustering coefficient of the Watts–Strogatz model and the Newman model are

$$C_{WS} = \frac{3(k-1)}{2(2k-1)}(1-p)^3 \quad C_N = \frac{3(k-1)}{2(2k-1)+4kp(p+2)}$$

respectively. This class of networks displays a high clustering coefficient for small values of p as we start with a regular lattice. However, even for small values of p, the average path length falls rapidly due to the few long-range connections. This coexistence of high clustering coefficient and small average path length is in excellent agreement with the characteristics of many real networks [8,72]. The degree distribution of both models depends on the parameter p, evolving from a univalued peak corresponding to the initial degree k to a

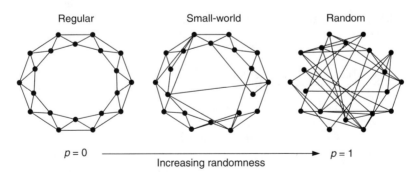

FIGURE 11.12 Illustration of the random rewiring process for the Watts–Strogatz model. This model interpolates between a regular ring lattice and a random network, without changing the number of vertices ($N=20$) or edges ($E=40$) in the graph. When $p=0$ the graph is regular (each node has four edges); as p increases, the graph becomes increasingly disordered until $p=1$; all the edges are rewired randomly. After Watts and Strogatz, 1998 [8].

somewhat broader but still peaked distribution. Thus, small-world models are even more homogeneous than random graphs, which is not the case with real networks.

11.3.3 Scale-Free Networks

As mentioned earlier, many real networks, including the World Wide Web [27–29], the Internet [5], peer-to-peer networks [23], metabolic networks [30], phone call networks [7,31], and movie actor collaboration networks [33–35], are scale-free, that is, their degree distribution follows a power-law, $p(k) \sim k^{-\gamma}$. Barabási and Albert [35] addressed the origin of this power-law degree distribution in many real networks. They argued that a static random graph or Watts–Strogatz model fails to capture two important features of large-scale networks: their constant growth and the inherent selectivity in edge creation. Complex networks like the World Wide Web, collaboration networks, and even biological networks are growing continuously by the creation of new Web pages, start of new researches, and by gene duplication and evolution. Moreover, unlike random networks where each node has the same probability of acquiring a new edge, new nodes entering the network do not connect uniformly to existing nodes, but attach preferentially to nodes of a higher degree. This reasoning led them to define the following mechanism:

- Growth: Start with a small number of connected nodes say m_0 and assume that every time a node enters the system, m edges are pointing from it, where $m < m_0$.
- Preferential attachment: Every time a new node enters the system, each edge of the newly entered node preferentially attaches to an already existing node i with degree k_i with the following probability:

$$\Pi_i = \frac{k_i}{\sum_j k_j}$$

It was shown that such a mechanism leads to a network with power-law degree distribution $p(k) = k^{-\gamma}$ with exponent $\gamma = 3$. These networks were called scale-free networks because of the lack of a characteristic degree and the broad tail of the degree distribution. The average path length of this network scales as $\frac{\log(N)}{\log(\log(N))}$ and thus displays small-world property. The clustering coefficient of a scale-free network is approximately $C \sim \frac{(\log N)^2}{N}$, which is a slower decay than $C = <k> N^{-1}$ decay observed in random graphs [73]. In the years following the proposal of the first scale-free model a large number of more refined models were introduced, leading to a well-developed theory of evolving networks [17–20].

11.4 Why "Complex" Networks

In this section, we will discuss why these large-scale networks are termed as "complex" networks. The reason is not merely because of the large size of the network, though the complexity does arise due to the size of the network. One must also distinguish "complex systems" from "complicated systems" [74]. Consider an airplane as an example. Even though it is a complicated system, we know its components and the rules governing its functioning. However, this is not the case with complex systems. Complex systems are characterized by diverse behaviors that emerge as a result of nonlinear spatio-temporal interactions among a large number of components [75]. These emergent behaviors cannot be fully explained by just understanding the properties of the individual components/constituents. Examples of such complex systems include ecosystems, economies, various organizations/societies,

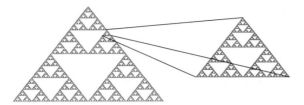

FIGURE 11.13 Illustration of self-similarity in the Sierpinski triangle. When we look at a small part of the triangle at a different scale, then it looks similar to the original triangle. Moreover, at each scale we look at the triangle, it is self-similar. This is a typical behavior of a scale invariant system.

the nervous system, the human brain, ant hills... the list goes on. Some of the behaviors exhibited by complex systems are discussed below:

- *Scale invariance or self-similarity*: A system is scale invariant if the structure of the system is similar regardless of the scale. Typical examples of scale invariant systems are fractals. For example, consider the Sierpinski triangle in Figure 11.13. Note that if we look at a small part of the triangle at a different scale, then it still looks similar to the original triangle. Similarly, at whichever scale we look at the triangle, it is self-similar and hence scale invariant.
- *Infinite susceptibility/response*: Most of the complex systems are highly sensitive or susceptive to changes in certain conditions. A small change in the system conditions or parameters may lead to a huge change in the global behavior. This is similar to the percolation threshold where a small change in connection probability induces the emergence of a giant connected cluster. Another good example of such a system is a sand pile. When we are adding more sand particles to a sand pile, they keep accumulating. But after reaching a certain point, an addition of one more small particle may lead to an avalanche, demonstrating that a sand pile is highly sensitive.
- *Self-organization and emergence*: Self-organization is the characteristic of a system by which it can evolve itself into a particular structure based on interactions between the constituents and without any external influence. Self-organization typically leads to an emergent behavior. Emergent behavior is a phenomenon in which the system global property is not evident from those of its individual parts. A completely new property arises from the interactions between the different constituents of the system. For example, consider an ant colony. Although a single ant (a constituent of an ant colony) can perform a very limited number of tasks in its lifetime, a large number of ants interact in an ant colony, which leads to more complex emergent behaviors.

Now let us consider the real large-scale networks such as the Internet, the WWW, and other networks mentioned in Section 11.1. Most of these networks have power-law degree distribution that does not have any specific scale [35]. This implies that the networks do not have any characteristic degree and an average behavior of the system is not typical (see Figure 11.11b). Due to these reasons, they are called scale-free networks. This heavy-tailed degree distribution induces a high level of heterogeneity in the degrees of the vertices. The heterogeneity makes the network highly sensitive to external disturbances. For example, consider the network shown in Figure 11.14a. This network is highly sensitive when we remove just two nodes in the network. It completely disintegrates into small components. On the other hand, the network shown in Figure 11.14b, having the same number of nodes

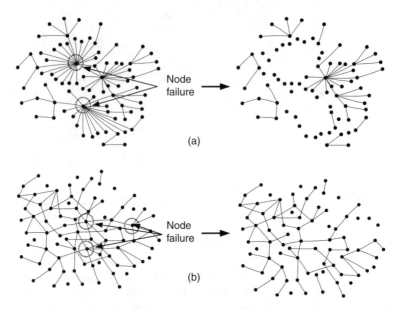

FIGURE 11.14 Illustration of high sensitivity phenomena in complex networks. (a) Observe that when we remove the two highest degree nodes from the network, it disintegrates into small parts. The network is highly sensitive to node removals. (b) Example of a network with the same number of nodes and edges which is not sensitive. This network is not affected much when we remove the three highest degree nodes. The network in (a) is highly sensitive due to the presence of high heterogeneity in node degree.

and edges, is not very sensitive. Most real networks are found to have a structure similar to the network shown in Figure 11.14a, with a huge heterogeneity in node degree. Also, studies [37–41] have shown that the presence of heterogeneity has a huge impact on epidemiological processes such as disease spreading. They have shown that in networks that do not have a heavy-tailed degree distribution, if the disease transmission rate is less than a certain threshold, it will not cause an epidemic or a major outbreak. However, if the network has power-law or scale-free distribution, it becomes highly sensitive to disease propagation. They further show that no matter what the transmission rate is, there exists a finite probability that the infection will cause a major outbreak. Hence, we clearly see that these real large-scale networks are highly sensitive or infinitely susceptible. Further, all these networks have evolved over time with new nodes joining the network (and some leaving) according to some self-organizing or evolutionary rules. There is no external influence that controls the evolution process or structure of the network. Nevertheless, these networks have evolved in such a manner that they exhibit complex behaviors such as power-law degree distributions and many others. Hence, they are called "complex" networks [76].

The above discussion on complexity is an intuitive explanation rather than one of technical details. More rigorous mathematical definitions of complexity can be found in Refs. [77] and [78].

11.5 Optimization in Complex Networks

The models discussed in Section 11.3 are focused on explaining the evolution and growth process of many large real networks. They mainly concentrate on the statistical properties of real networks and network modeling. *But the ultimate goal in studying and modeling the* structure of complex networks is to understand and optimize the processes taking place on these networks. For example, one would like to understand how the structure of the Internet affects its survivability against random failures or intentional attacks, how the structure of the WWW helps in efficient surfing or search on the Web, how the structure of social networks affects the spread of diseases, and so on. In other words, to design rules for optimization, one has to understand the interactions between the structure of the network and the processes taking place on the network. These principles will certainly help in redesigning or restructuring the existing networks and perhaps even help in designing a network from scratch. In the past few years, there has been a tremendous amount of effort by the research communities of different disciplines to understand the processes taking place on networks [17–20]. In this chapter, we concentrate on two processes, namely node failures and local search, because of their high relevance to engineering systems, and discuss a few other topics briefly.

11.5.1 Network Resilience to Node Failures

All real networks are regularly subject to node/edge failures due to either normal malfunctions (random failures) or intentional attacks (targeted attacks) [9,36]. Hence, it is extremely important for the network to be robust against such failures for proper functioning. Albert et al. [36] demonstrated that the topological structure of the network plays a major role in its response to node/edge removal. They showed that most real networks are extremely resilient to random failures. On the other hand, they are very sensitive to targeted attacks. They attribute this to the fact that most of these networks are scale-free networks, which are highly heterogenous in node degree. As a large fraction of nodes have small degree, random failures do not have any effect on the structure of the network. On the other hand, the removal of a few highly connected nodes that maintain the connectivity of the network drastically changes the topology of the network. For example, consider the Internet: despite frequent router problems in the network, we rarely experience global effects. However, if a few critical nodes in the Internet are removed then it would lead to a devastating effect. Figure 11.15 shows the decrease in the size of the largest connected component for both scale-free networks and ER graphs due to random failures and targeted attacks. ER graphs are homogenous in node degree; that is, all the nodes in the network have approximately the same degree. Hence, they behave almost similarly for both random failures and targeted attacks (see Figure 11.15a). In contrast, for scale-free networks, the size of the largest connected component decreases slowly for random failures and drastically for targeted attacks (see Figure 11.15b).

Ideally, we would like to have a network that is as resilient as scale-free networks to random failures and as resilient as random graphs to targeted attacks. To determine the feasibility of modeling such a network, Valente et al. [79] and Paul et al. [80] have studied the following optimization problem: "What is the optimal degree distribution of a network of size N nodes that maximizes the robustness of the network to both random failures and targeted attacks with the constraint that the number of edges remain the same?"

Note that we can always improve the robustness by increasing the number of edges in the network (for instance, a completely connected network will be the most robust network for both random failures and targeted attacks). Hence the problem has a constraint on the number of edges. Valente et al. [79] showed that the optimal network configuration is very different from both scale-free networks and random graphs. They showed that the optimal networks that maximize robustness for both random failures and targeted attacks have at most three distinct node degrees and hence the degree distribution is three-peaked. Similar results were demonstrated by Paul et al. [80]. They showed that the optimal network

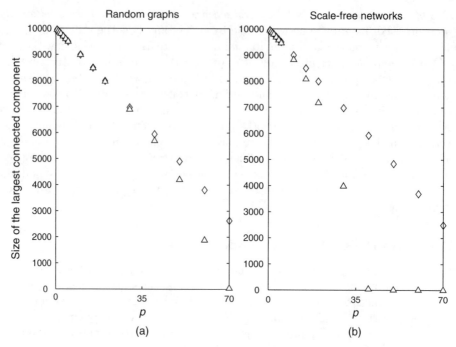

FIGURE 11.15 The size of the largest connected component as the percentage number of nodes (p) removed from the networks due to random failures ($\diamond$) and targeted attacks ($\triangle$). (a) ER graph with number of nodes (N) = 10,000 and mean degree $<k>$ = 4; (b) scale-free networks generated by the Barabási–Albert model with N = 10,000 and $<k>$ = 4. The behavior with respective to random failures and targeted attacks is similar for random graphs. Scale-free networks are highly sensitive to targeted attacks and robust to random failures.

design is one in which all the nodes in the network except one have the same degree, k_1 (which is close to the average degree), and one node has a very large degree, $k_2 \sim N^{2/3}$, where N is the number of nodes. However, these optimal networks may not be practically feasible because of the requirement that each node has a limited repertoire of degrees.

Many different evolutionary algorithms have also been proposed to design an optimal network configuration that is robust to both random failures and targeted attacks [81–85]. In particular, Thadakamalla et al. [84] consider two other measures, *responsiveness* and *flexibility*, along with robustness for random failures and targeted attacks, specifically for supply-chain networks. They define responsiveness as the ability of a network to provide timely services with effective navigation and measure it in terms of the average path length of the network. The lower the average path length, the better is the responsiveness of the network. Flexibility is the ability of the network to have alternate paths for dynamic rerouting. Good clustering properties ensure the presence of alternate paths, and the flexibility of a network is measured in terms of the clustering coefficient. They designed a parameterized evolutionary algorithm for supply-chain networks and analyzed the performance with respect to these three measures. Through simulation they have shown that there exist tradeoffs between these measures and proposed different ways to improve these properties. However, it is still unclear as to what would be the optimal configuration of such survivable networks. The research question would be "what is the optimal configuration of a network of N nodes that maximizes the robustness to random failures, targeted attacks, flexibility, and responsiveness, with the constraint that the number of edges remain the same?"

Until now, we have focused on the effects of node removal on the static properties of a network. However, in many real networks, the removal of nodes will also have dynamic effects on the network as it leads to avalanches of breakdowns also called *cascading failures*. For instance, in a power transmission grid, the removal of nodes (power stations) changes the balance of flows and leads to a global redistribution of loads over all the network. In some cases, this may not be tolerated and might trigger a cascade of overload failures [86], as happened on August 10, 1996, in eleven U.S. states and two Canadian provinces [87]. Models of cascades of irreversible [88] or reversible [89] overload failures have demonstrated that the removal of even a small fraction of highly loaded nodes can trigger global cascades if the load distribution of the nodes is heterogenous. Hence, cascade-based attacks can be much more destructive than any other strategies considered in Refs. [36,55]. Later, Motter [90] showed that a defense strategy based on a selective further removal of nodes and edges, right after the initial attack or failure, can drastically reduce the size of the cascade. Other studies on cascading failures include Refs. [91–95].

11.5.2 Local Search

One of the important research problems that has many applications in engineering systems is searching in complex networks. Local search is the process in which a node tries to find a network path to a target node using only local information. By local information, we mean that each node has information only about its first or perhaps second neighbors and it is not aware of nodes at a larger distance and how they are connected in the network. This is an intriguing and relatively little studied problem that has many practical applications. Let us suppose some required information such as computer files or sensor data is stored at the nodes of a distributed network or database. Then, to quickly determine the location of particular information, one should have efficient local (decentralized) search strategies. Note that this is different from neighborhood search strategies used for solving combinatorial optimization problems [96]. For example, consider the networks shown in Figure 11.16a and 11.16b. The objective is for node 1 to send a message to node 30 in the shortest possible path. In the network shown in Figure 11.16a, each node has global connectivity information

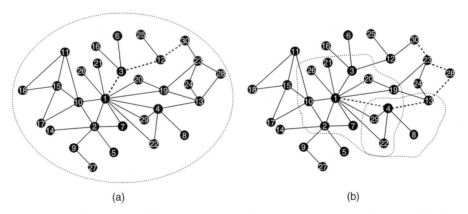

FIGURE 11.16 Illustration of different ways of sending message from node 1 to node 30. (a) In this case, each node has global connectivity information about the whole network. Hence, node 1 calculates the optimal path and send the message through this path. (b) In this case, each node has only information about its neighbors (as shown by the dotted curve). Using this local information, node 1 tries to send the message to node 30. The path obtained is longer than the optimal path.

about the network (i.e., how each and every node is connected in the network). In such a case, node 1 can calculate the optimal path using traditional algorithms [1] and send the message through this path (1 - 3 - 12 - 30, depicted by the dotted line). Next, consider the network shown in Figure 11.16b, in which each node knows only about its immediate neighbors. Node 1, based on some search algorithm, chooses to send the message to one of its neighbors: in this case, node 4. Similarly, node 4 also has only local information, and uses the same search algorithm to send the message to node 13. This process continues until the message reaches the target node. We can clearly see that the search path obtained (1 - 4 - 13 - 28 - 23 - 30) is not optimal. However, given that we have only local information available, the problem tries to design optimal search algorithms in complex networks. The algorithms discussed in this section may look similar to "distributed routing algorithms" that are abundant in wireless ad hoc and sensor networks [97,98]. However, the main difference is that the former try to exploit the statistical properties of the network topology, whereas the latter do not. Most of the algorithms in wireless sensor networks literature find a path to the target node either by broadcasting or random walk and then concentrate on efficient routing of the data from the start node to the end node [97,99]. As we will see in this section, the statistical properties of the networks have a significant effect on the search process. Hence, the algorithms in wireless sensor networks could be integrated with these results for better performance.

We discuss this problem for two types of networks. In the first type of network, the global position of the target node can be quantified and each node has this information. This information will guide the search process in reaching the target node. For example, if we look at the network considered in Milgram's experiment each person has the geographical and professional information about the target node. All the intermediary people (or nodes) use this information as a guide for passing the messages, whereas, in the second type of network, we cannot quantify the global position of the target node. In this case, during the search process, we would not know whether a step in the search process is going toward the target node or away from it. This makes the local search process even more difficult. One such kind of network is the peer-to-peer network, *Gnutella* [100], where the network structure is such that one may know very little information about the location of the target node. Here, when a user is searching for a file, the user does not know the global position of the node that has the file. Further, when the user sends a request to one of its neighbors, it is difficult to find out whether this step is toward the target node or away from it. For lack of a more suitable name, we call the networks of the first type spatial networks and the networks of the second type nonspatial networks. In this chapter, we focus more on search in non-spatial networks.

Search in Spatial Networks

The problem of local search goes back to the famous experiment by Stanley Milgram [24] (discussed in Section 11.2) illustrating the short distances in social networks. Another important observation of the experiment, which is even more surprising, is the ability of these nodes to find these short paths using just local information. As pointed out by Kleinberg [101–103], this is not a trivial statement because most of the time people have only local information in the network. That is the information about their immediate friends or perhaps their friends' friends. They do not have global information about the acquaintances of all people in the network. Even in Milgram's experiment, the people to whom he gave the letters had only local information about the entire social network. Still, from the results of the experiment, we can see that arbitrary pairs of strangers are able to find short chains of acquaintances between them by using only local information alone. Many models have been proposed to explain the existence of such short paths [8,17–20,72]. However, these models are not sufficient to explain the second phenomenon. The observations from Milgram's experiment suggest that there is something more embedded in the underlying social network that guides the message implicitly from the source to the target. Such networks which are inherently easy to search are called *searchable networks*. Mathematically, a network is searchable if the length of the search path obtained scales logarithmically with the number of nodes N ($\sim \log N$) or lesser. Kleinberg demonstrated that the emergence of such a phenomenon requires special topological features [101–103]. Considering a family of network models on an n-dimensional lattice that generalizes the Watts–Strogatz model, he showed that only one particular model among this infinite family can support efficient decentralized algorithms. Unfortunately, the model given by Kleinberg is highly constrained and represents a very small subset of complex networks. Watts et al. [104] presented another model that is based upon plausible hierarchical social structures and contentions regarding social networks. This model defines a class of searchable networks and offers an explanation for the searchability of social networks.

Search in Nonspatial Networks

The traditional search methods in nonspatial networks are broadcasting or random walk. In broadcasting, each node sends the message to all its neighbors. The neighbors in turn broadcast the message to all their neighbors, and the process continues. Effectively, all the nodes in the network would have received the message at least once or maybe even more. This could have devastating effects on the performance of the network. A hint of the potential damages of broadcasting can be viewed by looking at the Taylorsville, NC, elementary school project [105]. Sixth-grade students and their teacher sent out a sweet email to all the people they knew. They requested the recipients to forward the email to everyone they knew and notify the students by email so that they could plot their locations on a map. A few weeks later, the project had to be canceled because they had received about 450,000 responses from all over the world [105]. A good way to avoid such a huge exchange of messages is by doing a walk. In a walk, each node sends the message to one of its neighbors until it reaches the target node. The neighbor can be chosen in different ways depending on the algorithm. If the neighbor is chosen randomly with equal probability, then it is called *random search*, while in a *high-degree search* the highest degree neighbor is chosen. Adamic et al. [32] have demonstrated that a high-degree search is more efficient than random search in networks with a power-law degree distribution (scale-free networks). A high-degree search sends the message to a more connected neighbor that has a higher probability of reaching the target node and thus exploiting the presence of heterogeneity in node degree to perform better. They showed that the number of steps (s) required for the random search until the whole graph is revealed is $s \sim N^{3(1-2/\gamma)}$ and for the high-degree search it is $s \sim N^{(2-4/\gamma)}$. Clearly, for $\gamma > 2.0$, the number of steps taken by high-degree search scales with a smaller exponent than the random walk search. As most real networks have power-law degree distribution with exponent (γ) between 2.1 and 3.0, a high-degree search would be more effective in these networks.

All the algorithms discussed until now [32,101–104] have assumed that the edges in the network are equivalent. But the assumption of equal edge weights (which may represent the cost, bandwidth, distance, or power consumption associated with the process described by the edge) usually does not hold in real networks. Many researchers [11,15,106–116], have pointed out that it is incomplete to assume that all the edges are equivalent. Recently, Thadakamalla et al. [117] have proposed a new search algorithm based on a network measure called local betweenness centrality (LBC) that utilizes the heterogeneities in node degrees

TABLE 11.2 Comparison of Different Search Strategies in Power-Law Networks with Exponent 2.1 and 2000 Nodes with Different Edge-Weight Distributions

Search Strategy	Beta $\sigma^2 = 2.3$	Uniform $\sigma^2 = 8.3$	Exp. $\sigma^2 = 25$	Power-law $\sigma^2 = 4653.8$
Random walk	1107.71 (202%)	1097.72 (241%)	1108.70 (272%)	1011.21 (344%)
Minimum edge weight	704.47 (92%)	414.71 (29%)	318.95 (7%)	358.54 (44%)
Highest degree	379.98 (4%)	368.43 (14%)	375.83 (26%)	394.99 (59%)
Minimum average node weight	1228.68 (235%)	788.15 (145%)	605.41 (103%)	466.18 (88%)
Highest LBC	**366.26**	**322.30**	**298.06**	**247.77**

Note: The mean for all the edge-weight distributions is 5 and the variance is σ^2. The values in the table are the average distances obtained for each search strategy in these networks. The values in the parentheses show the relative difference between the average distance for each strategy with respect to the average distance obtained by the LBC strategy. LBC search, which reflects both the heterogeneities in edge weights and node degree, performed the best for all edge-weight distributions.

and edge weights. The LBC of a neighbor node i, $L(i)$, is given by

$$L(i) = \sum_{\substack{s \neq n \neq t \\ s,t \in \text{local network}}} \frac{\sigma_{st}(i)}{\sigma_{st}}$$

where σ_{st} is the total number of shortest paths (shortest path means the path over which the sum of weights is minimal) from node s to t. $\sigma_{st}(i)$ is the number of these shortest paths passing through i. If the LBC of a node is high, it implies that this node is critical in the local network. Thadakamalla et al. assume that each node in the network has information about its first and second neighbors and using this information the node calculates the LBC of each neighbor and passes the message to the neighbor with the highest LBC. They demonstrated that this search algorithm utilizes the heterogeneities in node degree and edge-weights to perform well in power-law networks with exponent between 2.0 and 2.9 for a variety of edge-weight distributions. Table 11.2 compares the performance of different search algorithms for scale-free networks with different edge-weight distributions. The values in the parentheses show the relative difference between the average distance for each algorithm with respect to the average distance obtained by the LBC algorithm. In specific, they observed that as the heterogeneity in the edge weights increase, the difference between the high-degree search and LBC search increase. This implies that it is critical to consider the edge weights in the local search algorithms. Moreover, given that many real networks are heterogeneous in edge weights, it becomes important to consider an LBC-based search rather than a high-degree search, as shown by Adamic et al. [32].

11.5.3 Other Topics

There are various other applications to real networks that include the issues related to the structure of the networks and their dynamics. In this section, we briefly summarize these applications and give some references for further study.

Detecting Community Structures

As mentioned earlier, community structures are typically found in many real networks. Finding these communities is extremely helpful in understanding the structure and function of the network. Sometimes the statistical properties of the community alone may be very different from the whole network and hence these may be critical in understanding the dynamics in the community. The following are some of the examples:

- *The World Wide Web*: Identification of communities in the Web is helpful for implementation of search engines, content filtering, automatic classification, automatic realization of ontologies, and focused crawlers [51,118].
- *Social networks*: Community structures are a typical feature of a social network. The behavior of an individual is highly influenced by the community that the individual belongs to. Communities often have their own norms, subcultures that are an important source of a person's identity [42,52].
- *Biological networks*: Community structures are found in cellular [45,119], metabolic [46], and genetic networks [120]. Identifying them helps in finding the functional modules that correspond to specific biological functions.

Algorithmically, the community detection problem is the same as the cluster analysis problem studied extensively by the OR community, computer scientists, statisticians, and mathematicians [121]. One of the major classes of algorithms for clustering is hierarchical algorithms that fall into two broad types, agglomerative and divisive. In an agglomerative method, an empty network (n nodes with no edges) is considered and edges are added based on some similarity measure between nodes (e.g., similarity based on the number of common neighbors) starting with the edge between the pairs with the highest similarity. This procedure can be stopped at any step and the distinct components of the network are taken to be the communities. On the other hand, in divisive methods the edges are removed from the network based on a certain measure (e.g., the edge with the highest betweenness centrality [52]). As this process continues the network disintegrates into different communities. Recently, many such algorithms have been proposed and applied to complex networks [18,122]. A comprehensive list of algorithms to identify community structures in complex networks can be found in Ref. [122], where Danon et al. have compared them in terms of sensitivity and computational cost.

Another interesting problem in community detection is to find a clique of maximum cardinality in the network. A clique is a complete subgraph in the network. In the network $G(V, E)$, let $G(S)$ denote the subgraph induced by a subset $S \subseteq V$. A network $G(V, E)$ is complete if each node in the network is connected to every other node, that is, $\forall i, j \in V$, $\{i, j\} \in E$. A clique C is a subset of V such that the induced graph $G(C)$ is complete. The maximum clique problem has many practical applications such as project selection, coding theory, computer vision, economics, and integration of genome mapping data [123–125]. For instance, Boginski et al. [13] solve this problem for finding maximal independent set in the market graph which can form a base for forming a diversified portfolio. The maximum clique problem is known to be NP-hard [47], and details on the various algorithms and heuristics can be found in Refs. [125,126]. Further, if the network size is large, then the data may not fit completely inside the computer's internal memory. Then we need to use external memory algorithms and data structures [127] for solving the optimization problems in such networks. These algorithms use slower external memory (such as disks) and the resulting communication between internal memory and external memory can be a major performance bottleneck. Using external memory algorithms, Abello et al. [7] proposed decomposition schemes that make large sparse graphs suitable for processing by graph optimization algorithms.

Spreading Processes

Diffusion of an infectious disease, computer virus, or information on a network constitute examples of spreading processes. In particular, the spread of infectious diseases in a population is called *epidemic spreading*. The study of epidemiological modeling has been an active research area for a long time and is heavily used in planning and implementing various prevention and control programs [128]. Recently, there has been a burst of activities on understanding the effects of the network properties on the rate and dynamics of disease propagation [17–20]. Most of the earlier methods used the *homogenous mixing hypothesis* [129], which implies that the individuals who are in contact with susceptible individuals are uniformly distributed throughout the entire population. However, recent findings (Section 11.2) such as heterogeneities in node degree, presence of high clustering coefficients, and community structures indicate that this assumption is far from reality. Later, many models have been proposed [17–20,38,41,130] which consider these properties of the network. In particular, many researchers have shown that incorporating these properties in the model radically changes the results previously established for random graphs. Other spreading processes that are of interest include spread of computer viruses [131–133], data dissemination on the Internet [134,135], and strategies for marketing campaigns [136].

Congestion

The transport of packets or materials ranging from packet transfer in the Internet to the mass transfer in chemical reactions in the cell is one of the fundamental processes occurring on many real networks. Due to limitations in resources (bandwidth), the increase in the number of packets (packet generation rate) may lead to overload at the node and unusually long delivering times, in other words, congestion in networks. Considering a basic model, Ohira and Sawatari [137] have shown that there exists a phase transition from a free flow to a congested phase as a function of the packet generation rate. This critical rate is commonly called "congestion threshold" and the higher the threshold, the better is the network performance with respect to congestion.

Many studies have shown that an important role is played by the topology and routing algorithms in the congestion of networks [138–146]. Toroczkai et al. [146] have shown that on large networks on which flows are influenced by gradients of a scalar distributed on the nodes, scale-free topologies are less prone to congestion than random graphs. Routing algorithms also influence congestion at nodes. For example, in scale-free networks, if the packets are routed through the shortest paths then most of the packets pass through the hubs and hence cause higher loads on the hubs [43]. Singh and Gupte [144] discuss strategies to manipulate hub capacity and hub connections to relieve congestion in the network. Similarly many congestion-aware routing algorithms [138,140,141,145] have been proposed to improve performance. Sreenivasan et al. [145] introduced a novel static routing protocol that is superior to the shortest path routing under intense packet-generation rates. They propose a mechanism in which packets are routed through hub avoidance paths unless the hubs are required to establish the route. Sometimes when global information is not available, routing is done using local search algorithms. Congestion due to such local search algorithms and optimal network configurations is studied in Ref. [147].

11.6 Conclusions

Complex networks are pervasive in today's world and are continuously evolving. The sheer size and complexity of these networks pose unique challenges in their design and analysis. Such complex networks are so pervasive that there is an immediate need to develop new analytical approaches. In this chapter, we presented significant findings and developments in recent years that led to a new field of interdisciplinary research, Network Science. We discussed how network approaches and optimization problems are different in network science from traditional OR algorithms and addressed the need and opportunity for the OR community to contribute to this fast-growing research field. The fundamental difference is that large-scale networks are characterized based on macroscopic properties such as degree distribution and clustering coefficient rather than the individual properties of the nodes and edges. Importantly, these macroscopic or statistical properties have a huge influence on the dynamic processes taking place on the network. Therefore, to optimize a process on a given configuration, it is important to understand the interactions between the macroscopic properties and the process. This will further help in the design of optimal network configurations for various processes. Due to the growing scale of many engineered systems, a macroscopic network approach is necessary for the design and analysis of such systems. Moreover, the macroscopic properties and structure of networks across different disciplines are found to be similar. Hence the results of this research can easily be migrated to applications as diverse as social networks to telecommunication networks.

Acknowledgments

The authors acknowledge the National Science Foundation (NSF, Grant # DMI 0537992) and a Sloan Research Fellowship to one of the authors (R.A.) for making this work feasible. In addition, the authors thank the anonymous reviewer for helpful comments and suggestions. Any opinions, findings and conclusions, or recommendations expressed in this material are those of the author(s) and do not necessarily reflect the views of the NSF.

References

1. R.K. Ahuja, T.L. Magnanti, and J.B. Orlin. *Network Flows: Theory, Algorithms, and Applications*. Prentice-Hall, Englewood Cliffs, NJ, 1993.
2. Committee on Network Science for Future Army Applications. *Network Science*. National Academies Press, Washington, D.C., 2005.
3. S. Lawrence and C.L. Giles. Accessibility of information on the web. *Nature*, 400:107–109, 1999.
4. A. Gulli and A. Signorini. The indexable web is more than 11.5 billion pages. In *WWW '05: Special Interest Tracks and Posters of the 14th International Conference on World Wide Web*, pp. 902–903. ACM Press, New York, 2005.
5. M. Faloutsos, P. Faloutsos, and C. Faloutsos. On power-law relationships of the internet topology. *Comput. Commun. Rev.*, 29:251–262, 1999.
6. R. Govindan and H. Tangmunarunkit. Heuristics for internet map discovery. *IEEE INFOCOM*, 3:1371–1380, 2000.
7. J. Abello, P.M. Pardalos, and M.G.C. Resende. On maximum clique problems in very large graphs. In: *External Memory Algorithms: DIMACS Series in Discrete Mathematics and Theoretical Computer Science*, vol. 50, pp. 119–130. American Mathematical Society, Boston, MA, 1999.
8. D.J. Watts and S.H. Strogatz. Collective dynamics of "small-world" networks. *Nature*, 393:440–442, 1998.
9. R. Albert, I. Albert, and G.L. Nakarado. Structural vulnerability of the North American power grid. *Phys. Rev. E*, 69(2):025103, 2004.

10. M. Barthelemy, A. Barrat, R. Pastor-Satorras, and A. Vespignani. Characterization and modeling of weighted networks. *Physica A*, 346:34–43, 2005.
11. R. Guimera, S. Mossa, A. Turtschi, and L.A.N. Amaral. The worldwide air transportation network: Anomalous centrality, community structure, and cities' global roles. *Proc. Nat. Acad. Sci.*, 102:7794–7799, 2005.
12. V. Boginski, S. Butenko, and P. Pardalos. Statistical analysis of financial networks. *Comput. Stat. Data Anal.*, 48:431–443, 2005.
13. V. Boginski, S. Butenko, and P. Pardalos. Mining market data: A network approach. *Comput. Oper. Res.*, 33:3171–3184, 2006.
14. M.E.J. Newman. Scientific collaboration networks: I. Network construction and fundamental results. *Phys. Rev. E*, 64(1):016131, 2001.
15. M.E.J. Newman. Scientific collaboration networks: II. Shortest paths, weighted networks, and centrality. *Phys. Rev. E*, 64(1):016132, 2001.
16. The Internet Movie Database can be found on the WWW at http://www.imdb.com/.
17. R. Albert and A.L. Barabási. Statistical mechanics of complex networks. *Rev. Mod. Phys.*, 74(1):47–97, 2002.
18. S. Boccaletti, V. Latora, Y. Moreno, M. Chavez, and D.U. Hwang. Complex networks: Structure and dynamics. *Phys. Reports*, 424:175–308, 2006.
19. S.N. Dorogovtsev and J.F.F. Mendes. Evolution of networks. *Adv. Phys.*, 51: 1079–1187, 2002.
20. M.E.J. Newman. The structure and function of complex networks. *SIAM Review*, 45: 167–256, 2003.
21. A. Broder, R. Kumar, F. Maghoul, P. Raghavan, S. Rajagopalan, R. Stata, A. Tomkins, and J. Wiener. Graph structure in the web. *Comput. Netw.*, 33:309–320, 2000.
22. R. Ferrer i Cancho, C. Janssen, and R.V. Solé. Topology of technology graphs: Small world patterns in electronic circuits. *Phys. Rev. E*, 64(4):046119, 2001.
23. M. Ripeanu, I. Foster, and A. Iamnitchi. Mapping the Gnutella network: Properties of large-scale peer-to-peer systems and implications for system design. *IEEE Internet Comput. J.*, 6:50–57, 2002.
24. S. Milgram. The small world problem. *Psychol. Today*, 2:60–67, 1967.
25. D.J. Watts, P.S. Dodds, and R. Muhamad. http://smallworld.columbia.edu/index.html, date accessed: March 22, 2006.
26. N. Contractor, S. Wasserman, and K. Faust. Testing multi-theoretical multilevel hypotheses about organizational networks: An analytic framework and empirical example. *Acad. Mgmt. Rev.*, 31(3):681–703, 2006.
27. L.A. Adamic and B.A. Huberman. Growth dynamics of the world-wide web. *Nature*, 401(6749):131, 1999.
28. R. Albert, H. Jeong, and A.L. Barabási. Diameter of the world wide web. *Nature*, 401(6749):130–131, 1999.
29. R. Kumar, P. Raghavan, S. Rajalopagan, D. Sivakumar, A. Tomkins, and E. Upfal. The web as a graph. *Proceedings of the Nineteenth ACM SIGMOD-SIGACT-SIGART Symposium on Principles of Database Systems*, pp. 1–10, 2000.
30. H. Jeong, B. Tombor, R. Albert, Z.N. Oltvai, and A.-L. Barabási. The large-scale organization of metabolic networks. *Nature*, 407:651–654, 2000.
31. W. Aiello, F. Chung, and L. Lu. A random graph model for massive graphs. *Proceedings of the Thirty-Second Annual ACM Symposium on Theory of Computing*, pp. 171–180, 2000.
32. L.A. Adamic, R.M. Lukose, A.R. Puniyani, and B.A. Huberman. Search in power-law networks. *Phys. Rev. E*, 64(4):046135, 2001.
33. R. Albert and A.L. Barabási. Topology of evolving networks: Local events and universality. *Phys. Rev. Lett.*, 85(24):5234–5237, 2000.

34. L.A.N. Amaral, A. Scala, M. Barthelemy, and H.E. Stanley. Classes of small-world networks. *Proc. Natl. Acad. Sci.*, 97(21):11149–11152, 2000.
35. A. L Barabási and R. Albert. Emergence of scaling in random networks. *Science*, 286 (5439):509–512, 1999.
36. R. Albert, H. Jeong, and A.L. Barabási. Attack and error tolerance of complex networks. *Nature*, 406(6794):378–382, 2000.
37. R. Pastor-Satorras and A. Vespignani. Epidemic dynamics and endemic states in complex networks. *Phys. Rev. E*, 63(6):066117, 2001.
38. R. Pastor-Satorras and A. Vespignani. Epidemic spreading in scale-free networks. *Phys. Rev. Lett.*, 86:3200–3203, 2001.
39. R. Pastor-Satorras and A. Vespignani. Epidemic dynamics in finite size scale-free networks. *Phys. Rev. E*, 65(3):035108, 2002.
40. R. Pastor-Satorras and A. Vespignani. Immunization of complex networks. *Phys. Rev. E*, 65(3):036104, 2002.
41. R. Pastor-Satorras and A. Vespignani. Epidemics and immunization in scale-free networks. In: *Handbook of Graphs and Networks*, Wiley-VCH, Berlin, 2003.
42. S. Wasserman and K. Faust. *Social Network Analysis*. Cambridge University Press, New York, 1994.
43. K.I. Goh, B. Kahng, and D. Kim. Universal behavior of load distribution in scale-free networks. *Phys. Rev. Lett.*, 87(27), 278701, 2001.
44. D. Koschützki, K.A. Lehmann, L. Peeters, S. Richter, D. Tenfelde-Podehl, and O. Zlotowski. Centrality indices. In: *Network Analysis*, pp. 16–61. Springer-Verlag, Berlin, 2005.
45. P. Holme, M. Huss, and H. Jeong. Subnetwork hierarchies of biochemical pathways. *Bioinformatics*, 19:532–538, 2003.
46. E. Ravasz, A.L. Somera, D.A. Mongru, Z.N. Oltvai, and A.L. Barabási. Hierarchical organization of modularity in metabolic networks. *Science*, 297:1551–1555, 2002.
47. M.R. Garey and D.S. Johnson. *Computers and Intractability, A Guide to the Theory of NP-Completeness*. W.H. Freeman, New York, 1979.
48. B. Hendrickson and R.W. Leland. A multilevel algorithm for partitioning graphs. In *Supercomputing '95: Proceedings of the 1995 ACM/IEEE Conference on Supercomputing*, p. 28. ACM Press, New York, 1995.
49. B.W. Kernighan and S. Lin. An efficient heuristic procedure for partitioning graphs. *Bell System Tech. J.*, 49:291–307, 1970.
50. A. Pothen, H. Simon, and K. Liou. Partitioning sparse matrices with eigenvectors of graphs. *SIAM J. Matrix Anal.*, 11(3):430–452, 1990.
51. G. Flake, S. Lawrence, and C. Lee Giles. Efficient identification of web communities. In: *Sixth ACM SIGKDD International Conference on Knowledge Discovery and Data Mining*, pp. 150–160, 2000.
52. M.E.J. Newman and M. Girvan. Finding and evaluating community structure in networks. *Phys. Rev. E*, 69(2):026113, 2004.
53. G. Palla, I. Derenyi, I. Farkas, and T. Vicsek. Uncovering the overlapping community structure of complex networks in nature and society. *Nature*, 435:814–818, 2005.
54. F. Radicchi, C. Castellano, F. Cecconi, V. Loreto, and D. Parisi. Defining and identifying communities in networks. *Proc. Natl. Acad. Sci.*, 101:2658–2663, 2004.
55. P. Holme and B.J. Kim. Attack vulnerability of complex networks. *Phys. Rev. E*, 65(5): 056109, 2002.
56. P. Erdős and A. Rényi. On random graphs. *Publicationes Mathematicae*, 6:290–297, 1959.
57. P. Erdős and A. Rényi. On the evolution of random graphs. *Magyar Tud. Mat. Kutato Int. Kozl.*, 5:17–61, 1960.

58. P. Erdős and A. Rényi. On the strength of connectedness of a random graph. *Acta Math. Acad. Sci. Hungar.*, 12:261–267, 1961.
59. B. Bollobas. *Random Graphs*. Academic, London, 1985.
60. W. Aiello, F. Chung, and L. Lu. A random graph model for power law graphs. *Exp. Math.*, 10(1):53–66, 2001.
61. F. Chung and L. Lu. Connected components in random graphs with given degree sequences. *Ann. Combinatorics*, 6:125–145, 2002.
62. M. Molloy and B. Reed. A critical point for random graphs with a given degree sequence. *Random Structures Algorithms*, 6:161–179, 1995.
63. M.E.J. Newman. Random graphs as models of networks. In:*Handbook of Graphs and Networks*, pp. 35–68. Wiley-VCH, Berlin, 2003.
64. M.E.J. Newman, S.H. Strogatz, and D.J. Watts. Random graphs with arbitrary degree distributions and their applications. *Phys. Rev. E*, 64(2):026118, 2001.
65. H.S. Wilf. *Generating Functionology*. Academic, Boston, 1990.
66. C. Anderson, S. Wasserman, and B. Crouch. A p* primer: Logit models for social networks. *Social Networks*, 21(1):37–66, 1999.
67. O. Frank and D. Strauss. Markov graphs. *J. Am. Stat. Assoc.*, 81:832–842, 1986.
68. P.W. Holland and S. Leinhardt. An exponential family of probability distributions for directed graphs. *J. Am. Stat. Assoc.*, 76:33–65, 1981.
69. D. Strauss. On a general class of models for interaction. *SIAM Rev.*, 28:513–527, 1986.
70. S. Wasserman and P. Pattison. Logit models and logistic regressions for social networks 1: An introduction to Markov random graphs and p*. *Psychometrika*, 61:401–426, 1996.
71. T.A.B. Snijders. Markov chain Monte Carlo estimation of exponential random graph models. *J. Soc. Struct.*, 3(2):1–40, 2002.
72. M.E.J. Newman. Models of small world. *J. Stat. Phys.*, 101:819–841, 2000.
73. B. Bollobas and O. Riordan. Mathematical results on scale-free graphs. In:*Handbook of Graphs and Networks*. Wiley-VCH, Berlin, 2003.
74. A. Vespignani. Complex networks: Ubiquity, importance, and implications. In: *Frontiers of Engineering: Reports on Leading-Edge Engineering from the 2005 Symposium*, pp. 75–81. National Academies Press, Washington, DC, 2006.
75. V. Honavar. Complex Adaptive Systems Group at Iowa State University, http://www.cs.iastate.edu/~honavar/cas.html, date accessed: March 22, 2006.
76. A. Vespignani. Epidemic modeling: Dealing with complexity. http://www.indiana.edu/talks-fall04/, 2004. Date accessed: July 6, 2006.
77. R. Badii and A. Politi. *Complexity: Hierarchical Structures and Scaling in Physics*. Cambridge University Press, Cambridge, UK, 1997.
78. C.H. Bennett. How to define complexity in physics, and why. In: *From Complexity to Life*, pp. 34–43. Oxford University Press, New York, 2003.
79. A.X.C.N. Valente, A. Sarkar, and H.A. Stone. Two-peak and three-peak optimal complex networks. *Phys. Rev. Lett.*, 92(11):118702, 2004.
80. G. Paul, T. Tanizawa, S. Havlin, and H.E. Stanley. Optimization of robustness of complex networks. *Eur. Phys. J. B*, 38:187–191, 2004.
81. L.F. Costa. Reinforcing the resilience of complex networks. *Phys. Rev. E*, 69(6):066127, 2004.
82. R. Ferrer i Cancho and R.V. Solé. Optimization in complex networks. In:*Statistical Mechanics of Complex Networks*, pp. 114–126. Springer-Verlag, Berlin, 2003.
83. B. Shargel, H. Sayama, I.R. Epstein, and Y. Bar-Yam. Optimization of robustness and connectivity in complex networks. *Phys. Rev. Lett.*, 90(6):068701, 2003.

84. H.P. Thadakamalla, U.N. Raghavan, S.R.T. Kumara, and R. Albert. Survivability of multi-agent based supply networks: A topological perspective. *IEEE Intell. Syst.*, 19:24–31, 2004.
85. V. Venkatasubramanian, S. Katare, P.R. Patkar, and F. Mu. Spontaneous emergence of complex optimal networks through evolutionary adaptation. *Comput. & Chem. Eng.*, 28(9):1789–1798, 2004.
86. R. Kinney, P. Crucitti, R. Albert, and V. Latora. Modeling cascading failures in the North American power grid. *Euro. Phys. J. B*, 46:101–107, 2005.
87. M.L. Sachtjen, B.A. Carreras, and V.E. Lynch. Disturbances in a power transmission system. *Phys. Rev. E*, 61(5):4877–4882, 2000.
88. A.E. Motter and Y. Lai. Cascade-based attacks on complex networks. *Phys. Rev. E*, 66(6):065102, 2002.
89. P. Crucitti, V. Latora, and M. Marchiori. Model for cascading failures in complex networks. *Phys. Rev. E*, 69(4):045104, 2004.
90. A.E. Motter. Cascade control and defense in complex networks. *Phys. Rev. Lett.*, 93(9):098701, 2004.
91. B.A. Carreras, V.E. Lynch, I. Dobson, and D.E. Newman. Critical points and transitions in an electric power transmission model for cascading failure blackouts. *Chaos*, 12(4):985–994, 2002.
92. Y. Moreno, J.B. Gomez, and A.F. Pacheco. Instability of scale-free networks under node-breaking avalanches. *Europhys. Lett.*, 58(4):630–636, 2002.
93. Y. Moreno, R. Pastor-Satorras, A. Vazquez, and A. Vespignani. Critical load and congestion instabilities in scale-free networks. *Europhys. Lett.*, 62(2):292–298, 2003.
94. X.F. Wang and J. Xu. Cascading failures in coupled map lattices. *Phys. Rev. E*, 70(5):056113, 2004.
95. D.J. Watts. A simple model of global cascades on random networks. *Proc. Natl. Acad. Sci.*, 99(9):5766–5771, 2002.
96. E. Aarts and J.K. Lenstra (eds.). *Local Search in Combinatorial Optimization*. John Wiley & Sons, Chichester, UK, 1997.
97. I.F. Akyildiz, W. Su, Y. Sankarasubramaniam, and E. Cayirci. Wireless sensor networks: A survey. *Comput. Netw.*, 38(4):393–422, 2002.
98. J.N. Al-Karaki and A.E. Kamal. Routing techniques in wireless sensor networks: A survey. *IEEE Wireless Communications*, 11(6):6–28, 2004.
99. C. Intanagonwiwat, R. Govindan, and D. Estrin. Directed diffusion: A scalable and robust communication paradigm for sensor networks. Proceedings of ACM MobiCom '00, Boston, MA, pp. 174–185, 2000.
100. G. Kan. Gnutella. In: *Peer-to-Peer Harnessing the Power of Disruptive Technologies*, edited by A. Oram, O'Reilly, Beijing, 2001.
101. J. Kleinberg. Navigation in a small world. *Nature*, 406:845, 2000.
102. J. Kleinberg. The small-world phenomenon: An algorithmic perspective. *Proc. 32nd ACM Symposium on Theory of Computing*, ACM Press, New York, pp. 163–170, 2000.
103. J. Kleinberg. Small-world phenomena and the dynamics of information. *Adv. Neural Inform. Process. Syst.*, 14:431–438, 2001.
104. D.J. Watts, P.S. Dodds, and M.E.J. Newman. Identity and search in social networks. *Science*, 296:1302–1305, 2002.
105. D.J. Watts. *Six Degrees: The Science of a Connected Age*. W.W. Norton & Company, New York, 2003.
106. E. Almaas, B. Kovacs, T. Viscek, Z.N. Oltval, and A.L. Barabási. Global organization of metabolic fluxes in the bacterium *Escherichia coli*. *Nature*, 427(6977):839–843, 2004.

107. A. Barrat, M. Barthelemy, R. Pastor-Satorras, and A. Vespignani. The architecture of complex weighted networks. *Proc. Natl. Acad. Sci.*, 101(11):3747, 2004.
108. A. Barrat, M. Barthelemy, and A. Vespignani. Modeling the evolution of weighted networks. *Phys. Rev. E*, 70(6):066149, 2004.
109. L.A. Braunstein, S. V. Buldyrev, R. Cohen, S. Havlin, and H.E. Stanley. Optimal paths in disordered complex networks. *Phys. Rev. Lett.*, 91(16):168701, 2003.
110. K.I. Goh, J.D. Noh, B. Kahng, and D. Kim. Load distribution in weighted complex networks. *Phys. Rev. E*, 72(1):017102, 2005.
111. M. Granovetter. The strength of weak ties. *Am. J. Sociol.*, 78(6):1360–1380, 1973.
112. A.E. Krause, K.A. Frank, D.M. Mason, R.E. Ulanowicz, and W.W. Taylor. Compartments revealed in food-web structure. *Nature*, 426:282–285, 2003.
113. J.D. Noh and H. Rieger. Stability of shortest paths in complex networks with random edge weights. *Phys. Rev. E*, 66(6):066127, 2002.
114. R. Pastor-Satorras and A. Vespignani. *Evolution and Structure of the Internet: A Statistical Physics Approach*. Cambridge University Press, Cambridge, UK, 2004.
115. S.L. Pimm. *Food Webs*. University of Chicago Press, Chicago, 2nd edition, 2002.
116. S.H. Yook, H. Jeong, A.L. Barabási, and Y. Tu. Weighted evolving networks. *Phys. Rev. Lett.*, 2001:5835–5838, 86.
117. H.P. Thadakamalla, R. Albert, and S.R.T. Kumara. Search in weighted complex networks. *Phys. Rev. E*, 72(6):066128, 2005.
118. R.B. Almeida and V.A.F. Almeida. A community-aware search engine. In: *Proceedings of the 13th International Conference on World Wide Web*, ACM Press, New York, 2004.
119. A.W. Rives and T. Galitskidagger. Modular organization of cellular networks. *Proc. Natl. Acad. Sci.*, 100(3):1128–1133, 2003.
120. D.M. Wilkinson and B.A. Huberman. A method for finding communities of related genes. *Proc. Natl. Acad. Sci.*, 101:5241–5248, 2004.
121. P. Hansen and B. Jaumard. Cluster analysis and mathematical programming. *Math. Program.*, 79:191–215, 1997.
122. L. Danon, A. Diaz-Guilera, J. Duch, and A. Arenas. Comparing community structure identification. *J. Stat. Mech.*, p. P09008, 2005.
123. S. Butenko and W.E. Wilhelm. Clique-detection models in computational biochemistry and genomics. *Eur. J. Oper. Res.*, 173:1–17, 2006.
124. J. Hasselberg, P.M. Pardalos, and G. Vairaktarakis. Test case generators and computational results for the maximum clique problem. *J. Global Optim.*, 3: 463–482, 1993.
125. P.M. Pardalos and J. Xue. The maximum clique problem. *J. Global Optim.*, 4:301–328, 1994.
126. D.J. Johnson and M.A. Trick (eds.). *Cliques, Coloring, and Satisfiability: Second DIMACS Implementation Challenge, Workshop, October 11–13, 1993*. American Mathematical Society, Boston, 1996.
127. J. Abello and J. Vitter (eds.). *External Memory Algorithms: DIMACS Series in Discrete Mathematics and Theoretical Computer Science*, vol. 50. American Mathematical Society, Boston, MA, 1999.
128. O. Diekmann and J. Heesterbeek. *Mathematical Epidemiology of Infectious Diseases: Model Building, Analysis and Interpretation*. Wiley, New York, 2000.
129. R.M. Anderson and R.M. May. *Infectious Diseases in Humans*. Oxford University Press, Oxford, UK, 1992.
130. V. Colizza, A. Barrat, M. Barthlemy, and A. Vespignani. The role of the airline transportation network in the prediction and predictability of global epidemics. *PNAS*, 103(7):2015–2020, 2006.

131. J. Balthrop, S. Forrest, M.E.J. Newman, and M.M. Williamson. Technological networks and the spread of computer viruses. *Science*, 304(5670):527–529, 2004.
132. A.L. Lloyd and R.M. May. How viruses spread among computers and people. *Science*, 292:1316–1317, 2001.
133. M.E.J. Newman, S. Forrest, and J. Balthrop. Email networks and the spread of computer viruses. *Phys. Rev. E*, 66(3):035101, 2002.
134. A.-M. Kermarrec, L. Massoulie, and A. J. Ganesh. Probabilistic reliable dissemination in large-scale systems. *IEEE Trans. on Parallel and Distributed Sys.*, 14(3):248–258, 2003.
135. W. Vogels, R. van Renesse, and K. Birman. The power of epidemics: Robust communication for large-scale distributed systems. *SIGCOMM Comput. Commun. Rev.*, 33(1):131–135, 2003.
136. J. Leskovec, L.A. Adamic, and B.A. Huberman. The dynamics of viral marketing, *ACM Trans. on the Web*, 1(1):5, 2007.
137. T. Ohira and R. Sawatari. Phase transition in a computer network traffic model. *Phys. Rev. E*, 58(1):193–195, 1998.
138. Z.Y. Chen and X.F. Wang. Effects of network structure and routing strategy on network capacity. *Phys. Rev. E*, 73(3):036107, 2006.
139. M. Argollo de Menezes and A.-L. Barabási. Fluctuations in network dynamics. *Phys. Rev. Lett.*, 92(2):028701, 2004.
140. P. Echenique, J. Gomez-Gardenes, and Y. Moreno. Improved routing strategies for internet traffic delivery. *Phys. Rev. E*, 70(5):056105, 2004.
141. P. Echenique, J. Gomez-Gardenes, and Y. Moreno. Dynamics of jamming transitions in complex networks. *Europhys. Lett.*, 71(2):325–331, 2005.
142. R. Guimera, A. Arenas, A. Díaz-Guilera, and F. Giralt. Dynamical properties of model communication networks. *Phys. Rev. E*, 66(2):026704, 2002.
143. R. Guimera, A. Díaz-Guilera, F. Vega-Redondo, A. Cabrales, and A. Arenas. Optimal network topologies for local search with congestion. *Phys. Rev. Lett.*, 89(24):248701, 2002.
144. B.K. Singh and N. Gupte. Congestion and decongestion in a communication network. *Phys. Rev. E*, 71(5):055103, 2005.
145. S. Sreenivasan, R. Cohen, E. Lopez, Z. Toroczkai, and H.E. Stanley. Communication bottlenecks in scale-free networks, *Phys. Rev. E*, 75(3):036105, 2007.
146. Z. Toroczkai and K.E. Bassler. Network dynamics: Jamming is limited in scale-free systems. *Nature*, 428:716, 2004.
147. A. Arenas, A. Cabrales, A. Diaz-Guilera, R. Guimera, and F. Vega. Search and congestion in complex networks. In: *Statistical Mechanics of Complex Networks*, pp. 175–194. Springer-Verlag, Berlin, 2003.
148. A.L. Barabási, H. Jeong, Z. Neda, E. Ravasz, A. Schubert, and T. Vicsek. Evolution of the social network of scientific collaborations. *Physica A*, 311:590–614, 2002.

12
Simulation

12.1	Introduction	**12**-1
	When Should Simulation Be Used? • Examples of Successful Applications	
12.2	Basics of Simulation	**12**-3
	Basic Definitions • Need for Randomness • Output Analysis for Discrete-Event Simulation	
12.3	Simulation Languages and Software	**12**-15
	A Sampling of Discrete-Event Simulation Software • Summary	
12.4	Simulation Projects—The Bigger Picture	**12**-20
12.5	Summary	**12**-22
	References	**12**-23

Catherine M. Harmonosky
Pennsylvania State University

12.1 Introduction

The word "simulation" can invoke many different images. Wind tunnels provide a physical simulation environment for testing the aerodynamic properties of wings and cars. Water tunnels also provide a physical simulation environment for testing the hydraulic and acoustical properties of submerged vehicles. Pilots train in flight simulators. There are Monte Carlo simulation methods that allow for statistical analysis of many trials of mathematical equations. Computer models allow simulation of electrical circuits during their design to ensure completeness. In the foundry industry, there is mold solidification simulation software, which allows analysis of how the molten metal will solidify for a particular mold design.

This chapter focuses on discrete-event simulation. Discrete-event simulation is a computer-based methodology that allows you to model an existing or proposed system, capturing key characteristics and parameters of that system, such that the model emulates the behavior and performance of the system as events take place over time. Examples of systems that can be simulated include factories, warehouses, hospitals, amusement parks, military operations, airports, convenience stores, or supply chains. Analyzing the simulation output can allow the user to:

- Better understand the behavior and performance of existing systems, including determining the cause of different system responses.
- Predict the effect upon system performance due to a proposed change in the system.
- Predict if a system design will have a performance that meets a company's specified goals.

Simulation allows the analyst to draw inferences about new systems without building them or about changes to existing systems without disturbing them. Consequently, it can provide a very cost-effective means of analyzing multiple "what-if" scenarios before any investment is made in capital, resources, or materials. Some examples of the types of system changes, either in existing systems or in a system design, that simulation can evaluate include:

- Introducing a lean manufacturing philosophy into an existing system.
- Modifying a supply chain by deleting or adding suppliers.
- Adding, replacing, or deleting resources (equipment, workers).
- Changing process/customer flow patterns.
- Changing scheduling or control rules.
- Modifying material handling methods.

12.1.1 When Should Simulation Be Used?

Of course, there are many other operations research modeling techniques that are described in other chapters of this handbook. Techniques such as linear and integer programming are analytical optimal methods—for the given input values, the mathematically optimal answer will be returned. It is important to understand that discrete-event simulation is a heuristic—although it seeks the optimal answer, it can not guarantee that it will deliver the optimal answer. It uses numerical methods to track the changing state of the system over time. Therefore, we say that simulation models are "run" rather than solved, generating a dynamic series of system changing events while collecting observations of system performance that are analyzed to estimate the true performance of the actual system being modeled [1]. Further, due to the stochastic input, each simulation run, or replication, of the same model will have a different specific series of events yielding slightly different output measures.

So, why should you use simulation, when it does not guarantee the optimal? In many complex manufacturing and service systems, for example, factories, hospitals, and amusement parks, it is very difficult (some may say impossible) to exactly capture the relationships and interactions of the system parameters in strict mathematical equations, as is required of mathematical modeling techniques. Additionally, complex systems often have significant sources of variability, for example, variable processing or service times, variable walking times, probabilities of rework or re-entry into the line, nonstationary arrival rates, breakdowns or worker breaks, and the like. Variability is particularly evident in service systems where customer behavior introduces another source of variability in addition to resource variability. Simulation can directly incorporate most types of system variability into the model, which is often difficult to incorporate in optimal techniques. Also, unlike queueing models, which have certain underlying assumptions regarding arrival time and processing time distributions, in simulation you can specify the distributions that most directly match your data. Therefore, in complex, stochastic environments we often do not have a directly applicable optimal technique that will return an answer in a reasonable amount of time. We must rely on heuristics, such as simulation.

12.1.2 Examples of Successful Applications

There are many reported examples of successful application of discrete-event simulation in a variety of environments. Evaluating different material control policies in a warehouse allowed a company to avoid purchasing new equipment that they previously thought was necessary. Simulation is routinely used in the automotive industry to investigate bottlenecks, evaluate different possible layouts and equipment, or to model the design plans for a completely new facility. Customer service call centers have been simulated, leading to better and faster call routing policies to improve response time. The introduction of lean manufacturing concepts can be investigated with simulation to determine which concepts work best for a particular company before investing in any physical changes. In the banking industry, simulation is used to model the processing of transactions, identify bottlenecks, and develop new processing strategies that allow transactions to be processed faster, thus saving the bank money or increasing revenues. The benefit to the companies may be a one-time benefit of hundreds of thousands of dollars for a larger project, or some companies use simulation extensively throughout the year, yielding $50,000–$100,000 benefit per project.

Many more examples of successful applications can be found in papers in the *Proceedings of the Winter Simulation Conferences* (WSC), which can be accessed via the WSC Web site, www.wintersim.org. Some examples of tracks at the WSC 2006 conference that have associated papers focused on specific application areas include manufacturing, aerospace and military applications, logistics, transportation and distribution, risk analysis, homeland security/emergency response, and biotechnology/health care. Additionally, four case study tracks dedicated to presentations of simulation applications in a multitude of industries were part of the conference. Although only abstracts are available for the case study presentations, they provide another perspective of the types of simulation applications.

The Society for Modeling and Simulation International sponsors a Summer Computer Simulation Conference, which also publishes a proceedings, as part of its annual Summer Simulation Multiconference. Topical areas are similar to those mentioned above. More information is available at the Society's Web site, www.scs.org.

12.2 Basics of Simulation

Although there are many discrete-event simulation languages and a list of some languages will follow later in this chapter, there are fundamentals about simulation and the simulation modeling process that transcend software platforms. Fundamentally, a discrete-event simulation model is stochastic (there is some type of randomness in the system), dynamic (the system state changes over time), and system changes occur at discrete points in time [2].

12.2.1 Basic Definitions

Figure 12.1 is a graphical representation of an overview of the main ideas of a simulation model. First, the model boundary must be established. For example, will your model include the entire factory or only one department; will your model include the transportation method for patients arriving to the hospital or just start with the registration desk? Setting the boundary defines the scope of the model. Within that boundary, entities move through the system interacting with each other and triggering events. *Entities* may be parts, people, or things and an *event* is an instantaneous occurrence at a discrete point in time (hence the term "discrete-event"), typically associated with a change in the state of the system. A discrete-event simulation tracks, over time, the *state of the system*, which is like a snapshot of the system at any given time. Defining the system and in some ways governing the entity flow are *system parameters*, for example, number of resources, queue space, and so on. Affecting the system are *external factors*, such as supplier schedules and customer orders, which provide some of the input parameters for the model.

Ultimately, your goal is to analyze the *output* from the model, which is the performance measures of interest, for example, average time in the system, work in process, meeting due dates, and so on. There is output at the end of each model *run or replication*. A run is

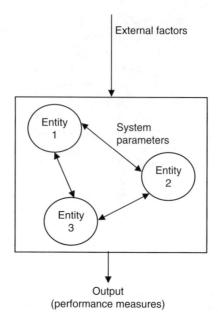

FIGURE 12.1 Basic simulation modeling process.

typically the simulation executing over some length of simulation time or entity count, and the output is based on all the observations that have been made as the state of the system has changed over simulated time. When you repeat the simulation run for the same model multiple times, you are making replications of that model, which differ slightly from each other due to the stochastic inputs.

Entities have *attributes*, which are descriptors of the entity characteristics, such as part or customer type, or routing through the system. Attributes are somewhat like a simulated version of a route sheet in a factory—they contain information specific to a particular entity that is attached to it for the duration of its time in the system. Attributes for an entity may change their specific values during the simulation; for example, a part's priority may get higher as it gets closer to its due date. The system has *parameters*, which are descriptors of the system characteristic, for example, the number of machines or workers, queue sizes, and so on. Although some system parameters may change during one simulation run (for example the number of workers may change according to a specified schedule), they typically remain the same value throughout one run. There are also *system variables*, sometimes called *global variables*, which take on different values throughout a simulation run based on system conditions, for example, the number of entities in a queue, resource status (busy or idle), or the number that have left the system. System variables are often used to track time-persistent performance measures in the system.

Most popular simulation languages take a *process-oriented* approach to modeling entity movement through a system, thinking of a series of events, delays, and activities for a specific entity as a "process" [1,3]. Figure 12.2 illustrates a very simple process for one entity at a single server or a single machine, with events, an activity, and a queue delay noted. First, there is the arrival event for the entity. Then, as entities move through the system, they try to access *resources*, for example, machine, worker, nurse, which are temporarily allocated to a particular entity and made available again later to other entities. *Queues* are the waiting lines for resources where entity movement is temporarily suspended if the resource is busy. This type of entity delay is a status-dependent delay because the length of the delay is

Simulation

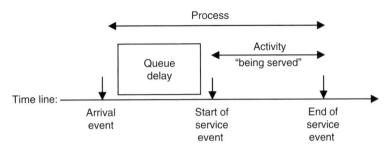

FIGURE 12.2 Process-oriented view of entity movement.

based on the status of factors outside of that entity's control. For example, the length of time an entity spends in a queue is dependent upon the processing or service times of all the entities before it in the queue. When the resource becomes available and is allocated to the entity, the entity begins its service *activity*, which begins with a "start" event and ends with an "end" event. Thus, activities, such as this "being served" activity, are triggered by instantaneous events and have a specific duration, making them deterministic delays.

It should be noted, however, that although most discrete-event simulation languages adopt a process orientation, behind the scenes they are still executed one event at a time. This requires coordination between the *event calendar* and the *simulation clock*. The event calendar can be thought of as the "to do" list of the simulation model, and processing events from the event calendar moves the simulation clock forward in time. Future events are scheduled on the event calendar. Similar to our personal "to do" list, after the simulation has processed an event, it is deleted from the list. There are three pieces of information associated with each event: (1) the entity that is associated with that event, (2) the time the event is to occur, and (3) the type of event. For a single-server single-queue system, a snapshot of the event calendar is shown below:

Entity	Time	Type of event
2	5.3	arrive to system
3	7.1	arrive to system
1	10.7	end processing

The event calendar is always sorted in chronological order. Figure 12.3 shows how the event calendar operates. The simulation progresses by taking events one at a time from the event calendar in chronological order and updating the simulation clock to be equal to the event time. This means that the simulation clock time jumps forward in discrete amounts instead of having continuous time progression. Then, the simulation processes the model logic triggered by that event and its associated entity until the entity hits a delay. All processing of the associated model logic, until the delay starts, occurs in zero simulated time. Once that entity is in a delay, the simulation model must look for something else to do. So, it returns to the event calendar, takes the next event off the calendar, and repeats the procedure.

Our small event calendar above can illustrate the event calendar basic operation, assuming a single-server, single-queue system. The simulation finds the next event is entity 2 arriving at time 5.3. The simulation clock is immediately updated to time 5.3, and a new arrival event is scheduled onto the event calendar, assuming entity 4 will arrive at time 11.2. The simulation then attempts to move entity 2 through the logic of checking to see if the server

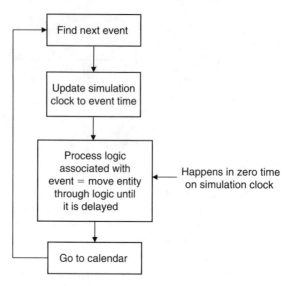

FIGURE 12.3 How a simulation progresses through time.

is available. Since it is busy serving entity 1, entity 2 is placed in the queue and is now in a queue delay. The simulation clock time remains 5.3 because the logic processing took zero time. Consequently, the simulation goes back to the event calendar to find the most imminent event to process. Because the event associated with entity 2 has been processed, it is removed from the event calendar, resulting in the following updated event calendar:

Entity	Time	Type of event
3	7.1	arrive to system
1	10.7	end processing
4	11.2	arrive to system

The next event is entity 3 arriving at time 7.1. The simulation clock is immediately updated from time 5.3 to time 7.1, and another arrival is scheduled on the event calendar, assuming entity 5 will arrive at time 17.5. When the simulation processes entity 3 through the logic of checking server availability, the server is still busy with entity 1. So, entity 3 is also placed in the queue in a delay, and the updated event calendar is below:

Entity	Time	Type of event
1	10.7	end processing
4	11.2	arrive to system
5	17.5	arrive to system

When the simulation returns to the event calendar this time, it finds the next event is "end processing" for entity 1 at time 10.7. The simulation clock updates from time 7.1 to time 10.7. The logic for entity 1 is to end processing and leave the system. This triggers the next entity in the queue, entity 2 in our example, to start its service. Entity 2 is now in a delay for a service activity, which is a deterministic delay, and the end of service activity for

Simulation

entity 2—let's assume it will occur at time 15.7—is put onto the event calendar in proper chronological order as follows:

Entity	Time	Type of event
4	11.2	arrive to system
2	15.7	end processing
5	17.5	arrive to system

Our simulation again returns to the event calendar in search of the next event to process, and this cycle repeats continuously until the simulation ends. There are two ways a simulation run will end: either there is a specific "end of run" event scheduled on the event calendar at a specified time or there are no events on the event calendar, causing the simulation to simply run out of things to do and end.

Although this is a very short example, it illustrates the frequency with which the event calendar is accessed and must consequently be re-sorted into chronological order. This is one reason why the execution time for a discrete-event simulation model is directly related to the number of events and how the lists are managed. Although most simulation languages had traditionally used linked-lists to manage the event calendar and other lists in the simulation, binary heaps, used for example in SIMAN, or Henriksen's Algorithm, used for example in SLAM and GPSS/H, provide faster execution times [4].

12.2.2 Need for Randomness

Because systems that are appropriate to analyze with discrete-event simulation are stochastic, the ability to include randomness in the model is essential. Reliable random number and random variate generation are at the heart of discrete-event simulation languages. A random number is used in a simulation model anytime a sample is generated from a distribution or when a probability must be checked, for example, determining if a part passes or fails inspection. One random number generator characteristic that is of particular importance for simulation languages is the need for a long cycle length, which is the number of random values generated before the random number string starts to repeat. When modeling large systems, many random numbers are needed during each run. Just a few years ago, a well-researched linear congruential generator that could provide a cycle length of about 2.1 billion values was considered adequate and the norm. Recently, however, a combined multiple recursive generator (e.g., used in Arena) provides a cycle length of 3.1×10^{57} [3]. Having a long cycle length and the additional ability to specify different streams for different distributions is needed for variance reduction techniques for output analysis, which is discussed later in this chapter.

When the random number is used to generate a sample from a distribution, generating the random number is the first step. If the distribution has a continuous cumulative distribution function (CDF), then the inverse transform method is used. In this method, the CDF is set equal to the random number and then solved for the sample value. This is the preferred method of generating a variate. If the CDF is not continuous, then other mathematical methods are used, but they still rely on having a random number.

12.2.3 Output Analysis for Discrete-Event Simulation

The main goal of modeling is to analyze output data regarding the performance measures of interest to your system and make decisions or recommendations based on that analysis.

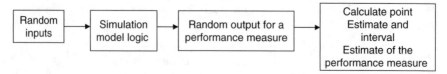

FIGURE 12.4 Randomness in inputs and outputs.

For discrete-event simulation models, which are stochastic and dynamic, the randomness inherent in the model inputs means the model outputs also exhibit randomness (see Figure 12.4). As previously mentioned, the stochastic input causes each simulation run, or replication, of the same model to have slightly different output. Also, as simulation models are run rather than solved, accurate output data analysis depends upon both an accurate translation of the physical system into simulation model logic and setting the correct simulation run control parameters so that the output will have appropriate statistical properties. If this is done correctly, you can then use standard statistical analysis techniques, for example, confidence intervals and paired-t tests. Therefore, you must know the type of output analysis that you want to accomplish as you develop your model.

The first step for proper output analysis is classifying the type of system you are modeling and the type of analysis needed as either a terminating system or a nonterminating system. A terminating system has fixed starting conditions that are usually the same for every system start, for example, empty (no parts or customers) and resources idle, and some event defines the end time or closing time of the system, for example, store closing time or the time the last diner leaves. Many service systems that are not open 24 hours a day are terminating systems, for example, banks, restaurants, retail stores. A nonterminating system might not have the same starting conditions each day and there is not a clear ending event. Any 24-hour operation is a nonterminating system, for example, emergency rooms, 24-hour manufacturing plants, convenience stores. Additionally, any system that carries over work from the previous day should also be considered nonterminating because the starting conditions each day have a different level of work in process, and thus they are not the same. Most manufacturing facilities exhibit this characteristic.

For systems classified as terminating systems, all the data collected during the simulation run is included in the output analysis. These systems exhibit transient behavior when the system starts up, which is important to include in the analysis. For nonterminating systems, we are usually interested in the long-term, steady-state behavior of the system after the transient or start-up conditions are gone. In these systems, only the steady-state data should be included in the analysis. Therefore, we need slightly different approaches to properly analyze each case.

Regardless of the type of system, the output performance measure of interest could be an observation-based statistic (sometimes called discrete-time or tally statistic), a time-persistent statistic (sometimes called discrete-change or continuous-time), or a simple running count. Examples of observation-based statistics include average customer or part time in the system or average time in the queue. Letting X_j be a single entity's observed value with J total entities, then the average for replication i is calculated in Equation 12.1.

$$\overline{x_i} = \frac{\sum_{j=1}^{J} X_j}{J} \tag{12.1}$$

Examples of time-persistent statistics include average work in process, average number in the queue, and utilization. To find the average of a time-persistent statistic, a system or global variable tracks the value of the specific measure (e.g., the number in queue) over the entire time of the simulation replication, as in Figure 12.5. To calculate the average, the

Simulation

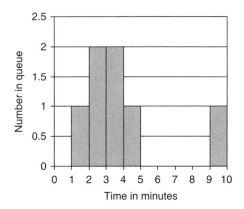

FIGURE 12.5 Tracking the number in queue.

area under the curve is divided by the total time of the simulation replication from time 0 to T, as in Equation 12.2.

$$\bar{y}_i = \frac{1}{T}\int_0^T Y(t)\mathrm{d}t \tag{12.2}$$

Examples of simple running counts include the total number leaving the system, the total number of bad parts, and total number of customers that cannot enter a queue that is full. You simply increment a counter variable throughout the total time of the simulation replication each time an entity matches the criteria for the count.

The next two sections discuss how to properly analyze these performance measures for terminating and nonterminating systems.

Terminating System Analysis

As previously mentioned, stochastic input causes each simulation replication of the same model to have a slightly different output. Therefore, the output from one simulation replication represents one sample from the population of all possible replications, paralleling a physical experiment that takes sample observations from an entire population. In physical experiments, we need to have population samples exhibit independence and be normally distributed to apply many of the common statistical analysis techniques. The same is true for simulation, where we commonly calculate a confidence interval about the mean value and use that confidence interval to reach conclusions and make recommendations.

An advantage of statistical analysis in our simulated environment is that it is very easy to insure independence of the output data between replications. To obtain independent output data from multiple replications, when running your simulation, you specify (1) that the statistics be cleared (set back to 0) between replications, (2) that the model be reinitialized to the same starting conditions (usually empty and idle) between replications, and (3) that a different random number seed is used for each replication to generate the model inputs. Because these factors are easy to specify in simulation languages, it is easy to insure that the assumption of independence will hold.

It is not quite as easy to demonstrate that the normality assumption will hold for any simulated system. Fortunately, the performance measure of interest is typically an average over all the observations or all the time of each replication. For example, the output "average time in the system" is the average over all the individual observations for each entity that has exited the simulated system. Then, to find a point estimate of the mean over multiple replications, take the average of values that are themselves averages. Simulation practitioners tend to agree that based on both practical experience with empirical testing and theoretical general versions of the central limit theorem, it is valid to use standard statistical techniques, which have an assumption of normality, with output data from discrete-event simulation models [3].

Suppose that we have modeled a terminating system and we are interested in the average time that a part or customer spends in our system. Our typical analysis proceeds according to the following steps:

Step 1: Run multiple independent replications.
Step 2: Record the output value for average time spent in our system, calculated by Equation 12.1, from each replication (1 value per replication).
Step 3: Using the average time in system values from each replication as calculated in Step 2, calculate the mean and standard deviation of these values.
Step 4: Calculate a confidence interval about the mean calculated in Step 3.
Step 5: Make recommendations based on your statistical analysis and knowledge of the system and company goals.

For example, suppose 20 replications (n) had end of replication average time in the system values (in minutes) of 12, 15, 10, 11, 12, 11, 16, 12, 17, 14, 13, 15, 11, 9, 10, 13, 12, 8, 12, 15. The mean ($\bar{x}$) is 12.4 min with standard deviation (s) of 2.37 min. If we want a 99% confidence interval (CI) ($\alpha = 0.01$), then the appropriate critical value from Student's t distribution is $t_{\alpha/2, n-1} = 2.86$. From the general confidence interval Equation 12.3, the 99% CI for time in the system is 12.4 ± 1.516 min or [10.884, 13.916].

$$\bar{x} \pm t_{\alpha/2, n-1} \frac{s}{\sqrt{n}} \tag{12.3}$$

This confidence interval means that we have an approximately 99% confidence that the true mean of the average time for parts in this system falls between 10.884 and 13.916 min [1]. The range of the confidence interval also gives you a sense of the variability in the system— the wider the range at a given confidence level the more the uncertainty concerning average time in the system.

Ultimately, it is still the responsibility of the analyst to appropriately use these data values to reach a conclusion or make recommendations. Good, accurate statistical analysis as described above is a key component for making appropriate recommendations. But you must also consider the type of system, the type of company, and the goals the company has for the specific system being modeled.

Nonterminating System Analysis

As previously mentioned, with nonterminating systems we are usually interested in the long-term, steady-state behavior of the system after the transient or start-up conditions are gone. Steady-state behavior does *not* mean that the system has *no* variability and that performance measures have become a constant value. Steady-state behavior is when the variability in the system has reached a more consistent level. This is the case in nonterminating physical systems that have been in operation for a long time. When a system is brand new, it typically starts in an empty and idle state. So, initially there is little competition for resources, utilizations are low, and there are no queues at many resources. Consequently, parts or customers have low time in the system values and very little of this time is nonvalue added. As time goes on, however, queues start to build because there is more competition for resources, time in the system values rise with more nonvalue added time (e.g., queue time)

Simulation

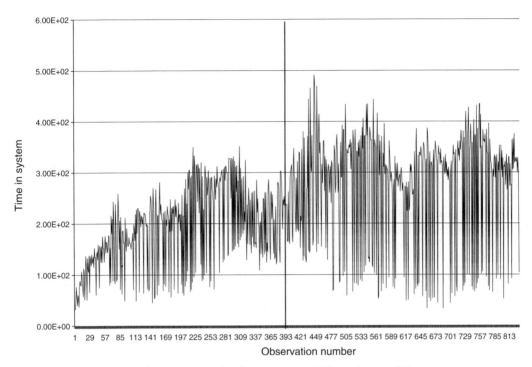

FIGURE 12.6 Time in the system starting from empty and idle system conditions.

and so on, to a certain point where the system starts to "stabilize" at some level of variability about some mean values for performance measures of interest, as shown in Figure 12.6.

Looking at Figure 12.6, you might estimate that the initial start-up or transient conditions end around observation 393. The data after this point could then be considered to be reflecting the steady-state behavior of the physical system. When we develop a simulation model of a nonterminating system to analyze its steady-state behavior, our goal is to only analyze simulation output data from the time where our model is also exhibiting steady-state behavior. If we start our simulation model with empty and idle conditions, as we did in terminating system analysis, a plot of the output data would look similar to Figure 12.6. If all of data from time 0 is included in the output data calculations, our values will be biased and will not accurately reflect only the steady-state system behavior. Therefore, since we only want to analyze data after the start-up transient conditions have ended, we must have a way of identifying the point at which the start-up transient conditions end and eliminate the biasing effect of initial transient data upon our steady-state statistical output data analysis.

There are a few methods to eliminate the effect of initial transient data. One method is to select starting conditions for your simulation model other than empty and idle. The idea is to "pre-load" the simulation to make it look more like an ongoing system at steady state right from time 0. Then, all the data from each replication can be used for output analysis since all the data is from a time where the simulation is emulating the steady-state behavior of the physical system being modeled. Pre-loading the system involves creating entities with attribute values assigned that would be placed in different queues throughout the system; thus, when the simulation begins, resources throughout the system will immediately have work to do and entities will soon be exiting the system. An advantage of this method is that it is easy to explain to anyone, even if they are not familiar with simulation. A disadvantage of this method is that if the system is large, you may need to pre-load many entities. This means (1) you need to collect data from the physical system to approximate the typical number of entities in each queue and the part mix at each queue and (2) actually loading all those entities into the simulation model can be tedious. For example, with 10 resources and approximately 5 entities waiting at each there are 50 entities that must be entered. Also, if the system being modeled is in the design phase, then this is not a feasible approach.

A second method is to allow the simulation to start in its typical empty and idle condition, then delete initial transient phase data from the output analysis. Deleting early data is easily accomplished in most simulation languages by simply specifying a *warm-up period* time in the run control parameters for a model. With a warm-up period specified, only data collected after that time is included in the output statistics provided by the simulation. This requires estimating the point where the initial transient phase ends and steady state begins. Typically, this is done by plotting the output performance measure of interest over time for the output from one long replication of your model [4]. Then, by looking at the plot, visually estimate when the data appears to stabilize and select that point as your warm-up period length. To assist this process, plotting a moving average of the data instead of the raw data is sometimes helpful, as shown in Figure 12.7. Figure 12.7 is the same raw data for time in the system as Figure 12.6, but it has a moving average with a window of width 50 superimposed. Looking at this graph, we might estimate that steady-state behavior begins at approximately 500 observations. An advantage of this method is that it can be applied when modeling either an existing or proposed system as it does not need data from the physical system. A disadvantage is that determining the point at which the steady-state behavior starts can still be difficult. Also, there is some amount of wasted data and run time during the warm-up period. In some large systems, the warm-up period can be significant.

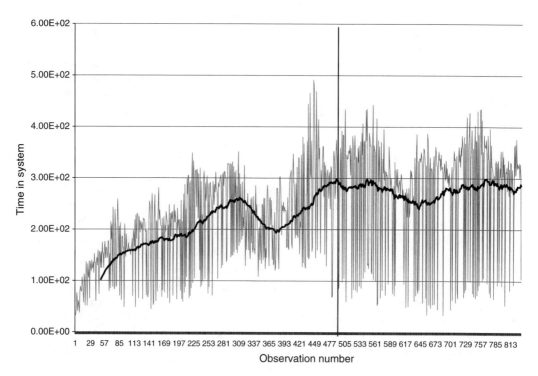

FIGURE 12.7 Time in the system starting from empty and idle system conditions, with a moving average line of window width 50 imposed.

A third method is to simply run the model for an extremely long time so that the initial transient data is dominated by the steady-state data. With computer power and speed continually improving, this method becomes more feasible. An advantage of this method is that it is simple to implement. However, you still need to get a rough estimate of the point in time that steady-state behavior begins to be certain you have a long enough run length to truly dominate the initial transient data. This could be done using the second method above. Another disadvantage is that some larger systems may have a long initial transient phase, which requires a very long total run length to dominate the data.

After you have solved the problem of the initial transient data with one of the methods above, then you need to get independent and normally distributed data so you can use the standard statistical techniques just as we did with terminating system analysis. Perhaps the easiest approach is to combine any of the above three methods with making multiple independent replications. In other words, you implement a method that insures that the effect of initial transient data has been eliminated, such that the output statistics in each replication from the simulation reflect the steady-state behavior of the system. Then, when multiple independent replications are run, by making the same specifications as discussed in the previous section on terminating systems, we will have independent samples of the steady-state performance measures of interest. Next, apply the same standard statistical analysis techniques for mean, standard deviation, and confidence interval as shown in Equations 12.1–12.3. So, once the initial transient issue has been removed, simply follow the same 5-step analysis procedure used with terminating systems as described in Section 12.2.3.1. One of the most common approaches is the truncated replication approach, because it is easy to specify a warm-up period and to specify the running of independent replications. This combines the second method above, eliminating the initial transient data by specifying a warm-up period, with running multiple independent replications.

Another method for analyzing steady-state performance is batch means analysis, which is quite different from the above techniques. Batch means analysis requires that the initial transient data be truncated from only one long replication, and then it uses the remaining steady-state data from that one replication for analysis. However, if all the steady-state individual observations or time-persistent values from within that one long replication are used to calculate a sample variance, there is a serious problem with bias in the variance calculation due to the significant correlation of data points from within one simulation replication. Therefore, this does not allow for statistically accurate confidence interval construction. Batch means analysis acknowledges this correlation and attempts to create large enough batches of individual data points from within the one replication, such that the calculated means of these batches will be statistically independent from each other [3]. These batch means may then be treated as independent observations to be used in standard statistical calculations. Steps in batch means analysis are:

Step 1: Truncate the initial transient data.
Step 2: Group the remaining data into batches of observations that are independent. N batches of size m.
Step 3: Let $\overline{X_j}(j = 1, 2, \ldots, N)$ be the mean of batch j.
Step 4: Treat $\overline{X_j}$ like independent observations of a steady-state performance measure, then use standard statistical analysis techniques, such as confidence intervals.

Step 2 may need to be repeated with increasingly larger batch sizes (m) and a corresponding reduction in the number of batches (N) until statistical testing indicates that the batch means are no longer correlated.

An advantage of the batch means technique is that you only need one very long simulation run. In a very complex system where one replication may take several hours of computing time, this can save valuable time. A disadvantage is that you must perform a statistical analysis on the batch means testing for independence, and the proper batch size must be determined. This technique may be more difficult to explain to someone not familiar with simulation, and it is a bit more involved with respect to statistical tests than the previous methods. However, some simulation languages have a built-in capability for analyzing simulation output via batch means.

Comparing Two Systems

Many times the reason that a simulation model is developed is to evaluate a change to the current system. In this case, you need to do an analysis of the output performance measures from two simulation models, one of the current system configuration and one of the changed/modified system, to estimate if there is indeed a significant difference in performance due to the system modification. A common way to do this is to perform a paired-t test using the output performance measures from the two simulation models, with $H_0 =$ there is no difference between the mean performance and with $H_1 =$ there is a difference. This is coupled with a simulation variance reduction technique called the common random numbers (CRN) method. CRN requires that within each simulation model different random number streams are assigned to each distribution used, but between the two models, the same random number stream should be used for the same type of distribution. For example, if random number stream 1 is assigned to the interarrival distribution in the model of the current system, then random number stream 1 is assigned to the interarrival distribution in the model of the modified system. This is similar to what we do when conducting physical experiments when we try to control all factors in the experiment other than the one factor that we are modifying. Then, if a change is observed in the performance measure between the experiments, we feel confident that the difference is due to the modification that we made and not due to some chance variation from another system factor. The way to control the other factors in a simulation experiment is to control the random number streams.

CRN is a variance reduction technique because it induces a positive correlation between the outputs from the two simulations. When using paired-t tests to statistically compare the output from the two systems, the difference between the performance measure for both systems for each replication becomes the new performance measure of interest, as follows:

Output current system (A)	Output modified system (B)	Difference $X_{Ai} - X_{Bi}$
X_{A1}	X_{B1}	D_1
X_{A2}	X_{B2}	D_2
.	.	.
X_{An}	X_{Bn}	D_n

Then, use Equation 12.3 to calculate a confidence interval about the mean of the differences. If 0 is contained in the interval, then we fail to reject the null hypothesis and assume the systems have no statistically significant difference at the specified confidence level. If 0 is not contained in the interval, then we reject the null hypothesis and assume there is a statistically significant difference between the systems. The variance reduction occurs because we are taking the difference of positively correlated output, and then the standard deviation of those differences, which is the square root of the variance calculation

Simulation

shown in Equation 12.4, is used in the confidence interval. Consequently, in Equation 12.4 the final covariance term is now greater than 0, making the variance of the difference less than what it would be if the random variables in the difference were independent making the covariance equal 0.

$$Var(X_A - X_B) = Var(X_A) + Var(X_B) - 2Cov(X_A X_B) \qquad (12.4)$$

If you reject the null hypothesis and determine there is a difference between the system performances, then you must decide if it is a difference that would be a benefit if the modification occurred. Finally, a cost analysis should be done to insure that it would be a cost effective change to implement.

Evaluating Multiple Potential Modifications to a System

At times there may be more than one system parameter that we are considering changing. In this case, you should develop a design of experiments with those parameters. Once you have identified the design points, you can modify the simulation model to emulate each design point and run multiple replications at each point. The output from each simulation then becomes the input to further statistical analysis, such as analysis of variance (ANOVA).

12.3 Simulation Languages and Software

There are many discrete-event simulation software products available. Although they all have some differences to distinguish themselves in the marketplace, there are many common features among all [1,4]:

- Graphical user interface to make model building easier.
- Sampling from several different distributions.
- Animation to visualize the model, possibly 3-D—this allows you to observe short-term variability in the performance of the model and can help quickly identify bottlenecks or quickly identify errors in the simulation model logic.
- Collect a variety of performance measure statistics and provide summary output statistics, such as means or confidence intervals, in an output report.
- Provide data in tables and in graphical form.
- Easily run multiple independent replications.

Additionally, there are features that many, although not all, simulation environments share:

- Extended statistical analysis, either as an add-on or integrated into the package—this may include both input data analysis (e.g., fitting data to a distribution) and output data analysis (e.g., comparing systems for statistical differences, batch means analysis).
- Import CAD drawings as a foundation for developing animation.
- Interfacing with other software, such as spreadsheets.
- Creating Visual Basic interfaces to ease changing the parameter specification for the model to create different scenarios.

Some characteristics that will set simulation languages apart include the level and sophistication of material handling or warehousing logic and the price of the package. Although some simulation software companies offer a bare-bones version of their packages for under

$1000, prices for versions that allow you to model a reasonably sized system range from approximately $1300 on the low end to mid-range of about $15,000 to $30,000 and up for specialized versions.

There are two main categories of simulation languages: general-purpose and simulator. A *general-purpose language* can be used to model any environment, manufacturing, service, business processes, and so on. The terminology used in the modeling constructs is very generic, for example, entity, resource, and process. A *simulator* is a language that is customized for a particular type of industry, for example, electronic chip manufacturing, hospital, call center. The terminology used in a simulator is specific to the environment, for example, wave solder machine, nurse, phone line. There is a tradeoff between ease and quickness of building a model and flexibility to capture a wide variety of real-world complex system logic. A general-purpose language offers a lot of flexibility regarding the structure of the physical system logic that can be properly accommodated in the simulation model. However, it may take longer to develop a model to make sure that the generic logic constructs are put together to properly emulate the logic of a specific type of environment. On the other hand, a simulator has logic constructs that are more structured to match typical physical system logic for a particular industry. Therefore, it has less flexibility to model anything and everything that might be encountered in all systems. But a model can usually be developed faster using a simulator compared to a general-purpose language, assuming the simulator is properly matched to an environment for which it was intended.

One additional comment regarding simulators: some simulators can still seem like a very general-purpose tool within an industry. For example, a manufacturing simulator, such as ProModel, is applicable in a wide variety of manufacturing environments. However, some simulators for more specific environments, for example, electronics manufacturing, may not be perfectly applicable to all types of systems within that larger category. Therefore, if you are considering a simulator, you must make sure that the structured logic of the simulator is truly applicable to your specific system.

12.3.1 A Sampling of Discrete-Event Simulation Software

This section contains a sampling of the many discrete-event simulation software packages available. All the following software run on a PC, with a typical operating system requirement of either Windows 98, Windows 2000, or Windows XP. They also take the process-oriented approach to modeling. Software is discussed in alphabetical order.

Arena

Arena is a general-purpose simulation software that can model manufacturing or service systems, and it is a product of Rockwell Automation [3,5]. The underlying simulation language is SIMAN. The Basic version is focused on business process modeling; however, Standard and Professional versions can model any environment, including continuous systems. It has a graphical interface for model building with logic modules, whose parameters you specify, connected with arrows designating entity flow. It includes input data analysis with distribution fitting through its built-in input analyzer and it allows more detailed statistical analysis of output data through its built-in output analyzer along with graphical chart capability. Standard output statistics, such as means and confidence intervals, are provided in the standard output report in numerical tables. Arena can model a variety of material handling devices including fork lifts, automated guided vehicles (AGVs), and different types of conveyors. Version 10.0 also includes specialized logic to more easily model continuous manufacturing that uses tanks, regulators, and pipes. It allows interfacing with various other software applications. Arena has built-in capability to easily run and analyze multiple scenarios with its process analyzer, and you can add-on OptQuest for Arena to perform optimization. OptQuest was developed by Dr. Fred Glover and is offered by OptTek Systems, Inc., as an add-on to several simulation packages [6]. It is based on several methods, including tabu search, scatter search, and neural networks.

AutoMod

AutoMod, available through Applied Materials, Inc. [7,8], can be used to model a wide range of applications; however, it is primarily applied to manufacturing and it includes significant material handling capabilities. It has built-in templates for bridge cranes, conveyors, power and free systems, automated storage and retrieval systems (AS/RS), and path-based movement (automated guided vehicles (AGVs), fork lifts, humans). It has 3-D graphics and template models of some specific environments to speed model development. However, it also is a simulation language to be used in a variety of manufacturing environments. In Version 11.0 you can create "composite systems," which allow the reuse of model objects through the encapsulation of submodels and models that are saved as library objects. It can also link to other software through ActiveX. Output may be viewed as numerical tables or a variety of graphical charts. They also have the product AutoStat, which can help with determining the warm-up period, sensitivity analysis, design of experiments, and optimization using evolutionary strategies. Another product, AutoView, provides postprocessed animation capability, allowing playback and creation of presentation graphics.

Extend

Extend simulation software refers to a family of products by Imagine That, Inc., that covers a range of applications, including continuous systems, manufacturing, and service systems [9,10]. All products have similar structure and capability. Models are built by combining different iconic blocks, representing a logic step or calculation, according to the proper logic flow for entities. During a simulation run, the user may change simulation parameters on the fly to observe the near-term effect of changes. Data relating to a specific block, for example, a queue block, may be viewed via plots and tables by double clicking the block. Output from a replication may be displayed in a table or plot, exported to another program or copied to a different area of the worksheet. An evolutionary optimization tool is included in the software. Extend also has the capability of creating a hierarchical block by grouping together blocks defining the logic of a submodel representing part of the larger system. These hierarchical blocks may be stored for future use. Extend supports the COM model and ODBC without any additional programming by the user. The Extend blocks are open source, so they may be customized to a specific need. Also, Extend has been chosen as a simulation engine by third-party applications. The Extend Suite product includes Proof Animation from Wolverine Software [11].

Flexsim

Flexsim, developed by Flexsim Software Products, Inc., is an object-oriented simulator aimed at all types of manufacturing environments, including both discrete-event systems and continuous flow systems [12,13]. It allows a hierarchical approach to modeling, making use of inheritance. Everything in Flexsim is open to customization and users can create their own new objects using C++. All models are built and run in 3-D. Input data can be imported from databases and spreadsheets using ODBC or Dynamic Data Exchange connections, and output can be exported to word processors or spreadsheets. The output can be viewed in reports and graphs.

GPSS/H

GPSS/H is a widely used version of the GPSS simulation language originally released by IBM in 1962, and it is available through Wolverine Software Corporation [14]. It is a general-purpose language that can model any manufacturing or service system that can be thought of as a queueing system. Models are built using a text-based interface, as opposed to a graphical interface commonly used in most other simulation languages. GPSS/H has built-in file and screen input and output, allowing data to be read into and out of standard text files that can be interfaced with standard spreadsheets. Customized output can also be created. GPSS/H has an interactive debugger to assist verification, and it uses Proof Animation, another Wolverine Software product [11], which can be viewed postprocessed after the simulation completes or while it runs. Because of GPSS/H's flexibility and speed of execution, it is sometimes used as the simulation engine for simulators developed for specific environments. With GPSS/H's structure and text-based model building interface, it tends to be more frequently used by experienced modelers with more complex systems.

Micro Saint Sharp

Micro Saint Sharp is a general-purpose object-oriented simulation language by Micro Analysis and Design, Inc. [15,16]. It was developed specifically to increase model execution speed, and allows users to turn off all interfaces for significant speed-up. Micro Saint Sharp can be used to model any system that can be represented by a flowchart. It uses task network modeling with activities represented as nodes and arrows connecting nodes into sequences of activities. The user defines the structure of the task network and then defines the objects in the network. It has a symbolic animation that animates the network diagram. This can be helpful for quickly gaining insight into the entity flow and can help with debugging. Charts showing dynamically changing outputs can also be presented. Alternately, you can also use their animation tool called Animator, which gives a more realistic view of the system. Micro Saint Sharp has a plug-in interface allowing for some modularity to customize how you use the software. Its options include OptQuest for optimization [6], the Animator tool discussed above, and interoperability using COM services.

ProModel

ProModel Corporation has several simulation products available, including ProModel, MedModel, and ServiceModel [17,18]. These are simulators because they are written for a specific industry. ProModel is applicable in manufacturing and warehousing environments, MedModel is specifically for modeling healthcare systems, and ServiceModel is designed for service industries. However, within these broad industry categories, they are very generally applicable to many specific environments. These products use object-oriented modeling constructs. Although the rest of this section focuses on ProModel, the capabilities are similar across these three products.

ProModel is a simulator aimed at general manufacturing environments. Models can usually be built graphically with the set of modeling elements provided (e.g., resources); however, complete programming is available if necessary for very unique physical system logic. Animation is developed when the model logic is being specified, and you can create both 2-D and 3-D perspectives. In ProModel you can also specify costing information that can be associated with a specific resource, location, or entity, which is reported in the output report. The representation of output data can be in tabular form or in graphics and charts. There is a run time interface that can be used at the start of a run to make changes to system parameters for just that run, or it can be used to define and store multiple scenarios of the system for experimentation. Two optimization capabilities are available. SimRunner, which was developed by ProModel Corp. specifically for their products, uses genetic algorithms and evolutionary strategies. It can be used to aid model analysis by determining the number of replications needed to estimate performance measure mean values to a specific precision, or to aid with steady-state analysis by estimating the proper warm-up period to eliminate transient data and estimating run time. It can also be used to optimize a user-specified performance measure by evaluating various combinations of user-defined input factors. The OptQuest for ProModel add-on is another optimization alternative.

QUEST

QUEST from Delmia Corporation is a manufacturing environment simulator that has a graphical model building interface [1,19]. It is a 3-D simulation environment that can integrate 2-D and 3-D CAD drawings to more accurately capture the layout of the factory being modeled. It can also interface with spreadsheets and production software, such as manufacturing execution systems (MES), material requirements planning (MRP), enterprise requirements planning (ERP), and scheduling packages. QUEST has an emphasis in its capability to model a variety of material handling types, including fork lifts, conveyors, power and free, humans, automated guided vehicles (AGVs), kinematic devices (robots), and automated storage and retrieval systems (AS/RS). Output can be displayed numerically in tables or graphically using various charts, or exported to a spreadsheet.

SIMUL8

SIMUL8 is a product of SIMUL8 Corporation and has a Standard and Professional version [1,20]. It has a graphical model-building interface, allowing icons representing logic to be placed and connected by arrows showing entity flow. It also allows user-defined icons that can be saved and reused. SIMUL8 is a general-purpose language; however, it has been frequently used to model service industries or business processes, and it also has some specialized logic capability for modeling tanks and pipes found in continuous fluid manufacturing processes. On the input side, it can interface with Excel and in the Professional version it can also interface with databases. Output can be viewed as either numbers in tables or graphically via plots, pie charts, and the like. Output can also be exported directly to Excel or MiniTab, or a file option allows you to send the date to other file types. The professional version has additional input data fitting help with STAT::FIT, a product of Geer Mountain Software Corp. [21], and optimization capability through OptQuest for SIMUL8.

WITNESS

WITNESS 2006 is a product of Lanner Group, Inc., and has a manufacturing performance edition (MPE), a manufacturing simulator, and a service and process performance edition (SAPPE), a service industry simulator [22]. Its manufacturing version can model both discrete manufacturing and continuous flow manufacturing. Models are built using a building block design, having a modular and hierarchical structure. It can interface with databases, spreadsheets, and other applications. It also has a Virtual Reality module that can integrate with the standard system to provide 3-D modeling capability. It has a simulation scenario manager to assist with experimentation, and the results can be exported to other applications. WITNESS has an optimizer integrated with its standard system, which is based on a variant of simulated annealing.

12.3.2 Summary

There is a wide variety of discrete-event simulation software available. With the exception of GPSS/H, all the other software mentioned here have graphical interfaces for model building. Most have user-friendly modules that typically represent several lines of simulation language code put together in a particularly structured way to capture a small logic element of the physical system, for example, an entity accessing a resource for service. In software that has a hierarchical structure, modules at higher levels have more lines of code within each module with a specific structure, and they consequently allow less flexibility to capture complex or unusual entity flows or control methods that may be present in the physical systems being modeled. Then, it is necessary to go lower in the hierarchy to modules that represent fewer lines of code with more elementary logic statements to buy the user more modeling flexibility. Although the graphical module approach tends to blur the lines of "modeling" and "programming," it is important to remember that underlying all the software mentioned above is a simulation language and as the "picture" of the system is developed on the screen, actual code is being created to capture that picture. So, remember that when building a simulation model, all rules of programming apply, including garbage-in-garbage-out.

Additional information about the simulation software discussed above is available at the following Web sites:

Arena: www.arenasimulation.com
AutoMod: www.automod.com
Extend: www.imaginethatinc.com
Flexsim: www.flexsim.com
GPSS/H: www.wolverinesoftware.com
Micro Saint: www.maad.com
ProModel: www.promodel.com
QUEST: www.delmia.com
SIMUL8: www.simul8.com
WITNESS: www.witness-for-simulation.com or www.lanner.com

12.4 Simulation Projects—The Bigger Picture

Any time you are developing a model of a physical system, there are several steps in the process that will help to make the modeling project a success, and simulation is no different. The steps are nothing more than the scientific process learned in grade school, but with a few added simulation details. Some slightly different procedures can be found in Refs. [1–3].

Step 1: Identify the problem: A statement is needed of the problem that is initiating the model development. For example, the manager of a facility may be concerned that the amount of work in process (WIP) is exceeding the maximum level that the corporation desires.

Step 2: State your objectives: Based on the problem statement, exactly what questions will be addressed by modeling this system? Specify what performance measures will be evaluated, what system changes will be studied, and what type of analysis of output data is required. For example, with the problem identified in Step 1, we may need to evaluate the utilization of resources and time spent in the system in addition to our main performance measure of interest WIP. Possible system changes might focus on first identifying if a single bottleneck station exists, and specifying possible changes that could be investigated, such as adding a resource, modifying the set-up operation, or changing the scheduling or control policies in the system. The type of analysis for this problem may be comparing the steady-state WIP performance of different modifications of the system to determine which best adheres to the corporate maximum WIP level. This may also be the best time to decide if simulation is the correct modeling tool for this situation.

Step 3: Identify, collect, and prepare input data: Identify the input needed to develop the model. What are the entities, the resources, and other system parameters? Are processing times, interarrival rates, and other system parameters known or must a data collection scheme be developed? If equipment failure is important to incorporate, does the company have reasonable records of time between failures and downtime? If data must be collected, it may take a few weeks or months to get enough data points for analysis, particularly for events/activities that occur less regularly, such as breakdowns. Therefore, it is a good idea to develop the collection plan and begin that process as soon as possible, even before you start model development.

Step 4: Formulate the model: Creating a logic flow diagram of the physical system being modeled is the first step in formulating the model. Further, it is independent of the particular simulation software that will be used for the project. Developing the logic flow diagram requires visiting the existing physical system or studying the plans for a proposed physical system and talking to all parties involved with the system to ensure that the modeler understands the way that system works or how it is proposed to work. The logic flow diagram also becomes an effective communication tool between the modeler and the system experts to help all parties have confidence that the appropriate level of understanding and detail of system operations will be properly included in the model. Once the logic flow diagram is complete, it can be translated into a simulation model in a specific simulation language.

Step 5: Verify and validate the model: First, verify that the model is behaving as you think it should according to your logic flow diagram. This includes debugging the model and doing some preliminary output data analysis to determine if it is reasonable given the model inputs and parameters. Second, validate that the model is accurately representing the physical system, that is, did we really understand the system correctly and build the right model? Validation can be a challenging part of the simulation modeling process. For both verification and validation, simple animation can be helpful to identify bottlenecks or basic logic problems in the model.

Some verification strategies include:

1. Test a large model in stages. Break the bigger problem into logical subproblems, for example, individual work stations or one department.
2. Send a single entity through the model at a time. If there are multiple possible routes, force a single entity to take each route to ensure you have the proper routing in your model.
3. Replace the random times with constant values, for example, the means of the distributions, so that you can predict with certainty how long it should take an entity to complete a specific route (combine with number 2).
4. Make a single replication during verification, first with constant times, then with the randomness put back into the model. Look at the output data in each case,

and use your engineering judgment. Are these the performance measure output values you expect?

5. Ask an expert—show this output to a person familiar with the physical system and get their opinion. You can also ask an outside expert familiar with simulation modeling but not necessarily with the specific system for their opinion on your modeling approach as well as the output data.
6. Perform some sensitivity analysis. Change an input parameter in a certain direction and see if the output data responds as expected. For example, in a moderately loaded system, if you increase the number of arrivals per unit time, you would expect higher utilization, longer queues, and longer time in the system.

Validation can occur simultaneously with verification. In particular, numbers 5 and 6 above are also validation strategies. Number 5, ask the expert, can be taken one step further if you are modeling an existing system. You can put the data from the model beside actual performance measure data from the physical system and ask the expert if they are reasonably close.

Step 6: Experiment and analyze the results: Once the model is verified and validated, it can be used for its intended purpose—allow you to experiment with system modifications. Depending upon the type and number of modifications, an appropriate experimental design is developed. The proper statistical analysis is then done with means, standard deviations, confidence intervals, paired t-tests, and so on.

Step 7: Conclusions and recommendations: Using the data from Step 6, reach your conclusions based on the simulation output results, and specify what you recommend for the modification or design of that system to address the objective identified in Step 2.

12.5 Summary

Discrete-event simulation is an important heuristic tool in the OR tool kit. It is applicable when the performance of a large, complex system needs to be evaluated, but we cannot properly capture the system dynamics in straightforward equations that could be used in an optimal technique. This chapter has presented an overview of discrete-event simulation, including discussion of basic definitions and operations of simulation, output analysis, some examples of where simulation has been applied, simulation languages, and the steps of a simulation project.

In addition to the software Web sites previously listed, there are some other readily accessible sources for simulation information. Attending simulation conferences gives you quick access to the latest information in both technical language developments and how people in other companies are using simulation. Conferences also provide an excellent networking opportunity to meet others facing similar problems. In a small- or medium-sized company, you might be the only person engaged in simulation modeling, and you may feel isolated with no one to bounce off ideas. Making contacts at a conference allows you to develop a support network for your ideas or problems in the future. Further, most conferences have a vendor area, which allows you to speak with many vendors in a short period of time and make some comparisons of products. Three examples of conferences are the Winter Simulation Conference (WSC) held in December at www.wintersim.org, the Summer Computer Simulation Conference typically held in July and sponsored by the Society for Modeling and Simulation International at www.scs.org, and the Simulation Solutions Conference usually held in the spring at www.simsol.org.

Another source of information is simulation software surveys that periodically appear in professional society journals. For example, OR/MS Today [23], a publication of INFORMS (Institute for Operations Research and the Management Sciences), reported a simulation software survey in December of 2005, which has since been updated on their Web site [24].

For further reading about simulation modeling in general, please consult some of the textbooks previously referenced in this chapter [1–4,25].

References

1. Banks, J. et al., *Discrete-Event System Simulation*, 4th ed., Pearson Prentice Hall, Upper Saddle River, 2005, chap.1–4, 11.
2. Leemis, L.M. and Park, S.K., *Discrete-Event Simulation: A First Course*, Pearson Prentice Hall, Upper Saddle River, 2006, chap. 1.
3. Kelton, W.D., Sadowski, R.P., and Sturrock, D.T., *Simulation with Arena*, 4th ed., McGraw-Hill, New York, 2007, chap. 2,3,6.
4. Fishman, G.S., *Discrete-Event Simulation: Modeling, Programming, and Analysis*, Springer-Verlag, New York, 2001, chap. 5.
5. Bapat, V. and Sturrock, D., The Arena product family: Enterprise modeling solutions, in *Proc. 2003 Winter Simulation Conference*, Chick, S. et al., Eds., IEEE, Piscataway, 2003, 210.
6. April, J. et al., Practical introduction to simulation optimization, in *Proc. 2003 Winter Simulation Conference*, Chick, S. et al., Eds., IEEE, Piscataway, 2003, 71.
7. Rohrer, M.W., Maximizing simulation ROI with AutoMod, in *Proc. 2003 Winter Simulation Conference*, Chick, S. et al., Eds., IEEE, Piscataway, 2003, 201.
8. Applied Materials, Inc., AutoMod Web site, Available at www.automod.com, accessed 2/17/06.
9. Krahl, D., Extend: an interactive simulation environment, in *Proc. 2003 Winter Simulation Conference*, Chick, S. et al., Eds., IEEE, Piscataway, 2003, 188.
10. Imagine That, Inc., Extend Web site, available at www.imaginthatinc.com, accessed 2/17/06.
11. Henriksen, J.O., General-purpose concurrent and post-processed animation with Proof, in *Proc. 1999 Winter Simulation Conference*, Farrington, P.A. et al., Eds., IEEE, Piscataway, 1999, 176.
12. Norgren, W.B., Flexsim simulation environment, in *Proc. 2003 Winter Simulation Conference*, Chick, S. et al., Eds., IEEE, Piscataway, 2003, 197.
13. Flexsim Software Products, Inc., Flexsim Web site, Available at www.flexsim.com, accessed 2/17/06.
14. Henriksen, J.O. and Crain, R.C., GPSS/H: a 23-year retrospective view, in *Proc. 2000 Winter Simulation Conference*, Joines, J.A. et al., Eds., IEEE, Piscataway, 2000, 177.
15. Bloechle, W.K. and Schunk. D., Micro Saint Sharp simulation software, in *Proc. 2003 Winter Simulation Conference*, Chick, S. et al., Eds., IEEE, Piscataway, 2003, 182.
16. Micro Analysis and Design, Inc., Micro Saint Web site, available at www.maad.com, accessed 2/17/06.
17. Harrell, C.R. and Price, R.N., Simulation modeling using ProModel, in *Proc. 2003 Winter Simulation Conference*, Chick, S. et al., Eds., IEEE, Piscataway, 2003, 175.
18. PROMODEL Corporation, ProModel Web site, available at www.promodel.com, accessed 2/17/06.
19. Delmia Corporation, QUEST Web site, available at www.delmia.com, accessed 2/17/06.
20. SIMUL8 Corporation, SIMUL8 Web site, available at www.simul8.com, accessed 2/17/06.

21. Geer Mountain Software, Corp., STAT::FIT Web site, available at www.geerms.com, accessed 2/17/06.
22. Lanner Group, WITNESS Web site, available at www.witness-for-simulation.com accessed 2/17/06.
23. Lionheart Publishing, OR/MS Today Web site, available at www.lionhrtpub.com/ORMS.shtml, accessed 8/24/06.
24. OR/MS Today, Simulation software survey Web site, available at www.lionhrtpub.com/orms/surveys/Simulation/Simulation.html, accessed 8/24/06.
25. Law, A.M. and Kelton, D.W., *Simulation Modeling and Analysis*, 3rd Ed., McGraw-Hill, Boston, 2000.

13
Metaheuristics for Discrete Optimization Problems

13.1	Mathematical Framework for Single Solution Metaheuristics	**13**-3
13.2	Network Location Problems	**13**-3
13.3	Multistart Local Search	**13**-5
13.4	Simulated Annealing	**13**-6
13.5	Plain Vanilla Tabu Search	**13**-8
13.6	Active Structural Acoustic Control (ASAC) ...	**13**-10
13.7	Nature Reserve Site Selection	**13**-13
	Problem Formulations and Data Details • MCSP Interchange Heuristics • Tabu Search for MCSP • Heuristics for MECP and LMECP	
13.8	Damper Placement in Flexible Truss Structures	**13**-21
	Optimization Models • Tabu Search for DPP • Computational Results	
13.9	Reactive Tabu Search	**13**-29
	Basic Features of Reactive Tabu Search • Computational Experiments	
13.10	Discussion	**13**-35
	References	**13**-36

Rex K. Kincaid
College of William and Mary

A metaheuristic is a general algorithmic framework for finding solutions to optimization problems. Within this framework underlying local heuristics are guided and adapted to effectively explore the solution space. Metaheuristics are designed to be robust. That is, solutions to a wide variety of optimization problems can be found with relatively few modifications. Examples of metaheuristics include simulated annealing (SA), tabu search (TS), iterated local search (ILS), evolutionary algorithms (EA), evolutionary programs (EP), greedy randomized adaptive search procedure (GRASP), memetic algorithms (MA), variable neighborhood descent (VND), genetic algorithms (GA), scatter search (SS) and ant colony optimization (ACO). In addition, there are many hybrid approaches combining features of several of these techniques simultaneously.

Standard references are available for each particular metaheuristic as well as reference and textbooks that compare and contrast a wider variety of metaheuristic approaches. My aim here is not to do a complete survey, but to address how to attack a practical discrete optimization problem (DOP). Metaheuristics are one popular algorithmic approach for generating solutions to DOPs. Consequently, the primary focus of this chapter is how to develop a particular metaheuristic for a DOP.

Metaheuristics attempt to find better solutions to a DOP by either maintaining a collection of solutions, a *population*, or a single solution. Examples of population-based metaheuristics include EA, EP, GA, ACO, and MA. The interested reader is referred to Goldberg [1], Back [2], Dasgupta and Michalewicz [3], and Moscato [4] for more information on these and other population-based techniques. Single solution metaheuristics include SA, TS, ILS, GRASP, VND, and SS. The books by Glover and Laguna [5] and van Laarhoven and Aarts [6] cover tabu search and simulated annealing, respectively. The books by Michalewicz and Fogel [7] and Hoos and Stutzle [8] compare and contrast a wide variety of metaheuristics. In particular, Ref. [8] provides a unified framework for 12 metaheuristics. Throughout this chapter the discussion is limited to single solution metaheuristics.

The focus of this chapter is on how to find high-quality solutions for a DOP. The intended audience is an individual with a DOP in need of answers to the following questions. If I have a feasible solution how do I know if it is a good one? Which solution strategy should I start with if I don't want to invest too much of my time? How do I improve upon these solutions to develop better ones? Further, the DOP of interest is assumed to be nonstandard in some way so that traditional approaches (branch and bound or linear programming relaxations) are not applicable or do not provide tight enough bounds on the optimal value. The contents of this chapter reflect my own preferences for how to go about answering the above questions. I begin by developing a neighborhood structure for traversing the space of feasible solutions and embed this within a multistart local search heuristic (see Sections 13.3 and 13.7) with a random initial feasible solution. (It may not always be possible to generate random feasible solutions, but for the problems I have faced it has been.) After implementing multistart, I now have some idea of the diversity of local optima both with respect to solution value and placement within the feasible space. Next, I embed this same neighborhood structure within a simulated annealing heuristic. I use a standard set of parameter values (see Section 13.4). I am not interested in fine-tuning the search parameters. I simply want to get an idea if the search terrain is difficult. If I am *not* happy with the solutions I have found with SA, I develop a tabu search heuristic. A basic tabu search (see Sections 13.5 and 13.6) is straightforward to implement given a previously defined neighborhood structure. In my experience, tabu search is able to find the same quality (typically with respect to objective value but may include other desirable features as well) of solutions as simulated annealing but in significantly less time and with significantly fewer function evaluations (typically an order of magnitude decrease). If basic tabu search is unable to improve upon the solutions found by multistart and simulated annealing, then I begin adding more advanced features to tabu search (see Sections 13.8 and 13.9).

The chapter is organized as follows. First, a framework for single solution metaheuristics is given. A description of a prototypical class of DOPs (network location problems) is provided next. Following this section are three sections describing multistart heuristics, simulated annealing, and tabu search, respectively. Section 13.3 on multistart heuristics utilizes the notation developed in Section 13.2 to explain first-improving and best-improving two interchange heuristics. Section 13.4 on simulated annealing provides an outline of two basic implementations of one approach to SA and provides recommendations on parameter selections. Section 13.5 presents the basic features of tabu search. The following three sections contain applications of metaheuristics in the location of actuators for noise control (Section 13.6), selecting nature reserve sites (Section 13.7), and the placement of vibrational dampers in large flexible space structures (Section 13.8). Section 13.6 is a nonstandard location problem with a complex performance measure. Traditional approaches are not applicable and the utility of a basic tabu search is demonstrated. The nature reserve site selection problem in Section 13.7 applies two known location models in an attempt to find meaningful solutions. The performance of multistart heuristics, tabu search, and branch and bound are compared. Section 13.8 describes an unusual location problem. A traditional optimization model is developed. LP-relaxations, branch and bound, simulated annealing, and tabu search are highlighted as solution techniques. A recency-based diversification scheme, an advanced feature of tabu search, is shown to be necessary so that high quality solutions can be uncovered. Section 13.9 presents additional features to augment the basic tabu search described earlier in Section 13.5. We close with a short discussion in Section 13.10.

13.1 Mathematical Framework for Single Solution Metaheuristics

Metaheuristics can be divided into those that work with a collection of solutions and those that are guided by a single solution. For those that work with a single solution consider the following definitions and notation. Without loss of generality, we assume that the performance measure is to be minimized.

A *state space*, $\sum$, is the set of all feasible states (solutions). A *state* S denotes an element of $\sum$. To qualitatively distinguish between the states we define a function c. The domain of c is $\sum$ and its range is typically $\mathbb{R}$ (the set of real numbers). $c(S)$ is the cost of state S. A move set Δ defines the set of moves from one state to another. The entries in Δ are denoted by δ. The set of allowable moves from a state S is denoted $\Delta(S)$. Note that $\delta \in \Delta(S)$ maps $\sum \to \sum$. The outcome of applying all moves $\delta \in \Delta(S)$ to S defines the set of states reachable from S. The set of reachable states is typically referred to as the *neighborhood* of S. The value of a move $\delta(S)$, $c(\delta(S)) - c(S)$, is the difference between the cost of the new state and the cost of S. Thus, moves are improving if their value is negative (downhill).

Each metaheuristic begins with an initial state $S_0 \in \sum$, chosen at random or constructed algorithmically. Each metaheuristic generates a sequence of moves, $\delta_1, \delta_2, \ldots, \delta_n$, which determines a sequence of states through which the search proceeds. The mechanism by which a move is selected is one of the critical differences between competing metaheuristics. For example, if the search mechanism is a greedy local search, then at state S_t the next move selected is δ_t where

$$c(\delta_t(S_t)) = \min_{\delta \in \Delta(S_t)} c(\delta(S_t))$$

13.2 Network Location Problems

We will use network location problems as a platform for illustrating the development of metaheuristics throughout this chapter. For additional information on network location theory the interested reader is referred to Mirchandani and Francis [9] and Nickel and Albandoz [10]. Network location theory typically addresses the question of where to locate desirable facilities to interact with customers who must travel to the facility to obtain a particular good or service. In these instances, a customer is assumed to prefer nearby facilities. The prototypical problems for desirable facilities are the p-median and p-center problems. In the p-median problem, the location of p facilities is determined so that the average travel distance for all customers is minimized. However, there are certain types of service for which all customers desire to be within a minimum threshold distance. Emergency medical services is one such example. The performance measure in the p-center problem is to minimize the maximum distance (or time) traveled by a customer to the nearest facility.

Although both the p-median and p-center problem are NP-complete, in practice each of these problems has many high quality local optima that are not difficult to uncover.

Facilities that customers prefer *not* to have nearby are called undesirable or obnoxious. Examples of undesirable facilities include garbage incinerators, hazardous waste dumps, and nuclear power plants. Both the scarcity of available land for siting undesirable facilities and an increase in public awareness of the potential dangers of living near such facilities have made locating undesirable facilities a difficult and sensitive problem. Many disparate factors—costs, politics, and social perceptions—affect the location choice. In addition, undesirable facilities may be mutually obnoxious as in the case of missile silos.

Formulating an optimization problem associated with the location of desirable or undesirable facilities varies with respect to several criteria including the feasible region—discrete or continuous; the number of facilities to be located—one or many; the performance measure—single or multiobjective; the feasible solution space—a tree (acyclic) network or a network that allows cycles; and the interactions considered—the distances between facilities or the distances between facilities and customers. For the desirable facility location problem we consider *discrete, multiple facility, single objective* models in which only the interactions between the facilities and the customers are of importance. For the undesirable facility location problem we consider *discrete, multiple facility, single objective* models in which only the interactions *between the facilities* are of importance. All six models appear in the literature and are summarized below. In each model, we are given an undirected network and the corresponding shortest path distance matrix. For the undesirable facility problems the performance measure seeks a set of facilities X of size p that maximizes some function of the distance between the members of X while for the desirable facility problems we seek a set of facilities X of size p that minimizes some function of the distance between the members of X and the members of the (discrete) feasible region denoted, without loss of generality, as the vertices of the network, V.

There is not complete agreement in the literature on the names for these problems. The names given above are consistent with most of the literature. The *p-dispersion* problem maximizes the minimum distance between any two facilities. For example, missile silos should be placed as far from one another as possible to reduce the number of facilities destroyed in a single attack. The *p-defense-sum* maximizes the average distance between facilities. These two models, perhaps because of their applicability, have received the most attention in the literature. See Kincaid and Yellin [11] for additional information on the undesirable location problems. Although there are no reported applications of the p-dispersion-sum problem to date, a closely related problem formulation in Kincaid and Berger [12] has been used to locate structural dampers on large, flexible, space, truss structures (see Section 13.8 of this chapter). As is seen in Table 13.1, three of these four problems are strongly NP-hard. Optimization problems (more precisely the decision problem associated with an optimization problem), which are still NP-hard even when all numbers in the input are bounded by some

TABLE 13.1 Discrete Network Location Problems

Name	Performance Measure	Complexity
p-median	$\min_{X \subseteq V} \sum_{j \in V} \min_{i \in X} d_{ij}$	NP-Complete
p-center	$\min_{X \subseteq V} \max_{j \in V} \min_{i \in X} d_{ij}$	NP-Complete
p-dispersion	$\max_{X \subseteq V} \min_{i \in X} \min_{j \in X} d_{ij}$	NP-Complete
p-dispersion-sum	$\max_{X \subseteq V} \min_{i \in X} \sum_{j \in X} d_{ij}$	Str. NP-Hard
p-defense	$\max_{X \subseteq V} \sum_{i \in X} \min_{j \in X} d_{ij}$	Str. NP-Hard
p-defense-sum	$\max_{X \subseteq V} \sum_{i \in X} \sum_{j \in X} d_{ij}$	Str. NP-Hard

polynomial in the length of the input, are labeled strongly NP-hard. The interested reader is referred to the seminal reference on complexity theory [13].

13.3 Multistart Local Search

Local searches are the centerpiece of many metaheuristics for DOPs. A typical local search is an interchange heuristic. For network location problems an interchange heuristic begins with a feasible solution—a set of p locations—and a set of potential site locations, V. Although we use V, the set of vertices in the underlying network for our set of potential site locations, the set can be any predetermined finite set of points on the network. Next, we define a set of moves Δ so that we can examine other nearby solutions. Here, we focus on interchange heuristics that swap one site currently in the solution with one site not in the solution, although other move sets are clearly possible. Consequently, a move $\delta_{ij} \in \Delta$ interchanges i, one of the p chosen sites, and j, one of the potential sites not yet chosen ($j \in V \backslash \{k\}$; often we will abuse notation and simply write $j \in V - k$). Algorithm 13.1 below describes a rudimentary two interchange search. At Step 2, if the first move (interchange) that leads to an improvement in the objective value is chosen then we have a *first-improving* two interchange search. We note that in our version of a first-improving search one complete iteration searches all of Δ with, possibly, multiple improving moves accepted. An alternative at Step 2 is to search among all possible interchanges and select an interchange that yields the largest improvement in the objective value. This latter version is called a *best-improving* two interchange search. Since we will consider no other move sets beyond the two interchange we can drop, without confusion, the two interchange identifier and refer to these heuristics simply as first-improving and best-improving. Further comparisons of these two approaches are made in Section 13.7. Algorithm 13.1 stops when no more improvements can be made, that is, when $X^{(t)}$ is a local optima with respect to the two interchange neighborhood. Ideally, the solution, $X^{(t)}$, at which the search stops is a global optima. However, there is no guarantee of this for any of the location problems we have discussed nor is it true for most DOPs of interest.

Algorithm 13.1: Rudimentary Two Interchange Search

Step 0: Choose $X^{(0)}$ by selecting p entries in V and set $t \leftarrow 0$.
Step 1: If no move $\delta_{i,j} \in \Delta^t$ is improving, stop.
        $X^{(t)}$ is a local optima with respect to Δ^t.
Step 2: Choose an improving $\delta_{i,j}^* \in \Delta^t$.
Step 3: Apply $\delta_{i,j}^*$ to $X^{(t)}$ yielding $X^{(t+1)}$. Increment t if no moves remain in Δ^t.
        Return to Step 1.

Algorithm 13.1 becomes a multistart local search by wrapping a loop around it that generates distinct starting solutions for Step 0. These solutions can be randomly generated or determined by another heuristic. In the latter case, the heuristics are often constructive. Greedily adding a site that improves the performance measure the most until p sites are selected is one such example. Multistart local searches are easy to implement and provides a metric for testing the quality of solutions generated by any competing metaheuristic. One way to make an effective comparison is to assume a budget for the number of solutions to be examined. If the metaheuristic cannot produce better solutions than a multistart local search within the same budget the approach should be abandoned for that problem.

13.4 Simulated Annealing

Simulated annealing (SA) was first proposed and used in statistical mechanics by Metropolis et al. [14]. Not until Kirkpatrick et al. [15] and Cerny [16], however, was simulated annealing implemented as a heuristic for an NP-Hard combinatorial optimization problem, the traveling salesman problem. There are two basic approaches to implementing SA. The first is championed by van Laarhoven and Aarts [6], while the second is well represented by Johnson et al. [17]. It is the latter approach that we will take. SA consists of a loop over a block of code that generates a random move $\delta \in \Delta(S)$. If $\delta(S)$, where S is the current state of the procedure, yields a decrease in the objective function value, $c(\delta(S)) < c(S)$, then $\delta(S)$ is immediately accepted as the new current state. However, if $\delta(S)$ yields an increase in the objective function value, then $\delta(S)$ still may be accepted with some positive probability as the new current state. Typically (and in our implementation) this probability remains fixed for a predetermined number of random moves and is then decreased in a uniform manner.

Markov chain theory provides a basis for determining when simulated annealing produces good results. Lundy and Mees [18] as well as Aarts and Van Laarhoven [19] provide detailed descriptions as to how Markov chain theory is used to prove convergence of simulated annealing. In particular, they show that at each temperature (each probability of acceptance of a non-improving state) a homogeneous Markov chain is generated; the transition matrix associated with the Markov chain is irreducible; and the state space is connected. Unfortunately, theoretical convergence to a global optimum is only guaranteed when the number of states visited is exponentially long. Consequently, practical implementations require a knowledge of how to determine problem parameters so that the number of examined solutions is not too many and the final solution quality is high. For an extensive survey of results obtained by simulated annealing algorithms see Collins et al. [20] and Johnson et al. [17].

The paradigm for simulated annealing is physical annealing. If a metal or a crystal is cooled too quickly (a greedy search), then the end product is poor. The metal may be brittle and break easily or the crystal structure may be irregular. Slower cooling is the preferred method. As a result of this physical analogy the problem parameters in SA are referred to as the initial temperature, T_0, the final temperature, T_f, and the cooling rate, *Tempfactor*. However, as nice as this analogy is there are limitations. Johnson et al. [17] provide an extensive set of experiments with SA's problem parameters for the graph partitioning problem (GPP). Selecting the values for these problem parameters can be a daunting task. We will discuss two readily available implementations. One is given in *Numerical Recipes* [21] and the other is given in Johnson et al. [17].

Although the physical annealing paradigm is helpful in explaining the process to scientists, it does not really explain how the algorithm works in practice. I like to think of simulated annealing as a game similar to one I played in the car on long trips. There are a set of holes and a ball bearing in an enclosed hand-held game. The goal is to get the ball bearing into a particular hole without getting stuck in any of the other holes. There are plastic ridges that form a maze and, as you tilt the game, you are able to move through the maze. In simulated annealing's version of this game the objective function is traced out by the plastic ridges and you try to get the ball bearing to end up in the hole associated with the lowest valley (global minimum). The only control parameter is how vigorously you shake the game to move the ball bearing out of undesirable holes. However, you are not allowed to look at the game to see if you are in the correct hole (valley). The game proceeds by successively decreasing the amount of shaking. Eventually the shaking is unable to allow an escape from a hole. Hopefully, you are in the hole that is the global minimum. Of course this oversimplifies simulated annealing but the visual image is a good one.

The parameter T_0 drives the probability of accepting nonimproving states—$exp(-\Delta obj/T)$. For a fixed change in the objective function, T_0 controls the amount of shaking you are allowed to do. Moreover, if T_0 is large enough you can shake your way out of any local optima. *Tempfactor* denotes the rate at which T is to be decreased, while *sample_size* denotes the maximum number of moves sampled at each T. As there is no formal stopping condition, a final value for the parameter T, T_f, must be specified. In Ref. [21], additional parameters (*nsucc* and *nover*) are included for exiting the local improvement phase of the algorithm early if a pre-specified (*nover*) number of successful (*nsucc*) moves has been achieved. If these parameters are not chosen appropriately, simulated annealing will produce poor results and exceedingly long execution times. A pseudocode for this version of simulated annealing is given below.

```
T = T_0
nsucc = 0
while (T > T_f) do
    do i = 1, sample_size
        [generate random move, compute Δobj]
        if (Δobj < 0 or uniform(0,1) < exp(−Δobj/T))then
            [update system, obj = obj + Δobj]
            nsucc = nsucc + 1
        endif
        if (nsucc ≥ nover) exit loop
    continue
        T = T · Tempfactor
end do
```

The parameters *sample_size*, *Tempfactor*, *nover*, T_0, and T_f are often problem specific. As was shown by Lundy and Mees [18], the value of T_0 must be much larger than the difference between the worst objective function value and the best objective function value to guarantee convergence to a globally optimal solution. Rather than specify a value for T_f a maximum iteration bound as suggested in Ref. [18] may be used. Some experimentation is involved in selecting the appropriate combination of *Tempfactor*, and *sample_size*, the maximum number of moves generated at each value of the parameter T (length of the Markov chain).

Johnson et al. [17] discuss the usefulness of these parameters for the GPP. Moreover, rather than use the somewhat unnatural parameters *nsucc* and *nover* they show that a lower initial value of T_0 achieves the same effect. In particular they found that rather than selecting a value of T_0 at which nearly every move (uphill or downhill) is selected, a value of T_0 at which 30%–60% of the moves examined are accepted is more effective. They recommend a value of 40% for GPP and call this parameter *initprob*. In addition, experiments with T_f led them to conclude that simulated annealing for the GPP should be stopped when five consecutive temperatures led to less than 2% (labeled *minpercent* in their parameter list) of the moves being accepted. Borrowing a term from physical annealing, the authors refer to this as the SA process being *frozen*. In the pseudocode below *frozen* is a counter used to determine when the frozen state occurs. Similar experiments led to recommended values of *Tempfactor* = 0.95 and *sample_size* = 16 N (here $N = \Delta(S)$ is the neighborhood size). Lastly, the recommended choices for the above parameter values are such that a suite of replications should be performed and the best solution found during the *num_reps* replications recorded. Below is the pseudocode for this version of simulated annealing.

Randomly select or construct an initial solution S.
Select T_0 so that $initprob = 0.40$.
while ($frozen < 5$) do
    do $i = 1$, $sample\_size$
        [generate move, compute Δobj]
        if ($\Delta obj < 0$ or $uniform(0,1) < exp(-\Delta obj/T)$)then
            [update system, $obj = obj + \Delta obj$]
            $nsucc = nsucc + 1$
            if current solution is best save it.
        endif
    continue
    $T = T \cdot Tempfactor$
    if $nucc/sample\_size < minpercent$
        $frozen = frozen + 1$
    else $frozen = 0$
end do

The robustness of the parameter choices given by Ref. [17] for the GPP is difficult to prove. However, independent of Ref. [17], Kincaid [22] found that similar values worked well for two location problems. Moreover, in Kincaid and Yellin [11] this version of simulated annealing was shown to work well for yet another location problem. Of course there is no guarantee that these parameter choices are best for every DOP, but they have proven robust enough to serve as an excellent starting choice. In addition to the selection of these parameter values, Ref. [17] also gives critical recommendations on how to improve simulated annealing.

For example, rather than randomly select a move $\delta \in \Delta(S)$ use a random permutation of $\Delta(S)$ and examine each move in order. Second, as we are starting at a value of T_0 that does not randomly reorder the entire structure of an initial solution, starting from a better than random solution is preferred. In particular, if the starting solution is constructed by a process that is distinct from the mechanism used to improve solutions in simulated annealing. Third, computing $exp(-\Delta obj/T)$ is expensive. The recommended approach is to implement a table lookup scheme, which cuts the computation time of $exp(-\Delta obj/T)$ by a third. For further details about these speedups see Ref. [17].

13.5 Plain Vanilla Tabu Search

Tabu search and, more generally, adaptive memory programming continues to have startling success in efficiently generating high-quality solutions to difficult practical optimization problems. Glover and Laguna [5] provides 42 vignettes, each of which describes a different application of tabu search by researchers and practitioners. The applications include standard problems such as the p-median problem, job shop scheduling, and the quadratic assignment problem, as well as unusual applications including polymer chemistry, forest management, and the control of flexible space structures (see Section 13.8).

If the topology of the solution space is such that there are no steep valleys or ridges then only the most basic features of tabu search may be needed. We describe such a basic tabu search in this section and call it plain vanilla tabu search (PVTS) to clearly distinguish it from the more general methodology that tabu search encompasses.

The definitions and notations that follow are taken from Glover [23,24] and Kincaid and Berger [12]. For network location problems PVTS begins with an initial state S_0, p entries of V, chosen either at random or by some constructive heuristic. Next PVTS generates a sequence of moves $\delta_0, \delta_1, \ldots$ which determines a sequence of states through which the search proceeds. The mechanism by which a move is selected is crucial. PVTS, at state S_t, selects the greatest available one-move improvement. That is, a next move δ_t is defined to be

$$c(\delta_t(S_t)) = \min_{\delta \in \Delta - \tau + \alpha} c(\delta(S_t))$$

As will be described below, τ restricts the move set and α extends the move set. Without the addition of τ and α the above equation would describe the move selection criterion in a greedy local improvement search and the search would conclude with a local optima. In effect, PVTS selects the largest downhill move, if one is available, or the least uphill move. Hence, there is no natural stopping criteria and the length of the sequence of generated moves must be specified.

PVTS attempts to avoid entrapment in local optima by keeping a list of previously selected moves and deleting them from the move set Δ for a state S in the hope of avoiding a return to a previously observed state. A list of move *attributes* of length *tabusize*, called a *tabu list* and designated by τ, is constructed and updated during each iteration of the search. An example of a move for a location problem is to swap a pair of vertices $(i, j) \in V$, where i is currently selected and j is not. The attributes of this move are the indices i and j, and the designations *drop* for i and *add* for j are often applied.

An *admissible move* in PVTS is a move $\delta \in \Delta$ that is either not on the tabu list τ or one that meets an aspiration level criterion. A *best* admissible move is one that yields the greatest improvement or the least degradation in the objective function value (or any other performance measure). The best admissible move is appended to the tabu list after the examination of all moves $\delta \in \Delta$. Once the tabu list becomes full, which occurs after tabu size iterations, the *oldest* move is removed. Moreover, it is possible for a move in τ to be selected provided that it meets one (or more) *aspiration level criterion*. The purpose of an aspiration level criterion is to choose *good* moves by allowing the tabu status of a move to be overridden if the move satisfies a particular condition. We label the set of tabu moves that meet an aspiration criterion α. The goal is to do this in a manner that retains the search's ability to avoid returning to a previously generated state (*cycling*).

Tabu restrictions and aspiration level criteria play a dual role in constraining and guiding the search process. Tabu restrictions allow a move to be regarded as admissible if they *do not* apply, while aspiration criteria allow a move to be regarded as admissible if they *do* apply. There are at least three ways to avoid cycling (see Ref. [23]), which prevent the move from S to $\delta(S)$:

1. $\delta(S)$ has been visited previously.
2. δ has been applied to S previously. [retracing]
3. δ^{-1} has been applied to S previously.

The first criterion is the strongest and is the only one of the three that fully assures cycling will not occur. This criterion would require the search to store every state visited. Typically, this requires too much memory and too large a computational effort to perform the necessary comparisons. If a tabu move is deemed admissible, provided that only condition (2) is met, then it is possible to reverse the move as soon as it is made. Hence, it makes sense to test condition (3) in addition to (2). By using conditions (2) and (3) to avoid cycling and allowing a tabu move that leads from state x to $S(x)$ to be made—unless the move from $S(x)$ to x had occurred previously—the attempt to prevent a return to an earlier location will be more strongly supported than relying solely on condition (2). Conditions (2) and (3) together are a close approximation of (1).

As it is impractical to store a large number of states, we need to find a way to store the same information in a compressed form. One possibility for compressed storage is to use the objective function value. The standard aspiration level criterion is an example of storing information in such a compressed form. This criterion has the property that if the tabu move (i, j) yields a state with the best objective function value encountered in any previous iteration, then the move is allowed. If $\delta \in \tau \subseteq \Delta(S)$ and $c(\delta(S)) < c_{\text{best}}$, then δ is admissible. The tabu status is overridden, because the move δ meets the aspiration level criterion. Cycling is avoided here since if $\delta(S)$ had appeared previously, then $c(\delta(S)) < c_{\text{best}}$ would not be possible. This particular aspiration criterion is easy to implement but treats all states with the same objective function value as if they were the same state. However, Kincaid and Berger [12] found this simple aspiration level criterion to perform as well as more complicated ones for a damper placement problem (see problem DPP1, described in Section 13.8).

13.6 Active Structural Acoustic Control (ASAC)

The optimization model in this section has a distinct network location flavor (Section 13.2), but the performance measure is nonstandard. Once a set of candidate locations is chosen, the solution of a complex valued least squares problem must be solved to provide the metric for comparing sets of candidate locations. In addition, the utility of plain vanilla tabu search (Section 13.5) is demonstrated.

Most air carriers use turboprop aircraft to transport passengers between local or nearby airports and their hubs. Turboprop aircraft are preferred for short hops due to their superior fuel efficiency at low speeds. However, the benefits of decreased fuel cost and increased travel range come at the price of increased levels of cabin noise and vibration. Turboprop noise levels are typically 10–30 dB louder than commercial jet noise levels.

Turboprop noise is fundamentally different from jet noise. The former is caused by periodic sources such as rotating engine parts, rotating props, and by vortices shed from the propellers. Turboprop noise spectrum is dominated by a few low-frequency tones. On the other hand, jet noise is caused by random sources such as the mixing of high speed, high temperature gases. Jet noise spectrum is dominated by high frequency broad band noise. Jet noise can be controlled by putting sound absorbing material in the walls of the fuselage. This passive treatment is not effective for propeller noise.

Interior noise levels are of concern to commercial airlines for three reasons. First, the airlines must provide a comfortable working environment for their employees. Occupational Safety and Health Administration (OSHA) mandates ear protection or restricted working hours for employees who are exposed to high levels of noise. Second, the airlines must attract customers. Uncomfortable noise levels may encourage the public to seek alternate travel options. Third, the airlines must consider safety issues. High noise and vibration levels cause fatigue in the airframe as well as fatigue in the passengers and crew.

Active control methods attempt to counteract propeller noise by introducing a secondary (or multiple secondary) sound source that is of the same frequency and amplitude as the propeller noise, but is 180° out of phase. Thus, in theory, the result is a complete silencing of the propeller noise source. We consider an acoustic control model in which the control inputs, which are used to reduce interior noise, are applied directly to a vibrating structural acoustic system (e.g., an aircraft fuselage). The feasibility of this approach has been demonstrated by using measured data from the aft cabin of a Douglas DC-9 fuselage. A simplified model of the ASAC problem is used to explain the process. Assume that an aircraft fuselage is represented as a cylinder with rigid end caps and that a turboprop propeller is represented as an acoustic point source with a frequency related to the propeller blade passage frequency. Piezoelectric actuators bonded to the fuselage skin are represented as line force distributions in the axial and azimuthal directions. With this simplified model, the point source produces predictable pressure waves that are exterior to the cylinder. These periodic pressure waves cause predictable structural vibrations in the cylinder wall and predictable noise levels in the interior space. The noise level at any interior microphone location depends on the control forces applied at each piezoelectric actuator location. For a given set of sensor and actuator locations, the control forces that minimize the average acoustic response are easy to calculate. However, methods for choosing good locations for the sensors and actuators are needed.

Here we will fix p microphone locations (sensors) and seek to determine the best set of k actuator locations chosen from a set of $q \gg k$ potential locations for the active controls. The goal is to determine the force inputs and sites for the piezoceramic actuators so that (1) the interior noise is effectively damped; (2) the level of vibration of the cylinder shell is not increased; and (3) the power requirements needed to drive the actuators are not excessive.

To model objective (1) a p by k complex transfer matrix H is determined. Each entry of H provides the contribution of each of the potential k actuators sites to interior noise reduction at each of the fixed p error microphone locations whenever a unit amplitude input is applied to actuator site k. Objective (2) requires more information which will not be described here. Lastly, since power is directly related to the magnitude of the force inputs required, objective (3) can be measured by considering either the Euclidean or max norm of the actuator force input vector.

A complex transfer matrix H can be generated (either experimentally or with a mathematical model) for the complete set of q actuator sites. A column j of H yields the contribution of a unit amplitude input at actuator site j in controlling the noise at each of the p sensors. Hence, given a particular set of k columns (actuators) the submatrix of H induced by these columns, H^k, is the associated transfer matrix. Once a particular set of k actuator sites have been selected then a force (amplitude and phase) must be selected for each actuator so that the resultant force cancels out $\vec{p}$ as nearly as possible. In the acoustic literature the force vector $\vec{f}$ is chosen as the solution of the complex least squares problem

$$\min_{f} \|H^k \vec{f} + \vec{p}\|_2 \tag{13.1}$$

We note that although the usual convention is to have a subtraction operator in Equation 13.1, the acoustic literature uniformly displays the plus operator. Thus, we adhere to acoustic literature convention. The solution to Equation 13.1 is found by solving

$$H^{k*} H^k \vec{f} = -H^{k*} \vec{p} \tag{13.2}$$

for $\vec{f}$ (* denotes the complex conjugate transpose). This means that each time a different set of k columns of H is selected a new complex least squares problem must be solved.

If there exists a column of H which is collinear with $\vec{p}$, then we would need only pick this single actuator site, driven at the appropriate amplitude, and total silencing would be achieved. Typically this situation does not exist. However, a likely set of columns for a starting solution are the k columns of H most nearly collinear with $\vec{p}$. To find these columns, we determine the angle between each actuator j site and $\vec{p}$. This is done by calculating the dot product of each column of the H matrix, $\vec{h}_j$, with the vector p and dividing that product by the product of the norms of each. That is, the expression:

$$\frac{\vec{h}_j \cdot \vec{p}}{\|\vec{h}_j\|_2 \|\vec{p}\|_2}$$

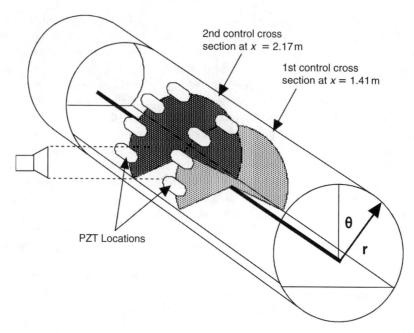

FIGURE 13.1 Laboratory cylinder for ASAC test.

would calculate the cosine of the angle between the two vectors. As the actuators that would be most effective, theoretically, in silencing the outside noise source would be those either in phase or 180° out of phase, we seek cosine values close to either 1 or −1. Thus, the k initial actuators are chosen that have the absolute value of the cosine closest to one. This greedy heuristic for constructing initial solutions performed better than randomly selecting actuator sites.

The performance of this set of k columns (or any set of k columns for that matter) is measured on a decibel scale given by the expression below.

$$10 \log_{10} \frac{\|H^k \vec{f} + \vec{p}\|_2}{\|\vec{p}\|_2} \tag{13.3}$$

The decibel value computed in Equation 13.3 compares the interior noise norm with actuator controls in place with the norm of the uncontrolled interior noise. Thus, a negative decibel (dB) level signifies a decrease in interior cylinder noise due to the control effects of the actuators. An analogous decibel expression computes the vibration level on the cylinder shell.

As we mentioned earlier, a set of experiments with a laboratory test article (see Figure 13.1) have been performed and reported on in Palumbo et al. [25]. For these experiments the transfer matrices were generated experimentally. The predicted performance of the best actuator sites found for the mathematical model by tabu search are shown to correlate well with the actual measured performance.

The cylinder in Figure 13.1 is a cartoon of the laboratory test article. The cylinder in the laboratory tests is 3.6 m long and 1.68 m in diameter. The outer shell is a nine-layer filament wound graphite epoxy composite. Total skin thickness is 1.7 mm. In an effort to make the cylinder as similar as possible to a commercial fuselage the cylinder is stiffened with composite stringers and ring frames. To complete the fuselage effect, three inner trim panel sections are attached. Each trim panel has a honeycomb core sandwiched between two graphite epoxy laminate sheets. The interested reader is referred to Lyle and Silcox [26]

TABLE 13.2 Experimental Results with Actuator/Sensor Location

Freq. (Hz)	PVTS Predicted (dB)	PVTS Measured (dB)	Modal Analysis Measured (dB)
210	−5.3	−4.4	−2.1
230	−3.8	−2.5	−1.2
275	−5.7	−3.9	−2.7

for a detailed description of the laboratory test article. The monopole noise source was a 100 W electrodynamic loudspeaker. An array of eight piezoceramic patches was installed on one of the trim panels. A set of six microphones was swept along a boom throughout the interior of the cylinder (near the shell) and data were taken at 462 locations. The 462 transfer functions at these locations were used to construct transfer matrices (462 by 8) for three frequencies—210, 230, and 275 Hz—of interest. The 230 Hz case has weak acoustic modes but strong structural modes.

The laboratory test was done with only eight potential actuator sites of which four were to be selected. Table 13.2 catalogs the performance of PVTS versus standard acoustical techniques in the selection of the best set of eight sensors. The acoustical techniques (cf. [25]) involve engineers examining visual and numerical displays of the modal decomposition of the interior pressure field $\vec{p}$ and the individual actuator responses. Dominant modes in the interior pressure field are matched with dominant modes in the actuator responses. The columns marked PVTS compare the predicted noise reduction versus the actual noise reduction measured in the laboratory. The column marked modal analysis shows noise reduction levels achieved in laboratory tests run prior to the PVTS selection of actuator and sensor locations.

The good news in Table 13.2 is that the measured values for PVTS are nearly 1 dB better than the modal decomposition analysis solutions. Hence, even for this relatively easy case (only four actuators) the optimization approach proved superior. A second positive feature is that the predicted performance, although always at least a 1 dB overestimate, is still a reasonable predictor of the measured performance. This provides evidence that the model assumptions made in the optimization approach are not too far off. A third positive result of these experiments is that the acoustic engineers now believe that the optimization approach is an important contributor in finding good solutions to the ASAC problems.

Much additional work has been done on this problem. The methodology in this section was used to select actuator locations for an actual aircraft that was flight tested. Upon completion of the flight test experiments the technology was transferred for development to a private corporation. For additional information see Refs. [25] and [27–30].

13.7 Nature Reserve Site Selection

A variety of heuristic solution techniques are highlighted in this section for two distinct network location models—the maximal covering species problem (MCSP) and the maximal expected coverage problem (MECP). For MCSP, a set of computational experiments evaluating the performance of competing multistart searches (Section 13.3) with tabu search (Section 13.5) is provided. For MECP, results for a tabu search are presented and are contrasted with results for a linear programming approximation for MECP. In both MCSP and MECP experiments with two types of tabu list structures are given.

In recent decades, human population growth and land development patterns have threatened to destroy many ecosystems and, as a direct result, the species that are part of and depend upon these ecosystems for survival. Consequently, nature reserves and parks are being identified and established as much for species preservation and the maintenance of biological diversity as they are for human enjoyment. As selection of nature reserve sites has become more complex, methods drawn from operations research have been introduced to aid in the solution of this problem. One approach has been the species set cover problem (SSCP) [31], which seeks to minimize investment while protecting all target species. A related model is the selection of a network of reserve sites so as to maximize preservation for a specified level of investment. This latter model is closely related to the maximal covering location problem. Consequently it is referred to as the MCSP [32,33]. In both the MCSP and the SSCP, a species is said to be covered if the species is present in at least one site selected for the reserve system. Further, these models assume that the presence or absence of a species in a site can be known with certainty. If, instead, we assume that the information as to where a species resides in an ecosystem is not known with certainty or if the foraging range of a species makes it difficult to pinpoint a home site, then probabilistic models are needed. One such model is the MECP. To be able to formulate an MECP the probability that a species is present in a given site must be specified. These probabilities are assumed to be independent. That is, the probability that one species is located in a particular site is independent of any other species' presence or absence.

We develop and test the performance of heuristics for both MCSP and MECP. Following the literature [34,35] we use the Oregon Terrestrial Vertebrate dataset [36] as a testbed. This dataset has location information for 426 terrestrial vertebrate species for the state of Oregon. The data is organized by partitioning the state into 441 regular hexagons (see Figure 13.2). The bulk of the material found in this section appears in [37].

Both MCSP and MECP are NP-Hard [34] and thus solution by exact methods (e.g., branch and bound) can be troublesome for problem sizes of interest to the ecological community. Hence, heuristics are popular alternatives and include greedy adding and interchange heuristics, which require significantly less time, but may yield suboptimal solutions. (See Refs. [35] and [38] for a more complete summary and comparison of previous computational

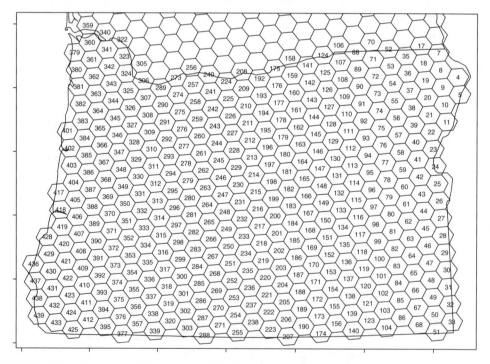

FIGURE 13.2 Hexagon partition and numbers for the state of Oregon.

Metaheuristics for Discrete Optimization Problems

experience.) In addition to heuristics, relaxation and approximation techniques are available. Of particular interest here is the linearized maximal expected covering problem (LMECP) developed by Camm et al. [34] as an approximation to MECP.

An additional issue is the utility of the information provided by the optimization model to the ecological community. Both MECP and MCSP are, at best, approximations of the true optimization model. Hence, the provision of a single solution is problematic. As in many public sector decision problems, tradeoffs are essential and the ability to provide a decision maker with a collection of high-quality solutions from which a network of reserve sites can be selected is a worthy goal.

13.7.1 Problem Formulations and Data Details

The MCSP can be formulated as the following binary integer program [39,40].

$$\max \sum_{i \in I} y_i$$

$$\text{subject to } y_i \leq \sum_{j \in M_i} x_j, \quad \forall i \in I$$

$$\sum_{j \in J} x_j = k$$

$$y_i, x_j \in \{0, 1\}, \quad \forall i \in I, \forall j \in J$$

where y_i and x_j are binary decision variables, defined such that

$$y_i = \begin{cases} 1 & \text{if species } i \text{ is covered in the reserve network} \\ 0 & \text{otherwise} \end{cases}$$

$$x_j = \begin{cases} 1 & \text{if site } j \text{ is selected for the reserve} \\ 0 & \text{otherwise} \end{cases}$$

In addition, I is the set of all species, J is the set of all potential reserve sites, and M_i is the set of sites in which species i resides.

The MCSP objective function maximizes the total number of species covered by the reserve system. The first constraint states that for a species to be covered, it must be present in one or more of the sites selected for the reserve system. The next constraint restricts the number of sites selected to k. Finally, the last constraint requires that all decision variables are binary. Extensions of this model are possible [41] and include budget constraints and objective function coefficients to reflect the conservation value of a species.

The MECP can be formulated as a binary nonlinear integer program [34]

$$\max \sum_{i \in I} \left[1 - \prod_{j \in J} (1 - p_{ij} x_j) \right]$$

$$\text{subject to } \sum_{j \in J} x_j = k$$

$$x_j \in \{0, 1\}, \quad \forall i \in I, \forall j \in J$$

where x_j are binary decision variables, defined such that

$$x_j = \begin{cases} 1 & \text{if site } j \text{ is selected for the reserve} \\ 0 & \text{otherwise} \end{cases}$$

In addition, I is the set of all species, J is the set of all potential reserve sites, p_{ij} is the probability that species i exists and will survive at site j, and k denotes the number of sites to be selected for the reserve. The MECP objective function maximizes the expected number of species covered by the reserve system. The first constraint restricts the number of sites selected to k. The remaining constraints require that all decision variables are binary.

The final model is the LMECP established in Ref. [34]. The development of LMECP is not duplicated here as the final model provides little insight without the complete derivation. LMECP is a binary integer linear program. The error in the approximation of LMECP to MECP can be controlled by the number of breakpoints. A more accurate approximation of the MECP is obtained with more breakpoints, but at the expense of additional decision variables. Following Ref. [34], the number of breakpoints is set at 10 for each run of the model for all values of k tested. Similar extensions for MECP exist as for MCSP [41].

The Oregon Terrestrial Vertebrate dataset [36] has location information on 426 species spread over a grid made up of 441 hexagons (see Figure 13.2). For each hexagon, a probability is assigned to each species existence. The probabilities are categorized as follows:

- 0.00: Not present—The habitat is unsuitable for the species.
- 0.10: Possible—The habitat is of questionable suitability for the species, but expert local opinion is that the species might occur in the site.
- 0.80: Probable—The site contains suitable habitat for the species and there have been verified sightings in nearby sites; local expert opinion is that it is highly probable that the species occurs in the site.
- 0.95: Confident—A verified sighting of the species has occurred in the last two decades.

Notice that the time span for the most confident classification is 20 years, while survival data on the species is not included.

For the MCSP the probability that species i is found in hexagon (site) j must be either 0 or 1. Following Ref. [35] let entries of 0.00 and 0.10 be mapped into 0 and entries of 0.80 and 0.95 be mapped into 1. Unfortunately, such certainty is not guaranteed, leading to the designation of *soft* for these datasets. There are a variety of reasons for this lack of certainty. Terrestrial vertebrates are ambulatory and the location in which they are observed may not be their primary residence. In addition, the certainty of the observation itself may be in question if the species is viewed from a distance or if it is based upon an interview with a time lag. Lastly, population dynamics may affect the certainty of a species location.

13.7.2 MCSP Interchange Heuristics

We test the performance of both first-improving and best-improving heuristics for MCSP on the terrestrial vertebrate species survey data from the Biodiversity Research Consortium for the state of Oregon [36]. We note that a similar set of experiments was done in Ref. [35] for the same dataset. These authors found that when compared to the 18 heuristics tested in Ref. [38], the number of species covered by a first-improving heuristic was always equal to or better than the number covered by the best solution found across all 18 heuristics, for each value of k from 1 to 24. We provide a similar set of experiments as in Ref. [35]. However, we also compare first-improving's performance against best-improving as well as embed first-improving within a tabu search in Section 13.4. For comparative purposes, MCSP is solved exactly. The results are recorded in Tables 13.3 through 13.7.

The first two columns in Table 13.3 list the number of sites to be selected and the corresponding optimal objective function value. Only 23 sites are needed to cover all 426 species.

Metaheuristics for Discrete Optimization Problems

TABLE 13.3 Comparative Results of Optimal Solution versus Interchange Heuristics

Sites	Opt	Best-Imp. 1 Start	Best-Imp. Multi-Start	First-Imp. Multi-Start
1	254	254	254	254
2	318	318	318	318
3	356	356	356	356
4	374	374	374	374
5	384	383	384	384
6	390	388	390	390
7	395	393	395	395
8	400	397	400	400
9	403	400	403	403
10	406	401	**406**	405
11	408	403	408	408
12	410	406	410	410
13	412	407	412	412
14	414	410	414	414
15	416	411	416	416
16	418	412	417	417
17	419	413	**419**	418
18	420	414	419	**420**
19	422	415	421	421
20	423	416	422	422
21	424	418	**423**	422
22	425	419	423	423
23	426	420	424	424
24	426	421	425	425
25	426	421	426	426

TABLE 13.4 Frequency of Best Observed Solutions for First-Improving

Sites	# Reps	Freq.
1	1	50
2	3	13
3	1	48
4	1	41
5	1	43
6	2	25
7	6	9
8	21	1
9	11	3
10	1	9
11	12	3
12	12	3
13	3	3
14	1	1
15	21	1
16	3	5
17	3	8
18	25	2
19	47	1
20	47	1
21	17	4
22	19	5
23	7	8
24	7	5
25	7	5

TABLE 13.5 Comparison of First-Improving and Tabu Search Solutions

		First-Imp	TabuA		TabuB	
Sites	Opt	50 Reps	2k	Asp	2k	Asp
16	418	417	418	0	418	1
17	419	418	419	0	419	1
18	420	420	420	0	420	0
19	422	421	421	0	421	0
20	423	422	422	0	422	0
21	424	422	423	0	423	0
22	425	423	424	0	424	1
23	426	424	425	0	425	1
24	426	425	426	0	426	2

TABLE 13.6 Tabu Search Results for MECP

k	MECP	Locations of Sites
1	217.4	319
2	288.5	135, 395
3	330.7	55, 135, 438
4	354.2	55, 135, 375, 438
5	368.1	55, 135, 274, 375, 438
6	376.7	55, **121**, 135, **274**, 375, 438
7	384.2	47, 55, 135, 289, 319, **395**, 438
10	398.4	9, 46, 135, 268, 274, 345, 357, 375, 428, 440
13	407.2	9, 20, 27 121, 135, 268, 274, 314, 345, 357, 375, 428, 440
15	411.9	9, 46, 75, 121, 135, 141, 268, 273, 314, 324, 345, 357, 375, 428, 440

The third column of Table 13.3 records the objective for a single replication of the best-improving heuristic for $k = 1, \ldots, 18$ when the first k sites are chosen as the initial starting solution. The progress of the search is even more surprising as this is such a poor starting solution. All of the first k sites are clumped together in the northeast corner of Oregon (see Figure 13.2). The maximum relative error between the best-improving solution and the optimal solution is 1.67%.

Columns 4 and 5 of Table 13.3 record the best solutions found with 50 restarts of the best-improving and first-improving heuristics, respectively. Starting solutions for each of the multiple starts are obtained by randomly permuting the 441 sites and then selecting the first k indices as the initial set of selected sites. In column 5 the optimal solution was

TABLE 13.7 Tabu Search Results for MECP—with Replications

k	Previous Best	Avg MECP	Tabu Tenure	Avg # Asp	Avg Itrs
7	384.2	384.11	20	0.6	18.7
		383.95	10	0.1	18.1
		384.22	4	3.1	19.5
10	398.4	398.22	20	0.7	18.2
		398.36	10	1.5	21.7
		398.36	5	5.7	22.0
13	407.2	407.23	20	0.9	27.6
		407.20	10	0.9	27.3
		407.22	7	6.8	25.6
15	411.9	411.67	20	1.8	26.6
		411.67	10	1.3	26.5
		411.84	8	5.2	24.3

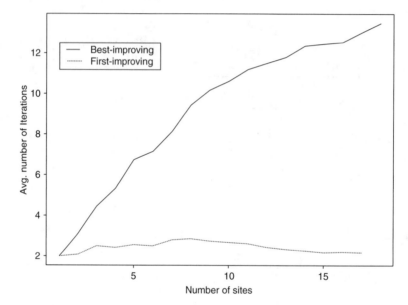

FIGURE 13.3 Average number of iterations: first-improving versus best-improving.

found at least once (see Table 13.4) out of the 50 starts for all values of k from 1 to 15. For $k > 15$ the best solution differs by at most two sites from the optimal solution. For $k > 18$ neither heuristic is able to uncover the optimal solution. Table 13.4 records the frequency with which these best solutions are found for first-improving search over the 50 replications.

How do the performances of first-improving and best-improving compare? The solution quality is nearly identical. See columns 4 and 5 of Table 13.3 for the best solutions found. The boldface values highlight the differences between the two heuristics. Out of the 25 experiments there are 21 ties, three wins for best-improving, and one win for first-improving. However, both the average CPU time and the average number of iterations decreases dramatically for first-improving when compared to best-improving as k increases. Instead of increasing with the problem size parameter k, as in the case for best-improving, the average number of iterations remains nearly constant between two and three for the first-improving heuristic. Figure 13.3 plots the average number of iterations required for both best-improving and first-improving, thus illustrating a dramatic difference between the two approaches. Please recall that one iteration of first-improving requires a complete perusal of all possible two interchanges, and that multiple moves can be accepted during a single iteration—for example, when $k=10$ each iteration of first-improving (or best-improving) examines $k(n-k)=10(431)$ moves (interchanges). The difference is that best-improving selects only one move in Δ while first-improving may select many improving moves. The CPU times for best-improving are growing much faster than for first-improving, with nearly an order of magnitude difference when $k=25$. As both approaches provide solutions of similar quality, first-improving is the clear winner for MCSP.

13.7.3 Tabu Search for MCSP

Next we develop a simple tabu search for MCSP. We embed first-improving within tabu search. We note, however, that after the first two to three iterations very few improving interchanges exist. Consequently, after iteration three first-improving behaves like best-improving. The entire neighborhood is searched for every iteration to find an improving or at least a nonimproving solution. The question to resolve is whether the ability of tabu search to move away from local optima and continue searching for other local optima will outperform the first-improving approach of the previous section [24,42].

One of the decisions that must be made to implement a tabu search is when to terminate the search. We follow the typical practice and set an iteration limit. As the first-improving heuristic required between two and three iterations for each of the 50 random starts, we let the maximum number of iterations be 100. Other decisions required to implement tabu search include how long a move should remain tabu (also called the tabu tenure of a move) and whether both the site dropped and the site added should be tabu or only one of the pair. We experiment with two different types of tabu moves. First, the complete move (drop and add) is made tabu—*TabuA*. Second, only the dropped sites are tabu—*TabuB*. That is, adding a dropped site back into the solution is tabu. Finally, we must decide what the aspiration criterion will be and if it should be invoked. Again, following standard practice [42] our aspiration criterion allows the tabu status of a move to be ignored if its objective value is better than the current best objective value.

The first-improving heuristic finds optimal solutions for all $k \leq 15$ and the frequency with which optimal solutions are uncovered is large (see Table 13.4) for $k \leq 6$. Consequently, we compare tabu search with first-improving for all values of $k>15$. Table 13.5 records the objective value for both types of tabu lists, with a tabu tenure of $2k$. These values are compared to the optimal objective value (column 2) and to the best values found by the first-improving heuristic of the previous section (column 3). Both implementations of tabu search find the same or better objective values than 50 replications of first-improving. TabuA and TabuB yield the same objective values for all k. TabuA was relatively insensitive to the tabu tenure parameter and values between k and $3k$ typically yield the same solutions. The tabu tenure value was less robust for TabuB. We report values found with a tabu tenure of $2k$ for both TabuA and TabuB. The only exceptions are for $k=22, 23$, and 24 for TabuB. For $k=22$ and 23 a tabu tenure of size k was best, while for $k=24$ a tabu tenure of $1.5k$ yielded the best result. As can be seen in column 5 and 7 of Table 13.5, no tabu moves in TabuA met the aspiration criteria, while the aspiration criteria were met several times for TabuB. In particular, the aspiration criterion for TabuB always led to the best observed solution for the entire run. It is not surprising that TabuB utilizes the aspiration criteria since the percentage of available moves held tabu by this approach is much greater than for TabuA.

How does the computational effort for tabu search compare with our earlier winner—first-improving multistart? The average number of iterations needed by first-improving is nearly constant at 2.5 across all values of k. The average number of restarts, before the best observed value out of 50 restarts is found, is 15.7 for k between 11 and 25. For all $k \leq 15$ the best observed value over the 50 replications is optimal. But for $k \geq 16$ it is not. Moreover, for $k \geq 16$ the number of replications needed to uncover the best observed solutions (see Table 13.4) varies between 1 and 47 ($k = 19, 20$). We argue that, *a priori*, it is impossible to know how many restarts will be needed and since at least 47 must be done, 50 is a reasonable choice for the number or restarts. How does this compare with tabu search? For TabuA the average iteration at which the best solution is found is 19.8 with a range from 3 to 42. For TabuB the average is 24.0 and the range is from 3 to 59. Since at least 59 iterations are needed, 60 iterations is a reasonable choice as an iteration limit for TabuB, while a 50 iteration limit suffices for TabuA. (We note that if a best-improving search is embedded within our tabu search 100 iterations are needed to ensure high quality solutions.) The CPU times for TabuA (50 iterations) increases linearly from a time of 9.1 s when $k = 16$ to 13 s when $k = 24$. TabuB follows a similar pattern with 15 s needed to complete 60 iterations of TabuB for $k = 24$. Letting the tabu search CPU times serves as a constraint for first-improving then 25 restarts of first-improving (on average) can be completed in 13 s (28 restarts in 15 s) when $k = 24$. Only three restarts for best-improving can be completed in 15 s for $k = 24$. Thus, TabuA and TabuB easily dominate best-improving when a CPU budget is imposed. When restricted to 25 (or 28) restarts the results for first-improving (column 3, Table 13.5) remain the same with the exception of $k = 19$ and $k = 20$. The best observed solution has an objective value of 420 for $k = 19$ and 421 for $k = 20$ when restricted to 25 (or 28) restarts. We note as well that, like the multistart approach, tabu search uncovers a number of local optima that may be recorded as the search progresses.

13.7.4 Heuristics for MECP and LMECP

Following Ref. [34], the resulting LMECP with 10 breakpoints for the Oregon Terrestrial Vertebrate dataset has 2256 real variables, 441 binary variables, and 1279 constraints. The AMPL model and data were obtained from the authors of Ref. [34]. The LMECP model was run for various values of k. As was mentioned in Ref. [34], running the model for $k = 7$ sites was quite lengthy. This was true for our implementation as well with over four wall clock hours required for $k = 7$ (computing environment was a network of Pentium 4s).

Next, we develop and test the performance of a tabu search for MECP. As we will see, the solutions generated by these approaches are competitive with the approximate solutions generated by LMECP and require far fewer computational resources. We begin with a simple tabu search approach. The embedded heuristic is a best-improving two-interchange heuristic described earlier for MCSP. Unlike MCSP, first-improving found inferior solutions (on average) when compared with best-improving for MECP. For example, 10 replications of a first-improving tabu search for $k = 10$ had an average best objective value of 397.97 while 10 replications of a best-improving tabu search for $k = 10$ had an average best objective value of 398.36. While it is still true that a first-improving tabu search requires only two to three iterations to find its best solution, we chose the higher solution quality for our reported results in this section. As was the case for TabuA for MCSP, the solution quality found by our tabu search for MECP is not particularly sensitive to the tenure of the tabu moves. That is, tabu tenures from 6 up to 20 yield the same best solutions. We note, however, that for shorter tabu tenures the aspiration criterion is invoked more frequently. The stopping condition is as before—an iteration limit. For all $k \leq 15$ the maximum iterations allowed is 50. The same simple aspiration criterion as before is also used—the tabu status of a move is ignored if it leads to a solution whose value is better than the current best. Initial solutions are generated as we did for MCSP. Randomly permute the 441 sites and select the first k as the initial set of sites. Table 13.6 records the best solutions found by our tabu search.

The results for $k=1$ to $k=7$ in Table 13.6 are nearly identical to the solution reported in Ref. [34] for LMECP. In Table 13.6 the sites that are different than those chosen in Ref. [34] are in boldface. The solution values in column 2 are identical to those in Ref. [34] except for $k=6$ where there is a difference of 0.2. One pair of sites that are different (289 and 274) is adjacent, so this is a small change. But the other pair (47 to 121) has a spatial separation of three hexagons, which is a significant change. Unlike LMECP computation time is not restrictive so we report solutions on larger values of k as well. In general, heuristics for the larger values of k run in minutes while LMECP took more than 4 wall clock hours for $k=7$.

Table 13.7 records experiments with three different tabu restrictions, one row for each type of tabu move in columns 3 through 6. In each case 10 replications are performed. Each set of 10 replications takes roughly 15 min on a Pentium IV computer for $k=15$. Computing times are shorter for smaller values of k. The first row, for each k, is the approach recorded in Table 13.6—both the site added and the site removed are tabu. The second row, for each k, makes it tabu to add a site that has been removed. The third row, for each k, makes it tabu to remove a site that has been added. Column 3 gives the average best objective value over the replications. Column 5 records the average number of tabu interchanges that meet the aspiration criterion, while column 6 records the average iteration at which the best solution was observed. Clearly the third method makes the greatest use of the aspiration criterion, which can be partly explained by the restriction that the tabu tenure can be no larger than k. If the tabu tenure is larger than k then no interchanges would be allowed. We found that a tabu tenure of $\lceil k/2 \rceil$ worked best. All implementations of tabu search were able to uncover the best observed solution. Although all three versions of tabu search performed well our recommendation is that the tabu restriction associated with row 3 of Table 13.7 is the best (tabu to add a site that has been removed). Moreover, we note that for all three tabu restrictions the solution quality is worse if the aspiration criterion is omitted.

13.8 Damper Placement in Flexible Truss Structures

The damper placement problem described in this section is similar to the p-dispersion-sum problem (Section 13.2). A tabu search is developed for this problem that includes more advanced features than those described in Section 13.5. In particular a recency-based diversification scheme is shown to yield high-quality solutions. A comparison is made between starting tabu search from a random set of damper locations versus using the result of a linear programming relaxation to generate an initial solution. In addition, tabu search results are compared to an *off the shelf* branch and bound code as well as simulated annealing.

The demand for larger-sized space structures with lower mass has led to the development of highly flexible structures where, in effect, every point can move relative to the next. Traditionally, structural motion is viewed more simply in terms of a sum of several dozen or more independent motions called *natural motions*. The problem of controlling the motion of a flexible structure is then reduced to controlling the natural motions. Associated with each natural motion are three parameters: a *mode* that is a natural spatial shape, a *natural frequency* that expresses the rate of oscillation, and a *natural decay* rate that is a measure of the time required for the motion to decay. The contribution of each natural motion to the overall motion depends on the degree to which it is excited by external forces.

The overall structural motion of a flexible truss structure can be reduced by the use of structural dampers that both sense and dissipate vibrations. We focus on where to locate these dampers so that vibrations arising from the control or operation of the structure and its payloads, or from cyclic thermal expansion and contraction of the space structure, can

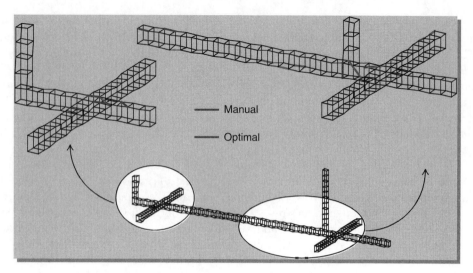

FIGURE 13.4 CSI Phase I Evolutionary Model (CEM).

be damped as effectively as possible. There are several mechanisms available for vibrational damping. We consider the replacement of some of the truss members by active dampers that sense axial displacement (strain) and induce a compensating displacement. A related option is to replace some of the truss members with passive dampers which dissipate strain energy due to their material properties. Each of these techniques for damping increases the weight and cost of the truss structure, as well as changing the stiffness properties of the structure. Hence, structural designers are required to locate as few dampers as possible and still maintain an appropriate level of vibrational damping. The results in the following sections are taken primarily from Kincaid [43].

13.8.1 Optimization Models

The CSI Phase I Evolutionary Model (CEM) in Figure 13.4 is an example of a large flexible space truss structure. (Please ignore the manual/optimal labeling in Figure 13.4 as it only shows up in color. Moreover, the labeling is not referenced in the ensuing discussion.) Let $nmode$ denote the number of modes and $nmemb$ denote the number of truss members in the structure. A normal modes analysis of a finite element model of this structure yielded a 10 ($nmode$) by 1507 ($nmemb$) modal strain energy matrix. Let D denote this matrix with row index set I and column index set J. The entries in the matrix have been normalized so that each d_{ij} denotes the percentage of the total modal strain energy imparted in mode i to truss member j.

In this section, we formulate two DOPs. The goal in each of the problems is to select p truss members to be replaced by active (or passive) dampers so that the *modal damping ratio* is maximized for all significant modes. Maximizing the modal damping ratio is a widely accepted goal in damper placement problems (cf. [44]). However, the modal damping ratio is difficult to determine explicitly and, consequently, the placement of active (or passive) dampers has proved difficult (cf. [45] and [46]). Both active and passive dampers dissipate forces that are internal to the structure and are most effective replacing truss members with maximum extension or compression. The truss elements with maximum internal displacement are those with the largest strain energy over all modes. Given a finite element model and the results of a normal modes analysis, the modal strain energy in each candidate location (truss member) for each significant normal vibration mode can be estimated quite accurately. The damping achieved with active dampers depends on the properties of the damper and the control law that is implemented. We use a force-feedback control law (cf. [46]) yielding damping ratios that are directly proportional to the fraction of modal strain energy.

The first damper placement problem (DPP1) assumes that the damping effectiveness of each active (or passive) damper is the same over all vibrational frequencies of interest. Hence, the maximization of the modal damping ratio for all modes can be accomplished by selecting the p damper locations that maximize the minimum sum of modal strain energy over the p chosen locations. Padula and Sandridge [45] formulate DPP1 as a mixed binary integer linear program. Their formulation is given below.

$$\max \quad \beta$$
$$\text{subject to} \quad \sum_{j=1}^{nmemb} d_{ij} x_j \geq \beta \quad \forall i = 1, \ldots, nmode$$
$$\sum_{j=1}^{nmemb} x_j = p$$
$$\beta \in \mathbb{R}, \ x_j \in \{0,1\} \quad \forall j = 1, \ldots, nmemb$$

Alternatively, DPP1 may be viewed as a combinatorial optimization problem. That is, given $\mathbf{D}$ we seek to find the $nmode$ by p submatrix whose smallest row sum is as large as possible. Let $Z(X) = \min_{i \in I} \sum_{j \in X} d_{ij}$ and let $|\cdot|$ denote the cardinality of the set X. Then DPP1 becomes

$$\max_{X \subseteq J} \quad Z(X)$$
$$\text{subject to} \quad |X| = p$$

Heuristic algorithms for DPP1 were first introduced by Kincaid and Berger [12] for a tetrahedral truss structure developed for NASA at the MIT Space Engineering Research Center (cf. [51]). This symmetric structure had 600 truss members and 34 vibrational modes of interest. The heuristics tested in Ref. [12] included a variety of tabu search heuristics for DPP1. DPP1 was studied as a *maxminsum* location problem on tree graphs in Ref. [47]. Characterizations of the optimal solutions for small values of p were obtained.

Unlike active dampers, each passive damper has a particular frequency, $\hat{\omega}$, at which its damping effectiveness is best. The damping effectiveness, $e(\omega)$, deteriorates the further the modal frequency, ω, is from $\hat{\omega}$. A typical graph of $e(\omega)$, where the y-axis $e(\omega)$, designates the fraction of mean strain energy dissipated by the passive damper, is roughly ski-sloped shaped. Let ω_i for all $i = 1, \ldots, nmode$ denote the frequencies of the nmode significant modes of vibration for CEM. We note that in the special case when $\hat{\omega}$ is identical for each of the passive dampers, and we replace each entry d_{ij} of $\mathbf{D}$ with $e(\omega_i) \cdot d_{ij}$ (where $\mathbf{e} = (e(\omega_1), \ldots, e(\omega_{nmode})))$, then DPP1 is an appropriate model.

Although there are some manufacturing restrictions, passive dampers can be designed so that $\hat{\omega}$ occurs at nearly any frequency of interest. Table 13.8 lists the 10 modes considered in the CEM model and their frequency ω_i in Hz. In particular, we assume that passive dampers can be designed so that $\hat{\omega}$ is any one of the $nmode$ frequencies (vibrational modes) of interest. Further, we assume there is no advantage in designing the passive dampers to have $\hat{\omega}$ at a value other than $\omega_1, \ldots, \omega_{nmode}$.

TABLE 13.8 Ten Modes for CEM and Their Frequencies

Mode	1	2	3	4	5	6	7	8	9	10
ωi, Hz	1.498	2.434	2.590	4.994	5.648	5.740	6.150	6.726	6.801	7.830

To extend model DPP1, we make *nmode* copies of **D** and pre-multiply each row i of each copy of **D** by $e_k(\omega_i)$—the damping effectiveness associated with mode i when the peak design frequency is ω_k. Each of the resulting *nmode* strain energy matrices corresponds to one of the potential design variable values of $\hat{\omega}$. In addition, a new set of constraints must be added to allow at most one of the x_{kj} variables to be 1 for each j. The resulting mixed binary integer linear program is given below. For CEM, there will be 15,070 binary variables, 1 continuous variable, and 1518 constraints. We note that in practice this model would require *nmode* prototype passive dampers to be built and tested, so that the damping effectiveness values are available. This formulation, DPP2, is new and is given below.

$$\max \quad \beta$$

$$\text{subject to} \quad \sum_{k=1}^{nmode} \sum_{j=1}^{nmemb} e_k(\omega_i) d_{ij} x_{kj} \geq \beta \quad \forall i = 1, \ldots, nmode$$

$$\sum_{k=1}^{nmode} x_{kj} \leq 1 \quad \forall j = 1, \ldots, nmemb$$

$$\sum_{k=1}^{nmode} \sum_{j=1}^{nmemb} x_{kj} = p$$

$$\beta \in \mathbb{R}, x_{kj} \in \{0, 1\} \quad \forall j = 1, \ldots, nmemb;\ k = 1, \ldots, nmode$$

We illustrate DPP2 with the following small example. In this example, there are two modes to be damped, and four truss members that may be replaced with $p=2$ passive dampers. The mean strain energy matrix **D** is

$$\begin{pmatrix} .25 & .30 & .20 & .25 \\ .10 & .20 & .50 & .20 \end{pmatrix}$$

Each of the p passive dampers may be manufactured so that its peak damping effectiveness coincides with ω_1 (mode 1) or ω_2 (mode 2). Assume that $e_1(\omega_1) = 0.25$, $e_1(\omega_2) = 0.20$ $e_2(\omega_1) = 0.15$, and $e_2(\omega_2) = 0.30$. There are $\binom{4}{2} = 6$ combinations of two truss members to be replaced with passive dampers. Coupled with the two manufacturing design choices for each passive damper, we get a total 24 possible solutions. Consider the solution that replaces truss members (columns) 2 and 3 with passive dampers. The passive dampers replacing truss members 2 and 3 are designed to damp most effectively for mode 1 and mode 2, respectively. The corresponding objective function value is

$$\min\{e_1(\omega_1)d_{12} + e_2(\omega_1)d_{13}, e_1(\omega_2)d_{22} + e_2(\omega_2)d_2\}$$

$$= \min\{(.25)(.30) + (.15)(.20), (.20)(.20) + (.30)(.50)\}$$

$$= .105$$

13.8.2 Tabu Search for DPP

The basic solution approach for TS consists of a *construction* phase that generates a starting solution and an *improvement* phase that seeks to iteratively improve upon the starting solution. After max_it iterations of the improvement phase we do one of the following; *intensify* the search by restarting the improvement phase at the current best solution: or *diversify* the search by restarting the improvement phase in an unexplored region of the solution space; or stop and display the best solution found.

TS begins with an initial state $S_0 \in \sum$ that is either chosen at random or constructed algorithmically. For our dataset, **D**, random solutions are easily generated. The p columns are randomly chosen from the $nmemb = 1507$ available entries. Several possibilities exist for constructing an initial state. One heuristic is to pick the p column indices corresponding to the p largest entries of **D**. This heuristic did not perform any better than randomly selecting p column indices of **D**. A second heuristic is based on solving the linear programming (LP) relaxation of the integer programming formulation presented in Section 13.2. Solve the LP-relaxation and pick the p largest of the nonzero decision variables in the LP optimal solution as our initial state X_0. Tables 13.9 through 13.11 catalog the performance of this second heuristic.

The improvement phase is a greedy local improvement scheme. That is, from a state S_t, we select a move that generates the greatest one move improvement in the objective function. That is, we select a move δ_t from among all $p \cdot (nmemb - p)$ pairwise interchanges that satisfies $c(\delta_t(S'_t)) = \max_{\delta \in \Delta(S_t)} c(\delta(S_t))$. Once the interchange is made, the set of all pairwise interchanges is generated again, and a move that improves the objective function the most is selected. This process continues until there are no more local changes available that lead to an improved solution. An important step in defining a local optimization scheme is specifying a method for perturbing the current solution to obtain other solutions. The neighborhood structure we use in our implementation of TS is the $p(nmemb - p)$ neighborhood structure. The $nmemb$ column indices of **D** are partitioned into two sets, set X of size p and set X' of size $(nmemb - p)$ (X' is the complement of X in the column index set J). To generate neighbors, we swap one entry from the set X with one entry from the set X'. As an example, suppose we let $J = \{1, 2, 3, 4, 5\}$ and $p = 2$. If $X = \{1, 3\}$ and $X' = \{2, 4, 5\}$, then one neighbor would be $X = \{2, 3\}$ and $X' = \{1, 4, 5\}$. The size of this neighborhood is $O(p(nmemb - p))$. Every member of the neighborhood is examined before the best admissible move is selected.

The primary goal of the short-term memory component of TS is to permit the heuristic to go beyond points of local optimality, while still making high-quality moves. Choosing the best admissible move is a critical step at each iteration of the improvement scheme. It is

TABLE 13.9 Best Objective Function Value Comparisons

P	LP	U. Bnd	IP/BB (%)	Tabu (%)
8	1.6144	1.6131	81.9	94.1
16	3.1629	3.1110	89.3	95.3
32	5.8867	5.8838	96.4	98.5

TABLE 13.10 Greedy Local Improvement Schemes with Different Initial Solutions

P	Random %	Time (min)	IP/BB (%)	Time (min)	LP (%)	Time (min)
8	81.3	3	84.7	1	88.6	1
16	86.9	18.5	92.4	4.5	94.3	1
32	90.1	37	97.9	2	98.4	4

TABLE 13.11 Performance of Tabu Search with Different Initial Solutions

P	Random (%)	Time (min)	IP/BB (%)	Time (min)	LP (%)	Time (min)
8	88.6	14	84.7	14	94.1	14
16	95.3	210	3.3	56	95.3	56
32	98.2	270	97.9	57	98.5	57

this step that embodies the aggressive orientation of the short-term memory. Each of the candidate moves is evaluated in turn. Initially, the evaluation of the candidate moves can be based on the change produced in the objective function value. As the search progresses, however, the form of evaluation employed becomes more adaptive and may incorporate references to intensification and diversification.

Search methods that are based on local optimization often rely on diversification strategies to increase their effectiveness in exploring the solution space. Some of the strategies are designed with the chief purpose of preventing cycling, while others attempt to impart additional robustness or vigor to the search. Kelly et al. [48] discuss two designs for diversification strategies—recency-based and frequency-based. We use a recency-based diversification scheme in our implementation of TS.

A recency-based diversification scheme is intended to direct the search to a solution that is "maximally diverse" with respect to the local minimum most recently visited. There are two important concepts in this scheme—separation and quality. First, two solutions are considered to be increasingly diverse as their separation increases. Separation is measured as the minimum number of moves (pairwise interchanges) necessary to get from one to the other. Next we want an estimate of the probability that a particular solution will be encountered in a set of solutions equally separated from a given solution. We assume that the probability of traversing between any two solutions separated by a fixed number of swaps decreases with the difference between the two objective function values. Hence, among the set of solutions separated from a given solution at a fixed distance, we are interested in those with the best (highest quality) objective function value. Therefore, a new solution is considered to be maximally diverse from the current best solution if it is maximally separated and the objective function value is close to, or better than, the objective function value of our current solution.

For the damper placement problem associated with **D**, we let the separation between two solutions A and B, $d(A, B)$, be the minimum number of two interchanges needed to produce solution B from solution A. Furthermore, we assume that highest quality objective function values are more difficult to obtain at a fixed distance than other solutions. A state S is described by (X, X'). Given two states S_A and S_B, if $X_A(i) \neq X_B(j)$ for all $i, j = 1 \ldots p$ then $d(A, B) = p$. For **D** there are $\binom{nmemb - p}{p}$ solutions at a distance p from any state X, which are too many to consider. To reduce the number of solutions examined, we implement a greedy search that interchanges each $X(i)$ for all $i = 1, \ldots, p$ with each of the $nmemb - p$ indices in X'. For each i the interchange that improves the objective function value the most or degrades it the least is selected.

In summary, our version of TS has a construction phase that generates an initial feasible solution to DPP1 and DPP2. The improvement phase searches the entire $O(nmemb (nmemb - p))$ move set (neighborhood) and selects the best admissible move. The tabu size $= 25$ and the aspiration criterion is met if the new objective function value is larger than the best objective function value yet generated. After repeating the improvement phase max_it times, a recency-based diversification scheme generates a new starting solution and the process is repeated.

13.8.3 Computational Results

Metaheuristic search strategies such as simulated annealing and tabu search can be used to generate solutions to the damper placement problem. However, by themselves these metaheuristics provide little information concerning the quality of the solution found. A traditional method for obtaining a bound on the optimal objective function value of a integer linear program is to solve the associated linear programming relaxation. That is, if x_j is restricted to be either 0 or 1, then it is relaxed by replacing the 0, 1 restriction with $0 \leq x_j \leq 1$. The objective function value associated with the optimal solution to the linear programming relaxation then serves as an upper bound on the optimal objective function value for the original integer linear program if the objective was to maximize some linear function.

The optimal objective function value of the linear programming (LP) relaxation of DPP1 and DPP2, then, provides an upper bound on the optimal objective function value for DPP1 and DPP2. In addition, if the optimal solution to the LP relaxation resulted in all x_j having values of 0 or 1, then we would have the true optimum to DPP1 or DPP2. It is unusual for this to occur, but it does show the strength of using an LP relaxation strategy. We make use of the upper bound obtained via the LP relaxation of DPP1 and DPP2 to evaluate the performance of our metaheuristics. Generally, there is no guarantee that this will be a tight upper bound but, as can be seen in Table 13.9, for DPP1 associated with the CEM, the LP relaxation does indeed provide a tight upper bound.

A traditional solution strategy for binary integer linear programs that makes use of linear programming relaxations is branch and bound. Given a solution to an LP relaxation, a typical branch and bound scheme fixes one or more of the fractional design variables to either 0 or 1, resolves the LP relaxation, and repeats this process until an all binary solution is found (a feasible solution to the original binary integer linear program). At this point the branch and bound process may continue fixing fractional design variables on other unexplored branches until the optimal binary solution is found, or stop with the current feasible (suboptimal) solution. For more details on branch and bound the interested reader is referred to Parker and Rardin [49].

We solved the LP relaxation of DPP1 using a commercially available branch and bound code for linear integer programs. This code provides (slightly) improved upper bounds (column 3 of Table 13.9) as well as a feasible solution to DPP1. The objective function values listed in Table 13.9 under the IP/BB column were generated in this manner. All objective function values in Tables 13.9 through 13.11 are given as a percentage of the best-known upper bound (column 3 of Table 13.9). Table 13.9 compares the quality of solutions generated by solving DPP1 with this branch and bound scheme (with a limit of 10,000 iterations) to tabu search. The solutions generated in column 4, labeled Integer Program/Branch and Bound (IP/BB) took roughly 3000 times as long to generate as those in column 6 for tabu search. Column 1 in Table 13.9 (as well as Tables 13.10 and 13.11) designates the number of active dampers to be placed. Column 2 of Table 13.9 lists the optimal objective function value of the LP relaxation of DPP1 (an upper bound on the optimal value of DPP1), while column 3 is the upper bound generated during the branch and bound scheme (a slight improvement over the LP relaxation value). Columns 4 and 5 of Table 13.11 provide the best objective function values generated by the branch and bound scheme and tabu search, respectively. When $p=32$ Table 13.11 illustrates that the optimal objective function value of DPP1 lies between 5.7943 (98.5%) and 5.8838, and that the solution corresponding to 5.7943 is feasible to DPP1 (all entries are 0 or 1). As can be seen from Table 13.11, tabu search consistently outperforms the branch and bound approach.

Tabu search can be used to further improve the branch and bound solution of DPP1 or the LP relaxation solution of DPP1. In the latter case, fractional solutions will be present and a mechanism for choosing a subset of the optimal decision variables must be found. We picked the p (where $p = 8$, 16, or 32) decision variables with the largest nonzero value (closest to 1). For example, when $p = 8$ the LP solution had 12 nonzero decision variables in the optimal solution. Of these 12, 5 had a value of 1. When $p = 32$ there were even fewer choices to be made. The optimal solution of the LP relaxation of DPP1 had only 35 nonzero decision variables of which 29 had a value of 1.

Tables 13.10 and 13.11 summarize the performance of tabu search under three different initial (starting) solutions—random (columns 2 and 3), the DPP1 solution generated by branch and bound (column 4 and 5), and the LP relaxation of DPP1 (columns 6 and 7). Reported timings are for a 16 MHz Intel 386-class microcomputer. The solutions generated by branch and bound for DPP1 in Table 13.9 were computed on a CONVEX computer in about 4 min. This corresponds to approximately 200 h of computational effort on the 386 microcomputer. Table 13.10 gives the best objective function values (as a percent of the best known upper bound) obtained using a greedy local improvement search (tabu size $= 0$, no aspiration, and no diversification), as well as the time required for convergence to a local optimum. Table 13.11 catalogs the additional improvement gained by implementing tabu search versus a greedy local improvement search. When $p = 16$ and $p = 32$ ($p = 8$) and one of the hot starts (LP relaxation or IP/BB) is used our implementation of TS sets max_it $= 100$ (50). In addition, when $p = 8$ and $p = 16$ we allow one restart using the recency-based diversification scheme of Section 13.8.2. The quality of solutions when $p = 32$ is so good after 100 iterations that the recency-based diversification and restart are not implemented. For the randomly generated starting solutions we allowed TS to run longer—7 restarts when $p = 16$ and 5 restarts when $p = 32$. The diversification scheme is most effective with the random starting solution version of TS. For example, when $p = 8$, the best solution found in the first set of max_it $= 50$ iterations was achieved at iteration 23 with a value of 1.313 (81.4%); the diversification scheme generated a new starting solution with a value of 1.326 (82.2 %); and the best solution found during the final set of max_it iterations was achieved at iteration 5 with a value of 1.429 (88.6%).

Table 13.12 records the performance of simulated annealing for DPP1. Column 4 is the average (over the number of replications given in column 2) of the final objective function value as a percentage of the best-known upper bound (column 3 of Table 13.9). For simulated annealing this typically is not the best objective function value generated, and column 3 records the best value ever generated over the given number of replications. Although Ref. [17] reports that starting SA with a good initial solution produced better results than using randomly generated starting solutions, this was not the case for DPP1. We used the p largest nonzero entries from the optimal solution of the LP relaxation of DPP1 and better results were not obtained. Simulated annealing is predicated upon a certain amount of randomness and as observed in Ref. [17] either many replications or one extremely long run should be performed. We have opted for 10 replications (except for $p = 32$ for which a single run took 10 h) instead of one long run. In addition, we have used the set of parameters given in Ref. [17] for our implementation of simulated annealing. Johnson et al.'s [17] experiments are extensive and the resulting set of parameters is quite robust. Column 5 gives the average CPU times for the simulated annealing runs. When multiplied by the number of replications this makes simulated annealing 10 to 100 times slower than tabu search with nearly the same objective function values.

Both TS and SA solutions select truss members to replace with dampers in the same general area of the CEM truss. The TS solution contains four longeron truss members near the tower, two diagonal truss members near the cross piece in front of the tail, and two

Metaheuristics for Discrete Optimization Problems

TABLE 13.12 Simulated Annealing for CSI-Phase I Design

P	Reps	Best Obj (%)	Final Obj (%)	Avg CPU (min)	TS Best (%)
8	10	89.2	88.0	169	94.1
16	10	95.4	94.1	310	95.3
32	1	98.5	97.8	626	98.5

diagonal truss members in or near the tail. The SA solution contains two longeron and one diagonal truss members near the tower, three diagonal truss members near the cross piece in front of the tail, and two diagonal truss members in or near the tail. Hence, the SA solution places more dampers (five) in the tail end of the structure and uses more diagonal truss members (six) than the TS solution.

We do not report on the computational experiments for DPP2 here. However, the LP relaxation becomes an even better starting solution as p increases. (It takes roughly five passive dampers to get the same performance as one active damper resulting in larger values of p.) In fact, a greedy local improvement scheme would have produced the same results as TS when $p = 80$.

As a result of the computational experiments, several conclusions may be drawn. The LP relaxation of DPP1 and DPP2 provide good, low cost (computationally) upper bound on the optimal value of the DPP problems for CEM. The combination of the LP relaxation *hot start* and tabu search yields the highest quality solutions and the smallest run-times. Simulated annealing produces nearly as high of quality solutions as tabu search, but, for DPP1, is about 100 times slower.

13.9 Reactive Tabu Search

The utility of several advanced features of tabu search is highlighted in this section. A dynamic memory structure is explained and tested as well as a frequency-based diversification scheme. The main goal of this section is to illustrate the salient features of reactive tabu search. We do this by examining the terrain of a function $f : [-10, 10] \times [-10, 10] \to \mathbb{R}$. As we can plot the domain and range of f together in $\mathbb{R}^3$ we are able to visualize key features of the search mechanism. The function we chose is a frequent test problem for global (continuous) optimization routines (see Hansen [50], Jansson and Knuppel [51], and Levy et al. [52]). The function is

$$f(x,y) = \left(\sum_{i=1}^{5} i(\cos((i-1)x + i)) \right) \left(\sum_{j=1}^{5} j(\cos((j+1)y + j)) \right); \quad x, y \in [-10, 10]$$

and is referred to as Levy No. 3 in the literature.

The Maple command `plot3d(sum(i*cos((i-1)*x+i),i=1..5)*sum(j*cos((j+1)*y+j), j=1..5), x=-10..10, y=-10..10, grid=[50,50]);` generates the plot in Figure 13.5 for $f(x, y)$ over the interval $[-10, 10]$ for both x and y. The resolution is a grid of 50 points along each axis (every 0.4 units) for a total of 2500 plotted points.

For our computational experiments we require a DOP. To discretize the domain of $f(x, y)$ for $x, y \in [-10, 10]$ we impose a regular grid of 800 points along each axis (every 0.025 units) for a total of 640,000 points. We were unable to plot $f(x, y)$ at this resolution but Figure 13.5 provides a reasonable picture of the discrete valued function at 2500 points. It is known that Levy No. 3 has 9 global minima and 760 local minima. The nearest points

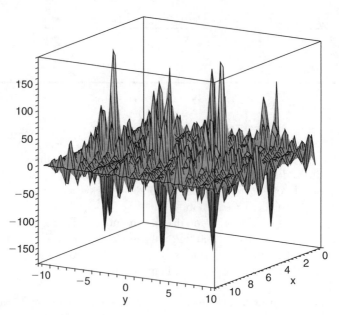

FIGURE 13.5 Levy No.3 evaluated over 50 by 50 grid.

TABLE 13.13 9 Points on Grid Closest to 9 Global Optima

x	y	$f(x,y)$
4.975	−1.425	−176.540
−1.300	−1.425	−176.505
−7.600	−1.425	−176.458
4.975	4.850	−176.397
−1.300	4.850	−176.363
−1.300	−7.700	−176.354
−7.600	4.850	−176.316
−7.600	−7.700	−176.307

to the 9 global optima on the 800 by 800 grid are given in Table 13.13. We don't know how many of the 640,000 points are local minima but we suspect that it is at least 760. Nearly all the results in this section are taken from Kincaid and Laba [29].

13.9.1 Basic Features of Reactive Tabu Search

We begin by describing the neighborhood structure we imposed on the 800 by 800 grid version of Levy No. 3. Given an interior point (x, y) of the grid the neighbors of this point are all of the compass points—North, Northeast, East, Southeast, South, Southwest, West, and Northwest—generated by adding or subtracting the grid size (0.025) from (x, y). For example, the grid point $(x, y + 0.025)$ is the North neighbor of (x, y). At each iteration of tabu search the eight neighbors of the current point are evaluated and the best neighbor among the non-tabu and tabu neighbors that satisfy the aspiration criteria is selected (where best means lowest $f(x, y)$ value). The aspiration criteria checks to see if the function value at a tabu neighbor is better than the best observed function value. If so, the tabu neighbor satisfies the aspiration criteria. Once a best neighbor is selected it becomes the new current point and the (x, y) coordinates of the best neighbor are placed on the tabu list.

The tabu list is a collection of (x, y) coordinates of previously selected best neighbors. To detect whether it is necessary to increase (or decrease) the number of points that are stored on the tabu list we must keep track of points that are visited more than once. The idea is that if the search continues to return to the same point several times then it is likely that the search is cycling and one of the functions of a tabu list is to avoid cycling. Here, instead of maintaining the frequency with which each point (640,000 of them) is uncovered by the search, we maintain the frequency distribution of collections of points. In particular, we place a coarser 50 by 50 grid over the finer 800 by 800 grid. Each point of the coarse grid is mapped to a 16 by 16 set of grid points (256 point in total) on the fine grid. We maintain the frequency data for each of the 50 by 50 grid points (2500 total points). We call each of these coarse grid points a *zone*. Moreover, we make further use of these zones in conjunction with the tabu list. A pass through reactive tabu search may terminate for a variety of reasons (to be described later) and restart at a new point. Upon completion of a single pass the frequency data on the 2500 zones are used to update a matrix describing which zones are to remain tabu in subsequent passes. In particular if the frequency of a zone on a pass is greater than 16 (the number of grid points on a horizontal or vertical line through a zone) then that zone will be labeled tabu. So, in addition to the tabu list the search also checks to see if any of the neighbors is in a tabu zone.

The name reactive tabu search (RTS) was coined by Battiti and Tecchiolli [53], who first described the procedure. The basic features include a tabu list whose length is dynamic and a mechanism for detecting and escaping from a basin of local minima. We initialize the length of the tabu list, *memsize* to 1000. If the frequency of a zone is larger than 256, indicating that each of the 256 points in the zone are visited once on average, then we conclude that we are likely to cycle and *memsize* is increased. That is, $memsize_{new} = 1.2\ memsize_{old}$. Conversely, if *memsize* has not increased for memsize iterations then we decrease *memsize*. That is, $memsize_{new} = 1.0/1.2\ memsize_{old}$.

Starting points from nontabu zones are generated randomly on the 800 by 800 grid. Tabu zones are determined as described earlier, but with one additional feature. Upon completion of a single pass of the search if the frequency of a zone is greater than 16 then that zone, along with its eight adjacent neighboring zones, are labeled tabu with regard to finding new starting points. The idea is to keep the random starting point away from zones that have been previously searched. The x and y coordinates of a potential starting point are each generated from a uniform $[-10, 10]$ distribution. If the point (x, y) is a member of a tabu zone (including the eight neighboring zones) then the point is discarded. This process is repeated until a new starting point from a non-tabu zone is found. We could have been a bit more sophisticated by sampling from only the tabu zones, but in our computational experiments no more than ten tries were required to find a new starting point from a nontabu zone even after 50 passes of the search.

There are two ways for one pass of RTS to terminate—if a basin is found or if the frequency of a zone is large. A basin is a collection of local minimums of similar magnitude from which it is difficult for a tabu search armed only with a tabu list and an aspiration criteria to escape. For Levy No. 3, we say that the search has entered a basin if the difference between the best value of $f(x, y)$ and the worst value of $f(x, y)$ is less than 30.0 for more than 1024 iterations. The number 1024 was selected with the idea that we would traverse at least four zones and 30.0 was selected as a rough estimate. We knew that the range of $f(x, y)$ was roughly 350 and we wanted to avoid the flat regions of the solution space. In practice, the range of f would not be known and the allowable range for basin detection would be a percentage of the best known maximum and minimum values. When a basin is detected the search restarts at a point generated from two features of the basin—the coordinates of the best point in the basin and the distance between the best and worst points in the basin. We call this latter value the diameter of the basin. By distance we mean the maximum of the difference between the x and y coordinates of the two points. Then we generate all eight neighbors of the best point at a distance of six times the diameter of the basin. We used a value of six so that we would be certain to evaluate points outside the basin. Among these eight points we select the non-tabu point with lowest function value. If all eight points are tabu we generate a new point randomly.

The second means of terminating a single pass of RTS is to avoid repetition of a particular point even though the search has been increasing the value of *memsize*, that is, detecting when the search is likely to be stuck in a cycle from which the simple increase in the length of the tabu list will not let the search escape. In other words, the search is stalled in a deep valley or a collection of valleys. Here again, our search relies on zones rather than on individual points. If the frequency of any zone is greater than 512 then we restart reactive tabu search at a new (randomly generated) grid point. The choice of 512 follows from deciding that if each point in a zone is visited (on average) twice then the search is most likely cycling. Clearly this number must be larger than 256, the zone frequency at which *memsize* is increased.

13.9.2 Computational Experiments

One of the goals of this section is to provide visual examples of the features of RTS. To do this we compare RTS with a more traditional tabu search. The traditional scheme has the same neighborhood structure, tabu zones, and aspiration criteria as RTS, but has a fixed tabu list size of 1500 and no capability to detect or escape from basins. The tabu list size of 1500 was determined by experimentation. Indeed the performance of the traditional tabu search is sensitive to this parameters value. In our experiment 1500 produced the best results. Thus, the traditional tabu search does not have the ability to alter the length of the tabu list, nor does it make any attempt to discover basins and escape from them. The only mechanism for terminating a pass of the search is the detection of cycling associated with stalling in a deep valley or a collection of valleys. As in RTS, if the frequency of a zone is greater than 512 then we restart the search at a new randomly generated grid point. To avoid extremely long passes we set a maximum of 5000 iterations per pass. Finally, the traditional tabu search generates its random starting points just as in RTS.

Figure 13.6 summarizes the performance of the traditional tabu search on Levy No. 3. Figure 13.6 plots two data items, the $f(x,y)$ values on the z-axis and the (x,y) values. The numbers from 1 to 7 denote a single pass of the search started at a randomly generated grid point. Clearly the fixed memory size of 1500 and the aspiration criteria are able to force the search out of small local minima and in most cases avoid cycling. Five of the seven passes terminate due to the maximum iteration limit. Only passes 1 and 2 detect the presence of cycling associated with stalling in a deep valley before 5000 iterations are complete. Notice that three of the nine best solutions are uncovered by the search and that to get to these three solutions relatively deep valleys were explored and escaped. Besides these nine valleys with minima on the order of -176 there are many valleys with minima on the order of -145 and -116 in the landscape (see Figure 13.5). Although not illustrated here, we observed that both the traditional search and RTS have great difficulty escaping from the -145 and -116 valleys using only a tabu list and aspiration criteria. The lowest point in pass 4 in Figure 13.6 has an objective value on the order of -116. The total number of iterations (neighborhood searches) generated by the seven passes was 31,658.

Figures 13.7 and 13.8 summarize the performance of RTS on Levy No. 3. Figure 13.8 adds the $f(x,y)$ values as the z-axis to the (x,y) values recorded in Figure 13.7. Each number (in both figures) denotes a single pass of the search. The checkerboard pattern in Figure 13.7 is most easily explained by Figure 13.8. Here we see that each checkerboard maps into a series of adjacent valleys (local minima). The unexamined spaces of the checkerboard

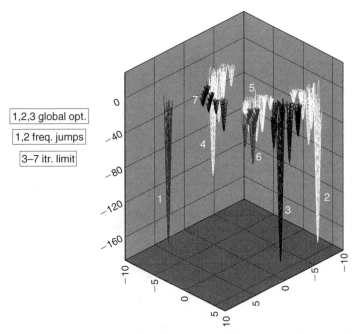

FIGURE 13.6 Static TS Levy No.3—31,568 iterations.

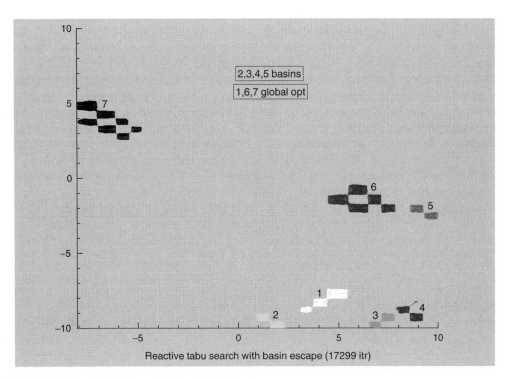

FIGURE 13.7 RTS for Levy No.3—17,299 iterations.

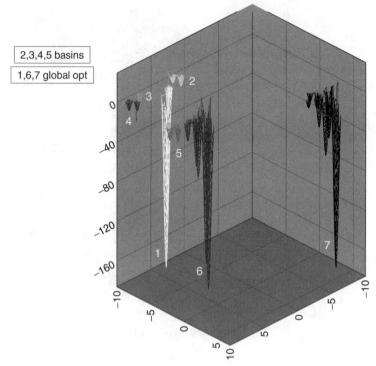

Reactive tabu search with basin escape (17299 itr)

FIGURE 13.8 RTS for Levy No.3—17,299 iterations.

most likely correspond to hills (local maximum) and were avoided by the search. Pass 1 is nearly the same as the one generated by the traditional tabu search. The difference is that in Figure 13.6 the search terminated due to a repeated solution (iteration 1725), but in Figure 13.8 RTS terminates due to the detection of a basin (iteration 2943). The variable length of the tabu list accounts for RTS not terminating at the same iteration (1725) as the traditional tabu search scheme. Clearly the valley for pass 1 in Figure 13.8 is not a basin (a collection of local minima). What must be happening is that the search is circling around the valley at nearly a constant height. Hence, since the change in the objective value is less than 30.0 (points at nearly the same depth of the valley) for more than 1024 iterations, the basin escape mechanism of RTS is activated.

As a result the starting point of the search for pass 2 is generated from the characteristics of the detected basin (best point and diameter). Starting solutions for passes 3–6 are generated in the same manner. That is, passes 3, 4, and 5 all terminate due to the detection of a basin. In each of these cases RTS detects what we initially had in mind for a basin—a collection of local minima (valleys). Each of the passes (2–5) contains a pair of small valleys that fulfill the basin criteria. Pass 6 terminates with a repeated zone and, therefore, pass 7 is a randomly generated starting point. Pass 7 finds a point with function value −176 at iteration 1145 and then spends the next 3200 iterations escaping from this valley and uncovering several small nearby valleys. These valleys trigger the basin escape mechanism and pass 7 concludes. As in the traditional search results, three of the seven passes find −176 solutions. Two of these −176 solutions (passes 1 and 7) were uncovered by the traditional method as well.

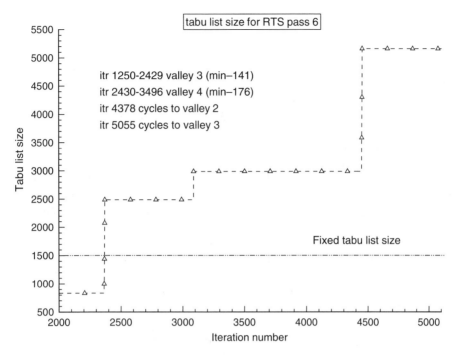

FIGURE 13.9 Size history of tabu list in RTS for pass 6.

Lastly we examine the length of the tabu list—*memsize*—in RTS. Figure 13.9 catalogs the value of *memsize* over the search history (iterations) for passes 6 and 7 of RTS summarized in Figures 13.7 and 13.8. What is the impact of allowing *memsize* to change? The value of *memsize* increases from 833 at iteration 2363 to 2489 at iteration 2369. The growth of *memsize* in iterations 2363–2369 comes just as the search is trying to escape from valley 3. Valley 4 is a deep local minima. Its lowest point has a value on the order of −141. The value of *memsize* increases again at iteration 3087 while in valley 4. Presumably this assists the search in escaping from the −176.540 local minima at iteration 3496. After iteration 4377 the search falls back into valley 2, from which it escapes for a second time on iteration 5055. Next the search falls into valley 3 again, from which it is unable to escape prior to the search meeting its repeated solution condition. Thus, pass 6 stops at iteration 5917 with a *memsize* of 6193. These latter increases of *memsize* were not large enough to drive it away from the previously discovered valleys. The search is restarted for pass 7 and *memsize* is reset to 1000. Unlike pass 6 there is only one deep valley traversed (with −176.316 at its minimum). The other six valleys are all relatively shallow, with the last two causing the basin escape mechanism to be activated and the search restarted. The −176.316 solution is uncovered at iteration 1145 in the third valley traversed. The value of *memsize* is increased from 833 on iteration 2153 to 2489 on iteration 2159. This enables the search to escape from the −176.316 valley on iteration 2222.

13.10 Discussion

The thrust of this chapter has been on the implementation details and performance of local search heuristics (Sections 13.3 and 13.7), tabu search (Sections 13.5 through 13.9) and simulated annealing (Sections 13.4 and 13.8). The focus on tabu search represents my own bias. I believe that the tabu search framework allows greater flexibility and allows for more insight about the problem structure to be built into the search than do most other metaheuristics. Plain vanilla tabu search (Section 13.5) is straightforward to implement with only one parameter to set—the length of time moves should remain on the tabu list. In addition there are a variety of advanced features that can significantly improve the performance of tabu search if the search terrain proves difficult (Sections 13.8 and 13.9). The material presented in Sections 13.1 through 13.5 is amplified by the results presented in Sections 13.6 through 13.9.

Returning to the questions asked at the outset: If I have a feasible solution, how do I know if it is a good one? My recommendation here would be to begin with a plain vanilla tabu search (Section 13.5). If the best solutions found by PVTS are not significantly better than your current feasible solution then you need not look further. However, having other high-quality solutions available may provide additional insights into the optimization model you are trying to solve and you may wish to continue the search. Which solution strategy should I start with if I don't want to invest too much of my time? The easiest to implement are multistart local searches (Section 13.4). However, if you have been solving other problems with simulated annealing and tabu search it is typically not difficult to adapt these codes to new problems. The section of the code dedicated to evaluating the quality of a given solution must change, but often much of the remaining code requires only minor changes.

The contents of this chapter reflect my own preferences for how to go about answering the above questions. As I suggested at the outset, when faced with a nonstandard DOP, I begin by developing a neighborhood structure for traversing the space of feasible solutions and embed this within a multistart local search heuristic (see Sections 13.3 and 13.7) with a random, constructed, or known initial feasible solution. Section 13.6 provides an example of a construction scheme for generating initial solutions while in Section 13.8 rounded linear programming relaxation solutions are used. Multistart provides a benchmark for comparing the performance of any other solution approach chosen for implementation as well as information about the search terrain. Next, I suggest embedding the neighborhood structure within a simulated annealing heuristic. I use the standard set of parameter values (see Section 13.4) without any fine-tuning. If simulated annealing is unable to improve upon the solutions found by multistart, I often stop any further computational tests and conclude that the search terrain is not difficult and that the solutions I have found with multistart are good enough. If simulated annealing is able to uncover better solutions than multistart, or if I have a competing solution found by a different approach that is better (e.g., in Section 13.6 the modal analysis approach for actuator placement problem), then I proceed to develop a tabu search for the problem. If basic tabu search (see Sections 13.5 and 13.6) is unable to improve upon the solutions found by multistart and simulated annealing then I begin adding advanced features to my tabu search (see Sections 13.8 and 13.9). In my experience, tabu search is able to find the same quality of solutions as simulated annealing but in significantly less time and with significantly fewer function evaluations (typically an order of magnitude decrease).

References

1. Goldberg, D.E., *Genetic Algorithms in Search, Optimization and Machine Learning*, Addison-Wesley, Reading, MA, 1989, ISBN: 0201157675.
2. Back, T., *Evolutionary Algorithms in Theory and Practice*, Oxford University Press, New York, 1996.

3. Dasgupta, D. and Z. Michalewicz, Evolutionary Algorithms—An Overview. In: D. Dasgupta and Z. Michalewicz (eds.), *Evolutionary Algorithms in Engineering Applications*, pp. 3–28, Springer-Verlag, Heidelberg, 1997.
4. Moscato, P., Memetic Algorithms: A Short Introduction. In: D. Corne, M. Dorigo, and F. Glover (eds.), *New Ideas in Optimization,* pp. 219–234, McGraw Hill, London, 1999.
5. Glover, F. and M. Laguna, *Tabu Search*, Kluwer Academic Publishers, Boston, 1997.
6. van Laarhoven, P.J.M. and E.H.L. Aarts, *Simulated Annealing: Theory and Applications*, D. Reidel Publishing Company, Kluwer, Dordrecht, 1987.
7. Michalewicz, Z. and D.B. Fogel, *How to Solve It: Modern Heuristics*. Springer-Verlag, Berlin, 2000.
8. Hoos, H.H. and T. Stutzle, *Stochastic Local Search: Foundations and Applications*, Morgan Kaufmann Publishers, San Francisco, 2005.
9. Mirchandani, P.B. and R.L. Francis (eds.), *Discrete Location Theory*, Wiley-Interscience, New York, 1990.
10. Nickel, S. and J.P. Albandoz, *Location Theory: A Unified Approach*, Springer-Verlag, Berlin, 2005.
11. Kincaid, R.K. and L.G. Yellin, "The P-dispersion-sum problem: Results on trees and graphs," *Location Science*, 1:171–186, 1993.
12. Kincaid, R.K. and R.T. Berger, "The damper placement problem on space truss structures," *Location Science*, 1:219–234, 1993.
13. Garey, M.R. and D.S. Johnson, *Computers and Intractibility: A Guide to the Theory of NP-Completeness*, W.H. Freeman, New York, 1979.
14. Metropolis, N., A. Rosenbluth, M. Rosenbluth, A. Teller, and E. Teller, "Equation of state calculations by fast computing machines," *Journal of Chemical Physics*, 21:1087–1092, 1953.
15. Kirkpatrick, S., C.D. Gelatt, and M.P. Vecchi, "Optimization by simulated annealing," *Science*, 220:671–680, 1983.
16. Cerny, V., "Thermodynamical approach to the traveling salesman problem: An efficient simulation algorithm," *Journal of Optimization Theory and Applications*, 45:41–52, 1985.
17. Johnson, D.S., C.R. Aragon, L.A. McGeoch, and C. Schevon, "Optimization by simulated annealing: An experimental evaluation. Part I, Graph partitioning," *Operations Research*, 37:865–893, 1989.
18. Lundy, M. and A. Mees, "Convergence of an annealing algorithm," *Mathematical Programming*, 34:111–124, 1986.
19. Aarts, E.H.L. and P.J.M. Van Laarhoven, "Statistical cooling: A general approach to combinatorial optimization problems," *Phillips Journal of Research*, 40:193–226, 1985.
20. Collins, N.E., R.W. Eglese, and B.L. Golden, "Simulated annealing—an annotated bibliography," *American Journal of Mathematical and Management Sciences*, 8:209–308, 1988.
21. Press, H., B.P. Flannery, S.A. Teukolsky, and W.T. Vetterling, *Numerical Recipes: The Art of Scientific Computing*, Cambridge University Press, Cambridge, UK, 1986.
22. Kincaid, R.K., "Good solutions to discrete noxious location problems via metaheuristics," *Annals of Operations Research*, 40:265–281, 1992.
23. Glover, F., "Tabu search, Part I," *ORSA Journal on Computing*, 1:190–206, 1989.
24. Glover, F.,"Tabu search: A tutorial," *INTERFACES*, 20:74–94, 1990.
25. Palumbo, D., S.L. Padula, K.H. Lyle, J.H. Cline, and R.H. Cabell, "Performance of Optimized Actuator and Sensor Arrays in an Active Noise Control System," NASA TM-110281, 1996.
26. Lyle, K.H. and R.J. Silcox, "A study of active trim panels for noise reduction in an aircraft fuselage," paper no. 951179 in SAE General, Corporate and Regional Aviation Meeting and Exposition, May, 1995.

27. Palumbo, D.L., "Flight test of ASAC interior noise control system," AIAA Paper 99-1933, 1999.
28. Kincaid, R., K. Laba, and S. Padula, "Quelling cabin noise in turboprop aircraft via active control," *Journal of Combinatorial Optimization*, 1(3):1–22, 1997.
29. Kincaid, R. and K. Laba, "Reactive tabu search and sensor selection in active structural acoustic control problems," *Journal of Heuristics*, 4:199–220, 1998.
30. Kincaid, R. and S. Padula, Actuator Selection for the Control of Multi-Frequency Noise in Aircraft Interiors. In: S. Voss, S. Martello, I. Osman, and C. Roucairol (eds.), *Meta-Heuristics: Advances and Trends in Local Search Paradigms for Optimization*, Kluwer, Dordrecht, pp. 111–124, 1999.
31. Underhill, L.G., "Optimal and suboptimal reserve selection algorithms," *Biological Conservation*, 70:85–87, 1994.
32. Church, R.L. and C.S. ReVelle, "The maximum covering location problem," *Papers in Regional Science*, 32:101–118, 1974.
33. Schilling, D.A., C.S. ReVelle, C. Cohon, and D.J. Elzinga, "Some models for fire protection location decisions," *European Journal of Operational Research*, 5:1–7, 1980.
34. Camm, J.D., S.K. Norman, S. Polasky, and A.R. Solow, "Nature reserve site selection to maximize expected species covered," *Operations Research*, 50:946–955, 2002.
35. Rosing, K.E., C.S. ReVelle, and J.C. Williams, "Maximizing species representation under limited resources: a new and efficient heuristic," *Environmental Modeling and Assessment*, 7:91–98, 2002.
36. Masters, L., N. Clupper, E. Gaines, C. Bogert, R. Solomon, and M. Ornes, *Biodiversity Research Consortium Species Database Manual*, The Nature Conservancy, Boston, 1995.
37. Kincaid, R.K., C. Easterling, and M. Jeske, "Computational experiments with heuristics for two nature reserve site selection problems," *Computers and Operations Research*, 35(2):499–512, 2008.
38. Csuti, B., S. Polasky, P.H. Williams, R.L. Pressey, J.D. Camm, M. Kershaw, A.R. Kiester, B. Downs, R. Hamilton, M. Huso, and K. Sahr, "A comparison of reserve selection algorithms using data on terrestrial vertebrates in Oregon," *Biological Conservation*, 80:83–97, 1997.
39. Camm, J.D., S. Polasky, A.R. Solow, and B. Csuti, "A note on optimal algorithms for reserve site selection," *Biological Conservation*, 78:353–355, 1996.
40. Church, R.L., D.M. Stoms, and F.W. Davis, "Reserve selection as a maximal covering location problem," *Biological Conservation*, 76:105–112, 1996.
41. ReVelle, C.S., J.C. Williams, and J.J. Boland, "Counterpart models in facility location science and reserve selection science," *Environmental Modeling and Assessment*, 7:71–80, 2002.
42. Ronald L. Rardin. *Optimization in Operations Research*, Prentice Hall, Upper Saddle River, NJ, 1998.
43. Kincaid, R.K., "Solving the damper placement problem using local search heuristics," special issue of *OR Spektrum* on Applied Local Search, 17:149–158, 1995.
44. Anderson, E., M. Trubert, and J. Fanson, "Testing and application of a viscous passive damper for use in precision truss structures," *Proceedings of 32nd Structures, Structural Dynamics and Materials Conference*, Baltimore, MD, April, pp. 2795–2807, 1991.
45. Padula, S. and C.A. Sandridge, Passive/Active Strut Placement by Integer Programming, In M.P. Bendsoe and C.A. Morta Soares (eds.), *Topology Design of Structures*, Kluwer Academic Publishers, Dordrecht, pp. 145–156, 1993.
46. Preumont, A., "Active damping by a local force feedback with piezoelectric actuators," *Proceedings of 32nd Structures, Structural Dynamics and Materials Conference*, Baltimore, MD, April, pp. 1879–1887, 1991.

47. Kincaid, R.K. and R.T. Berger, "The MaxMin sum problem on trees," *Location Science*, 2:1–10, 1994.
48. Kelly, J.P., M. Laguna, and F. Glover, "A study of the diversification strategies for the quadratic assignment problem," *Computers and OR*, 21(8):885–893, 1994.
49. Parker, R.G. and R.L. Rardin, *Discrete Optimization*, Academic Press, San Diego, CA, 1988.
50. Hansen, E. *Global Optimization Using Interval Analysis*, Marcel Dekker, New York, 1992.
51. Jansson, C. and O. Knuppel, "Numerical results for a self-validating global optimization method," working paper No. 94.1, Technische Universität Hamburg-Harburg, Technische Informatik III, D-21071 Hamburg, Germany, 1994.
52. Levy, A.V., A. Montalvo, S. Gomez, and A. Calderon, *Topics in Global Optimization*, Lecture Notes in Mathematics No. 909, Springer-Verlag, New York, 1981.
53. Battiti, R. and G. Tecchiolli, "The reactive tabu search," *ORSA Journal on Computing*, 6:126–140, 1994.

14
Robust Optimization

H. J. Greenberg
University of Colorado at Denver and Health Sciences Center

Tod Morrison
University of Colorado at Denver and Health Sciences Center

14.1	Introduction	14-1
14.2	Classical Models	14-2
	Mean-Risk Model • Recourse Model • Chance-Constrained Model	
14.3	Robust Optimization Models	14-10
	Worst-Case Hedge • Simple Case of Interval Uncertainty • Minimax Regret • Uncertainty Sets	
14.4	More Applications	14-16
14.5	Summary	14-30
	Acknowledgments	14-32
	References	14-32

14.1 Introduction

We all want to maximize our gains and minimize our losses, but decisions have uncertain outcomes. What if you could choose between an expected return of $1000 with no chance of losing any amount, or an expected return of $5000 with a chance of losing $50,000. Which would you choose? The answer depends upon how risk-averse you are. Many would happily take the nearly certain $1000, and some would take the risk with hope of greater profit. This chapter is concerned with how to extend mathematical programming models to deal with such uncertainty.

Sources of uncertainty could be due to at least three different conditions [1]:

- *ignorance*, such as not knowing exactly how much oil is in a reserve;
- *noise*, such as measurement errors, or incomplete data;
- *events that have not yet occurred*, such as future product demand.

The *effects* of uncertainty on decision-making include variations in actual returns or resources consumed, but there could be a *catastrophic effect* that one seeks to avoid completely. Losing money in an investment is not good, but losing so much that a firm goes bankrupt is catastrophic. Losing additional lives due to a mis-estimate of where and when some tragedy will strike is bad, but losing an ability to recover is catastrophic. Uncertainty creates a range of concerns about the *volatility* of one policy versus another, and the one with better expected value may miss essential points of concern. One hedge against uncertainty is to *diversify*, keeping a greater level of *flexibility* in what recourses are available.

In any problem that requires optimization under uncertainty we must first answer some basic questions: "How do we represent uncertainty?" "What are our attitudes towards risk?" In addition, robust optimization raises the question, "What is a *robust* solution?" A naive approach to robust optimization is to define a robust solution as one that remains optimal, or feasible, under any realization of the data. Unfortunately, such a definition is very restrictive and may lead to infeasible models for reasonable problems. In the following sections we define some alternative robust optimization models, each of which captures some aspect of volatility or flexibility.

One useful perspective is to ask how much degradation we are willing to accept in our objective to reduce the risk that our solution is infeasible. Suppose we have two candidate solutions, x^1 and x^2, and we can measure their performance (such as by computing the expected value of each). Based on this performance, we prefer x^1 to x^2. Suppose, however, that there is uncertainty in the constraint $Dx \geq d$ such that less than 1% variation in uncertain values of D causes x^1 to violate the constraints by as much as 50%—that is, $1.01Dx^1 < \frac{1}{2}d$ [2]. One must consider the *price of robustness* [3] but avoid being overly conservative. In particular, one typically does not want to use the most pessimistic model, where each uncertain coefficient is replaced by its most pessimistic value [4]:

$$\underline{D}x \geq \overline{d}$$

where $\underline{D}_{ij} = \min_D \{D_{ij}\}$ and $\overline{d}_i = \max_d \{d_i\}$. Replacing the stochastic constraint, $Dx \geq d$, with this pessimistic one rules out x^1 and leaves only those x that absolutely guarantee feasibility. It also ignores correlations among the coefficients in D and between D and d, and it is not influenced by their probability of occurrence.

The rest of this chapter is organized as follows. Section 14.2 presents a succinct description of each model that was proposed in the early development of operations research (1950s). Section 14.3 presents robust optimization from its beginnings to now. Section 14.4 presents some additional applications. The last section is a summary and an indication of avenues for further research. Most terms are defined here, but see the *Mathematical Programming Glossary* [5] for extended definitions.

14.2 Classical Models

The *fallacy of averages* [6] is the misconception resulting from the replacement of each uncertain parameter with its expected value. This can lead to poor results in supporting decision making for at least two reasons: (1) the parameters may be correlated—namely, $\mathrm{E}[XY] \neq \mathrm{E}[X]\,\mathrm{E}[Y]$, and (2) the functions may not be linear, so $f(\mathrm{E}[X]) \neq \mathrm{E}[f(X)]$.

Consider the simple two-asset investment problem using only the expected returns:

$$\max\ 1000\,x_1 + 50{,}000\,x_2 : x \geq 0,\, x_1 + x_2 = 1$$

where x_i is the portion of the total budget, as a percentage, invested in the i-th asset. This is a totally deterministic problem; the solution is to invest everything into asset 2, giving an expected return of $50,000 per amount invested. Suppose there is a correlation between investment outcomes such that the return from asset 1 could be much greater than the return from asset 2. In particular, suppose we have three scenarios with their probabilities of occurrence and the returns they generate, shown in Table 14.1.

Notice that the properties of asset returns do not depend upon the actual levels of investment. The total investment could be $1,000,000 to reap the net returns, or it could be $1. This is what is implied by the constraint $x_1 + x_2 = 1$, where x_i is the *portion* of the capital invested in asset i.

We shall revisit this finance application as we explain modeling approaches to deal with the uncertainty, but this chapter is concerned with the broad view, beyond this application.

Robust Optimization

TABLE 14.1 Scenarios That Are Uncertain

Scenario	Probability	Asset 1 Return	Asset 2 Return
1	0.2	1050	−2,000,000
2	0.4	975	1,000,000
3	0.4	1000	125,000
Expected Return:		1000	50,000

It is thus imperative that you examine Section 14.4 to learn how robust optimization applies to a variety of applications and how it compares with other approaches. Seeing how different models apply will help to clarify the basic terms and concepts, particularly those developed in Section 14.3.

There are three classical models that are still used today: mean-risk, recourse, and chance-constrained. A brief discussion of each will give us a base from which to present robust optimization.

14.2.1 Mean-Risk Model

The first mathematical programming model of uncertainty was Markowitz's *mean-risk model* [7,8] for the portfolio selection problem. Markowitz rejected the hypothesis that an investor should simply maximize expected returns. Instead, he suggested that investors consider expected return a desirable objective to maximize, but only while *also* considering risk undesirable, to be minimized. Recognizing that these two objectives are usually in conflict, he considered them in a bi-objective model.

Consider the example in Table 14.1. One measure of risk is the expected square of how much the actual return differs from the expected return. Let R_1 and R_2 denote the (random) returns from the two assets, and let $p = \text{E}[R]$ denote their expected values. We have already noted that $\text{E}[R_1] = 1000$ and $\text{E}[R_2] = 50{,}000$. However, the potential for a \$2,000,000 loss is daunting, so let us assess the risk of some arbitrary mixture of investments.

The total return is $R_1 x_1 + R_2 x_2$, so the square deviation from the mean is:

$$\text{E}\left[(R_1 x_1 + R_2 x_2 - (p_1 x_1 + p_2 x_2))^2\right]$$

$$= \text{E}\left[((R_1 - p_1)x_1 + (R_2 - p_2)x_2)^2\right]$$

$$= \text{E}\left[(R_1 - p_1)^2\right] x_1^2 + \text{E}\left[(R_2 - p_2)^2\right] x_2^2 + 2 x_1 x_2 \text{E}\left[(R_1 - p_1)(R_2 - p_2)\right]$$

We write this as the quadratic form: $(x_1, x_2) V \begin{pmatrix} x_1 \\ x_2 \end{pmatrix}$, where V is the *variance-covariance matrix*:

$$V = \begin{bmatrix} \text{E}\left[(R_1 - p_1)^2\right] & \text{E}\left[(R_1 - p_1)(R_2 - p_2)\right] \\ \text{E}\left[(R_1 - p_1)(R_2 - p_2)\right] & \text{E}\left[(R_2 - p_2)^2\right] \end{bmatrix} = \begin{bmatrix} \sigma_1^2 & \sigma_{12} \\ \sigma_{12} & \sigma_2^2 \end{bmatrix}$$

To obtain the variances and covariance, we consider each scenario's contribution.

$$\sigma_1^2 = 0.2(2500) \qquad\qquad + 0.4(625) \qquad\qquad + 0.4(0)$$
$$\sigma_2^2 = 0.2(4.2025 \times 10^{12}) + 0.4(9.025 \times 10^{11}) + 0.4(5.625 \times 10^9)$$
$$\sigma_{12} = 0.2(-1.025 \times 10^8) + 0.4(-2.375 \times 10^7) + 0.4(0)$$

Thus,

$$V = \begin{bmatrix} 750 & -30{,}000{,}000 \\ -30{,}000{,}000 & 1{,}203{,}750{,}000{,}000 \end{bmatrix}$$

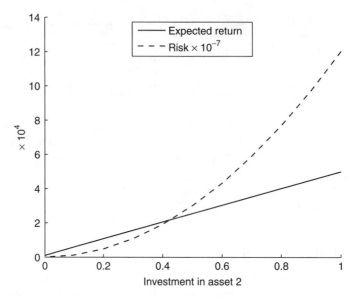

FIGURE 14.1 Expected return and risk as functions of x_2.

and the risk is measured by $x'Vx = 750x_1^2 - 2x_1x_2(3.00 \times 10^6) + 1.20375 \times 10^{12} x_2^2$ (approximating the coefficient of x_2^2). If we invest everything into asset 2, this means $x_1 = 0$ and $x_2 = 1$. In that case, the risk is 1.20375×10^{12}, reflecting not only a chance of much greater return than what is "expected," but also a risk of a great loss. At the other extreme, if we invest everything into asset 1, $x = (1,0)$ and the risk is only 750, reflecting a much more conservative investment. Figure 14.1 plots the expected return as a (linear) function of how much is invested into asset 2 and the risk as a (quadratic) function. (To fit on the same plot, the risk is scaled by 10^{-7}.)

Markowitz's mean-risk model is

$$\max_{x \in X} px - \lambda x'Vx$$

where $X = \{x \geq 0 : \sum_i x_i = 1\}$. We sometimes allow no investment by replacing the equation with the inequality $\sum_i x_i \leq 1$. These are equivalent as we could add a risk-free asset with zero return, also called the *slack variable*.

Depending upon the choice of λ, this also represents the constrained mathematical program that maximizes expected return subject to a limit on risk:

$$\text{P: } \max_{x \in X} px : x'Vx \leq r$$

Reciprocally, one could view the mean-risk model as one of minimizing risk subject to a required expected return:

$$\text{Q: } \min_{x \in X} x'Vx : px \geq \mu$$

All three mathematical programs are related; for each λ, there is a corresponding r and μ for which $x^*(\lambda)$ is the optimal solution in P and Q, respectively.

THEOREM 14.1 *Define $X = \{x \in \mathbb{R}_+^n : \sum_{i=1}^n x_i \leq 1\}$, and let $x^* \in \text{argmax}\{px - \lambda x'Vx : x \in X\}$ for $\lambda > 0$, with $\mu = px^*$ and $r = x^{*'}Vx^*$. Then,*

$$x^* \in \text{argmax}\{px : x'Vx \leq r, x \in X\} \text{ and } x^* \in \text{argmin}\{x'Vx : px \geq \mu, x \in X\}$$

PROOF 14.1   We have

$$px^* - \lambda x^{*\prime}Vx^* \geq px - \lambda x'Vx \qquad \text{for all } x \in X$$
$$\Rightarrow px^* \geq px - \lambda(x'Vx - r) \geq px \qquad \text{for all } x \in X : x'Vx \leq r$$

Similarly,

$$px^* - \lambda x^{*\prime}Vx^* \geq px - \lambda x'Vx \qquad \text{for all } x \in X$$
$$\Rightarrow x^{*\prime}Vx^* \leq x'Vx + \tfrac{1}{\lambda}(\mu - px) \leq x'Vx \qquad \text{for all } x \in X : px \geq \mu \qquad \blacksquare$$

There is also a converse, due to the Lagrange multiplier rule and the fact that the quadratic risk function is convex (i.e., V is positive semi-definite).

THEOREM 14.2 *If x^* solves P or Q, there exists $\lambda \geq 0$ such that x^* solves the mean-risk model.*

PROOF 14.2   Suppose x^* solves P. Applying the Lagrange multiplier rule, there exist α, μ, $\mu_0 \geq 0$ such that

$$p - 2\alpha x^{*\prime}V + \mu - \mu_0 \mathbf{1} = 0, \quad \alpha x^{*\prime}Vx^* = \alpha r, \quad \mu x^* = 0, \quad \mu_0 \sum_{i=1}^{n} x_i^* = \mu_0$$

Defining $\lambda = 2\alpha$, these are the Karush–Kuhn–Tucker conditions for the optimality of x^* in the mean-risk model. A similar proof applies if we suppose x^* solves Q. $\blacksquare$

Example 14.1

Suppose we have three independent assets with $p = (3, 2, 1)$, plus a risk-free choice with zero return. Consider the following mean-risk model:

$$\max 3x_1 + 2x_2 + x_3 - \lambda(x_1^2 + x_2^2 + x_3^2) : x \geq 0, x_1 + x_2 + x_3 \leq 1$$

where $1 - (x_1 + x_2 + x_3)$ is the level of investment in the risk-free asset. Letting λ vary, the parametric solutions yield the efficient frontier, shown in Figure 14.2.

Setting $\lambda = 0$ ignores risk and gives the maximum expected return. This is achieved by investing all capital into asset 1 (i.e., $x_1 = 1$) because it has the greatest expected return. This remains true for $0 \leq \lambda \leq \tfrac{1}{2}$. As we increase λ past $\tfrac{1}{2}$, we penalize the risk of putting all capital into one asset and begin to invest into the next one, which is asset 2, having an expected return of 2. Splitting the investment between assets 1 and 2 gives a greater overall objective value:

$$px - \lambda x'Vx \quad \text{for } x = (x_1, 1 - x_1, 0) = 3x_1 + 2(1 - x_1) - \lambda(x_1^2 + (1 - x_1)^2)$$
$$= x_1 + 2 - \lambda(2x_1^2 - 2x_1 + 1)$$

which is maximized at $x_1 = \frac{1 + 2\lambda}{4\lambda}$.

For λ sufficiently great, splitting the risk is better than investing in just one asset because the square of x_1 outweighs the linear expected return. The greater the price of risk (λ), the

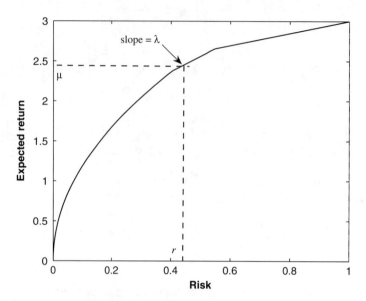

FIGURE 14.2 Efficient frontier of example portfolio model.

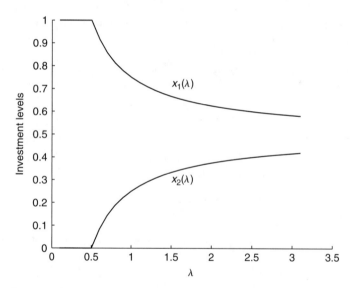

FIGURE 14.3 Splitting the investment to reduce risk for $\lambda > 0$.

greater the split. Figure 14.3 shows how the optimal split, $x_1(\lambda)$, $x_2(\lambda) = 1 - x_1(\lambda)$, varies as functions of λ ($x_2(\lambda) = 0$ for $0 \leq \lambda \leq \frac{1}{2}$).

At $\lambda = 1\frac{1}{2}$, asset 3 comes into play, and the optimality conditions are:

$$p_i - 2\lambda x_i - \frac{1}{3}\left(\sum_k p_k - 2\lambda\right) = 0 \quad \text{for } i = 1, 2, 3$$

Thus,

$$x_i = \frac{p_i - (2 + \frac{2}{3}\lambda)}{2\lambda}$$

TABLE 14.2 Optimal Asset Investments as Functions of λ

λ	x_1^*	x_2^*	x_3^*	px^*	$x^{*\prime}Vx^*$
$0 - \frac{1}{2}$	1	0	0	3	1
$\frac{1}{2} - 1\frac{1}{2}$	$\frac{1}{2} + \frac{1}{4\lambda}$	$\frac{1}{2} - \frac{1}{4\lambda}$	0	$2\frac{1}{2} + \frac{1}{4\lambda}$	$2\frac{1}{2} - \frac{1}{4}\lambda - \frac{1}{8\lambda}$
$1\frac{1}{2} - 3$	$\frac{1}{3} + \frac{1}{2\lambda}$	$\frac{1}{3}$	$\frac{1}{3} - \frac{1}{2\lambda}$	$2 + \frac{1}{\lambda}$	$\frac{1}{3} + \frac{1}{2\lambda^2}$
≥ 3	$\frac{3}{2\lambda}$	$\frac{1}{\lambda}$	$\frac{1}{2\lambda}$	$\frac{3}{\lambda}$	$\frac{7}{2\lambda^2}$

which remains valid for $1\frac{1}{2} \leq \lambda \leq 3$. At $\lambda > 3$, the price of risk is so great that $\sum_i x_i < 1$. The levels are $x(\lambda) = \frac{1}{2\lambda}(3, 2, 1)$—that is, the portion of investment in an asset is proportional to its expected return (Table 14.2).

If you draw a vertical line in Figure 14.2, the expected return is the maximum possible if the risk is limited to be on or to the left of that line. That is the solution to problem P. If you draw a horizontal line, the risk is the minimum possible if the expected return is required to be on or above that line. That is the solution to problem Q. So, we can generate the efficient frontier in Figure 14.2 by letting the price of risk, λ, vary between 0 and ∞. In practice, a decision maker would choose a point on the efficient frontier based on risk tolerance.

14.2.2 Recourse Model

The second mathematical programming model of uncertainty was Dantzig's recourse [9]. The two-stage recourse model has two vectors of decision variables: x must be specified before uncertain parameters become known, and y is chosen after the uncertain parameters become known. A classical example is to have a linear program (LP) with uncertain demand. The first-stage variables (x) specify production and process operation levels. Once the demand is known, the second-stage variables (y) take recourse in deciding what to do about any excess or shortage.

The two-stage recourse model is thus:

$$\min_{x \in X} \left\{ f(x) + \mathrm{E}\left[\min_{y \in Y(x)} \{\Phi(y; x)\} \right] \right\}$$

When the underlying mathematical program is linear, we have the form:

$$\min \{cx + \mathrm{E}[\min \{\Phi(y; x) : y = d - Dx\}] : Ax \geq b\}$$

where Φ penalizes over- and under-satisfaction of joint constraints, $Dx + y = d$.

When the number of possible realizations is finite, we call each realization a scenario and define a second-stage variable for each scenario. Let y^k denote the second-stage variable for scenario k. Then, the entire recourse model can be put into one large mathematical program:

$$\min cx + \sum_{k=1}^{K} s_k C^k |y^k| : Ax \geq b, y^k + D^k x = d^k, \quad \text{for } k = 1, \ldots, K$$

where s_k is the probability that scenario k prevails and K is the number of scenarios. Note that if there are additional constraints on y, such as $y^k \geq 0$ for all k, there could be no feasible solution. Otherwise, C^k is a linear penalty on the magnitude of y^k, and every x that satisfies $Ax \geq b$ is a feasible candidate since it is feasible to let $y^k = d^k - D^k x$ for all k.

To turn this into a linear program, we replace the absolute value with the difference of the positive and negative part: $y^k = u^k - v^k$. Then, the LP recourse model is given by:

$$\min cx + \sum_k s_k C^k (u^k + v^k) : Ax \geq b$$

$$y^k + D^k x = d^k, (u^k, v^k) \geq 0, u^k \geq y^k, v^k \geq -y^k, \quad \text{for } k = 1, \ldots, K$$

Every optimal solution has $u_i^k = \max\{0, y_i^k\}$ and $v_i^k = \max\{0, -y_i^k\}$, with $y^k = u^k - v^k$. Having partitioned each y_i^k, we could have different costs for slack ($u_i^k = y_i^k > 0$) and surplus ($u_i^k = -y_i^k > 0$).

As an example, consider the following transportation problem:

$$\min \sum_{i=1}^m \sum_{j=1}^n c_{ij} x_{ij} : x \geq 0$$

$$\sum_{j=1}^n x_{ij} \leq S_i \qquad \text{for } i = 1, \ldots, m$$

$$\sum_{i=1}^m x_{ij} \geq D_j \qquad \text{for } j = 1, \ldots, n$$

where x_{ij} is the amount shipped from source i to destination j.

Suppose demand is uncertain and there is recourse in the face of shortage or excess. Let f_j denote the unit cost of a shortage (demand exceeds the amount shipped) for demand market j, and let g_j denote the unit cost of excess (demand is less than the amount shipped). For example, f could be a spot market price or a cost of loss of good will, and g could be a salvage cost (net of the production cost). Then, $\Phi(y) = \sum_j \Phi_j(y)$, where

$$\Phi_j(y) = \begin{cases} f_j y & \text{if } y \geq 0 \\ -g_j y & \text{if } y \leq 0 \end{cases}$$

We extend the transportation model to include our recourse function by introducing for each market j in scenario k two additional non-negative variables, $\hat{y}_{jk}$ and $\check{y}_{jk}$, reflecting over-supply and unmet demand, respectively. Letting s_k be the probability of scenario k, we have $E[\Phi_j(y)] = \sum_k s_k (f_j \hat{y}_{jk} + g_j \check{y}_{jk})$, so the recourse model is

$$\min \sum_{i=1}^m \sum_{j=1}^n c_{ij} x_{ij} + \sum_k s_k (f_j \hat{y}_j^k + g_j \check{y}_j^k) : x \geq 0$$

$$\sum_{j=1}^n x_{ij} \leq S_i \qquad \text{for } i = 1, \ldots, m$$

$$\sum_{i=1}^m x_{ij} + \hat{y}_j^k - \check{y}_j^k = D_j^k \quad \text{for } j = 1, \ldots, n, k = 1, \ldots, K$$

This modeling approach extends to the *n-stage recourse model* in a natural fashion. Besides production, the finance community found applications in setting up portfolios of investments [10]. The stages correspond to time, and the uncertainty is with the asset prices. Each recourse variable adjusts for the decisions of the previous stages. Deficits in capital can be made up by borrowing and surpluses can be added to reserve capital.

14.2.3 Chance-Constrained Model

The third model to appear as a certainty equivalent was the *chance-constrained model*, by Charnes, Cooper, and Symonds [11]. Their motivating application was planning production levels in the face of uncertain sales. The general model has the form:

$$\max \ cx : x \in X, P(Dx \geq d) \geq \alpha$$

where $\alpha \in [0,1]$.

For the portfolio problem indicated by Table 14.1, consider the chance constraint: $P(R_1 x_1 + R_2 x_2 \geq \mu) \geq \alpha$, where μ is a specified return and α is an acceptable level of its probability.

Let s_k denote the probability of scenario k and apply Bayes' Law to obtain:

$$P(R_1 x_1 + R_2 x_2 \geq \mu) = \sum_{k=1}^{3} P\left((R_1^k - R_2^k) x_1 \geq \mu - R_2^k \,|\, k\right) s_k$$

where we substituted $x_2 = 1 - x_1$. We determine which scenarios (if any) satisfy the condition $(R_1^k - R_2^k) x_1 \geq \mu - R_2^k$. Let $\mathcal{K}(x)$ be this set of scenarios, so

$$P(R_1 x_1 + R_2 x_2 \geq \mu) = \sum_{k \in \mathcal{K}(x)} s_k$$

In particular, consider $\mu = 0$, so the chance constraint bounds the probability that the investment will not lose money. This chance constraint will hold for any $\alpha \in [0, 1]$ if we set $x_1 = 1$; it will hold for $\alpha \leq 0.8$ if we set $x_1 = 0$. To determine thresholds between these extremes consider the three scenarios shown in Table 14.3.

For $\mu = 0$, the chance constraint is the probability that there be no loss. In this case, the condition to include scenario 1 is $x_1 \geq 0.995$; scenarios 2 and 3 always hold because $x_1 \leq 1$.

Thus, we have:

$$P(R_1 x_1 + R_2 x_2 \geq 0) = \begin{cases} 1 & \text{if } x_1 \geq 0.995 \\ 0.8 & \text{otherwise} \end{cases}$$

The chance-constrained model is

$$\max \ px : x \in X, P(R(x) \geq \mu) \geq \alpha$$

In our example, $X = \{(x_1, x_2) \geq 0: x_1 + x_2 = 1\}$ and $R(x) = R_1 x_1 + R_2 x_2 = R_2 + (R_1 - R_2) x_1$. Hence, $px = 50{,}000 - 49{,}000 x_1$, and, for $\mu = 0$, our model is

$$\max -49{,}000 x_1 : 0 \leq x_1 \leq 1, x_1 \geq 0.995$$

The solution is $x_1^* = 0.995$. For $\mu > 1050$, the solution is $x^* = (0, 1)$ if $\alpha \leq 0.8$, and the problem is infeasible if $\alpha > 0.8$.

TABLE 14.3 Conditions to Compute Probability of Minimum Return

Scenario (k)	Condition for k Included in $\mathcal{K}(x)$
1	$x_1 \geq \dfrac{2{,}000{,}000 + \mu}{2{,}001{,}050}$
2	$x_1 \leq \dfrac{1{,}000{,}000 - \mu}{999{,}025}$
3	$x_1 \leq \dfrac{125{,}000 - \mu}{124{,}000}$

A class of models for which the chance-constrained paradigm is especially well suited is when a government regulation stipulates that some limit cannot be violated more than some specified portion of time. One example is sulfur emissions: $P(Dx \leq d) \geq \alpha$ requires that the probability that the sulfur emitted (Dx) is within a prescribed limit (d) and must be at least some specified level (α). Similar chance-constrained models for water quality control are described in Ref. [12].

14.3 Robust Optimization Models

The modeling paradigm called "robust optimization" emerged from dissatisfaction with limitations of the three classical models, which do not quite capture major effects of relatively small variations. Gupta and Rosenhead [13] first introduced the notion of "robustness" into an optimal solution, where they meant it to favor *flexibility* in what recourses are subsequently available. Given two policies, x^1 and x^2, with x^1 preferred on the basis of expected return, they considered the relative flexibility of how many and what recourses are available in each case. Unlike the mean-risk model, they need not have variances available. Unlike the recourse model, they need not have some estimated cost structure precisely defined. Unlike the chance-constrained model, they need not know any probability distributions. They can choose based on some qualitative information. For example, if x represents levels of production, they can prefer x^2 if it admits a greater number of recourse alternatives. The key in their definition of a robust solution is the flexibility it allows after the uncertain values become known.

A second view of robustness is the degree to which a solution is *sensitive* to the underlying assumptions (beyond data values, such as functional relations). Drawing from the concept of robust statistics, Mulvey et al. [14] introduced the concept of model and solution robustness using a penalty function much like the recourse model. (An example in statistics is the use of the median, rather than the mean, to estimate a parameter. The median is robust because it does not change with small fluctuations in the data on each side of the median, while the mean does.) Unlike the recourse model, the robust model could penalize risk as in the Markowitz model without requiring scenarios to be defined. This has evolved into a collection of robust optimization models, which we will now describe in a general way, followed by a section of specific applications that should help clarify how these general models may arise.

14.3.1 Worst-Case Hedge

A form of robustness is to choose a policy that does as well as possible under worst possible outcomes, regardless of the probability of occurrence. This view is that of a two-person game, where the choice of policy x is made first, then some "player" chooses an outcome c that is worst for that x. This is the *worst-case model*:

$$\min_{x \in X} \max_{c} cx : Dx \geq d \quad \text{for all } [D\ d]$$

To avoid having an infinite number of constraints, and to correlate the objective coefficients with the constraint data, this is modelled by discretizing the set of uncertain values into a finite set of scenarios:

$$\min_{z, x \in X} z : z \geq c^k x, D^k x \geq d^k \quad \text{for } k = 1, \ldots, K$$

The solution to this problem is sometimes called the *absolute robust decision* [15] (also see [16]). Note that z is introduced to equal $\max_k c^k x^*$ in every optimal solution, x^*.

Robust Optimization

This is pessimistic, and it may be unrealistic. The constraints require x to be feasible in every possible realization of $[D\ d]$, regardless of its probability of occurrence. The worst objective value (over possible c-values) may be appropriate when all constraints are deterministic, which we consider in the special case of interval uncertainty in the next section.

14.3.2 Simple Case of Interval Uncertainty

Consider the worst-case robust optimization model:

$$\min_{x \in X} \max_{c \in \mathcal{C}} cx$$

(Only the objective coefficients are uncertain.) Let $\hat{c}$ denote a *central value* in the box $[\underline{c}, \overline{c}]$, such as its expected, or most likely, value. Define a *central value solution*:

$$\hat{x} \in \operatorname{argmin}\{\hat{c}x : x \in X\}$$

In this section, we consider the restricted case where we have a box of uncertainty whose endpoints are proportional to the central value vector: $\underline{c} = (1-\varepsilon)\hat{c}$ and $\overline{c} = (1+\varepsilon)\hat{c}$ for some $\varepsilon \in [0,1)$. Let $\mathcal{P}(\hat{c}, \varepsilon) = \{c \in [\underline{c}, \overline{c}] : \sum_{j=1}^{n} c_j = K\}$, where $K = \sum_i \hat{c}_i$. We call the equation the *constant-sum constraint*.

Let $Q(\varepsilon)$ denote the robust optimization problem:

$$\min_{x \in X} \max_{c \in \mathcal{P}(\hat{c}, \varepsilon)} cx$$

where X is any subset of binary vectors. The following theorem demonstrates that the central solution also solves every robust formulation that allows percentage deviations within a constant proportion of its central value. The proof is based on the fact that the constant-sum constraint restricts the uncertain parameter to balance coefficients at their upper bounds with those at their lower bounds. Our method of proof is based on the duality theorem of linear programming.

THEOREM 14.3 *Let $\varepsilon \in [0,1)$. Then, x^* is an optimal solution to $Q(0)$ if, and only if, x^* is an optimal solution to $Q(\varepsilon)$.*

PROOF 14.3 We begin with some notation and general observations. Let $\sigma(x) = \{j : x_j \neq 0\}$ (called the "support set" of x). Also, let $\mathbf{1}$ denote the vector of all ones: $(1, 1, \ldots, 1)'$. The following identities follow from the definitions of K and $\sigma : \hat{c}x = \sum_{j \in \sigma(x)} \hat{c}_j = K - \sum_{j \notin \sigma(x)} \hat{c}_j$, and $\hat{c}(\mathbf{1}-x) = K - \hat{c}x = \sum_{j \notin \sigma(x)} \hat{c}_j$. Let $L = (1-\varepsilon)$ and $U = (1+\varepsilon)$. The dual of the linear program, $\max_{c \in \mathcal{P}(\hat{c}, \varepsilon)} cx$, is

$$\min \pi K + U\mu\hat{c} - L\lambda\hat{c} : \lambda, \mu \geq 0, \text{ and } \pi + \mu_j - \lambda_j = x_j \text{ for all } j = 1, \ldots, n$$

The dual variable π is associated with the constant-sum constraint, and λ, μ are associated with the lower and upper bound constraints on c, respectively.

Let x^ε be an optimal solution to $Q(\varepsilon)$. Our proof divides into two cases, depending on whether $\hat{c}x^0$ is greater or less than $\frac{1}{2}K$.

Case 1: $\hat{c}x^0 \geq \frac{1}{2}K$.

Consider the dual solution $\pi = 1$, $\mu = 0$, and $\lambda' = 1 - x^0$. This is dual-feasible, where $\lambda \geq 0$ because $x^0 \leq 1$. The dual objective value satisfies

$$\pi K + U\mu\hat{c} - L\lambda\hat{c} = K - L\hat{c}(1 - x^0) = K - L(K - \hat{c}x^0) = \varepsilon K + L\hat{c}x^0$$

Therefore,
$$\max_{c \in \mathcal{P}(\hat{c},\varepsilon)} cx^0 \leq \varepsilon K + L\hat{c}x^0 \tag{14.1}$$

Now we define $c_j^\varepsilon = L\hat{c}_j$ for $j \notin \sigma(x^\varepsilon)$. As we assume that $\hat{c}x^0 \geq \frac{1}{2}K$, it follows that $\hat{c}x^\varepsilon \geq \frac{1}{2}K$, which implies that $\hat{c}(1-x^\varepsilon) \leq \frac{1}{2}K$. Consequently, we have

$$c^\varepsilon x^\varepsilon = K - \sum_{j \notin \sigma(x^\varepsilon)} c_j^\varepsilon$$
$$= K - L \sum_{j \notin \sigma(x^\varepsilon)} \hat{c}_j$$
$$= K - L(K - \hat{c}x^\varepsilon) = \varepsilon K + L\hat{c}x^\varepsilon$$

which gives us the bound:
$$\max_{c \in \mathcal{P}(\hat{c},\varepsilon)} cx^\varepsilon \geq \varepsilon K + L\hat{c}x^\varepsilon \tag{14.2}$$

Using Equations 14.1 and 14.2, we then obtain the following chain of inequalities:
$$\max_{c \in \mathcal{P}(\hat{c},\varepsilon)} cx^\varepsilon \geq \varepsilon K + L\hat{c}x^\varepsilon \geq \varepsilon K + L\hat{c}x^0 \geq \max_{c \in \mathcal{P}(\hat{c},\varepsilon)} cx^0 \geq \max_{c \in \mathcal{P}(\hat{c},\varepsilon)} cx^\varepsilon$$

Thus, equality must hold throughout. This establishes the following two results:

$$\max_{c \in \mathcal{P}(\hat{c},\varepsilon)} cx^0 = \max_{c \in \mathcal{P}(\hat{c},\varepsilon)} cx^\varepsilon \quad \text{(first = last expression)}$$
$$\hat{c}x^0 = \hat{c}x^\varepsilon \quad \text{(second = third expression and } L > 0\text{)}$$

which completes this case.

Case 2: $\hat{c}x^0 \leq \frac{1}{2}K$.

The dual objective value of any dual-feasible solution is an upper bound on the primal value, cx^0. Choose $\pi = 0$, $\mu' = x^0$, and $\lambda = 0$. This is clearly dual-feasible, and its dual objective value is $U\hat{c}x^0$. Therefore,

$$\max_{c \in \mathcal{P}(\hat{c}\varepsilon)} cx^0 \leq U\hat{c}x^0 \tag{14.3}$$

Now consider the value of $\hat{c}x^\varepsilon$. Suppose $\hat{c}x^\varepsilon \leq \frac{1}{2}K$. Then define $c_j^\varepsilon = U\hat{c}_j$ for $j \in \sigma(x^\varepsilon)$, and note that $c^\varepsilon \in \mathcal{P}(\hat{c},\varepsilon)$. This is feasible (i.e., $c^\varepsilon \in \mathcal{P}(\hat{c},\varepsilon)$) because $c^\varepsilon x^\varepsilon \leq \frac{1}{2}K$. It follows that $c^\varepsilon x^\varepsilon = U\hat{c}x^\varepsilon$, so we have $\max_{c \in \mathcal{P}(\hat{c},\varepsilon)} cx^\varepsilon \geq U\hat{c}x^\varepsilon$. On the other hand, suppose $\hat{c}x^\varepsilon > \frac{1}{2}K$. Then, define $c_j^\varepsilon = L\hat{c}_j$ for $j \notin \sigma(x^\varepsilon)$, and note that $c^\varepsilon \in \mathcal{P}(\hat{c},\varepsilon)$. It follows from our analysis in Case 1 that $\max_{c \in \mathcal{P}(\hat{c},\varepsilon)} cx^\varepsilon \geq \varepsilon K + L\hat{c}x^\varepsilon$. Taken together, this gives us the bound:

$$\max_{c \in \mathcal{P}(\hat{c},\varepsilon)} cx^\varepsilon \geq \min\{U\hat{c}x^\varepsilon, \varepsilon K + L\hat{c}x^\varepsilon\} \tag{14.4}$$

Using Equations 14.3 and 14.4, we then obtain the following chain of inequalities:

$$\max_{c \in \mathcal{P}(\hat{c},\varepsilon)} cx^\varepsilon \geq \min\{U\hat{c}x^\varepsilon, \varepsilon K + L\hat{c}x^\varepsilon\} \geq \min\{U\hat{c}x^0, \varepsilon K + L\hat{c}x^0\}$$
$$= U\hat{c}x^0 \geq \max_{c \in \mathcal{P}(\hat{c},\varepsilon)} cx^0 \geq \max_{c \in \mathcal{P}(\hat{c},\varepsilon)} cx^\varepsilon$$

The equality in this chain follows from our assumption that $\hat{c}x^0 \leq \frac{1}{2}K$. We conclude that equality must hold throughout, and $\max_{c \in \mathcal{P}(\hat{c},\varepsilon)} cx^0 = \max_{c \in \mathcal{P}(\hat{c},\varepsilon)} cx^\varepsilon$. Furthermore, this shows that $U\hat{c}x^0 = \min\{U\hat{c}x^\varepsilon, \varepsilon K + L\hat{c}x^\varepsilon\}$ (fourth expression = second), so either

Robust Optimization

$U\hat{c}x^0 = U\hat{c}x^\varepsilon$ or $U\hat{c}x^0 = \varepsilon K + L\hat{c}x^\varepsilon$. In the former case, we have immediately that $\hat{c}x^0 = \hat{c}x^\varepsilon$. In the latter case, we have the following chain of inequalities:

$$U\hat{c}x^0 \leq \varepsilon K + L\hat{c}x^0 \leq \varepsilon K + L\hat{c}x^\varepsilon = U\hat{c}x^0$$

As equality must hold throughout, we conclude $\hat{c}x^0 = \hat{c}x^\varepsilon$. ∎

Example 14.2

Let $X = \{(1,0),(0,1),(1,1)\}$ and $\hat{c} = (1,2)$. The central solution is $\hat{x} = (1,0)$, and the worst-case model is

$$\min_{x \in X} \max_{c} c_1 x_1 + c_2 x_2 :$$
$$1 - \varepsilon \leq c_1 \leq 1 + \varepsilon$$
$$2 - 2\varepsilon \leq c_2 \leq 2 + 2\varepsilon$$
$$c_1 + c_2 = 3$$

For each x, the maximization problem is a linear program with one equality constraint. Therefore, an optimal c occurs at one of the four basic solutions. Of these, only two are feasible: $c = (1 \pm \varepsilon, 3 - c_1)$. The robust optimization problem is thus:

$$\min_{x \in X} \max\{(1+\varepsilon)x_1 + (2-\varepsilon)x_2, (1-\varepsilon)x_1 + (2+\varepsilon)x_2\}$$

We have

$$\max\{(1+\varepsilon)x_1 + (2-\varepsilon)x_2, (1-\varepsilon)x_1 + (2+\varepsilon)x_2\} = \hat{c}x + \max\{\varepsilon(x_1 - x_2), \varepsilon(x_2 - x_1)\}$$

The robust optimization model is thus equivalent to

$$\min_{x \in X} \{\hat{c}x + \varepsilon|x_1 - x_2|\}$$

The central solution yields $\hat{c}_1 + \varepsilon$, which dominates the value of $x = (0,1)$. The value of $x = (1,1)$ is $\hat{c}_1 + \hat{c}_2$, which is 3. This is greater than the central solution value for $\varepsilon < 1$. The relations that make this example result in the optimality of the central solution are used in the proof. In particular, if $c_2 > \hat{c}_2$, we must have $c_1 = (1 - \varepsilon)\hat{c}_1$, in which case $c_2 = K - c_1 = \hat{c}_2 - \varepsilon\hat{c}_1$. Thus, $cx = \hat{c}x - 2\varepsilon\hat{c}_1 x_2 \leq \hat{c}x$, so $c = \hat{c}$ is a better value than c for any x, whence $\hat{c}$ is a maximum of the rival "player," for any x.

Now consider the following variation of the portfolio problem. Instead of allowing capital to be split among assets in arbitrary fashion, let x_j be a binary variable that equals 1 if we select asset j (0 otherwise). Then, let $X = \{x \in \{0,1\}^n : ax \leq b\}$, where a_j is the amount of capital needed to invest in asset j and b is the total budget.

The interval uncertainty model of Theorem 14.3 assumes that the return is known to be within the bounds, $(1 - \varepsilon)\hat{c}$ to $(1 + \varepsilon)\hat{c}$ for some $\varepsilon \in [0,1)$, and that the sum of the returns is the same for all possible realizations. Theorem 14.3 tells us that the assets selected are those that solve the central value problem:

$$\max_{x \in X} \hat{c}x$$

which is a knapsack problem.

Our example in Table 14.1 does not satisfy these interval assumptions. However, if we keep the same expected returns, with $\hat{c} = (1000, 50000)$, the solution for $a_2 \leq b < a_1 + a_2$ is $x^* = (0,1)$, assuming the actual returns satisfy

$$(1-\varepsilon)(1000, 50000) \leq R \leq (1+\varepsilon)(1000, 50000)$$

for some $\varepsilon \in [0,1)$ and $R_1 + R_2 = 51000$.

The motivating application for Theorem 14.3 was the following sensor placement problem [17]. We are given a water distribution network that could be contaminated by some entry at a node. Sensors can be placed on the pipes (arcs) to detect the contaminant when it flows by, and we wish to minimize the expected number of nodes that the contaminant reaches undetected. Let x_{ij} denote a binary variable that is 1 if, and only if, node j is contaminated without detection, given the entry is at node i. (These are actually auxiliary variables in a model where the primary variables are where to place the sensors.) The problem is

$$\min_{x \in X} \sum_i \sum_{j \in R_i} \alpha_i x_{ij}$$

where $\mathcal{R}_i$ is the set of reachable nodes from node i.

The contaminant entry probabilities (α) are estimated using data and expert judgment, and they are subject to error. We know $\sum_i \alpha_i = 1$, and we assume $(1-\varepsilon)\hat{\alpha} \leq \alpha \leq (1+\varepsilon)\hat{\alpha}$, where $\hat{\alpha}$ is the central value (viz., the original estimate). The situation meets the assumptions of Theorem 14.3, so we can solve the worst-case robust optimization model by solving the central problem:

$$\min_{x \in X} \sum_i \sum_{j \in \mathcal{R}_i} \hat{\alpha}_i x_{ij}$$

Yaman et al. [18] proved a similar theorem for the minimum spanning tree problem, where they defined the central solution to be a *permanent solution*, as it remains optimal under perturbations of the data.

14.3.3 Minimax Regret

Another robust optimization model is to minimize the regret, which we now define. For each scenario (k), let z^k be the optimal objective value if the data were known with certainty—that is,

$$z^k = \max_{x \in X} \{c^k x : D^k x \geq d^k\}$$

(The problem could be minimization, but the regret value given below is the same—the greatest amount that a candidate solution deviates from z^k.)

For the robust deviation decision, define the objective value of $x \in X : D^k x \geq d^k$ to be the *maximum regret*:

$$f(x) = \max_k \{z^k - c^k x\}$$

For the *relative* robust decision, assume $z^k > 0$ for all k, and define the objective value of $x \in X : D^k x \geq d^k$ to be the maximum regret normalized by z^k:

$$f(x) = \max_k \frac{z^k - c^k x}{z^k} = 1 - \min_k \frac{c^k x}{z^k}$$

Then, the certainty equivalent models are, respectively:

$$\min_{x \in X} \max_k \{z^k - c^k x\} : D^k x \geq d^k \quad \text{for all } k = 1, \ldots, K$$

and

$$\max_{x \in X} \min_k \left\{ \frac{c^k x}{z^k} \right\} : D^k x \geq d^k \quad \text{for all } k = 1, \ldots, K$$

Robust Optimization

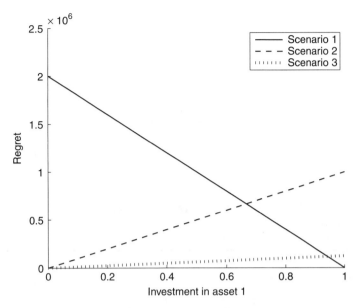

FIGURE 14.4 Regret functions for each scenario in example.

For the investment problem (Table 14.1), we have $z^1 = 1050$, $z^2 = 1{,}000{,}000$, and $z^3 = 125{,}000$. Figure 14.4 shows the regret functions, $z^k - R^k x$ (with $x_2 = 1 - x_1$) for $k = 1, 2, 3$. The minimum regret solution is $x^* = (\tfrac{2}{3}, \tfrac{1}{3})$.

As long as the solution is obtained for each scenario, it is useful to present how that solution fares in the other scenarios. This gives rise to a square array of the form:

Policy	1	2	$\cdots$	K
x^1	z^1	$c^2 x^1$	$\cdots$	$c^K x^1$
x^2	$c^1 x^2$	z^2	$\cdots$	$c^K x^2$
$\vdots$	$\vdots$		$\ddots$	$\vdots$
x^K	$c^1 x^K$	$c^2 x^K$	$\cdots$	z^K

The off-diagonal terms are well defined when the policy is feasible (for example, x^1 is feasible in scenario 2). Otherwise, the table could show the infeasibility, or there can be some recourse function to penalize the infeasibility consistently across scenarios.

One might add other information, such as the probability of each scenario, if available. One can then compare these solutions with the minimax robust solution for additional judgments about its quality. For the example, we have the array shown in Table 14.4.

The normalized regret problem is

$$\max_{0 \le x_1 \le 1} \min \left\{ \frac{1050 x_1 - 2{,}000{,}000(1-x_1)}{1050}, \frac{975 x_1 + 1{,}000{,}000(1-x_1)}{1{,}000{,}000}, \frac{1000 x_1 + 125{,}000(1-x_1)}{125{,}000} \right\}$$

The solution is $x^* = (0.9995, 0.0005)$.

TABLE 14.4 Scenario Solutions for Example

	Scenario				
Policy	1	2	3	E[x]	$x'Vx$
$x^1 = (1, 0)$	1050*	975	1000	1000	750
$x^2 = (0, 1)$	−2,000,000	1,000,000*	125,000	50,000	1,203,750,000,000
$x^3 = (0, 1)$	−2,000,000	1,000,000	125,000*	50,000	1,203,750,000,000
$x^* = (\frac{2}{3}, \frac{1}{3})$	−1,331,233	333,983	42,333	17,333	133,736,667,000

*Optimal.

TABLE 14.5 Uncertainty Sets for Example Investment Problem

ω	$\{R : |R_1 - 1000| + |R_2 - 50{,}000| \leq \omega\}$
<75000	$\emptyset$
[75000, 950025)	$\{(1000, 125000)\}$
[950025, 2050050)	$\{(1000, 125000), (975, 1000000)\}$
≥ 2050050	$\{(1000, 125000), (975, 1000000), (1050, -2000000)\}$

14.3.4 Uncertainty Sets

Ben-Tal and Nemirovski [2,19] introduced the model:

$$\max_{(z,x)} z : x \in X$$

$$z \leq c^k x, \quad y^k = \max\{0, d^k - D^k x\}, \quad \|y^k\| \leq r_k \quad \text{for } k = 1, \ldots, K$$

The key here is the representation of uncertainty by limiting the violation (y) by r. For $r_k = 0$, we require $y^k = 0$, which means we require that the constraint for scenario k hold: $D^k x \geq d^k$. Thus, this includes the worst-case model with all scenarios required to hold for x to be feasible. More generally, with $r > 0$, the feasible set is *ellipsoidal*, and it includes the Markowitz mean-risk model by defining the norm to be $\sqrt{y'Vy}$. Furthermore, the objective constraints can be merged with the others, so the uncertainty of c can be included in the uncertainty of $[D\ d]$.

We use the extended framework by Chen et al. [20]. Let $(\hat{D}, \hat{d})$ be some central (or reference) value, and define the *uncertain set* of data: $\mathcal{U}_\omega = \{(D, d) : \|(D, d) - (\hat{D}, \hat{d})\| \leq \omega\}$, where $\omega \geq 0$ is called the *budget of uncertainty*. Then, the feasible region is defined as

$$X(\omega) = \{x \in X : Dx \geq d \quad \text{for all } [D\ d] \in \mathcal{U}_\omega\}$$

An important observation by Chen et al. [20] is that under mild assumptions the set of x satisfying the chance constraint, $P(Dx \geq d) \geq \alpha$, contains $X(\omega)$ for some ω.

For the investment problem in Table 14.1, let the central value of the returns be their expected values. Then, using the absolute value norm, we have the uncertainty sets shown in Table 14.5.

This particular approach is among the most active areas of robust optimization research. See Ref. [21] for an overview and recent results in theory, algorithm design, and applications.

14.4 More Applications

In this section, we describe more applications and apply different robust optimization models to them.

Job Sequencing. A batch of n jobs arrives into a system and each must be processed on a machine. We can decide the order. Job j takes t_j minutes to be completed. If we sequence

Robust Optimization

the jobs as $1, 2, \ldots, n$, job 1 will be in the system t_1 minutes and job 2 will be in the system $t_1 + t_2$ minutes; in general, job j will be in the system $t_1 + \cdots + t_j$ minutes. The total job-time in the system is the sum:

$$T = \sum_{i=1}^{n}(n-i+1)t_i$$

If each job's time is known with certainty, an optimal sequence, one which minimizes total job-time T, is by descending order: $t_1 \le t_2 \le \cdots \le t_n$. This is called the shortest processing time (SPT) rule.

In general, a sequence is represented by a permutation, $\pi = (\pi_1, \ldots, \pi_n)$, so the total job-time in the system is

$$T(\pi) = \sum_{i=1}^{n}(n-i+1)t_{\pi_i}$$

Suppose the job-times are not known with certainty. One rule is to order them by their mean values. In particular, if t_i is uniformly distributed on $[a_i, b_i]$, the mean value is $m_i = \frac{1}{2}(a_i + b_i)$ and the variance is $\frac{1}{12}(b_i - a_i)^2$. To illustrate, consider four jobs with intervals [5, 50], [15, 45], [22, 40], and [31, 35]. Sequencing by their means, the expected total job-time is given by:

$$\mathrm{E}[T(\pi)] = 4m_1 + 3m_2 + 2m_3 + m_4 = 4(27.5) + 3(30) + 2(31) + 33 = 295$$

A worst-case robust solution sorts the jobs by their max times. That sequence is $\pi = (4, 3, 2, 1)$, and the expected total job-time is

$$\mathrm{E}[T(\pi)] = 4b_4 + 3b_3 + 2b_2 + b_1 = 4(35) + 3(40) + 2(45) + 50 = 400$$

Suppose the job-times are independent. Their variances are $\sigma^2 = \frac{1}{12}(45^2, 30^2, 18^2, 4^2) = (168\frac{3}{4}, 75, 27, 1\frac{1}{3})$. Thus, for $\pi = (1, 2, 3, 4)$,

$$V(T(\pi)) = 16(168\tfrac{3}{4}) + 9(75) + 4(27) + 1\tfrac{1}{3} = 3{,}484\tfrac{1}{3}$$

For $\pi = (4, 3, 2, 1)$, $V(T(\pi)) = 733.1$, which is much less.

More generally, Figure 14.5 shows the efficient frontier of $\mathrm{E}[T]$ versus $V(T)$; the mean-value ordering is at one end, and the max-value ordering is at the other.

Newsboy Problem. This is a classical problem in operations research, which formed the basis for the early inventory models. A newspaper is concerned with controlling the number of papers to be distributed to newsstands. The cost of a paper varies, and the demand is a random variable, d, with probability function $P(d)$. Unsold papers are returned, where millions of dollars are lost in the production cost. It is possible, however, for a newsstand to order more papers the same day. There are holding and shortage costs. The profit function for ordering Q papers with demand d is given by:

$$f(Q, d) = \begin{cases} (p-c)Q - g(d-Q) & \text{if } Q \le d \\ (p-c)d - (c-s)(Q-d) & \text{if } Q \ge d \end{cases}$$

where p = sales price, c = cost, g = shortage cost, and s = salvage value of unsold papers. ($s = 0$ in the original newsboy problem, but we include it here, as it appears in the more general extensions.)

The decision variable is Q, the amount ordered. An expected-value model seeks to maximize $\mathrm{E}[f(Q, d)] = \sum_d f(Q, d) P(d)$. If demand has a significant variance, the newsboy could miss opportunities for sales, or he could be stuck with a lot of rather worthless newspapers.

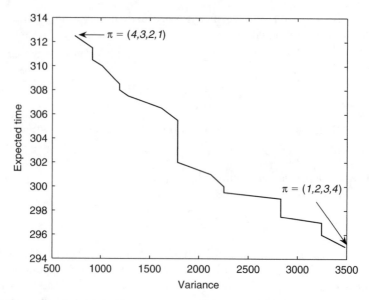

FIGURE 14.5 Efficient frontier for mean-variance model of scheduling example.

The worst-case robust optimization model seeks to maximize $\min_d f(Q,d)$. For known demand, the optimal ordering quantity is the demand and the maximum profit is $z(d) = (p-c)d$. Thus, the minimax regret model is:

$$\min_Q \max_d \{z(d) - f(Q,d)\}$$

For example, suppose d has only two values: $\underline{d} < \bar{d}$. Then, the worst-case robust optimization model is:

$$\max_{\underline{d} \leq Q \leq \bar{d}} F(Q)$$

where

$$\begin{aligned} F(Q) &= \min_d \{f(Q,d)\} \\ &= \min\{f(Q,\underline{d}), f(Q,\bar{d})\} \\ &= \min\{(p-c)\underline{d} - (c-s)(Q-\underline{d}), (p-c)Q - g(\bar{d}-Q)\} \\ &= \min\{(p-s)\underline{d} - (c-s)Q, (p-c+g)Q - g\bar{d}\} \end{aligned}$$

The solution occurs where $(p-s)\underline{d} - (c-s)Q = (p-c+g)Q - g\bar{d}$, which is at $Q^* = \frac{p-s}{p-s+g}\underline{d} + \frac{g}{p-s+g}\bar{d}$. This is independent of the cost of a newspaper (c) because both outcomes incur the cost cQ.

The absolute minimax regret model is given by:

$$\max\{z(\underline{d}) - f(Q,\underline{d}), z(\bar{d}) - f(Q,\bar{d})\} = \max\{(c-s)(Q-\underline{d}), g(\bar{d}-Q)\}$$
$$\Rightarrow Q^* = \frac{c-s}{c-s+g}\underline{d} + \frac{g}{c-s+g}\bar{d}$$

This says that the minimax regret ordering quantity is in the interval bounded by the two possible demands, where the weight on the lower bound is the cost minus salvage value and the weight on the upper bound is the shortage cost. Thus, if the salvage value or the

Robust Optimization

shortage cost is nearly zero, the minimax absolute regret solution is near the lower bound, accepting the possible shortage $(d - \underline{d})$. If the shortage cost dominates, the solution is near the upper bound, accepting the possible excess $(\bar{d} - d)$. If $c - s = g$, the solution is at the midpoint because the shortage and excess recourse costs are the same.

Assuming $z(d) > 0$ for all d, the normalized regret model is

$$\max_Q \min_d \frac{f(Q, d)}{z(d)}$$

For the two-demand case,

$$\min \left\{ \frac{f(Q, \underline{d})}{z(\underline{d})}, \frac{f(Q, \bar{d})}{z(\bar{d})} \right\} = \min \left\{ 1 - \frac{(c-s)(Q - \underline{d})}{(p-c)\underline{d}}, 1 - \frac{g(\bar{d} - Q)}{(p-c)\bar{d}} \right\}$$

$$\Rightarrow Q^* = \frac{\underline{d}\bar{d}(c - s + g)}{(c-s)\bar{d} + g\underline{d}}$$

This is also in the interval $[\underline{d}, \bar{d}]$, with coefficients

$$Q^* = \frac{(c-s)\bar{d}}{(c-s)\bar{d} + g\underline{d}} \underline{d} + \frac{g\underline{d}}{(c-s)\bar{d} + g\underline{d}} \bar{d}$$

Unlike the absolute regret solution, these coefficients depend upon the bounds. In particular, as $\bar{d} \to \infty$, $Q^* \to \underline{d}(1 + \frac{g}{c-s})$, whereas in the absolute regret solution, $Q^* \to \infty$ (assuming $g\underline{d} > 0$), as does the worst-case robust optimization solution.

Revenue Management [22]. A hotel can sell a room for a low price of $100, or it can hold the room in hope of a customer who will pay the regular price of $160. The regular demand is uncertain, but the hotel estimates that its demand will have one room available with probability 0.25; the probability of no surplus rooms is 0.75. The decision is whether to sell now at the lower price.

Table 14.6 shows the decision process and the outcome possibilities. The maximum expected revenue is $120 by holding the room. However, it has a high variance, compared to a zero variance for selling. A worst-case robust solution is to sell because we are guaranteed $100, while holding the room could result in zero revenue.

The maximum revenue for each possible outcome is $z^0 = 160$ (hold) and $z^1 = 100$ (sell). The minimax absolute regret solution is thus:

$$\min\{\underbrace{\max\{z^0 - 100, z^1 - 100\}}_{\text{Sell}}, \underbrace{\max\{z^0 - 160, z^1 - 0\}}_{\text{Hold}}\}$$

$$= \min\{60, 100\} = 60 \text{ (sell)}$$

The relative regret solution is:

$$\min\left\{\max\left\{\frac{z^0 - 100}{z^0}, \frac{z^1 - 100}{z^1}\right\}, \max\left\{\frac{z^0 - 160}{z^0}, \frac{z^1 - 0}{z^1}\right\}\right\}$$

$$= \min\{\tfrac{3}{8}, 1\} = \tfrac{3}{8} \text{ (sell)}$$

TABLE 14.6 Outcomes for Sell or Hold Decision

Decision	Demand 0	Surplus ≥ 1	Expected Revenue	Variance
Sell	100	100	100	0
Hold	160	0	120	4800
Probability	0.75	0.25		

In this example, the worst-case robust optimization model has the same decision as the two minimax regret models, which is to sell the room at the discount price. These run counter to the expected-value solution, but a risk term could be added, as in the Markowitz modeling paradigm, in which case there is a price of risk that renders the optimal solution to sell.

Transportation Problem. Another classical OR problem is a min-cost shipment from sources to destinations. Let x_{ij} = amount shipped from source i to destination j. The data are unit shipment costs (c_{ij}), supplies (s_i), and demands (d_j). Here is the standard model:

$$\min \sum_{i=1}^{m} \sum_{j=1}^{n} c_{ij} x_{ij} : x \geq 0$$

$$\sum_{j=1}^{n} x_{ij} \leq s_i \quad \text{(supply limit)}$$

$$\sum_{i=1}^{m} x_{ij} \geq d_j \quad \text{(demand requirement)}$$

When $m = n$ and $s_i = d_j = 1$ for all i, j, this is the *standard assignment problem*, and $x_{ij} = 1$ if person i is assigned to job j (otherwise, $x_{ij} = 0$).

Let us first assume that s and d are known, so the uncertainty is only in the costs. A worst-case robust optimization model has the form:

$$\min_{x \in X} \max_{c \in \mathcal{C}} \sum_{i=1}^{m} \sum_{j=1}^{n} c_{ij} x_{ij}$$

where $X = \{x \geq 0 : \sum_{j=1}^{n} x_{ij} \leq s_i, \sum_{i=1}^{m} x_{ij} \geq d_j\}$, and $\mathcal{C}$ is the set of possible cost matrices.

The assignment problem satisfies the conditions of Theorem 14.3 if there exists a central value ($\hat{c}$) and

$$\mathcal{C} = \left\{ c : (1-\varepsilon)\hat{c} \leq c \leq (1+\varepsilon)\hat{c}, \sum_{i,j} c_{ij} = \sum_{i,j} \hat{c}_{ij} \right\}$$

In that case, the central solution is all we need to solve this for all $\varepsilon \in (0, 1)$.

If this interval uncertainty is not the case, or if this is not the assignment problem, suppose there is a finite set of scenarios, indexed by $k = 1, \ldots, K$. The worst-case robust optimization model is given by:

$$\min z : x \in X, z \geq \sum_{i=1}^{m} \sum_{j=1}^{n} c_{ij}^{k} x_{ij} \quad \text{for } k = 1, \ldots, K$$

Define the minimum cost for each scenario:

$$z^k = \min_{x \in X} \sum_{i=1}^{m} \sum_{j=1}^{n} c_{ij}^{k} x_{ij}$$

The absolute minimax regret model is the linear program:

$$\min z : x \in X, z \geq \sum_{i=1}^{m} \sum_{j=1}^{n} c_{ij}^{k} x_{ij} - z^k \quad \text{for } k = 1, \ldots, K$$

Robust Optimization

The relative minimax regret model is also a linear program:

$$\min z : x \in X, z \geq \sum_{i=1}^{m}\sum_{j=1}^{n} c_{ij}^k x_{ij} - z^k \quad \text{for } k = 1, \ldots, K$$

Now consider uncertain supplies or demands. The recourse model puts additional variables associated with each scenario:

$$\min \sum_{i,j} \mathrm{E}[c_{ij}] x_{ij} + \Phi(u,v) :$$

$$\sum_j x_{ij} - u_i^k = s_i^k, \; \forall i$$

$$\sum_i x_{ij} - v_j^k = d_j^k, \; \forall j$$

$$\text{for } k = 1, \ldots, K$$

where Φ penalizes supply shortage ($u_i^k > 0$) and demand shortage ($v_j^k > 0$).

The robust optimization model introduced by Mulvey et al. [14] includes penalty functions akin to the Markowitz mean-variance model:

$$\min \sum_{i,j} \mathrm{E}[c_{ij}] x_{ij} + \sum_k \lambda_k \left(\sum_i U_{ik}^2 + \sum_j V_{jk}^2 \right) : x \geq 0$$

$$\sum_j x_{ij} - u_i^k = s_i^k, U_{ik} \geq 0, U_{ik} \geq u_i^k \; \forall i$$

$$\sum_i x_{ij} + v_j^k = d_j^k, V_{jk} \geq 0, V_{jk} \geq v_j^k \; \forall j$$

$$\text{for } k = 1, \ldots, K$$

where $\lambda > 0$ is a vector of parameters that represents a cost of shortage. The penalty costs can vary by scenario, which could include the probability of the scenario's occurrence. If $\sum_i x_{ij} \leq s_i^k, U_{ik} = 0$ is feasible, so there is no penalty. Similarly, if $\sum_j x_{ij} \geq d_j^k, V_{jk} = 0$ is feasible, so there is no penalty. If either $U_{ik} > 0$ for some i or $V_{jk} > 0$ for some j, the penalty is λ_k times the sum-squares of those shortages. If we follow the Markowitz modeling paradigm, we have only one risk parameter, with $\lambda_k = \lambda$ for all k.

Example 14.3

Figure 14.6 shows a 2×3 transportation network with supply and demand data for two scenarios.

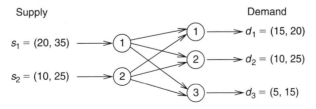

FIGURE 14.6 Transportation network example.

TABLE 14.7 Optimal Flows for Transportation Example

Policy	Flow			Cost	Shortage
Scenario 1	5	10	5	365	0
	10	0	0		
Scenario 2	0	20	15	720	0
	20	5	0		
Recourse	0	20	15	750*	15*
	20	5	0		

*Expected value.

The unit costs are given by:

$$c^1 = \begin{bmatrix} 10 & 12 & 15 \\ 12 & 15 & 18 \end{bmatrix} \quad \text{and} \quad c^2 = \begin{bmatrix} 15 & 12 & 10 \\ 13 & 14 & 15 \end{bmatrix}$$

The optimal solution to each scenario is shown in Table 14.7, along with a recourse solution that requires demand feasibility in each scenario.

The minimax regret models are infeasible because they require a flow to satisfy supply and demand constraints for each scenario. This means x is restricted by:

$$\sum_j x_{ij} \leq \min_k s_i^k, \quad \forall i$$

$$\sum_i x_{ij} \geq \max_k d_j^k, \quad \forall j$$

This is infeasible whenever $\sum_i \min_k s_i^k < \sum_j \max_k d_j^k$, as in the example.

We consider a robust optimization model that penalizes shortages but does not require supply and demand feasibility in all scenarios. Further, we use a common risk penalty parameter (λ). For this problem, we thus have the robust optimization model:

$$\min \text{E}[c]x + \lambda \sum_k \left(\sum_i u_i^k + \sum_j v_j^k \right):$$

$$\sum_j x_{ij} - u_i^k \leq s_i^k, \forall i, k$$

$$\sum_i x_{ij} + v_j^k \geq d_j^k, \forall j, k$$

$$x, u, v \geq 0$$

With $\lambda = 0$ and the added constraints $u_i^k = 0$ for all i, k, we obtain the recourse model solution shown in Table 14.7. Here we allow shortages to occur at a supply node ($u_i^k > 0$) or at a demand node ($v_j^k > 0$); the penalty is the same, namely λ for each unit. Figure 14.7 shows the expected cost versus shortage as λ varies.

The curves are mostly flat, except that there is a sharp ramp from $\lambda = 5.8$ to $\lambda = 6.25$. Table 14.8 shows the optimal flows; note the rise in x_{12}, then suddenly to the flows that comprise the optimal solution for scenario 1.

Robust Optimization

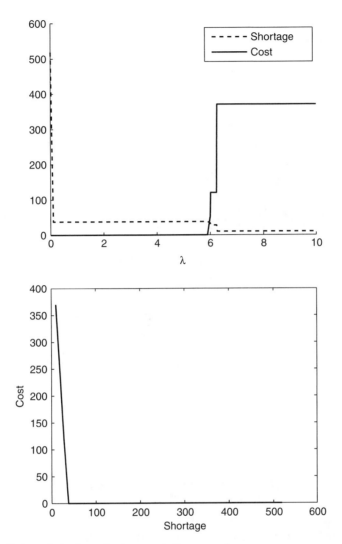

FIGURE 14.7 Costs and shortages for transportation example.

TABLE 14.8 Optimal Flows as Functions of λ

λ	x^*
0–5.9	$\begin{bmatrix} 0 & 0 & 0 \\ 0 & 0 & 0 \end{bmatrix}$
6	$\begin{bmatrix} 0 & 4.4 & 0 \\ 0 & 0 & 0 \end{bmatrix}$
6.02–6.24	$\begin{bmatrix} 0 & 10 & 0 \\ 0 & 0 & 0 \end{bmatrix}$
≥ 6.29	$\begin{bmatrix} 5 & 10 & 5 \\ 10 & 0 & 0 \end{bmatrix}$

As an alternative, consider an uncertainty-set robust optimization model, where the supplies and demands are correlated to be feasible:

$$\mathcal{D} = \left\{ (s,d) \geq 0 : \sum_i s_i \geq \sum_j d_j \right\}$$

In particular, we could have a central value and maintain constant-sum supplies and constant-sum demands, allowing individual supplies and demands to vary independently within the two degrees of freedom lost for the constant-sum equations. Then,

$$\mathcal{U}_\omega = \{ (s,d) \in \mathcal{D} : \|(s,d) - (\hat{s}, \hat{d})\| \leq \omega \}$$

The robust optimization model is

$$\min \sum_{i,j} c_{ij} x_{ij} : x \geq 0, \left\{ \sum_j x_{ij} \leq s_i, \sum_i x_{ij} \geq d_j \right\} \forall (s,d) \in \mathcal{U}_w$$

The robust models for cost uncertainty can be included as well. The net result is that we can find a shipment (or assignment) that is feasible in all *admissible* scenarios (i.e., in $\mathcal{U}_\omega$), and for which we minimize the maximum cost (or maximum regret).

A problem with this uncertainty-set model is the likelihood of infeasibility. As in the example, we could wind up with the impossibility:

$$\min_{(s,d)\in \mathcal{U}_\omega} \left(\sum_i s_i - \sum_j d_j \right) < 0$$

If the central value has surplus supply (i.e., $\sum_i \hat{s}_i > \sum_j \hat{d}_j$), this infeasibility can be mitigated by choosing ω sufficiently small. This is appropriate when the robustness sought is with respect to insensitivity to small perturbations in the data.

The standard transportation problem can be modified in several ways. One is to suppose the network is *sparse*—not every supplier can ship to every consumer. This is realistic and makes the deterministic problem much more tractable for very large networks. However, the topology of the network could then become a source of uncertainty, which renders the robust optimization models difficult to solve. Another extension is to add uncertain bounds on the flows: $x \leq U$, where U is the *capacity* of arc $<i,j>$. If major variations are allowed, the bounds could control the topology by having $U_{ij} = 0$ mean that arc x_{ij} is absent. There could also be *gains* or *losses* during shipment, and those values could be uncertain. Again, if large variations are possible, a 100% loss value essentially removes the arc from the network. In general, robust optimization models can deal effectively with insensitivity to small fluctuations in the data, but allowing the topology to change is a hard problem.

Capacity Expansion. Consider the following power system problem [23,24]. Electricity can be generated by different types of plants, like coal-fired boilers, oil-fired turbines, and nuclear power plants. During a cycle, like 24 hours, the demand for electricity varies, and that *load* is sorted into the *load–duration* curve, as shown in Figure 14.8. The curve is approximated by three steps, called *modes*: (1) *base*, which is ongoing demand for electricity throughout the region being serviced; (2) *peak*, which is the greatest amount, but for a short duration, usually during mid-day for temperature control in business and residence; and (3) *intermediate*, which is between the base and peak loads.

Robust Optimization

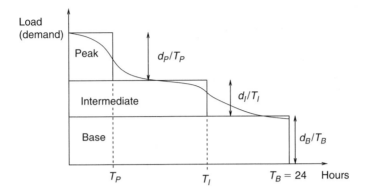

FIGURE 14.8 Load–duration curve for example electric power system.

The approximation allows us to model electricity generation and demand satisfaction as a linear program:

$$\min \sum_i f_i x_i + \sum_j T_j \sum_i c_{ij} y_{ij} : x, y \geq 0$$

$$\sum_j y_{ij} - x_i \leq 0 \quad \text{(capacity limit)}$$

$$T_j \sum_i y_{ij} \geq d_j \quad \text{(demand)}$$

where x_i is the amount of capacity of type i that is made available (f_i = annualized capital cost), and y_{ij} is the level of operation of plant i to fulfill demand during mode j (c_{ij} is its unit operating cost). The demand, d_j, is by mode (base, intermediate, and peak).

The capacity decisions (x) must be made before the demands are known. Then, y must operate within those capacity limits, which could result in a shortage (where d_j is greater than expected) or excess capacity (where d_j is less than expected).

One robust optimization model is to prepare for the worst case—that is, the greatest demand in each mode. This is likely to be overly pessimistic because demands might shift, rather than move in parallel across modes, in which case we would inevitably incur excess capacity.

Here is a scenario-based robust optimization model:

$$\min fx + E[\zeta] + \lambda_1 E[(\zeta - E[\zeta])^2] + \lambda_2 E[\|\kappa^k\|^2] + \lambda_3 E[\|\delta^k\|^2]$$

$$\kappa_i^k \geq x_i - \sum_j y_{ij}^k \quad \forall i, k \quad \text{(excess capacity)}$$

$$\delta_j^k \geq d_j^k - T_j^k \sum_i y_{ij}^k \quad \forall j, k \quad \text{(shortage)}$$

$$x, y, \kappa, \delta \geq 0,$$

where κ^k is the excess capacity vector and δ^k is the shortage vector for scenario k. (Note: $\kappa_i^k = \max\{0, x_i - \sum_j y_{ij}^k\}$ and $\delta_j^k = \max\{0, d_j^k - T_j^k \sum_i y_{ji}^k\}$.) The random variable, ζ, is defined over scenarios:

$$P\left(\zeta = \sum_j T_j^k \sum_i c_{ij} y_{ij}^k\right) = s_k$$

The multipliers are
λ_1 = usual price of risk, as in the Markowitz mean-risk model
λ_2 = price of having excess capacity
λ_3 = price of having shortage.

To illustrate, Table 14.9 gives data for four types of plants (taken from Refs. [16] and [21]), and Table 14.10 gives the demands for each of four scenarios.

Now consider the regret models. We have

$$z^k = \min \sum_i f_i x_i + \sum_j T_j \sum_i c_{ij} y_{ij} : x, y \geq 0$$

$$\sum_j y_{ij} - x_i \leq 0 \quad \text{(capacity limit)}$$

$$T_j \sum_i y_{ij} \geq d_j^k \quad \text{(demand)}$$

(Note: T and c are the same for all scenarios.)

Table 14.11 shows the table introduced in Section 14.3.3. The diagonal values are the scenario optima, $z^1, \ldots, z^4$. The off-diagonal entries reflect the cost of (x^p, y^{kp}), the best possible response to scenario k given that the capacity decisions (x^p) were fixed assuming scenario (p). We assume each scenario is equally-likely, so $s_k = \frac{1}{4}, \forall k$.

Table 14.12 shows the optimal capacities for each of the seven policies—$x^1, \ldots, x^4$ are the scenario optima; x^A, x^R are the absolute and relative minimax regret solutions, and x^E is the expected-cost optimum.

TABLE 14.9 Supply Options and Associated Costs for Example

Plant	Annualized Capital Cost	Operating Cost (same for all modes)
A	200	30
B	500	10
C	300	20
D	0	200

TABLE 14.10 Demand Loads and Durations for Scenarios

	Base		Intermediate		Peak	
Scenario	Load	Duration	Load	Duration	Load	Duration
1	8	24	6	12	4	6
2	8.67	24	7	10	1.33	5
3	9	24	7.33	10	1.67	4
4	8.25	24	5	12	2.5	4

TABLE 14.11 Scenario Optima over Scenarios

	Scenario				Expected
Policy	1	2	3	4	Cost
1	10,680.0*	10,608.1	10,899.4	10,110.0	10,574.38
2	11,620.4	10,381.3*	11,398.8	9,980.2	10,855.13
3	10,860.0	10,614.3	10,859.4*	10,200.0	10,633.43
4	12,560.0	11,210.6	12,561.0	9,605.0*	11,484.15
E[Cost]	10,715.0	10,898.1	10,889.4	10,055.0	10,055.00*
Minimax	Regret Solutions				
Absolute	11,048.6	10,749.9	11,228.0	9,973.6	10,750.05
Relative	11,081.1	10,771.2	11,267.2	9,965.7	10,771.30

*Optimal.

TABLE 14.12 Optimal Capacities

	Plant			
Policy	A	B	C	D
x^1	4.00	8.00	6.00	0
x^2	8.33	8.67	0.00	1.00
x^3	9.00	9.00	0.00	0
x^4	2.50	8.25	5.00	2.25
x^A	4.34	8.25	5.00	0.41
x^R	4.30	8.25	5.00	0.45
x^E	4.75	8.25	5.00	0

The absolute regret solution uses slightly more capacity of plant A, thus reducing the amount of plant D needed to fulfill demand; otherwise, the two regret solutions are essentially the same. The expected value solution is constrained to satisfy demand for each scenario, and it does so without plant D. The difference with the regret solutions is the amount of plant A capacity made available.

The total excess capacity in each case is 0, 1, 1.25, or 2.25, depending on the scenario and policy. The expected value solution has 2.25 excess capacity of plant A in scenario 1, while the regret solutions have only 2.05. All three solutions have no excess capacity in the other plants.

Facility Location. A standard facility location problem is to choose the coordinates of a point that minimizes the total (weighted) distance to each of a set of given points. The points could represent population centers in a region and the weights are the populations. The facility is one of service, like a hospital, and the objective is to provide an overall best service by being centrally located. Here is a basic model:

$$\min_{x,y} \sum_{i=1}^{m} w_i d_i(x,y)$$

where $w \geq 0$ and d_i is the distance from (x_i, y_i) to (x, y). The distance is typically given by some norm: $d_i(x,y) = \|(x_i, y_i) - (x,y)\|$.

Suppose the problem is unweighted and we use the city-block distance (L_1 norm):

$$\min_{x,y} \sum_{i=1}^{m} (|x_i - x| + |y_i - y|)$$

First, we note that we can optimize each coordinate separately. Second, the minimum L_1 placement is the median, given as follows. Let π be a permutation of the x values such that $x_{\pi_1} \leq x_{\pi_2} \leq \cdots \leq x_{\pi_m}$. If m is odd, the median is the mid-point, $x_{\pi_{\frac{m+1}{2}}}$. Thus, 5 is the median of 1,2,5,9,50. If m is even, the median can be any point in the interval such that half the points are less and half are greater. Thus, any value in the interval (5, 9) is a median of 1,2,5,9,50,90. We do the same thing to compute the median of $\{y_i\}$ to obtain the median coordinate in the plane. We leave as an exercise that a median is the location problem solution when using the L_1 norm.

Now consider the Euclidean norm (L_2). The location problem is given by:

$$\min_{x,y} \sum_{i=1}^{m} \sqrt{(x_i - x)^2 + (y_i - y)^2}$$

The solution to this problem is the mean: $(x^*, y^*) = \frac{1}{m} \sum_{i=1}^{m} (x_i, y_i)$. Compared to the median, this is very sensitive to the given coordinates, especially to outliers—a few points far from the others. For that reason, the median is a more robust solution than the mean.

The family of norms, $L_p(v) = (\sum_j |v_j|^p)^{\frac{1}{p}}$, approaches $L_\infty = \max_j\{|v_j|\}$. The location problem for the L_p norm is given by:

$$\min_{x,y} \sum_{i=1}^{m} w_i(|x_i - x|^p + |y_i - y|^p)^{\frac{1}{p}}$$

and as $p \to \infty$

$$\min_{x,y} \sum_{i=1}^{m} w_i \max\{|x_i - x|, |y_i - y|\}$$

In statistical estimation, L_p is more robust than L_{p+1} because it is less sensitive to the data (both the weights and the given coordinates).

However, a robust optimization model does not sum the distances at all. It defines the problem as:

$$\min_{x,y} \max_i \{w_i d_i(x,y)\}$$

We call this solution the center of the convex hull of the given points. It is insensitive to small changes in the coordinates or weights of non-extreme points, and in that sense the center is a robust solution. While there are two concepts of robustness, median vs. mean and center vs. L_∞, both pertain to insensitivity to the given data, but they have different kinds of sensitivities. The median, which is optimal with summing the L_1 norms, is insensitive to perturbations of all data values; it depends upon only their relative values. The center depends upon the data of the outliers, which are extreme points of the convex hull. (The converse is not true—not all extreme points are outliers.)

Figure 14.9 illustrates this for 9 points. Table 14.13 shows the objective values for each of the three location points.

Engineering Design. The engineering design problem specifies outputs (y) as a function of inputs that consist of decision variables (x) and (non-controllable) parameters (p):

$$y = F(x, p)$$

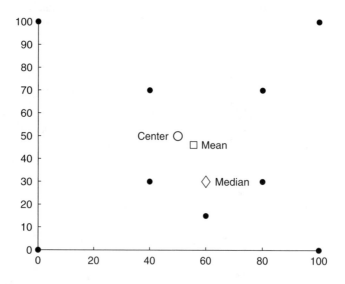

FIGURE 14.9 Median, mean, and center are respective solutions for facility location using $\sum L_1, \sum L_2$, and max L_1 objectives.

Robust Optimization

TABLE 14.13 Median, Mean, and Center Distances Using $\sum L_1, \sum L_2$, and max L_1 Objectives

	x	y	$\sum L_1$	$\sum L_2$	Max L_1
Median	60	30	575*	164.39	130
Mean	55.56	46.11	595.56	156.56*	109.44
Center	50	50	605	157.88	100*

*Optimal.

TABLE 14.14 Variations in Output Current for $(R, L) \in \mathcal{R} \times \mathcal{L}(y^{\min}, y^{\max})L$

R	L	$y^{\min}$	$y^{\max}$	$\max\{|y^{\min} - 10|, |y^{\max} - 10|\}$	$\bar{y}$	s	$|\bar{y} - 10|$
1	1	21.5	38.4	28.4	29.95	11.95	19.95
1	2	10.8	19.4	9.4	15.10	6.08	5.10
1	3	7.2	13.0	3.0	10.10	4.38	0.10
2	1	13.1	20.7	10.7	16.90	5.37	6.90
2	2	9.0	15.2	5.2	12.10	4.38	2.10
2	3	6.6	11.5	3.4	9.05	3.46	0.95
3	1	8.0	12.2	2.2	10.10	2.97	0.10
3	2	6.8	10.7	3.2	8.75	2.76	1.25
3	3	5.5	9.1	4.5	7.30	2.54	2.70

Some parameters are random variables. If the actual value of a decision variable is random, the optimization model uses the mean of its distribution as the decision variable and accounts for its variation in some way.

There is a target value, denoted T, and a robust design is one that chooses x (or its mean) to minimize some combination of the distance between y and the target, $\mathrm{E}\left[\|y - T\|\right]$, and variation around that, typically with a standard deviation risk metric. The model is similar to the Markowitz mean-risk model, but not quite the same. We illustrate with some examples.

Electric circuit example [25,26]. We want to build an AC circuit with output current given by

$$y = \sqrt{\frac{V}{R^2 + (2\pi f L)^2}}$$

where V is the input voltage, R is the resistance, f is the frequency, and L is the self-inductance. Resistance and self-inductance are decision variables, restricted to the sets $\mathcal{R} = \{1\,\text{ohm}, 2\,\text{ohms}, 3\,\text{ohms}\}$ and $\mathcal{L} = \{1\,\text{henry}, 2\,\text{henries}, 3\,\text{henries}\}$, respectively. Voltage and frequency are not decision variables, but they have some noise—that is, V and f are random variables with known means and variances.

For each $(R, L) \in \mathcal{R} \times \mathcal{L}$, there are observed deviations in y with $y^{\min}, y^{\max}$ being the minimum and maximum noise values, shown in Table 14.14. The $\bar{y}$ is the observed mean value of y, and s is its observed standard deviation. The last column is the distance of the mean from the target of 10 amps.

The design parameters that minimize the distance from the target is $(R, L) = (1, 3), (3, 1)$. The former has a deviation of 4.38, while the latter has a deviation of 2.97. Thus, the latter would be preferred. However, the overall min-deviation solution is with $(R, L) = (3, 3)$, though its distance from the target is 2.70. One worst-case robust optimization model is given by:

$$\min_{\substack{R \in \mathcal{R} \\ L \in \mathcal{L}}} \max\{|y^{\min}(R, L) - 10|, |y^{\max}(R, L) - 10|\}$$

The solution to this model is $(R, L) = (3, 1)$.

TABLE 14.15 Design Variables and Parameters Data

Variable/Parameter	Mean	Standard Deviation	Range
FPR	μ_{FPR}	0.1	$1.25 \leq \mu_{\text{FPR}} \leq 1.6$
JVR	μ_{JVR}	100	$0.6 \leq \mu_{\text{JVR}} \leq 0.9$
CET	μ_{CET}	1.5	$2400 \leq \mu_{\text{CET}} \leq 4000$
HCPR	μ_{HCPR}	0.5	$10.2 \leq \mu_{\text{HCPR}} \leq 25$
LCPR	μ_{LCPR}	0.1	$1.15 \leq \mu_{\text{LCPR}} \leq 4.9$
htce	0.891	0.0033	n/a
hte	0.933	0.0033	n/a
ltc	0.9	0.0033	n/a

Using the extreme values $(y^{\min}, y^{\max})$ is overly conservative, as they are obtained by a large number of experiments. Instead, consider a mean-risk model, where two objectives are to minimize the deviation from the target, $|\bar{y}(R, L) - 10|$, and the standard deviation about the mean, $s(R, L)$. We define a solution (R, L) to be robust if there does not exist another solution (R', L') for which $|\bar{y}(R', L') - 10| \leq |\bar{y}(R, L) - 10|$ and $s(R', L') \leq s(R, L)$. Here we have three robust solutions: (1, 3), (3, 2), and (3, 3).

Engine design example [27]. We have five design variables: fan pressure ratio (FPR), exhaust jet velocity ratio (JVR), combustor exit temperature (CET), high compressor pressure ratio (HCPR), and low compressor pressure ratio (LCPR). In addition, there are three parameters: high turbine compressor efficiency (htce), high turbine efficiency (hte), and low turbine efficiency (lte). The design variables and the parameters are each normally distributed with given standard deviations, shown in Table 14.15.

The aircraft's range is a nonlinear function of the design variables and parameters. The robust optimization model has the form of Markowitz's mean-risk model, except that the design engineer uses standard deviation instead of variance:

$$\max \mu_{\text{Range}} - \lambda \sigma_{\text{Range}}$$

where $\lambda > 0$ is chosen to reflect the relative importance of robustness.

Figure 14.10 gives four solutions. One is the expected-value solution ($\lambda = 0$), one is the worst-case robust solution ($\lambda = \infty$), and two are in between. The maximum expected range has a chance of having the aircraft's range be less than that of the worst-case robust solution. The optimal robust solution results with $\mu_{\text{Range}} = 2200.52$ and $\sigma_{\text{Range}} = 54.1189$, so $\mu_{\text{Range}} \pm 3\sigma_{\text{Range}} = [2038.1658, 2362.8792]$. The optimal expected range results with $\mu_{\text{Range}} = 2671.1454$ and $\sigma_{\text{Range}} = 225.7006$, so $\mu_{\text{Range}} \pm 3\sigma_{\text{Range}} = [1994.0436, 3348.2472]$. Which solution is best (among the four) depends on what value we place on the range. If being less than 2050 has dire consequences, the robust solution is preferred over the expected-value solution.

14.5 Summary

We have seen that robust optimization can be a worst-case model, minimax regret, or a model of uncertainty that simply restricts the solution space in a way that avoids "too much" deviation from some central value. Collectively, these capture much of the essence of the classical models within their scope, except that robust optimization avoids the explicit need for probability distribution information within the model. This difference has been key, but current research is building a more inclusive framework, addressing the desire to use whatever information is available.

In this view, robust optimization falls firmly in the tradition of stochastic programming paradigms. Historically, robust optimization has emerged from the three classical paradigms, as shown in Figure 14.11. The diagram represents fundamental concepts, like risk, flexibility,

Robust Optimization

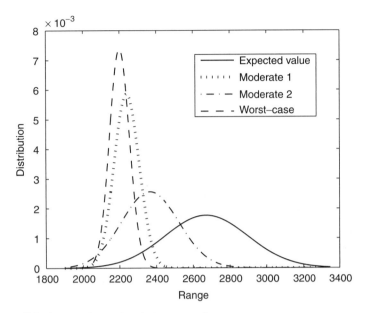

FIGURE 14.10 Solutions to the engine design example.

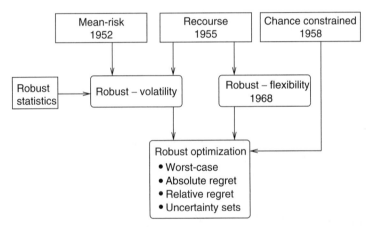

FIGURE 14.11 Historical perspective of stochastic programming modeling paradigms.

recourse, and the meaning of constraining for robustness. An algorithmic view would add divisions by underlying structures, like continuous versus discrete.

The key elements of robust optimization are volatility and flexibility. The former asks for a solution that is relatively insensitive to data variations and hedges against catastrophic outcomes. The latter is concerned with keeping options open in a sequential decision process having recourses for the effects of earlier decisions.

An area of concern, which we did not have space to present fully, is the computational difficulty to solve instances of a robust optimization model. Kouvelis and Yu [15] address this with complexity theory; the uncertainty set model by Ben-Tall and Nemirovski [2,19] exploits the tractability of semi-definite conic programming; and, Carr et al. [17] show a progression of incremental computational difficulty for a particular sensor placement problem, starting with the simple case expressed in Theorem 14.3.

Acknowledgments

We thank colleagues who reviewed an earlier version of this chapter: Ignacio Grossman, Istvan Maros, Frederic H. Murphy, Bora Tarhan, and H. Paul Williams.

References

1. J. Rosenhead. Robustness analysis: Keeping your options open. In J. Rosenhead (ed.), *Rational Analysis for a Problematic World: Problem Structuring Methods for Complexity, Uncertainty and Conflict*, pages 193–218, John Wiley & Sons, Chichester, UK, 1989.
2. A. Ben-Tal and A. Nemirovski. Robust optimization—methodology and applications. *Mathematical Programming, Series B*, 92:453–480, 2002.
3. D. Bertsimas and M. Sim. The price of robustness. *Operations Research*, 52(1):35–53, 2004.
4. A.L. Soyster. Convex programming with set-inclusive constraints and applications to inexact linear programming. *Operations Research*, 21(5):1154–1157, 1973.
5. A. Holder (ed.). *Mathematical Programming Glossary*, INFORMS Computing Society, http://glossary.computing.society.informs.org, 2006–2007.
6. S.E. Elmaghraby. On the fallacy of averages in project risk management. *European Journal of Operational Research*, 165(2):307–313, 2005.
7. H. Markowitz. Portfolio selection. *Journal of Finance*, 7(1):77–91, 1952.
8. H.M. Markowitz. *Portfolio Selection: Efficient Diversification of Investments*. John Wiley & Sons, New York, 1959.
9. G.B. Dantzig. Linear programming under uncertainty. *Management Science*, 1(3/4):197–206, 1955.
10. W.T. Ziemba and J.M. Mulvey (eds.). *Worldwide Asset and Liability Modeling*. Cambridge University Press, Cambridge, UK, 2003.
11. A. Charnes, W.W. Cooper, and G.H. Symonds. Cost horizons and certainty equivalents: An approach to stochastic programming of heating oil. *Management Science*, 4(3):235–263, 1958.
12. H.J. Greenberg. Mathematical programming models for environmental quality control. *Operations Research*, 43(4):578–622, 1995.
13. S.K. Gupta and J. Rosenhead. Robustness in sequential investment decisions. *Management Science*, 15(2):B-18–29, 1968.
14. J.M. Mulvey, R.J. Vanderbei, and S.A. Zenios. Robust optimization of large-scale systems. *Operations Research*, 43(2):264–281, 1995.
15. P. Kouvelis and G. Yu. Robust Discrete Optimization and Its Applications. *Nonconvex Optimization and Its Applications*. Vol. 14, Kluwer Academic Press, Dordrecht, 1997.
16. G. Cornuejols and R. Tutiincu. Optimization methods in finance. Course notes, Carnegie-Mellon University, Pittsburgh, PA, 2005.
17. R.D. Carr, H.J. Greenberg, W.E. Hart, G. Konjevod, E. Lauer, H. Lin, T. Morrison, and C.A. Phillips. Robust optimization of contaminant sensor placement for community water systems. *Mathematical Programming, Series B*, 107(1–2):337–356, 2006.
18. H. Yaman, O.E. Karasan, and M.C. Pinar. The robust minimum spanning tree problem with interval data. *Operations Research Letters*, 29(1):31–40, 2001.
19. A. Ben-Tal and A. Nemirovski. Robust solutions of uncertain linear programs. *Operations Research Letters*, 25(1):1–13, 1999.
20. X. Chen, M. Sim, and P. Sun. A robust optimization perspective of stochastic programming. *Optimization Online*, June 2005.

21. A. Ben-Tal, L.E. Ghaoui, and A. Nemirovski. Forward: Special issue on robust optimization. *Mathematical Programming, Series B*, 107(1–2):1–3, 2006.
22. S. Netessine and R. Shumsky. Introduction to the theory and practice of yield management. *INFORMS Transactions on Education*, 3(1):34–44, 2002.
23. S.A. Malcolm and S.A. Zenios. Robust optimization for power systems capacity expansion under uncertainty. *Journal of the Operational Research Society*, 45(9):1040–1049, 1994.
24. F.M. Murphy, S. Sen, and A.L. Soyster. Electric utility capacity expansion planning with uncertain load forecasts. *IIE Transactions*, 1:452–459, 1982.
25. G. Taguchi and S. Konishi. *Taguchi Methods—Research and Development*. ASI Press, Dearborn, MI, 1992.
26. T.S. Arthanari. A game theory application in robust design. *Quality Engineering*, 17:291–300, 2005.
27. CD. McAllister and T.W. Simpson. Multidisciplinary robust design optimization of an internal combustion engine. *Transactions of the American Society of Mechanical Engineers*, 125:124–130, 2003.

II

OR/MS Applications

15 Project Management *Adedeji B. Badiru* **15**-1
Introduction • Critical Path Method • PERT Network Analysis • Statistical Analysis of Project Duration • Precedence Diagramming Method • Software Tools for Project Management • Conclusion

16 Quality Control *Qianmei Feng and Kailash C. Kapur* **16**-1
Introduction • Quality Control and Product Life Cycle • New Trends and Relationship to Six Sigma • Statistical Process Control • Process Capability Studies • Advanced Control Charts • Limitations of Acceptance Sampling • Conclusions

17 Reliability *Lawrence M. Leemis* **17**-1
Introduction • Reliability in System Design • Lifetime Distributions • Parametric Models • Parameter Estimation in Survival Analysis • Nonparametric Methods • Assessing Model Adequacy • Summary

18 Production Systems *Bobbie L. Foote and Katta G. Murty* **18**-1
Production Planning Problem • Demand Forecasting • Models for Production Layout Design • Scheduling of Production and Service Systems

19 Energy Systems *C. Randy Hudson and Adedeji B. Badiru* **19**-1
Introduction • Definition of Energy • Harnessing Natural Energy • Mathematical Modeling of Energy Systems • Linear Programming Model of Energy Resource Combination • Integer Programming Model for Energy Investment Options • Simulation and Optimization of Distributed Energy Systems • Point-of-Use Energy Generation • Modeling of CHP Systems • Economic Optimization Methods • Design of a Model for Optimization of CHP System Capacities • Capacity Optimization • Implementation of the Computer Model • Other Scenarios

20 Airline Optimization *Jane L. Snowdon and Giuseppe Paleologo* **20**-1
Introduction • Schedule Planning • Revenue Management • Aircraft Load Planning • Future Research Directions and Conclusions

21 Financial Engineering *Aliza R. Heching and Alan J. King* **21**-1
Introduction • Return • Estimating an Asset's Mean and Variance • Diversification • Efficient Frontier • Utility Analysis • Black–Litterman Asset Allocation Model • Risk Management • Options • Valuing Options • Dynamic Programming • Pricing American Options Using Dynamic Programming • Comparison of Monte Carlo Simulation and Dynamic Programming • Multi-Period Asset Liability Management • Conclusions

22 Supply Chain Management *Donald P. Warsing* **22**-1
Introduction • Managing Inventories in the Supply Chain • Managing Transportation in the Supply Chain • Managing Locations in the Supply Chain • Managing Dyads in the Supply Chain • Discussion and Conclusions

23 E-Commerce   *Sowmyanarayanan Sadagopan* **23**-1
Introduction • Evolution of E-Commerce • OR/MS and E-Commerce • OR Applications in E-Commerce • Tools–Applications Matrix • Way Forward • Summary

24 Water Resources   *G.V. Loganathan* **24**-1
Introduction • Optimal Operating Policy for Reservoir Systems • Water Distribution Systems Optimization • Preferences in Choosing Domestic Plumbing Materials • Stormwater Management • Groundwater Management • Summary

25 Military Applications   *Jeffery D. Weir and Marlin U. Thomas* **25**-1
Introduction • Background on Military OR • Current Military Applications of OR • Concluding Remarks

26 Future of OR/MS Applications: A Practitioner's Perspective
P. Balasubramanian **26**-1
Past as a Guide to the Future • Impact of the Internet • Emerging Opportunities

15
Project Management

15.1	Introduction	**15**-1
15.2	Critical Path Method	**15**-3
	CPM Example • Forward Pass • Backward Pass • Determination of Critical Activities • Using Forward Pass to Determine the Critical Path • Subcritical Paths • Gantt Charts • Gantt Chart Variations • Activity Crashing and Schedule Compression	
15.3	PERT Network Analysis	**15**-18
	PERT Estimates and Formulas • Modeling of Activity Times • Beta Distribution • Triangular Distribution • Uniform Distribution	
15.4	Statistical Analysis of Project Duration	**15**-23
	Central Limit Theorem • Probability Calculation • PERT Network Example	
15.5	Precedence Diagramming Method	**15**-26
	Reverse Criticality in PDM Networks	
15.6	Software Tools for Project Management	**15**-34
	Computer Simulation Software	
15.7	Conclusion	**15**-37
	Further Reading	**15**-37

Adedeji B. Badiru
Air Force Institute of Technology

15.1 Introduction

Project management techniques continue to be a major avenue to accomplishing goals and objectives in various organizations ranging from government, business, and industry to academia. The techniques of project management can be divided into three major tracks as summarized below:

- Qualitative managerial principles
- Computational decision models
- Computer implementation tools.

This chapter focuses on computational network techniques for project management. Network techniques emerged as a formal body of knowledge for project management during World War II, which ushered in a golden era for operations research and its quantitative modeling techniques.

Badiru (1996) defines project management as the process of managing, allocating, and timing resources to achieve a given objective expeditiously. The phases of project management are:

1. Planning
2. Organizing
3. Scheduling
4. Controlling.

The network techniques covered in this chapter are primarily for the scheduling phase, although network planning is sometimes included in the project planning phase too. Everyone in every organization needs project management to accomplish objectives. Consequently, the need for project management will continue to grow as organizations seek better ways to satisfy the constraints on the following:

- Schedule constraints (time limitation)
- Cost constraints (budget limitation)
- Performance constraints (quality limitation).

The different tools of project management may change over time. Some tools will come and go over time. But the basic need of using network analysis to manage projects will always be high. The network of activities in a project forms the basis for scheduling the project. The critical path method (CPM) and the program evaluation and review technique (PERT) are the two most popular techniques for project network analysis. The precedence diagramming method (PDM) has gained popularity in recent years because of the move toward concurrent engineering. A project network is the graphical representation of the contents and objectives of the project. The basic project network analysis of CPM and PERT is typically implemented in three phases (network planning phase, network scheduling phase, and network control phase), which has the following advantages:

- Advantages for communication
  It clarifies project objectives.
  It establishes the specifications for project performance.
  It provides a starting point for more detailed task analysis.
  It presents a documentation of the project plan.
  It serves as a visual communication tool.
- Advantages for control
  It presents a measure for evaluating project performance.
  It helps determine what corrective actions are needed.
  It gives a clear message of what is expected.
  It encourages team interactions.
- Advantages for team interaction
  It offers a mechanism for a quick introduction to the project.
  It specifies functional interfaces on the project.
  It facilitates ease of application.
  It creates synergy between elements of the project.

Network planning is sometimes referred to as activity planning. This involves the identification of the relevant activities for the project. The required activities and their precedence relationships are determined in the planning phase. Precedence requirements may be determined on the basis of technological, procedural, or imposed constraints. The activities are then represented in the form of a network diagram. The two popular models for network drawing are the activity-on-arrow (AOA) and the activity-on-node (AON) conventions. In the AOA approach, arrows are used to represent activities, whereas nodes represent starting and ending points of activities. In the AON approach, nodes represent activities, whereas arrows represent precedence relationships. Time, cost, and resource requirement estimates are developed for each activity during the network planning phase. Time estimates may be based on the following:

1. Historical records
2. Time standards
3. Forecasting
4. Regression functions
5. Experiential estimates.

Network scheduling is performed by using forward pass and backward pass computational procedures. These computations give the earliest and latest starting and finishing times for each activity. The slack time or float associated with each activity is determined in the computations. The activity path with the minimum slack in the network is used to determine the critical activities. This path also determines the duration of the project. Resource allocation and time-cost tradeoffs are other functions performed during network scheduling.

Network control involves tracking the progress of a project on the basis of the network schedule and taking corrective actions when needed; an evaluation of actual performance versus expected performance determines deficiencies in the project.

15.2 Critical Path Method

Precedence relationships in a CPM network fall into the three major categories listed below:

1. Technical precedence
2. Procedural precedence
3. Imposed precedence

Technical precedence requirements are caused by the technical relationships among activities in a project. For example, in conventional construction, walls must be erected before the roof can be installed. Procedural precedence requirements are determined by policies and procedures. Such policies and procedures are often subjective, with no concrete justification. Imposed precedence requirements can be classified as resource-imposed, project-imposed, or environment-imposed. For example, resource shortages may require that one task be scheduled before another. The current status of a project (e.g., percent completion) may determine that one activity be performed before another. The environment of a project, for example, weather changes or the effects of concurrent projects, may determine the precedence relationships of the activities in a project.

The primary goal of a CPM analysis of a project is the determination of the "critical path." The critical path determines the minimum completion time for a project. The computational analysis involves forward pass and backward pass procedures. The forward pass determines the earliest start time and the earliest completion time for each activity in the network. The backward pass determines the latest start time and the latest completion time for each activity. Figure 15.1 shows an example of an activity network using the activity-on-node convention. Conventionally the network is drawn from left to right. If this convention is followed, there is no need to use arrows to indicate the directional flow in the activity network. The notations used for activity A in the network are explained below:

- A: Activity identification
- ES: Earliest starting time
- EC: Earliest completion time
- LS: Latest starting time
- LC: Latest completion time
- t: Activity duration

During the forward pass analysis of the network, it is assumed that each activity will begin at its earliest starting time. An activity can begin as soon as the last of its predecessors is finished. The completion of the forward pass determines the earliest completion time of the project. The backward pass analysis is a reverse of the forward pass. It begins at the latest project completion time and ends at the latest starting time of the first activity in the project network. The rules for implementing the forward pass and backward pass analyses in CPM are presented below. These rules are implemented iteratively until the ES, EC, LS, and LC have been calculated for all nodes in the network.

Rule 1: Unless otherwise stated, the starting time of a project is set equal to time zero. That is, the first node in the network diagram has an earliest start time of zero.

Rule 2: The earliest start time (ES) for any activity is equal to the maximum of the earliest completion times (EC) of the immediate predecessors of the activity. That is,

$$ES = Maximum\ \{Immediately\ Preceding\ ECs\}$$

Rule 3: The earliest completion time (EC) of an activity is the activity's earliest start time plus its estimated duration. That is,

$$ES = ES + (Activity\ Time)$$

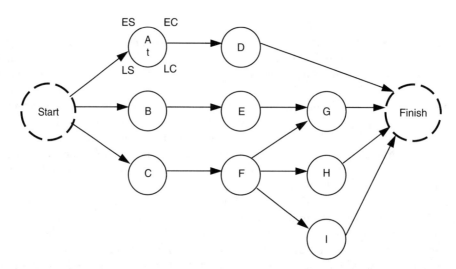

FIGURE 15.1 Example of activity network.

Rule 4: The earliest completion time of a project is equal to the earliest completion time of the very last node in the network. That is,

$$EC\ of\ Project = EC\ of\ last\ activity$$

Rule 5: Unless the latest completion time (LC) of a project is explicitly specified, it is set equal to the earliest completion time of the project. This is called the *zero-project-slack* assumption. That is,

$$LC\ of\ Project = EC\ of\ Project$$

Rule 6: If a desired deadline is specified for the project, then

$$LC\ of\ Project = Specified\ Deadline$$

It should be noted that a latest completion time or deadline may sometimes be specified for a project based on contractual agreements.

Rule 7: The latest completion time (LC) for an activity is the smallest of the latest start times of the activity's immediate successors. That is,

$$LC = Minimum\ \{Immediately\ Succeeding\ LS's\}$$

Rule 8: The latest start time for an activity is the latest completion time minus the activity time. That is,

$$LS = LC - (Activity\ Time)$$

15.2.1 CPM Example

Table 15.1 presents the data for an illustrative project. This network and its extensions will be used for other computational examples in this chapter. The AON network for the example is given in Figure 15.2. Dummy activities are included in the network to designate single starting and ending points for the project.

TABLE 15.1 Data for Sample Project for CPM Analysis

Activity	Predecessor	Duration (Days)
A	—	2
B	—	6
C	—	4
D	A	3
E	C	5
F	A	4
G	B,D,E	2

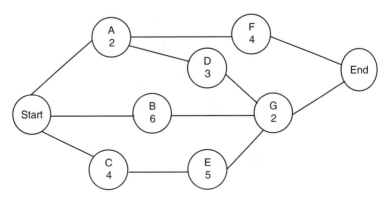

FIGURE 15.2 Project network for illustrative example.

15.2.2 Forward Pass

The forward pass calculations are shown in Figure 15.3. Zero is entered as the ES for the initial node. As the initial node for the example is a dummy node, its duration is zero. Thus, EC for the starting node is equal to its ES. The ES values for the immediate successors of the starting node are set equal to the EC of the START node and the resulting EC values are computed. Each node is treated as the "start" node for its successor or successors. However, if an activity has more than one predecessor, the maximum of the ECs of the preceding activities is used as the activity's starting time. This happens in the case of activity G, whose ES is determined as Max $\{6,5,9\} = 9$. The earliest project completion time for the example is 11 days. Note that this is the maximum of the immediately preceding earliest completion times: Max $\{6,11\} = 11$. As the dummy ending node has no duration, its earliest completion time is set equal to its earliest start time of 11 days.

15.2.3 Backward Pass

The backward pass computations establish the latest start time (LS) and latest completion time (LC) for each node in the network. The results of the backward pass computations are shown in Figure 15.4. As no deadline is specified, the latest completion time of the project is set equal to the earliest completion time. By backtracking and using the network analysis rules presented earlier, the latest completion and start times are determined for each node. Note that in the case of activity A with two successors, the latest completion time is determined as the minimum of the immediately succeeding latest start times. That

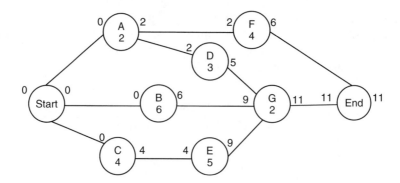

FIGURE 15.3 Forward pass analysis for CPM example.

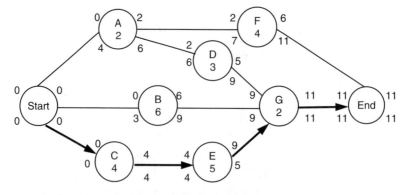

FIGURE 15.4 Backward pass analysis for CPM example.

is, Min{6,7} = 6. A similar situation occurs for the dummy starting node. In that case, the latest completion time of the dummy start node is Min {0,3,4} = 0. As this dummy node has no duration, the latest starting time of the project is set equal to the node's latest completion time. Thus, the project starts at time 0 and is expected to be completed by time 11.

Within a project network, there are usually several possible paths and a number of activities that must be performed sequentially and some activities that may be performed concurrently. If an activity has ES and LS times that are not equal, then the actual start and completion times of that activity may be flexible. The amount of flexibility an activity possesses is called a slack. The slack time is used to determine the critical activities in the network as discussed below.

15.2.4 Determination of Critical Activities

The critical path is defined as the path with the least slack in the network diagram. All the activities on the critical path are said to be critical activities. These activities can create bottlenecks in the network if they are delayed. The critical path is also the longest path in the network diagram. In some networks, particularly large ones, it is possible to have multiple critical paths. If there is a large number of paths in the network, it may be very difficult to visually identify all the critical paths.

The slack time of an activity is also referred to as its *float*. There are four basic types of activity slack as described below:

- *Total Slack* (TS). Total slack is defined as the amount of time an activity may be delayed from its earliest starting time without delaying the latest completion time of the project. The total slack time of an activity is the difference between the latest completion time and the earliest completion time of the activity, or the difference between the latest starting time and the earliest starting time of the activity.

$$TS = LC - EC$$

Or

$$TS = LS - ES$$

Total slack is the measure that is used to determine the critical activities in a project network. The critical activities are identified as those having the minimum total slack in the network diagram. If there is only one critical path in the network, then all the critical activities will be on that one path.

- *Free Slack* (FS). Free slack is the amount of time an activity may be delayed from its earliest starting time without delaying the starting time of any of its immediate successors. Activity free slack is calculated as the difference between the minimum earliest starting time of the activity's successors and the earliest completion time of the activity.

$$FS = \text{Min}\{\text{Succeeding ES's}\} - EC$$

- *Interfering Slack* (IS). Interfering slack or interfering float is the amount of time by which an activity interferes with (or obstructs) its successors when its total slack is fully used. This is rarely used in practice. The interfering float is computed as the difference between the total slack and the free slack.

$$IS = TS - FS$$

- *Independent Float* (IF). Independent float or independent slack is the amount of float that an activity will always have regardless of the completion times of its predecessors or the starting times of its successors. Independent float is computed as:

$$IF = \text{Max}\{0, (ES_j - LC_i - t)\}$$

where ES_j is the earliest starting time of the preceding activity, LC_i is the latest completion time of the succeeding activity, and t is the duration of the activity whose independent float is being calculated. Independent float takes a pessimistic view of the situation of an activity. It evaluates the situation whereby the activity is pressured from either side, that is, when its predecessors are delayed as late as possible while its successors are to be started as early as possible. Independent float is useful for conservative planning purposes, but it is not used much in practice. Despite its low level of use, independent float does have practical implications for better project management. Activities can be buffered with independent floats as a way to handle contingencies.

In Figure 15.4 the total slack and the free slack for activity A are calculated, respectively, as:

$$TS = 6 - 2 = 4 \text{ days}$$
$$FS = \text{Min}\{2, 2\} - 2 = 2 - 2 = 0$$

Similarly, the total slack and the free slack for activity F are:

$$TS = 11 - 6 = 5 \text{ days}$$
$$FS = \text{Min}\{11\} - 6 = 11 - 6 = 5 \text{ days}$$

Table 15.2 presents a tabulation of the results of the CPM example. The table contains the earliest and latest times for each activity as well as the total and free slacks. The results indicate that the minimum total slack in the network is zero. Thus, activities C, E, and G are identified as the critical activities. The critical path is highlighted in Figure 15.4 and consists of the following sequence of activities:

$$\text{Start} \rightarrow C \rightarrow E \rightarrow G \rightarrow \text{End}$$

The total slack for the overall project itself is equal to the total slack observed on the critical path. The minimum slack in most networks will be zero as the ending LC is set equal to the ending EC. If a deadline is specified for a project, then we would set the project's latest completion time to the specified deadline. In that case, the minimum total slack in the network would be given by:

$$TS_{\text{Min}} = \text{Project Deadline} - EC \text{ of the last node}$$

TABLE 15.2 Result of CPM Analysis for Sample Project

Activity	Duration	ES	EC	LS	LC	TS	FS	Criticality
A	2	0	2	4	6	4	0	—
B	6	0	6	3	9	3	3	—
C	4	0	4	0	4	0	0	Critical
D	3	2	5	6	9	4	4	—
E	5	4	9	4	9	0	0	Critical
F	4	2	6	7	11	5	5	—
G	2	9	11	9	11	0	0	Critical

Project Management

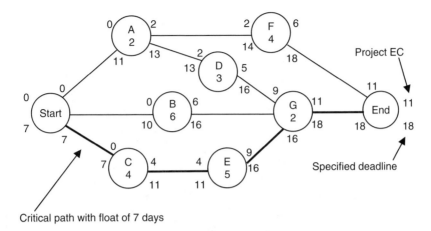

FIGURE 15.5 CPM network with deadline.

This minimum total slack will then appear as the total slack for each activity on the critical path. If a specified deadline is lower than the EC at the finish node, then the project will start out with a negative slack. That means that it will be behind schedule before it even starts. It may then become necessary to expedite some activities (i.e., crashing) to overcome the negative slack. Figure 15.5 shows an example with a specified project deadline. In this case, the deadline of 18 days comes after the earliest completion time of the last node in the network.

15.2.5 Using Forward Pass to Determine the Critical Path

The critical path in CPM analysis can be determined from the forward pass only. This can be helpful in cases where it is desired to quickly identify the critical activities without performing all the other calculations needed to obtain the latest starting times, the latest completion times, and total slacks. The steps for determining the critical path from the forward pass only are:

1. Complete the forward pass in the usual manner.
2. Identify the last node in the network as a critical activity.
3. Work backward from the last node. Whenever a merge node occurs, the critical path will be along the path where the earliest completion time (EC) of the predecessor is equal to the earliest start time (ES) of the current node.
4. Continue the backtracking from each critical activity until the project starting node is reached. Note that if there is a single starting node or a single ending node in the network, then that node will always be on the critical path.

15.2.6 Subcritical Paths

In a large network, there may be paths that are near critical. Such paths require almost as much attention as the critical path as they have a high potential of becoming critical when changes occur in the network. Analysis of subcritical paths may help in the classification of tasks into ABC categories on the basis of Pareto analysis. Pareto analysis separates the "vital" few activities from the "trivial" many activities. This permits a more efficient allocation of resources. The principle of Pareto analysis originated from the work of Italian economist Vilfredo Pareto (1848–1923). Pareto discovered from his studies that most of the wealth in his country was held by a few individuals.

TABLE 15.3 Analysis of Sub-Critical Paths

Path No.	Activities on Path	Total Slack	λ (%)	λ'
1	A,C,G,H	0	100	10
2	B,D,E	1	97.56	9.78
3	F,I	5	87.81	8.90
4	J,K,L	9	78.05	8.03
5	O,P,Q,R	10	75.61	7.81
6	M,S,T	25	39.02	4.51
7	N,AA,BB,U	30	26.83	3.42
8	V,W,X	32	21.95	2.98
9	Y,CC,EE	35	17.14	2.54
10	DD,Z,FF	41	0	1.00

For project control purposes, the Pareto principle states that 80% of the bottlenecks are caused by only 20% of the tasks. This principle is applicable to many management processes. For example, in cost analysis, one can infer that 80% of the total cost is associated with only 20% of the cost items. Similarly, 20% of an automobile's parts cause 80% of the maintenance problems. In personnel management, about 20% of the employees account for about 80% of the absenteeism. For critical path analysis, 20% of the network activities will take up 80% of our control efforts. The ABC classification based on Pareto analysis divides items into three priority categories: A (most important), B (moderately important), and C (least important). Appropriate percentages (e.g., 20%, 25%, 55%) may be assigned to the categories.

With Pareto analysis, attention can be shifted from focusing only on the critical path to managing critical and near-critical tasks. The level of criticality of each path may be assessed by the steps below:

1. Sort paths in increasing order of total slack.
2. Partition the sorted paths into groups based on the magnitudes of their total slacks.
3. Sort the activities within each group in increasing order of their earliest starting times.
4. Assign the highest level of criticality to the first group of activities (e.g., 100%). This first group represents the usual critical path.
5. Calculate the relative criticality indices for the other groups in decreasing order of criticality.

Define the following variables:
α_1 = the minimum total slack in the network
α_2 = the maximum total slack in the network
β = total slack for the path whose criticality is to be calculated.

Compute the path's criticality as:

$$\lambda = \frac{\alpha_2 - \beta}{\alpha_2 - \alpha_1}(100\%)$$

The above procedure yields relative criticality levels between 0% and 100%. Table 15.3 presents a hypothetical example of path criticality indices. The criticality level may be converted to a scale between 1 (least critical) and 10 (most critical) by the expression below:

$$\lambda' = 1 + 0.09\lambda$$

15.2.7 Gantt Charts

When the results of a CPM analysis are fitted to a calendar time, the project plan becomes a schedule. The Gantt chart is one of the most widely used tools for presenting a project

Project Management

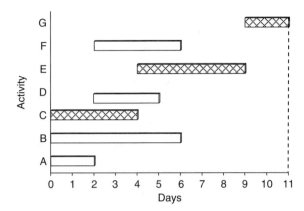

FIGURE 15.6 Gantt chart based on earliest starting times.

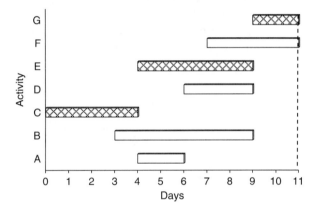

FIGURE 15.7 Gantt chart based on latest starting times.

schedule. A Gantt chart can show planned and actual progress of activities. The time scale is indicated along the horizontal axis, while horizontal bars or lines representing activities are ordered along the vertical axis. As a project progresses, markers are made on the activity bars to indicate actual work accomplished. Gantt charts must be updated periodically to indicate project status. Figure 15.6 presents the Gantt chart using the earliest starting (ES) times from Table 15.2. Figure 15.7 presents the Gantt chart for the example based on the latest starting (LS) times. Critical activities are indicated by the shaded bars.

Figure 15.6 shows the Gantt chart for the example based on earliest starting time. From the CPM network computations, it is noted that activity F can be delayed from day two until day seven (i.e., TS = 5) without delaying the overall project. Likewise, A, D, or both may be delayed by a combined total of 4 days (TS = 4) without delaying the overall project. If all the 4 days of slack are used up by A, then D cannot be delayed. If A is delayed by 1 day, then D can only be delayed by up to 3 days without causing a delay of G, which determines project completion. CPM computations also reveal that activity B may be delayed up to 3 days without affecting the project completion time.

In Figure 15.7, the activities are scheduled by their latest completion times. This represents a pessimistic case where activity slack times are fully used. No activity in this schedule can be delayed without delaying the project. In Figure 15.7, only one activity is scheduled over the first 3 days. This may be compared to the schedule in Figure 15.6, which has three starting activities. The schedule in Figure 15.7 may be useful if there is a situation that permits only a few activities to be scheduled in the early stages of the project. Such situations may involve shortage of project personnel, lack of initial budget, time for project initiation, time for personnel training, allowance for learning period, or general resource constraints. Scheduling of activities based on ES times indicates an optimistic view. Scheduling on the basis of LS times represents a pessimistic approach.

15.2.8 Gantt Chart Variations

The basic Gantt chart does not show the precedence relationships among activities. The chart can be modified to show these relationships by coding appropriate bars, as shown by the cross-hatched bars in Figure 15.8. Other simple legends can be added to show which bars are related by precedence linking. Figure 15.9 shows a Gantt chart that presents a comparison of planned and actual schedules. Note that two tasks are in progress at the current time indicated in the figure. One of the ongoing tasks is an unplanned task. Figure 15.10 shows a Gantt chart on which important milestones have been indicated. Figure 15.11 shows a Gantt chart in which bars represent a combination of related tasks. Tasks may be combined for scheduling purposes or for conveying functional relationships required in a project. Figure 15.12 presents a Gantt chart of project phases. Each phase is further divided into

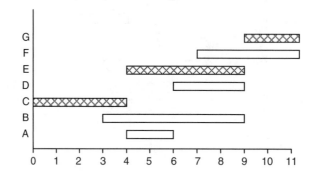

FIGURE 15.8 Coding of bars that are related by precedence linking.

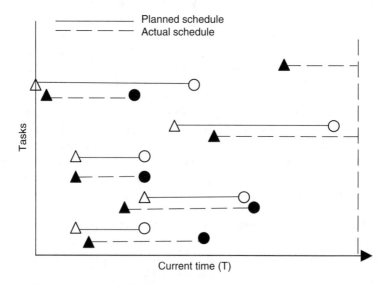

FIGURE 15.9 Progress monitoring Gantt chart.

Project Management

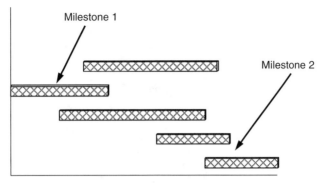

FIGURE 15.10 Milestone Gantt chart.

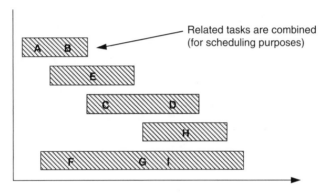

FIGURE 15.11 Task combination Gantt chart.

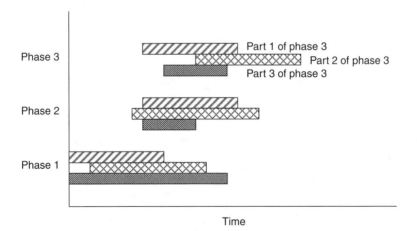

FIGURE 15.12 Phase-based Gantt chart.

parts. Figure 15.13 shows a Gantt chart for multiple projects. Multiple project charts are useful for evaluating resource allocation strategies. Resource loading over multiple projects may be needed for capital budgeting and cash flow analysis decisions. Figure 15.14 shows a project slippage chart that is useful for project tracking and control. Other variations of the basic Gantt chart may be developed for specific needs.

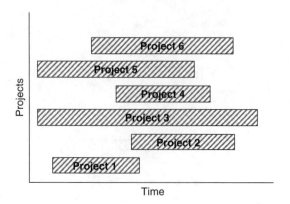

FIGURE 15.13 Multiple projects Gantt chart.

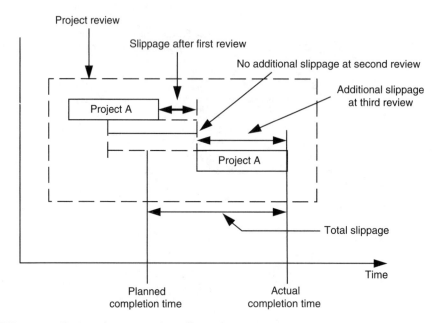

FIGURE 15.14 Project slippage tracking Gantt chart.

15.2.9 Activity Crashing and Schedule Compression

Schedule compression refers to reducing the length of a project network. This is often accomplished by crashing activities. Crashing, sometimes referred to as expediting, reduces activity durations, thereby reducing project duration. Crashing is done as a tradeoff between shorter task duration and higher task cost. It must be determined whether the total cost savings realized from reducing the project duration is enough to justify the higher costs associated with reducing individual task durations. If there is a delay penalty associated with a project, it may be possible to reduce the total project cost even though individual task costs are increased by crashing. If the cost savings on delay penalty is higher than the incremental cost of reducing the project duration, then crashing is justified. Under conventional crashing, the more the duration of a project is compressed, the higher the total cost of the project. The objective is to determine at what point to terminate further crashing in a network. Normal task duration refers to the time required to perform a task under normal

Project Management

TABLE 15.4 Normal and Crash Time and Cost Data

Activity	Normal Duration	Normal Cost	Crash Duration	Crash Cost	Crashing Ratio
A	2	$210	2	$210	0
B	6	400	4	600	100
C	4	500	3	750	250
D	3	540	2	600	60
E	5	750	3	950	100
F	4	275	3	310	35
G	2	100	1	125	25
		$2775		$3545	

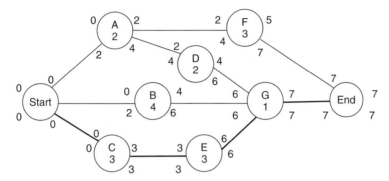

FIGURE 15.15 Example of fully crashed CPM network.

circumstances. Crash task duration refers to the reduced time required to perform a task when additional resources are allocated to it.

If each activity is assigned a range of time and cost estimates, then several combinations of time and cost values will be associated with the overall project. Iterative procedures are used to determine the best time and cost combination for a project. Time-cost trade-off analysis may be conducted, for example, to determine the marginal cost of reducing the duration of the project by one time unit. Table 15.4 presents an extension of the data for the earlier example to include normal and crash times as well as normal and crash costs for each activity. The normal duration of the project is 11 days, as seen earlier, and the normal cost is $2775.

If all the activities are reduced to their respective crash durations, the total crash cost of the project will be $3545. In that case, the crash time is found by CPM analysis to be 7 days. The CPM network for the fully crashed project is shown in Figure 15.15. Note that activities C, E, and G remain critical. Sometimes, the crashing of activities may result in a new critical path. The Gantt chart in Figure 15.16 shows a schedule of the crashed project using the ES times. In practice, one would not crash all activities in a network. Rather, some heuristic would be used to determine which activity should be crashed and by how much. One approach is to crash only the critical activities or those activities with the best ratios of incremental cost versus time reduction. The last column in Table 15.4 presents the respective ratios for the activities in our example. The crashing ratios are computed as:

$$r = \frac{\text{Crash Cost} - \text{Normal Cost}}{\text{Normal Duration} - \text{Crash Duration}}$$

This method of computing the crashing ratio gives crashing priority to the activity with the lowest cost slope. It is a commonly used approach in CPM networks.

Activity G offers the lowest cost per unit time reduction of $25. If our approach is to crash only one activity at a time, we may decide to crash activity G first and evaluate the increase in project cost versus the reduction in project duration. The process can then be

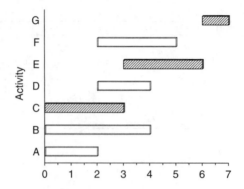

FIGURE 15.16 Gantt chart of fully crashed CPM network.

TABLE 15.5 Selected Crashing Options for CPM Example

Option No.	Activities Crashed	Network Duration	Time Reduction	Incremental Cost	Total Cost
1.	None	11	–	–	2775
2.	G	10	1	25	2800
3.	G,F	10	0	35	2835
4.	G,F,D	10	0	60	2895
5.	G,F,D,B	10	0	200	3095
6.	G,F,D,B,E	8	2	200	3295
7.	G,F,D,B,E,C	7	1	250	3545

repeated for the next best candidate for crashing, which is activity F in this case. After F has been crashed, activity D can then be crashed.

This approach is repeated iteratively in order of activity preference until no further reduction in project duration can be achieved or until the total project cost exceeds a specified limit.

A more comprehensive analysis is to evaluate all possible combinations of the activities that can be crashed. However, such a complete enumeration would be prohibitive, as there would be a total of 2^c crashed networks to evaluate, where c is the number of activities that can be crashed out of the n activities in the network ($c <= n$). For our example, only 6 out of the 7 activities in the sample network can be crashed. Thus, a complete enumeration will involve $2^6 = 64$ alternate networks. Table 15.5 shows 7 of the 64 crashing options. Activity G, which offers the best crashing ratio, reduces the project duration by only 1 day. Even though activities F, D, and B are crashed by a total of 4 days at an incremental cost of $295, they do not generate any reduction in project duration. Activity E is crashed by 2 days and it generates a reduction of 2 days in project duration. Activity C, which is crashed by 1 day, generates a further reduction of 1 day in the project duration. It should be noted that the activities that generate reductions in project duration are the ones that were identified earlier as the critical activities.

Figure 15.17 shows the crashed project duration versus the crashing options, while Figure 15.18 shows a plot of the total project cost after crashing versus the selected crashing options. As more activities are crashed, the project duration decreases while the total project cost increases. If full enumeration were performed, Figure 15.17 would contain additional points between the minimum possible project duration of 7 days (fully crashed) and the normal project duration of 11 days (no crashing). Similarly, the plot for total project cost (Figure 15.18) would contain additional points between the normal cost of $2775 and the crash cost of $3545.

Project Management   15-17

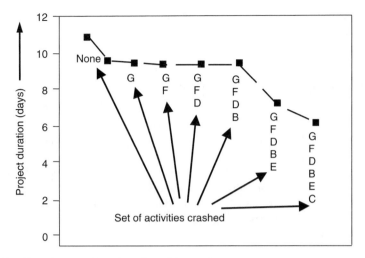

FIGURE 15.17 Duration as a function of crashing options.

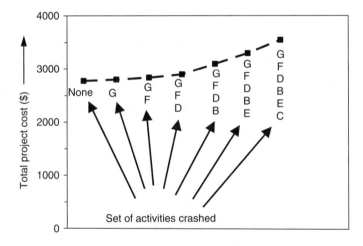

FIGURE 15.18 Project cost as a function of crashing options.

Several other approaches exist for determining which activities to crash in a project network. Two alternate approaches are presented below for computing the crashing ratio, r.

Let $r = $ Criticality Index

or

Let $r = \dfrac{\text{Crash Cost} - \text{Normal Cost}}{(\text{Normal Duration} - \text{Crash Duration})(\text{Criticality Index})}.$

The first approach uses a *critical index* criterion, giving crashing priority to the activity with the highest probability of being on the critical path. In deterministic networks, this refers to the critical activities. In stochastic networks, an activity is expected to fall on the critical path only a percentage of the time. The second approach is a combination of the approach used for the illustrative example and the criticality index approach. It reflects the process of selecting the least-cost expected value. The denominator of the expression represents the expected number of days by which the critical path can be shortened. For different project networks, different crashing approaches should be considered, and the one that best fits the nature of the network should be selected.

15.3 PERT Network Analysis

Program evaluation review technique (PERT) is an extension of CPM that incorporates variability in activity durations into project network analysis. PERT has been used extensively and successfully in practice. In real life, activities are often prone to uncertainties that determine the actual durations of the activities. In CPM, activity durations are assumed to be deterministic. In PERT, the potential uncertainties in activity durations are accounted for by using three time estimates for each activity. The three time estimates represent the spread of the estimated activity duration. The greater the uncertainty of an activity the wider the range of the estimates.

15.3.1 PERT Estimates and Formulas

PERT uses three time estimates (optimistic, most likely, and pessimistic) to compute the expected duration and variance for each activity. The PERT formulas are based on a simplification of the expressions for the mean and variance of a beta distribution. The approximation formula for the mean is a simple weighted average of the three time estimates, with the end points assumed to be equally likely and the mode four times as likely. The approximation formula for PERT is based on the recognition that most of the observations from a distribution will lie within plus or minus three standard deviations, or a spread of six standard deviations. This leads to the simple method of setting the PERT formula for standard deviation equal to one-sixth of the estimated duration range. While there is no theoretical validation for these approximation approaches, the PERT formulas do facilitate ease of use. The formulas are presented below:

$$t_e = \frac{a + 4m + b}{6}$$

$$s = \frac{(b - a)}{6}$$

where
a = optimistic time estimate
m = most likely time estimate
b = pessimistic time estimate $a < m < b$
t_e = expected time for the activity
s^2 = variance of the duration of the activity

After obtaining the estimate of the duration for each activity, the network analysis is carried out in the same manner previously illustrated for the CPM approach. The major steps in PERT analysis are summarized below:

1. Obtain three time estimates $a, m,$ and b for each activity.
2. Compute the expected duration for each activity by using the formula for t_e.
3. Compute the variance of the duration of each activity from the formula for s^2.
4. Compute the expected project duration, T_e. As in the case of CPM, the duration of a project in PERT analysis is the sum of the durations of the activities on the critical path.

Project Management

5. Compute the variance of the project duration as the sum of the variances of the activities on the critical path. The variance of the project duration is denoted by S^2. It should be recalled that CPM cannot compute the variance of the project duration, as variances of activity durations are not computed.
6. If there are two or more critical paths in the network, choose the one with the largest variance to determine the project duration and the variance of the project duration. Thus, PERT is pessimistic with respect to the variance of project duration when there are multiple critical paths in the network. For some networks, it may be necessary to perform a mean-variance analysis to determine the relative importance of the multiple paths by plotting the expected project duration versus the path duration variance.
7. If desired, compute the probability of completing the project within a specified time period. This is not possible under CPM.

15.3.2 Modeling of Activity Times

In practice, a question often arises as to how to obtain good estimates of a, m, and b. Several approaches can be used to obtain the time estimates for PERT. Some of the approaches are:

- Estimates furnished by an experienced person
- Estimates extracted from standard time data
- Estimates obtained from historical data
- Estimates obtained from simple regression or forecasting
- Estimates generated by simulation
- Estimates derived from heuristic assumptions
- Estimates dictated by customer requirements.

The pitfall of using estimates furnished by an individual is that they may be inconsistent, as they are limited by the experience and personal bias of the person providing them. Individuals responsible for furnishing time estimates are usually not experts in estimation, and they generally have difficulty in providing accurate PERT time estimates. There is often a tendency to select values of $a, m,$ and b that are optimistically skewed. This is because a conservatively large value is typically assigned to b by inexperienced individuals.

The use of time standards, on the other hand, may not reflect the changes occurring in the current operating environment due to new technology, work simplification, new personnel, and so on. The use of historical data and forecasting is very popular because estimates can be verified and validated by actual records. In the case of regression and forecasting, there is the danger of extrapolation beyond the data range used for fitting the regression and forecasting models. If the sample size in a historical data set is sufficient and the data can be assumed to reasonably represent prevailing operating conditions, the three PERT estimates can be computed as follows:

$$\hat{a} = \bar{t} - kR$$
$$\hat{m} = \bar{t}$$
$$\hat{b} = \bar{t} + kR$$

where
R = range of the sample data
$\bar{t}$ = arithmetic average of the sample data

$k = 3/d_2$
$d_2 =$ an adjustment factor for estimating the standard deviation of a population

If $kR > \bar{t}$, then set $a = 0$ and $b = 2\bar{t}$. The factor d_2 is widely tabulated in the quality control literature as a function of the number of sample points, n. Selected values of d_2 are presented below.

n	5	10	15	20	25	30	40	50	75	100
d_2	2.326	3.078	3.472	3.735	3.931	4.086	4.322	4.498	4.806	5.015

As mentioned earlier, activity times can be determined from historical data. The procedure involves three steps:

1. Appropriate organization of the historical data into histograms.
2. Determination of a distribution that reasonably fits the shape of the histogram.
3. Testing of the goodness-of-fit of the hypothesized distribution by using an appropriate statistical model. The chi-square test and the Kolmogrov-Smirnov (K-S) test are two popular methods for testing goodness-of-fit. Most statistical texts present the details of how to carry out goodness-of-fit tests.

15.3.3 Beta Distribution

PERT analysis assumes that the probabilistic properties of activity duration can be modeled by the beta probability density function. The beta distribution is defined by two end points and two shape parameters. The beta distribution was chosen by the original developers of PERT as a reasonable distribution to model activity times because it has finite end points and can assume a variety of shapes based on different shape parameters. While the true distribution of activity time will rarely ever be known, the beta distribution serves as an acceptable model. Figure 15.19 shows examples of alternate shapes of the standard beta

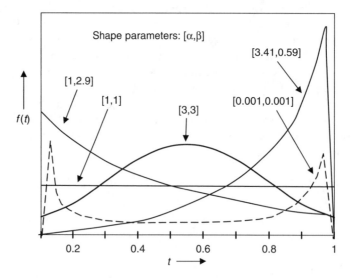

FIGURE 15.19 Alternate shapes of the beta distribution.

distribution between zero and one. The uniform distribution between 0 and 1 is a special case of the beta distribution with both shape parameters equal to one.

The standard beta distribution is defined over the interval 0 to 1, while the general beta distribution is defined over any interval a to b. The general beta probability density function is given by:

$$f(t) = \frac{\Gamma(\alpha+\beta)}{\Gamma(\alpha)\Gamma(\beta)} \cdot \frac{1}{(b-a)^{\alpha+\beta-1}} \cdot (t-a)^{\alpha-1}(b-t)^{\beta-1}$$

for $a \leq t \leq b$ and $\alpha > 0, \beta > 0$

where
$a =$ lower end point of the distribution
$b =$ upper end point of the distribution
α and β are the shape parameters for the distribution.

The mean, variance, and mode of the general beta distribution are defined as:

$$\mu = a + (b-a)\frac{\alpha}{\alpha+\beta}$$

$$\sigma^2 = (b-a)^2 \frac{\alpha\beta}{(\alpha+\beta+1)(\alpha+\beta)^2}$$

$$m = \frac{a(\beta-1) + b(\alpha-1)}{\alpha+\beta-2}$$

The general beta distribution can be transformed into a standardized distribution by changing its domain from $[a, b]$ to the unit interval $[0, 1]$. This is accomplished by using the relationship $t_s = a + (b-a)t_s$, where t_s is the standard beta random variable between 0 and 1. This yields the standardized beta distribution, given by:

$$f(t) = \frac{\Gamma(\alpha+\beta)}{\Gamma(\alpha)\Gamma(\beta)} t^{\alpha-1}(1-t)^{\beta-1}; \quad 0 < t < 1; \quad \alpha, \beta > 0$$

$$= 0; \quad \text{elsewhere}$$

with mean, variance, and mode defined as:

$$\mu = \frac{\alpha}{\alpha+\beta}$$

$$\sigma^2 = \frac{\alpha\beta}{(\alpha+\beta+1)(\alpha+\beta)^2}$$

$$m = \frac{a(\beta-1) + b(\alpha-1)}{\alpha+\beta-2}$$

15.3.4 Triangular Distribution

The triangular probability density function has been used as an alternative to the beta distribution for modeling activity times. The triangular density has three essential parameters: a

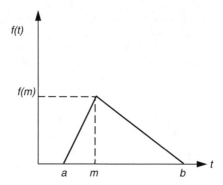

FIGURE 15.20 Triangular probability density function.

minimum value (a), a mode (m) and a maximum (b). It is defined mathematically as:

$$f(t) = \frac{2(t-a)}{(m-a)(b-a)}; \quad a \leq t \leq m$$

$$= \frac{2(b-t)}{(b-m)(b-a)}; \quad m \leq t \leq b$$

with mean and variance defined, respectively, as:

$$\mu = \frac{a+m+b}{3}$$

$$\sigma^2 = \frac{a(a-m) + b(b-a) + m(m-b)}{18}$$

Figure 15.20 presents a graphical representation of the triangular density function. The three time estimates of PERT can be inserted into the expression for the mean of the triangular distribution to obtain an estimate of the expected activity duration. Note that in the conventional PERT formula, the mode (m) is assumed to carry four times as much weight as either a or b when calculating the expected activity duration. By contrast, under the triangular distribution, the three time estimates are assumed to carry equal weights.

15.3.5 Uniform Distribution

For cases where only two time estimates instead of three are to be used for network analysis, the uniform density function may be assumed for activity times. This is acceptable for situations where extreme limits of an activity duration can be estimated and it can be assumed that the intermediate values are equally likely to occur. The uniform distribution is defined mathematically as:

$$f(t) = \frac{1}{b-a}; \quad a \leq t \leq b$$

$$= 0; \quad \text{otherwise}$$

with mean and variance defined, respectively, as:

$$\mu = \frac{a+b}{2}$$

$$\sigma^2 = \frac{(b-a)^2}{12}$$

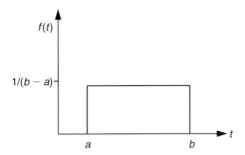

FIGURE 15.21 Uniform probability density function.

Figure 15.21 presents a graphical representation of the uniform distribution, for which the expected activity duration is computed as the average of the upper and lower limits of the distribution. The appeal of using only two time estimates a and b is that the estimation error due to subjectivity can be reduced and the estimation task simplified.

Other distributions that have been explored for activity time modeling include the normal distribution, lognormal distribution, truncated exponential distribution, and Weibull distribution. Once the expected activity durations have been computed, the analysis of the activity network is carried out just as in the case of single-estimate CPM network analysis.

15.4 Statistical Analysis of Project Duration

Regardless of the distribution assumed for activity durations, the central limit theorem suggests that the distribution of the project duration will be approximately normally distributed. The theorem states that the distribution of averages obtained from any probability density function will be approximately normally distributed if the sample size is large and the averages are independent. In mathematical terms, the theorem is stated as follows.

15.4.1 Central Limit Theorem

Let $X_1, X_2, \ldots, X_N$ be independent and identically distributed random variables. Then the sum of the random variables is normally distributed for large values of N. The sum is defined as:

$$T = X_1 + X_2 + \cdots + X_N$$

In activity network analysis, T represents the total project length as determined by the sum of the durations of the activities of the critical path. The mean and variance of T are expressed as:

$$\mu = \sum_{i=1}^{N} E[X_i]$$

$$\sigma^2 = \sum_{i=1}^{N} V[X_i]$$

where
$E[X_i]$ = expected value of random variable X_i
$V[X_i]$ = variance of random variable X_i.

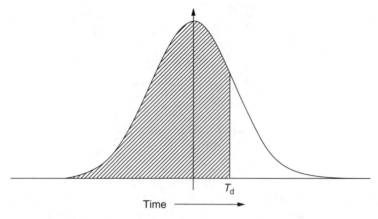

FIGURE 15.22 Area under the normal curve.

When applying the central limit theorem to activity networks, one should note that the assumption of independent activity times may not always be satisfied. Because of precedence relationships and other interdependencies of activities, some activity durations may not be independent.

15.4.2 Probability Calculation

If the project duration T_e can be assumed to be approximately normally distributed based on the central limit theorem, then the probability of meeting a specified deadline T_d can be computed by finding the area under the standard normal curve to the left of T_d. Figure 15.22 shows an example of a normal distribution describing the project duration.

Using the familiar transformation formula below, a relationship between the standard normal random variable z and the project duration variable can be obtained:

$$z = \frac{T_d - T_e}{S}$$

where
T_d = specified deadline
T_e = expected project duration based on network analysis
S = standard deviation of the project duration.

The probability of completing a project by the deadline T_d is then computed as:

$$P(T \leq T_d) = P\left(z \leq \frac{T_d - T_e}{S}\right)$$

The probability is obtained from the standard normal table. Examples presented below illustrate the procedure for probability calculations in PERT.

15.4.3 PERT Network Example

Suppose we have the project data presented in Table 15.6. The expected activity durations and variances as calculated by the PERT formulas are shown in the last two columns of the table. Figure 15.23 shows the PERT network. Activities C, E, and G are shown to be critical, and the project completion time is 11 time units.

Project Management

TABLE 15.6 Data for PERT Network Example

Activity	Predecessors	a	m	b	t_e	s^2
A	–	1	2	4	2.17	0.2500
B	–	5	6	7	6.00	0.1111
C	–	2	4	5	3.83	0.2500
D	A	1	3	4	2.83	0.2500
E	C	4	5	7	5.17	0.2500
F	A	3	4	5	4.00	0.1111
G	B,D,E	1	2	3	2.00	0.1111

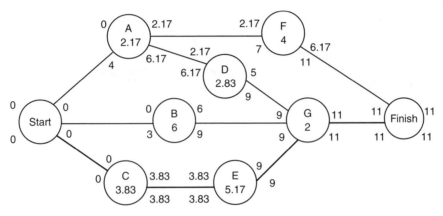

FIGURE 15.23 PERT network example.

The probability of completing the project on or before a deadline of 10 time units (i.e., $T_d = 10$) is calculated as shown below:

$$T_e = 11$$

$$S^2 = V[C] + V[E] + V[G]$$

$$= 0.25 + 0.25 + 0.1111$$

$$= 0.6111$$

$$S = \sqrt{0.6111}$$

$$= 0.7817$$

$$P(T \leq T_d) = P(T \leq 10)$$

$$= P\left(z \leq \frac{10 - T_e}{S}\right)$$

$$= P\left(z \leq \frac{10 - 11}{0.7817}\right)$$

$$= P(z \leq -1.2793)$$

$$= 1 - P(z \leq 1.2793)$$

$$= 1 - 0.8997$$

$$= 0.1003$$

Thus, there is just over 10% probability of finishing the project within 10 days. By contrast, the probability of finishing the project in 13 days is calculated as:

$$P(T \leq 13) = P\left(z \leq \frac{13-11}{0.7817}\right)$$

$$= P(z \leq 2.5585)$$

$$= 0.9948$$

This implies that there is over 99% probability of finishing the project within 13 days. Note that the probability of finishing the project in exactly 13 days will be zero. That is, $P(T = T_d) = 0$. If we desire the probability that the project can be completed within a certain lower limit (T_L) and a certain upper limit (T_U), the computation will proceed as follows: Let $T_L = 9$ and $T_U = 11.5$. Then,

$$P(T_L \leq T \leq T_U) = P(9 \leq T \leq 11.5)$$

$$= P(T \leq 11.5) - P(T \leq 9)$$

$$= P\left(z \leq \frac{11.5-11}{0.7817}\right) - P\left(z \leq \frac{9-11}{0.7817}\right)$$

$$= P(z \leq 0.6396) - P(z \leq -2.5585)$$

$$= P(z \leq 0.6396) - [1 - P(z \leq 2.5585)]$$

$$= 0.7389 - [1 - 0.9948]$$

$$= 0.7389 - 0.0052$$

$$= 0.7337$$

That is, there is 73.4% chance of finishing the project within the specified range of duration.

15.5 Precedence Diagramming Method

The precedence diagramming method (PDM) was developed in the early 1960s as an extension of PERT/CPM network analysis. PDM permits mutually dependent activities to be performed partially in parallel instead of serially. The usual finish-to-start dependencies between activities are relaxed to allow activities to overlap. This facilitates schedule compression. An example is the requirement that concrete should be allowed to dry for a number of days before drilling holes for handrails. That is, drilling cannot start until so many days after the completion of concrete work. This is a finish-to-start constraint. The time between the finishing time of the first activity and the starting time of the second activity is called the lead–lag requirement between the two activities. Figure 15.24 shows the basic lead–lag relationships between activity A and activity B. The terminology presented in Figure 15.24 is explained as follows.

SS_{AB} (Start-to-Start) lead: Activity B cannot start until activity A has been in progress for at least SS time units.

FF_{AB} (Finish-to-Finish) lead: Activity B cannot finish until at least FF time units after the completion of activity A.

Project Management

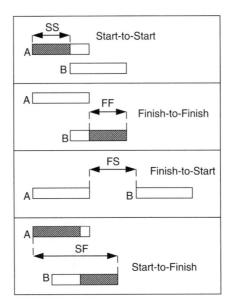

FIGURE 15.24 Lead–lag relationships in PDM.

FS_{AB} **(Finish-to-Start) lead:** Activity B cannot start until at least FS time units after the completion of activity A. Note that PERT/CPM approaches use $FS_{AB} = 0$ for network analysis.

SF_{AB} **(Start-to-Finish) lead:** This specifies that there must be at least SF time units between the start of activity A and the completion of activity B.

The leads or lags may, alternately, be expressed in percentages rather than time units. For example, we may specify that 25% of the work content of activity A must be completed before activity B can start. If the percentage of work completed is used for determining lead–lag constraints, then a reliable procedure must be used for estimating the percent completion. If the project work is broken up properly using work breakdown structure (WBS), it will be much easier to estimate percent completion by evaluating the work completed at the elementary task level. The lead–lag relationships may also be specified in terms of at most relationships instead of at least relationships. For example, we may have at most FF lag requirement between the finishing time of one activity and the finishing time of another activity. Splitting of activities often simplifies the implementation of PDM, as will be shown later with some examples. Some of the factors that will determine whether or not an activity can be split are technical limitations affecting splitting of a task, morale of the person working on the split task, set-up times required to restart split tasks, difficulty involved in managing resources for split tasks, loss of consistency of work, and management policy about splitting jobs.

Figure 15.25 presents a simple CPM network consisting of three activities. The activities are to be performed serially and each has an expected duration of 10 days. The conventional CPM network analysis indicates that the duration of the network is 30 days. The earliest times and the latest times are as shown in the figure.

The Gantt chart for the example is shown in Figure 15.26. For a comparison, Figure 15.27 shows the same network but with some lead–lag constraints. For example, there is an SS constraint of 2 days and an FF constraint of 2 days between activities A and B. Thus, activity B can start as early as 2 days after activity A starts, but it cannot finish until 2 days after the completion of A. In other words, at least 2 days must separate the finishing

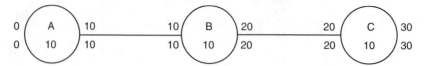

FIGURE 15.25 Serial activities in CPM network.

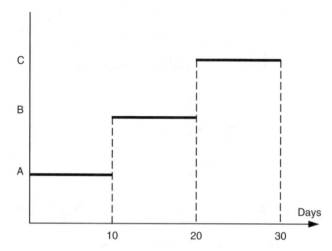

FIGURE 15.26 Gantt chart of serial activities in CPM example.

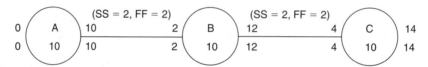

FIGURE 15.27 PDM network example.

time of A and the finishing time of B. A similar precedence relationship exists between activity B and activity C. The earliest and latest times obtained by considering the lag constraints are indicated in Figure 15.27.

The calculations show that if B is started just 2 days after A is started, it can be completed as early as 12 days as opposed to the 20 days obtained in the case of conventional CPM. Similarly, activity C is completed at time 14, which is considerably less than the 30 days calculated by conventional CPM. The lead–lag constraints allow us to compress or overlap activities. Depending on the nature of the tasks involved, an activity does not have to wait until its predecessor finishes before it can start. Figure 15.28 shows the Gantt chart for the example incorporating the lead–lag constraints. It should be noted that a portion of a succeeding activity can be performed simultaneously with a portion of the preceding activity.

A portion of an activity that overlaps with a portion of another activity may be viewed as a distinct portion of the required work. Thus, partial completion of an activity may be evaluated. Figure 15.29 shows how each of the three activities is partitioned into contiguous parts. Even though there is no physical break or termination of work in any activity, the distinct parts (beginning and ending) can still be identified. This means that there is no physical splitting of the work content of any activity. The distinct parts are determined on the basis of the amount of work that must be completed before or after another activity, as dictated by the lead–lag relationships. In Figure 15.29, activity A is partitioned into parts A_1 and A_2. The duration of A_1 is 2 days because there is an $SS = 2$ relationship between

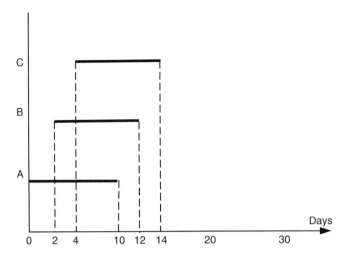

FIGURE 15.28 Gantt chart for PDM example.

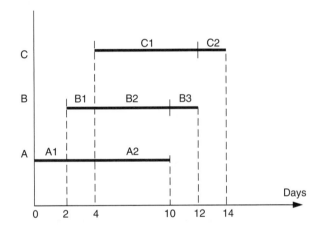

FIGURE 15.29 Partitioning of activities in PDM example.

activity A and activity B. As the original duration of A is 10 days, the duration of A_2 is then calculated to be $10 - 2 = 8$ days.

Likewise, activity B is partitioned into parts B_1, B_2, and B_3. The duration of B_1 is 2 days because there is an $SS = 2$ relationship between activity B and activity C. The duration of B_3 is also 2 days because there is an $FF = 2$ relationship between activity A and activity B. As the original duration of B is 10 days, the duration of B_2 is calculated to be $10 - (2 + 2) = 6$ days. In a similar fashion, activity c is partitioned into C_1 and C_2. The duration of C_2 is 2 days because there is an $FF = 2$ relationship between activity B and activity C. As the original duration of C is 10 days, the duration of C_1 is then calculated to be $10 - 2 = 8$ days. Figure 15.30 shows a conventional CPM network drawn for the three activities after they are partitioned into distinct parts. The conventional forward and backward passes reveal that all the activity parts are performed serially and no physical splitting of activities has been performed. Note that there are three critical paths in Figure 15.30, each with a length of 14 days. It should also be noted that the distinct parts of each activity are performed contiguously.

Figure 15.31 shows an alternate example of three serial activities. The conventional CPM analysis shows that the duration of the network is 30 days. When lead–lag constraints are

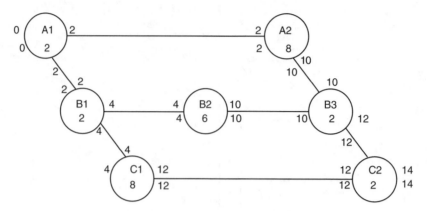

FIGURE 15.30 CPM network of partitioned activities.

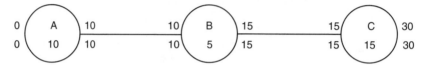

FIGURE 15.31 Another CPM example of serial activities.

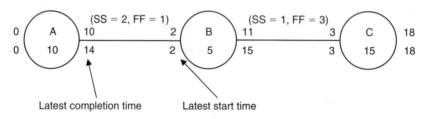

FIGURE 15.32 Compressed PDM network.

introduced into the network as shown in Figure 15.32, the network duration is compressed to 18 days.

In the forward pass computations in Figure 15.32, note that the earliest completion time of B is time 11, because there is an FF = 1 restriction between activity A and activity B. As A finishes at time 10, B cannot finish until at least time 11. Even though the earliest starting time of B is time 2 and its duration is 5 days, its earliest completion time cannot be earlier than time 11. Also note that C can start as early as time 3 because there is an SS = 1 relationship between B and C. Thus, given a duration of 15 days for C, the earliest completion time of the network is $3 + 15 = 18$ days. The difference between the earliest completion time of C and the earliest completion time of B is $18 - 11 = 7$ days, which satisfies the FF = 3 relationship between B and C.

In the backward pass, the latest completion time of B is 15 (i.e., $18 - 3 = 15$), as there is an FF = 3 relationship between activity B and activity C. The latest start time for B is time 2 (i.e., $3 - 1 = 2$), as there is an SS = 1 relationship between activity B and activity C. If we are not careful, we may erroneously set the latest start time of B to 10 (i.e., $15 - 5 = 10$). But that would violate the SS = 1 restriction between B and C. The latest completion time of A is found to be 14 (i.e., $15 - 1 = 14$), as there is an FF = 1 relationship between A and B. All the earliest times and latest times at each node must be evaluated to ensure that they conform to all the lead–lag constraints. When computing earliest start or earliest completion times, the smallest possible value that satisfies the lead–lag constraints should

Project Management

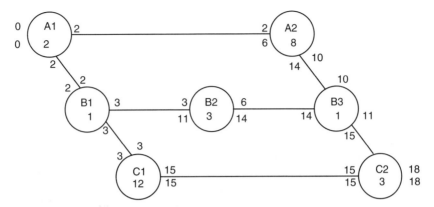

FIGURE 15.33 CPM expansion of second PDM example.

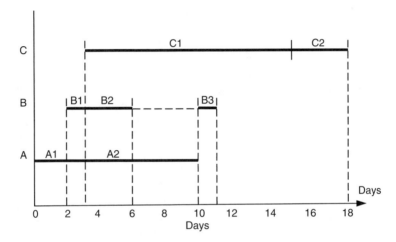

FIGURE 15.34 Compressed PDM schedule based on ES times.

be used. By the same reasoning, when computing the latest start or latest completion times, the largest possible value that satisfies the lead–lag constraints should be used.

Manual evaluations of the lead–lag precedence network analysis can become very tedious for large networks. A computer tool may be needed to implement PDM. If manual analysis must be done for PDM computations, it is suggested that the network be partitioned into more manageable segments. The segments may then be linked after the computations are completed. The expanded CPM network in Figure 15.33 was developed on the basis of the precedence network in Figure 15.32. It is seen that activity A is partitioned into two parts, activity B is partitioned into three parts, and activity C is partitioned into two parts. The forward and backward passes show that only the first parts of activities A and B are on the critical path, whereas both parts of activity C are on the critical path.

Figure 15.34 shows the corresponding earliest-start Gantt chart for the expanded network. Looking at the earliest start times, one can see that activity B is physically split at the boundary of B_2 and B_3 in such a way that B_3 is separated from B_2 by 4 days. This implies that work on activity B is temporarily stopped at time 6 after B_2 is finished and is not started again until time 10. Note that despite the 4-day delay in starting B_3, the entire project is not delayed. This is because B_3, the last part of activity B, is not on the critical path. In fact, B_3 has a total slack of 4 days. In a situation like this, the duration of activity

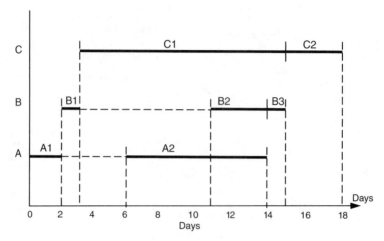

FIGURE 15.35 Compressed PDM schedule based on LS times.

B can actually be increased from 5 days to 9 days without any adverse effect on the project duration. It should be recognized, however, that increasing the duration of an activity may have negative implications for project cost, personnel productivity, and morale.

If physical splitting of activities is not permitted, then the best option available in Figure 15.34 is to stretch the duration of B_3 so as to fill up the gap from time 6 to time 10. An alternative is to delay the starting time of B_1 until time 4 so as to use up the 4-day slack right at the beginning of activity B. Unfortunately, delaying the starting time of B_1 by 4 days will delay the overall project by 4 days, as B_1 is on the critical path as shown in Figure 15.33. The project analyst will need to evaluate the appropriate tradeoffs between splitting of activities, delaying activities, increasing activity durations, and incurring higher project costs. The prevailing project scenario should be considered when making such trade-off decisions. Figure 15.35 shows the Gantt chart for the compressed PDM schedule based on latest start times. In this case, it will be necessary to split both activities A and B even though the total project duration remains the same at 18 days. If activity splitting is to be avoided, then we can increase the duration of activity A from 10 to 14 days and the duration of B from 5 to 13 days without adversely affecting the entire project duration. The benefit of precedence diagramming is that the ability to overlap activities facilitates flexibility in manipulating individual activity times and reducing project duration.

15.5.1 Reverse Criticality in PDM Networks

Care must be exercised when working with PDM networks because of the potential for misuse or misinterpretation. Because of the lead and lag requirements, activities that do not have any slacks may appear to have generous slacks. Also, "reverse critical" activities may occur in PDM. Reverse critical activities are activities that can cause a decrease in project duration when their durations are increased. This may happen when the critical path enters the completion of an activity through a finish lead–lag constraint. Also, if a "finish-to-finish" dependency and a "start-to-start" dependency are connected to a reverse critical task, a reduction in the duration of the task may actually lead to an increase in the project duration. Figure 15.36 illustrates this anomalous situation. The finish-to-finish constraint between A and B requires that B should finish no earlier than 20 days. If the duration of task B is reduced from 10 days to 5 days, the start-to-start constraint between

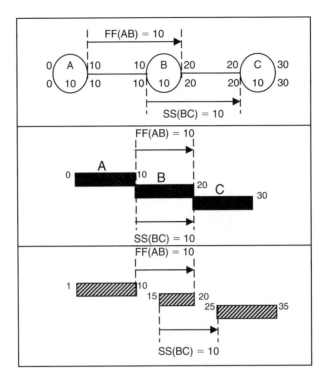

FIGURE 15.36 Reverse critical activity in PDM network.

B and C forces the starting time of C to be shifted forward by 5 days, thereby resulting in a 5-day increase in the project duration.

The preceding anomalies can occur without being noticed in large PDM networks. One safeguard against their adverse effects is to make only one activity change at a time and document the resulting effect on the network structure and duration. The following categorizations are used for the unusual characteristics of activities in PDM networks.

Normal Critical (NC): This refers to an activity for which the project duration shifts in the same direction as the shift in the duration of the activity.

Reverse Critical (RC): This refers to an activity for which the project duration shifts in the reverse direction to the shift in the duration of the activity.

Bi-Critical (BC): This refers to an activity for which the project duration increases as a result of any shift in the duration of the activity.

Start Critical (SC): This refers to an activity for which the project duration shifts in the direction of the shift in the start time of the activity, but is unaffected (neutral) by a shift in the overall duration of the activity.

Finish Critical (FC): This refers to an activity for which the project duration shifts in the direction of the shift in the finish time of the activity, but is unaffected (neutral) by a shift in the overall duration of the activity.

Mid Normal Critical (MNC): This refers to an activity whose mid-portion is normal critical.

Mid Reverse Critical (MRC): This refers to an activity whose mid-portion is reverse critical.

Mid Bi-Critical (MBC): This refers to an activity whose mid-portion is bi-critical.

A computer-based decision support system can facilitate the integration and consistent usage of all the relevant information in a complex scheduling environment. Task precedence relaxation assessment and resource-constrained heuristic scheduling constitute an example of a problem suitable for computer implementation. Examples of pseudocoded heuristic rules for computer implementation are shown below:

IF: *Logistical conditions are satisfied*
THEN: *Perform the selected scheduling action*

IF: condition A is satisfied and
         condition B is false and
         evidence C is present or
         observation D is available
THEN: precedence belongs in class X

IF: *precedence belongs to class X*
THEN: *activate heuristic scheduling procedure Y*

The function of the computer model will be to aid a decision maker in developing a task sequence that fits the needs of concurrent scheduling. Based on user input, the model will determine the type of task precedence, establish precedence relaxation strategy, implement task scheduling heuristic, match the schedule to resource availability, and present a recommended task sequence to the user. The user can perform "what-if" analysis by making changes in the input data and conducting sensitivity analysis. The computer implementation can be achieved in an interactive environment as shown in Figure 15.37. The user will provide task definitions and resource availabilities with appropriate precedence requirements. At each stage, the user is prompted to consider potential points for precedence relaxation. Wherever schedule options exist, they will be presented to the user for consideration and approval. The user will have the opportunity to make final decisions about task sequence.

15.6 Software Tools for Project Management

There are numerous commercial software packages available for project management. Because of the dynamic changes in software choices on the market, it will not be effective to include a survey of the available software in an archival publication of this nature. Any such review may be outdated before the book is even published. To get the latest on commercial software capabilities and choices, it will be necessary to consult one of the frequently published trade magazines that carry software reviews. Examples of such magazines are *PC Week*, *PC World*, and *Software*. Other professional publications also carry project management software review occasionally. Examples of such publications are *Industrial Engineering Magazine*, *OR/MS Today Magazine*, *Manufacturing Engineering Magazine*, *PM Network Magazine*, and so on. Practitioners can easily get the most current software information through the Internet. Prospective buyers are often overwhelmed by the range of products available. However, there are important factors to consider when selecting a software package. Some of the factors are presented in this section.

The proliferation of project management software has created an atmosphere whereby every project analyst wants to have and use a software package for every project situation. However, not every project situation deserves the use of project management software. An analyst should first determine whether or not the use of software is justified. If this

Project Management   15-35

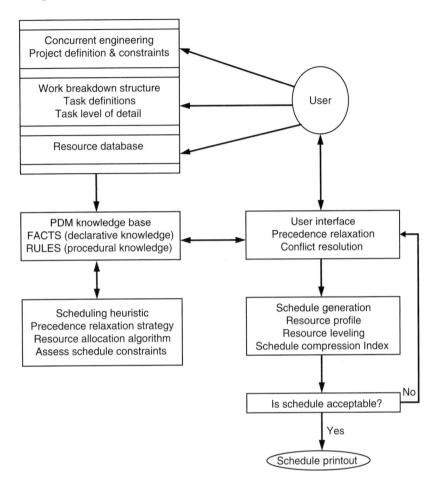

FIGURE 15.37 Decision support model for PDM.

evaluation is affirmative, then the analyst will need to determine which specific package out of the many that are available should be used. Some of the important factors that may indicate the need for project management software are:

1. Multiple projects are to be managed concurrently.
2. A typical project contains more than 20 tasks.
3. The project scheduling environment is very complex.
4. More than five resource types are involved in each project.
5. There is a need to perform complementing numerical analysis to support project control functions.
6. The generation of graphics (e.g., Gantt, PERT charts) are needed to facilitate project communication.
7. Cost analysis is to be performed on a frequent basis.
8. It is necessary to generate forecasts from historical project data.
9. Automated reporting is important to the organization.
10. Computerization is one of the goals of the organization.

A close examination of a project will reveal whether it fits the criteria for using project management software. Only very small and narrowly focused projects will not need the help of software for effective management. Some of the important factors for software selection are summarized below:

1. Cost
2. Need
3. Complexity of project scope and plan
4. Diversity of resource types
5. Frequency of progress tracking
6. Need for versatile report generation
7. Ease of use and learning curve requirement
8. Supporting analytical tools (e.g., optimization add-on, economic analysis)
9. Hardware requirements
10. General characteristics (e.g., software version, vendor accessibility, technical support).

Some of the above factors may be more important than others in specific project situations. A careful overall analysis should be done by the project analyst. With more and more new programs and updates appearing in the market, a crucial aspect of the project management function is keeping up with what is available and making a good judgment in selecting a software tool.

The project management software market continues to be very competitive. Many packages that were originally developed for specific and narrow applications, such as data analysis and conventional decision support, now offer project management capabilities. For example, SAS/OR, a statistical analysis software package that is popular on large computers, has a PC-based version that handles project management. Similarly, AutoCAD, the popular computer-aided design software, now has a project management option within it. The option, called AutoProject, has very good graphics and integrated drafting capabilities.

Some of the most popular project management software packages include InstaPlan 5000, Artemis Project, Microsoft Project for Windows, Plantrac II, Advanced Project Workbench, Qwiknet Professional, Super Project Plus, Time Line, Project Scheduler, Primavera, Texim, ViewPoint, PROMIS, Topdown Project Planner, Harvard Project Manager, PCS (Project Control System), PAC III, VISION, Control Project, SAS/OR, Autoproject, Visual Planner, Project/2, Timesheet, Task Monitor, Quick Schedule Plus, Suretrak Project Scheduler, Supertime, On Target, Great Gantt, Pro Tracs, Autoplan, AMS Time Machine, Mac-Project II, Micro Trak, Checkpoint, Maestro II, Cascade, OpenPlan, Vue, and Cosmos. So prolific are the software offerings that the developers are running out of innovative productive names. At this moment, new products are being introduced while some are being phased out. Prospective buyers of project management software should consult vendors for the latest products. No blanket software recommendation can be offered in this chapter as product profiles change quickly and frequently.

15.6.1 Computer Simulation Software

Special purpose software tools have found a place in project management. Some of these tools are simulation packages, statistical analysis packages, optimization programs, report writers, and others. Computer simulation is a versatile tool that has a potential for enhancing project planning and control analysis. Computer simulation is a tool that can be effectively utilized to enhance project planning, scheduling, and control. At any given time, only a small segment of a project network will be available for direct observation and analysis. The major portion of the project either will have been in the past or will be expected in the future. Such unobservable portions of the project can be studied by simulation.

Using the historical information from previous segments of a project and the prevailing events in the project environment, projections can be made about future expectations of the project. Outputs of simulation can alert management to real and potential problems. The information provided by simulation can be very helpful in projecting selection decisions. Simulation-based project analysis may involve the following components:

- Activity time modeling
- Simulation of project schedule
- What-if analysis and statistical modeling
- Management decisions and sensitivity analysis.

15.7 Conclusion

This chapter has presented the basic techniques of activity network analysis for project management. In business and industry, project management is rapidly becoming one of the major tools used to accomplish goals. Engineers, managers, and OR professionals are increasingly required to participate on teams in complex projects. The technique of network analysis is frequently utilized as a part of the quantitative assessment of such projects. The computational approaches contained in this chapter can aid project analysts in developing effective project schedules and determining the best way to exercise project control whenever needed.

Further Reading

1. Badiru, A.B., *Project Management in Manufacturing and High Technology Operations*, 2nd edition, John Wiley & Sons, New York, 1996.
2. Badiru, A.B. and P. S. Pulat, *Comprehensive Project Management: Integrating Optimization Models, Management Principles, and Computers*, Prentice-Hall, Englewood Cliffs, NJ, 1995.

16
Quality Control

16.1	Introduction	**16**-1
16.2	Quality Control and Product Life Cycle	**16**-2
16.3	New Trends and Relationship to Six Sigma	**16**-5
16.4	Statistical Process Control	**16**-7
	Process Control System • Sources of Variation • Use of Control Charts for Problem Identification • Statistical Control • Control Charts • Benefits of Control Charts	
16.5	Process Capability Studies	**16**-16
16.6	Advanced Control Charts	**16**-18
	Cumulative Sum Control Charts • Exponentially Weighted Moving Average Control Charts • Other Advanced Control Charts	
16.7	Limitations of Acceptance Sampling	**16**-20
16.8	Conclusions	**16**-20
	References	**16**-21
	Appendix	**16**-23

Qianmei Feng
University of Houston

Kailash C. Kapur
University of Washington

16.1 Introduction

Quality has been defined in different ways by various experts and the operational definition has even changed over time. The best way is to start from the beginning and look at the origin and meaning of the word. Quality, in Latin *qualitas*, comes from the word *qualis*, meaning "how constituted" and signifying "such as a thing really is." The Merriam-Webster dictionary defines quality as "... peculiar and essential character ... a distinguishing attribute...." Thus, a product has several or infinite qualities. Juran and Gryna [1] looked at multiple elements of fitness of use based on various quality characteristics (or qualities), such as technological characteristics (strength, dimensions, current, weight, ph values), psychological characteristics (beauty, taste, and many other sensory characteristics), time-oriented characteristics (reliability, availability, maintainability, safety, and security), cost (purchase price, life cycle cost), and product development cycle. Deming also discussed several faces of quality; the three corners of quality relate to various quality characteristics and focus on evaluation of quality from the viewpoint of the customer [2]. The American Society for Quality defines quality as the "totality of features and characteristics of a product or service that bear on its ability to satisfy a user's given needs" [3]. Thus the quality of a process or product is defined and evaluated by the customer. Any process has many processes before it that are typically called the suppliers and has many processes after it that are its customers. Thus, anything (the next process, environment, user) the present process affects is its customer.

```
                    Quality
                      ⇓ ⇑
             Customer satisfaction
                      ⇓ ⇑
              Voice of customer
                      ⇓ ⇑
            Substitute characteristics
                      ⇓ ⇑
                 Target values
```

FIGURE 16.1 Quality, customer satisfaction, and target values.

One of the most important tasks in any quality program is to understand and evaluate the needs and expectations of the customer and to then provide products and services that meet or exceed those needs and expectations. Shewhart states this as follows: "The first step of the engineer in trying to satisfy these wants is, therefore, that of translating as nearly as possible these wants into the physical characteristics of the thing manufactured to satisfy these wants. In taking this step, intuition and judgment play an important role as well as the broad knowledge of human element involved in the wants of individuals. The second step of the engineer is to set up ways and means of obtaining a product which will differ from the arbitrary set standards for these quality characteristics by no more than may be left to chance" [4]. One of the objectives of quality function deployment (QFD) is exactly to achieve this first step proposed by Shewhart. Mizuno and Akao have developed the necessary philosophy, system, and methodology to achieve this step [5]. QFD is a means to translate the "voice of the customer" into substitute quality characteristics, design configurations, design parameters, and technological characteristics that can be deployed (horizontally) through the whole organization: marketing, product planning, design, engineering, purchasing, manufacturing, assembly, sales, and service [5,6]. Products have several characteristics, and an "ideal" state or value of these characteristics must be determined from the customer's viewpoint. This ideal state is called the target value (Figure 16.1). QFD is a methodology to develop target values for substitute quality characteristics that satisfy the requirements of the customer. The purpose of statistical process control is to accomplish the second step mentioned by Shewhart, and his pioneering book [4] developed the methodology for this purpose.

16.2 Quality Control and Product Life Cycle

Dynamic competition has become a key concept in world-class design and manufacturing. Competition is forcing us to provide products and services with less variation than our competitors. Quality effort in many organizations relies on a combination of audits, process and product inspections, and statistical methods using control charts and sampling inspection. These techniques are used to control the manufacturing process and meet specifications in the production environment. We must now concentrate on achieving high quality by starting at the beginning of the product life cycle. It is not enough for a product to work well when manufactured according to engineering specifications. A product must be designed for manufacturability and must be insensitive to variability present in the production environment and in the field when used by the customer. Reduced variation translates into greater

Quality Control

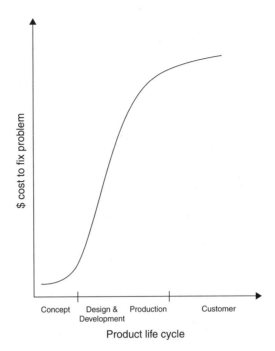

FIGURE 16.2 Cost to fix problems vs. product life cycle.

repeatability, reliability, and ultimately cost savings to both the producer and the consumer and thus the whole society.

Figure 16.2 shows how the cost to fix or solve problems increases as we move downstream in the product life cycle. Another way to emphasize the early and proactive activities related to quality is to evaluate Figure 16.3, which shows that approximately 90% of the life cycle cost is determined by the concept and development phases of the life cycle.

Different quality improvement methods should be applied for the different phases of the product life cycle. During the phase of design and development of products, design of experiments should be utilized to select the optimal combination of input components and minimize the variation of the output quality characteristic, as shown in Figure 16.4. Sometimes, we use offline quality engineering to refer to the quality improvement efforts including experiment design and robust parameter design, because these efforts are made off the production line.

As opposed to offline quality engineering, online quality control refers to the techniques employed during the manufacturing process of products. Statistical quality control (SQC) is a primary online control technique for monitoring the manufacturing process or any other process with key quality characteristics of interest. Before we use statistical quality control to monitor the manufacturing process, the optimal values of the mean and standard deviation of the quality characteristics are determined by minimizing the variability of the quality characteristics through experimental design and process adjustment techniques. Consequently, the major goal of SQC (SPC) is to monitor the manufacturing process, keep the values of mean and standard deviation stable, and finally reduce variability.

The manufactured products need to be inspected or tested before they reach the customer. Closely related to the inspection of output products, acceptance sampling is defined as the inspection and classification of samples from a lot randomly and decision about disposition of the lot.

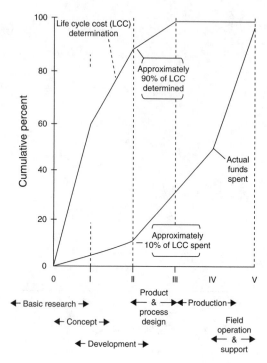

FIGURE 16.3 Life cycle cost and production life cycle.

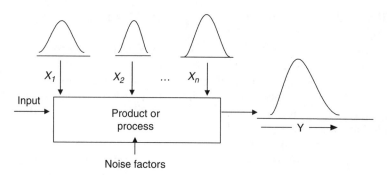

FIGURE 16.4 Offline quality engineering.

At different phases of the product life cycle, the modeling approaches in operations research and management science (OR/MS) described in the first part of this handbook are widely used. For example, as a collection of statistical and optimization methods, the response surface methodology (RSM) is a specialized experimental design technique in offline quality engineering. The basic idea of RSM is to fit a response model for the output variable and then explore various settings of the input design variables with the purpose to maximize or minimize the response [7]. For online quality control and inspection efforts, various optimization modeling approaches are involved for process adjustment, economic design of control charts, and optimization of specifications, and so on [8,9].

Six Sigma methodology is one of the most distinctive applications of quality improvement effort with the involvement of OR/MS approaches. As a collection of optimization, statistical, engineering, and management methods, Six Sigma methodology emerged in manufacturing industry, and has been widely practiced in many industries including energy,

16.3 New Trends and Relationship to Six Sigma

Industrial, manufacturing, and service organizations are interested in improving their processes by decreasing the process variation because the competitive environment leaves little room for error. Based on the ideal or target value of the quality characteristic from the viewpoint of the customer, the traditional evaluation of quality is based on average measures of the process/product and their deviation from the target value. But customers judge the quality of process/product not only on the average measure, but also by the variance in each transaction with the process or use of the product. Customers want consistent, reliable, and predictable processes that deliver best-in-class levels of quality. This is what the Six Sigma process strives to achieve. Six Sigma has been applied by many manufacturing companies such as GE, Motorola, and service industries including health care systems.

Six Sigma is a customer-focused, data-driven, and robust methodology, which is well rooted in mathematics and statistics [10–12]. A typical process for Six Sigma quality improvement has six phases: Define, Measure, Analyze, Improve, Control, and Technology Transfer, denoted by (D)MAIC(T). Traditionally, a four-phase process, MAIC is often referred in the literature [13]. We extend it to the six-phase process, (D)MAIC(T). We want to emphasize the importance of the define (D) phase as the first phase for problem definition and project selection and technology transfer (T) as the never-ending phase for continuous applications of the Six Sigma technology to other parts of the organization to maximize the rate of return on the investment in developing this technology.

The process of (D)MAIC(T) stays on track by establishing deliverables at each phase, and by creating engineering models over time to reduce the process variation. Each of the six phases answers some target questions, and this continuously improves the implementation and the effectiveness of the methodology [8].

Define—What problem needs to be solved?
Measure—What is the current capability of the process?
Analyze—What are the root causes for the process variability?
Improve—How to improve the process capability?
Control—What controls can be put in place to sustain the improvement?
Technology Transfer—Where else can these improvements be applied?

In each phase, there are several steps that need to be implemented. For each step, many quality improvement methods, tools, and techniques are used. We next describe the six phases in more detail.

Phase 0: Define (D)
Once an organization decides to launch a Six Sigma process improvement project, they need to first define the improvement activities. Usually the following two steps are taken in the define phase:

Step 0.1: *Identify and prioritize customer requirements.* Methods such as benchmarking surveys, spider charts, and customer needs mapping must be put in place to ensure that the customer requirements are properly identified. The critical to quality (CTQ) characteristics are defined from the viewpoint of customers, which are also called external CTQs. We need to translate the external CTQs into internal CTQs that are key process requirements. This translation is the foremost step in the measure phase.

Step 0.2: *Select projects.* Based on customer requirements, the target project is selected by analyzing the gap between the current process performance and the requirement of customers. Specifically, we need to develop a charter for the project, including project scope, expectations, resources, milestones, and the core processes.

Phase 1: Measure (M)

Six Sigma is a data-driven approach that requires quantifying and benchmarking the process using actual data. In this phase, the performance or process capability of the process for the CTQ characteristics are evaluated.

Step 1.1: *Select CTQ characteristics.* This step uses tools such as QFD and FMECA to translate the external CTQs established in the define phase into internal requirements denoted by Y's. Some of the objectives for this step are:

- Define, construct, and interpret the QFDs.
- Participate in a customer needs mapping session.
- Apply failure mode, effect, and criticality analysis (FMECA) to the process of selecting CTQ characteristics.
- Identify CTQs and internal Y's.

Step 1.2: *Define performance standards.* After identifying the product requirements, Y's, measurement standards for the Y's are defined in this step. QFD, FMECA, as well as process mapping can be used to establish internal measurement standards.

Step 1.3: *Validate measurement system.* We need to learn how to validate measurement systems and determine the repeatability and reproducibility of these systems using tools such as gage R&R. This determination provides for separation of variability into components and thus into targeted improvement actions.

Phase 2: Analyze (A)

Once the project is understood and the baseline performance is documented, it is time to do an analysis of the process. In this phase, the Six Sigma approach applies statistical tools to validate the root causes of problems. The objective is to understand the process in sufficient detail so that we are able to formulate options for improvement.

Step 2.1: *Establish product capability.* This step determines the current product capability, associated confidence levels, and sample size by process capability analysis, described in Section 16.5. The typical definition for process capability index, C_{pk}, is $C_{\text{pk}} = \min\left\{\frac{USL - \hat{\mu}}{3\hat{\sigma}}, \frac{\hat{\mu} - LSL}{3\hat{\sigma}}\right\}$, where USL is the upper specification limit, LSL is the lower specification limit, $\hat{\mu}$ is the point estimator of the mean, and $\hat{\sigma}$ is the point estimator of the standard deviation. If the process is centered at the middle of the specifications, which is also interpreted as the target value, that is, $\hat{\mu} = \frac{USL + LSL}{2} = y_0$, then 6σ process means that $C_{\text{pk}} = 2$.

Step 2.2: *Define performance objectives.* The performance objectives are defined to establish a balance between improving customer satisfaction and available resources. We should distinguish between courses of actions necessary to improve process capability versus technology capability.

Step 2.3: *Identify variation sources.* This step begins to identify the causal variables that affect the product requirements, or the responses of the process. Some of these causal variables might be used to control the responses Y's. Experimental design and analysis should be applied for the identification of variation sources.

Phase 3: Improve (I)
In the improvement phase, ideas and solutions are implemented to initialize the change. Experiments are designed and analyzed to find the best solution using optimization approaches.

Step 3.1: *Discover variable relationships.* In the previous step, the causal variables X's are identified with a possible prioritization as to their importance in controlling the Y's. In this step, we explore the impact of each vital X on the responses Y's. A system transfer function (STF) is developed as an empirical model relating the Y's and the vital X's.

Step 3.2: *Establish operating tolerances.* After understanding the functional relationship between the vital X's and the responses Y's, we need to establish the operating tolerances of the X's that optimize the performance of the Y's. Mathematically, we develop a variance transmission equation (VTE) that transfers the variances of the vital X's to variances of Y's.

Step 3.3: *Optimize variable settings.* The STF and VTE will be used to determine the key operating parameters and tolerances to achieve the desired performance of the Y's. Optimization models are developed to determine the optimum values for both means and variances for these vital X's.

Phase 4: Control (C)
The key to the overall success of the Six Sigma methodology is its sustainability. Performance tracking mechanisms and measurements are put in place to assure that the process remains on the new course.

Step 4.1: *Validate measurement system.* The measurement system tools first applied in Step 1.3 will now be used for the X's.

Step 4.2: *Implement process controls.* Statistical process control is a critical element in maintaining a Six Sigma level. Control charting is the major tool used to control the vital few X's. Special causes of process variations are identified through the use of control charts, and corrective actions are implemented to reduce variations.

Step 4.3: *Documentation of the improvement.* We should understand that the project is not complete until the changes are documented in the appropriate quality management system, such as QS9000/ISO9000. A translation package and plan should be developed for possible technology transfer.

Phase ∞: Technology Transfer (T)
Using the infinity number, we convey the meaning that transferring technology is a never-ending phase for achieving Six Sigma quality. Ideas and knowledge developed in one part of the organization can be transferred to other parts of the organization. In addition, the methods and solutions developed for one product or process can be applied to other similar products or processes. With technology transfer, the Six Sigma approach starts to create phenomenal returns.

16.4 Statistical Process Control

A traditional approach to manufacturing and addressing quality is to depend on production to make the product and on quality control to inspect the final product and screen out the items that do not meet the requirements of the next customer. This detection strategy related to after-the-fact inspection is mostly uneconomical, as the wasteful production has

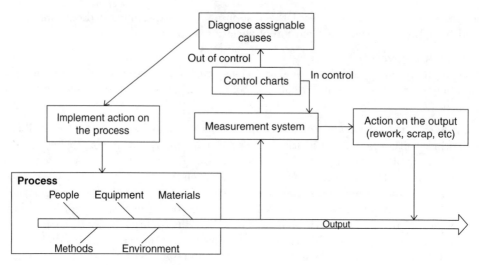

FIGURE 16.5 A process control system.

already been produced. A better strategy is to avoid waste by not producing the unacceptable output in the first place and thus focus more on prevention. Statistical process control (SPC) is an effective prevention strategy to manufacture products that will meet the requirements of the customer [14]. We will discuss the following topics in this section.

1. What is a process control system?
2. What are different types of variation and how do they affect the process output? We will discuss two types of variation based on common or system causes and special or assignable causes.
3. How can a control chart or statistical methods tell us whether a problem is due to special causes or common causes?
4. What is meant by a process being in statistical control?
5. What are control charts and how are they used?
6. What benefits can be expected from using control charts?

16.4.1 Process Control System

A process control system can be described as a feedback system as shown in Figure 16.5. Four elements of that system are important to the discussions that will follow:

1. The process—by the process, we mean the whole combination of people, equipment, input materials, methods, and environment that work together to produce output. The total performance of the process—the quality of its output and its productive efficiency—depends on the way the process has been designed and built, and on the way it is operated. The rest of the process control system is useful only if it contributes to improved performance of the process.
2. Information about performance—much information about the actual performance of the process can be learned by studying the process output. In a broad sense, process output includes not only the products that are produced, but also any intermediate "outputs" that describe the operating state of the process, such as

temperatures, cycle times, and the like. If this information is gathered and interpreted correctly, it can show whether action is necessary to correct the process or the just-produced product. If timely and appropriate actions are not taken, however, any information-gathering effort is wasted.

3. Action on the process—action on the process is future-oriented, as it is taken when necessary to prevent the production of nonconforming products. This action might consist of changes in the operations (e.g., operator training, changes to the incoming materials, etc.) or the more basic elements of the process itself (e.g., the equipment—which may need rehabilitation, or the design of the process as a whole—which may be vulnerable to changes in shop temperature or humidity).

4. Action on the output—action on the output is past-oriented, because it involves detecting out-of-specification output already produced. Unfortunately, if current output does not consistently meet customer requirements, it may be necessary to sort all products and to scrap or rework any nonconforming items. This must continue until the necessary corrective action on the process has been taken and verified, or until the product specifications have been changed.

It is obvious that inspection followed by action only on the output is a poor substitute for effective first-time process performance. Therefore, the discussions that follow focus on gathering process information and analyzing it so that action can be taken to correct the process itself.

Process control plays a very important role during the effort for process improvement. When we try to control a process, analysis and improvement naturally result; and when we try to make an improvement, we naturally come to understand the importance of control. We can only make a breakthrough when we have achieved control. Without process control, we do not know where to improve, and we cannot have standards and use control charts. Improvement can only be achieved through process analysis.

16.4.2 Sources of Variation

Usually, the sources of variability in a process are classified into two types: chance causes and assignable causes of variation. Chance causes, or common causes, are the sources of inherent variability, which cannot be removed easily from the process without fundamental changes in the process itself. Assignable causes, or special causes, arise in somewhat unpredictable fashion, such as operator error, material defects, or machine failure. The variability due to assignable causes is comparatively larger than chance causes, and can cause the process to go out of control. Table 16.1 compares the two sources of variation, including some examples.

TABLE 16.1 Sources of Variation

Common or Chance Causes	Special or Assignable Causes
1. Consist of many individual causes	1. Consist of one or just a few individual causes
2. Any one chance cause results in only a minute amount of variation. (However, many of chance causes together result in a substantial amount of variation)	2. Any one assignable cause can result in a large amount of variation
3. As a practical matter, chance variation cannot be economically eliminated—the process may have to be changed to reduce variability	3. The presence of assignable variation can be detected (by control charts) and action to eliminate the causes is usually economically justified
4. Examples: • Slight variations in raw materials • Slight vibrations of a machine • Lack of human perfection in reading instruments or setting controls	4. Examples: • Batch of defective raw materials • Faulty setup • Untrained operator

16.4.3 Use of Control Charts for Problem Identification

Control charts by themselves do not correct problems. They indicate that something is wrong and it is up to you to take the corrective action. Assignable causes are the factors that cause the process to go out of control. They are due to change in the condition of manpower, materials, machines, or methods or a combination of all of these.

Assignable causes relating to manpower:
- New or wrong man on the job
- Careless workmanship and attitudes
- Improper instructions
- Domestic, personal problems

Assignable causes relating to materials:
- Improper work handling
- Stock too hard or too soft
- Wrong dimensions
- Contamination, dirt, etc.
- Improper flow of materials

Assignable causes relating to machines or methods:
- Dull tools
- Poor housekeeping
- Machine adjustment
- Improper machine tools, jigs, fixtures
- Improper speeds, feeds, etc.
- Improper adjustments and maintenance
- Worn or improperly located locators

When assignable causes are present, as shown in Figure 16.6, the probability of nonconformance may increase, and the process quality deteriorates significantly. The eventual goal of SPC is to improve the process quality by reducing variability in the process. As one of the primary SPC techniques, the control chart can effectively detect the variation due to the assignable causes and reduce process variability if the identified assignable causes can be eliminated from the process.

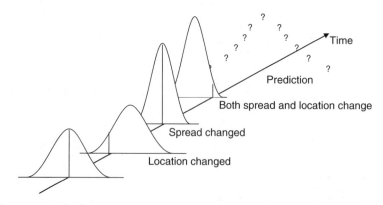

FIGURE 16.6 Unstable and unpredictable process with presence of assignable causes.

16.4.4 Statistical Control

Statistically, SPC techniques aim to detect changes over time in the parameters (e.g., mean and standard deviation) of the underlying distribution for the process. In general, the statistical process control problem can be described as below [15]. Let X denote a random variable for a quality characteristic with the probability density function, $f(x; \theta)$, where θ is a set of parameters. If the process is operating with $\theta = \theta_0$, it is said to be in statistical control; otherwise, it is out of control. The value of θ_0 is not necessarily equal to the target (or ideal) value of the process. Due to the effort of experimental design and process adjustment techniques, a process is assumed to start with the in-control state [7,16,17]. After a random length of time, variability in the process will possibly cause deterioration or shift of the process. This shift can be reflected by a change in θ from the value of θ_0, and the process is said to be out of control. Therefore, the basic goal of control charts is to detect changes in θ that can occur over time.

A process is said to be operating in statistical control when the only source of variation is common causes. The status of statistical control is obtained by eliminating special causes of excessive variation one by one.

Process capability is determined by the total variation that comes from common causes. A process must first be brought into statistical control, and then its capability to meet specifications can be assessed. We will discuss the details of process capability analysis in the next section.

16.4.5 Control Charts

The basic concept of control charts was proposed by Walter A. Shewhart of the Bell Telephone Laboratories in the 1920s, which indicates the formal beginning of statistical quality control. The effective use of the control chart involves a series of process improvement activities. For a process variable of interest, one must observe data from a process over time, or monitor the process, and apply a control chart to detect process changes. When the control chart signals the possible presence of an assignable cause, efforts should be made to diagnose the assignable causes and implement corrective actions to remove the assignable causes so as to reduce variability and improve the process quality. The long history of control charting application in many industries has proven its effectiveness for improving productivity, preventing defects and providing information about diagnostic and process capability.

A typical control chart is given in Figure 16.7. The basic model for Shewhart control charts consist of a center line, an upper control limit (UCL) and a lower control limit (LCL) [18].

$$\text{UCL} = \mu_s + L\sigma_s$$
$$\text{Center line} = \mu_s \tag{16.1}$$
$$\text{LCL} = \mu_s - L\sigma_s$$

where μ_s and σ_s are the mean and standard deviation of the sample statistic, such as sample mean (X-bar chart), sample range (R chart), and sample proportion defective (p chart). $L\sigma_s$ is the distance of the control limits from the center line and L is most often set at three. To construct a control chart, one also needs to specify the sample size and sampling frequency. The common wisdom is to take smaller samples at short intervals or larger samples at longer intervals, so that the sampling effort can be allocated economically. An important concept related to the sampling scheme is the rational subgroup approach, recommended by Shewhart. To maximize detection of assignable causes between samples, the rational subgroup approach takes samples in a way that the within-sample variability is only due to common causes, while the between-sample variability should indicate assignable causes in

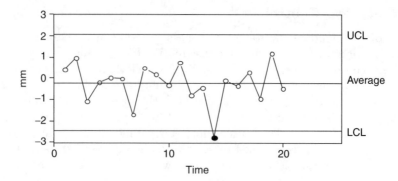

FIGURE 16.7 A typical control chart (x-bar chart).

TABLE 16.2 The Most Commonly Used Shewhart Control Charts

Symbol	Description	Sample Size
	Variable Charts	
X-bar and R	The average (mean) and range of measurements in a sample	Must be constant
X-bar and S	The average (mean) and standard deviation of measurements in a sample	May be variable
	Attributes Charts	
p	The percent of defective (nonconforming) units in a sample	May be variable
np	The number of defective (nonconforming) units in a sample	Must be constant
c	The number of defects in a sample	Must be constant
u	The number of defects per unit	May be variable

the process. Further discussion of the rational subgroup approach can be found in Refs. [19] and [20].

An out-of-control signal is given when a sample statistic falls beyond the control limits, or a nonrandom pattern presents. Western Electric rules are used to identify the nonrandom pattern in the process. According to Western Electric rules [21], a process is considered out of control if any of the following occur:

1. One or more points outside 3σ limits.
2. Two of three consecutive points outside 2σ limits.
3. Four of five consecutive points beyond the 1σ limits.
4. A run of eight consecutive points on one side of the center line.

More decision rules or sensitizing rules can be found in the textbook by Montgomery.

The measurements of quality characteristics are typically classified as attributes or variables. Continuous measurements, such as length, thickness, or voltage, are variable data. Discrete measurements, such as number of defective units, number of nonconformities per unit, are attributes. The most commonly used Shewhart control charts for both attributes and variables are summarized in Table 16.2.

Control Charts for Variables

When a quality characteristic is measured as variables, it is necessary to monitor the process mean and standard deviation. For grouped data, we use $\overline{X}$ charts to detect the process mean shift (between-group variability), and R charts or S charts to monitor the process variation (within-group variability). The control limits of each chart are constructed based on the Shewhart model in Equation 16.1. When we use $\overline{X}$, R, and S charts, we assume that the underlying distribution of the quality characteristic is normal and the observations exhibit no correlation over time. If the quality characteristic is extremely nonnormal or the observations are autocorrelated, other control charts such as the exponentially weighted moving average control charts (EWMA) or time series model (ARIMA) may be used instead.

In practice, the parameters of the underlying distribution of a quality characteristic are not known. We need to estimate the process mean and standard deviation based on the preliminary data. It can be shown that an unbiased estimate of the standard deviation is $\hat{\sigma} = s/c_4$, where s is the sample standard deviation. A more convenient approach in quality control application is the range method, where the range of the sample, R, is used to estimate the standard deviation, and it is obtained as $\hat{\sigma} = R/d_2$. The resulting control charts using different estimators of standard deviation are the R chart and the S chart, respectively.

$\overline{X}$ and R Charts

When the sample size is not very large ($n < 10$), $\overline{X}$ and R charts are widely used to monitor variable quality characteristics due to their simplicity of application. To use the basic Shewhart model in Equation 16.1 for $\overline{X}$ and R charts, we need to estimate $\mu_{\overline{x}}$ and $\sigma_{\overline{x}}$, μ_R, and σ_R first.

It is obvious that we can use the grand average to estimate $\mu_{\overline{x}}$ and μ_R, that is, $\hat{\mu}_{\overline{x}} = \overline{\overline{x}}$ and $\hat{\mu}_R = \overline{R}$. Using the range method, we have $\hat{\sigma}_{\overline{x}} = \hat{\sigma}/\sqrt{n} = \overline{R}/(d_2\sqrt{n})$ and $\hat{\sigma}_R = d_3\hat{\sigma} = d_3\overline{R}/d_2$. The control limits for $\overline{X}$ and R charts are

$$\text{LCL} = \overline{\overline{x}} - A_2\overline{R} \qquad \text{LCL} = D_3\overline{R}$$
$$\text{CL} = \overline{\overline{x}} \qquad \text{and} \qquad \text{CL} = \overline{R}$$
$$\text{ULC} = \overline{\overline{x}} + A_2\overline{R} \qquad \text{UCL} = D_4\overline{R}$$

respectively, where $A_2 = \frac{3}{d_2\sqrt{n}}$, $D_3 = 1 - \frac{3d_3}{d_2}$, and $D_4 = 1 + \frac{3d_3}{d_2}$. The values of d_2, d_3, A_2, D_3, and D_4 can be obtained from most books on control charts for n up to 25 [22,20]. Normally, the preliminary data used to establish the control limits are 20–25 samples with sample size 3–5. The established control limits are then used to check if the preliminary samples are in control. The R chart (or S chart) should be checked first to ensure the process variability is in statistical control, and then the $\overline{X}$ chart is checked for the process mean shift. Once a set of reliable control limits is constructed, they can be used for process monitoring.

$\overline{X}$ and S Charts

When the sample size is relatively large ($n > 10$), or the sample size is variable, the $\overline{X}$ and S charts are preferred to $\overline{X}$ and R charts. To construct the control limits, we need to estimate the mean and standard deviation of $\overline{X}$ and S, that is, $\mu_{\overline{x}}$ and $\sigma_{\overline{x}}, \mu_S$ and σ_S first. We have $\hat{\mu}_{\overline{x}} = \overline{\overline{x}}$ and $\hat{\mu}_S = \overline{S}$. Using $\hat{\sigma} = s/c_4$, we have $\hat{\sigma}_{\overline{x}} = \hat{\sigma}/\sqrt{n} = \overline{S}/(c_4\sqrt{n})$ and $\hat{\sigma}_s = \overline{S}\sqrt{1-c_4^2}/c_4$. Therefore, the control limits for $\overline{X}$ and S charts are

$$\text{LCL} = \overline{\overline{x}} - A_3\overline{S} \qquad \text{LCL} = B_3\overline{S}$$
$$\text{CL} = \overline{\overline{x}} \qquad \text{and} \qquad \text{CL} = \overline{S}$$
$$\text{ULC} = \overline{\overline{x}} + A_3\overline{S} \qquad \text{UCL} = B_4\overline{S}$$

respectively, where $A_3 = \frac{3}{c_4\sqrt{n}}$, $B_3 = 1 - \frac{3}{c_4}\sqrt{1-c_4^2}$, and $B_4 = 1 + \frac{3}{c_4}\sqrt{1-c_4^2}$. The values of c_4, A_3, B_3, and B_4 can be obtained from most books on control charts for n up to 25.

Control Charts for Attributes

When quality characteristics are expressed as attribute data, such as defective or conforming items, control charts for attributes are established. Attribute charts can handle multiple quality characteristics jointly because the unit is classified as defective if it fails to meet the specification on one or more characteristics. The inspection of samples for attribute charts is usually cheaper due to less precision requirement. Attribute charts are particularly useful in quality improvement efforts where numerical data are not easily obtained, such as service, industrial, and health care systems. In the context of quality control, the attribute data include a proportion of defective items and a number of defects on items. A defective unit may have one or more defects that are a result of nonconformance to standard on one or more quality characteristics. Nevertheless, a unit with several defects may not necessarily be classified as a defective unit. It requires two different types of attribute charts: control charts for proportion defective (p chart and np chart), and control charts for number of defects (c chart and u chart).

p Chart and np Chart

The proportion that is defective is defined as the ratio of the number of defective units to the total number of units in a population. We usually assume that the number of defective units in a sample is a binomial variable; that is, each unit in the sample is produced independently and the probability that a unit is defective is constant, p. Using preliminary samples, we can estimate the defective rate, that is, $\bar{p} = \sum_{i=1}^{m} D_i/mn$, where D_i is the number of defective units in sample i, n is the sample size, and m is the number of samples taken. The formulas used to calculate control limits are then given as

$$\text{UCL}_{\hat{p}} = \bar{p} + 3\sqrt{\frac{\bar{p}(1-\bar{p})}{n}}$$

$$\text{Centerline} = \bar{p}$$

$$\text{LCL}_{\hat{p}} = \bar{p} - 3\sqrt{\frac{\bar{p}(1-\bar{p})}{n}}.$$

Sometimes, it may be easier to interpret the number that is defective instead of the proportion that is defective. That is why the np chart came into use:

$$\text{UCL} = n\bar{p} + 3\sqrt{n\bar{p}(1-\bar{p})}$$

$$\text{Centerline} = n\bar{p}$$

$$\text{LCL} = n\bar{p} - 3\sqrt{n\bar{p}(1-\bar{p})}.$$

The developed trial control limits are then used to check if the preliminary data are in statistical control, and the assignable causes may be identified and removed if a point is out of control. As the process improves, we expect a downward trend in the p or np control chart.

c Chart and u Chart

Control charts for monitoring the number of defects per sample are constructed based on Poisson distribution. With this assumption of reference distribution, the probability of occurrence of a defect at any area is small and constant, the potential area for defects is

Quality Control

infinitely large, and defects occurs randomly and independently. If the average occurrence rate per sample is a constant c, we know that both the mean and variance of the Poisson distribution are the constant c. Therefore, the parameters in the c chart for the number of defects are

$$\text{LCL} = c - 3\sqrt{c}$$

$$\text{CL} = c$$

$$\text{UCL} = c + 3\sqrt{c}$$

where c can be estimated by the average number of defects in a preliminary sample. To satisfy the assumption of constant rate of occurrence, the sample size is required to be constant.

For variable sample size, the u chart should be used instead of the c chart. Compared to the c chart that is used to monitor the number of defects per sample, the u chart is designed to check the average number of defects per inspection unit. Usually, a sample may contain one or more inspection units. For example, in a textile finishing plant, dyed cloth is inspected for defects per $50\,\text{m}^2$, which is one inspection unit. A roll of cloth of $500\,\text{m}^2$ is one sample with 10 inspection units. Different rolls of cloth may have various areas, hence variable sample sizes. As a result, it is not appropriate to use the c chart, because the occurrence rate of defects in each sample is not a constant. The alternative is to monitor the average number of defects per inspection unit in a sample, $u_i = c_i/n_i$. In this way, the parameters in the u chart are given as

$$\text{LCL} = \bar{u} - 3\sqrt{\frac{\bar{u}}{n}}$$

$$\text{CL} = \bar{u}$$

$$\text{UCL} = \bar{u} + 3\sqrt{\frac{\bar{u}}{n}}$$

where $\bar{u} = \sum_{i=1}^{m} u_i/m$, is an estimation of the average number of defects in an inspection unit. For variable sample size, the upper and lower control limits vary for different n.

To effectively detect small process shifts (on the order of 1.5σ or less), the cumulative sum (CUSUM) control chart and the exponentially weighted moving average (EWMA) control chart may be used instead of Shewhart control charts. In addition, there are many situations where we need to simultaneously monitor two or more correlated quality characteristics. The control charts for multivariate quality characteristics will also be discussed in the following.

16.4.6 Benefits of Control Charts

In this section, we summarize some of the important benefits that can come from using control charts.

- Control charts are simple and effective tools to achieve statistical control. They lend themselves to being maintained at the job station by the operator. They give the people closest to the operation reliable information on when action should be taken—and on when action should not be taken.
- When a process is in statistical control, its performance to specification will be predictable. Thus, both producer and customer can rely on consistent quality levels, and both can rely on stable costs of achieving that quality level.

- After a process is in statistical control, its performance can be further improved to reduce variation. The expected effects of proposed improvements in the system can be anticipated, and the actual effects of even relatively subtle changes can be identified through the control chart data. Such process improvements will:
  - Increase the percentage of output that meets customer expectations (improve quality),
  - Decrease the output requiring scrap or rework (improve cost per good unit produced), and
  - Increase the total yield of acceptable output through the process (improve effective capacity).
- Control charts provide a common language for communications about the performance of a process—between the two or three shifts that operate a process; between line production (operator, supervisor) and support activities (maintenance, material control, process engineering, quality control); between different stations in the process; between supplier and user; between the manufacturing/assembly plant and the design engineering activity.
- Control charts, by distinguishing special from common causes of variation, give a good indication of whether any problems are likely to be correctable locally or will require management action. This minimizes the confusion, frustration, and excessive cost of misdirected problem-solving efforts.

16.5 Process Capability Studies

As discussed earlier, statistical control of a process is arrived at by eliminating special causes of excessive variation one by one. Process capability is determined by the total variation that comes from common causes. Therefore, a process must first be brought into statistical control and then its capability to meet specifications can be assessed.

The process capability to meet specifications is usually measured by process capability indices that link process parameters to product design specifications. Using a single number, process capability indices measure the degree to which the stable process can meet the specifications [23,24]. If we denote the lower specification limit as LSL and the upper specification limit as USL, the process capability index C_p is defined as:

$$C_p = \frac{\text{USL} - \text{LSL}}{6\sigma}$$

which measures the potential process capability. To measure the actual process capability, we use C_{pk} that is defined as:

$$C_{pk} = \min\left(\frac{\text{USL} - \mu}{3\sigma}, \frac{\mu - \text{LSL}}{3\sigma}\right)$$

The measure of C_{pk} takes the process centering into account by choosing the one side C_p for the specification limit closest to the process mean. The estimations of C_p and C_{pk} are obtained by replacing μ and σ using the estimates $\hat{\mu}$ and $\hat{\sigma}$. To consider the variability in terms of both standard deviation and mean, another process capability index C_{pm} is defined as

$$\hat{C}_{pm} = \frac{\text{USL} - \text{LSL}}{6\hat{\tau}}$$

where $\hat{\tau}$ is an estimator of the expected square deviation from the target, T, and is given by

$$\tau^2 = E\big[(x-T)^2\big] = \sigma^2 + (\mu-T)^2$$

Therefore, if we know the estimate of C_p, we can estimate C_{pm} as:

$$\hat{C}_{pm} = \frac{\hat{C}_p}{\sqrt{1+\left(\frac{\hat{\mu}-T}{\hat{\sigma}}\right)^2}}$$

In addition to process capability indices, capability can also be described in terms of the distance of the process mean from the specification limits in standard deviation units, Z, that is

$$Z_U = \frac{\text{USL}-\hat{\mu}}{\hat{\sigma}}, \quad \text{and} \quad Z_L = \frac{\hat{\mu}-\text{LSL}}{\hat{\sigma}}$$

Z values can be used with a table of standard normal distribution to estimate the proportion of process fallout for a normally distributed and statistically controlled process. The Z value can also be converted to the capability index, C_{pk}:

$$C_{pk} = \frac{Z_{\min}}{3} = \frac{1}{3}\min(Z_U, Z_L)$$

A process with $Z_{\min}=3$, which could be described as having $\hat{\mu}\pm 3\hat{\sigma}$ capability, would have $C_{pk}=1.00$. If $Z_{\min}=4$, the process would have $\hat{\mu}\pm 4\hat{\sigma}$ capability and $C_{pk}=1.33$.

Example 16.1

For a process with $\hat{\mu}=0.738$, $\hat{\sigma}=0.0725$, USL $=0.9$, and LSL $=0.5$,

- Since the process has two-sided specification limits,

$$Z_{\min} = \min\left(\frac{\text{USL}-\hat{\mu}}{\hat{\sigma}}, \frac{\hat{\mu}-\text{LSL}}{\hat{\sigma}}\right)$$

$$= \min\left(\frac{0.9-0.738}{0.0725}, \frac{0.738-0.5}{0.0725}\right) = \min(2.23, 3.28) = 2.23$$

and the proportion of process fallout would be:

$$p = 1 - \Phi(2.23) + \Phi(-3.28) = 0.0129 + 0.0005 = 0.0134$$

The process capability index would be:

$$C_{pk} = \frac{Z_{\min}}{3} = 0.74$$

- If the process could be adjusted toward the center of the specification, the proportion of process fallout might be reduced, even with no change in σ:

$$Z_{\min} = \min\left(\frac{\text{USL}-\hat{\mu}}{\hat{\sigma}}, \frac{\hat{\mu}-\text{LSL}}{\hat{\sigma}}\right) = \min\left(\frac{0.9-0.7}{0.0725}, \frac{0.7-0.5}{0.0725}\right) = 2.76$$

and the proportion of process fallout would be:

$$p = 2\Phi(-2.76) = 0.0058$$

The process capability index would be:

$$C_{pk} = \frac{Z_{\min}}{3} = 0.92$$

- To improve the actual process performance in the long run, the variation from common causes must be reduced. If the capability criterion is $\hat{\mu} \pm 4\hat{\sigma}$ ($Z_{\min} \geq 4$), the process standard deviation for a centered process would be:

$$\sigma_{new} = \frac{\text{USL} - \hat{\mu}}{Z_{\min}} = \frac{0.9 - 0.7}{4} = 0.05$$

Therefore, actions should be taken to reduce the process standard deviation from 0.0725 to 0.05, about 31%. ∎

At this point, the process has been brought into statistical control and its capability has been described in terms of process capability index or $Z_{\min}$. The next step is to evaluate the process capability in terms of meeting customer requirements. The fundamental goal is never-ending improvement in process performance. In the near term, however, priorities must be set as to which processes should receive attention first. This is essentially an economic decision. The circumstances vary from case to case, depending on the nature of the particular process in question. While each such decision could be resolved individually, it is often helpful to use broader guidelines to set priorities and promote consistency of improvement efforts. For instance, certain procedures require $C_{pk} > 1.33$, and further specify $C_{pk} = 1.50$ for new processes. These requirements are intended to assure a minimum performance level that is consistent among characteristics, products, and manufacturing sources.

Whether in response to a capability criterion that has not been met, or to the continuing need for improvement of cost and quality performance even beyond minimum capability requirement, the action required is the same: improve the process performance by reducing the variation that comes from common causes. This means taking management action to improve the system.

16.6 Advanced Control Charts

The major disadvantage of the Shewhart control chart is that it uses the information in the last plotted point and ignores information given by the sequence of points. This makes it insensitive to small shifts. One effective way is to use

- cumulative sum (CUSUM) control charts
- exponentially weighted moving average (EWMA) control charts

16.6.1 Cumulative Sum Control Charts

CUSUM charts incorporate all the information in the sequence of sample values by plotting the CUSUM of deviations of the sample values from a target value, defined as

$$C_i = \sum_{j=1}^{i} (\bar{x}_j - T)$$

A significant trend developed in C_i is an indication of the process mean shift. Therefore, CUSUM control charts would be more effective than Shewhart charts to detect small process shifts. Two statistics are used to accumulate deviations from the target T:

$$C_i^+ = \max[0, \ x_i - (T + K) + C_{i-1}^+]$$
$$C_i^- = \max[0, \ (T - K) - x_i + C_{i-1}^-]$$

where $C_0^+ = C_0^- = 0$, and K is the slack value, and it is often chosen about halfway between the target value and the process mean after shift. If either C^+ or C^- exceeds the decision interval H (a common choice is $H = 5\sigma$), the process is considered to be out of control.

16.6.2 Exponentially Weighted Moving Average Control Charts

As discussed earlier, we use Western Electric rules to increase the sensitivity of Shewhart control charts to detect nonrandom patterns or small shifts in a process. A different approach to highlight small shifts is to use a time average over past and present data values as an indicator of recent performance. Roberts [25] introduced the EWMA as such an indicator, that is, past data values are remembered with geometrically decreasing weight. For example, we denote the present and past values of a quality characteristic x by x_t, x_{t-1}, x_{t-2},...; then the EWMA y_t with discount factor q is

$$y_t = a(x_t + qx_{t-1} + q^2 x_{t-2} + \cdots)$$

where a is a constant that makes the weights add up to 1 and it equals to $1-q$. In the practice of process monitoring, the constant $1-q$ is given the distinguishing symbol λ. Using λ, the EWMA can be expressed as $y_t = \lambda x_t + (1-\lambda)y_{t-1}$, which is a more convenient formula for updating the value of EWMA at each new observation. It is observed from the formula that a larger value of λ results in weights that die out more quickly and place more emphasis on recent observations. Therefore, a smaller value of λ is recommended to detect small process shifts, usually $\lambda = 0.05$, 0.10, or 0.20.

An EWMA control chart with appropriate limits is used to monitor the value of EWMA. If the process is in statistical control with a process mean of μ and a standard deviation of σ, the mean of the EWMA would be μ, and the standard deviation of the EWMA would be $\sigma \left(\frac{\lambda}{2-\lambda}\right)^{1/2}$. Thus, given a value of λ, three-sigma or other appropriate limits can be constructed to monitor the value of EWMA.

16.6.3 Other Advanced Control Charts

The successful use of Shewhart control charts and the CUSUM and EWMA control charts have led to the development of many new techniques over the last 20 years. A brief summary of these techniques and references to more complete descriptions are provided in this section.

The competitive global market expects smaller defect rates and higher quality level that requires 100% inspection of output products. The recent advancement of sensing techniques and computer capacity makes 100% inspection more feasible. Due to the reduced intervals between sampling of the 100% inspection, the complete observations will be correlated over time. However, one of the assumptions for Shewhart control charts is the independence between observations over time. When the observations are autocorrelated, Shewhart control charts will give misleading results in the form of many false alarms. ARIMA are used to remove autocorrelation from data, and then control charts are applied to the residuals. Further discussion on SPC with autocorrelated process data can be found in Refs. [16,20].

It is often necessary to simultaneously monitor or control two or more related quality characteristics. Using individual control charts to monitor the independent variables separately can be very misleading. Multivariate SPC control charts were developed based on multivariate normal distribution by Hotelling [26]. More discussion on multivariate SPC can be found in Refs. [20,26].

The use of control charts requires the selection of sample size, sampling frequency or interval between samples, and the control limits for the charts. The selection of these parameters has economic consequences in that the cost of sampling, the cost of false alarms, and the cost of removing assignable causes will affect the choice of the parameters. Therefore, the economic design of control charts has received attention in research and practice. Related discussion can be found in Refs. [8,27]. Other research issues and ideas in SPC can be found in a review paper by Woodall and Montgomery [28].

16.7 Limitations of Acceptance Sampling

As one of the earliest methods of quality control, acceptance sampling is closely related to inspection of output of a process, or testing of a product. Acceptance sampling is defined as the inspection and classification of samples from a lot randomly and decision about disposition of the lot. At the beginning of the concept of quality conformance back in the 1930s, the acceptance sample took the whole effort of quality improvement. The most widely used plans are given by the Military Standard tables (MIL STD 105A), which were developed during World War II. The last revision (MIL STD 105E) was issued in 1989, but cancelled in 1991. The standard was adopted by the American Society for Quality as ANSI/ASQ A1.4.

Due to its less proactive nature in terms of quality improvement, acceptance sampling is less emphasized in current quality control systems. Usually, methods of lot sentencing include no inspection, 100% inspection, and acceptance sampling. Some of the problems with acceptance sampling were articulated by Dr. W. Edwards Deming [2], who pointed out that this procedure, while minimizing the inspection cost, does not minimize the total cost to the producer. To minimize the total cost to the producer, Deming indicated that inspection should be performed either 100% or not at all, which is called Deming's "All or None Rule." In addition, acceptance sampling has several disadvantages compared to 100% inspection [20]:

- There are risks of accepting "bad" lots and rejecting "good" lots.
- Less information is usually generated about the product or process.
- Acceptance sampling requires planning and documentation of the acceptance sampling procedure.

16.8 Conclusions

The quality of a system is defined and evaluated by the customer [29]. A system has many qualities and we can develop a utility or customer satisfaction measure based on all of these qualities. The design process substitutes the voice of the customer with engineering or technological characteristics. Quality function deployment (QFD) plays an important role in the development of those characteristics. Quality engineering principles can be used to develop ideal values or targets for these characteristics. The methodology of robust design is an integral part of the quality process [30–32]. Quality should be an integral part of all the elements of the enterprise, which means that it is distributed throughout the enterprise and also all of these elements must be integrated together as shown in Figure 16.8.

Statistical quality control is a primary technique for monitoring the manufacturing process or any other process with key quality characteristics of interests. Before we use statistical quality control to monitor the manufacturing process, the optimal values of the mean and

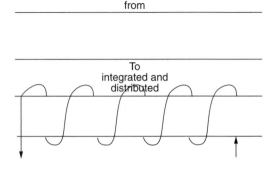

FIGURE 16.8 Integrated and distributed quality management.

standard deviation of the quality characteristic are determined by minimizing the variability of the quality characteristic through experimental design and process adjustment techniques. Consequently, the major goal of SQC (SPC) is to monitor the manufacturing process, keep the values of mean and standard deviation stable, and finally reduce variability. When the process is in control, all the assignable causes are not present and consequently the probability to produce a nonconforming unit is very small. When the process changes to out of control, the probability of nonconformance may increase, and the process quality deteriorates significantly. As one of the primary SPC techniques, the control charts we discussed in this chapter can effectively detect the variation due to the assignable causes and reduce process variability if the identified assignable causes can be eliminated from the process.

The philosophy of Deming, Juran, and other quality gurus implies that the responsibility for quality spans the entire organization. It is critical that the management in any enterprise recognize that quality improvement must be a total, company-wide activity, and that every organizational unit must actively participate. Statistical quality control techniques are the common language of communication about quality problems that enables all organizational units to solve problems rapidly and efficiently.

References

1. Juran, J.M. and Gryna, F.M. *Quality Planning and Analysis: From Product Development Through Usage*, 3rd edition. McGraw-Hill, New York, 1993.
2. Deming, W.E., *Quality, Productivity, and Competitive Position*, Massachusetts Institute of Technology, Center for Advanced Engineering Study, Cambridge, MA, 1982.
3. American Society for Quality Control, *ASQC Glossary and Tables for Statistical Quality Control*, Milwaukee, 1983.
4. Shewhart, W.A., *Economic Control of Quality of a Manufactured Product*, D. Van Nostrand Company, New York, 1931.
5. Mizuno, S. and Akao, Y. (Eds), *Quality Function Deployment: A Company-Wide Quality Approach*, JUSE Press, Tokyo, 1978.
6. Akao, Y. (Ed), *Quality Function Deployment: Integrating Customer Requirements into Product Design*, translated by Mazur, G.H., Productivity Press, Cambridge, MA, 1990.
7. Montgomery, D.C., *Design and Analysis of Experiments*, 5th edition, John Wiley & Sons, New York, 2001.

8. Kapur, K.C. and Feng, Q., Integrated optimization models and strategies for the improvement of the Six Sigma process. *International Journal of Six Sigma and Competitive Advantage*, 1(2), 2005.
9. Montgomery, D.C., The economic design of control charts: A review and literature survey, *Journal of Quality Technology*, 14, 75–87, 1980.
10. Breyfogle, F.W., *Implementing Six Sigma: Smarter Solutions Using Statistical Methods*, 2nd edition, John Wiley & Sons, New York, 2003.
11. Pyzdek, T., *The Six Sigma Handbook, Revised and Expanded: The Complete Guide for Greenbelts, Blackbelts, and Managers at All Levels*, 2nd edition, McGraw-Hill, New York, 2003.
12. Yang, K. and El-Haik, B., *Design for Six Sigma: A Roadmap for Product Development*, McGraw-Hill, New York, 2003.
13. De Feo, J.A. and Barnard, W.W., *Juran Institute's Six Sigma Breakthrough and Beyond: Quality Performance Breakthrough Methods*, McGraw-Hill, New York, 2004.
14. Duncan, A.J., *Quality Control and Industrial Statistics*, 5th edition, Irwin, Homewood, IL, 1986.
15. Stoumbos, Z.G., Reynolds, M.R., Ryan, T.P., and Woodall, W.H., The state of statistical process control as we proceed into the 21st century, *Journal of the American Statistical Association*, 95, 992–998, 2000.
16. Box, G. and Luceno, A., *Statistical Control by Monitoring and Feedback Adjustment*, John Wiley & Sons, New York, 1997.
17. Hicks, C.R. and Turner, K.V., *Fundamental Concepts in the Design of Experiments*, 5th edition, Oxford University Press, New York, 1999.
18. ASTM Publication STP-15D, *Manual on the Presentation of Data and Control Chart Analysis*, 1916 Race Street, Philadelphia, PA, 1976.
19. Alwan, L.C., Statistical Process Analysis, McGraw-Hill, New York, 2000.
20. Montgomery, D.C., *Introduction to Statistical Quality Control*, 5th edition, John Wiley & Sons, New York, 2005.
21. Western Electric, *Statistical Quality Control Handbook*, Western Electric Corporation, Indianapolis, IN, 1956.
22. Chandra, M.J., *Statistical Quality Control*, CRC Press LLC, Boca Raton, FL, 2001.
23. Kane, V.E., Process capability indices, *Journal of Quality Technology*, 18, 1986.
24. Kotz, S. and Lovelace, C.R., *Process Capability Indices in Theory and Practice*, Arnold, London, 1998.
25. Roberts, S.W., *Control Chart Tests Based on Geometric Moving Averages*, Technometrics, 42(1), 97–102 1959.
26. Hotelling, H., Multivariate Quality Control, In: *Techniques of Statistical Analysis*, Edited by Eisenhart, C., Hastay, M.W., and Wallis, W.A., McGraw-Hill, New York, 1947.
27. Duncan, A.J., The economic design of charts used to maintain current control of a process, *Journal of the American Statistical Association*, 51, 228–242, 1956.
28. Woodall, W.H. and Montgomery, D.C., Research issues and ideas in statistical process control, *Journal of Quality Technology*, 31(4), 376–386, 1999.
29. Kapur, K.C., An integrated customer-focused approach for quality and reliability, *International Journal of Reliability, Quality and Safety Engineering*, 5(2), 101–113, 1998.
30. Phadke, M.S., *Quality Engineering Using Robust Design*, Prentice-Hall, Englewood Cliffs, NJ, 1989.
31. Taguchi, G., *Introduction to Quality Engineering*, Asia Productivity Organization, Tokyo, 1986.
32. Taguchi, G., *System of Experimental Design, Volume I and II*, UNIPUB/Krauss International, White Plains, New York, 1987.

Appendix

Constants and Formulas for Constructing Control Charts

	$\overline{X}$ and R Charts				$\overline{X}$ and S Charts			
	Chart for Averages $\overline{X}$	Chart for Ranges (R)			Chart for Averages $\overline{X}$	Chart for Standard Deviations (S)		
Subgroup Size n	Factors for Control Limits A_2	Devisors for Estimate of Standard Deviation d_2	Factors for Control Limits D_3	Factors for Control Limits D_4	Factors for Control Limits A_3	Divisors for Estimator of Standard Deviation c_4	Factors for Control Limits B_3	Factors for Control Limits B_4
2	1.880	1.128	–	3.267	2.659	0.7979	–	3.267
3	1.023	1.693	–	2.574	1.954	0.8862	–	2.568
4	0.729	2.059	–	2.282	1.628	0.9213	–	2.266
5	0.577	2.326	–	2.114	1.427	0.9400	–	2.089
6	0.483	2.534	–	2.004	1.287	0.9515	0.030	1.970
7	0.419	2.704	0.076	1.924	1.182	0.9594	0.118	1.882
8	0.373	2.847	0.136	1.864	1.099	0.9650	0.185	1.815
9	0.337	2.970	0.184	1.816	1.032	0.9693	0.239	1.761
10	0.308	3.078	0.223	1.777	0.975	0.9727	0.284	1.716

$$\text{LCL} = \overline{\overline{x}} - A_2\overline{R} \qquad \text{LCL} = D_3\overline{R} \qquad \text{LCL} = \overline{\overline{x}} - A_3\overline{S} \qquad \text{LCL} = B_3\overline{S}$$

$$\text{CL} = \overline{\overline{x}} \quad \text{and} \quad \text{CL} = \overline{R} \qquad \text{CL} = \overline{\overline{x}} \quad \text{and} \quad \text{CL} = \overline{S}$$

$$\text{ULC} = \overline{\overline{x}} + A_2\overline{R} \qquad \text{UCL} = D_4\overline{R} \qquad \text{ULC} = \overline{\overline{x}} + A_3\overline{S} \qquad \text{UCL} = B_4\overline{S}$$

$$\hat{\sigma} = R/d_2 \qquad\qquad \hat{\sigma} = s/c_4$$

Guide for Selection of Charts for Attributes:

	Nonconforming Units	Nonconformities
Number of Nonconformities (Simple, but needs constant sample size)	np	c
Proportion (More complex, but adjusts to understandable proportion, and can cope with varying sample sizes)	p	u

- **p chart** for proportion of units nonconforming, from samples not necessarily of constant size: (If n varies, use $\overline{n}$ or individual n_i.)

$$\text{UCL}_{\hat{p}} = \overline{p} + 3\sqrt{\frac{\overline{p}(1-\overline{p})}{n}}$$

$$\text{Centerline} = \overline{p}$$

$$\text{LCL}_{\hat{p}} = \overline{p} - 3\sqrt{\frac{\overline{p}(1-\overline{p})}{n}}$$

- **np chart** for number of units nonconforming, from samples of constant size:

$$\text{UCL} = n\overline{p} + 3\sqrt{n\overline{p}(1-\overline{p})}$$

$$\text{Centerline} = n\overline{p}$$

$$\text{LCL} = n\overline{p} - 3\sqrt{n\overline{p}(1-\overline{p})}$$

- *c chart* for number of nonconformities, from samples of constant size:

$$\text{LCL} = c - 3\sqrt{c}$$

$$\text{CL} = c$$

$$\text{UCL} = c + 3\sqrt{c}$$

- *u chart* for number of nonconformities per unit, from samples not necessarily of constant size: (If n varies, use $\bar{n}$ or individual n_i.)

$$\text{LCL} = \bar{u} - 3\sqrt{\frac{\bar{u}}{n}}$$

$$\text{CL} = \bar{u}$$

$$\text{UCL} = \bar{u} + 3\sqrt{\frac{\bar{u}}{n}}$$

17
Reliability

17.1	Introduction	17-1
17.2	Reliability in System Design	17-3
	Structure Functions • Reliability Functions	
17.3	Lifetime Distributions	17-10
	Survivor Function • Probability Density Function • Hazard Function • Cumulative Hazard Function • Expected Values and Fractiles • System Lifetime Distributions	
17.4	Parametric Models	17-17
	Parameters • Exponential Distribution • Weibull Distribution	
17.5	Parameter Estimation in Survival Analysis	17-22
	Data Sets • Exponential Distribution • Weibull Distribution	
17.6	Nonparametric Methods	17-32
17.7	Assessing Model Adequacy	17-36
17.8	Summary	17-40
	References	17-41

Lawrence M. Leemis
The College of William and Mary

17.1 Introduction

Reliability theory involves the mathematical modeling of systems, typically comprised of components, with respect to their ability to perform their intended function over time. Reliability theory can be used as a *predictive* tool, as in the case of a new product introduction; it can also be used as a *descriptive* tool, as in the case of finding a weakness in an existing system design.

Reliability theory is based on probability theory. The reliability of a component or system at one particular point in time is a real number between 0 and 1 that represents the probability that the component or system is functioning at that time. Reliability theory also involves statistical methods. Estimating component and system reliability is often performed by analyzing a data set of lifetimes.

Recent tragedies, such as the space shuttle accidents, nuclear power plant accidents, and aircraft catastrophes, highlight the importance of reliability in design. This chapter describes probabilistic models for reliability in design and statistical techniques that can be applied to a data set of lifetimes. Although the majority of the illustrations given here come from engineering problems, the techniques described here may also be applied to problems in actuarial science and biostatistics.

Reliability engineers concern themselves primarily with lifetimes of inanimate objects, such as switches, microprocessors, or gears. They usually regard a complex system as a collection of components when performing an analysis. These components are arranged in a structure that allows the system state to be determined as a function of the component states. Interest in reliability and quality control has been revived by a more competitive international market and increased consumer expectations. A product or service that has a reputation for high reliability will have consumer goodwill and, if appropriately priced, will gain in market share.

Literature on reliability tends to use *failure*, actuarial literature tends to use *death*, and point process literature tends to use *epoch*, to describe the event at the termination of a lifetime. Likewise, reliability literature tends to use a *system*, *component*, or *item*, actuarial literature uses an *individual*, and biostatistical literature tends to use an *organism* as the object of a study. To avoid switching terms, *failure* of an *item* will be used as much as possible throughout this chapter, as the emphasis is on reliability. The concept of failure time (or lifetime or survival time) is quite generic, and the models and statistical methods presented here apply to any nonnegative random variable (e.g., the response time at a computer terminal).

The remainder of this chapter is organized as follows. Sections 17.2 through 17.4 contain probability models for lifetimes, and the subsequent remaining sections contain methods related to data collection and inference.

Mathematical models for describing the arrangement of components in a system are introduced in Section 17.2. Two of the simplest arrangements of components are series and parallel systems. The notion of the reliability of a component and system at a particular time is also introduced in this section. As shown in Section 17.3, the concept of reliability generalizes to a survivor function when time dependence is introduced. In particular, four different representations for the distribution of the failure time of an item are considered: the survivor, density, hazard, and cumulative hazard functions.

Several popular parametric models for the lifetime distribution of an item are investigated in Section 17.4. The *exponential distribution* is examined first due to its importance as the only continuous distribution with the memoryless property, which implies that a used item that is functioning has the same conditional failure distribution as a new item. Just as the normal distribution plays a central role in classical statistics due to the central limit theorem, the exponential distribution is central to the study of the distribution of lifetimes, as it is the only continuous distribution with a constant hazard function. The more flexible Weibull distribution is also outlined in this section.

The emphasis changes from developing probabilistic models for lifetimes to analyzing lifetime data sets in Section 17.5. One problem associated with these data sets is that of censored data. Data are censored when only a bound on the lifetime is known. This would be the case, for example, when conducting an experiment with light bulbs, and half of the light bulbs are still operating at the end of the experiment. This section surveys methods for fitting parametric distributions to data sets. Maximum likelihood parameter estimates are emphasized because they have certain desirable statistical properties. Section 17.6 reviews a nonparametric method for estimating the survivor function of an item from a censored data set: the Kaplan–Meier product-limit estimate. Once a parametric model has been chosen to represent the failure time for a particular item, the adequacy of the model should be assessed. Section 17.7 considers the Kolmogorov–Smirnov goodness-of-fit test for assessing how well a fitted lifetime distribution models the lifetime of the item. The test uses the largest vertical distance between the fitted and empirical survivor functions as the test statistic.

We have avoided references throughout the chapter to improve readability. There are thousands of journal articles and over 100 textbooks on reliability theory and applications. As this is not a review of the current state-of-the-art in reliability theory, we cite only a few key comprehensive texts for further reading. A classic, early reference is Barlow and

Proschan (1981). Meeker and Escobar (1998) is a more recent comprehensive textbook on reliability. The analysis of survival data is also considered by Kalbfleisch and Prentice (2002) and Lawless (2003). This chapter assumes that the reader has a familiarity with calculus-based probability and statistical inference techniques.

17.2 Reliability in System Design

This section introduces mathematical techniques for expressing the arrangement of components in a system and for determining the reliability of the associated system. We assume that an item consists of n components, arranged into a system. We first consider a system's *structural properties* associated with the arrangement of the components into a system, and then consider the system's *probabilistic properties* associated with determining the system reliability. *Structure functions* are used to map the states of the individual components to the state of the system. *Reliability functions* are used to determine the system reliability at a particular point in time, given the component reliabilities at that time.

17.2.1 Structure Functions

A *structure function* is used to describe the way that the n components are related to form a system. The structure function defines the system state as a function of the component states. In addition, it is assumed that both the components and the system can either be functioning or failed. Although this *binary assumption* may be unrealistic for certain types of components or systems, it makes the mathematics involved more tractable. The functioning and failed states for both components and systems will be denoted by 1 and 0, respectively, as in the following definition.

DEFINITION 17.1 The state of component i, denoted by x_i is

$$x_i = \begin{cases} 0 & \text{if component } i \text{ has failed} \\ 1 & \text{if component } i \text{ is functioning} \end{cases}$$

for $i = 1, 2, \ldots, n$.

These n values can be written as a system state vector, $\boldsymbol{x} = (x_1, x_2, \ldots, x_n)$. As there are n components, there are 2^n different values that the system state vector can assume, and $\binom{n}{j}$ of these vectors correspond to exactly j functioning components, $j = 0, 1, \ldots, n$. The structure function, $\phi(\boldsymbol{x})$, maps the system state vector $\boldsymbol{x}$ to 0 or 1, yielding the state of the system.

DEFINITION 17.2 The *structure function* ϕ is

$$\phi(\boldsymbol{x}) = \begin{cases} 0 & \text{if the system has failed when the state vector is } \boldsymbol{x} \\ 1 & \text{if the system is functioning when the state vector is } \boldsymbol{x} \end{cases}$$

The most common system structures are the series and parallel systems, which are defined in the examples that follow. The series system is the worst possible way to arrange components in a system; the parallel system is the best possible way to arrange components in a system.

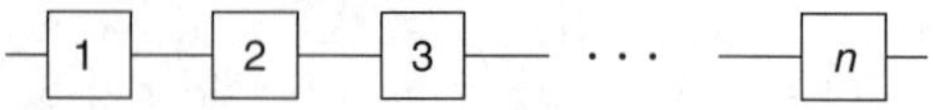

FIGURE 17.1 A series system.

Example 17.1

A *series system* functions when all its components function. Thus $\phi(\boldsymbol{x})$ assumes the value 1 when $x_1 = x_2 = \cdots = x_n = 1$, and 0 otherwise. Therefore,

$$\phi(\boldsymbol{x}) = \begin{cases} 0 & \text{if there exists an } i \text{ such that } x_i = 0 \\ 1 & \text{if } x_i = 1 \text{ for all } i = 1, 2, \ldots, n \end{cases}$$

$$= \min\{x_i, x_2, \ldots, x_n\}$$

$$= \prod_{i=1}^{n} x_i$$

These three different ways of expressing the value of the structure function are equivalent, although the third is preferred because of its compactness. Systems that function only when all their components function should be modeled as series systems.

Block diagrams are useful for visualizing a system of components. The block diagram corresponding to a series system of n components is shown in Figure 17.1. A block diagram is a graphic device for expressing the arrangement of the components to form a system. If a path can be traced through functioning components from left to right on a block diagram, then the system functions. The boxes represent the system components, and either component numbers or reliabilities are placed inside the boxes. ∎

Example 17.2

A *parallel system* functions when one or more of the components function. Thus $\phi(\boldsymbol{x})$ assumes the value 0 when $x_1 = x_2 = \cdots = x_n = 0$, and 1 otherwise.

$$\phi(\boldsymbol{x}) = \begin{cases} 0 & \text{if } x_i = 0 \text{ for all } i = 1, 2, \ldots, n \\ 1 & \text{if there exists an } i \text{ such that } x_i = 1 \end{cases}$$

$$= \max\{x_1, x_2, \ldots, x_n\}$$

$$= 1 - \prod_{i=1}^{n}(1 - x_i)$$

As in the case of the series system, the three ways of defining $\phi(\boldsymbol{x})$ are equivalent. The block diagram of a parallel arrangement of n components is shown in Figure 17.2. A parallel arrangement of components is appropriate when all components must fail for the system to fail. A two-component parallel system, for instance, is the brake system on an automobile that contains two reservoirs for brake fluid. Arranging components in parallel is also known as *redundancy*. ∎

Series and parallel systems are special cases of k-out-of-n systems, which function if k or more of the n components function. A series system is an n-out-of-n system, and a parallel system is a 1-out-of-n system. A suspension bridge that needs only k of its n cables to support the bridge or an automobile engine that needs only k of its n cylinders to run are examples of k-out-of-n systems.

Reliability

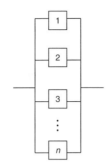

FIGURE 17.2 A parallel system.

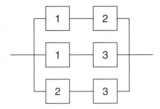

FIGURE 17.3 A 2-out-of-3 system.

Example 17.3

The structure function of a k-out-of-n system is

$$\phi(\boldsymbol{x}) = \begin{cases} 0 & \text{if } \sum_{i=1}^{n} x_i < k \\ 1 & \text{if } \sum_{i=1}^{n} x_i \geq k \end{cases}$$

The block diagram for a k-out-of-n system is difficult to draw in general, but for specific values of k and n it can be drawn by repeating components in the block diagram. The block diagram for a 2-out-of-3 system, for example, is shown in Figure 17.3. The block diagram indicates that if all three, or exactly two out of three components (in particular 1 and 2, 1 and 3, or 2 and 3) function, then the system functions. The structure function for a 2-out-of-3 system is

$$\phi(\boldsymbol{x}) = 1 - (1 - x_1 x_2)(1 - x_1 x_3)(1 - x_2 x_3)$$
$$= x_1 x_2 + x_1 x_3 + x_2 x_3 - x_1^2 x_2 x_3 - x_1 x_2^2 x_3 - x_1 x_2 x_3^2 + (x_1 x_2 x_3)^2 \quad \blacksquare$$

Most real-world systems have a more complex arrangement of components than a k-out-of-n arrangement. The next example illustrates how to combine series and parallel arrangements to determine the appropriate structure function for a more complex system.

Example 17.4

An airplane has four propellers, two on each wing. The airplane will fly (function) if at least one propeller on each wing functions. In this case, the four propellers are denoted by components 1, 2, 3, and 4, with 1 and 2 being on the left wing and 3 and 4 on the right wing. For the moment, if the plane is considered to consist of two wings (not considering individual propellers), then the wings are arranged in series, as failure of the propulsion on either wing results in system failure. Each wing can be modeled as a two-component parallel subsystem of propellers, as only one propeller on each wing is required to function. The appropriate

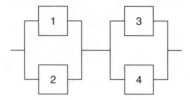

FIGURE 17.4 A four-propeller system.

block diagram for the system is shown in Figure 17.4. The structure function is the product of the structure functions of the two parallel subsystems:

$$\phi(\boldsymbol{x}) = [1-(1-x_1)(1-x_2)][1-(1-x_3)(1-x_4)] \qquad \blacksquare$$

To avoid studying structure functions that are unreasonable, a subset of all possible systems of n components, namely *coherent systems*, has been defined. A system is coherent if $\phi(\boldsymbol{x})$ is nondecreasing in $\boldsymbol{x}$ [e.g., $\phi(x_1,\ldots,x_{i-1},0,x_{i+1},\ldots,x_n) \leq \phi(x_1,\ldots,x_{i-1},1,x_{i+1},\ldots,x_n)$ for all i] and there are no irrelevant components. The condition that $\phi(\boldsymbol{x})$ be nondecreasing in $\boldsymbol{x}$ implies that the system will not degrade if a component upgrades. A component is irrelevant if its state has no impact on the structure function. Many theorems related to coherent systems have been proven; one of the more useful is that redundancy at the component level is more effective than redundancy at the system level. This is an important consideration in *reliability design*, where a reliability engineer decides the components to choose (reliability allocation) at appropriate positions in the system (reliability optimization).

17.2.2 Reliability Functions

The discussion of structure functions so far has been completely deterministic in nature. We now introduce probability into the mix by first defining reliability. The paragraphs following the definition expand on the italicized words in the definition.

DEFINITION 17.3 The *reliability* of an *item* is the *probability* that it will *adequately perform* its specified *purpose* for a specified *period of time* under specified *environmental conditions*.

The definition implies that the object of interest is an *item*. The definition of the item depends on the purpose of the study. In some situations, we will consider an item to be an interacting arrangement of components; in other situations, the component level of detail in the model is not of interest.

Reliability is defined as a *probability*. Thus, the axioms of probability apply to reliability calculations. In particular, this means that all reliabilities must be between 0 and 1 inclusive, and that the results derived from the probability axioms must hold. For example, if two independent components have 1000-hour reliabilities of p_1 and p_2, and system failure occurs when either component fails (i.e., a two-component series system), then the 1000-hour system reliability is $p_1 p_2$.

Adequate performance for an item must be stated unambiguously. A *standard* is often used to determine what is considered adequate performance. A mechanical part may require tolerances that delineate adequate performance from inadequate performance. The performance of an item is related to the mathematical model used to represent the condition of the item. The simplest model for an item is a binary model, which was introduced earlier, in which the item is in either the functioning or failed state. This model is easily applied to

Reliability

a light bulb; it is more difficult to apply to items that gradually degrade over time, such as a machine tool. To apply a binary model to an item that degrades gradually, a threshold value must be determined to separate the functioning and failed states.

The definition of reliability also implies that the *purpose* or intended use of the item must be specified. Machine tool manufacturers, for example, often produce two grades of an item: one for professional use and another for consumer use.

The definition of reliability also indicates that *time* is involved in reliability, which implies five consequences. First, the units for time need to be specified (e.g., minutes, hours, years) by the modeler to perform any analysis. Second, many lifetime models use the random variable T (rather than X, which is common in probability theory) to represent the failure time of the item. Third, time need not be taken literally. The number of miles may represent time for an automobile tire; the number of cycles may represent time for a light switch. Fourth, a time duration associated with a reliability must be specified. The reliability of a component, for example, should not be stated as simply 0.98, as no time is specified. It is equally ambiguous for a component to have a 1000-hour life without indicating a reliability for that time. Instead, it should be stated that the 1000-hour reliability is 0.98. This requirement of stating a time along with a reliability applies to systems as well as components. Finally, determining what should be used to measure the lifetime of an item may not be obvious. Reliability analysts must consider whether continuous operation or on/off cycling is more effective for items such as motors or computers.

The last aspect of the definition of reliability is that *environmental conditions* must be specified. Conditions such as temperature, humidity, and turning speed all affect the lifetime of a machine tool. Likewise, the driving conditions for an automobile will influence its reliability. Included in environmental conditions is the preventive maintenance to be performed on the item.

We now return to the mathematical models for determining the reliability of a system. Two additional assumptions need to be made for the models developed here. First, the n components comprising a system must be *nonrepairable*. Once a component changes from the functioning to the failed state, it cannot return to the functioning state. This assumption was not necessary when structure functions were introduced, as a structure function simply maps the component states to the system state. The structure function can be applied to a system with nonrepairable or repairable components. The second assumption is that the components are independent. Thus, failure of one component does not influence the probability of failure of other components. This assumption is not appropriate if the components operate in a common environment where there may be common-cause failures. Although the independence assumption makes the mathematics for modeling a system simpler, the assumption should not be automatically applied.

Previously, x_i was defined to be the state of component i. Now X_i is a random variable with the same meaning.

DEFINITION 17.4 The random variable denoting the state of component i, denoted by X_i, is

$$X_i = \begin{cases} 0 & \text{if component } i \text{ has failed} \\ 1 & \text{if component } i \text{ is functioning} \end{cases}$$

for $i = 1, 2, \ldots, n$.

These n values can be written as a random system state vector $\boldsymbol{X}$. The probability that component i is functioning at a certain time is given by $p_i = P(X_i = 1)$, which is often called the *reliability* of the ith component, for $i = 1, 2, \ldots, n$. The P function denotes probability. These n values can be written as a reliability vector $\boldsymbol{p} = (p_1, p_2, \ldots, p_n)$.

The *system reliability*, denoted by r, is defined by

$$r = P[\phi(\boldsymbol{X}) = 1]$$

where r is a quantity that can be calculated from the vector $\boldsymbol{p}$, so $r = r(\boldsymbol{p})$. The function $r(\boldsymbol{p})$ is called the *reliability function*. In some of the examples in this section, the components have identical reliabilities (that is, $p_1 = p_2 = \cdots = p_n = p$), which is indicated by the notation $r(p)$.

Several techniques are used to calculate system reliability. We will illustrate two of the simplest techniques: definition and expectation. The first technique for finding the reliability of a coherent system of n independent components is to use the definition of system reliability directly, as illustrated in the example.

Example 17.5

The system reliability of a *series* system of n components is easily found using the definition of $r(\boldsymbol{p})$ and the independence assumption.

$$r(\boldsymbol{p}) = P[\phi(\boldsymbol{X}) = 1]$$

$$= P\left[\prod_{i=1}^{n} X_i = 1\right]$$

$$= \prod_{i=1}^{n} P[X_i = 1]$$

$$= \prod_{i=1}^{n} p_i$$

The product in this formula indicates that system reliability is always less than the reliability of the least reliable component. This "chain is only as strong as its weakest link" result indicates that improving the weakest component causes the largest increase in the reliability of a series system. ∎

In the special case when all components are identical, the reliability function reduces to $r(p) = p^n$, where $p_1 = p_2 = \cdots = p_n = p$. The plot in Figure 17.5 of component reliability versus system reliability for several values of n shows that highly reliable components are necessary to achieve reasonable system reliability, even for small values of n.

The second technique, *expectation*, is based on the fact that $P[\phi(\boldsymbol{X}) = 1]$ is equal to $E[\phi(\boldsymbol{X})]$, because $\phi(\boldsymbol{X})$ is a Bernoulli random variable. Consequently, the expected value of $\phi(\boldsymbol{X})$ is the system reliability $r(\boldsymbol{p})$, as illustrated in the next example.

Example 17.6

Since the components are assumed to be independent, the system reliability for a parallel system of n components using the expectation technique is

$$r(\boldsymbol{p}) = E[\phi(\boldsymbol{X})]$$

$$= E\left[1 - \prod_{i=1}^{n}(1 - X_i)\right]$$

Reliability

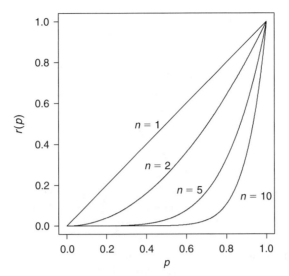

FIGURE 17.5 Reliability of a series system of n components.

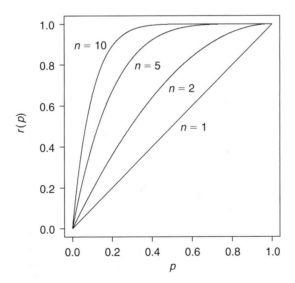

FIGURE 17.6 Reliability of a parallel system of n components.

$$= 1 - E\left[\prod_{i=1}^{n}(1-X_i)\right]$$

$$= 1 - \prod_{i=1}^{n} E[1-X_i]$$

$$= 1 - \prod_{i=1}^{n}(1-p_i)$$

In the special case of identical components, this expression reduces to $r(p) = 1 - (1-p)^n$. Figure 17.6 shows component reliability versus system reliability for a parallel system of n identical components. The law of diminishing returns is apparent from the graph when a fixed component reliability is considered. The marginal gain in reliability decreases dramatically as more components are added to the system. ∎

There are two systems that appear to be similar to parallel systems on the surface, but they are not true parallel systems such as the one considered in the previous example. The first such system is a *standby system*. In a standby system, not all the components function simultaneously, and components are switched to standby components upon failure. Examples of standby systems include a spare tire for an automobile and having three power sources (utility company, backup generator, and batteries) for a hospital. In contrast, all components are functioning simultaneously in a true parallel system.

The second such system is a *shared-parallel system*. In a shared-parallel system, all components are online, but the component reliabilities change when one component fails. The lug nuts that attach a wheel to an automobile are an example of a five-component shared-parallel system. When one lug nut fails (i.e., loosens or falls off), the load on the remaining functioning lug nuts increases. Thus, the static reliability calculations presented in this section are not appropriate for a wheel attachment system. In contrast, the failure of a component in a true parallel system does not affect the reliabilities of any of the other components in the system.

17.3 Lifetime Distributions

Reliability has only been considered at one particular instant of time. Reliability is generalized to be a function of time in this section, and various lifetime distribution representations that are helpful in describing the evolution of the risks to which an item is subjected over time are introduced. In particular, four lifetime distribution representations are presented: the *survivor* function, the *probability density* function, the *hazard* function, and the *cumulative hazard* function. These four distribution representations apply to both continuous (e.g., a fuse) and discrete (e.g., software executed daily) lifetimes, although the focus here is on continuous lifetimes. These four representations are not the only ways to define the distribution of the continuous, nonnegative random variable T, referred to generically here as a "lifetime." Other methods include the moment generating function $E[e^{sT}]$, the characteristic function $E[e^{isT}]$, the Mellin transform $E[T^s]$, and the mean residual life function $E[T - t | T \geq t]$. The four representations used here have been chosen because of their intuitive appeal, their usefulness in problem solving, and their popularity in the literature.

17.3.1 Survivor Function

The first lifetime distribution representation is the *survivor function*, $S(t)$. The survivor function is a generalization of reliability. Whereas reliability is defined as the probability that an item is functioning at one particular time, the survivor function is the probability that an item is functioning at any time t:

$$S(t) = P[T \geq t] \quad t \geq 0$$

It is assumed that $S(t) = 1$ for all $t < 0$. A survivor function is also known as the reliability function [since $S(t)$ is the reliability at time t] and the complementary cumulative distribution function [since $S(t) = 1 - F(t)$ for continuous random variables, where $F(t) = P[T \leq t]$ is the cumulative distribution function]. All survivor functions must satisfy three conditions:

$$S(0) = 1 \quad \lim_{t \to \infty} S(t) = 0 \quad S(t) \text{ is nonincreasing}$$

Reliability

There are two interpretations of the survivor function. First, $S(t)$ is the probability that an individual item is functioning at time t. This is important, as will be seen later, in determining the lifetime distribution of a system from the distribution of the lifetimes of its individual components. Second, if there is a large population of items with identically distributed lifetimes, $S(t)$ is the expected fraction of the population that is functioning at time t.

The survivor function is useful for comparing the survival patterns of several populations of items. If $S_1(t) \geq S_2(t)$, for all t values, for example, it can be concluded that the items in population 1 are superior to those in population 2 with regard to reliability.

17.3.2 Probability Density Function

The second lifetime distribution representation, the *probability density function*, is defined by $f(t) = -S'(t)$, where the derivative exists. It has the probabilistic interpretation

$$f(t)\Delta t = P[t \leq T \leq t + \Delta t]$$

for small values of Δt. Although the probability density function is not as effective as the survivor function in comparing the survival patterns of two populations, a graph of $f(t)$ indicates the likelihood of failure for a new item over the course of its lifetime. The probability of failure between times a and b is calculated by an integral:

$$P[a \leq T \leq b] = \int_a^b f(t) \mathrm{d}t$$

All probability density functions for lifetimes must satisfy two conditions:

$$\int_0^\infty f(t) \mathrm{d}t = 1 \quad f(t) \geq 0 \text{ for all } t \geq 0$$

It is assumed that $f(t) = 0$ for all $t < 0$.

17.3.3 Hazard Function

The *hazard function*, $h(t)$, is perhaps the most popular of the five representations for lifetime modeling due to its intuitive interpretation as the amount of *risk* associated with an item at time t. The hazard function goes by several aliases: in reliability it is also known as the hazard rate or failure rate; in actuarial science it is known as the force of mortality or force of decrement; in point process and extreme value theory it is known as the rate or intensity function; in vital statistics it is known as the age-specific death rate; and in economics its reciprocal is known as Mill's ratio.

The hazard function can be derived using conditional probability. First, consider the probability of failure between t and $t + \Delta t$:

$$P[t \leq T \leq t + \Delta t] = \int_t^{t+\Delta t} f(\tau) \mathrm{d}\tau = S(t) - S(t + \Delta t)$$

Conditioning on the event that the item is working at time t yields

$$P[t \leq T \leq t + \Delta t | T \geq t] = \frac{P[t \leq T \leq t + \Delta t]}{P[T \geq t]} = \frac{S(t) - S(t + \Delta t)}{S(t)}$$

If this conditional probability is averaged over the interval $[t, t + \Delta t]$ by dividing by Δt, an average rate of failure is obtained:

$$\frac{S(t) - S(t + \Delta t)}{S(t)\Delta t}$$

As $\Delta t \to 0$, this average failure rate becomes the instantaneous failure rate, which is the hazard function

$$\begin{aligned} h(t) &= \lim_{\Delta t \to 0} \frac{S(t) - S(t + \Delta t)}{S(t)\Delta t} \\ &= -\frac{S'(t)}{S(t)} \\ &= \frac{f(t)}{S(t)} \quad t \geq 0 \end{aligned}$$

Thus, the hazard function is the ratio of the probability density function to the survivor function. Using the previous derivation, a probabilistic interpretation of the hazard function is

$$h(t)\Delta t = P[t \leq T \leq t + \Delta t | T > t]$$

for small values of Δt, which is a conditional version of the interpretation for the probability density function. All hazard functions must satisfy two conditions:

$$\int_0^\infty h(t)\,dt = \infty \quad h(t) \geq 0 \text{ for all } t \geq 0$$

The *units* on a hazard function are typically given in failures per unit time, for example, $h(t) = 0.03$ failures per hour. Since the magnitude of hazard functions can often be quite small, they are often expressed in scientific notation, for example, $h(t) = 3.8$ failures per 10^6 hours, or the time units are chosen to keep hazard functions from getting too small, for example, $h(t) = 8.4$ failures per year.

The shape of the hazard function indicates how an item ages. The intuitive interpretation as the amount of *risk* an item is subjected to at time t indicates that when the hazard function is large the item is under greater risk, and when the hazard function is small the item is under less risk. The three hazard functions plotted in Figure 17.7 correspond to an increasing hazard function (labeled IFR for increasing failure rate), a decreasing hazard function (labeled DFR for decreasing failure rate), and a bathtub-shaped hazard function (labeled BT for bathtub-shaped failure rate).

The increasing hazard function is probably the most likely situation of the three. In this case, items are more likely to fail as time passes. In other words, items wear out or degrade with time. This is almost certainly the case with mechanical items that undergo wear or fatigue. The second situation, the decreasing hazard function, is less common. In this case, the item is less likely to fail as time passes. Items with this type of hazard function improve with time. Some metals, for example, work-harden through use and thus have increased strength as time passes. Another situation for which a decreasing hazard function might be appropriate for modeling is in working the bugs out of computer programs. Bugs are more likely to appear initially, but the likelihood of them appearing decreases as time passes.

The third situation, a bathtub-shaped hazard function, occurs when the hazard function decreases initially and then increases as items age. Items improve initially and then degrade

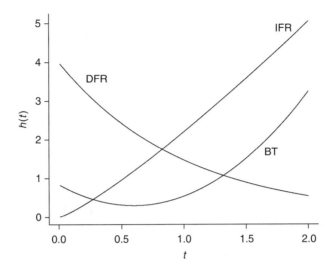

FIGURE 17.7 Hazard functions.

as time passes. One instance where the bathtub-shaped hazard function arises is in the lifetimes of manufactured items. Often manufacturing, design, or component defects cause early failures. The period in which these failures occur is sometimes called the *burn-in* period. If failure is particularly catastrophic, this part of the lifetime will often be consumed by the manufacturer in a controlled environment. The time value during which early failures have been eliminated may be valuable to a producer who is determining an appropriate warranty period. Once items pass through this early part of their lifetime, they have a fairly constant hazard function, and failures are equally likely to occur at any point in time. Finally, as items continue to age, the hazard function increases without limit, resulting in *wear-out* failures. Bathtub-shaped hazard functions are also used for modeling human lifetimes.

Care must be taken to differentiate between the hazard function for a population and the hazard function for an individual *item* under consideration. Consider the lifetimes of a certain model of a laptop computer as an illustration. Consider the following question: do two laptops operating in identical environments necessarily have the same hazard function? The answer is no. The laptops have their own individual hazard functions as they might have been manufactured at different facilities that may have included differing conditions (e.g., temperature, raw materials, parts suppliers). So, although a hazard function could be drawn for all laptop computers, it would be an aggregate hazard function representing the population, and individual laptops may be at increased or decreased risk.

17.3.4 Cumulative Hazard Function

The fourth lifetime distribution representation, the *cumulative hazard function* $H(t)$, can be defined by

$$H(t) = \int_0^t h(\tau) d\tau \quad t \geq 0$$

The cumulative hazard function is also known as the integrated hazard function. All cumulative hazard functions must satisfy three conditions:

$$H(0) = 0 \quad \lim_{t \to \infty} H(t) = \infty \quad H(t) \text{ is nondecreasing}$$

The cumulative hazard function is valuable for variate generation in Monte Carlo simulation, implementing certain procedures in statistical inference, and defining certain distribution classes.

The four distribution representations presented here are equivalent in the sense that each completely specifies a lifetime distribution. Any one lifetime distribution representation implies the other four. Algebra and calculus can be used to find one lifetime distribution representation given that another is known. For example, if the survivor function is known, the cumulative hazard function can be determined by

$$H(t) = \int_0^t h(\tau)d\tau = \int_0^t \frac{f(\tau)}{S(\tau)}d\tau = -\log S(t)$$

where log is the natural logarithm (log base e).

17.3.5 Expected Values and Fractiles

Once a lifetime distribution representation for a particular item (which may be a component or an entire system) is known, it may be of interest to compute a moment or a fractile of the distribution. Moments and fractiles contain less information than a lifetime distribution representation, but they are often useful ways to summarize the distribution of a random lifetime. Examples of these performance measures include the mean time to failure, $E[T]$, the median, $t_{0.50}$, and the 99th fractile of a distribution, $t_{0.99}$.

A formula for the expectation of some function of the random variable T, say $u(T)$, is

$$E[u(T)] = \int_0^\infty u(t)f(t)dt$$

The most common measure associated with a distribution is its *mean*, or first moment,

$$\mu = E[T] = \int_0^\infty tf(t)dt = \int_0^\infty S(t)dt$$

where the last equality is proved using integration by parts and is based on the assumption that $\lim_{t\to\infty} tS(t) = 0$. The mean is a measure of the central tendency or average value that a lifetime distribution assumes and is known as the center of gravity in physics. It is often abbreviated by MTTF (mean time to failure) for nonrepairable items. For repairable items that can be completely renewed by repair, it is often abbreviated by MTBF (mean time between failures). Another value associated with a distribution is its *variance*, or second moment about the mean,

$$\sigma^2 = V[T] = E[(T-\mu)^2] = E[T^2] - (E[T])^2$$

which is a measure of the dispersion of a lifetime distribution about its mean. The positive square root of the variance is known as the *standard deviation*, which has the same units as the random variable T.

Fractiles of a distribution are the times to which a specified proportion of the items survives. The definition of the pth fractile of a distribution, t_p, (often called the pth quantile or $100p$th percentile) satisfies

$$F(t_p) = P[T \leq t_p] = p$$

or, equivalently,

$$t_p = F^{-1}(p)$$

Example 17.7

The *exponential* distribution has survivor function

$$S(t) = e^{-\lambda t} \quad t \geq 0$$

where λ is a positive parameter known as the failure rate. Find the mean, variance, and the pth fractile of the distribution. The mean can be found by integrating the survivor function from 0 to infinity:

$$\mu = E[T] = \int_0^\infty S(t)\,dt = \int_0^\infty e^{-\lambda t}\,dt = \frac{1}{\lambda}$$

Since the probability density function is $f(t) = -S'(t) = \lambda e^{-\lambda t}$ for $t > 0$, the second moment about the origin is

$$E[T^2] = \int_0^\infty t^2 f(t)\,dt = \int_0^\infty t^2 \lambda e^{-\lambda t}\,dt = \frac{2}{\lambda^2}$$

using integration by parts twice. Therefore, the variance is

$$\sigma^2 = V[T] = E[T^2] - (E[T])^2 = \frac{2}{\lambda^2} - \frac{1}{\lambda^2} = \frac{1}{\lambda^2}$$

Finally, the pth fractile of the distribution, t_p, is found by solving

$$1 - e^{-\lambda t_p} = p$$

for t_p, yielding $t_p = -\frac{1}{\lambda}\log(1-p)$. ∎

17.3.6 System Lifetime Distributions

To this point, the discussion concerning the four lifetime representations $S(t), f(t), h(t)$, and $H(t)$ has assumed that the variable of interest is the *lifetime* of an *item*. For systems of components, both the individual components and the system have random lifetimes whose lifetime distributions can be defined by any of the four lifetime distributions. We now integrate reliability functions from Section 17.2 and the lifetime distribution representations from this section, which allows a modeler to find the distribution of the system lifetime, given the distributions of the component lifetimes. The component lifetime representations are denoted by $S_i(t), f_i(t), h_i(t)$, and $H_i(t)$, for $i = 1, 2, \ldots, n$, and the system lifetime representations are denoted by $S(t), f(t), h(t)$, and $H(t)$.

The survivor function is a time-dependent generalization of reliability. Whereas reliability always needs an associated time value (e.g., the 4000-hour reliability is 0.96), the survivor function is the reliability at any time t.

To find the reliability of a system at any time t, the component survivor functions should be used as arguments in the reliability function, that is,

$$S(t) = r(S_1(t), S_2(t), \ldots, S_n(t))$$

Once $S(t)$ is known, it is straightforward to determine any of the other four lifetime representations, moments, or fractiles, as illustrated in the following examples.

Example 17.8

Two independent components with survivor functions

$$S_1(t) = e^{-t} \quad \text{and} \quad S_2(t) = e^{-2t}$$

for $t \geq 0$, are arranged in series. Find the survivor and hazard functions for the system lifetime. Since the reliability function for a two-component series system is $r(\mathbf{p}) = p_1 p_2$, the system survivor function is

$$\begin{aligned} S(t) &= S_1(t) S_2(t) \\ &= e^{-t} e^{-2t} \\ &= e^{-3t} \quad t \geq 0 \end{aligned}$$

which can be recognized as the survivor function for an exponential distribution with $\lambda = 3$. The hazard function for the system is

$$h(t) = -\frac{S'(t)}{S(t)} = 3 \quad t \geq 0$$

Thus, if two independent components with exponential times to failure are arranged in series, the time to system failure is also exponentially distributed with a failure rate that is the sum of the failure rates of the individual components. This result can be generalized to series systems with more than two components. If the lifetime of component i in a series system of n independent components has an exponential distribution with failure rate λ_i, then the system lifetime is exponentially distributed with failure rate $\sum_{i=1}^{n} \lambda_i$. ∎

The next example considers a parallel system of components having exponential lifetimes.

Example 17.9

Two independent components have hazard functions

$$h_1(t) = 1 \quad \text{and} \quad h_2(t) = 2$$

for $t \geq 0$. If the components are arranged in parallel, find the hazard function of the time to system failure and the mean time to system failure.

The survivor functions of the components are

$$S_1(t) = e^{-H_1(t)} = e^{-\int_0^t h_1(\tau) d\tau} = e^{-t}$$

for $t \geq 0$. Likewise, $S_2(t) = e^{-2t}$ for $t \geq 0$. Since the reliability function for a two-component parallel system is $r(\mathbf{p}) = 1 - (1 - p_1)(1 - p_2)$, the system survivor function is

$$\begin{aligned} S(t) &= 1 - (1 - S_1(t))(1 - S_2(t)) \\ &= 1 - (1 - e^{-t})(1 - e^{-2t}) \\ &= e^{-t} + e^{-2t} - e^{-3t} \quad t \geq 0 \end{aligned}$$

The hazard function is

$$h(t) = -\frac{S'(t)}{S(t)} = \frac{e^{-t} + 2e^{-2t} - 3e^{-3t}}{e^{-t} + e^{-2t} - e^{-3t}} \quad t \geq 0$$

Reliability

To find the mean time to system failure, the system survivor function is integrated from 0 to infinity:

$$\mu = \int_0^\infty S(t)\mathrm{d}t = \int_0^\infty (\mathrm{e}^{-t} + \mathrm{e}^{-2t} - \mathrm{e}^{-3t})\mathrm{d}t = 1 + \frac{1}{2} - \frac{1}{3} = \frac{7}{6}$$

The mean time to failure of the stronger component is 1 and the mean time to failure of the weaker component is $1/2$. The addition of the weaker component in parallel with the stronger only increases the mean time to system failure by $1/6$. This is yet another illustration of the law of diminishing returns for parallel systems. ∎

17.4 Parametric Models

The survival patterns of a machine tool, a fuse, and an aircraft are vastly different. One would certainly not want to use the same failure time distribution with identical parameters to model these diverse lifetimes. This section introduces two distributions that are commonly used to model lifetimes. To adequately survey all the distributions currently in existence would require an entire textbook, so detailed discussion here is limited to the exponential and Weibull distributions.

17.4.1 Parameters

We begin by describing parameters, which are common to all lifetime distributions. The three most common types of parameters used in lifetime distributions are location, scale, and shape. Parameters in a lifetime distribution allow modeling of such diverse applications as light bulb failure time, patient postsurgery survival time, and the failure time of a muffler on an automobile by a single lifetime distribution (e.g., the Weibull distribution).

Location (or *shift*) parameters are used to shift the distribution to the left or right along the time axis. If c_1 and c_2 are two values of a location parameter for a lifetime distribution with survivor function $S(t; c)$, then there exists a real constant α such that $S(t; c_1) = S(\alpha + t; c_2)$. A familiar example of a location parameter is the mean of the normal distribution.

Scale parameters are used to expand or contract the time axis by a factor of α. If λ_1 and λ_2 are two values for a scale parameter for a lifetime distribution with survivor function $S(t; \lambda)$, then there exists a real constant α such that $S(\alpha t; \lambda_1) = S(t; \lambda_2)$. A familiar example of a scale parameter is λ in the exponential distribution. The probability density function always has the same shape, and the units on the time axis are determined by the value of λ.

Shape parameters are appropriately named because they affect the shape of the probability density function. Shape parameter values might also determine whether a distribution belongs to a particular distribution class such as IFR or DFR. A familiar example of a shape parameter is κ in the Weibull distribution.

In summary, location parameters *translate* survival distributions along the time axis, scale parameters *expand* or *contract* the time scale for survival distributions, and all other parameters are shape parameters.

17.4.2 Exponential Distribution

Just as the normal distribution plays an important role in classical statistics because of the central limit theorem, the exponential distribution plays an important role in reliability and lifetime modeling because it is the only continuous distribution with a constant hazard function. The exponential distribution has often been used to model the lifetime of electronic components and is appropriate when a used component that has not failed is statistically as good as a new component. This is a rather restrictive assumption. The exponential distribution is presented first because of its simplicity. The Weibull distribution, a more complex two-parameter distribution that can model a wider variety of situations, is presented subsequently. The exponential distribution has a single positive scale parameter λ, often called the *failure rate*, and the four lifetime distribution representations are

$$S(t) = e^{-\lambda t} \quad f(t) = \lambda e^{-\lambda t} \quad h(t) = \lambda \quad H(t) = \lambda t \text{ for } t \geq 0$$

There are several probabilistic properties of the exponential distribution that are useful in understanding how it is unique and when it should be applied. In the properties to be outlined below, the nonnegative lifetime T typically has the exponential distribution with parameter λ, which denotes the number of failures per unit time. The symbol $\sim$ means "is distributed as." Proofs of these results are given in most reliability textbooks.

THEOREM 17.1 (*Memoryless Property*) *If $T \sim exponential(\lambda)$, then*

$$P[T \geq t] = P[T \geq t + s | T \geq s] \quad t \geq 0; s \geq 0$$

As shown in Figure 17.8 for $\lambda = 1$ and $s = 0.5$, the memoryless property indicates that the conditional survivor function for the lifetime of an item that has survived to time s is identical to the survivor function for the lifetime of a brand new item. This used-as-good-as-new assumption is very strong. The exponential lifetime model should not be applied to mechanical components that undergo wear (e.g., bearings) or fatigue (e.g., structural supports) or electrical components that contain an element that burns away (e.g., filaments) or degrades with time (e.g., batteries). An electrical component for which the exponential lifetime assumption may be justified is a fuse. A fuse is designed to fail when there is a power surge that causes the fuse to burn out. Assuming that the fuse does not undergo any weakening or degradation over time and that power surges that cause failure occur with

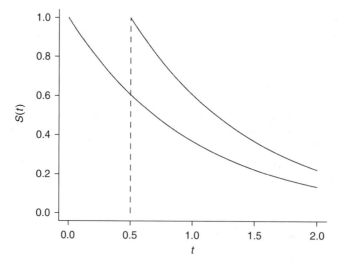

FIGURE 17.8 The memoryless property.

Reliability

equal likelihood over time, the exponential lifetime assumption is appropriate, and a used fuse that has not failed is as good as a new one.

The exponential distribution should be applied judiciously as the memoryless property restricts its applicability. It is usually misapplied for the sake of simplicity as the statistical techniques for the exponential distribution are particularly tractable, or small sample sizes do not support more than a one-parameter distribution.

THEOREM 17.2 *The exponential distribution is the only continuous distribution with the memoryless property.*

This result indicates that the exponential distribution is the only continuous lifetime distribution for which the conditional lifetime distribution of a used item is identical to the original lifetime distribution. The only discrete distribution with the memoryless property is the geometric distribution.

THEOREM 17.3 *If $T \sim exponential(\lambda)$, then*

$$E[T^s] = \frac{\Gamma(s+1)}{\lambda^s} \quad s > -1$$

where the gamma function *is defined by*

$$\Gamma(\alpha) = \int_0^\infty x^{\alpha-1} e^{-x} dx$$

When s is a nonnegative integer, this expression reduces to $E[T^s] = s!/\lambda^s$. By setting $s = 1$ and 2, the mean and variance can be obtained:

$$E[T] = \frac{1}{\lambda} \quad V[T] = \frac{1}{\lambda^2}$$

THEOREM 17.4 *(Self-Reproducing) If $T_1, T_2, \ldots, T_n$ are independent, $T_i \sim$ exponential(λ_i), for $i = 1, 2, \ldots, n$, and $T = \min\{T_1, T_2, \ldots, T_n\}$, then*

$$T \sim exponential\left(\sum_{i=1}^n \lambda_i\right)$$

This result indicates that the minimum of n exponential random lifetimes also has the exponential distribution, as alluded to in the previous section. This is important in two applications. First, if n independent exponential components, each with exponential times to failure, are arranged in series, the distribution of the system failure time is also exponential with a failure rate equal to the sum of the component failure rates. When the n components have the same failure rate λ, the system lifetime is exponential with failure rate $n\lambda$. Second, when there are several independent, exponentially distributed *causes* of failure competing for the lifetime of an item (e.g., failing by open or short circuit for an electronic item or death by various risks for a human being), the lifetime can be modeled as the minimum of the individual lifetimes associated with each cause of failure.

THEOREM 17.5 If $T_1, T_2, \ldots, T_n$ are independent and identically distributed exponential(λ) random variables, then

$$2\lambda \sum_{i=1}^{n} T_i \sim \chi^2(2n)$$

where $\chi^2(2n)$ denotes the chi-square distribution with $2n$ degrees of freedom.

This property is useful for determining a confidence interval for a λ based on a data set of n independent exponential lifetimes. For instance, with probability $1-\alpha$,

$$\chi^2_{2n,1-\alpha/2} < 2\lambda \sum_{i=1}^{n} T_i < \chi^2_{2n,\alpha/2}$$

where the left- and right-hand sides of this inequality are the $\alpha/2$ and $1-\alpha/2$ fractiles of the chi-square distribution with $2n$ degrees of freedom, that is, the second subscript denotes *right*-hand tail areas. Rearranging this expression yields a $100(1-\alpha)\%$ confidence interval for λ:

$$\frac{\chi^2_{2n,1-\alpha/2}}{2\sum_{i=1}^{n} T_i} < \lambda < \frac{\chi^2_{2n,\alpha/2}}{2\sum_{i=1}^{n} T_i}$$

THEOREM 17.6 If $T_1, T_2, \ldots, T_n$ are independent and identically distributed exponential(λ) random variables, $T_{(1)}, T_{(2)}, \ldots, T_{(n)}$ are the corresponding order statistics (the observations sorted in ascending order), $G_k = T_{(k)} - T_{(k-1)}$ for $k = 1, 2, \ldots, n$, and if $T_{(0)} = 0$, then

- $P[G_k \geq t] = e^{-(n-k+1)\lambda t}$; $t \geq 0$; $k = 1, 2, \ldots, n$.
- $G_1, G_2, \ldots, G_n$ are independent.

This property is most easily interpreted in terms of a life test of n items with exponential(λ) lifetimes. Assume that the items placed on the life test are *not* replaced with new items when they fail. The ith item fails at time $t_{(i)}$, and $G_i = t_{(i)} - t_{(i-1)}$ is the time between the $(i-1)$st and ith failure, for $i = 1, 2, \ldots, n$, as indicated in Figure 17.9 for $n = 4$. The result states that these gaps (G_i's) between the failure times are independent and exponentially distributed. The proof of this theorem relies on the memoryless property and the self-reproducing property of the exponential distribution, which implies that when the ith failure occurs the time until the next failure is the minimum of $n - i$ independent exponential random variables.

THEOREM 17.7 If $T_1, T_2, \ldots, T_n$ are independent and identically distributed exponential(λ) random variables and $T_{(r)}$ is the rth order statistic, then

$$E[T_{(r)}] = \sum_{k=1}^{r} \frac{1}{(n-k+1)\lambda}$$

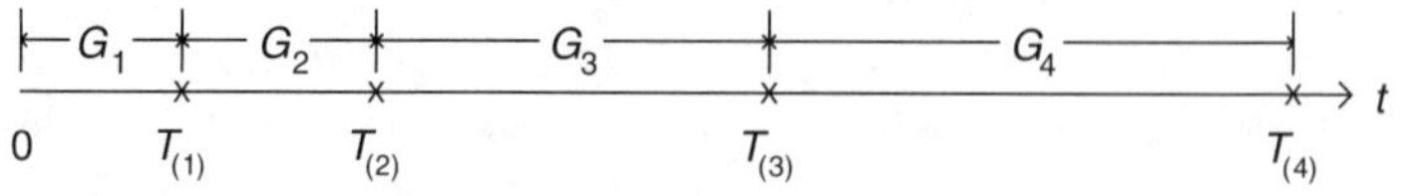

FIGURE 17.9 Order statistics and gap statistics.

Reliability

and

$$V[T_{(r)}] = \sum_{k=1}^{r} \frac{1}{[(n-k+1)\lambda]^2}$$

for $r = 1, 2, \ldots, n$.

The expected value and variance of the rth-ordered failure are simple functions of n, λ, and r. The proof of this result is straightforward, as the gaps between order statistics are independent exponential random variables from the previous theorem, and the rth ordered failure on a life test of n items is the sum of the first r gaps. This result is useful in determining the expected time to complete a life test that is discontinued after r of the n items on test fail.

THEOREM 17.8 *If $T_1, T_2, \ldots$ are independent and identically distributed exponential(λ) random variables denoting the interevent times for a point process, then the number of events in the interval $[0,t]$ has the Poisson distribution with parameter λt.*

This property is related to the memoryless property and can be applied to a component that is subjected to shocks occurring randomly over time. It states that if the time between shocks is exponential(λ) then the number of shocks occurring by time t has the Poisson distribution with parameter λt. This result also applies to the failure time of a cold standby system of n identical exponential components in which nonoperating units do not fail and sensing and switching are perfect. The probability of fewer than n failures by time t (the system reliability) is

$$\sum_{k=0}^{n-1} \frac{(\lambda t)^k}{k!} e^{-\lambda t}$$

The exponential distribution, for which the item under study does not age in a probabilistic sense, is the simplest possible lifetime model. Another popular distribution that arises in many reliability applications is the Weibull distribution, which is presented next.

17.4.3 Weibull Distribution

The exponential distribution is limited in applicability because of the memoryless property. The assumption that a lifetime has a constant failure rate is often too restrictive or inappropriate. Mechanical items typically degrade over time and hence are more likely to follow a distribution with a strictly increasing hazard function. The Weibull distribution is a generalization of the exponential distribution that is appropriate for modeling lifetimes having constant, strictly increasing, and strictly decreasing hazard functions. The four lifetime distribution representations for the Weibull distribution are

$$S(t) = e^{-(\lambda t)^\kappa} \quad f(t) = \kappa \lambda^\kappa t^{\kappa-1} e^{-(\lambda t)^\kappa} \quad h(t) = \kappa \lambda^\kappa t^{\kappa-1} \quad H(t) = (\lambda t)^\kappa$$

for all $t \geq 0$, where $\lambda > 0$ and $\kappa > 0$ are the scale and shape parameters of the distribution. The hazard function approaches zero from infinity for $\kappa < 1$, is constant for $\kappa = 1$, the exponential case, and increases from zero when $\kappa > 1$. Hence, the Weibull distribution can attain hazard function shapes in both the IFR and DFR classes and includes the exponential distribution as a special case. One other special case occurs when $\kappa = 2$, commonly known as the Rayleigh distribution, which has a linear hazard function with slope $2\lambda^2$. When $3 < \kappa < 4$, the shape of the probability density function resembles that of a normal probability density function.

Using the expression

$$E[T^r] = \frac{r}{\kappa \lambda^r} \Gamma\left(\frac{r}{\kappa}\right)$$

for $r = 1, 2, \ldots$, the mean and variance for the Weibull distribution are

$$\mu = \frac{1}{\lambda \kappa} \Gamma\left(\frac{1}{\kappa}\right)$$

and

$$\sigma^2 = \frac{1}{\lambda^2} \left\{ \frac{2}{\kappa} \Gamma\left(\frac{2}{\kappa}\right) - \left[\frac{1}{\kappa} \Gamma\left(\frac{1}{\kappa}\right)\right]^2 \right\}$$

Example 17.10

The lifetime of a machine used continuously under known operating conditions has the Weibull distribution with $\lambda = 0.00027$ and $\kappa = 1.55$, where time is measured in hours. (Estimating the parameters for the Weibull distribution from a data set will be addressed subsequently, but the parameters are assumed to be known for this example.) What is the mean time to failure and the probability the machine will operate for 5000 hours?

The mean time to failure is

$$\mu = E[T] = \frac{1}{(0.00027)(1.55)} \Gamma\left(\frac{1}{1.55}\right) = 3331 \text{ hours}$$

The probability that the machine will operate for 5000 hours is

$$S(5000) = e^{-[(0.00027)(5000)]^{1.55}} = 0.203 \qquad \blacksquare$$

The Weibull distribution also has the self-reproducing property, although the conditions are slightly more restrictive than for the exponential distribution. If $T_1, T_2, \ldots, T_n$ are independent component lifetimes having the Weibull distribution with identical shape parameters, then the minimum of these values has the Weibull distribution. More specifically, if $T_i \sim \text{Weibull}(\lambda_i, \kappa)$ for $i = 1, 2, \ldots, n$, then

$$\min\{T_1, T_2, \ldots, T_n\} \sim \text{Weibull}\left(\left(\sum_{i=1}^n \lambda_i^\kappa\right)^{1/\kappa}, \kappa\right)$$

Although the exponential and Weibull distributions are popular lifetime models, they are limited in their modeling capability. For example, if it were determined that an item had a bathtub-shaped hazard function, neither of these two models would be appropriate. Dozens of other models have been developed over the years, such as the gamma, lognormal, inverse Gaussian, exponential power, and log logistic distributions. These distributions provide further modeling flexibility beyond the exponential and Weibull distributions.

17.5 Parameter Estimation in Survival Analysis

This section investigates fitting the two distributions presented in Section 17.4, the exponential and Weibull distributions, to a data set of failure times. Other distributions, such as

Reliability

the gamma distribution or the exponential power distribution, have analogous methods of parameter estimation. Two sample data sets are introduced and are used throughout this section.

The analysis in this section assumes that a random sample of n items from a population has been placed on a test and subjected to typical field operating conditions. The data values are assumed to be independent and identically distributed random lifetimes from a particular population distribution. As with all statistical inference, care must be taken to ensure that a random sample of lifetimes is collected. Consequently, random numbers should be used to determine which n items to place on test. Laboratory conditions should adequately mimic field conditions. Only representative items should be placed on test because items manufactured using a previous design may have a different failure pattern than those with the current design.

A data set for which all failure times are known is called a complete data set. Figure 17.10 illustrates a complete data set of $n=5$ items placed on test, where the X's denote failure times. (The term "items placed on test" is used instead of "sample size" or "number of observations" because of potential confusion when right censoring is introduced.) The likelihood function for a complete data set of n items on test is given by

$$L(\boldsymbol{\theta}) = \prod_{i=1}^{n} f(t_i)$$

where $t_1, t_2, \ldots, t_n$ are the failure times and $\boldsymbol{\theta}$ is a vector of unknown parameters. (Although lowercase letters are used to denote the failure times here to be consistent with the notation for censoring times, the failure times are nonnegative random variables.)

Censoring occurs frequently in lifetime data because it is often impossible or impractical to observe the lifetimes of all the items on test. A censored observation occurs when only a bound is known on the time of failure. If a data set contains one or more censored observations, it is called a *censored data set*. The most frequent type of censoring is known as *right censoring*. In a right-censored data set, one or more items have only a lower bound known on the lifetime. The number of items placed on test is still denoted by n and the number of observed failures is denoted by r.

One special case of right censoring is considered here. Type II or *order statistic* censoring corresponds to terminating a study upon one of the ordered failures. Figure 17.11 shows the case of $n=5$ items are placed on a test that is terminated when $r=3$ failures are observed. The third and fourth items on test had their failure times right censored.

Writing the likelihood function for a censored data set requires some additional notation. As before, let $t_1, t_2, \ldots, t_n$ be lifetimes sampled randomly from a population. The corresponding right-censoring times are denoted by $c_1, c_2, \ldots, c_n$. The set U contains the indexes

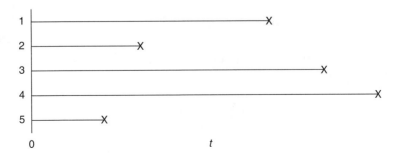

FIGURE 17.10 A complete data set with $n=5$.

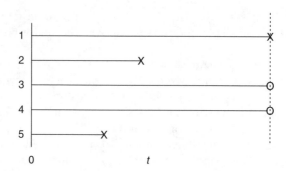

FIGURE 17.11 A right-censored data set with $n=5$ and $r=3$.

of the items that are observed to fail during the test (the uncensored observations):

$$U = \{i | t_i \leq c_i\}$$

The set C contains the indexes of the items whose failure time exceeds the corresponding censoring time (they are right censored):

$$C = \{i | t_i > c_i\}$$

Let $x_i = \min\{t_i, c_i\}$. The likelihood function is:

$$L(\boldsymbol{\theta}) = \prod_{i \in U} f(t_i) \prod_{i \in C} S(c_i) = \prod_{i \in U} f(x_i) \prod_{i \in C} S(x_i)$$

The reason that the survivor function is the appropriate term in the likelihood function for a right-censored observation is that $S(c_i)$ is the probability that item i survives to censoring time c_i. The log likelihood function is

$$\log L(\boldsymbol{\theta}) = \sum_{i \in U} \log f(x_i) + \sum_{i \in C} \log S(x_i)$$

As the density function is the product of the hazard function and the survivor function, the log likelihood function can be simplified to

$$\log L(\boldsymbol{\theta}) = \sum_{i \in U} \log h(x_i) + \sum_{i \in U} \log S(x_i) + \sum_{i \in C} \log S(x_i)$$

or

$$\log L(\boldsymbol{\theta}) = \sum_{i \in U} \log h(x_i) + \sum_{i=1}^{n} \log S(x_i)$$

Finally, as $H(t) = -\log S(t)$, the log likelihood can be written in terms of the hazard and cumulative hazard functions only as

$$\log L(\boldsymbol{\theta}) = \sum_{i \in U} \log h(x_i) - \sum_{i=1}^{n} H(x_i)$$

The choice of which of these three expressions for the log likelihood may be used for a particular distribution depends on the particular forms of $S(t), f(t), h(t)$, and $H(t)$.

Reliability

17.5.1 Data Sets

Two lifetime data sets are presented here that are used to illustrate inferential techniques for survivor (reliability) data. The two types of lifetime data sets presented here are a complete data set, where all failure times are observed, and a Type II right-censored data set (order statistic right censoring).

Example 17.11

A complete data set of $n = 23$ ball bearing failure times to test the endurance of deep-groove ball bearings has been extensively studied (e.g., Meeker and Escobar, 1998, page 4). The ordered set of failure times measured in 10^6 revolutions is

$$17.88 \quad 28.92 \quad 33.00 \quad 41.52 \quad 42.12 \quad 45.60 \quad 48.48 \quad 51.84 \quad 51.96$$
$$54.12 \quad 55.56 \quad 67.80 \quad 68.64 \quad 68.64 \quad 68.88 \quad 84.12 \quad 93.12 \quad 98.64$$
$$105.12 \quad 105.84 \quad 127.92 \quad 128.04 \quad 173.40 \quad \blacksquare$$

Example 17.12

A Type II right-censored data set of $n = 15$ automotive a/c switches is given by Kapur and Lamberson (1977, pages 253–254). The test was terminated when the fifth failure occurred. The $r = 5$ ordered observed failure times measured in number of cycles are

$$1410 \quad 1872 \quad 3138 \quad 4218 \quad 6971$$

Although the choice of "time" as cycles in this case is discrete, the data will be analyzed as continuous data. $\blacksquare$

17.5.2 Exponential Distribution

The exponential distribution is popular due to its tractability for parameter estimation and inference. Using the failure rate λ to parameterize the distribution, recall that the survivor, density, hazard, and cumulative hazard functions are

$$S(t) = e^{-\lambda t} \quad f(t) = \lambda e^{-\lambda t} \quad h(t) = \lambda \quad H(t) = \lambda t \quad \text{for all } t \geq 0$$

We begin with the analysis of a complete data set consisting of failure times $t_1, t_2, \ldots, t_n$. Since all of the observations belong to the index set U, the log likelihood function derived earlier becomes

$$\log L(\lambda) = \sum_{i=1}^{n} \log h(x_i) - \sum_{i=1}^{n} H(x_i)$$

$$= \sum_{i=1}^{n} \log \lambda - \sum_{i=1}^{n} \lambda t_i$$

$$= n \log \lambda - \lambda \sum_{i=1}^{n} t_i$$

The maximum likelihood estimator for λ is found by maximizing the log likelihood function. To determine the maximum likelihood estimator for λ, the single element "score vector"

$$U(\lambda) = \frac{\partial \log L(\lambda)}{\partial \lambda} = \frac{n}{\lambda} - \sum_{i=1}^{n} t_i$$

often called the score statistic, is equated to zero, yielding

$$\hat{\lambda} = \frac{n}{\sum_{i=1}^{n} t_i}$$

Example 17.13

Consider the complete data set of $n = 23$ ball bearing failure times. For this particular data set, the total time on test is $\sum_{i=1}^{n} t_i = 1661.16$, yielding a maximum likelihood estimator

$$\hat{\lambda} = \frac{n}{\sum_{i=1}^{n} t_i} = \frac{23}{1661.16} = 0.0138$$

failure per 10^6 revolutions.

As the data set is complete, an exact 95% confidence interval for the failure rate of the distribution can be determined. Since $\chi^2_{46,0.975} = 29.16$ and $\chi^2_{46,0.025} = 66.62$, the confidence interval derived in Section 17.4:

$$\frac{\chi^2_{2n,1-\alpha/2}}{2\sum_{i=1}^{n} T_i} < \lambda < \frac{\chi^2_{2n,\alpha/2}}{2\sum_{i=1}^{n} T_i}$$

becomes

$$\frac{(0.0138)(29.16)}{46} < \lambda < \frac{(0.0138)(66.62)}{46}$$

or

$$0.00878 < \lambda < 0.0201$$

Note that, due to the use of the chi-square distribution for this confidence interval, the interval is not symmetric about the maximum likelihood estimator. For this and subsequent examples, care has been taken to perform intermediate calculations involving numeric quantities such as critical values or total time on test values to as much precision as possible; then final values are reported using only significant digits.

Figure 17.12 shows the empirical survivor function, which takes a downward step of $1/n = 1/23$ at each data point, along with the survivor function for the fitted exponential distribution. It is apparent from this figure that the exponential distribution is a very poor fit. This particular data set was chosen for this example to illustrate one of the shortcomings of using the exponential distribution to model any data set without assessing the adequacy of the fit. Extreme caution must be exercised when using the exponential distribution since, as indicated in Figure 17.12, the exponential distribution might be a poor fit. The appropriate distribution is probably in the IFR class, as the ball bearings are wearing out. As shown subsequently, the Weibull distribution is a much better approximation to this particular data set. As the exponential distribution can be fitted to any data set that has at least one observed failure, the adequacy of the model must always be assessed. The point and interval estimators associated with the exponential distribution are meaningful only if the data set is a random sample from an exponential population. ∎

The importance of model adequacy assessments, such as those indicated in the previous example, applies to all fitted distributions, not just the exponential distribution. Furthermore, if a modeler knows the failure physics (e.g., fatigue crack growth) underlying a process, then an appropriate model consistent with the failure physics should be chosen.

We now turn to the analysis of right-censored data sets drawn from exponential populations. The previous discussion concerning complete data sets is a special case of Type II

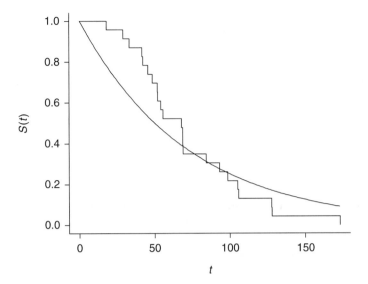

FIGURE 17.12 Empirical and exponential fitted survivor functions for the ball bearing data.

censoring when $r = n$. As before, assume that the failure times are $t_1, t_2, \ldots, t_n$, the test is terminated upon the rth ordered failure, the censoring times are $c_1 = c_2 = \cdots = c_n = t_{(r)}$ for all items, and $x_i = \min\{t_i, c_i\}$ for $i = 1, 2, \ldots, n$.

The log likelihood function is

$$\log L(\lambda) = \sum_{i \in U} \log h(x_i) - \sum_{i=1}^{n} H(x_i)$$

$$= \sum_{i \in U} \log \lambda - \sum_{i=1}^{n} \lambda x_i$$

$$= r \log \lambda - \lambda \sum_{i=1}^{n} x_i$$

Since there are r observed failures, the expression

$$\sum_{i=1}^{n} x_i$$

is often called the *total time on test* as it represents the total accumulated time that the n items accrue while on test. To determine the maximum likelihood estimator, the log likelihood function is differentiated with respect to λ,

$$U(\lambda) = \frac{\partial \log L(\lambda)}{\partial \lambda} = \frac{r}{\lambda} - \sum_{i=1}^{n} x_i$$

and is equated to zero, yielding the maximum likelihood estimator

$$\hat{\lambda} = \frac{r}{\sum_{i=1}^{n} x_i}$$

Exact confidence intervals and hypothesis tests concerning λ can also be derived in the Type II censoring case by using the result

$$2\lambda \sum_{i=1}^{n} x_i = \frac{2r\lambda}{\hat{\lambda}} \sim \chi^2(2r)$$

where $\chi^2(2r)$ is the chi-square distribution with $2r$ degrees of freedom. This result can be proved in an analogous fashion to the case of a complete data set. Using this fact, it can be stated with probability $1 - \alpha$ that

$$\chi^2_{2r, 1-\alpha/2} < \frac{2r\lambda}{\hat{\lambda}} < \chi^2_{2r, \alpha/2}$$

Rearranging terms yields an exact $100(1-\alpha)\%$ confidence interval for the failure rate λ:

$$\frac{\hat{\lambda} \chi^2_{2r, 1-\alpha/2}}{2r} < \lambda < \frac{\hat{\lambda} \chi^2_{2r, \alpha/2}}{2r}$$

Example 17.14

Consider the Type II right-censored data set of automotive switches, where $n = 15$ and there are $r = 5$ observed failures, which are

$$t_{(1)} = 1410, \ t_{(2)} = 1872, \ t_{(3)} = 3138, \ t_{(4)} = 4218, \ t_{(5)} = 6971$$

For this particular data set, the total time on test is $\sum_{i=1}^{n} x_i = 87{,}319$ cycles, yielding a maximum likelihood estimator

$$\hat{\lambda} = \frac{r}{\sum_{i=1}^{n} x_i} = \frac{5}{87{,}319} = 0.00005726$$

failure per cycle. Equivalently, the maximum likelihood estimator for the mean of the distribution is

$$\hat{\mu} = \frac{\sum_{i=1}^{n} x_i}{r} = \frac{87{,}319}{5} = 17{,}464$$

cycles to failure. As the data set is Type II right censored, an exact 95% confidence interval for the failure rate of the distribution can be determined. Using the chi-square critical values, $\chi^2_{10, 0.975} = 3.247$ and $\chi^2_{10, 0.025} = 20.49$, the formula for the confidence interval

$$\frac{\hat{\lambda} \chi^2_{2r, 1-\alpha/2}}{2r} < \lambda < \frac{\hat{\lambda} \chi^2_{2r, \alpha/2}}{2r}$$

becomes

$$\frac{(0.00005726)(3.247)}{10} < \lambda < \frac{(0.00005726)(20.49)}{10}$$

or

$$0.00001859 < \lambda < 0.0001173$$

Taking reciprocals, this is equivalent to a 95% confidence interval for the mean number of cycles to failure of

$$8525 < \mu < 53{,}785$$

Reliability

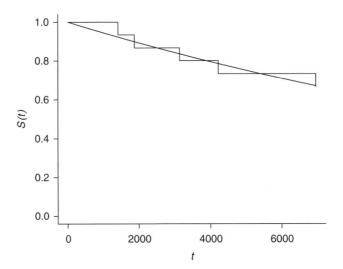

FIGURE 17.13 Empirical and exponential fitted survivor functions for the automotive a/c switch data.

Not surprisingly, with only five observed failures, this is a rather wide confidence interval for μ, and hence there is not as much precision as in the ball bearing example, where there were 23 observed failures. Assessing the adequacy of the fit is more difficult in the case of censoring, as it is impossible to determine what the lifetime distribution looks like after the last observed failure time, which is 6971 cycles in this case. Figure 17.13 shows the empirical survivor function and the associated fitted exponential survivor function. In this case, the exponential distribution appears to adequately model the lifetimes through 6971 cycles. ∎

Situations often arise when it is useful to compare the failure rates of two exponential populations based on data collected from each of the two populations. Examples include comparing the survival of one brand of integrated circuit versus another and comparing the survival of a single item at two levels of an environmental variable. Let the failure rate in the first population be λ_1 and the failure rate in the second population be λ_2.

As in the previous subsections, x denotes the minimum of the lifetime t and the censoring time c. The two data sets are denoted by

$$x_{11}, x_{12}, \ldots, x_{1n_1}$$

the n_1 values from the first population, and

$$x_{21}, x_{22}, \ldots, x_{2n_2}$$

the n_2 values from the second population. Thus x_{ji} is observation i (failure or right-censoring time) from population j. Assume further that $r_1 > 0$ failures are observed in the first population and that $r_2 > 0$ failures are observed in the second population. For tractability, it is assumed that the tests performed on both populations use Type II censoring. This assumption allows exact confidence intervals for λ_1/λ_2 to be derived. Approximate methods exist when other types of censoring are used. In addition, these methods generalize to the case of comparing more than two populations.

Since $2\lambda_1 \sum_{i=1}^{n_1} x_{1i}$ has the chi-square distribution with $2r_1$ degrees of freedom and $2\lambda_2 \sum_{i=1}^{n_2} x_{2i}$ has the chi-square distribution with $2r_2$ degrees of freedom, the statistic

$$\frac{2\lambda_1 \sum_{i=1}^{n_1} x_{1i}/(2r_1)}{2\lambda_2 \sum_{i=1}^{n_2} x_{2i}/(2r_2)} = \frac{r_2 \lambda_1 \sum_{i=1}^{n_1} x_{1i}}{r_1 \lambda_2 \sum_{i=1}^{n_2} x_{2i}} = \frac{\lambda_1 \hat{\lambda}_2}{\lambda_2 \hat{\lambda}_1}$$

has the F distribution with $2r_1$ and $2r_2$ degrees of freedom. This is true because the ratio of two independent chi-square random variables divided by their respective degrees of freedom results in an F random variable. So with probability $1 - \alpha$

$$F_{2r_1, 2r_2, 1-\alpha/2} < \frac{\lambda_1 \hat{\lambda}_2}{\lambda_2 \hat{\lambda}_1} < F_{2r_1, 2r_2, \alpha/2}$$

or

$$\frac{\hat{\lambda}_1}{\hat{\lambda}_2} F_{2r_1, 2r_2, 1-\alpha/2} < \frac{\lambda_1}{\lambda_2} < \frac{\hat{\lambda}_1}{\hat{\lambda}_2} F_{2r_1, 2r_2, \alpha/2}$$

Two points are important to keep in mind with respect to this confidence interval. First, it is typically of interest to see whether this confidence interval contains 1, which indicates that there is no statistical evidence to conclude that the failure rates of the two populations are different. Second, if the null hypothesis

$$H_0 : \lambda_1 = \lambda_2$$

is to be tested directly, then there is cancellation in $\lambda_1 \hat{\lambda}_2 / \lambda_2 \hat{\lambda}_1$ under H_0, so that the test statistic $\hat{\lambda}_2 / \hat{\lambda}_1$ has the F distribution with $2r_1$ and $2r_2$ degrees of freedom.

17.5.3 Weibull Distribution

As mentioned earlier, the Weibull distribution is typically more appropriate for modeling the lifetimes of items with increasing and decreasing failure rates, such as mechanical items. We present the most general case of random censoring, rather than looking at each censoring mechanism individually.

As before, let $t_1, t_2, \ldots, t_n$ be the failure times, $c_1, c_2, \ldots, c_n$ be the censoring times, and $x_i = \min\{t_i, c_i\}$ for $i = 1, 2, \ldots, n$. Recall that the Weibull distribution has hazard and cumulative hazard functions

$$h(t) = \kappa \lambda (\lambda t)^{\kappa - 1} \quad \text{and} \quad H(t) = (\lambda t)^{\kappa}$$

for $t \geq 0$. When there are r observed failures, the log likelihood function is

$$\log L(\lambda, \kappa) = \sum_{i \in U} \log h(x_i) - \sum_{i=1}^{n} H(x_i)$$

$$= \sum_{i \in U} (\log \kappa + \kappa \log \lambda + (\kappa - 1) \log x_i) - \sum_{i=1}^{n} (\lambda x_i)^{\kappa}$$

$$= r \log \kappa + \kappa r \log \lambda + (\kappa - 1) \sum_{i \in U} \log x_i - \lambda^{\kappa} \sum_{i=1}^{n} x_i^{\kappa}$$

and the 2×1 score vector has elements

$$U_1(\lambda, \kappa) = \frac{\partial \log L(\lambda, \kappa)}{\partial \lambda} = \frac{\kappa r}{\lambda} - \kappa \lambda^{\kappa - 1} \sum_{i=1}^{n} x_i^{\kappa}$$

and

$$U_2(\lambda, \kappa) = \frac{\partial \log L(\lambda, \kappa)}{\partial \kappa} = \frac{r}{\kappa} + r \log \lambda + \sum_{i \in U} \log x_i - \sum_{i=1}^{n} (\lambda x_i)^\kappa \log(\lambda x_i)$$

When these equations are equated to zero, the simultaneous equations

$$\frac{\kappa r}{\lambda} - \kappa \lambda^{\kappa-1} \sum_{i=1}^{n} x_i^\kappa = 0$$

and

$$\frac{r}{\kappa} + r \log \lambda + \sum_{i \in U} \log x_i - \sum_{i=1}^{n} (\lambda x_i)^\kappa \log(\lambda x_i) = 0$$

have no closed-form solution for $\hat{\lambda}$ and $\hat{\kappa}$. One piece of good fortune, however, to avoid solving a 2×2 set of nonlinear equations, is that this first equation can be solved for λ in terms of κ as follows:

$$\lambda = \left(\frac{r}{\sum_{i=1}^{n} x_i^\kappa} \right)^{1/\kappa}$$

Using this expression for λ in terms of κ in the second element of the score vector yields a single, albeit more complicated, expression with κ as the only unknown. Applying some algebra, this equation reduces to

$$g(\kappa) = \frac{r}{\kappa} + \sum_{i \in U} \log x_i - \frac{r \sum_{i=1}^{n} x_i^\kappa \log x_i}{\sum_{i=1}^{n} x_i^\kappa} = 0$$

which must be solved iteratively using the Newton–Raphson technique or a fixed point method.

Example 17.15

It was seen in a previous example that the exponential distribution poorly approximated the ball bearing data set. The Weibull distribution is fit to the ball bearing failure times yielding maximum likelihood estimators $\hat{\lambda} = 0.0122$ and $\hat{\kappa} = 2.10$. Figure 17.14 shows the empirical survival function along with the exponential and Weibull fits to the data. It is clear that the Weibull distribution is far superior to the exponential for modeling the ball bearing failure times. This is due to the fact that the Weibull distribution is capable of modeling wear out for $\kappa > 1$.

In the case of the exponential distribution, we found a confidence *interval* for the parameter λ. In the case of the Weibull distribution, we desire a confidence *region* for the parameters λ and κ. Using the fact that the likelihood ratio statistic, $2[\log L(\hat{\lambda}, \hat{\kappa}) - \log L(\lambda, \kappa)]$, is asymptotically $\chi^2(2)$, a 95% confidence region for the parameters is all λ and κ satisfying

$$2[-113.691 - \log L(\lambda, k)] < 5.99$$

where $\log L(\hat{\lambda}, \hat{\kappa}) = -113.691$ and $\chi^2_{2, 0.05} = 5.99$. The two degrees of freedom for the χ^2 distribution come from the fact that the Weibull distribution has two unknown parameters λ and κ. The 95% confidence region is shown in Figure 17.15, and, not surprisingly, the line $\kappa = 1$ is not interior to the region. This is further proof that the exponential distribution is not an appropriate model for this particular data set. The entire confidence region is in the $\kappa > 1$ region of the graph; this is statistically significant evidence provided by the data that the ball bearings are indeed wearing out. ∎

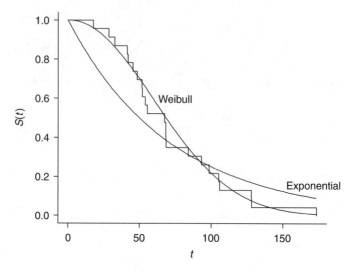

FIGURE 17.14 Empirical, exponential fitted, and Weibull fitted survivor functions for the ball bearing data.

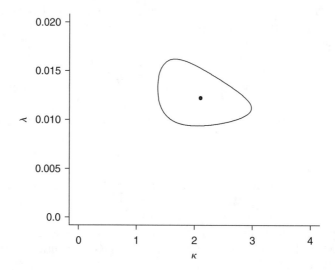

FIGURE 17.15 Confidence region for λ and κ for the ball bearing data.

17.6 Nonparametric Methods

In the previous sections, the focus was on developing parametric models for lifetimes and fitting them to a data set. The emphasis switches here to letting the data speak for itself, rather than approximating the lifetime distribution by one of the parametric models. There are several reasons to take this approach. First, it is not always possible to find a parametric model that adequately describes the lifetime distribution. This is particularly true of data arising from populations with nonmonotonic hazard functions. The most popular parametric models, such as the Weibull distribution, have monotonic hazard functions. A nonparametric analysis might provide more accurate estimates. Second, data sets are often so small that fitting a parametric model results in parameter estimators with confidence intervals that are so wide that the models are of little practical use. To cover all nonparametric methods used

Reliability

in reliability is not possible due to space constraints. We focus on just one such method here: the Kaplan–Meier product-limit survivor function estimate and the associated Greenwood's formula for assessing the precision of the estimate.

Let $y_1 < y_2 < \cdots < y_k$ be the k distinct failure times, and let d_j denote the number of observed failures at y_j, $j = 1, 2, \ldots, k$. Let $n_j = n(y_j)$ denote the number of items on test just before time y_j, $j = 1, 2, \ldots, k$, and it is customary to include any values that are censored at y_j in this count. Also, let $R(y_j)$ be the set of all indexes of items that are at risk just before time y_j, $j = 1, 2, \ldots, k$.

The search for a nonparametric survivor function estimator begins by assuming that the data were drawn from a discrete distribution with mass values at $y_1, y_2, \ldots, y_k$. For a discrete distribution, $h(y_j)$ is a conditional probability with interpretation $h(y_j) = P[T = y_j | T \geq y_j]$. For a discrete distribution, the survivor function can be written in terms of the hazard function at the mass values:

$$S(t) = \prod_{j \in R(t)'} [1 - h(y_j)] \quad t \geq 0$$

where $R(t)'$ is the complement of the risk set at time t. Thus a reasonable estimator for $S(t)$ is $\prod_{j \in R(t)'} [1 - \hat{h}(y_j)]$, which reduces the problem of estimating the survivor function to that of estimating the hazard function at each mass value. An appropriate element for the likelihood function at mass value y_j is

$$h(y_j)^{d_j} [1 - h(y_j)]^{n_j - d_j}$$

for $j = 1, 2, \ldots, k$. This expression is correct because d_j is the number of failures at y_j, $h(y_j)$ is the conditional probability of failure at y_j, $n_j - d_j$ is the number of items on test not failing at y_j, and $1 - h(y_j)$ is the probability of failing after time y_j conditioned on survival to time y_j. Thus the likelihood function for $h(y_1), h(y_2), \ldots, h(y_k)$ is

$$L(h(y_1), h(y_2), \ldots, h(y_k)) = \prod_{j=1}^{k} h(y_j)^{d_j} [1 - h(y_j)]^{n_j - d_j}$$

and the log likelihood function is

$$\log L(h(y_1), h(y_2), \ldots, h(y_k)) = \sum_{j=1}^{k} \{d_j \log h(y_j) + (n_j - d_j) \log [1 - h(y_j)]\}$$

The ith element of the score vector is

$$\frac{\partial \log L(h(y_1), h(y_2), \ldots, h(y_k))}{\partial h(y_i)} = \frac{d_j}{h(y_i)} - \frac{n_i - d_i}{1 - h(y_i)}$$

for $i = 1, 2, \ldots, k$. Equating this vector to zero and solving for $h(y_i)$ yields the maximum likelihood estimate:

$$\hat{h}(y_i) = \frac{d_i}{n_i}$$

This estimate for $\hat{h}(y_i)$ is sensible, since d_i of the n_i items on test at time y_i fail, so the ratio of d_i to n_i is an appropriate estimate of the conditional probability of failure at time y_i. This derivation may strike a familiar chord since, at each time y_i, estimating $h(y_i)$ with d_i divided by n_i is equivalent to estimating the probability of success, that is, failing at time y_i, for each of the n_i items on test. Thus, this derivation is equivalent to finding the maximum likelihood estimators for the probability of success for k binomial random variables.

Using this particular estimate for the hazard function at y_i, the survivor function estimate becomes

$$\hat{S}(t) = \prod_{j \in R(t)'} [1 - \hat{h}(y_j)]$$

$$= \prod_{j \in R(t)'} \left[1 - \frac{d_j}{n_j}\right]$$

commonly known as the Kaplan–Meier or product-limit estimate. One problem that arises with the product-limit estimate is that it is not defined past the last observed failure time. The usual way to handle this problem is to cut the estimator off at the last observed failure time y_k. The following example illustrates the product-limit estimate.

Example 17.16

An experiment is conducted to determine the effect of the drug 6-mercaptopurine (6-MP) on leukemia remission times (Lawless, 2003, page 5). A sample of $n = 21$ leukemia patients is treated with 6-MP, and the remission times are recorded. There are $r = 9$ individuals for whom the remission time is observed, and the remission times for the remaining 12 individuals are randomly censored on the right. There are $k = 7$ distinct observed failure times. Letting an asterisk denote a censored observation, the remission times (in weeks) are

$$6 \quad 6 \quad 6 \quad 6^* \quad 7 \quad 9^* \quad 10 \quad 10^* \quad 11^* \quad 13 \quad 16$$
$$17^* \quad 19^* \quad 20^* \quad 22 \quad 23 \quad 25^* \quad 32^* \quad 32^* \quad 34^* \quad 35^*$$

Find an estimate for $S(14)$.

Table 17.1 gives the values of y_j, d_j, n_j, and $1 - d_j/n_j$ for $j = 1, 2, \ldots, 7$. In particular, the product-limit survivor function estimate at $t = 14$ weeks is

$$\hat{S}(14) = \prod_{j \in R(14)'} \left[1 - \frac{d_j}{n_j}\right]$$

$$= \left[1 - \frac{3}{21}\right]\left[1 - \frac{1}{17}\right]\left[1 - \frac{1}{15}\right]\left[1 - \frac{1}{12}\right]$$

$$= \frac{176}{255}$$

$$\cong 0.69$$

TABLE 17.1 Product-Limit Calculations for the 6-MP Data

j	y_j	d_j	n_j	$1 - \dfrac{d_j}{n_j}$
1	6	3	21	$1 - \dfrac{3}{21}$
2	7	1	17	$1 - \dfrac{1}{17}$
3	10	1	15	$1 - \dfrac{1}{15}$
4	13	1	12	$1 - \dfrac{1}{12}$
5	16	1	11	$1 - \dfrac{1}{11}$
6	22	1	7	$1 - \dfrac{1}{7}$
7	23	1	6	$1 - \dfrac{1}{6}$

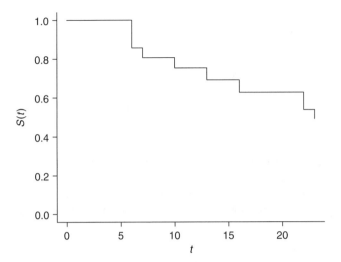

FIGURE 17.16 Product-limit survivor function estimate for the 6-MP data.

The product-limit survivor function estimate for all t values is plotted in Figure 17.16. Downward steps occur only at observed failure times. The effect of censored observations in the survivor function estimate is a larger downward step at the next subsequent failure time. If there are ties between the observations and a censoring time, as there is at time 6, our convention of including the censored values in the risk set means that there will be a larger downward step following this tied value. Note that the estimate is truncated at time 23, the last observed failure time. ∎

To find an estimate for the variance of the product-limit estimate is significantly more difficult than for the uncensored case. The Fisher and observed information matrices require a derivative of the score vector:

$$-\frac{\partial^2 \log L(h(y_1), h(y_2), \ldots, h(y_k))}{\partial h(y_i) \partial h(y_j)} = \frac{d_i}{h(y_i)^2} + \frac{n_i - d_i}{(1 - h(y_i))^2}$$

when $i = j$ and 0 otherwise, for $i = 1, 2, \ldots, k$, $j = 1, 2, \ldots, k$. Both the Fisher and observed information matrices are diagonal. Replacing $h(y_i)$ by its maximum likelihood estimate, the diagonal elements of the observed information matrix are

$$\left[-\frac{\partial^2 \log L(h(y_1), h(y_2), \ldots, h(y_k))}{\partial h(y_i)^2}\right]_{h(y_i) = d_i/n_i} = \frac{n_i^3}{d_i(n_i - d_i)}$$

for $i = 1, 2, \ldots, k$. Using this fact and some additional approximations, an estimate for the variance of the survivor function is

$$\hat{V}[\hat{S}(t)] = [\hat{S}(t)]^2 \sum_{j \in R(t)'} \frac{d_j}{n_j(n_j - d_j)}$$

commonly known as "Greenwood's formula." The formula can be used to find asymptotically valid confidence intervals for $S(t)$ by using the normal critical values as in the uncensored case:

$$\hat{S}(t) - z_{\alpha/2}\sqrt{\hat{V}[\hat{S}(t)]} < S(t) < \hat{S}(t) + z_{\alpha/2}\sqrt{\hat{V}[\hat{S}(t)]}$$

where $z_{\alpha/2}$ is the $1-\alpha/2$ fractile of the standard normal distribution.

Example 17.17

For the 6-MP treatment group in the previous example, give a 95% confidence interval for the probability of survival to time 14.

The point estimator for the probability of survival to time 14 from the previous example is $\hat{S}(14) \cong 0.69$. Greenwood's formula is used to estimate the variance of the survivor function estimator at time 14:

$$\hat{V}[\hat{S}(14)] = [\hat{S}(14)]^2 \sum_{j \in R(14)'} \frac{d_j}{n_j(n_j - d_j)}$$

$$= (0.69)^2 \left[\frac{3}{21(21-3)} + \frac{1}{17(17-1)} + \frac{1}{15(15-1)} + \frac{1}{12(12-1)} \right]$$

$$\cong 0.011$$

Thus an estimate for the standard deviation of the survivor function estimate at $t = 14$ is $\sqrt{0.011} = 0.11$. A 95% confidence interval for $S(14)$ is

$$\hat{S}(14) - z_{0.025}\sqrt{\hat{V}[\hat{S}(14)]} < S(14) < \hat{S}(14) + z_{0.025}\sqrt{\hat{V}[\hat{S}(14)]}$$

$$0.69 - 1.96\sqrt{0.011} < S(14) < 0.69 + 1.96\sqrt{0.011}$$

$$0.48 < S(14) < 0.90$$

Figure 17.17 shows the 95% confidence bands for the survivor function for all t values. These have also been cut off at the last observed failure time, $t = 23$. The bounds are particularly wide as there are only $r = 9$ observed failure times. ∎

17.7 Assessing Model Adequacy

As there has been an emphasis on continuous lifetime distributions thus far in the chapter, the discussion here is limited to model adequacy tests for continuous distributions. The popular chi-square goodness-of-fit test can be applied to both continuous and discrete distributions, but suffers from the limitations of arbitrary interval widths and application only to large data sets. This section focuses on the Kolmogorov–Smirnov (KS) goodness-of-fit test for assessing model adequacy.

A notational difficulty arises in presenting the KS test. The survivor function $S(t)$ has been emphasized to this point in the chapter, but the cumulative distribution function, where $F(t) = P[T \leq t] = 1 - S(t)$ for continuous distributions, has traditionally been used to define the KS test statistic. To keep with this tradition, $F(t)$ is used in the definitions in this section.

The KS goodness-of-fit test is typically used to compare an empirical cumulative distribution function with a fitted or hypothesized parametric cumulative distribution function

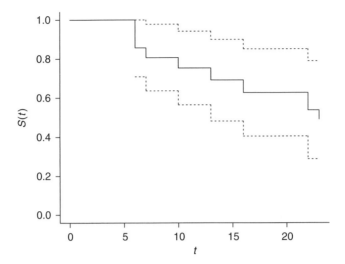

FIGURE 17.17 Confidence bands for the product-limit survivor function estimate for the 6-MP data.

for a continuous model. The KS test statistic is the maximum vertical difference between the empirical cumulative distribution function $\hat{F}(t)$ and a hypothesized or fitted cumulative distribution function $F_0(t)$. The null and alternative hypotheses for the test are

$$H_0 : F(t) = F_0(t)$$

$$H_1 : F(t) \neq F_0(t)$$

where $F(t)$ is the true underlying population cumulative distribution function. In other words, the null hypothesis is that data set of random lifetimes has been drawn from a population with cumulative distribution function $F_0(t)$. For a complete data set, the defining formula for the test statistic is

$$D_n = \sup_t |\hat{F}(t) - F_0(t)|$$

where *sup* is an abbreviation for *supremum*. This test statistic has intuitive appeal since larger values of D_n indicate a greater difference between $\hat{F}(t)$ and $F_0(t)$ and hence a poorer fit. In addition, D_n is independent of the parametric form of $F_0(t)$ when the cumulative distribution function is hypothesized. From a practical standpoint, computing the KS test statistic requires only a single loop through the n data values. This simplification occurs because $\hat{F}(t)$ is a nondecreasing step function and $F_0(t)$ is a nondecreasing continuous function, so the maximum difference must occur at a data value.

The usual computational formulas for computing D_n require a single pass through the data values. Let

$$D_n^+ = \max_{i=1,2,\ldots,n} \left(\frac{i}{n} - F_0(t_{(i)}) \right)$$

$$D_n^- = \max_{i=1,2,\ldots,n} \left(F_0(t_{(i)}) - \frac{i-1}{n} \right)$$

so that $D_n = \max\{D_n^+, D_n^-\}$. These computational formulas are typically easier to translate into computer code for implementation than the defining formula.

We consider only hypothesized (as opposed to fitted) cumulative distribution functions $F_0(t)$ here because the distribution of D_n is free of the hypothesized distribution specified.

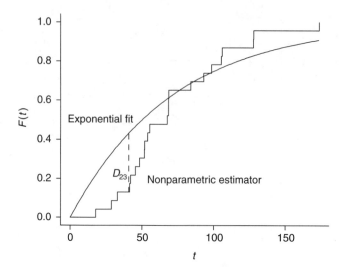

FIGURE 17.18 KS statistic for the ball bearing data set (exponential fit).

To illustrate the geometric aspects of the KS test statistic D_n, however, the fitted exponential distribution is compared to the empirical cumulative distribution function for the ball bearing data set. Figure 17.18 shows the empirical step cumulative distribution function $F(t)$ associated with the failure times of the $n = 23$ ball bearing failure times, along with the exponential fit $F_0(t)$. The maximum difference between these two cumulative distribution functions occurs just to the left of $t_{(4)} = 41.52$ and is $D_{23} = 0.301$ as indicated on the figure.

The test statistic for the KS test is nonparametric in the sense that it has the same distribution regardless of the distribution of the parent population under H_0 when all the parameters in the hypothesized distribution are known. The reason for this is that $F_0(t_{(1)}), F_0(t_{(2)}), \ldots, F_0(t_{(n)})$ have the same joint distribution as $U(0,1)$-order statistics under H_0 regardless of the functional form of F_0. These are often called *omnibus tests* as they are not tied to one particular distribution (e.g., the Weibull) and apply equally well to any hypothesized distribution $F_0(t)$. This also means that fractiles of the distribution of D_n depend on n only.

The rows in Table 17.2 denote the sample sizes and the columns denote several levels of significance. The values in the table are estimates of the $1 - \alpha$ fractiles of the distribution of D_n under H_0 in the all-parameters-known case (hypothesized, rather than fitted distribution) and have been determined by Monte Carlo simulation with one million replications. Not surprisingly, the fractiles are a decreasing function of n, as increased sample sizes will have lower sampling variability. Test statistics that exceed the appropriate critical value lead to rejecting H_0.

Example 17.18

Run the KS test (at $\alpha = 0.10$) to assess whether the ball bearing data set was drawn from a Weibull population with $\lambda = 0.01$ and $\kappa = 2$.

Note that the Weibull distribution in this example is a *hypothesized*, rather than *fitted* distribution, so the all-parameters-known case for determining critical values is appropriate. The goodness-of-fit test

$$H_0 : F(t) = 1 - e^{-(0.01t)^2}$$

$$H_1 : F(t) \neq 1 - e^{-(0.01t)^2}$$

TABLE 17.2 Selected Approximate KS Percentiles for Small Sample Sizes

n	$\alpha = 0.20$	$\alpha = 0.10$	$\alpha = 0.05$	$\alpha = 0.01$
1	0.900	0.950	0.975	0.995
2	0.683	0.776	0.842	0.930
3	0.565	0.636	0.708	0.829
4	0.493	0.565	0.624	0.733
5	0.447	0.509	0.563	0.668
6	0.410	0.468	0.519	0.617
7	0.381	0.436	0.483	0.576
8	0.358	0.409	0.454	0.542
9	0.339	0.388	0.430	0.513
10	0.323	0.369	0.409	0.489
11	0.308	0.352	0.391	0.468
12	0.296	0.338	0.376	0.449
13	0.285	0.325	0.361	0.433
14	0.275	0.314	0.349	0.418
15	0.266	0.304	0.338	0.404
16	0.258	0.295	0.327	0.392
17	0.250	0.286	0.318	0.381
18	0.243	0.278	0.309	0.370
19	0.237	0.271	0.302	0.361
20	0.232	0.265	0.294	0.352
21	0.226	0.259	0.287	0.345
22	0.221	0.253	0.281	0.337
23	0.217	0.248	0.275	0.330
24	0.212	0.242	0.269	0.323
25	0.208	0.237	0.264	0.317

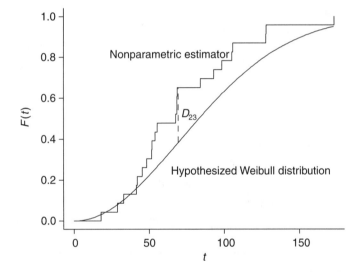

FIGURE 17.19 KS statistic for the ball bearing data set (Weibull fit).

does not involve any parameters estimated from data. The test statistic is $D_{23} = 0.274$. The empirical cumulative distribution function, the Weibull(0.01, 2) cumulative distribution function and the maximum difference between the two [which occurs at $t_{(15)} = 68.88$] are shown in Figure 17.19. At $\alpha = 0.10$, the critical value is 0.248, so H_0 is rejected. The test statistic is very close to the critical value for $\alpha = 0.05$, so the attained p-value for the test is approximately $p = 0.05$. ∎

The KS test can be extended in several directions. First, it can be adapted for the case the parameters are estimated from the data. Unfortunately, a separate table of critical values must be given for each fitted distribution. Second, the KS test can be adapted for right-censored data sets. Many researchers have devised approximate methods for determining the critical values for the KS test with random right censoring and parameters estimated from data. Finally, there are several variants of the KS test, such as the Anderson–Darling and Cramer–von Mises test, which improve on the power of the test.

17.8 Summary

The purpose of this chapter has been to introduce the mathematics associated with the design and assessment of systems with respect to their reliability. In specific, this chapter has:

- outlined basic techniques for describing the arrangement of components in a system by defining the structure function $\phi(\boldsymbol{x})$ that maps the states of the components to the state of the system;
- defined reliability as the probability that a nonrepairable item (component or system) is functioning at a specified time;
- introduced two techniques, definition and expectation, for determining the system reliability from component reliabilities;
- defined four functions, the survivor function $S(t)$, the probability density function $f(t)$, the hazard function $h(t)$, and the cumulative hazard function $H(t)$, which describe the distribution of a nonnegative random variable T, which denotes the lifetime of a component or system;
- reviewed formulas for calculating the mean, variance, and a fractile (percentile) of T;
- illustrated how to determine the system survivor function as a function of the component survivor functions;
- introduced two parametric lifetime distributions, the exponential and Weibull distributions, and outlined some of their properties;
- surveyed characteristics (e.g., right-censoring) of lifetime data sets;
- outlined point and interval estimation techniques for the exponential and distributions;
- derived a technique for comparing the failure rates of items with lifetimes drawn from two populations;
- derived and illustrated the nonparametric Kaplan–Meier product-limit estimate for the survivor function;
- introduced the Kolmogorov–Smirnov goodness-of-fit test for assessing model adequacy.

All of these topics are covered in more detail in the references. In addition, there are many topics that have not been covered at all, such as repairable systems, incorporating covariates into a survival model, competing risks, reliability growth, mixture models, failure modes and effects analysis, accelerated testing, fault trees, Markov models, and life testing. These topics and others are considered in the reliability literature, highlighted by the textbooks cited below. Software for reliability analysis has been written by several vendors and incorporated into existing statistical packages, such as SAS, S-Plus, and R.

References

1. Barlow, R. and Proschan, F. (1981), *Statistical Theory of Reliability and Life Testing Probability Models*, To Begin With, Silver Spring, MD.
2. Kalbfleisch, J.D. and Prentice, R.L. (2002), *The Statistical Analysis of Failure Time Data*, Second Edition, John Wiley & Sons, New York.
3. Kapur, K.C. and Lamberson, L.R. (1977), *Reliability in Engineering Design*, John Wiley & Sons, New York.
4. Lawless, J.F. (2003), *Statistical Models and Methods for Lifetime Data*, Second Edition, John Wiley & Sons, New York.
5. Meeker, W.Q. and Escobar, L.A. (1998), *Statistical Methods for Reliability Data*, John Wiley & Sons, New York.

18
Production Systems

18.1	Production Planning Problem	18-1
18.2	Demand Forecasting	18-2
	Commonly Used Techniques for Forecasting Demand • Parametric Methods for Forecasting Demand Distributions • A Nonparametric Method for Updating and Forecasting the Entire Demand Distribution • An Application of the Forecasting Method for Computing Optimal Order Quantities • How to Incorporate Seasonality in Demand into the Model	
18.3	Models for Production Layout Design	18-12
	An Example Problem • Optimal Plant Layouts with Practical Considerations • The Tretheway Algorithm • A Spanning Tree Network Model for Layout Design • A Cut Tree Network Model for Layout Design • Facility Shapes	
18.4	Scheduling of Production and Service Systems	18-20
	Definition of the Scheduling Decision • Single Machines • Flow Shops • Job Shops	
	References	18-29

Bobbie L. Foote
U.S. Military Academy

Katta G. Murty
University of Michigan

18.1 Production Planning Problem

The total problem of planning production consists of the following decisions: demand forecasting, the total floor space needed, the number and type of equipment, their aggregation into groups, the floor space needed for each group, the spatial relationship of each group relative to one another, the way the material and work pieces move within groups, the equipment and methods to move work between groups, the material to be ordered, the material to be produced in-house, the assembly layout and process, the quantities and timing of stock purchases, and the manner of storage of purchases and the items that are to be part of the information support system. There are no methods that determine the answers to these questions simultaneously. There are numerous models that consider each of these decisions as a subproblem to be optimized. When all the solutions are pieced together sometimes the whole will be suboptimal.

The production planning problem can also be looked at as a system of systems: forecasting, material handling, personnel, purchasing, quality assurance, production, assembly, marketing, design, finance, and other appropriate systems. At an advanced level one hopes that these systems integrate: that is, the design is such that it is easy to produce by using snap-in fasteners; materials easy to form; financial planning provides appropriate working

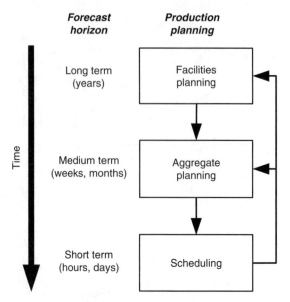

FIGURE 18.1 Planning and forecast horizons in production planning.

capital; purchases arrive on time and have appropriate quality. In this context, the interfaces of the systems become the appropriate design issue.

Forecasts form the basis of production planning. Strategic plans (e.g., where to locate factories, etc.) are based on long-term forecasts whereas aggregate production plans (allocating labor and capital resources for the next quarter's operations, say) are usually made a few months in advance, and production scheduling at the shop floor level may occur a few hours or days in advance of actual production. This also allows better information and a more detailed history to tailor production to demand or desired inventory levels. Figure 18.1 shows this graphically.

From the above comments, clearly forecasting is as much an art as a science. Given its importance, forecasting is introduced in Section 18.2. In production planning, demand forecasts are used for aggregate planning and more detailed materials requirement plans. These plans provide details on the components required and the relationships between the processing steps involved in production. This information is then used to design the facility (i.e., positioning of various machines and processing equipment within the facility) and schedule production. We assume that aggregate plans and more detailed material requirements plans are available and do not discuss these stages of production planning; the interested reader may consult Nahmias (1993) for details. Instead, we focus on the facility layout and scheduling issues in Sections 18.3 and 18.4, respectively.

18.2 Demand Forecasting

Forecasting is a key activity that influences many aspects of business planning and operation. In simple terms, a forecast is a prediction or estimate of the future. Forecasting is common at multiple levels, from macroeconomic forecasts (e.g., forecasts of oil prices) to microlevel estimates (of component requirements, say). We focus on forecasting in support of the production planning function. The *forecast horizon* refers to how far into the future we would like to make a prediction. Some common features of forecasts are as follows (Nahmias, 1993):

- *Forecasts are usually wrong*—Forecasts are *estimates* of a random variable (demand, price, etc.) and actual outcomes may be significantly different.
- *Aggregate forecasts are better*—Demand forecasts for an entire product family are more accurate (on a percentage basis) than those for each member of the product family.
- *Forecasting error increases with the forecast horizon*—Demand forecasts for the coming week are likely to be much more accurate than forecasts of next year's demand.
- *Environmental evidence*—The environment is very important. For example, if you are forecasting demand for spare parts for a military aircraft, then it is important to account for the type of operations in progress. For regular training, past history may be valid, but if the aircraft are being used in war, then peacetime history is likely to be invalid.

18.2.1 Commonly Used Techniques for Forecasting Demand

Models for controlling and replenishing of inventories in production have the aim of determining order quantities to minimize the sum of total overage costs (costs of excess inventory remaining at the end of the planning period) and underage costs (shortage costs, or costs of having less than the desired amount of stock at the end of the planning period). In production planning literature, the total overage (underage) cost is usually assumed to be proportional to the overage (shortage) amount or quantity, to make the analysis easier. Some companies find that a piecewise linear (PL) function provides a much closer representation of the true overage and underage costs. In these companies, there is a buffer with limited space in which excess inventory at the end of the planning period can be stored and retrieved later at a low cost (i.e., with minimum requirements of human-hours needed) per unit. Once this buffer is filled up, any remaining excess quantity has to be held at a location farther away that requires greater number of man hours for storing or retrieval/unit. A similar situation exists for underage cost as a function of the shortage amount. This clearly implies that the overage and underage costs are PL functions of the excess, shortage quantities. Determining optimum order quantities to minimize such unusual overage and underage cost functions is much harder with inventory control models using forecasting techniques in current literature. After reviewing the forecasting methods commonly used in production applications at present (Section 18.2.2), we will discuss a new nonparametric forecasting method (Section 18.2.3), which has the advantage of being able to accommodate such unusual overage and underage cost functions easily (Murty, 2006).

Almost all production management problems in practice are characterized by the uncertainty of demand during a future planning period; that is, this demand is a random variable. Usually, we do not have knowledge about its exact probability distribution, and the models for these problems have the objective of minimizing the sum of expected overage and underage costs. Successful production management systems depend heavily on good demand forecasts to provide data for inventory replenishment decisions. The output of forecasting is usually presented in the literature as the *forecasted demand quantity*; in reality it is an estimate of the expected demand during the planning period. Because of this, the purpose of forecasting is often misunderstood to be that of generating this single number, even though sometimes the standard deviation of demand is also estimated.

All commonly used methods for demand forecasting are parametric methods; they usually assume that demand is normally distributed, and they update its distribution by updating the parameters of the distribution, the mean μ, and the standard deviation σ. The most commonly used methods for updating the values of the parameters are the method of moving averages, and the exponential smoothing method.

The method of moving averages uses the average of n most recent observations on demand as the forecast for the expected demand for the next period. n is a parameter known as the order of the *moving average method* being used; typically it is between 3 to 6 or larger.

The other method, perhaps the most popular method in practice, is the exponential smoothing method introduced and popularized by Brown (1959). It takes $\hat{D}_{t+1}$, the forecast of expected demand during next period $t+1$, to be $\alpha x_t + (1-\alpha)\hat{D}_t$, where x_t is the observed demand during current period t, $\hat{D}_t$ is the forecasted expected demand for current period t, and $0 < \alpha \leq 1$ is a *smoothing constant*, which is the relative weight placed on the current observed demand. Typically, values of α between 0.1 and 0.4 are used, and normally the value of α is increased whenever the absolute value of the deviation between the forecast and observed demand exceeds a tolerance times the standard deviation. Smaller values of α (like 0.1) yield predicted values of expected demand that have a relatively smooth pattern, whereas higher values of α (like 0.4) lead to predicted values exhibiting significantly greater variation, but doing a better job of tracking the demand series. Thus using larger α makes forecasts more responsive to changes in the demand process, but will result in forecast errors with higher variance.

One disadvantage of both the method of moving averages and the exponential smoothing method is that when there is a definite trend in the demand process (either growing or falling), the forecasts obtained by them lag behind the trend. Variations of the exponential smoothing method to track trend linear in time in the demand process have been proposed (see Holt, 1957), but these have not proved very popular.

There are many more sophisticated methods for forecasting the expected values of random variables, for example, the Box–Jenkins ARIMA models (Box and Jenkins, 1970), but these methods are not popular for production applications, in which forecasts for many items are required.

18.2.2 Parametric Methods for Forecasting Demand Distributions

Using Normal Distribution with Updating of Expected Value and Standard Deviation in Each Period

As discussed in the previous section, all forecasting methods in the literature only provide an estimate of the expected demand during the planning period. The optimum order quantity to be computed depends of course on the entire probability distribution of demand, not just its expected value. So, almost everyone assumes that the distribution of demand is the normal distribution because of its convenience. One of the advantages that the normality assumption confers is that the distribution is fully characterized by only two parameters, the mean and the standard deviation, both of which can be very conveniently updated by the exponential smoothing or the moving average methods.

Let t be the current period, x_r the observed demand in period r for $r \leq t$, $\hat{D}_t$ the forecast (i.e., estimate) of expected demand in current period t (by either the exponential smoothing or the moving average methods, whichever is being used), and $\hat{D}_{t+1}, \hat{\sigma}_{t+1}$ the forecasts for

Production Systems

expected demand, standard deviation of demand for the planning period which is the next period $t+1$. Then these forecasts are:

Method	Forecast
Method of moving averages of order n	$\hat{D}_{t+1} = \frac{1}{n}\sum_{r=t-n+1}^{t} x_r$
Exponential smoothing method with smoothing constant α	$\hat{D}_{t+1} = \alpha x_t + (1-\alpha)\hat{D}_t$
Method of moving averages of order n	$\hat{\sigma}_{t+1} = +\sqrt{(\sum_{r=t-n+1}^{t}(x_r - \hat{D}_{t+1})^2)/n}$

To get $\hat{\sigma}_{t+1}$ by the exponential smoothing method, it is convenient to use the mean absolute deviation (MAD), and use the formula: standard deviation $\sigma \approx (1.25)\text{MAD}$ when the distribution is the normal distribution. Let MAD_t denote the estimate of MAD for current period t. Then the forecasts obtained by the exponential smoothing method with smoothing parameter α for the next period $t+1$ are:

$$\text{MAD}_{t+1} = \alpha|x_t - \hat{D}_t| + (1-\alpha)\text{MAD}_t$$
$$\hat{\sigma}_{t+1} = (1.25)\text{MAD}_{t+1}$$

Usually $\alpha = 0.1$ is used to ensure stability of the estimates. And the normal distribution with mean $\hat{D}_{t+1}$ and standard deviation $\hat{\sigma}_{t+1}$ is taken as the forecast for the distribution of demand during the next period $t+1$ for making any planning decisions under this procedure.

Using Normal Distribution with Updating of Expected Value and Standard Deviation Only when There Is Evidence of Change

In some applications, the distribution of demand is assumed to be the normal distribution, but estimates of its expected value and standard deviation are left unchanged until there is evidence that their values have changed. Foote (1995) discusses several statistical control tests on demand data being generated over time to decide when to re-estimate these parameters. Under this scheme, the method of moving averages is commonly used to estimate the expected value and the standard deviation from recent data whenever the control tests indicate that a change may have occurred.

Using Distributions Other Than Normal

In a few special applications in which the expected demand is low (i.e., the item is a slow-moving item), other distributions like the Poisson distribution are sometimes used, but by far the most popular distribution for making inventory management decisions is the normal distribution because of its convenience, and because using it has become a common practice historically.

For the normal distribution the mean is the mode (i.e., the value associated with the highest probability), and the distribution is symmetric around this value. If histograms of observed demand data of an item do not share these properties, it may indicate that the normal distribution is a poor approximation for the actual distribution of demand, in this case order quantities determined using the normality assumption may be far from being optimal.

These days, the industrial environment is very competitive with new products replacing the old periodically due to rapid advancements in technology. In this dynamic environment, the life cycles of components and end products are becoming shorter. Beginning with the introduction of the product, its life cycle starts with a growth period due to gradual market penetration of the product. This is followed by a stable period of steady demand. It is then followed by a final decline period of steadily declining demand, at the end of which the item disappears from the market. Also, the middle stable period seems to be getting shorter for many major components. Because of this constant rapid change, it is necessary to periodically update demand distributions based on recent data.

The distributions of demand for some components are far from being symmetric around the mean, and the skewness and shapes of their distributions also seem to be changing over time. Using a probability distribution like the normal defined by a mathematical formula, involving only a few parameters, it is not possible to capture changes taking place in the shapes of distributions of demand for such components. This is the disadvantage of existing forecasting methods based on an assumed probability distribution. Our conclusions can be erroneous if the true probability distribution of demand is very different from the assumed distribution.

Nonparametric methods use statistical learning, and base their conclusions on knowledge derived directly from data without any unwarranted assumptions. In the next section, we discuss a nonparametric method for forecasting the entire demand distribution (Murty, 2006) that uses the classical empirical probability distribution derived from the relative frequency histogram of time series data on demand. It has the advantage of being capable of updating all changes occurring in the probability distribution of demand, including those in the shape of this distribution.

Then, in the following section, we illustrate how optimal order quantities that optimize piecewise linear and other unusual cost functions discussed in the previous section can be easily computed using these empirical distributions.

18.2.3 A Nonparametric Method for Updating and Forecasting the Entire Demand Distribution

In production systems, the important random variables are daily or weekly (or whatever planning period is being used) demands of various items (raw materials, components, subassemblies, finished goods, spare parts, etc.) that companies either buy from suppliers, or sell to their customers. Observed values of these random variables in each period are generated automatically as a time series in the production process, and are usually available in the production databases of companies. In this section, we discuss a simple nonparametric method for updating changes in the probability distributions of these random variables using these data directly.

Empirical Distributions and Probability Density Functions

The concept of the probability distribution of a random variable evolved from the ancient practice of drawing histograms for the observed values of the random variable. The observed range of variation of the random variable is usually divided into a convenient number of value intervals (in practice about 10 to 25) of equal length, and the relative frequency of each interval is defined to be the proportion of observed values of the random variable that lie in that interval. The chart obtained by marking the value intervals on the horizontal axis, and erecting a rectangle on each interval with its height along the vertical axis equal to the relative frequency, is known as the relative frequency histogram of the random variable, or its discretized probability distribution. The relative frequency in each value interval I_i is the estimate of the probability p_i that the random variable lies in that interval; see Figure 18.2 for an example.

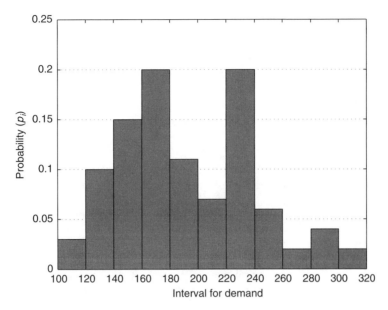

FIGURE 18.2 Relative frequency histogram for daily demand for a major component at a PC assembling plant in California.

Let $I_1, \ldots, I_n$ be the value intervals with $u_1, \ldots, u_n$ as their midpoints, and $p = (p_1, \ldots, p_n)$, the probability vector in the discretized probability distribution of the random variable. Let

$$\overline{\mu} = \sum_{i=1}^{n} u_i p_i, \quad \overline{\sigma} = \sqrt{\sum_{i=1}^{n} p_i (u_i - \overline{\mu})^2}$$

Then, $\overline{\mu}$, $\overline{\sigma}$ are estimates of the expected value μ and standard deviation σ of the random variable, respectively.

We will use the phrase *empirical distribution* to denote such a discretized probability distribution of a random variable, obtained either through drawing the histogram, or by updating a previously known discretized probability distribution based on recent data.

When mathematicians began studying random variables from the sixteenth century onwards, they found it convenient to represent the probability distribution of the random variable by the *probability density function*, which is the mathematical formula for the curve defined by the upper boundary of the relative frequency histogram in the limit as the length of the value interval is made to approach 0, and the number of observed values of the random variable goes to infinity. So the probability density function provides a mathematical formula for the height along the vertical axis of this curve as a function of the variable represented on the horizontal axis. Because it is a mathematically stated function, the probability density function lends itself much more nicely into mathematical derivations than the somewhat crude relative frequency histogram.

It is rare to see empirical distributions used in decision making models these days. Almost everyone uses mathematically defined density functions characterized by a small number of parameters (typically two or less) to represent probability distributions. In these decision making models, the only freedom we have in incorporating changes is to change the values of those parameters. This may be inadequate to capture all the dynamic changes occurring in the shapes of probability distributions from time to time.

Extending the Exponential Smoothing Method to Update the Empirical Probability Distribution of a Random Variable

We will now see that representing the probability distributions of random variables by their empirical distributions gives us unlimited freedom in making any type of change including changes in shape (Murty, 2002).

Let $I_1, \ldots, I_n$ be the value intervals, and $p_1, \ldots, p_n$ the probabilities associated with them in the present empirical distribution of a random variable. In updating this distribution, we have the freedom to change the values of all the p_i; this makes it possible to capture any change in the shape of the distribution.

Changes, if any, will reflect in recent observations on the random variable. The following table gives the present empirical distribution, histogram based on most recent observations on the random variable (e.g., the most recent k observations where k could be about 30), and x_i to denote the probabilities in the updated empirical distribution to be determined.

Value interval	Probability vector in the		
	Present empirical distribution	Recent histogram	Updated empirical distribution
I_1	p_1	f_1	x_1
$\vdots$	$\vdots$	$\vdots$	$\vdots$
I_n	p_n	f_n	x_n

$f = (f_1, \ldots, f_n)$ represents the estimate of the probability vector in the recent histogram, but it is based on too few observations. $p = (p_1, \ldots, p_n)$ is the probability vector in the empirical distribution at the previous updating. $x = (x_1, \ldots, x_n)$, the updated probability vector, should be obtained by incorporating the changing trend reflected in f into p. In the theory of statistics the most commonly used method for this incorporation is the weighted least squares method, which provides the following model (Murty, 2002) to compute x from p and f.

$$\text{Minimize} \quad (1-\beta)\sum_{i=1}^{n}(p_i - x_i)^2 + \beta \sum_{i=1}^{n}(f_i - x_i)^2$$
$$\text{Subject to} \quad \sum_{i=1}^{n} x_i = 1 \quad (18.1)$$
$$x_i \geq 0, \quad i = 1, \ldots, n$$

where β is a weight between 0 and 1, similar to the smoothing constant α in the exponential smoothing method for updating the expected value (like α there, here β is the relative weight placed on the probability vector from the histogram composed from recent observations). x is taken as the optimum solution of this convex quadratic program. $\beta = 0.1$ to 0.4 works well; the reason for choosing the weight for the second term in the objective function to be small is because the vector f is based on only a small number of observations. As the quadratic model minimizes the weighted sum of squared forecast errors over all value intervals, when used periodically, it has the effect of tracking gradual changes in the probability distribution of the random variable.

The above quadratic program has a unique optimum solution given by the following explicit formula.

$$x = (1-\beta)p + \beta f \quad (18.2)$$

Production Systems

So we take the updated empirical distribution to be the one with the probability vector given by Equation 18.2.

The formula for updating the probability vector in Equation 18.2 is exactly analogous to the formula for forecasting the expected value of a random variable using the latest observation, in exponential smoothing. Hence, the above formula can be thought of as the extension of the exponential smoothing method to update the probability vector in the empirical distribution of the random variable.

When there is a significant increase or decrease in the mean value of the random variable, new value intervals may have to be opened up at the left or right end. In this case, the probabilities associated with value intervals at the other end may become very close to 0, and these intervals may have to be dropped from further consideration at that time.

This procedure can be used to update the discretized demand distribution either at every ordering point, or periodically at every rth ordering point for some convenient r, using the most recent observations on demand.

18.2.4 An Application of the Forecasting Method for Computing Optimal Order Quantities

Given the empirical distribution of demand for the next period based on the forecasting method given in Section 18.2.3, the well-known newsvendor model (Murty, 2002) can be used to determine the optimal order quantity for that period that minimizes the sum of expected overage and underage costs very efficiently numerically. We will illustrate with a numerical example. Let the empirical distribution of demand (in units) for the next period be

I_i = interval for demand	Probability p_i	u_i = mid-point of interval i
100–120	0.03	110
120–140	0.10	130
140–160	0.15	150
160–180	0.20	170
180–200	0.11	190
200–220	0.07	210
220–240	0.20	230
240–260	0.06	250
260–280	0.02	270
280–300	0.04	290
300–320	0.02	310

The expected value of this distribution $\bar{\mu} = \sum_i u_i p_i = 192.6$ units, and its standard deviation $\bar{\sigma} = \sqrt{\sum_i (u_i - \bar{\mu})^2 p_i} = 47.4$ units.

Let us denote the ordering quantity for that period, to be determined, by Q, and let d denote the random variable that is the demand during that period. Then

- y = overage quantity in this period = amount remaining after the demand is completely fulfilled = $(Q-d)^+$ = maximum$\{0, Q-d\}$
- z = underage quantity during this period = unfulfilled demand during this period = $(Q-d)^-$ = maximum$\{0, d-Q\}$.

Suppose the overage cost $f(y)$, is the following piecewise linear function of y:

Overage amount $=y$	Overage cost $f(y)$ in $	Slope
$0 \leq y \leq 30$	$3y$	3
$30 \leq y$	$90 + 10(y-30)$	10

Suppose the underage cost $g(z)$ in $, is the fixed cost depending on the amount given below:

Underage amount $=y$	Underage cost $g(z)$ in $
$0 \leq z \leq 10$	50
$10 < z$	150

To compute $E(Q) =$ the expected sum of overage and underage costs when the order quantity is Q, we assume that the demand value d is equally likely to be anywhere in the interval I_i with probability p_i. This implies, for example, that the probability that the demand d is in the interval 120–125 is = (probability that d lies in the interval 120–140)/4 = (0.10)/4 = 0.025.

Let $Q = 185$. When the demand d lies in the interval 120–140, the overage amount varies from 65 to 45 and the overage cost varies from $440 to $240 linearly. So the contribution to the expected overage cost from this interval is $0.10(440 + 240)/2$.

Demand lies in the interval 140–160 with probability 0.15. In this interval, the overage cost is not linear, but it can be partitioned into two intervals 140–155 (with probability 0.1125), and 155–160 (with probability 0.0375) in each of which the overage cost is linear. In the interval $140 \leq d \leq 155$ the overage cost varies linearly from $240 to 90; and in $155 \leq d \leq 160$ the overage cost varies linearly from $90 to 75. So, the contribution to the expected overage cost from this interval is $\$(0.115(240+90)/2) + (0.0375(90+75)/2)$.

Proceeding this way, we see that $E(Q)$ for $Q = 185$ is: $\$(0.03(640+440)/2) + (0.10(440+240)/2) + [(0.115(240+90)/2) + (0.0375(90+75)/2)] + (0.20(75+15)/2) + [0.0275(15+0)/2) + 0.055(50) + 0.0275(150)] + (0.07 + 0.20 + 0.06 + 0.02 + 0.04 + 0.02)150 = \$140.87$.

Q	$E(Q)$
195	178.00
190	162.27
185	143.82
180	139.15
175	130.11
170	124.20
165	120.40
160	121.95
155	122.60
150	124.40
145	139.70

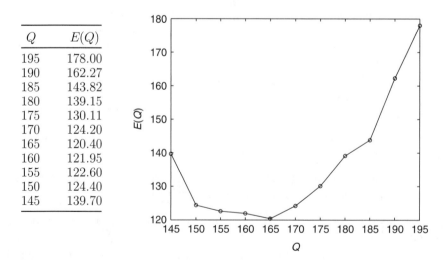

FIGURE 18.3 Plot of $E(Q)$ for various values of Q.

In the same way, we computed the values of $E(Q)$ for different values of Q spaced 5 units apart, given by the side of Figure 18.3. The graph of Figure 18.3 is a plot of these values of $E(Q)$. Here, we computed $E(Q)$ at values of Q, which are multiples of 5 units, and it can be seen that $Q = 165$ is the optimum order quantity correct to the nearest multiple of 5. If the optimum is required to greater precision, the above calculation can be carried out for values of Q at integer (or closer) values between 150 and 170 and the best value of Q there chosen as the optimum order quantity.

The optimum value of Q can then be translated into the actual order quantity for the next period by subtracting the expected on-hand inventory at the end of the present period from it.

For each i, assuming that demand d is equally likely to be anywhere in the interval I_i with probability p_i makes the value of $E(Q)$ computed accurate for each Q. However, in many applications, people make the simpler assumption that p_i is the probability of demand being equal to u_i, the midpoint of the interval I_i. The values of $E(Q)$ obtained with this assumption will be approximate, particularly when the overage and underage costs are not linear (i.e., when they are piecewise linear etc.); but this assumption makes the computation of $E(Q)$ much simpler; that's why people use this simpler assumption.

18.2.5 How to Incorporate Seasonality in Demand into the Model

The discussion so far has dealt with the case when the values of demand in the various periods form a stationary time series. In some applications this series may be seasonal; that is, it has a pattern that repeats every N periods for some known value of N. The number of periods N, before the pattern begins to repeat, is known as the length of the season. To use seasonal models, the length of the season must be known.

For example, in the computer industry, the majority of the sales are arranged by sales agents who operate on quarterly sales goals. That's why the demand for components in the computer industry and demand for their own products tend to be seasonal with the quarter of the year as the season. The sales agents usually work much harder in the last month of the quarter to meet their quarterly goals; so demand for products in the computer industry tends to be higher in the third month of each quarter than in the beginning two months. As most of the companies are building to order nowadays, weekly production levels and demands for components inherit the same kind of seasonality.

At one company in this industry each quarter is divided into three homogeneous intervals. Weeks 1–4 of the quarter are slack periods; each of these weeks accounts a fraction of about 0.045 of the total demand in the quarter. Weeks 5–8 are medium periods; each of these weeks accounts for a fraction of about 0.074 of the total demand in the quarter. Weeks 9–13 are peak periods; each of these weeks accounts for a fraction of about 0.105 of the total demand in the quarter. This fraction of demand in each week of the season is called the seasonal factor of that week.

In the same way, in the paper industry, demand for products exhibits seasonality with each month of the year as the season. Demand for their products in the 2nd fortnight in each month tends to be much higher than in the 1st fortnight.

There are several ways of handling seasonality. One way is for each $i = 1$ to $N (=$ length of the season), consider demand data for the ith period in each season as a time series by itself, and make the decisions for this period in each season using this series based on methods discussed in earlier sections.

Another method that is more popular is based on the assumption that there exists a set of indices c_i, $i = 1$ to N called seasonal factors or seasonal indices (see Meybodi and Foote, 1995), where c_i represents the demand in the ith period of the season as a fraction of the demand during the whole season (as an example, see the seasonal factors given for the computer company described above). Once these seasonal factors are estimated, we divide each observation of demand in the original demand time series by the appropriate seasonal factor to obtain the de-seasonalized demand series. The time series of de-seasonalized demand amounts still contains all components of information of the original series except for seasonality. Forecasting is carried out using the methods discussed in the earlier sections, on the de-seasonalized demand series. Then estimates of the expected demand, standard deviation, and the optimal order quantities obtained for each period must be re-seasonalized by multiplying by the appropriate seasonal factor before being used.

18.3 Models for Production Layout Design

The basic engineering tasks involved in producing a product are

1. Developing a process plan,
2. Deciding whether to make or buy each component,
3. Deciding on the production doctrine: group technology, process oriented, or a mix,
4. Designing the layout, and
5. Creating the aisles and choosing the material handling system.

In this section, we will focus on tasks 3–5, assuming tasks 1 and 2 have been done. We will then have available the following data: a matrix that shows the flows between processes.

18.3.1 An Example Problem

Our first problem is how to judge between layouts. A common metric is to minimize $\sum d_i * f_i$. This metric uses the distance from the centers of the process area multiplied by the flow. The idea is that if this is minimized then the processes with the largest interactions will be closest together. This has an unrealistic side in that flow usually is door to door. We will look at two approaches. The first approach is the spanning tree concept and then an approximation for the above metric.

As a practical matter, we always want to look at an ideal layout. We would like for the material to be delivered and moved continuously through processes and be packaged in motion and moved right into the truck or railway car that carries the product to the store. Some examples already exist as logs of wood go to a facility that moves them directly through a few processes that cut them into 2 × 4's or 4 × 4's and capture the sawdust and create particle board, all of which with few stationary periods and go right into the trucks that move them to lumber yards. In this case, the product is simple; there are only a small number of processes that are easy to control by computers armed with expert system and neural net rules that control the settings of the processes. Consider Example 1, where the basic data on the first level is given in Table 18.1.

The matrix illustrates a factory with seven process areas and lists the number of containers that will flow between them. Let us first assume that each department is of the same unit size, say 100 sq. feet. We can draw a network to represent the flows between process areas as in Figure 18.4. We can act as if the flows of material are from center to center of departments by a conveyor. We can estimate a layout as in Figure 18.5. Note that we now look at the size of the flows and put P2 and P3 adjacent and the center for P4 is slightly further away. A similar reasoning locates the other departments. Note also the room left for expansion. This is not a mathematical principle but is based on experience. Just as traffic on a freeway always expands, production needs to expand if the business succeeds.

Production Systems

TABLE 18.1 Flow Values for Example 1

	P1	P2	P3	P4	P5	P6	P7
P1	0	20	30	15			
P2		0			20		
P3			0		28		
P4				0		16	
P5		5			0	28	
P6				4		0	48
P7							0

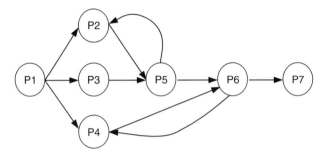

FIGURE 18.4 A process network for Example 1.

P1	P2	P6	P7
P3	P4	P5	Expansion area

FIGURE 18.5 Possible layout for Example 1.

The layout in Figure 18.5 has subjective virtues as department one is "door-to-door" close to the departments that it flows material to. An overhead conveyor down the middle has a straight line layout that is very convenient and is of minimum length. This is not necessarily an optimal layout but it provides a great starting point. If the departments are of different sizes, the layout problem is much more complex and natural adjacencies are sometimes infeasible.

Another heuristic approach to the layout adjacency problem is the use of the cut tree. If you look at Figure 18.4 and assume the arcs are conveyors, then you try to find the set of conveyors that, if inactive, will cut flow from 1 to 7. If the conveyors from 5 to 6 and 4 to 6 are inactive then flow from 1 to 7 cannot occur. This flow has value 44 and is the minimum of such cuts. A common sense interpretation is that P6 and P7 should be on one end and P1–P5 grouped on the other end. The flow between ends is the minimum it can be, which is 44. A cut can be made in the set P1–P5 and new smaller groups can be seen. Other cut trees can be computed and can provide a basis for several trials. The cut tree approach is discussed further in Section 18.3.5.

18.3.2 Optimal Plant Layouts with Practical Considerations

One of the difficulties in defining "optimal" plant layouts is choosing the metric that defines "optimal." Most of the metrics try to minimize the product of distance and flow. The problem is what is flowing. The weight and shape of what is moved varies tremendously. What is moved can be a box of parts, a chassis 7' by 4', a container of waste, or a bundle of electrical wire. At some point in time one needs to just find how many moves are made. Thirty boxes can cause a move or one waste container can cause a move. The movement

TABLE 18.2 Flow Values for Example 2

i/j	1	2	3	4	5
1		5	20		
2	5		20		
3	20	30			
4					100
5				100	

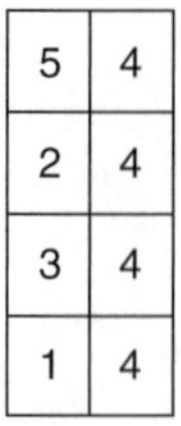

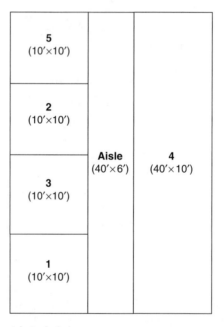

FIGURE 18.6 Conceptual layout without aisles.

FIGURE 18.7 Layout with aisle included.

can be along aisles or on overhead conveyors. If the move is along aisles then it comes out of doors. If it is moved by overhead conveyors, then the move is from a point inside the area that produces it and the area that needs it. If the move is on overhead conveyors, then the distance is Euclidean. If the move is along aisles, then distance is measured by the Manhattan metric, which is the sum of N-S and E-W moves. The problem is clearly NP complete, so that optimizing means optimizing in an ideal situation and then engineering the solution into a practical satisficing layout. An approach by Tretheway and Foote (1994) illustrates this approach.

Consider Example 2, where there are five departments with areas 100, 100, 100, 100, and 400. Let the flow values be as given in Table 18.2.

Let us assume a single aisle structure. Figures 18.6 and 18.7 illustrate an actual optimal layout for this aisle structure and set of areas. Each block is 100 sq. ft. Based on the flow and

Production Systems

the aisle down the middle of the plant, this is optimal regardless of the material handling system. Notice that the aisle takes some area. This means that the overall structure must contain perhaps 1024 sq. feet if the aisle is 40′ by 6′, or that each department donates some area to the aisle.

18.3.3 The Tretheway Algorithm

Tretheway and Foote (1994) suggest a way to actually compute the two-dimensional coordinates of the location of departments in a layout. Their approach works for buildings that are rectangular, and variants of rectangular such as U, T, or L shaped structures. They adapt a concept by Drezner (1987) to develop a technique to actually draw the layout with aisles. The basic idea is to get a good adjacency layout on a line, then get a second layout on another line, use the cross product to obtain a two-dimensional relative position, then draw in aisles, account for loss of space to aisles, and create the drawing. "Rotations" of the aisles structure create alternative structures that can be evaluated in terms of practical criteria. An example is that a rotation which puts a test facility with a risk of explosions occurring during test in a corner would be the layout chosen if the aisle structure can support the material handling design.

NLP problem: Drezner's nonlinear program for n departments is given by

$$\min \sum_{ij} f_{ij} d_{ij} \quad i = 1, \ldots, n; \ j = 1, \ldots, m \tag{18.3}$$

where f_{ij}'s are flows (given data) and d_{ij} (decision variables) are distances between departments i and j. Utilizing a squared Euclidean distance metric, the nonlinear program is transformed to

$$\min \frac{\sum_{ij} f_{ij} \left[(x_i - x_j)^2 + (y_i - y_j)^2 \right]}{\sum_{i,j} (x_i - x_j)^2 + (y_i - y_j)^2} \tag{18.4}$$

The objective in one dimension becomes

$$\min \frac{\sum_{i,j} f_{ij} x_{ij}^2}{\sum_{i,j} x_{ij}^2} \tag{18.5}$$

Equation 18.5 can be optimized. The minimizing solution is the eigen vector associated with the second least eigen value of a matrix derived from the flow matrix.

To get a solution to the layout problem in a plane, we use the second and third least eigen values to create two eigen vectors $[x]$ and $[y]$, each with n values. The Cartesian product of these vectors creates a *scatter diagram* (a set of points in $\mathbb{R}^2$) that suggests an optimal spatial relationship in two dimensions for the n departments that optimizes the squared distance ∗ flow relationship between the departments. Notice this relationship is not optimized based on practical flow measures. However, it may give us a set of alternatives that can be compared based on a practical measure. Further, this set of points can be expressed as a layout based on two practical considerations: the areas of the departments and the aisle structure selected. The Tretheway algorithm will illustrate how to do this. The algorithm can actually be done easily by hand, especially if we remember that the space requirements usually have some elasticity. The selection of an aisle structure first allows the departments to be painted in a simple way.

Example 3: Consider the input data for 12 departments shown in Table 18.3. For this input, consider an aisle structure with three vertical aisles. The scatter diagram with the

TABLE 18.3 Input Data for Example 3

	1	2	3	4	5	6	7	8	9	10	11	12
1	0	5	2	4	1	0	0	6	2	1	1	1
2	5	0	3	0	2	2	2	0	4	5	0	0
3	2	3	0	0	0	0	0	5	5	2	2	2
4	4	0	0	0	5	2	2	10	0	0	5	5
5	1	2	0	5	0	10	0	0	0	5	1	1
6	0	2	0	2	10	0	5	1	1	5	4	0
7	0	2	0	2	0	5	0	10	5	2	3	3
8	6	0	5	10	0	1	10	0	0	0	5	0
9	2	4	5	0	0	1	5	0	0	0	10	10
10	1	5	2	0	5	5	2	0	0	0	5	0
11	1	0	2	5	1	4	3	5	10	5	0	2
12	1	0	2	5	1	0	3	0	10	0	2	0

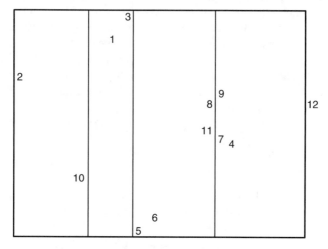

FIGURE 18.8 Example 3 scatter diagram for vertical aisles.

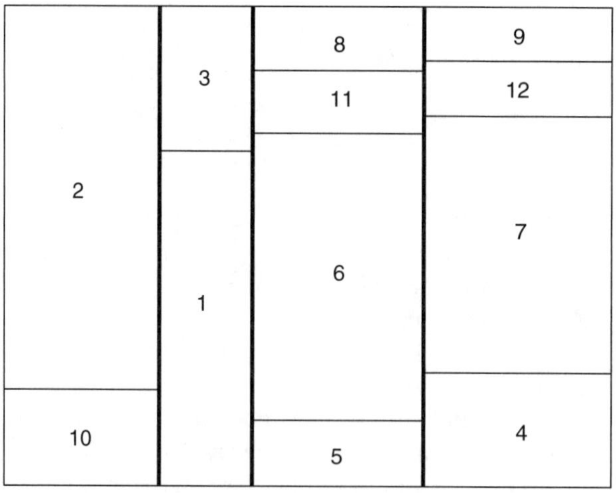

FIGURE 18.9 Example 3 final solution with vertical aisles.

overlaid *relative* position of the aisles is shown in Figure 18.8. Figure 18.9 shows the area proportioned final layout, with the heavy lines representing the aisles. For the same data input of Table 18.3, Figures 18.10 and 18.11 demonstrate the technique for one main aisle and two sub-aisles with the subaisles located between the aisle and the top wall. Note the different proximities for the two different aisle structure layouts.

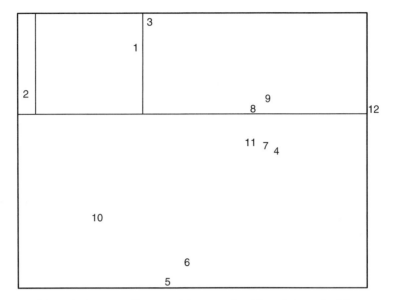

FIGURE 18.10 Example 3 scatter diagram with one main aisle and two sub-aisles.

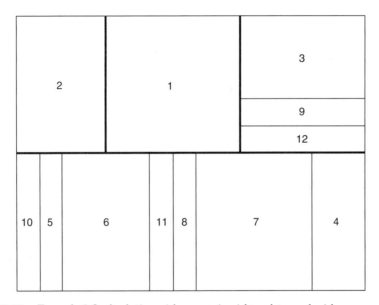

FIGURE 18.11 Example 3 final solution with one main aisle and two sub-aisles.

18.3.4 A Spanning Tree Network Model for Layout Design

Consider the flow network in Figure 18.12. The base example is from Ravindran et al. (1987).

The flows are $f_{12} = 3$, $f_{13} = 7$, $f_{14} = 4$, $f_{26} = 9$, $f_{36} = 6$, $f_{35} = 3$, $f_{46} = 3$, $f_{23} = 2$, and $f_{34} = 1$. These represent costs, not the flow of containers, if the associated departments are neighbors. Let us generate a spanning tree. This is a tree with no cycles such that all nodes can be reached by a path. According to Ravindran et al. (1987), we arbitrarily select node 1 to generate a tree such that the cost of each path segment is minimal. Thus, from 1 we go

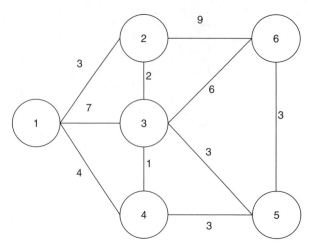

FIGURE 18.12 A material flow network.

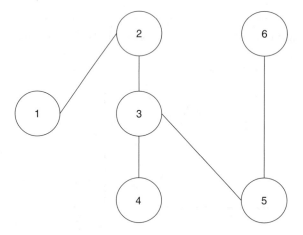

FIGURE 18.13 The minimum cost spanning tree.

5	6
3	4
2	1

FIGURE 18.14 A layout for equal area departments.

to 2 as 3 = min (3, 7, 4). We now add node 3 as its cost of 2 is less than any other cost from nodes 1 and 2. Now we look at arcs connecting from nodes 1, 2 and 3. Arc (3,4) is the best.

So we have [(1,2),(2,3),(3,4)]. Now (3,5) and (4,5) are tied, so we arbitrarily pick (3,5). (5,6) is left so we have [(1,2),(2,3),(3,4),(3,5),(5,6)]. This is the minimal spanning tree illustrated in Figure 18.13. We have five arcs which is one less than the number of nodes and hence the smallest number of paths. If all departments have the same area an optimal layout would be as given in Figure 18.14.

What if the departments do not have equal areas? Then an appropriate aisle structure could help. If you are lucky with areas and 1 has the biggest area and 2, 3, and 4 are not

Production Systems   18-19

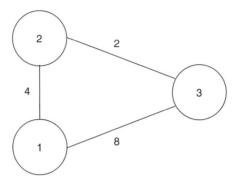

FIGURE 18.15 A layout adjusted for areas.

FIGURE 18.16 A network model of container flows.

too large and 5 and 6 are intermediate then we could have a layout as given in Figure 18.15. The aisle structure in Figure 18.15 could be two parallel aisles. This would work if the areas are roughly proportional to 1:300, 2,3,4:100, 5:200, and 6:100. Plant layout will always be NP complete relative to optimality in general with any metric, but in special cases apparent optimality can be achieved.

Optimizing (flow × distance) metrics must be constrained by unique circumstances. If department 3 in Figure 18.15 has some dangerous equipment that might explode, it has to go to a corner. This reduces the number of outer walls that can be blown out to two. It gets the area away from the center where all directions are in jeopardy. Buildings do not have to be in a rectangular shape. If space permits and modules are feasible, then the shape of the building can follow the spanning tree or the cut tree. U shapes can be solved by adding dummy departments to the Tretheway set and setting their artificial flows such that they will be in the middle at the end. We can also not have aisles and let the material handling system overhead be a minimum spanning tree with nodes at the appropriate points in departments or cells. See also Ravindran et al. (1989) for the use of shortest route network model to minimize overhead conveyor distances along with material movement control.

18.3.5 A Cut Tree Network Model for Layout Design

If you consider the numbers on arcs representing the flow of containers, then the cut tree provides a great basis for a layout that simply overlays a cut tree derived from the network flow model on a plant shape. Kim et al. (1995) give an algorithm for generating the cut tree and show an example layout for a nine area department cut tree. A cut tree divides the network model into a final spanning tree ($n-1$ arcs) that assumes flows go along the arcs and that flows to other nodes pass through predecessor nodes. The following is a simple example.

Consider the example network with three departments D1, D2, and D3 given in Figure 18.16. We treat the arcs as conveyors. If we cut the arcs (2,1) and (2,3) we interrupt 6 units of flow to D1 and D3. If D1 is isolated, then a flow of 12 is cut. If D3 is isolated then a flow of 10 is cut. The minimum cut is 6. If we cut arc (2,3), the 2 units from D2 to D3 must move to D1 and flow from D1 to D3 increases to 10 as seen in Figure 18.17. If we

FIGURE 18.17 The derived cut tree.

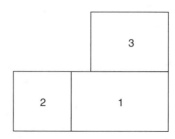

FIGURE 18.18 Layout of L shaped area using the cut tree solution.

move material by conveyors, then we have the minimum flow on the conveyors and a layout of 2-1-3 pasted into a building shape.

In Figure 18.18 the building shape is deliberately L-shaped, as occurs often when areas inside a plant are re-engineered. The areas are different in size, but the material handling costs will be minimized. See Kim et al. (1995) and also Benson and Foote (1997) for other ways to use cut trees and door to door metrics to rate layouts.

18.3.6 Facility Shapes

The shapes of facilities can be designed with a few basic principles. For example, the shape with the minimum perimeter for a given area is a circle. This means that this simple shape should be explored for nursing wards (minimum walking per day and maximum visual surveillance of patients). This would also work for sleeping areas and jails of abused children. A circular shape for production cells gives the best visual and response for the operator. If walls are needed, they will have a minimum length. Of course wall construction costs also depend on other factors such as materials (brick is more costly for circular walls) and insulation (minimum perimeter means less insulation, and so on). Shopping mall designers want longer walks to expose more goods, so shapes and paths will meet this requirement. Long materials require long straight shapes. Towers work if work can trickle down by gravity.

18.4 Scheduling of Production and Service Systems

18.4.1 Definition of the Scheduling Decision

The basic function of scheduling is the decision to start a job on a given process and predict when it will finish. This decision is determined by the objective desired, the data known, and the constraints. The basic data required to make a mathematical decision are the process plan, the time of operation projected for each process, the quality criteria, the bill of materials (which may determine start dates if materials are not immediately available), the due date, if any, and the objectives that must be met, if any. A process plan is a list of the operations needed to complete the job and any sequence constraints. It is sometimes thought that there is always a strict precedence requirement, but that is not true. Many times one can paint or cut to shape in either order. A quality requirement may impose a precedence, but this is to be determined. The basic plan can be listed as cut, punch, trim, and smooth (emery wheel or other). This list is accompanied by instructions as to how to execute each process (usually in the form of a drawing or pictures) and standards of quality to be achieved; as well, there will be an estimate of the time to set up, perform the process, do quality checks, record required information for the information and tracking systems, and move. A decision is required when there are two or more jobs at a process and the decision has to be made as to which job to start next. Some jobs require only a single machine. The process could be welding, creating a photo from film, X-ray, creating an IC board, or performing surgery. The study of single machines is very important because this theory forms the basis of many heuristics for more complex manufacturing systems.

18.4.2 Single Machines

We are assuming here a job that only requires one process, and that the time to process is given as a deterministic number or a random value from a given distribution. The standard objectives are (1) to Min $\overline{F}$ the average flow time, (2) Min Max lateness $L_{\max}$, (3) Min n_t, the number of tardy jobs, and combinations of earliness and lateness. The flow time F is the time to completion from the time the job is ready. If the job is ready at time equal to 0, flow time and completion time are the same. The function (finish time-due date) is the basic calculation. If the value is positive the job is late, if it is negative the job is early. The number of tardy jobs is the number of late jobs. Finish time is simply computed in the deterministic case as the time of start + processing time. There are surprisingly few objectives that have an optimal policy. This fact implies that for most systems we must rely on good heuristics. The following is a list of objectives and the policy that optimizes the objective. The list is not exclusive, as there are many types of problems with unusual assumptions.

Deterministic Processing Time

The objective is followed by the policy that achieves that objective:

1. Min $\overline{F}$: Schedule job with shortest processing time first (SPT). Break ties randomly or based on a second objective criterion. The basic idea is this: if the shortest job is worked first, that is the fastest a job can come out. Every job's completion time is the smallest possible (Figure 18.19).
2. Min Max lateness ($L_{\max}$): Schedule the job with the earliest due first (EDD). Break ties randomly or based on a second objective criterion (Figure 18.20).
3. Min n_t: Execute Moore's algorithm. This algorithm is explained to illustrate a basic sequencing algorithm. Some notation is required (Figure 18.21).

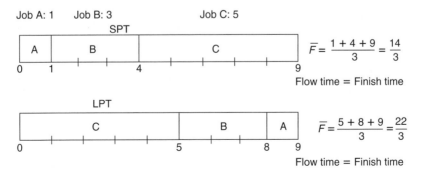

FIGURE 18.19 Optimality of SPT for min $\overline{F}$ (single machine Gantt charts comparing optimal with nonoptimal).

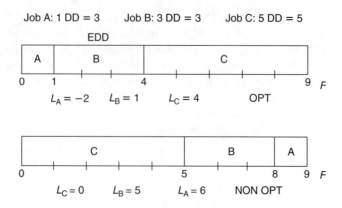

FIGURE 18.20 Optimality of EDD to Min Max lateness (single machine problem).

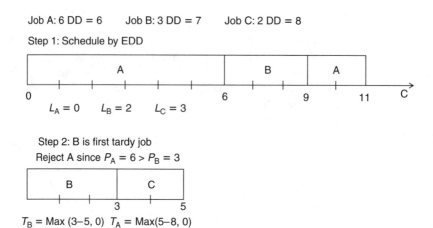

FIGURE 18.21 Use of Moore's algorithm.

Notation

Classification of problems: $n/m/A/B$, where n is the number of jobs, m the number of machines, A the flow pattern (blank when $m=1$, otherwise F for flow-shop, P for permutation job shop (jobs in same order on all machines), G the general job shop (where jobs can have different priorities at different process plan order), B the performance measure (objective), C the completion time of a job and equal to the ready time plus sum of waiting and processing times of jobs ahead of it, and D the due date of the job

L (lateness of job) $= C - D$

T (tardiness of job) $= \text{Max}(L, 0)$

Production Systems

E (earliness of job) = Max$(-L,0)$

P_i (processing time of job i) n_t = number of tardy jobs

Moore's algorithm solves the $n/1//n_t$ problem. The idea of the algorithm is simple. Schedule by EDD to min max tardiness. Then find the first tardy job and look at it and all the jobs in front of it. The problem job is the one with the largest processing time. It keeps other jobs from finishing; remove it. It could be a job that is not tardy. Now repeat the process. When no tardy jobs are found in the set of jobs that remain, schedule them in EDD order and place the rejected jobs at the end in any order.

Figure 18.21 illustrates Moore's algorithm and basic common sense ideas. For other single-machine problems that can be solved optimally, see French (1982) or any current text on scheduling (a fruitful path is to web-search Pinedo or Baker).

18.4.3 Flow Shops

More than one-third of production systems are flow shops. This means that all jobs have a processing plan that goes through the processes in the same order (some may take 0 time at a process). Only the two-process case has an optimal solution for min $F_{\max}$. A special case of the three-machine flow shop has an optimal solution for the same criteria. All other problems for m(number of processes) >2 do not have optimal solutions that are computable in a time proportional to some quadratic function of a parameter of the problem, such as n for any criteria. This will be addressed in more detail later.

Two-Process Flow Shops

It has been proved for flow shops that only schedules that schedule jobs through each process (machine) in the same sequence need be considered. This does not mean that deviations from this form will not sometimes get a criterion value that is optimal. When $m=2$, an optimal solution is possible using Johnson's algorithm. The idea of the algorithm is to schedule jobs that have low processing times on the first machine and thus get them out of the way. Then schedule jobs later that have low processing times on the last machine, so that when they get on they are finished quickly and can make way for jobs coming up. In this way, the maximum flow time of all jobs is minimized (the time all jobs are finished).

Job	Time on machine A	Time on machine B
1	8	12
2	3	7
3	10	2
4	5	5
5	11	4

The algorithm is easy. Look at all the processing times and find the smallest. If it is on the first machine, schedule it as soon as possible, and if it is on the second machine, schedule it as late as possible. Then remove this job's data and repeat. The solution here has five positions. The following shows the sequence construction step by step. Schedule is represented by S.

$$S = [,,,,3]$$
$$S = [2,,,,3]$$
$$S = [2,,,5,3]$$
$$S = [2,4,,5,3] \text{ or } [2,,4,5,3]$$

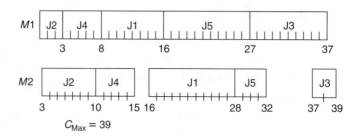

FIGURE 18.22 Gantt chart of an optimal schedule. For the example in two C_{max} problem schedule, using Johnson's algorithm.

Job 1 now goes in either sequence in the only position left: $S = [2,4,1,5,3]$ or $S = [2,1,4,5,3]$ (see Figure 18.22).

Heuristics for General Flow Shops for Some Selected Criteria

$$n/m/F/F_{max}$$

The Campbell–Dudek heuristic makes sense and gives good solutions. There are better heuristics if one is not solving by hand. The idea is common sense. Assume we have six machines. If we find a job that gets through the six machines quickly, schedule it first so it gets out of the way. If there is a job that gets through the first five quickly, schedule it as soon as you can. This follows up to a job that gets off the first machine quickly. Alternatively, if a job gets off the last machine quickly, schedule it last, or if it gets off the last two quickly, or the last three, and so on. This leads to solving five constructed two-machine flow shop problems (surrogates), finding the optimal sequences, and then picking the one with the best maximum flow. A sixth sequence can be tested by sequencing in order of the least total (on all machines) processing time. Here is a five-job, four-machine problem to illustrate the surrogate problems. There will be $m - 1$ or three of them.

Job	$m1$	$m2$	$m3$	$m4$	Total
1	5	7	4	11	27
2	2	3	6	7	18
3	6	10	1	3	20
4	7	4	2	4	17
5	1	1	I	2	5

$S = [5,4,2,3,1]$ when looking at total processing time.

Three Surrogate Problems

Surrogate one: times on first and last machines only (solution by Johnson's algorithm: [5,2,3,4,1])

Job	$m1'$	$m2''$
1	5	11
2	2	7
3	6	3
4	7	4
5	1	2

Surrogate two: times on first two and last two machines (solution by Johnson's algorithm: [5,2,1,4,3])

Job	m1'	m2''
1	12	15
2	5	13
3	16	4
4	11	6
5	2	3

Surrogate three: times on first three machines and last three machines (solution by Johnson's algorithm: [5,4,3,2,1])

Job	m1'	m2''
1	16	22
2	11	16
3	17	14
4	13	10
5	3	4

These four sequences are then Gantt-charted, and the one with the lowest F max is used. If we want to min average flow time, then these would also be good *solutions to try*. If we want to min max lateness, then try EDD, and also try these sequences. We could also put job 1 last and schedule by EDD as a commonsense approach. Later we will talk about computer heuristics for these types of problems.

The Concept of Fit in Flow Shops

Consider a case where job (i) precedes job (j), with a six-machine flow shop. Let their processing times be (i): [5 4 2 6 8 10] and (j): [3 2 5 8 9 15]. In the Gantt chart (see Figure 18.23) you will see that job (j) is always ready to work when job (i) is finished on a machine. No idle time is created (see Figure 18.23). If computer-processing time is available, then a matrix of job fit can be created to be used in heuristics.

Fit is computed with the following notation: t_{ij} is the processing time of job (i) on machine (j). Then let job (j) precede job (k):

$$\text{Fit} = \sum_{s=2}^{s=m} \max[t_{js} - t_{k,s-1}, 0]$$

Heuristics will try to make job (j) ahead of job (k) if fit is the largest positive found, that is, $[\ldots k, j, \ldots]$. By removing the max operator, one can get a measure of the overall fit or one can count the number of positive terms and use this to pick a pair that should be in sequence. If there are six machines, pick out all the pairs that have a count of 5, then 4, then 3, and so on, and try to make a common sense Gantt chart. Consider the following data: J1: [4 5 4 7], J2: [5 6 9 5], J3: [6 5 6 10], J4: [3 8 7 6], J5: [4 4 6 11]. The largest number of fits is 3.

$4 \to 3$ $3 \to 2$ $2 \to 1$ $1 \to 5$ have three positive terms in the fit formula. Hence, an optimal sequence is $S = [4\ 3\ 2\ 1\ 5]$ (Figure 18.24).

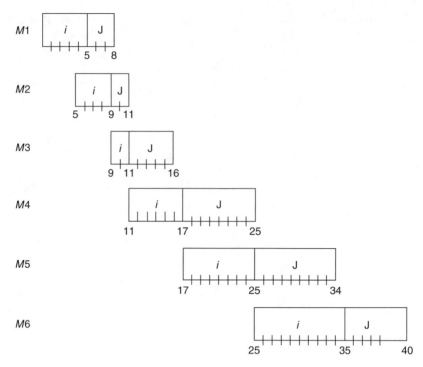

FIGURE 18.23 Job J fits behind job i.

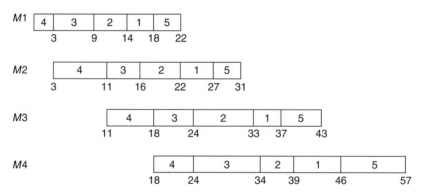

FIGURE 18.24 An optimal sequence since we have perfect fit.

18.4.4 Job Shops

Job shops are notoriously tough (see later discussion on NP-complete problems). In job shops, a job may have processing plans that have a variety of sequences. For large job shops some simulation studies have shown that job shops are dominated by their bottlenecks. A bottleneck is the machine that has the largest total work to do when all jobs are completed.

In practice, it has been effective to schedule the jobs on each machine by using the sequence determined by solving a one-machine problem with the bottleneck as the machine. For large job shops, jobs are constantly entering. Thus, the structure constantly changes. This means that at the beginning of the period, each machine is treated as a single machine, and single machine theory is used for scheduling. At the beginning of the next period, a new set of priorities is determined. The determination of period length is a matter still under

Production Systems

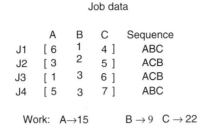

Job data

	A	B	C	Sequence
J1	[6	1	4]	ABC
J2	[3	2	5]	ACB
J3	[1	3	6]	ACB
J4	[5	3	7]	ABC

Work: A→15   B → 9   C → 22

A and C are top two bottlenecks

Johnson's table

	A	C
1	6	4
2	3	5
3	1	6
4	5	7

Johnson's solution → [3,2,4,1] = 5

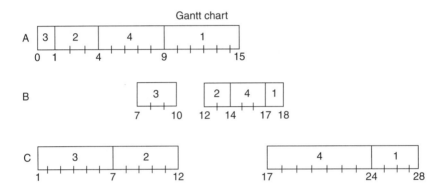

FIGURE 18.25 A heuristic approach to scheduling a job shop. Compare with other guesses and pick lowest C_{max}.

research. A period length of a multiple of total average job-processing time is a good start. A multiple such as 10 is a good middle ground to try. Perform some experiments and try going up if you have a lot of setup time and down if setup time is minimal.

Some other approaches are to take the top two bottlenecks and schedule the jobs as a two-machine flow shop, using only their times on the two bottlenecks. If you are using min average flow time, use flow shop heuristics designed for that purpose. If you are using min max flow time, use Johnson's algorithm. If a job does not go through the two bottlenecks, schedule it later, if you can. In the real world sometimes jobs can be worked faster than predicted. Most schedulers try to find such jobs. They then look for small processing time jobs and insert them. Schedulers also look for jobs that have remaining processes that are empty and pull them out if average flow time is an issue.

It is never good to operate at 100% capacity. Quality goes down under this kind of pressure. Studies have shown that around 90% is good. At times, random variation will put the shop at 100% even then. But a little gap is good for preventive maintenance and for training. Most maintenance will be off shift if unplanned, but off shift can be expensive. Figure 18.25 is an example using Johnson's algorithm on the two bottlenecks and then scheduling the three machines with that sequence.

Scheduling and Systems Analysis

If output from one subsystem goes to another, then there is a timing issue and hence a due date issue. Basic queuing issues come into play. If service centers are in a flow mode, then the jobs should be in SPT order. This gets them out of the early systems quickly and allows the other jobs get going. If work has a large variance, then low variance jobs go first. Simulation software analysis is a must for large systems. The biggest aid is total system knowledge. If a truck coming in has an accident and this is known by the warehouse, some healing action can be taken. Some borrowing from other trucks might happen. Work that will not be completed due to the accident can be put off. Top-level knowledge of demand from the primary data is a must. Inferring it from orders is a major mistake. This subsystem knowledge allows coordination.

Control

What does control mean for a system? If a policy is simulated, there will be a mean and standard deviation of the queue length. When a queue exceeds the mean plus two standard deviations something is likely wrong. So, what questions should you ask? Are the processing times different from predictions? Is a large amount of maintenance going on? Do we have critical material shortages? Have we instituted just-in-time scheduling when the variance in system parameters is too high? If your trucks go completely through urban areas, you may not be able to arrive in a just-in-time window.

It is now possible with computers to predict a range of conditions that are normal. Readers should refer to the literature for current information, journals, and books.

Solvability Problems

We now know of many areas that have no solution. The normal distribution function does not have a closed-form integral. It must be computed numerically. Most scheduling problems are such that to guarantee optimality, total enumeration or implied total enumeration (branch and bound) must be used. However, it has been shown over and over that heuristics can get optimal solutions or solutions that are only off by 1 or 2%. When data quality is considered, this will be satisfactory. In linear programming, any point in the vicinity of a basic solution will be near optimal and, in fact, due to round off errors in computing, the basis may even be better than the basis presented by the computation. Problems that require total enumeration to prove optimality are interesting, but in practice are solved and the solutions are good. Practically speaking, problems such as $n/m/F/n_t$ are what are called NP complete. They must have implied total enumeration to prove the answer is optimal. However, in practice, one can obtain good solutions. Moreover, for some problems, the optimal solution is obvious (for example, in flow shops where all jobs have unit processing times on all machines for the objective of min average flow time).

Bootstrapping

With the advent of modern computers and Monte Carlo simulation packages, it is possible to obtain statistical predictions of systems. By using heuristics and genetic algorithms to generate policies to simulate, a good statistical prediction can be made to set control standards with only three or four replications. Bootstrapping is the technique in which actual data are used for lead times, processing times, etc. These data can be arranged as a histogram from which values can be drawn randomly to determine how long a job will be processed, how long it will take to repair a machine, and so on. This uses the best evidence, the data, and has no chance of making an error in distribution assumption. Good data collection systems are a must. Distribution assumptions do have their virtues, in that we know repair times can occur that have not previously been observed. Running a Monte Carlo simulation both ways can throw much light on the matter.

Practical Issues

In both build-to-stock and build-to-order policies, the unexpected may happen. Machines break down, trucks get caught in traffic delays, materials are found to be defective, and workers do not show up. The plan and schedule must be repaired. A basic plan to illustrate repair is to assign workers to the bottleneck one at a time, then find the new bottleneck and assign the next worker there. Repairing, regaining feasibility, and extending the plan when new work arrives requires software if the system is large ("large" meaning that hand computation is simply not doable). An illustration of repair occurs when some jobs I are not ready in an $n/m/G$ or $F/C_{\max}$ problem at time 0. One then goes ahead and solves, assuming all i ready times are 0. If the solution requires a job to start before it is ready, repair must be done. These needs form the criteria for selecting commercial software from an available list. The software must be algorithmically correct and computationally efficient with good structures for search and information retrieval. Professional magazines such as *IE Solutions* provide lists of vendors to assess.

References

1. Baker, K.R. *Elements of Sequencing and Scheduling*. John Wiley & Sons, New York, revised and updated edition, 1995. URL http://www.dartmouth.edu/~kbaker/.
2. Baker, K. http://www.pearson.ch/pageid/34/artikel/28138PH/PrenticeHall/0130281387/ Scheduling Theory Algorithms.aspx, 1995 (also self published).
3. Benson, B. and B.L. Foote. DoorFAST: A constructive procedure to optimally layout a facility including aisles and door locations based on an aisle flow distance metric. *International Journal of Production Research*, 35(7):1825–1842, 1997.
4. Box, G.E.P. and G.M. Jenkins. *Time Series Analysis, Forecasting and Control*. Holden Day, San Francisco, 1970.
5. Brown, R.G. *Statistical Forecasting for Inventory Control*. McGraw-Hill, New York, 1959.
6. Drezner, Z. A heuristic procedure for the layout of a large number of facilities. *Management Science*, 33:907–915, 1987.
7. Foote, B.L. On the implementation of a control-based forecasting system for aircraft spare parts procurement. *IIE Transactions*, 27:210–216, 1995.
8. French, S. *Sequencing and Scheduling: An Introduction to the Mathematics of the Job-Shop*. John Wiley & Sons, New York, 1982.
9. Holt, C.C. Forecasting seasonals and trends by exponentially weighted moving averages. ONR Memo No. 52, 1957.
10. Kim, C-B, B.L. Foote, and P.S. Pulat. Cut-tree construction for facility layout. *Computers & Industrial Engineering*, 28(4):721–730, 1995.
11. Meybodi, M.Z. and B.L. Foote. Hierarchical production planning and scheduling with random demand and production failure. *Annals of Operations Research*, 59:259–280, 1995.
12. Montgomory, D.D. and L.A. Johnson. *Forecasting and Time Series Analysis*. McGraw-Hill, St. Louis, 1976.
13. Murty, K.G. Histogram, an ancient tool and the art of forecasting. Technical report, IOE Dept., University of Michigan, Ann Arbor, 2002.
14. Murty, K.G. Forcasting for supply chain management. Technical report, IOE dept., University of Michigan, Ann Arbor, 2006. Paper for a keynote talk at ICLS (International

Congress on Logistics and SCM Systems), Taiwan 2006, and appeared in the Proceedings of that Congress.
15. Nahmias, S. *Production and Operations Analysis*. Irwin, Boston, 2nd edition, 1993.
16. Pinedo, M. *Scheduling: Theory, Algorithms, and Systems*. Prentice Hall, Upper Saddle River, NJ, 2nd edition, 2001.
17. Pinedo, M.L. http://www.amazon.co.uk/exec/obidos/ASIN/0963974610/tenericbusine-21/026-2125298-8551636, Prentice Hall, Upper Saddle River, NJ, 2002.
18. Ravindran, A., D.T. Phillips, and J.J. Solberg. *Operations Research: Principles and Practice*. John Wiley & Sons, New York, 2nd edition, 1987.
19. Ravindran, A., B.L. Foote, A.B. Badiru, L.M. Leemis, and L. Williams. An application of simulation and network analysis to capacity planning and material handling systems at Tinker Air Force Base. *Interfaces*, 19(1):102–115, 1989.
20. Silver, E.A. and R. Peterson. *Decision Systems for Inventory Management and Production Planning*. John Wiley & Sons, New York, 2nd edition, 1979.
21. Tretheway, S.J. and B.L. Foote. Automatic computation and drawing of facility layouts with logical aisle structures. *International Journal of Production Research*, 32(7):1545–1555, 1994.

19
Energy Systems

19.1	Introduction..	**19**-1
19.2	Definition of Energy	**19**-2
19.3	Harnessing Natural Energy	**19**-3
19.4	Mathematical Modeling of Energy Systems	**19**-3
19.5	Linear Programming Model of Energy Resource Combination..	**19**-4
19.6	Integer Programming Model for Energy Investment Options .... Energy Option 1 • Option 2 • Option 3 • Option 4	**19**-5
19.7	Simulation and Optimization of Distributed Energy Systems..	**19**-11
19.8	Point-of-Use Energy Generation	**19**-11
19.9	Modeling of CHP Systems	**19**-12
19.10	Economic Optimization Methods	**19**-13
19.11	Design of a Model for Optimization of CHP System Capacities	**19**-16
19.12	Capacity Optimization	**19**-21
19.13	Implementation of the Computer Model.........	**19**-24
	Example Calculation	
19.14	Other Scenarios....................................	**19**-27
	References	**19**-27

C. Randy Hudson
Oak Ridge National Laboratory

Adedeji B. Badiru
Air Force Institute of Technology

19.1 Introduction

Energy is an all-encompassing commodity that touches the life of everyone. Managing energy effectively is of paramount importance in every organization and every nation. Energy issues have led to political disputes and wars. This chapter presents examples of the application of operations research (OR) to optimizing energy decisions. Kruger (2006) presents a comprehensive assessment of energy resources from different perspectives, which include historical accounts, fossil fuel, energy sustainability, consumption patterns, exponentially increasing demand for energy, environmental impact, depletion of energy reserves, renewable energy, nuclear energy, economic aspects, industrialization and energy consumption, and energy transportation systems. Badiru (1982) presents the application of linear programming (LP) technique of OR to energy management with a specific case example of deciding between alternate energy sources. Badiru and Pulat (1995) present an OR model for making investment decisions for energy futures. Hudson (2005) developed an OR-based spreadsheet model for energy cogeneration. All these and other previous applications of OR to energy systems make this chapter very important for both researchers and practitioners.

19.2 Definition of Energy

The word energy comes from the Greek word for "work." Indeed, energy is what makes everything *work*. Everything that we do is dependent on the availability of energy. Newton's law of conservation of momentum and energy states that energy cannot be created or destroyed. It can, however, be converted from one form to other. Recent energy-related events around the world have heightened the need to have full understanding of energy issues, from basic definitions to consumption and conservation practices. Tragic examples can be seen in fuel-scavenging practices that turn deadly in many energy impoverish parts of the world. In May 2006, more than 200 people died when a gasoline pipeline exploded in Nigeria while poor villagers were illegally tapping into the pipeline to obtain the much-needed energy source. This illustrates a major lack of understanding of the volatility of many energy sources.

Before we can develop a mathematical model of energy systems, we must understand the inherent characteristics of energy. There are two basic types of energy:

- Kinetic energy
- Potential energy

All other forms of energy are derived from the above two fundamental forms. Energy that is stored (i.e., not being used) is potential energy. *Kinetic energy* is found in anything that moves (e.g., waves, electrons, atoms, molecules, and physical objects). Electrical energy is the movement of electrical charges. Radiant energy is electromagnetic energy traveling in waves. Radiant energy includes light, X-rays, gamma rays, and radio waves. Solar energy is an example of radiant energy. Motion energy is the movement of objects and substances from one place to another. Wind is an example of motion energy. Thermal or heat energy is the vibration and movement of matter (atoms and molecules inside a substance). Sound is a form of energy that moves in waves through a material. Sound is produced when a force causes an object to vibrate. The perception of sound is the sensing (picking up) of the vibration of an object.

Potential energy represents stored energy as well as energy of position (e.g., energy due to fuel, food, and gravity). Chemical energy is energy derived from atoms and molecules contained in materials. Petroleum, natural gas, and propane are examples of chemical energy. Mechanical energy is the energy stored in a material by the application of force. Compressed springs and stretched rubber bands are examples of stored mechanical energy. Nuclear energy is stored in the nucleus of an atom. Gravitational energy is the energy of position and place. Water retained behind the wall of a dam is a demonstration of gravitational potential energy. Light is a form of energy that travels in waves. The light we see is referred to as visible light. However, there is also an invisible spectrum. Infrared or ultraviolet rays cannot be seen, but can be felt as heat. Getting a sunburn is an example of the effect of infrared. The difference between visible and invisible light is the length of the radiation wave, known as wavelengths. Radio waves have the longest rays while gamma rays have the shortest rays.

When we ordinarily talk about conserving energy, we are referring to adjusting the thermostat (for cooling or heating) to save energy. When scientists talk of conserving energy, they are referring to the law of physics, which states that energy cannot be created or destroyed. When energy is used (consumed), it does not cease to exist; it simply turns from one form to another. As examples, solar energy cells change radiant energy into electrical energy. As an automobile engine burns gasoline (a form of chemical energy), it is transformed from the chemical form to a mechanical form.

When energy is converted from one form to another, a useful portion of it is always lost because no conversion process is 100% efficient. It is the objective of energy analysts to minimize that loss, or convert the loss into another useful form. Therein lies the need to use OR techniques to mathematically model the interaction of variables in an energy system to achieve optimized combination of energy resources. The process of combined heat and power (CHP) to achieve these objectives can benefit from OR modeling of energy systems.

19.3 Harnessing Natural Energy

There is abundant energy in our world. It is just a matter of meeting technical requirements to convert it into useful and manageable forms. For example, every second, the sun converts 600 million tons of hydrogen into 596 million tons of helium through nuclear fusion. The balance of 4 million tons of hydrogen is converted into energy in accordance with Einstein's theory of relativity,

$$E = mc^2$$

This is a lot of energy that equates to 40,000 W per square inch on the visible surface of the sun. Can this be effectively harnessed for use on Earth? Although the Earth receives only one-half of a billionth of the sun's energy, this still offers sufficient potential for harnessing. Comprehensive technical, quantitative, and qualitative analysis will be required to achieve the harnessing goal. OR can play an importance role in that endeavor. The future of energy will involve several integrative decision scenarios involving technical and managerial issues such as:

- Micropower generation systems
- Negawatt systems
- Energy supply transitions
- Coordination of energy alternatives
- Global energy competition
- Green-power generation systems
- Integrative harnessing of sun, wind, and water energy sources
- Energy generation, transformation, transmission, distribution, storage, and consumption across global boundaries.

19.4 Mathematical Modeling of Energy Systems

The rapid industrialization of society, coupled with drastic population growth, has fueled increasingly complex consumption patterns that require optimization techniques to manage. It is essential to first develop mathematical representations of energy consumption patterns to apply OR methods of optimization. Some of the mathematical modeling options for energy profiles are linear growth model, exponential growth model, logistic curve growth model, regression model, and logarithmic functions. There is often a long history of energy data to develop appropriate mathematical models. The huge amount of data involved and the diverse decision options preclude the use of simple enumeration approaches. Thus, LP and other nonlinear OR techniques offer proven solution techniques. The example in the

19.5 Linear Programming Model of Energy Resource Combination

This example illustrates the use of LP for energy resource allocation (Badiru, 1982). Suppose an industrial establishment uses energy for heating, cooling, and power. The required amount of energy is presently being obtained from conventional electric power and natural gas. In recent years, there have been frequent shortages of gas, and there is a pressing need to reduce the consumption of conventional electric power. The director of the energy management department is considering a solar energy system as an alternate source of energy. The objective is to find an optimal mix of three different sources of energy to meet the plant's energy requirements. The three energy sources are

- Natural gas
- Commercial electric power grid
- Solar power

It is required that the energy mix yield the lowest possible total annual cost of energy for the plant. Suppose a forecasting analysis indicates that the minimum kilowatt-hour (kwh) needed per year for heating, cooling, and power are 1,800,000, 1,200,000, and 900,000 kwh, respectively. The solar energy system is expected to supply at least 1,075,000 kwh annually. The annual use of commercial electric grid must be at least 1,900,000 kwh due to a prevailing contractual agreement for energy supply. The annual consumption of the contracted supply of gas must be at least 950,000 kwh. The cubic foot unit for natural gas has been converted to kwh (1 cu. ft. of gas = 0.3024 kwh).

The respective rates of $6/kwh, $3/kwh, and $2/kwh are applicable to the three sources of energy. The minimum individual annual conservation credits desired are $600,000 from solar power, $800,000 from commercial electricity, and $375,000 from natural gas. The conservation credits are associated with the operating and maintenance costs. The energy cost per kilowatt-hour is $0.30 for commercial electricity, $0.05 for natural gas, and $0.40 for solar power. The initial cost of the solar energy system has been spread over its useful life of 10 years with appropriate cost adjustments to obtain the rate per kilowatt-hour. The sample data is summarized in Table 19.1. If we let x_{ij} be the kilowatt-hour used from source i for purpose j, then we would have the data organized as shown in Table 19.2. Note that the energy sources (solar, commercial electricity grid, and natural gas) are used to power devices to meet energy needs for cooling, heating, and power.

TABLE 19.1 Energy Resource Combination Data

Energy Source	Minimum Supply (1000's kwh)	Minimum Conservation Credit (1000's $)	Conservation Credit Rate ($/kwh)	Unit Cost ($/kwh)
Solar power	1075	600	6	0.40
Electricity grid	1900	800	3	0.30
Natural gas	950	375	2	0.05

TABLE 19.2 Tabulation of Data for LP Model

Energy Source	Heating	Type of Use Cooling	Power	Constraint
Solar power	x_{11}	x_{12}	x_{13}	≥ 1075K
Electric power	x_{21}	x_{22}	x_{23}	≥ 1900K
Natural gas	x_{31}	x_{32}	x_{33}	≥ 950K
Constraint	≥ 1800	≥ 1200	≥ 900	

TABLE 19.3 LP Solution to the Resource Combination Example (in kwh)

Energy Source	Heating	Type of Use Cooling	Power
Solar power	0	1075	0
Commercial Electricity	975	0	925
Natural gas	825	125	0

The optimization problem involves the minimization of the total cost function, Z. The mathematical formulation of the problem is presented below.

$$\text{Minimize:} \quad Z = 0.4 \sum_{j=1}^{3} x_{1j} + 0.3 \sum_{j=1}^{3} x_{2j} + 0.05 \sum_{j=1}^{3} x_{3j}$$

Subject to:
$$x_{11} + x_{21} + x_{31} \geq 1800$$
$$x_{12} + x_{22} + x_{32} \geq 1200$$
$$x_{13} + x_{23} + x_{33} \geq 900$$
$$6(x_{11} + x_{12} + x_{13}) \geq 600$$
$$3(x_{21} + x_{22} + x_{23}) \geq 800$$
$$2(x_{31} + x_{32} + x_{33}) \geq 375$$
$$x_{11} + x_{12} + x_{13} \geq 1075$$
$$x_{21} + x_{22} + x_{23} \geq 1900$$
$$x_{31} + x_{32} + x_{33} \geq 950$$
$$x_{ij} \geq 0, \qquad i,j = 1,2,3$$

Using the LINDO LP software, the solution presented in Table 19.3 was obtained. The table shows that solar power should not be used for heating or power; commercial electricity should not be used for cooling, and natural gas should not be used for power. In pragmatic terms, this LP solution may have to be modified before being implemented on the basis of the prevailing operating scenarios and the technical aspects of the facilities involved. The minimized value of the objective function is $1047.50 (in thousands).

19.6 Integer Programming Model for Energy Investment Options

This section presents an integer programming formulation as another type of OR modeling for energy decision making. Planning a portfolio of energy investments is essential in resource-limited operations. The capital rationing example presented here (Badiru and Pulat, 1995) involves the determination of the optimal combination of energy investments so as to maximize present worth of total return on investment. Suppose an energy analyst is given N energy investment options, $X_1, X_2, X_3, \ldots, X_N$, with the requirement to determine the level of investment in each option so that present worth of total investment return is maximized subject to a specified limit on available budget. The options are not mutually exclusive.

The investment in each option starts at a base level b_i $(i = 1, 2, \ldots, N)$ and increases by variable increments k_{ij} $(j = 1, 2, 3, \ldots, K_i)$, where K_i is the number of increments used for option i. Consequently, the level of investment in option X_i is defined as

$$x_i = b_i + \sum_{j=1}^{K_i} k_{ij}$$

where

$$x_i \geq 0 \quad \forall i$$

For most cases, the base investment will be 0. In those cases, we will have $b_i = 0$. In the modeling procedure used for this example, we have:

$$X_i = \begin{cases} 1 & \text{if the investment in option } i \text{ is greater than zero} \\ 0 & \text{otherwise} \end{cases}$$

and

$$Y_{ij} = \begin{cases} 1 & \text{if the increment of alternative } i \text{ is used} \\ 0 & \text{otherwise.} \end{cases}$$

The variable x_i is the actual level of investment in option i, while X_i is an indicator variable indicating whether or not option i is one of the options selected for investment. Similarly, k_{ij} is the actual magnitude of the jth increment, while Y_{ij} is an indicator variable that indicates whether or not the jth increment is used for option i. The maximum possible investment in each option is defined as M_i such that

$$b_i \leq x_i \leq M_i$$

There is a specified limit, B, on the total budget available to invest such that

$$\sum_i x_i \leq B$$

There is a known relationship between the level of investment, x_i, in each option and the expected return, $R(x_i)$. This relationship is referred to as the *utility function*, $f(.)$, for the option. The utility function may be developed through historical data, regression analysis, and forecasting models. For a given energy investment option, the utility function is used to determine the expected return, $R(x_i)$, for a specified level of investment in that option. That is,

$$R(x_i) = f(x_i)$$
$$= \sum_{j=1}^{K_i} r_{ij} Y_{ij}$$

where r_{ij} is the incremental return obtained when the investment in option i is increased by k_{ij}. If the incremental return decreases as the level of investment increases, the utility function will be concave. In that case, we will have the following relationship:

$$r_{ij} - r_{i,j+1} \geq 0$$

Thus,

$$Y_{ij} \geq Y_{i,j+1}$$

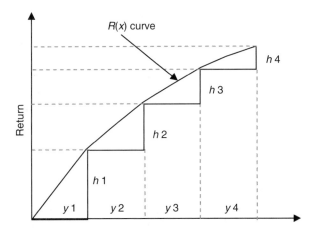

FIGURE 19.1 Utility curve for investment yield.

or

$$Y_{ij} - Y_{i,j+1} \geq 0$$

so that only the first n increments ($j = 1, 2, \ldots, n$) that produce the highest returns are used for project i. Figure 19.1 shows an example of a concave investment utility function.

If the incremental returns do not define a concave function, $f(x_i)$, then one has to introduce the inequality constraints presented above into the optimization model. Otherwise, the inequality constraints may be left out of the model, as the first inequality, $Y_{ij} \geq Y_{i,j+1}$, is always implicitly satisfied for concave functions. Our objective is to maximize the total investment return. That is,

$$\text{Maximize: } Z = \sum_i \sum_j r_{ij} Y_{ij}$$

Subject to the following constraints:

$$x_i = b_i + \sum_j k_{ij} Y_{ij} \quad \forall i$$

$$b_i \leq x_i \leq M_i \quad \forall i$$

$$Y_{ij} \geq Y_{i,j+1} \quad \forall i,j$$

$$\sum_i x_i \leq B$$

$$x_i \geq 0 \quad \forall i$$

$$Y_{ij} = 0 \text{ or } 1 \quad \forall i,j$$

Now suppose we are given four options (i.e., $N = 4$) and a budget limit of \$10 million. The respective investments and returns are shown in Tables 19.4 through 19.7.

All the values are in millions of dollars. For example, in Table 19.7, if an incremental investment of \$0.20 million from stage 2 to stage 3 is made in option 1, the expected incremental return from the project will be \$0.30 million. Thus, a total investment of \$1.20 million in option 1 will yield present worth of total return of \$1.90 million. The question addressed by the optimization model is to determine how many investment increments should be used for each option. That is, when should we stop increasing the investments

TABLE 19.4 Investment Data for Energy Option 1

Stage (j)	Incremental Investment y_{1j}	Level of Investment x_1	Incremental Return r_{1j}	Total Return $R(x_1)$
0	—	0	—	0
1	0.80	0.80	1.40	1.40
2	0.20	1.00	0.20	1.60
3	0.20	1.20	0.30	1.90
4	0.20	1.40	0.10	2.00
5	0.20	1.60	0.10	2.10

TABLE 19.5 Investment Data for Energy Option 2

Stage (j)	Incremental Investment y_{2j}	Level of Investment x_2	Incremental Return r_{2j}	Total Return $R(x_2)$
0	—	0	—	0
1	3.20	3.20	6.00	6.00
2	0.20	3.40	0.30	6.30
3	0.20	3.60	0.30	6.60
4	0.20	3.80	0.20	6.80
5	0.20	4.00	0.10	6.90
6	0.20	4.20	0.05	6.95
7	0.20	4.40	0.05	7.00

TABLE 19.6 Investment Data for Energy Option 3

Stage (j)	Incremental Investment y_{3j}	Level of Investment x_3	Incremental Return r_{3j}	Total Return $R(x_3)$
0	0	—	—	0
1	2.00	2.00	4.90	4.90
2	0.20	2.20	0.30	5.20
3	0.20	2.40	0.40	5.60
4	0.20	2.60	0.30	5.90
5	0.20	2.80	0.20	6.10
6	0.20	3.00	0.10	6.20
7	0.20	3.20	0.10	6.30
8	0.20	3.40	0.10	6.40

TABLE 19.7 Investment Data for Energy Option 4

Stage (j)	Incremental Investment y_{4j}	Level of Investment x_4	Incremental Return r_{4j}	Total Return $R(x_4)$
0	—	0	—	0
1	1.95	1.95	3.00	3.00
2	0.20	2.15	0.50	3.50
3	0.20	2.35	0.20	3.70
4	0.20	2.55	0.10	3.80
5	0.20	2.75	0.05	3.85
6	0.20	2.95	0.15	4.00
7	0.20	3.15	0.00	4.00

in a given option? Obviously, for a single option we would continue to invest as long as the incremental returns are larger than the incremental investments. However, for multiple investment options, investment interactions complicate the decision so that investment in one project cannot be independent of the other projects. The IP model of the capital rationing example was solved with LINDO software. The model is

Maximize: $Z = 1.4Y11 + .2Y12 + .3Y13 + .1Y14 + .1Y15 + 6Y21 + .3Y22 + .3Y23$
$+ .2Y24 + .1Y25 + .05Y26 + .05Y27 + 4.9Y31 + .3Y32 + .4Y33 + .3Y34$
$+ .2Y35 + .1Y36 + .1Y37 + .1Y38 + 3Y41 + .5Y42 + .2Y43 + .1Y44$
$+ .05Y45 + .15Y46$

Energy Systems

Subject to:

$.8Y11 + .2Y12 + .2Y13 + .2Y14 + .2Y15 - X1 = 0$

$3.2Y21 + .2Y22 + .2Y23 + .2Y24 + .2Y25 + .2Y26 + .2Y27 - X2 = 0$

$2.0Y31 + .2Y32 + .2Y33 + .2Y334 + .2Y35 + .2Y36 + .2Y37 + .2Y38 - X3 = 0$

$1.95Y41 + .2Y42 + .2Y43 + .2Y44 + .2Y45 + .2Y46 + .2Y47 - X4 = 0$

$X1 + X2 + X3 + X4 <= 10$

$Y12 - Y13 >= 0$

$Y13 - Y14 >= 0$

$Y14 - Y15 >= 0$

$Y22 - Y23 >= 0$

. . . . . .

$Y26 - Y27 >= 0$

$Y32 - Y33 >= 0$

$Y33 - Y34 >= 0$

$Y35 - Y36 >= 0$

$Y36 - Y37 >= 0$

$Y37 - Y38 >= 0$

$Y43 - Y44 >= 0$

$Y44 - Y45 >= 0$

$Y45 - Y46 >= 0$

$X_i >= 0 \quad$ for $i = 1, 2, \ldots, 4$

$Y_{ij} = 0, 1 \quad$ for all i and j

The solution indicates the following values for Y_{ij}.

19.6.1 Energy Option 1

Y11 = 1, Y12 = 1, Y13 = 1, Y14 = 0, Y15 = 0
Thus, the investment in option 1 is $X1 = \$1.20$ million. The corresponding return is $1.90 million.

19.6.2 Option 2

Y21 = 1, Y22 = 1, Y23 = 1, Y24 = 1, Y25 = 0, Y26 = 0, Y27 = 0
Thus, the investment in option 2 is $X2 = \$3.80$ million. The corresponding return is $6.80 million.

19.6.3 Option 3

Y31 = 1, Y32 = 1, Y33 = 1, Y34 = 1, Y35 = 0, Y36 = 0, Y37 = 0
Thus, the investment in option 3 is $X3 = \$2.60$ million. The corresponding return is $5.90 million.

19.6.4 Option 4

Y41 = 1, Y42 = 1, Y43 = 1
Thus, the investment in option 4 is $X4 = \$2.35$ million. The corresponding return is $3.70 million.

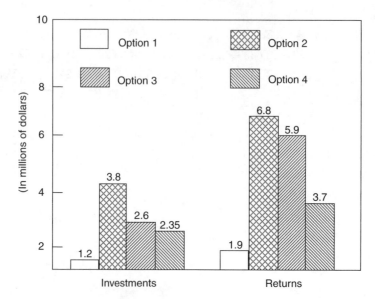

FIGURE 19.2 Comparison of investment levels.

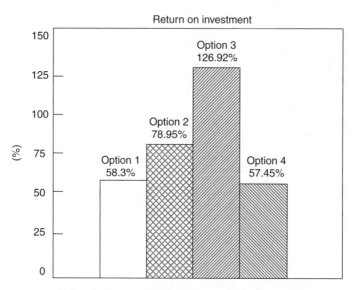

FIGURE 19.3 Histogram of returns on investments.

The total investment in all four options is $9,950,000. Thus, the optimal solution indicates that not all of the $10,000,000 available should be invested. The expected present worth of return from the total investment is $18,300,000. This translates into 83.92% return on investment. Figure 19.2 presents histograms of the investments and the returns for the four options. The individual returns on investment from the options are shown graphically in Figure 19.3.

The optimal solution indicates an unusually large return on total investment. In a practical setting, expectations may need to be scaled down to fit the realities of the investment environment. Not all optimization results will be directly applicable to real-world scenarios. Possible extensions of the above model of capital rationing include the incorporation of risk and time value of money into the solution procedure. Risk analysis would be relevant, particularly for cases where the levels of returns for the various levels of investment are not known with certainty. The incorporation of time value of money would be useful if the investment analysis is to be performed for a given planning horizon. For example, we might need to make investment decisions to cover the next 5 years rather than just the current time.

19.7 Simulation and Optimization of Distributed Energy Systems

Hudson (2005) has developed a spreadsheet-based adaptive nonlinear optimization model that utilizes the OR methods of simulation and nonlinear optimization to determine the optimal capacities of equipment in combined heat and power (CHP) applications, often called cogeneration. Evaluation of potential CHP applications requires an assessment of the operations and economics of a particular system in meeting the electric and thermal demands of a specific end-use facility. Given the electrical and thermal load behavior of a facility, the tariff structure for grid-supplied electricity, the price of primary fuel (e.g., natural gas), the operating strategy and characteristics of the CHP system, and an assumed set of installed CHP system capacities (e.g., installed capacity of prime mover and absorption chiller), one can determine the cost of such a system as compared to reliance solely on traditional, grid-supplied electricity and on-site boilers. Selecting the optimal capacities for a CHP system will help to produce optimal cost benefits and potentially avoid economic losses. The following material describes the methodology of the approach and provides an example of the results obtained.

19.8 Point-of-Use Energy Generation

Distributed energy is the provision of energy services at or near the point of use. It can take many forms, but a central element is the existence of a prime mover for generating electricity. Typical prime movers for current distributed energy applications are gas or light oil-fired turbines, fuel cells, or reciprocating engines fired with natural gas or diesel fuel. Such prime movers are only able to utilize roughly 30% of the input fuel energy in the production of electricity. The remaining energy can either be utilized as a thermal resource stream or must be exhausted to the atmosphere. When the waste heat is used to satisfy heating needs, the system is typically termed a cogeneration or combined heat and power system. Through the use of an absorption chiller, waste heat can also be utilized to provide useful cooling, in which case the system is considered a CHP application.

Generally, CHP systems are not the sole source of electricity and thermal resource for a facility. In most cases, these systems are merely alternatives to utility grid-supplied electricity, electric chillers, and electric or gas-fired on-site water heating. As a result, CHP systems are characteristic of the classic "make-or-buy" decision, and economic viability is relative to grid-based electricity and on-site boiler heating. An assessment of the economic viability of a particular CHP system requires an assumption regarding the installed equipment capacities of the system. As costs are a direct function of the installed capacities of these systems, the challenge is to determine the most economically optimal capacities of the equipment.

An important consideration in assessing the potential for CHP systems is recognition of the noncoincident behavior of the electric and thermal (i.e., heating and cooling) loads of a facility. That is, the load patterns for the three load streams are not perfectly correlated with each other through time. As a result, the peak of electrical demand will most likely not occur at the same point in time as either the heating or cooling demand peak. Absence of means to store electrical or thermal energy on-site, producing electricity with a distributed generator to track electrical demand (i.e., electric load following) may produce recovered thermal energy that cannot be used due to lack of thermal demand at that moment. Similarly, operating a CHP system in a thermal load following mode (i.e., tracking thermal demand), combined with limits on the sale of electricity into the grid, may also impact the degree to which the three demand streams can be satisfied by distributed energy.

19.9 Modeling of CHP Systems

To optimize a CHP system, the operational behavior of the system and the loads that it serves must be taken into account. One of the best ways to do this is by simulating the interaction of the system with its loads. With a properly detailed simulation, the variabilities of the load streams, the time-related prices of grid-based electricity, and the performance limitations of the CHP equipment can be recognized.

In addition to the use of the OR method of simulation, application of optimization methods is also required. An important distinction is the optimization of a system's *operation* as compared to the optimization of the system's installed *capacity*. A number of early works address the optimization of the operation of a given system. Baughman et al. (1989) developed a cogeneration simulation model in Microsoft Excel that sought optimal operation of industrial cogeneration systems over a 15-year planning horizon using the minimization of net present value of operating costs as the objective function. Consonni et al. (1989) developed an operations optimization simulator based on 36 separate sample day patterns to represent annual operations. Using a binary representation of equipment being either on or off, the model was a mixed integer linear program with an objective function of maximizing hourly profits from operation.

Regarding the optimization of installed system *capacity*, Yokoyama et al. (1991) introduced a coupled, "hierarchical" modeling concept, whereby the optimization of a system's installed capacity was an outer shell or layer serving to drive a separate inner operations optimization model based on mixed integer linear programming. Similar to Consonni, Yokoyama et al. used 36 sample day patterns to describe annual load behavior. Utilizing the hierarchical optimization process described by Yokoyama et al., Asano et al. (1992) considered the impact of time-of-use rates on optimal sizing and operations of cogeneration systems. Using 14 sample day patterns to represent the load behavior, Asano evaluated three commercial applications (hotel, hospital, and office building) and calculated optimal capacities ranging from 50% to 70% of peak electricity demand. Contemporaneously, a set of closed form equations for calculating the optimal generation capacity of an industrial cogeneration plant with stochastic input data was developed by Wong-Kcomt (1992). Wong-Kcomt's approach relied upon single unit prices for electricity (i.e., no separate demand charges) and assumed independent Gaussian distributions to describe aggregate thermal and electrical demand. The effects of hourly non-coincidence of loads were not addressed. Wong-Kcomt showed, however, that the solution space of the objective function (cost minimization) was convex in nature.

As an extension of the hierarchical model proposed by Yokoyama et al. in 1991, Gamou et al. (2002) investigated the impact that variation in end-use energy demands had on optimization results. Modeling demand (i.e., load) variation as a continuous random variable, probability distributions of electrical and thermal demands were developed. Dividing the problem into discrete elements, a piecewise LP approach was used to find the minimum cost objective function. It was observed that the capacity deemed as optimal using average data was, in fact, suboptimal when load variations were introduced. In characterizing the variability of electrical and thermal demands, the non-coincident behavior of the electrical and thermal loads led to the determination of a lower optimal capacity value when variability was recognized, relative to the value obtained when considering only average demands. A key finding from this work was that use of average demand data (e.g., sample day patterns) may not accurately determine the true optimal system capacity. Orlando (1996) states a similar conclusion in that "any averaging technique, even multiple load-duration curves, by definition, cannot fully model the interaction between thermal and electrical loads."

Within the last 10 years, cogeneration technology has evolved to include systems with smaller electric generation unit capacities in uses other than large, industrial applications. Termed "distributed energy" or "distributed generation," these systems are now being applied in commercial markets such as hospitals, hotels, schools, and retail stores. In addition, traditional cogeneration (i.e., the production of electricity and useful heat) has been expanded to include trigeneration, that is, the use of waste heat from electrical production to produce both useful heat and cooling. A number of works are focused on this recent development. Marantan (2002) developed procedures to evaluate a predefined list of candidate system capacities for an office building application, selecting the CHP system with the minimum net annual cost. Campanari et al. (2002) used a simulation model with 21 sample day patterns and a predefined list of operating scenarios to select the least cost operating strategy for a CHP system in commercial buildings. They did a somewhat reverse approach in investigating capacity-related optimization by varying the building size for a CHP system of fixed capacity. An important conclusion of their manual, trial and error optimization was that "due to the inherent large variability of heating, cooling, and electric demand typical of commercial buildings, the optimum size of a cogeneration plant is significantly lower than peak demand." A similar conclusion was found by Czachorski et al. (2002) while investigating the energy cost savings resulting from the use of CHP systems in hospitals, hotels, offices, retail, and educational facilities in the Chicago area. Through manual capacity enumeration, they found that, based on maximum annual energy cost savings, "the corresponding size of the power generator was between 60% and 80% of the maximum electric demand for CHP systems" in the applications considered. The study by Czachorski et al. also showed that annual energy cost savings exhibit a concave behavior with respect to generator capacity. While their work did not reflect life-cycle cost savings by including investment cost as a function of generator capacity, the inclusion of generation equipment capital cost should not eliminate the general concave behavior produced by the annual energy economics.

19.10 Economic Optimization Methods

Based on the modeling efforts described in the previous section, consideration is now given to the question of an appropriate method to apply in seeking an optimum of an economic objective function. A discussion of relevant optimization techniques cannot be made without some knowledge of the structure of the model in which the optimization will be conducted. Therefore, rather than providing a pedagogic recitation of the wide variety of optimization methods and algorithms that exist, this section will focus on the specific methods that are applicable to the problem at hand, which will then be further developed in the following section. There are, however, a number of good texts on optimization methods. Two examples are Bazaraa et al. (1993) and Gill et al. (1986).

As mentioned above, to determine an appropriate optimization method (i.e., to select the correct tool for the job), one must have some understanding of the system or model upon which the optimization will be applied. One approach to this selection is to consider the attributes of the system or model and proceed through somewhat of a classification process. A good initial step in the classification is to determine if the system or model is linear or nonlinear in either its objective function or its constraints. If the objective function and all constraints are linear, then linear optimization methods (e.g., linear programming) should be applied. If either the objective function or any constraint is nonlinear, then the nonlinear class of methods may be required. A further distinction is whether the independent variables are constrained. If the feasible region is defined by constraints, constrained optimization methods generally should be applied. In addition, if one or more of the independent variables can only take on integer values, specialized integer programming methods may be required.

With respect to the economic modeling of CHP systems, life-cycle cost, or alternatively, the life-cycle savings relative to some non-CHP alternative, as a function of installed equipment capacity, has been shown to be convex and concave, respectively (Wong-Kcomt, 1992; Czachorski et al., 2002). Therefore, using either life-cycle cost or savings as the objective function necessitates a nonlinear optimization approach. Beyond this, consideration must be given as to whether the current problem has independent variables that are constrained to certain sets of values (i.e., equality constraints) or somehow bounded (i.e., inequality constraints). In either case, constrained nonlinear optimization is generally performed by converting the problem in such a way that it can be solved using unconstrained methods (e.g., via Lagrangian multipliers or penalty methods) (Bazaraa et al., 1993). In this study, the independent variables are installed equipment capacities, which are assumed to be continuous and non-negative. Thus, the only constraints are simple bounds, defined as $x_i \geq 0$. Fletcher (1987) suggests a number of ways to handle such constraints, including variable transformation (e.g., $x = y^2$) and introduction of slack variables. With slack variables, a problem of the type Maximize $F(x)$ subject to $x_i \geq 0$ can be rewritten using slack variables as Maximize $F(x)$ subject to $x_i - w_i^2 = 0$, where w_i^2 is a squared slack variable. The solution can then follow through the development of the Karush–Kuhn–Tucker (KKT) conditions and the Lagrangian function. It has been shown that for linear constraints and a concave objective function, as in this study, the global optimum will be at a point satisfying the KKT conditions (Bazaraa et al., 1993; Winston, 1994).

Another method to extend equality-constraint methods to inequalities is through the use of a generalized reduced gradient (GRG) approach. The reduced gradient method seeks to reduce the number of degrees of freedom, and therefore, free variables, that a problem has by recognizing the constraints that are active (i.e., at their bounds) during each iteration. If a variable is at an active bound, it is excluded from calculations related to the determination of the incremental solution step. If no variables are at active constraints, the GRG method is very similar to the standard quasi-Newton method for unconstrained variables.

There are a number of methods available to perform unconstrained nonlinear optimization. A central distinction is whether the method relies on derivatives of the objective function. If derivatives are not available or are computationally difficult to obtain, nonderivative methods can be employed. Such methods are also needed when the objective function or gradient vector is not continuous. Methods that rely solely on function comparison are considered direct search methods (Gill et al., 1986). A common direct search method is the polytope or Nelder–Mead method in which prior functional evaluations are ordered such that the next iteration is a step in the direction away from the worst point in the current set of points. Another nonderivative method is the Hook and Jeeves method, which performs exploratory searches along each of the coordinate directions followed by pattern searches defined by the two most recent input vectors. The main disadvantage of these direct search methods is that they can be very slow to converge.

The two direct search methods mentioned above are considered sequential methods in that new trial inputs are the product of the previous result. Another class of the direct search method is the simultaneous direct search in which the trial points are defined a priori (Bazaraa et al., 1993). For variables in two dimensions, an example of this method would be a grid-pattern search, which is employed in this particular study.

If objective function derivatives are available, other, more efficient, methods can be brought to bear. One of the most fundamental procedures for optimizing a differentiable function is the method of steepest descent, also called the gradient method. In this method, the search direction is always the negative gradient, and the step size is calculated to minimize the objective function (assuming the function is convex). This is repeated until a stopping criterion, such as the gradient norm, is sufficiently small. However, it has been shown that following the direction of steepest descent does not necessarily produce rapid convergence, particularly near a stationary point, and that other derivative methods perform better (Bazaraa et al., 1993; Bartholomew-Biggs, 2005). For large problems (i.e., those with more than 100 decision variables), the conjugate gradient method is useful as it does not require storage of large matrices (Bazaraa et al., 1993). As this method is typically less efficient and less robust than other methods, and as the current problem concerns a small number of independent variables, the conjugate gradient method was not used for this application.

The remaining methods of interest are the Newton method and the related quasi-Newton method. The Newton method has been shown to be very efficient at unconstrained nonlinear optimization if the objective function has continuous first and second derivatives. If first and second derivatives are available, a Taylor-series expansion in the first three terms of the objective function yields a quadratic model of objective function that can subsequently be used to define a Newton direction for function minimization. As long as the Hessian is positive definite and the initial input values are in the neighborhood of the optimum, Newton's method converges to the optimum quadratically (Gill et al., 1986).

Due to the discrete form of the model in this study, analytical expressions for first and second derivatives are not available. In these situations, derivatives can be approximated using finite difference techniques. The lack of an exact expression for second derivatives means that curvature information, typically provided by calculating the Hessian matrix, is not directly available for use in a Newton method. The solution to this problem is to utilize the well-known quasi-Newton method, in which an approximation to the inverse Hessian is successively built-up during the iteration process. While typically expecting the first derivative to be analytically available in a quasi-Newton method, the additional lack of explicit first derivatives to form the gradient does not appear to be a fatal impediment. Van der Lee et al. (2001) successfully used this approach in studying the optimization of thermodynamic efficiency in power plant steam cycles. As Gill et al. (1986) state, "when properly implemented, finite-difference quasi-Newton methods are extremely efficient, and display the same robustness and rapid convergence as their counterparts with exact gradients."

With respect to the iterative update of the Hessian matrix, a number of Hessian update methods have been proposed over the years, including the Davidson–Fletcher–Powell (DFP) method, the Powell–Symmetic–Broyden (PSB) update, and the Broyden–Fletcher–Goldfarb–Shanno (BFGS) method. The literature indicates that the BFGS method is clearly accepted as the most effective update method currently available (Gill et al., 1986; Nocedal, 1992; Zhang and Xu, 2001; Bertsekas, 2004; Yongyou et al., 2004). Details of the BFGS algorithm will be provided in the following section.

A final element related to the quasi-Newton method is the use of line search methods when the full quasi-Newton step produces an objective function response that does not make satisfactory progress relative to the previous iteration, thus possibly indicating the passing of a local optimum. In that case, a "backtracking" process along the step direction is needed. As discussed by Dennis and Schnabel (1983), the backtracking approach should conform to the Armijo and Goldstein (AG) conditions to ensure satisfactory convergence. Dennis and Schnabel provide the classic quadratic fit using three previously calculated function values to solve for the optimum quasi-Newton step multiplier, followed by the cubic spline fit, should the new quadratic step not meet AG conditions.

It should be noted that the quadratic/cubic fit method is but one method to determine an appropriate step value. While less efficient in terms of computational requirements, line search methods that do not rely on the gradient, or in this case, an approximation to the gradient, can also be used. Sequential search methods such as the Fibonacci and related Golden section methods can be utilized to determine an acceptable step multiplier (Bazaraa et al., 1993).

19.11 Design of a Model for Optimization of CHP System Capacities

This section provides a detailed explanation of the simulation model as well as the approach used to determine an optimum set of equipment capacities for a CHP system. Similar to the approach used by Edirisinghe et al. (2000) and Yokoyama et al. (1991), the model consists of two nested sections: an outer, controlling optimization algorithm and an inner operation simulation routine. The overall flow of the optimization model is shown in Figure 19.4. Starting with an initial "guess" for the installed electrical generator and absorption chiller capacities, an hour-by-hour operation simulation is performed to develop a value of the objective function for the given generator and chiller capacities. Within the optimization algorithm, a stopping criterion is used to control the updating of the optimization routine and subsequent iterative looping back to the operation simulation with a new set of candidate installed capacities. The optimization algorithm seeks to maximize the net present value (NPV) savings produced by using the CHP system relative to a non-CHP scenario (where electricity is obtained solely from the grid and heating loads are met by an on-site boiler). The maximization of NPV savings (i.e., maximization of overall profitability) is an appropriate method for evaluating mutually exclusive alternatives (Sullivan et al., 2006).

In recognition of the problems identified earlier regarding the use of average or aggregated demand data (Gamou et al., 2002; Hudson and Badiru, 2004), this approach utilizes demand data expressed on an hourly basis, spanning a 1-year period. Use of hourly data has the advantage of explicitly capturing the seasonal and diurnal variations, as well as non-coincident behaviors, of electrical and thermal loads for a given application. In many cases, actual hourly demand data for an entire year may not be available for a specific site.

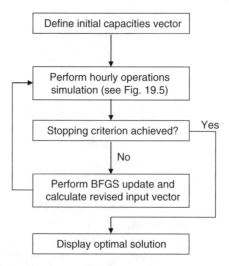

FIGURE 19.4 Overview flow chart for optimization model.

TABLE 19.8 Input Variables Used in Operation Simulation Model

Variable	Typical Units
Facility loads	
Hourly electrical demand (non-cooling related)	kW
Hourly heating demand	Btu/h
Hourly cooling demand	Btu/h
Electric utility prices	
Demand charge	$/kW-month
Energy charge	$/kWh
Standby charge	$/kW-month
On-site fuel price (LHV basis)	$/MMBtu
Equipment parameters	
Boiler efficiency (LHV)	Percent
Conventional chiller COP	Without units
Absorption chiller (AC) COP	Without units
Absorption chiller (AC) capacity	RT
AC minimum output level	Percent
AC system parasitic electrical load	kW/RT
Distributed generation (DG) capacity, net	kW
DG electric efficiency (LHV) at full output	Percent
DG minimum output level	Percent
DG power/heat ratio	Without units
Operating and maintenance (O&M) cost	$/kWh
Number of DG units	Units
DG capital cost	$/kW installed
AC capital cost	$/RT installed
General economic parameters	
Planning horizon	Years
Discount rate	Percent/year
Effective income tax rate	Percent

In these situations, building energy simulation programs, such as Building Energy Analyzer or BCHP Screening Tool, are available that can develop projected hourly loads for electricity, heating, and cooling on the basis of building application, size, location, and building design attributes (e.g., dimensions, insulation amounts, glazing treatments) (InterEnergy/GTI, 2005; Oak Ridge National Laboratory, 2005).

The data needed to simulate the operation of a CHP system are shown in Table 19.8. The input for the hourly facility electrical demand should include all facility electrical demand *except* for cooling-related demand. As cooling may be provided by an absorption chiller under CHP operation, electrical demand related to cooling is calculated explicitly within the simulation model. For the hourly heating and cooling demands, the input values are expressed on an end-use, as-consumed thermal basis.

The prices for utility-supplied electricity typically have a price component related to the amount of energy consumed (i.e., an energy charge) as well as a component proportional to the monthly peak rate of energy consumed (i.e., a demand charge). Some utilities will price their electricity at different rates to those who self-generate a portion of their electrical needs. In addition, some electric utilities charge a monthly standby fee for the availability of power that may be called upon should the distributed generation not be available. Utility tariff structures can also have unit prices that vary both seasonally and diurnally. Similar to electricity rates, the unit price for on-site fuel may be different for those who operate a CHP system.

The fuel assumed for on-site distributed generation and on-site water/steam heating in this study is natural gas, expressed on a $/million Btu (MMBtu) lower heating value (LHV) basis. The heating value of natural gas refers to the thermal energy content in the fuel, which can be expressed on a higher heating value (HHV) or lower heating value basis. The difference in the two heating values relates to the water formed as a product of combustion. The higher heating or gross value includes the latent heat of vaporization of the water vapor.

The lower heating or net value excludes the heat that would be released if the water vapor in the combustion products were condensed to a liquid. As DG/CHP systems try to limit exhaust vapor condensation due to corrosion effects, the usable heat from natural gas is typically the LHV. In the United States, natural gas is typically priced on a HHV basis, so care should be used in entering the proper value. For natural gas, the conversion between HHV and LHV is

$$\text{heat content}_{HHV} = \text{heat content}_{LHV} \times 1.11 \text{ (Petchers, 2003)}$$

The definitions for the equipment and economic parameters listed in Table 19.8 are as follows:

Boiler efficiency—The thermal efficiency of the assumed on-site source of thermal hot water/steam (e.g., boiler) for the baseline (non-CHP) scenario, expressed on an LHV basis.

Conventional chiller COP—The coefficient of performance (COP) for a conventional electricity-driven chiller. It is determined by dividing the useful cooling output by the electrical energy required to produce the cooling, adjusted to consistent units.

Absorption chiller COP—The coefficient of performance for the CHP system absorption chiller. It is determined by dividing the useful cooling output by the thermal energy required to produce the cooling, adjusted to consistent units. Parasitic electrical support loads (e.g., pump and fan loads) are addressed separately.

Absorption chiller capacity—The installed capacity of the absorption chiller in refrigeration tons (RT). This is an independent variable in the model.

AC minimum output level—The minimum percent operating level, relative to full output, for the absorption chiller. This is also known as the minimum turndown value.

AC system parasitic electrical load—The electrical load required to support the absorption chiller. The chiller load should include the chiller solution pump, the AC cooling water pump, and any cooling tower or induced draft fan loads related to the AC.

Distributed generation (DG) capacity—The installed capacity of the distributed electrical generator (i.e., prime mover), expressed in net kilowatts. This is an independent variable in the model.

DG electric efficiency (LHV) at full output—The electricity production efficiency of the DG prime mover at full output. This efficiency can be determined by dividing the electricity produced at full output by the fuel used on a LHV basis, adjusted to consistent units.

DG minimum output level—The minimum percent operating level, relative to full output, for the DG unit. Also known as the minimum economic turndown value.

DG power/heat ratio—The ratio of net electrical power produced to useful thermal energy available from waste heat, adjusted to consistent units.

O&M cost—The operating and maintenance cost of the total cooling, heating, and power system, expressed on a $/kWh of electricity generated basis.

Number of DG units—The number of prime mover units comprising the system. Currently, the model is limited to no more than two units, each identical in size and performance. The optimum capacity determined by the model is the total capacity of the CHP system, and for a two-unit system, that capacity is split equally between the units.

DG capital cost—The fully installed capital cost of the distributed generation system, expressed on a \$/net kW basis.

AC capital cost—The fully installed capital cost of the absorption chiller system, expressed on a \$/RT basis.

Planning horizon—The assumed economic operating life of the CHP system. The default value is 16 years to be consistent with U.S. tax depreciation schedules for a 15-year property. Currently, 16 years is the maximum allowed planning horizon in the model.

Discount rate—The rate used to discount cash flows with respect to the time-value of money.

Effective income tax rate—The income tax rate used in income tax-related calculations such as depreciation and expense deductions. The effective rate reflects any relevant state income tax and its deductibility from federal taxes.

The general flow of calculations within the operation simulation is shown in Figure 19.5. Once the electrical and thermal loads and general equipment/economic parameters are defined for each iteration of the optimization routine, a trial set of distributed generator and absorption chiller capacities are provided to the operations simulator. Two separate simulations must be performed. First, the hour-by-hour costs for satisfying the thermal and electric loads solely by a traditional utility grid/on-site boiler arrangement must be

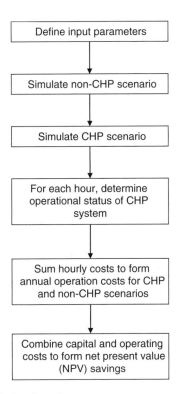

FIGURE 19.5 Operation simulation flow chart.

calculated. This is referred to as the non-CHP or grid-only scenario. In the non-CHP scenario, grid-supplied electricity is assumed to support all facility loads, including cooling but excluding heating. Heating is assumed to be provided by an on-site fossil-fired boiler. A second, separate calculation develops the hour-by-hour costs of meeting at least some part of the specified loads with a CHP system. The degree of contribution of the CHP system is determined hourly by the relative cost of using the CHP system (i.e., making) to satisfy the loads versus the traditional grid-purchase/on-site boiler operation (i.e., buying). If the operation of the CHP system is less expensive in a given hour than the grid/boiler approach, the CHP system will supply the loads, with the grid/boiler providing any supplemental energy needed to satisfy that hour's load. If the operation of the CHP system is more expensive, loads are satisfied in the traditional grid-purchase manner for that hour. As the price of grid-based electricity can change depending upon the time-of-day, this test is performed for each hour of the year. With typical time-of-day rates, the CHP system generally operates during the workday hours and is offline during the night.

Relative to the non-CHP scenario, developing the annual cost for a CHP-based system is substantially more complicated. There can be utility surcharges (e.g., standby fees) that are imposed as a result of operating self-generation equipment. In addition, the unit pricing for electricity may be different for customers using a CHP system than for those buying all their supply solely from the utility. The operational considerations related to the CHP system are of considerable influence as well. As an example, the fuel efficiency of electrical generation equipment is directly proportional to the relative output level. Typically, the highest efficiency (i.e., most electricity produced for the least fuel consumed) is at or near full-rated output. Depending upon the type of prime mover, electrical efficiencies at low part-load can be 65%–75% of full-load efficiency. As a result, there is a general lower limit on part-load operations. A typical minimum operating value is 50% of rated unit capacity. The limit becomes influential when the electrical demand is less than 50% of the rated unit capacity, requiring that electricity be purchased from the grid. Thus, there is an economic trade-off related to the size of the CHP generation capacity. A CHP system sized to meet peak electrical or thermal loads will incur higher utility standby charges and will have less ability to operate during periods of low demand. Conversely, a smaller sized system may be able to operate a larger fraction of time, but may result in a higher fraction of unmet load for the facility (resulting in higher utility purchases, typically at peak pricing). The economics are further influenced by the direct relationship of CHP electrical generation capacity and useful thermal energy available. Smaller electrical capacity means less useful thermal byproduct, which might then require more supplemental gas-boiler or electric chiller operation.

At the CHP/non-CHP scenario level, two sets of annual operating costs are then determined by summing the relevant hourly costs of meeting thermal and electric demands from either the grid and on-site boiler solely (i.e., the non-CHP scenario) or from CHP operations. A differential annual operating cost (or net annual savings, if the CHP scenario is less costly than the non-CHP scenario) is determined based on the annual cost difference between the non-CHP scenario and the CHP-available scenario. A net present value is then determined by calculating the present worth of the net annual savings over the number of years defined by the planning horizon at the defined discount rate and adding the installed capital costs of the CHP system, adjusted for income tax effects (e.g., depreciation).

As stated earlier, the planning horizon for this model can be up to 16 years. Unit prices for electricity and gas, as well as O&M unit costs, are not likely to remain constant over such a long period. Similarly, it is possible that electrical or thermal loads may change over such a period. As a result, the ability to reflect escalating unit prices and possible load changes is needed. As annual operations are calculated on an hourly basis, performing an explicit calculation for every hour within a 16-year duration would require 140,160 hourly calculations. It was felt that explicitly performing 140,160 hourly calculations would be intractable in the Excel platform. A solution to this dilemma was to express the variables that are subject to escalation as levelized values. A common method used in public utility economic analysis, a levelized value is determined by calculating the present value of the escalating annual stream and then applying an annual capital recovery factor to produce the levelized annual equivalent value (Park and Sharp-Bette, 1990). The levelized values are then used in the operation simulation calculations. Thus, in the material that follows, unless explicitly stated, values for electricity and gas unit prices, thermal and electric loads, unit O&M costs, and the resulting annual costs should be considered annual levelized values, spanning the duration of the planning horizon.

19.12 Capacity Optimization

As mentioned, the optimization goal is to maximize NPV cost savings by determining the optimum installed capacities for the electricity generation system and the absorption chiller. Given that only objective function values are directly available in this computational model (i.e., no analytical expressions for first or second derivatives), it is felt that, based on a review of current literature, the use of a quasi-Newton method with BFGS updates of the inverse Hessian is the most appropriate approach.

The quasi-Newton method is a variant of the Newton method and can be found in any good nonlinear optimization textbook (Gill et al., 1986; Fletcher, 1987; Bazaraa et al., 1993; Bertsekas, 2004; Bartholomew-Biggs, 2005). The Newton method relies on a three-term Taylor approximation of an assumed quadratic behavior of the objective function. As such, the quadratic model of the objective function, F, can be expressed as

$$F(x_k + p) \approx F_k + g_k^T p + \frac{1}{2} p^T G_k p$$

where g, p, and G are the gradient (Jacobian) in x, step direction, and Hessian in x, respectively. As we seek to find a stationary point of the function with respect to the step direction p, the objective function can be rewritten in p as

$$F(p) = g_k^T p + \frac{1}{2} p^T G_k p$$

A necessary condition for a stationary point is that the gradient vector vanishes at that point. Thus,

$$\nabla F(p) = g_k + G_k p = 0 \quad \text{or} \quad p = -G_k^{-1} g_k$$

If G is positive definite, then conditions are sufficient to state that p can be a minimum stationary point (Gill et al., 1986). In the case of maximization, G should be negative definite. The Newton method requires, however, that the Hessian of the objective function be known or determinable. In the current problem, the Hessian cannot be determined analytically. Thus, we rely on a sequential approximation to the Hessian as defined by the quasi-Newton method.

The typical quasi-Newton method assumes that the gradient of the objective function is available. The model used in this study has no analytic representation of either first or second derivatives. In this situation, a forward-difference approximation must be used to estimate the gradient vector. For the ith independent variable, x_i, the gradient is estimated by

$$g_i = \frac{1}{h}(F(x_i + h) - F(x_i))$$

where h is the finite-difference interval. For this study, a finite-difference interval of 10^{-4} was selected after evaluating choices ranging from 10^{-2} to 10^{-6}.

The general outline of the quasi-Newton method for maximization (Bartholomew-Biggs, 2005) is as follows:

- Choose some x_o as an initial estimate of the maximum of $F(x)$.
- Set the initial inverse Hessian, H_0, equal to the negative identity matrix (an arbitrary symmetric negative definite matrix).
- Repeat for $k = 0, 1, 2, \ldots$.
  - Determine $g_k = \nabla F(x_k)$ by forward-difference approximation.
  - Set the step length scalar, λ, equal to 1.
  - Calculate the full step direction $p_k = -H_k g_k$.
  - Evaluate whether the full step is appropriate by comparing $F(x_k + \lambda p_k)$ to $F(x_k)$. If $F(x_k + \lambda p_k) < F(x_k) + \rho \lambda g_k^T p_k$, solve for the step length λ that produces a univariate maximum $F(\lambda)$ for $0 \leq \lambda \leq 1$.
  - Set $x_{k+1} = x_k + \lambda p_k$, $y_k = g_{k+1} - g_k$, $d_k = x_{k+1} - x_k$.
  - Evaluate stopping criteria, and if not achieved,
  - Update the approximate inverse Hessian such that $H_{k+1} y_k = d_k$.
  - Increment k.

The stopping criteria used in this model are consistent with prior work by Edirisinghe et al. (2000) and Kao et al. (1997), in which the algorithm is terminated when either the change (i.e., improvement) in the objective function is less than a prescribed threshold amount or when the gradients of the objective function at a particular input vector are zero. The setting of the termination threshold value is a matter of engineering judgment. If a value is chosen that requires very small changes in the objective function before termination, the algorithm can cycle for a large number of iterations with very little overall improvement in the objective function. Conversely, a more relaxed threshold value can terminate the optimization algorithm prematurely, producing a suboptimal solution. A balance must therefore be struck between long execution times and less than total maximization of the objective function. As the objective function in this study is NPV cost savings over a multiyear period, one must select a value at which iterative improvements in NPV cost savings are considered negligible. There are two approaches used in setting this termination threshold. First, on an absolute basis, if the iterative improvement of the NPV cost savings is less than $50.00, it is considered reasonable to terminate the algorithm. In some cases, however, this absolute value can be a very small percentage of the overall savings, thus leading to long execution times with little relative gain. The second termination approach is based upon a relative measure on the objective function. If the change in NPV cost savings between iterations is greater than $50.00, but less than 0.00001 times the objective function value, then the algorithm terminates under the assumption that a change of less than 0.001% is insignificant.

In some situations, the objective function can exhibit multiple local optima of low magnitude relative to the average value within a neighborhood around the stationary point (i.e., low-level noise of the objective function). When this occurs, a means to help avoid getting "trapped" in a *near* optimum response space, particularly when the response surface is relatively flat, is to require two or three consecutive iterative achievements of the stopping criterion (Kim, D., personal communication, 2005). For this study, two consecutive achievements of the stopping criterion detailed in the previous paragraph were required to end the optimization process. In some cases with multiple local optima, the model may find a local optimum rather than the global optimum. A useful technique to improve the solution is to try different starting points for the optimization (Fylstra et al., 1998).

The updating of the matrix H, representing a sequential approximation of the inverse Hessian, is done using the BFGS method. As mentioned earlier, the BFGS update method is clearly considered to be the most efficient and robust approach available at this time. The BFGS formula for H_{k+1}, as presented by Zhang and Xu (2001) and Bartholomew-Biggs (2005), is:

$$H_{k+1} = H_k - \frac{H_k y_k d_k^T + d_k y_k^T H_k}{d_k^T y_k} + \left[1 + \frac{y_k^T H_k y_k}{d_k^T y_k}\right] \frac{d_k d_k^T}{d_k^T y_k}$$

There are a number of methods that can be employed in the backtracking search for the Newton step length λ that produces a maximum in the objective function. In this study, a quadratic and cubic spline fit was evaluated, but the method was not stable under some input conditions or required a large number of iterations before reaching the stopping criterion. This appears to be due to the lack of strict concavity of the objective function. As a result, the Golden sequential line search method was selected for its accuracy and stability. The Golden search was terminated when the interval of uncertainty (IOU) for the step length λ became less than 0.025. It should be noted that the step length can be unique to each variable rather than being a single scalar value. Such an approach was explored, but the additional computations did not seem to produce sufficiently improved results (i.e., faster optimization) to merit incorporating the approach in the final model.

To provide visual guidance regarding the surface behavior of the objective function within the overall solution space, a simultaneous uniform line search method was utilized as well. Using a 21×7 (electric generator capacity $\times$ absorption chiller capacity) grid, grid step sizes were selected to evaluate the complete range of possible CHP equipment capacities (i.e., $0 \leq \text{size} \leq \text{max load}$) for both the distributed generator and the absorption chiller. For each of the 147 cells, the capacity combination was used as an explicit input to the simulation model to determine the corresponding NPV cost savings. A contour plot of the NPV cost savings was produced to graphically display the overall solution space.

As mentioned earlier, there are simple lower bound constraints that require the capacities of the distributed generator and absorption chiller to be greater than or equal to zero. In an unconstrained method, it is possible that the direction vector could propose a solution that would violate the lower bound. This model checks for this condition, and if present, sets the capacity value to zero. As an added element to improving the efficiency of the algorithm, if the capacity of the distributed generation is set to zero, the capacity of the absorption chiller is also set to zero, as DG capacity is the energy source to operate the absorption chiller. This approach does not violate the quasi-Newton method as the effect of zeroing the capacity when a negative capacity is suggested is equivalent to reducing the Newton step size for that iteration. In this situation, the new x_{k+1} point is set to zero, and gradients are calculated at the new input vector for use in the quasi-Newton algorithm. Should the economic conditions of the problem be such that the maximum objective function (given the lower bound constraints) is truly at zero capacity for the distributed generator, the next iteration will yield the same adjusted input vector (owing to a direction vector pointing into the negative capacity space) and same NPV cost savings, which will appropriately terminate the optimization on the basis of similar NPV results, as discussed above.

19.13 Implementation of the Computer Model

To provide useful transparency of the calculations, the methods were implemented using Microsoft Excel. Excel spreadsheets allow others to view the computational formulae, which enhances understanding and confidence in the modeling approach. In addition, Microsoft Excel is a ubiquitous platform found on most personal computer (PC) systems. The model in this study, named the CHP Capacity Optimizer, was developed using Microsoft Office Excel 2003 on a PC running the Microsoft Windows XP Professional operating system (version 2002). The model makes use of Excel's Visual Basic for Applications (VBA) macro language to control movement to various sheets within the overall spreadsheet file and to initiate the optimization procedure. The Excel model and User's Manual are available from the author or can be downloaded from the Internet (search "CHP capacity optimizer").

19.13.1 Example Calculation

As an example of the use of the optimization tool, the potential use of CHP at a hospital in Boston, Massachusetts, will be evaluated. The hospital consists of five stories with a total floor area of 500,000 square feet. The maximum electrical load is 2275 kW; the maximum heating load is 17 million Btu/h; and the maximum cooling load is 808 refrigeration tons (RT). Hourly electrical and thermal demands for the facility were obtained using the building simulator program, Building Energy Analyzer (InterEnergy/GTI, 2005). Grid-based electricity prices were based on the Boston Edison T-2 time-of-use tariff. The price of natural gas in the initial year of operation was assumed to be $11.00/million Btu. Escalation assumptions, expressed in percent change from the previous year, for this example are provided in Table 19.9.

Other data needed to calculate the optimum capacity relate to equipment cost and performance and general modeling behavior (e.g., discount rate, planning horizon). The data assumed for the hospital in Boston are shown in Table 19.10.

TABLE 19.9 Sample Escalation Assumptions

Year	Fuel Price (%)	Elec Price (%)	O&M Cost (%)	Heat Load (%)	Cool Load (%)	Elec Load (%)
2	−0.5	0.5	0.5	0.0	0.0	0.0
3	0.0	1.0	0.5	0.0	0.0	0.0
4	0.0	1.0	0.5	0.0	0.0	0.0
5	0.0	1.0	0.5	0.0	0.0	0.0
6	0.0	1.0	0.5	0.0	0.0	0.0
7	0.5	0.5	0.5	0.0	0.0	0.0
8	0.5	0.5	1.0	0.0	0.0	0.0
9	0.5	0.5	1.0	0.0	0.0	0.0
10	0.5	0.5	1.0	0.0	0.0	0.0
11	0.5	0.5	1.0	0.0	0.0	0.0
12	1.0	1.0	1.0	0.0	0.0	0.0
13	1.0	1.0	1.0	0.0	0.0	0.0
14	1.0	1.0	2.0	0.0	0.0	0.0
15	1.0	1.0	2.0	0.0	0.0	0.0
16	1.0	1.0	2.0	0.0	0.0	0.0

TABLE 19.10 General Data for the Boston Hospital Case

On-site boiler efficiency	82.0%
Conventional chiller COP	3.54
DG electric efficiency (full output)	29.0%
DG unit minimum output	50%
Absorption chiller COP	0.70
Absorption chiller min. output	25%
Abs chiller sys elec req (kW/RT)	0.20
CHP O&M cost ($/kWh)	0.011
DG power/heat ratio	0.65
Number of DG units	1
Type of prime mover	Recip
Discount rate	8.0%
Effective income tax rate	38.0%
DG capital cost ($/net kW installed)	1500
AC capital cost ($RT installed)	1000
Planning horizon (years)	16

Demands	Electricity	Heating	Cooling
Annual	12,406,742 kWh	37,074 MMBtu	1,617,306 RT-hr
Maximum	2275 kW	17.0 MMBtu/hr	808 RT
Minimum	934 kW	0.51 MMBtu/hr	0 RT

Installed DG capacity:	1130.1	kW (net)
Installed AC capacity:	210.5	RT
Installed capital cost:	$1,905,607	
Hours of DG operation	6,717	hours/year
DG generated electricity	7,422,145	kWh/year
DG supplied heating	27,839	MMBtu/year
AC supplied cooling	535,793	RT-hr/year

Annual costs (before tax)	With CHP	No CHP
CHP system	$1,056,847	$0
Utility elec	$661,305	$1,785,547
Non-CHP fuel	$125,137	$502,367
Total	$1,843,290	$2,287,913

Annual operating savings (after tax):	$275,666
NPV savings:	$954,175

Optimum DG capacity:	1130.1 kW
Optimum AC capacity:	210.5 RT
NPV savings:	$954,175

FIGURE 19.6 Optimization results for a Boston hospital.

The numeric results of the optimization are shown in Figure 19.6. At the top of the figure is a summary of the electric and thermal loads, as estimated by the building simulation program mentioned above. The optimal capacities for a reciprocating engine prime mover and an absorption chiller are 1130 kW and 210.5 RT, respectively. As shown, the CHP system operates for 6717 h each year, producing 60% of the total electricity and over 75% of the total heating required by the facility. Owing to the relative economics of gas and electricity, it is preferable that waste heat first go to satisfying heating demands before contributing to cooling demands. As a result, only one-third of the total cooling demand is provided by the CHP system. The levelized total annual operating cost savings from the CHP system is $275,666/year. The resulting NPV cost savings, including capital investment, over the 16-year planning horizon is $954,175.

A summary of the operating frequency by hour of the day is provided in Figure 19.7. It can be observed from the figure that the frequency of the CHP system operation is influenced

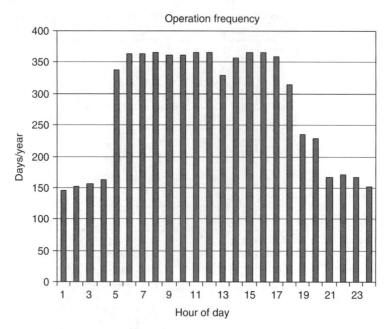

FIGURE 19.7 Operation frequency by time of day.

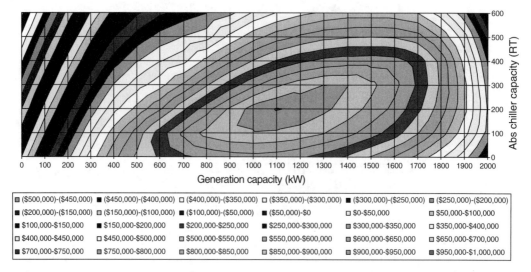

FIGURE 19.8 Optimization contour plot for a Boston hospital.

heavily by the time-of-use electricity rates, which are higher during the normal workday hours. Higher grid-based electricity rates increase the likelihood that the CHP system will be the less expensive alternative, and therefore the one to be selected, during those hours.

A contour plot of the objective function surface for the hospital in Boston is shown in Figure 19.8. Each iso-savings line represents a $50,000 increment. As shown, the surface behavior near the optimum is relatively flat. From a practical standpoint, this is a serendipitous result. Equipment capacities are offered in discrete sizes, and having a flat objective function surface in the neighborhood of the optimum gives a degree of flexibility in matching the calculated optimum set of capacities to near-values consistent with manufactured equipment sizes. As an example, vendors currently offer a reciprocating engine

TABLE 19.11 Summary Results

City	Parameter	Hospital	Hotel	Retail Store	Supermarket
Boston	Peak elec load (kW)	2275	1185	337	260
	Peak chiller load (RT)	808	492	147	40
	Optimal DG capacity (kW)	1130.1	417.2	118.3	117.7
	Optimal AC capacity (RT)	210.5	88.3	35.9	6.9
	NPV savings	$954,175	$341,877	$57,945	$67,611
	DG capacity % of peak load	50	35	35	45
	AC capacity % of peak load	26	18	24	17
San Francisco	Peak elec load (kW)	1949	1004	433	248
	Peak chiller load (RT)	452	334	131	17
	Optimal DG capacity (kW)	513.8	257.2	63.9	54.6
	Optimal AC capacity (RT)	30.9	36.2	19.1	5.0
	NPV savings	$431,123	$212,726	$15,954	$42,666
	DG capacity % of peak load	26	26	15	22
	AC capacity % of peak load	7	11	15	29

prime mover at 1100 kW and an absorption chiller at 210 RT. Manually substituting these capacities into the model produces a NPV cost savings of $951,861, which is a negligible difference relative to the value determined for the optimum capacity. Not all cases may have such a close match, but the flat gradient of the objective function near the optimum provides a reasonably wide range for matching actual equipment.

19.14 Other Scenarios

Due to the variation in fuel and electricity prices, site weather, and electrical and thermal loads of a specific facility, each potential CHP application will have a unique optimal solution that maximizes economic benefit. To illustrate, similar optimization runs were made for hotel, retail store, and supermarket applications in Boston as well as in San Francisco. It is instructive to consider the results of these eight cases together, looking in particular at the percent of peak load that the DG and AC optimum capacities represent. As shown in Table 19.11, the percent of peak load represented by the optimum capacities varies tremendously both by application and location. Thus, to maximize the economic benefit of a CHP system, each application should be individually evaluated by the methods described in this chapter.

References

1. Asano, H., S. Sagai, E. Imamura, K. Ito, and R. Yokoyama (1992). "Impacts of Time-of-Use Rates on the Optimal Sizing and Operation of Cogeneration Systems." *Power Systems, IEEE Transactions* **7**(4): 1444.
2. Badiru, A. B. (1982). "An Illustrative Example of the Application of Linear Programming to Energy Management," presented at the *4th Annual Conference on Computers & Industrial Engineering*, Orlando, Florida, March 1982.
3. Badiru, A. B. and P. S. Pulat (1995). *Comprehensive Project Management: Integrating Optimization Models, Management Principles, and Computers.* Prentice-Hall, Englewood Cliffs, NJ.
4. Bartholomew-Biggs, M. (2005). *Nonlinear Optimization with Financial Applications.* Kluwer, Boston.
5. Baughman, M. L., N. A. Eisner, and P. S. Merrill (1989). "Optimizing Combined Cogeneration and Thermal Storage Systems: An Engineering Economics Approach." *Power Systems, IEEE Transactions* **4**(3): 974.

6. Bazaraa, M. S., H. D. Sherali, and C. M. Shetty (1993). *Nonlinear Programming: Theory and Algorithms.* Wiley, New York.
7. Bertsekas, D. P. (2004). *Nonlinear Programming.* Athena Scientific, Belmont, MA.
8. Campanari, S., L. Boncompagni, and E. Macchi (2002). *GT-2002-30417 Microturbines and Trigeneration: Optimization Strategies and Multiple Engine Configuration Effects*, v. 4A & 4B, ASME Turbo Expo 2002, Amsterdam, the Netherlands. ASME, 2002.
9. Consonni, S., G. Lozza, and E. Macchi (1989). *Optimization of Cogeneration Systems Operation. Part B. Solution Algorithm and Examples of Optimum Operating Strategies.* ASME, New York.
10. Czachorski, M., W. Ryan, and J. Kelly (2002). *Building Load Profiles and Optimal CHP Systems.* American Society of Heating, Refrigerating and Air-Conditioning Engineers, Honolulu, HI.
11. Dennis, J. E. and R. B. Schnabel (1983). *Numerical Methods for Unconstrained Optimization and Nonlinear Equations.* Prentice-Hall, Englewood Cliffs, NJ.
12. Edirisinghe, N. C. P., E. I. Patterson, and N. Saadouli (2000). "Capacity Planning Model for a Multipurpose Water Reservoir with Target-Priority Operation." *Annals of Operations Research* **100**: 273–303.
13. Fletcher, R. (1987). *Practical Methods of Optimization.* Wiley, Chichester, UK.
14. Fylstra, D., L. Lasdon, J. Watson, and A. Waren (1998). "Design and Use of the Microsoft Excel Solver." *Informs Interfaces* **28**(5): 29–55.
15. Gamou, S., R. Yokoyama, and K. Ito (2002). "Optimal Unit Sizing of Cogeneration Systems in Consideration of Uncertain Energy Demands as Continuous Random Variables." *Energy Conversion and Management* **43**(9–12): 1349.
16. Gill, P. E., W. Murray, and M. H. Wright (1986). *Practical Optimization.* Elsevier Academic Press.
17. Hudson, Carl R. (2005). "Adaptive Nonlinear Optimization Methodology for Installed Capacity Decisions in Distributed Energy/Cooling Heat and Power Applications." http://etd.utk.edu/2005/HudsonCarlRandolph.pdf, Knoxville, The University of Tennessee. PhD: 133.
18. Hudson, C. R. and A. B. Badiru (2004). "Use of Time-Aggregated Data in Economic Screening Analyses of Combined Heat and Power Systems." *Cogeneration and Distributed Generation Journal* **19**(3): 7.
19. InterEnergy/GTI (2005). Building Energy Analyzer. http://www.interenergysoftware.com/BEA/BEAPROAbout.htm. Chicago, IL.
20. Kao, C., W. T. Song, and S.-P. Chen (1997). "A Modified Quasi-Newton Method for Optimization in Simulation." *International Transactions in Operational Research* **4**(3): 223.
21. Kruger, P. (2006). *Alternative Energy Resources: The Quest for Sustainable Energy.* John Wiley & Sons, New York.
22. Marantan, A. P. (2002). "Optimization of Integrated Microturbine and Absorption Chiller Systems in CHP for Buildings Applications." *DAI* **64**: 112.
23. Nocedal, J. (1992). "Theory of Algorithms for Unconstrained Optimization." *Acta Numerica*, 1992: 199–242.
24. Oak Ridge National Laboratory (2005). *BCHP Screening Tool.* Oak Ridge, TN.
25. Orlando, J. A. (1996). *Cogeneration Design Guide.* American Society of Heating, Refrigerating and Air-Conditioning Engineers, Atlanta, GA.
26. Park, C. S. and G. P. Sharp-Bette (1990). *Advanced Engineering Economics.* Wiley, New York.
27. Petchers, N. (2003). *Combined Heating, Cooling & Power Handbook: Technologies & Applications: An Integrated Approach to Energy Resource Optimization.* Fairmont Press, Lilburn, GA.

28. Sullivan, W. G., E. M. Wicks, and J. T. Luxhoj (2006). *Engineering Economy.* Pearson/Prentice Hall, Upper Saddle River, NJ.
29. Van der Lee, P. E. A., T. Terlaky, and T. Woudstra (2001). "A New Approach to Optimizing Energy Systems." *Computer Methods in Applied Mechanics and Engineering* **190** (40–41): 5297.
30. Winston, W. L. (1994). *Operations Research: Applications and Algorithms.* Belmont, CA: Duxbury Press.
31. Wong-Kcomt, J. B. (1992). "A Robust and Responsive Methodology for Economically Based Design of Industrial Cogeneration Systems." *DAI* **54**: 218.
32. Yokoyama, R., K. Ito, and Y. Matsumoto (1991). *Optimal Sizing of a Gas Turbine Cogeneration Plant in Consideration of Its Operational Strategy.* Budapest, Hungary, published by ASME, New York.
33. Yongyou, H., S. Hongye, and C. Jian (2004). "A New Algorithm for Unconstrained Optimization Problem with the Form of Sum of Squares Minimization." 2004 IEEE International Conference on Systems, Man and Cybernetics, IEEE.
34. Zhang, J. and C. Xu (2001). "Properties and Numerical Performance of Quasi-Newton Methods with Modified Quasi-Newton Equations." *Journal of Computational and Applied Mathematics* **137**(2): 269.

20
Airline Optimization

	20.1	Introduction... 20-1
		Airline Industry Environment • Airline Planning and Operations Business Processes • Airline Optimization Solutions • Benefits of Airline Optimization
	20.2	Schedule Planning................................. 20-5
		Schedule Design • Fleet Assignment • Aircraft Routing • Crew Scheduling
	20.3	Revenue Management 20-14
		Forecasting • Overbooking • Seat Inventory Control • Pricing
Jane L. Snowdon and	20.4	Aircraft Load Planning............................ 20-21
Giuseppe Paleologo	20.5	Future Research Directions and Conclusions.... 20-23
IBM T. J. Watson Research Center		References 20-24

20.1 Introduction

The main objective of this chapter is to describe and review airline optimization from an applied point of view. The airline industry was one of the first to apply operations research methods to commercial optimization problems. The combination of advancements in computer hardware and software technologies with clever mathematical algorithms and heuristics has dramatically transformed the ability of operations researchers to solve large-scale, sophisticated airline optimization problems over the past 60 years. As an illustration, consider United Airlines' eight fleet problem in Figure 20.1, which took nearly 9 h to solve in 1975 on an IBM System 370 running Mathematical Programming System Extended (MPSX), and only 18 s to solve in 2001 on an IBM Thinkpad running IBM's Optimization Subroutine Library (OSL) V3. Better models and algorithms supported by faster hardware achieved more than four orders of magnitude improvement in solution time for this problem instance over the course of 26 years.

In the subsequent sections, the state-of-the-art in the application of optimization to the airline industry is given for solving traditional airline planning problems and is arranged by functional area. The remainder of Section 20.1 presents an overview of the major historical events in the airline industry environment, gives a description of the high-level business processes for airline planning and operations, provides a representative list of major airline optimization vendors, and illustrates select cases where airlines have adopted and embraced optimization to realize productivity and efficiency gains and cost reductions. Section 20.2 covers schedule planning including the four steps of schedule design, fleet assignment, aircraft routing, and crew scheduling. Section 20.3 describes the four steps of revenue

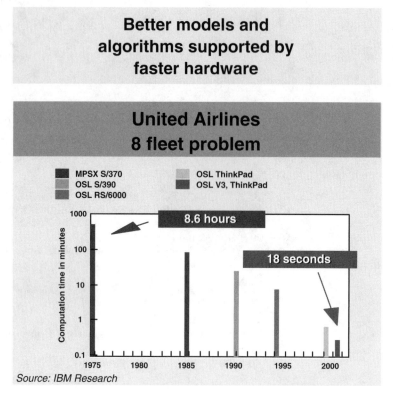

FIGURE 20.1 United Airlines' eight fleet problem.

management, which consist of forecasting, overbooking, seat inventory control, and pricing. Section 20.4 reviews aircraft load planning. Finally, Section 20.5 provides directions for future research.

20.1.1 Airline Industry Environment

The Chicago Convention was signed in December 1944 by Franklin Roosevelt and his peers to establish the basic rules for civil aviation. Over 60 years later, their vision of international civil aviation as a means for fostering friendship and understanding among nations and peoples worldwide has been realized. Air travel has not only become a common, affordable means of transportation for developed countries, but it has also brought enormous benefits to developing countries with potential for social progress and economic prosperity through trade and tourism.

According to the Air Transport Action Group (ATAG), aviation transported, in 2004, nearly 1.9 billion passengers and 38 million tons of freight worth $1.75 billion, which is over 40% of the world trade of goods by value [1]. Giovanni Bisignani, director general and CEO of the International Air Transport Association (IATA), predicted that 2.2 billion passengers will travel in 2006 during his speech at the State of the Industry Annual General Meeting and World Air Transport Summit. International passenger traffic increased by 15.6% in 2004, driven primarily by a strong rebound in Asia from SARS-affected levels in 2003, and by substantial increases in route capacity in the Middle East. However, international passenger traffic is forecast to grow at an annual average rate of 5.6% from 2005 to 2009, but may decline due to high oil prices [2]. Likewise, the 15.8% growth in international freight volumes in 2004 is expected to slow with an estimated forecast of a 6.3% annual average growth rate between 2005 and 2009 [2].

Aviation generated nearly $3 trillion, equivalent to 8% of the world's gross domestic product (GDP), in 2004. The air transport industry employs 5 million people directly and an additional 24 million indirectly or induced through spending by industry employees globally. The world's 900 airlines have a total fleet of nearly 22,000 aircraft [3], which serves an estimated 1670 airports [4] using a route network of several million kilometers managed by approximately 160 air navigation service providers [5]. Airbus forecasts that over 17,000 new planes will be needed by 2024 [6] and Boeing forecasts that over 27,000 new planes will be needed by 2025 [7] to meet growing demand for air transport, to replace older fleet, and to introduce more fuel-efficient planes.

Between 1944, when 9 million passengers traveled on the world's airlines, and 1949, when the first jet airliner flew, the modern era of airline optimization was born. George Dantzig designed the simplex method in 1947 for solving linear programming formulations of U.S. Air Force deployment and logistical planning problems. Applications to solve large-scale, real-world problems, such as those found in military operations, enabled optimization to blossom in the 1950s with the advent of dynamic programming (Bellman), network flows (Ford and Fulkerson), nonlinear programming (Kuhn and Tucker), integer programming (Gomory), decomposition (Dantzig and Wolfe), and the first commercial linear programming code (Orchard-Hays) as described in Nemhauser [8].

The Airline Group of the International Federation of Operational Research Societies (AGIFORS) was formed in 1961 and evolved from informal discussions between six airline operational research practitioners from Trans Canada, Air France, Sabena, BEA, and Swissair. Today AGIFORS is a professional society comprised of more than 1200 members representing over 200 airlines, aircraft manufacturers, and aviation associations dedicated to the advancement and application of operational research within the airline industry. AGIFORS conducts four active study groups in the areas of cargo, crew management, strategic and scheduling planning, and reservations and yield management, and holds an annual symposium.

In the late 1960s and early 1970s, United Airlines, American Airlines, British Airways, and Air France, recognizing the competitive advantage that decision technologies could provide, formed operations research groups. As described by Barnhart, Belobaba, and Odoni, these groups grew rapidly, developing decision support tools for a variety of airline applications, and in some cases offering their services to other airlines [9]. Most major airlines worldwide have launched similar groups.

A dramatic change in the nature of airline operations occurred with the Airline Deregulation Act of 1978 led by the United States. This Act is widely credited for having stimulated competition in the airline industry. For example, the year 1986, which was marked by economic growth and stable oil prices, saw the poor financial results of major airlines. This was just a symptom of the increased competitive pressures faced by the carriers. Effective 1997, the European Union's final stage of deregulation allows an airline from one member state to fly passengers within another member's domestic market. Since 1978, other important forces such as the rising cost of fuel, the entrance of low-cost carriers into the market, the abundance of promotional and discount fares, and consolidations have significantly shaped the competitive landscape of the airline industry. Hundreds of airlines have entered into alliances and partnerships, ranging from marketing agreements and code-shares to franchises and equity transfers, resulting in globalization of the industry. According to IATA, the three largest airline alliances combined today fly 58% of all passengers traveling each year, including Star Alliance (425 million passengers/year, 23.6% market share), Skyteam Alliance (372.9 million passengers/year, 20.7% market share), and oneworld (242.6 million passengers/year, 13.5% market share). Alliances between cargo airlines are also taking place—for example, WOW Alliance and SkyTeam Cargo.

20.1.2 Airline Planning and Operations Business Processes

Business processes for airline planning and operations may be described by four high-level processes as shown in Figure 20.2. First, resource planning consists of making strategic decisions with respect to aircraft acquisition, crew manpower planning, and airport resources such as slots. Resource planning is typically performed over a 1–2-year time horizon prior to the day of operations. Due to the competitive and proprietary nature of the data, little is published about strategic planning. Second, market planning and control consists of schedule design, fleet assignment, and yield management decisions. Market planning and control is typically performed over a 6 months' time horizon. Third, operations planning consists of making operational decisions with respect to aircraft routing and maintenance, crew resources, and airport operations. Operations planning is typically performed over a 1-month time horizon. Fourth, operations control consists of making tactical decisions with respect to aircraft rerouting, crew trip recovery, airport operations, and customer service. Operations control is typically performed over a 2-week time horizon and includes the day of operations.

20.1.3 Airline Optimization Solutions

Any discussion of airline optimization would not be complete without mentioning some of the major vendors in this space. While not exhaustive, this list is representative. First, global distribution systems (GDS) are operated by airlines to store and retrieve information and conduct transactions related to travel, such as making reservations and generating tickets. Four GDS dominate the marketplace: Amadeus (founded in 1987 by Air France, Iberia, Lufthansa, and SAS and now owned by InterActive Corporation, which also owns Expedia), Galileo International (founded in 1993 by 11 major North American and European airlines and acquired in 2001 by Cendant Corporation, which also owns Orbitz), Sabre (founded in 1960, it also owns Travelocity), and Worldspan (founded in 1990 and currently owned by affiliates of Delta Air Lines, Northwest Airlines, and American Airlines). Sabre also offers solutions for pricing and revenue management, flight and crew scheduling, operations control, among others.

Second, revenue management, also referred to as yield management, is the process of comprehending, predicting, and reacting to consumer behavior to maximize revenue. PROS Revenue Management is a market leader in revenue optimization. Their product suite includes forecasting demand, optimizing the allocation of inventory, providing dynamic

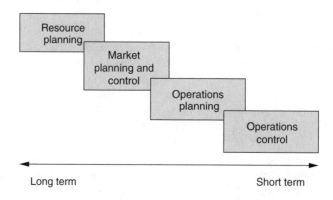

FIGURE 20.2 Airline planning and operations business processes.

packaging, and offering dynamic pricing. Recently JDA Software Group acquired Manugistics, another leading provider of solutions for demand management, pricing, and revenue management.

Third, the airline industry has seen a wave of consolidation of schedule planning providers. Jeppesen, a wholly owned subsidiary of Boeing Commercial Aviation Services, acquired Carmen Systems and its portfolio of manpower planning, crew pairing, crew rostering, crew tracking, crew control, fleet control, passenger recovery, and integrated operations control modules in 2006. Kronos purchased AD OPT Technologies' suite of customized crew planning, management, and optimization solutions for the airline industry in 2004. Navitaire, a wholly owned Accenture business that hosts technology solutions for reservations, direct distribution, operations recovery, decision-support and passenger revenue accounting, acquired Caleb Technologies (operations recovery decision-support and resource planning systems) in 2004 and Forte Solutions (operations and crew management control) in 2005.

Finally, a number of general purpose modeling and optimization tools are available including Dash Optimization's Xpress-MP, the IBM Optimization Solutions and Library, and ILOG's CPLEX.

20.1.4 Benefits of Airline Optimization

The practice of airline optimization has yielded significant business benefits. For example, the fractional aircraft operations system designed and developed by AD OPT Technologies, Bombardier, and GERAD to simultaneously maximize the use of aircraft, crew, and facilities generated an initial savings of $54 million with projected annual savings of $27 million as reported by Hicks et al. [10]. UPS claims savings of more than $87 million over 3 years for a system co-developed with MIT to optimize the design of service networks for delivering express packages by simultaneously determining aircraft routes, fleet assignments, and package routings as given in Armacost et al. [11]. The CrewSolver system developed by CALEB Technologies for recovering crew schedules in real-time when disruptions occur has saved Continental an estimated $40 million for major disruptions alone, as presented in Yu et al. [12]. Air New Zealand and Auckland University teamed to develop a suite of systems to optimize the airline's crew schedules, which has resulted in an annual savings of $10 million as explained in Butchers et al. [13]. Subrahmanian et al. [14] provide an account of solving a large-scale fleet assignment problem with projected savings to Delta Air Lines of $300 million over 3 years. According to Robert Crandell, the former president and chairman of American Airlines, yield management was the single most important technical development in transportation management since airline deregulation in 1979 and is estimated to have generated $1.4 billion in incremental revenue in 3 years, with projected annual savings of more than $500 million for the airline as quoted in Smith et al. [15]. These are but a few of many examples demonstrating the tremendous business impact and potential of airline optimization.

20.2 Schedule Planning

The problem of airline scheduling is among the largest and most complex of any industry. Airline scheduling includes determining the concurrent flows of passengers, cargo, aircraft, and flight crews through a space-time network, performing aircraft maintenance, and deploying ground-based resources such as gates. The problem is further complicated by air traffic control restrictions, noise curfews, company/union regulations, market structure (density, volume, and elasticity of demand), fare classes, and safety and security concerns. The overarching objective in airline schedule planning is to maximize profitability.

Etschmaier and Mathaisel [16], Rushmeier and Kontogiorgis [17], Barnhart and Talluri [18], Clarke and Smith [19], Klabjan [20], and Barnhart and Cohn [21] supply excellent overviews of the airline schedule development process. Airline schedule planning is traditionally broken down into four steps in such a way as to make each step computationally viable. No single optimization model has been formulated or solved to address the complex schedule planning task in its entirety, although attempts have been made to solve combinations of two or more of the steps in a single model. The steps in sequential order are:

1. Schedule Design—defining the markets to serve and how frequently, and determining the origin, destination, and departure times of flights.
2. Fleet Assignment—designating the aircraft size for each flight.
3. Aircraft Routing—selecting the sequence of flights to be flown by each aircraft during the course of every day and satisfying maintenance requirements.
4. Crew Scheduling—assigning crews to flights.

Next, each of these steps will be described in more detail.

20.2.1 Schedule Design

The scheduling process starts with the schedule design. The schedule design constructs the markets and how frequently to serve them, and determines the origin, destination, and departure times of flights. In general, schedules are dictated largely by differences in seasonal demand and to a smaller extent by month-to-month differences. Profitability and feasibility are two important criteria that affect how a flight schedule is designed. Schedule profitability is determined by the degree to which the origin–destination city pairs attract revenue from passengers and cargo compared to the operational costs of transporting the people and goods. Schedule feasibility is determined by the ability of the airline to assign sufficient resources, such as airplanes, crew, and airport facilities, to the flights.

Construction of the schedule typically begins 1 year in advance of its operation and 6 months in advance of its publication. The schedule construction process encompasses three steps:

1. Establishing the service plan (a list of routes and frequencies where factors such as demand forecasts, competitive information, and marketing initiatives are considered).
2. Developing an initial feasible schedule (establish flight departure and arrival times while satisfying maintenance planning and resource constraints, such as the number of aircraft).
3. Iteratively reviewing and modifying the schedule based upon evaluations of new routes, aircraft, or frequencies, and connection variations.

Decision support systems based upon operations research have been slow to develop for schedule design, primarily because of the inherent complexity and problem size, the lack of sufficient and timely data from the airline revenue accounting process, the method for collecting and forecasting market demand, and the difficulty of predicting competing airlines' responses to market conditions. Published literature on airline schedule design is sparse. One of the earliest descriptions of the schedule construction process can be found in Rushmeier et al. [22]. Erdmann et al. [23] address a special case of the schedule generation problem for charter airlines, model the problem as a capacitated network design problem using a path-based mixed integer programming formulation, and solve it using a branch-and-cut approach. Armacost et al. [24,25] apply a composite variable formulation approach by removing aircraft- and package-flow decisions as explicit decisions, and solve a set-covering instance to the next-day-air network-design problem for United Parcel Service (UPS).

20.2.2 Fleet Assignment

Given a flight schedule and a set of aircraft, the fleet assignment problem determines which type of aircraft, each having a different capacity, should fly each flight segment to maximize profitability, while complying with a large number of operational constraints. The objective of maximizing profitability can be achieved both by increasing expected revenue, for example, assigning aircraft with larger seating capacity to flight legs with high passenger demand, and by decreasing expected costs, for example, fuel, personnel, and maintenance. Operational constraints include the availability of maintenance at arrival and departure stations (i.e., airports), gate availability, and aircraft noise, among others. Assigning a smaller aircraft than needed on a flight results in spilled (i.e., lost) customers due to insufficient capacity. Assigning a larger aircraft than needed on a flight results in spoiled (i.e., unsold) seats and possibly higher operational costs.

Literature on the fleet assignment problem spans nearly 20 years. For interested readers, Sherali et al. [26] provide a comprehensive tutorial on airline fleet assignment concepts, models, and algorithms. They suggest future research directions including the consideration of path-based demands, network and recapture effects, and exploring the interactions between initial fleeting and re-fleeting. Additional review articles on fleet assignment include Gopalan and Talluri [27], Yu and Yang [28], Barnhart et al. [9], Clarke and Smith [19], and Klabjan [20].

Several of the earliest published accounts are by Abara [29], who describes a mixed integer programming implementation for American Airlines; Daskin and Panayotopoulos [30], who present an integer program that assigns aircraft to routes in a single hub and spoke network; and Subramanian et al. [14], who develop a cold-start solution (i.e., a valid initial assignment does not exist) approach for Delta Airlines that solves the fleet assignment problem initially as a linear program using an interior point method, fixes certain variables, and then solves the resulting problem as a mixed integer program. Talluri [31] proposes a warm-start solution (i.e., a valid initial assignment does exist) that performs swaps based on the number of overnighting aircraft for instances when it becomes necessary to change the assignment on a particular flight leg to another specified aircraft type. Jarrah et al. [32] propose a re-fleeting approach for the incremental modification of planned fleet assignments. Hane et al. [33] employ an interior point algorithm and dual steepest edge simplex, cost perturbation, model aggregation, and branching on set-partitioning constraints with prioritized branching order. Rushmeier and Kontogiorgis [17] focus on connect time rules and incorporate profit implications of connections to assign USAir's fleets resulting in an annual benefit of at least $15 million. Their approach uses a combination of dual simplex and a fixing heuristic to solve a linear programming relaxation of the problem to obtain an initial solution, which in turn is fed into a depth-first branch-and-bound process. Berge and Hopperstad [34] propose a model called Demand Driven Dispatch for dynamically assigning aircraft to flights to leverage the increased accuracy of the flight's demand forecast as the actual flight departure time approaches.

Attempts have been made to integrate the fleet assignment model with other airline processes such as schedule design, maintenance routing, and crew scheduling. Because of the interdependencies of these processes, the optimal solution for processes considered separately may not yield a solution that is optimal for the combined processes.

For example, integrated fleet assignment and schedule design models have the potential to increase revenues through improved flight connection opportunities. Desaulniers et al. [35] introduce time windows on flight departures for the fleet assignment problem and solve the multicommodity network by branch-and-bound and column generation, where the column generator is a specialized time-constrained shortest path problem. Rexing et al. [36] discretize time windows and create copies of each flight in the underlying graph to represent different departure time possibilities and then solve using a column generator that is a shortest path problem on an acyclic graph. Lohatepanont and Barnhart [37] present an integrated schedule design and fleet assignment solution approach that determines incremental changes to existing flight schedules to maximize incremental profits. The integrated schedule design and fleet assignment model of Yan and Tseng [38] includes path-based demand considerations and uses Lagrangian relaxation, where the Lagrangian multipliers are revised using a subgradient optimization method.

Integrated fleet assignment and maintenance, routing, and crew considerations have the potential for considerable cost savings and productivity improvements. Clarke et al. [39] capture maintenance and crew constraints to generalize the approach of Hane et al. [33] and solve the resulting formulation using a dual steepest-edge simplex method with a customized branch-and-bound strategy. Barnhart et al. [40] explicitly model maintenance issues using ground arcs and solve the integrated fleet assignment, maintenance, and routing problem using a branch-and-price approach where the column generator is a resource-constrained shortest path problem over the maintenance connection network. Rosenberger et al. [41] develop a fleet assignment model with hub isolation and short rotation cycles (i.e., a sequence of legs assigned to each aircraft) so that flight cancellations or delays will have a lower risk of impacting subsequent stations or hubs. Belanger et al. [42] present both a mixed-integer linear programming model and a heuristic solution approach for the weekly fleet assignment problem for Air Canada in the case where homogeneity of aircraft type is desired over legs sharing the same flight number, which enables easier ground service planning.

Models to integrate fleet assignment with passenger flows and fare classes adopt an origin–destination-based approach compared to the more traditional flight-leg approach. Examples of fleet assignment formulations that incorporate passenger considerations include Farkas [43], Kniker [44], Jacobs et al. [45], and Barnhart et al. [46].

The basic fleet assignment model (FAM) can be described as a multicommodity flow problem with side constraints defined on a time-space network and solved as an integer program using branch-and-bound. The time-space network has a circular time line representing a 24-hour period, or daily schedule, for each aircraft fleet at each city. Along a given time line, a node represents an event: either a flight departure or a flight arrival. Each departure (arrival) from the city splits an edge and adds a node to the time line at the departure (arrival + ground servicing) time. A decision variable connects the two nodes created at the arrival and departure cities and represents the assignment of that fleet to that flight.

Mathematically, the fleet assignment model may be stated as follows, which is an adaptation of Hane et al. [33] without required through-flights (i.e., one-stops). The objective is to minimize the cost of assigning aircraft types to flight legs as given in Equation 20.1. To be feasible, the fleet assignment must be done in such a way that each flight in the schedule is assigned exactly one aircraft type (i.e., cover constraints as given in Equation 20.2), the itineraries of all aircraft are circulations through the network of flights that can be repeated cyclically over multiple scheduling horizons (i.e., balance constraints as given in Equation 20.3), and the total number of aircraft assigned cannot exceed the number available in the fleet (i.e., plane count constraints as given in Equation 20.4). Additional inequality

Airline Optimization

constraints may be incorporated to address such issues as through-flight assignments, maintenance, crew, slot allocation, and other issues.

$$\min \sum_{j \in J} \sum_{i \in I} c_{ij} X_{ij} \tag{20.1}$$

$$\text{subject to:} \sum_i X_{ij} = 1, \quad \forall j \in J \tag{20.2}$$

$$\sum_d X_{idot} + Y_{iot^-t} - \sum_d X_{iodt} - Y_{iott^+} = 0, \quad \forall \{iot\} \in N \tag{20.3}$$

$$\sum_{j \in O(i)} X_{ij} + \sum_{o \in C} Y_{iot_n t_1} \leq S(i), \quad \forall i \in I \tag{20.4}$$

$$Y_{iott^+} \geq 0, \quad \forall \{iott^+\} \in N \tag{20.5}$$

$$X_{ij} \in \{0, 1\}, \quad \forall i \in I, \forall j \in J \tag{20.6}$$

The mathematical formulation requires the following notation with parameters:

C = set of stations (cities) serviced by the schedule,
I = set of available fleets,
$S(i)$ = number of aircraft in each fleet for $i \in I$,
J = set of flights in the schedule,
$O(i)$ = set of flight arcs, for $i \in I$, that contains an arbitrary early morning time (i.e., 4 AM, overnight),
N = set of nodes in the network, which are enumerated by the ordered triple $\{iot\}$ consisting of fleet $i \in I$, station $o \in C$, and t = takeoff time or landing time at o.
t^- = time preceding t,
t^+ = time following t,
$\{i_{ot_n}\}$ = last node in a time line, or equivalently, the node that precedes 4 AM,
$\{i_{ot_1}\}$ = successor node to the last node in a time line, and decision variables:
$X_{iodt} = X_{ij} = 1$ if fleet i is assigned to the flight leg from o to d departing at time t, and 0 otherwise;
Y_{iott^+} = number of aircraft of fleet $i \in I$ on the ground at station $o \in C$ from time t to t^+.

20.2.3 Aircraft Routing

The Federal Aviation Administration (FAA) mandates some safety requirements for aircraft. Those are of four types:

- A-checks are routine visual inspections of major systems, performed every 65 block-hours or less. A-checks take 3–10 h, and are usually performed at night at the gate.
- B-checks are detailed visual inspections, and are performed every 3 months.
- C-checks are performed every 12–18 months depending on aircraft type and operational circumstances, and consist of in-depth inspections of many systems. C-checks require disassembly of parts of the aircraft, and must be performed in specially equipped spaces.
- D-checks are performed every 4–5 years, perform extensive disassembly and structural, chemical, and functional analyses of each subsystem, and can take more than 2 months.

Of these checks, A and B are performed on the typical planning horizon of fleet planning and crew scheduling, and must therefore be incorporated into the problem.

Aircraft routing determines the allocation of candidate flight segments to a specific aircraft tail number within a given fleet-type while satisfying all operational constraints, including maintenance. Levin [47] is the first author to analyze maintenance scheduling. One simplified formulation of the maintenance problem, following Barnhart et al. [40], is the following. We define a string as a sequence of connected flights (i.e., from airport a to airport b, then from b to c, until a final airport) performed by an individual aircraft. Moreover, a string satisfies the following requirements: the number of block-hours satisfies maintenance requirements, and the first and last nodes in the sequence of airports are maintenance stations. A string k has an associated cost c_k. Let S be the set of all augmented strings. For every node v, $I(v)$ and $O(v)$ denote the set of incoming and outgoing links, respectively. The problem has two types of decision variables: x_s is set to 1 if string s is selected as a route for the associated fleet, and 0 otherwise. y_m is the number of aircraft being serviced at service station m. The problem then becomes:

$$\min \sum_{s \in S} c_s x_s \tag{20.7}$$

$$\text{subject to:} \sum_{i \in S} x_s = 1 \quad \text{for all flights } i \tag{20.8}$$

$$\sum_{\substack{j \in O(v) \\ j \in s}} x_s - \sum_{\substack{j \in I(v) \\ j \in s}} x_s + y_{O(v)} - y_{I(v)} = 0 \quad \text{for all stations } v \tag{20.9}$$

where $y \geq 0$, $x = 0/1$.

The problem can be formulated as a multicommodity flow problem (e.g., Cordeau et al. [48]), as a set partitioning problem (Feo and Bard [49]; Daskin and Papadopoulos [30]), and employing eulerian tours (Clarke et al. [50], Talluri [31], and Gopalan and Talluri [51]). Recent work includes that of Gabteni and Gronkvist [52], who provide a hybrid column generation and constraint programming optimization solution approach, Li and Wang [53], who present a path-based integrated fleet assignment and aircraft routing heuristic, Mercier et al. [54], who solve the integrated aircraft routing and crew scheduling model using Benders decomposition, and Cohn and Barnhart [55], who propose an extended crew pairing model that integrates crew scheduling and maintenance routing decisions.

20.2.4 Crew Scheduling

The area of crew scheduling involves the optimal allocation of crews to flights and can be partitioned into three phases: crew pairing, crew assignment based upon crew rostering or preferential bidding, and recovery from irregular operations. Each of these three phases will be described in more detail in this section. The literature on crew scheduling is plentiful over its nearly 40-year history with descriptions in comprehensive survey papers by Etschmaier and Mathaisel [16], Clarke and Smith [19], Klabjan [20], Barnhart and Cohn [21], Arabeyre et al. [56], Bornemann [57], Barutt and Hull [58], Desaulniers et al. [59], Barnhart et al. [60], and Gopalakrishnan and Johnson [61].

Crew Pairing

In crew pairing optimization, the objective is to find a minimum cost set of legal duties and pairings that covers every flight leg. An airline generally starts with the schedule of flight segments and their corresponding fleet assignments and decomposes the problem for different crew types (pilots, flight attendants) and for different fleet types. The flight legs are joined together to form duties, which are essentially work shifts. These duties are then combined into trips/pairings that range from 1 to 5 days in length and which start and end at a crew's home base. There are many FAA and union regulations governing the legality of duties and pairings. Some typical ones may include: a limit on the number of hours of flying time in a duty, a limit on the total length of a duty (possibly stricter for a nighttime duty), a limit on the number of duties in a pairing, and upper and lower bounds on the rest time between two duties. The generation of good crew pairings is complicated by the fact that a crew's pay is determined by several guarantees that are based on the structure of the duties in the pairing. Some common rules governing crew compensation may include: a crew is paid overtime for more than 5 h of flying in a duty, a crew is guaranteed a minimum level of pay per duty period, a crew is guaranteed that a certain percentage of the duty time will count as flying time, and a crew receives a bonus for overnight flying. Some companies also include additional constraints and objective function penalties to avoid marginally legal pairings, tight connections, and excessive ground transportation.

The area of crew pairing optimization is complex and challenging for two reasons. First, it is not practical to generate the complete set of feasible pairings, which can number in the billions for major carriers, for a problem. Second, the cost of a pairing usually incorporates linear components for hotel and per-diem charges and several nonlinear components based upon actual flying time, total elapsed work time, and total time away from the home base.

The crew pairing problem is normally solved in three stages for North American carriers, as described in Anbil et al. [62]. First, a daily problem is solved where all flights operate every day. Next, a weekly problem is solved, which reuses as much of the daily problem as possible, and also handles weekly exceptions. Finally, a dated problem is solved, which reuses as much of the weekly problem as possible, and also produces a monthly schedule that handles holidays and weekly transitions.

Several differences exist for European carriers. First, European crews typically receive fixed salaries, so the objective is simply to minimize the number of crews needed. Second, European airlines do not tend to run on a daily, repeating schedule or have hub-and-spoke networks. Third, European carriers must comply with government regulations and work with somewhat variable union rules, while for North American carriers, crew pairing is primarily determined by FAA regulations.

The crew pairing problem can be modeled as a set partitioning problem where the rows represent flights to be covered and the columns represent the candidate crew pairings as follows:

$$\min \quad \sum_{j \in P} c_j x_j \tag{20.10}$$

$$\text{subject to:} \quad \sum_{j \in P} a_{ij} x_j = 1, \quad \forall i \in F \tag{20.11}$$

$$x_j \in \{0, 1\} \tag{20.12}$$

The mathematical formulation requires the following notation with parameters:

$P =$ set of all feasible pairings,
$c_j =$ cost of pairing j,
$a_{ij} = 1$ if pairing j includes (covers) flight i, and 0 otherwise,
$F =$ set of all flights that must be covered in the period of time under consideration,

and decision variable:

$x_j = 1$ if the pairing j is used, and 0 otherwise.

The objective is to minimize the total cost as shown in Equation 20.10. Equation 20.11 ensures that each flight leg is covered once and only once. In practice, side constraints are often added to reflect the use of crew resources; for example, upper and lower bounds on the number of available crew at a particular base, and balancing pairings across crew bases with respect to cost, the number of days of pairings, or the number of duties.

A variety of solution methodologies have been suggested for the crew pairing problem and the majority are based upon pairing generation and pairing selection strategies. Pairings are frequently generated using either an enumerative or shortest path approximation approach on a graph network that can be represented either as a flight segment network or a duty network. Pairing generation approaches and selected illustrations include:

- *Row approach*: Start with a feasible (or artificial) initial solution. Choose a subset of columns and exhaustively generate all possible pairings from the flight segments covered by these columns, and solve the resulting set partitioning problem. If the solution to this sub-problem provides an improvement, replace the appropriate pairings in the current solution. Repeat until no further improvements are found or until a limit on the execution time is reached. Selected illustrations include Gershkoff [63], Rubin [64], Anbil et al. [65,66], and Graves et al. [67].
- *Column approach*: Consider all the rows simultaneously, explicitly generate all possible legal pairings by selecting a relatively small set of pairings that have small reduced cost from a relatively large set of legal pairings, and solve the sub-problem until optimality. Continue until the solution to the sub-problem no longer provides an improvement. A selected illustration includes Hu and Johnson [68].
- *Network approach*: The network approach can utilize one of two strategies: the column approach where the columns are generated using the flight segment or duty network, or column generation based on shortest paths. Selected illustrations include Minoux [69], Baker et al. [70], Lavoie et al. [71], Barnhart et al. [72], Desaulniers et al. [73], Stojković et al. [74], and Galia and Hjorring [75].

Strategies for selecting pairings and solving the crew pairing optimization problem include TPACS [64], TRIP [65,66], SPRINT (Anbil et al. [62]), volume algorithm [76,77], branch-and-cut [78], branch-and-price [79–82], and branch-and-cut-and-price [62]. Others have proposed strategies using genetic algorithms [58,83–86].

Crew Assignment

In crew assignment, pairings are assembled into monthly work schedules and assigned to individual crew members. Depending upon the airline's approach, either crew rostering or preferential bidding is used for crew assignment. In crew rostering, a common process outside of North America, the objective is to assign pairings to individual crew members based upon the individual's preferences while minimizing overall costs. In preferential bidding, a common process used within North America, individual crew members bid on a set of cost-minimized schedules according to their relative preferences. The airline assigns specific schedules to the crew members based upon their bids and seniority.

The crew rostering and preferential bidding problems can be formulated as set partitioning problems. The problem decomposes based upon the aircraft equipment type and the certification of the crew member (e.g., Captain, First Officer, Flight Attendant). The crew assignment problem is complex, with FAA, union and company rules, and is compounded by additional rules involving several rosters, including language restrictions and married crew members preferring to fly together. The airline strives to produce equitable crew assignments in terms of equal flying time and number of days off.

The crew assignment problem can be modeled simply as follows:

$$\min \quad c_r^k x_r^k \tag{20.13}$$

$$\text{subject to:} \quad \sum_{\substack{k \in K \\ i \in r}} x_r^k \geq n_i, \quad \forall i \tag{20.14}$$

$$\sum_r x_r^k = 1, \quad \forall k \tag{20.15}$$

$$x \in \{0, 1\} \tag{20.16}$$

The mathematical formulation requires the following notation with parameters:

K = set of crew members,
c_r^k = cost of assigning roster r to crew member k,
n_i = number of crew members that are required for task i,

and decision variable:

$x_r^k = 1$ if roster r is assigned to crew member k, and 0 otherwise.

The objective is to minimize the total cost as shown in Equation 20.13. Equation 20.14 ensures that each task is covered. Equation 20.15 assigns a roster to every crew member.

Recent work in the area of crew assignment includes Kohl and Karisch [87], Cappanera and Gallo [88], Caprara et al. [89], Christou et al. [90], Dawid et al. [91], Day and Ryan [92], Gamache and Soumis [93], Gamache et al. [94,95], Shebalov and Klabjan [96], and Sohoni et al. [97]. Recent solution approaches to the crew assignment problem focus on branch-and-price where sub-problems are solved based upon constrained shortest paths, and on numerous heuristic approaches, for instance, simulated annealing as described in Lučić and Teodorvić [98] and Campbell et al. [99].

Recovery from Irregular Operations

In the face of disruptions such as bad weather, unscheduled aircraft maintenance, air traffic congestion, security problems, flight cancellations or delays, passenger delays, and sick or illegal crews on the day of operations, the three aspects of aircraft, crew, and passengers are impacted, often causing the airline to incur significant costs. Clarke [100] states that a major U.S. airline can incur more than $400 million annually in lost revenue, crew overtime pay, and passenger hospitality costs due to irregular operations. When any of these disruptions happen, operations personnel must make real-time decisions that return the aircraft and crew to their original schedule and deliver the passengers to their destinations as soon and as cost effectively as possible.

Aircraft recovery usually follows groundings and delays where the objective is to minimize the cost of reassigning aircraft and re-timing flights while taking into account available resources and other system constraints. Interested readers are referred to detailed reviews in Clarke and Smith [19], Rosenberger et al. [101], and Kohl et al. [102].

Crew recovery usually follows flight delay, flight cancellation, and aircraft rerouting decisions, but often flight dispatchers and crew schedulers coordinate to explore the best options. In crew recovery, the objective is to reassign crew to scheduled flights. It is desirable to return to the original crew schedule as soon as possible. Crew recovery is a hard problem because of complex crew work rules, complex crew cost computations, changing objectives, and having to explore many options. Often, airlines are forced to implement the first solution found due to time pressures.

Airline investment in technologies to expedite crew recovery and reduce crew costs has increased in recent years. Anbil et al. [103] develop a trip repair decision support system with a solution approach that involves iterating between column generation and optimization, based on a set partitioning problem formulation where the columns are trip proposals and the rows refer to both flights and non-reserve crews. The solution strategy is to determine whether the problem can be fixed by swapping flights among only the impacted crew, and bringing in reserves and other nondisrupted crews as needed. Heuristics are used to localize the search during column generation. Teodorović and Stojković [104] present a greedy heuristic based upon dynamic programming and employ a first-in-first-out approach for minimizing crew ground time. Wei et al. [105] and Song et al. [106] describe a heuristic-based framework and algorithms for crew management during irregular operations. Yu et al. [12] describe the CrewSolver system developed by CALEB Technologies for recovering crew schedules at Continental Airlines in real-time when disruptions occur. Abdelghany et al. [107] design a decision support tool that automates crew recovery for an airline with a hub-and-spoke network structure. Nissen and Haase [108] present a duty-period-based formulation for the airline crew rescheduling problem that is tailored to European carriers and labor rules. Schaefer et al. [109] develop approximate solution methods for crew scheduling under uncertainty. Once a crew schedule has been determined, Schaefer and Nemhauser [110] define a framework for perturbing the departure and arrival times that keeps the schedule legal without increasing the planned crew schedule cost. Yen and Birge [111] describe a stochastic integer programming model and devise a solution methodology for integrating disruptions in the evaluation of crew schedules.

In the past decade, there has been a trend toward integrated aircraft, crew, and passenger recovery from irregular operations. Clarke [100] provides a framework that simultaneously addresses flight delays and flight cancellations, and solves the mixed-integer linear programming aircraft recovery problem incorporating crew availability using an optimization-based solution procedure. Lettovský [112] and Lettovský et al. [113] discuss an integrated crew assignment, aircraft routing, and passenger flow mixed-integer linear programming problem, and solve it using Bender's decomposition. Stojković and Soumis [114,115] solve a nonlinear multicommodity flow model that considers both flight and crew schedule recovery using branch-and-price. Rosenberger et al. [101] develop a heuristic for rerouting aircraft and then revise the model to minimize crew and passenger disruptions. Kohl et al. [102] develop a disruption management system called Descartes (Decision Support for integrated Crew and AiRcrafT recovery), which is a joint effort between British Airways, Carmen Systems, and the Technical University of Denmark. Lan et al. [116] propose two new approaches for minimizing passenger disruptions and achieving more robust airline schedule plans by (a) intelligently routing aircraft to reduce the propagation of delays using a mixed-integer programming formulation with stochastically generated inputs and (b) minimizing the number of passenger misconnections by adjusting flight departure times within a small time window. Bratu and Barnhart [117] minimize both airline operating costs and estimated passenger delay and disruption costs to simultaneously determine aircraft, crew, and passenger recovery plans.

20.3 Revenue Management

The Airline Deregulation Act of 1978 is widely credited for having stimulated competition in the airline industry. For example, the year 1986, which was marked by economic growth and stable oil prices, saw the poor financial results of major airlines. This was just a symptom

Airline Optimization

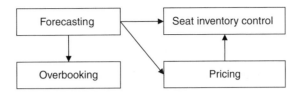

FIGURE 20.3 Interacting revenue management sub-problems.

of the increased competitive pressure faced by the carriers. In search of greater efficiencies, airlines began exploring ways to increase the average aircraft load and the revenue per seat. The discipline that has developed around this issue is usually referred to as "revenue management" (RM) although other terms are sometime used, such as "perishable asset revenue management" (PARM), "yield management," or "demand management." Given the complex and evolving nature of airline services, it does not consist of a single problem, with an unequivocal formulation, but rather of a collection of interacting sub-problems. The first such problem to be analyzed (before deregulation took place) was *overbooking*, which addresses the decision to sell more seats than currently available. The overbooking level depends on the likelihood that a customer will not show up at the gate, and on other factors, such as the lost revenue due to an unsold seat and the reimbursement to an eligible customer who is refused boarding. The former input to the overbooking problem is provided by forecasting methods applied to the reservation process and to the actual conversion rates of reservations in "go shows." The period prior to the 1978 Act saw another innovation: the promotional fare. "Early birds" would receive a discount over regular customers. From these early instances, *pricing* has evolved into a sophisticated discipline, with a deep impact in other industries as well. This simple yet effective innovation exploited the diversity of the customer population, namely the willingness of certain customers to trade off dollars for flexibility. However, the development of time-dependent fares posed a new challenge for airlines: *inventory seat control*, which consists of assigning inventory (number of seats) to a certain fare class. This is perhaps among the most important problems in the practice of revenue management. The earliest known implementation of inventory seat control is that of American Airlines in the first half of the 1980s. Since then, great progress has been made in improving the accuracy of inventory control systems. The relationship between the four sub-problems is shown in Figure 20.3.

The literature on revenue management is vast and growing rapidly. A complete bibliography is beyond the scope of this survey. Recent comprehensive treatments of revenue management are given by Phillips [118] and Van Ryzin and Talluri [119]. Yeoman and McMahon-Beattie [120] offer several real-world examples. The surveys of Weatherford and Bodily [121], McGill and Van Ryzin [122], and Barnhart et al. [9] contain comprehensive bibliographies. Several journals regularly publish RM articles, notably *Operations Research, Management Science, Interfaces,* the *European Journal of Operations Research,* and the recently founded *Journal of Pricing and Revenue Management,* which is entirely dedicated to this subject.

The rest of this section is organized as follows. First, the forecasting issue in revenue management is described. Then overbooking, inventory control, and pricing are reviewed. The section concludes with speculations on trends of the field.

20.3.1 Forecasting

Demand forecasting is a fundamental step in the RM exercise, as it influences all the others. Nevertheless, there are fewer published articles in this area than in any other area of RM. There are several possible reasons behind this relative scarcity. Actual booking data are not usually made available to academic researchers, and industry practitioners have no incentive in revealing internal methods or data. Another disincentive for researchers is that a majority of the articles published in flagship journals in operations research are of a theoretical, rather than empirical, nature. Finally, forecasting for RM purposes presents a formidable set of challenges. Next, the current approaches to this problem are reviewed, and the reader is pointed to the relevant literature.

Airlines sell a nonstorable product: a seat on an origin–destination itinerary fare. To maximize the financial metric of interest, it is vital for the airline not only to forecast the final demand of a given product, but also the process through which the demand accumulated over time. The former quantity is sometimes called the *historical booking forecast* and is a necessary input to overbooking decisions, while the latter, *the advanced booking forecast*, is used for pricing and inventory control, and aims at forecasting the number of bookings in a certain time interval prior to departure. Some of the relevant factors determining advanced booking behavior are:

1. Origin–destination pair
2. Class
3. Price
4. Time of the year
5. Time to departure
6. Presence of substitute products (e.g., a seat offered by the same company in a different class, or products offered by competing airlines).

Moreover, *group bookings* may be modeled explicitly, as their impact on inventory control policies is significant. Many methods have been proposed to explain observed reservation patterns; for a survey, see the book of Van Ryzin and Talluri [119] and the experimental comparison by Weatherford and Kimes [123]. Linear methods currently in place include autoregressive, moving-average models, Bayes and Hierarchical Bayes, and linear or multiplicative regressive models (sometime termed "pick-up" models). Model selection depends also on the chosen level of aggregation of the forecast. By choosing more aggregate models (e.g., aggregating similar classes) the modeler trades off greater bias for smaller variance; there is no "rule of thumb" or consensus on the ideal level of aggregation, although more sophisticated and application-specific models implemented in industrial applications usually provide very detailed forecasts.

It is important to remark that the data used in these statistical analyses are usually *censored*: the observed reservations are always bounded by the inventory made available on that day. Unconstraining demand is either based on mathematical methods (e.g., projection detruncation, or expectation maximization), or recorded reservation denials. For a comparison of popular unconstraining methods, see Weatherford and Pölt [124]. The availability of direct reservation channels (internet, airline-operated contact centers) has made reservation denials data both more abundant and reliable; demand correction based on denial rates is preferable, as it is usually more accurate and requires less computational effort. Figure 20.4 illustrates the relationships among the different components of the forecasting stage.

20.3.2 Overbooking

As mentioned before, research on overbooking predates deregulation. The original objective of overbooking was to maximize average load, subject to internally or externally

Airline Optimization

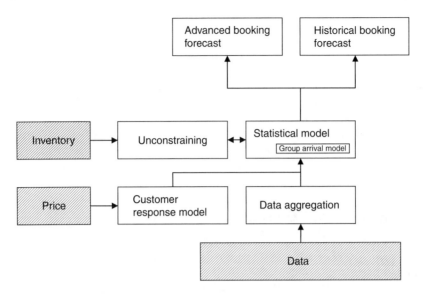

FIGURE 20.4 Forecasting components and relationships.

mandated maximum denial rates. More recently, airlines have been implementing revenue-maximizing overbooking policies. Overbooking can be roughly partitioned into three subproblems: (1) identification of optimal policies (overbooking limits); (2) identification of "fair bumping" policies (i.e., to whom should one refuse boarding); and (3) estimation of no-show rates. Of the three problems, (1) is the better studied and understood. In its simplest, static, form (see, e.g., Beckmann [125]), it is akin to a newsvendor problem: the availability of a product in the face of underage and overage costs, and of uncertain demand, must be chosen. To better understand the similarity with this problem, consider a simple one-period model, in which random demand is observed in the first period, and departure occurs in the second period. The available capacity is denoted by C, and the overbooking limit must be set prior to the observation of demand to maximize expected revenue. Let p be the unit price per seat and a be the penalty incurred for bumping a passenger. If the number of reservations is r, a random number $Z(r)$, with cumulative distribution function $F_r(z)$, of passengers will show up. A popular behavioral model for "show" demand posits that customers show up with probability q, and that each individual behaves independently of each other; this "binomial" model is defined by:

$$F_r(x) = \binom{r}{x} q^x (1-q)^{r-x} \tag{20.17}$$

The problem faced by the airline is given by:

$$\max_r E(pZ(r) - a(Z(r) - C)^+) \tag{20.18}$$

It can be shown through simple algebra that the solution is given by:

$$1 - F_r(C) = \frac{p}{aq} \tag{20.19}$$

which closely resembles the solution of a newsvendor problem.

The static policy is robust and effective in practical applications; however, using dynamical models could yield revenue increases that, albeit small, can be of crucial importance to achieve profitability. Commonly used models are in discrete time, and assume Markov or independent and identically distributed (iid) demand, and binomial no-show rates, allowing a stochastic dynamic programming formulation. Rothstein [126] is an early contribution to this approach; Alstrup et al. [127] and Alstrup [128] present both theoretical and empirical results.

The problem of equitable bumping has been studied, among others, by Simon [129,130] and Vickrey [131]. Interestingly, the solution proposed in these early days—to conduct an ascending auction to bump passengers with the lowest disutility from boarding denial, has been implemented some three decades later.

Finally, the problem of estimating no-show rates from historical data has received relatively little attention in the published literature. Lawrence et al. [132] present a data-driven no-show forecasting rule based on customer/fare attributes.

20.3.3 Seat Inventory Control

One of the new entrants following the 1978 Act was PEOPLExpress. A nonunionized, barebone service airline carrier with very low fares, it grew phenomenally in the first half of the 1980s, and had become a serious threat to the incumbents. One of them, American Airlines, was competing directly with PEOPLExpress on its principal routes. Partly as a response to the threat posed by the new entrant, American Airlines introduced a seat inventory control system. Under this system, the availability of seats belonging to a certain class could be dynamically controlled. For example, a certain low-price class (e.g., non-refundable tickets) could be "closed" (i.e., made unavailable) at a certain date because the same seat had a high probability of being reserved under a higher-price class (e.g., refundable tickets). The first rule for inventory control dates back to the seminal work of Littlewood [133], which derives it for a two-class inventory. Some assumptions are necessary to obtain closed-form rules:

1. Over time classes are booked in a certain order
2. Low-valuation-before-high-valuation arrival pattern
3. Independent arrivals for different booking classes
4. Absence of no-shows
5. Single-leg reservations
6. No group reservations.

The formulation is reviewed here, as it is the basis for widely used revenue management heuristics in real-world applications. Let the classes be ranked from low valuation to high valuation $p_1 \leq p_2 \leq \cdots \leq p_N$, and denote their uncertain demands D_i, based on assumptions 1 and 2, where i also denotes the period during which the demand is observed. The decision of interest is the capacity x_i that should be allocated to class i. Let $V_i(y)$ denote the expected revenue from period i to period N when the available capacity at the beginning of that period is y. At stage $i-1$, the revenue is given by:

$$p_{i-1}x_{i-1} + V(C_{i-1} - x_{i-1}) \qquad (20.20)$$

where x_i cannot exceed the available capacity in period $i-1$, C_{i-1}, and D_{i-1} the observed demand in that period. The Bellman equation is then

$$V_i(y) = E(\max\{p_{i-1}x_{i-1} + V_i(y - x_{i-1}) : x_{i-1} \in [0, \min\{D_{i-1}, x\}]\}), \quad 0 < i < N \quad (20.21)$$

$$V_N(y) = 0 \qquad (20.22)$$

It can be shown that the optimal policy takes a simple form as shown in Van Ryzin and Talluri [119]: the optimal solution $\{x_i^*, i = 1, \ldots, N\}$ is such that $x_1^* \leq x_2^* \leq \cdots \leq x_N^*$; the value x_i^* does not depend on D_i, that is, the optimal inventory decision in a certain period can be taken *before* observing the demand in that period. According to the optimal policy, more and more capacity is reserved for classes with higher marginal value. This policy is sometimes termed *theft nesting*: at a given period, the set of higher-value classes is allocated at least a predetermined amount of inventory, but those classes can "steal" inventory from the lower-value class if seats are not sold by the end of the period; the opposite can never happen. An equivalent formulation of the problem uses as the relevant decision variables the booking limits for classes $i, \ldots, N$, defined as the sum of inventory allocations reserved to these classes. In the case of two classes (and two periods) only one decision is taken, and the solution to the recursive equation (20.1) takes the closed form for the booking limit for class 2: $C - F^{-1}(1 - p_1/p_2)$, where F is the cumulative distribution function of demand for class 2. This formula, also known as Littlewood's rule, has been used to formulate heuristics. Their popularity stems from their simplicity; from the fact that they predate the general formulation for the n-class problem; and finally because their sub-optimality, as observed from numerical experiments, is in the range 0–2%, which is considered low when compared to the vastly simplifying assumptions underlying the model. Two such heuristics are particularly popular in practice: expected marginal seat revenue version a (EMSR-a), and b (EMSR-b). Both originate from the work of Belobaba [134,135]. Both heuristics reduce the n-class problem to a 2-class problem. In the case of EMSR-a, pairwise comparisons of class i with higher-value classes are performed, and the booking limit for each such class in isolation is computed. The overall booking limit for classes $i+1, \ldots, N$ is then set equal to the sum of the individual booking limits. By considering booking limits in isolation, EMSR-a ignores pooling effects among the demand of various classes, and as a result it is often more conservative than the optimal policy. EMSR-b alleviates this shortcoming by aggregating the demand of classes $i+1, \ldots, N$ into an aggregated class, with demand equal to the sum of demands and marginal revenue equal to the weighted sum of marginal revenues: $\bar{p} = \sum_{k=i+1}^{N} p_k E[D_k] / \sum_{k=i+1}^{N} E[D_k]$. The booking limit in period i is then set by applying Littlewood's rule to class i and the aggregated class. EMSR-b is especially popular among practitioners, and has been extended in various directions. Notably, the case in which there is interaction between classes has been analyzed. In this setting, closing inventory of lower-value classes results in higher demand for higher-value classes (buy up); references on this subject are Belobaba and Weatherford [136] and Weatherford et al. [137].

Before concluding this section, we briefly describe the extension of single-leg capacity control mechanisms to a network setting. Indeed, the appropriate framework of analysis of RM is on an itinerary basis. However, the complexity of the problem increases significantly with size; moreover, network RM also poses organizational and data collection challenges, as responsibilities for revenue targets are assigned on a geographical basis, and data requirements for network RM are more demanding than for single-leg instances. Consequently, several exact formulations and heuristics are available. The heuristics rely on the concept of marginal network revenue value: when a seat is sold at a certain price, it generates revenue, but also results in lost revenues, as certain other itinerary fares will have one less seat available for sale. The network revenue value reflects this tradeoff, and can be effectively employed to make allocation decisions on a network. The advantage of such an approach lies in its robustness, and in the fact that it employs leg-based sub-problems, for which well-tested heuristics and tractable optimization formulations exist. As mentioned above, several network optimizations have been proposed as well. The topology of the network is captured through an incidence matrix, and the decision variables are the booking limits associated to an origin–destination class. The models differ in the way they capture uncertainty, in the intertemporal modeling of the decisions, and finally in the algorithmic approaches adopted to find a solution, exact or approximated. A comprehensive and current description of models is provided in Van Ryzin and Talluri [119]. There are no available data quantifying the financial benefits of network RM. Estimates rely on simulations, as those provided in the industry-standard platform PODS (Passenger Origin–Destination Simulator) and on the assessment of industry experts. The agreement [9] is that network RM can yield an improvement of 1% above and beyond single-leg RM.

20.3.4 Pricing

Most airlines condition passenger demand through seat inventory control. An alternative is to condition demand through pricing. The two practices are conceptually similar: if the only differentiating attribute among a set of classes is price, then closing lower-value classes is equivalent to posting a bid price for tickets equal to the cheapest class that is still open. Price controls, however, are more granular than inventory controls, as price points are not limited to the finite number of classes; moreover, dynamic price controls can effectively be used to limit the access to available seats. In its early days, commercial flight did not have the instruments to change prices dynamically, and to effectively advertise these changes. Moreover, customers were not used to observing changing prices for the same service; finally, a suitable framework of analysis was not available. In recent years, however, we are witnessing the evolution of revenue management systems in the direction of price-based controls. Low-cost carriers have grown in number and have thrived: Southwest, JetBlue in the U.S., EasyJet, RyanAir in Europe, TAM, and Go in South America. These carriers often offer a single product: an economy, nonrefundable ticket, and this product's availability is controlled through price. Despite the simplicity of the setting, dynamic pricing is a complex problem that has been attacked from many different angles, and is far from being solved. To obtain insightful results, researchers are forced to make simplifying assumptions that are either unrealistic or are difficult to test. Perhaps the single biggest challenge in dynamic pricing is understanding consumer behavior. Among the questions that have not received a definite answer are: how sensitive are the customers to a product's price changes, and to price changes of substitute products? Do customers take into account expected future prices when making their purchase decisions, and if they do, what is their learning process? How can we define the potential market for a certain product, and what is the impact of market size on optimal pricing? Some of these issues are studied in the marketing literature (for example, Nagle and Hogan [138]), but empirical validation in the airline industry is still missing. Perhaps the simplest model that captures the essential features of the airline industry is given by the following deterministic formulation for a single-product pricing problem with myopic customers:

$$\max \quad \sum_{i=1}^{N} \pi_i(p_i, d_i(p_i)) \qquad (20.23)$$

$$\text{subject to:} \quad \sum_{i=1}^{N} d_i(p_i) \leq C \qquad (20.24)$$

$$p_i \geq 0, \quad i = 1, \ldots, N \qquad (20.25)$$

The decision variables are the prices p_i of the product in period i, the market response is modeled by a deterministic demand function $d_i(\cdot)$, and profit is function of both price and demand. The above model assumes customers' myopic behavior, as demand in a certain period does not depend on the sequence of observed prices. Extensions of the model are possible in several directions. Gallego and Van Ryzin [139] considered a continuous-time, stochastic version of the pricing problem, and proved monotonicity and structural properties of the solution. The same authors analyze the multiperiod, multiproduct case (Gallego and Van Ryzin[140]). The case of non-myopic behavior has been analyzed by Stokey [141], Gale and Holmes [142], Gallego and Sahin [143], and Gallego et al. [144], who consider consumers with intertemporal valuations.

20.4 Aircraft Load Planning

The general problem of aircraft load planning can be described as transporting equipment (vehicles, palletized cargo, helicopters, ammunitions, etc.) and personnel from a set of departure points to a set of destinations with the objective of minimizing the number of loads used. The arrangement of the equipment and personnel affects the position of the center of gravity of the aircraft, which in turn has an impact on aircraft drag, which in turn has an impact on fuel consumption. The problem is further complicated by additional constraints such as hazardous material incompatibilities. The aircraft load planner must balance the load to ensure the safe and efficient use of the aircraft by complying with aircraft safety, weight and balance, and floor load restrictions for takeoff, in-flight, and landing. The speed and ease of on-loads and off-loads are of utmost importance too.

Aircraft load planning is both an important civilian and military application. The load planning process was done by manual procedures until 1977 when the first account of the use of computer-assisted approaches began within the U.S. Air Force, according to Martin-Vega [145], who provides a comprehensive review of work prior to 1985. Early semi-automated systems include CARLO (Cargo Loading system used by Scandinavian Airlines) by Larsen and Mikkelsen [146], AALPS (Automated Air Load Planning System by SRI International for the U.S. Army) as described in Frankel et al. [147,148] and Anderson and Ortiz [149], and DMES in Cochard and Yost [150]. Today AALPS is used at more than 300 sites throughout the continental United States Army community and by operations *Just Cause, Desert Shield, Desert Storm,* and *Restore Democracy.* Lockheed Corporation and the U.S. Air Force have utilized AALPS for designing cargo aircraft and configuring the C-17, respectively. Additional work in the area of aircraft load planning includes that of Eilon and Christofides [151], Hane et al. [152], Ng [153], Amiouny et al. [154], Heidelberg et al. [155], Thomas et al. [156], Gueret et al. [157], and Mongeau and Bes [158].

The aircraft loading problem belongs to the broad class of cutting and packing problems [159]. Even one-dimensional packing problems, such as the knapsack and bin-packing problems, are NP-hard. Typically 2-D bin-packing with the length and width of both the cargo and the aircraft's cargo hold are used. Height is not as significant because usually cargo is not stacked. Higher dimensional problems are much more combinatorial and cannot be solved exactly for large instances. A heuristic approach must be taken to achieve an acceptable result in a short amount of time, as discussed in Bischoff and Marriott [160] and Pisinger [161].

The challenge of balancing is generally focused on one dimension: the length of the plane, that is, positioning the center of gravity along the fore and aft axis of the aircraft. Balancing the center of gravity side-to-side is generally considered not significant.

Mongeau and Bes [158] take a mathematical programming approach to the aircraft loading problem for Airbus France, which can be solved optimally in a reasonable time.

To illustrate, the formulation is given as:

$$\max \quad M(x) \tag{20.26}$$

$$\text{subject to:} \quad CG(x) \le X_{\text{stab}} \tag{20.27}$$

$$M(x) \le M_{\max} \tag{20.28}$$

$$\sum_{i=1}^{N_{\text{cont}}} \sum_{j \in H_k} M_i x_{ij} \le M_{\max}^k, \quad k = 1, 2, \ldots, N_{\text{hold}} \tag{20.29}$$

$$\sum_{j=1}^{N_{\text{comp}}} x_{ij} \le 1, \qquad \forall i = 1, 2, \ldots, N_{\text{cont}} \tag{20.30}$$

$$\sum_{j=1}^{N_{\text{comp}}} x_{ij} = 1, \qquad \forall i \in I \tag{20.31}$$

$$x_{ik} = 1, \qquad \text{for any given container } i \text{ required} \tag{20.32}$$
$$\text{to be in a specific compartment } k$$

$$X_{\text{target}} - \varepsilon \le CG(x) \le X_{\text{target}} + \varepsilon \tag{20.33}$$

The mathematical formulation requires the following notation with parameters:

N_{cont} = number of containers on the ground,
N_{comp} = number of compartments,
N_{hold} = number of holds,
M_i = mass of container i,
$CG(x)$ = center of gravity of the aircraft after loading,
X_{stab} = maximal (longitudinal) position of the center of gravity of the aircraft after loading to satisfy stability requirements,
$M_{\max}$ = maximal mass of freight that can be loaded,
$M_{\max}^k$ = maximal mass of freight that can be loaded in hold k and $k = 1, 2, \ldots, N_{\text{hold}}$,
H_k where $H_k \subseteq \{1, 2, \ldots, N_{\text{comp}}\}$ are the compartments in hold k and $k = 1, 2, \ldots, N_{\text{hold}}$,
I = a subset of the container list that is required to be loaded,
X_{target} = ideal (longitudinal) position of the center of gravity of the aircraft after loading,
ε = some positive allowable displacement of the center of gravity of the aircraft from its ideal position,
$x_{ij} \in \{0, 1\}$ where $x_{ij} = 1$, if container i is to be placed in container j, and 0 otherwise; and $i = 1, 2, \ldots, N_{\text{cont}}$; $j = 1, 2, \ldots, N_{\text{comp}}$.

The objective given in Equation 20.26 is to maximize the mass loaded $M(x)$ where

$$M(x) = \sum_{i=1}^{N_{\text{cont}}} \sum_{j=1}^{N_{\text{comp}}} M_i x_{ij} \tag{20.34}$$

Equation 20.27 ensures that the aircraft stability requirement is satisfied. Equations 20.28 and 20.29 limit the stress/mass capacity overall and for each hold, respectively. Equations 20.30 state that each container must be loaded at most once. Equations 20.31 enforce

Airline Optimization

that a subset I of the container list is loaded. Equations 20.32 ensure, in the event that a specific container must be placed in a specific compartment, that the requirement is met. Equation 20.33 keeps the center of gravity within an allowable displacement from the ideal position. In the formulation, Mongeau and Bes [158] also include volume capacity constraints, which involve nonconvex, piecewise-linear functions that are transformed into simple linear constraints. These constraints are quite complex and vary according to both the aircraft and the container types. The interested reader may refer to the details in the article.

20.5 Future Research Directions and Conclusions

The application of Operations Research to airline planning processes has resulted in significant benefits to the industry through cost reductions, revenue improvements, operational efficiencies, and automated tools. The combination of advancements in computer hardware and software technologies with clever mathematical algorithms and heuristics has dramatically transformed the ability of operations researchers to solve large-scale, sophisticated airline optimization problems over the past 60 years. This trend will continue in the future; airline planning problems are rich with complexities that can lead to breakthroughs in methods and decision-support tools.

Many promising areas for future research exist in the airline industry. Some examples include:

1. Incorporating new forecasting techniques, overbooking strategies, and pricing strategies into revenue management decisions to address the increased market transparency of fares and fare classes on the Web.
2. Developing accurate consumer response models that take into account the multiple sales channels available to customers (e.g., travel agency, phone, online agencies, online airline Web sites), and their ability to compare fares.
3. Integrating schedule development to incorporate aircraft and crew planning decisions; for example, simultaneous consideration of aircraft routing, fleet assignment, and crew pairing.
4. Dynamically reassigning aircraft capacities to the flight network when improved passenger demand forecasts become available to maximize total revenue. This line of research of demand driven re-fleeting/demand driven dispatching has been recently explored in Sherali et al. [162] and Wang and Regan [163].
5. Exploiting the availability of real-time data to enable real-time decision-making; for example, in the area of recovery from irregular operations.
6. Developing real-time policies to assign locations and decide ordering quantities for spare parts. The problem formulation must take into account savings obtained from sharing the same warehouse for multiple service centers, and the service time requirements. Similar systems have already been implemented in military and industrial production environments.
7. Developing loyalty programs and customer contact programs that take into account the financial value associated to a customer, as well as his/her potential future evolution. This line of research, traditionally associated to the field of relationship marketing, has been recently explored in the context of airline campaign management [164].

8. Developing decision-support tools for fractional fleet services, which are growing in popularity, to address the stochastic nature of demand.
9. Developing new algorithms for aircraft fuel operational optimization and fuel hedging as fuel represents the second largest cost component of an airline's operations after labor [165].

References

1. The economic & social benefits of air transport, Air Transport Action Group brochure, 2004.
2. IATA Passenger and Freight Forecast 2005–2009, October 2005.
3. ICAO Annual Report of the Council, 2004.
4. Airports Council International (ACI) figure, 2005.
5. CANSO estimation, 2005.
6. Airbus: Global market forecast 2004–2023, http://www.airbus.com/en/myairbus/global_market_forcast.html.
7. Boeing: 2006 current market outlook, http://www.boeing.com/commercial/cmo.
8. Nemhauser, G. L., The age of optimization: solving large-scale real-world problems, *Operations Research*, 42, 5, 1994.
9. Barnhart, C., Belobaba, P., and Odoni, A. R., Applications of operations research in the air transport industry, *Transportation Science*, 37, 368, 2003.
10. Hicks, R. et al., Bombardier Flexjet significantly improves its fractional aircraft ownership operations, *Interfaces*, 35, 49, 2005.
11. Armacost, A. P. et al., UPS optimizes its air network, *Interfaces*, 34, 15, 2004.
12. Yu, G. et al., A new era for crew recovery at Continental Airlines, *Interfaces*, 33, 5, 2003.
13. Butchers, E. R. et al., Optimized crew scheduling at Air New Zealand, *Interfaces*, 31, 30, 2001.
14. Subramanian, R. et al., Coldstart: fleet assignment at Delta Air Lines, *Interfaces*, 24, 104, 1994.
15. Smith, B. C., Leimkuhler, J. F., and Darrow, R. M., Revenue management at American Airlines, *Interfaces*, 22, 8, 1992.
16. Etschmaier, M. and Mathaisel, D., Airline scheduling: an overview, *Transportation Science*, 19, 127, 1985.
17. Rushmeier, R. and Kontogiorgis, S. A., Advances in the optimization of airline fleet assignment, *Transportation Science*, 31, 159, 1997.
18. Barnhart, C. and Talluri, K., Airlines operations research, in C. ReVelle and A. E. McGarity (eds.), *Design and Operation of Civil and Environmental Engineering Systems*, John Wiley & Sons, Inc., New York, 435, 1997.
19. Clarke, M. and Smith, B., Impact of operations research on the evolution of the airline industry, *Journal of Aircraft*, 41, 62, 2004.
20. Klabjan, D., Large-scale models in the airline industry, in G. Desaulniers, J. Desrosiers, and M. M. Solomon (eds.), *Column Generation*, Springer, New York, 2005.
21. Barnhart, C. and Cohn, A., Airline schedule planning: accomplishments and opportunities, *Manufacturing & Service Operations Management*, 6, 3, 2004.
22. Rushmeier, R. A., Hoffman, K. L., and Padberg, M., Recent advances in exact optimization of airline scheduling problems, Technical Report, 1995.
23. Erdmann, A. et al., Modeling and solving an airline schedule generation problem, *Annals of Operations Research*, 107, 117, 2001.

24. Armacost, A., Barnhart, C., and Ware, K., Composite variable formulations for express shipment service network design, *Transportation Science*, 36, 1, 2002.
25. Armacost, A., Barnhart, C., Ware, K., and Wilson, A., UPS optimizes its air network, *Interfaces*, 34, 15, 2004.
26. Sherali, H. D., Bish, E. K., and Zhu, X., Airline fleet assignment concepts, models, and algorithms, *European Journal of Operational Research*, 172, 1, 2006.
27. Gopalan, R. and Talluri, K. T., Mathematical models in airline schedule planning: a survey, *Annals of Operations Research*, 76, 155, 1998.
28. Yu, G. and Yang, J., Optimization applications in the airline industry, in D. Z. Du and P. M. Pardalos (eds.), *Handbook of Combinatorial Optimization*, Kluwer Academic Publishers, Norwell, MA, 635, 1998.
29. Abara, J., Applying integer linear programming to the fleet assignment problem, *Interfaces*, 19, 20, 1989.
30. Daskin, M. S. and Panayotopoulos, N. D., A Lagrangian relaxation approach to assigning aircraft to routes in hub and spoke networks, *Transportation Science*, 23, 91, 1989.
31. Talluri, K. T., Swapping applications in a daily airline fleet assignment, *Transportation Science*, 30, 237, 1996.
32. Jarrah, A., Goodstein, J., and Narasimhan, R., An efficient airline re-fleeting model for the incremental modification of planned fleet assignments, *Transportation Science*, 34, 349, 2000.
33. Hane, C. A. et al., The fleet assignment problem: solving a large-scale integer program, *Mathematical Programming*, 70, 211, 1995.
34. Berge, M. A. and Hopperstad, C. A., Demand driven dispatch: a method for dynamic aircraft capacity assignment, models, and algorithms, *Operations Research*, 41, 153, 1993.
35. Desaulniers, G. et al., Daily aircraft routing and scheduling, *Management Science*, 43, 841, 1997.
36. Rexing, B. et al., Airline fleet assignment with time windows, *Transportation Science*, 34, 1, 2000.
37. Lohatepanont, M. and Barnhart, C., Airline scheduling planning: integrated models and algorithms for schedule design and fleet assignment, *Transportation Science*, 38, 19, 2004.
38. Yan, S. and Tseng, C. H., A passenger demand model for airline flight scheduling and fleet routing, *Computers and Operations Research*, 29, 1559, 2002.
39. Clarke, L. W. et al., Maintenance and crew considerations in fleet assignment, *Transportation Science*, 30, 249, 1996.
40. Barnhart, C. et al., Flight string models for aircraft fleeting and routing, *Transportation Science*, 32, 208, 1998.
41. Rosenberger, J. M., Johnson, E. L., and Nemhauser, G. L., A robust fleet-assignment model with hub isolation and short cycles, *Transportation Science*, 38, 357, 2004.
42. Belanger, N. et al., Weekly airline fleet assignment with homogeneity, *Transportation Research Part B—Methodological*, 40, 306, 2006.
43. Farkas, A., The influence of network effects and yield management on airline fleet assignment decisions, Ph.D. dissertation, MIT, Cambridge, MA, 1995.
44. Kniker, T. S., Itinerary-based airline fleet assignment, Ph.D. dissertation, Massachusetts Institute of Technology, Cambridge, MA, 1998.
45. Jacobs, T. L., Smith, B. C., and Johnson, E., O&D FAM: Incorporating passenger flows into the fleeting process, In: *Proceedings of the AGIFORS Symposium*, New Orleans, LA, 128, 1999.
46. Barnhart, C., Kniker, T. S., and Lohatepanont, M., Itinerary-based airline fleet assignment, *Transportation Science*, 36, 199, 2002.

47. Levin, A., Scheduling and fleet routing models for transportation system, *Transportation Science*, 5, 232, 1971.
48. Cordeau, J., Stojković, G., Soumis, F., and Desrosiers, J., Benders decomposition for simultaneous aircraft routing and crew scheduling, *Transportation Science*, 35, 375, 2001.
49. Feo, T. and Bard, J., Flight scheduling and maintenance base planning, *Management Science*, 35, 1415, 1989.
50. Clarke, L., Johnson, E., Nemhauser, G., and Zhu, Z., The aircraft rotation problem, in R.E. Burkard, T. Ibaraki, and M. Queyranne (eds.), *Annals of OR: Mathematics of Industrial Systems*, II, 33, 1997.
51. Gopalan, R. and Talluri, K., The aircraft maintenance routing problem, *Operations Research*, 46, 260, 1998.
52. Gabteni, S. and Gronkvist, M., A hybrid column generation and constraint programming optimizer for the tail assignment problem, *Computer Science*, 3990, 89, 2006.
53. Li, Y. H. and Wang, X. B., Integration of fleet assignment and aircraft routing, *Airports, Airspace, and Passenger Management Transportation Research Record*, 1915, 79, 2005.
54. Mercier, A., Cordeau, J. F., and Soumis, F., A computational study of Benders decomposition for the integrated aircraft routing and crew scheduling problem, *Computers & Operations Research*, 32, 1451, 2005.
55. Cohn, A. M. and Barnhart, C., Improving crew scheduling by incorporating key maintenance routing decisions, *Operations Research*, 51, 387, 2003.
56. Arabeyre, J. P. et al., The airline crew scheduling problem: a survey, *Transportation Science*, 3, 140, 1969.
57. Bornemann, D. R., The evolution of airline crew pairing optimization, *The 22nd AGIFORS Crew Management Study Group Proceeding*, Paris, France, 1982.
58. Barutt, J., and Hull, T., Airline crew scheduling: supercomputers and algorithms, *SIAM News*, 23, 20, 1990.
59. Desaulniers, G. et al., Crew scheduling in air transportation, in T. Cranic and G. Laporte (eds.), *Fleet Management and Logistics*, Kluwer Academic Publishers, Boston, MA, 1998.
60. Barnhart, C. et al., Airline crew scheduling, in Randolph W. Hall (ed.), *Handbook of Transportation Science*, 2nd ed., Kluwer Academic Publishers, Norwell, MA, 2003.
61. Gopalakrishnan, B. and Johnson, E. L., Airline crew scheduling: state-of-the-art, *Annals of Operations Research*, 140, 305, 2005.
62. Anbil, R., Forrest, J. J., and Pulleyblank, W. R., Column generation and the airline crew pairing problem, *Documenta Mathematica Journal der Deutschen Mathematiker, Extra Volume ICM*, 677, 1998.
63. Gershkoff, I., Optimizing flight crew schedules, *Interfaces*, 19, 29, 1989.
64. Rubin, J., A technique for the solution of massive set covering problems, with applications to airline crew scheduling, *Transportation Science*, 7, 34, 1973.
65. Anbil, R. et al., Recent advances in crew pairing optimization at American Airlines, *Interfaces*, 21, 62, 1991.
66. Anbil, R. et al., Crew-pairing optimization at American Airlines Decision Technologies, *Optimization in Industry*, in T. A. Ciriani and R. C. Leachman (eds.), John Wiley & Sons, 31, 1993.
67. Graves, G. W. et al., Flight crew scheduling, *Management Science*, 39, 736, 1993.
68. Hu, J. and Johnson, E., Computational results with a primal-dual subproblem simplex method, *Operations Research Letters*, 25, 149, 1999.
69. Minoux, M., Column generation techniques in combinatorial optimization: a new application to the crew pairing problems, *Proceedings of the XXIV AGIFORS Symposium*, France, 1984.

70. Baker, E. K., Bodin, L. D., and Fisher, M., The development of a heuristic set covering based system for air crew scheduling, *Transportation Policy Decision Making*, 3, 95, 1985.
71. Lavoie, S., Minoux, M., and Odier, E., A new approach of crew pairing problems by column generation and application to air transport, *European Journal of Operational Research*, 35, 45, 1988.
72. Barnhart, C. et al., A column generation technique for the long-haul crew assignment problem, in T. A. Ciriani and R. Leachman (eds.), *Optimization in Industry 2: Mathematical Programming and Modeling Techniques in Practice*, John Wiley & Sons, New York, 7, 1994.
73. Desaulniers, G. et al., Crew scheduling in air transportation, *Les Cahiers du GERAD*, G97-26, GERAD Inst. of Montreal, Quebec, Canada, 1997.
74. Stojković, M., Soumis, F., and Desrosiers, J., The operational airline crew scheduling problem, *Transportation Science*, 32, 232, 1998.
75. Galia, R. and Hjorring, C., Modeling of complex costs and rules in a crew pairing column generator, *Technical Report CRTR-0304*, Carmen Systems, 2003.
76. Barahona, F. and Anbil, R., The volume algorithm: producing primal solutions with a subgradient algorithm, *Mathematical Programming*, 87, 385, 2000.
77. Barahona, F. and Anbil, R., On some difficult linear programs coming from set partitioning, *Discrete Applied Mathematics*, 118, 3, 2002.
78. Hoffman, K. L. and Padberg, M., Solving airline crew scheduling problems by branch-and-cut, *Management Science*, 39, 657, 1993.
79. Vance, P. H., Barnhart, C., Johnson, E. L., and Nemhauser, G. L., Airline crew scheduling: a new formulation and decomposition algorithm, *Operations Research*, 45, 188, 1997.
80. Klabjan, D. et al., Solving large airline crew scheduling problems: random pairing generation and strong branching, *Computational Optimization and Applications*, 20, 73, 2001.
81. Makri, A. and Klabjan, D., Efficient column generation techniques for airline crew scheduling, *INFORMS Journal on Computing*, 16, 56, 2004.
82. Tran, V. H., Reinelt, G., and Bock, H. G., BoxStep methods for crew pairing problems, *Optimization and Engineering*, 7, 33, 2006.
83. Marchiori, E. and Steenbeck, A., An evolutionary algorithm for large scale set covering problems with application to airline crew scheduling, in *Real-World Applications of Evolutionary Computing*, LNCS1803, 367, 2000.
84. Ozdemir, H. T. and Mohan, C., Flight graph based genetic algorithm for crew scheduling in airlines, *Information Sciences*, 133, 165, 2001.
85. Kornilakis, H. and Stamatopoulos, P., Crew pairing optimization with genetic algorithms, *Methods and Applications of Artificial Intelligence, Lecture Notes in Artificial Intelligence*, 2308, 109, 2002.
86. Lagerholm, M., Peterson, C., and Söderberg, B., Airline crew scheduling using Potts mean field techniques, *European Journal of Operational Research*, 120, 81, 2000.
87. Kohl, N. and Karisch, S. E., Airline crew rostering: problem types, modeling, and optimization, *Annals of Operations Research*, 127, 223, 2004.
88. Cappanera, P. and Gallo, G., A multicommodity flow approach to the crew rostering problem, *Operations Research*, 52, 583, 2004.
89. Caprara, A. et al., Modeling and solving the crew rostering problem, *Operations Research*, 46, 820, 1998.
90. Christou, I. T. et al., A two-phase genetic algorithm for large-scale bidline-generation problems at Delta Air Lines, *Interfaces*, 29, 51, 1999.
91. Dawid, H., König, J., and Strauss, C., An enhanced rostering model for airline crews, *Computers and Operations Research*, 28, 671, 2001.

92. Day, P. R. and Ryan, D. M., Flight attendant rostering for short-haul airline operations, *Operations Research*, 45, 649, 1997.
93. Gamache, M. and Soumis, F., A method for optimally solving the rostering problem, in G. Yu (ed.), *Operations Research in the Airline Industry*, Kluwer Academic Publishers, Norwell, MA, 1998, 124.
94. Gamache, M. et al., The preferential bidding system at Air Canada, *Transportation Science*, 32, 246, 1998.
95. Gamache, M. et al., A column generation approach for large-scale aircrew rostering problems, *Operations Research*, 47, 247, 1999.
96. Shebalov, S. and Klabjan, D., Robust airline crew pairing: move-up crews, *Transportation Science*, 40, 300, 2006.
97. Sohoni, M. G., Johnson, E. L., and Bailey, T. G., Operational airline reserve crew planning, *Journal of Scheduling*, 9, 203, 2006.
98. Lučić, P. and Teodorvić, D., Simulated annealing for the multi-objective aircrew rostering problem, *Transportation Research Part A*, 33, 19, 1999.
99. Campbell, K., Durfee, R., and Hines, G., FedEx generates bid lines using simulated annealing, *Interfaces*, 27, 1, 1997.
100. Clarke, M., Development of heuristic procedures for flight rescheduling in the aftermath of irregular airline operations, Ph.D. dissertation, Massachusetts Institute of Technology, International Center for Air Transportation, Cambridge, MA, 1997.
101. Rosenberger, J. M., Johnson, E. L., and Nemhauser, G. L., Rerouting aircraft for airline recovery, *Transportation Science*, 37, 408, 2003.
102. Kohl, N. et al., Airline disruption management—perspectives, experiences, and outlook, forthcoming in *Journal of Air Transport Management*.
103. Anbil, R., Barahona, F., Ladanyi, L., Rushmeier, R., and Snowdon, J., Airline optimization, *OR/MS Today*, 26, 26, 1999.
104. Teodorvić, D. and Stojković, G., Model for operational daily airline scheduling, *Transportation Planning and Technology*, 14, 273, 1990.
105. Wei, G., Yu, G., and Song, M., Optimization model and algorithm for crew management during airline irregular operations, *Journal of Combinatorial Optimization*, 1, 305, 1997.
106. Song, M., Wei, G., and Yu, G., A decision support framework for crew management during airline irregular operations, in G. Yu (ed.), *Operations Research in the Airline Industry*, Kluwer Academic Publishers, Norwell, MA, 1998, 260.
107. Abdelghany, A., Ekollu, G., Narasimhan, R., and Abdelghany, K., A proactive crew recovery decision support tool for commercial airlines during irregular operations, *Annals of Operations Research*, 127, 309, 2004.
108. Nissen, R. and Haase, K., Duty-period-based network model of crew rescheduling in European airlines, *Journal of Scheduling*, 9, 255, 2006.
109. Schaefer, A. J. et al., Airline crew scheduling under uncertainty, *Transportation Science*, 39, 340, 2005.
110. Schaefer, A. J. and Nemhauser, G. L., Improving airline operational performance through schedule perturbation, *Annals of Operations Research*, 144, 3, 2006.
111. Yen, J. W. and Birge, J. R., A stochastic programming approach to the airline crew scheduling problem, *Transportation Science*, 40, 3, 2006.
112. Lettovský, L., Airline operations recovery: an optimization approach, Ph.D. thesis, Georgia Institute of Technology, 1997.
113. Lettovský, L., Johnson, E., and Nemhauser, G., Airline crew recovery, *Transportation Science*, 34, 337, 2000.
114. Stojković, M. and Soumis, F., An optimization model for the simultaneous operational flight and pilot scheduling problem, *Management Science*, 47, 1290, 2001.

115. Stojković, M. and Soumis, F., The operational flight and multi-crew scheduling problem, Technical Report G-2000-27, GERAD, 2003.
116. Lan, S., Clarke, J. P., and Barnhart, C., Planning for robust airline operations: optimizing aircraft routings and flight departure times to minimize passenger disruptions, *Transportation Science*, 40, 15, 2006.
117. Bratu, S. and Barnhart, C., Flight operations recovery: new approaches considering passenger recovery, *Journal of Scheduling*, 9, 279, 2006.
118. Phillips, R. L., *Pricing and Revenue Optimization*, Stanford University Press, Stanford, CA, 2005.
119. Van Ryzin, G. J., and Talluri, K. T., *The Theory and Practice of Revenue Management*, Kluwer Academic Publishers, Norwell, MA, 2004.
120. Yeoman, I. and McMahon-Beattie, U. (eds.), *Revenue Management and Pricing: Case Studies and Applications*, Int. Thomson Business Press, London, 2004.
121. Weatherford, L. R. and Bodily, S. E., A taxonomy and research overview of perishable-asset revenue management: yield management, overbooking, and pricing, *Operations Research*, 40, 831, 1992.
122. McGill, J. I. and Van Ryzin, G. J., Revenue management: research overview and prospects, *Transportation Science*, 33, 233, 1999.
123. Weatherford, L. R. and Kimes, S. E., A comparison of forecasting methods for hotel revenue management, *International Journal of Forecasting*, 19, 401, 2003.
124. Weatherford, L. R. and Pölt, S., Better unconstraining of airline demand data in revenue management systems for improved forecast accuracy and greater revenues, *Journal of Revenue Management and Pricing*, 1, 234, 2001.
125. Beckmann, J. M., Decision and team problems in airline reservations, *Econometrica*, 26, 134, 1958.
126. Rothstein, M., Stochastic models for airline booking policies, Ph.D. thesis, Graduate School of Engineering and Science, New York University, New York, 1968.
127. Alstrup, J., Boas, S., Madsen, O. B. G., Vidal, R., and Victor, V., Booking policy for flights with two types of passengers, *European Journal of Operational Research*, 27, 274, 1986.
128. Alstrup, J., Booking control increases profit at Scandinavian Airlines, *Interfaces*, 19, 10, 1989.
129. Simon, J., An almost practical solution to airline overbooking, *Journal of Transportation and Economic Policy*, 2, 201, 1968.
130. Simon, J., Airline overbooking: the state of the art—a reply, *Journal of Transportation and Economic Policy*, 6, 254, 1972.
131. Vickrey, W., Airline overbooking: some further solutions, *Journal of Transportation and Economic Policy*, 6, 257, 1972.
132. Lawrence, R. D., Hong, S. J., and Cherrier, J., Passenger-based predictive modeling of airline no-show rates, *ACM SIGKDD '03*, 2003.
133. Littlewood, K., Forecasting and control of passenger bookings, *AGIFORS Symposium Proceedings*, 1972, 95.
134. Belobaba, P., Air travel demand and airline seat inventory management, Ph.D. thesis, Flight Transportation Laboratory, Massachusetts Institute of Technology, Cambridge, MA, 1987.
135. Belobaba, P., Application of a probabilistic decision model to airline seat inventory control, *Operations Research*, 37, 183, 1989.
136. Belobaba, P., and Weatherford, L. R., Comparing decision rules that incorporate customer diversion in perishable asset revenue management situations, *Decision Sciences*, 27, 343, 1996.

137. Weatherford, L., Bodily, S. E., and Pfeifer, P. E., Modeling the customer arrival process and comparing decision rules in perishable asset revenue management situations, *Transportation Science*, 27, 239, 1993.
138. Nagle, T. T. and Hogan, J., *The Strategy and Tactics of Pricing: A Guide to Growing More Profitably* (4th ed.), Prentice Hall, Upper Saddle River, NJ, 2005.
139. Gallego, G. and Van Ryzin, G., Optimal dynamic pricing of inventories with stochastic demand over finite horizons, *Management Science*, 40, 999, 1994.
140. Gallego, G. and Van Ryzin, G., A multi-product dynamic pricing problem and its applications to network yield management, *Operations Research*, 45, 24, 1997.
141. Stokey, N., Intertemporal price discrimination, *Quarterly Journal of Economics*, 94, 355, 1979.
142. Gale, I. L. and Holmes, T. J., The efficiency of advance-purchase discounts in the presence of aggregate demand uncertainty, *International Journal of Industrial Organization*, Elsevier, 10, 413, 1992.
143. Gallego, G. and Sahin, O., Inter-temporal valuations, product design and revenue management, Working paper, Columbia University Department of Industrial Engineering and Operations Research, New York, 2006.
144. Gallego, G., Phillips, R., and Sahin, O., Strategic management of distressed inventory, Working paper, Columbia University Department of Industrial Engineering and Operations Research, New York, NY, 2004.
145. Martin-Vega, L., Aircraft load planning and the computer, *Computers & Industrial Engineering*, 9, 357, 1985.
146. Larsen, O. and Mikkelsen, G., An interactive system for the loading of cargo aircraft, *European Journal of Operational Research*, 4, 367, 1980.
147. Frankel, M. S. et al., Army data distributions system/packet radio testbed, Contract MDA903-80-C-0217, Quarterly Technical Reports 1–3, SRI International, Menlo Park, California, 1980.
148. Frankel, M. S. et al., Army data distributions system/packet radio testbed, Contract MDA903-80-C-0217, Final Technical Report, SRI International, Menlo Park, California, 1981.
149. Anderson, D. and Ortiz, C., AALPS: a knowledge-based system for aircraft loading, *IEEE Expert*, 1987, 71.
150. Cochard, D. D. and Yost, K. A., Improving utilization of air force cargo aircraft, *Interfaces*, 15, 53, 1985.
151. Eilon, S. and Christofides, N., The loading problem, *Management Science*, 17, 259, 1971.
152. Hane, C. et al., Plane loading algorithms, presented at the ORSA/TIMS Joint Meeting, Denver, 1988.
153. Ng, K. Y. K., A multicriteria optimization approach to aircraft loading, *Operations Research Technical Note*, 40, 1200, 1992.
154. Amiouny, S. V. et al., Balanced loading, *Operations Research*, 40, 238, 1992.
155. Heidelberg, K. R. et al., Automated air load planning, *Naval Research Logistics*, 45, 751, 1998.
156. Thomas, C. et al., Airbus packing at Federal Express, *Interfaces*, 28, 21, 1998.
157. Gueret, C. et al., Loading aircraft for military operations, *Journal of the Operational Research Society*, 54, 458, 2003.
158. Mongeau, M. and Bes, C., Optimization of aircraft container loading, *IEEE Transactions on Aerospace and Electronic Systems*, 39, 140, 2003.
159. Sweeney, P. E. and Paternoster, E. R., Cutting and packing problems: a categorized, application-oriented research bibliography, *Journal of the Operational Research Society*, 43, 691, 1992.

160. Bischoff, E. E. and Marriott, M. E., A comparative evaluation of heuristics for container loading, *European Journal of Operational Research*, 44, 267, 1990.
161. Pisinger, D., Heuristics for the container loading problem, *European Journal of Operational Research*, 141, 382, 2002.
162. Sherali, H. D., Bish, E. K., and Zhu, X. M., Polyhedral analysis and algorithms for a demand-driven refleeting model for aircraft assignment, *Transportation Science*, 39, 349, 2005.
163. Wang, X. B. and Regan, A., Dynamic yield management when aircraft assignments are subject to swap, *Transportation Research Part B—Methodological*, 40, 563, 2006.
164. Labbi, A., Tirenni, G., Berrospi, C., and Elisseeff A., Customer equity and lifetime management (CELM): Finnair case study, *Marketing Science*, submitted.
165. Subcommittee on Aviation Hearing on Commercial Jet Fuel Supply: impact and cost on the U.S. airline industry, http://www.house.gov/transportation/aviation/02-15-06/02-15-06memo.html, 2006.

21
Financial Engineering

21.1	Introduction..	**21**-1
21.2	Return..	**21**-3
	Expected Portfolio Return • Portfolio Variance	
21.3	Estimating an Asset's Mean and Variance.......	**21**-4
	Estimating Statistics Using Historical Data • The Scenario Approach to Estimating Statistics	
21.4	Diversification	**21**-6
21.5	Efficient Frontier	**21**-8
	Risk-Reward Portfolio Optimization	
21.6	Utility Analysis.....................................	**21**-10
	Utility Functions • Utility in Practice	
21.7	Black–Litterman Asset Allocation Model........	**21**-13
	Market Equilibrium Returns • Investor Views • An Example of an Investor View • Combining Equilibrium Returns with Investor Views • Application of the Black–Litterman Model	
21.8	Risk Management	**21**-18
	Value at Risk • Coherent Risk Measures • Risk Management in Practice	
21.9	Options	**21**-22
	Call Option Payoffs • Put Option Payoffs	
21.10	Valuing Options	**21**-24
	Risk Neutral Valuation in Efficient and Complete Markets • Brownian Motion • Simulating Risk-Neutral Paths for Options Pricing • A Numerical Example	
21.11	Dynamic Programming.............................	**21**-28
21.12	Pricing American Options Using Dynamic Programming.............................	**21**-29
	Building the T Stage Tree • Pricing the Option Using the Binomial Tree • A Numerical Example	
21.13	Comparison of Monte Carlo Simulation and Dynamic Programming	**21**-33
21.14	Multi-Period Asset Liability Management.......	**21**-33
	Scenarios • Multi-Period Stochastic Programming	
21.15	Conclusions	**21**-36
	References	**21**-36

Aliza R. Heching and
Alan J. King
IBM T. J. Watson Research Center

21.1 Introduction

Operations research provides a rich set of tools and techniques that are applied to financial decision making. The first topic that likely comes to mind for most readers is Markowitz's

Nobel Prize–winning treatment of the problem of portfolio diversification using quadratic programming techniques. This treatment, which first appeared in 1952, underlies almost all of the subsequent research into the pricing of risk in financial markets. Linear programming, of course, has been applied in many financial planning problems, from the management of working capital to formulating a bid for the underwriting of a bond issue. Less well known is the fundamental role that duality theory plays in the pricing of options and contingent claims, both in its discrete state and time formulation using linear programming and in its continuous time counterparts. This duality leads directly to the Monte Carlo simulation method for pricing and evaluating the risk of options portfolios for investment banks; this activity probably comprises the single greatest use of computing resources in any industry.

This chapter does not cover every possible topic in the applications of operations research (OR) to finance. We have chosen to highlight the main topics in investment theory and to give an elementary, mostly self-contained, exposition of each. A comprehensive perspective of the application of OR techniques to financial markets along with an excellent bibliography of the recent literature in this area can be found in the survey by Board et al. (2003). In this chapter, we chose not to cover the more traditional applications of OR to financial management for firms, such as the management of working capital, capital investment, taxation, and financial planning. For these, we direct the reader to consult Ashford et al. (1988). We also excluded financial forecasting models; the reader may refer to Campbell et al. (1997) and Mills (1999) for treatments of these topics. Finally, Board et al. (2003) provide a survey of the application of OR techniques for the allocation of investment budgets between a set of projects. Complete and up-to-date coverage of finance and financial engineering topics for readers in operations research and management science may be found in the handbooks of Jarrow et al. (1995) and the forthcoming volume of Birge and Linetsky (2007).

We begin this chapter by introducing some basic concepts in investment theory. In Section 21.2, we present the formulas for computing the return and variance of return on a portfolio. The formulas for a portfolio's mean and variance presume that these parameters are known for the individual assets in the portfolio. In Section 21.3, we discuss two methods for estimating these parameters when they are not known.

Section 21.4 explains how a portfolio's overall risk can be reduced by including a diverse set of assets in the portfolio. In Section 21.5, we introduce the risk–reward tradeoff efficient frontier and the Markowitz problem. Up to this point, we have assumed that the investor is able to specify a mathematical function describing his attitude toward risk. In Section 21.6, we consider utility theory that does not require an explicit specification of a risk function. Instead, utility theory assumes that investors specify a utility, or satisfaction, with any cash payout. The associated optimal portfolio selection problem will seek to maximize the investor's expected utility.

Section 21.7 discusses the Black–Litterman model for asset allocation. Black and Litterman use Bayesian updating to combine historical asset returns with individual investor views to determine a posterior distribution on asset returns that is used to make asset allocation decisions. Section 21.8 considers the challenges of risk management. We introduce the notion of coherent risk measures and conditional value-at-risk ($CVaR$), and show how a portfolio selection problem with a constraint on $CVaR$ can be formulated as a stochastic program.

In Sections 21.9 through 21.13, we turn to the problem of options valuation. Options valuation combines a mathematical model for the behavior of the underlying uncertain market factors with simulation or dynamic programming (or combinations thereof) to determine

21.2 Return

Suppose that an investor invests in asset i at time 0 and sells the asset at time t. The rate of return (more simply referred to as the return) on asset i over time period t is given by:

$$r_i = \frac{\text{amount received at time } t - \text{amount invested in asset } i \text{ at time } 0}{\text{amount invested in asset } i \text{ at time } 0} \qquad (21.1)$$

Now suppose that an investor invests in a portfolio of N assets. Let f_i denote the fraction of the portfolio that is comprised of asset i. Assuming that no short selling is allowed, $f_i \geq 0$. Clearly, $\sum_{i=1}^{N} f_i = 1$.

The portfolio return is given by the weighted sum of the returns on the individual assets in the portfolio:

$$r_p = \sum_{i=1}^{N} f_i r_i \qquad (21.2)$$

We have described asset returns as if they are known with certainty. However, there is typically uncertainty surrounding the amount that will be received at the time that an asset is sold. We can use a probability distribution to describe this unknown rate of return. If return is normally distributed, then only two parameters—its expected return and its standard deviation (or variance)—are needed to describe this distribution. The expected return is the return around which the probability distribution is centered; it is the expected value of the probability distribution of possible returns. The standard deviation describes the dispersion of the distribution of possible returns.

21.2.1 Expected Portfolio Return

Suppose there are N assets with random returns $r_1, \ldots, r_N$. The corresponding expected returns are $E(r_1), \ldots, E(r_N)$. An investor wishes to create a portfolio of these N assets, by investing a fraction f_i of his wealth in asset i.

Using the properties of expectation, we may compute the expected portfolio return using Equation 21.2:

$$E(r_p) = \sum_{i=1}^{N} f_i E(r_i)$$

That is, an expected portfolio return is equal to the weighted sum of the expected returns of its individual asset components.

21.2.2 Portfolio Variance

The volatility of an asset's return can be measured by its variance. Variance is often adopted as a measure of an asset's risk. If σ_i^2 denotes the variance of asset i's return, then the variance of the portfolio's return is given by:

$$\sigma_p^2 = E\left[\left(\sum_{i=1}^N f_i r_i - \sum_{i=1}^N f_i \bar{r}_i\right)^2\right]$$

$$= E\left[\left(\sum_{i=1}^N f_i(r_i - \bar{r}_i)\right)\left(\sum_{j=1}^N f_j(r_j - \bar{r}_j)\right)\right]$$

$$= E\left[\sum_{i,j=1}^N f_i f_j (r_i - \bar{r}_i)(r_j - \bar{r}_j)\right] \qquad (21.3)$$

$$= \sum_{i,j=1}^N f_i f_j \sigma_{ij}$$

$$= \sum_{i=1}^N f_i^2 \sigma_i^2 + \sum_{i=1}^N \sum_{\substack{j=1 \\ j \neq i}}^N f_i f_j \sigma_{ij}$$

where $\bar{r}_i = E(r_i)$. Note that portfolio variance is a combination of the variance of the returns of each individual asset in the portfolio plus their covariance.

21.3 Estimating an Asset's Mean and Variance

Of course, asset i's rate of return and variance are not known and must be estimated. These values can be estimated based upon historical data using standard statistical methods. Alternatively, one can use a scenario-based approach. We describe the two methods below.

21.3.1 Estimating Statistics Using Historical Data

To estimate statistics using historical data, one must collect several periods of historical returns on the assets. The estimated average return on asset i is computed as the sample average of returns on asset i, $\overline{X}_i$, given by:

$$\overline{X}_i = \frac{\sum_t x_{it}}{T}$$

where x_{it} is the historical return on asset i in period t and there are T periods of historical data.

The variance of return on asset i is estimated by s_i^2, the historical sample variance of returns on investment i:

$$s_i^2 = \frac{\sum_t (x_{it} - \overline{X}_i)^2}{T-1}$$

For example, Table 21.1 contains the monthly closing stock prices and monthly returns for Sun Microsystems and Continental Airlines for the months January 2004 through February 2006. The first column of this table indicates the month, the second and third columns contain the closing stock prices for SUN and Continental Airlines, respectively. The fourth and fifth columns contain the monthly stock returns for SUN ($X_{SUN,t}$) and Continental

TABLE 21.1 Monthly Closing Stock Prices and Returns for Sun Microsystems and Continental Airlines

Month	SUN Stock Price ($)	CAL Stock Price ($)	SUN Return (%)	CAL Return (%)
Jan-04	55.55	15.65	8.12	−5.04
Feb-04	61.66	14.97	11.00	−4.35
Mar-04	62.38	12.58	1.17	−15.97
Apr-04	63.00	10.66	0.99	−15.26
May-04	62.04	10.50	−1.52	−1.50
Jun-04	63.63	11.37	2.56	8.29
Jul-04	68.17	8.75	7.13	−23.04
Aug-04	61.60	9.65	−9.64	10.29
Sep-04	73.98	8.60	20.10	−10.88
Oct-04	74.76	9.24	1.05	7.44
Nov-04	82.45	11.16	10.29	20.78
Dec-04	81.71	13.75	−0.90	23.21
Jan-05	88.04	10.50	7.75	−23.64
Feb-05	98.80	11.14	12.22	6.10
Mar-05	105.00	12.00	6.28	7.72
Apr-05	98.74	12.00	−5.96	0.00
May-05	103.00	13.70	4.31	14.17
Jun-05	114.14	13.32	10.82	−2.77
Jul-05	126.33	15.81	10.68	18.69
Aug-05	75.35	13.28	−40.35	−16.00
Sep-05	78.20	9.70	3.78	−26.96
Oct-05	73.76	13.20	−5.68	36.08
Nov-05	78.10	15.59	5.88	18.11
Dec-05	79.34	21.27	1.59	36.43
Jan-06	95.20	20.50	19.99	−3.62
Feb-06	74.35	23.79	−21.90	16.05

TABLE 21.2 Expected Historical Monthly Returns, Variances, and Standard Deviations of Returns

	Expected Monthly Return	Variance of Return	Standard Deviation
SUN	2.30%	1.54	12.40%
Continental Airlines	2.86%	3.04	17.43%

Airlines ($X_{CAL,t}$), respectively; these columns were populated using Equation 21.1. In this example, $T = 26$.

Table 21.2 shows the mean and standard deviation of returns for these two stocks, based upon the 26 months of historical data. The average monthly return for SUN, $\overline{X}_{SUN}$, and the average monthly return for Continental, $\overline{X}_{CAL}$, is computed as the arithmetic average of the monthly returns in the fourth and fifth columns, respectively. An estimate of the variance of monthly return on SUN's (Continental's) stock is computed as the variance of the returns in the fourth (fifth) column of Table 21.1. If variance is used as a measure of risk, then Continental is a riskier investment as it has a higher volatility (its variance is higher).

21.3.2 The Scenario Approach to Estimating Statistics

Sometimes, historical market conditions are not considered a good predictor of future market conditions. In this case, historical data may not be a good source for estimating expected returns or risk. When historical estimates are determined to be poor predictors of the future, one can consider a scenario approach.

The scenario approach proceeds as follows:

Define a set of S future economic scenarios and assign likelihood $p(s)$ that scenario s will occur. $\sum_{s \in S} p(s) = 1$, since in the future the economy must be in exactly one of these

TABLE 21.3 Definition of Future Possible Scenarios for States of the Economy

Scenario (s)	Likelihood ($p(s)$)	Return ($r_i(s)$)	$p(s)^* r_i(s)$	$p(s)^* (r_i(s) - r_i)^2$
Weak economy	0.30	-15.00%	-4.50%	1.03
Stable economy	0.45	9.00%	4.05%	0.13
Strong economy	0.25	16.00%	4.00%	0.39
			3.55%	1.55

economic conditions. Next, define each asset's behavior (its return) under each of the defined economic scenarios. Asset i's expected return is computed as:

$$r_i = \sum_s p(s) r_i(s) \tag{21.4}$$

where $r_i(s)$ is asset i's return under scenario s.

Similarly, we compute the variance of return on asset i as:

$$v_i = \sum_s p(s)(r_i(s) - r_i)^2 \tag{21.5}$$

For example, suppose we use the scenario approach to predict expected monthly return on SUN stock. We have determined that the economy may be in one of three states: weak, stable, or strong, with a likelihood of 0.3, 0.45, and 0.25, respectively. Table 21.3 indicates the forecasted monthly stock returns under each of these future economic conditions.

The first and second columns in Table 21.3 indicate the economic scenario and likelihood that the scenario will occur, respectively. The third column contains the expected return under each of the defined future scenarios. The fourth and fifth columns contain intermediate computations needed to calculate the expected return and standard deviation of returns on SUN stock, based upon Equations 21.3 and 21.4. Using these equations we find that the expected monthly return on SUN stock is 3.55%, variance of monthly return is 1.55, with corresponding standard deviation of 12.46%.

Comparing the estimates of mean and standard deviation of SUN's monthly return using the historical data approach versus the scenario-based approach we find that while the estimates of volatility of return are close in value, the estimates of monthly return differ significantly (2.3% versus 3.55%). In Section 21.7, we discuss the negative impact that can result from portfolio allocation based upon incorrect parameter estimation. Thus, care must be taken to determine the correct method and assumptions when estimating these values.

21.4 Diversification

We now explore how a portfolio's risk, as measured by the variance of the portfolio return, can be reduced when stocks are added to the portfolio. This phenomenon, whereby a portfolio's risk is reduced when assets are added to the portfolio, is known as diversification.

Portfolio return is the weighted average of the returns of the assets in the portfolio, weighted by their appearance in the portfolio. However, portfolio variance (as derived in Equation 21.3) is given by:

$$\sum_{i=1}^{N} f_i^2 \sigma_i^2 + \sum_{i=1}^{N} \sum_{\substack{j=1 \\ j \neq i}}^{N} f_i f_j \sigma_{ij}$$

Namely, portfolio variance is comprised of two components. One component is the variances of the individual assets in the portfolio and the other component is the covariance between the returns on the different assets in the portfolio. Covariance of return between asset i and j is the expected value of the product of the deviations of each of the assets from their respective means. If the two assets deviate from their respective means in identical fashions (i.e., both are above their means or below their means at the same time) then the covariance is highly positive, and its contribution to portfolio variance is highly positive. If the return on one asset deviates below its mean at the time that the return on the other asset deviates above its mean, then the covariance is highly negative, which reduces the overall portfolio variance.

Let us explore the impact of the covariance term on overall portfolio variance. If all of the assets are independent, then the covariance terms equal zero and only the variance of the individual assets in the portfolio contribute to overall portfolio variance. In this case, the variance formula is:

$$\sum_{i=1}^{N} f_i^2 \sigma_i^2$$

If the investor invests an equal amount in each of the N independent assets, then the portfolio variance is:

$$\sum_{i=1}^{N} \left(\frac{1}{N}\right)^2 \sigma_i^2 = \frac{1}{N} \sum_{i=1}^{N} \frac{1}{N} \sigma_i^2 = \frac{1}{N} \overline{\sigma}_i^2$$

where $\overline{\sigma}_i^2$ is the average variance of the assets in the portfolio. As N gets larger, the portfolio variance goes to zero. Thus, for a portfolio comprised only of independent assets, when the number of assets in the portfolio is large enough the variance of the portfolio return is zero.

Now consider N assets that are not independent. Without loss of generality, assume the assets appear in the portfolio with equal weight. Then, variance of portfolio return is:

$$\frac{1}{N}\overline{\sigma}_i^2 + \sum_{i=1}^{N} \sum_{\substack{j=1 \\ j \neq i}}^{N} \left(\frac{1}{N}\right)^2 \sigma_{ij}$$

$$= \frac{1}{N}\overline{\sigma}_i^2 + \frac{N-1}{N} \sum_{i=1}^{N} \sum_{\substack{j=1 \\ j \neq i}}^{N} \frac{\sigma_{ij}}{N(N-1)} \qquad (21.6)$$

$$= \frac{1}{N}\overline{\sigma}_i^2 + \frac{N-1}{N} \overline{\sigma}_{ij}$$

The final equality in Equation 21.6 reveals that as the number of assets in the portfolio increases, the contribution of the first component (variance) becomes negligible—it is diversified away—and the second term (covariance) approaches average covariance. Thus, while increasing the number of assets in a portfolio will diversify away the individual risk of the assets, the risk attributed to the covariance terms cannot be diversified away.

As a numerical example, consider the SUN and Continental Airlines monthly stock returns from Section 21.3.1. Average monthly return on Sun Microsystems (Continental Airlines) stock from January 2004 through February 2006 was 2.3% (2.86%). A portfolio comprised of equal investments in Sun Microsystems and Continental Airlines yields an average monthly return of 2.58%. The variance of the monthly returns on Sun Microsystems stock over the 26 months considered is 1.54. The variance of the monthly returns on Continental Airlines stock over that same time period is 3.04. However, for the portfolio consisting of equal investments in Sun Microsystems and Continental Airlines, the variance of portfolio returns is 1.09. This represents a significant reduction in risk from the risk associated with either of the stocks alone. The explanation lies in the value of the covariance between the monthly returns on these two stocks; the value of the covariance is −0.12. Because the returns on the stocks have a negative covariance, diversification reduces the portfolio risk.

21.5 Efficient Frontier

The discussion in Section 21.4 illustrates the potential benefits of combining assets in a portfolio. For a *risk averse* investor, diversification provides the opportunity to reduce portfolio risk while maintaining a minimum level of return. An investor can consider different combinations of assets, each of which has an associated risk and return.

This naturally leads one to question whether there is an *optimal* way to combine assets. We address this question within the context of assuming that: (i) for a fixed return investors prefer the lowest possible risk, that is, investors are risk averse, (ii) for a given level of risk, investors always prefer the highest possible return, (this property is referred to as nonsatiation), and (iii) the first two moments of the distribution of an asset's return are sufficient to describe the asset's character; there is no need to gather information about higher moments such as skew.

Given these assumptions, the risk-return trade-off of portfolios of assets can be graphically displayed by constructing a plot with risk (as measured by standard deviation) on the horizontal axis and return on the vertical axis. We can plot every possible portfolio on this risk–return space. The set of all portfolios plotted form a *feasible region*.

By the risk averse assumption, for a given level of return investors prefer portfolios that lie as far to the left as possible, as these have the lowest risk. Similarly, by the nonsatiation property for a given level of risk investors prefer portfolios that lie higher on the graph as these yield a greater return. The upper left perimeter of the feasible region is called the *efficient frontier*. It represents the least risk combination for a given level of return. The efficient frontier is concave.

Figure 21.1 shows a mean-standard deviation plot for 10,000 random portfolios created from the 30 stocks in the Dow from 1986 through 1991. It represents the feasible region of portfolios. Figure 21.2 is the efficient frontier for the stocks from the Dow Industrials from 1985 through 1991. (Both figures were taken from the NEOS Server for Optimization website.)

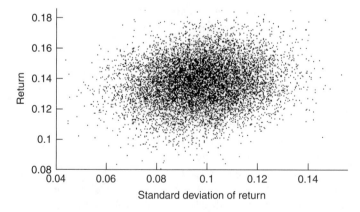

FIGURE 21.1 Return vs. standard deviation for 10,000 random portfolios from the Dow Industrials.

Financial Engineering

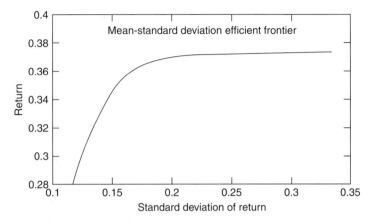

FIGURE 21.2 Efficient frontier for Dow Industrials from 1985–1991.

21.5.1 Risk-Reward Portfolio Optimization

Markowitz (1952) developed a single period portfolio choice problem where the objective is to minimize portfolio risk (variance) for a specified return on the portfolio. For this model, it is assumed that all relevant information required by investors to make portfolio decisions is captured in the mean, standard deviation, and correlation of assets. This method for portfolio selection is often referred to as mean-variance optimization as it trades off an investor's desire for higher mean return against an aversion to greater risk as measured by portfolio variance.

The Markowitz model for portfolio optimization is given by:

$$\text{Minimize} \quad \sum_{i,j} f_i f_j \sigma_{ij}$$

$$\text{Subject to:} \quad \sum_{i=1}^{N} f_i r_i = \underline{R} \tag{21.7}$$

$$\sum_{i=1}^{N} f_i = 1$$

The objective is to minimize variance subject to two constraints: (i) portfolio return must equal the targeted return $\underline{R}$ and (ii) total allocation must equal 1. Negative values for f_i correspond to short selling.

One method that can be employed to solve this constrained optimization problem is to form an auxiliary function L called the Lagrangian by (i) rearranging each constraint so that the right hand side equals to zero and (ii) introducing a single Lagrange multiplier for each constraint in the problem as follows:

$$L = \sum_{i,j} f_i f_j \sigma_{ij} - \lambda \left(\sum_{i=1}^{N} f_i r_i - \underline{R} \right) - \mu \left(\sum_{i=1}^{N} f_i - 1 \right) \tag{21.8}$$

We now treat the Lagrangian Equation 21.8 as an unconstrained minimization problem. A necessary condition for a point to be optimal is that the partial derivative of the Lagrangian with respect to each of the variables must equal zero. Thus, we take the partial derivative of L with respect to each of the i asset weights f_i, μ, and λ and set each partial derivative equal to zero. (Notice that the partial derivative of L with respect to λ yields the portfolio return constraint, and the partial derivative of L with respect to μ yields the constraint on the asset weights.) The result is a system of $i+2$ constraints. We use this set of constraints to solve for the $i+2$ unknowns: f_i, μ, and λ.

The Lagrangian method can often only be successfully implemented for small problems. For larger problems with many variables it may be virtually impossible to solve the set of equations for the $i+2$ unknowns. Instead, one can find the solution to Problem 21.7 using methods developed for optimizing quadratic programs. A *quadratic program* is a mathematical optimization model where the objective function is quadratic and all constraints are linear equalities or inequalities. The constraints of the optimization problem define the *feasible region* within which the optimal solution must lie. Quadratic programs are, in general, difficult to solve. Quadratic programming solution methods work in two phases. In the first phase a feasible solution is found. In the second phase, the method searches along the edges and surfaces of the feasible region for another solution that improves upon the current feasible solution. Unless the objective function is convex, the method will often identify a local optimal solution.

The solution to the Markowitz problem yields a point that lies on the efficient frontier. By varying the value of $\underline{R}$ in Problem 21.7, one can map out the entire efficient frontier.

21.6 Utility Analysis

The solution to the Markowitz problem provides one means for making investment decisions in the mean-variance space. It requires the investor to define a measure of risk and a measure of value and then utilizes an explicitly defined trade-off between these two measures to determine the investor's preferences. Utility theory provides an alternative way to establish an investor's preferences without explicitly defining risk functions.

Utility describes an investor's attitude toward risk by translating the investor's satisfaction associated with different cash payouts into a utility value. The application of utility to uncertain financial situations was first introduced by von Neumann and Morgenstern (1944). Utility functions can be used to explain how investors make choices between different portfolios.

Utility theory is often introduced by way of the concept of *certainty equivalent*. Certainty equivalent is the amount of wealth that is equally preferred to an uncertain alternative. Risk averse investors will prefer a lower certain cash payout to a higher risky cash payout. That is, their certainty equivalent is lower than the expected value of uncertain alternatives. Risk seeking individuals have a certainty equivalent that is higher than the expected value of the uncertain alternatives. A utility function assigns different weights to different outcomes according to the risk profile of the individual investor. The shape of a utility function is defined by the risk profile of the investor.

Each individual investor may have a different utility function, as each investor may have a different attitude toward risk. However, all utility functions U satisfy the following:

- *Nonsatiation*: Utility functions must be consistent with more being preferred to less. If x and y are two cash payouts and $x > y$, then $U(x) > U(y)$. This property is equivalent to stating that the first derivative of the utility function, with respect to cash payout, is positive.
- *Risk preference*: Economic theory presumes that an investor will seek to maximize the utility of his investment. However, all investors will not make identical

investment decisions because they will not all share the same attitudes toward risk. Investors can be classified into three classes, according to their willingness to accept risk: risk averse, risk neutral, or risk taking. Risk-averse investors invest in investments where the utility of the return exceeds the risk-free rate. If no such investment exists, the investor invests in risk-free investments. The utility function of a risk-averse investor is increasing and concave in the cash payout ($U''(x) < 0$); the value assigned to each additional dollar received decreases due to the risk-averse nature of the individual. Risk-neutral investors ignore risk when making investment decisions. They seek investments with a maximum return, irrespective of the risk involved. The utility function of a risk-neutral individual is linear increasing in the cash payout; the same value is assigned to each additional dollar. Risk-taking investors are more likely to invest in investments with a higher risk involved. The utility function of a risk-taking individual is convex ($U''(x) > 0$).

- *Changes in wealth*: Utility functions also define how an investor's investment decisions will be affected by changes in his wealth. Specifically, if an investor has a larger amount of capital will this change his willingness to accept risk? The Arrow–Pratt absolute risk aversion coefficient, given by

$$A(x) = -\frac{U''(x)}{U'(x)}$$

is a measure of an investor's absolute risk aversion. $A'(x)$ measures changes in an investor's absolute risk aversion as a function of changes in his wealth. If $A'(x) > 0$ ($A'(x) < 0$), the investor has increasing (decreasing) absolute risk aversion and he will hold fewer (more) dollars in risky assets as his wealth increases. If $A'(x) = 0$, the investor has constant absolute risk aversion and he will hold the same dollar amount of risky assets as his wealth increases.

A utility function only *ranks* alternatives according to risk preferences; its numerical value has no real meaning. Thus, utility functions are unique up to a positive affine transformation. More specifically, if $U(x)$ is a utility function then $V(x) = a + bU(x)$ ($b > 0$) will yield the same rankings (and hence the same investment decisions) as $U(x)$.

21.6.1 Utility Functions

Although each investor can define his own utility function, there are a number of predefined utility functions that are commonly used in the finance and economics literature. In this section, we describe the exponential utility function. We also reconcile between mean-variance optimization and utility optimization.

The exponential utility function is often adopted for financial analysis. The exponential utility function is defined as:

$$U(x) = 1 - e^{-x/R}, \quad \text{for all } x \tag{21.9}$$

where x is the wealth and $R > 0$ represents the investor's risk tolerance. Greater values of R mean that the investor is less risk averse. (In fact, as $R \to \infty$ the investor becomes risk neutral.) $U'(x) = (1/R)e^{-x/R}$ and $U''(x) = -(1/R)^2 e^{-x/R}$. The second derivative of Equation 21.9 is strictly negative so the exponential utility function is concave; the exponential utility function describes the behavior of a risk-averse investor. The absolute risk aversion coefficient is $(1/R)$, which is constant with wealth; the investor invests constant dollar amounts in risky investments as his wealth increases.

We now demonstrate the relationship between utility and mean-variance for concave (i.e., risk-averse) investors, following King (1993). Suppose $U(X)$ is a strictly concave utility function. The Taylor expansion of $U(X)$ is an approximation of the function at a particular point, say M, using its derivatives. The second-order Taylor expansion of $U(X)$ about point M is given by:

$$U(X) = U(M) + U'(M)(X - M) + \frac{1}{2}U''(M)(X - M)^2 \qquad (21.10)$$

Now suppose that the point M is equal to expected wealth, that is, $E(X) = M$. Then the expected value of the second-order Taylor expansion expression (Equation 21.10) is equal to

$$E[U(X)] = U(M) + \frac{1}{2}U''(M)E(X - M)^2$$
$$= U(M) + \frac{1}{2}U''(M)\sigma^2$$

(The middle term drops out as $E(X) = M$.) The second derivative is negative for a strictly concave utility function. This implies that maximizing the second-order Taylor expansion is equivalent to

$$\min \quad E(X - M)^2 = \sigma^2$$

for all X with $E(X) = M$, which is a mean-variance problem with a given mean return M.

It follows that mean-variance is a second-order approximation to utility maximization. Of course, due to the two-sided nature of the variance, eventually this approximation will become negatively-sloped—and hence not really valid as a utility function—as X increases in value. The range over which the approximation is valid can be pretty wide. The upper bound of the range is the point where $U'(X) + U''(X)(X - M) = 0$; which is the maximum point of the approximating quadratic. For the logarithmic utility $U(X) = \log(X)$, the upper bound of the range where the mean-variance approximation remains quadratic is $X = 2M$. In other words, the mean-variance approximation for the logarithm is valid for a range that includes twice the mean value of the return!

21.6.2 Utility in Practice

In practice, utility is not often used as an objective criterion for investment decision making because utility curves are difficult to estimate. However, Holmer (1998) reports that Fannie Mae uses expected utility to optimize its portfolio of assets and liabilities. Fannie Mae faces a somewhat unique set of risks due to the specific nature of its business. Fannie Mae buys mortgages on the secondary market, pools them, and then sells them on the open market to investors as mortgage backed securities. Fannie Mae faces many risks such as: prepayment risk, risks due to potential gaps between interest due and interest owed, and long-term asset and liability risks due to interest rate movements. Utility maximization allows Fannie Mae to explicitly consider degrees of risk aversion against expected return to determine its risk-adjusted optimal investment portfolio.

21.7 Black–Litterman Asset Allocation Model

Markowitz mean-variance portfolio optimization requires mean and covariance as input and outputs optimal portfolio weights. The method has been criticized because:

i. The optimal portfolio weights are highly dependent upon the input values. However, it is difficult to accurately estimate these input values. Chopra and Ziemba (1993), Kallberg and Ziemba (1981, 1984), and Michaud (1989) use simulation to demonstrate the significant cash-equivalent losses due to incorrect estimates of the mean. Bengtsson (2004) showed that incorrect estimates of variance and covariance also have a significant negative impact on cash returns.

ii. Markowitz mean-variance optimization requires the investor to specify the universe of return values. It is unreasonable to expect an investor to know the universe of returns. On the other hand, mean-variance optimization is sensitive to the input values so incorrect estimation can significantly skew the results.

iii. Black and Litterman (1992) and He and Litterman (1999) have studied the optimal Markowitz model portfolio weights assuming different methods for estimating the assets' means and found that the resulting portfolio weights were unnatural. Unconstrained mean-variance optimization typically yields an optimal portfolio that takes many large long and short positions. Constrained mean-variance optimization often results in an extreme portfolio that is highly concentrated in a small number of assets. Neither of these portfolio profiles is typically considered acceptable to investors.

iv. Due to the intricate interaction between mean and variance, the optimal weights determined by Markowitz's mean-variance estimation are often non-intuitive. A small change in an estimated mean of a single asset can drastically change the weights of many assets in the optimal portfolio.

Black and Litterman observed the potential benefit of using mathematical optimization for portfolio decision making, yet understood an investment manager's hesitations in implementing Markowitz's mean-variance optimization model. Black and Litterman (1992) developed a Bayesian method for combining individual investor subjective views on asset performance with market equilibrium returns to create a mixed estimate of expected returns. The Bayes approach works by combining a prior belief with additional information to create an updated "posterior" distribution of expected asset returns. In the Black–Litterman framework the equilibrium returns are the prior and investor subjective views are the additional information. Together, these form a posterior distribution on expected asset returns. These expected returns can then be used to make asset allocation decisions. If the investor has no subjective views on asset performance, then the optimal allocation decision is determined solely according to the market equilibrium returns. Only if the investor expresses opinions on specific assets will the weights for those assets shift away from the market equilibrium weights in the direction of the investor's beliefs.

The Black–Litterman model is based on taking a market equilibrium perspective on asset returns. Asset "prior" returns are derived from the market capitalization weights of the optimal holdings of a mean-variance investor, given historical variance. Then, if the investor has specific views on the performance of any assets, the model combines the equilibrium returns with these views, taking into consideration the level of confidence that the investor associates with each of the views. The model then yields a set of updated expected asset returns as well as updated optimal portfolio weights, updated according to the views expressed by the investor.

The key inputs to the Black–Litterman model are market equilibrium returns and the investor views. We now consider these inputs in more detail.

21.7.1 Market Equilibrium Returns

Black and Litterman use the market equilibrium expected returns, or capital asset pricing model (CAPM) returns, as a neutral starting point in their model. (See, e.g., Sharpe, 1964.) The basic assumptions are that (i) security markets are frictionless, (ii) investors have full information relevant to security prices, and (iii) all investors process the information as if they were mean-variance investors. The starting point for the development of the CAPM is to form the efficient frontier for the market portfolios and to draw the capital market line (CML). The CML begins at the risk-free rate on the vertical axis (which has risk 0) and is exactly tangent to the efficient frontier. The point where the CML touches the efficient frontier is the pair (σ_m, r_m), which is defined to be the "market" standard deviation and "market" expected return. By changing the relative proportions of riskless asset and market portfolio, an investor can obtain any combination of risk and return that lies on the CML. Because the CML is tangent to the efficient frontier at the market point, there are no other combinations of risky and riskless assets that can provide better expected returns for a given level of risk. Now, consider a particular investor portfolio i with expected return $E(r_i)$ and standard deviation σ_i. For an investor to choose to hold this portfolio it must have returns comparable to the returns that lie on the CML. Thus, the following must hold:

$$E(r_i) = r_f + \frac{\sigma_i}{\sigma_m}(r_m - r_f)$$

where r_f is the risk-free rate. This equation is called the CAPM.

The interpretation of the CAPM is that investors' portfolios have an expected return that includes a reward for taking on risk. This reward, by the CAPM hypothesis, must be equal to the return that would be obtained from holding a portfolio on the CML that has an equivalent risk. Any remaining risk in the portfolio can be diversified away (by, for example, holding the market portfolio) so the investor does not gain any reward for the non-systematic, or diversifiable, risk of the portfolio. The CAPM argument applied to individual securities implies that the holders of individual securities will be compensated only for that part of the risk that is correlated with the market, or the so-called systematic risk. For an individual security j the CAPM relationship is

$$E(r_j) = r_f + \beta_j(r_m - r_f)$$

where $\beta_j = (\sigma_j/\sigma_m)\rho_{mj}$ and ρ_{mj} is the correlation between asset j returns and the market returns.

The Black–Litterman approach uses the CAPM in reverse. It assumes that in equilibrium the market portfolio is held by mean-variance investors and it uses optimization to back out the expected returns that such investors would require given their observed holdings of risky assets. Let N denote the number of assets and let the excess equilibrium market returns (above the risk-free rate) be defined by:

$$\prod = \lambda \sum w \qquad (21.11)$$

where
$\prod = N \times 1$ vector of implied excess returns
$\sum = N \times N$ covariance matrix of returns

Financial Engineering

$w = N \times 1$ vector of market capitalization weights of the assets
λ = risk aversion coefficient that characterizes the expected risk-reward tradeoff.

λ is the price of risk as it measures how risk and reward can be traded off when making portfolio choices. It measures the rate at which an investor will forego expected return for less variance. λ is calculated as $\lambda = (r_m - r_f)/\sigma_m^2$, where σ_m^2 is the variance of the market return. The elements of the covariance matrix are computed using historical correlations and standard deviations. Market capitalization weights are determined by measuring the dollar value of the global holdings of all equity investors in the large public stock exchanges. The capitalization weight of a single equity name is the dollar-weighted market-closing value of its equity share times the outstanding shares issued. Later we will show that Equation 21.11 is used to determine the optimal portfolio weights in the Black–Litterman model.

21.7.2 Investor Views

The second key input to the Black–Litterman model is individual investor views.

Assume that an investor has K views, denoted by a $K \times 1$ vector Q. Uncertainty regarding these views is denoted by an error term ε, where ε is normally distributed with mean zero and $K \times K$ covariance matrix Ω. Thus, a view has the form:

$$Q + \varepsilon = \begin{bmatrix} Q_1 \\ \vdots \\ Q_K \end{bmatrix} + \begin{bmatrix} \varepsilon_1 \\ \vdots \\ \varepsilon_K \end{bmatrix}$$

$\varepsilon = 0$ means that the investor is 100% confident about his view; in the more likely case that the investor is uncertain about his view, ε takes on some positive or negative value.

ω denotes the variance of each error term. We assume that the error terms are independent of each other. (This assumption can be relaxed.) Thus, the covariance matrix Ω is a diagonal matrix where the elements on the diagonal are ω, the variances of each error term. A higher variance indicates greater investor uncertainty with the associated view. The error terms ε do not enter directly into the Black–Litterman formula; only their variances enter via the covariance matrix Ω.

$$\Omega = \begin{bmatrix} \omega_1 & 0 & \cdots & 0 \\ 0 & \ddots & & \vdots \\ \vdots & & & 0 \\ 0 & \cdots & 0 & \omega_K \end{bmatrix}$$

The Black–Litterman model allows investors to express views such as:

View 1: Asset A will have an excess return of 5.5% with 40% confidence.
View 2: Asset B will outperform asset C by 3% with 15% confidence.
View 3: Asset D will outperform assets E and F by 1% with 20% confidence.

The first view is called an *absolute* view, while the second and third views are called *relative* views. Notice that the investor assigns a level of confidence to each view.

Each view can be seen as a portfolio of long and short positions. If the view is an absolute view then the portfolio position will be long. If the view is a relative view, then the portfolio will take a long position in the asset that is expected to "overperform" and a short position in the asset that is expected to "underperform." In general, the impact on the optimal portfolio weights is determined by comparing the equilibrium difference in the performance of these assets to the performance expressed by the investor view. If the performance expressed in the view is better than the equilibrium performance, the model will tilt the portfolio toward the outperforming asset. More specifically, consider View 2, which states that asset B will outperform asset C by 3%. If the equilibrium returns indicate that asset B will outperform asset C by more than 3% then the view represents a weakening view in the performance of asset B and the model will tilt the portfolio away from asset B.

One of the most challenging questions in applying the Black–Litterman model is how to populate the covariance matrix Ω and how to translate the user specified expressions of confidence into uncertainty in the views. We will discuss this further below.

21.7.3 An Example of an Investor View

We now illustrate how one builds the inputs for the Black–Litterman model, given the three views expressed. Suppose that there are $N = 7$ assets: Assets A – G. The Q matrix is given by:

$$Q = \begin{bmatrix} 5.5 \\ 3 \\ 1 \end{bmatrix}$$

Note that the investor only has views on six of the seven assets. We use the matrix P to match the views to the individual assets. Each view results in a $1 \times N$ vector so that P is a $K \times N$ matrix. In our case, where there are seven assets and three views, P is a 3×7 matrix. Each column corresponds to one of the assets; column 1 corresponds to Asset A, column 2 corresponds to Asset B, and so on. In the case of absolute views, the sum of the elements in the row equals 1. In our case, View 1 yields the vector:

$$P_1 = \begin{bmatrix} 1 & 0 & 0 & 0 & 0 & 0 & 0 \end{bmatrix}$$

In the case of relative views, the sum of the elements equals zero. Elements corresponding to relatively outperforming assets have positive values; elements corresponding to relatively underperforming assets take negative values. We determine the values of the individual elements by dividing 1 by the number of outperforming and underperforming assets, respectively. For View 2, we have one outperforming asset and one underperforming asset. Thus, View 2 yields the vector:

$$P_2 = \begin{bmatrix} 0 & 1 & -1 & 0 & 0 & 0 & 0 \end{bmatrix}$$

View 3 has one outperforming asset (Asset D) and two relatively underperforming assets (Assets E and F). Thus, Asset D is assigned a value of +1 and Assets E and F are assigned values of −0.5 each. View 3 yields the vector:

$$P_3 = \begin{bmatrix} 0 & 0 & 0 & 1 & -0.5 & -0.5 & 0 \end{bmatrix}$$

Matrix P is given by:

$$\begin{bmatrix} 1 & 0 & 0 & 0 & 0 & 0 & 0 \\ 0 & 1 & -1 & 0 & 0 & 0 & 0 \\ 0 & 0 & 0 & 1 & -0.5 & -0.5 & 0 \end{bmatrix}$$

The variance of the kth view portfolio can be calculated according to the formula $p_k \sum p_k'$, where p_k is the kth row of the P matrix and $\sum$ is the covariance matrix of the excess

Financial Engineering

equilibrium market returns. (Recall, these form the neutral starting point of the Black–Litterman model.) The variance of each view portfolio is an important source of information regarding the confidence that should be placed in the investor's view k.

21.7.4 Combining Equilibrium Returns with Investor Views

We now state the Black–Litterman equation for combining equilibrium returns with investor views to determine a vector of expected asset returns that will be used to determine optimal portfolio weights. The vector of combined asset returns is given by:

$$E[R] = \left[\left(\tau \sum\right)^{-1} + P'\Omega^{-1}P\right]^{-1}\left[\left(\tau \sum\right)^{-1}\Pi + P'\Omega^{-1}Q\right] \quad (21.12)$$

where:
$E[R] = N \times 1$ vector of combined returns
$\tau =$ scalar, indicating uncertainty of the CAPM prior
$\sum = N \times N$ covariance matrix of equilibrium excess returns
$P = K \times N$ matrix of investor views
$\Omega = K \times K$ diagonal covariance matrix of view error terms (uncorrelated view uncertainty)
$\Pi = N \times 1$ vector of equilibrium excess returns
$Q = K \times 1$ vector of investor views.

Examining this formula, we have yet to describe how the value of τ should be set and how the matrix Ω should be populated. Recall that if an investor has no views, the Black–Litterman model suggests that the investor does not deviate from the market equilibrium portfolio. Only weights on assets for which the investor has views should change from their market equilibrium weights. The amount of change depends upon τ, the investor's confidence in the CAPM prior, and ω, the uncertainty in the views expressed.

The literature does not have a single view regarding how the value of τ should be set. Black and Litterman (1992) suggest a value close to zero. He and Litterman (1999) set τ equal to 0.025 and populate the covariance matrix Ω so that

$$\Omega = \begin{bmatrix} \left(p_1 \sum p_1'\right)\tau & 0 & \cdots & 0 \\ 0 & \ddots & & \vdots \\ \vdots & & & 0 \\ 0 & \cdots & 0 & \left(p_K \sum p_K'\right)\tau \end{bmatrix}$$

We note that the implied assumption is that the variance of the view portfolio is the information that determines an investor's confidence in his view. There may be other information that contributes to the level of an investor's confidence but it is not accounted for in this method for populating Ω.

Formula 21.12 uses Bayes approach to yield posterior estimates of asset returns that reflect a combination of the market equilibrium returns and the investor views. These updated returns are now used to compute updated optimal portfolio weights.

In the case that the investor is unconstrained, we use Formula 21.11. Using Formula 21.11 w^*, the optimal portfolio weights, are given by:

$$w^* = \left(\lambda \sum\right)^{-1}\mu \quad (21.13)$$

where μ is the vector of combined returns given by Equation 21.12. Equation 21.13 is the solution to the unconstrained maximization problem $\max_w w'\mu - \lambda w' \sum w/2$.

In the presence of constraints (e.g., risk, short selling, etc.) Black and Litterman suggest that the vector of combined returns be input into a mean-variance portfolio optimization problem.

We note two additional comments on the updated weights w^*:

i. Not all view portfolios necessarily have equal impact on the optimal portfolio weights derived using the Black–Litterman model. A view with a higher level of uncertainty is given less weight. Similarly, a view portfolio that has a covariance with the market equilibrium portfolio is given less weight. This is because such a view represents less new information and hence should have a smaller impact in moving the optimal portfolio weights away from the market equilibrium weights. Finally, following the same reasoning, a view portfolio that has a covariance with another view portfolio has less weight.

ii. A single view causes all returns to change, because all returns are linked via the covariance matrix $\sum$. However, only the weights for assets for which views were expressed change from their original market capitalization weights. Thus, the Black–Litterman model yields a portfolio that is intuitively understandable to the investor. The optimal portfolio represents a combination of the market portfolio and a weighted sum of the view portfolios expressed by the investor.

21.7.5 Application of the Black–Litterman Model

The Black–Litterman model was developed at Goldman Sachs in the early 1990s and is used by the Quantitative Strategies group at Goldman Sachs Asset Management. This group develops quantitative models to manage portfolios. The group creates views and then uses the Black–Litterman approach to transform these views into expected asset returns. These expected returns are used to make optimal asset allocation decisions for all of the different portfolios managed by the group. Different objectives or requirements (such as liquidity requirements, risk aversion, etc.) are incorporated via constraints on the portfolio. The Black–Litterman model has gained widespread use in other financial institutions.

21.8 Risk Management

Risk exists when more than one outcome is possible from the investment. Sources of risk may include business risk, market risk, liquidity risk, and the like. Variance or standard deviation of return is often used as a measure of the risk associated with an asset's return. If variance is small, there is little chance that the asset return will differ from what is expected; if variance is large then the asset returns will be highly variable.

Financial institutions manage their risk on a regular basis both to meet regulatory requirements as well as for internal performance measurement purposes. However, while variance is a traditional measure of risk in economics and finance, in practice it is typically not the risk measure of choice. Variance assumes symmetric deviations above and below expected return. In practice, one does not observe deviations below expected return as often as deviations above expected return due to positions in options and options-like instruments in portfolios. Moreover, variance assigns equal penalty to deviations above and below the mean return. However, investors typically are not averse to receiving higher than anticipated returns. Investors are more interested in shortfall risk measures. These are risk measures that measure either the distance of return below a target or measure the likelihood that return will fall below a threshold.

One measure of shortfall risk is downside risk. Downside risk measures the expected amount by which the return falls short of a target. Specifically, if z is the realized return and X is the target then downside risk is given by $E[\max(X-z,0)]$. Semivariance is another measure of shortfall risk. Semivariance measures the variance of the returns that fall below a target value. Semivariance is given by $E[\max(X-z,0)^2]$.

21.8.1 Value at Risk

Value at risk (VaR) is a risk measure that is used for regulatory reporting. Rather than measuring risk as deviation from a target return, VaR is a quantile of the loss distribution of a portfolio. Let $L(f,\tilde{r})$ be the random loss on a portfolio with allocation vector f and random return vector $\tilde{r}$. Let F be its distribution function so that $F(f,u) = \Pr\{L(f,\tilde{r}) \leq u\}$. VaR is the α-quantile of the loss distribution and is defined by:

$$VaR_\alpha(f) = \min\{u : F(f,u) \geq \alpha\} \quad (21.14)$$

Thus, VaR is the smallest amount u such that with probability α the loss will not exceed u.

The first step in computing VaR is to determine the underlying market factors that contribute to the risk (uncertainty) in the portfolio. The next step is to simulate these sources of uncertainty and the resulting portfolio loss. Monte Carlo simulation is used largely because many of the portfolios under management by the more sophisticated banks include a preponderance of instruments that have optional features. As we shall see in Section 21.10, the price changes of these instruments can best be approximated by simulation. VaR can then be calculated by determining the distribution of the portfolio losses. The key question is what assumptions to make about the distributions of these uncertain market factors. Similar to the two methods that we discuss for estimating asset returns and variances, one can use historical data or a scenario approach to build the distributions.

Using historical data, one assumes that past market behavior is a good indicator of future market behavior. Take T periods of historical data. For each period, simulate the change in the portfolio value using the actual historical data. Use these T data points of portfolio profit/loss to compute the loss distribution and hence VaR. The benefit of this method is that there is no need for artificial assumptions about the distribution of the uncertainty of the underlying factors that impact the value of the portfolio. On the other hand, this method assumes that future behavior will be identical to historical behavior.

An alternative approach is to specify a probability distribution for each of the sources of market uncertainty and to then randomly generate events from those distributions. The events translate into behavior of the uncertain factors, which result in a change in the portfolio value. One would then simulate the portfolio profit/loss assuming that these randomly generated events occur and construct the loss distribution. This distribution is used to compute VaR.

21.8.2 Coherent Risk Measures

VaR is a popular risk measure. However, VaR does not satisfy one of the basic requirements of a good risk measure: VaR is not subadditive for all distributions (i.e., it is not always the case that $VaR(A+B) < VaR(A) + VaR(B)$), a property one would hope to hold true if risk is reduced by adding assets to a portfolio. This means that the VaR of a diversified portfolio may exceed the sum of the VaR of its component assets.

Artzner, Delbaen, Eber, and Heath (1999) specify a set of axioms satisfied by all *coherent* risk measures. These are:

- Subadditivity: $\rho(A+B) \leq \rho(A) + \rho(B)$
- Positive homogeneity: $\rho(\lambda A) = \lambda \rho(A)$ for $\lambda \geq 0$
- Translation invariance: $\rho(A+c) = \rho(A) - c$ for all c
- Monotonicity: $A \leq B$ then $\rho(B) \leq \rho(A)$

Subadditivity implies that the risk of a combined position of assets does not exceed the combined risk of the individual assets. This allows for risk reduction via diversification, as we discuss in Section 21.4.

Conditional value-at-risk (*CVaR*), also known as expected tail loss, is a coherent risk measure. *CVaR* measures the expected losses conditioned on the fact that the losses exceed *VaR*. Following the definition of *VaR* in Equation 21.14, if F is continuous then *CVaR* is defined as:

$$CVaR(f,\alpha) = E\{L(f,\tilde{r})|L(f,\tilde{r}) \geq VaR(f,\alpha)\} \quad (21.15)$$

An investment strategy that minimizes *CVaR* will minimize *VaR* as well.

An investor wishing to maximize portfolio return subject to a constraint on maximum *CVaR* would solve the following mathematical program:

$$\begin{aligned}
\max \quad & E\left(\sum_{i=1}^{N} f_i r_i\right) \\
\text{subject to:} \quad & CVaR_\alpha(f_1, \ldots, f_N) \leq C \\
& \sum_{i=1}^{N} f_i = 1 \\
& 0 \leq f_i \leq 1
\end{aligned} \quad (21.16)$$

where f_i is the fraction of wealth allocated to asset i, r_i is the return on asset i, and C is the maximum acceptable risk. Formulation 21.16 is a nonlinear formulation due to the constraint on *CVaR*, and is a hard problem to solve.

Suppose, instead of assuming that the loss distribution F is continuous, we discretize the asset returns by considering a finite set of $s = 1, \ldots, S$ scenarios of the portfolio performance. Let $p(s)$ denote the likelihood that scenario s occurs; $0 \leq p(s) \leq 1$; $\sum_{s \in S} p(s) = 1$. $\varphi(s)$ denotes the vector of asset returns under scenario s and $\rho(f, \varphi(s))$ denotes the portfolio return under scenario s assuming asset allocation vector f. Then, using this discrete set of asset returns, the expected portfolio return is given by

$$\sum_{s \in S} p(s) \rho(f, \varphi(s)) \quad (21.17)$$

The investor will wish to maximize Equation 21.17. Definition 21.15 applies to continuous loss distributions. Rockafellar and Uryasev (2002) have proposed an alternative definition of *CVaR* for any general loss distribution, where $u = VaR_{\alpha(f)}$:

$$CVaR(f,\alpha) = \frac{1}{1-\alpha}\left(\sum_{\{s \in S | \rho(f,\varphi(s)) \leq u\}} p(s) - \alpha\right)u + \frac{1}{1-\alpha}\sum_{\{s \in S | \rho(f,\varphi(s)) > u\}} p(s)\rho(f,\varphi(s)) \quad (21.18)$$

Using Definition 21.18, we can solve Problem 21.16 using a stochastic programming approach. First, we define an auxiliary variable $z(s)$ for each scenario s, which denotes the shortfall in portfolio return from the target u:

$$z(s) = \max(0, u - \rho(f, \varphi(s))) \qquad (21.19)$$

Following Rockafellar and Uryasev (2002), $CVaR$ can be expressed using the shortfall variables (Equation 21.19) as:

$$CVaR(f, \alpha) = \min \left[u - \frac{1}{1-\alpha} \sum_{s=1}^{S} p(s) z(s) \right]$$

The linear program is given by:

$$\begin{aligned}
\max \quad & \sum_{s \in S} p(s) \rho(f, \varphi(s)) \\
\text{subject to:} \quad & \sum_{i=1}^{N} f_i = 1 \\
& 0 \leq f_i \leq 1 \\
& u - \frac{1}{1-\alpha} \sum_{s=1}^{S} p(s) z(s) \geq C \\
& z(s) \geq u - \rho(f, \varphi(s)) \quad s = 1, \ldots, S \\
& z(s) \geq 0 \qquad\qquad\qquad\ \ s = 1, \ldots, S
\end{aligned} \qquad (21.20)$$

where the last two inequalities follow from the definition of the shortfall variables (Equation 21.19) and the maximization is taken over the variables (f, u, z).

Formulation 21.20 is a linear stochastic programming formulation of the $CVaR$ problem. To solve this problem requires an estimate of the distribution of the asset returns, which will be used to build the scenarios. If historical data are used to develop the scenarios then it is recommended that as time passes and more information is available, the investor reoptimize (Equation 21.20) using these additional scenarios. We direct the reader to Rockafellar and Uryasev (2000, 2002) for additional information on this subject.

21.8.3 Risk Management in Practice

VaR and *CVaR* are popular risk measures. *VaR* is used in regulatory reporting and to determine the minimum capital requirements to hedge against market, operational, and credit risk. Financial institutions may be required to report portfolio risks such as the 30-day 95% *VaR* or the 5% quantile of 30-day portfolio returns and to hold reserve accounts in proportion to these calculated amounts. *VaR* is used in these contexts for historical reasons. But even though as we saw above it is not a coherent risk measure, there is possibly some justification in continuing to use it. *VaR* is a frequency measure, so regulators can easily track whether the bank is reporting *VaR* accurately; *CVaR* is an integral over the tail probabilities and is likely not as easy for regulators to track.

In addition to meeting regulatory requirements, financial institutions may use *VaR* or *CVaR* to measure the performance of its business units that control financial portfolios by comparing the profit generated by the portfolio actions versus the risk of the portfolio itself.

For purposes of generating risk management statistics, banks will simulate from distributions that reflect views of the market and the economy. Banks will also incorporate some probability of extreme events such as the wild swings in correlations and liquidity that occur in market crashes.

21.9 Options

In this section, we discuss options. An option is a derivative security, which means that its value is derived from the value of an underlying variable. The underlying variable may or may not be a traded security. Stocks or bonds are examples of traded securities; interest rates or the weather conditions are examples of variables upon which the value of an option may be contingent but that are not traded securities. Derivative securities are sometimes referred to as contingent claims, as the derivative represents a claim whose payoff is contingent on the history of the underlying security.

The two least complex types of option contracts for individual stocks are calls and puts. Call options give the holder the right to buy a specified number of shares (typically, 100 shares) of the stock at the specified price (known as the exercise or strike price) by the expiration date (known as the exercise date or maturity). A put option gives the holder the right to sell a specified number of shares of the stock at the strike price by maturity. American options allow the holder to exercise the option at any time until maturity; European options can only be exercised at maturity. The holder of an option contract may choose whether or not he wishes to exercise his option contract. However, if the holder chooses to exercise, the seller is obligated to deliver (for call options) or purchase (for put options) the underlying securities.

When two investors enter into an options contract, the buyer pays the seller the option price and takes a *long* position; the seller takes a *short* position. The buyer has large potential upside from the option, but his downside loss is limited by the price that he paid for the option. The seller's profit or loss is the reverse of that of the buyer's. The seller receives cash upfront (the price of the option) but has a potential future liability in the case that the buyer exercises the option.

21.9.1 Call Option Payoffs

We first consider European options. We will define European option payoffs at their expiration date.

Let C represent the option cost, S_T denote the stock price at maturity, and K denote the strike price. An investor will only exercise the option if the stock price exceeds the strike price, that is, $S_T > K$. If $S_T > K$, the investor will exercise his option to buy the stock at price K and gain $(S_T - K)$ for each share of stock that he purchases. If $S_T < K$, then the investor will not exercise the option to purchase the stock at price K. Thus, the option payoff for each share of stock is

$$\max(S_T - K, 0) \qquad (21.21)$$

The payoff for the option writer (who has a short position in a call option) is the opposite of the payoff for the long position. If, at expiration, the stock price is below the strike price the holder (owner) will not exercise the option. However, if the stock price is above the strike price the owner will exercise his option. The writer must sell the stock to the owner at the strike price. For each share of stock that he sells, the writer must purchase the stock in the open market at a price per share of S_T and then sell it to the owner at a price of K. Thus, the writer loses $(S_T - K)$ for each share that he is obligated to sell to the owner.

Financial Engineering

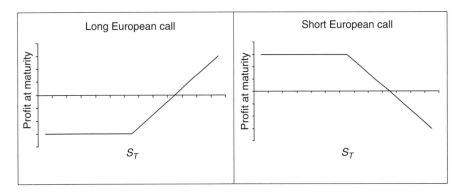

FIGURE 21.3 Profit from European call options.

The writer thus has an unlimited potential loss depending upon the final stock price of the underlying asset; the writer's payoff is

$$-\max(S_T - K, 0) = \min(K - S_T, 0)$$

Thus, an investor with a short position in a European call option has potentially unlimited loss depending upon the final stock price of the underlying stock. This risk must be compensated by the price of the option C.

The graph on the left in Figure 21.3 shows the profit at maturity for the owner of a call option. The point of inflection occurs when the ending stock price equals the strike price. The negative payoff is the price the investor paid for the option. As mentioned, the option holder's downside loss is limited by the price paid for the option. The payoff for the call option writer is shown in the graph on the right in Figure 21.3 and is the reverse of the payoff to the call option buyer.

When the price of the underlying stock is above the strike price, we say that the option is "in the money." If the stock price is below the strike price we say that the option is "out of the money." If the stock price is exactly equal to the strike price we say that the call option is "at the money."

American options can be exercised at any time prior to maturity. The decision of whether to exercise hinges upon a comparison of the value of exercising immediately (the *intrinsic value* of the option) against the expected future value of the option if the investor continues to hold it. We will discuss this in further detail in Section 21.12, where we discuss pricing American options.

21.9.2 Put Option Payoffs

An investor with a long position in a put option profits if the price of the underlying stock drops below the option's strike price. Similar to the definitions for a call option, we say that a put option is "in the money" if the stock price at maturity is lower than the strike price. The put option is "out of the money" if the stock price exceeds the strike price. The option is "at the money" if the strike price equals the stock price at maturity.

Let P denote the cost of the put option, K is its strike price, and S_T the stock price at expiration. The holder of a put option will only sell if the stock price at expiration is lower than the strike price. In this case, the owner can sell the shares to the writer at the strike price and will gain $(K - S_T)$ per share. Thus, payoff on a put option is $\max(K - S_T, 0)$; we note that positive payoff on a put is limited at K. If the stock price is higher than the strike price at maturity then the holder will not exercise the put option as he can sell his shares on the open market at a higher price. In this case, the option will expire worthless.

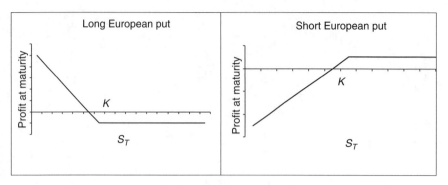

FIGURE 21.4 Profit from European put options.

A put writer has opposite payoffs. If the stock price exceeds the strike price, the put holder will never exercise his option; however, if the stock price declines, the writer will lose money. The price P paid by the owner must compensate the writer for this risk.

Figure 21.4 shows the option payoff for a put holder and writer.

21.10 Valuing Options

The exact value of a stock option is easy to define at maturity. Valuing options prior to expiration is more difficult and depends upon the distribution of the underlying stock price, among other factors. Black and Scholes (1973) derived a differential equation that can be used to price options on non-dividend paying stocks. We discuss the Black–Scholes formula in Section 21.10.1. However, in general, exact formulas are not available for valuing options. In most cases, we rely on numerical methods and approximations for options valuation. In Sections 21.10.3 and 21.12, we discuss two numerical methods for pricing options: Monte Carlo simulation and dynamic programming. Dynamic programming is useful for pricing American options, where the holder has the ability to exercise prior to the expiration date. Monte Carlo simulation is useful for pricing a European option, where the option holder cannot exercise the option prior to the maturity date. In the following sections, we describe Monte Carlo simulation and dynamic programming and show how they can be used to price options. We first will lay out some background and basic assumptions that are required.

21.10.1 Risk Neutral Valuation in Efficient and Complete Markets

We base our discussion of options pricing on the assumption that markets are efficient and complete. Efficient markets are arbitrage-free. Arbitrage provides an investor with a riskless investment opportunity with unlimited return, without having to put up any money. We assume that if any such opportunities exist there would be infinite demand for such assets. This would immediately raise the price of the investments and the arbitrage opportunity would disappear.

Black and Scholes derived a differential equation that describes the value of a trader's portfolio who has a short position in the option and who is trading in the underlying asset and a cash-like instrument. Efficiency of markets is one of the key assumptions required in their derivation. In addition, they assume that instantaneous and continuous trading opportunities exist, no dividends, transaction costs, or taxes are paid, and that short selling is permitted. Finally, they assume that the price of the underlying stock follows a specific stochastic process called Brownian motion. (See Section 21.10.2 for discussion of Brownian motion.) In this Black–Scholes framework it turns out that there is a trading strategy (called "delta-hedging") that makes the portfolio's return completely riskless. In an efficient market, a riskless portfolio will return the risk-free rate. This arbitrage reasoning combined with the delta-hedging strategy leads to a partial differential equation that resembles the heat equation of classical physics. Its solution provides the option's value at any point in time. In the case of European-style options (those that have a fixed exercise date) the solution can be achieved in closed form—this is the famous Black–Scholes formula. The Black–Scholes formulas for the values of a European call C or put P are:

$$C = S\Phi(d_1) - Ke^{-rT}\Phi(d_2)$$

$$P = Ke^{-rT}\Phi(-d_2) - S\Phi(-d_1)$$

where:

$$d_1 = [\log(S/K) + T(r + \sigma^2/2)]/\sigma\sqrt{T}$$

$$d_2 = d_1 - \sigma\sqrt{T}$$

Here, r is the risk-free rate (the rate of return of a riskless security such as a U.S. Treasury security over time T), log denotes the natural logarithm, and $\Phi()$ is the cumulative distribution function for the standard normal distribution $N(0, 1)$.

Exotic option contracts, especially those with exercise rules that give the owner the discretion of when to exercise, or options with underlying assets that are more complicated than equity stocks with no dividends, or options that depend on multiple assets, are very difficult to solve using the Black–Scholes partial differential equation.

Harrison and Pliska (1981) developed a more general perspective on options pricing that leads to a useful approach for these more general categories of options. The existence of the riskless trading strategy in the Black–Scholes framework is mathematically equivalent to the existence of a dual object called a "risk-neutral measure," and the options price is the integral of the option payouts with this risk-neutral measure. When the risk-neutral measure is unique, the market is called "complete." This assumption means that there is a single risk-neutral measure that can be used to price all the options.

This perspective leads to the following methodology for options pricing. Observe the prices of the traded options. Usually these are of a fairly simple type (European or American calls and puts), for which closed-form expressions like the Black–Scholes formula can be used. Then invert the formula to find the parameters of the risk-neutral distribution. This distribution can then be used to simulate values of any option—under the assumption that the market is efficient (arbitrage-free) and complete.

21.10.2 Brownian Motion

A key component of valuing stock options is a model of the price process of the underlying stock. In this section, we describe the Brownian motion model for the stock prices.

The efficient market hypothesis, which states that stock prices reflect all history and that any new information is immediately reflected in the stock prices, ensures that stock prices follow a Markov process so the next stock price depends only upon the current stock price and does not depend upon the historical stock price process. A Markov process is a stochastic process with the property that only the current value of the random variable is relevant for the purposes of determining the next value of the variable. A Wiener process is a type of Markov process. A Wiener process $Z(t)$ has normal and independent increments with variance proportional to the square root of time, that is, $Z(t) - Z(s)$ has a normal distribution with mean zero and variance $\sqrt{t-s}$. It turns out that $Z(t), t > 0$ will be a continuous function of t. If Δt represents an increment in time and ΔZ represents the change in Z over that increment in time then the relationship between ΔZ and Δt can be expressed by:

$$\Delta Z(t) = \varepsilon \sqrt{\Delta t} \tag{21.22}$$

where ε is drawn from a standard normal distribution. A Wiener process is the limit of the above stochastic process as the time increments get infinitesimally small, that is, as $\Delta t \to 0$. Equation 21.22 is expressed as

$$dZ(t) = \varepsilon \sqrt{dt} \tag{21.23}$$

If $x(t)$ is a random variable and Z is a Wiener process, then a generalized Wiener process is defined as

$$dx(t) = a\, dt + b\, dZ$$

where a and b are constants. An Ito process is a further generalization of a generalized Wiener process. In an Ito process, a and b are not constants rather, they can be functions of x and t. An Ito process is defined as

$$dx(t) = a(x,t)\, dt + b(x,t)\, dZ$$

Investors are typically interested in the rate of return on a stock, rather than the absolute change in stock price. Let S be the stock price and consider the change in stock price dS over a small period of time dt. The rate of return on a stock, dS/S, is often modeled as being comprised of a deterministic and stochastic component. The deterministic component, μdt, represents the contribution of the average growth rate of the stock. The stochastic component captures random changes in stock price due to unanticipated news. This component is often modeled by σdZ, where Z is a Brownian motion. Combining the deterministic growth rate (also known as drift) with the stochastic contribution to rate of change in stock price yields the equation:

$$\frac{dS}{S} = \mu\, dt + \sigma\, dZ \tag{21.24}$$

an Ito process. μ and σ can be estimated using the methods described in Section 21.3.

The risk-neutral measure Q of Harrison and Pliska as applied to the Black–Scholes framework is induced by the risk-neutral process X that satisfies the modified Brownian motion

$$\frac{dX}{X} = (r - \sigma^2/2)\, dt + \sigma\, dZ \tag{21.25}$$

It is important to note that this process is not the same as the original process followed by the stock—the drift has been adjusted. This adjustment is required to generate the probability measure that makes the delta-hedging portfolio process into a martingale. According to the theory discussed in Section 21.10.1, we price options in the Black–Scholes framework by integrating their cash flows under the risk-neutral measure generated by Equation 21.25. In the following section, we discuss how efficient markets and risk neutral valuation are used to compute an option's value using Monte Carlo simulation.

21.10.3 Simulating Risk-Neutral Paths for Options Pricing

In this section, we discuss how simulation can be used to price options on a stock by simulating the stock price under the risk-neutral measure over T periods, each of length $\Delta t = 1/52$. At each time interval, we simulate the current stock price and then step the process forward so that there are a total of T steps in the simulation. To simulate the path of the stock price over the T week period, we consider the discrete time version of Equation 21.25: $\Delta S/S = (r - \sigma^2/2)\Delta t + \sigma dZ = (r - \sigma^2/2)\Delta t + \sigma \varepsilon \sqrt{\Delta t}$. As ε is distributed like a standard normal random variable, $\Delta S/S \sim N((r - \sigma^2/2)\Delta t, \sigma\sqrt{\Delta t})$.

Each week, use the following steps to determine the simulated stock price:

i. Set $i = 0$.
ii. Generate a random value v_1 from a standard normal distribution. (Standard spreadsheet tools include this capability.)
iii. Convert v_1 to a sample v_2 from a $N((r-\sigma^2/2)\Delta t, \sigma\sqrt{\Delta t})$ by setting $v_2 = (r-\sigma^2/2)\Delta t + \sigma\sqrt{\Delta t}v_1$.
iv. Set $\Delta S = v_2 S$. ΔS represents the incremental change in stock price from the prior period to the current period.
v. $S' = S + \Delta S$, where S' is the simulated updated value of the stock price after one period.
vi. Set $S = S'$, $i = i + 1$.
vii. If $i = T$ then stop. S is the simulated stock price at the end of 6 months. If $i < T$, return to step (i).

Note that randomness only enters in step (ii) when generating a random value v_1. All other steps are mere transformations or calculations and are deterministic. The payoff of a call option at expiration is given by Equation 21.21. Further, in the absence of arbitrage opportunities (i.e., assuming efficient markets) and by applying the theory of risk neutral valuation we know that the value of the option is equal to its expected payoff discounted by the risk-free rate. Using these facts, we apply the Monte Carlo simulation method to price the option. The overall methodology is as follows:

i. Define the number of time periods until maturity, T.
ii. Use Monte Carlo simulation to simulate a path of length T describing the evolution of the underlying stock price, as described above. Denote the final stock price at the end of this simulation by S^F.
iii. Determine the option payoff according to Equation 21.21, assuming S^F, the final period T stock price determined in step (ii).
iv. Discount the option payoff from step (iii) assuming the risk-free rate. The resulting value is the current value of the option.
v. Repeat steps (ii)–(iv) until the confidence bound on the estimated value of the option is within an acceptable range.

21.10.4 A Numerical Example

A stock has expected annual return of $\mu = 15\%$ per year and standard deviation of $\sigma = 20\%$ per year. The current stock price is $S = \$42$. An investor wishes to determine the value of a European call option with a strike price of \$40 that matures in 6 months. The risk-free rate is 8%.

We will use Monte Carlo simulation to simulate the path followed by the stock price and hence the stock price at expiration which determines the option payoff. We consider weekly time intervals, that is, $\Delta t = 1/52$. Thus $T = 24$ assuming, for the sake of simplicity, that there are 4 weeks in each month.

To simulate the path of the stock price over the 24-week period, we follow the algorithm described in Section 21.10.3. We first compute the risk-neutral drift $(r - \sigma^2/2)\Delta t$, which with these parameter settings works out to be 0.0012. The random quantity ε is distributed like a standard normal random variable, so $\Delta S/S \sim N(.0012, .0277)$.

The starting price of the stock is \$42. Each week, use the following steps to determine the simulated updated stock price:

i. Generate a random value v_1 from a standard normal distribution.
ii. Convert v_1 to a sample v_2 from a $N(.0012, .0277)$ by setting $v_2 = .0012 + .0277 v_1$.
iii. Set $\Delta S = v_2 S$.
iv. Set $S = S + \Delta S$

Steps (i)–(iv) yield the simulated updated stock price after 1 week. Repeat this process $T = 24$ times to determine S^F, the stock price at the end of 6 months. Then, the option payoff equals $P = \max(S^F - 40, 0)$. P is the option payoff based upon a single simulation of the ending stock price after 6 months, that is, based upon a single simulation run. Perform many simulation runs and after each run compute the arithmetic average and confidence bounds of the simulated values of P. When enough simulation runs have been performed so that the confidence bounds are acceptable, the value of the option can be computed based upon the expected value of P: $V = e^{-.08(.5)} E(P)$.

21.11 Dynamic Programming

Dynamic programming is a formal method for performing optimization over time. The algorithm involves breaking a problem into a number of subproblems, solving the smaller subproblems, and then using those solutions to help solve the larger problem. Similar to stochastic programming with recourse, dynamic programming involves sequential decision making where decisions are made, information is revealed, and then new decisions are made. More formally, the dynamic programming approach solves a problem in *stages*. Each stage is comprised of a number of possible *states*. The optimal solution is given in the form of a policy that defines the optimal action for each stage. Taking action causes the system to transition from one stage to a new state in the next stage.

There are two types of dynamic programming settings: deterministic and stochastic. In a deterministic setting, there is no system uncertainty. Given the current state and the action taken, the future state is known with certainty. In a stochastic setting, taking action will select the probability distribution for the next state. For the remainder we restrict our discussion to a stochastic dynamic programming setting, as finance problems are generally not deterministic. If the current state is the value of a portfolio, and possible actions are allocations to different assets, the value of the portfolio in the next stage is not known with certainty (assuming that some of the assets under consideration contain risk).

A dynamic program typically has the following features:

i. The problem is divided into $t = 1, \ldots, T$ stages. x_t denotes the state at the beginning of stage t and $a_t(x_t)$ denotes the action taken during stage t given state x_t.

Financial Engineering

Taking action transitions the system to a new state in the next stage so that $x_{t+1} = f(x_t, a_t(x_t), \varepsilon_t)$, where ε_t is a random noise term. The initial state x_0 is known.

ii. The cost (or profit) function in period t is given by $g_t(x_t, a_t(x_t), \varepsilon_t)$. This cost function is additive in the sense that the total cost (or profit) over the entire T stages is given by:

$$g_T(x_T, a_T(x_T), \varepsilon_T) + \sum_{t=1}^{T-1} g_t(x_t, a_t(x_t), \varepsilon_t) \tag{21.26}$$

The objective is to optimize the expected value of Equation 21.26.

iii. Given the current state, the optimal solution for the remaining states is independent of any previous decisions or states. The optimal solution can be found by backward recursion. Namely, the optimal solution is found for the period T subproblem, then for the periods $T-1$ and T subproblem, and so on. The final period T subproblem must be solvable.

The features of dynamic programming that define the options pricing problem differ somewhat from the features described here. In Section 21.12.1 we note these differences.

21.12 Pricing American Options Using Dynamic Programming

Monte Carlo simulation is a powerful tool for options pricing. It performs well even in the presence of a large number of underlying stochastic factors. However, at each step simulation progresses forward in time. On the other hand, options that allow for early exercise must be evaluated backward in time where in each period the option holder must compare the intrinsic value of the option against the expected future value of the option.

Dealing with early exercise requires one to go backward in time, as at each decision point the option holder must compare the value of exercising immediately against the value of holding the option. The value of holding the option is simply the price of the option at that point.

In this section we will show how one can use dynamic programming to price an American option. The method involves two steps:

i. Build a T stage tree of possible states. The states correspond to points visited by the underlying stock price process.

ii. Use dynamic programming and backward recursion to determine the current value of the option.

21.12.1 Building the T Stage Tree

Cox et al. (1979) derived an exact options pricing formula under a discrete time setting. Following their analysis, we model stock price as following a multiplicative binomial distribution: if the stock price at the beginning of stage t is S then at the beginning of stage $t+1$

the stock price will be either uS or dS with probability q and $(1-q)$, respectively. Each stage has length Δt. We will build a "tree" of possible stock prices in any stage. When there is only one stage remaining, the tree looks like:

$$S \begin{cases} uS & \text{with probability } q \\ dS & \text{with probability } 1-q \end{cases}$$

Suppose there are two stages. In each stage, the stock price will move "up" by a factor of u with probability q and "down" by a factor of d with probability $(1-q)$. In this case, there are three possible final stock prices and the tree is given by:

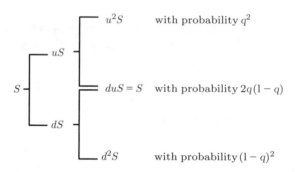

The tree continues to grow according to this method. In general, at stage t there are $t+1$ possible stock prices (states) that will appear on the tree. These are given by:

$$u^j d^{t-j} S, \quad \text{for } j = 0, \ldots, t$$

The values of u, d, and q are determined based upon the assumptions of efficient markets, risk neutral valuation, and the fact that the variance of the change in stock price is given by $\sigma^2 \Delta t$ (from Section 21.10.3). These values are:

$$u = e^{\sigma \sqrt{\Delta t}}$$

$$d = e^{-\sigma \sqrt{\Delta t}}$$

$$q = \frac{a - d}{u - d}$$

where $a = e^{r \Delta t}$.

21.12.2 Pricing the Option Using the Binomial Tree

We now use backward enumeration through the binomial tree to determine the current stage 0 value of the option. We will illustrate the concept using the trees developed in

Section 12.12.1. Let K denote the strike price. With one period remaining, the binomial tree had the form:

$$S \begin{cases} uS & \text{with probability } q \\ dS & \text{with probability } 1-q \end{cases}$$

The corresponding values of the call at the terminal points of the tree are $C_u = \max(0, uS - K)$ with probability q and $C_d = \max(0, dS - K)$ with probability $(1-q)$. The current value of the call is given by the present value (using the risk-free rate) of $qC_u + (1-q)C_d$.

When there is more than one period remaining, each node in the tree must be evaluated by comparing the intrinsic value of the option against its continuation value. The intrinsic value is the value of the option if it is exercised immediately; this value is determined by comparing the current stock price to the option strike price. The continuation value is the discounted value of the expected cash payout of the option under the risk neutral measure, assuming that the optimal exercise policy is followed in the future. Thus, the decision is given by:

$$g_t = \max\left\{\max(0, x_t - K), E[e^{-r\Delta t} g_{t+1}(x_{t+1})|x_t]\right\} \qquad (21.27)$$

The expectation is taken over the risk neutral probability measure. x_t, the current state, is the current stock price. Notice that the action taken in the current state (whether to exercise) does not affect the future stock price. Further, this value function is not additive. However, its recursive nature makes a dynamic programming solution method useful.

21.12.3 A Numerical Example

We illustrate the approach using the identical setting as that used to illustrate the Monte Carlo simulation approach to options pricing. However, here we will assume that we are pricing an American option. The investor wishes to determine the value of an American call option with a strike price of $40 that matures in 1 month. (We consider only 1 month to limit the size of the tree that we build.) The current stock price is $S = \$42$. The stock return has a standard deviation of $\sigma = 20\%$ per year. The risk-free rate is 8%.

We first build the binomial tree and then use the tree to determine the current value of the option. We consider weekly time intervals, that is, $\Delta t = 1/52$. Thus $T = 4$ assuming, for the sake of simplicity, that there are 4 weeks in each month.

$$u = e^{\sigma\sqrt{\Delta t}} = 1.028$$

$$d = e^{-\sigma\sqrt{\Delta t}} = 0.973$$

$$q = \frac{e^{r\Delta t} - d}{u - d} = 0.493$$

The tree of stock movements over the 4-week period looks like:

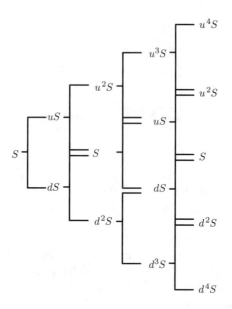

We will evaluate each node in the tree by backward evaluation starting at the fourth time period and moving backward in time. For each node we will use the Equation 21.27 to compare the intrinsic value of the option against its continuation value to determine the value of that node. The binomial tree for the value of the American call option is:

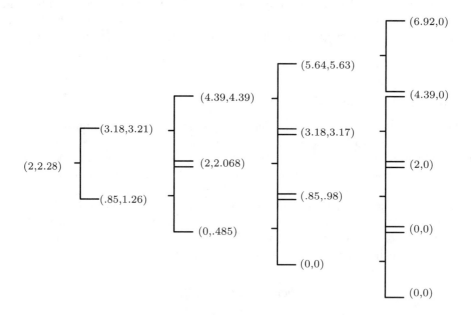

Every node in the tree contains two numbers in parenthesis. The first number is the intrinsic value of the option. The second number is the discounted expected continuation value, assuming that optimal action is followed in future time periods. The option value at time zero (current time) is 2.28. Note that although the option is currently in the money, it is not optimal to exercise even though it is an American option and early exercise is allowed. By using the binomial tree to evaluate the option we find that the expected continuation value of the option is higher than its current exercise value.

21.13 Comparison of Monte Carlo Simulation and Dynamic Programming

Dynamic programming is a powerful tool that can be used for pricing options with early exercise features. However, dynamic programming suffers from the so-called curse of dimensionality. As the number of underlying variables increases, the time required to solve the problem grows significantly. This reduces the practical use of dynamic programming as a solution methodology. The performance of Monte Carlo simulation is better in the sense that its convergence is independent of the state dimension. On the other hand, as we have discussed, simulation has traditionally been viewed as inappropriate for pricing options with early exercise decisions as these require estimates of future values of the option and simulation only moves forward in time. Recent research has focused on combining simulation and dynamic programming approaches to pricing American options to gain the benefits of both techniques. See, for example, Broadie and Glasserman (1997).

21.14 Multi-Period Asset Liability Management

The management of liability portfolios of relatively long-term products, like pensions, variable annuities, and some insurance products requires a perspective that goes beyond a single investment period. The portfolio optimization models of Sections 21.5 through 21.7 are short-term models. Simply rolling these models over into the next time horizon can lead to suboptimal results. First, the models may make an excessive number of transactions. Transactions are not free, and realistic portfolio management models must take trading costs into consideration. Second, the models depend only on second moments. Large market moves, such as during a market crash, are not part of the model assumptions. Finally, policies and regulations place conditions on the composition of portfolios.

Academic and finance industry researchers have, over the past few decades, been exploring the viability of using multi-period balance sheet modeling to address the issues of long-term asset liability management. A multi-period balance sheet model treats the assets and liabilities as generating highly aggregated cash flows over multiple time periods. The aggregations are across asset classes, so that individual securities in an asset class, say stocks, are accumulated into a single asset class, say S&P 500. Other asset classes are corporate bonds of various maturity classes, and so forth. The asset class cash flows are aggregated over time periods, typically 3 or 6 months, so that cash flows occurring within a time interval, say, $(t-1, t]$, are treated as if they all occur at the end-point t. The model treats the aggregate positions in each asset category as variables in the model. There is a single decision to be made for each asset class at the beginning of each time period, which is the change in the holdings of each asset class. The asset holdings and liabilities generate cash flows, which then flow into account balances. These account balances are assigned targets, and the objective function records the deviation from the targets. The objective of the model is to maximize the sum of the expected net profits and the expected penalties for missing the targets over the time horizon of the model.

A simplified application of such a model to a property-casualty insurance problem is as follows. Let A_t denote a vector of asset class values at time t and i_t denote their cash flows (e.g., interest payment, dividends, etc.) at time t. Let x_t denote the portfolio of holdings in the asset classes. Cash flows are generated by the holdings and by asset sales:

$$C_t = A_t \Delta x_t + i_t x_{t-1}$$

where $\Delta x_t := x_t - x_{t-1}$. The cash flows are subject to market and economic variability over the time horizons of interest, say $t = 1, \ldots, T$.

Liability flows from the property-casualty portfolio are modeled by aggregating and then forecasting the aggregated losses minus premium income. Loss events are classified by frequency of occurrence and intensity given loss. These can be simulated over the time horizon T using actuarial models. The evolution of the liability portfolio composition (say, by new sales, lapses of coverage, and so forth) can also be modeled. The key correlation to capture in the liability model is the relationship between the liability flows and the state of the economy. For example, high employment is indicative of strong economic activity, which can lead to increases in the number and size of the policies in the insurance portfolio; high inflation will lead to higher loss payouts given losses; and so forth.

Various performance, accounting, tax, and regulatory measurements are computed from the net cash flows. For example, one measurement could be shareholder's equity at the time horizon S_T. Another could be annual net income N_t. Yet another could be premium-surplus ratio P_t—a quantity used in the property-casualty industry as a proxy for the rating of the insurance company.

In these aggregated models, we try to model the change in performance measurements as linear computations from the cash and liability flows and the previous period quantities. For example, net income is computed as cash flow minus operating expenses. If $O_t \Delta x_t$ is a proxy for the contribution of portfolio management to expenses, for example, the cost of trading, then net income can be modeled by the following equation

$$N_t = C_t - L_t - O_t \Delta x_t$$

Shareholder's equity can change due to a number of influences; here we just capture the change due to the addition of net income:

$$S_t = S_{t-1} + N_t$$

Finally, premium-surplus ratio can be approximated by fixing premium income to a level L and assuming (this is a major simplification!) that the surplus is equivalent to shareholders equity:

$$P_t = \frac{L}{S_t}$$

A typical objective for a multi-period balance sheet model/an asset-liability matching problem is to create performance targets for each quantity and to penalize the shortfall. Suppose that the targets are $\overline{N}_t$ for annual net income, $\overline{S}_T$ for shareholder's equity, and $\overline{P}_t$ for premium-surplus ratio. Then the model to be optimized is:

$$\begin{aligned}
\text{Maximize} \quad & E\{S_T - \lambda \sum_t [\overline{N}_t - N_t]^+ - \mu \sum [L - S_t \overline{P}_t]^+\} \\
\text{Subject to} \quad & N_t = A_t \Delta x_t + i_i x_{t-1} - L_t - O_t \Delta x_t \\
& S_t = N_t + S_{t-1}
\end{aligned} \quad (21.28)$$

Financial Engineering

where the parameters λ and μ are used to balance the various contributions in the objective function, the premium-surplus ratio target relationship has been multiplied through by the denominator to make the objective linear in the decision variables, and the objective is integrated over the probability space represented by the discrete scenarios.

The objective function in formulation 21.28 can be viewed as a variation of the Markowitz style, where we are modeling "expected return" through the shareholder's equity at the end of the horizon, and "risks" through the shortfall penalties relative to the targets for net income and premium-surplus ratio.

21.14.1 Scenarios

In multi-period asset liability management the probability distribution is modeled by discrete scenarios. These scenarios indicate the values, or states, taken by the random quantities at each period in time. The scenarios can branch so that conditional distributions given a future state can be modeled. The resulting structure is called a "scenario tree." Typically there is no recombining of states in a scenario tree, so the size of the tree grows exponentially in the number of time periods. For example, in the property-casualty model, the scenarios are the values and cash flows of the assets and the cash flows of the liabilities. The values of these quantities at each time point t and scenario s is represented by the triple (A_t^s, i_t^s, L_t^s). The pair (s,t) is sometimes called a "node" of the scenario tree. The scenario tree may branch at this node, in which case the conditional distribution for the triple $(A_{t+1}, i_{t+1}, L_{t+1})$ given node (s,t) is the values of the triples on the nodes that branch from this node.

It is important to model the correlation between the asset values and returns and the liability cash flows in these conditional distributions. Without the correlations, the model will not be able to find positions in the asset classes that hedge the variability of the liabilities. In property-casualty insurance, for example, it is common to correlate the returns of the S&P 500 and bonds with inflation and economic activity. These correlations can be obtained from historical scenarios, and conditioned on views as discussed in Section 21.7.

The scenario modeling framework allows users to explicitly model the probability and intensity of extreme market movements and events from the liability side. One can also incorporate "market crash" scenarios in which the historical correlations are changed for some length of time that reflects unusual market or economic circumstances—such as a stock market crash or a recession. Finally, in these models it is usual to incorporate the loss event scenarios explicitly rather than follow standard financial valuation methodology, which would tend to analyze the expected value of loss distributions conditional on financial return variability. Such methodology would ignore the year-to-year impact of loss distribution variability on net income and shareholder's equity. However, from the asset liability management (ALM) perspective, the variability of the liability cash flows is very important for understanding the impact of the hedging program on the viability of the firm.

21.14.2 Multi-Period Stochastic Programming

The technology employed in solving an asset liability management problem such as this is multi-period stochastic linear programming. For a recent survey of stochastic programming, see Shapiro and Ruszczynski (2003).

The computational intensity for these models increases exponentially in the number of time periods, so the models must be highly aggregated and strategic in their recommendations. Nevertheless, the models do perform reasonably well in practice, usually generating 300 basis points of excess return over the myopic formulations based on the repetitive application of one period formulations, primarily through controlling transaction costs and because the solution can be made more robust by explicitly modeling market crash scenarios. A recent collection of this activity is in the volume edited by Ziemba and Mulvey (1998). See also Ziemba's monograph (Ziemba, 2003) for an excellent survey of issues in asset liability management.

21.15 Conclusions

In this chapter, we saw the profound influence of applications of Operations Research to the area of finance and financial engineering. Portfolio optimization by investors, Monte Carlo simulation for risk management, options pricing, and asset liability management, are all techniques that originated in OR and found deep application in finance. Even the foundations of options pricing are based on deep applications of duality theory. As the name financial engineering suggests, there is a growing part of the body of financial practice that is regarded as a subdiscipline of engineering—in which techniques of applied mathematics and OR are applied to the understanding of the behavior and the management of the financial portfolios underpinning critical parts of our economy: in capital formation, economic growth, insurance, and economic-environmental management.

References

1. Artzner, P., F. Delbaen, J.-M. Eber, D. Heath, 1999, Coherent measures of risk, *Mathematical Finance*, 203–228.
2. Ashford, R.W., R.H. Berry, and R.G. Dyson, 1988, "Operational Research and Financial Management," *European Journal of Operational Research*, 36(2), 143–152.
3. Bengtsson, C., 2004, "The Impact of Estimation Error on Portfolio Selection for Investors with Constant Relative Risk Aversion." Working paper, Department of Economics, Lund University.
4. Bertsekas, D.P. and J.N. Tsitsiklis, 1996, *Neuro-Dynamic Programming*, Athena Scientific, Belmont, MA, 1996.
5. Bevan, A. and K. Winkelmann, 1998, "Using the Black Litterman Global Asset Allocation Model: Three Years of Practical Experience," Fixed Income Research, Goldman Sachs & Company, December.
6. Birge, J. and V. Linetsky, "Financial Engineering," in: *Handbook in Operations Research and Management Science*, Elsevier, Amsterdam, In press.
7. Black, F. and R. Litterman, 1992, "Global Portfolio Optimization," *Financial Analysts Journal*, September/October, 28–43.
8. Black, F. and M. Scholes, 1973, "The Price of Options and Corporate Liabilities," *Journal of Political Economy*, 81, 637–654.
9. Board, J., C. Sutcliffe, and W.T. Ziemba, 2003, "Applying Operations Research Techniques to Financial Markets," *Interfaces*, 33, 12–24.
10. Broadie, M. and P. Glasserman, 1997, "Pricing American-Style Securities Using Simulation," *Journal of Economic Dynamics and Control*, 21, 1323–1352.
11. Campbell, J.Y., A.W. Lo, and A.C. MacKinlay, 1997, *The Econometrics of Financial Markets*, Princeton University Press, Princeton, NJ.
12. Chopra, V.K. and W.T. Ziemba, 1993, "The Effect of Errors in Mean, Variance, and Covariance Estimates on Optimal Portfolio Choice," *Journal of Portfolio Management*, 19(2), 6–11.
13. Cox, J.C. and S.A. Ross, 1976, "The Valuation of Options for Alternative Stochastic Processes," *Journal of Financial Economics*, 3, 145–166.

14. Cox, J.C., S.A. Ross, and M. Rubinstein, (1979), "Option Pricing: A Simplified Approach," *Journal of Financial Economics*, 7, 229–263.
15. Fama, E.F., 1965, "The Behavior of Stock Prices," *Journal of Business*, 38, 34–105.
16. Fusai, G. and A. Meucci, 2003, "Assessing Views," *Risk*, March 2003, s18–s21.
17. Glasserman, P., 2004, *Monte Carlo Methods in Financial Engineering*, Springer-Verlag, New York.
18. Harrison, J.M. and S.R. Pliska, 1981, "Martingales and Stochastic Integrals in the Theory of Continuous Trading," *Stochastic Processes and their Applications*, 11, 215–260.
19. He, G. and R. Litterman, 1999, "The Intuition Behind Black-Litterman Model Portfolios," *Investment Management Research*, Goldman Sachs & Company, December.
20. Holmer, M.R., 1998, "Integrated Asset-Liability Management: An Implementation Case Study," in *Worldwide Asset and Liability Modeling*, W.T. Ziemba and J.R. Mulvey (eds.), Cambridge University Press.
21. Jarrow, R.V. Maksimovic and W. Ziemba (eds.), 1995, "Finance," In: *Handbooks in Operations Research and Management Science*, vol. 9, North-Holland, Amsterdam.
22. Kallberg, J.G. and W.T. Ziemba, 1981, "Remarks on Optimal Portfolio Selection," In: *Methods of Operations Research*, G. Bamberg, O. Optin, Oelgeschlager, Gunn and Hain (eds.), Cambridge, MA, pp. 507–520.
23. Kallberg, J.G. and W.T. Ziemba, 1984, "Mis-Specifications in Portfolio Selection Problems," In: *Risk and Capital*, G. Bamberg and K. Spremann (eds.), Springer-Verlag, New York, pp. 74–87.
24. King, A.J. 1993, "Asymmetric Risk Measures and Tracking Models for Portfolio Optimization Under Uncertainty, *Annals of Operations Research*, 45, 165–177.
25. Michaud, R.O., 1989, "The Markowitz Optimization Enigma: is 'Optimized' Optimal?" *Financial Analysts Journal*, 45(1), 31–42.
26. Mills, T.C., 1999, *The Econometric Modelling of Financial Time Series*, 2nd ed., Cambridge University Press, Cambridge, UK.
27. Pedersen, C.S. and S.E. Satchell, 1998, "An Extended Family of Financial Risk Measures," *Geneva Papers on Risk and Insurance Theory*, 23, 89–117.
28. Rockafellar, R.T. and S. Uryasev, 2000, "Optimization of Conditional Value-at-Risk," *The Journal of Risk*, 2(3), 21–41.
29. Rockafellar, R.T. and S. Uryasev, 2002, "Conditional Value-at-Risk for General Distributions," *Journal of Banking and Finance*, 26(7), 1443–1471.
30. Samuelson, P.A., 1965, "Rational Theory of Warrant Pricing," *Industrial Management Review*, 6, 13–31
31. Shapiro, A. and A. Ruszczynski, 2003, "Stochastic Programming," In: *Handbooks in Operations Research and Management Science*, vol. 10, Elsevier.
32. Sharpe, W.F., 1964, "Capital Asset Prices: A Theory of Market Equilibrium," *Journal of Finance*, September, 425–442.
33. Uryasev, S. 2001, "Conditional Value-at-Risk: Optimization Algorithms and Applications," *Financial Engineering News*, 14, 1–5.
34. von Neumann, J. and O. Morgenstern, 1944, *Theory of Games and Economic Behaviour*, Princeton University Press, Princeton, NJ.
35. Ziemba, W.T. 2003, *The Stochastic Programming Approach to Asset, Liability, and Wealth Management*, AIMR Research Foundation Publications, No. 3: 1–192.
36. Ziemba, W.T. and J.R. Mulvey (eds.), 1998, *Worldwide Asset and Liability Modeling*, Cambridge University Press, Cambridge, UK.

22

Supply Chain Management

22.1	Introduction...	**22**-1
22.2	Managing Inventories in the Supply Chain......	**22**-6
	Newsvendor and Base-Stock Models • Reorder Point—Order Quantity Models • Distribution Requirements Planning	
22.3	Managing Transportation in the Supply Chain .	**22**-24
	Transportation Basics • Estimating Transportation Costs • Transportation-Inclusive Inventory Decision Making • A More General Transportation Cost Function	
22.4	Managing Locations in the Supply Chain	**22**-38
	Centralized Versus Decentralized Supply Chains • Aggregating Inventories and Risk Pooling • Continuous-Location Models	
22.5	Managing Dyads in the Supply Chain............	**22**-48
22.6	Discussion and Conclusions.......................	**22**-58
References		**22**-59

Donald P. Warsing
North Carolina State University

22.1 Introduction

Perhaps the best way to set the tone for this chapter is to consider a hypothetical encounter that an industrial engineer or operations manager is likely to experience. It's not quite the "guy walks into a bar" setup, but it's close. Consider the following scenario: A software salesperson walks into the office of an industrial engineer and says, "I have some software to sell you that will optimize your supply chain." Unfortunately, the punch line—which I'll leave to the reader's imagination—is not as funny as many of the "guy walks into a bar" jokes you're likely to hear. In fact, there are some decidedly not-funny examples of what could happen when "optimizing" the supply chain is pursued via software implementations without careful consideration of the structure and data requirements of the software system. Hershey Foods' enterprise resource planning (ERP) system implementation (Stedman, 1999) and Nike's advanced planning system (APS) implementation (Koch, 2004) are frequently cited examples of what could go wrong. Indeed, without some careful thought and the appropriate perspective, the industrial engineer (IE) in our hypothetical scenario might as well be asked to invest in a bridge in Brooklyn, to paraphrase another old joke.

Encounters similar to the one described above surely occur in reality countless times every day as of this writing in the early twenty-first century, and in this author's opinion it happens all too often because those magical words "optimize your supply chain" seem to be used frivolously and in a very hollow sense. The question is, what could it possibly mean to optimize the supply chain? Immediately, an astute IE or operations manager must ask himself or herself a whole host of questions, such as: What part of my supply chain? All of it? What does "all of it" mean? Where does my supply chain begin and where does it end? What is the objective function being optimized?...minimize cost?...which costs?...maximize profit?...whose profit? What are the constraints? Where will the input data behind the formulation come from? Is it even conceivable to formulate the problem? Provided I could formulate the problem, could I solve it?...to optimality?

Now, in fairness, this chapter is not technically about supply chain *optimization*, but supply chain *management*. Still, we should continue our line of critical questioning. What does it mean to "manage the supply chain"? Taken at face value, it's an *enormous* task. In the author's view, the term "manage the supply chain," like "optimize your supply chain," is thrown around much too loosely. A better approach, and the one taken in this chapter, is to consider the various aspects of the supply chain that must be managed. Our challenge at hand, therefore, is to figure out how to identify and use the appropriate tools and concepts to lay a good foundation for the hard work that comprises "managing the supply chain."

In the first edition of their widely-cited text book on supply chain management (SCM), Chopra and Meindl (2001) define a *supply chain* as "all stages involved, directly or indirectly, in fulfilling a customer request" (p. 3). (In the more recent edition of the text, Chopra and Meindl 2004, the authors substitute the word "parties" for stages. Since this is inconsistent with the definition of SCM that is to follow, I have chosen the earlier definition.) This definition is certainly concise, but perhaps a bit too concise. A more comprehensive definition of the term "supply chain" is given by Bozarth and Handfield (2006), as follows: "A network of manufacturers and service providers that work together to convert and move goods from the raw materials stage through to the end user" (p. 4). In a more recent edition of their text, Chopra and Meindl (2004) go on to define *supply chain management* as "the management of flows between and among supply chain stages to maximize total supply chain profitability" (p. 6). The nice aspects of this definition are that it is, once again, concise and that it clearly emphasizes managing the *flows* among the stages in the chain. The problematic aspect of the definition, for this writer, is that its objective, while worthy and theoretically the right one (maximizing the net difference between the revenue generated by the chain and the total costs of running the chain), would clearly be difficult to measure in reality. Would all parties in the supply chain willfully share their chain-specific costs with all others in the chain? This brings us back to the criticisms lodged at the outset of this chapter related to "optimizing the supply chain"—a noble goal, but perhaps not the most practical one.

The author is familiar with a third-party logistics (3PL) company that derives a portion of its revenue from actually carrying out the goals of optimizing—or at least *improving*—its clients' supply chains. Figure 22.1 is taken from an overview presentation by this 3PL regarding its services and represents a cyclic perspective on what it means to manage the supply chain—through design and redesign, implementation, measurement, and design (re-)assessment. Figure 22.2 provides an idea of the kinds of tools this 3PL uses in carrying out its SCM design, evaluation, and redesign activities. What is striking about this company's approach to doing this is that it involves neither a single software application nor a simple series of software applets integrated into a larger "master program," but a whole host of discrete tools applied to various portions of the overarching problem of "managing the supply chain." The software tools listed are a mix of off-the-shelf software and proprietary code, customized in an attempt to solve more integrated problems. The result is an approach that views the overall challenge of managing the supply chain as one of understanding how the various problems that comprise "supply chain management" interact with one another, even as they are solved discretely and independently, many times for the sheer purpose of tractability.

Supply Chain Management

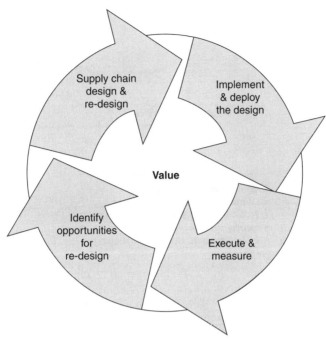

FIGURE 22.1 Supply chain design cycle. Presentation by unnamed 3PL company.

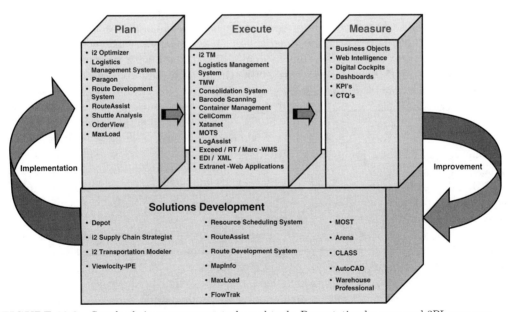

FIGURE 22.2 Supply chain management tasks and tools. Presentation by unnamed 3PL company.

Coming back to definitions of SCM, the definition from Bozarth and Handfield (2006) offers an objective different from Chopra and Meindl's profit-maximizing one—more customer focused, but perhaps no less problematic in practice. Their definition is as follows: "The *active* management of supply chain activities and relationships in order to maximize

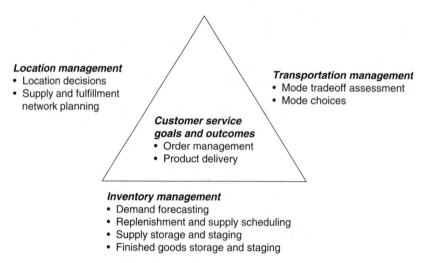

FIGURE 22.3 Drivers of supply chain management strategy. (Adapted from Ballou, 2004.)

customer value and achieve a sustainable competitive advantage" (p. 8). The idea of a sustainable competitive advantage through the management of SCM-related activities is also emphasized by Ballou (2004), who does a commendable job basing the whole of his textbook on logistics and SCM on the various management activities that must be understood and integrated to derive positive customer outcomes and thereby achieve a sustainable advantage. This will be the approach taken in the remainder of this chapter.

Figure 22.3 shows an adaptation of the conceptual model that serves as the structural basis for Ballou's (2004) text. Ballou describes SCM as a triangle, the legs of which are the inventory, transportation, and location strategies of the chain, with customer service goals located in the middle of the triangle, representing the focus of these strategies. This approach is consistent with that of the Chopra and Meindl (2004) text, whose *drivers* of SCM are inventory (*what* is being moved through the chain), transportation (*how* the inventory is moved through the chain), and facilities (*where* inventories are transformed in the chain). These SCM drivers form the essence of a supply chain strategy, laying out the extent to which a firm chooses approaches to fulfilling demand that achieve higher service, but almost necessarily higher cost, or lower cost, but almost necessarily lower service. (Chopra and Meindl also list *information* as an SCM driver, but in the author's view, information is more appropriately viewed as an *enabler* that may be used to better manage inventory, transportation, and facilities. More discussion on *enablers* of SCM will follow.)

The idea of SCM drivers is useful in thinking about another theme in SCM emphasized by many authors and first proposed by Fisher (1997) in an important article that advanced the notion that "one size fits all" is not an effective approach to SCM. Fisher (1997) cogently lays out a matrix that matches *product* characteristics—what he describes as a dichotomy between *innovative* products like technology-based products and *functional* products like toothpaste or other staple goods—and *supply chain* characteristics—another dichotomy between *efficient* (cost-focused) supply chains and *responsive* (customer service-focused) supply chains. Chopra and Meindl (2004) take this conceptual model a step further, first by pointing out that Fisher's product characteristics and supply chain strategies are really continuous spectrums, and then by superimposing the Fisher model, as it were, on a frontier that represents the natural tradeoff between responsiveness and efficiency. Clearly, it stands to reason that a firm, or a supply chain, *cannot* maximize cost efficiency and customer responsiveness simultaneously; some aspects of each of these objectives necessarily

Supply Chain Management

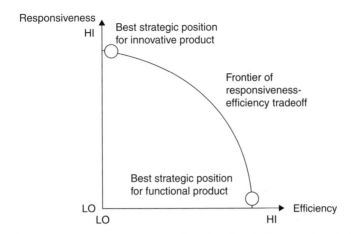

FIGURE 22.4 Overlay of conceptual models of Fisher (1997) and Chopra and Meindl (2004).

work at cross purposes. Chopra and Meindl's frontier and Fisher's product dichotomy are presented in Figure 22.4. The value of this perspective is that it clearly identifies a *market-driven* basis for strategic choices regarding the three SCM drivers: Should our inventory management decisions be focused more on efficiency—minimizing inventory levels—or on responsiveness—maximizing product availability? Should our transportation choices be focused more on efficiency—minimizing transportation costs, perhaps through more extensive economies of scale—or on responsiveness—minimizing delivery lead times and maximizing reliability? Should our facilities (network design) decisions be focused more on efficiency—minimizing the number of locations and maximizing their size and scale—or on responsiveness—seeking high levels of customer service by choosing many, focused locations closer to customers?

Clearly, the legs of the Ballou (2004) triangle frame the SCM strategy for any firm. What is it, however, that allows a firm to implement that strategy and execute those driver-related decisions, effectively? Marien (2000) presents the results of a survey of approximately 200 past attendees of executive education seminars, asking these managers first to provide a rank-ordering of four *enablers* of effective SCM, and then to rank the various attributes that support each enabler. The enablers in Marien's survey were identified through an extensive search of the academic and trade literature on SCM, and the four resulting enablers were (1) technology, (2) strategic alliances, (3) organizational infrastructure, and (4) human resources management. Marien's motivation for the survey was similar to the criticism laid out at the outset of this chapter—the contrast between what he was hearing from consultants and software salespeople, that effective SCM somehow flowed naturally from information technology implementations, and from practicing managers, who, at best, were skeptical of such technology-focused approaches and, at worst, still bore the scars of failed, overzealous IT implementations. Indeed, the result of Marien's (2000) survey indicate that managers, circa 1998, viewed organizational infrastructure as, far and away, the most important enabler of effective SCM, with the others—technology, alliances, and human resources—essentially in a three-way tie for second place. The most important attributes of organizational infrastructure, according to Marien's survey results, were a coherent business strategy, formal process flows, commitment to cross-functional process management, and effective process metrics.

The process management theme in Marien's (2000) results is striking. It is also consistent with what has arguably become the most widely-accepted structural view of supply chain management among practitioners, the supply chain operations reference model, or

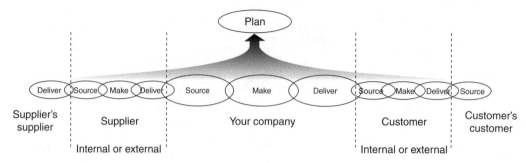

FIGURE 22.5 The Supply Chain Operations Reference (SCOR®) model. (From Supply-Chain Council, 2006.)

SCOR* (Supply-Chain Council, 2006). Figure 22.5 presents the structural framework for the SCOR model, based on the primary business processes that comprise SCM according to the model—plan, source, make, deliver, and return. Underlying these primary processes is an extensive checklist, as it were, of the sub-processes and tasks that, according to the model, should be in place to enable effective SCM. Bozarth and Warsing (2006) offer an academic perspective that attempts to tie the underlying detail of the SCOR model back to company performance and offers a research agenda for understanding this link; in fact, this theme has already begun to emerge in the literature (Lockamy and McCormack, 2004).

An important theme that must be addressed in this chapter, however—and one that is clear from Figure 22.5—is that supply chain management is encapsulated in managing *internal* business processes and their interfaces with the firms *immediately* upstream and downstream from your company. While the common lingo claiming to manage the chain from the "supplier's suppliers to the customer's customers" is catchy, it is not clear how this is to be done. What is more important, from an external perspective, is managing the most important *dyadic* relationships in the chain, the links immediately upstream and downstream. These dyads form the business relationships that a company can directly influence, and *direct* influence is what really matters in dealing with a system as complex and dynamic as a supply chain. Vollmann et al. (2005, Chapter 17) offer an excellent discussion of this key idea, and one that, hopefully, will help bring practitioners of SCM down from the clouds and back to the hard work of making their links in the chain more effective.

In the remainder of this chapter, therefore, I discuss each of the drivers of SCM separately—managing inventories, managing transportation, and managing locations (i.e., the network design problem)—identifying important enablers along the way. Then, I present the important issues in managing dyads in the chain. Finally, a discussion follows to tie these themes together, and the chapter ends with a few concluding remarks.

22.2 Managing Inventories in the Supply Chain

Inventory management can be concisely captured in the following policy, of sorts: "Before you run out, get more."† Though concise, the statement captures the essence of the basic problem of inventory management, and immediately begs for an answer to two

---

*SCOR® is a registered trademark of the Supply-Chain Council in the United States and Canada.

†I owe this concise statement, with only slight alterations applied, to my good friend and colleague Douglas J. Thomas.

	Independent demand	Dependent demand
Stationary demand	Reorder point – order quantity systems	Kanban
Non-stationary demand	Distribution requirements planning (DRP)	Manufacturing requirements planning (MRP)

FIGURE 22.6 Framework for choice of inventory management tools. (Based on Bozarth, 2005.)

questions—"How long before you run out (i.e., *when*)?" and "*How much* more do you get?" Thus, managing inventory boils down to answering those two questions, "when?" and "how much?"

Before I dive into tools and techniques for managing inventories—and consistent with the theme I laid out in the Introduction—let us first step back and make sure that we are clear about the characteristics of the items to be managed. These characteristics, it stands to reason, should have a significant impact on the kinds of tools that are appropriate to the management task. Again, my overarching theme is to beware of (buying or) applying a single, elaborate, sophisticated tool to every problem. Sometimes you really might be better off with the proverbial "hammer" to solve the simple, "nail-like" aspects of your problem. Sometimes, I will admit, an advanced power-tool might be warranted.

Bozarth (2005) offers an effective framework for placing the inventory management problem in the context of two important aspects of demand for the products being managed. This framework has been reproduced in Figure 22.6. The first of the two context-shaping aspects of demand is whether demand is driven externally by forces outside the firm's direct control or whether demand is driven by internal management decisions. The former situation describes *independent demand* items, those items whose demand is driven by customer tastes, preferences, and purchasing patterns; typically, these are finished goods. Inventory management in these cases is influenced greatly by the extent to which the firm can effectively describe the random variations in these customer purchasing patterns. The latter situation describes *dependent demand* items, those items whose demand is driven by the demand of other items; a good example of such items would be component parts, whose demand is driven by the production schedules for the finished goods of which they are components. Note that the production schedule for finished items is solely under the control of the manufacturing firm; its management chooses the production plan in response to projected consumer demand, and therefore, inventory management for dependent demand items is largely an issue of ensuring that component parts are available in sufficient quantities to execute the production plan.

The second important aspect of demand that helps determine which inventory management tools to apply is the stability of demand over time. Though Bozarth (2005) describes the dichotomy as "stable" versus "variable," I have chosen to characterize demand as *stationary* or *non-stationary*—the former being the case in which the expected value of demand remains constant over time, or at least over the planning horizon, and the latter being the case in which it does not. In either case, stationary or non-stationary, demand could be uncertain or known with certainty. Obviously, then, the simplest case would be one in which demand is stationary and known with certainty, and the most challenging would be one in which demand is non-stationary and uncertain.

As indicated by Bozarth (2005), another idea embedded in his framework is that the stationary demand approaches are *pull systems*, whereas the non-stationary approaches are *push systems*. Although definitions tend to vary (see Hopp and Spearman, 2004, for an excellent discussion), pull systems are those that generate orders only in response to actual demand, while push systems drive replenishments from the schedule of projected future demands, which, consistent with Bozarth's framework, will vary over time. Finally, I should point out that the matrix in Figure 22.6 presents what would appear to be *reasonable* approaches to managing demand in these four demand contexts, but not necessarily the *only* approaches. For example, non-stationary reorder point-order quantity systems are certainly possible, but probably not applied very often in practice due to the significant computational analysis that would be required. The effort required to manage the system must clearly be considered in deciding which system to apply. While the "optimal" solution might indeed be a non-stationary reorder point-order quantity system, the data required to systematically update the policy parameters of such a system and the analytical skill required to carry out—and *interpret*—the updates may not be warranted in all cases.

As this chapter deals with managing the supply chain, I will focus only on managing those items that move between firms in the chain—that is, the independent demand items. Management of dependent demand items is an issue of production planning, more of a "microfocus" than is warranted in this chapter. An excellent discussion of production planning problems and comparisons between MRP and Kanban systems can be found in Vollmann et al. (2005). Thus, I will spend the next few sections describing the important issues in managing independent demand inventories in the supply chain. First, however, I will present an inventory management model that does not appear, *per se*, in the framework of Figure 22.6. This model is presented, however, to give the reader an appreciation for the important tradeoffs in *any* inventory management system—stationary or non-stationary, independent or dependent demand.

22.2.1 Newsvendor and Base-Stock Models

An important starting point with all inventory management decision making is an understanding of the nature of demand. At one extreme, demand for an item can be a "spike," a single point-in-time, as it were, at which there is demand for the item in question, after which the item is never again demanded by the marketplace. This is the situation of the classic "newsvendor problem." Clearly, demand for today's newspaper is concentrated solely in the current day; there will be, effectively, no demand for today's newspaper tomorrow (it's "yesterday's news," after all) or at any point in future. Thus, if I run a newsstand, I will stock papers only to satisfy this spike in demand. If I don't buy enough papers, I will run out and lose sales—and potentially also lose the "goodwill" of those customers who wanted to buy, but could not. If I buy too many papers, then I will have to return them or sell them to a recycler at the end of the day, in either case at pennies on the dollar, if I'm lucky. The other extreme for demand is for it to occur at a stationary rate per unit time infinitely into the future. That situation will be presented in the section that follows.

I present the newsvendor model in this chapter for the purpose of helping the reader understand the basic tension in inventory decision making—namely, balancing the cost of having too much inventory versus the cost of having too little.[*] It does not take too much

---

[*] For an excellent treatment of the newsvendor problem, the reader is referred to Chapter 9 in Cachon and Terwiesch (2006).

imagination to see that this too much–too little knife edge can be affected significantly by transportation and network design decisions. The important issue before us at this point, however, is to understand how to evaluate the tradeoff. Note also that in the newsvendor case, one only needs to answer the question of "how much" as the answer to the "when" question, in the case of a demand spike, is simply "sufficiently in advance of the first sale of the selling season." The newsvendor model, however, is also the basis for a base-stock inventory policy, which applies to the case of recurring demand with no cost to reorder, in which case, the answer to "when" is "at every review of the inventory level."

To illustrate the costs and tradeoffs in the newsvendor model, let us imagine a fictional retailer—we'll call them "The Unlimited"—who foresees a market for hot-pink, faux-leather biker jackets in the Northeast region for the winter 2007–08 season. Let's also assume that The Unlimited's sales of similar products in Northeastern stores in recent winter seasons— for example, flaming-red, fake-fur, full-length coats in 2005–06 and baby-blue, faux-suede bomber jackets in 2004–05—were centered around 10,000 units. The Unlimited's marketing team, however, plans a stronger ad campaign this season and estimates that there is a 90% chance that demand will be at least 14,000 units. The Unlimited also faces fierce competition in the Northeast from (also fictional) catalog retailer Cliff's Edge, who has recently launched an aggressive direct-mail and internet ad campaign. Thus, there is concern that sales might come in three to four thousand units below historical averages for similar products. Jackets must be ordered from an Asian apparel producer in March for delivery in October; only one order can be placed and it cannot be altered once it is placed. The landed cost of each jacket is $100 (the cost to purchase the item, delivered). They sell for $200 each. In March 2008, any remaining jackets will be deeply discounted for sale at $25 each.

Thus, this case gives us all the basics we will need to determine a good order quantity: a distribution of demand (or at least enough information to infer a distribution), unit sales price, unit cost, and end-of-season unit salvage value. In general, if an item sells for p, can be purchased for c, and has end-of-season salvage value of s, then the cost of missing a sale is $p-c$, the profit margin that we would have made on that sale, and the cost of having a unit in excess at the end of the selling season is $c-s$, the cost we paid to procure the item less its salvage value. If demand is a random variable D with probability distribution function (pdf) f, then the optimal order quantity maximizes expected profit, given by

$$\pi(Q) = \int_{-\infty}^{Q} [(p-c)D - (c-s)(Q-D)]f(D)\mathrm{d}D + \int_{Q}^{\infty} Q(p-c)f(D)\mathrm{d}D \qquad (22.1)$$

(i.e., if the order quantity exceeds demand, we get the unit profit margin on the D units we sell, but we suffer a loss of $c-s$ on the $Q-D$ units in excess of demand; however, if demand exceeds the order quantity, then we simply get the unit profit margin on the Q units we are able to sell).

One approach to finding the best order quantity would be to minimize Equation 22.1 directly. Another approach, and one that is better in drawing out the intuition of the model, is to find the best order quantity through a marginal analysis. Assume that we have ordered Q units. What are the benefits and the risks of ordering another unit versus the risks and benefits of not ordering another unit? If we order another unit, the probability that we will not sell it is $\Pr\{D \leq Q\} = F(Q)$, where F is obviously the cdf of D. The cost in this case is $c-s$, meaning that the expected cost of ordering another unit beyond Q is $(c-s)F(Q)$. If we do not order another unit, the probability that we could have sold it is $\Pr\{D > Q\} = 1 - F(Q)$ (provided that D is a continuous random variable). The cost in this case is $p-c$, the opportunity cost of the profit margin foregone by our under-ordering, meaning that the expected cost of not ordering another unit beyond Q is $(p-c)[1-F(Q)]$.

Another way of expressing the costs is to consider $p - c$ to be the *underage cost*, or $c_u = p - c$, and to consider $c - s$ the *overage cost*, or $c_o = c - s$. A marginal analysis says that we should, starting from $Q = 1$, continue ordering more units until that action is no longer profitable. This occurs where the net difference between the expected cost of ordering another unit—the cost of overstocking—and the expected cost of not ordering another unit—the cost of understocking—is zero, or where $c_o F(Q) = c_u[1 - F(Q)]$. Solving this equation, we obtain the optimal order quantity,

$$Q^* = F^{-1}\left(\frac{c_u}{c_u + c_o}\right) \tag{22.2}$$

where the ratio $c_u/(c_u + c_o)$ is called the *critical ratio*, which we denote by CR, and which specifies the optimal order quantity at a *critical fractile* of the demand distribution.

Returning to our fictional retailer example, we have $c_u = \$100$ and $c_o = \$75$; therefore, $CR = 0.571$. The issue now is to translate this critical ratio into a critical fractile of the demand distribution. From the information given above for "The Unlimited," let's assume that the upside and downside information regarding demand implies that the probabilities are well-centered around the mean of 10,000. Moreover, let's assume that demand follows a normal distribution. Thus, we can use a normal look-up table (perhaps the one embedded in the popular Excel spreadsheet software) and use the 90th-percentile estimate from Marketing to compute the standard deviation of our demand distribution. Specifically, $z_{0.90} = 1.2816 = (14,000 - 10,000)/\sigma$, meaning that $\sigma = 3,121$. Thus, another table look-up (or Excel computation) gives us $Q^* = 10,558$, the value that accumulates a total probability of 0.571 under the normal distribution pdf with mean 10,000 and standard deviation 3,121.

As it turns out, the optimal order quantity in this example is not far from the expected value of demand. This is perhaps to be expected as the cost of overage is only 25% less than the cost of underage. One might reasonably wonder what kind of service outcomes this order quantity would produce. Let us denote α as the *in-stock probability*, or the probability of not stocking out of items in the selling season, and β as the *fill rate*, or the percentage of overall demand that we are able to fulfill. For the newsvendor model, let us further denote $\alpha(Q)$ as the in-stock probability given that Q units are ordered, and let $\beta(Q)$ denote the fill rate if Q units are ordered. It should be immediately obvious that $\alpha(Q^*) = CR$. Computing the fill rate is a little more involved. As fill rate is given by (Total demand − Unfulfilled demand) ÷ (Total demand), it follows that

$$\beta(Q) = 1 - \frac{S(Q)}{\mu} \tag{22.3}$$

where $S(Q)$ is the expected number of units short (i.e., the expected unfulfilled demand) and μ is the expected demand. Computing $S(Q)$ requires a *loss function*, which also is typically available in look-up table form for the standard normal distribution.* Thus, for normally distributed demand

$$\beta(Q) = 1 - \frac{\sigma L(z)}{\mu} \tag{22.4}$$

where $L(z)$ is the standard normal loss function[†] and $z = (Q - \mu)/\sigma$ is the standardized value of Q. Returning once again to our example, we can compute $\beta(Q^*) = 0.9016$, meaning

---

*Cachon and Terwiesch (2006) also provide loss function look-up tables for Erlang and Poisson distributions.

[†] That is, $L(z) = \int_z^\infty (x - z)\phi(x)dx$, where $x \sim N(0, 1)$ and ϕ is the standard normal pdf. This can also be computed directly as $L(z) = \phi(z) - z[1 - \Phi(z)]$, where Φ is the standard normal cdf.

TABLE 22.1 Comparative Outcomes of Newsvendor Example for Increasing $g = k(p-c)$

k	$\alpha(Q^*)$	$\beta(Q^*)$	Q^*	Q^*/μ	z^*	Expected Overage
0	0.571	0.901	10,558	1.06	0.179	1543.9
0.2	0.615	0.916	10,913	1.09	0.293	1754.5
0.4	0.651	0.927	11,211	1.12	0.388	1943.2
0.6	0.681	0.935	11,468	1.15	0.470	2114.3
0.8	0.706	0.942	11,691	1.17	0.542	2269.0
1	0.727	0.948	11,884	1.19	0.604	2407.3
2	0.800	0.965	12,627	1.26	0.842	2975.4
4	0.870	0.980	13,515	1.35	1.126	3718.3
6	0.903	0.986	14,054	1.41	1.299	4196.4
8	0.923	0.989	14,449	1.44	1.426	4557.2
10	0.936	0.991	14,750	1.48	1.522	4837.0

Note: Q^* is rounded to the nearest integer, based on a critical ratio rounded to three decimal places.

that we would expect to fulfill about 90.2% of the demand for biker jackets next winter season if we order 10,558 of them.

A reasonable question to ask at this point is whether those service levels, a 57% chance of not stocking out and a 90% fill rate are "good" service levels. One answer to that question is that they are neither "good" nor "bad"; they are *optimal* for the overage and underage costs given. Another way of looking at this is to say that, if the decision maker believes that those service levels are too low, then the cost of underage must be understated, for if this cost were larger, then optimal order quantity would increase and the in-stock probability and fill rate would increase commensurately. Therefore, newsvendor models like our example commonly employ another cost factor, what we will call g, the cost of losing customer *goodwill* due to stocking out. This cost inflates the underage cost, making it $c_u = p - c + g$. The problem with goodwill cost, however, is that it is not a cost that one could pull from the accounting books of any company. Although one could clearly argue that it is "real"—customers do indeed care, in many instances, about the inconvenience of not finding items they came to purchase—it would indeed be hard to evaluate for the "average" customer. The way around this is to allow the service outcome to serve as a proxy for the actual goodwill cost. The decision maker probably has a service target in mind, one that represents the in-stock probability or fill rate that the firm believes is a good representation of their commitment to their customers, but does not "break the bank" in expected overstocking costs. Again, back to our example, Table 22.1 shows the comparative values of Q, α, and β for goodwill costs $g = k(p - c)$, with k ranging from 0 to 10. Thus, from Table 22.1, as g ranges from 0 to relatively large values, service—measured both by α and β—clearly improves, as does the expected unsold inventory.

For comparison, Table 22.1 displays a ratio of the optimal order quantity to the expected demand, Q^*/μ, and also shows the standardized value of the optimal order quantity, $z^* = (Q^* - \mu)/\sigma$. These values are, of course, specific to our example and would obviously change as the various cost factors and demand distribution parameters change. It may be more instructive to consider a "parameter-free" view of the service level, which is possible if we focus on the in-stock probability. Figure 22.7 shows a graph similar to Figure 12.2 from Chopra and Meindl (2004), which plots $\alpha(Q^*) = CR$, the optimal in-stock probability, versus the ratio of overage and underage costs, c_o/c_u.[*] This graph shows clearly how service levels increase as c_u increases relative to c_0, in theory increasing to 100% in-stock probability at $c_0 = 0$ (obviously an impossible level for a demand distribution that is unbounded in the positive direction). Moreover, for high costs of overstocking, the optimal in-stock probability could be well below 50%.

---

[*]Note that $CR = c_u/(c_u + c_o)$ can also be expressed as $CR = 1/(1 + c_o/c_u)$.

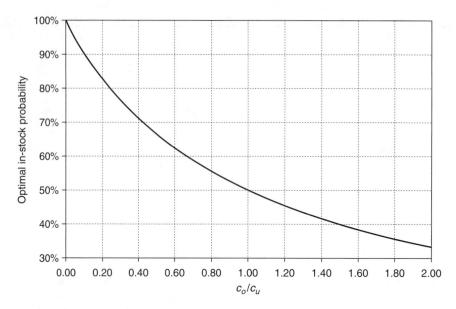

FIGURE 22.7 Optimal service level versus ratio of overage and underage costs.

These ideas lead to a natural perspective on *safety stock*, which is the amount of inventory held in excess of expected demand to protect against stockouts. In the case of the newsvendor model, where demand is a spike, the safety stock is simply given by $Q - \mu$. Note also that at $Q = \mu$, the safety stock is zero and the expected in-stock probability is 50%.

We also pointed out above that the newsvendor model is the basis for the simplest inventory system to manage items with repeating demands, namely a base-stock policy. A base-stock policy is appropriate when the cost of placing orders is zero (or negligible). In a *periodic review* setting, one in which the inventory level is reviewed on a regular cycle, say every T days, a base-stock policy is one in which, at each review, an order is placed for $B - x$ units, where B is the base-stock level and x is the *inventory position* (inventory on-hand plus outstanding orders). In a *continuous review* setting, where the inventory level is reviewed continuously (e.g., via a computerized inventory records system), a base-stock policy is sometimes called a "sell one-buy one" system as an order will be placed every time an item is sold (i.e., as soon as the inventory position dips below B). One can see how a newsvendor model serves as the basis for setting a base-stock level. If one can describe the distribution of demand over the replenishment lead time, with expected value $E[D_{DLT}]$ and variance $\text{var}[D_{DLT}]$, then the base-stock level could be set by choosing a fractile of this distribution, in essence, by setting the desired in-stock probability. For normally distributed demand over the lead time, with mean μ_{DLT} and standard deviation σ_{DLT}, this gives $B = \mu_{DLT} + z_\alpha \sigma_{DLT}$, where z_α is the fractile of the standard normal distribution that accumulates α probability under the pdf. Moreover, if demand is stationary, this base stock level is as well.

A Supply-Chain Wide Base-Stock Model

In a supply-chain context, one that considers managing base-stock levels across multiple sites in the chain, Graves and Willems (2003) present a relatively simple-to-compute approximation for the lead times that tend to result from the congestion effects in the network caused by demand uncertainty and occasional stockouts at various points in the network.

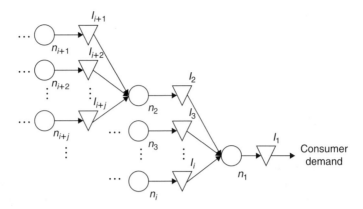

FIGURE 22.8 A general production network.

We will focus on what is called the *stochastic service formulation*, in which the *actual* time to replenish supply at any node in the network is stochastic, even if *nominal* replenishment times are deterministic. (Graves and Willems cite Lee and Billington, 1993, as the motivation for this approach). This results from each node occasionally stocking out, causing congestion in the system as other orders build up while supply is unavailable at some point in the network. The other alternative is a *guaranteed-service formulation*, in which each node holds sufficient inventory to guarantee meeting its service time commitment; this, however, requires an assumption of bounded demand. Graves and Willems trace this model back to the work of Kimball, from 1955, but published in Kimball (1988). For the stochastic service formulation, the congestion-based adjustment to lead time employed by Graves and Willems comes from Ettl et al. (2000), and leads to a fairly straightforward non-linear optimization problem to minimize the cost of holding inventories throughout the network, subject to an in-stock probability constraint at the network node farthest downstream.

In this case, the supply chain network is defined as a graph $G = (N, E)$, comprised of a set of nodes $N = \{1, \ldots, n\}$ and edges (arcs) $E = \{(i,j) : i, j \in N, i \neq j\}$. Figure 22.8 shows a pure production network, in which all nodes have exactly one successor, but could have multiple predecessors. (This is done for simplicity of exposition. The optimization method applies to more general arborescent networks as well. Graves and Willems give examples of networks for battery packaging and distribution and for bulldozer assembly.) Node 1 (n_1) is the node farthest downstream (i.e., the node that fulfills demand for the finished item), and downstream inventory I_i is controlled by node n_i. The *fulfillment time* of any node j in this network, which we denote by L_j, is the nominal amount of time required for node j to fulfill an order from a successor node. We denote the *replenishment time* of node j by τ_j. Thus, as indicated by Graves and Willems (following the model of Ettl et al., 2000), the worst-case replenishment time of node j is

$$\tau_j^{\max} = L_j + \max_{i:(i,j)\in E} \{\tau_i\} \qquad (22.5)$$

but the expected replenishment time can be expressed as

$$E[\tau_j] = L_j + \sum_{i:(i,j)\in E} \pi_{ij} L_i \qquad (22.6)$$

where π_{ij} is the probability that supply node i causes a stockout at j. The values of π_{ij} can be estimated using equation 3.4 in Graves and Willems (2003), derived from Ettl et al. (2000), which is

$$\pi_{ij} = \frac{1-\Phi(z_i)}{\Phi(z_i)} \left[1 + \sum_{h:(h,j)\in E} \frac{1-\Phi(z_h)}{\Phi(z_h)}\right]^{-1} \quad (22.7)$$

where z_i is the safety stock factor at node i, as in our discussion above.

Using the equations above—and assuming that demand over the replenishment lead time at each node j is normally distributed with mean $\mu_j E[\tau_j]$ and standard deviation $\sigma_j \sqrt{E[\tau_j]}$—leads to a relatively straightforward non-linear optimization problem to minimize the total investment in safety stock across the network. Specifically, that optimization problem is to

$$\min C = \sum_{j=1}^{N} h_j \sigma_j \sqrt{E[\tau_j]} \left(z_j + \int_{z_j}^{\infty} (x - z_j)\phi(x)dx \right) \quad (22.8)$$

subject to $z_1 = \Phi^{-1}(\alpha)$, where α is the target service level at the final inventory location, Φ is the standard normal cdf, ϕ is the standard normal pdf, and h_i is the annual inventory holding cost at node i. The safety stock factors z_i ($i=2,\ldots,n$) are the decision variables and are also used in computing π_{ij} in Equation 22.6 via Equation 22.7. Feigin (1999) provides a similar approach to setting base-stock levels in a network, but one that uses fill rate targets—based on a fill rate approximation given by Ettl et al. (1997), an earlier version of Ettl et al. (2000)—as constraints in a non-linear optimization.

Base-Stock Model with Supply Uncertainty

Based on the work of Bryksina (2005), Warsing, Helmer, and Blackhurst (2006) extend the model of Graves and Willems (2003) to study the impact of disruptions or uncertainty in supply on stocking levels and service outcomes in a supply chain. Warsing et al. consider two possible fulfillment times, a normal fulfillment time L_j, as in Equation 22.6, and worst possible fulfillment time K_j, the latter of which accounts for supply uncertainty, or disruptions in supply. Using p_j to denote the probability that node j is disrupted in any given period, Warsing et al. extend the estimate of Equation 22.6 further by using $(1-p_j)L_j + p_j K_j$ in place of L_j. They then compare the performance of systems using the original Graves and Willems base-stock computations, base-stock levels adjusted for supply uncertainty, and a simulation-based adjustment to safety stock levels that they develop.

Table 22.2 shows the base case parameters from a small, three-node example from Warsing et al. (2006), and Table 22.3 shows the results, comparing the Graves and Willems (2003) solution (GW), the modification of this solution that accounts for supply uncertainty (Mod-GW), and the algorithmic adjustment to the safety stock levels developed by Warsing et al. (WHB). An interesting insight that results from this work is the moderating effect of safety stock placement on the probability of disruptions. Specifically, as the ratio of downstream inventory holding costs become less expensive relative to upstream holding costs, the system places more safety stock downstream, which buffers the shock that results from a 10-fold increase in the likelihood of supply disruptions at an upstream node in the network.

Supply Chain Management

TABLE 22.2 Base Case Parameters for Example from Warsing et al. (2006)

Node	L_i	K_i	p_i	h_i	μ_i	σ_i
1	1	2	0.01	10	200	30
2	3	10	0.01	1	200	30
3	3	10	0.01	1	200	30

TABLE 22.3 Results from Warsing et al. (2006) with Target In-Stock Probability of 95%

		Base Stock Levels			In-Stock Probability		Fill Rate	
Scenario	Model	Node 1	Node 2	Node 3	Mean (%)	Std Dev (%)	Mean (%)	Std Dev (%)
Base case	GW	296	694	694	91.5	2.1	93.1	2.0
	Mod-GW	299	709	709	92.9	1.8	94.3	1.8
	WHB	299	1233	709	94.5	2.2	95.8	2.0
$h_3 = 5$	GW	549	655	613	94.9	2.8	95.7	2.6
	Mod-GW	556	670	628	95.4	1.9	96.3	1.7
	WHB	556	670	628	95.4	1.9	96.3	1.7
$p_2 = 0.10$	GW	296	694	694	71.1	4.9	75.4	4.3
	Mod-GW	301	847	709	77.5	5.4	81.3	4.8
	WHB	301	1488	800	91.6	2.8	93.8	2.3
$h_3 = 5$	GW	549	655	613	80.2	3.2	83.6	2.7
$p_2 = 0.10$	Mod-GW	554	810	629	84.3	4.4	87.3	3.9
	WHB	554	1296	629	94.4	2.4	95.9	1.8

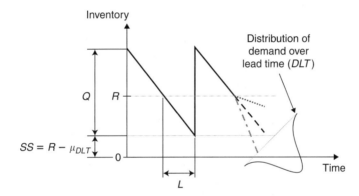

FIGURE 22.9 Continuous-review, perpetual-demand inventory model.

22.2.2 Reorder Point—Order Quantity Models

As discussed above, when there is no cost to place a replenishment order, a base-stock model that reorders at every periodic review—or whenever inventory is depleted by one or more units in a continuous review system—is appropriate. With costly ordering, however, this is likely not to be the cost-minimizing policy. Therefore, in this case, the question of *when* to get more is now relevant in addition to the question of *how much* to get.

In addition to costly ordering, let us now consider a *perpetual demand* setting, one in which demand is assumed to continue infinitely far into the future, and clearly one that is the antithesis of the spike demand that characterized the newsvendor setting. The easiest case to analyze in this setting would be a continuous-review inventory model. A pictorial view of a continuous-review, perpetual-demand model with stationary demand is shown in Figure 22.9. In this setting, an order of size Q is placed whenever the inventory level reaches, or dips below, reorder point R. The picture also provides a visual representation of safety stock for this model. From the visual we see that safety stock SS is not only the amount of inventory held in excess of demand—in this case demand over the replenishment lead time—to protect against stockouts, but it can also be seen to be the amount of inventory that is expected to be on hand when a replenishment order arrives. As indicated by the three dashed lines in the second cycle of the figure, if demand over the replenishment lead time is uncertain, the demand rate could track to expectations over the replenishment lead time (middle line), or it could experience a slow down (upper line) or an increase (lower line), in the latter case leaving the system at risk of a stockout. Setting the reorder point sufficiently high is the means of preventing such stockouts, and therefore, the reorder point, as indicated by the figure is given by

$$R = \mu_{DLT} + SS \tag{22.9}$$

The problem in this case, then, is to find good values of two decision variables, the reorder point R and the order quantity Q, or the "when" and "how much" decisions we discussed above. If D now denotes the expected value of annual demand, A denotes the fixed cost to place a replenishment order, and h denotes the annual cost of holding a unit in inventory, then the total annual costs of ordering and holding in this model are given by

$$TAC(R, Q) = \frac{AD}{Q} + h\left(\frac{Q}{2} + R - \mu_{DLT}\right) \tag{22.10}$$

since D/Q obviously gives the number of orders placed annually, $Q/2$ gives the cycle stock (or the average inventory—over and above any safety stock—given a stationary demand rate over time), and $R - \mu_{DLT}$ gives the safety stock.* Some time-honored texts—for example, Hax and Candea (1984) and Silver et al. (1998)—include explicit stockout costs in the annual cost equation. Our approach, however, will be to consider a service constraint in finding the (Q, R) solution that minimizes Equation 22.10.

Thus, an important issue in our formulation will be how we specify the service constraint on our optimization problem. The most general case is the one in which *both* demand per unit time period (typically days or weeks) *and* replenishment lead time (correspondingly expressed in days or weeks) are random variables. Let us, therefore, assume that the lead time L follows a normal distribution with mean μ_L and standard deviation σ_L. Further, let us assume that the distribution of the demand rate per unit time d also is normal, and let μ_d and σ_d denote its mean and standard deviation, respectively.[†] The normal-distribution assumption allows a well-defined convolution of demand over the lead time, which is therefore also normally distributed, with mean $\mu_{DLT} = \mu_d \mu_L$ and standard deviation[‡] $\sigma_{DLT} = \sqrt{\mu_L \sigma_d^2 + \mu_d^2 \sigma_L^2}$. Moreover, given a value of the reorder point R, the expected number of units short in an order cycle can be expressed as $S(R) = \sigma_{DLT} L(z)$, where $L(z)$, as above, is the standard normal loss function evaluated at z, which in this case is given by $z = (R - \mu_{DLT})/\sigma_{DLT}$. If unmet demand in a replenishment cycle is fully backlogged, it should be apparent that the expected demand met in a replenishment cycle is Q, meaning

---

*Note that this model implies that expected inventory is $Q/2 + SS$. Technically, however, we should be careful about how demand is handled in the event of a stockout. If demand is backlogged, then expected inventory at any point in time is actually $Q/2 + SS + E[BO]$, where $E[BO]$ is the expected number of backorders. Silver et al. (1998, p. 258), however, indicate that since $E[BO]$ is typically assumed to be small relative to inventory, it is reasonable to use $Q/2 + SS$ for expected inventory. If demand is lost in the event of a stockout, then expected inventory is exactly $Q/2 + SS$.

[†]As indicated by Silver et al. (1998), the assumption of the normal distribution for lead time demand is common for a number of reasons, particularly that it is "convenient from an analytic standpoint" and "the impact of using other distributions is usually quite small" (p. 272).

[‡]The general form of this equation, from the variance of a random-sized sum of random variables, can be found in Example 4 in Appendix E of Kulkarni (1995, pp. 577–578).

that the fill rate in this continuous-review, perpetual demand model is given by

$$\beta(R,Q) = 1 - \frac{S(R)}{Q} = 1 - \frac{\sigma_{DLT}L(z)}{Q}* \tag{22.11}$$

Note, therefore, that the problem to minimize Equation 22.10 subject to a fill rate constraint based on Equation 22.11 is non-linear in both the objective function and the constraint. The objective (Equation 22.10), however, is convex in both Q and R,[†] so the solution is fairly straightforward.

A less rigorous approach to finding a (Q,R) solution would be to solve for Q and R separately. Note that $z = (R - \mu_{DLT})/\sigma_{DLT}$ gives a fractile of the distribution of demand over the lead time. Thus, we could set R to achieve a desired in-stock probability, along the lines of the newsvendor problem solution discussed above (i.e., to accumulate a given amount of probability under the lead-time demand distribution). In this setting, the in-stock probability is typically referred to as the *cycle service level* (CSL), or the expected in-stock probability in each replenishment cycle. With R fixed to achieve a desired CSL, we then use the first-order condition on Q in Equation 22.10,

$$\frac{\partial TAC(R,Q)}{\partial Q} = -\frac{AD}{Q^2} + \frac{h}{2} = 0 \tag{22.12}$$

to solve for

$$Q^*_{EOQ} = \sqrt{\frac{2AD}{h}} \tag{22.13}$$

the familiar *economic order quantity* (EOQ). Note also that if demand uncertainty is removed from our model altogether, then the safety stock term is dropped from Equation 22.10 and EOQ via Equation 22.13 provides the optimal order quantity.

Reorder point–order quantity (ROP-OQ) models are also relevant to periodic review systems, in which the system inventory is reviewed at some regular period, say every T days, and an order is placed if the inventory position is below a given level. Analysis of the periodic-review system is more complex and requires a dynamic programming solution (see, e.g., Clark and Scarf, 1960), with the result being an (s, S) policy, wherein an order of size $S - s$ is placed if the inventory position is below s at the current review. As pointed out by Silver et al. (1998), however, a reasonably good, approximate (s, S) policy can be generated from continuous-review-type parameters. Specifically, one could set s equal to an extension, of sorts, of the continuous-review reorder point R, set to cover demand over the lead time *plus* the review period and achieve a desired cycle service level. The order-up-to level S would then be given by $S = R + Q$, with Q given by EOQ (or a some slight adjustment—see Silver et al., 1998, pp. 331–336, for more details).

Multi-Item Models

Our treatment of ROP-OQ models up to this point has concerned only a single item at a single location. It would be the rare company, however, that managed only a single item at a single location. As one complication introduced with multiple items, consider a situation in

---

*Again, the backlogging assumption is important. If unmet demand is lost, then the replenishment quantity Q is never used to meet backlogged demand, and therefore, the total demand experienced in an average cycle is $Q + E[BO] = Q + S(R)$, yielding $\beta(R,Q) = 1 - \sigma_{DLT}L(z)/[Q + \sigma_{DLT}L(z)]$.

[†]See Hax and Candea (1984, p. 206) for a proof of the convexity of a related model.

which k items can be ordered jointly at a discount in the ordering cost—that is, where the cost of ordering k items independently is kA, but ordering these items jointly incurs only an incremental cost, $a_i \geq 0$, for each item i ordered beyond the single base cost of ordering, A, such that $A + \sum_{i=1}^{k} a_i < kA$.

A straightforward way to address this situation is to consider ordering all items jointly on a common replenishment cycle. In essence this changes the decision from "how much" to order—the individual order quantities, Q_i—to "how often" to order, which is given by n, defined as the number of joint orders placed per year. Then since $n = D_i/Q_i \Rightarrow Q_i = D_i/n$ for each item i, we can build a joint-ordering TAC model with

$$TAC(n) = \left(A + \sum_{i=1}^{k} a_i\right) n + \sum_{i=1}^{k} h_i \left(\frac{D_i}{2n} + SS_i\right) \quad (22.14)$$

where h_i is the annual cost of holding one unit of item i in inventory, $D_i/(2n) = Q_i/2$ is the cycle stock of item i, and SS_i is the safety stock of item i. For fixed values of SS_i (or if demand is certain and $SS_i = 0$, $i = 1, \ldots, k$), the optimal solution to Equation 22.14 is found by setting

$$\frac{\partial TAC(n)}{\partial n} = A + \sum_{i=1}^{k} a_i - \frac{1}{2n^2} \sum_{i=1}^{k} h_i D_i = 0 \quad (22.15)$$

yielding

$$n^* = \sqrt{\frac{\sum_{i=1}^{k} h_i D_i}{2\left(A + \sum_{i=1}^{k} a_i\right)}} \quad (22.16)$$

Introducing safety stock and reorder points into the decision complicates the idea of joint replenishments. Clearly, it would negate the assumption that all items would appear in each replenishment since the inventory levels of the various items are unlikely to hit their respective reorder points at exactly the same point in time. Silver et al. (1998) discuss an interesting idea first proposed by Balintfy (1964), a "can-order" system. In this type of inventory control system, two reorder points are specified, a "can-order" point, at which an order *could* be placed, particularly if it would allow a joint-ordering discount, and a "must-order" point, at which an order *must* be placed to guard against a stockout. Although I will not discuss the details here, Silver et al. (1998: p. 435) provide a list of references for computing "must-order" and "can-order" levels in such a system. Thus, one could propose a review period for the joint-ordering system and set reorder points to cover a certain cumulative probability of the distribution of demand over the lead time plus the review period.

Let's consider an example that compares individual orders versus joint orders for two items. Floor-Mart—"We set the floor on prices" (and, apparently, on advertising copy)—is a large, (fictional) discount retailer. Floor-Mart stocks two models of 20-inch, LCD-panel TVs, Toshiba and LG, that it buys from two different distributors. Annual demand for the Toshiba 20-inch TV is $D_1 = 1600$ units, and the unit cost to Floor-Mart is \$400. Annual demand for the LG 20-inch TV is $D_2 = 2800$ units, and the unit cost to Floor-Mart is \$350. Assuming 365 sales days per year, this results in an expected daily demand of $\mu_{d,1} = 4.38$ units for the Toshiba TV and $\mu_{d,2} = 7.67$ units for the LG TV. Assume that Floor-Mart also has data on demand uncertainty such that $\sigma_{d,1} = 1.50$ and $\sigma_{d,2} = 2.50$. Annual holding costs at Floor-Mart are estimated to be 20% on each dollar held in inventory, and Floor-Mart's target fill rate on high-margin items like TVs is 99%. Floor-Mart uses the same contract carrier to ship TVs from each of the distributors. Let's put a slight twist on the problem formulation as it was stated above. Assume that Floor-Mart's only fixed cost of placing a replenishment order with its TV distributors is the cost the carrier charges to move the goods to Floor-Mart. The carrier charges $600 for each shipment from the Toshiba distributor, and the mean and standard deviation of the replenishment lead time are $\mu_{L,1} = 5$ and $\sigma_{L,1} = 1.5$ days. The carrier charges $500 for each shipment from the LG distributor, and the mean and standard deviation of the replenishment lead time are $\mu_{L,2} = 4$ and $\sigma_{L,2} = 1.2$ days. However, the carrier also offers a discounted "stop-off" charge to pick up TVs from *each* distributor on a *single* truck, resulting in a charge of $700.

With the parameters above, we find that $\mu_{DLT,1} = 21.92$, $\sigma_{DLT,1} = 7.38$, $\mu_{DLT,2} = 30.68$, and $\sigma_{DLT,2} = 10.60$. Solving Equation 22.10 for each TV to find the optimal independent values of Q and R (via Excel Solver), we obtain $(Q_1, R_1) = (166, 25)$ and $(Q_2, R_2) = (207, 36)$. By contrast, independent EOQ solutions would be $Q_{EOQ,1} = 155$ and $Q_{EOQ,2} = 200$. Using Equation 22.16 for the joint solution, we obtain $n^* = \sqrt{(0.2)(400 \cdot 1600 + 350 \cdot 2800)/(2 \cdot 700)} = 15.21$, yielding $Q_1 = 105$ and $Q_2 = 184$. The joint solution—assuming that it does not exceed the truck capacity—saves approximately $5100 in ordering and holding costs over either independent solution. The issue on the table, then, as we discuss above, is to set safety stock levels in this joint solution. As orders would be placed only every 3.4 weeks with $n = 15.21$, a review period in line with the order cycle would probably inflate safety stock too dramatically. Thus, the reader should be able to see the benefit of a "can-order" system in this case, perhaps with a one-week review period, a high CSL—that is, relatively large value of $z = (R - \mu_{DLT})/\sigma_{DLT}$—on the "can-order" reorder point, and a lower CSL on the "must-order" reorder point. In closing the example, we should point out that we have assumed a constant rate of demand over time, which may be reasonable for TVs, but clearly ignores the effects of intermittent advertising and sales promotions. Later in this section, we discuss distribution requirements planning, which addresses situations where demand is not stationary over time.

Finally, the more general case for joint ordering is to consider what Silver et al. (1998) call *coordinated replenishment*, where each item i is ordered on a cycle that is a multiple, m_i, of a base order cycle of T days. Thus, each item is ordered every $m_i T$ days. Some subset, possibly a single item, of the k items has $m_i = 1$, meaning that those items appear in every replenishment order. Silver et al. (1998, pp. 425–430) offer an algorithmic solution to compute m_i $(i = 1, \ldots, k)$ and T for this case. Jackson et al. (1985) consider the restrictive assumption that $m_i \in \{2^0, 2^1, 2^2, 2^3, \ldots\} = \{1, 2, 4, 8, \ldots\}$, which they call a "powers-of-two" policy, and show that this approach results in an easy-to-compute solution whose cost is no more than 6% above the cost of an optimal policy with unrestricted values of m_i. Another situation that could link item order quantities is an aggregate quantity constraint or a budget constraint. Hax and Candea (1984) present this problem and a Lagranian-relaxation approach to solving it; they refer to Holt et al. (1960) for several methods to solve for the Lagrange multiplier that provides near-optimal order quantities.

Multi-Echelon Models

In a *multi-echelon* supply chain setting, Clark and Scarf (1960) show that the supply-chain optimal solution at each echelon is an "echelon-stock policy" of (s, S) form. Clark and Scarf are able to solve their stochastic dynamic programming formulation of the problem by redefining the manner in which inventory is accounted for at successive upstream echelons in the chain. Clark and Scarf's definition of *echelon stock* is "the stock at any given installation plus stock in transit to or on hand at a lower installation" (p. 479). Thus, the echelon stock at an upstream site includes stock at that site *plus* all stock downstream that has not yet been sold to a customer. Using an echelon-stock approach coordinates the ordering processes across the chain, or at least across the echelons that can be coordinated. This last point refers to the complication that a supply-chain-wide, echelon-stock policy would require all firms in the chain (except the one farthest upstream) to provide access to their inventory data. Such supply-chain-wide transparency is perhaps unlikely to occur, but coordination between company-owned production sites and company-owned distribution sites is eminently reasonable. Silver et al. (1998, 477–480), for example, provide computational procedures to determine coordinated warehouse (upstream) and retailer (downstream) order quantities using an echelon-stock approach. In addition, I refer the reader to Axsäter (2000) for an excellent overview of echelon-inventory methods and computational procedures.

The astute reader should note that we have already discussed an approach to setting safety stocks in a multi-echelon system, namely the non-linear optimization suggested by Graves and Willems (2003). In the author's view, this approach is more intuitive and more directly solved than an echelon inventory approach, but perhaps only slightly less problematic in terms of obtaining data about upstream and downstream partners in the supply chain. An approach that uses the GW optimization (see the earlier section in this chapter on "A Supply-Chain Wide Base-Stock Model") to set network-wide safety stock levels, coupled with simple EOQ computations as starting points for order quantities, and possibly "tweaked" by simulation to find improved solutions, may be a promising approach to solving network-wide inventory management problems.

22.2.3 Distribution Requirements Planning

In an environment in which independent-item demands vary over time, distribution requirements planning (DRP) provides a method for determining the time-phased inventory levels that must be present in the distribution system to fulfill projected customer demands on time. As Vollmann et al. (2005) point out, "DRP's role is to provide the necessary data for matching customer demand with the supply of products being produced by manufacturing" (p. 262). Note that in any make-to-stock product setting, the linchpin between product manufacturing and customer demand is the distribution system, and indeed, it is distribution system inventory that coordinates customer demand with producer supply. While distribution system inventories could be set using the ROP-OQ control methods as described above, the critical difference between DRP and ROP-OQ is that DRP plans *forward* in time, whereas ROP-OQ systems react to actual demands whenever those demands drive inventory to a level that warrants an order being placed. This is not to say that the two approaches are inconsistent. Indeed, it is quite easy to show that, in a stationary-demand environment with sufficiently small planning "time buckets," DRP and ROP-OQ produce equivalent results. Where DRP provides a clear benefit, however, is in situations where demand is *not* constant over time, particularly those situations in which surges and drops in demand can be anticipated reasonably accurately.

The basic logic of DRP is to compute the time-phased replenishment schedule for each stock-keeping unit (SKU) in distribution that keeps the inventory level of that SKU at or above a specified safety stock level (which, of course, could be set to zero). Let the DRP horizon span planning periods $t = 1, \ldots, T$, and assume that safety stock SS, order lot size Q, and lead time L are given for the SKU in question. In addition, assume that we are given

Supply Chain Management

a period-by-period forecast of demand for this SKU, D_t, $t=1,\ldots,T$. Then, we can compute the *planned receipts* for each period t, namely the quantity—as an integer multiple of the lot size—required to keep inventory at or above SS.

To illustrate these concepts, let us consider an example, altered from Bozarth and Handfield (2006). Assume that MeltoMatic Company manufactures and distributes snow blowers. Sales of MeltoMatic snow blowers are concentrated in the Midwestern and Northeastern states of the United States. Thus, the company has distribution centers (DCs) located in Minneapolis and Buffalo, both of which are supplied by MeltoMatic's manufacturing plant in Cleveland. Table 22.4 shows the DRP records for the Minneapolis and Buffalo DCs for MeltoMatic's two SKUs, Model SB-15 and Model SBX-25. The first three lines in the table sections devoted to each SKU at each DC provide the forecasted demand (requirements), D_t; the scheduled receipts, SR_t, already expected to arrive in future periods; and the projected ending inventory in each period, I_t. This last quantity, I_t, is computed on a period-by-period basis as follows:

1. For period t, compute *net requirements* $NR_t = \max\{0, D_t + SS - (I_{t-1} + SR_t)\}$.
2. Compute planned receipts $PR_t = \lceil NR_t/Q \rceil \cdot Q$, where $\lceil x \rceil$ gives the smallest integer greater than or equal to x. (Note that this implies that if $NR_t = 0$, then $PR_t = 0$ as well.)
3. Compute $I_t = I_{t-1} + SR_t + PR_t - D_t$. Set $t \leftarrow t+1$.
4. If $t \leq T$, go to step 1; else, done.

Then, planned receipts are offset by the lead time for that SKU, resulting in a stream of *planned orders*, PO_t, to be placed so that supply arrives in the DC in time to fulfill the projected demands (i.e., $PO_t = PR_{t+L}$ for $t=1,\ldots,T-L$, and any planned receipts in periods $t=1,\ldots,L$ result in orders that are, by definition, *past due*).

Note that an important aspect of DRP is that it *requires human intervention* to turn planned orders into *actual* orders, the last line in Table 22.4 for each SKU. To further illustrate this point, let's extend our example to consider a situation where the human being charged with the task of converting planned orders into actual orders might alter the planned order stream to achieve some other objectives. For example, let us assume that the carrier that ships snow blowers into MeltoMatic's warehouses in Minneapolis and Buffalo provides a discount if the shipment size exceeds a truckload quantity of 160 snow blowers, relevant to shipments of either SKU or to combined shipments containing a mix of both SKUs. Table 22.5 shows an altered actual order stream that generates orders across both SKUs of at least 160 units in all periods $t=1,\ldots,T-L$ where $PO_t > 0$ for either SKU. In addition, Table 22.5 also shows the temporary overstock created by that shift in orders, with some units arriving in advance of the original plan. The question would be whether the projected savings in freight transportation exceeds the expense of temporarily carrying excess inventories. Interestingly, we should also note that the shift to larger shipment quantities, somewhat surprisingly, creates a smoother production schedule at the Cleveland plant (after an initial bump in Week 46).

Once again, therefore, we revisit the enormity of the problem of "optimizing the supply chain," noting the complexity of the above example for just two SKUs at a single firm. Granted, jointly minimizing inventory cost, transportation cost, and possibly also "production-change" costs might be a tractable problem at a single firm for a small set of SKUs, but the challenge of formulating and solving such problems clearly grows as the number of component portions of the objective function grows, as the number of SKUs being planned grows, and as the planning time scale shrinks (e.g., from weeks to days).

TABLE 22.4 DRP Records for MeltoMatic Snow Blowers Example with Actual Orders Equal to Planned Orders

	Week		45	46	47	48	49	50	51	52	1	2	3	4
MINNEAPOLIS DC														
Model SB-15		Forecasted requirements	60	60	70	70	80	80	80	80	90	90	95	95
LT (weeks):	2	Scheduled receipts	120											
Safety stock (units):	10	Projected ending inventory 80	140	80	10	60	100	20	60	100	10	40	65	90
Lot size (units):	120	Net requirements	0	0	0	70	30	0	70	30	0	90	65	40
		Planned receipts	0	0	0	120	120	0	120	120	0	120	120	120
		Planned orders	**0**	**120**	**120**	**120**	**120**	**120**	**120**	**120**	**120**	**120**	**0**	**0**
		Actual orders 0	0	120	120	0	120	120	0	120	120	120	0	0
Model SBX-25		Forecasted requirements	40	40	50	60	80	90	100	100	110	110	100	100
LT (weeks):	2	Scheduled receipts	40											
Safety stock (units):	10	Projected ending inventory 100	100	60	10	30	30	20	80	60	30	80	60	40
Lot size (units):	80	Net requirements	0	0	0	60	60	70	90	30	60	90	30	50
		Planned receipts	0	0	0	80	80	80	160	80	80	160	80	80
		Planned orders	**0**	**80**	**80**	**80**	**160**	**80**	**80**	**160**	**80**	**80**	**0**	**0**
		Actual orders 0	0	80	80	80	160	80	80	160	80	80	0	0
BUFFALO DC														
Model SB-15		Forecasted requirements	70	70	80	80	90	90	100	100	120	120	140	140
LT (weeks):	1	Scheduled receipts	100											
Safety stock (units):	10	Projected ending inventory 30	60	110	30	70	100	10	30	50	50	50	30	10
Lot size (units):	120	Net requirements	0	20	0	60	30	0	100	30	30	80	100	120
		Planned receipts	0	120	0	120	120	0	120	120	120	120	120	120
		Planned orders	**120**	**0**	**120**	**120**	**0**	**120**	**120**	**120**	**120**	**120**	**120**	**0**
		Actual orders 0	120	0	120	120	0	120	120	120	120	120	120	0
Model SBX-25		Forecasted requirements	50	50	60	70	80	80	80	110	130	120	100	100
LT (weeks):	1	Scheduled receipts	60											
Safety stock (units):	15	Projected ending inventory 30	40	70	90	20	20	20	80	50	80	30	90	70
Lot size (units):	80	Net requirements	0	25	5	0	75	75	95	45	95	65	85	25
		Planned receipts	0	80	80	0	80	80	160	80	160	80	160	80
		Planned orders	**80**	**80**	**0**	**80**	**80**	**160**	**80**	**160**	**80**	**160**	**80**	**0**
		Actual orders 0	80	80	0	80	80	160	80	160	80	160	80	0
CLEVELAND PLANT		Master production schedule (Gross requirements)												
		SB-15	120	120	240	120	120	240	120	240	240	240		
		SBX-25	80	160	80	160	240	240	160	320	160	240		
		Total	200	280	320	280	360	480	280	560	400	480		

TABLE 22.5 DRP Records for MeltoMatic Snow Blowers Example with Actual Orders Altered for Transportation Discount

		Week	45	46	47	48	49	50	51	52	1	2	3	4
MINNEAPOLIS DC														
Model SB-15	80	Forecasted requirements	60	60	70	70	80	80	80	80	90	90	95	95
		Projected ending inventory	140	80	10	60	100	20	60	100	10	40	65	90
		Planned orders	0	120	120	0	120	120	0	120	120	120	0	0
		Actual orders	0	120	120	0	120	120	0	120	120	120	0	0
Model SBX-25	100	Forecasted requirements	40	40	50	60	80	90	100	100	110	110	100	100
		Projected ending inventory	100	60	10	30	30	20	80	60	30	80	60	40
		Planned orders	0	80	80	80	160	80	80	160	80	80	0	0
		Actual orders	0	80	80	160	80	80	160	80	80	80	0	0
Total for DC		Planned orders	0	200	200	80	280	200	80	280	200	200	0	0
		Actual orders	0	200	200	160	200	200	160	200	200	200	0	0
		Temporary overstock	**0**	**0**	**0**	**80**	**0**	**0**	**80**	**0**	**0**	**0**	**0**	**0**
BUFFALO DC														
Model SB-15	30	Forecasted requirements	70	70	80	80	90	90	100	100	120	120	140	140
		Projected ending inventory	60	110	30	70	100	10	30	50	50	50	30	10
		Planned orders	120	0	120	120	0	120	120	120	120	120	120	0
		Actual orders	120	0	120	120	0	120	120	120	120	120	120	0
Model SBX-25	30	Forecasted requirements	50	50	60	70	80	80	100	110	130	130	100	100
		Projected ending inventory	40	70	90	20	20	20	80	50	80	30	90	70
		Planned orders	80	80	0	80	80	160	80	160	80	160	80	0
		Actual orders	80	160	80	80	160	80	80	80	80	80	80	0
Total for DC		Planned orders	200	80	120	200	80	280	200	280	200	280	200	0
		Actual orders	200	160	200	200	160	200	200	200	200	200	200	0
		Temporary overstock	**0**	**80**	**160**	**160**	**240**	**160**	**160**	**80**	**80**	**0**	**0**	**0**
CLEVELAND PLANT		Master production schedule (Gross requirements)												
	SB-15		120	120	240	120	120	240	120	240	240	240	200	0
	SBX-25		80	240	160	240	240	160	240	160	160	160	200	0
	Total		200	360	400	360	360	400	360	400	400	400	0	0

The examples above, however, give us some insights into how inventory decisions could be affected by transportation considerations, a topic we consider further in the section that follows.

22.3 Managing Transportation in the Supply Chain

22.3.1 Transportation Basics

Ballou (2004) defines a *transport service* "as a set of transportation performance characteristics purchased at a given price" (p. 167). Thus, in general, we can describe and measure a freight transport mode by the combination of its cost and its service characteristics. Moreover, we can break down the service characteristics of a transport service into *speed* (average transit time), *reliability* (transit time variability), and risk of *loss/damage*. Freight transit modes are water, rail, truck, air, and pipeline; interestingly, Chopra and Meindl (2004) also include an *electronic* mode, which may play an increasing role in the future, and already has for some goods like business documents (e.g., e-mail and fax services) and recorded music (e.g., MP3 file services like Napster and Apple's iTunes). I will therefore distinguish *physical* modes of freight transit from this nascent electronic mode and focus our discussion exclusively on the physical modes. Ballou's characterization of a freight transport service leads nicely to a means of comparing physical modes—with respect to their relative cost, speed of delivery, reliability of delivery, and risk of loss and damage. Ballou (2004) builds a tabular comparison of the relative performance of the modes in terms of cost, speed, reliability, and risk of loss/damage, but I would suggest that there is some danger in trying to do this irrespective of the nature of the cargo or the transportation *lane* (i.e., origin–destination pair) on which the goods are traveling. Some comparisons are fairly obvious, however: Air is clearly the fastest and most expensive of the physical modes, whereas water is the slowest and perhaps the least expensive, although pipeline could probably rival water in terms of cost. In terms of "in-country" surface modes of travel, rail is much cheaper than truck, but also much slower—partially because of its lack of point-to-point flexibility—and also more likely to result in damage to or, loss of, the cargo due to the significant number of transfers of goods from train to train as a load makes its way to its destination via rail.

A Harvard Business School (Harvard Business School, 1998) note on freight transportation begins with a discussion of the deregulation of the rail and truck freight transport markets in the United States and states that "... an unanticipated impact of regulatory reform was a significant consolidation in the rail and LTL [less-than-truckload] trucking sectors" (p. 3). This stemmed from the carriers in these markets seeking what the HBS note calls "economies of flow," consolidating fixed assets by acquiring weaker carriers to gain more service volume, against which they could then more extensively allocate the costs of managing those fixed assets. This provides interesting insights to the cost structures that define essentially any transport service, *line-haul costs*—the variable costs of moving the goods—versus *terminal/accessorial costs*—the fixed costs of owning or accessing the physical assets and facilities where shipments can be staged, consolidated, and routed.

Even without the benefit of detailed economic analysis, one can infer that modes like pipeline, water, and rail have relatively large terminal/accessorial costs; that air and truck via LTL have less substantial terminal costs; and that truck via truckload (TL), as a point-to-point service, has essentially no terminal costs. Greater levels of fixed costs for the carrier create a stronger incentive for the carrier to offer per-shipment volume discounts on its services. Not only does this have a direct impact on the relative cost of these modes, but it also has implications for the relative lead times of freight transit modes, as shown in Figure 22.10, reproduced (approximately) from Piercy (1977). In this case, more substantial terminal operations—that is, rail yards to build trains or LTL terminals to build full trailer

Supply Chain Management

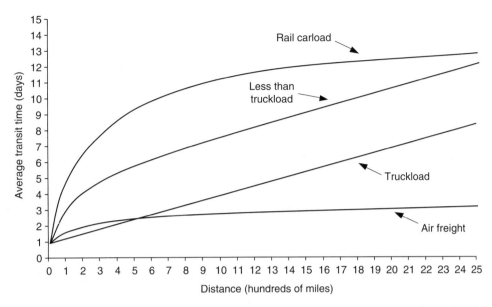

FIGURE 22.10 Comparison of shipping distance versus lead time for various modes. (Reproduced from Piercy, 1977.)

loads—result in more non-linearity in the distance-versus-transit time relationship. Note from the figure that for TL, this relationship is essentially linear as this mode does not require terminal operations to mix and consolidate loads.

In building a model of transportation cost as a function of the shipment quantity, the remainder of this section will focus on truck transit, as that is the mode for which the most data is available and for which the full-load versus less-than-full-load analysis is most interesting. The basic ideas of the results, however, apply to other transit modes as well. The basic tension for the decision maker is between using smaller shipments more frequently, but at a potentially higher per-unit shipping cost, versus making larger, but less expensive, shipments less frequently. Clearly, this has implications for transportation and inventory costs, and potentially for customer service as well. The analysis that follows, therefore, begins to round out the triangle of SCM strategy—inventory, transportation, locations, and customer service—that was presented earlier in this chapter.

In the trade press, Speigel (2002) picks up on exactly this tension in the TL versus LTL decision. As Speigel indicates, and as we draw out in an example later in this section, the answer to the TL–LTL question depends on a number of factors and explains why many shippers use transportation management system (TMS) software to attempt to generate minimum cost shipment decisions. Moreover, Speigel points out that it is getting increasingly difficult to fill truckloads in practice when one considers the many factors driving small shipment sizes: an increasing number of stock-keeping-units driven by customers' demands for greater customization, lean philosophies and the resulting push to more frequent shipments, and customer service considerations that are driving more decentralized distribution networks serving fewer customers per distribution center (this last theme returns in the succeeding section of this chapter on "Managing Locations"). As I point out after our example below, such trends probably lead to a need to consider mixed loads as a possible means of building larger, more economical shipments. This section closes with a general cost model that is well-positioned for carrying out such an analysis, even without the help of a costly TMS.

TABLE 22.6 Breakdown of United States Physical Distribution Costs, 1988 Versus 2002

	1988*		2002†	
	% of Sales	$/cwt‡	% of Sales	$/cwt
Transportation	3.02	12.09	3.34	26.52
Warehousing	1.90	11.99	2.02	18.06
Customer service/order entry	0.79	7.82	0.43	4.58
Administration	0.35	2.35	0.41	2.79
Inventory carrying cost @ 18%/year	1.76	14.02	1.72	22.25
Other	0.88	5.94	NR	NR
Total distribution cost	7.53	45.79	7.65	67.71

* From Davis, H. W., *Annual Conference Proceedings of the Council of Logistics Management*, 1988. With permission.
† From Davis, H. W. and W. H. Drumm, *Annual Conference Proceedings of the Council of Logistics Management*, 2002. With permission.
‡ Weight measure "cwt" stands for "hundred-weight," or 100 lbs.

22.3.2 Estimating Transportation Costs

Table 22.6 shows a breakdown of total logistics cost as a percent of sales as given in Ballou (1992) and in Ballou (2004). From the table, one can note the slight growth in transportation cost as a fraction of sales and the slight decline in inventory cost over time. Thus, it would appear that effectively managing transportation costs is an important issue in controlling overall logistics costs, in addition to managing the impact of transportation decisions on warehousing and inventory carrying costs, the other large cost components in the table.

Nearly all operations management or logistics textbooks that incorporate quantitative methods (e.g., Nahmias, 1997; Silver et al., 1998; Chopra and Meindl, 2004; Ballou, 2004) present methods for computing reorder points and order quantities for various inventory models and, in some cases, also discuss accounting for transportation costs in managing inventory replenishment. To the author's knowledge, however, no textbook currently presents any models that effectively tie these interrelated issues together by incorporating a general model of transportation costs in the computation of optimal or near-optimal inventory policy parameters. Moreover, although many academics have previously proposed or studied joint inventory–transportation optimization models—extending back to the work of Baumol and Vinod (1970)—the author's experience is that the use of previously published approaches has not become commonplace among practitioners.

Given the recent estimate of Swenseth and Godfrey (2002) that transportation costs could account for as much as 50% of the total annual logistics cost of a product (consistent with Table 22.6, which shows aggregate data), it seems clear that failing to incorporate these costs in inventory management decisions could lead to significantly sub-optimal solutions—that is, those with much higher than minimal total logistics costs. Perhaps the most common model for transportation costs in inventory management studies is simply to assume a fixed cost to ship a truckload of goods (see, e.g., Waller et al., 1999, for use of such a model in studying vendor-managed inventory). This model, however, is insensitive to the effect of the shipment quantity on the per-shipment cost of transportation and seems unrealistic for situations in which goods are moved in smaller-sized, less-than-truckload shipments. In contrast, several authors have considered discrete rate-discount models, using quantity breaks to represent the shipment weights at which the unit cost changes. Tersine and Barman (1991) provide a review of some earlier modeling work. A recent example is the work of Çetinkaya and Lee (2002), who modify a cost model introduced by Lee (1986), which represents transportation costs as a stepwise function to model economies of scale by explicitly reflecting the lower per-unit rates for larger shipments.

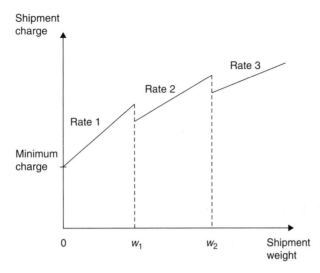

FIGURE 22.11 LTL rate structure as typically stated.

TABLE 22.7 Example Class 100 LTL Rates for NYC Dallas Lane

Minimum Weight (lbs)	Rate ($/cwt)
1	107.75
500	92.26
1000	71.14
2000	64.14
5000	52.21
10,000	40.11
20,000	27.48
30,000	23.43
40,000	20.52

In contrast to explicit, discrete rate-discount models of freight rates, Swenseth and Godfrey (1996) review a series of models that specify the freight rate as a continuous, non-linear function of the shipment weight, building off of the work of Ballou (1991), who estimated the accuracy of a linear approximation of trucking rates. One advantage of continuous functions is that they do not require the explicit specification of rate breakpoints for varying shipment sizes nor do they require any embedded analysis to determine if it is economical to increase—or over-declare—the shipping weight on a given route. A second advantage is that a continuous function may be parsimonious with respect to specifying its parameters and can therefore be used in a wide variety of optimization models.

The idea that a shipper might benefit from "over-declaring" a shipment, or simply shipping more in an attempt to reduce the overall shipping cost, stems from the manner in which LTL rates are published by carriers. Figure 22.11 shows a pictorial view of the typical LTL rate structure as it is published to potential shippers. This rate structure uses a series of weight levels at which the per-unit freight rate—that is, the slope of the total shipment charge in Figure 22.11—decreases. A sample of such published rates is shown in Table 22.7, which contains rates for the New York City-to-Dallas lane for Class 100 freight,[*]

---

[*]Coyle et al. (2006) indicate that the factors that determine the freight class of a product are its density, its storability and handling, and its value. Class 100 freight, for example, would have a density between 6 and 12 lbs/ft^3 for LTL freight.

as given in Ballou (2004, Table 6.5). Modeling per-unit rates for shipments of an arbitrary size requires one to account for *weight breaks*, which are used by carriers to determine the shipment weight at which the customer receives a lower rate, and which are typically *less* than the specific cutoff weight for the next lower rate to avoid creating incentives for a shipper to over-declare, as discussed above. For example, using the rate data in Table 22.7 for a Class 100 LTL shipment from New York to Dallas, if one were to ship 8000 pounds (80 cwt) on this lane, the charge without considering a possible weight break would be $80 \times 52.21 = \$4176.80$. If, however, one were to add 2000 pounds of "dummy weight," the resulting charge would be only $100 \times 40.11 = \$4011$. In Figure 22.11, this is the situation for a shipment weight just slightly less than w_2, for example. Thus, to discourage shippers from over-declaring or "over-shipping"—and also leaving space for more goods on the carrier's trailer and, therefore, providing the carrier with an opportunity to generate more revenue— the carrier actually bills any shipment at or above 7682 lbs (the weight break, or the weight at which the \$52.21 rate results in a total charge of \$4011) as if it were 10,000 lbs. Therefore, an 8000-lb shipment will actually result in carrier charges of only \$4011. The *effective rate* for this 8000-lb shipment is $\$4011 \div 80 = \$50.14/\text{cwt}$. Figure 22.12 shows a pictorial view of the rates *as actually charged*, with all shipment weights between a given weight break at the next rate break point (i.e., w_1 or w_2) being charged the same amount.

Tyworth and Ruiz-Torres (2000) present a method to generate a continuous function from rate tables like those in Table 22.7. The functional form is $r(W) = C_w W^b$, where W is the shipment weight; $r(W)$ is the rate charged to ship this amount, expressed in dollars per hundred pounds (i.e., \$/cwt); and C_w and b are constants found using non-linear regression analysis. With knowledge of the weight of the item shipped, this rate function can easily be expressed as $r(Q) = CQ^b$ (in \$/unit), where Q is the number of units shipped and C is the "quantity-adjusted" equivalent of C_w. Appending the rate function $r(Q)$ to the annual cost equation for a continuous-review, ROP-OQ inventory system like the one presented earlier in this chapter results in a relatively straightforward non-linear optimization problem, as we will demonstrate below.

Thus, to generate the rate function parameters, we can, for a given origin–destination pair and a given set of LTL rates, compute effective rates for a series of shipment weights. The resulting paired data can be mapped to the continuous function $r(W)$ presented

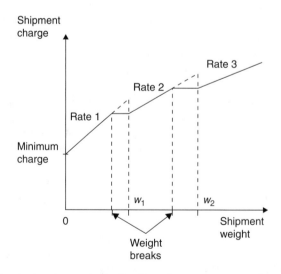

FIGURE 22.12 LTL rate structure as typically charged.

above. Although Tyworth and Ruiz-Torres (2000) present this mapping as a non-linear curve-fitting process, let us consider a computationally simpler linear regression approach. Since $r(W) = C_w W^b$, then $\ln r(W) = \ln[C_w W^b] = \ln C_w + b \ln W$, and therefore, the parameters of the continuous function that approximates the step-function rates can be found by performing simple linear regression analyses on logarithms of the effective rate–shipment weight pairs that result from computations like those described above to find the effective rate. Sample plots of the results of this process for the New York-Dallas rate data in Table 22.7 are shown in Figure 22.13, resulting in $C_w = 702.67$ and $b = -0.3189$. Thus, shipping Q units of a Class 100 item that weighs w lbs on this lane would result in a per-unit rate of $r(Qw) = 702.67(Qw)^{-0.3189}$ \$/cwt. Converting to \$/unit by multiplying this by

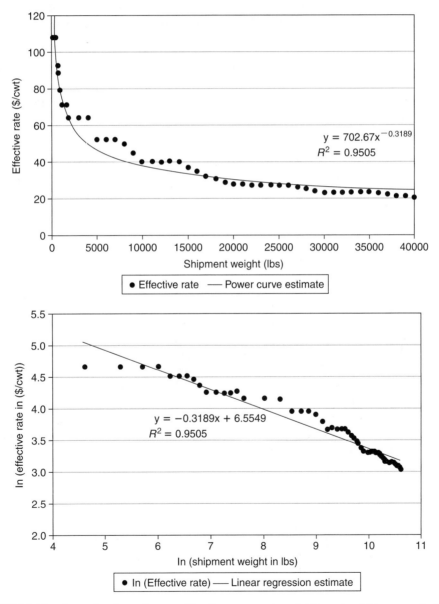

FIGURE 22.13 Examples of non-linear and log-linear regression of shipment weight versus effective rate.

$w/100$ cwt/unit, one obtains $r(Q) = 7.0267 w^{0.6811} Q^{-0.3189}$ \$/unit. Depending on the source of the rates, it might also be necessary to assume a discount factor that would reflect what the shipper would actually contract to pay the LTL carrier for its services. For example, Tyworth and Ruiz-Torres (2000) apply a 50% discount, which they claim to be typical of contract-based discounts in practice.*

22.3.3 Transportation-Inclusive Inventory Decision Making

Let us return to the total annual cost equation developed earlier in the chapter, Equation 22.10. This equation accounts for the cost of placing orders and the costs of holding cycle stock and safety stock inventories. In the preceding subsections, we indicated that transportation costs tend to exhibit economies of scale, and we demonstrated, using LTL freight rates, a method for building a continuous function for transportation costs, $r(Q) = CQ^b$. Note also that, using this model of scale economies in transportation costs, the annual cost of freight transportation for shipments of size Q to satisfy demand of D units is $Q \cdot CQ^b \cdot D/Q = DCQ^b$. Although it seems immediately clear that this rate function *could* be incorporated into an annual cost equation, the question is, first, *whether* it should be incorporated into the TAC expression, and second, whether it implies any other inventory-related costs.

Marien (1996) provides a helpful overview of what are called *freight terms of sale* between a supplier shipping goods and the customer that ultimately will receive them from the carrier that is hired to move the goods. Two questions are answered by these negotiated freight terms, namely, "*Who owns the goods in transit?*" and "*Who pays the freight charges?*" Note that legally, the carrier that moves the goods is only the *consignee* and never takes ownership of the goods. The question of ownership of goods in transit is answered by the "F.O.B." terms,† indicating whether the transfer of ownership from supplier (shipper) to customer (receiver) occurs at the point of *origin* or the point of *destination*, hence the terms "F.O.B. origin" and "F.O.B. destination." The next issue is which party, shipper or receiver, pays the freight charges, and here there is a wider array of possibilities. Marien (1996), however, points out that three arrangements are common: (1) F.O.B. origin, freight collect—in which goods in-transit are owned by the receiver, who also pays the freight charges; (2) F.O.B. destination, freight prepaid—in which the shipper retains ownership of goods in transit and also pays the freight bill; and (3) F.O.B. origin, freight prepaid and charged back, similar to (1), except that the shipper pays the freight bill up-front and then charges it to the receiver upon delivery of the goods. These three sets of terms are described pictorially in Figure 22.14. The distinction of who pays the bill, and the interesting twist implied in (3), is important in that the party that pays the freight charges is the one who has the right, therefore, to hire the carrier and choose the transport service and routing.

Note that our TAC expression for inventory management is a decision model for ordering; thus, it takes the perspective of the *buyer*, the downstream party that will receive the goods being ordered. Therefore, the freight transportation terms that involve "freight collect" or "freight prepaid and charged" back imply that the decision maker should incorporate

---

*Using empirical LTL rate data from 2004, Kay and Warsing (2006) estimate that the average discount from published rates is approximately 46%.

†According to Coyle et al. (2006, pp, 473, 490), "F.O.B." stands for "free on board," shortened from "free on board ship." This distinguishes it from the water-borne transportation terms "free alongside ship" (FAS), under which the goods transfer ownership from shipper to receiver when the shipper delivers the goods to the port (i.e., "alongside ship"), leaving the receiver to pay the cost of lifting the goods onto the ship.

Supply Chain Management

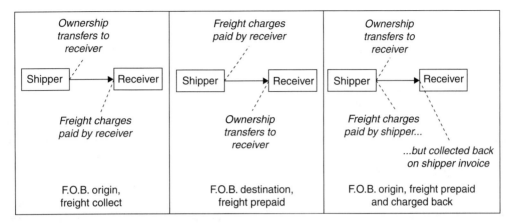

FIGURE 22.14 Common freight transportation terms.

a transportation cost term in the TAC expression that is the objective function for the order-placement decision. If the terms are freight prepaid, then the order-placing decision maker would not include the transportation cost in its objective function. At this point, one might object, pointing out that supply chain management must concern itself with a "supply-chain-wide" objective to minimize costs and maximize profits. My counterargument is that the Securities and Exchange Commission has not yet required, and likely never will require, any *supply chain* to report its revenues and costs; moreover, the Wall Street analysts who are so important in determining the market valuation of a firm care only about that firm's individual profits, not those of its supply chain. Thus, decision models that take the perspective of the decision-making firm and incorporate only those costs relevant to the firm are eminently reasonable. (An analysis of complementary, and to some extent competing, objective functions between a supplier and its customer is presented in Section 22.5.)

The next question we should address is whether the freight terms have any bearing on inventory holding cost. The answer is—as it often is—"It depends." Specifically, it depends on where the ownership of the goods transfers to the receiver. If the receiver is responsible for the goods in-transit, then it stands to reason that this liability should be reflected in annual inventory-related costs. From a conservative standpoint, those goods are "on the books" of the receiver, since, from the shipper's perspective, the goods were "delivered" as soon as the carrier took them under consignment. Although typical invoice terms dictate that money will not change hands between the supplier and customer for between 10 and 30 days, the goods are technically now part of the payable accounts of the customer. Thus, a conservative perspective would consider them to be "money tied up in inventory," in this case, inventory in-transit.

Using an approach suggested by Coyle et al. (2003, pp. 270–274), let us assume that the lead time L is composed of two parts, a random transit time T and a constant (and obviously non-negative) order processing time. If the expected transit time is μ_T, then every unit shipped spends a fraction μ_T/Y of a year in transit, where Y is the number of days per year. If h_p is the annual cost of holding inventory in transit, then annual *pipeline inventory* cost is given by $Dh_p\mu_T/Y$. The resulting total annual cost equation,* under F.O.B. origin

---

*It is straightforward to show that this cost equation is convex in Q and R. First, note that we are merely adding a convex term to an equation we already know to be convex. Moreover, one can modify the convexity proof of the transportation-exclusive equation from Hax and Candea (1984) to formalize the result, although the details are not included here.

terms that also require the receiver to pay the freight charges is

$$TAC(R,Q) = A\frac{D}{Q} + h\left(\frac{Q}{2} + R - \mu_{DLT}\right) + h_p D\frac{\mu_T}{Y} + DCQ^b \qquad (22.17)$$

Note that the pipeline inventory cost does not depend on the order quantity Q or the reorder point R. Thus, while it will not affect the optimal values of the decision variables for a given mode, it may affect the ultimate choice *between* modes through the effect of different values of μ_T on pipeline inventory. Note also that this approach to computing pipeline inventory costs ignores the time value of money over the transit time; one would assume, however, that such time value effects are insignificant in almost all cases.

Putting all of these pieces together, let's consider an example that compares cases with significant differences in item weight—affecting transportation costs—and item value—affecting holding costs. Assume that our company currently imports the item in question from a company based in Oakland, CA, and that our company's southeast regional distribution center is located in Atlanta, GA. We compare a reorder point–order quantity (Q, R) solution that excludes freight cost and pipeline inventory carrying costs—that is, from "F.O.B. destination, freight prepaid" terms—to solutions that incorporate these costs, both for LTL and TL shipments—using "F.O.B. origin, freight collect" terms—from Oakland to Atlanta. Assume that our target fill rate at the Atlanta DC is 99%. Our approach to computing the order quantity and reorder point for each of the cases—without transportation costs, using TL, and using LTL—is as follows:

- For "F.O.B. destination, freight prepaid" terms, compute (Q^*, R^*) using Equation 22.10 as the objective function, constrained by the desired fill rate.

- For "F.O.B. origin, freight collect" terms using LTL, compute (Q^*_{LTL}, R^*_{LTL}) using Equation 22.17 as the objective function, with the appropriate, empirically derived values of C and b, and with the solution constrained by the fill rate and by $Q_{LTL} \cdot w \leq 40{,}000$, where w is the item weight, which therefore assumes a truck trailer weight capacity of 40,000 lbs.*

- For "F.O.B. origin, freight collect" terms using TL, solve Equation 22.17 for (Q^*_{TL}, R^*_{TL}) as follows:

  1. $TAC_{\min} \leftarrow \infty$; $TAC_{TL} \leftarrow 0$; number of truckloads $= i = 1$
  2. Minimize Equation 22.17 with respect to R_{TL} with $Q_{TL} = i(40{,}000/w)$, C equal to the cost of i truckloads, $b = -1$, and constrained by the fill rate to obtain $TAC(Q_{TL}, R_{TL})$.
  3. $TAC_{TL} \leftarrow TAC(Q_{TL}, R_{TL})$
  4. If $TAC_{TL} < TAC_{\min}$,

     $TAC_{\min} \leftarrow TAC_{TL}$

     $i \leftarrow i + 1$; go to step 2.
  5. $(Q^*_{TL}, R^*_{TL}) = (Q_{TL}, R_{TL})$

---

*Actually, an LTL carrier would probably not take a shipment of 40,000 lbs, and the shipper would clearly get a better rate from a TL carrier for this full-TL weight. This is, however, the logical upper bound on the size of an LTL shipment. In addition, in using this value, we assume that a TL quantity will hit the trailer *weight* capacity before the trailer *space* (volume) capacity—i.e., that we "weigh-out" before we "cube-out."

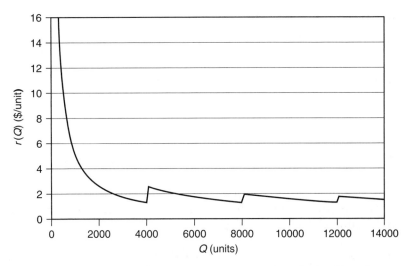

FIGURE 22.15 Example of truckload cost per unit versus number of units shipped.

TABLE 22.8 Example Class 100 LTL Rates for Oakland-Atlanta Lane

Minimum Weight (lbs)	Rate ($/cwt)
1	136.26
500	109.87
1000	91.61
2000	79.45
5000	69.91
10,000	54.61
20,000	48.12
30,000	41.85
40,000	39.12

Before continuing with our example, note that the procedure laid out above to solve for (Q^*_{TL}, R^*_{TL}) is not necessarily guaranteed to generate a global optimum due to the non-smooth nature of the $r(Q)$ function for TL shipments. For example, Figure 22.15 shows a plot of $r(Q)$ versus Q for a scenario in which the cost per TL is \$5180, the TL capacity is 40,000 lbs, and the item weight is $w = 10$ lbs. As Q increases, and therefore the number of truckloads increases, the freight rate per unit falls, but not in a smooth fashion. Thus, if $TAC_{TL,i}$ is the total cost of shipping i truckloads, depending on the problem parameters, even if $TAC_{TL,i} < TAC_{TL,i-1}$ and $TAC_{TL,i+1} > TAC_{TL,i}$, it does not immediately follow that $TAC_{TL,i+2} > TAC_{TL,i+1}$, leaving it unclear as to how $TAC_{TL,i}$ compares to $TAC_{TL,i+2}$. Thus, returning to our discussion at the outset of this section, one can see the possible value of a TMS in performing the analysis that takes these kinds of parametric complexities of the problem into account. Nonetheless, for our illustrative example, let us assume that our approach for computing (Q^*_{TL}, R^*_{TL}) is sufficient.

Continuing with our example, the distance from Oakland to Atlanta is 2471 miles.[*] Using a TL price of \$0.105/ton-mi (adjusted upward from the TL revenue per mile of \$0.0914/ton-mi from Wilson, 2003), the TL cost is \$259/ton. For the LTL rates, we will assume that we are shipping Class 100 goods with a published rate table as shown in Table 22.8, resulting in $b = -0.2314$ and $C = 2.435 w^{0.7686}$, after applying an assumed 50%

---

[*]The distances used in this example were obtained from MapQuest, www.mapquest.com.

discount from these published rates.* Further, we assume a cost of $80 to place a replenishment order, an on-site holding cost rate at the Atlanta DC of 15%, and an in-transit holding cost rate of 12%. Finally, we assume that the LTL shipment lead time is 5 days with a standard deviation of 1.5 days, that the TL lead time is 4 days with standard deviation of 0.8 days, that mean daily demand is 30 units with a standard deviation of 8 units, and that there are 365 selling days per year, such that $D = 10{,}950$. Table 22.9 displays the results, with item weights of $w \in \{10, 100\}$ (pounds) and item costs of $v \in \{\$20, \$500\}$. We report TAC only for the transportation-inclusive cases as this is the only relevant point of comparison. From the table, one can see that for this case, if our company pays transportation costs, TL is preferred in three of the four weight-cost combinations, with LTL being preferred only in the low-weight, high-item-cost case. This is reasonable, given the ability of LTL to allow small shipments sizes that keep inventory holding costs in check.

Let us now assume that we have the option to switch to an importer located in New York City (NYC), dramatically decreasing the transit distance for replenishment shipments to the Atlanta DC to 884 mi. Assume further that this importer charges a premium over the current importer located in Oakland so that the item cost will be 10% higher than in the Oakland-to-Atlanta case above. Given the distance from NYC to Atlanta, using the same approach as above, the TL cost is $92.82/ton. For an LTL shipment, the published rates are given in Table 22.10, resulting in $b = -0.3704$ and $C = 3.993w^{0.6296}$. Let us assume that the LTL lead time from NYC to Atlanta is 2 days, with a standard deviation of 0.6 days, and that the TL lead time is 1 day, with a standard deviation of 0.2 days. Therefore, although our annual materials cost will increase by 10% if we switch to the NYC importer, our transportation costs, where applicable, and inventory holding costs—safety stock and, where applicable, pipeline inventory—will decline. The comparative results are shown in Table 22.11. As the annual increase in materials cost is $21,900 in the low-item-cost case and $547,500 in the high-item-cost case, these values serve as the hurdles for shifting to the NYC source. For the high-item-cost cases, this increase in annual materials costs swamps the transportation and holding cost savings, even though the transportation costs are cut dramatically for the cases in which $w = 100$ lbs. The analysis shows, however, that the switch is viable in the case in which our company pays for freight transportation with a high-weight, low-cost item. In this case, TL is the preferred freight transit option, and the switch to the NYC source saves over $88,000 in annual transportation costs and over $91,000 overall.

22.3.4 A More General Transportation Cost Function

In most commercial transportation management and planning systems, LTL rates are determined using tariff tables, as in the examples above. The major limitations of this approach are that access to the tariff tables must be purchased and that it is necessary to know what discount to apply to the tariff rates in determining transportation costs. Kay and Warsing (2006), however, present a more general model of LTL rates, one that is not specific to a transportation lane or to the class of the items being shipped, with the item's density serving as a proxy for its class rating. This generality allows the rate model to be used in the early stages of logistics network design, when location decisions are being made and when the most appropriate shipment size for each lane in the network is being determined. Thus, the model developed by Kay and Warsing is important for the sort of analysis presented in Section 22.4.

In their model, Kay and Warsing (2006) use average, industry-wide LTL and TL rates empirically to develop an LTL rate model that is scaled to economic conditions, using the

---

*In general, $C = [\delta C_w w^{(b+1)}]/100$, in $/unit, where δ is the discount factor.

TABLE 22.9 Comparative Results from Inventory-Transportation Analysis

Oakland–Atlanta	Freight Terms	Mode	Q	R	Ordering Cost($)	Holding Cost($)	Pipeline Cost($)	Transp'n Cost($)	TAC ($)
$w=10, v=\$20$	Origin–collect	TL	4000	120	219	6000	288	14,180	20,687
	Origin–collect	LTL	3750.99	150	234	5626	360	23,306	29,526
	Destination–prepaid	—	794.35	179.88	1103	1281	—	—	—
$w=10, v=\$500$	Origin–collect	TL	4000	120	219	150,000	7200	14,180	171,599
	Origin–collect	LTL	361.52	201.19	2423	17,396	9000	40,047	68,866
	Destination–prepaid	—	176.38	217.94	4966	11,710	—	—	—
$w=100, v=\$20$	Origin–collect	TL	800	127.86	1095	1224	288	141,803	144,409
	Origin–collect	LTL	400	198.65	2190	746	360	229,616	232,912
	Destination–prepaid	—	794.35	179.88	1103	1281	—	—	—
$w=100, v=\$500$	Origin–collect	TL	400	140.70	2190	16,552	7200	141,803	167,745
	Origin–collect	LTL	400	198.65	2190	18,648	9000	229,616	259,454
	Destination–prepaid	—	176.38	217.94	4966	11,710	—	—	—

TABLE 22.10 Example Class 100 LTL Rates for NYC-Atlanta Lane

Minimum Weight (lbs)	Rate ($/cwt)
1	81.96
500	74.94
1000	61.14
2000	49.65
5000	39.73
10,000	33.44
20,000	18.36
30,000	14.90
40,000	13.36

producer price index for LTL services provided by the U.S. Bureau of Labor Statistics. The LTL rate model requires inputs of W, the shipment weight (in tons); s, the density (in lbs/ft^3) of the item being shipped; and d, the distance of the move (in mi). From a non-linear regression analysis on CzarLite tariff rates* for 100 random O-D pairs (5-digit zip codes), with weights ranging from 150 to 10,000 lbs (corresponding to the midpoints between successive rate breaks in the LTL tariff), densities from approximately 0.5 to 50 lbs/ft^3 (serving as proxies for class ratings 500 to 50), and distances from 37 to 3354 mi, the resulting generalized LTL rate function is

$$r_{LTL}(W,s,d) = PPI_{LTL} \left[\frac{\frac{s^2}{8} + 14}{\left(W^{(1/7)} d^{(15/29)} - \frac{7}{2}\right)(s^2 + 2s + 14)} \right] \quad (22.18)$$

where, as suggested above, PPI_{LTL} is the producer price index for LTL transportation, reported by the U.S. Bureau of Labor Statistics as the index for "General freight trucking, long-distance, LTL."† Kay and Warsing (2006) report a weighted average residual error of approximately 11.9% for this functional estimate of LTL rates as compared to actual CzarLite tariff rates. Moreover, as industry-wide revenues are used to scale the functional estimate, the model reflects the average discount provided by LTL carriers to their customers, estimated by Kay and Warsing to be approximately 46%.

Thus, for a given lane, with d therefore fixed, the model given by Equation 22.18 provides estimates of LTL rates for various shipment weights and item densities. Kay and Warsing (2006) demonstrate the use of the model to compare LTL and TL shipment decisions under different item value conditions. Moreover, the functional estimate could easily be used to evaluate the comparative annual cost of sending truckloads of mixed goods with different weights and densities versus individual LTL shipments of these goods. On the other hand, for a given item, with a given density, Equation 22.18 provides a means of analyzing the effect of varying the origin point to serve a given destination, possibly considering different shipment weights as well. These two analytical perspectives are, as we suggest above, important in the early stages of supply chain network design, the subject of the section that follows.

---

*CzarLite is an LTL rating application developed by SMC^3 Company, and it is commonly used as the basis for carrier rates. The rates used in the study by Kay and Warsing (2006) were obtained in November–December 2005 from http://www.smc3.com/applications/webczarlite/entry.asp.

†See the Bureau of Labor Statistics website at http://www.bls.gov/ppi/home.htm, Series ID=PCU484122484122.

TABLE 22.11 Comparative Results from Extended Inventory-Transportation Analysis

NYC–Atlanta	Freight Terms	Mode	Q	R	Ordering Cost($)	Holding Cost($)	Pipeline Cost($)	Transp'n Cost($)	TAC($)	Change from OAK-ATL($)
$w=10, v=\$22$	Origin–collect	TL	4000	30	219	6600	79	5082	11,980	−8707
	Origin–collect	LTL	2512.92	60	349	4146	158	10,253	14,906	−14,619
	Destination–prepaid	—	745.02	62.15	1176	1236	—	—	—	28
$w=10, v=\$550$	Origin–collect	TL	4000	30	219	165,000	1980	5082	172,281	682
	Origin–collect	LTL	298.71	75.09	2933	13,567	3960	22,565	43,025	−25,841
	Destination–prepaid	—	157.04	82.59	5578	8341	—	—	—	−2757
$w=100, v=\$22$	Origin–collect	TL	800	30.00	1095	1320	79	50,819	53,313	−91,096
	Origin–collect	LTL	400	71.31	2190	697	158	86,313	89,358	−143,554
	Destination–prepaid	—	745.02	62.15	1176	1236	—	—	—	28
$w=100, v=\$550$	Origin–collect	TL	400	30.00	2190	16,500	1980	50,819	71,489	−96,256
	Origin–collect	LTL	400	71.31	2190	17,433	3960	86,313	109,895	−149,559
	Destination–prepaid	—	157.04	82.59	5578	8341	—	—	—	−2757

22.4 Managing Locations in the Supply Chain

At this point, we have discussed the inventory and transportation legs of Ballou's triangle, and along the way, had some brief discussion of customer service as given by α, the cycle service level, and β, the fill rate. Therefore, we now turn our attention to "location strategy," or designing and operating the supply chain network. Key questions to answer in this effort are as follows: How do we find a good (or perhaps the *best*) network? What are the key objectives for the network? What network alternatives should we consider? Once again, we need to consider the perspective of the decision maker charged with making the design decisions. Typically, the perspective will be that of a single firm. The objectives may be diverse. We clearly wish to minimize the cost of operating the network, and as product prices typically will not change as a function of our network design, this objective should be sufficient from a financial perspective. Customer service may also need to be incorporated in the decision, but this may pose a challenge in any explicit modeling efforts. Assuming an objective to minimize the costs of operating the network for a single firm, the key decision variables will be the number and locations of various types of facilities—spanning as far as supplier sites, manufacturing sites, and distribution sites—and the quantities shipped from upstream sites to downstream sites and ultimately out to customers.

To get a sense of the cost structure of these kinds of models, let us consider a single-echelon location-allocation problem. This problem is used to determine which sites to include in a network from a set of n production or distribution sites that serve a set of m customers. The fixed cost of including site i in the network and operating it across the planning period (say, a year) is given by f_i, and each unit of flow from site i to customer j results in variable cost c_{ij}. Site i has capacity K_i, and the total planning period (e.g., annual) demand at customer j is given by D_j. Therefore, our objective is to

$$\min \quad \sum_{i=1}^{n} f_i y_i + \sum_{i=1}^{n}\sum_{j=1}^{m} c_{ij} x_{ij} \qquad (22.19)$$

$$\text{subject to} \quad \sum_{i=1}^{n} x_{ij} = D_j, \quad j=1,\ldots,m \qquad (22.20)$$

$$\sum_{j=1}^{m} x_{ij} \leq K_i y_i, \quad i=1,\ldots,n \qquad (22.21)$$

$$y_i \in \{0,1\}, \quad i=1,\ldots,n \qquad (22.22)$$

$$x_{ij} \geq 0, \quad i=1,\ldots,n;\ j=1,\ldots,m \qquad (22.23)$$

where decision variables y_i $(i=1,\ldots,n)$ determine whether site i is included in the network $(y_i=1)$ or not $(y_i=0)$ and x_{ij} $(i=1,\ldots,n;\ j=1,\ldots,m)$ determine the number of units shipped from site i to customer j. As parameters f_i and c_{ij} are constants, this problem is a mixed-integer program (MIP). Depending on the values of n and m, solving it may be challenging, even by twenty-first-century computing standards, given the combinatorial nature of the problem.

Moreover, the problem represented by Equations 22.19 to 22.23 above is only a small slice of the overall network design formulation. If we do indeed wish to consider supplier location, production facility locations, *and* distribution facility locations, then we need location variables (y) and costs (f) at each of three echelons and flow variables (x) and costs (c) to represent the levels and costs of supplier-to-factory, factory-to-distribution, and distribution-to-customer flows. In addition, we need to add *conservation of flow* constraints that ensure that all units shipped *into* the factory and distribution echelons are also shipped back *out*—unless we consider the possibility of carrying inventories at the various echelons, another complication that would also possibly entail adding time periods to the model to reflect varying inventory levels over time. Further, we should keep in mind the data requirements and restrictions for formulating this problem. One must have a reasonably good estimate of the fixed cost of opening and operating each site, the capacity of each site (or at least a reasonable upper bound on this value), and the variable cost of producing or shipping goods from this site to other network locations or to customers. In this latter case, one must assume that these variable costs are linear in the quantities produced or shipped. Earlier in this chapter, however, we indicated that shipping costs typically exhibit economies of scale; production costs may do so as well. Moreover, the *discrete-location* problem formulated above assumes that a set of candidate locations has been precisely determined in advance of formulating the problem.

In this author's mind, the typical IE or operations management treatment of the facility location problem rushes to a description of the mathematical techniques without first considering *all* of the various inputs to the decision. As Ballou (2004, pp. 569–570) rightfully reminds his readers, "... It should be remembered that the optimum solution to the real-world location problem is no better than the model's description of the problem realities."

In addition, decisions regarding facility locations have significant *qualitative* aspects to them as well. As with customer service, these factors may be challenging to incorporate into a quantitative decision model. Indeed, in the author's opinion, the typical academic treatment of the facility location problem is one that could benefit from a serious consideration of what practitioners faced with facility location decisions actually *do*. This author's reading of the trade press indicates that those practitioners, rightly or wrongly, devote only a small portion—if *any*—of their decision-making process to solving traditional facility location models. Consider this list of the "ten most powerful factors in location decisions" from a 1999 survey of readers of *Transportation and Distribution* magazine (Schwartz, 1999a), in order of their appearance in the article: reasonable cost for property, roadway access for trucks, nearness to customers, cost of labor, low taxes, tax exemptions, tax credits, low union profile, ample room for expansion, community disposition to industry. In fact, only two, or perhaps three, of these ten—property and labor costs, and perhaps low taxes—could be directly incorporated into one of the classic facility location models. The others could be grouped into what another *T&D* article (Schwartz, 1999b) referred to as *service-oriented issues* (roadway access, nearness to customers) and *local factors* (tax exemptions, tax credits, union profile, expansion opportunity, and community disposition). That *T&D* article goes on to state, mostly from consulting sources, that service-oriented issues now "dominate," although this is not, in this author's mind, a "slam-dunk" conclusion. For example, 2004–2005 saw significant public debate in North Carolina as to the public benefit—or public detriment—of using state and local tax incentives to lure businesses to locate operations in a given community (Reuters, 2004; Cox, 2005; Craver, 2005). Given the coverage devoted to the effects of tax incentives on these location decisions, it would be hard to conclude as the *T&D* article did, that "local issues" are "least consequential" (p. 44). A more thoughtful perspective can be found in another trade publication discussion (Atkinson, 2002, p. S65), which states, "While everyone can identify the critical elements, not all of them agree on the order of priority."

From a broader perspective, Chopra and Meindl (2004) suggest that supply chain network design encompasses four phases: Phase I—SC Strategy, Phase II—Regional Facility Configuration, Phase III—Desirable Sites, Phase IV—Location Choices. They also suggest a number of factors that enter into these decisions. In our discussion, we will indeed start from supply chain strategy. The primary strategic choice in the treatment of supply chain network design that follows will be how *centralized* or *decentralized* the network design should be. Given the complexity of the network design process, it is reasonable to frame it as an iterative one, perhaps evaluating several network options that range from highly centralized to highly decentralized. Ultimately, a choice on this centralized-decentralized spectrum will lead to a sense of how many facilities there should be at each echelon of the network. Once the decision maker has an idea of the number of facilities and a rough regional configuration, he or she can answer the question of where good candidate locations would be using a *continuous-location* model, along the lines of that studied by Cooper (1963, 1964). This "ballpark" set of locations can then be subjected to qualitative analysis that considers service-oriented issues and local factors. Therefore, let us first turn our attention to understanding the differences between centralized and decentralized networks.

22.4.1 Centralized Versus Decentralized Supply Chains

Let us focus on the design of a distribution network and address the following questions: How centralized should the network be? What are the advantages of a more centralized or a more decentralized network? This issue can clearly be addressed in the context of the legs and center of Ballou's strategic triangle—inventory, locations, and transportation on the legs and customer service in the center. Consider the two DC networks laid out in Figure 22.16. Network 1 is more centralized, with only four distribution points, laid out essentially on the compass—west, central, northeast, and southeast. Network 2 is more decentralized, with a hub-spoke-type structure that has twenty additional DCs located near major population centers and ostensibly fed by the regional DCs that encompassed all of Network 1. Starting with transportation, we can break the tradeoffs down if we first consider transportation *outbound* from DCs to customers versus transportation *inbound* to the DCs, or essentially *within* the network. In the context of the facility location problem (Equations 22.19 to 22.23) laid out above, these would be costs c_{ij} for shipments from DC i to customer j and costs c_{0i} for shipments from the factory (location 0) to DC i.

As shipments to customers—often called the "last mile" of distribution—typically involve smaller quantities, perhaps only a single unit, it is often difficult to garner significant scale-driven discounts for these shipments. However, it is clear from our discussion above that distance, in addition to weight, is a significant factor in the cost of freight transportation. More weight allows greater economies in unit costs, but more distance increases the variable costs of the carrier and thereby increases the cost of the transportation service to the

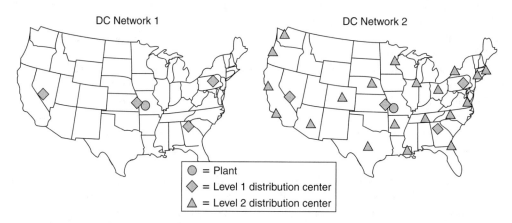

FIGURE 22.16 Comparative distribution networks.

Supply Chain Management

shipper. Note that the centralized network will clearly involve longer "last-mile" shipments to customers than the decentralized network. Although it is possible that larger loads may be moved on vehicles outbound to a set of customers from a centralized site, it is quite possible that neither network affords much chance for significant bulk shipping activity outbound, particularly if the outbound shipments must be made to customers that are dispersed widely. Thus, one could argue persuasively that outbound distribution costs are larger in centralized networks.

On the inbound side of the DCs, however, the situation flips. With only a few DCs in the centralized network, the shipments from the supply points (e.g., plants, suppliers) into the final distribution points are more likely to garner significant economies of scale in transportation, and therefore, "within-network" freight costs should be lower in a centralized distribution network. In addition, Figure 22.16 clearly shows that there are fewer miles to travel in moving goods to the final distribution points in a centralized network, another source of lower within-network freight transportation costs in a centralized network. However, as shipments inbound will tend to be packed in bulk, more dense, and perhaps less time-sensitive, inbound transportation costs for a given network are likely to be much lower than outbound costs. Thus, it stands to reason that transportation costs will net out to be lower in a decentralized network, wherein relatively expensive "last-mile" transportation covers shorter distances. Moreover, it should be immediately clear that customers would prefer the decentralized network for its ability to reduce order fulfillment lead times.

From the standpoint of locations, there are obviously fewer locations in a centralized DC network, and therefore, this network will result in lower fixed and operating costs driven by facilities (e.g., amortization of loans, payment of leases, operating equipment, labor). In the context of our facility location problem since fewer DC locations (y_i) would be involved in the centralized problem (or more of them would be forced to be $y_i = 0$), the sum $\sum_{i=1}^{n} f_i y_i$ in the objective function would necessarily be smaller in a more centralized solution.

Thus, we have addressed two legs of the SC strategy triangle, transportation, and locations. We are left with inventory. The effects of centralizing, or not, on this leg are richer, and therefore, I will devote the section that follows to this topic.

22.4.2 Aggregating Inventories and Risk Pooling

Vollmann et al. (2005, p. 275) point out that "... if the uncertainty from several field locations could be aggregated, it should require less safety stock than having stock at each field location." Indeed, in one's mind's eye, it is relatively easy to see that the ups and downs of uncertain demand over time in several independent field locations are likely to cancel each other out, to some extent, when those demands are added together and served from an aggregated pool of inventory. This *risk pooling* result is the same as the one that forms the basis of financial portfolio theory, in which the risk—that is, the variance in the returns—of a portfolio of investments is less than the risk of any single investment in the portfolio. Thus, an investor can cushion the ups and downs, as it were, or reduce the variability in his investments by allocating his funds across a portfolio as opposed to concentrating all of his funds in a single investment.

Looking at this risk pooling effect from the standpoint of inventory management, since safety stock is given by $SS = z\sigma_{DLT}$, one would reasonably infer that the risk pooling effects of aggregating inventories must center on the standard deviation of demand over the replenishment lead time, σ_{DLT}. Indeed, this is the case. If we have, for example, descriptions of lead-time demand for N distribution regions, with a mean lead-time demand in region i of $\mu_{DLT,i}$ and standard deviation of lead-time demand in region i of $\sigma_{DLT,i}$, then aggregating demand across all regions results in a pooled lead-time demand with mean

$$\mu_{DLT,\text{pooled}} = \sum_{i=1}^{N} \mu_{DLT,i} \qquad (22.24)$$

and standard deviation

$$\sigma_{DLT,\text{pooled}} = \sqrt{\sum_{i=1}^{N} \sigma_{DLT,i}^2 + 2 \sum_{i=1}^{N-1} \sum_{j=i+1}^{N} \rho_{ij} \sigma_{DLT,i} \sigma_{DLT,j}} \qquad (22.25)$$

where ρ_{ij} is the correlation coefficient of the demand between regions i and j. If demand across all regions is independent, then $\rho_{ij} = 0$ for all $i, j = 1, \ldots, N$ $(i \neq j)$, and Equation 22.25 reduces to

$$\sigma_{DLT,\text{pooled}} = \sqrt{\sum_{i=1}^{N} \sigma_{DLT,i}^2} \qquad (22.26)$$

By the triangle inequality, $\sqrt{\sum_{i=1}^{N} \sigma_{DLT,i}^2} \leq \sum_{i=1}^{N} \sigma_{DLT,i}$, and therefore,

$$SS_{\text{pooled}} = z\sigma_{DLT,\text{pooled}} \leq z\sum_{i=1}^{N} \sigma_{DLT,i} = SS_{\text{unpooled}} \qquad (22.27)$$

a result that holds for any values of ρ_{ij}, since perfect positive correlation ($\rho_{ij} = 1$ for all $i, j = 1, \ldots, N$) makes Equation 22.27 an equality ($SS_{\text{pooled}} = SS_{\text{unpooled}}$) and *negative* correlation actually *increases* the pooling-driven reduction in safety stock beyond the independent case.

Let's consider an example. Assume we are interested in designing a distribution network to serve 40 customers. An important decision is the number of distribution sites we need to serve those 40 customers. Our choices range from just a single distribution center (DC) up to, theoretically, 40 DCs, one dedicated to serving each customer. In our simplified system, let's assume that there is no fixed cost to place an order to replenish inventory at any distribution site and that demand over the replenishment lead time for any distribution site and any customer served is normally distributed with mean $\mu = 50$ and standard deviation $\sigma = 15$. Therefore, at any DC in the network, since there is no ordering cost, we can set the stocking level by using a base-stock policy, with a stocking level at DC i equal to $B_i = \mu_i + z_\alpha \sigma_i$, where μ_i is the mean demand over the replenishment lead time for the set of customers served by DC i, σ_i is the standard deviation of this demand, and z_α is the safety stock factor that yields a CSL of α. If DC i serves n_i $(1 \leq n_i \leq 40)$ customers, then using Equations 22.24 and 22.25 above,

$$\mu_i = n_i \mu \qquad (22.28)$$

and

$$\sigma_i = \sqrt{\sum_{j=1}^{n_i} \sigma^2 + 2 \sum_{j=1}^{n_i-1} \sum_{k=j+1}^{n_i} \rho_{jk} \sigma^2} \qquad (22.29)$$

where ρ_{jk} is the correlation coefficient of the demand between customers j and k. If $\rho_{jk} = \rho$ for all $j, k = 1, \ldots, 40$ $(j \neq k)$, then it is easy to show that Equation 22.29 reduces to

$$\sigma_i = \sqrt{[n_i(1-\rho) + \rho n_i^2]\sigma^2} \qquad (22.30)$$

Supply Chain Management

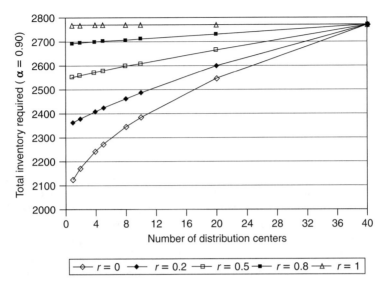

FIGURE 22.17 Example of risk-pooling-driven reduction in inventory.

TABLE 22.12 Risk Pooling-Based Reduction in Safety Stock with $n = 40$ Customers

Number of Customers per DC	Number of DCs in Network	SS at $\rho = 0$ Units	% of $E[D] = n\mu$
1	40	768.9	38.45
2	20	543.7	27.19
4	10	384.5	19.22
5	8	343.9	17.19
8	5	271.9	13.59
10	4	243.2	12.16
20	2	171.9	8.60
40	1	121.6	6.08

Note that when $\rho = 0$ (i.e., the aggregated customer demands are independent), $\sigma_i = \sigma\sqrt{n_i}$, meaning that safety stock increases with the square root of the number of sites aggregated—that is, sub-linearly. In contrast, when $\rho = 1$ (i.e., the aggregated customer demands are perfectly correlated), $\sigma_i = n_i \sigma$, meaning that—as we pointed out in the more general discussion above—the standard deviation of the aggregated demands is simply the sum of the standard deviations of the disaggregated demands. Figure 22.17 shows the total system inventory required in the distribution system to achieve $\alpha = 0.90$ as n_i ranges from 1 to 40 at various levels of demand correlation (where ρ is expressed as r in the legend of the graph). Table 22.12 shows the reduction in system safety stock as a percentage of overall lead-time demand as the number of DCs in the network declines, assuming independent demands across customers.

The example above highlights the notion of postponement in the supply chain. In our network configuration, risk pooling through aggregation is a means of time (or place) postponement, where the forward movement of the goods to (or near) the final points of consumption is postponed until more is known about demand at those points of consumption. The opposite approach, in a decentralized network, is to speculate and move the goods to points closer to customers in advance of demand. As Chopra and Meindl (2004) point out, an approach that attempts to attain the best of both worlds—safety stock reduction through aggregation while still retaining the customer service benefits of a decentralized network—would be to virtually aggregate inventory in the network. The idea is to allow all DCs in the network to have visibility to and access to inventory at all other DCs in the network, such that when a particular DC's inventory approaches zero before a replenishment can arrive from the supplier, that DC can request a transshipment of inventory in excess at another DC. Chopra and Meindl (2004) point out that Wal-Mart has used this strategy effectively. With such an approach, demand at all locations is virtually pooled, with the reductions in safety stock and improvements in customer service coming at the expense of the transshipments that rebalance supply when necessary.

Rounding out this discussion, it is worthwhile to point out that another type of postponement is form postponement, whereby the final physical conversion of a product is delayed until more is known about demand. The two now-classic examples of form postponement in the operations management literature come from Hewlett-Packard (HP), the computing equipment manufacturer, and Benetton, a major apparel company. As described by Feitzinger and Lee (1997), the HP Deskjet printer was redesigned such that the power supply was converted to a drop-in item that could be added to a generic printer in distribution centers. The previous approach was to manufacture printers with country-specific power supplies—due to the different electrical power systems in various countries—in the manufacturing process. This change in product and packaging design allowed HP to aggregate all demand around the globe into demand for a single, generic printer and hold only the differentiating items, the power supply, and various cables, at in-country distribution centers. In the case of Benetton, it changed the order of manufacturing operations to re-sequence the dyeing and sewing processes for sweaters. The previous approach, time-honored in the apparel manufacturing industry, was first to dye bolts of "greige" (un-colored) fabric and then cut that dyed fabric into its final form to be sewn into apparel items. Benetton swapped this sequence to first cut and sew "greige" sweaters that would be dyed after more information could be collected about demand for color-specific items. By doing so, demand for all colors can be aggregated into demand for a single item until a point in time when the inherent uncertainty in demand might be reduced, closer to the point of consumption.

One final point regarding risk pooling effects on inventory is that these effects change significantly with changes in the coefficient of variation, $cv = \sigma/\mu$, of demand. This gets to the notion of classifying inventory items as slow movers versus fast movers. Fast movers have expected demands that are relatively large and predictable, thereby resulting in lower cv values. On the other hand, slow movers have expected demands that are relatively small and unpredictable, resulting in higher values of cv. Thus, the relative benefit of centralizing slow movers, in terms of safety stock reduction, will tend to be quite a bit larger than the benefit of centralizing fast movers. This effect is demonstrated in Figure 22.18, which uses our 40-customer example, still with mean lead-time demand of $\mu = 50$ for each customer, but now with standard deviations varying from $\sigma = 2.5$ ($cv = 0.05$) to $\sigma = 17.5$ ($cv = 0.35$—beyond which the normal distribution will exhibit more than a 1 in 1000 chance of a negative value, inappropriate for modeling customer demands). A simple rule of thumb that results from this data would be to centralize the slow movers to maximize the benefits of risk pooling-driven reductions in inventory and to stock the fast movers closer to points of consumption to maximize the benefits of rapid customer service for these high-demand items.

22.4.3 Continuous-Location Models

Note that the discrete-location MIP model for facility location does not, in the form we stated it above, include any inventory effects of the decision variables. Moreover, it is clear

Supply Chain Management

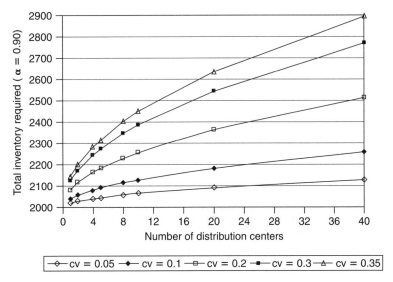

FIGURE 22.18 Effect of variability on risk-pooling impact on inventory levels.

from our discussion above that the effect of the centralizing decision on inventory levels in the network is non-linear. An MIP formulation could handle such non-linear effects via a piecewise linear approximation, but this would only add more integer variables to the formulation.* A better approach, however, might be to consider a facility location model that allows a non-linear objective function, as the most general formulation of the problem is likely to have a number of sources of non-linearity in costs: non-linear production costs driven by economies of scale in production, non-linear transportation costs driven by economies of scale in transportation, and non-linear inventory holding costs driven by the risk-pooling effects of centralizing inventory in the network. Moreover, in the opinion of the author, this approach is probably more consistent with the manner in which facility location decisions are made in practice. The resulting approach would be iterative and would not feed a specific set of candidate locations into the modeling effort a priori, but rather would use some preliminary modeling efforts to generate a rough geographic idea of what might be good candidate locations, perhaps guided by the customer service implications of fewer distribution locations farther from customers, on average, or vice versa. The resulting candidate locations from the cost-service tradeoffs could be subjected to a more qualitative analysis based on the types of "local factors" discussed earlier in this section.

The basis for this formulation of the facility location problem is commonly referred to as the "gravity model," typically formulated to minimize the costs of moving goods *through* the facility, in general, taking receipts from sources and ultimately satisfying demand at customers. In this formulation, the decision variables are x and y, the coordinates of the facility location on a Cartesian plane. The input parameters are D_i, the total volume—in

---

*Most introductory operations research textbooks include such formulations; see, for example, Winston (1994).

TABLE 22.13 Input Data for Gravity Location Example

	i		F_i ($/ton-mi)	D_i (Tons)	x_i	y_i
Source Locations	1	Buffalo	0.90	500	700	1200
	2	Memphis	0.95	300	250	600
	3	St. Louis	0.85	700	225	825
Customer Locations	4	Atlanta	1.50	225	600	500
	5	Boston	1.50	150	1050	1200
	6	Jacksonville	1.50	250	800	300
	7	Philadelphia	1.50	175	925	975
	8	New York	1.50	300	1000	1080

tons, for example—moved into (out of) the facility from source (to customer) $i \in \{1,\ldots,m\}$ and F_i, the cost—in $/ton-mi, for example—to move goods to (from) the facility from source (to customer) $i \in \{1,\ldots,m\}$. Then, the problem is to

$$\min \sum_{i=1}^{m} F_i D_i d_i \qquad (22.31)$$

where $d_i = \sqrt{(x_i - x)^2 + (y_i - y)^2}$, the Euclidean distance between the location (x, y) of the facility and the locations (x_i, y_i) $(i = 1, \ldots, m)$ of the sources and customers that interact with the facility being located. The "gravity" aspect of the problem stems from the fact that the combined cost and distance of the moves into and out of the facility drive the solution, with a high-volume or high-cost source or customer exerting significant "gravity" in pulling the ultimate location toward it to reduce the cost of moving goods between the facility and that source or customer.

Consider a small example, based on an example from Chopra and Meindl (2004), with three sources and five customers. The input data for this example is shown in Table 22.13. The optimal solution, found easily using the Solver tool embedded in Excel, is to locate the facility that interacts with these sources and customers at $(x, y) = (681.3, 882.0)$. A graphical view of the problem data and the solution (labeled as "initial solution") are shown in Figure 22.19.

To illustrate the "gravity" aspects of this model, consider an update to the original problem data. In this situation, assume that it is cheaper to move goods from the source in Buffalo ($F_1 = 0.75$) and more expensive to move them from the sources in Memphis ($F_2 = 1.00$) and St. Louis ($F_3 = 1.40$), and assume that it is also cheaper to move goods to the customers in Boston, New York, and Philadelphia ($F_5 = F_7 = F_8 = 1.30$), but that the costs of moving goods to Atlanta and Jacksonville stay the same. The combined effects of these changes in transportation costs is to exert some "gravity" on the optimal solution and shift it west and south, to attempt to reduce the mileage of the moves that are now more expensive. Indeed, the updated optimal solution is $(x, y) = (487.2, 796.8)$, shown as "updated solution" in Figure 22.19.

Of course, the example above is a simplified version of the larger-scale problem. A typical problem would involve a larger number of customers, and possibly also a larger number of sources. In addition, the cost of moving goods to/from the facility location in our example is a constant term, irrespective of the distance the goods are moved. In general, this cost could be some non-linear function of the distance between the source/customer and the facility, as discussed earlier in the chapter (e.g., the general function for LTL rates derived by Kay and Warsing, 2006). In addition, a more general formulation would allocate sources

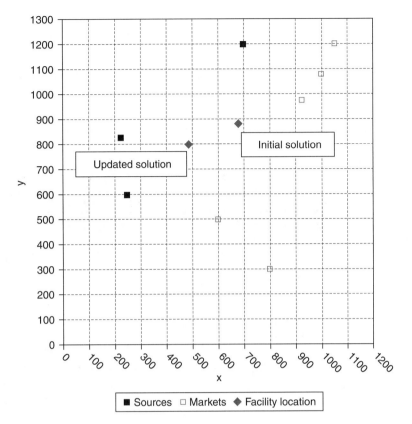

FIGURE 22.19 Gravity location data and solution.

and customers to n candidate locations. The general formulation would be to

$$\min \sum_{i=1}^{n}\sum_{j=1}^{m_S} \delta_{ij}^S F_{ij} D_j d_{ij} + \sum_{i=1}^{n}\sum_{k=1}^{m_C} \delta_{ik}^C F_{ik} D_k d_{ik} \tag{22.32}$$

$$\text{subject to} \quad \sum_{j=1}^{m_S} \delta_{ij}^S = 1 \quad i=1,\ldots,n \tag{22.33}$$

$$\sum_{k=1}^{m_C} \delta_{ik}^C = 1 \quad i=1,\ldots,n \tag{22.34}$$

$$\delta_{ij}^S \in \{0,1\} \quad i=1,\ldots,n; \quad j=1,\ldots,m_S \tag{22.35}$$

$$\delta_{ik}^C \in \{0,1\} \quad i=1,\ldots,n; \quad k=1,\ldots,m_C \tag{22.36}$$

In addition to the generalized location decision variables (x_i, y_i) for each candidate facility location $i = 1, \ldots, n$, this formulation also introduces allocation decisions δ_{ij}^S and δ_{ik}^C, binary variables that allocate each source and each customer, respectively, to a candidate facility. Constraints (22.33) and (22.34) ensure that each source and each customer is allocated to exactly one candidate facility.

Given the nature of this problem—a constrained, non-linear optimization—it would appear that this is no easier to solve, and perhaps significantly more challenging to solve, than the MIP facility location problem we introduced at the outset of this section. Indeed, solving Equations 22.32 to 22.36 to optimality would be challenging. Note, however, that for a given allocation of sources and customers to candidate facilities—that is, if we *fix* the values of δ_{ij}^S and δ_{ik}^C—the problem decomposes into n continuous-location problems of the form of Equation 22.31. In fact, this approach has been suggested by Cooper (1964), in what he described as an "alternate location-allocation" (ALA) algorithm. Starting from some initial allocation of sources and customers to sites, one solves the n separate location problems exactly. Then, if some source or customer in this solution could be allocated to another candidate location at a net reduction in total cost, that source or customer is reallocated and the overall solution is updated. The process continues until no further cost-reducing reallocations are found. This algorithm results in a local minimum to the problem, but not necessarily a global minimum. Cooper shows, however, that the algorithm is fast and generates reasonably good solutions. More recent work (Houck et al., 1996) has shown that the ALA procedure is actually quite efficient in reaching an optimal solution in networks with fewer than 25 candidate facilities.

In addition, the ALA approach appears to be promising in allowing for a more generalized functional form of the cost to move goods and fulfill demands in the network. Bucci and Kay (2006) present a formulation similar to Equations 22.32 to 22.36 above for a large-scale distribution problem—limited to customer allocations only—applied to the continental United States. They use U.S. Census Bureau data to generate 877 representative "retail locations" from 3-digit ZIP codes, with the relative demand values based on the populations of those ZIP codes. They consider locating six fulfillment centers to serve these 877 customers. The fulfillment centers exhibit economies of scale in production such that the cost of producing a unit of output decreases by an exponential rate with increasing size of the facility. Facility size is determined by the amount of demand allocated to the facility, with the relative costs given by $c_2/c_1 = (S_2/S_1)^b$, where c_i is the production cost for facility i, S_i is the size of facility i, and b is a constant value provided as an input parameter. Bucci and Kay (2006) cite Rummelt (2001), who in turn cites past studies to estimate that $b \approx -0.35$ for U.S. manufacturers. In addition, Bucci and Kay consider the ratio of production and transportation costs as an input parameter to their problem formulation. Thus, their formulation minimizes the sum of the production and transportation costs that result from the allocation of candidate facilities to customers. Figure 22.20 shows a series of solutions that result from their analysis, using $b = -0.4$ and production costs at 16% of transportation costs. The initial solution at the top of the figure is a static solution that does not incorporate production economies; the middle solution is intermediate in the ALA process; and the bottom solution is the final ALA solution. This solution process provides a graphical sense of the manner in which Bucci and Kay's modified ALA procedure balances production and transportation costs, with the final solution aggregating the demands of the densely populated eastern United States to be served by a single large-scale facility and consolidating what initially were three western facilities into two, ultimately resulting in a five-facility solution.

22.5 Managing Dyads in the Supply Chain

Much of the discussion in previous sections in this chapter, particularly in the immediately preceding section, considers issues related to the *design* of the supply chain. Our emphasis up to this point has been to carefully consider and understand the interaction of inventory, transportation, and location decisions on SC design. In our inventory discussion, we got some sense of how *operating policies* can affect the performance of the SC. The perspective of the entire chapter thus far has really been from the standpoint of a single decision maker.

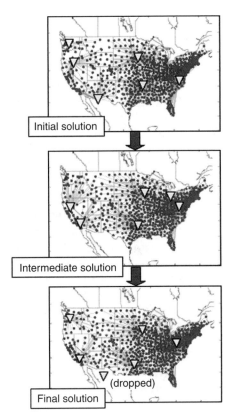

FIGURE 22.20 Series of ALA solutions (from Bucci and Kay, 2006).

An important issue in managing the supply chain, however, is how firms in the supply chain interact with one another. From a supply and demand perspective, the firms in any supply chain interact through orders placed upstream to replenish supply and shipments sent downstream to fulfill the demand represented by those orders. Vollmann et al. (2005), however, cogently point out that understanding how to manage the supply chain really requires a focus on firm-to-firm interactions in *dyads*. While it is indeed the complex network of these dyads that truly comprises the supply chain, one cannot possibly hope to understand all of the complexity of those many interactions without first stepping back to try and understand the factors that drive each dyad relationship individually.

In the past decade, a number of practitioners and supply chain management researchers have lent a significant attention to the level of variability in the stream of orders placed over time. Empirical evidence (Blinder, 1982, 1986; Lee et al., 1997b; Chen et al., 2000; Callioni and Billington, 2001) shows that replenishment orders demonstrate increasing variability at successive upstream echelons of a supply chain. Lee et al. (1997b) report that Procter & Gamble noted this phenomenon, called the *bullwhip effect*, in its supply chain for Pampers diapers. Underlying demand for diapers over a relatively short period of time, say a few months, tends to be quite stable, as one would predict—babies clearly go through them at a pretty steady rate, as any parent knows first-hand! However, P&G noticed that, although underlying consumer demand for Pampers was stable, the orders from retailers to their wholesalers were more volatile than consumer demand, while orders from wholesalers to P&G were more volatile still, and orders from P&G to its suppliers were more volatile still.

Indeed, the bullwhip effect is formally defined by this phenomenon, an increasing level of variability in orders at successive upstream echelons of the supply chain.

Although the effect has been the subject of much discussion in the supply chain management literature since, roughly, the mid-1990s, the effect has been studied quite a bit longer than this. As we indicated above, economists began to study the effect and its origins in earnest in the mid-1980s (Blinder, 1982, 1986; Caplin, 1985; Khan, 1987). Moreover, the famous systems dynamics studies of Forrester (1961) ultimately have had a tremendous impact on supply chain management education, with scores of students each academic year playing the famous "Beer Game" that resulted from follow-on work to Forrester's by Sterman (1989). In playing the Beer Game, students run a four-echelon supply chain for "beer" (typically, only coins that represent cases of beer—to the great relief of many university administrators, and to the great dismay of many students), ultimately trying to fulfill consumer demand at the retailer by placing orders with upstream partners, eventually culminating in the brewing of "beer" by a factory at the farthest upstream echelon in the chain. Indeed, the author's experience in playing the game with business students at all levels—undergraduates, MBAs, and seasoned managers—has never failed to result in bullwhip effects in the chains playing the game, sometimes in a quite pronounced fashion.

A natural question, and one that researchers continue to debate, is what causes this effect. If the underlying demand is relatively stable, and orders are meant ultimately to replenish this underlying demand, why should orders not also be relatively stable? The astute reader may quickly identify an issue from our discussion of the inventory above, namely that the economics of the ordering process might dictate that orders be batched in order to, reduce the administrative costs of placing orders or the scale-driven costs of transporting order fulfillment quantities from the supplier to the customer. Indeed, Lee et al. (1997a,b) identify order batching as one of four specific causes of the bullwhip effect, and the economic studies of the 1980s (Blinder, 1982, 1986; Caplin, 1985) clearly note this cause. The other causes as described by Lee et al. are demand forecast updating, price fluctuations, and supply rationing and shortage gaming. The last two effects are fairly straightforward to understand, namely that a customer might respond to volume discounts or short-term "special" prices by ordering out-of-sync with demand, or that a customer might respond to allocations of scarce supply by placing artificially-inflated orders to attempt to secure more supply than other customers. For the purposes of understanding reasonably simple dyad relationships, we will focus only on order batching and demand forecast updating, leaving the complexities of pricing and shortage gaming outside the scope of our discussion.

Indeed, the method used to generate the demand forecast at each echelon in the supply chain can have a significant impact on the stream of orders placed upstream. Vollmann et al. (2005) provide a simple example to demonstrate the existence of the bullwhip effect; this example is repeated and extended in Table 22.14 and Figure 22.21. We extend what is a 10-period example from Vollmann et al. to 20 periods that cover two repeating cycles of a demand stream with slight seasonality. In addition, we institute a few policy rules: (1) that orders must be non-negative, (2) that shortages at the manufacturer result in lost consumer demand, (3) that shortages at the supplier are met by the manufacturer with a supplemental source of perfectly reliable, immediately available supply. The example of Vollman et al. employs a relatively simple, and essentially *ad hoc*, forecast updating and ordering policy at each level in the chain: in any period t, order $2D_t - I_t$, where D_t is actual demand in period t and I_t is the ending inventory in period t, computed after fulfilling as much of D_t as possible from the beginning inventory. Thus, the current demand is used as the forecast of future demand, and as the replenishment lead time is one period (i.e., an order placed at the end of the current period arrives at the beginning of the next period—one could obviously

Supply Chain Management

TABLE 22.14 Bullwhip Example from Vollmann et al. (2005) Altered for Minimum Order Size of 0 and Horizon of 20 Periods

Period	Consumer Sales	Manufacturer				Supplier			
		Beg Inv	End Inv	Lost Dmd	Order	Beg Inv	End Inv	Lost Dmd	Order
1	50	100	50	0	50	100	50	0	50
2	55	100	45	0	65	100	35	0	95
3	61	110	49	0	73	130	57	0	89
4	67	122	55	0	79	146	67	0	91
5	74	134	60	0	88	158	70	0	106
6	67	148	81	0	53	176	123	0	0
7	60	134	74	0	46	123	77	0	15
8	54	120	66	0	42	92	50	0	34
9	49	108	59	0	39	84	45	0	33
10	44	98	54	0	34	78	44	0	24
11	50	88	38	0	62	68	6	0	118
12	55	100	45	0	65	124	59	0	71
13	61	110	49	0	73	130	57	0	89
14	67	122	55	0	79	146	67	0	91
15	74	134	60	0	88	158	70	0	106
16	67	148	81	0	53	176	123	0	0
17	60	134	74	0	46	123	77	0	15
18	54	120	66	0	42	92	50	0	34
19	49	108	59	0	39	84	45	0	33
20	44	98	54	0	34	78	44	0	24
Min	44		38		34		6		0
Max	74		81		88		123		118
Avg	58.1		58.7		57.5		60.8		55.9
Fill rate				100.0%				100.0%	
Range	30				54				118
Std Dev	9.16				17.81				39.04
				Ratio	1.94			Ratio	4.26

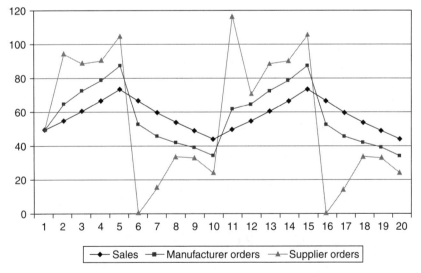

FIGURE 22.21 Bullwhip example from Vollmann et al. (2005) with horizon of 20 periods.

debate whether this a lead time of zero), the policy amounts to carrying a period's worth of safety stock.

We measure the severity of the bullwhip effect by taking the ratio of the standard deviation of the orders placed at each echelon to the standard deviation of consumer sales. As one can see, the example in Table 22.14 and Figure 22.21 demonstrates a clear bullwhip effect, with manufacturer orders exhibiting about twice the volatility of consumer sales and supplier orders exhibiting more than four times the volatility. One conclusion that we could

TABLE 22.15 Bullwhip Example with Two-Period Lead Time

Period	Consumer Sales	Manufacturer				Supplier			
		Beg Inv	End Inv	Lost Dmd	Order	Beg Inv	End Inv	Lost Dmd	Order
1	50	100	50	0	50	100	100	–	0
2	55	50	0	5	110	100	50	0	170
3	61	50	0	11	122	50	0	60	244
4	67	110	43	0	91	170	48	0	134
5	74	165	91	0	57	292	201	0	0
6	67	182	115	0	19	335	278	0	0
7	60	172	112	0	8	278	259	0	0
8	54	131	77	0	31	259	251	0	0
9	49	85	36	0	62	251	220	0	0
10	44	67	23	0	65	220	158	0	0
11	50	85	35	0	65	158	93	0	37
12	55	100	45	0	65	93	28	0	102
13	61	110	49	0	73	65	0	0	146
14	67	114	47	0	87	102	29	0	145
15	74	120	46	0	102	175	88	0	116
16	67	133	66	0	68	233	131	0	5
17	60	168	108	0	12	247	179	0	0
18	54	176	122	0	0	184	172	0	0
19	49	134	85	0	13	172	172	0	0
20	44	85	41	0	47	172	159	0	0
Min	44		0		0		0		0
Max	74		122		122		278		244
Avg	58.1		59.55		57.4		130.8		55.0
Fill rate				98.6%				94.8%	
Range	30				122				244
Std Dev	9.16				35.23				77.32
				Ratio	3.85			Ratio	8.44

draw from this example is that an ad hoc ordering policy can cause a bullwhip effect. Fair enough—but an important aspect of this conclusion is that many supply chains are, in practice, driven by similarly ad hoc ordering policies, an observation noted by the author in conversations with practicing managers and also supported by Chen et al. (2000).

Thus, even without the type of complexities in the demand stream such as the serial correlation in demand studied by Khan (1987), Lee et al. (1997a), and Chen et al. (2000), our simple example shows the bullwhip effect in action. Chen et al. (2000), however, demonstrate a key point—that the magnitude of the bullwhip effect can be shown to be a function, indeed in their case a *super-linear* function, of the replenishment lead time. Thus, let us alter the Vollmann et al. (2005) example to consider a situation in which the replenishment lead time is *two* periods—that is, where an order placed at the end of period t arrives at the beginning of period $t+2$. Using the same ad hoc ordering policy as above, the results are shown in Table 22.15 and Figure 22.22. Clearly, the bullwhip effect gets dramatically worse in this case, essentially doubling the variability in orders at both echelons with respect to consumer demand.

Now let's try some simple things to mitigate the bullwhip effect in this example. First, we will use a base-stock ordering policy at both the manufacturer and the supplier. As the original example did not specify an ordering cost, we will assume that this cost is zero or negligible. Thus, for *stationary* demand, a stationary base-stock policy would be warranted; however, in our example, demand is not stationary. Let us therefore use an "updated base-stock" policy, where the base-stock level will be updated periodically, in this case every five periods. We will assume that the manufacturer bases its base stock policy on a forecast of demand, with a forecasted mean and standard deviation of per-period demand of $\mu_1^M = 65$ and $\sigma_1^M = 10$ in the periods 1–5 and $\mu_2^M = 55$ and $\sigma_2^M = 10$ in the periods 6–10. These parameters are also used in the second demand cycle, with the higher mean being used as the demand rises in periods 11–15 and the lower mean as demand falls in periods 16–20. We will assume that the supplier uses a forecast of the manufacturer's orders to compute

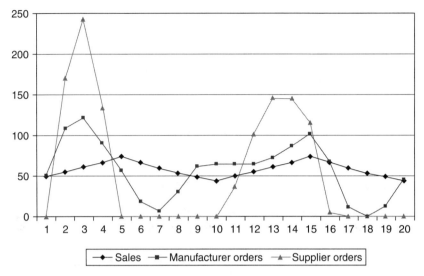

FIGURE 22.22   Bullwhip example with two-period lead time.

TABLE 22.16   Bullwhip Example with Updated Base-Stock Orders

Period	Consumer Sales	Manufacturer					Supplier				
		Beg Inv	End Inv	Lost Dmd	B	Order	Beg Inv	End Inv	Lost Dmd	B	Order
1	50	100	50	0	83	33	100	100	—	96	0
2	55	50	0	5	83	83	100	67	0	96	29
3	61	33	0	28	83	83	67	0	16	96	96
4	67	83	16	0	83	67	29	0	54	96	96
5	74	99	25	0	83	58	96	29	0	96	67
6	67	92	25	0	73	48	125	67	0	102	35
7	60	83	23	0	73	50	134	86	0	102	16
8	54	71	17	0	73	56	121	71	0	102	31
9	49	67	18	0	73	55	87	31	0	102	71
10	44	74	30	0	73	43	62	7	0	102	95
11	50	85	35	0	83	48	78	35	0	60	25
12	55	78	23	0	83	60	130	82	0	60	0
13	61	71	10	0	83	73	107	47	0	60	13
14	67	70	3	0	83	80	47	0	26	60	60
15	74	76	2	0	83	81	13	0	67	60	60
16	67	82	15	0	73	58	60	0	21	94	94
17	60	96	36	0	73	37	60	2	0	94	92
18	54	94	40	0	73	33	96	59	0	94	35
19	49	77	28	0	73	45	151	118	0	94	0
20	44	61	17	0	73	56	153	108	0	94	0
Min	44		0			33		0			0
Max	74		50			83		118			96
Avg	58.1		20.65			57.35		45.45			45.8
Fill rate				97.2%					84.0%		
Range	30					50					96
Std Dev	9.16					16.11					36.22
				Ratio		1.76			Ratio		3.95

its base-stock level in the first five periods, with $\mu_1^S = 60$ and $\sigma_1^S = 20$. In the next three five-period blocks, the supplier uses actual orders from the manufacturer to compute the mean and standard deviation of demand, so that $(\mu_i^S, \sigma_i^S) = \{(64.8, 20.8), (50.4, 5.3), (68.4, 14.2)\}$, $i = 2, 3, 4$. At both echelons, therefore, the base-stock policy is computed as $B_i = \mu_i + z\sigma_i\sqrt{2}$ ($i = 2, 3, 4$), with $z = 1.282$, yielding a target cycle service level of 90%, and $\sigma_{DLT,i} = \sigma_i\sqrt{2}$. As one can see from Table 22.16 and Figure 22.23, the base-stock policies work quite well in reducing the bullwhip effect, but at the expense of lower fill rates at both the manufacturer

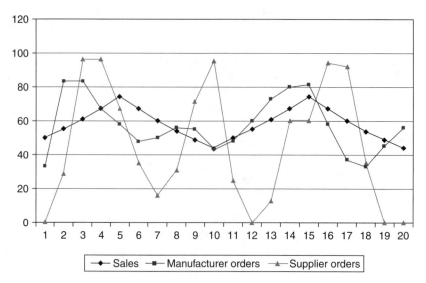

FIGURE 22.23 Bullwhip example with updated base-stock orders.

and the supplier. (Note that a higher CSL target does not have much impact on this outcome.)

One key prescription of the literature on the bullwhip effect is to centralize demand information so that demand forecasts are not inferred from the orders of the echelon immediately downstream—the essence of the "demand forecast updating" (Lee et al., 1997b) cause of the bullwhip effect—but instead come directly from consumer sales data. Therefore, let us update our example to account for this. We will assume the same base-stock policy at the manufacturer as above, based on its forecasts of rising and falling consumer demand. At the supplier, however, we will assume that the initial base stock level is computed directly from the manufacturer's forecast so that $\mu_1^S = 65$ and $\sigma_1^S = 10$. In the succeeding five-period blocks, the supplier updates its base-stock levels using actual consumer sales data, such that $(\mu_i^S, \sigma_i^S) = \{(61.4, 9.5), (54.8, 9.0), (61.4, 9.5)\}$, $i = 2, 3, 4$. As proved in the literature (Chen et al., 2000), this will mitigate the bullwhip effect at the supplier, as our results in Table 22.17 and Figure 22.24 demonstrate, but only slightly, and it certainly does not eliminate the effect. Also, the supplier fill rate in this latter case improves from the previous base-stock case, but only slightly.

Finally, let us consider the effect of batch ordering at the manufacturer. Let us assume that the manufacturer has occasion to order in consistent lots, with $Q = 150$. We further assume that the manufacturer follows a (Q, R) policy with R set to the base-stock levels from the prior two examples. In contrast, we assume that the supplier has no need to batch orders and follows a base-stock policy, with $\mu_1^S = 100$ and $\sigma_1^S = 30$, set in anticipation of the large orders from the manufacturer. The results are shown in Table 22.18 and Figure 22.25, in which the supplier's base-stock levels in the remaining five-period blocks are updated using the manufacturer orders, resulting in $(\mu_i^S, \sigma_i^S) = (60.0, 82.2)$, $i = 2, 3, 4$, as each block contains exactly two orders from the manufacturer. (Note that the assumption of a normal distribution for lead-time demand probably breaks down with such a large standard deviation, due to the high probability of negative values, but we retain this assumption nonetheless.) From the table and figure, one can see that the variability in the order stream increases dramatically at the manufacturer. Interestingly, however, the significant increase in variability is not further amplified at the supplier when it uses a base-stock policy.

Supply Chain Management

TABLE 22.17 Bullwhip Example with Updated Base-Stock Orders Using Consumer Demand

Period	Consumer Sales	Manufacturer					Supplier				
		Beg Inv	End Inv	Lost Dmd	B	Order	Beg Inv	End Inv	Lost Dmd	B	Order
1	50	100	50	0	83	33	100	100	–	83	0
2	55	50	0	5	83	83	100	67	0	83	16
3	61	33	0	28	83	83	67	0	16	83	83
4	67	83	16	0	83	67	16	0	67	83	83
5	74	99	25	0	83	58	83	16	0	83	67
6	67	92	25	0	73	48	99	41	0	79	38
7	60	83	23	0	73	50	108	60	0	79	19
8	54	71	17	0	73	56	98	48	0	79	31
9	49	67	18	0	73	55	67	11	0	79	68
10	44	74	30	0	73	43	42	0	13	79	79
11	50	85	35	0	83	48	68	25	0	71	46
12	55	78	23	0	83	60	104	56	0	71	15
13	61	71	10	0	83	73	102	42	0	71	29
14	67	70	3	0	83	80	57	0	16	71	71
15	74	76	2	0	83	81	29	0	51	71	71
16	67	82	15	0	73	58	71	0	10	79	79
17	60	96	36	0	73	37	71	13	0	79	66
18	54	94	40	0	73	33	92	55	0	79	24
19	49	77	28	0	73	45	121	88	0	79	0
20	44	61	17	0	73	56	112	67	0	79	12
Min	44		0			33		0			0
Max	74		50			83		100			83
Avg	58.1		20.65			57.35		34.45			44.85
Fill rate				97.2%					84.9%		
Range	30					50					83
Std Dev	9.16					16.11					29.47
				Ratio		1.76			Ratio		3.22

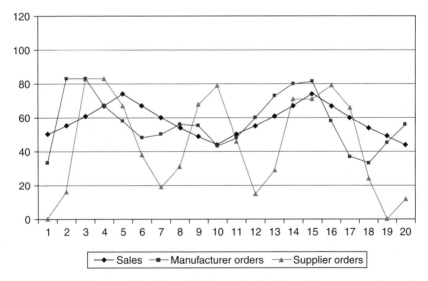

FIGURE 22.24 Bullwhip example with updated base-stock orders using consumer demand.

These examples are not intended to prove any general results, per se, but merely to demonstrate the effect of ordering policies on supply chain performance and to emphasize the challenge of identifying policy prescriptions to mitigate these effects. Note that Lee et al. (1997a,b) lay out a more extensive set of prescriptions that may be used in practice to attempt to address the full set of causes for the bullwhip effect—order batching, forecast updating, price fluctuations, and shortage gaming. From our examples, interestingly, the fill rate performance is actually best with the ad hoc policy from the original example

TABLE 22.18 Bullwhip Example with (Q, R) Orders at Manufacturer and Base-Stock Orders at Supplier

	Consumer	Manufacturer					Supplier				
Period	Sales	Beg Inv	End Inv	Lost Dmd	R	Order	Beg Inv	End Inv	Lost Dmd	B	Order
1	50	100	50	0	83	150	100	100	–	154	54
2	55	50	0	5	83	150	100	0	50	154	154
3	61	150	89	0	83	0	54	0	96	154	154
4	67	239	172	0	83	0	154	154	0	154	0
5	74	172	98	0	83	0	308	308	0	154	0
6	67	98	31	0	73	150	308	308	0	209	0
7	60	31	0	29	73	150	308	158	0	209	51
8	54	150	96	0	73	0	158	8	0	209	201
9	49	246	197	0	73	0	59	59	0	209	150
10	44	197	153	0	73	0	260	260	0	209	0
11	50	153	103	0	83	0	410	410	0	209	0
12	55	103	48	0	83	150	410	410	0	209	0
13	61	48	0	13	83	150	410	260	0	209	0
14	67	150	83	0	83	0	260	110	0	209	99
15	74	233	159	0	83	0	110	110	0	209	99
16	67	159	92	0	73	0	209	209	0	209	0
17	60	92	32	0	73	150	308	308	0	209	0
18	54	32	0	22	73	150	308	158	0	209	51
19	49	150	101	0	73	0	158	8	0	209	201
20	44	251	207	0	73	0	59	59	0	209	150
Min	44		0			0		0			0
Max	74		207			150		410			201
Avg	58.1		85.55			60.0		169.85			68.2
Fill rate				94.1%					87.8%		
Range	30					150					201
Std Dev	9.16					75.39					75.37
				Ratio		8.23			Ratio		8.23

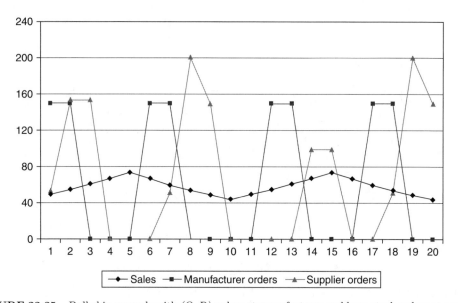

FIGURE 22.25 Bullwhip example with (Q, R) orders at manufacturer and base-stock orders at supplier.

in Vollmann et al. (2005). This is important, in that lost demand at the supplier in our example is a proxy for additional cost at the manufacturer to supplement supply. (A more complex assumption would be to allow backorders at the supplier, thus deflating the service performance at the manufacturer further and likely exacerbating the bullwhip effect at the supplier.) Our example might really speak to a need for a richer model of the interaction between customer and supplier, especially in a case where there is an underlying pattern to demand, and particularly in the case where the customer has access to more information about consumer demand than the supplier could infer on its own simply from the customer's stream of orders. Indeed, in practice, this has led to burgeoning efforts such as collaborative planning, forecasting, and replenishment (CPFR).*

Lest one conclude that sharing forecast information *always* improves the outcome for all parties in the supply chain, let us briefly consider some other research in progress that identifies cases when forecast sharing is helpful, and when it might actually make things worse. Thomas et al. (2005) offer a richer model of a specific dyad relationship than the simple ordering relationships we consider above. Their model considers a contract manufacturer (CM) that builds a product for an original equipment manufacturer (OEM). The product is composed of two sets of components, a long lead time set and a shorter lead time set, such that orders for the long lead time components must be committed by the CM to its supplier some time before orders are placed for the short lead time components. In the intervening period between these lead times, the OEM may be able to provide an updated forecast of the single-period consumer demand to the CM. In the base model, the only cost to the OEM is the opportunity cost of lost sales if the CM does not order sufficient quantities of both component sets to assemble complete products and fulfill consumer demand for the product. The CM, however, bears both a lost-sales cost, based on its share of the product profit margin, *and* the overage loss related to having components left over in excess of demand. Since the CM bears the costs of both underage and overage, it may tend to be risk averse in its ordering, particularly when its profit margin is relatively small, meaning that the OEM could benefit by offering contract terms to share a portion of the overage cost with the CM.

With D denoting demand, ξ the demand-related information update, and γ_1 and γ_2 the fraction of overage loss the OEM bears on the long lead time and short lead time components, respectively, the OEM's cost function in Thomas et al. (2005) is

$$TC_{\text{fs}}^{\text{OEM}}(\gamma_1, \gamma_2) = E_\xi[E_{D|\xi}[m_o(D - Q_2^*)^+ + \gamma_1 l_1(Q_1^* - \min\{D, Q_2^*\})^+ + \gamma_2 l_2(Q_2^* - D)^+]] \tag{22.37}$$

where m_o is the OEM's share of the product margin, l_1 and l_2 are the overage losses (unit cost minus salvage value) on the long lead time and short lead time components, respectively, and Q_1^* and Q_2^* are the CM's optimal ordering decisions (for long and short lead time components, respectively), given the forecast sharing and risk sharing decisions of the OEM. The CM's optimal ordering decisions come from solving

$$TC_{\text{fs}}^{\text{CM}}(Q_1) = E_\xi[E_{D|\xi}[m_c(D - Q_2^*)^+ + (1 - \gamma_1)l_1(Q_1 - \min\{D, Q_2^*\})^+ + (1 - \gamma_2)l_2(Q_2^* - D)^+]] \tag{22.38}$$

where m_c is the CM's share of the product margin and

$$Q_2^* = \min\{F_{D|\xi}^{-1}(\alpha_2), Q_1\} \tag{22.39}$$

with $F_{D|\xi}$ being the cdf of D given ξ and with α_2 given by

$$\alpha_2 = \frac{m_c + (1 - \gamma_1)l_1}{m_c + (1 - \gamma_1)l + (1 - \gamma_1)l_1} \tag{22.40}$$

If the demand forecast update is not shared, Equations 22.37 and 22.38 are simplified, with the expectation taken over the random variable D alone. Using a model that employs nested

---

*See the Web site of the Voluntary Interindustry Commerce Standards (VICS) organization at http://www.vics.org/committees/cpfr/ for more details on this set of practices.

TABLE 22.19 OEM-CM Cost Reductions from Forecast and Risk Sharing, Thomas et al. (2005)

			No Risk Sharing (%)	Limited Risk Sharing (%)	Complete Risk Sharing (%)
Reduction in TC^{OEM}	No forecast sharing	Mean	–	4.60	6.87
		Max	–	55.23	68.90
	Forecast sharing	Mean	8.69	12.83	15.80
		Max	43.75	50.17	72.97
Reduction in TC^{CM}	No forecast sharing	Mean	–	6.63	10.90
		Max	–	58.97	78.30
	Forecast sharing	Mean	3.16	9.19	14.28
		Max	17.49	49.05	79.87

uniform distributions for demand and demand-related information, Thomas et al. are able to express solutions for Q_1^* and Q_2^* in closed-form. They also demonstrate that this nested-uniform demand model is a very good approximation to the case where the demand forecast and the related updating information form a bivariate normal pair.

Thomas et al. consider different contracts for sharing overage risk, with the extremes being no risk sharing ($\gamma_1 = \gamma_2 \equiv 0$), limited risk sharing (only on the long lead time components—i.e., $\gamma_2 \equiv 0$), and complete risk sharing (on both long and short lead time components). The interaction of forecast sharing and risk sharing creates some interesting results in this relatively complex situation involving components with a complementary relationship—noting that the CM's order for short lead time components is bounded above by its order for long-lead time components, per Equation 22.39 above. Table 22.19 shows the comparative reductions in OEM and CM costs, on average, of sharing forecast information and sharing overage risk from the base case of "no forecast sharing—no risk sharing" across 18,225 computational experiments with m_o and m_c ranging from 0.1 to 1.5 and l_1 and l_2 ranging from 10% to 90% of unit cost. Interestingly, 12.5% of the computational experiments under limited risk sharing exhibit *increased* cost for the OEM—by an average of about 4% and as much as 12%—when the forecast update is shared with the CM. This effect is completely mitigated, however, when the OEM shares the overage risk on both sets of components.

22.6 Discussion and Conclusions

Let us now step back and tie together the themes and ideas presented in this chapter. An important theme, and one that appeared throughout the chapter, is that supply chain management strategy is framed by three important decision areas—inventory, transportation, and locations—and their joint impact on customer service. Moreover, we can now turn that statement around and state that it is the *interaction* of the inventory, transportation, and location decisions—and their joint impact on customer service—that frame the design of the supply chain itself. Indeed, it is the author's view that SC design initiatives that are undertaken without a clear understanding and appreciation of those interactions will generate poor results, at best, and may be doomed to spectacular failure, at worst.

Another theme, albeit one that appeared late in the chapter and to which a much smaller number of pages has been devoted, is the effect of SCM *policies* on the operation of the supply chain. Policies in SCM take many forms, and many IE and operations management researchers have devoted significant time and effort to the study of inventory management policies. Indeed, a relatively wide-ranging chapter on SCM such as this one could not do justice to the study of inventory management, about which many books have been written. However, my hope in this chapter was to take at least a small step beyond inventory policies and also consider coordinating mechanisms like the sharing of overage risk that help to better align the incentives of the parties in a supply chain. This theme focuses on the *dyad relationships* in the supply chain, the point at which any two parties in the chain interact, and, in the author's mind, the *only* point at which one can hope to study supply chain management—as opposed to making the bold and rather far-fetched claim that one could, for example, "optimize the management of the supply chain," whatever that might mean. Similar to the manner in which the decisions and interactions of the "triangle-leg decisions" (inventory, transportation, and location) frame SC design, it should be clear that an understanding and appreciation of the interaction of dyad partners frames the design of SCM policies.

So, we return to the "guy walks into a bar" scenario—or at least the "software salesperson walks into the IE's office" scenario—that opened the chapter. Could one "optimize the supply chain" with software? Certainly not with a single piece of software solving a single problem. After all of the discussion presented in this chapter, my sincere hope is that the reader has an appreciation for the significant scope, complexity, and subtlety of the types of problems underlying the relatively large domain that is supply chain management. Could one, however, optimize portions of the supply chain?... individually optimize the various decision problems that together frame SCM?... find policies that minimize or jointly improve the independent objective functions of dyad partners? Certainly, all of those things can be done. Will they, together, represent a "supply chain optimal" solution? That seems like a loaded question, and one that I'll avoid answering.

Perhaps the discussion at the outset of the chapter was a bit too critical, however. Softening that criticism somewhat, I will close by noting the relatively recent advent of advanced planning systems (APS) software, which seems to indicate that the marketplace for SCM software recognizes that "optimizing" the supply chain in one fell swoop would be a Herculean, if not impossible, task. Indeed, referring back to Figure 22.2 in the Introduction to this chapter, one is reminded that at least one company that does business by attempting to create and implement supply chain (re)design initiatives does so in modular fashion, employing tools that solve a variety of discrete problems in the process of building an overall SC design solution. An excellent reference on this subject is the book chapter by Fleischmann and Meyr (2003), which includes a mapping of their conceptual model of APS activities to the APS software offerings of the major software vendors in this market—namely Baan, i2 Technologies, J.D. Edwards, Manugistics, and SAP. Each company offers its software in a series of modules that focus on discrete problem areas—for example, forecasting and demand management, transportation management, inventory management—very much along the lines of the discussion in this chapter.

Indeed, supply chain management is a wide-ranging and challenging subject. Solving the decision problems that frame SCM requires a similarly wide-ranging knowledge and an appreciation of the manner in which the various aspects of SC design solutions and SCM policy solutions interact to impact company objectives like minimizing operating costs at acceptably high levels of customer service. However, armed by this chapter with an appreciation of the challenges implied by designing and managing—let alone *optimizing*—the supply chain, my sincere hope is that you, the reader, won't be tempted to rush into buying that bridge in Brooklyn.

References

1. Atkinson, W. (2002), DC siting—What makes most sense?, *Logistics Management and Distribution Report*, May issue, 41(5), S63–S65.
2. Axsäter, S. (2000), *Inventory Control*, Boston: Kluwer Academic.

3. Balintfy, J. (1964), On a basic class of multi-item inventory problems, *Management Science*, 10(2), 287–297.
4. Ballou, R.H. (1991), The accuracy in estimating truck class rates for logistical planning, *Transportation Research—Part A*, 25A(6), 327–337.
5. Ballou, R. H. (1992), *Business Logistics Management*, 3rd ed., Englewood Cliffs, NJ: Prentice Hall.
6. Ballou, R. H. (2004), *Business Logistics/Supply Chain Management*, 5th ed., Upper Saddle River, NJ: Pearson Prentice Hall.
7. Baumol, W. J. and H. D. Vinod (1970), An inventory theoretic model of freight transport demand, *Management Science*, 16(7), 413–421.
8. Blinder, A. S. (1982), Inventories and sticky prices: More on the microfoundations of macroeconomics, *American Economic Review*, 72(3), 334–348.
9. Blinder, A. S. (1986), Can the production smoothing model of inventory behavior be saved?, *Quarterly Journal of Economics*, 101(3), 431–453.
10. Bozarth C. C. and D. P. Warsing (2006), *The Evolving Role of Manufacturing Planning and Control Systems in Supply Chains*, College of Management working paper, North Carolina State University, Raleigh, NC.
11. Bozarth, C. C. and R. B. Handfield (2006), *Introduction to Operations and Supply Chain Management*, Upper Saddle River, NJ: Pearson Prentice Hall.
12. Bozarth. C. C. (2005), Executive education session notes on *Inventory Operations*, North Carolina State University, Raleigh, NC.
13. Bryksina, E. A. (2005), *Assessing the impact of strategic safety stock placement in a multi-echelon supply chain*, M.S. thesis, North Carolina State University, Raleigh, NC.
14. Bucci, M. J. and M. G. Kay (2006), A modified ALA procedure for logistic network designs with scale economies, *Proceedings of the Industrial Engineering Research Conference*, Orlando, FL.
15. Cachon, G. and C. Terwiesch (2006), *Matching Supply with Demand: An Introduction to Operations Management*, New York: McGraw-Hill/Irwin.
16. Callioni, G. and C. Billington (2001), Effective collaboration, *OR/MS Today*, October issue, 28(5), 4–39.
17. Caplin, A. S. (1985), The variability of aggregate demand with (S,s) inventory policies, *Econometrica*, 53(6), 1395–1410.
18. Çentinkaya, S. and C. Y. Lee (2002), Optimal outbound dispatch policies: Modeling inventory and cargo capacity, *Naval Research Logistics*, 49(6), 531–556.
19. Chen, F., Z. Drezner, J. K. Ryan, and D. Simchi-Levi (2000), Quantifying the bullwhip effect in a simple supply chain: The impact of forecasting, lead times, and information, *Management Science*, 46(3), 436–443.
20. Chopra, S. and P. Meindl (2001),*Supply Chain Management: Strategy, Planning, and Operation*, Upper Saddle River, NJ: Pearson Prentice Hall.
21. Chopra, S. and P. Meindl (2004), *Supply Chain Management: Strategy, Planning, and Operation*, 2nd ed., Upper Saddle River, NJ: Pearson Prentice Hall.
22. Clark, A. J. and H. Scarf (1960), Optimal policies for a multi-echelon inventory problem, *Management Science*, 6(4), 475–490.
23. Cooper, L. (1963), Location-allocation problems, *Operations Research*, 11(3), 331–343.
24. Cooper, L. (1964), Heuristic methods for location-allocation problems, *SIAM Review*, 6(1), 37–53.
25. Cox, J. B. (2005), They got Dell. Now what?, *The News & Observer*, October 9 edition, Raleigh, NC.

26. Coyle J. J., E. J. Bardi, and C. J. Langley, Jr. (2003), *The Management of Business Logistics*, 7th ed., Mason, OH: South-Western.
27. Coyle, J. J., E. J. Bardi, and R. A. Novack (2006), *Transportation*, 6th ed., Mason, OH: Thomson–South-Western.
28. Craver, R. (2005), Dell computer-assembly plant opens in Forsyth County, N.C., *Winston-Salem Journal*, October 5 edition, Winston-Salem, NC.
29. Ettl, M., G. E. Feigin, G. Y. Lin, and D. D. Yao (1997), *A supply network model with base-stock control and service requirements*, IBM Corporation technical report RC 20473, Armonk, NY.
30. Ettl, M., G. E. Feigin, G. Y. Lin, and D. D. Yao (2000), A supply network model with base-stock control and service requirements, *Operations Research*, 48(2), 216–232.
31. Feigin, G. E. (1999), Inventory planning in large assembly supply chains, in *Quantitative Models for Supply Chain Management*, S. Tayur, R. Ganeshan, and M. Magazine, Eds., Boston: Kluwer Academic Publishers.
32. Feitzinger, E. and H. L. Lee (1997), Mass customization at Hewlett-Packard: The power of postponement, *Harvard Business Review*, 75(1), 116–121.
33. Fisher, M. L. (1997), What is the right supply chain for your product?, *Harvard Business Review*, 75(2), 105–116.
34. Fleischmann, B. and H. Meyr (2003), Planning hierarchy, modeling and advanced planning systems, in *Handbooks in OR & MS, Vol. 11*, A. de Kok and S. Graves, Eds., Amsterdam: Elsevier.
35. Forrester, J. W. (1961), *Industrial Dynamics*, Cambridge, MA: MIT Press.
36. Graves, S. and S. Willems (2003), Supply chain design: Safety stock placement and supply chain configuration, in *Handbooks in OR & MS, Vol. 11*, A. de Kok and S. Graves, Eds., Amsterdam: Elsevier.
37. Harvard Business School (1998), *Note on the U.S. Freight Transportation Industry*, # 9-688-080, Boston, MA: Harvard Business School Publishing.
38. Hax, A. C. and D. Candea (1984), *Production and Inventory Management*, Englewood Cliffs, NJ: Prentice-Hall.
39. Holt, C. C., F. Modigliani, J. F. Muth, and H. A. Simon (1960), *Planning Production, Inventories, and Work Force*, Englewood Cliffs, NJ: Prentice-Hall.
40. Hopp, W. J. and M. L. Spearman (2004), To pull or not to pull: What is the question?, *Manufacturing & Service Operations Management*, 6(2), 133–148.
41. Houck, C. R., J. A. Joines, and M. G. Kay (1996), Comparison of genetic algorithms, random restart, and two-opt switching for solving large location-allocation problems, *Computers & Operations Research*, 23(6): 587–596.
42. Jackson, P., W. Maxwell, and J. Muckstadt (1985), The joint replenishment problem with a powers-of-two restriction, *IIE Transactions*, 17(1), 25–32.
43. Kay, M. G. and D. P. Warsing (2006), *Modeling Truck Rates Using Publicly Available Empirical Data*, North Carolina State University working paper, Raleigh, NC.
44. Khan, J. A. (1987), Inventories and the volatility of production, *American Economic Review*, 77(4), 667–679.
45. Kimball, G. E. (1988), General principles of inventory control, *Journal of Manufacturing and Operations Management*, 1(1), 119–130.
46. Koch, C. (2004), Nike rebounds: How (and why) Nike recovered from its supply chain disaster, *CIO*, 17 (Jun 15 issue).
47. Kulkarni, V. G. (1995), *Modeling and Analysis of Stochastic Systems*, Boca Raton, FL: Chapman & Hall/CRC.

48. Lee, C. Y. (1986), The economic order quantity for freight discount costs, *IIE Transactions*, 18(3), 318–320.
49. Lee, H. L. and C. Billington (1993), Material management in decentralized supply chains, *Operations Research*, 41(5), 835–847.
50. Lee, H. L., V. Padmanabhan, and S. Whang (1997a), Information distortion in a supply chain: The bullwhip effect, *Management Science*, 43(4), 546–558.
51. Lee, H. L., V. Padmanabhan, and S. Whang (1997b), The bullwhip effect in supply chains, *Sloan Management Review*, 38(3), 93–102.
52. Lockamy, A. and K. McCormack (2004), Linking SCOR planning practices to supply chain performance, *International Journal of Operations and Production Management*, 24(11/12), 1192–1218.
53. Marien, E. J. (1996), Making sense of freight terms of sale, *Transportation & Distribution*, 37(9), 84–86.
54. Marien, E. J. (2000), The four supply chain enablers, *Supply Chain Management Review*, 4(1), 60–68.
55. Nahmias, S. (1997), *Production and Operations Analysis*, 3rd ed., Chicago: Irwin.
56. Piercy, J. E. (1977), A Performance Profile of Several Transportation Freight Services, unpublished Ph.D. dissertation, Case Western Reserve University, Cleveland, OH.
57. Reuters (2004), N. Carolina lures new Merck plant with incentives, on-line *Update*, April 6 edition.
58. Rumelt, R. P. (2001), *Note on Strategic Cost Dynamics*, POL 2001-1.2, Anderson School of Management, UCLA, Los Angeles, CA.
59. Schwartz, B. M. (1999a), Map out a site route, *Transportation & Distribution*, 40(11), 67–69.
60. Schwartz, B. M. (1999b), New rules for hot spots, *Transportation & Distribution*, 40(11), 43–45.
61. Silver, E. A., D. F. Pyke, and R. Peterson (1998), *Inventory Management and Production Planning and Scheduling*, 3rd ed., New York: John Wiley & Sons.
62. Speigel, R. (2002), Truckload vs. LTL, *Logistics Management*, 47(7), 54–57.
63. Stedman, C. (1999), Failed ERP gamble haunts Hershey, *Computerworld*, 33(44), 1–2.
64. Sterman, J. D. (1989), Modeling managerial behavior misperceptions of feedback in a dynamic decision making experiment, *Management Science*, 35(3), 321–329.
65. Supply-Chain Council (2006), *Supply-Chain Operations Reference-Model—SCOR Version 8.0 Overview*, Washington, D.C. (available via http://www.supply-chain.org/).
66. Swenseth, S. R. and M. R. Godfrey (1996), Estimating freight rates for logistics decisions, *Journal of Business Logistics*, 17(1), 213–231.
67. Swenseth, S. R. and M. R. Godfrey (2002), Incorporating transportation costs into inventory replenishment decisions, *International Journal of Production Economics*, 77(2), 113–130.
68. Tersine, R. J. and S. Barman (1991), Lot size optimization with quantity and freight rate discounts, *Logistics and Transportation Review*, 27(4), 319–332.
69. Thomas, D. J., D. P. Warsing, and X. Zhang (2005), *Forecast Updating and Supplier Coordination for Complementary Component Purchases*, Smeal College of Business working paper, Pennsylvania State University, University Park, PA.
70. Tyworth, J. E. and A. Ruiz-Torres (2000), Transportation's role in the sole- versus dual-sourcing decision, *International Journal of Physical Distribution and Logistics Management*, 30(2), 128–144.
71. Vollmann, T., W. Berry, D. C. Whybark, and R. Jacobs (2005), *Manufacturing Planning & Control for Supply Chain Management*, 5th ed., New York: McGraw-Hill/Irwin.

72. Waller, M., M. E. Johnson, and T. Davis (1999), Vendor-managed inventory in the retail supply chain, *Journal of Business Logistics*, 20(1), 183–203.
73. Warsing, D. P., E. A. Helmer, and J. V. Blackhurst (2006), Strategic Safety Stock Placement in Production Networks with Supply Risk, *Proceedings of the Industrial Engineering Research Conference*, Orlando, FL.
74. Wilson, R. A. (2003), *Transportation in America*, Washington, D.C.: Eno Transportation Foundation.
75. Winston, W. L. (1994), *Operations Research: Applications and Algorithms*, 3rd ed., Belmont, CA: Duxbury Press.

23

E-Commerce

23.1 Introduction... **23**-1
23.2 Evolution of E-Commerce........................ **23**-3
23.3 OR/MS and E-Commerce......................... **23**-5
23.4 OR Applications in E-Commerce **23**-7
  E-Commerce Relating to Digital Goods •
  E-Commerce Relating to Physical Goods • Electronic
  Delivery of Services • Support for E-Commerce
  Infrastructure
23.5 Tools–Applications Matrix........................ **23**-20
23.6 Way Forward....................................... **23**-21
23.7 Summary **23**-21
References **23**-21
Further Reading **23**-23

Sowmyanarayanan Sadagopan
Indian Institute of Information Technology

23.1 Introduction

Commerce is the exchange of goods and services between a buyer and a seller for an agreed price. The commerce ecosystem starts with a customer who interacts with a *salesperson*, typically at a *storefront*; the *store* gets the supply from a *distributor*; the distribution is typically from a *warehouse* or a *factory*; if the customer decides to buy the goods, the seller *invoices* the buyer for the *price* of goods, including taxes and shipping charges, if any; and finally the invoiced amount is *collected* either immediately or through an intermediary such as a bank or credit company. This commerce ecosystem has evolved over centuries. With the explosion of the Internet since 1995, every aspect of commerce could be supplemented by electronic support leading to an e-commerce ecosystem similar to the commerce eco-system. This includes systems to support customer acquisition, price negotiation, order processing, contracting, order fulfillment, logistics and delivery, bill presentation and payment, and post-sales support [1–3].

> E-commerce or electronic commerce is commerce conducted electronically, typically over the Internet. E-commerce includes support to all stages of commerce.

E-commerce initially was limited to "digital goods"—software, e-books, and information. Thanks to "digitization," today digital goods encompass a large set of items including

- *Computer and telecom software*;
- *Music* with analog recording on tapes evolving to digital music—first as CD audio and currently digital audio in the form of MP3 and other forms like RealAudio;

- *Video* with the evolution of digital cinematography and MPEG compression scheme;
- *Images* including still images captured by digital camera, video images captured by camcorders, document images and medical images like ultrasound image, digital X-ray image, CAT scan, and MRI image;
- *Documents* including books, journals, tech reports, magazines, handbooks, manuals, catalogs, standards documentation, brochures, newspapers, phone books, and a multitude of other "forms" often created using DTP; and
- *Drawings* including building plans, circuit diagrams, and flow diagrams typically created using CAD tools.

Digital goods could be completely delivered electronically, and formed the first focus of e-commerce. Soon it was clear that a whole range of information services lend themselves to electronic delivery. This includes

- *Financial services* that include banking, insurance, stock trading, and advisory services;
- *E-governance* including electronic delivery of justice and legal services, electronic support to legislation and law making, issue of identities like passport and social service numbers, visa processing, licensing for vehicles, toll collection, and tax collection;
- *Education* that includes classroom support, instructional support, library, laboratories, textbooks, conducting examinations, evaluation, and issue of transcripts; and
- *Healthcare delivery* including diagnostics using imaging, record keeping, and prescription handling.

Thanks to the success of digital goods delivery and electronic delivery of information services, e-commerce could address the support issues for buying and selling of physical goods as well. For example,

- *Customer acquisition* through search engine-based advertisements, online promotion, and e-marketing;
- *E-ticketing* for airline seats, hotel rooms, and movie tickets;
- *E-procurement* including e-tendering, e-auction, reverse auction, and electronic order processing;
- *Electronic payment* that includes electronic shopping cart and online payments; and
- *Electronic post-sales support* through status monitoring, tracking, health-checking, and online support.

To support digital goods delivery, delivery of information services, and support commerce related to physical goods, a huge Internet infrastructure had to be built (and continues to get upgraded). Optimizing this infrastructure that converged data communications with voice communications on the one hand and wire-line with wireless communications on the other hand is leading to several interesting challenges that are being addressed today. Behind those challenges are the key techniques and algorithms perfected by the OR/MS community over the past five decades [4–6]. In this chapter, we will address the key OR/MS applications relating to the distribution of digital goods, electronic delivery of information services, and optimizing the supporting Internet infrastructure.

23.2 Evolution of E-Commerce

Electronic commerce predates the widespread use of the Internet. Electronic data interchange (EDI) used from the early 1960s did facilitate transfer of commercial data between seller and buyer. The focus, however, of EDI was mostly to clear congestion of goods movement at seaports and airports due to rapid globalization of manufacturing, improved transportation, and the emergence of global logistics hubs such as Rotterdam and Singapore. EDI helped many governments in the process of monitoring of export and import of goods between countries and also control the movement of undesired goods such as drugs and narcotics across borders [1,2,7–15].

Less widely known e-commerce applications in the pre-Internet era include

- *Online databases* like *MEDLARS* for medical professionals and *Lexis-Nexis* for legal professionals;
- *Electronic funds transfer* such as SWIFT network; and
- *Online airline tickets reservations* such as SABRE from American Airlines.

In all these cases electronic data transfer was used to facilitate business between *partners*:

- Pre-determined intermediaries such as port authorities, customs inspectors, and export-import agencies in the case of EDI;
- Librarians and publishing houses in the case of online databases;
- Banks and banking networks in the case of electronic funds transfer; and
- Airlines and travel agents in the case of online reservations systems.

The ordering of books online on Amazon.com in 1994 brought e-commerce to the end users. One could order books electronically, and by the year 2000 make payments also electronically through a secure payment gateway that connects the buyer to a bank or a credit company. All this could be achieved from the desktop computer of an end user, a key differentiator compared to the earlier generation of electronic commerce. Such a development that permitted individual buyers to complete end-to-end buying (and later selling through eBay) made e-commerce such a revolutionary idea in the past decade.

The rise of digital goods (software, music, and e-books), the growth of the Internet, and the emergence of the "Internet-economy" companies—Yahoo and eBay—led to the *second stage* of e-commerce. Software was the next commodity to embrace e-commerce, thanks to iconic companies like Microsoft, Adobe, and Macromedia. Selling software over the Internet was far more efficient than any other means, particularly for many small software companies that were writing software to run on Wintel (Windows + Intel) platform (as PCs powered by Intel microprocessors were a dominant hardware platform and Windows dominated the desktop software platform). The outsourcing of software services to many countries including India gave further impetus to e-commerce.

The spectacular growth of the Internet and World Wide Web (WWW) fueled the demand for widespread access and very large network-bandwidth, leading to phenomenal investment in wire-line (fiber optic networks and undersea cables) and wireless networks in the past decade (the trend is continuing even today). With consumers demanding broadband Internet access both at home and in the office, developed countries (USA, Western Europe, Japan, and Australia) saw near ubiquitous access to Internet at least in urban areas. Widespread Internet reach became a natural choice to extend e-commerce to support the sale of physical goods as well.

- E-procurement (e.g., CommerceOne)
- Electronic catalogs (e.g., Library of Congress)

- Electronic shopping cart (e.g., Amazon)
- Payment gateway (e.g., PayPal), and
- On-line status tracking (e.g., FedEx)

are tools used by ordinary citizens today. Practically every industry dealing with physical goods—steel, automotive, chemicals, retail stores, logistics companies, apparel stores, and transportation companies—could benefit from this *third stage* of e-commerce.

With the widespread adoption of e-commerce by thousands of corporations and millions of consumers by the turn of the century, the equivalent of Yellow Pages emerged in the form of online catalogs like Yahoo. The very size of the Yellow Pages and the need for speed in exploration of the desired items by the buyer from online catalogs led to the arrival of the search engines. The algorithmic approach to search (in place of database search) perfected by Google was a watershed in the history of e-commerce, leading to a *fourth stage* in e-commerce.

The next stage of growth in e-commerce was the end-to-end delivery of information services. Banks and financial institutions moved to online banking and online delivery of insurance booking. NASDAQ in USA and National Stock Exchange (NSE) in India moved to all-electronic stock trading. Airlines moved beyond online reservations to e-ticketing. Hotels moved to online reservations, even linking to online reservations of airline tickets to offer seamless travel services. Several movie theaters, auditoriums, and games/stadiums moved to online ticketing. Even governments in Australia and Canada started delivering citizen services online. Many libraries and professional societies such as IEEE shifted to electronic libraries and started serving their global customers electronically from the year 2000, representing the *fifth stage* of e-commerce.

The *sixth stage* is the maturing of the Internet into a highly reliable, available, and scaleable infrastructure that can be depended upon for 24×7 delivery of services worldwide. Services like Napster selling music to millions of users (before its unfortunate closure) and the recent success of Apple iTunes puts enormous pressure on network bandwidth and availability. In turn, this demand led to significant improvements in content delivery mechanisms, pioneered by companies like Akamai.

A whole range of issues including next generation network infrastructure, such as

- Internet II
- IPv6
- Security issues
- Privacy issues
- Legal issues, and
- Social issues

are getting addressed today to take e-commerce to the next stage.

During the dotcom boom, e-commerce created a lot of hype. It was predicted that all commerce will become e-commerce and eventually "brick and mortar" stores will be engulfed by "click and conquer" electronic storefront. Commerce in its conventional form is far from dead (it may never happen); yet e-commerce has made considerable progress [1,16].

- By December 2005 the value of goods and services transacted electronically exceeded $1 trillion, as per Forrester Research.
- Stock trading, ticketing (travel, cinema, and hospitality), and music are the "top 3" items traded electronically today.
- Online trading accounts for 90% of all stock trading.

E-Commerce

- Online ticketing accounts for 40% of the total ticketing business.
- Online music accounts for 30% of all "label" music sales is increasing dramatically, thanks to iPod and iTunes from Apple.
- By the year 2010, it is expected that nearly 50% of all commercial transactions globally will happen electronically.

23.3 OR/MS and E-Commerce

"Operations Research (OR) is the discipline of applying analytical methods to help make better decisions using the key techniques of Simulation, Optimization and the principles of Probability and Statistics," according to Institute of Management Science and Operations Research (INFORMS) [17]. Starting with the innovative application of sophisticated mathematical tools to solve logistics support issues for the Allied Forces during the Second World War, OR has moved into every key industry segment. For example, OR has been successfully used in the

- Airline industry to improve "yield" (percentage of "paid" seats);
- Automotive industry to optimize production planning;
- Consumer goods industry for supply chain optimization;
- Oil industry for optimizing product mix;
- Financial services for portfolio optimization; and
- Transportation for traffic simulation.

Practically all Fortune 500 companies use one or the other of the well-developed tools of OR, namely,

- Linear, integer, nonlinear, and dynamic programming
- Network optimization
- Decision analysis and multi-criteria decision making
- Stochastic processes and queuing theory
- Inventory control
- Simulation, and
- Heuristics, AI, genetic algorithms, and neural networks.

OR tools have been extensively used in several aspects of e-commerce in the past decade. The success behind many iconic companies of the past decade can be traced to OR embedded within their patented algorithms—Google's Page Ranking algorithm, Amazon's Affiliate program, and i2's Supply Chain Execution Engine, for example [4–6,18–20].

During the dotcom boom a number of e-commerce companies started using OR tools extensively to offer a range of services and support a variety of business models, though many of them collapsed during the dotcom bust. As early as the year 1998 Geoffrion went to the extent of stating "OR professionals should prepare for a future in which most businesses will be e-business" [5,6] and enumerated numerous OR applications to e-commerce including

- *OptiBid* using integer programming for Internet-based auction;
- *Trajectra* using stochastic optimization for optimizing credit card portfolio planning;
- *SmartSettle* using integer programming for automated settling of claims;

- Optimal Retirement Planner that used linear programming to provide advisory services;
- Home Depot using integer programming-based bidding for truck transportation planning;

indicating the importance of OR to e-commerce. Over the past 8 years the applications have increased manyfold and the area has matured.

In this chapter, we will outline a number of OR applications that are widely used in the different stages of e-commerce. This being a handbook, we will limit ourselves to outlining the interesting applications, indicating the possible use of OR tools to address the selected applications, and the benefits generated. The actual formulation and the algorithmic implementation details of specific applications are beyond the scope of this chapter, though interested readers can refer to literature for further details. There are dozens of OR tools, hundreds of algorithms, and thousands of e-commerce companies, often with their own patented implementation of algorithms. Instead of making a "shopping list" of the algorithms and companies, we will limit our scope to nine key applications that provide a broad sense of the applications and their richness.

We limit our scope to Internet-based e-commerce in this chapter. With the spectacular growth of wireless networks and mobile phones—particularly in India and China with lower PC penetration and land-line phones—e-commerce might take different forms including

- Mobile commerce (m-commerce)
- Smart cards, and
- Internet kiosks.

This is an evolving area with not-so-clear patterns of e-commerce and best captured after the market matures.

In the next section, we will outline the OR tools that are useful in the different stages of e-commerce, namely,

- Customer acquisition
- Order processing
- Order fulfillment
- Payment, and
- Post-sales support.

Section 23.4.1 will be devoted to the case of e-commerce related to digital goods. Section 23.4.2 will address the issues relating to e-commerce for physical goods. Section 23.4.3 will be devoted to the case of electronic delivery of services. Section 23.4.4 will focus on the optimization of the infrastructure to support e-commerce.

A very large number of e-commerce applications and business models have emerged over the past decade. For brevity, we have identified a set of nine key tools/applications that form the core set of tools to support e-commerce today. These nine candidates have been so chosen that they are representative of the dozens of issues involved and *comprehensive* enough to capture the essence of contemporary e-commerce.

We will discuss the chosen nine core tools/applications under one of the four categories mentioned earlier, where the specific tool/application is most relevant to the particular category. Accordingly,

 i. Internet search
 ii. Recommendation algorithms

E-Commerce

   iii. Internet affiliate programs

will be discussed in Section 23.4.1.
   Section 23.4.2 will address

   iv. Supply chain optimization
   v. Auction and reverse auction engines

   Section 23.4.3 will be devoted to

   vi. Pricing engines

And finally Section 23.4.4 will focus on

   vii. Content delivery network (CDN)
   viii. Web-site performance tuning
   ix. Web analytics.

23.4 OR Applications in E-Commerce

OR applications predate Internet e-commerce, as demonstrated by the airline industry that used seat reservation information to optimize route planning; in the process, innovative concepts like "planned overbooking" were developed that dramatically improved "yield." In the past decade (1995–2005) a lot more has been achieved in terms of the use of algorithms to significantly improve user experience, decrease transactions cost, reach a larger customer base, or improve service quality. We will outline the major applications of OR to the key aspects of e-commerce in this section.

23.4.1 E-Commerce Relating to Digital Goods

Goods that are inherently "soft" are referred to as "digital goods" as they are likely to be created by digital technology. Packaged software such as Microsoft Office and Adobe Acrobat that are used by millions of users globally are among the first recognizable digital goods. Until the mid-1990s shrink-wrapped software was packaged in boxes; with the widespread availability of the Internet and high-speed access at least in schools and offices, if not homes, software could be distributed electronically. It is common these days even for lay people, let alone IT professionals, to routinely "download" and "install" software on to their desktop or notebook computers. With the arrival of MP3 standard digital music and freely downloadable digital music players for PCs, such as Real Networks' RealAudio player, and Microsoft Media Player, digital music became a widely traded digital good, particularly during the Napster days. In recent years, with the phenomenal success of the Apple iPod MP3 player and the digital music stores Apple iTunes, music has become the fastest growing digital goods. With the world's largest professional society IEEE launching its Digital Library in the year 2000, online books from Safari Online have become popular; and with many global stock exchanges going online, e-books, digital libraries, and stock trading became popular digital goods in the past 5 years.

i. Internet Search
The first stage in commerce is the identification and acquisition of customers by the seller. In traditional commerce this is accomplished by advertising, promotion, catalog printing and distribution, and creation of intermediaries like agents. In e-commerce, particularly

in the case of digital goods, the customer acquisition process itself often happens on the Internet. Typically, a customer looking for a digital good is likely to "Google" (a term that has become synonymous with "searching" thanks to the overwhelming success of Google over the past 7 years) for the item over the Internet.

Search engines like Google and portals like Yahoo, MSN, eBay, and AOL have become the focal points of customer identification today and many corporations, big and small, are fine-tuning their strategies to reach potential customers' eyeballs over the portal sites. Internet advertising is only next to TV advertising today. The portal sites Yahoo and MSN and online bookstore Amazon have their own search engines today; eBay has started using Yahoo's search engine recently.

Search engines help the customer acquisition process by way of helping potential customers to explore alternatives. The reason for the phenomenal success of Google is its speed and relevance, as well as the fact that it is free. Very few users would use Google if every search took 5 minutes, however good the search is; similarly, Google usage would be miniscule if users have to wade through a large file with millions of "hits" to find the item they are looking for, even if the million-long list is delivered within seconds.

The secret behind the speed–relevance combination of Google (and other search engines today) is the sophisticated use of algorithms, many of them being OR algorithms.

Search engines broke the tradition followed by librarians for centuries; instead of using databases (catalogs in the language of librarians), search engines use indexes to store the links to the information (often contained in Web pages); they use automated tools to index, update the index, and most importantly, use algorithms to rank the content. For the ranking of the content another principle perfected by information scientists in the form of "citation index" is used; it is quite natural that important scientific papers will be cited by many scientific authors and the count of citation can be used as a surrogate for the quality of the paper. Scientific literature goes through peer review and fake citations are exceptions. In the "free-for-all" world of WWW publishing, a simple count of citation will not do; also the sheer size of the content on the WWW and the instantaneous response expected by Internet users (unlike the scientific community that is used to waiting for a year for the citations to be "compiled") makes it imperative to use fast algorithms to aid search. Google manages speed through its innovative use of server farms with nearly 1 million inexpensive servers across multiple locations and distributed algorithms to provide availability. Search engine vendors like Google also use some of the optimization techniques discussed in Section 23.4.4 (multicommodity flows over networks, graph algorithms, scheduling and sequencing algorithms) to optimize Internet infrastructure. The success behind the "relevance" of search engines is contributed by another set of algorithms, the famous one being Google's patented "page ranking" algorithm [21–23].

A typical user looking for an item on the Internet would enter a couple of keywords on the search window (another successful idea by Google of a simple text-box on a browser screen as user interface). Google uses the keywords to match; instead of presenting all the items that would match (often several millions), Google presents a list of items that are likely to be the most relevant items for the keywords input by the user. Often the search is further refined using local information (location of the computer and the network from where the search is originating and user preferences typically stored in cookies). Google is able to order the relevant items using a rank (patented Page Rank); the rank is computed using some simple graph theoretic algorithm.

A typical Web page has several pieces of content (text, graphics, and multimedia); often the contents are hyperlinked to other elements in the same Web page or to other Web pages. Naturally, a Web page can be reduced to a graph with Web pages as nodes and hyperlinks as edges. There are outward edges that point to other Web pages and inward edges that

E-Commerce

link other pages. Inward edges are equivalent to citations and outward edges are similar to references. By counting the number of edges one can compute a surrogate for citation; this can be a starting point for computing the rank of a Web page; the more the citations the better the rank. What is special about Google's page ranking algorithm is that it recognizes that there are important Web pages that must carry a lot of weight compared to a stray Web page. While a page with several inward links must get a higher importance, a page with a single inward link from an important page must get its due share of importance. However, the importance of a page would be known only after all the pages are ranked. Using a recursive approach, Google's page ranking algorithm starts with the assumption that the sum of ranks of all Web pages will be a constant (normalized to unity). If page A has inward links (citations) T1, T2,..., Tn and C(A) is the number of outward links of page A, the page rank (PR) could be defined as

$$PR(A) = (1-d) + d\{PR(T1)/C(T1) + PR(T2)/C(T2) + \cdots + PR(Tn)/C(Tn)\} \quad (23.1)$$

In a sense, instead of adding the citations linearly (as in the case of citation index), page rank algorithm takes the weighted rank approach with citation from higher-ranked page getting higher weight. "d" is used as a "dampening" constant with an initial value of 0.85 in the original Google proposal. In a sense, page rank is used as a probability distribution over the collection of Web pages. Defining page rank for the millions of Web pages leads to a very large system of linear equations; fortunately, OR professionals have the experience of solving such large system of linear equations using linear programming for years. The special structure of the set of equations admits even a faster algorithm. According to the original paper by Google co-founders Sergey and Page, their collection had 150 million pages (nodes) and 1.7 billion links (edges) way back in 1998. Their paper reports the following performance: 24 million pages had links to 75 million unique URLs; with an average 11 links per page and 50 Web pages per second it took 5 days to update the content. Distributing the load to several workstations they could get convergence to the set of equations (Equation 23.1) within 52 iterations; the total work distributed across several workstations took 5 hours (about 6 min/workstation). Based on this empirical performance the algorithm is extremely efficient (with complexity of O(log n) and can scale to Web scale [21]. That explains why Google search performs so well even today, though Web page collection has increased severalfold over the period 1998 to 2006 and Google is still serving results real fast. There are a number of special issues such as dangling links that have been addressed well; in the year 2005 Google introduced the no-follow link to address spam control, too.

To constantly keep pace with the changing content of the Web pages, search engines use a crawler to collect new information from the sites all over the world. The background infrastructure constantly updates the content and keeps an index of links to documents for millions of possible keyword combinations as a sorted list. After matching the user-provided keywords, Google collects the links to the documents, formats them as HTML page, and gives it to the browser that renders it over the user screen.

Google is not the only search engine; there are many other search engines including specialized search engines used by eBay for searching price lists. Many of them use specialized algorithms to improve ranking [20]; there are others like Vivisimo [24] and Kartoo [25] that use clustering techniques and visual display to improve user experience with search; yet Google remains the supreme search engine today.

The key OR algorithms used by search engines include linear programming to solve large-scale linear equations, neural networks for updating new information (learning), and heuristics to provide a good starting solution for the rank.

ii. Recommendation Algorithms

When a customer visits an online store to order an item there is a huge opportunity to up sell and cross sell similar items. Recommending such potential buy items to prospective buyers is a highly researched area. While such direct selling is not new, online direct selling is new. Common sense suggests that the online store recommends items similar to those bought by the customer during the current interaction or based on prior purchases (using database of prior purchases); a more refined common sense is to use the customer profile (based on cookies or customer-declared profile information) and match items that suit the customer profile. A more sophisticated approach is to use similarity between customers with preferences that match the current customer. Many users are familiar with Amazon Recommends [26,27] that talks of "people who ordered *this* book also ordered *these* books." Knowing customer profiles and customer interests it makes sense to suggest items that are likely to appeal "most" to the customer (taking care not to hurt any customer sensitivities). All such approaches form the core of recommendation algorithms [28].

Recommendation needs a lot of computation; both associative prediction and statistical prediction have been widely used by market researchers for ages, but at the back-end; tons of junk promotional materials and free credit card offers are the results of such recommendation. Recently recommendation has invaded e-mail too. But recommendation in e-commerce space needs superfast computation, as online visitors will not wait beyond seconds; this poses special challenges. Considering the fact that online stores like eBay or Amazon stock tens of thousands of items, millions of customers, and billions of past sales transaction data, the computational challenge is immense even with super computers and high-speed networks. Brute force method will *not* work! There are other constraints too. A typical user browsing the Internet will typically see the results on a PC screen that at best accommodates a handful of products' information; even if the screen size is large the customer is unlikely to view hundreds of products. The challenge is to present just a handful of the most relevant products out of hundreds or thousands.

Starting with crude technology that simply recommended items others have purchased, recommendation technology has evolved to use complex algorithms, use large volume of data, incorporate tastes, demographics, and user profiles to improve customer satisfaction and increased sales for the e-commerce site. The Group Lens project of the University of Minnesota was an early pioneer in recommendation algorithms. E-commerce sites today have much larger data sets leading to additional challenges. For example, Amazon had 29 million customers in 2003 and a million catalog items. While the earlier generation of OR analysts had the problem of powerful methods and scanty data, today's recommendation algorithms have to contend with excessive data!

Common approaches to recommendation include [27,28]

- *Simple "similarity" search* that treats recommendation as a search problem over the user's purchased item and related items stocked by the e-commerce seller.
- *Collaborative filtering* uses a representation of a typical customer as an N-dimensional vector of items where N is the number of distinct catalog items with positive vector for items purchased or rated positively and negative vector for negatively rated products. The vector is scaled by the number of customers who purchased the item or recommended the item. Typically the vector is quite sparse as most customers purchase just a few items. Collaborative filtering algorithm measures the similarity between two customers (cosine of the angle between the two vectors) and uses this similarity measure to recommend items.
- *Item-to-item collaborative filtering* goes beyond matching the purchased item to each of the user's purchased items and items rated by the user to similar

E-Commerce

items and combines them into a recommendation list. The Amazon site "Your recommendations" leads to the customer's earlier recommendations categorized by product lines and provides feedback as to why an item was recommended.
- *Cluster methods* that divide the customer base into many segments using a classification scheme. A prospective user is classified into one of the groups and a recommendation list is based on the buying pattern of the users in that segment.

Recommendation algorithms have been fairly successful; with unplanned online purchases amounting to just 25%, it is important to recommend the right items—those the customer likes but did not plan to buy, a sure way to increase sales.

Major recommendation technology vendors include *AgentArts, ChoiceStream, Expert-Maker,* and *Movice* in addition to Google.

The tools of OR that recommendation algorithms use include graphs—to capture similarity and identify clusters—and in turn implement collaborative filtering. Simulation and heuristics are used to cut down computation (dimensionality reduction).

iii. Internet Affiliate Programs

The Internet search techniques help in reaching the eyeballs of potential customers; the recommendation algorithms address the issue of recommending the right products to the customers. Obviously the customer might purchase some or all of the items recommended. The next stage is to retain the customer and attempt to convert the recommendation into an actual sale. As the Internet is a platform available to billions of people and a great equalizer that makes the platform accessible even to a small vendor in a remote area, it is not possible for all vendors to be able to reach potential customers through search engines or portals; an alternative is to ride on the major e-commerce sites like Amazon; the small vendors act as intermediaries or referrals that bring customers to the major sites and get a small commission in return. These are generally known as affiliate programs in marketing literature. The Internet gave a new meaning to the entire area of affiliate programs [9,29].

Intermediation is not new. For decades the print advertisement media have been engaging agents who design and get the advertisement for their agents displayed in the newspapers and get paid a small commission from the publishers for bringing advertisement revenue and a nominal fee from clients toward the labor of designing the advertisement. The Internet, however, is more complex in the sense that the intermediary has no precise control on the real estate, being a personalizeable medium rather than a mass medium, the advertisement—typically "banner ad"—gets displayed differently on different user screens.

Amazon is credited to be a pioneer in affiliate programs. Starting in 1994, Amazon let small booksellers link their site to Amazon site; if a prospective buyer who visits the affiliate site is directed to the Amazon site and buys an item from Amazon, the small bookstore vendor is entitled to receive a commission (a small percentage of the sale value). At a later date, it was extended to any associate that brings the buyer to the Amazon site. Today Amazon has more than 600,000 affiliates (one of the largest Internet affiliate programs). Google started its affiliate program in 2003; initially it was *AdWords* meant for large Web sites and extended later as *AdSense* in May 2005 for all sites. Google and others took the program to further sophistication through cost per lead, cost per acquisition, and cost per action and also extended it to cost per thousand that is more like the print advertisements. Dealing with such large intermediaries (affiliates) calls for significant management resources (track earnings, accounting, reporting, quality of service monitoring, and filter sensitive content) leading to Internet-based solutions to manage affiliate programs from third party companies. LinkShare [29] is one of the most successful agencies to manage Internet-based affiliate programs.

Yahoo Search Marketing and Microsoft *adCenter* are other large-size affiliate programs to watch.

Managing the banner ad that is typically 460 pixels wide × 60 pixels high is itself a sophisticated space optimization; the banner ad must be placed within the small real estate of the space earmarked for banner ad within the display page of the browser window on the user screen—an interesting integer programming problem (Knapsack problem).

Internet affiliate programs use optimization techniques to optimize banner ads and estimate initial cost per click; they use genetic algorithms and neural networks to adaptively optimize and improve estimates; and graph theory to capture associations.

23.4.2 E-Commerce Relating to Physical Goods

The success of e-commerce in digital goods sales prompted the brick and mortar industry to utilize the power of the Internet. One of the key challenges of many physical goods companies embracing e-commerce is the speed mismatch between electronic information delivery and the physical movement of goods that must be shipped to the customer who has ordered electronically. It is one thing to be able to order groceries over the Internet and yet another thing to receive it at your home by the time you reach home.

Obviously, logistics and supply chain optimization are two key issues that would make or mar the success of e-commerce for physical goods.

iv. Supply Chain Optimization

Traditional OR models focused on the optimization of inventory, typically in a factory. Typically one would forecast sales as accurately as possible, including variations, build optimal level of safety stocks suiting a particular replenishment policy. Supply chain optimization refers to the optimization of inventory across the entire supply chain—factory to warehouse to distributors to final stores. Supply chain optimization must ensure that at every stage of the supply chain there is the *right material* in the *right quantity* and *quality* at the *right time* and supplied at the *right price*. Often there will be uncontrollability in predicting the demand; this must be protected against through the cushion of safety stocks at different levels. One must optimize the safety stocks so that they are neither too small leading to nonavailability when needed, nor too large leading to unnecessary locking up of capital. Naturally, supply chain optimization would involve multistage optimization and the forecasting of inventory must evolve from a simple probability distribution to a sophisticated stochastic process.

A key success behind today's supply chain optimization is the power of online databases to forecast at low granularity involving hundreds, if not thousands of items and to aggregate the forecasts; today's distributed computing power permits statistical forecasting to arrive at the best fit using past demand pattern, even if it involves thousands of items.

Today's supply chain optimization engines [30,31] use academically credible algorithms and commercially effective implementations. Supply chain optimization leads to

- Lower inventories,
- Increased manufacturing throughput,
- Better return on investment,
- Reduction in overall supply chain costs.

Pioneering companies like i2 and Manugistics took many of the well-studied algorithms in multistage optimization, inventory control, network optimization, and mixed-integer programming, and combining them with recent concepts like third party logistics and

E-Commerce

comanaged inventory along with exceptional computing power available to corporations, they created supply chain optimization techniques that could significantly save costs. The simultaneous development of enterprise resource planning (ERP) software—SAP R/3 and PeopleSoft (now part of Oracle)—helped supply chain optimization to be integrated with the core enterprise functions; together it was possible to integrate

- Supply chain network design
- Product families
- Suppliers/customers integration
- Multiple planning processes
- Demand planning
- Sales and operations planning
- Inventory planning
- Manufacturing planning
- Shipment planning

with

- Execution systems

paving the way for optimizing across the supply chain (compared to single-stage optimization earlier).

Today's supply chain optimizing engines use established OR tools like

- Mixed integer programming
- Network optimization
- Simulation
- Theory of constraints
- Simulated annealing
- Genetic algorithms, and
- Heuristics

routinely; and solvers like iLog are part of such supply chain optimization engines.

Dell and Cisco are two vendors who used supply chain optimization extensively and built highly successful extranets that enabled suppliers and customers to conduct end-to-end business online—ordering, status monitoring, delivery, and payment—leading to unprecedented improvements in managing the supply chain. For example, Dell reported the following metrics for its online business [30]

- User configurable ordering with maximum of 3 days for feasibility checking
- Online order routing to the plant with a maximum of 8 h
- Assembly within 8 h
- Shipping within 5 days.

With the estimated value of 0.5–2.0% value depletion per week in the PC industry, optimizing supply chain makes a lot of sense; the leadership position that Dell enjoyed for several years is attributed to its success with supply chain optimization—a classic success story in e-commerce for physical goods.

Supply chain uses multistage optimization and inventory control theory to reduce the overall inventory and hence the costs; it uses networks to model multistage material flows; queuing theory and stochastic processes to model demand uncertainty; simulation to model scenarios; and heuristics to reduce computation.

v. Internet Auction

In e-commerce relating to physical goods, supply chain optimization dramatically improves the delivery side of many manufacturing enterprises. Material inputs (raw materials, subassemblies) account for a significant part of the overall business of most manufacturing enterprises and procuring the inputs, if optimized, will lead to significant savings in costs. Another area where e-commerce made a big difference to physical goods sales is that of auction over the Internet pioneered by eBay. Another interesting twist pioneered by Freemarket Online is that of reverse auction used by many in the procurement space. Internet-based reverse auction brings unprecedented efficiency and transparency in addition to billions of dollars of savings to large corporations like GE (though not always sustainable over a longer period).

Auction is not new; antiques, second-hand goods, flower auctions, or tea auctions are centuries old. The technology industry is familiar with spectrum auction. Electricity auction is an idea that is being experimented with in different parts of the world, though with limited success [32].

Auctioning is an idea that has been well researched. Simply stated, a seller (generally a single seller) puts out an item for sale; many buyers bid for it by announcing a bid price; the process is transparent with all bidders knowing the bid price of other bidders; the bidders can quote a higher price to annul the earlier bid; generally, an auction is for a limited time and at the end of the time limit, the seller sells the item to the highest bidder; the seller reserves the right to fix a floor price or reserve price below which bids are not accepted. If there are no bids at a price higher than the reserve price, the sale does not go through, though the seller might bring down the reserve price.

The Internet brings a number of efficiencies to the process of auctioning; unlike physical auction electronic auction can happen across time zones globally and on a 24×7 basis; with geographic constraints removed electronic auction (e-auction) may bring larger numbers of bidders leading to network economics and improving the efficiency of the auction process [33].

To conduct e-auction an auction engine has to be built either by the service provider (eBay) or the seller can use third-party e-auction engines from a number of vendors OverStock, TenderSystem, and TradeMe, for example.

The auction engine has modules to publish the tender, process queries, accept quotations from the bidders, present the highest bid at any time (call out) electronically to any of the bidders over the Net, provide time for the next higher bidder to bid, close the bid at the end of the time, collect commission charges for conducting the auction, and declare the results—all in a transparent manner. Though the process is fairly simple, sophisticated algorithms have to work behind the screen as tens of thousands of auctions would be taking place simultaneously; even the New Zealand-based TradeMe.com reports 565,000 auctions running at any day, while eBay talks of millions of auctions running simultaneously [34,35].

A more interesting Internet auction is reverse auction (also called procurement auction) pioneered by an ex-GE employee, who founded Freemarkets Online (later acquired by Ariba). In reverse auction, instead of the seller putting out the auction for the buyers to bid, the buyer puts out the auction for all the sellers to bid; a typical application is the procurement manager wanting to buy items from multiple suppliers.

In the conventional purchase, the specifications for an item, that the buyer is interested to buy, is put out as tender. Typically, sellers put out a sealed bid; after ensuring that all the bids have quoted for the same item the sealed bids are opened and the seller who quotes the minimum price is awarded the contract. In reverse auction, there is no sealed bid; everyone

E-Commerce

would know the bid price, leading to higher levels of transparency and efficiency. Such buying is a huge opportunity globally—*Fortune* magazine estimated that industrial goods buying alone was $50 billion in year 2000—and even a single percentage improvement could translate to millions of dollars of savings.

A typical reverse auction engine facilitates the process of auctioning [33–35] by way of

- Preparation (help in preparing standardized quotations that makes supply/ payment schedule, payment terms, and inventory arrangements common across suppliers to get everyone a "level playing field"). In the process, many industrial items with complex specifications that are vaguely spelt out get so standardized that they become almost a "commodity" that can be "auctioned";
- Finding new suppliers (being an intermediary and a "market maker," reverse auction vendors bring in a larger range of suppliers to choose from);
- Training the suppliers (reverse auction engine vendors train the suppliers so that they take part more effectively in the process, leading to efficiency and time savings);
- Organizing the auction (including the necessary electronic support, server capacity, network capacity, interface software, special display to indicate the various stages of bidding);
- Providing auction data (analysis for future auctions).

Several large buyers including General Electric and General Motors have found reverse auctions extremely valuable. There are also consortiums like GM, Ford, and Daimler Chrysler who run their own "specialized" reverse auction sites. A whole range of industry-specific portals such as *MetalJunction* also offer reverse auction as a standard option today.

Internet auctions use economic theory (often modeled through simultaneous equations that are solved through LP Solver), multicriteria decision making to model conflicting requirements between bidder and buyer, and neural networks to model learning.

23.4.3 Electronic Delivery of Services

Internet search helps in the identification of potential customers; recommendation algorithms help in suggesting the most relevant items for possible buy by a visitor to the specific site or portal; affiliate programs bring in intermediaries who help in converting the potential customer visit on the Internet to a possible sale; supply chain optimization helps in supporting the sales of physical good; auction and reverse auction help in bringing efficiency of online markets to business-to-business commerce, particularly in the case of physical goods [36].

There are a whole range of services that are delivered electronically; these include information services—hotel room booking, airline booking, e-trading, banking, online stores, e-governance services, and e-learning. The delivery is completely online in the case of digital goods and information services while all but the physical shipment (seeking, price negotiation, ordering, order tracking, and payment) happen over the Net. For example [37],

- Hotel rooms are physical that the visitor physically occupies during the visit, but all supporting services are delivered online.
- University classes are held in the classrooms of a University, but registration for the course, payment of fees, reading materials, assignment submission, and award of the grades happen electronically.
- In electronic trading if the stock exchange is online (like NASDAQ) the entire process happens electronically.

- Many banks provide full service-banking electronically, with the customer using the "physical world" (the branch of a bank or ATM) only for cash withdrawal/payment.
- Many governments provide electronic access to their services (marriage registration, passport, tax payment, drivers license) where all but physical documents (actual passports, driver license, license plates) are delivered electronically.
- Many stores (bookstores, computers, electronic gadgets, video stores, merchandise, and even groceries) support electronic access for all but except the actual delivery.

vi. Price Engines

A key advantage of e-commerce is a far superior "comparison shopping" that the electronic stores can offer. Normal paper-based comparison shopping is limited to a few stores, a few items, and a few merchants/agents, typically in the same area. Thanks to the global footprint of the Internet and the 24×7 availability, "pricing engines" [38] can offer a far richer information source, electronic support to compare "real" prices (calculating the tax, freight charges, and accounting for minor feature variations all within seconds) and displaying the results in a far more user-friendly manner. Special discounts that are in place in specific time zones or geographies for specific customer segments (age groups, profession, and sex) can be centrally served from a common database with the user not having to "scout" for all possible promotions for which he/she is eligible on that day in that geography.

A number of price engines are available today. Google offers "Froogle"; the early pioneer *NexTag* started with a focus on electronics and computer products and recently expanded to cars, mortgages, and interestingly online education programs! PriceGrabber is an innovator in comparative shopping with 22 channels and brings the entire ecosystem together that includes merchants, sellers, and other intermediaries; *BestBuy*, Office Depot, and Wal-Mart are *PriceGrabber* customers.

Comparison shopping can be viewed as a specialized search engine, specialized e-marketplace, or as a shopping advisory service firm. Yahoo introduced *YahooShopping*, another comparative shopping service.

Comparative shopping sites organize structured data—price lists, product catalog, feature lists—unlike ordinary search engines like Google that deal with unstructured Web pages. Specialized techniques are needed (table comparison) as well as "intelligence" to check which accessory goes with what main item. Often many of these higher level processes are done manually by experts and combined with automated search, though attempts to automate structured data are underway by many comparative shopping engines. Specialized comparative shopping engines focusing on travel would address issues like connectivity of different segments, multimodal availability checking like room availability, transportation facilities, along with airline booking.

Comparative shopping engines use expert systems, ideas from AI, heuristics, and specialized algorithms like assignment and matching to get the speed and richness of the recommendations (in that sense a comparative shopping engine can be viewed as another "recommendation engine").

Price engines use simulation to study millions of alternatives to generate initial alternatives, linear and nonlinear programming to optimize costs, and multiattribute utility theory to balance the conflicting requirements of buyers and sellers.

23.4.4 Support for E-Commerce Infrastructure

The sale of digital goods, electronic support for sale of physical goods, or electronic delivery of other services need a reliable, highly available, and scaleable Internet infrastructure with

E-Commerce

a range of challenges. We address the challenges relating to the distribution of content by building a specialized infrastructure for mission critical sites; we address the issues relating to content distribution of very demanding sites like the World Cup games site or Olympic games site during the peak period using Web-caching.

Finally, e-commerce should address post-sales support also. While there are many issues, we address the specific technology of "Web analytics" that provides insights into customer behavior in the e-commerce context.

vii. Content Delivery Network

Mission critical servers that distribute large content (software downloads, music, video) must plan for sustained high performance. Downloading of software, if it takes too much time due to network congestion, would lead to "lost customers" as the customer would "give up" after some time; dynamic information like Web-casting (for Webinar support) would need guaranteed steady network availability to sustain delivery of audio or video content; any significant delay would lead to poor quality sessions that the users would give up. Companies like Akamai [39] have pioneered a way of distributing content by equipping the "edge" with sufficient content delivery for high availability, replicating the content at multiple locations for faster delivery, and running an independent managed network for sustained high performance, in addition to tools to manage the whole process without human intervention.

A content delivery network (CDN) consists of servers, storage, and network working in a cooperative manner to deliver the right content to the right user as fast as possible, typically serving the user from the nearest possible server location, and if possible using the cheapest network. Obviously, it is a continuously adaptive optimization problem for which multicommodity flow algorithms have been extensively applied.

The key issues that must be optimized are the location of the servers and the choice of multiple backbones that improves performance (user gets the content delivered real fast) and decreases cost (minimal replication, least number of servers/least expensive servers, minimal usage of bandwidth/least cost bandwidth). Naturally the objective is to provide consistent experience to the users without their having to direct their search to specific locations; working behind the infrastructure must be a smart system that constantly watches the network, copies content at multiple locations, replicates servers at multiple networks/locations, and negotiates bandwidth.

Typically, content delivery networks are built as "overlay network" over the public Internet and use DNS re-direct to re-direct traffic to the overlay network. The key elements of CDN include

- High-performance servers
- Large storage
- Backup/storage equipment and "smart" policies to manage backup/storage
- Automated content distribution/replication
- Content-specific (audio/video, file transfer) delivery
- Content routing
- Performance measurement, and
- Accounting for charging the clients.

A content delivery network can be viewed as "higher layer" routing [39]; "Ethernet routing" happens at Layer 2 (as per OSI protocol); IP routing happens at Layer 3; port-based routing takes place at Layer 4; and content-based routing can be viewed as Layer 4 to Layer 7 routing. New protocols that permit query of cache have led to the maintenance of "cache farms" to improve content delivery.

Akamai, Speedera (now part of Akamai), *Nexus, Clearway*, and *iBeam* Broadcasting (that pioneered the use of satellite for replication) are the pioneers in CDN technology development. This is an evolving area of research currently.

The importance of CDN can be gauged by the fact that the leader in CDN, Akamai, has Google, Microsoft, AOL, Symantec, American Express, and FedEx as its customers today.

Google uses commodity servers and high availability algorithms that use a distributed computing model to optimize search; content delivery is optimized using sophisticated caching servers; as search results are less demanding (compared to video streaming, Webinars, and Webcasts) Google is able to provide high performance at lower costs; with Google getting into video, some of the strategies used by other CDN vendors might be pursued by Google also.

CDN uses a distributed computing model to improve availability, networks to model the underlying infrastructure, linear and dynamic programming to optimize flows, stochastic processes to model user behavior, and heuristics to decide routing at high-speed for real-time high performance.

viii. Web Site Performance Tuning

With Internet penetration on the rise, end users depending on Google, Yahoo, MSN, AOL, and eBay for their day-to-day operations (search, mail, buy), the demand on the Internet infrastructure is increasing both quantitatively and qualitatively. Users expect better performance and ISPs want more customers. Users want screen refresh to happen within a couple of seconds irrespective of network congestion or the number of users online at a time. The users also expect the same response irrespective of the nature of the content—text, graphics, audio, or video. Several researchers have addressed this issue. Google uses several server farms with tens of thousands of servers to make its site available all the time and the site contents refresh fairly fast. A whole area of "load-balancing" content delivery has emerged over the past decade with several novel attempts [40,41].

One attempt to improve Web-site performance by load balancing is to use content distribution strategies that use differentiated content delivery strategies for different content [41]. In this scheme, the differential content is distributed to servers of varying capability:

- High-end mainframe database servers for critical content
- Medium-end Unix box running low-end DBMS for less-critical content
- Low-end commodity server for storing other information as text files.

For example, in an airline reservation system,

- Fast-changing and critical content like "seat matrix" could be stored on a mainframe/high-performance server,
- Less frequently changing content like "fare data" that is less critical for access performance could be stored on a mid-range database server, and
- "Time table" information that does not change for a year can be kept as a flat file on a commodity server.

By storing differentiated content on servers of different capability, criticality, and costs, it is possible to cut down costs without compromising on quality of service; of course, sophisticated replication schemes must be in place to ensure that the data across the three levels of servers are consistent. The queries would arrive at an intermediate server and be directed to the right server based on the content, yet another content delivery distribution scheme.

With the rising popularity of some sites, hit rates are reaching staggering proportions—access rates of up to million hits per minute on some sites are not uncommon, posing great challenges to servers, software, storage, and network. Content-based networking is a

E-Commerce   **23**-19

solution for some premier sites (Microsoft and FedEx), but ordinary sites too need enhanced performance.

One approach to improve Web-site performance is to create concurrency through

- Process-based server (e.g., Apache server)
- Thread-based server (e.g., Sun Java server)
- Event-driven server (e.g., Zeus server), or
- In-kernel servers (very fast though not portable).

OR algorithms have been used to combine the various schemes; in addition, the "popularity index" of the content is used in a prioritization algorithm to improve Web site performance of critical application infrastructure.

Such strategies have been used successfully to optimize Web servers for critical applications like the Olympic Games site during the days of the Olympic Games with telling performance. For example, the Guinness Book of World Records announced on July 14, 1998, has two notable records:

- Most popular Internet record of 634.7 million requests over the 16 days of the Olympic Games.
- Most hits on an Internet site in a minute—110,114 hits—occurred at the time of women's freestyle figure skating.

Using special techniques that use embedded O/S to improve router performance by optimizing TCP stack, such strategies helped IBM achieve superior performance—serving 5000 pages/second on a single PowerPC-processor-based PC class machines running at 200 MHz [41].

Yet another technique widely used for Web-caching uses "greedy algorithms" to improve server throughput and client response; typical strategies include

1. Least recently used (LRV) caching and
2. Greedy dual size that uses cost of object cashing (how expensive it is to fetch the object) and size of the object (Knapsack problem)

to fill the cache.

Web site performance tuning uses simulation to get broad patterns of usage, optimization, and decision analysis for arriving at strategies, stochastic processes and queuing theory to model network delays, and inventory control to decide on network storage needs.

ix. Web Analytics

The final stage of e-commerce is post-sales support. In the context of the Web it is important to analyze the consumer behavior in terms of the number of visits, how much time it took for the user to find the information that was being searched, the quality of the recommendations, the time the user spent on the various activities of the buyer–seller interaction that takes place on the Internet. In turn, such insights would help the seller to offer better buying experience and offer more focused products in the next buying cycle. Web analytics is the technology that addresses this issue [42].

From the early days, Web site administrators have used logs to study the visit pattern of the visitors visiting the site—number of visitors, distribution of visitors over time, popular Web pages of the site, most frequently visited Web page, least frequently visited Web page, most downloaded Web page, and so on. With the proliferation of the Web and the sheer size of the Web content, many Web masters could not cope up with the large numbers thrown up by the logs to get any key insights. It is the marketing and marketing communications professionals along with tool vendors that have helped the ideas of Web analytics mature over the past decade.

The technique of page tagging that permitted a client to write an entry on an external server paved the way for third-party Web analytics vendors who could use the "logs" and "page tag databases" to provide a comprehensive analysis of the way a particular e-commerce Web site was being used. JavaScript technology permitted a script file to be copied (say, for example, a "counter" application) that would let third party vendors remotely "manage" the site and provide analytics.

Google started providing "Google Analytics" free of charge on an invitation basis since 2005. SAS Web Analytics, *Webtrends, WebsideStory, Coremetrics*, Site Catalyst, and *HitBix* are other Web analytics vendors who pioneered many ideas in this emerging area [42]. A comprehensive Web analytics report goes beyond simple page counts, unique visitors, or page impressions to provide insights into usability, segmentation, campaigns to reach specific customers, and identification of potential customers, in the process becoming a powerful marketing tool.

Web analytics uses data mining including optimization and simulation for pattern generation and pattern matching, stochastic processes, and simulation for traffic analysis and neural networks, and genetic algorithms for postulating patterns and hypotheses.

23.5 Tools–Applications Matrix

We have outlined a range of OR applications for e-commerce in this section. Instead of listing the tools and techniques we have discussed the different OR algorithms within the life-cycle of typical e-commerce—customer acquisition, promotion, price negotiation, order processing, delivery, payment, and after-sales support. We have indicated at the end of every subsection describing the nine tools and applications (Internet search to Web analytics), the key OR algorithms that the researchers have found useful. To provide a better perspective of the interplay of e-commerce applications and tools of OR, the following tools–applications matrix indicates the applicability of all relevant tools for each of the nine key e-commerce tools and applications discussed in this chapter. The chart is more indicative and includes the tools most likely to be used; after all, most tools can be used on every occasion.

OR Techniques used in	1	2	3	4	5	6	7
Internet search	☑						☑
Recommendation algorithms	☑	☑				☑	☑
Internet affiliate programs	☑						☑
Supply chain optimization	☑	☑		☑	☑	☑	☑
Internet auction	☑		☑				☑
Price engines	☑		☑			☑	☑
Content delivery network	☑	☑		☑		☑	☑
Web performance tuning	☑	☑	☑	☑	☑	☑	☑
Web analytics	☑	☑		☑		☑	☑

1. Linear, integer, nonlinear, and dynamic programming
2. Network optimization
3. Decision analysis and multi-criteria decision making
4. Stochastic processes and queuing theory
5. Inventory control

6. Simulation, and
7. Heuristics, AI, genetic algorithms, and neural networks.

23.6 Way Forward

E-commerce is fundamentally transforming every business—corporations, universities, NGOs, media, and governments. With Internet still reaching only 2 billion of the 6+ billion people in the world, e-commerce has the potential to grow; the existing users of e-commerce will see further improvements in the services they see today and a whole new set of services that have not even been touched. We are still in the growth phase of global e-commerce. OR algorithms have influenced significantly the various business models and services offered by e-commerce today. Several of the e-commerce applications pose special challenges; OR professionals would find new algorithms and techniques to solve the challenges arising out of e-commerce powered by WWW 2.0 and Internet II. It will be exciting to see OR influencing the future of e-commerce and OR getting influenced by e-commerce applications.

23.7 Summary

In this chapter, we have outlined a number of OR applications in the context of e-commerce. With e-commerce growing phenomenally, it is impossible to document all the e-commerce applications and the OR tools used in those applications. We have taken a more pragmatic view of outlining the key OR tools used in the different stages of the life-cycle of a typical e-commerce transaction, namely, customer acquisition, promotion, order processing, logistics, payment, and after-sales support. A set of nine key e-commerce applications, namely,

1. Internet search
2. Recommendation algorithms
3. Internet affiliate programs
4. Supply chain optimization
5. Internet auction
6. Price engines
7. Content delivery network
8. Web-site performance tuning
9. Web analytics

have been discussed as a way to provide insight into the role of OR models in e-commerce. It is not possible to list the thousands of e-commerce applications and the use of hundreds of OR tools; what is indicated in this chapter is the immense potential of OR applications to e-commerce through a set of nine core tools/applications within the world of e-commerce. A tools–applications matrix provides the e-commerce applications versus OR tools match. While a lot has happened in the past decade, one expects a much larger application of OR to the next stage of growth in global e-commerce.

References

1. E-commerce-times www.ecommercetimes.com/.
2. Kalakota, Ravi and Andrew Whinston, *Electronic Commerce: A Manager's Guide*, Addison-Wesley (1997).
3. Wikipedia (www.wikipedia.com).

4. e-Optimization site http://www.e-optimization.com/.
5. Geoffrion, A.M., "A new horizon for OR/MS," *INFORMS Journal of Computing*, 10(4) (1998).
6. Geoffrion, A.M. and Ramayya Krishnan, "Prospects of Operation Research in the E-Business era," *Interfaces*, 31(2): 6–36 (March–April 2001).
7. Chaudhury, Abhijit and Jean-Pierre Kuilboer, *E-Business and E-Commerce Infrastructure: Technologies Supporting the E-Business Initiative*, McGraw-Hill College Publishing (January 2002).
8. Kalakota, Ravi and Andrew B. Whinston, *Frontiers of Electronic Commerce* (1st edition), Addison-Wesley (January 1996).
9. Krishnamurthy, S., *E-Commerce Management: Text and Case*, South Western Thomson Learning (2002).
10. Laudon, Kenneth C. and Carol Guercio Traver, *E-Commerce: Business, Technology, Society* (2nd edition), Prentice Hall (2004).
11. Robinson, Marcia, Don Tapscott, Ravi Kalakota, *e-Business 2.0: Roadmap for Success* (2nd edition), Addison-Wesley Professional (December 2000).
12. MIT E-commerce resource center, http://ebusiness.mit.edu/.
13. Penn State e-biz center, Smeal.psu.edu/ebrc.
14. Smith, Michael et al., "Understanding digital markets," in *Understanding the Digital Economy*, Erik Brynjolfsson (ed.) MIT Press (2000).
15. University of Texas at Austin Center for Research in Electronic Commerce, http://crec.mccombs.utexas.edu/.
16. e-marketer site http://www.emarketer.com/.
17. INFORMS (The Institute of Operations Research and Management Science) www.informs.org.
18. Bhargava, H. et al., "WWW and implications for OR/MS," *INFORMS Journal of Computing*, 10(4): 359–382 (1998).
19. Bhargava, H. et al., "Beyond spreadsheets," *IEEE Computer*, 32(3): 31–39 (1999).
20. Willinger, W. et al., "Where mathematics meets the Internet," *American Math Society*, 45(8): 961–970 (1998).
21. Brin, Sergey and Larry Page, "The anatomy of a large-scale hyper-textual Web search engine," Working Paper, CS Department, Stanford University (1998).
22. Google Page Rank Algorithm, "Method for node ranking in a linked database," U.S. Patent 6,285,999, dated September 4, 2001.
23. Page, L., S. Brin, R. Motwani, and T. Winogard, "Page rank citation ranking—bringing order to the Web," Technical Report, Stanford University (1999).
24. Vivisimo search engine (www.vivisimo.com).
25. Kartoo search engine (www.kartoo.com).
26. Amazon one-click patent, "Method and system for placing a purchase order via a communications network," U.S. Patent 5,960,411 dated Sep 28, 1999.
27. Linden, Greg et al., "Amazon.com recommendation," *IEEE Internet Computing*, pp. 76–80 (Jan–Feb 2003).
28. Leavitt, Neal, "Recommendation technology: will it boost e-commerce?," *IEEE Computer*, May 2006, pp. 13–16.
29. Linkshare patent, "System method and article of manufacture for internet based affiliate pooling, U.S. Patent 6.804,660 dated October 12, 2004.
30. Kapuscinski, R. et al., "Inventory decisions in Dell's supply chain," *Interfaces*, 34(3): 191–205 (May–June 2004).
31. Lapide, Larry, "Supply chain planning optimization: Just the facts," *AMR Research* (1998).

32. Beam, C., Segev, A., and Shantikumar, G. "Electronic Negotiation Through Internet-Based Auctions," Working Paper 96-WP-1019, University of California, Berkeley.
33. *Fortune* magazine, "Web auctions," March 20, 2000.
34. Nissanoff, Daniel, *Future Shop: How the New Auction Culture Will Revolutionize the Way We Buy, Sell and Get the Things We Really Want*, The Penguin Press (2006) ISBN 1-594-20077-7.
35. Sawhney, Monhanbir, "Reverse auctions," *CIO Magazine*, June 1, 2003.
36. Alvarez, Edurado et al., "Beyond shared services—E-enabled service delivery," Booze Allen & Hamilton Report (1994).
37. Seybold, P., *Customers.com: How to Create a Profitable Business Strategy for the Internet and Beyond*, Time Books (1998).
38. Priceline patent "Method and apparatus for a cryptographically assisted commercial network system designed to facilitate buyer-driven conditional purchase offers," U.S. Patent 5,794,207 dated Aug 11, 1998.
39. Akamai patent, "Method and apparatus for testing request-response service using live connection traffic, U.S. Patent 6,981,180 dated December 27, 2005.
40. Chen, H. and A. Iyengar, "A tiered system for serving differentiated content," *World-Wide Web: Internet and Web Information Systems Journal*, no. 6, Kluwer Academic Publishers, 2003.
41. Iyengar, A., Nahum, E., Shaikh, A., and Tewari, R., "Improving web-site performance," in *Practical Handbook of Internet Computing*, Munindar, P. Singh (ed.) CRC Press (2004).
42. Peterson, Eric, "Web analytics demystified," Celilo Group Media (2004), ISBN 0974358428.

Further Reading

1. Awad, E., *Electronic Commerce* (2nd edition), Prentice Hall (2003).
2. Chan, H. et al., *E-Commerce: Fundamentals and Applications*, 0-471-49303-1, Wiley (January 2002).
3. Deitel, Harvey M. and Paul J. Deitel, *e-Business & e-Commerce for Managers* (1st edition), Prentice Hall (December 2000).
4. Hillier, Frederick S. and Gerald J. Lieberman, *Introduction to Operations Research*, (8th edition), McGraw Hill, ISBN 0-07-732-1114-1 (2005).
5. Jeffrey F. Rayport et al., *Introduction to e-Commerce*, McGraw-Hill/Irwin (August 2001).
6. Carroll, Jim, Rick Broadhead, *Selling Online: How to Become a Successful E-Commerce Merchant*, Kaplan Business (2001).
7. Murty, Katta, *Operations Research: Deterministic Optimization Models*, Prentice Hall (1994).
8. Ravindran, A., D. T. Phillips, and J. Solberg, *Operations Research: Principles and Practice* (2nd edition), John Wiley and Sons, Inc., New York, 1987.
9. Schneider, G., *Electronic Commerce* (4th edition), Canada: Thompson Course Technology (2003).
10. Schneiderjans, Marc J. and Qing Cao, *E-commerce Operations Management*, Singapore: FuIsland Offset Printing (S) Pte. Ltd. (2002).
11. Taha, Hamdy, *Operations Research—An Introduction* (4th edition), Macmillan (1987).
12. Turban, E. and David King, *E-Commerce Essentials*, Prentice Hall (2002).
13. Wagner, H. M., *Principles of Operations Research*, Prentice Hall (1969).
14. Winston, Wayne, *Operations Research: Applications & Algorithms* (4th edition), Duxbury (2004).

24
Water Resources

24.1	Introduction.. **24**-1	
	Sustainability • Minor Systems • Chapter Organization	
24.2	Optimal Operating Policy for Reservoir Systems **24**-4	
	Introduction • Reservoir Operation Rules • Problem Formulation	
24.3	Water Distribution Systems Optimization....... **24**-10	
	Introduction • Global Optimization • Two-Stage Decomposition Scheme • Multistart-Local Search • Simulated Annealing • Extension to Parallel Expansion of Existing Networks • Analysis of an Example Network • Genetic Algorithm	
24.4	Preferences in Choosing Domestic Plumbing Materials **24**-18	
	Introduction • Analytical Hierarchy Process	
24.5	Stormwater Management........................... **24**-21	
	Introduction • Problem Formulation	
24.6	Groundwater Management......................... **24**-23	
	Introduction • Problem Formulation	
24.7	Summary **24**-25	

G.V. Loganathan
Virginia Tech

Acknowledgments **24**-25
References **24**-26

24.1 Introduction

24.1.1 Sustainability

There has been an increased awareness and concern in sustaining precious natural resources. Water resources planning is rising to the top of the international agenda (Berry, 1996). In the following, we present some of the problems that are affecting water resources. Chiras et al. (2002) list the following water-related problems in the United States: increasing demand in states such as Florida, Colorado, Utah, and Arizona; water used for food has tripled in the last 30 years to about $24(10^6)$ hectares straining one of the largest aquifers, Ogallala aquifer, ranging from Nebraska to Texas with a water table decline of 1.5 m per year in some regions; high industrial and personal use; unequal distribution; and water pollution. In the United States, the estimated water use is about 408 billion $[408(10^9)]$ gallons per day with a variation of less than 3% since 1985. The water use by categories is as follows: 48% for thermoelectric power, 34% for irrigation, 11% for public supply, and 5% for other industrial use (Hutson et al., 2004).

Sustaining this resource is attracting significant attention (Smith and Lant, 2004). Sustainability is defined as "meeting current needs without compromising the opportunities of future generations to meet their need" (World Commission, 1987). According to the American Society of Civil Engineers (1998) and UNESCO, "Sustainable water systems are designed and managed to fully contribute to the objective of the society now and in the future, while maintaining their ecological, environmental and hydrologic integrity." Heintz (2004) points to economics to provide a framework for sustainability as not spending the "principal." He also elucidates the use of technology for resource substitution especially for nonrenewable resources. Dellapenna (2004) argues that if the replenishment time is too long, such as for iron ore or oil cycles, the resource should not be considered sustainable. Replenishing groundwater in general is a long-term process. Kranz et al. (2004), Smith (2004), Loucks (2002), and Loucks and Gladwell (1999) provide indicators for sustaining water resources. An indicator used for assessing the receiving water quality is the total maximum daily load (TMDL), which is the maximum mass of contaminant per time that a waterway can receive from a watershed without violating water quality standards; an allocation of that mass to pollutant's sources is also required. In 2004, about 11,384 miles of Virginia's 49,220 miles of rivers and streams were monitored; of these, 6948 miles have been designated as impaired under Virginia's water quality standards (Younos, 2005). Reinstating water resources to acceptable standards requires a holistic engineering approach. Delleur (2003) emphasizes alternative technological solutions that should be assessed for (i) scientific and technical reliability, (ii) economic effectiveness, (iii) environmental impact, and (iv) social equity.

24.1.2 Minor Systems

In the following, we draw attention to a minor system, namely the home plumbing system. Domestic copper plumbing pipes (for drinking water use) are experiencing pinhole leaks. Figure 24.1 shows the distribution of locations reporting pinhole leaks around the country. We define the major system as the public utility water distribution system that brings drinking water to houses. While the major system is readily recognized as a vast infrastructure system of nearly 1,409,800 km of piping within the United States (Material

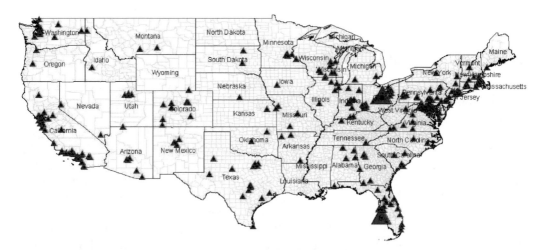

FIGURE 24.1 Plumbing pipe pinhole leak distribution (2000–2004) (small triangles show number of leaks less than 10). (From Lee and Loganathan, *Virginia Water Central*, Virginia Water Resources Research Center, 2006. With permission.)

Performance, 2002), the minor system that is at least 5–10 times larger is generally not well addressed. When a pipe has a leak, the homeowner is faced with the following issues: water damage cost, repair cost, service disruption, possible lowering of home value, home insurance/premium increase/nonrenewal, and health consequences, resulting from brown mold growth and mental stress. The interaction between hydraulics, water quality parameters, and pipe material has to be understood. There is a need for a decision model recommending whether to continue to repair or replace the system.

Another example is related to urbanization, which increases imperviousness along with quick draining channelization for runoff. These modifications result in increases in runoff volume and peak flow and decreases in time to peak and recharge to groundwater storage. It is required by many local governments that the post development peak flows of certain designated frequencies must not exceed those of the predevelopment conditions and proper flood controlling detention ponds must be installed to contain the increased post development peak flows. An associated problem is that the falling rain can capture airborne pollutants with it as well as dislodge particulate matter settled on the ground. The surface runoff can dissolve as well as carry the adsorbed pollutants downstream. The water quality impacts of stormwater discharges on receiving lakes and rivers are quite significant, as indicated by the Environmental Protection Agency's (EPA) guidelines for stormwater discharge permits in the National Pollutant Discharge Elimination System (NPDES) program. Civil action may be brought by the EPA against large municipalities that do not comply with the permit requirements.

The above description addresses the most frequent drainage problem within an urban area. The control measures put in place together to alleviate the problem constitute the *minor drainage system* consisting of street gutters, inlets, culverts, roadside ditches, swales, small channels, and pipes. A less frequent but significant flood damage problem arises when an urban area is located along a water course draining a large basin. The issue here is controlling the flood plain use and providing for proper drainage of flood waters. The Federal Emergency Management Agency (FEMA) issues guidelines with regard to flood plain management under the National Flood Insurance Program (NFIP). The components involved in this latter problem are the natural waterways, large man-made conduits including the trunkline sewer system that receives the stormwater from the minor drainage system, large water impoundments, and other flood protection devices. These components together are designated as the *major drainage system*. Distributing, dimensioning, maintaining, and operating these devices for maximum benefit is part of efficient water resources management.

24.1.3 Chapter Organization

Even though the foregoing discussion seems to delineate large and small-scale systems, in reality they all are intertwined. For example, phosphate is typically added as a corrosion inhibitor at the treatment plant but may have to be removed at the wastewater plant before it reaches the receiving water body. It has become clearer that water systems cannot be engineered in isolation. This introduction emphasizes the need for a holistic approach. In the following, we present a set of water resources problem formulations with the understanding that by no means are these exhaustive. All real life water resources problems require an examination of alternatives. These problems typically require a combined simulation-optimization approach highly suited to evolutionary optimization techniques.

There are also specific structures that can be exploited for many of the problems. They seem to break into the following categories, namely (1) project planning including sequencing of multiple projects, (2) design, (3) operation, and (4) replacement. The goals and data needs under these categories differ. Project planning involves national and regional economic development, damage reduction, protection of ecosystems, and improvement of quality of life. *Planning* process requires a reconnaissance study followed by feasibility assessment on a system-wide basis (National Research Council, 1999). Typically, a design life is used. It includes factors such as period of cost recovery, meeting the intended function, technological obsolescence, and the life of the weakest component. The *design* aspect mainly focuses on dimensioning of the components subject to regulatory, physical, societal, and technical constraints. The *operational* phase addresses system behavior of an already built system and its coexistence under changing needs. McMahon and Farmer (2004) provide a lucid account of the need for reallocation of federal multipurpose reservoirs because of changing social preferences and purposes that were originally intended. The *replacement* facet assumes that the facility's continued service is essential and it fails periodically. Here the analysis focuses on failure patterns and alternative components to provide improved service.

In this chapter, we cover two major systems, namely, reservoirs and water distribution networks. The minor system, namely, the home plumbing system, is actually much larger than the public water distribution system and is addressed next. We present outcomes of a focus study group on consumer preferences toward different attributes of plumbing pipes. Urban stormwater management related to nonpoint source pollution is a major problem. In addition to regulations, financial considerations due to limited space and price of land steer the engineering design. Groundwater contamination is widespread and remains a significant cause of concern. All these problems require a process simulation-optimization approach. In this chapter, mathematical formulations for the efficient utilization of water resources are presented.

24.2 Optimal Operating Policy for Reservoir Systems

24.2.1 Introduction

In this section, we consider one of the largest and most complex water resource systems, namely, the reservoir. Labadie (2004) provides a comprehensive updated review of optimization of reservoir systems. He also includes a discussion on available software. He points out that with construction of no new large-scale projects there is a need for methods that can guide in efficient operation of these systems. It is not clear whether the increased demands on fossil fuels will result in a re-examination of policy toward building dams. A comprehensive analysis of reallocation of reservoir storage for changing uses and social preferences is given in McMahon and Farmer (2004) (also see National Research Council, 1999). A reservoir system captures, stores, and distributes water in sufficient quantities at sufficient head when and where needed. Such a system is typically subjected to three kinds of analysis. At preconstruction planning stage, determine the required capacity using the lowest inflows called the critical duration sequence that shifts burden to storage in meeting the demand; or compute the maximum yield that is obtainable for a given capacity. At post-construction planning stage, reevaluate performance with multiple inflow sequences in meeting the demand.

The United States Army Corps of Engineers (USACE) is the primary agency responsible for flood control for major reservoir systems. In the following, USACE's storage classification (Feldman, 1981; Wurbs, 1991) is adopted. In broad terms, reservoir storage is divided into three zones. The bottom zone, called *inactive pool*, is located beneath the low-level outlet so that no release is possible. This storage is meant to trap sediment and may provide for fish habitat. The top zone, called *flood control storage*, is empty space and is used only during periods of high flows so that releases will not exceed downstream channel capacity. When the top of the flood control zone is exceeded, the reservoir is out of control and an emergency spillway is used to divert the bulk of the flow. Major dams are designed for large flood events called *probable maximum floods* based on meteorological factors. The moisture maximization procedure employed in computing *probable maximum precipitation* provides the upperbound flood and it is highly unlikely this design event can ever be exceeded (Wurbs, 1991). The middle zone, called *conservation storage*, is utilized to satisfy day to day demands.

The USACE employs an *index level* associated with the top of each storage zone from the inactive pool, conservation pool to the flood pool. For example, a reservoir of capacity 700,000 ac-ft may have an index level of 1 assigned to a storage of 50,000 ac-ft at the top of the inactive pool; index 2 may correspond to 200,000 ac-ft at the top of the conservation pool; and index 3 corresponds to 700,000 ac-ft at the top of the flood control pool. The number of index levels can be increased by further subdividing the storage zones, mainly in the conservation pool. When the index levels at all reservoirs coincide, the entire system is considered to be in equilibrium and deviations from the equilibrium should be minimized. The equilibrium maintenance of index levels sets up a framework for a mathematical formulation in determining the reservoir operations schedule (Loganathan, 1996).

In real-time operations, forecasted inflows are used. The problem is solved with a rolling horizon of a few days with the results being implemented for the current day only. For the next day, the problem is re-solved with the updated forecast information from the current day observations. To aid the operator in day to day operations a *rule curve* is used. It specifies where the water level (storage) should be as a function of a particular month or week. These rule curves are based on the river flow sequence and the storage capacity that are available to smooth out operations. Because they are typically based on long-term simulations subjecting the system to a variety of inflow sequences, inherent in a rule curve is the anticipatory nature with regard to the stream flow. Hedging is the ability to shift storage to later times by providing only a portion of the demand when in fact it is possible to meet the entire demand. It is done to minimize drastic deficits over prolonged low flow periods.

The maximum flow that can be guaranteed during the most adverse stream flow period, also known as *critical period*, is the *firm yield*. The energy that can be produced under the most adverse flow conditions is called *firm energy* (Mays and Tung, 1992). The firm yield and firm energy set the lower bounds on flow and energy that are available for all times from a reservoir. A *flow duration curve*, which is a plot of flow versus percentage of times flow that is equaled or exceeded, is utilized to select the firm yield and firm energy at the near 100% exceedance level. This curve is important because it shows the inflow potential for a reservoir. The larger the reservoir, the higher the firm yield and energy that can be sustained for a prolonged time period. Preliminary power plant capacity is determined based on the average flow between selected threshold percentage exceedance values (American Society of Civil Engineers, 1989; Warnick et al., 1984). These limits help to impose a realistic demand pattern over the system.

The demand requirements are as follows. The water supply demands require specific amounts of flow at various locations and time; hydropower demands require both flow and hydraulic head at the turbine; navigation operations require sufficient flow depths in the channels and flows for the locks; recreation aspects require minimum water level fluctuations in the reservoir and sufficient depth for boats in rivers; ecological considerations require maintaining water quality, selective withdrawal of water from certain zones of storage for temperature maintenance downstream, providing sufficient amounts of aeration for required quantity of dissolved oxygen and at the same time avoiding supersaturation of water with gases such as nitrogen, which can cause gas bubble disease in fish and other biota (Mattice, 1991). These requirements in general become constraints on flow and reservoir storage. In the following some general guidelines as recommended by the U.S. Army Corps of Engineers (USACE) in meeting the demands are outlined.

24.2.2 Reservoir Operation Rules

The USACE's (2003) HEC-ResSim computer program [HEC5 (USACE, 1982)] is designed to simulate system behavior including hydropower under a set of operating rules. The storage index levels for each reservoir along with downstream flow conditions, anticipated inflows, and a view toward reaching equilibrium for the system as a whole dictate how the releases should be made. This approach eliminates much of the leeway in determining the releases; however, this strategy helps in minimizing flooding and empties the system as quickly as possible. Such a strategy should be reconsidered in a situation wherein flood control is not a major issue. In the following the operating rules are presented.

Reservoir Releases

Reservoir releases can be based on the following criteria: (1) channel capacity at the dam; (2) rate of change of release so that current period release cannot deviate from previous period release by more than a specified amount unless the reservoir is in flood surcharge operation; (3) storage not exceeding the top of conservation pool; (4) downstream control point flooding potential; (5) meeting specified target levels; (6) exceeding required low flows; (7) reaching various storage values; and (8) hydropower requirements.

Reservoir Filling/Drawing Down

Reservoirs are operated to satisfy constraints at the individual reservoirs to maintain specified flows at downstream control points and to keep the system as a whole in balance.

Constraints at individual reservoirs:

1. When the level of a reservoir is between the top of the conservation pool and top of the flood pool, releases are made to attempt to draw the reservoir to the top of the conservation pool without exceeding the channel capacity.
2. Releases are made compatible with the storage divisions within the conservation pool.
3. Channel capacity releases (or greater) are to be made prior to the time the reservoir storage reaches the top of the flood pool if forecasted inflows are excessive. The excess flood water is dumped if sufficient outlet capacity is available. If insufficient capacity exists, a surcharge routing is made.
4. Rate of change criterion specifies the maximum difference between consecutive period releases.

Constraints for downstream control points:

1. Releases are not made (as long as flood storage remains) which would contribute to flooding during a predetermined number of future periods. During flooding at a downstream location, there may not be any release for power requirements.
2. Releases are made, where possible, to exactly maintain downstream releases at channel capacity for flood operation or for minimum desired or required flows for conservation operation.

System Balancing

To keep the entire system in balance, priority is given to making releases from the reservoir with the highest index level; if one of two parallel reservoirs has one or more reservoirs upstream, whose storage should be considered in making the releases, an upstream reservoir release is permitted only if its index level is greater than both the levels of the downstream reservoir and the combined reservoir equivalent index level. The combined equivalent index

Water Resources

level corresponds to the sum of the reservoirs' storage at the appropriate index levels. For example, if reservoirs A, B, and C have index level 5 at storage 400, 300, and 200 (in kiloacre feet say) the combined equivalent index 5 corresponds to the storage of 900. For an elaborate discussion the reader is referred to the HEC5 user's manual (USACE, 1982). In the following, a general problem formulation that accommodates many of the facets of the above operating policy guidelines is presented.

24.2.3 Problem Formulation

A general mathematical programming formulation for the reservoir-operations problem may be stated as
Problem 24.1:

$$\text{minimize: deviations} = F(Sd_{11}^+, Qd_{11}^+, Ed_{11}^+, Sd_{11}^-, Qd_{11}^-, Ed_{11}^-, \ldots) \quad (24.1)$$

Subject to:

Real Constraints

Reservoir continuity

$$S(i, t+1) + R(i,t) = S(i,t) + I(i,t) \quad \text{for } i = 1, \ldots, \text{NRES}; \ t = 1, \ldots, T \quad (24.2)$$

Reach routing

$$Q(i,t) = g[P(i,t), \text{TF}(i,t), Q(j_i), R(j_i)] \quad \text{for } i = 1, \ldots, \text{NREACH}; \ t = 1, \ldots, T \quad (24.3)$$

Hydropower

$$R(i,t) = A(i,t) + B(i,t) \quad \text{for } i = 1, \ldots, \text{NPLANT}; \ t = 1, \ldots, T \quad (24.4)$$

$$E(i,t) = \eta(i)\gamma A(i,t) H[S(i,t)] \quad \text{for } i = 1, \ldots, \text{NPLANT}; \ A = 1, \ldots, T \quad (24.5)$$

Bounds on flows

$$\text{LQ}(i) \leq Q(i,t) \leq \text{UQ}(i) \quad \text{for } i = 1, \ldots, \text{NREACH}; \ t = 1, \ldots, T \quad (24.6)$$

Bounds on storages

$$\text{LS}(i) \leq S(i,t) \leq \text{US}(i) \quad \text{for } i = 1, \ldots, \text{NRES}; \ t = 1, \ldots, T \quad (24.7)$$

Powerplant capacity

$$E(i,t) \leq \text{EP}(i) \quad \text{for } i = 1, \ldots, \text{NPLANT} \quad (24.8)$$

Nonnegativity of variables

$$S(i,t), R(i,t), Q(i,t), Sd_{i,t}^-, Sd_{i,t}^+, Qd_{i,t}^-, Qd_{i,t}^+, Ed_{i,t}^-, Ed_{i,t}^+ \geq 0 \quad (24.9)$$

Goal Constraints

$$S(i,t) + Sd_{i,t}^- - Sd_{i,t}^+ = \text{TS}(i,t) \quad \text{for } i = 1, \ldots, \text{NRES}; \ t = 1, \ldots, T \quad (24.10)$$

$$Q(i,t) + Qd_{i,t}^- - Qd_{i,t}^+ = \text{TQ}(i,t) \quad \text{for } i = 1, \ldots, \text{NREACH}; \ t = 1, \ldots, T \quad (24.11)$$

$$E(i,t) + Ed_{i,t}^- - Ed_{i,t}^+ = \text{TE}(i,t) \quad \text{for } i = 1, \ldots, \text{NPLANT}; \ t = 1, \ldots, T \quad (24.12)$$

in which $S(i,t)$ = the storage at the beginning of period; $I(i,t)$ and $R(i,t)$ = inflow and release during period t for reservoir i, respectively; NRES is the number of reservoirs;

T = the operating horizon; $Q(i,t)$ = the flow in reach i written as some function g; $P(i,t)$ and TF (i,t) = precipitation and tributary flow to reach i for period t, respectively; $A(i,t)$ = flow for hydropower from reservoir i for period t; $B(i,t)$ = nonpower release from reservoir i for period t; $\eta(i)$ = efficiency of plant i; γ = specific weight of water; H$[S(i,t)]$ = head over the turbine, a function of reservoir storage and tailwater level; EP(i) = plant capacity for the ith power plant; NPLANT = the number of power plants; ji = the set of control stations contributing flow for reach i; NREACH = the number of reaches; LQ(i) and UQ(i) = lower and upper bounds for flow in reach i, respectively; LS(i) and US(i) = lower and upper bounds for storage in reservoir i, respectively; Sd_i^-, Sd_i^+, and Qd_i^-, Qd_i^+, and Ed_i^-, Ed_i^+ = the slack and surplus deviational variables for storage, flow, and power respectively; TS(i,t) = the storage target for reservoir i at time t; TQ(i,t) = the flow target for reach i at time t; and TE(i,t) = power target for plant i at time t. The storage goal constraints provide for recreation and hydropower; the flow goal constraints provide for water supply, navigation, instream flow and irrigation; the hydropower goal constraints provide for target power production at each plant.

Hydropower

Eschenbach et al. (2001) provide a detailed description of Riverware optimization decision support software based on a goal programming formulation. Linearization is used to solve it as a linear program. They also include a description of Tennessee Valley Authority's use of the software in managing their reservoirs to determine optimal hydropower schedules. The hydropower aspect has been considered in various forms, such as assigning a target storage for power production; maintaining a fixed head; linearization by Taylor series about an iterate; induced separability/subsequent linearization; division of storage into known intervals and choosing constant heads for each interval with the selection of intervals aided by integer variables or by dynamic programming state transition; optimal control strategy and direct nonlinear optimization. Comprehensive reviews are given in Labadie (2004), Yeh (1985), Wurbs (1991), and Wunderlich (1991). Martin (1987) and Barritt-Flatt and Cormie (1988) provide detailed formulations. Successive linear programming (Palacios-Gomez et al., 1982; Martin, 1987; Tao and Lennox 1991) and separable programming (Can et al., 1982; Ellis and ReVelle, 1988) are widely adopted for solution. Pardalos et al. (1987) have offered a strategy to convert an indefinite quadratic objective into tight lower bounding separable convex objective which can be solved efficiently. If the hydropower constraints are dropped or linearized, then using linear routing schemes in (Equation 24.3) Problem 24.1 becomes a linear program (Loganathan and Bhattacharya, 1990; Changchit and Terrell, 1989). Martin (1995) presents a well-detailed real system application to the Lower Colorado River Authority district. Moore and Loganathan (2002) present an application to a pumped storage system.

Trigger Volumes for Rationing

Thus far the discussion has paid more attention to floods. However, water shortages do occur and how to cope with them is a critical issue. Lohani and Loganathan (1997) offer a procedure for predicting droughts on a regional scale. In the case of a reservoir one should know at what low storage levels the regular releases should be curtailed toward an impending drought (Shih and ReVelle, 1995). If the available storage plus forecasted inflow is less than V_{1p}, the first-level trigger volume, level 1 rationing is initiated. If the anticipated storage for the next period is less than V_{2p}, the second level trigger volume, level 2 rationing, which is severer than level 1 rationing, is initiated. The objective is not to have any rationing at all by prudently utilizing the storage to satisfy the regular demands. With this objective, the V_{1p} and V_{2p} will be chosen optimally by preserving storage whenever possible to hedge against future shortfalls.

Shih and ReVelle (1995) propose optimal trigger volumes V_{1p} and V_{2p} to initiate two levels of rationing of water during drought periods as described by the following constraints.

$$S_{t-1} + I_t \leq V_{1p} + M\delta_{1t} \tag{24.13}$$

$$S_{t-1} + I_t \geq V_{1p} - M(1 - \delta_{1t}) \tag{24.14}$$

$$S_{t-1} + I_t \leq V_{2p} + M\delta_{2t} \tag{24.15}$$

$$S_{t-1} + I_t \geq V_{2p} - M(1 - \delta_{2t}) \tag{24.16}$$

$$R_t = (1 - \alpha_1)D^*\delta_{1t} + (\alpha_1 - \alpha_2)D*\delta_{2t} + \alpha_2 D \tag{24.17}$$

$$V_{1p} \geq (1 + \beta_1)V_{2p} \tag{24.18}$$

$$\delta_{1t} \leq \delta_{2,t+1} \text{ with } \delta_{1t}, \delta_{2t} \in 0 \text{ or } 1 \text{ binary} \tag{24.19}$$

in which V_{1p}, V_{2p} = cutoff storage values for levels 1 and 2 rationing; α_1, α_2 = percentages of demand D provided during rationing levels 1 and 2, respectively; β_1 = suitably chosen percentage level, say 20%. When $\delta_{1t} = 1$, from Equation 24.14 there is no rationing; when $\delta_{1t} = 0$, and $\delta_{2t} = 1$ from Equations 24.13 and 24.16 there is level 1 rationing. When $\delta_{2t} = 0$, there is level 2 rationing and $R_t = \alpha_2 D$. Note that when $\delta_{2t} = 0$, we must have $\delta_{1t} = 0$ by Equation 24.14. When $\delta_{1t} = 1$, δ_{2t} has to be 1 and $R_t = D$. Constraint 24.19 says that level 1 rationing must precede level 2 rationing. The objective is to maximize $\sum \delta_{1t}$ over t.

Linear Release Rule

There have been attempts to follow the rule curve in the sense of relating a fixed amount of storage to a specific time period. For example, the release rule given by ReVelle et al. (1969), also called linear decision rule, is

$$R_t = S_{t-1} - b_t \tag{24.20}$$

in which S_{t-1} = storage at the end of $t - 1$; b_t = decision parameter for period t; R_t = release for period t. This rule says that from the end of $t - 1$ period storage S_{t-1}, save b_t amount and release the remaining during period t as R_t regardless of inflow, I_t, which will add to b_t. It is seen from Equation 24.20 that the end period of period t storage S_t is the sum of the left-over storage b_t from the previous period plus the inflow. That is,

$$S_t = b_t + I_t \tag{24.21}$$

With the aid of Equation 24.21 each storage variable S_t can be replaced by the corresponding decision variable b_t which will be optimally determined for each t. If t represents a month, $b_1 = b$ (January) is fixed and repeats in an annual cycle in replacing S_1, S_{13}, S_{25}, and so on. From Equation 24.21 it is seen that the single decision variable b_1 along with the known I_1, I_{13}, and I_{25} replaces the three variables S_1, S_{13}, and S_{25}. Therefore, this substitution leads to a few decision variables. These twelve b_t values help to regulate a reservoir. Refinements to this type of release rule are given in Loucks et al. (1981) (also see Karamouz et al., 2003, and Jain and Singh, 2003).

Chance Constraint

Instead of using the predicted value of the inflow, its quantile may be used at a particular probability level. For example, consider the probability that storage at the end of time period t denoted by S_t exceeding some maximum storage SMAX must be less than or equal to a small value, say 0.1. That is,

$$P[S_t \geq \text{SMAX}] \leq 0.1 \tag{24.22}$$

Substituting Equation 24.21 in Equation 24.22 we obtain

$$P[I_t \geq \text{SMAX} - b_t] \leq 0.1 \tag{24.23}$$

with its deterministic equivalent given by

$$\text{SMAX} - b_t \geq i_{0.9}(t) \tag{24.24}$$

in which $i_{0.9}(t)$ = cutoff value at the 90% cumulative probability level for period t.

Firm Yield, Storage Capacity, and Maximization of Benefits

As mentioned before, firm yield and storage capacity to meet fixed demands are important parameters. The determination of firm yield requires maximize [minimum $R(i, t)$] for specified reservoir capacity. The required storage capacity is obtained by minimizing capacity to satisfy the demands. The optimal capacity therefore will be determined by the critical period with the most adverse flow situation requiring maximum cumulative withdrawal as dictated by the continuity constraint (Equation 24.2). Of course, a large demand that cannot be supported by the inflow sequence would trigger infeasibility. While the objective in Equation 24.1 minimizes the deviations from targets, one may choose to maximize the benefits from the releases including hydropower, and recreation. Such a formulation can yield a problem with nonlinear objective and linear constraint region. Can et al. (1982) discuss strategies to linearize the objective function so that the problem can be solved as a linear program.

In the simulation program HEC5, by employing the index level discretization scheme for storage capacity, releases compatible with the storage are made, which enhances operations within the conservation zone. This scheme also minimizes flood risk at a reservoir by having its storage compatible with the others. The program attempts to find near optimal policies by cycling through the allocation routines several times. This approach has a tendency to place the releases at their bounds. Practical considerations in reservoir operations include a well spelt out emergency preparedness plan and the operation and maintenance of electrical, mechanical, structural, and dam instrumentation facilities by the operating personnel located at or nearest to the dam. These practical operating procedures at the dam site are covered in the standing operating procedures guide (U.S. Department of Interior, 1985). Because loss of lives and property damages are associated with any faulty decisions with regard to reservoir operations, all model results should be subjected to practical scrutiny.

24.3 Water Distribution Systems Optimization

24.3.1 Introduction

Water distribution systems constitute one of the largest public utilities. Optimal water distribution system design remains intriguing because of its complexity and utility. Comprehensive reviews of optimization of water distribution systems are given in Boulos et al. (2004) and Mays (2000). Bhave (1991) and Walski (1984) provide thorough details on formulating water distribution network problems. The pipe network problems have feasible regions that are nonconvex. Also, the objective function is multimodal. These two aspects make the conventional (convex) optimization methods to result in a local optimum sensitive to the starting point of the search. In this section, a standard test problem from the literature is considered. The pipe network is judiciously subjected to an outer search scheme that chooses alternative flow configurations to find an optimal flow division among pipes. An inner linear program is employed for the design of least cost diameters for the pipes.

The algorithm can also be employed for the optimal design of parallel expansion of existing networks. Three global search schemes, multistart-local search, simulated annealing, and genetic algorithm, are discussed. Multistart-local search selectively saturates portions of the feasible region to identify the local minima. Simulated annealing iteratively improves the objective function by finding successive better points and to escape out of a local minimum it exercises the Metropolis step, which requires an occasional acceptance of a worse point. Genetic algorithm employs a generation of improving solutions as opposed to the other two methods that use a single solution as an iterate. Spall (2003) provides comprehensive details on search methods.

24.3.2 Global Optimization

It is not uncommon to find real life problems that have cost or profit functions defined over a nonconvex feasible region involving multiple local optima and the pipe network optimization problem falls into this group. While classical optimization methods find only a local optimum, global optimization schemes adapt the local optimum seeking methods to migrate among local optima to find the best one. Of course, without presupposing the nature of local optima, a global optimum cannot be guaranteed except to declare a relatively best optimum. Detailed reviews are given in Torn and Zilinskas (1987) and Rinnooy Kan and Timmer (1989). Consider Problem P0 given by:

Problem P0:

$$\text{Minimize} \quad f(x)$$
$$\text{Subject to:} \quad g_i(x) \geq 0 \quad \text{for } i = 1, 2, \ldots, m$$

Let the feasible region be $X = \{x | g_i(x) \geq 0\}$. A solution x^1 is said to be a local optimum if there exists a neighborhood B around x^1 such that $f(x^1) \leq f(x)$ for all $x \in B(x^1)$. A solution x^g is said to be a global optimum if $f(x^g) \leq f(x)$ for all $x \in X$.

24.3.3 Two-Stage Decomposition Scheme

In this section, a two-stage decomposition with the outer search being conducted among feasible flows, and an inner linear programming to optimally select pipe diameters and hydraulic heads is employed. The following formulation retains the same general inner linear programming framework when addressing multiple loadings, pumps, and storage tanks as given in Loganathan et al. (1990). However, for the sake of clarity and the nature of the example problem these elements are suppressed. Consider a pipe network comprised of N nodes. Let S be the set of fixed head nodes. Let $\{N-S\}$ be the set of junction nodes and $\mathbf{L}$ be the set of links. $Q_{(i,j)}$ is the steady state flow rate through link $(i,j) \in \mathbf{L}$. Let $L_{(i,j)}$ denote the length of link $(i,j) \in \mathbf{L}$ and $D_{(i,j)}$ be its diameter which must be selected from a standard set of discrete diameters $D = \{d_1, d_2, \ldots, d_M\}$. In the present formulation it is assumed that each link (i,j) is made up of M segments of *unknown* lengths $x_{(i,j)m}$ (decision variable) but of *known* diameter d_m for $m = 1, 2, \ldots, M$ (Karmeli et al., 1968). Let $C_{(i,j)m}$ be the cost per unit length of a pipe of diameter d_m. Let r_k be the path from a fixed head node (source) to demand node, k. Let $\mathbf{P}$ be the set of paths connecting fixed head nodes and basic loops. Let there be P_l loops and b_p denote the head difference between the fixed head nodes for path p connecting them; and it is zero corresponding to loops. Let H_s be the fixed head and $H_k^{\min}$ be the minimum head at node $k \in \{N-S\}$. Let q_i be the supply at node i which is positive; if it is demand, it is negative. Let $J_{(i,j)m}$ be the hydraulic gradient for segment m (partial length of a link with diameter d_m) which is given by

$$J_{(i,j)m} = K[Q_{(i,j)}/C]^{1.85} d_m^{-4.87} \qquad (24.25)$$

in which $K = 8.515(10^5)$ for $Q_{(i,j)}$ in cfs and d_m in inches; C = Hazen-Williams Coefficient. The pipe network problem may be stated as follows:

Problem P1:

$$\text{Minimize} \quad f(x) = \sum_{(i,j)} \sum_{m=1}^{M} C_{(i,j)m} x_{(i,j)m} \tag{24.26}$$

$$\text{Subject to:} \quad \sum_j Q_{(i,j)} - \sum_j Q_{(j,i)} = q_i \quad \text{for } i \in \{N\text{--}S\} \tag{24.27}$$

$$H_s - H_k^{\min} - \sum_{(i,j) \in r_k} \pm \sum_m J_{(i,j)m} x_{(i,j)m} \geq 0 \quad \text{for } s \in S \text{ and } k \in \{N\text{--}S\} \tag{24.28}$$

$$\sum_{(i,j) \in p} \pm \sum_m J_{(i,j)m} x_{(i,j)m} = b_p \quad \text{for } p \in P \tag{24.29}$$

$$\sum_m x_{(i,j)m} = L_{(i,j)} \quad \text{for } (i,j) \in L \tag{24.30}$$

$$x_{(i,j)} \geq 0 \tag{24.31}$$

In Problem P1 pipe cost objective function (Equation 24.26) is minimized; constraint 24.27 represents steady state flow continuity; constraint 24.28 is the minimum head restriction; constraint 24.29 represents the sum of head losses in a path which is zero for loops; constraint 24.30 dictates that sum of segment lengths must equal link length; constraint 24.31 is the nonnegativity on segment lengths. The decision variables are: $Q_{(i,j)}$, $J_{(i,j)m}$, and $x_{(i,j)m}$. The following two-stage strategy Problem P2 is suggested for the solution of Problem P1.

Problem P2:

$$\min_{Q_{(i,j)}} \left[\min_{x \in X} f(x) \right] \tag{24.32}$$

in which $Q_{(i,j)}$ are selected as perturbations of the flows of an underlying near optimal spanning tree of the looped layout satisfying constraint 24.27 and X is the feasible region made up of constraints 24.28 through 24.31. It is worth noting that the optimal layout tends to be a tree layout. Deb (1973) shows that a typical pipe cost objective is a concave function that attains its minimum on a boundary point resulting in a tree solution (also see Deb and Sarkar, 1971). Gessler (1982) and Templeman (1982) also argue that because optimization has the tendency to remove any redundancy in the system the optimal layout should be a tree. However, a tree gets disconnected even when one link fails. Loganathan et al. (1990) have proposed a procedure that yields a set of near optimal trees that are to be augmented by loop-forming pipes so that each node has two distinct paths to source. It is suggested that the flows of the optimal tree be considered as initial flows that are to be perturbed to obtain flows in all pipes of the looped network. It is observed that the inner Problem P3 of Problem P2 given by

Problem P3:

$$\min_{x \in X} f(x), \text{ for fixed flows} \tag{24.33}$$

is a linear program that can be solved efficiently. To choose the flows in the outer problem of P3, multistart-local search and simulated annealing are adopted. To make the search efficient, first a set of flows corresponding to a near optimal spanning tree of the network is found. The flows in the looped network are taken as the perturbed tree link flows by

$$Q_{(i,j)}(\text{loop}) = Q_{(i,j)}(\text{tree}) + \sum \pm \Delta Q_{p(i,j)} \qquad (24.34)$$

in which the sum is taken over loops "p" which contain link (i, j) with positive loop change flow ΔQ used for clockwise flow.

24.3.4 Multistart-Local Search

Multistart-local search covers a large number of local minima by saturating the feasible region with randomly generated starting points and applying a local minimization procedure to identify the local optima. A suitable generating function is needed. Because flows in the looped network are generated by perturbing the optimal tree flows, a probability density function for loop change flows with mean zero is a good choice. However, other parameters of the density function must be adjusted by experimentation for the problem under consideration. The algorithm is summarized as follows:

Procedure MULTISTART

   Step 0: (*Initialization*) Select a suitable density function (pdf) with known parameters and assign number of seed points to be generated, NMAX; determine an estimate of global optimal value f_g; assign a large value for the incumbent best objective function value, BEST; set $n = 1$.

   Step 1: (*Iterative Local Minimization*) Until $n = $ NMAX
   do: Generate a point using the pdf that serves as perturbation flow in Equation 24.34. Apply the minimizer Problem P3. Update incumbent BEST.

   Step 2: (*Termination*) If $|f_g - BEST| < tolerance$, report BEST and the optimal solution and stop. Otherwise, revise parameter values for the generating function, and/or global value, f_g and tolerance. Set $n = 1$; go to Step 1.

   END Procedure.

24.3.5 Simulated Annealing

Simulated annealing is an iterative improvement algorithm in which the cost of the current design is compared with the cost of the new iterate. The new iterate is used as the starting point for the subsequent iteration if the cost difference is favorable. Otherwise, the new iterate is discarded. It is clear that such an algorithm needs help to get out of a local optimum to wander among local minima in search of the global minimum. It uses the Metropolis step to accomplish this. The Metropolis step dictates that a worse point (higher objective value for a minimization problem) be accepted from the current point with a user specified probability "pR." The algorithm accepts the next iterate from the current iterate with probability one if it is a better point (lower objective value).

From the given initial point, the corresponding objective function value, f_0, is calculated. Then, a point is randomly generated from a unit hypersphere with the current iterate at its center. This random point specifies a random direction along which a new point is generated by taking a user specified step size of α from the current iterate. A new objective function value (f_1) is evaluated at the point. The new point is accepted with the probability

p given by:

$$pR = 1 \quad \text{if } \Delta f = f_1 - f_0 \leq 0 \qquad (24.35)$$
$$= \exp(-\beta \Delta f / f_0^\theta) \quad \text{if } \Delta f > 0$$

where β and θ are user-specified parameters. From the first condition of Equation 24.35 it is seen that the method readily accepts an improving solution if it is found. From the second condition, it is apparent that an inferior point (i.e., with worse objective function value) becomes an acceptable solution with probability $\exp(-\beta \Delta f / f_0^\theta)$ when an improving solution is not available. Bohachevsky et al. (1986) recommend selecting the parameter β such that the inequality $0.5 < \exp(-\beta \Delta f / f_0^\theta) < 0.9$ holds, which implies that 50–90% of the detrimental steps are accepted. The procedure continues until a solution that is within a *tolerance* level from the user-specified (estimated) global optimal value is obtained. The algorithm is stated as follows (Loganathan et al., 1995).

Procedure ANNEALING

Step 0: (*Initialization*) Let f_g be the user-specified (estimated) global optimal value (which may be unrealistic): α a step size, and β and θ acceptance probability parameters. Let the vector of loop change flows be $\boldsymbol{\varepsilon} = \{\Delta Q_1, \Delta Q_2, \ldots, \Delta Q_{P_l}\}$. Let $\boldsymbol{\varepsilon}^0$ be the arbitrary feasible starting point of dimension P_l.

Step 1: (*Local Minimization*) Set $f_0 = f(\boldsymbol{\varepsilon}^0)$. If $|f_0 - f_g| <$ *tolerance*, stop.

Step 2: (*Random Direction*) Generate P_l (number of loops) independent standard normal variates, $Y_1, \ldots, Y_{P_l}$ and compute components of unit vector $\mathbf{U}$: $U_i = Y_i/(Y_1^2 + \cdots Y_{P_l}^2)^{1/2}$, for $i = 1, \ldots, P_l$.

Step 3: (*Iterative Local Minimum*) Set $\boldsymbol{\varepsilon}^1 = \boldsymbol{\varepsilon}^0 + \alpha \mathbf{U}$. Apply minimizer Problem P3. If $\boldsymbol{\varepsilon}^1$ is infeasible, return to Step 2. Otherwise, set $f_1 = f(\boldsymbol{\varepsilon}^1)$ and $\Delta f = f_1 - f_0$.

Step 4: (*Termination*) If $f_1 > f_0$, go to step 5. Otherwise, set $\boldsymbol{\varepsilon}^0 = \boldsymbol{\varepsilon}^1$ and $f_0 = f_1$. If $|f_0 - f_g| <$ *tolerance*, save $\boldsymbol{\varepsilon}^1$ and stop. Otherwise go to Step 2.

Step 5: (*Acceptance–Rejection*) ($f_1 > f_0$): Set $pR = \exp(-\beta \Delta f / f_0^\theta)$. Generate a uniform 0–1 variate v. If $v \geq pR$, go to Step 2. Else $v < pR$ and set $\boldsymbol{\varepsilon}^0 = \boldsymbol{\varepsilon}^1$, $f_0 = f_1$, and go to Step 2.
END Procedure.

The success of the procedure depends upon the selection of the parameters α (step size), β (exponential probability parameter), and θ (exponent in the exponential probability function), which must be adjusted as necessary. If β is too large (pR is too small) too many function evaluations are needed to escape from a local minimum; if β is too small (pR is close to unity), an inefficient search results by accepting almost every inferior point generated. The parameter α controls the search radius around $\boldsymbol{\varepsilon}^0$: α should be chosen such that the distance is sufficiently long to prevent from falling back to the same local minimum. Also, estimates for f_g and *tolerance* should be updated based on the performance of the algorithm. Goldman and Mays (2005) provide additional applications of simulated annealing to water distribution systems.

24.3.6 Extension to Parallel Expansion of Existing Networks

Another important aspect that must be considered is the expansion of existing networks. In the present formulation parallel expansion of existing pipes is considered. Adaptation of Problem P2 for existing networks is as follows. For existing pipes, the pipe diameters are fixed and parallel links may be required for carrying additional flow in order not to increase the head loss. The flow from node i to node j, for a parallel system, is denoted by $Q_{(i,j)} = Q_{(i,j),O} + Q_{(i,j),N}$, in which subscripts O and N indicate Old and New, respectively.

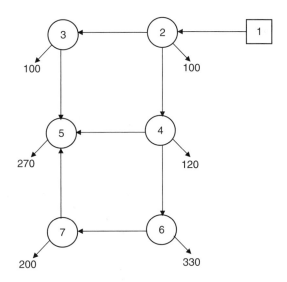

FIGURE 24.2 Two-loop example network.

TABLE 24.1 Node Data

Node i	Elevation (m)	Minimum Head (m)	Demand (m³/h)
1	210	0	−1120
2	150	30	100
3	160	30	100
4	155	30	120
5	150	30	270
6	165	30	330
7	160	30	200

It is suggested that Problem P2 be solved with the restriction that an existing pipe should not be undersized. From the solution to Problem P2 the head loss for each pipe can be obtained by multiplying the optimal segment lengths, $x_{(i,j)m}$'s and the corresponding head gradients $J_{(i,j)m}$'s. Because optimal head loss, friction parameter, and existing pipe size are known, the flow in an existing link and therefore the flow for the parallel pipe can be computed (Wylie and Streeter, 1985). The parallel pipe in turn is decomposed into discrete diameter pipes in series.

24.3.7 Analysis of an Example Network

In this section, an example network (Alperovits and Shamir, 1977) using the data given therein is considered (see Figure 24.2). The data for the minimization problem P2 are the minimum and maximum heads and demand at each node, length and Hazen-Williams coefficient C for each link, and the unit cost for different pipe diameters. These data for Alperovits and Shamir network are given in Tables 24.1 and 24.2. Each link has length 1000 m and a Hazen-Williams coefficient of 130. For known flow rates, Problem P3 is a linear program. The decision variables are the unknown segment lengths of known diameters. The network consists of seven nodes, eight links (see Figure 24.2), and fourteen different candidate diameters as shown in Table 24.2. Because there are only fifteen possible spanning trees for the network, the global optimal tree solution is easily obtained (Loganathan et al., 1990). These flows are {1120, 370, 650, 0, 530, 200, 270, 0} in links {(1,2), (2,3), (2,4), (4,5), (4,6), (6,7), (3,5), (7,5)}, respectively, which are systematically perturbed to obtain the flows for the looped network.

TABLE 24.2 Diameter/Cost Data

Diameter (in.)	Unit Cost ($/m)	Diameter (in.)	Unit Cost ($/m)
1	2	12	50
2	5	14	60
3	8	16	90
4	11	18	130
6	16	20	170
8	23	22	300
10	32	24	550

TABLE 24.3 Optimal Flows for Different Methods (Flow Rate m^3/h)

Link	Loganathan et al. (1990)	Global Search	Global Search with Min. Flow
(1,2)	1120.0	1120.0	1120.0
(2,3)	360.0	368.33	368.0
(2,4)	660.0	651.67	652.0
(3,5)	260.0	268.33	268.0
(4,6)	534.0	530.69	531.0
(6,7)	204.0	200.69	201.0
(7,5)	4.0	0.97	1.0
(4,5)	6.0	0.69	1.0
Total Cost	412,931	403,657	405,381

TABLE 24.4 Optimal Solution without Minimum Flow

Link	Diameter (in.)	Length (m)	Hydraulic Gradient (m/m)	Head Loss (m)
(1,2)	18	1000.00	0.0068	6.76
(2,3)	10	792.54	0.0150	11.95
	12	207.46	0.0062	1.29
(2,4)	16	1000.00	0.0044	4.40
(4,5)	1	1000.00	0.0190	18.84
(4,6)	14	303.46	0.0058	1.75
	16	696.54	0.0030	2.09
(6,7)	8	10.32	0.0150	0.15
	10	989.68	0.0049	4.85
(3,5)	8	97.70	0.0250	2.43
	10	902.30	0.0084	7.57
(7,5)	1	1000.00	0.0100	10.00

To implement the procedure MULTISTART, a probability distribution function with a high probability mass near the origin (loop change flow vector is zero) accounting for the core tree link flows, is needed. The generated random points are used within Problem P3 to obtain the optimal designs. Because the normal distribution allows for positive or negative perturbations and high probability mass close to the mean, it is chosen as the standard distribution for each loop flow change variable ΔQ_p. A standard deviation of 6.5 is used both for ΔQ_1 and ΔQ_2 and the maximum number of points generated NMAX is set at 200. The optimal point $\varepsilon = (\Delta Q_1, \Delta Q_2) = (1.95, 0.69)$ is further refined by gradient search to obtain $\varepsilon = (\Delta Q_1, \Delta Q_2) = (1.67, 0.69)$ producing a cost of 403,657.94 units. The optimal link flows after ΔQ change flows applied to the optimal tree link flows are given in Table 24.3. The procedure ANNEALING is implemented next. When the procedure locates the optimum, $\varepsilon = (\Delta Q_1, \Delta Q_2) = (2.25, 1.22)$, it circles around the neighborhood and does not escape. The parameters are $\beta = 1$, $\theta = 1$, $\alpha = 2.5$, $f_g = 405,000$, and NMAX = 200. Further refinement as well as the gradient search yields the same optimum point (1.67, 0.69).

The global optimal tree solution made up of links {(1, 2), (2, 3), (2, 4), (3, 5), (4, 6), (6, 7)} has a cost of $399,561 (Loganathan et al., 1990). Using the minimum possible diameter of 1 in. for two loop-forming links {(4, 5), (5, 7)} at a unit cost of $2/meter length we obtain the global minimum cost $403,561. The proposed algorithm yields $403,657, verifying global optimality. It is noted that for this solution a minimum diameter of 1 in. is imposed for

TABLE 24.5 Energy Heads without Minimum Flow

Node (i)	Elevation (m)	Minimum Head (m)	Head (m)
1	210.00	210.00	210.00
2	150.00	180.00	203.24
3	160.00	190.00	190.00
4	155.00	185.00	198.84
5	150.00	180.00	180.00
6	165.00	195.00	195.00
7	160.00	190.00	190.00

TABLE 24.6 Optimal Solution with Minimum Flow

Link	Diameter (in)	Length (m)	Hydraulic Gradient (m/m)	Head Loss (m)
(1,2)	18	1000.00	0.0068	6.76
(2,3)	10	795.05	0.0150	11.97
	12	204.95	0.0062	1.27
(2,4)	16	1000.00	0.0044	4.40
(4,5)	1	951.65	0.0200	18.81
	2	48.35	0.00068	0.03
(4,6)	14	300.46	0.0058	1.73
	16	699.54	0.0030	2.11
(6,7)	8	9.01	0.0150	0.13
	10	990.99	0.0049	4.87
(3,5)	8	99.12	0.0250	2.46
	10	900.88	0.0084	7.54
(7,5)	1	488.53	0.0200	9.65
	2	511.46	0.00068	0.35

TABLE 24.7 Energy Heads with Minimum Flow

Node (i)	Elevation (m)	Minimum Head (m)	Head (m)
1	210.00	210.00	210.00
2	150.00	180.00	203.24
3	160.00	190.00	190.00
4	155.00	185.00	198.84
5	150.00	180.00	180.00
6	165.00	195.00	195.00
7	160.00	190.00	190.00

all the pipes in the looped layout as previously done by the other authors. The results are given in Tables 24.3 through 24.5. A minimum flow of constraint $1\,\mathrm{m}^3/\mathrm{h}$, in addition to the minimum diameter of 1 in., was implemented and the results are given in Tables 24.6 and 24.7, with the cost of $405,301.

24.3.8 Genetic Algorithm

Savic and Walters (1997) applied genetic algorithms for the same problem. Deb (2001) presents an authoritative description of genetic algorithms (GA). In GA solutions are represented by strings called chromosomes. A locus is the location of a gene (bit in binary) on a chromosome. An allele represents the set of values that can be assigned to a gene. Good solutions in terms of their objective function value are identified from an initial set of solutions. Multiple copies of good solutions are made while eliminating the weak solutions keeping the population size constant (Selection). The selected good solutions are combined to create new solutions by exchanging portions of strings (Crossover). Note this process retains the population size. To enhance diversity in the populations a bit may be changed (Mutation). Using the new population from crossover and mutation operators, the cycle is restarted with the selection operator. Deb (2001) makes the following key points. The three operators progressively result in strings with similarities at certain string positions (schema). A GA with its selection operator alone increases the density of solutions with above average value while reducing the variance; mutation increases the variance. The crossover operator has the tendency to decorrelate the decision variables. The combined effect of selection and crossover is to enable hyperplanes of higher average fitness values to be sampled more frequently.

Handling constraints is an issue in direct search methods. Variable elimination by constraint substitution for equality constraints, penalty function methods including biasing feasible solutions over infeasible solutions, hybrid methods, and methods based on decoders are procedures for accommodating constraints. Hybrid methods combine the classical pattern search methods within the GA. Bazaraa et al. (2006) and Reklaitis et al. (1983) provide a thorough analysis of penalty function methods. Savic and Walters (1997) use a penalty for constraint violation. They report a cost of \$419,000 with a minimum diameter of 1 in. and without using split pipes.

24.4 Preferences in Choosing Domestic Plumbing Materials

24.4.1 Introduction

In the United States, about 90% of drinking water home plumbing systems are copper pipes. Pinhole leaks in copper plumbing pipes are reported in several parts of the country. Figure 24.1 shows the distribution of pinhole leaks over a five-year period, from 2000 to 2004. For a majority of homeowners, their home is the most valuable asset. A decision has to be made on whether to continue to repair or replace the system. Loganathan and Lee (2005) have proposed an optimal replacement time for the home plumbing system exploiting an economically sustainable optimality criterion. Dietrich et al. (2006) present results of their study related to material propensity to leach organic chemicals, metals, and odorants; ability to promote growth of nitrifying bacteria; and their impact on residual disinfectant for five polymeric and four metallic pipes. Typically, copper, PEX (cross-linked polyethylene), and CPVC (chlorinated polyvinyl chloride) pipes are used. Stainless steel is also being considered. The material-related attributes of price, corrosion resistance, fire retardance, health effects, longevity, re-sale value of home, and taste and odor are considered in making a selection. Consumer preferences toward these attributes are assessed with the aid of a focus group.

24.4.2 Analytical Hierarchy Process

Marshutz (2001) reports that copper accounts for 90% of new homes, followed by PEX (cross linked polyethylene) at 7%, and CPVC (chlorinated polyvinyl chloride) at 2%. All materials have relative advantages and disadvantages. One material may be easy to install but may not be reliable while another material may be corrosion resistant but may have health or taste and odor problems. It is generally recognized that comparing several attributes simultaneously is complicated and people tend to compare two attributes at a time. The analytic hierarchy process (AHP) is used to determine the preference for attributes by pair-wise comparison. Assessing pair-wise preferences is easier as it enables to concentrate judgment on taking a pair of elements and compare them on a single property without thinking about other properties or elements (Saaty, 1990). It is noted that elicited preferences may be based on the standards already established in memory through a person's experience or education. Based on Saaty (1980) the following steps are adopted in performing the analytical hierarchy process.

Step 0: (*Identify attributes*) For plumbing material the following seven attributes are considered: Price—includes cost of materials and labor for installation and repair; Corrosion

TABLE 24.8 Standard Numerical Scores

Preference Level	Numerical Score, $a(i,j)$ 1–9 Scale
Equally preferred	1
Equally to moderately preferred	2
Moderately preferred	3
Moderately to strongly preferred	4
Strongly preferred	5
Strongly to very strongly preferred	6
Very strongly preferred	7
Very strongly to extremely preferred	8
Extremely preferred	9

TABLE 24.9 Pair-Wise Preference Weight Matrix [General]

	Attribute 1	Attribute 2	...	Attribute n
Attribute 1	$w1/w1$	$w1/w2$	...	$w1/wn$
Attribute 2	$w2/w1$	$w2/w2$	...	$w2/wn$
...	...	...	...	...
Attribute n	$wn/w1$	$wn/w2$	...	wn/wn
Sum	$X/w1$	$X/w2$	...	X/wn

$[X = (w1 + w2 + \cdots + wn)]$.

TABLE 24.10 Pair-Wise Preference Weight Matrix

	P	C	F	H	L	R	T
P	1	0.20	0.25	0.14	0.20	0.50	0.33
C	5	1	5	0.33	1	6	1
F	4	0.20	1	0.17	0.33	1	0.25
H	7	3	6	1	4	8	3
L	5	1	3	0.25	1	2	0.33
R	2	0.17	1	0.13	0.50	1	0.50
T	3	1	4	0.33	3	2	1
Sum	27.00	6.57	20.25	2.35	10.03	20.50	6.42

P: price, C: corrosion resistance, F: fire retardance, H: health effects, L: longevity, R: resale value of home, and T: taste and odor.

resistance—dependability of material to remain free of corrosion; Fire retardance—ability of material to remain functional at high temperatures and not to cause additional dangers such as toxic fumes; Health effects—ability of material to remain inert in delivering water without threatening human health; Longevity—length of time material remains functional; Resale value of home—people's preference for a particular material including aesthetics; and Taste and odor—ability of material to deliver water without imparting odor or taste.

Step 1: (*Use standard preference scores*) A scale (1–9) of pair-wise preference weights is given in Table 24.8 (Saaty, 1980).

Step 2: (*Develop pair-wise preference matrix*) In AHP, instead of directly assessing the weight for attribute, i, we assess the relative weight $a_{ij} = wi/wj$ between attribute i and j. As shown in Tables 24.9 and 24.10, each participant is asked to fill in a 7×7 attribute matrix of pair-wise preferential weights. An example is given in Table 24.10. In Table 24.10, row H and column P, the entry of 7 implies that health effects are very strongly preferred in comparison to price in the ratio of 7:1. Row H for health effects overwhelms all other attributes with the entries staying well above 1. In row P and column C, the cell value of 0.2 indicates corrosion resistance is strongly preferred to price in the ratio of 5:1.

Step 3: (*Evaluate re-scaled pair-wise preference matrix*) A rescaled preference matrix is generated by dividing each column entry in Table 24.10 by that column's sum, yielding Table 24.11. The last column "Average" contains average weights for each row and shows the ranking of the attributes. Table 24.12 shows the ordered relative ranking of the attributes.

TABLE 24.11 Rescaled Pair-Wise Weight Matrix

Attribute	P	C	F	H	L	R	T	Average
P	0.04	0.03	0.01	0.06	0.02	0.02	0.05	0.03
C	0.19	0.15	0.25	0.14	0.10	0.29	0.16	0.18
F	0.15	0.03	0.05	0.07	0.03	0.05	0.04	0.06
H	0.26	0.46	0.30	0.43	0.40	0.39	0.47	0.38
L	0.19	0.15	0.15	0.11	0.10	0.10	0.05	0.12
R	0.07	0.03	0.05	0.05	0.05	0.05	0.08	0.05
T	0.11	0.15	0.20	0.14	0.30	0.10	0.16	0.17
Sum	1.00	1.00	1.00	1.00	1.00	1.00	1.00	

TABLE 24.12 Relative Ranking of Attributes

Attribute	Weight
Health effects	0.38
Corrosion resistance	0.18
Taste and odor	0.17
Longevity	0.12
Fire resistance	0.06
Resale value	0.05
Price	0.03

TABLE 24.13 Material Pair-Wise Matrix and the Associated Rescaled Matrix for the Attribute Price

Material	Mat. A	Mat. B	Mat. C		Mat. A	Mat. B	Mat. C	Average
Mat. A	1.000	0.333	2.000	Mat. A	0.222	0.200	0.333	0.252
Mat. B	3.000	1.000	3.000	Mat. B	0.667	0.600	0.500	0.589
Mat. C	0.500	0.333	1.000	Mat. C	0.111	0.200	0.167	0.159
Sum	4.500	1.667	6.000	Sum	1.000	1.000	1.000	1.000

Step 4: (*Evaluate preferences*) Corresponding to each attribute, pair-wise weight matrix and the associated rescaled matrix for three hypothetical pipe materials A, B, and C are obtained. Results for the price attribute are shown as Table 24.13. This procedure is repeated for the other attributes for the three materials. Table 24.14 shows the results. We obtain the final ranking of the three materials by multiplying the pipe material preference matrix given in Table 24.14 and the attribute preference vector given as the last column of Table 24.11. The results are shown in Table 24.15. Mat. C is the most preferred material with a preference score of 0.460, followed by Mat. A with a score of 0.279. Mat. B and Mat. A have rather close scores of 0.261 and 0.279, respectively. A consistency check can be performed following Saaty (1980) as given in Step 5. Participants were advised to reassess the pair-wise weights if the consistency check failed.

Step 5: (*Perform consistency check*) The calculated maximum eigenvalue for the pair-wise weight matrix, $\lambda \max = n$. If it is different from n, we have inconsistencies in our weight assignments. Saaty (1980) defines a consistency index as C.I. $= (\lambda \max - n)/(n-1)$. Instead of the eigenvalue, it is possible to manipulate the matrices to get a different form of the consistency index. Let the average ratio AR $= 1/n \sum_{i=1}^{n} (i\text{th row of }[A]\{\text{ave W}\})/(i\text{th row }\{\text{ave W}\})$ in which $[A] =$ matrix of pair-wise preference weights in Table 24.10, $\{\text{ave W}\} =$ vector of average rescaled weights in the last column of Table 24.11. The newly formed consistency index is C.I. $= [AR - n]/(n-1)$. If the ratio of C.I. to R.I. (random index given in Table 24.16) is less than 0.1, the weights should be taken as consistent. Table 24.16 contains the random index values calculated from randomly generated weights as a function of the pair-wise matrix size (number of criteria).

Table 24.17 contains the attribute ranking for 10 participants. From Table 24.17, it is seen that the participants rank health effects the highest, followed by taste and odor, corrosion

TABLE 24.14 Average Ranking for the Materials and Attributes

	P	C	F	H	L	R	T
Mat. A	0.252	0.164	0.444	0.328	0.250	0.297	0.250
Mat. B	0.589	0.297	0.111	0.261	0.250	0.164	0.250
Mat. C	0.159	0.539	0.444	0.411	0.500	0.539	0.500
Average	0.03	0.18	0.06	0.38	0.12	0.05	0.17

TABLE 24.15 Final Preference Matrix

Material	Preference
Mat. A	0.279
Mat. B	0.261
Mat. C	0.460

TABLE 24.16 Random Index (R.I.)

Matrix Size	1	2	3	4	5	6	7	8	9	10	11	12	13	14	15
R.I.	0	0	0.58	0.90	1.12	1.24	1.32	1.41	1.45	1.49	1.51	1.48	1.56	1.57	1.59

TABLE 24.17 Participants' Ranking of Attributes

Participant/Attribute	1	2	3	4	5	6	7	8	9	10
P	0.034	0.161	0.025	0.040	0.200	0.063	0.310	0.202	0.048	0.087
C	0.182	0.037	0.107	0.061	0.090	0.091	0.158	0.111	0.112	0.256
F	0.060	0.056	0.078	0.085	0.070	0.119	0.044	0.057	0.188	0.044
H	0.385	0.211	0.395	0.304	0.270	0.450	0.154	0.369	0.188	0.367
L	0.120	0.097	0.101	0.132	0.090	0.120	0.154	0.067	0.112	0.062
R	0.054	0.020	0.027	0.127	0.050	0.098	0.103	0.064	0.033	0.029
T	0.165	0.419	0.267	0.250	0.230	0.059	0.078	0.131	0.318	0.155

P: price, C: corrosion resistance, F: fire retardance, H: health effects, L: longevity, R: resale value of home, and T: taste and odor.

resistance, and longevity; price, resale value, and fire resistance are showing the lower ranks of the list. Participants 2 and 9 rate taste and odor above health. Participant 5 has the closest preferences among price, health, and taste and odor. Participant 7 has the highest preference for price. These results indicate that health, and taste, and odor may be a surrogate for the purity of water that dominates preferences for a plumbing material. The mental stress resulting from frequent leaks is also an issue. While copper may have remained a relatively inert carrier of water, the corrosion leaks have forced consumers to consider alternatives including other materials, the use of corrosion inhibitors such as phosphate, and lining the interior of the pipe.

24.5 Stormwater Management

24.5.1 Introduction

Urban runoff is a carrier of contaminants. As discussed in Section 24.1.1, the total maximum daily load (TMDL) study should not only estimate the maximum permissible load per time but also should allocate it. Runoff contributes to nonpoint source pollution and a distributed pollutant control strategy is needed. The best management practices (BMPs) are both nonstructural measures and structural controls employed to minimize contaminants carried by the surface runoff. The BMPs also help to serve to control the peak flows due to development. Therefore, the decision should be comprised of optimal location and level of treatment at a selected location.

24.5.2 Problem Formulation

Zhen et al. (2004) suggest the following general framework.

$$\text{Minimize total annual cost} = \sum_{i=1}^{I} C(\mathbf{d_i})$$

$$\text{subject to:} \quad L_j \leq L\max_j \quad \text{for } j = 1, 2, \ldots, J$$

$$\mathbf{d_i} \in S_i$$

in which $C(\mathbf{d_i})$ = annual cost of implementing a BMP at level d_i at location i; $\mathbf{d_i}$ = decision vector pertaining to the dimensions of the BMP; L_j = annual contaminant load at the designated check point j; I = total number of potential BMP locations; J = total number of check points; and S_i = feasible set of BMPs applicable at location i. The decision vector is given by $\{\mathbf{T_i}; \mathbf{H_i}\}$ in which $\mathbf{T_i}$ is the vector of possible detention times compatible with the BMP dimension vector (height) $\mathbf{H_i}$ at location i. Typically, for a choice of the decision vectors the load, L_j at location j will have to be obtained from a simulation model. Zhen et al. (2004) use the scatter search (Glover, 1999) to find near optimal solutions. Kuo et al. (1987) use the complex search (Box, 1965; Reklaitis et al., 1983) to obtain near optimal solutions.

Loganathan and Sherali (1987) consider the probability of failure of a BMP (also see Loganathan et al., 1994). They consider the risk of failure and cost of the BMP as two objectives within a mutiobjective optimization framework. They generate a cutting plane based on pair-wise tradeoff between objective functions provided by the decision maker at each iterate. It is shown that the iterates are on the efficient frontier and any accumulation point generated by the algorithm is a best compromise solution. The method is applicable even when the feasible region is nonconvex. Lee et al. (2005) consider a production function of release rate and storage for a certain percent pollutant removal. They recommend using an evolutionary optimizer for the solution. They also provide a review of progress in optimizing the storm water control systems. Adams and Papa (2000) provide analytical probabilistic models for estimating the performance of the storm water control systems. They include optimization strategies for these systems.

Mujumdar and Vemula (2004) present an optimization-simulation model for waste load allocation. They apply the genetic algorithm along with an appealing constraint handling strategy due to (Koziel and Michalewicz, 1998, 1999). The decoder method proposed by Koziel and Michalewicz (1999) is based on homomorphous mapping between the representation space of $[-1, 1]^n$ n-dimensional cube and the feasible region, X in the decision space. Because the feasible region may be nonconvex, it may have several points of intersection between a line segment $\mathbf{L}$ and the boundary of the feasible region X. The method permits selecting points from the broken feasible parts of the line segment $\mathbf{L}$. To identify these break points, the following procedure is used. A feasible point $\mathbf{r}$ is chosen. A line segment $\mathbf{L} = \mathbf{r} + \alpha\ (\mathbf{s} - \mathbf{r})$ for $0 \leq \alpha \leq 1$, starting with $\mathbf{r}$ and ending with a boundary point $\mathbf{s}$ of the search space S encompassing the feasible region is considered. Such a defined line segment $\mathbf{L}$ involves a single variable α when substituted into each of the constraints, $g_i(x) \leq 0$. The domain of α is partitioned into v subintervals of length $(1/v)$. Each subinterval is explored for the root $g_i(\alpha) = 0$. The subintervals are chosen in such a manner that there can be at most one root within each subinterval. The intersection points between the line segment and the constraints $g_i(x) \leq 0$ can thus be found.

The problem is as follows. A discharger m (source) removes a fractional amount $x(i, m, n)$ of pollutant n to control water quality indicator i at check point l. For discharger m after removing a fractional load $x(i, m, n)$ of pollutant n, the concentration is $c(m, n)$. This input concentration $c(m, n)$ for $m = 1, 2, \ldots, M$ and $n = 1, 2, \ldots, N$ and the effluent flow at m

denoted as $q(m)$ for $m = 1, 2, \ldots, M$ serve as forcing functions in the water quality simulation model QUAL2E (Brown and Barnwell, 1987) to determine the concentration $c(i, l)$ of water quality indicator i (dissolved oxygen) at the check point l. The discharger would like to limit $x(i, m, n)$ and the pollutant controlling agency would like to control $c(i, l)$ for all check points l. The problem is given below:

Maximize Level of satisfaction $= \lambda$

subject to: $[c(i,l) - c_\text{L}(i,l)]/[c_\text{D}(i,l) - c_\text{L}(i,l)]^{a(i,l)} \geq \lambda$

$$[x_\text{M}(i,m,n) - x(i,m,n)]/[x_\text{M}(i,m,n) - x_\text{L}(i,m,n)]^{b(i,m,n)} \geq \lambda$$

$$c_\text{L}(i,l) \leq c(i,l) \leq c_\text{D}(i,l)$$

$$x_\text{L}(i,m,n) \leq x(i,m,n) \leq x_\text{M}(i,m,n)$$

$$0 \leq \lambda \leq 1$$

in which $c_\text{D}(i,l) =$ desirable concentration level such as the dissolved oxygen, $c_\text{L}(i,l) =$ permissible safe level concentration, $x_\text{M}(i,m,n) =$ technologically possible maximum pollutant removal, $x_\text{L}(i,m,n) =$ minimum pollutant removal, $a(i,l)$ and $b(i,m,n) =$ positive parameters, and $\lambda =$ compromise satisfactory level.

The decision variables $x(i,m,n)$ are determined with the aid of the genetic algorithm. They use QUAL2E and genetic algorithm optimization packages PGAPack (Levine, 1996) and GENOCOP (Koziel and Michalewicz, 1999) to solve the problem. Using the GA generated population of pollutant reduction values $x(i,m,n)$, QUAL2E determines the corresponding concentration of water quality indicator i at check point l, $c(i,l)$. The feasibility is checked by GENOCOP. They point out that the selection of the initial feasible point in constraint handling could affect the final solution. They recommend using the incumbent best solution as the initial point.

24.6 Groundwater Management

24.6.1 Introduction

As rain falls over land, a part of it runs over land and a part of it infiltrates into the ground. There is void space between soil grains underground and water and air occupy that space. Groundwater flows through the tortuous paths between the solid grains. Because of the extremely small diameter, the tortuous nature of the path, and the nature of the hydraulic head, the groundwater velocity is very small. Therefore, if the groundwater storage is mined, it cannot be replenished in a short time. Driscoll (1986) reports that there has been a 400-ft decline in groundwater table in Phoenix, Arizona, in the last 50 years. Perennial rivers carry the base flow resulting from groundwater storage even when there is no rain. In the United States 83.3 billion [$83.3(10^9)$] gallons of fresh groundwater and 1.26 billion gallons of saline groundwater are used, whereas 262 billion gallons of fresh and 61 billion gallons of saline surface water are used (Hutson et al., 2004). Declining groundwater storage and contaminated groundwater are two major problems.

Willis and Yeh (1987), Bear and Verruijt (1987), and Bedient et al. (1999) provide comprehensive details on groundwater flow and contaminant transport modeling. The groundwater problem involves detailed numerical modeling. There is an added complexity because of the complex nature of the underground domain. The groundwater velocity is determined by hydraulic conductivity which is not known deterministically. Chan Hilton and Culver

(2005) offer a comprehensive review of groundwater remediation problem. They also propose a strategy within the framework of simulation-optimization to consider the uncertainty in the hydraulic conductivity field. Following Bedient et al. (1999) the groundwater remediation may involve one of the following: (1) removal of source by excavation, (2) containment of source by barriers and hydraulic control, (3) reducing the mass of the source by pump and treat, bioremediation, soil vapor extraction, and natural attenuation.

Chan Hilton and Culver (2005) propose the following simulation-optimization approach to groundwater remediation problems. The procedure differs from the previous studies in accommodating the uncertainty in the hydraulic conductivity field. There are three ways in which the hydraulic conductivity field is incorporated: (1) Keep the same field throughout the simulation-optimization study. In the genetic algorithm framework all strings are subjected to the same field in every generation. (2) Use multiple fields within each generation and each such realization is applied to all strings for that generation. Calculate the average fitness value over all these realizations for each string. (3) Obtain a hydraulic conductivity field for a generation. Apply the same field to all the strings in that generation. For the next generation, obtain a new hydraulic conductivity field. The fitness value is calculated applicable to that conductivity field. Chan Hilton and Culver use the third type and call it the robust genetic algorithm.

24.6.2 Problem Formulation

The groundwater remediation problem formulation is as follows. There are N sampling points for checking contaminant concentration (computational nodes) and W potential extraction wells. The problem is to optimally decide which extraction wells should be used such that the contaminant concentration, $C(j)$ can be controlled; hydraulic head, $h(j)$, above a minimum can be maintained; extraction flow rates can be set within chosen bounds. The mathematical formulation is

Minimize   Total Cost = Pumping cost + Treatment cost + Well cost
Subject to: $C(j) \leq C\max$ for $j = 1, 2, \ldots, N$
    $h(j) \geq h(\min)$ for $j = 1, 2, \ldots, N$
    $Q(\min) \leq Q(i) \leq Q(\max)$ for $I = 1, 2, \ldots, W$

in which Pumping cost = Cpump $(c1/\eta)$ $Q(i)$ [Hdatum $- h(i) + \mu$ Pads] Tdur, Cpump = energy cost, $c1$ = unit conversion, η = pump efficiency, $Q(i)$ = extracted flow at well i, Hdatum = depth to datum, μ = technological coefficient for adsorber, Pads = pressure required for the adsorber, Tdur = remediation period in appropriate units; Treatment cost = f[Cave, Qtot] + Cads Nads, Qtot = $\sum_{i=1}^{W} Q(i)$, Cave = $\sum_{i=1}^{W} \frac{Q(i)}{Q\text{tot}} \left[\frac{C(i,T\text{beg}) + C(i,T\text{end})}{2} \right]$, $C(i,T$beg$)$ = concentration of contaminant at well i at the beginning time, Tbeg, $C(i,T$end$)$ = concentration of contaminant at well i at the ending time, Tend, Tdur = Tend $- T$beg, and Cave = average influent concentration for the adsorber, Qtot = total flow, Cads = cost of an adsorber, Nads = number of adsorbers = ceiling[Qtot(Tcont)/Vpore], ceiling(.) = rounds up to nearest multiple of 1, Vpore = pore volume of an adsorber, Tcont = required contact time.

The above stated problem is solved using simulation–optimization approach by the genetic algorithm. The concentration $c(j)$, and head $h(j)$ constraints are incorporated through a multiplicative penalty function as

Fitness = Total cost $*$ [$1 + $ pen$C$ Viol$C$ + pen$H$ Viol$H$]

in which pen$C$ = penalty for violation in concentration $C(j)$ and pen$H$ = penalty for violation in head $h(j)$. For the flow $Q(i)$ constraint, it is assumed that if a well is active, it will extract flow at a fixed rate. Therefore, it becomes a binary variable and is incorporated into the decision variable string in the genetic algorithm.

The entire procedure is as follows. (1) *Initialize:* Generate the initial population of strings with age$(i) = 1$ for the ith string. (2) *Selection:* For a realization of the hydraulic conductivity field, obtain the fitness value for each string, Fitness(i). Rank order the strings by their Fitness values with the highest rank assigned to the string with the best Fitness value. Let $r(i)$ be the rank for string i. Define modified rank fitness measure

$$r\text{fit}(i) = \frac{[\min\{\text{age}(i), \text{age}T\} - 1]r\text{prev}(i) + r(i)}{\min\{\text{age}(i), \text{age}T\}} \quad \text{if age}(i) > 1$$

$$r\text{fit}(i) = 0.9\ r(i) \quad \text{if age}(i) = 1$$

in which age$T$ = chosen value greater than 1 (set at 11), $r\text{prev}(i) = r\text{fit}(i)$ from the previous generation. List the strings by their $r\text{fit}(i)$ values; assign new $r\text{new}(i)$ ranks to these strings. These become $r\text{prev}(i)$ in the next generation. Retain the highest rnew(i) strings as the *elite group*. These are directly passed to the next generation without crossover and mutation. Apply *tournament selection* to the population in which two strings are compared and the string with better rank fitness is retained for crossover. The tournament selection is repeated with replacement until the desired population size is reached. (3) *Crossover:* Crossover is performed between two strings to create a new string. This new string's age$(i) = 0$. If there is no crossover due to specified probability, the first string proceeds into the next generation with its age unchanged. (4) *Mutation:* If mutation causes a new string to attain the same bits as its parent string, its age is reset to the original string's age. If not, its age$(i) = 0$. (5) *New generation:* The strings passed on to the next generation are assigned age$(i) = $ age$(i) + 1$. (6) *Termination check:* If there is convergence to lower costs, stop. If not, go to Step (2), selection. The robust GA retains strings that perform well over a number of different hydraulic conductivity fields. In the noisy GA, a number hydraulic conductivity fields are used within the same generation and the fitness value is taken as the average over the realizations. The comparison indicates that the robust GA yields comparable results to the noisy GA while using a smaller number of fitness evaluations.

24.7 Summary

In this chapter, selected mathematical programming formulations related to water resources systems are presented. The overarching theme has been to expose certain underlying problem structures such as in reservoir operations and water distribution systems optimization, the need to assess public preference toward material attributes as in plumbing systems, and the necessity to combine process simulation and optimization such as in stormwater and groundwater management problems. The evolutionary optimization techniques provide a good approach. The constraint handling in these techniques is an important issue; guaranteeing the nature of optimality is also an issue. Often, a zero-one decision has to be made. This decision requires consideration of political and public preferences and implications over a long term. Both the analyst and the decision maker are interested in the alternatives and the solution behavior with changing constraints as the result of incorporating regulatory issues and societal preferences.

Acknowledgments

The writer gratefully acknowledges the support provided by the National Science Foundation under the grant DMI-0329474, American Water Works Association Research Foundation,

Weston Solutions, Inc., Virginia Department of Environmental Quality, and Virginia Water Resources Research Center. He also wishes to thank Professors Marc Edwards, Andrea Dietrich, Darrell Bosch, Sharon Dwyer, Mr. Bob Buglass, Dr. Paolo Scardina, Dr. A. K. Deb, Dr. Hwandon Jun, Mr. Juneseok Lee, and Mr. Owais Farooqi.

References

1. Adams, B.J. and Papa, F., *Urban Stormwater Management Planning with Analytical Probabilistic Models*, John Wiley, New York, 2000.
2. Alperovits, E. and Shamir, U., Design of Optimal Water Distribution Systems, *Water Resources Research*, 13, 885–900, 1977.
3. American Society of Civil Engineers, *Civil Engineering Guidelines for Planning and Designing Hydroelectric Developments*, ASCE, New York, 1989.
4. American Society of Civil Engineers, Sustainability Criteria for Water Resources Systems, *Project M-4.3*, WRPM Division (Task Committee Sustainability Criteria) and UNESCO International Hydrological Programme, 1998.
5. Barritt-Flatt, P.E. and Cormie, A.D., Comprehensive Optimization Model for Hydroelectric Reservoir Operations, in Labadie, L.W., et al. (eds.) *Computerized Decision Support System for Water Managers*, ASCE, New York, 1988.
6. Bazaraa, M.S., Sherali, H.D., and Shetty, C.M., *Nonlinear Programming: Theory and Algorithms*, John Wiley, New York, 2006.
7. Bear, J. and Verruijt, A., *Modeling Groundwater Flow and Pollution*, D. Reidel, Boston, 1987.
8. Bedient, P.B., Rifai, H.S., and Newell, C.J., *Groundwater Contamination*, Prentice Hall, Upper Saddle River, 1999.
9. Berry, L., Water: The Emerging Crisis? L. Berry (Issue editor) *Water Resources Update*, Universities Council on Water Resources, 102, 7–9, Winter, 1996.
10. Bhave, P.R., *Analysis of Flow in Water Distribution Networks*, Technomic Publishers, Lancaster, 1991.
11. Bohachevsky, I.O., Johnson, M.E., and Stain, M.L., Generalized Simulated Annealing for Optimization, *Technometrics*, 28(3), 209–217, 1986.
12. Boulos, P.F., Lansey, K.E., and Karney, B.W., *Comprehensive Water Distribution Analysis Handbook for Engineers and Planners*, MWH Soft, Inc., Pasadena, 2004.
13. Box, M.J., A New Method of Constrained Optimization and a Comparison with Other Methods, *Computer Journal*, 8, 42–51, 1965.
14. Brown, L.C. and Barnwell, T.O., Jr., The Enhanced Stream Water Quality Models QUAL2E and QUAL2E-UNCAS: Documentation and User Manual, *Report No. EPA/600/3/87/007*, U.S. Environmental Protection Agency, Athens, GA, 1987.
15. Can, E.K., Houck, M.H., and Toebes, G.H., Optimal Real-time Reservoir Systems Operation: Innovative Objectives and Implementation Problems, *Technical Report 150*, Purdue University Water Resources Research Center, 1982.
16. Chan Hilton, A.B. and Culver, T., Groundwater Remediation Design under Uncertainty Using Genetic Algorithms, *Journal of Water Resources Planning and Management*, 131(1), 25–34, January/February 2005.
17. Changchit, C. and Terrell, M.P., CCGP Model for Multiobjective Reservoir Systems, *Journal of Water Resources Planning and Management*, 115(5), 658–670, 1989.
18. Chiras, D.D., Reganold, J.P., and Owen, O.S., *Natural Resources Conservation*, 8th ed., Prentice Hall, Upper Saddle River, 2002.

19. Deb, A.K. and Sarkar, A.K., Optimization in Design of Hydraulic Network, *Journal of the Sanitary Engineering Division*, 97(SA2), 141–159, 1971.
20. Deb, A.K., Least Cost Pipe Network Derivation, *Water and Water Engineering*, 77, 18–21, 1973.
21. Deb, K., *Multi-Objective Optimization Using Evolutionary Algorithms*, John Wiley, New York, 2001.
22. Dellapenna, J.W., Is Sustainable Development a Serviceable Legal Standard in the Management of Water? *Water Resources Update*, Universities Council on Water Resources, 127, 87–93, February 2004.
23. Delleur, J.W., The Evolution of Urban Hydrology: Past, Present, and Future, *Journal of Hydraulic Engineering*, 129(8), 563–573, August 2003.
24. Dietrich, A. et al., Plumbing Materials: Costs, Impacts on Drinking Water Quality, and Consumer Willingness to Pay, *Proceedings of 2006 NSF Design, Service, and Manufacturing Grantees Conference*, St. Louis, Missouri, 2006.
25. Driscoll, F.G., *Groundwater and Wells*, Johnson Filtration Systems, St. Paul, MN, 1986.
26. Eschenbach, E., Magee, T., Zagona, E., Goranflo, M., and Shane, R., Goal Programming Decision Support System for Multiobjective Operation of Reservoir Systems, *Journal of Water Resources Planning and Management*, 127(2), 108–120, March/April 2001.
27. Ellis, J.H. and ReVelle, C.S., A Separable Linear Algorithm for Hydropower Optimization, *Water Resources Bulletin*, 24(2), 435–447, 1988.
28. Feldman, A.D., HEC Models for Water Resources System Simulation: Theory and Experience, *Advances in Hydroscience*, 12, 297–423, 1981.
29. Gessler, J., Optimization of Pipe Networks, *International Symposium on Urban Hydrology, Hydraulics and Sediment Control*, University of Kentucky, Lexington, KY, pp. 165–171, 1982.
30. Glover, F., Scatter Search and Path Relinking, *HCES-01-99, Working paper series*, Hearin Center for Enterprise Science, The University of Mississippi, 1999.
31. Goldman, F.E. and Mays, L.W., Water Distribution System Operation: Application of Simulated Annealing, in Mays, L.W. (ed.) *Water Resource Systems Management Tools*, McGraw Hill, New York, 2005.
32. Heintz, T.H., Applying the Concept of Sustainability to Water Resources Management, *Water Resources Update*, Universities Council on Water Resources, 127, 6–10, February 2004.
33. Hutson, S.S., Barber, N.L., Kenny, J.F., Linsey, K.S., Lumia, D.S., and Maupin, M., Estimated Use of Water in the United States in 2000, *U.S. Geological Survey Circular 1268*, U.S. Geological Survey, Reston, VA, 2004.
34. Jain, S.K. and Singh, V.P., *Water Resources Systems Planning and Management*, Elsevier, Boston, 2003.
35. Karamouz, M., Szidarovszky, F., and Zahraie, B., *Water Resources Systems Analysis*, Lewis Publishers, Boca Raton, FL, 2003.
36. Karmeli, D., Gadish, Y., and Myers, S., Design of Optimal Water Distribution Networks, *Journal of Pipeline Division*, 94, 1–10, 1968.
37. Koziel, S. and Michalewicz, Z., A Decoder-Based Evolutionary Algorithm for Constrained Parameter Optimization Problems, *Proceedings of the 5 Parallel Problem Solving from Nature*, T. Back, A.E. Eiben, M. Schoenauer, and H.-P. Schwefel (eds.), Amsterdam, September 27–30, Lecture Notes in Computer Science, Springer-Verlag, New York, 231–240, 1998.
38. Koziel, S. and Michalewicz, Z., Evolutionary Algorithms, Homomorphous Mappings, and Constrained Parameter Optimization, *Evolutionary Computation*, 7(1), 19–44, 1999.

39. Kranz, R., Gasteyer, S.P., Heintz, T., Shafer, R., and Steinman, A., Conceptual Foundations for the Sustainable Water Resources Roundtable, *Water Resources Update*, Universities Council on Water Resources, 127, 11–19, February 2004.
40. Kuo, C.Y., Loganathan, G.V., Cox, W.E., Shrestha, S., and Ying, K.J., *Effectiveness of BMPs for Stormwater Management in Urbanized Watersheds*, Bulletin 159, Virginia Water Resources Research Center, 1987.
41. Labadie, J.W., Optimal Operation of Multireservoir Systems: State of the Art Review, *Journal of Water Resources Planning and Management*, 130(2), 93–111, March/April 2004.
42. Lee, J. and Loganathan, G.V., Consumer Preferences Towards Plumbing Material Attributes, *Virginia Water Central*, Virginia Water Resources Research Center, August 2006.
43. Lee, J.G., Heaney, J.P., and Lai, F.-H., Optimization of Integrated Urban Wet-Weather Control Strategies, *Journal of Water Resources Planning and Management*, 131(4), 307–315, July/August 2005.
44. Levine, D., User's guide to the PGAPack parallel genetic algorithm library, Tech. Rep. ANL-95/18, Argonne National Lab, 1996.
45. Loganathan, G.V., Optimal Operating Policy for Reservoir Systems, in Kenneth D. Lawrence and Gary R. Reeves (eds.), *Applications of Management Science: Engineering Applications*, Vol. 9, JAI Press, Greenwich, CT, 1996.
46. Loganathan, G.V. and Bhattacharya, D., Goal-Programming Techniques for Optimal Reservoir Operations, *Journal of Water Resources Planning and Management*, 116(6), 820–838, 1990.
47. Loganathan, G.V., Greene, J.J., and Ahn, T., A Design Heuristic for Globally Minimum Cost Water Distribution Systems, *Journal of Water Resources Planning and Management*, 121(2), 182–192, 1995.
48. Loganathan, G.V., Sherali, H.D., and Shah, M.P., A Two-Phase Network Design Heuristic for the Minimum Cost Water Distribution Systems under Reliability Constraints, *Engineering Optimization*, 15, 311–336, 1990.
49. Loganathan, G.V. and Lee, J., Decision Tool for Optimal Replacement of Plumbing Systems, *Civil Engineering and Environmental Systems*, 22(4), 189–204, December 2005.
50. Loganathan, G.V. and Sherali, H.D., A Convergent Cutting-Plane Algorithm for Multiobjective Optimization, *Operations Research*, 35(3), 365–377, May–June 1987.
51. Loganathan, G.V., Watkins, E.W., and Kibler, D.F., Sizing Stormwater Detention Basins for Pollutant Removal, *Journal of Environmental Engineering*, 120(6), 1380–1399, 1994.
52. Lohani, V.K. and Loganathan, G.V., An Early Warning System for Drought Management Using the Palmer Drought Index, *Water Resources Bulletin*, 33(6), 1375–1386, 1997.
53. Loucks, D.P. and Gladwell, J., *Sustainability Criteria for Water Resources Systems*, Cambridge University Press, Cambridge, 1999.
54. Loucks, D.P., Quantifying System Sustainability Using Multiple Risk Criteria, in Bogardi, J.J., and Kundzewicz, Z. (eds.), *Risk, Reliability, Uncertainty, and Robustness of Water Resources Systems*, Cambridge University Press/UNESCO, Cambridge, 2002.
55. Loucks, D.P., Stedinger, J.R., and Haith, D.A., *Water Resources Systems Planning and Analysis*, Prentice-Hall, Englewood Cliffs, 1981.
56. Marshutz, S., Hooked on Copper, *Reeves Journal*, March 2001.
57. Martin, Q.W., Optimal Daily Operation of Surface-Water Systems, *Journal of Water Resources Planning and Management*, 113(4), 453–470, 1987.
58. Martin, Q.W., Optimal Reservoir Control for Hydropower on Colorado River, Texas, *Journal of Water Resources Planning and Management*, 121(6), 438–4446, 1995.
59. Material Performance (Supplement), *Cost of Corrosion Study Unveiled*, July 2002.

60. Mattice, J.S., Ecological Effects of Hydropower Facilities, in Gulliver, J.S. and Arndt, R.E.A. (eds.), *Hydropower Engineering Handbook*, McGraw Hill, New York, 1991.
61. Mays, L.W. and Tung, Y.K., *Hydrosystems Engineering and Management*, McGraw Hill, New York, 1992.
62. Mays, L.W., *Water Distribution Systems Handbook*, McGraw Hill, New York, 2000.
63. McMahon, G.F. and Farmer, M.C., Reallocation of Federal Multipurpose Reservoirs: Principles, Policy, and Practice, *Journal of Water Resources Planning and Management*, 130(3), 187–197, July/August 2004.
64. Moore, C. and Loganathan, G.V., Optimizing Hydropower Operation in a Pumped-Storage System, *ASCE Environmental and Water Resources Institute (EWRI) Annual Conference*, May 19–22, 2002, Roanoke, Virginia.
65. Mujumdar, P.P. and Vemula, V.R.S., Fuzzy Waste Load Allocation Model: Simulation-Optimization Approach, *Journal of Computing in Civil Engineering*, 18(2), 120–131, April 2004.
66. National Research Council, *New Directions in Water Resources: Planning for the U.S. Army Corps of Engineers*, National Academy Press, Washington, DC, 1999.
67. Palacios-Gomez, Lasdon, L., and Engquist, M., Nonlinear Optimization by Successive Linear Programming, *Management Science*, 28(10), 1106–1120, 1982.
68. Pardalos, P.M., Glick, J.H., and Rosen, J.B., Global Minimization of Indefinite Quadratic Problems, *Computing*, 39, 281–291, 1987.
69. Reklaitis, G.V., Ravindran, A., and Ragsdell, K.M., *Engineering Optimization*, John Wiley, New York, 1983.
70. ReVelle, C.S., Joeres, E., and Kirby, W., The Linear Decision Rule in Reservoir Management and Design 1: Development of the Stochastic Model, *Water Resources Research*, 5(4), 767–777, 1969.
71. Rinnooy Kan, A.H.G. and Timmer, G.T., Global Optimization, Chapter IX in G.L. Nemhauser et al. (eds.), *Handbooks in OR & MS*, Vol. I, Elsevier Science Publishers B.V. (North Holland), New York, 1989.
72. Saaty, T.L. *The Analytic Hierarchy Process*, McGraw-Hill, New York, 1980.
73. Saaty, T.L. How to Make a Decision: The Analytic Hierarchy Process, *European Journal of Operational Research*, 48, 9–26, 1990.
74. Savic, D.A. and Walters, G.A., Genetic Algorithms for Least-Cost Design of water Distribution Networks, *Journal of Water Resources Planning and Management*, 123(2), 67–77, 1997.
75. Shih, J.S. and ReVelle, C.S., Water Supply Operations During Drought: A Discrete Hedging Rule, *European Journal of Operational Research*, 82, 163–17, 1995.
76. Smith, E.T., Water Resources Criteria and Indicators, *Water Resources Update*, Universities Council on Water Resources, 127, 59–67, February 2004.
77. Smith, E.T. and Lant, C., Water Resources Sustainability, *Water Resources Update*, Universities Council on Water Resources, 127, 1–5, February 2004.
78. Spall. J.C., *Introduction to Stochastic Search and Optimization: Estimation, Simulation, and Control*, John Wiley, New York, 2003.
79. Tao, T. and Lennox, W.C., Reservoir Operations by Successive Linear Programming, *Journal of Water Resources Planning and Management*, 117(2), 274–280, 1991.
80. Templeman, A.B., Discussion of Optimization of Looped Water Distribution Systems by Quindry et al., *Journal of the Environmental Engineering Division*, 108(EE3), 599–602, 1982.
81. Torn, A. and Zilinkas, A., *Global Optimization*, Lecture Notes in Computer Science 350, Springer-Velag, New York, 1987.

82. U.S. Army Corps of Engineers, *HEC-5: Simulation of Flood Control and Conservation Systems*, User's Manual, Hydrologic Engineering Center, Davis, CA, 1982.
83. U.S. Army Corps of Engineers, *HEC-ResSim: Reservoir System Simulation*, User's Manual, Hydrologic Engineering Center, Davis, CA, 2003.
84. U.S. Dept. of Interior, *Standing Operating Procedures Guide for Dams and Reservoirs*, U.S. Government Printing Office, Denver, CO, 1985.
85. Walski, T., *Analysis of Water Distribution Systems*, Van Nostrand Reinhold Company, New York, 1984.
86. Warnick, C.C., Mayo, H.A., Carson, J.L., and Sheldon, L.H., *Hydropower Engineering*, Prentice-Hall, Englewood Cliffs, 1984.
87. Wylie, E.B. and Streeter, V.L., *Fluid Mechanics*, McGraw Hill, New York, 1985.
88. Willis, R. and Yeh, W.W., *Ground Water Systems Planning and Management*, Prentice Hall, Englewood Cliffs, 1987.
89. World Commission on Environment and Development, *Our Common Future*, Oxford University Press, Oxford, 1987.
90. Wunderlich, W.O., System Planning and Operation, in Gulliver, J.S. and Arndt, R.E.A. (eds.), *Hydropower Engineering Handbook*, McGraw Hill, New York, 1991.
91. Wurbs, R.A., Optimization of Multiple Purpose Reservoir Operations: A Review of Modeling and Analysis Approaches, *Research Document No. 34*, U.S. Army Corps of Engineers, Hydrologic Engineering Center, Davis, CA, 1991.
92. Yeh, W.W.-G., Reservoir Management and Operations Models: A State of the Art Review, *Water Resources Research*, 21(12), 1797–1818, 1985.
93. Younos, T., *Water Research Needs in Virginia*, Virginia Water Resources Research Center, 2005. Available at http://www.vwrrc.vt.edu/publications/recent.htm (accessed in April 2006).
94. Zhen, X.-Y., Yu, S.L., and Lin, J.-Y., Optimal Location and Sizing of Stormwater Basins at Watershed Scale, *Journal of Water Resources Planning and Management*, 130(4), 339–347, July/August 2004.

25

Military Applications

Jeffery D. Weir and
Marlin U. Thomas
Air Force Institute of Technology*

25.1 Introduction.. **25-1**
25.2 Background on Military OR........................ **25-2**
25.3 Current Military Applications of OR............. **25-3**
  Linear and Integer Programming • Network Optimization • Multiple Criteria Decision-Making • Decision Analysis • Stochastic Models • Simulation • Metrics
25.4 Concluding Remarks.................................**25-10**
References **25-11**

25.1 Introduction

Operations Research (OR) has evolved through the need for analytical and analysis methods for dealing with decisions involving the use and utilization of critical resources during military operations. One could argue that OR has been around as long as standing armies and navies. History is replete with examples of military "geniuses" who used mathematics and science to outwit and overcome their enemy. In more recent history, one need only look at Thomas Edison helping the United States Navy overcome the stifling effect of German U-boats during World War I or Lanchester's modeling of attrition rates to see OR being applied to the military. It was not until the late 1930s, however, that OR was officially recognized as a scientific discipline that could help the military. At that time, British scientists were called upon to apply their scientific methods not to improving the technical aspects of the recent discoveries in Radio Detection and Ranging (RADAR) but rather the best way to use it operationally within the military. The field of OR grew rapidly from this point and many of the advances in the field can be directly tied to their applications in the military. Evidence of this can be found in the articles that appeared in the early issues of OR, documenting the interest and applications of OR methods following World War II and throughout the three decades that followed. It is easy to argue that military OR is the beginning and actual foundation of the OR profession which has since contributed to successful efforts in gaining and sustaining world peace, and further established the foundation for the field of OR and management science as it is known today.

The purpose of this chapter is to introduce and provide an overview of military OR. We will start in Section 25.2 with a brief background on its history. There is a fairly rich literature on the field including some excellent surveys that have appeared in several books

---

* The views expressed in this chapter are those of the authors and do not reflect the official policy or position of the United States Air Force, the Department of Defense, or the United States Government.

during the past 10 years. Of particular note here is the chapter by Washburn (1994). He gives an excellent summary of the measures of effectiveness, game theory applications, and combat modeling applied in wartime applications. Our focus will be on the methodological aspects of military OR. In Section 25.3, we discuss current military applications. Although the first known applications of OR might be considered the application of experimental designs for establishing optimal machine gun firing methods during World War II by the British, the initial methodological developments were through mathematical programming (Dantzig, 1963). In Section 25.3, these and other methods and applications are summarized. In Section 25.4, we provide some discussion on the current trends that are prevalent in military OR.

25.2 Background on Military OR

Before and during World War II the military recognized the value of OR and began setting up OR groups in the different branches. The British were the first to set up these groups and in fact by 1941 they had OR groups in all three branches of their military: Army, Air Force, and Navy. P. M. S. Blackett, credited by most as the father of operations research, stood up the British Army's first group, which as mentioned earlier dealt with air defense and the best use of radar. Blackett also stood up the British Navy's first OR section at their Costal Command. The naval groups were concerned with U-boats, as they were in World War I, but also started looking at convoy operations and protection of ships from aircraft. The Air Force OR groups in Britain were more focused on air-to-air and bombing tactics. About this same time, OR was growing in the United States military.

Several incidents have been credited with transferring OR from Great Britain to the United States. In 1940, Henry Tizzard from Britain's Committee for the Scientific Study on Air Defense traveled to the United States to share some of the secret developments Great Britain had made in sonar, radar, and atomic energy. Although there was no direct mention of OR in this technical interchange, there was mention of assigning scientists to the various commands within the United States. Captain Wilder D. Baker, in charge of antisubmarine studies for the Atlantic Fleet, was influenced by "Scientists at the Operational Level," a paper written by P. M. S. Blackett, and asked John Tate of the National Defense Research Committee to provide some scientists for his studies. It was at this time that Phillip M. Morse was brought into operations research. Finally, the commander of the United States' Eighth Bomber Command requested scientists be assigned to his command after conferring with B. G. Dickens of the ORS Bomber Command of Great Britain. This led to the formation of what became known as operations analysis in the U.S. Army Air Forces. How much these events actually contributed to the formation of OR units in the United States is debatable, but shortly after its entry into World War II, the U.S. Navy had two OR groups and the Army Air Forces had its first of many (McCloskey, 1987).

Before and during World War II, operations researchers proved the value of applying the scientific method to the day-to-day operations of the military and not just to weapon system development and strategy. The exploits of these early researchers are chronicled in many articles and books, too numerous to name (see, for example, Koopman's (1980) *Search and Screening*, E. S. Quade's (1964) *Analysis for Military Decisions*, and any issue of *Operations Research* dated in the 1950s). There is one, however, that can be used to demonstrate how OR modeling during this time was quickly outpacing our ability to solve many of the mathematical formulations and another that can be used to highlight the differences between the beginnings of OR and how it is practiced today.

Saul Gass in his article "The First Linear-Programming Shoppe" (2002a) details how the development of the computer industry and OR were linked. Of importance to this chapter, however, is how that link came about. In 1947, Project SCOOP (Scientific Computation Of Optimal Programs) was formed by the U.S. Air Force. The members of Project SCOOP included at some time during its eight-year existence, George B. Dantzig, Saul I. Gass, Murray Geisler, Leon Goldstein, Walter Jacobs, Julian Holley, George O'Brien, Alex Orden, Thomas L. Saaty, Emil D. Schell, Philip Wolfe, and Marshall K. Wood. Its main objective was to develop better answers to the problem of programming Air Force requirements. Wood and Dantzig defined programming as "the construction of a schedule of actions by means of which an economy, organization, or other complex of activities may move from one defined state to another, or from a defined state toward some specifically defined objective" (Wood and Dantzig 1949, p. 15).

This definition led Dantzig to name his rectangular optimization model linear programming (LP). In addition to research into LP and the simplex method, Project SCOOP researched inequality systems by the relaxation method and zero-sum two-person games by the fictitious play method. Project Scoop was the impetus for the U.S. National Bureau of Standards contracting the first Standards Eastern Automated Computer (SEAC) and purchasing the second production unit of the UNIVAC. During its eight-year span, Project SCOOP went from solving a 9×77 Stigler's diet problem in 120 person-days on a hand-operated desk calculator to solving an 18×34 gasoline blending LP in 20 minutes on an IBM 701. While Project SCOOP and other works were outpacing technology, there was also a shift beginning to happen in military OR (Gass, 2002).

After World War II many of the scientists that had helped the war effort were returning to their original lives in academia and were being offered jobs in industry. Businesses, seeing the success of OR in the military, began using these techniques to improve their own operations and the field of management science was beginning to evolve. In 1951, Morse and Kimball published *Methods of Operations Research*, originally written in 1946 as a classified report for the U.S. Navy's Operations Evaluation Group, in hopes to share information with the scientific population at large about the successes of OR before and during World War II. Morse and Kimball saw that many of the successes in OR were not widely known either due to classification or just a general lack of free flowing information. Their hope was that by compiling this information the field of OR could be expanded in the military and be applied to nonmilitary operations as well. Of note in this book is how Morse and his peers performed OR during World War II.

In the early stages of OR in the military, practitioners were focused on improving tactics and the effective use of newly invented equipment developed by scientists to enable the military to counter or overcome enemy techniques and equipment. Military analysts used close observation and mathematical modeling of processes and their salient properties to improve desired outcomes. After the war, the size of the nation's militaries was decreasing both because the global war was over and many of the tactics had been so improved that smaller forces could have the same effect as larger ones. Beasley in his OR notes discusses how the probability to attack and kill a U-boat went from 2% to 3% in 1941 to over 40% by 1945. Bothers (1954) explains how the Eighth Air Force used operations analysis to increase the percentage of bombs that fell within 1000 feet of their aimpoint from less than 15% in 1942 to over 60% by 1944.

25.3 Current Military Applications of OR

The rest of this chapter will outline selected general application areas for OR within the military. The goal is to describe how military organizations use common OR techniques and to provide some specific examples where it is useful. It is not a complete review of current literature, as that would require an entire book. On their Web site, http://www.mors.org, the Military Operations Research Society of the United States has a list of several excellent references regarding OR and the military. When talking about the application of OR to military problems, it is useful to look at what level the tools or techniques are being applied. These levels are similar to those of businesses. At the topmost level are strategic (corporate) decisions that are made for the good of a nation. The bottom level is comprised of tactical decisions. These decisions influence how small units or individuals will engage for short durations. Linking these levels is the operational level. At this level, decisions are made to achieve the strategic goals. It determines the when, where, and for what purpose forces will be employed. While each level of war is unique, many of the OR tools talked about in Part I of this book can be applied at more than one of these levels.

25.3.1 Linear and Integer Programming

Linear and integer programming are used throughout the military in a variety of ways. Most of their uses are at the strategic and operational levels. At the operational level, LPs and integer programs (IP) have been used to allocate operating room space, route unmanned air vehicles, and even to optimize rehabilitation and restoration of damaged land at training facilities. Its most frequent use, however, is at the strategic level in the area that it was originally designed by Dantzig to help, that being the planning of large-scale operations.

Often called campaigns, these operations look at possible future scenarios for major wars, that is, country on country conflicts. These models provide decision-makers with an analytical tool for determining the impacts of budget changes, force structures, target allocations, munitions inventories, and combat attrition rates on a nation's war-fighting capability. Typically, the scenarios analyzed use all of a nation's military forces, air, ground, and sea, and are modeled over a large time-period, for example, months or years. These large-scale operations can also be humanitarian in nature such as natural disaster relief. Typically, the decision variables used reflect combinations of platform, weapon, target, profile, loadout, and weather. A platform refers to the type of system delivering the weapon, for example, a tank, aircraft, or naval vessel. The weapon refers to what type or types of munitions or aid are being delivered, for example, food pallets, large equipment, flares, bombs, guided munitions, or torpedoes. The profile refers to the manner in which the platform will attack a target or deliver its cargo. For aircraft, this might include altitude, angle of attack, or landing area. The loadout refers to the number of weapons used against or delivered to the target. A submarine might use a volley of two torpedoes against a single target, or a large cargo aircraft might deliver bags of grain and a truck to haul them to a different location. Finally, target and weather are self-explanatory and refer to the target to be attacked or location of aid delivery and the weather expected at the target area. Associated with each of the variables are several parameters.

It is beyond the scope of this chapter to outline all of the parameters that could be associated with these decision variables, but a few bear mentioning as they will be mentioned later in this chapter. With respect to war-like operations, one parameter of interest is how much damage to the target can be expected from choosing the decision variable. This is referred to as damage expectancy (DE) or probability of kill (PK) and takes into account all of the characteristics used to describe the decision variable as well as those of the target. Another is the probability that the platform making the attack described by the variable will survive and be available for subsequent attacks. The complement of this is called the attrition rate. How these are modeled is discussed later in this chapter and is usually done at the tactical level.

25.3.2 Network Optimization

Most of the applications for network optimization within the military are no different from those of major corporations or small cities. Like corporations, they are concerned with building reliable communication networks to include local area networks (LANs) and wide area networks (WANs), logistics networks for supply chain management, and the exploitation of network structures and formulations for LPs and IPs. As the military must build and maintain self-sufficient bases, like municipalities they work with electrical grids, water and sewer supply, and other public utility networks. There are, however, some aspects of network optimization that are unique to the military. One area of research that is unique to the military (and possibly law enforcement) deals with the interdiction of an enemy's network.

When dealing with the interdiction of an enemy's network, the military considers both physical networks, such as command and control networks, and more esoteric networks such as clandestine and terrorist networks. Regardless, the goal for these types of network operations is the same, disruption. Many of the techniques used in optimizing networks can also be used to identify vulnerabilities within networks. The most obvious of these techniques is the max-flow min-cut. By modeling the max flow of an enemy's network, bottlenecks can be identified for potential exploitation. Besides looking at arcs, one might also be interested in the nodes of a network. In this instance a facility location problem may be useful to identify key nodes that are in contact with many other nodes or the use of graph theoretic centrality measures might be useful to exploit.

25.3.3 Multiple Criteria Decision-Making

Like nonlinear programming, multicriteria decisionmaking within the military has followed from LP and IP development. While not the only techniques for multicriteria decisionmaking, goal programming and multiple objective linear programming are used in conjunction with large-scale campaign models to account for various objectives. Two examples of these techniques follow from the U.S. Strategic Command's Weapon Allocation Model (WAM) and the U.S. Air Force Space Command's (AFSPC) Aerospace Integrated Investment Study (ASIIS).

WAM is a large-scale integer goal-programming model. It allocates weapons to targets. Several goals need to be met by the allocation. These include maximizing damage expectancy, ensuring specific weapons cover specific targets, and goals for damage expectancy within varying target categories. These goals are user defined and can be specified either as a goal to be obtained or as a hard constraint. Once the user identifies the goals, they are given priority and the model uses preemptive goal programming. That is, the model runs with the highest priority goal remaining as the objective function and if the goal is obtained, that goal is added to the model as a hard constraint. If a goal cannot be achieved it will be adjusted to the level of best achievement and added to the model as a constraint. This technique was chosen over weighting the priorities because there was no single decision maker from whom to solicit weights. ASIIS does support a single decision maker and therefore uses a weighted goal programming approach.

AFSPC is responsible for training and equipping the U.S. space force. They use the ASIIS model to aid in making decisions with regard to how to balance their spending on space assets and maintain a capable force. ASIIS is a large-scale integer program that uses *elastic* constraints (Brown et al., 2003) and has a linear objective function that penalizes deviations in the *elastic* constraints. This is a long-range planning tool that looks at an epoch of time, usually 25–30 years. There are four *elastic* constraints or goals in ASIIS: (1) do not exceed the authorized funding levels in any given year, (2) do not exceed total funding over the epoch, (3) minimize capability gaps, and (4) minimize underspending.

The first goal ensures that AFSPC does not spend more money on its investments than is authorized by the government in any given year. The second goal ensures that if in any year more money is needed, that money can come from a future or past year in the epoch. The third goal tries to purchase or fund those programs that will ensure that AFSPC can meet its current and future mission requirements with as little capability gap as possible based on the money and technology available. The fourth goal was added to the model to let the model choose redundant systems for better capability coverage. This was needed due to the way the capability gaps are calculated in the third goal. The ASIIS model uses absolute deviations from the goals and therefore, weights for each goal both prioritize the goals and ensure that they are of equal scale. For example, the units for money and capability gaps are several orders of magnitude different. As there is a single decision maker in charge of AFPSC, these weights can be solicited directly from the decision maker to ensure their priorities are met.

25.3.4 Decision Analysis

There has been much debate in the last 5 years as to what is decision analysis (DA). Keefer et al. (2002) define DA as a normative, not descriptive, systematic quantitative approach to making better decisions. Specifically, for an application to be considered DA, it must "explicitly analyze alternatives for a decision problem using judgmental probabilities and/or subjectively assessed utility/value functions." This definition excludes multicriteria decisionmaking and the analytic hierarchy process. In the military, DA applications fall into two categories: decision trees and value focused thinking (VFT).

VFT has been used extensively in the military to help senior decision makers gain insight into decisions that have competing objectives. Of the ten articles summarized by Keefer et al. in their survey of DA literature from 1990 to 2001, eight used some form of VFT. The models described in these articles were usually very large in size and scope. The largest effort described was the U.S. Air Force's 2025 (Jackson et al., 1997; Parnell et al., 1998) study that included over 200 subject matter experts working over a year to develop a 134-attribute model to score 43 future concepts using weights based on six different scenarios. Not all VFT models are at the strategic level, however. The Croatian Army (Barković and Peharda, 2004) used a VFT model to determine what rifle to acquire to replace its aging stores and a VFT model was developed to help U.S. psychological operations (PSYOP) detachment commanders choose the best strategy for the use of PSYOP, where PSYOPs are defined as operations that use selected information to influence an audience's behavior (Kerchner et al., 2001). Along with the VFT model, the military also use decision trees.

The application of decision trees is most useful in the military when there is a single objective to consider. When faced with a variety of options, the decision maker can then choose the best option based on the expected value of the objective. An example of this in the military was the use of a decision tree to choose the best allocation of aircraft to targets by Griggs et al. (1997). In this example, an IP was used to build the allocations with varying weather states and the plan with the best expected value was found by solving a decision tree with the IP solutions at the varying weather states.

25.3.5 Stochastic Models

One of the earliest quantitative approaches to combat modeling was by F. W. Lanchester (1916) who proposed the study of the dynamics of two opposing forces, say Red versus Blue, through a set of differential equations. Let $R(t)$ and $B(t)$ be the number of combatants

Military Applications

remaining for each force at a time $t > 0$ and assume that:

$$\frac{d}{dt} R(t) = -bB(t), \quad R(t) > 0$$

$$\frac{d}{dt} B(t) = -aR(t), \quad B(t) > 0$$

where a and b are constants representing the respective lethality rates of kill against each other. It can be shown from these equations that

$$a[R(0)]^2 > b[B(0)]^2 \Rightarrow \text{Red wins}$$

$$b[B(0)]^2 > a[R(0)]^2 \Rightarrow \text{Blue wins}$$

implying that the force with the greatest strength will first deplete the number of combatants of the weaker force. Strength is given by the product of the attrition rate and the square of the initial force. This has become known as the *Lanchester square law*. A simpler linear law follows by assuming that the attrition rates are directly proportional to the number of opposing units. Many variations of the attrition conditions for this model have been studied including that by Brackney (1959) and probabilistic extensions by Morse and Kimball (1951) and further work by Smith (1965), Kisi and Hirose (1966), and Jain and Nagabhushanam (1974).

The earlier studies of the Lanchester models focused on the development of closed form results for force survival probabilities and expected duration of combat. A complete treatment of these models can be found in Taylor (1980, 1983). Bhat (1984) formulated a continuous parameter Markov chain model for this force-on-force combat scenario. We assume the two forces to have given fixed initial levels $R(0) = R$ and $B(0) = B$. The states of the process are outcomes of the 2-tuple $\langle R(t), B(t) \rangle$ with transitions among the finite number of states satisfying the Markov property. For example, for $R = 2$ and $B = 3$ the state space is

$$\Omega = \{(2,3), (2,2), (2,1), (2,0), (1,3), (1,2), (1,1), (1,0)\}$$

The survival probabilities can be derived from the Kolmogorov equations and for a Red win with j survivors it can be shown that,

$$P_{i,0} = b_{i,1} \int_0^\infty P_{i,1}(t) dt$$

for the duration of combat, T

$$P(T > t) = \sum \sum_{i,j > 0} P_{i,j}(t)$$

and

$$E[T] = \int_0^\infty P(T > t) dt.$$

Another area of important military applications of OR is in search and detection. The basic problem is to try to locate a hidden object in a region A through examination of an area $a \in A$. This problem appears in many military settings; for example, A could be a region of water for antisubmarine operations, a weapon site in a region of desert, or a failed component in a piece of equipment. Various solution approaches have been taken, the earliest, of course, being an exhaustive search which, of course, has limitations due to costs for searching, that is, in time, costs, or other effort, and the criticality of the objective. Another popular method applied during World War II is the random search that can be formulated as a nonlinear programming problem. Let p_i denote the probability of a target being in $A_i \subseteq A$, $i=1,\ldots,n$, and B is the amount of available search effort. The problem is then to find a solution to

$$\text{min:} \quad \sum_{i=1}^{n} p_i e^{-b_i/A_i}$$

$$\text{subject to:} \quad \sum_{i=1}^{n} b_i = B$$

More information on methods and approaches for dealing with search and detection can be found in Washburn (2002).

25.3.6 Simulation

Simulation has become one of the most widely used OR techniques in the military. Many of the processes that were modeled with mathematical programming or that had difficult to find closed formed solutions using probabilistic models are now being modeled with simulations. Today's more powerful computers allow practitioners to better model the chance outcomes associated with many real world events and help describe how systems with many such events react to various inputs. Within the military, Monte Carlo, discrete event, and agent-based simulations are being used at all levels.

The most recognizable type of simulation in the military is probably the "war game." The "war game" gets its fame from movie and television, but actually has the same purpose as mathematical programming at the strategic level. This type of discrete event simulation allows decision makers to try out different strategies, force structures, and distributions of lethality parameters and see their affect on the outcome of a given scenario. Unlike their deterministic counterparts, these simulations show decision makers a distribution on the output parameters of interest and not just the most likely. This provides senior military officials insight into the inherent risks associated with war. Like their deterministic counterparts, these models too are aggregated at the strategic level, but can also be used at the operational and tactical level when disaggregated.

While senior military officials do not need to know what every soldier, sailor, or airman is doing on the battlefield, there are leaders that do. At the operational level, simulations focus more on small groups and allow commanders to dry run an operation. This type of simulation is not used for planning purposes necessarily, but to provide insight to commanders so that they can see how different decisions might play out on the battlefield. By identifying decisions that will likely result in bad outcomes, commanders can design new techniques, tactics, and procedures to mitigate those bad outcomes. These same types of simulations when run with even smaller groups can be useful at the tactical level.

At the tactical level, leaders of small units can rehearse actions, trying out different strategies for a single scenario. An army unit might simulate patrolling a village and try different tactics for repelling an ambush. Airmen can try several different approaches for bombing a target and sailors for enforcing an embargo. In these cases, the simulation acts as a training tool allowing the military to gain experience in what can happen during an operation without the loss of life and with substantial monetary savings when compared to live exercises. There is more value to these simulations thought than just training.

As is discussed in the section on LPs and IPs, many models require parameter estimates. As the funding of militaries decreases and the cost of weapons increases, simulation can help to estimate these parameters. During World War II, OR sections within bombing units would gather data on thousands of bomb drops to try and estimate the effectiveness of a particular tactic and try to find ways to improve them. Today's militaries do not have thousands of bomb drops to gather data and have turned to simulation for help. With more powerful computers, simulations that mirror reality better can be built at the individual weapon level. These simulations model the physics behind the weapon and its delivery vehicle allowing researchers to generate the thousands of "bomb drops" they cannot afford. Like their predecessors, these researchers are able to uncover problems with tactics as well as design flaws and produce better effects. These simulations can also help militaries learn about their enemies.

The same simulations that improve one's own tactics and weapons can also be used to model an enemy's weapons. These simulations can be used to provide the distributions used in simulations at the tactical level or higher. For example, when an aircraft is approaching a guarded target, a simulation can be used to build a distribution on the probability that the aircraft will survive the attack. By using engineering data and modeling the known physics of the system guarding the target, the simulation provides both a training tool to try to maximize the likelihood of surviving a mission, and a parameter estimate of the enemy system that can be used in larger simulations. While all of these types of simulation are discrete event, they are not the only type of simulation being used by the military today.

Agent-based simulations (ABS) are the newest simulation tool for the military. Originally developed by those who study the evolution of (complex) systems, such as biologists, AI researchers, sociologists, and psychologists, agent-based simulation is used to predict future events as well as past processes. It is not a computational approach to rational action such as game theory, but rather attempts to model autonomous and heterogeneous agents capable of adapting and learning. Most military operations require that individuals interact with their environment and other individuals. Before ABS, these interactions were modeled based on physics vice first principles of behavior (assuming one knows what these are). That is to say, most simulations before ABS were based on kinetics. As the military's missions continue to grow to include more peacekeeping and humanitarian relief, kinetic models are less useful.

For these ever-increasing missions, militaries need to be able to model the reaction of populations to their presence. Like the tactical discrete event simulations, these ABS models can help units train for these new missions. For example, what is the best tactic for crowd control? Should you call in more reinforcements, fire a warning shot, use non-lethal weapons? ABS models can show how each of these choices affects different cultures and are affected by different environments. These types of insights help commanders identify changing situations before they get out of control. While this type of modeling is still in its infancy, it shows promise for future military operations other than war.

25.3.7 Metrics

Once a military operations researcher has learned the techniques and tools described in this chapter and others and applies them to a problem or process, they must then try to build what Morse and Kimball (1970) call a "quantitative picture" for the decision maker. This quantitative picture allows for comparisons both within an operation and between competing alternatives that may be used for an operation. Morse and Kimball (1970) called these comparators "constants of the operation." Today we know them as metrics.

The two most common categories of metrics are measures of performance (MOP) and measures of effectiveness (MOE). MOPs are used to answer the question, "Are we doing things right?" MOEs answer the question, "Are we doing the right things?" Examples of military MOPs are casualties, bombs dropped, ton miles of cargo moved, or square miles searched in a given time period. Observed actual values of MOPs can be compared to theoretical values if calculable and show how well a unit is doing or where possible improvements may be made. However, even when a unit is operating at or near its theoretical limits, it still may not be effective. MOEs provide the commander or senior decision maker insight into how well an operation or groups of operations are achieving the overall objective. Going back to the example of ton miles of cargo moved, while a unit may be moving goods at its capacity, if it is moving the wrong type of goods or the goods are consistently late, then that unit is not effective.

There are countless examples in the literature of the use of metrics. Morse and Kimball (1970) devote a whole chapter of their book to the use of metrics and give many examples from their time with the Navy. Dyson (2006) recalls examples of metrics he used while assigned to the Operational Research Section (ORS) of the British Royal Air Force's Bomber Command in 1943. While these are historical examples, they help provide today's military operations researchers with foundations for developing the "art" of building the quantitative picture of tomorrow's battles. To see that this "art" is still being practiced today one need only look at recent publications. In the book by Perry et al. (2002), the Navy develops measures of effectiveness to verify the hypothesis that network-centric operations will enhance the effectiveness of combat systems. Other examples can be found by searching the Defense Technical Information Center (www.dtic.mil), where articles and technical reports by Schamburg and Kwinn (2005) and West (2005) detail the use of MOPs and MOEs to build a quantitative picture of future Army weapon systems.

25.4 Concluding Remarks

The nature of warfare is changing throughout the world and military OR practitioners will need to change along with it. In the United States, this new generation of warfare is referred to as the 4th generation (Lind et al., 1989). The generations of warfare are concisely defined by Echevarria (2001) as (1) massed manpower, (2) firepower, (3) maneuver, and (4) insurgency. While there is not complete agreement about 4th generation warfare, one thing is clear: insurgency and terrorist attacks have presented the military with new problems to solve.

Since World War II military OR practitioners have been focused on planning and training and not the quick fielding of new technologies. Their modeling has been characterized by the realistic portrayal of combat in simulations. They were often dislocated from the actual problem being solved and have lost sight of their beginnings. The notion of 4th generation warfare is changing this. Increased access to technology, information, and finances has given terrorists and insurgents alike the ability to create chaos at will. The scientific community has been asked to develop and field new technologies rapidly to thwart these new tactics. Military OR practitioners are once again being called back to the field to help the military employ these new technologies effectively against their enemies.

The threat of terrorism is not only affecting military OR practitioners through rapid fielding initiatives, but also through how to plan for and respond to a terrorist attack. While the location and timing of these attacks are not known with certainty, it is still possible to plan for some of them. In the military, this is called contingency planning. This is in contrast to deliberate planning where the location and enemy are known with certainty. During any type of planning, logistics support for personnel and equipment is important. While the military has many systems for assessing their deliberate plans, contingency or crisis planning assessment is lacking (Thomas, 2004). The idea of contingency logistics has been brought to the forefront of research by the recent tsunamis in the Indian Ocean and hurricanes in the United States. While not typically the responsibility of the military, responding to these types of contingencies is becoming more and more common. The military can learn from these operations what is needed to support a contingency. In the future, military OR practitioners will need to develop models both to plan for and assess contingency operations whether they be for war or operations other than war.

Finally, a new field of study, system of systems (SoS), is emerging and OR practitioners seem well placed to help. While there are many definitions for SoS, this chapter will use one from the U.S. Defense Acquisition University's Defense Acquisition Guidebook: "[SoS] engineering deals with planning, analyzing, organizing, and integrating the capabilities of a mix of existing and new systems into a system of systems capability greater than the sum of the capabilities of the constituent parts." The notion is that there are few standalone systems and by considering during the design phase how one system will interact with others, we can get better results in the end. As these SoS cut across many disciplines and OR practitioners are used to interacting with multiple disciplines it seems logical that we can help in this new field. Since the military purchases many high-cost systems it seems prudent that military OR practitioners begin to embrace this new field and see how they can help save their countries money and improve military capabilities. For an example of how the military is already beginning to use SoS, see Pei (2000).

References

1. Barković, M. and I. Peharda, 2004. The value-focused thinking application: Selection of the most suitable 5.56 mm automatic rifle for the Croatian Armed Forces, presented at Croatian Operations Research Society 10th International Conference on Operations Research, Trogir Croatia, Sept. 22–24, 2004.
2. Bhat, U. N., 1984. *Elements of Applied Stochastic Processes*, 2nd ed., Wiley, New York.
3. Beasley, J. E., "OR Notes," Downloaded from http://people.brunel.ac.uk/~mastjjb/jeb/or/intro.html.
4. Blackett, P. M. S., 1942. "Scientists at the Operational Level," reprinted in Blackett, P. M. S. *Studies of War* (Oliver and Boyd, Edinburgh, 1962), Part II, Chapter 1, pp. 171–176.
5. Brackney, H., 1959. "The Dynamics of Military Combat," *Operations Research*, 7(1):30–44.
6. Brothers, L. A., 1954. "Operations Analysis in the United States Air force," *Operations Research Society of America*, 2(1):1–16.
7. Brown, G. G., R. F. Dell, H. Heath, and A. M. Newman, 2003. "How US Air Force Space Command Optimizes Long-Term Investment in Space Systems," *Interfaces*, 33(4):1–14.
8. Dantzig, G. B., 1963. *Linear Programming and Extensions*, Princeton University Press, Princeton, NJ.
9. Defense Acquisition Guidebook, Version 1.6 (07/24/2006), http://akss.dau.mil/dag/.
10. Dyson F., 2006. "A Failure of Intelligence," Essay in *Technology Review* (Nov.–Dec. 2006).
11. Echevarria II, A. J., 2001. Fourth-Generation War and Other Myths, monograph from the Strategic Studies Institute of the U.S. Army War College. Available at http://www.strategicstudiesinstitute.army.mil/pubs/display.cfm?pubID=632, 8 Dec 2006.
12. Gass, S. I., 2002. "The First Linear-Programming Shoppe," *Operations Research*, 50(1):61–68.

13. Griggs, B. J., G. S. Parnell, and L. J. Lehmkuhl, 1997. "An Air Mission Planning Algorithm Using Decision Analysis and Mixed Integer Programming," *Operations Research*, 45:662–676.
14. Jackson, J. A., G. S. Parnell, B. L. Jones, L. J. Lehmkuhl, H. W. Conley, and J. M. Andrew, 1997. "Air Force 2025 Operational Analysis," *Military Operations Research*, 3(4):5–21.
15. Jain, G. C. and A. Nagabhushanarn, 1974. "Two-State Markovian Correlated Combat," *Operations Research*, 22(2): 440–444.
16. Keefer, D. L., C. W. Kirkwood, and J. L. Corner, 2002. Summary of Decision Analysis Applications in the Operations Research Literature, 1990–2001, Technical Report, Department of Supply Chain Management, Arizona State University, Nov. 2002.
17. Kerchner, P. M., R. F. Deckro, and J. M. Kloeber, 2001. "Valuing Psychological Operations," *Military Operations Research*, 6(2):45–65.
18. Kisi, T. and T. Hirose, 1966. "Winning Probability in an Ambush Engagement," *Operations Research*, 14(6):1137–1138.
19. Koopman, B. O., 1980. *Search and Screening: General Principles with Historical Application*. Pergamon, New York.
20. Lanchester, F. W. 1916. *Aircraft in Warfare: The Dawn of the Future Arm*, Constable and Co., London.
21. Lind, W. S., K. Nightengale, J. F. Schmitt, J. W. Sutton, and G. I. Wilson, 1989. "The Changing Face of War: Into the Fourth Generation," *Marine Corps Gazette*, October 1989, pp. 22–26.
22. McCloskey, J. F., 1987. "U.S. Operations Research in World War II," *Operations Research*, 35(6):910–925.
23. Morse, P. M. and G. E. Kimball, 1970. *Methods of Operations Research*. Peninsula Publishing, San Francisco, CA, 1970.
24. Parnell, G. S., H. W. Conley, J. A. Jackson, L. J. Lehmkuhl, and J. M. Andrew, 1998. "Foundations 2025: A Value Model for Evaluating Future Air and Space Forces," *Management Science*, 44:1336–1350.
25. Pei, R.S., 2000. "Systems-of-Systems Integration (SoSI)—A Smart Way of Acquiring Army C4I2WS Systems," Proceedings of the Summer Computer Simulation Conference, (2000), pp. 574–579.
26. Perry, W., R. Button, J. Bracken, T. Sullivan, and J. Mitchell, 2002. Measures of Effectiveness for the Information-Age Navy: The Effects of Network-Centric Operations on Combat Outcomes, RAND. Available electronically at www.rand.org/publications/MR/MR1449/.
27. Quade, E. S., ed., 1964. *Analysis of Military Decisions*. Amsterdam, the Netherlands.
28. Schamburg, J. B. and M. J. Kwinn, Jr., 2005. *Future Force Warrior Integrated Analysis Planning*, Technical Report. United States Military Academy Operations Research Center of Excellence. DSE-TR-0542, DTIC#: ADA434914
29. Smith, D. G., 1965. "The Probability Distribution of the Number of Survivors in a Two-Sided Combat Situation," *OR*, 16(4):429–437.
30. Taylor, J., 1980. *Force on Force Attrition Modeling*, Military Applications Section of the Operations Research Society, Baltimore, MD.
31. Taylor, J., 1983. *Lanchester Models of Warfare*, Vol. 1 and 2, Operations Research Society of America, Baltimore, MD.
32. Thomas, M. U., 2004. "Assessing the Reliability of a Contingency Logistics Network," *Military Operations Research*, 9(1):33–41.
33. Washburn, A., 1994. Military Operations Research, in S. M. Pollock et al. (eds.), *Handbooks in OR & MS. Vol. 6, Operations Research and the Public Sector*, Elsevier Science, North Holland.

34. Washburn, A. R., 2002. *Topics in O.R.: Search and Detection*, 4th ed., Military Applications Section, INFORMS.
35. West, P. D., 2005. *Network-Centric System Implications for the Hypersonic Interceptor System*, Technical Report. United States Military Academy Operations Research Center of Excellence. DSE-TR-0547, DTIC#: ADA434078.
36. Wood, M. K. and G. B. Dantzig, 1949. Programming of interdependent activities I, general discussion. Project SCOOP Report Number 5, Headquarters, U.S. Air Force, Washington, D.C. Also published in T. C. Koopmans, ed. *Activity Analysis of Production and Allocation*. John Wiley & Sons, New York, 1951, 15–18, and *Econometrica* 17 (3&4) July–October, 1949, pp. 193–199.

26

Future of OR/MS Applications: A Practitioner's Perspective

P. Balasubramanian
Theme Work Analytics

26.1	Past as a Guide to the Future	**26**-2
	Leg 1: Problem Formulation in an Inclusive Mode • Leg 2: Evolving Appropriate Models • Leg 3: Ensuring Management Support • Leg 4: Innovative Solutions	
26.2	Impact of the Internet	**26**-6
26.3	Emerging Opportunities..........................	**26**-8
References		**26**-12

As per Maltz [1], Simeon-Denis Poisson developed and used the Poisson distribution in the 1830s to analyze inherent variations in jury decisions and to calculate the probability of conviction in French courts. There have been many other initiatives prior to the 1930s similar to Poisson's, highlighting the use of quantitative techniques for decision support. However, it is widely acknowledged that operations research (OR) started as a practice in the 1930s. It owes its origin to the needs of Great Britain during World War II to enhance the deployment effectiveness of combat resources and the support facilities. Its subsequent growth is credited to the achievements during the war period. The year 1937 can be considered to be the base year as this was the year the term operational research (OR) was coined [2]. It is not even a century old and compared to other functional areas (such as production, marketing, financial management, or human resource development) or to scientific disciplines (like mathematics, physics, or chemistry) it is a very young field. Yet, there are many practitioners of OR who began questioning its utility, applicability [3], and even its survivability within four decades of its existence. Fildes and Ranyard [4] have documented that 96% of Fortune 500 companies had in-house OR groups in 1970 but the numbers dwindled by the 1980s.

Ackoff [5] fueled this debate in 1979 with the assertion that "The future of Operational Research is Past." In 1997, Gass [6] proposed that in many lines of business, issues pertaining to fundamental activities have been modeled and OR has saturated. And this debate seems to be in a continuous present tense in spite of the growth in the number of OR practitioners (more than 100,000), applications, and OR/MS journals. (The 1998 estimates are 42 primary OR journals, 72 supplementary journals, 26 specialist journals, 66 models of common practices, 43 distinct arenas of application, and 77 types of related theory development, according to Miser [7].) The fundamental issue has been the gap between developments on the theoretical front versus the application of OR models and concepts in practical circumstances. Consequently, the career growth of OR professionals in corporations beyond middle management positions has become difficult.

Rapid growth in affordable computing power since the 1970s and the advent of Internet access on a large scale have rekindled this debate during the current decade, however, with one significant difference. While the earlier issues of theory versus application remain, the need for inventing newer algorithms to locate optima is also questioned now. If exhaustive enumeration and evaluation can be done in seconds and at a small cost, even for large or complex problems, why would one want to invent yet another superior search algorithm?

These issues can be embedded into the larger context of the relevance and utility of OR in the present century, considered to be the growth period of Information Age. Will OR as a discipline survive in this digital economy? Will there be newer contexts in the emerging market place where OR would find relevance? Does the discipline need to undergo a paradigm shift? What is the future of OR practitioners?

26.1 Past as a Guide to the Future

I propose to start with the early days of OR in the British armed forces during World War II [7]. The original team at Royal Coastal Command (RCC) consisted of Henry T. Tizard, Professor P. M. S. Blackett, and A. V. Hill. They were specialists in chemistry, physics, and anatomy, respectively. They worked with many others in the military operations with backgrounds in a variety of other disciplines. They were asked to study the deployment of military resources and come up with strategies to improve their effectiveness. They were to help RCC in protecting the royal merchant ships in the Atlantic Ocean from being decimated by the German U-Boats; in enhancing the ability of the royal airforce to sight U-Boats in time; and in increasing the effectiveness of patrol of the coast with a given aircraft fleet size.

Time was of the essence here. Solutions had to be found quickly, often in days or weeks but not in months. Proposed solutions could not be simulated for verification of effectiveness, yet any wrong solution could spell doom for the entire program.

Blackett and the team relied on available but limited data not only to study each of the stated problems but also to give it a focus. They were inventive enough to relate concepts from mathematics (the ratio of circumference to area of a circle decreases as the radius increases) to suggest that the size of a convoy of merchant ships be increased to minimize damage to them against enemy attacks. They were lateral thinkers who asked for painting the aircrafts white to reduce their visibility against a light sky so that the aircraft could sight the U-Boats before the latter detected their presence. They converted the coast patrol problem into one of effective maintenance of aircraft. They documented the before and after scenarios so that the effectiveness of the solutions proposed could be measured objectively. They interacted with all stakeholders throughout the project to ensure that their understanding was correct, meaningful solutions were considered, and final recommendations would be acceptable. It helped their cause that they had the ears of RCC leaders so that their proposals could be tried with commitment.

In other words, the team was interdisciplinary to facilitate innovative thinking. It consisted of scientists who could take an objective, data-based approach for situation analysis, solution construct, and evaluation of solution effectiveness. They were capable modelers as they knew the concepts and their applicability. Above all, they had the management support from the start, they understood the time criticality of the assignment, and they were tackling real world problems.

Should we ever wonder why they succeeded?

Compare this to

Mitchell and Tomlinson's [8] enunciation of six principles for effective use of OR in practice in 1978. The second principle calls for OR to play the role of a Change Agent.

and
Ackoff's assertion in 1979 [5] that "American Operations Research is dead even though it has yet to be buried" (for want of interdisciplinary approach and due to techniques chasing solutions and not the other way round).
or
Chamberlain's agony [9] even in 2004 that OR faced the challenge of "to get the decision maker to pay attention to us" and to "get the practitioner and academics to talk to each other."

My detailed study of innumerable OR projects (both successful and failed) reveals four discriminators of success, namely

1. Problem formulation in an inclusive mode with all stakeholders
2. An appropriate model (that recognizes time criticality, organization structure for decision making, and ease of availability of data)
3. Management support at all stages of the project
4. Innovative solution.

These are to be considered the Four Legs of OR. Absence of any one or more of these legs seem to be the prime cause for any failed OR project. Presence of all four legs gives an almost complete assurance of the project's success. Besides, they seem to be time invariant. They hold good when applied to the pre-Internet era as well as the current period. (A review of the Franz Edelman Prize winning articles from 1972 to 2006 lends strength to this assertion.) Let us explore these in detail.

26.1.1 Leg 1: Problem Formulation in an Inclusive Mode

This is a lesson learned early in my career as an OR practitioner.

The governor of the state is the chancellor of the state-run university in India. The then governor of Maharashtra called the director-in-charge of my company, Mr. F. C. Kohli, to discuss a confidential assignment. As a young consultant I went with Mr. Kohli and met the governor in private. The governor, extremely upset about the malpractices in the university examination system wanted us to take up a consulting assignment and to submit a report suggesting remedial action. He wished to keep the study confidential and not to involve any of the university officials whose complicity was suspected. Mr. Kohli intervened to state that it would be impossible to do such a study focusing on malpractices without involving the key stakeholders. Much against his belief, the governor finally relented to let Mr. Kohli try.

We went and met the vice chancellor of the university and briefed him about the assignment. It was the turn of the VC to express his anguish. He berated the government in general for not giving the university an appropriate budget to meet the growing admission and examination needs of the university during the past decade. Mr. Kohli then suggested that the scope of the study be enhanced to cover the functioning of the examination system in entirety (and not the malpractices alone) and was able to get the VC to agree to participation.

As a consequence we went back to the governor and impressed upon him the need to enlarge the scope. He consented.

Identifying all stakeholders and getting their commitment to participate from the beginning is an essential prerequisite for the success of any OR assignment. That this has to be a part of the project process is often lost among the OR practitioners.

Stakeholder involvement enables scoping the project appropriately, facilitates problem construction in a comprehensive manner, and eases the path for solution implementation.

Above all it recognizes and accepts the decision making role assigned to stakeholders and creates an inclusive environment. It is a natural route to forming a multidisciplinary team.

26.1.2 Leg 2: Evolving Appropriate Models

This has turned out to be the Achilles' heel of OR professionals.

Stated simply, a model is not reality but a simplified and partial representation of it. A model solution needs to be adapted to reality. The right approach calls for qualitative description, quantification, and model building as a subset and finally qualitative evaluation and adaptation of the model solution. This realization is reflected in the approach of Professor Blackett in 1938. Yet it has been lost in many a subsequent effort by others.

Many a problem of OR practice, in the pre-Internet era, can be traced to the practitioner lamenting about the quality or reliability of data and nonavailability of data at the desired microlevel at the right time as per model needs. This is ironical as the role of the practitioner is to have a firm grip on data quality, availability, and reliability at the early stages so that an appropriate model is constructed. In other words, the relationship between model and data is not independent but interdependent.

Shrinking time horizons for decisions is the order of the day. Model-based solutions need to be operationalized in current time. Prevention of spread of communicable diseases before they assume epidemic proportions is critical to societal welfare when dealing with AIDS, SARS, or avian flu. Models that depend on knowing the spread rate of a current episode cannot do justice to the above proposition (while they are good enough to deal with communicable diseases that are endemic in a society and do flare up every now and then).

I was involved in a traffic management study in suburban commuter trains in Mumbai in the early 1980s. The intent was to minimize the peak load and flatten it across different commuting time periods. We evolved an ingenious way of using a linear programming (LP) model with peak load minimization as the objective and scheduling the work start time for different firms as decision variables. And we obtained optimality that showed reduction of peak load by 15%. This admirable solution, however, could never be implemented as it did not reflect the goal of thousands of decision makers in those firms. Instead their goals were to maximize interact time with their customer and supplier firms as well as with their branch offices spread out in the city.

Having learnt this lesson, I included the study of organization structure and decision making authority as an integral aspect of any subsequent OR project. When our group came up with the optimal product mix for a tire company, where I was the head of systems, we ensured that the objective function reflected the aspirations of stakeholders from sales, marketing, production, and finance functions; the constraint sets were as stated by them and the decision variables were realistic from their perspective. The optimal solution, along with sensitive analysis was given as an input to their fortnightly decision committee on production planning, resulting in a high level of acceptance of our work.

Models are context sensitive and contexts change over time. Hence data used to derive model parameters from a given period turn out to be poor representers with the passage of time. The model itself is considered unusable at this stage. The need to construct models where model parameters are recalculated and refreshed at regular intervals is evident. Assessing and establishing the effectiveness of a model continuously and revising the model at regular intervals depend on data handling capabilities inbuilt into the solution.

Data collection, analysis, hypothesis formulation, and evaluation in many instances is iterative (as opposed to the belief that they are strictly sequential). Andy Grove, then the CEO of Intel, was afflicted with prostate cancer in the early 1990s. He was advised to go for surgery by eminent urologists. His approach to understanding the situation and constructing a solution different from their suggestion is documented in Ref. [10]. The use of data and data-based approach and the iterative nature of finding a solution are exemplified here.

26.1.3 Leg 3: Ensuring Management Support

Operations research functionaries within a firm are usually housed in corporate planning, production planning and scheduling, management information systems (MIS), or industrial engineering (IE) departments or in an OR cell. They tend to be involved in staff functions. They are given the lead or coordination responsibility in productivity improvement assignments by the top management. As direct responsibility to manage resources (machines, materials, money, or workforce) rests with functional departments such as production, purchase, finance, and HRD the onus to form an all inclusive stakeholder team rests with the OR practitioner.

Most organizations are goal driven and these are generally short term. For listed companies the time horizon can be as short as a quarter and rarely extend beyond a year. The functional departments mirror this reality with monthly goals to be met.

Many OR assignments originate with top management support. As the assignment progresses, the OR practitioner has a challenging task of including the interest of all stakeholders and ensuring their continued participation vis-à-vis keeping the support and commitment of top management. These two interests can diverge at times as the former's goal is to protect the interest of a particular resource team while management, concerned with overall effectiveness of all resources, is willing for a tradeoff. Issues and conflicts arise at this stage resulting in inordinate delay in project execution. Management support soon evaporates as a timely solution is not found.

Mitchell and Tomlinson [8] have stressed the Change Agent role of OR. Playing this role effectively calls for many skills such as excellent articulation, unbiased representation of all interests, negotiation, and diplomacy. OR practitioners need to ensure that they are adequately skilled to do justice to these role requirements.

Apart from helping to ensure top management support, possession of these skills will facilitate long-term career growth of the OR professional by switching over to line functions. This is a career dilemma for many OR professionals. While their services are required at all times, their role rarely rises above that of a middle or senior management position. There are no equivalent top management positions such as Chief Finance Officer, Chief Operations Officer, or CEO for them. (Some with IT skills can become Chief Information Officers.) Hence, diversifying their skill base (beyond quantitative) is a precondition for the growth of any OR professional in most organizations.

26.1.4 Leg 4: Innovative Solutions

Andy Grove, in his article about his battle with prostate cancer [10], states that "The tenors always sang tenor, the baritones, baritone, and the basses, bass. As a patient whose life and well-being depended on a meeting of minds, I realized I would have to do some cross-disciplinary work on my own."

The single value most realized out of an interdisciplinary team is the innovativeness of the solution generated. Either the model or the process adopted should result in its emergence. Discussing the practice of OR, Murphy [11] asserts that the "OR tools in and of themselves are of no value to organizations." According to him, the ability to build a whole greater than the sum of the parts "is the missing link." Blackett and his team took full advantage of this. In spite of the use of models to structure and analyze the given problems, they went out of the box when finding solutions. (But returned to quantitative techniques to assess the impact of solutions.)

OR has become integral to the success of an organization whenever the solution turns out to be innovative. A case in point is the Revenue Management System of American Airlines [12]. Innovation occurred here when the value of attributes such as time to book and flexibility in travel planning were discovered and quantified in selling available seats. This has been a breakthrough idea that has subsequently been adopted by hospitality and other service-oriented industries extensively.

Indian Railways carries more than 14 million passengers every day. It is the second largest rail network in the world in terms of kilometers of track laid. Booking reservations through a manual system, catering to more than 14 different language requirements, prior to the mid-1980s, was a dreaded and time wasting experience for most citizens. One had to stand in multiple lines to book for different journeys; had to return in 2 days to know if one would get a reservation in a train originating from elsewhere; an entire afternoon could be wasted in this process. Meanwhile a gray market flourished for train reservations. The computerized reservation system installed in the mid-1980s has been a phenomenal success due to innovations introduced in the system using OR concepts. Citizens can stand in one line, instead of many, to book journeys from anywhere to anywhere in any class. (Now they can even do this online from their home.) Apart from saving valuable customer time, there has been a manifold increase in railway staff productivity. In addition, the malpractices have come down to a minimum.

Almost a decade ago, I led a team that was attempting to break into a segment of IT services, namely outsourced software maintenance through competitive bidding. We had superior quality and project management processes and a competent and dedicated team. We knew how to measure software quality and programmer productivity in multiple platforms on an ongoing basis. We had excellent knowledge management practices. But these were not adequate to ensure our win. Delving deep into my OR background and working with a team of software experts, I could evolve a model that would treat the arrival and service rates of software bugs in a queuing theoretical format. Our team could then submit an innovative proposal to reduce the size of the maintenance team gradually. This would result in continuous cost reduction and yet keep service quality undiluted. With this approach, starting from Nynex, we won multiple assignments with many customers around the world. Besides, post year 2000 (Y2K), when IT budgets were enhanced, we could obtain development contracts to replace many of these old systems as many of our team members now had in-depth knowledge of products and processes of the client organization. (True multiplier effect of an innovative solution.)

26.2 Impact of the Internet

Two distinct developments need to be discussed here: the spread of computing capabilities at an ever decreasing unit cost from the early 1950s to date and the exponential growth in computer to computer networking due to the Internet since the 1990s.

The introduction of computers in industry during the early 1950s played a seminal role in showcasing what OR can do. The field of OR flourished in this period. Statistical, linear and nonlinear optimization, and stochastic simulation applications grew. The next two decades turned out to be the most fertile age of growth in OR applications. Growth in computing power, presence of personal computers, and crashing cost of computing have impacted on management science practitioners in many ways. While they played a positive role in the spread and growth of OR in the early decades, there has been a role reversal, however, since the 1980s. Productivity gains can now be demonstrated easily by computerization projects rather than taking the route of model building and optimization. Worse still is the ability of the computer to do exhaustive enumeration in a few seconds, thus striking a death knell on many efforts to find the algorithms for optimal solution.

Personal productivity tools such as spreadsheets with programmable cells have enabled wide-scale use of simulation, particularly of financial scenarios. Some optimization routines are also embedded into functions such as internal rate of return (IRR) calculation resulting in extensive use of OR concepts by masses but without the need to know the concepts.

The second phenomenon, namely, the Internet with its ubiquitous connectivity since the mid-1980s, coupled with increase in computing power available per dollar, has impacted on every field of human endeavor dramatically. The field of OR is no exception.

Gone are the days of data issues such as adequacy, granularity, reliability, and timeliness. Instead it is a data avalanche now presenting a data management problem. Montgomery [13] highlights issues relating to extraction of value from the huge volume of data in the Internet era. The arrival of tools such as data warehousing and data mining has helped the storage and analysis aspects. Yet the interpretation and cause–effect establishment challenges remain.

Many practitioners have wondered if IT developments have overshadowed the practice of OR. They also believe that OR applications in industry have reached a saturation point with the modeling of all conceivable problems. Some other developments are considered even more profound.

For example, for decades, inventory management has been a business area that has received enormous attention from the OR community. Models have been built and successfully used to consider deterministic and stochastic demand patterns, supply and logistics constraints, price discounts, and the like. The focus has remained in determining the optimal inventory to hold. Contrast this with the developments in this era of supply chain. Supply chain management (SCM) has been largely facilitated by the relative ease and cost effective manner in which disjoint databases, processes, and decision makers can be linked in real time and by real time collection and processing of point of sale (POS) data. Coupled with the total quality management (TQM) principles it has also given rise to the new paradigm that inventory in any form is evil and hence needs to be minimized. No one is yet talking about the futility of inventory models, at least at the manufacturing plant level, but that managerial focus has shifted to just in time (JIT) and minimal finished goods (FG) inventory is real.

The Internet era has also squeezed out intermediaries between manufacturers and consumers and has usurped their roles. Issues such as capacity planning and product mix optimization (essential in made to stock scenario) have yielded place to direct order acceptance, delivery commitment, and efficient scheduling of available capacity (made to order scenario). Dell's success has been credited [14] to eliminating FG inventory as well as to minimizing inventory held with its suppliers. They treated inventory as a liability as many components lose 0.5% to 2.0% of value per week in the high tech industry.

OR expertise was called for in improving the forecasting techniques earlier. Forecasts were fundamental to many capacity planning and utilization decisions. With the ability to contact the end user directly, firms have embraced mass customization methodologies and choose to count each order than forecast. In other words, minimal forecasting is in.

As the Internet has shrunk distances and time lines and as it has facilitated exponential growth in networking, the planning horizon has shrunk from quarters to days or even hours. Decisions that were considered strategic once have become operational now. The revenue management principle that originated in the airline industry is a prime example. The allocation of seats to different price segments was strategic earlier. It is totally operational now and is embedded into the real time order processing systems. This practice has spread to many other service industries such as hospitality, cruise lines, and rental cars (a beneficial side effect of strategic decisions becoming operational is the reduced need for assumptions and worrying about the constancy aspects of assumptions).

When the look-ahead period was long, the need to unravel the cause and effect relationship between different variables was strong. Regression and other techniques used to determine this dependency have always been suspected of throwing up spurious correlations. This has resulted in low acceptance of such techniques for certain purposes, particularly in the field of social sciences. In the Internet age, the emphasis is not in unraveling but to accept the scenario in totality, build empirical relationships without probing for root causes, and move ahead with scenario management tasks.

Does it mean that many revered OR techniques of the past are becoming irrelevant in this ICE (information, communication, and entertainment) age? Is there a change in paradigm where the brute force method (applied at a low cost) of computers and the Internet is sweeping to dust all optimization algorithms? Do we see a trend to embed relevant OR concepts into many day to day systems that make every commoner an OR expert?

Or is there another side to this coin? Given that OR has the best set of tool kits to cope with complexity and uncertainty and facilitates virtual experimentation, as seen by Geoffrion and Krishnan [15], is this a door opening for vast new opportunities in the digital era?

I have chosen to highlight some situations already encountered in a limited way untill now but are likely to be widespread in future. These are forerunners of possible opportunities for OR professionals but are by no means exhaustive.

26.3 Emerging Opportunities

Gass [6] opined in 1997 that the future challenges for OR will come from real time decision support and control problems such as air traffic control, retail POS, inventory control, electricity generation and transmission, and highway traffic control. This prediction is in conformity with our findings that many strategic decisions have become operational in the Information Age. Data collection at source, instantaneous availability of data across the supply and service chains, and shrinking planning horizons have meant that fundamental tasks of OR such as cause–effect discovery, simulation of likely scenarios, evaluation of outcomes, and optimal utilization of resources, all need to be performed on the fly. It implies that for routine and repeatable situations OR faces the challenge of automating functions such as modeling, determining the coefficient values, and selecting the best solutions. In other words, static modeling needs to yield its place to dynamic modeling. I contend that this is a whole new paradigm. OR professionals, both academicians and practitioners alike, have to invent newer methodologies and processes, and construct templates to meet this challenge.

Gawande and Bohara [16], discussing the issue faced by U.S. Coast Guard, exemplify the dilemma faced by many developed nations. As economic development reaches new highs, so will concerns for ecology and preservation. Hence, many policy decisions have to wrestle with the complex issue of balancing between incentives versus penalties. According to Gawande and Bohara, maritime pollution laws dealing with oil spill prevention have to decide between penalties for safety (and design) violations of ships and penalties for pollution caused. This issue can be resolved only with OR models that reflect an excellent understanding of maritime laws, public policy, incentive economics, and organizational behavior. In other words, bringing domain knowledge along with the already identified interdisciplinary team into the solution process and enhancing the quantitative skill sets of the OR practitioner with appropriate soft skills is mission critical here. I assert that structuring of incentives vis-à-vis penalties will be a key issue in the future and OR practitioners with enhanced skill sets and domain knowledge will play a significant role in helping to solve such issues.

Another dilemma across many systems relates to investments made for prevention as opposed to breakdown management. Replacement models of yesteryear chose to balance between cost of prevention against cost of failure. The scope of application of these principles will expand vastly in this new age, from goods to services. They were difficult to quantify for the services sector earlier. One of the significant developments in the digital era has been the sharpening of our ability to quantify the cost of an opportunity lost or poor quality of service. Further, a class of situations has emerged where cost of failure is backbreaking and can lead to system shutdown. Aircraft and spaceship failures in flight, bugs in mission critical software systems, and spread of deadly communicable diseases among humans are good examples. Even a single failure has dramatic consequences. Innovative solutions need to be found. Redundancy and real time alternatives are possible solutions with cost implications. OR experts with innovation in their genes will find a great opportunity here.

The spread of avian flu is a global challenge threatening the lives of millions of birds, fowls, and humans. Within the past 5 years it has spread from the Southeast Asian subcontinent to Africa and Eastern Europe. The losses incurred run into millions of dollars whenever the H5N1 virus is spotted in any place. The United States alone has budgeted 7 billion dollars to stockpile drugs needed to tackle an epidemic. Policies to quarantine and to disinfect affected population, to create disjoint clusters of susceptible population, have been implemented in some countries. Innovative measures to monitor, test, and isolate birds and fowls that move across national borders through legitimate trade, illegitimate routes, and natural migration are needed in many countries [17]. Formulating detection, prevention, and treatment measures calls for spending millions of dollars judiciously. Cost effectiveness of each of these measures and their region-wise efficacy can be ensured by application of OR concepts at the highest levels of government in every country.

Almost a decade ago, Meyer [18] discussed the complexities in selecting employees for random drug tests at Union Pacific Railroad. The issue of selection is entwined with concerns for protection of employee privacy, confidentiality, as well as fairness. Post 9/11, similar situations have occurred in the airline industry with passengers being selected randomly for personal search. Modern society will face this dilemma more and more across many fields, from work places to public facilities to even individual homes. For the good of the public at large, a select few need to be subject to inconvenience, discomfort, and even temporary suspension of their fundamental rights. (The alternative of covering the entire population is cost prohibitive.) How does one be fair to all concerned and at the same time maximize the search effectiveness for a given budget? Or what is an optimal budget in such cases? OR expertise is in demand here.

Market researchers have struggled with the 4Ps and their interrelationships for four decades since McCarthy [19] identified them, namely, Product, Performance, Price, and Promotion. It is the application of OR that has helped to discover the value of each of the 4Ps and relate them in a commercially viable way. The 4Ps have been applied in a statistical sense in product and system design so far. Markets were segmented and products were designed to meet the needs of a given segment with features likely to appeal to either the majority or a mythical ideal customer. This meant that product attributes were subject to either the rule of least common denominator or with gaps in meeting specific customer needs. The Internet era has extended direct connectivity to each customer, and hence the 4Ps as they relate to any individual can be customized. The front end of mass customization, namely, configuring the product with relevant Ps for each customer, understanding his/her tradeoffs, can be done cost effectively only with innovative application of OR concepts.

Thompson et al. [20] have dealt with one aspect of the above challenge, namely the need for product design to reach a new level of sophistication. As costs plummet, every manufacturer wishes to cram the product with as many features as one can. Market research has shown that buyers assign a higher weightage to product capability over its usability at the time of purchase. This results in high expected utility. However, their experienced utility is always lower as many features remain unused or underutilized. This results in feature fatigue and lower customer satisfaction. Hence, manufacturers need to segment the market better, increase product variety, and invest to learn consumer behavior on an ongoing basis.

Similarly, consumer research can scale new heights in the emerging decades. Customer behavior can be tracked easily, whereas customer perception is hard to decipher. Correlating the two and deriving meaningful insights will be the focus of market researchers. OR when combined with market research and behavioral science concepts can play a vital role here.

Communicating with customers and potential customers on time and in a cost effective manner is the dream of every enterprise. As the Internet and phone-based customer reach/ distribution channels have emerged, as customers seek to interact in multiple modes with an organization, optimal channel design in terms of segments, activities within a segment, and balancing of direct versus indirect reach partners will assume criticality. Channel cannibalization or channel conflicts are issues to contend with. Channels such as the Internet, kiosks, and telephone have been around for a decade or more. However, data on channel efficacy is either sparse or too macro. Over time a higher level of sophistication in channel usage will emerge as a key competitive edge. OR has enough tools and concepts to help out in such decisions for different segments in the financial services sector.

Banks, savings and loan associations, insurance firms, and mutual funds constitute the financial services sector. Customer segmentation, product structuring, pricing, and credit worthiness assessment are issues tackled through sophisticated OR applications since the 1970s in this sector. The concept of multi-layering of risk and treating each layer as a distinct market segment is inherent to insurance and reinsurance lines of business [21]. Product portfolio optimization for a given level of capital and risk appetite is mission critical here. The Internet and online systems have enabled firms to gather product, customer segment, and channel specific data on costs, margins, and risks. Extending the product mix optimization applications from the manufacturing area to financial services is an exciting and emerging opportunity. Formulating cost effective programs to minimize identity theft and fraudulent claims without alienating genuine customers are industry challenges where OR professionals can play significant enabler roles.

With reference to physical goods, the back end of mass customization is intelligent manufacturing that can incorporate a feedback loop from the demand side into the manufacturing processes. Modularizing components and subassemblies, delaying final assembly, and linking final assembly line to specific customer orders as well as items on hand are tasks consequent to mass customization. As discussed earlier, the move now is toward minimal forecasting. Capacity management and order management are in focus rather than better forecasting and finished goods inventory control. Simulation techniques have found wide acceptance in shop floor layout, sequencing of production steps, scheduling and rescheduling of operations to meet customer commitments and to make delivery commitments. While optimization held sway when many decisions were strategic, operationalizing the strategic decisions has resulted in simulation systems replacing the optimizers.

FedEx, UPS, and DHL are early movers in the courier transportation industry to have used complex OR algorithms and sophisticated computer systems for routing and tracking packages [22]. Cell phones, RFID devices, and GPS technologies have immense potential to provide solutions to mobility-related issues of products and resources. Dynamic data captured through such means facilitate building optimizing online and real time solutions for logistic and transportation problems. When coupled with context-specific information from GIS databases as well as from enterprise-wide computers, higher levels of complex issues can be solved in an integrated mode. Constructing such solutions is a skill and capability building means for OR professionals, thus facilitating their career growth. Camm et al. report a pioneering GIS application at Proctor & Gamble carried out by OR practitioners [23].

OR being called upon to play the role of a fair judge recurs in the telecom sector too. Systems need to reach a new level of sophistication at the design stage itself. Many systems have facilities that are multifaceted (can carry data, voice, video, etc.). Hence, they end up serving multiple customer segments and multiple products. Then the critical issue is one of design to honor service level agreements (SLAs) with each customer segment and at the same time not to build overcapacity. This turns out to be both a facility design issue and operational policy issue for a given system capacity. Ramaswamy et al. [24] have discussed an Internet dial-up service clashing with the emergency phone services when provided over the same carriers and how OR models have been used to resolve attendant issues.

Law enforcement and criminal justice systems have had a long history of use of quantitative techniques, starting from Poisson in the 1830s [1]. From effective deployment of police forces, determining patrol patterns, inferring criminality, concluding guilt or innocence, deciding on appropriate level of punishment to scheduling of court resources, numerous instances are cited in the literature about the use of OR techniques. However, a virtually untapped arena of vast potential for OR practitioners is attacking cyber crime. This is fertile ground for correlating data from multiple and unconnected sources, drawing inference, and detection and prevention of criminal activities that take place through the Internet.

One of the telling effects of the Internet has been online auctions [25]. It is the virtual space where the buyer and seller meet disregarding time and place separations. One can imagine well the advantage of such auction sites when compared to physical sites. The emerging issues in this scenario call for ensuring fairness in transactions through an objective process while preserving user confidentiality. This is an exciting emerging application area for OR experts with in-depth understanding of market behavior as well as appropriate domain knowledge relating to the product or service in focus. Bidders in such auctions need to choose not only the best prices to offer but also sequence their offer effectively. The emergence of intelligent agents that can gather focused data, evolve user specific criteria for evaluation, help both parties to choose the best path of behavior (price to quote and when) are challenges awaiting the OR practitioner. Gregg and Walizak [26] are early entrants to this arena.

A conventional assumption in any system design has been to overlook the transient state and design for the steady state (in most queuing systems). Today's computing technology permits the analysis of the transient state within reasonable cost. Also it is true that many systems spend a large amount of time in the transient state and hence have the need to design them optimally in this state as well. OR theory has much to contribute in this area.

Developments in life sciences and related fields such as drug discovery, genetics, and telemedicine have opened doors to many opportunities for OR applications. OR experts are playing an increasingly assertive role in cancer diagnosis and treatment. 3D pattern recognition issues rely heavily on statistical and related OR concepts to identify the best fits. It is evident that such opportunities will grow manifold going forward.

The year 2003 is significant in the field of biology and bioinformatics. This was the 50th anniversary of the discovery of the double helical structure of DNA and the completion of the human genome project. While the first event gave birth to the field of genetics, the second is the beginning of the use of genomes in medicine. The current challenge is to utilize the genome data to its full extent and to develop tools that enhance our knowledge of biological pathways. This would lead to accelerated drug discovery. Abbas and Holmes [27] provide enormous insight into the opportunities for OR/MS expertise application in this area. Sequence alignment, phylogenetic tree, protein structure prediction, and genome rearrangement and the like are problems solvable with OR tools of probabilistic modeling, simulation, and Boolean networks.

Similarly, the next revolution in information technology itself awaits the invention of intelligent devices that recognize inputs beyond keyboard tapping and speech. Other means of inputs such as smell, vision, and touch require advancement in both hardware and software. And the software would rely heavily on OR to draw meaningful conclusions.

Industries in the services sector will dominate the economies of both developed and developing countries in the twenty-first century. To answer the challenges in the service sector, there is an emerging *science of service management* advocated by IBM and other global companies. This would open up many opportunities for the OR/MS practitioners (see the article by Dietrich and Harrison [28]). In Part II of this handbook, OR/MS applications to the service sector (airlines, energy, finance, military, and water resources) are discussed in detail.

Finally, one of the significant but unheralded contributions of OR to society at large has been the creation of means to handle variability. Previously, most systems and products were designed with either the average in mind or custom developed. OR with its embedded statistical tools and optimization algorithms has enabled a better understanding of variability at multiple levels (variance, skewness, and kurtosis), its associated costs, and has enabled meaningful and commercially viable market segmentation (as in the airlines industry). The twenty-first century demands that we accept diversity as an integral and inevitable aspect of nature in every aspect of our endeavor. We as a society are expected to build systems, evolve policies, and promote thoughts that accept diversity in totality. This can, however, be achieved only in an economic model that balances the associated costs with resultant benefits. OR is the only discipline that can quantify the diversity aspects anywhere and create the economic models to deal with the issues effectively. In that sense its longevity is assured.

References

1. Maltz, M.D., From Poisson to the present: Applying operations research to problems of crime and justice, *Journal of Quantitative Criminology*, 12(1), 3, 1996. http://tigger.uic.edu/~mikem//Poisson.PDF.
2. Haley, K.B., War and peace: The first 25 years of OR in Great Britain, *Operations Research*, 50(1), 82, Jan.–Feb. 2002.
3. Ledbetter, W. and Cox, J.F., Are OR techniques being used?, *Journal of Industrial Engineering*, 19, Feb. 1977.
4. Fildes, R. and Ranyard, J., Internal OR consulting: Effective practice in a changing environment, *Interfaces*, 30(5) 34, Sep.–Oct. 2000.
5. Ackoff, R.L., The future of operational research is past, *Journal of the Operational Research Society*, 30(2), 93, 1979.
6. Gass, S.I., Model world: OR is the bridge to the 21st century, *Interfaces*, 27(6), 65, Nov.–Dec. 1997.
7. Miser, H.J., The easy chair: What OR/MS workers should know about the early formative years of their profession, *Interfaces*, 30(2), 97, Mar.–Apr. 2000.
8. Mitchell, G.H. and Tomlinson, R.C., Six principles for effective OR—their basis in practice, Proceedings of the Eighth IFORS International Conference on Operational Research, Toronto, Canada, June 19–23, 1978, Edited by Haley, K.B., North Holland Publishing Company, 1979.
9. Chamberlain, R.G., 20/30 hindsight: What is OR?, *Interfaces*, 34(2), 123, Mar.–Apr. 2004.

10. Andy Grove, Taking on prostate cancer, *Fortune*, May 13, 1996 www.prostate-cancerfoundation.org/andygrove www.phoenix5.org/articles/Fortune96Grove.html.
11. Murphy, F.H., The practice of operations research and the role of practice and practitioners in INFORMS, *Interfaces*, 31(6), 98, Nov.–Dec. 2001.
12. Smith, B.C. et al., E-commerce and operations research in airline planning, marketing and distribution, *Interfaces*, 31(2), 37, Mar.–Apr. 2001.
13. Montgomery, A.L., Applying quantitative marketing techniques to the Internet, *Interfaces*, 31(2), 90, Mar.–Apr. 2001.
14. Kapuscinski, R. et al., Inventory decisions in Dell's supply chain, *Interfaces*, 34(3), 191, May–Jun. 2004.
15. Geoffrion, A.M. and Krishnan, R., Prospects for operations research in the e-business era, *Interfaces*, 31(2), 6, Mar.–Apr. 2001.
16. Gawande, K. and Bohara, A.K., Agency problems in law enforcement: Theory and applications to the U.S. Coast Guard, *Management Science*, 51(11), 1593, Nov. 2005.
17. Butler, D. and Ruttiman, J., Avian flu and the New World, *Nature*, 441, 137, 11 May 2006.
18. Meyer, J.L., Selecting employees for random drug tests at Union Pacific Railroad, *Interfaces*, 27(5), 58, Sep.–Oct. 1997.
19. McCarthy, E.J., *Basic Marketing: A Managerial Approach*, Homewood, IL: Richard D Irwin Inc., 1960.
20. Thompson, D.V., Hamilton, R.W., and Rust, R.T., Feature fatigue: When product capabilities become too much of a good thing, *Journal of Marketing Research*, XLII, 431, Nov. 2005.
21. Brackett, P.L. and Xia, X., OR in insurance: A review, *Transaction of Society of Actuaries*, 47, 1995.
22. Mason, R.O. et al., Absolutely positive OR: The Federal Express story, *Interfaces*, 27(2), 17, Mar.–Apr. 1997.
23. Camm, J.D. et al., Blending OR MS, judgement and GIS: restructuring P & G's supply chain, *Interfaces*, 27(1), 128, Jan.–Feb. 1997.
24. Ramaswami, V. et al., Ensuring access to emergency services in the presence of long Internet dial-up calls, *Interfaces*, 35(5), 411, Sept.–Oct. 2005.
25. Iyer, K.K. and Hult, G.T.M., Customer behavior in an online ordering application. A decision scoring model, *Decision Sciences*, 36(4), 569, Dec. 2005.
26. Gregg, D.G. and Walizak, S., Auction advisor: An agent based online auction decision support system, *Journal of Decision Support Systems*, 41, 449, 2006.
27. Abbas, A.E. and Holmes, S.P., Bio-informatics and management science: Some common tools and techniques, *Operations Research*, 52(2), 165, Mar.–Apr. 2004.
28. Dietrich, B. and Harrison, T., Serving the services industry, *OR/MS Today*, 33, 42–49, 2006.

Index

A

AALPS (Automated Air Load Planning System), **20**-21
ABC analysis, **10**-4, 35
absolute robust decision, **14**-10
absolute value, **1**-17–8, **14**-16, **19**-22
absorption chiller, **19**-11, 16, 18, 23
AC
    capital cost, **19**-19, 27
    minimum output level, **19**-18
    system parasitic electrical load, **19**-18
acceptance sampling, **5**-33, **16**-20
active structural acoustic control (ASAC), **13**-10–13
activity-on-arrow (AOA) convention, **15**-3
activity-on-node (AON) convention, **15**-3, 5
actuarial science, **17**-2
actuator placement, **13**-13
acyclic networks, **4**-2, 5
AD OPT Technologies, **20**-5
additivity assumptions, **1**-4
administrative costs, **22**-50
admissible move, **13**-9
Adobe, **23**-3
advanced booking forward, **20**-16
advanced planning systems (APS), **22**-1, 59
affine functions, **1**-14, **17**-8
affine scaling, **2**-13
age replacement policy, **8**-17, 40–2
AgentArts, **23**-11
agent-based simulation, military applications, **25**-8–9
aggregate forecasts, **18**-3
aggregate inventories, **22**-41–4
aircraft recovery problems, **20**-13–4
Airline Deregulation Act of 1978, **20**-3, 14–5, 18
Airline Group of the International Federation of Operational Research Societies (AGIFORS), **20**-3
airline network, **11**-4
airline optimization
    aircraft
        load planning, **20**-21–3
        recovery, **20**-13
    benefits of, **20**-1–2, 5
    charter airlines, **20**-7
    crew recovery, **20**-13–4
    crew scheduling
        crew assignment, **20**-12–3
        crew pairing, **20**-10–2, 23
        irregular operations, recovery from, **20**-13–4
    future research directions, **20**-23–4
    disruption management system, **20**-14
    fuel operational optimization/fuel hedging, **20**-24
    industry environment, **20**-2–4
    information resources, **20**-7, 15–6, 23
    irregular operations, **20**-13
    optimization solutions, **20**-4–5
    passenger recovery, **20**-14
    planning and operations business processes, **20**-2
    revenue management
        components of, **20**-4, 14–5
        forecasting, **20**-15–7
        overbooking, **20**-15–8, 23
        pricing, **20**-20–1
        seat inventory control, **20**-18–20
    schedule planning
        aircraft routing, **20**-6–7, 9–10
        fleet assignment, **20**-6–9
        problems with, **20**-5–6
        schedule design, **20**-6–7
    simulations, **20**-20
    software programs, **20**-5, 21
Akamai, **23**-4, 18
algorithms, types of
    decision analysis, **23**-5, 20
    dynamic programming, **23**-5, 20
    genetic algorithms, **23**-13
    integer programming, **23**-5–6
    inventory control, **23**-5, 20
    Knapsack problem, **23**-12
    linear programming, **23**-5, 9, 20
    mixed integer programming, **23**-13
    multi-criteria decision making (MCDM), **23**-5, 20
    network optimization, **23**-5, 13, 20
    nonlinear programming, **23**-5, 20
    queueing theory, **23**-5, 14, 20
    simulated annealing, **23**-13
    simulation, **23**-5, 13, 21
    stochastic processes, **23**-5, 12, 14, 20
Al-Khawarizmi, **1**-2

Al-Maqala fi Hisab al-jabr w'almuqabalah (Al-Khawarizmi), **1**-2
all pair shortest path problem, **4**-8
ALOHA protocols, **8**-3–4
alternate-location-alternate (ALA) algorithm, **22**-48–9
alternatives
  defined, **6**-2
  multiple criteria decision making (MCDM), **5**-3
Amazon, **23**-4–5, 10–1
American Express, **23**-18
American options, **21**-22–3, 29–33
American Society for Quality, **16**-1
AMP, **1**-31
analysis of variance (ANOA), **12**-15
analytic hierarchy process (AHP) method, **5**-11–3, 20, 35, **24**-18
Anderson-Darling test, **17**-40
animation, in simulation, **12**-15, 21
annealing, **2**-3, **3**-16
ant colony optimization, **13**-1–2
anti-ideal solution, **5**-4
ALL, **23**-8, 18
Apple, **23**-4–5, 7
approximations, **1**-5, **17**-35
arbitrage, **21**-25
arborescent distribution network, **10**-29
arc
  adjacency list, **4**-2
  defined, **4**-2
  directed, **4**-6
  undirected, **4**-3, **4**-6
Arena, **12**-16–7
Argonne National Laboratory, **1**-31
ANIMA, **16**-13, 19
Armijo and Goldstein (AG) conditions, energy system optimization, **19**-15
Armijo's conditions/rule, **2**-4, 7–8
Arrival Theorem, **9**-31–2
Arrow-Pratt absolute risk aversion coefficient, **21**-11
artificial intelligence (AI), **23**-5, 16, 21
aspiration criteria, **5**-30, **13**-9, 21, 28, 30
assembly network, **10**-29
asset allocation, **21**-13
asset classes, **21**-33–4
asset liability management (ALM), multi-period, **21**-33–6
assignable causes, **16**-9–10
assignment problem
  applications, **4**-4
  example of, **4**-14
  Hungarian algorithm, **4**-13–4

linear programming formulation, **4**-13
assumptions
  blending problem, **1**-7
  linear blending, **1**-6
  strict unimodality, **2**-4
  utility theory, **6**-14
at-the-money options, **21**-23
attributes
  multiple criteria decision making (MCDM), **5**-3
  tabu search, **13**-9
auction
  electricity, **23**-14
  internet, **23**-21
  reverse, **23**-7, **14**-5
augmenting path algorithm, **4**-9–13
automobile insurance, stochastics example, **8**-5, 2–3, **23**-4
AutoMod, **12**-17
automotive a/c switch functions, **17**-25–6
average path length, **11**-6–7, 14, 16, 20
aviation networks, minimum cost multicommodity flow problem, **4**-18
axioms, analytic hierarchy process (AHP), **5**-12

B

back ordering, **10**-16–8, **22**-56
backtracking
  energy systems optimization, **19**-15, 23
  search guidelines, **2**-7–8
backward recursion, **21**-29–30
balance equations, **8**-34
ball bearing failure time, **17**-30–2, 38–9
banner ads, **23**-11
Barabási-Albert complex network model, *see* scale-free networks
barrier functions, **2**-3–4, 8, 18
base-stock inventory
  characterized, **10**-27, **22**-8–14
  risk pooling, **22**-42
  supply-chain wide, **22**-12–4
  with supply uncertainty, **22**-14–5
basic feasible solution (BFS), **1**-27–8
basic vector, **1**-26–8
basin, **13**-31–2, 34
batch means analysis, **12**-13–4
batch-processing strategy, **1**-19
bathtub-shaped failure rate, **17**-12
Bayesian estimate, **21**-13
Bayes' Law, **14**-9
Bayes' probabilities, **6**-8, **20**-16

Index

Bayes' theorem, **6**-9
BCHP Screening Tool, **19**-17
beam search, **3**-12
Bellman equation, **20**-18
Bernoulli random variable, **17**-8
best-bound first search, integer
    programming, **3**-9
best compromise solution, **5**-18
"best" decision, determination factors, **6**-17
best-improving search, **13**-5, **17**–20
best management practices (BMPs),
    24-21–2
beta distribution, **15**-20–1
betweenness centrality, **11**-9
bi-criteria linear integer programs, **5**-33
bi-criteria linear programming, **5**-16, 25, 34
bi-criteria mathematical programming
    models, **5**-34
bi-criteria nonlinear integer programs, **5**-33
bi-critical, implications of, **15**-33
bimodal distribution, **9**-21
binary assumption, **17**-3
binary ILP (BILP), **3**-2
binary integer variables, integer
    programming, **3**-12
binary MILP (BMILP), **3**-2
binary pairwise comparison, **5**-29
binary variables, integer programming, **3**-6
binomial random variables, **17**-33
binomial tree, **21**-30–3
biological networks, **11**-25
biostatistics, **17**-2
bipartite graphs, **4**-2, **11**-14
bipartite networks, **4**-4, 13
birth and death process, **8**-29–30, 34–5,
    9-10–1
bisection search, **2**-5–7
Black-Litterman asset allocation model,
    21-13–8
Black-Scholes formula, **21**-24–6
blending problems, **1**-6–8
block diagrams, **17**-4–6
boiler efficiency, **19**-18
Bolzano search, **2**-6
Bombardier, **20**-5
Bonus Males table, **8**-2
bootstrapping, **18**-28–9
bottlenecks
    closed queueing networks, **9**-33–4
    job shops, **18**-27
    simulations, **12**-2, 15
boundary conditions, dynamic
    programming, **7**-2, 8, 10, **21**-3

bound-first search strategy, integer
    programming, **3**-10
Box-Jenkins ANIMA forecasting model,
    18-4
bracketing technique, **2**-4–6
brainstorming, **6**-24
branch and bound (B&B) ILP method
    airline optimization, **20**-7–8
    bound-first B&B algorithm for
        minimization/maximization, **3**-10
    branching, **3**-8–9
    components of, **3**-7–8, 16
    computer lower and upper bounds, **3**-9
    example of, **3**-10–12
    fathoming, **3**-8–12
    multiple criteria mathematical
        programming (MCM), **5**-33
    search strategies, **3**-9–10
branch and bound solution, **13**-21, 27–8
branch-and-cut, airline optimization,
    20-7, 12
branch-and-cut-and-price, airline
    optimization, **20**-12
branch-and-price, airline optimization, **20**-8,
    12–3
brand loyalty, **8**-15–6
brand switching, stochastics example, **8**-2, 5,
    22–4
breakeven price, **1**-25
brick and mortar stores, **23**-4
Brownian motion, **21**-24–6
Brownian processes, **8**-6
Broyden-Fletcher-Goldfarb-Shanno (BFGS)
    method
    energy systems optimization, **19**-1, 23
    updates, **2**-14
Building Energy Analyzer simulation,
    19-17, 24
build-to-order policy, **18**-29
build-to-stock policy, **18**-29
bullwhip effect, **22**-49–58
burn-in period, **17**-13

C

C++ code, multicommodity flow problems,
    4-19
CAD, **12**-15
Caleb Technologies, **20**-5, 14
call options, **21**-22–3, 31–2
can-order *vs.* must-order inventory policy,
    22-18–9
capacities, inventory control, **10**-8

capacity constraints, 4-19
capacity expansion
    dynamic programming models, 7-3–7
    problem, 14-24–7
capacity management, 26-10
capacity of a cut, 4-10–11
capacity optimization, energy systems,
    19-12, 16–24
capacity planning, 26-7
capital asset pricing model (CAPM),
    21-14, 17
capital budgeting problems, 3-3–4, 7
capital costs, energy systems optimization,
    19-19–20
capital market line (CML), 21-14
capital rationing, energy investment
    example, 19-4–11
capital recovery, energy savings
    optimization, 19-21
cardinal weights, 5-20
Carmen Systems, 20-5
cascading failures, 11-21
cash flow management, 4-4, 21-33–5
catastrophic effects, 14-1
c chart, 16-14–5, 24
censored data, 17-2, 23
centering step, 1-30–1
centrality, 11-23–4
central limit theorem (CLT), 15-23–24,
    17-2, 17
centralized supply chains
    characterized, 22-40–1
    inventory models, 10-7, 32–4
central value solution, 14-11–3
certainty equivalent models, 14-14–5, 21-10
chance-constrained robust optimization
    model, 14-9–10
channel design, optimal, 26-10
Chapman-Kolmogorov (CK) equations,
    8-19–20, 22, 28
chemical energy, 19-2
chi-square distribution, 17-20, 30
chi-square goodness-of-fit test, 17-36
ChoiceStream, 23-11
Choleski decomposition, 2-14
CHP Capacity Optimizer, 19-24
circulation problem, 4-9
classical statistics, 17-2, 17
class properties, 8-22
Clearway, 23-18
click and conquer electronic storefront, 23-4
clique, 11-25
closed-loop SC, 10-37
closed-queueing networks, 9-30–2

clustering coefficient, 11-7–8, 15, 20, 26–7
cluster methods, 23-11
coefficient matrices, 1-4
coefficient of variation, 22-44
coefficient vector, 1-3
coherent risk measure, 21-19–20
coherent systems, 17-6
cold standby reliability system, 8-29
collaborative filtering, 23-10–1
collaborative planning, forecasting, and
    replenishment (CPFR), 22-57
column approach, crew pairings, 20-12
column generation approaches, 4-1, 19
column generator, airline optimization,
    20-8, 10
combined heat and power (CHP) systems
    capacity optimization, design strategies,
        19-16–21
    characterized, 19-12
    "make or buy" decision, 19-11
    operation simulation, 19-16
    point-of-use energy generation, 19-11–2
    terminology, 19-19–20
    thermal load, 19-11–3, 20
CommerceOne, 23-3
commercial buildings, combined heat and
    power (CHP) systems, 19-11, 13
community structure, 11-9–10, 24–5
comparative trade-off method (CTM), 5-32
comparative trade-off ratio, 5-30
competitive analysis, 10-36
complementary slackness optimality
    solutions, 1-29–30
complex networks
    average path length, 11-6–7, 14, 16, 20
    betweenness centrality, 11-9
    characterized, 11-1–3, 26–7
    clustering coefficient, 11-7–8, 15, 20, 26–7
    community structure, 11-9–10, 24–5
    congestion, 11-26
    defined, 11-18
    degree distribution, 11-8–9, 14, 16,
        18–9, 27
    emergence, 11-17
    examples of, 11-4, 16–7
    flexibility, 11-20
    gas atoms analogy, 11-4–5
    high-sensitivity, 11-18
    infinite susceptibility/response, 11-17
    models, 11-11–6
    modularity, 11-9–10
    non-uniform random graphs, 11-13
    optimization, 11-18–26
    resilience, 11-9–11, 19–21

Index

responsiveness, **11**-20
robustness, **11**-19–20
scale-free networks, **11**-8, 16–7
scale invariance, **11**-17
self-organization, **11**-17
self-similarity, **11**-17
small-world effect, **11**-6–7, 15–6
spreading processes, **11**-26
statistical properties, **11**-6–11
complex system, reliability, **17**-4
compromise programming, **5**-9
computer(s), *see* software programs
   networks, **4**-9
   viruses, **11**-26
concordance index, **5**-10
conditional probability, **6**-9–10, **8**-15, 18, **17**-12
conditional value-at-risk (CVaR), **21**-20–1
confidence bounds, **21**-28
confidence intervals, **12**-8, 13, 22, **17**-20, 28, 32
confidence region, **17**-31
conflicts, multiobjective LP models, **1**-32
congestion, **11**-26
conjugate gradient, **2**-13
conservation of flow, **4**-5, **22**-39
conservative decisions, **6**-20
consistence index (CI), **5**-12–3
consistency ratio (CR), **5**-13
constant-sum constraint, **14**-11
constituents, blending problems, **1**-6–7
constrained nonlinear convex optimization problems, **2**-17–8
constrained optimization
   characterized, **2**-15
   direction finding methods, **2**-15–7
   transformation methods, **2**-15, 17–9
constrained shortest path problem, **4**-8
constraints
   airline optimization, **20**-9–10, 23
   blending problem, **1**-6
   diet problem, **1**-8
   dual, **1**-29
   energy system optimization, **19**-13–4
   integer programming, **3**-3
   linear, **1**-4–5, 12
   linear inequality, **1**-3
   material balance, **1**-12
   mutually contradictory, **1**-22
   nonnegativity, **1**-29–30
   primal, **1**-29
   product mix example, **1**-6
consumption rate, **10**-15
container shipping, **1**-18–22

content delivery networks (CDNs), **23**-7, 17–18, 21
continuous-location model, **22**-40, 44–8
continuous optimization problem, **2**-2
continuous review inventory policy, **8**-42–5, **22**-12, 15, 17
continuous-time Markov chain (CTMC)
   birth and death processes and applications, **8**-29–30
   defined, **8**-27–9
   limiting distribution, **8**-33–7
   queueing theory, **9**-8
   statistical inference, **8**-37–8
   transient analysis, **8**-31–3
continuous-time stochastic process, **8**-5
continuous variables, **1**-4, **3**-6
contracts, inventory management, **10**-36
contribution functions, defined, **7**-2
control charts
   advanced, **16**-18–20
   for attributes, **16**-14–5, 23
   benefits of, **16**-15–7
   construction, constants and formulas, **16**-23
   demand forecasting, **18**-5
   multivariate SPC, **16**-19
   for problem identification, **16**-10–6
   selection factors, **16**-20
   for variables, **16**-12–13
conventional chiller COP, **19**-18
convergence, **4**-2
convex functions, variable transformations, **1**-13
convex optimization problem, **2**-2
convex piecewise linear (PL) function
   defined, **1**-13
   separable, minimizing, **1**-14–6
convex polyhedra, **1**-4
coordinated replenishment, **22**-19
Coremetrics, **23**-20
correlation coefficient, **11**-4, **22**-42
cost(s)
   backorder, **10**-16–8
   coefficients, **1**-25
   holding, **10**-6, 10–2, 15, 17–8, 21, 23, 25, 31–2, 35, **22**-45
   life-cycle, **19**-13–4
   line-haul, **22**-24
   minimization, **19**-12
   ordering, **10**-5–6, 12, 20, 23–4, 32, 35
   overage, **10**-27, **18**-3, 10–1, **20**-17, **22**-12
   physical distribution, **22**-26
   set-up, **10**-5–6, 35
   shortage, **10**-21, 23–4, 35, **18**-3

cost(s) (*Contd.*)
    terminal/accesorial, **22**-24
    underage, **10**-27, **18**-3, 10–1, **20**-17, **22**-12
    variable, **10**-31–2
cost per unit, **1**-5
counting process, **8**-39
covariance
    implications of, **21**-6–7, **14**-5
    matrix, **21**-15–7
CPLEX, **1**-31, **3**-16, **20**-5
Cramer-von Mises test, **17**-40
Crandell, Robert, **20**-5
crew recovery, **20**-13–4
CrewSolver system, **20**-5, 14
criteria normalization, multiple criteria decision making (MCDM)
    characterized, **5**-3
    linear normalization, **5**-5, 8
    range equalization factor, **5**-6
    use of 10 raised to an appropriate power, **5**-6
    vector normalization, **5**-5–6
criteria weights, computation of
    analytic hierarchy process (AHP) method, **5**-11–3, 35
    compromise programming, **5**-9
    ELECTRE method, **5**-10–1, 35
    PROMETHEE method, **5**-13–5, 35
    from ranks, **5**-6–7
    rating method, **5**-7
    ratio weighting method, **5**-7
    TOPSIS method, **5**-9–10
CriteriumDecisionPlus, **5**-35
critical fractile, **10**-26, **22**-10
criticality index, **15**-17
critical path method (CPM)
    activity crashing, **15**-14–8
    backward pass analysis, **15**-3, 6–7
    characterized, **15**-3
    with deadline, **15**-8–9
    forward pass analysis, **15**-3, 6, 9
    goal of, **15**-3
    schedule compression, **15**-14
    subcritical paths, **15**-9–10
critical period, **24**-5
critical ratio, **22**-10
critical to quality (CTQ), **16**-5–6
critical values, determination of, **17**-38–9
CSI Phase I Evolutionary Model (CEM), **13**-22–4
cubic spline fit, **19**-15–6, 23
cumulative distribution function, **17**-36–9
cumulative hazard function, **17**-10, 13–4, 24, 40

cumulative process, **8**-41
cumulative sum (CUSUM) control charts, **16**-15, **18**-19
curse of dimensionality, **7**-24
curve fitting line search, **2**-7
customer loyalty programs, **20**-23
cuts, integer programming, *see* cutting plane ILP method
cutting plane
    ILP method
        cuts, defined, **3**-12
        components of, **3**-12, 16
        dual fractional cut, **3**-13–5
        dual fractional cutting-plane algorithm, **3**-14
        example of, **3**-14
        Gomory cuts, **3**-14
    multiple criteria mathematical programming (MCMP), **5**-32
cut tree network model, **18**-19–20
cyber crime, **26**-11
cycle service level (CSL), **22**-17, 19
cyclic coordinate method, **2**-13
cycling, **13**-9–10, 26, 31–2
cylinder, metaheuristic applications, **13**-12

D

Daimler Chrysler, **23**-15
damper placement problem
    computational results, **13**-27–9
    first DPP (DPP1), **13**-23
    optimization problems, **13**-22–4
    overview of, **13**-21–2
    tabu search, **13**-25–6, 29–35
dams, **24**-4–5, 10
Dantzig, George B., **1**-4, 26, **20**-3
Dantzig-Wolfe decomposition, **4**-19
DASH/XPRESS, **4**-19
data collection, **17**-2
data mining, **26**-7
data warehousing, **26**-7
Davidson-Fletcher-Powell (DFP) methods
    energy system optimization, **19**-15
    updates, **2**-14
decentralized networks, **11**-21
decentralized supply chains
    characterized, **22**-40–1
    inventory models, **10**-7, 31–2
decision analysis
    alternatives, **6**-2, 4–5, 21, 23–5, 28
    consequences, **6**-25–6, 28
    defined, **6**-1

Index

hospital problem example, **6**-21–9
Hurwicz criterion, **6**-19
importance of, **6**-1–2, 28–9
information resources, **6**-29
investment example, **6**-3–8, 14
linked decisions, **6**-27–29
military applications, **25**-6
objectives, **6**-22–3, 28
problem definition, **6**-2, 21–2
under risk, **6**-2–17
risk tolerance, **6**-17, 27, 29
software programs, **6**-6, 17
software survey, **6**-29
terminology, **6**-2–3
tradeoffs, **6**-25–6, 28
under uncertainty, **6**-2–6, 17–20, 26–8
Decision Analysis (Keeney/Raiffa), **5**-35
Decision Analysis Society of INFORMS, **6**-29
decision makers
    attitudes, **6**-17–8
    classification of, **6**-12
    defined, **6**-2
    functions in decision analysis, **6**-5
    inconsistency of, **5**-33
    preferences, **6**-17–8
Decision Lab, **5**-35–6
decision optimization strategies, **19**-1, 3
decision space, **2**-2, **5**-16–7
decision trees
    characterized, **6**-7–8
    lotteries, **6**-14
    military applications, **25**-6
    stock investment example, **6**-16
decision variables
    airline optimization, **20**-8–10, 12
    blending problem, **1**-6–7
    characterized, **1**-4
    continuous, **7**-10
    defined, **1**-2
    discrete optimization problems, **13**-15, 28
    dynamic programming, **7**-2, **20**-2
    financial engineering, **21**-35
    goal programming, **5**-19
    integer programming, **3**-3, 11
    intelligent modeling example, **1**-20
    linear constraints, **1**-4
    military applications, **25**-4
    partitioning algorithm, **5**-24
    production layout design, **18**-15
    product mix example, **1**-6
    robust optimization, **14**-17
    supply chain management, **22**-44, 47

decision vectors, **24**-22
decomposition
    analysis, **13**-13
    complex networks, **11**-25
    implications of, **4**-1, **20**-3
    queueing theory, **9**-34
decreasing failure rate (DFR), **17**-12, 17, 21
deflection techniques, **2**-11, 13–4
degree distribution, **11**-8–9, 14, 16, 18–9, 27
degree, of node, **4**-2
degree-constrained spanning tree problem, **4**-16
degrees of freedom, **8**-27, 38, **17**-28, 30
Dell, Inc., **10**-34, **23**-13
delta-hedging, **21**-25
demand centers/markets, **1**-10
demand distributions, updating, **18**-6
demand driven dispatch, **20**-7, 23
demand forecasting
    accuracy, **20**-7
    airline optimization, **20**-4, 6–7, 15–7
    characterized, **18**-2–3
    commonly used techniques, **18**-3–4
    distributions
        nonparametric methods for updating and forecasting, **18**-6–9
        parametric methods for forecasting, **18**-4–6
    information resources, **18**-29
    newsvendor model, **18**-9–11
    optimal order quantities, computing, **18**-9–12
    seasonality, **18**-11–2
    software programs, **18**-29
    supply chain management, **22**-54
demographics, significance of, **6**-26
dependencies, inventory management, **10**-28
depreciation, **10**-6, **19**-20
depth-first searches, integer programming, **3**-9
derivative securities, **21**-22
Descartes (Decision Support for integrated Crew and AiRcrafT recovery), **20**-14
descent algorithm, **2**-3
descent method, steepest, **2**-11–2, 14
descent step, **1**-31
de-seasonalized demand series, **18**-12
deterministic demand, **10**-8
deterministic dynamic programming models
    capacity expansion problem
        components of, **7**-3–6
        with discounting, **7**-6–7
    characterized, **21**-28

deterministic dynamic programming models (*Contd.*)
  dynamic inventory planning problem, **7**-13–6
  equipment replacement problem, **7**-7–9
  reliability system design, **7**-16–8
  simple production problem, **7**-9–13
deterministic inventory system
  characterized, **10**-9, 36
  economic order quantity (EOQ) inventory model, **10**-9–14, 16–8, 31
  maximum-minimum system, **10**-10
  time-varying demand, **10**-18–20
  uniform demand with replenishment, **10**-14–6
deterministic processing time, **18**-21–3
deterministic techniques, **2**-3
deviation variables, **1**-33–4, **5**-20
d-heap, **4**-6
DHL, **26**-10
diet problems, **1**-8–9
differential equations, **4**-4, **25**-6
diffusion approximation, queueing theory, **9**-32–4
Dijkstra's algorithm, **4**-5–8, **11**-2
directed cycle, **4**-2
directed graphs, **4**-2, **11**-14
directed network, **4**-2–3
directed path, **4**-2
directed walk, **4**-2
direction finding methods, **2**-15–7
direction finding step, **5**-31
discount, inventory control and, **10**-16
discounting, dynamic programming models, **7**-6–7
discount rate, **7**-6, **19**-19–20
discrete optimization problems, metaheuristics
  active structural acoustic control (ASAC), **13**-10–13
  damper placement in flexible truss structures, **13**-21–29
  implications of, **13**-1
  multistart local search, **13**-5, 36
  nature reserve site selection, **13**-13–21
  network location problems, **13**-3–5
  population-based, **13**-2
  simulated annealing, **13**-6–8, 27, 36
  single solution, **13**-2–3
  tabu search
    characterized, **13**-7, 36
    damper placement project, **13**-22–3
    for MCSP, **13**-19–20

  plain vanilla, **13**-8–10, 13, 36
  reactive, **13**-29–35
discrete-event simulation
  characterized, **12**-1–23
  information resources, **12**-30
  military applications, **25**-8
  software programs, **12**-16–30
discrete-location problem, **22**-39
discrete-time Markov chain (DTMC)
  automobile insurance, **8**-22–4
  brand switching model, **8**-15–6, 19
  classification of states, **8**-20–2
  defined, **8**-14
  gambler's ruin problem, **8**-24–5
  limiting probabilities, **8**-22–5
  random walk model, **8**-16–7
  slotted ALOHA model, **8**-18
  social mobility example, **8**-23
  statistical inference, **8**-25–7
  success runs, **8**-17–20
  transient analysis, **8**-18–20
discrete-time model, airline optimization, **20**-17
discrete-time statistic, **12**-8
discrete-time stochastic process, **8**-4–5
distance labels, shortest path problem, **4**-6, 8
distributed energy systems, **19**-11
distributed generation (DG)
  capacity, **19**-18–19, 23
  capital cost, **19**-19
  electric efficiency (LHV) at full output, **19**-18
  number of units, **19**-18–9
  peak load, **19**-27
  power/heat ratio, **19**-18
distributed routing algorithms, **11**-22
distribution networks, minimum cost multicommodity flow problems, **4**-18
distribution problems, impact of, **1**-11–2, **3**-6
distribution requirements planning (DRP), **22**-20–4
distribution system, **10**-29
diversifiable risk, **21**-14
diversification, **14**-1, **21**-6–8
Dmitri Bertsekas Network Optimization Codes, **4**-19
DNA sequencing, **4**-4
dominated solution, **5**-18
do nothing (DN) alternative, **6**-5
Dow Industrials, **21**-8–9
downside risk, **21**-19
downstream, in supply chain, **10**-29, 31
drinking water, **24**-2

Index

drug discovery, **26**-11
dual fractional cut
   for MILPs, **3**-15
   pure ILPs, **3**-13–4
dual fractional cutting plane algorithm
   for MILPs, **3**-15
   pure ILPs, **3**-14
dual problem, **1**-24–5
dual simplex method, **1**-9
dual variables, **1**-24–5
dyads, **22**-48–58
dynamic process, complex networks, **11**-27
dynamic programming (DP)
   advantages of, **7**-3
   airline optimization, **20**-18
   components of, **7**-1–2, **20**-3, **21**-28–9, 33
   defined, **7**-1
   deterministic models, **7**-3–18
   difficulties of, **7**-1
   formulation stages, **7**-3–8, 10, 13–4, 17–8, 20, 22, 24
   graphical solutions, **7**-15–6
   limitations of, **7**-3
   numerical solutions, **7**-23–4
   optimal solutions, **7**-5–6, 18, 24
   options trading, **21**-24
   state variables, **7**-3, 5
   stochastic models, **7**-19–24, **22**-19–20

E

earliest due first (EDD), **18**-21–3
eBay, **23**-3, 8–9, 14, 18
echelon inventory, **10**-32–3
echelon-stock policy, **22**-19–20
e-commerce
   airline industry, **23**-5, 15, 18–9
   auctions, *see* auctions
   comparative shopping, **23**-16
   characterized, **23**-1–2
   cluster methods, **23**-9, 11
   collaborative filtering, **23**-10–1
   content delivery network (CDN), **23**-7, 17–8, 21
   cookies, **23**-8
   delivery of services, **23**-15–6
   digital goods, **23**-2, 7–12
   dotcom boom, **23**-4–5
   evolution of, **23**-3–5
   future directions for, **2**-21
   global, **23**-21
   hyperlinks, **23**-9
   infrastructure, **23**-16–20

   life-cycle, **23**-20
   OR applications, **23**-7–20
   OR/MS and, **23**-5–7
   physical goods, **23**-12–5
   price engines, **23**-16, 21
   recommendation algorithms, **23**-10–1, 21
   routing, **23**-17
   servers, **23**-19
   software programs, **23**-3
   storefronts, **23**-4
   tools-applications matrix, **23**-20–1
   universities, **23**-15–6
   web analytics, **23**-17, 19–21
   web site performance tuning, **23**-18–9, 21
economic conditions, impact of, **21**-5–6
economic order quantity (EOQ) inventory model
   with back ordering, **10**-16–8
   characterized, **10**-9, 26, **22**-17, 19
   example of, **10**-9–10
   formulation of, **10**-10–3
   graphical presentation of, **10**-13
   lead time, **10**-13
   objective function, **10**-33
   reorder point (ROP), **10**-13–4, 16, 31, 33
   sawtooth pattern of, **10**-31, 33
economic production quantity (EPQ) inventory model, **10**-14
edge-weight distributions, **11**-24
effective income tax rate, **19**-19–20
efficient frontier, **5**-16, **14**-6–7, **21**-8–10, 14
efficient markets, **21**-24–6, 30
efficient point, global criterion method, **5**-28
efficient set, **5**-16
efficient solution
   multiple criteria decision making (MCDM), **5**-4
   multiple criteria mathematical programming (MCMP)
      defined, **5**-16
      determination of, **5**-17–8
      global criterion, **5**-28
      theorem, **5**-18
e-governance, **23**-2
eigenvectors, **5**-7, 12
either-or constraints, **3**-5–6
ELECTRE method, **5**-10–1, 35
electrical efficiency, **19**-20
electrical energy, **19**-2
electrical generator, **19**-11, 16
electric generator capacity, **19**-23
electricity
   grid-based, **19**-11–2, 24
   networks, **4**-9

electric power, grid-supplied, **19**-4–5
electric power system, robust optimization, **14**-24–7
electronic circuits, **11**-6
electronic data interchange (EDI), **23**-3
electronic trading, **23**-4, 15–6
elementary renewal theorem, **8**-40
elimination method, **1**-3
embedded DTMC, **8**-45
emergence, **11**-17
EMOO, **5**-36
empirical probability distributions
   characterized, **18**-6–7
   updating, **18**-7, 9
empirical survivor function, **17**-29, 32
employee drug testing, **26**-9
EMSR-a/EMSR-b, **20**-19
endpoints, of arc, **4**-2
energy systems
   capacity optimization
      CHP systems, **19**-16–21
      components of, **19**-21–4
   combined heat and power (CHP) systems modeling
      capacity optimization, design strategies, **19**-16–21
      characterized, **19**-12
      hiearchical, **19**-12
      "make or buy" decision, **19**-11
      thermal load, **19**-11–3
   computer model, implementation of
      calculation example, **19**-24–7
      scenarios, **19**-27
   cogeneration technology, **19**-11–3
   decision optimization, **19**-1, 3
   distributed, simulation and optimization, **19**-11
   economic optimization methods, **19**-13–16
   energy
      conservation, **19**-2
      conversion process, **19**-3
      defined, **19**-2
      end-use demands, **19**-12–3
      types of, **19**-2
   future directions for, **19**-3
   integer programming model, energy investment options, **19**-5–11
   linear programming model, energy resource combination, **19**-4–5
   mathematical models, **19**-3–4
   natural, harnessing strategies, **19**-3
   operating frequency, **19**-25–6
   point-of-use energy generation, **19**-11–2

enterprise resource planning (ERP), **10**-35, **22**-1, **23**-13
enumerative algorithms, **1**-3
environmental conditions, impact of, **17**-7
Environmental Protection Agency (EPA), **24**-3
e-payment, **23**-2
epidemic spreading, **11**-26
epoch, **17**-2
e-procurement, **23**-2–3
equality-constraint methods, **3**-2, **19**-14
equilibrium excess returns, **21**-17
equipment replacement
   decisions, **4**-4–5
   dynamic programming models, **7**-7–9, 19–22
   problem, **4**-4–5
Erdös-Rényi random graphs, **11**-11–13
Erlang, A.K., **8**-29–30
Erlang A formula, **8**-30, 36–7
Erlang C formula, **8**-29–30, 35–6
Erlang distribution, **8**-8
Erlang formula, **9**-3
Erlang loss system, **9**-12
e-ticketing, **23**-2
Euclidean distance metric, **18**-15
Euclidean norm, **14**-27
Eulerian circuit, **11**-1
European options, **21**-22–3, 25–8
EURO Working Group, **5**-36
evaluation criteria, multiple criteria decision making (MCDM), **5**-3
evolutionary algorithms, **13**-1–2
evolutionary optimization techniques, **2**-19
evolutionary programs, **13**-1–2
EXCEL, **1**-31
excess return, **21**-36–6
exercised options, **21**-22–3
exotic options, **21**-25
expectation
   maximization, **20**-16
   reliability and, **17**-8
expected opportunity lost (EOL), **6**-6–7
expected payoff, decision analysis, **6**-17
expected profit, **10**-26
expected regret, **6**-6
expected returns, **14**-5, 7, 10, **19**-6, **21**-6
expected utilities, decision analysis, **6**-14–7
expected value
   decision analysis
      decision trees, **6**-10–2
      implications of, **6**-4–5
      of perfect information, *see* expected value of perfect information (EVPI)

Index **I**-11

under uncertainty, **6**-5–6
implications of, **17**-14, **18**-4–5
of perfect information (EVPI)
   characterized, **6**-6–7
   expected opportunity lost (EOL), **6**-7
   expected value under uncertainty, **6**-7
ExpertChoices, **5**-35
ExpertMaker, **23**-11
expiration date, **21**-2–3
exponential distribution, **17**-2
exponentially weighted moving average (EWMA) control charts, **16**-13, 15, 19
exponential power distribution, **17**-21, 23, 40
exponential probability function, **24**-14
exponential random graphs models (ERGMs), **11**-14
exponential random variable, **8**-28
exponential smoothing method, **18**-4–5, 8–9
exponential survivor function, **17**-29, 32
Extend, **12**-17

F

facilities
  location problem, **14**-27–8
  shapes of, **18**-20
failure mode, effect, and criticality analysis (FMECA), **16**-6
failure physics, **17**-26
failure rate, defined, **17**-15
failures, common-cause, **17**-7
fallacy of averages, **14**-2
Fannie Mae, **21**-12
fathoming, **3**-8–12
feasibility
  airline optimization, **20**-6
  problem, **2**-2
  significance of, **14**-2, 7, 12, 15–6
feasible region, **21**-8, 10, **24**-22
feasible set, **2**-2
feasible solutions, **2**-2–3
Federal Aviation Administration (FAA) regulations
  aircraft safety requirements, **20**-9–10
  crew pairing, **20**-11
Federal Emergency Management System (FEMA), **24**-3
FedEx, **23**-4, 18–9, **26**-10
fertilizer product mix problem, **1**-5–6, **17**-8, 22, 24, 27–8, 34

Fibonacci heap, **4**-6
Fibonacci search, **2**-4–6, **19**-16
fill rate, **22**-11, 14, 32, 53, 55
fill-ratio equalization policy, **1**-20–1
financial analysts, roles in decision analysis, **6**-9–10, 12
financial engineering
  applications, **21**-1–2
  asset liability management, multi-period, **21**-33–6
  Black-Litterman asset allocation model, **21**-13–8
  diversification, **21**-6–8
  dynamic programming, **21**-28–9, 33
  efficient frontier, **21**-8–10
  investor perspectives, **21**-15–8
  mean, estimation of, **21**-4–6
  Monte Carlo simulation, dynamic programming distinguished from, **21**-33
  options, **21**-22–33
  pricing strategies, **21**-29–33
  return, **21**-3–4
  risk management, **21**-18–22
  utility analysis, **21**-10–2
  variance, estimation of, **21**-4–6
finish critical, **15**-33
finished goods, inventory control, **10**-9
"finish to finish" dependency, **15**-32
finite-difference interval, **19**-22
finite difference techniques, energy system optimization, **19**-15
finite population queues, **9**-13
firm energy, **24**-5
firm yield, **24**-5, 10
first come first served (FCFS), **9**-3–4, 10, 17, 22, 24–5, 27, 34–5, 38
first-improving search, **13**-5, 17–20
first moment, **17**-14
first-order approximations, **2**-16–7
Fisher information matrix, **17**-35
fixed-charge problem, **3**-4–5
fleet assignment model (FAM), airline optimization, **20**-8–9
flexibility, **14**-1, 10
Flexsim, **12**-17–8
flight simulators, **12**-1
flood control storage, **24**-4
flowcharts
  CHP operation simulation, **19**-19
  partitioning algorithm, **5**-22
flow duration curve, **24**-5
flow node, shortest path problem, **4**-5

flow shops
  "fit" in, **18**-25
  functions of, 18–23
  heuristics for selected criteria, **18**-24, 28
  three surrogate problems, **18**-24–5
  two-process, **18**-23–4
force-feedback control law, **13**-23
Ford, **23**-15
forecasted demand quantity, **18**-3
forecast horizon, **18**-2–3
forecasting errors, **18**-3
form postponement, **22**-44
forward-difference approximation, **19**-22
Fourier-Motzkin elimination method, **1**-3
fractiles, **17**-14–5
framing, **6**-17
Frank-Wolfe method, **5**-31
Freemarket Online, **23**-14
free slack, **15**-7–8
freight terms of sale, **22**-30
freight transportation, **22**-28–31
frequency
  histograms, demand forecasting, **18**-5–8, 28
  implications of, **13**-21
frequency-based diversification, **13**-26
fulfillment time, **22**-13
future directions for OR/MS applications
  background to, **26**-1-3
  emerging opportunities, **26**-8–12
  ensuring management support, **26**-5
  evolving appropriate models, **26**-4–5
  innovative solutions, **26**-5–6
  internet, **26**-6–8
  problem formulation in inclusive mode, **26**-3–4
future value, **21**-33
fuzzy goal programming, **5**-26–7

G

Galileo International, **20**-4
gambler's ruin problem, stochastics example, **8**-17, 24–5
game theory, **10**-36
gamma distribution, **8**-8, **17**-23
GAMS, **1**-31
Gantt charts
  applications, **15**-10–2
  variations, **15**-12–4
gap statistics, **17**-20
gas atoms, complex networks analogy, **11**-4–5

gas networks, **4**-9
Gauss-Jordan
  elimination method, **1**-3
  pivot step, **1**-27
Gaussian distribution, **5**-14, **19**-12
GDF procedure, **5**-31-3
general knapsack problem, **3**-4
General Electric, **23**-14–5
generalized reduced gradient (GRG) algorithm, **5**-24, 31, **19**-14
General Motors, **23**-15
general-purpose language, **12**-16
genetic algorithm, **3**-16, **13**-1–2, **20**-12, **23**-5, **24**-17–8, **22**-3
GENOCOP, **24**-23
GERAD, **20**-5
$G/G/1$ queue, **9**-23–4
$G/M/1$ queue, **9**-17
$G/M/s$ queue, **8**-47
Gibbs sampling, **11**-14
$GI/G/m$ queue, **9**-36
$GI/G/s$ queue, **9**-5–8, 35
global distribution systems (GDS), **20**-4
goal constraints, **5**-19, **21**-3
goal programming (GP)
  model, multiobjective LP models, **1**-32–4
  multiple criteria decision making applications, **5**-34
  formulation, **5**-19–20
  partitioning algorithm, **5**-21–5
  preemptive, **5**-20–1
  significance of, **5**-19
  software, **5**-35
  STEM (GPSTEM) method, **5**-31
goals, multiple criteria decision making (MCDM), **5**-3
GOBLIN-C++ Object, **4**-19
Goldberg, Andrew, codes for shortest path and minimum cost flow algorithms, **4**-19
Golden sequential line search method, energy systems optimization, **19**-23
Gomory cuts, **3**-14
goodness-of-fit test, **8**-14, **15**-20, **17**-2, 38–9
goodwill, cost, **22**-11
Google
  AdWords, **23**-11
  Analytics, **23**-20
  characterized, **23**-4–5, 8, 18
  Froogle, **23**-16
  page ranking algorithm, **23**-8–9
GPSS/H, **12**-7, 18, 20
gradient method, energy system optimization, **19**-13–5

Index

gradient vector, **19**-21
Gram-Schmidt orthogonalization procedure, **2**-10
graphical user interface, **12**-15
graphic-aided interactive approach, **5**-32
graph partitioning problem (GPP), **11**-10, **13**-6–8
graph theory, **11**-1–3
gravitational energy, **19**-2
gravity model, **22**-45–6
greedy heuristics, **13**-25–6, 28
greedy local improvement schemes, **13**-25
greedy randomized adaptive search procedures (GRASP), **13**-1–2
Greenwood's formula, **17**-36
GRG2, **5**-32
gross domestic product (GDP), **20**-3
groundwater management, **24**-23–5
groundwater remediation, **24**-24
group bookings, airline optimization, **20**-16
Group Lens project, **23**-10
group theoretic algorithms, **3**-16
guaranteed-service formulation, **22**-13

H

hazard function
   bathtub-shaped, **17**-12–3
   characterized, **17**-2, 10–3, 16, 18, 24, 40
   cumulative, **17**-13–4
   nonmonotonic, **17**-32
Hazen-Williams coefficient, **24**-12
head node, **4**-2
HEC-ResSim computer program, **24**-6–7
hedge, robust optimization, **14**-10–1
hedging, **21**-36
Henriksen's Algorithm, **12**-7
Hessian matrix, **2**-12–4, **19**-15, 21–22
heterogeneity, complex networks, **11**-17–8, 23–4
heuristic optimization strategies, **2**-3, 19
heuristic search algorithms, **3**-16
heuristics, e-commerce, **23**-5
Hewlett-Packard (HP)
   form postponement, **22**-44
   inventory management, **10**-34, 37
Hierarchical Bayes model, **20**-16
high-degree search, **11**-23
highest degree, **11**-24
highest LBC, **11**-24
high-sensitivity networks, **11**-18
histograms, energy investment options, **19**-10

historical booking forward, **20**-16
historical data, **21**-4–5, 19
HitBix, **23**-20
holding costs, **7**-13, **10**-2, 6, 10–2, 15, 17–8, 21, 23, 25, 31–2, 35, **22**-45
Home Depot, **23**-6
home plumbing system
   analytical hierarchy selection process, **24**-18–21
   piping system, **24**-2–3
HOM software, **9**-24
homogeneous mixing hypothesis, **11**-26
Hook and Jeeves approach, **19**-14
Hotelling, **1**-4
HTML, **23**-9
hub and spoke network, **20**-8
hub avoidance paths, **11**-26
Hungarian algorithm, **4**-13–4
hydraulic conductivity, **24**-23–5

I

iBeam Broadcasting, **23**-18
IBM, **10**-34, **20**-5, **26**-12
ICE (information, communication, and entertainment) age, **26**-8
ideal point, global criterion method, **5**-27
ideal solution, **5**-4, 16
identity theft, **26**-10
IEEE, **23**-4, 7
if-then constraints, **3**-6–7
if-then statements
   project management, **15**-34
   tabu search, **13**-10
iLog, **23**-13
ILOG/CPLEX, **4**-19
incidence matrix, **20**-19
increasing failure rate (IFR), **17**-12, 17, 21, 26
incumbent integer solution, **3**-9, 11
in-degree
   defined, **4**-2
   distribution, **11**-8
independent float, **15**-8
independent increments, **8**-5, 8–9
index specification and value trade-off, **5**-30
industrial engineer, functions of, **22**-1–2
inequality constraint function, **2**-2
infeasibility analysis, **1**-22–3
inferences, **17**-2, 25
infinite horizon case, **10**-27
infinitesimal generator, **8**-29, 31, 34
inflow, minimum cost flow problem, **4**-3

Information Age, **26**-8
information matrix, **17**-35
information networks, **4**-9
information resources
  conferences, **12**-3, 22
  databases, **11**-4
  literature review, **4**-19, **5**-35–6, **18**-23, 28, **25**-1–3, 10–1
  professional journals, xxii, **5**-36, **6**-29, **15**-34, **17**-2, **18**-29, **20**-7, 15–6, **22**-39, **25**-1–2
  textbooks, **17**-2–3, 40
  web sites, **1**-31, **2**-20, **4**-19, **5**-36, **6**-29, **12**-3, 22, 30, **25**-10
information technology, **1**-22
initial distribution, **8**-15
input-output analysis, **1**-5
input-output (I/O) coefficients, **1**-5
inspection interval, **2**-4
installation stock, **10**-31–2
Institute of Management Science and Operations Research (INFORM), **23**-5
Institute of Management Science (TIMS), xxii
Institute of Operations Research and Management Science (INFORMS)
  development of, xxii
  publications, **12**-23
in-stock probability, **22**-11–3, 15
integer goal program, **5**-24
integer linear programming (ILP)
  alternative solutions, **3**-15–6
  branch and bound (B&B) method, **3**-7–10, 16
  characterized, **3**-1–3
  computer solutions, **3**-15–6
  cutting plane method, **3**-12–6
  functions with N possible values, **3**-7
  model formulation, **3**-3–7
  NP-hardness, **3**-15
  pure, **3**-1, 13–4
  software systems, **3**-16
integer optimization problem, **2**-2
integer programming, **20**-3, 14, **25**-4
integrated SC models, **10**-37
intelligent design, **26**-10
intelligent modeling, container shipping example, **1**-18–22
interactive methods, multiple criteria mathematical programming (MCMP)
  branch-and-bound method, **5**-33
  criterion weight space methods, **5**-31–2
  feasible direction methods, **5**-31

feasible region reduction methods, **5**-30–1
  Lagrange multiplier methods, **5**-32
  trade-off cutting plane methods, **5**-32
  visual, **5**-32–3
interactive surrogate worth trade-off method (ISWTM), **5**-32
interarrival time
  queueing networks, **9**-36
  simulations, **12**-22
interchange heuristics, **13**-16–9
interfering slack, **15**-7–8
interior feasible solution, **1**-29–31
interior point methods, linear programming
  central path, **1**-29
  general step, **1**-29–30
  gravitational, **1**-30–1
  optimality conditions, **1**-29
  overview, **1**-28–9
internal rate of return (IRR), **26**-7
International Air Transport Association (IATA), **20**-2–3
International Society on Multiple Criteria Decision Making, **5**-35–6
internet
  affiliate programs, **23**-7, **11**-2, 21
  auction, **23**-7, **26**-11
  characterized, **4**-9, 15
  as complex network, **11**-3, 5–6, 8, 17
  internet service providers (ISPs), **23**-18
  kiosks, **23**-6
  protocol, IPv6, **23**-4
  search, **23**-8–9
interval of uncertainty (IOU), **19**-23
interval trade-off ratio, **5**-30
in-the-money options, **21**-23, 33
intranets, **4**-9
intrinsic value, **21**-23, 31–2
intuition, **6**-21
inventory control
  ABC classification, **10**-4, 35
  airline optimization, **20**-15, 18–20
  demand process, **10**-8
  deterministic inventory systems, **10**-9–20, 31, 35
  forecasting techniques, **18**-3
  inventory management, in practice, **10**-34–6
inventory system design
  graphic representations, **10**-10, 13, 19
  higher inventory levels, **10**-5–6
  implications of, **10**-4–5
  inventory management models, **10**-7–9
  lower inventory levels, **10**-6

Index

lead time, **10**-7–8, 13–4, 24
logic of, **10**-2–3
multiple locations, **10**-27–34
optimization tools, **10**-35
order quantity, **10**-7
policies
   base stock, **10**-27
   coordination, **10**-30
research directions, **10**-36–7
pure inventory management, **10**-36
reorder point (ROP), **10**-13–4, 22, 24–5, 31, 33
significance of, **10**-1–2
software programs, **10**-35
stochastic inventory systems, **10**-20–7, 36
types of inventory, **10**-3–4
inventory cycle, **10**-17
inventory equations, **1**-12
inventory management
   base-stock models, **22**-12–15, 52–5
   can-order vs. must-order inventory policy, **22**-18–9
   fill rate, **22**-11, 14, 32, 53, 55
   in-stock probability, **22**-11–3, 15
   newsvendor models, **22**-8–12
   risk pooling, **22**-41–4
   strategies overview, **22**-6–8
inventory planning, **4**-4, **7**-13–6
inventory-transportation analysis, **22**-34–6
inventory-transportation optimization models, **22**-26
investment decisions, financial engineering, **21**-11
investment planning, **7**-22–4
investment problems, integer programming, **3**-6
investment strategies, energy systems, **19**-5–11
investor views
   combining equilibrium returns with, **21**-17–8
   example of, **21**-16–7
   significance of, **21**-15–6
irreducible Markov chain, **8**-21, **8**-25
items, product mix example, **1**-6
iterated local search, **13**-1–2
Ito process, **21**-26

J

JDA Software Group, **20**-5
J E Beasley OR library, data sets for OR problems, **4**-19
Jeppesen, **20**-5

job sequencing, **3**-5, **14**-16–7
job shops, **18**-26–8
Johnson's algorithm, **18**-27
joint constraints, **14**-7
joint ordering, **22**-18
Jordan, Wilhelm, **1**-3
judgment values, **5**-12
just in time inventory, **26**-7
just-in-time scheduling, **18**-28

K

k, **4**-16, **5**-22
Kanban, **9**-30, **22**-8
Kantorovitch, L. V., **1**-9–10
Kaplan-Meier product-limit estimate, **17**-2, 34, 40
Karush-Kuhn-Tucker (KKT) conditions, **14**-5, **19**-14
Kendall notation, **9**-3–4
kernel, multiple criteria decision making, **5**-10–1
kinetic energy, **19**-2
knapsack problem, **3**-4
Kolmogorov equations
   applications, **25**-7
   backward equations, **8**-33
   differential equations, **8**-31–2
   forward equations, **8**-33
Kolmogorov-Smirnov (KS) goodness-of-fit test, **17**-2, 36–40
Königsberg bridge problem, **11**-1–2
Koopmans, T. C., **1**-10
k-out-of-n system, **8**-3, **17**-4–5
KRONOS, **20**-5
Kruskal's algorithm, **4**-16–8
$K \times K$ matrix, **21**-15
$K \times N$ matrix, **21**-16
k year planning horizon, **4**-5

L

label-correcting algorithms, **4**-6, 8
laboratory conditions, significance of, **17**-23
Lagrange multipliers, **2**-19, **5**-32, **14**-5, **20**-8, **21**-9–10, **22**-19
Lagrangian function, **19**-14
Lagrangian relaxation, **4**-1, 19, **20**-8, **22**-19
LAMSADE, **5**-35
Lanchester square law, **25**-7
Laplace criterion, in decision analysis, **6**-20
large-scale networks, see complex networks
large-scale problems, network optimization, **4**-19

last come first served (LCFS), **9**-3, 10, 22
lateral shipments, **10**-8
lateral supply, **10**-29
lattices, complex networks, **11**-15, 23
law of diminishing returns, **17**-10
lead-lag constraints, **15**-29–30
lead time
   demand, **22**-41–2
   inventory management, **10**-7–8, 13
   replenishment, **22**-16
   significance of, **18**-28
   supply chain management, **22**-12, 16, 50, 54
Leibniz's rule, **10**-26
less-than-truckload (LTL) shipment, **22**-24–5, 27–8, 30, 32, 34, 36, 46
Levenberg-Marquardt method, **2**-14–5
Library of Congress, **23**-3
lifetime distributions
   cumulative hazard function, **17**-13–4, 40
   expected values, **17**-14
   fractiles, **17**-14–5
   hazard function, **17**-10–3, 16, 18, 40
   probability density function, **17**-10–2, 15, 17, 40
   survivor function, **17**-10–1, 15–7, 40
   system, **17**-15–7
likelihood function, **17**-23–4
likelihood ratio test, **8**-27
limiting distributions, **8**-21, 24, 33–7, 45–6
LINDO LP software, **1**-31, **19**-5, 8
LindoSystems, **5**-35
linear algebra, **1**-3–4
linear constraints, dynamic programming, **7**-9
linear goal program, **5**-21–4
linear inequality
   constraints, **1**-3
   solving systems, **1**-4
linearity assumptions, **1**-4
linearized maximal expected covering problem (LEMCP)
   defined, **13**-15
   development of, **13**-16
   heuristics for, **13**-20–1
linear normalization, **5**-5
linear penalty function, **1**-32
linear programming (LP)
   algorithms
      applications, **1**-1–4
      solving LP models, **1**-26–31
   classical examples of direct applications
      blending problems, **1**-6–8
      diet problem, **1**-8–9

multiperiod production planning, storage, distribution problems, **1**-11–2
   overview, **1**-4–5
   product mix problem, **1**-5–6
   transportation problem, **1**-9–11
development of, **1**-4
duality theorem, **14**-11
energy systems, **19**-4–5
intelligent modeling, container shipping example, **1**-18–22
military applications, **25**-4
multiobjective models, **1**-32–3
Phase I problems, **1**-4, **1**-22–3
planning uses, **1**-22–6
relaxation, **3**-2, 4, 7–15, **13**-25, 27–8, **20**-7
simplex method, **1**-4
solving LP models
   algorithm applications, **1**-26–31
   software systems available for, **1**-9, 25–6, 31
utilization of, **26**-4
variable transformations
   absolute values of affine functions, **1**-17–8
   min-max, max-min problems, **1**-16–7
   types of, **1**-12–6
linear regression, **20**-16, **22**-29
line searches
   backtracking search, **2**-4, 7–8
   bisection search, **2**-4–7
   bracketing the minimum, **2**-4–6
   characterized, **2**-3–4
   energy system optimization, **19**-16
   Fibonacci search, **2**-4–6
   quadratic fit search, **2**-4, 7
   types of, **2**-4
line-haul costs, **22**-24
linked decisions, **6**-27–8
Linkshare, **23**-11
LIST data, **4**-5–6, 16
Little's law, **9**-6–7, 9, 23, 30
Littlewood's rule, **20**-19
load balancing, **23**-18
local area networks (LANs), **25**-5
local betweenness centrality (LBC), **11**-23–4
local search
   characterized, **2**-3, **11**-21–4
   multistart, **13**-5
local trade-off ratios, **5**-29, 31
location-allocation problem, **22**-38
location management
   aggregate inventories, **22**-41–4
   centralized vs. decentralized supply chains, **22**-40–1

Index

characterized, **22**-38–40
continuous-location model, **22**-44–8
risk pooling, **22**-41–5
location parameters, **17**-17
logic flow diagram, **12**-21
Logical Decisions, **5**-35
logistics network design, **22**-34
log likelihood function, **17**-24, 27, 33
log-linear regression, **22**-29
long position, **21**-15, 22, 24
LOQO, **1**-31
loss function, **22**-9
lotteries, **6**-14
lower bounds
 dynamic programming, **7**-4, 18
 exponential utility function, **21**-12
 goal programming, **5**-26
 implications of, **3**-7, 9, **4**-4
 multiple criteria decision making (MCDM), **5**-4
 robust optimization, **14**-11, 19
lower control limit (LCL), **16**-11
LP Solver, **23**-15
Lucent Technologies, **10**-34
Ludus Algebrae, **1**-2
lushlawn, **1**-25–6

M

Macromedia, **23**-3
maintenance, airline optimization, **20**-9–10
management information systems (MIS), **26**-5
management science (MS), xxii
Management Science, xxii
Manugistics, **20**-5, **23**-12
Maple, **8**-47
marginal analysis, **1**-25, **10**-27
marginal values, **1**-24–5
market analysis, **1**-17
market capitalization, **21**-15, 18
market crash scenarios, **21**-33, 35–6
market equilibrium returns, **21**-13–5
market graph, as complex network, **11**-4
marketing campaigns, **11**-26
marketing strategies, **20**-20, **26**-9
market risk, **5**-34
market share, **1**-34, **17**-2
Markov, A. A., **8**-15
Markov Analysis, **8**-47
Markov chain, military applications, **25**-7

MARkov Chain Analyzer (MARCA), **8**-47
Markov chain theory, **13**-6
Markov decision processes (MDPs), **8**-47
Markov process, **21**-26
Markov property, **7**-3, **8**-5, 15
Markowitz model, **21**-9–10
mass balance constraints, **4**-4–5
mass customization, **26**-10
material balance
 equations, **1**-12
 inequality, **1**-6
material flow network, **18**-18
Mathematica, **8**-47
Mathematical Methods in the Organization and Planning of Production (Kantorovich), **1**-9–10
mathematical modeling
 defined, **1**-2
 energy systems, **19**-3–4
 large-scale, **4**-1
Mathematical Programming System Extended (MPSX), **20**-1
MATLAB, **1**-31, **8**-47
max-flow min-cut, **25**-5
maximal covering species problem (MCSP)
 characterized, **13**-13–4
 data details, **13**-15–6
 interchange heuristics, **13**-16–9
 problem formulation, **13**-15–6
 tabu search for, **13**-19–20
maximal expected coverage problem (MECP), **13**-20–1
maximum flow problem
 augmenting path algorithm, **4**-9–13
 characterized, **4**-8–9
 linear programming formulation, **4**-9
maximum inventory, **10**-17
maximum leap spanning tree problem, **4**-16, **4**-18
maximum likelihood
 characterized, **6**-5, **13**-13–5
 estimate, **8**-13–4, **17**-2, 25–6, 28, 33, 35
 function, **8**-25–6, 38
 heuristics for, **13**-20–1
max-min
 method, **5**-8
 problem, **1**-16–7
maximum regret, **14**-14
maximum reliability path problem, **4**-4
maxmin (minimax) criterion, **6**-18
maxmin (minimin) criterion, **6**-19
MCQueue, **8**-47
mean absolute deviation (MAD), **18**-5
mean arrival rate, queueing networks, **9**-35

mean
  beta distribution, **15**-21
  estimation of, **21**-4–6
  exponential distribution, **17**-19
  inventory control, **10**-8
  lifetime distribution, **17**-14
  simulations, **12**-13, 22
  statistical process control, **16**-11
  supply chain management, **22**-12
  survival analysis, **17**-40
mean-risk constrained robust optimization model, **14**-3–7
mean time between failures (MTBF), **17**-14
mean time to failure (MTTF), **17**-14
mean-variance approximation, **21**-12
mean-variance model, **5**-34, **14**-18
mean-variance optimization, **21**-9, 11, 13
measures of effectiveness (MOEs), **25**-10
measures of performance (MOP), **25**-10
MEDLARS, **23**-3
Mellin transform, **17**-10
memetic algorithms, **13**-1–2
memoryless properties, **8**-5, 8–10, 41, **9**-3, **17**-18–20
memsize, **13**-31, 35
6-mercaptopurine (6-MP), **17**-34–6
merit successive quadratic programming method, **2**-17
metaheuristics
  applications, **13**-2
  defined, **13**-1–2
  single solution, **13**-2–3
MetalJunction, **23**-15
method of centers, **5**-32
method of multipliers (MOM), **2**-19–20
Metropolis, **24**-13
Metropolis-Hastings sampling method, **11**-14
$M/G/1/1$ queue, **8**-41, 47
$M/G/1$ queue, **9**-7, **14**-7, 22, **24**-6
$M/G/s$ queue, **9**-7
$M/G/s/s$ queue, **9**-12, **16**-7
Micro Saint Sharp, **12**-18
Microsoft, **23**-3, 7, 18–9
Microsoft Excel
  CallCenterDSS.xla, **8**-36
  CHP optimization simulation, **19**-12, 21, 24
  Solver, **3**-16, **22**-19, 46
  Stochastics programs, **8**-47
  supply chain management, **22**-10
mid bi-critical, **15**-33
mid normal critical, **15**-33
mid reverse critical, **15**-33

military applications
  airline optimization, **20**-3
  background on, **25**-2–3
  current, **25**-4–10
  decision analysis, **25**-6
  future directions for, **25**-11
  information resources, **25**-1–2, 10
  linear and integer programming, **25**-4, 9
  metrics, **25**-9–10
  multiple criteria decision-making (MCDM), **25**-5–6
  network optimization, **25**-5
  overview, **25**-1–2, 10
  simulation, **25**-8–9
  stochastic models, **25**-6–8
  system of systems (SoS), **25**-11
  terrorism and, **25**-10–1
Military Standard tables (MIL STD 105A), **16**-20
Mill's ratio, **17**-11
minimal finished goods inventory, **26**-7
minimax regret, **6**-20
Min-Max method, **5**-8–9, 26
min-max problem, **1**-16–7
minimax regret, robust optimization model, **14**-14–6, 18–22, 26
minimization of concave function, **10**-19
minimum average node weight, **11**-24
minimum cost flow problem, **4**-3–4, 9, 13
minimum cost multicommodity network flow (MCMNF) problem, **4**-18–9
minimum edge weight, **11**-24
minimum spanning tree problem
  applications, **4**-15
  characterized, **4**-14–5
  Kruskal's algorithm, **4**-16–8
  linear programming formulation, **4**-16
Minitab, **8**-47
mixed integer linear programs (MILPs), **3**-1–2, 15–6, **20**-7
mixed-integer optimization problem, **2**-2
mixed penalty-barrier method, **2**-19
mixed-integer problem (MIP), **3**-1, **20**-14, **22**-38, 45, 48
$M/M/1/K$ queue, **9**-11–2
$M/M/1$ queue, **8**-38, **9**-4, 7–10
$M/M/s/k + M$ queue, **8**-30, 36
$M/M/s$ queue, **8**-30, 35, 47, **9**-7, **10**-2, 29
$M/M/s/s+M$ queue, **8**-47
$M/M/\infty$, **9**-12–3, 16
mobile commerce, *see* e-commerce; mobile electronic commerce
mobile electronic commerce, **23**-6

Index

mode
  beta distribution, **15**-21
  implications of, **13**-21
modeling, *see specific types of models*
  defined, **1**-2
  software, **1**-31
modern portfolio theory, **5**-34
modularity, complex networks, **11**-9–10
mold solidification simulation software, **12**-1
monotonic slope, **1**-14
Monte Carlo simulation, **12**-1, **17**-14, 38, **18**-28–9, **21**-19, 24, 26–8, 31, 33, **25**-8
monthly returns, **21**-5, 7
Moore's algorithm, **18**-22–3
mortgage-backed securities, **21**-12
move set, **13**-5
Movice, **23**-11
movie actor collaboration network, **11**-4, 6
moving average method, **18**-4–5
MP3, **23**-7
MPL, **1**-31
MSN, **23**-8, 18
$m \times n$, **1**-4, 22, 26, 28
Multi Attribute Decision Making (Keeney/Raiffa), **5**-35
multiclass queues, **9**-21–7, 34–7
multicommodity flow problems, **1**-10, **4**-4, **20**-10
multicommodity network
  airline optimization, **20**-8
  flow problems, **4**-18
multicriteria shortest path problem, **4**-8
multidimensional optimization
  characterized, **2**-8
  Levenberg-Marquardt method, **2**-14–5
  method of
    Fletcher and Reeves, **2**-12–3
    Hooke and Jeeves, **2**-10–11
    Rosenbrock, **2**-10–11
    Zangwill, **2**-10–2
  quasi-Newton method, **2**-13–4, 17
  simplex method (S^2), **2**-9–10
  steepest descent method, **2**-11–2, 14
multi-echelon inventory
  controls, **10**-7, 31
  theory, **10**-28
multi-echelon supply
  chain network, **10**-28
multiobjective LP models, **1**-32–4
multiobjective optimization problem, **2**-2
multiperiod production planning problems, **1**-11–2

multiple attribute decision making (MADM) problem, **5**-2
multiple criteria decision making (MCDM)
  applications, **5**-34
  "best solutions," **5**-4–5
  characterized, **5**-1–2
  classification scheme, **5**-30
  compromise programming, **5**-29
  criteria normalization, **5**-5–6
  criteria weights, computation methods, **5**-6–7
  definitions, **5**-3
  goal programming, **5**-19–26, 34–5
  information resources, literature review, **5**-35–7
  interactive methods, **5**-29–34
  method of global criterion, **5**-27–8
  military applications, **25**-5–6
  multiple attribute decision making (MADM) problem, **5**-2
  multiple criteria mathematical programming (MCMPM) methods, **5**-2
  multiple criteria mathematical programming (MCMP) problems, **5**-2, 15–9
  multiple criteria methods for finite alternatives (MCMFA), **5**-2, 8–15
  multiple criteria selection problem (MCSP), **5**-2–3, 11
  multiple objective decision making (MODM), **5**-2
  software systems, **5**-35
multiple criteria mathematical programming methods (MCMPM), **5**-2
multiple criteria mathematical programming (MCMP) problems
  characterized, **5**-2, 15
  classification of, **5**-18–9
  definitions, **5**-16
  efficient solution, **5**-16–8
  test for efficiency, **5**-18
multiple criteria methods for finite alternatives (MCMFA), **5**-2, 8–15
multiple criteria selection problem (MCSP), **5**-2–3, 11, 15
multiple objective decision making (MODM), **5**-2
multiple objective linear programming (MOLP) problems, **5**-31
multiple sourcing, **10**-8
multiplicative regressive models, **20**-16
multiprogrammed computer system, **9**-30
multi-start local search, **13**-5, 36

multistation queues
  multiclass queues, **9**-34–7
  single-class queues, **9**-28–34
mutually exclusive subproblems, **3**-8

N

nadir solution, **5**-4
Napster, **23**-4
NASDAQ, **23**-4, 15
National Pollutant Discharge Elimination System (NPDES) program, **24**-3
National Stock Exchange (NSE), **23**-4
natural decay, **13**-21–22
natural energy systems, **19**-3
natural frequency, **13**-21
natural gas, **19**-4–5
natural motions, **13**-21
nature reserve site selection
  MCSP
    interchange heuristics, **13**-16–9
    tabu search for, **13**-19–20
  LMECP heuristics, **13**-20–1
  MECP heuristics, **13**-20–1
  overview of, **13**-13–5
  problem formulation and data details, **13**-15–6
Navitaire, **20**-5
negative cycle, shortest path problem, **4**-5
negative-ideal solution, **5**-4
negative inventory, **10**-16
negative-outranking flow, **5**-14–5
negative part, **1**-18
neighborhood structure, **13**-25, 30–2, 36
Nelder-Mead method, **19**-14
NEOS
  Guide, web site, **2**-20, **4**-19
  server, **1**-31
nested coordination policy, **10**-30
nested-uniform demand model, **22**-58
net present value (NPV)
  cost savings in energy systems optimization, **19**-20, **23**-5, 27
  energy system optimization, **19**-16, **21**-22
net profit, **1**-15–6, 32
network approach, crew pairings, **20**-12
network design problems, **4**-4
network flow models, **4**-2, **20**-3
network location theory, **13**-3
network optimization
  applications, generally, **4**-1–2
  assignment problem, **4**-4, **13**-4
  characterized, **4**-1–2, 19

information resources, **4**-19
maximum flow problem, **4**-8–13
military applications, **25**-5
minimum cost flow problem, **4**-3–4
minimum cost multicommodity network flow (MCMNF) problem, **4**-18–9
minimum spanning tree problem, **4**-14–18
notation, **4**-2–3
shortest path problem, **4**-4–8, 18
software systems, **4**-19
network revenue value, airline optimization, **20**-19
network simplex algorithm, **4**-4
Network Science, **11**-2–3, 5
neural networks, **23**-5, 9, 21
neutral branch, decision tree, **6**-10
new products, profitability evaluation, **1**-25–6
newsboy problem, robust optimization, **14**-17–19
newsvendor model
  forecasting method, **18**-9–11
  inventory management, **10**-26–7, **22**-8–12
Newton direction, **1**-30
Newton method, **2**-3, **19**-15
Newton-Raphson technique, **17**-31
NexTag, **23**-16
Nexus, **23**-18
node balance, **4**-5, 19
node set, defined, **4**-2
nonconvex, piecewise-linear functions, **20**-23
nondominated paths, **4**-8
non-dominated solution, **5**-4, 16
nonhomogeneous Poisson processes, **8**-12–4
nonlinear goal program, **5**-24–5
non-linear optimization
  problem, **2**-3, **22**-13–4, 28–9
  supply chain management, **22**-47
non-preemptive priority, **9**-24–6, 38
nonlinear programming
  applications, **1**-30, **20**-3, **21**-14
  constrained optimization, **2**-15–9
  defined, **2**-1
  dynamic programming, **7**-9
  problem statement, **2**-1–3
  production layout design, **18**-15
  techniques, **2**-19
  unconstrained optimization
    line searches, **2**-3–8
    multidimensional optimization, **2**-8–15
nonnegativity restriction, **1**-25–7
nonparametric survivor function, **17**-32–6

Index

nonproduct form-networks, diffusion approximation
   closed queueing networks, **9**-33–4
   open queueing networks, **9**-32–3
nonquadratic functions, **2**-14
nonrepairable systems, **17**-7
nonspatial networks, search in, **11**-23–4
nonstantiation, **21**-8, 10
nonstationary demands, **10**-8
non-systematic risk, **21**-14
normal critical, **15**-33
normal distribution, **5**-14, **17**-17, 36, **18**-4–5, **21**-15, **24**-16
notation, multistation queues, **9**-34–5
n-out-of-n system, **8**-3, **17**-4
np chart, **16**-14, 23
NP complete, **18**-28
n-step transition probabilities, **8**-18
$N \times N$ identity matrix, **9**-28, 32–3
null hypothesis, **8**-14, 38
null recurrent state, **8**-21–2, 25
Numerical Recipes, **13**-6

O

O&M cost, **19**-18, 20
objective functions
   damper placement, **13**-25
   energy system optimization, **19**-13–5, **21**-3
   integer programming, **3**-3, 7
   nonlinear programming, **2**-2
   quadratic, **7**-9
   shortest path problem, **4**-5
   water distribution systems, **24**-13–4, 22
objective problem, integer programming, **3**-5
objectives
   multiple, **6**-28–9
   multiple criteria decision making (MCDM), **5**-3
objective space, multiple criteria mathematical programming (MCMP), **5**-16–7
observed information matrix, **17**-35
obsolescence, **10**-6
off the shelf branch and bound code, **13**-21
offline quality engineering, **16**-3–4
omnibus tests, **17**-38
1-out-of-n system, **17**-4
one-step transition, **8**-15
one-to-one functional mapping, **5**-32
oneworld, **20**-3
online quality control, **16**-3–4

open-queueing networks, **9**-28–30
operational research teams, defined, xxi
operations research, generally
   historical perspective, xxi-xxii
   meaning of, xxii-xxiii
Operations Research, xxii
operations researchers, functions of, xxi
Operations Research Society of America (ORSA), xxi-xxiii
OptiBid, **23**-5
optimality, multiobjective LP models, **1**-32
optimal policy, dynamic programming, **7**-2, 8, 10–2, 14, 20, 22
Optimal Retirement Planner, **23**-6
optimal solution, **2**-2
optimal value, **2**-2, **7**-2, 8, **10**-1, 14, **21**-2
optimization models, **1**-5, **11**-18–26
optimization problem, branch and bound method, **3**-9
Optimization Software Guide Web site, **1**-31
optimization techniques, **1**-22
optimization vector, **2**-2
optimum solutions, **1**-22–4, 33–4
options
   American, **21**-22–3, 27–33
   European, **21**-22–3, 25
   intrinsic value, **21**-23, 31–2
   pricing, **21**-27, 29–33
   types of, **21**-22–4
   valuing, **21**-24–8, 31–3
option writer, **21**-22–4
ordering costs, **10**-5–6, 12, 20, 23–4, 32, 35
order management, **26**-10
order of magnitude, **5**-6, **13**-36
order quantity, **10**-7
order statistics, **17**-20, 23, 38
ordinal weights, **5**-20
Oregon Terrestrial Vertebrate dataset, **13**-14
origin-destination pair
   characterized, **20**-16
   network optimization, **4**-10, 18
OR/MS Today, **12**-23
OSL, **1**-31
outcomes, decision analysis, **6**-3, 8–9, 14
outdegree
   defined, **4**-2
   distribution, **11**-8
outflow, minimum cost flow problem, **4**-3
outliers, goal programming, **5**-26
out-of-control signals, **16**-12
out-of-the-money options, **21**-23
outperforming assets, **21**-16
outranking relationship, **5**-10–1
outsourcing, **10**-8

overage cost, **10**-27, **22**-12
overload failures, **11**-21
OverStock, **23**-14
overstocking, **22**-11

P

paired comparison method (PCM), **5**-32
paired-t test, **12**-8, 22
pairwise comparison, **5**-7, **10**–1, 15, 29, **24**-18–9
pair-wise preference matrix, **24**-19–20
parallel system, reliability of, **17**-4–6, 9–10
parameter-free SUMT, **2**-18
parametrized SUMT, **2**-18
Pareto analysis, **15**-9–10
Pareto optimal solution
   defined, **5**-16
   multiple criteria decision making (MCDM), **5**-4
Pareto optimality, **2**-2
Pareto Race, **5**-32–3
Part-Period Balancing method, **10**-20
partial differential equation, **21**-25
partitioning
   airline optimization, **20**-10
   algorithm, multiple criteria decision making, **5**-21–5
   branch and bound method, **3**-7–8
   gradient-based (PGB) algorithm, **5**-24
   implications of, **13**-25, **15**-29
   integer programming and, **3**-16
   network optimization, **4**-11
passive dampers, **13**-23–4
PASTA (Poisson Arrives See Time Averages), **9**-7–8
path, defined, **4**-2
payoff matrix, **5**-9–10, **6**-3–5, 14, 18, 26
pay-off table, **5**-2, 4–5
PayPal, **23**-4
p-center, **13**-4
p chart, **16**-14, 23
p-defense-sum, **13**-4
p-dispersion, **13**-4
p-dispersion-sum, **13**-4
peer-to-peer network, **11**-6
penalty coefficients, multiobjective LP models, **1**-33–4
penalty method, **2**-3
penalty successive linear programming (PSLP) method, **2**-16–17
penalty vectors, **1**-33
PeopleSoft, **23**-13

percolation transition, **11**-11–12
performance measurement, **12**-14–5, **13**-3–4
periodic review
   inventory policy, **22**-12, 15, 17
   models, **7**-13
permanent solution, **14**-14
perpetual demand, **22**-15
pessimism, **6**-19
petroleum industry, blending models, **1**-7–8
petroleum networks, **4**-9
PGAPack, **24**-23
Phase I linear programming problems, **1**-4, **22**-3
phone call network, **11**-3
physical annealing, **13**-6
physical distribution costs, **22**-26
"pick-up" models, **20**-16
piecewise linear approximation, **22**-45
piecewise linear (PL) function
   affine functions, **1**-14
   characterized, **1**-12–3, **18**-3
   variables and, **1**-13–4
piecewise linear programming, **19**-12
pipeline inventory, **22**-31–2, 34
pipelines, network optimization, **4**-8–9, 15
pivot step, **1**-26–8
plain vanilla tabu search (PVTS), **13**-8–10, 13, 36
planning horizon
   dynamic programming, **7**-4, 9
   energy investments, **19**-11
   energy systems optimization, **19**-19–20
   significance of, **3**-4, **4**-5
   supply chain management, **22**-7
p-median, **13**-4
PODS (Passenger Origin-Destination Simulator), **20**-20
point of sale (POS) data, **26**-7
Poisson, Simeon-Denis, **8**-6
Poisson arrival, **9**-28–9
Poisson distribution, **11**-13, **17**-21, **18**-5
Poisson processes
   compound, **8**-12–4
   defined, **8**-5, **8**-10
   exponential distribution and properties, **8**-7–8
   mark, defined, **8**-13
   nonhomogenous, **8**-12–4
   properties from, **8**-10–2
   testing for, **8**-14
polyhedra, mathematical theory of, **1**-4, **2**-9
population dynamics, **13**-16
population studies, **9**-30

Index

portfolio
   diversification, **11**-25, **21**-6–8
   management, **21**-33
   optimization, **21**-9–10, 33
   selection, **5**-34
   variance, **21**-3–4
positive-outranking flow, **5**-14–5
positive part, **1**-18
positive recurrent state, **8**-21–3, 44
posterior probabilities, **6**-8–12
potential energy, **19**-2
Powell-Symmetic-Broyden (PSB) method,
      energy system optimization, **19**-15
power grid network, **11**-3
power-law networks, **11**-8, 13, 23–4
powers-of-two policy, **22**-19
precedence diagramming method (PDM)
   activity characteristics, **15**-33
   characterized, **15**-2, 26
   compressed network, **15**-30
   computer implementation, **15**-34
   network example, **15**-27–8
   partitioning, **15**-29
   terminology, **15**-26–7
precise local trade-off ratio, **5**-29
preemptive goal programs, linear, **5**-21–4
preemptive priorities/goal programming,
      5-20–1
preemptive resume priority, **9**-26–7, 38
preference function, **5**-18, 33
preference index, **5**-14
preference judgment, **5**-11
preferential attachment, **11**-16
preferred solution, global criterion method,
      5-28
present value, **7**-6–7, **21**-31
prespecified weights, **5**-20
PriceGrabber, **23**-16
price engines, **23**-16
pricing strategies, **10**-36, **20**-5, 20, **21**-29–33
primal-dual path following interior point
      method, **1**-28
primal variables, **1**-25–6
Principle of Optimality, **7**-3
probabilistic models, **17**-2
probabilistic properties, **17**-3
probabilistic symbolic model checker
      (PRISM), **8**-46–7
probability
   applications, **4**-4, **15**-24
   distribution, **6**-17, **8**-3, **10**-21, 28, **14**-30,
      18-6, 8, **23**-12
   model, decision analysis, **6**-2, 4
   posterior, **6**-8–12

reliability and, **17**-6
theory, **17**-7
vector, **18**-7, 9
probability density function (PDF), **8**-7,
      17-10–2, 15, 17, 22, 40, **18**-6–7
process capability, **16**-11, **16**-8
process capability index, **16**-6, 17
processor sharing, **9**-10, 16
production allocation decisions, **1**-12
production layout design
   adjacency problem, **18**-13
   cut tree network model, **18**-19–20
   engineering tasks, **18**-12
   example problem, **18**-12–3
   facility shapes, **18**-20
   practical considerations, **18**-13–5
   reengineering, **18**-20
   spanning tree network model, **18**-17–9
   Tretheway algorithm, **18**-15–7
production planning, **4**-4
production problems
   dynamic programming models, **7**-9–13
   integer programming, **3**-6
production schedules
   bootstrapping, **18**-28–29
   flow shops, **18**-23–6
   job shops, **18**-26–9
   scheduling decision defined, **18**-20–1
   single machines, **18**-21–3
   solvability problems, **18**-28
   systems analysis, **18**-28
production systems
   demand forecasting, **18**-2–12
   planning problems, **18**-1–2
   production layout design models, **18**-12–20
   scheduling, **18**-20–9
product life cycle, **16**-2–5
product-limit
   estimate, **17**-2, 34, 37
   survivor function, **17**-34–5
product mix, **1**-5–6, 15, **23**-5, **26**-7
profitability, **1**-25–6, **10**-37, **20**-7
profit margin, **22**-9
profit maximization, **10**-26, **19**-12, **22**-2, 4
profit per unit, **1**-5
program evaluation and review technique
      (PERT) network analysis
   activity time models, **15**-19–20, 23
   beta distribution, **15**-20–1
   defined, **15**-18
   estimates and formulas, **15**-18–9
   network example, **15**-24–6
   triangular distribution, **15**-21–2
   uniform distribution, **15**-22–3

projection detruncation, **20**-16
project management
  backward pass, **15**-3, 6–7, 31
  constraints, **15**-2
  CPM example, **15**-5
  critical activities, determination of, **15**-7–9
  critical path method (CPM), **15**-2–5
  decision support model, **15**-35
  duration of project, **15**-32
  forward pass analysis, **15**-3, 6, 9, 29, 31
  Gantt charts, **15**-10–4
  implementation of, **15**-2
  information resources, **15**-34
  network planning, **15**-2–3
  network scheduling, **15**-3
  overview of, **4**-4, **15**-1
  phases of, **15**-2
  precedence diagramming method (PDM), **15**-2, 26–34
  program evaluation and review technique (PERT) network analysis, **15**-2, 18–23
  software tools
    characterized, **15**-34–6
    computer simulation software, **15**-36–7
  statistical analysis, project duration, **15**-23–6
  subcritical paths, **15**-9–10
project scheduling, **4**-4
Project SCOOP, **25**-3
project slippage chart, **15**-13–4
PROMETHEE method, **5**-13–5, 35
ProModel, **12**-18–9
proportionality assumptions, **1**-4, 14
prospect theory, **6**-17
pseudoconvex, **2**-6
pth fractile, **17**-14–5
$p \times n$, **1**-4
pull systems, **22**-8
push systems, **22**-8
put options, **21**-22–4

Q

QNA, **9**-36
(Q, R) inventory policy, **22**-32
QS9000/ISO9000, **16**-7
quadratically constrained quadratic optimization problem, **2**-2
quadratic fit search, **2**-7
quadratic optimization problem, **2**-2
quadratic program, **21**-10

quadratic programming problem, **2**-17
QUAL2E, **24**-23
quality control
  acceptance sampling, **16**-3, 20
  control charts
    advanced, **16**-18–20
    for problem identification, **16**-10–6
  online sources, **16**-3
  process capability studies, **16**-16–18
  product life cycle, **16**-2–5
  significance of, **16**-1–2, **20**-1
  Six Sigma methodology, **16**-4–7
  statistical (SQC), **16**-3
  statistical process control (SPC), **16**-7
  sources of variation, **16**-9
quality function deployment (QFD), **16**-2, 6, 20
quantity coordination, **10**-30
quantity discounts, **10**-6
quasi-Newton method, **2**-13–4,17, **19**-14–5, 21, 23
quasi-Newton step multiplier, **19**-15–6, 21–22
QUEST, **12**-19
queueing theory
  applications, **9**-1–2, 28, 37–8
  balking, **9**-9
  basics of, **8**-47, **9**-2–8
  bulk arrivals and service, **9**-13–4
  closed queueing network, **9**-30–4
  e-commerce, **23**-14
  finite capacity, **9**-5, 11
  finite population, **9**-13
  infinite capacity, **9**-5
  information resources, **9**-2
  Little's law, **9**-6–7, 9, 30
  multiserver systems, **9**-10–1
  multistation
    multiclass queues, **9**-34–7
    single-class queues, **9**-28–34, 37
  non-preemptive priority, **9**-25–6
  open-queueing networks, **9**-28–30, 32–3
  physical *vs.* psychology tradeoff, **9**-37
  preemptive resume priority, **9**-26–7
  queueing relations, **9**-4–5
  reneging, **9**-9
  in simulations, **12**-4–5, 8–10
  single-station queues
    closed-queueing networks, **9**-31–2
    and multiclass queues, **9**-21–7
    and single-class queues, **9**-8–21, 37
  state-dependent service, **9**-10, 29–30
queuing process, **8**-11

Index

R

radiant energy, **19**-2
radix heap, 4-6
random graphs, **11**-11–4, 16, 26
random index (RI), **5**-13
random order of service (ROS), **9**-3, 10
random search, **11**-23
random variable, reliability and, **17**-7–8
random walk, **8**-16–7, 29, **11**-22, 24
range equalization factor, **5**-6
rate of return, **6**-5
raw material, inventory control, **10**-9
Rayleigh distribution, **17**-21
$\mathbb{R}$, **2**-1, 4, 12
R, 17–40
R charts, **16**-13
reactive tabu search (RTS)
   characterized, **13**-29–32
   computational experiments, **13**-32–5
real constraints, **5**-19
Real Networks, **23**-7
recency-based diversification, **13**-3, 26, 28
recommendation engine, **23**-16
recommendations algorithms, **23**-10–1, 21
recourse robust optimization model, **14**-7–8, 21, 31
recurrence relation, **7**-2, 7–8, 10–1, 14, 17, 20–2
recurrent state, **8**-21–3
reduced cost matrix, **4**-13–14
redundancy, **3**-5, **17**-4, 6
refining process, blending problem example, **1**-7
refleeting, **20**-8
regenerative cycles, **8**-44
regret matrix, **6**-20
relationships, in linked decisions, **6**-28
relative cost
   coefficients, **1**-25, 27
   vector, **1**-28
relative value, **5**-12–3
Relex Markov, **8**-47
reliability
   defined, **17**-7, 40
   design, **17**-6–7
   failure rate, **17**-12, 15, 18, 28–30
   failure time, **17**-2, 17, 20–2
   fractiles, **17**-14–5
   functions, **17**-6–10
   hazard function, *see* hazard function
   information resources, **17**-2, 40
   of the ith component, **17**-8
   lifetime distributions, **17**-2, 10–7, 40

   literature review, **17**-2
   mathematical models, **17**-2, 6–7
   model adequacy, assessment of, **17**-36–40
   nonparametric methods, **17**-32–6
   parametric estimation, survival analysis, **17**-22–32
   parametric models, **17**-17–22
   probability density function, **17**-10–2, 15, 17, 40
   repairable system, **8**-32–3
   significance of, **4**-4, **17**-1–2
   software packages, **17**-40
   standards, **17**-6
   stochastic processes, **8**-29
   stochastics example, **8**-3
   structure functions, **17**-2–6
   survivor function, **17**-2, 10–11, 15–7, 29, 40
   system
      design, **17**-3–10
      machine maintenance example, **8**-45–6
   theory, characterized, **17**-1–2
reliability engineer, **17**-2–3, 6
renewal cycles, **8**-42
renewal function, **8**-39
renewal processes, statistical inference, **8**-46
renewal rate, **8**-40, 42
renewal reward process, **8**-43
renewal theory
   regenerative processes, **8**-39, 44–5
   renewal processes, **8**-39–41
   renewal reward processes, **8**-39, 41–4
   semi-Markov processes, **8**-39, 45–6
   statistical inference, **8**-46
reorder point (ROP)
   implications of, **10**-13–4, 22, 24–5, 31, 33
   order quantity model
      characterized, **22**-15–7
      multi-echelon models, **22**-19–20
      multi-item models, **22**-17–9
reorder point-order quantity (ROP-OQ) models, **22**-17, 28
re-seasonalized, **18**-12
reserve capital, **14**-8
reservoir systems
   dams, **24**-4–5, 10
   filling/drawing down, **24**-6
   flood control storage, **24**-4
   forecasted inflows, **24**-5
   hydropower, **24**-8
   inactive pool, **24**-4
   index level, **24**-5–6
   linear release rule, **24**-9
   maximization of benefits, **24**-10

reservoir systems (*Contd.*)
  operations rules, **24**-6
  optimal operating policy, **24**-4–10
  planning stages, **24**-4
  probable maximum floods, **24**-5
  program formulation, **24**-7–10
  releases, **24**-6
  software programs, **24**-6
  storage capacity, **24**-10
  system balancing, **24**-6–7
  trigger volumes, rationing, **24**-8–9
residual capacity, **4**-10
resilience, **11**-9–11, **19**-21
resource allocation, decision making process, **6**-2
resource availability, product mix problem, **1**-5, 17
response, in complex networks, **11**-17
response surface methodology (RSM), **16**-4
return on asset, **21**-4
returns
  expected portfolio return, **21**-3
  significance of, **21**-3–4
  weighted average, **21**-6
revenue management (RM)
  airline optimization, **20**-4, 14–21
  problem, **14**-19–20
reverse critical, **15**-33
reversible failures, **11**-21
RHS constants, **1**-23–4
right-censored data set, **17**-23–4, 26
risk analysis, energy investments, **19**-10–1
risk-averse decision makers, **6**-12, 27
risk-averse investors, **21**-8, 11–2
risk-averse utility function, **6**-12–3
risk-free rate, **21**-14, 25, 27, 31
riskless assets, **21**-14
riskless portfolio, **21**-25
riskless securities, **21**-25
risk management, **6**-3–17, 27, **21**-18–22
risk-neutral
  decision makers, **6**-12–3
  investors, **21**-11
  utility function, **6**-12–3
  valuation, **21**-25–7, 31
risk pooling, **22**-41–5
risk preference, **21**-10–1
risk-return tradeoff, **21**-8
risk-reward tradeoffs, **21**-15
risk-seeker decision makers, **6**-12–3, 27
risk-seeking utility function, **6**-12–3
risk tolerance, **6**-17
risky assets, **21**-14

robustness
  airline optimization, **20**-17, 19
  complex networks, **11**-19–20
  energy systems optimization, **19**-23
robust optimization
  capacity expansion problem, **14**-24–7
  classical models of
    chance-constrained model, **14**-9–10
    characterized, **14**-2–3
    fallacy of averages, **14**-2
    mean-risk model, **14**-3–7
    recourse model, **14**-7–8, 21, 31
  defined, **14**-2
  electric circuit example, **14**-29–30
  engineering design, **14**-28–31
  facility location problem, **14**-27–8
  job sequencing, **14**-16–7
  models of
    interval uncertainty, **14**-11–4, 20
    minimax regret, **14**-14–6, 18–22, 26
    uncertainty sets, **14**-16, 24–7
    worst-case hedge, **14**-10–1
  newsboy problem, **14**-17–19
  revenue management, **14**-19–20
  sensor placement problem, **14**-31
  stochastic programming modeling, **14**-30–1
  transportation problem, **14**-8, **20**-7
rounding up/down, **3**-2
round-off errors, **2**-20
routers, complex networks, **11**-3
Row approach, crew pairings, **20**-12

S

SABRE, **20**-4, **23**-3
Safari online, **23**-7
safety stock, **10**-21–5, 37, **22**-12, 14, 19, 42–3
salvage regret, **14**-18–9
S&P 500, **21**-33
SAP R/3, **23**-13
SAS, **8**-47, **17**-40
satisficing solution, **5**-4
satisfied constraints, integer programming, **3**-6
savage regret, **6**-20
scale-free networks, **11**-8, 16–7
scale invariance, **11**-17
scale parameters, **17**-17
scatter diagram, **18**-15–7
scatter search, **13**-1–2
scenario estimation, **21**-5–6
scenario tree, **21**-35

S charts, **16**-13–4
scheduling
  airline optimization, **20**-5–10
  components of, **5**-24, **18**-20–1
  optimal policy, **18**-21
  single machines, **18**-20–3
scientific collaboration networks, **11**-4
score vector, **17**-25–6, 33, 35
SCOV, **9**-32, 34–6
scrap metal blending problems, **1**-2–4
searchable networks, **11**-23
search engines, **23**-8
seasonality, **18**-11–2, **19**-16
second moment, **17**-14–5
second-order approximations, **2**-16–7
secord-order cone optimization problem, **2**-2
Securities and Exchange Commission (SEC), **22**-31
self-organization, **11**-17
self-reproducing property, **17**-19, 22
self-similarity, **11**-17
semidefinite optimization problem, **2**-2
semivariance, **21**-19
SensIt software, **6**-6
sensitivity analysis, **5**-11, **6**-5, 29, **12**-22, **26**-4
sensor placement problem, **13**-13, **14**-31
sequential search methods, **19**-16
sequential unconstrained minimization techniques (SUMT), **2**-17–18
serial supply chain inventory models
  centralized control, **10**-32–4
  decentralized control, **10**-31–2
serial system, **10**-29
series system, reliability of, **17**-4, 8–9, 16
set-covering problem, **3**-4
set partitioning problem, airline optimization, **20**-10
set-up costs, **10**-5–6, 35
shadow price, **5**-32
shape parameters, **17**-17
shared-parallel system, **17**-10
shareholder's equity, **21**-34–5
Shewhart, Walter A., **16**-11
Shewhart control charts, **16**-11–2, 18–9
shift parameters, **17**-17
shortage cost, **10**-21, 23–4
shortest path
  approximation, airline optimization, **20**-12
  distance matrix, metaheuristics, **13**-4
  dynamic programming compared with, **7**-16
  problem
    characterized, **4**-2, 5, 14–5, 18

    Dijkstra's algorithm, **4**-5–8
    large-scale networks, **11**-2
    linear program formulation, **4**-5
short position, **21**-15, 22, 24
shortest processing time first (SPTF), **9**-3, **18**-21, 8
shortest processing time (SPT) rule, **14**-17
shortest total path length spanning tree problem, **4**-18
shortfall risk, **21**-21
Sierpinski triangle, **11**-17
Silver-Meal inventory method, **10**-20
SIMAN, **12**-7
simplex method
  airline optimization, **20**-8
  characterized, **2**-9–10
  network optimization, **4**-9
simplex solution
  integer programming, **3**-13–5
  network optimization, **4**-4
simulated annealing, **2**-3, **3**-16, **13**-1–2, 6–8, 27, 35–6, **20**-13, **24**-13–4
simulation
  applications, **12**-1–2, 15
  attributes, **12**-4
  basics of, **12**-3–15
  batch means analysis, **12**-13–4
  characterized, **12**-1–2, 22–3
  clock, **12**-5
  common random numbers (CRN) method, **12**-14–5
  deterministic delay, **12**-6
  energy system optimization, **19**-11, 17
  event calendar, **12**-5–7
  examples of, **12**-2–3
  global variables, **12**-4
  information resources, **12**-3, 20, 22
  languages, **12**-7, 15–16, 22
  military applications, **25**-8–9
  modeling process, **12**-4
  nonterminating system analysis, **12**-8, 10–4
  output-analysis, **12**-7–15, 21–2
  preloading, **12**-11
  process-oriented, **12**-4–5
  queues, **12**-4
  randomness, **12**-7–8
  reasons for, **12**-2
  replications, **12**-3–4, 8, 10–1, **17**-38
  resources, **12**-4
  run, **12**-3, 10, 15
  server availability, **12**-5–6
  software, **12**-15–20
  state of the system, **12**-3–4

simulation (*Contd.*)
  system parameters, **12**-3–4
  system variables, **12**-4
  terminating system analysis, **12**-8–10
  work in process, **12**-20–2
SIMULS, **12**-19
single location inventory, **10**-7
single-product inventory, **10**-8
single-server, single-queue system, simulations, **12**-5
single solution metaheuristics, **13**-2
single sourcing, **10**-8
single-station queues
  closed-queueing networks, **9**-31–2
  and multiclass queues, **9**-21–7
  and single-class queues, **9**-8–21
single-variable quadratic function, dynamic programming, **7**-10
single warehouse, multiple retailer inventory system, **10**-30
sink nodes, network optimization, **4**-8–10, 15
sinks, transportation problem, **1**-10
six degrees of separation, **11**-7
Six Sigma methodology, **16**-4–7
skewness, **18**-6
Skyteam Alliance, **20**-3
slack variables, **1**-24, 27, **14**-1, **19**-14
SLAM, **12**-7
slope, **1**-13–4, **2**-4
slotted ALOHA, **8**-3, 18
small-world effect, **11**-6–7, **15**-6
smart card, **23**-6
SmartSettle, **23**-5
smoothing constant, **18**-4
social mobility, stochastics example, **8**-4–7, 23
social networks, **11**-25, 27
Society for Modeling and Simulation International, **12**-3
soft datasets, **13**-16
software programs
  airline optimization, **20**-5
  decision analysis, **6**-6, 17
  demand forecasting, **18**-29
  energy optimization, **19**-5, 8
  integer linear programming, **3**-16
  inventory management, **10**-35
  linear programming, **1**-9, 25–6, 31
  project management, **15**-34–7
  reliability, **17**-40
  simulations, **12**-1, 15–20
  supply chain management, **22**-2
sojourn time, **8**-28
solar energy, **19**-2

solar power, **19**-4–5
solvability problems, **18**-28
solver software, **1**-31
source nodes, network optimization, **4**-5, 8–10, 15
sources, transportation problem, **1**-10
spanning tree
  characterized, **4**-2, 4, 14
  network model, **18**-17–9
sparse networks, **14**-24
spatial networks, search in, **11**-22–3
species set cover problem (SSCP), **13**-14
Speedera, **23**-18
splitting
  activities, **14**-5–6, **15**-27, 32
  queueing theory, **9**-29, 35
S-Plus, **17**-40
spreading processes, **11**-18–9, 26
spreadsheet applications
  inventory control, **10**-35
  options pricing, **21**-27
  supply chain management, **22**-10
SPRINT, **20**-12
stage, defined, **7**-2
stage tree, **21**-29–30
standard deviation, implications of, **10**-8, **12**-13, **15**-18, **16**-11, **17**-14, **18**-4–5, **21**-6, 8–9, 27, 31, **22**-12, 41–2, **24**-16
standby system, **17**-10
StarAlliance, **20**-3
start critical, **15**-33
"start to start" dependency, **15**-32
state, defined, **7**-2
state-dependent service, **9**-10, 29–30
states of nature, defined, **6**-2–3, 5
state space, **8**-4–5, 17
stationary continuous-time Markov chains, **8**-28
stationary coordination policy, **10**-30
stationary demands, **10**-8, **22**-52
stationary discrete-time Markov chains, **8**-15
stationary distribution, **8**-24, 34, 45
stationary increments, **8**-5, 8–9
stationary nested coordination policy, **10**-30
statistical analysis, simulations, **12**-1, 22
statistical inference
  continuous-time Markov chain (CTMC), **8**-31–3
  discrete-time Markov chain (DTMC), **8**-25–7
  implications of, **8**-46
  renewal theory, **8**-46
statistical learning, **18**-6
statistical physics, **11**-4

Index

statistical process control (SPC)
  applications, **16**-7–8
  control charts
    characterized, **16**-11–2
    problem identification, **16**-10
  sources of variation, **16**-9
  statistical control, **16**-11
  system overview, **16**-8–9
statistical quality control (SQC), **16**-3, 21
steady-state probability, **9**-29, 31, 33
step length, **2**-4
STEP method (STEM), **5**-30–1, 33
step multiplier, energy system optimization, **19**-15
stochastic demand, **10**-8
stochastic dynamic programming
  applications, **21**-28, **22**-19–20
  models
    characterized, **7**-19
    equipment replacement problem, **7**-19–22
    investment planning, **7**-22–4
stochastic inventory systems
  characterized, **10**-20
  examples of, **10**-20–1
  maximum-minimum system, **10**-22–5
  multi-period inventory model (s, S) policy, **10**-27
  one-period system, **10**-26–7
  safety stock, **10**-21–5, 37
  simplified solution method, **10**-25–6
stochastic matrix, **8**-15, **9**-31
stochastic models, military applications, **25**-6–8
stochastic optimization problem, **2**-2
stochastic processes
  applications, **8**-1–7, **21**-26
  continuous-time Markov chains (CTMC), **8**-27–38, **9**-8
  defined, **8**-1
  discrete-time Markov chains (DTMC), **8**-5–6, 14–27
  Poisson processes, **8**-7–14
  renewal theory, **8**-39–46
  software products, **8**-46–7
stochastic programming
  airline optimization, **20**-18, 21
  impact of, **21**-21
  modeling, **14**-30–1
  multi-period, **21**-35–6
stochastics, CHP modeling, **19**-12
stochastic service formulation, **22**-13
stock exchanges, **21**-15

stock investment, decision analysis example, **6**-3–8, 14
stock-keeping unit (SKU), **22**-20–1
stockouts, **22**-12, 15–6
stopping criteria, energy systems optimization, **19**-22–3
stopping rules, **2**-4
storage problems, **1**-11–2
stormwater management
  best management practices (BMPs), **24**-21–2
  decoder method, **24**-22
  genetic algorithm, **24**-22
  problem formulation, **24**-22–3
  simulation, **24**-22
  total maximum daily load (TMDL), **24**-2, 21
  urban runoff, **24**-3, 21
strategic decisions, **26**-10
strike price, **21**-22–3, 31
subadditivity, **21**-20
subproblems
  dynamic programming, **7**-3, 5, 10–1
  integer linear programming, B&B method, **3**-7–12
  network optimization, **4**-1
  partitioning algorithm, **5**-21–4
  shortest path problem, **4**-4
substitute-objective-function technique, **1**-20
sum quota sampling, **8**-11–2
SUNSET SOFTWARE/XA, **4**-19
superposition, queueing theory, **9**-35
suppliers, unreliable, **10**-6
supply and demand, **4**-3–4, 9, **8**-30, **10**-13
supply chain
  common structures, **10**-29
  decentralized *vs.* centralized, **22**-40–1
  defined, **22**-2
  design, **22**-3, 48
  efficient, **22**-4
  inventory management, *see* inventory management
  inventory models, **10**-27
  management of, *see* supply chain management (SCM)
  market-driven, **22**-5
  network design, **22**-36
  optimization, **23**-5, 7, 12, 21
  responsive, **22**-4
  safety stock, **22**-12, 14, 19, 42–3
  transportation management, **22**-24–37
  vulnerability of, **10**-37

supply chain management (SCM)
  congestion effects, **22**-12
  decisions, types of, **22**-58
  defined, **22**-2
  dependent demand items, **22**-7
  drivers of, **22**-4
  distribution requirements planning (DRP), **22**-7, 20–4
  of dyads, **22**-48–58
  enablers, **22**-4–5
  importance of, **22**-1–2, **26**-7
  incentives for, **22**-1
  independent demand items, **22**-7
  information resources, **22**-39
  location management, **22**-38–48
  manufacturing requirements planning (MRP), **22**-7–8
  non-stationary demand items, **22**-7–8
  periodic review inventory policy, **22**-12, 15, 17
  pull systems *vs.* push systems, **22**-8
  reorder point, order quantity models, **22**-15–20
  software programs, **22**-1, 2, 10, 59
  stationary demand items, **22**-7–8, 12
  structural view of, **22**-6
  tasks and tools, **22**-3
  transportation
    basics of, **22**-24–5
    cost estimates, **22**-26–30
    cost function, **22**-34, 36
    inventory decision making, **22**-30–4
Supply Chain Operations Reference (SCOR®), **22**-6
surrogate criteria, **5**-3
surveys
  airline optimization, **20**-15
  decision analysis, **6**-29
  simulation software, **12**-23
survival analysis, parameter estimation
  characterized, **17**-22–4, 40
  data sets, **17**-25, 29
  exponential distribution, **17**-25–30, 40
  Weibull distribution, **17**-2, 17, 21, 30–2, 40
survivor function, **17**-10–1, 15–7, 32–6
susceptibility, complex networks, **11**-17
swarm intelligence, **2**-3
SWIFT, **23**-3
Swift & Co., inventory management, **10**-34
Symantec, **23**-18
system of linear equations, **1**-2
system of systems (SoS), **25**-11
systematic risk, **21**-14
systems analysis, **18**-28

T

table lookup scheme, **13**-8
tabu list, **13**-9, 32
tabu search, **2**-19, **13**-1–2, 8–10, 25–6
tabu size, **13**-9
tabu tenure, **13**-18–9, 21
tail node, **4**-2
tally statistic, **12**-8
target value, **1**-32, **16**-2, 11
tariff rate, in freight transportation, **22**-34, 36
Taylor expansion expression, **21**-12
Taylor-series expansion, **19**-15
Tchebycheff (Min-Max) goal programming, **5**-26
Tchebycheff Metric, **5**-29
technology coefficients, **1**-5
telecommunication networks, **4**-4, 9, **11**-27
Tempfactor, **13**-6–7
TenderSystem, **23**-14
terminal/accessorial costs, **22**-24
test statistic, **17**-2, 37
theft nesting, **20**-19
theorems
  multiple criteria mathematical programming (MCMP), **5**-18, 29
  optical policy and value function, **7**-11
  reliability, **17**-17–21
  robust optimization, **14**-4–5, 11
theory of constraints, **23**-13
third-party logistics (3PL) supply chain, **22**-2
time coordination, **10**-30
time-homogeneous continuous-time Markov chain, **8**-28
time-homogeneous discrete-time Markov chains, **8**-15
time horizon
  airline optimization, **20**-4
  financial engineering, **21**-33–5
  significance of, **26**-4
time series models, **16**-13
time value of money, **7**-7, **19**-11
Time Warner, inventory management, **10**-34
TOPSIS (technique for order preference by similarity to ideal solution), **5**-9–10
total quality management (TQM), **26**-7
total returns, energy investments, **19**-7
total slack, **15**-7–8, 10, 31
total time on test, **17**-27
TPACS, **20**-12
TradeMe.com, **23**-14

tradeoffs
  in decision analysis, **6**-25–6
  information, **1**-32
  ratio, **5**-29–30
traffic
  intensity, queueing theory, **9**-36
  management, **26**-4
Trajectra, **23**-5
transformation function, dynamic
    programming, **7**–2, 4, 8, 17
transformation methods, **2**-15, 17–9
transient analysis, **8**-18–20, 31–3
transient state, **8**-21, 25
transition matrix, **8**-20–1, 24–5, 28
transition probabilities, **8**-5, 17, 26–7
transportation
  costs, **22**-45, 48
  management system (TMS) software, **22**-25
  networks, minimum cost multicommodity flow problem, **4**-18
  problems, **1**-9–11, **4**-3–4, **14**-8, 20–7
transportation-distribution decisions, **1**-12
transshipment nodes, **4**-9
tree, *see specific types of trees*
  defined, **4**-2
  shortest path problem, **4**-6, 8
Tretheway algorithm, **18**-15–7
TRIP, **20**-12
truckload (TL) shipment, **22**-24–5, 33–4, 36
true criteria, **5**-3
truncation errors, **2**-20
turboprop, **13**-10
turboprop noise, **13**-14
T-year planning horizon, **4**-5
Type II right-censoring, **17**-25–8

U

u chart, **16**-14–5, 24
uncapacitated network flow problem, **4**-3–4
uncertainty
  airline optimization, **20**-20
  budget of, **14**-16
  in decision analysis, snack product example
    characterized, **6**-18
    expected value, **6**-20
    impact of, **6**-4, 17, 28
    Laplace criterion, **6**-20
    maxmin (minimax) criterion, **6**-18
    maxmin (minimin) criterion, **6**-19
    minmax regret (savage regret), **6**-20

  effects of, **14**-1
  financial engineering, **21**-19–20
  impact on decision making process, **6**-2–3, 17–20, 26–7
  interval, **14**-11–4
  inventory management and, **10**-22
  robust optimization models, **14**-16, 24–7
  sources of, **14**-1
unconstrained linear optimization, energy systems, **19**-14
underage cost, **10**-27, **22**-12
underlying assets, **21**-23
underlying security, **21**-22
understocking, **22**-10
undirected network, **4**-3
uniform distribution, **8**-11–12, 14, **13**-31
uniformization, **8**-31, 33
uniform line search method, energy systems optimization, **19**-23
uniform supply, **10**-14
United Parcel Service (UPS), **20**-7, **26**-10
U.S. Air Force Space Command (AFSPC), Aerospace Integrated Investment Study (ASIIS), **25**-5–6
U.S. Strategic Command, Weapon Allocation Model (WAM), **25**-5
Unix servers, **23**-18
unknowns, defined, **1**-2
unslotted ALOHA, **8**-3–5
upper bounds
  dynamic programming, **7**-18
  exponential utility function, **21**-12
  goal programming, **5**-26
  multiple criteria decision making (MCDM), **5**-4
  robust optimization, **14**-11–2
  significance of, **4**-10, **13**-27–8
  supply chain management, **22**-39
upper control limit (UCL), **16**-11
upstream, in supply chain, **10**-29
utility analysis, **21**-10–2
utility function
  approximate, **6**-16
  characterized, **6**-12–13
  classification of, **6**-13
  construction of, **6**-15
  energy investment options, **19**-6–7
  exponential, **21**-11
  financial engineering, **21**-10–2
  investment example, **6**-17
  lottery example, **6**-14–7
utility maximization, **21**-12
utility theory, **6**-14
utilization, queueing networks, **9**-35

V

validation, simulations, **12**-21
value at risk (VaR), **21**-19, 21–2
value focused thinking (VFT), **25**-6
variable neighborhood descent, **13**-1–2
variables, defined, **1**-2
variance
   of asset return, **21**-4–6
   beta distribution, **15**-21
   estimation of, **21**-4–6
   exponential distribution, **17**-19
   lifetime distribution, **17**-14
   survival analysis, **17**-40
variance-covariance matrix, **14**-3
variance transmission equation (VTE), **16**-7
vector(s), *see specific types of vectors*
   comparison, **5**-29
   optimization problem, **5**-2
vibrational damping, **13**-13
vibration control, **13**-21–3
volatility, impact of, **14**-1, **21**-5–6, **22**-51
volume algorithm, **20**-12
Von Neumann, **1**-4
voters' attitude changes problem, **8**-26–7

W

Wagner-Whitin inventory model, **10**-18–20
walk, defined, **4**-2
warehousing, inventory control and, **10**-6
warm-up period, **12**-12–3
warranty, prorated, **8**-43–4
water distribution systems optimization
   drinking water, **24**-2
   example network, **24**-15–7
   genetic algorithm (GA), **24**-17–8
   global optimization, **24**-11, 16
   multistart-local search, **24**-11, 13
   overview, **24**-10–1
   parallel expansion, **24**-14–5
   pipe networks, **24**-2, 11–2
   simulated annealing, **24**-11, 13–4
   two-stage decomposition scheme, **24**-11–3
water networks, **4**-9
water pollution, **24**-1
water quality, **24**-2, 23
water resources
   domestic plumbing materials, selection factors, **24**-18–21
   groundwater management, **24**-23–5
   major drainage system, **24**-3
   minor systems, **24**-2–3
   reservoir systems, *see* reservoir systems
   stormwater management, **24**-21–3
   sustainability, **24**-1–4
   total maximum daily load (TMDL), **24**-2
   urban runoff, **24**-3, 21
   water distribution systems optimization, **24**-10–8
Watts-Strogatz complex network model, *see* small-world networks
wear-out failures, **17**-13
web analytics, **23**-7, 17, 19–20
webcasts, **23**-17
webinars, **23**-17
web pages, contents of, **23**-8–9
web-site performance tuning, **23**-7. *See also* Information resources, web sites
WebSideStory, **23**-20
Webtrends, **23**-20
Weibull distribution, **17**-2, 17, 21–2, 30–2, 38, 40
Weibull survivor function, **17**-32
weight breaks, in freight transportation, **22**-28
weighted average, **5**-12
weighted least squares method, **18**-8
weighted sum of squared forecast errors, **18**-8
weighting, assignment problem, **4**-13
Western Electric, **16**-12, 19
what-if scenarios, **12**-2
Whitt, Ward, **9**-36
wide area networks (WANs), **25**-5
width-first searches, integer programming, **3**-9
Wiener process, **21**-25–6
window flow control, **9**-30
wireless sensor networks, **11**-22
WITNESS, **12**-19–20
Wolfe conditions, **2**-8, 14
work breakdown structure (WBS), **15**-27
workforce planning, **4**-4
work in process (WIP), **12**-20–1
workload estimates, **1**-19–20
World Wide Web, *see* internet
   as complex network, **11**-3, 5–6, 9, 16–7, 19, 25
   as information resource, **20**-23, **23**-3, 8
Worldspan, **20**-4
worst case analysis, **1**-17

X

x-bar chart, **16**-12
X charts, **16**-13–4

Index

Xpress-MP, **20**-5
x^2 statistic, **8**-27, 38

Y

Yahoo, **23**-3–4, 8, 12, 18
yield
    energy investments, **19**-7
    management, **20**-4

Z

zero inventory ordering property, **10**-19, 30–1
zero minimum inventory, **10**-17
zero-project-slack assumption, **15**-5
Zionts-Wallenius (ZW) method, **5**-31